D1159611

SMITHSONIAN MISCELLANEOUS COLLECTIONS.

—————— 850 ——————

A SELECT

BIBLIOGRAPHY

OF

CHEMISTRY

1492-1892.

BY

HENRY CARRINGTON BOLTON.

———————

Martino Publishing

Mansfield Centre, CT

2005

Martino Publishing
P.O. Box 373,
Mansfield Centre, CT 06250 USA

www.martinopublishing.com

ISBN 1-57898-553-6

© *2006 Martino Publishing*

All rights reserved. No new contribution to this publication may
be reproduced, stored in a retrieval system, or transmitted, in any form or
by any means, electronic, mechanical, photocopying, recording, or otherwise,
without the prior permission of the Publisher.

Library of Congress Cataloging-in-Publication Data

Bolton, Henry Carrington, 1843-1903.
   A select bibliography of chemistry / by Henry Carrington
Bolton.
      p. cm.
   Originally published: City of Washington: Smithsonian
Institution, 1893.
   Includes bibliographical references.
   ISBN 1-57898-553-6 (cloth: alk. paper)
   1. Chemistry--Bibliography. 2. Alchemy--Bibliography. I.
Title.

Z5522.B65 2006
[QD31]
016.54--dc22   2005045582

Printed in the United States of America On 100% Acid-Free Paper

Foreword

Henry Carrington Bolton (1843–1903) was truly a Renaissance man a chemist, world traveler, academic, alpine climber, folklorist, bibliographer, historian, and bibliophile. The Bolton Society, an organization of chemical bibliophiles sponsored by the Chemical Heritage Foundation, is named for him.

Of the many contributions which grew out of Bolton's many and varied interests, the most lasting has proved to be his exhaustive works in bibliography which are not likely to be surpassed. A Select Bibliography of Chemistry, 1452-1892 published In 1893, contains 12,031 titles. Supplements in 1899 and 1904 added 3803 titles. A fourth volume published in 1901 is devoted exclusively to cataloging more than 3000 academic dissertations in chemistry. His pen was seldom idle.

The first volume of the Bibliography was reprinted by Krause Reprint Corporation of New York in 1966, and again by The Bolton Society of Philadelphia in 2005. Member Steven Beare proposed the republication and selected the publisher, Maurizio Martino

A sampling of topics covered by Bolton's more than 300 monographs, journal articles and addresses shows the breadth of his interests: action of light on uranium, counting-out rhymes for children, Russian transliteration, Chinese alchemy, glaciers, the early medical practice of women, microscopic crystals found in the vertebrae of toads, Bolton family genealogy, humor in chemistry, index to the literature of uranium, fortune telling, Hawaiian pastimes, evolution of the thermometer, divination with mirrors, physics and faith, language used in talking to domesticated animals, and musical notes emitted by certain beach sands when stepped on. His collection of 600 early books on alchemy and chemistry was one of the finest of its kind at the time. It seems that he was equally at home in all subjects.

Born into an affluent family in New York City, Bolton had been afforded the luxuries of good breeding, culture, travel, and a superior education (Columbia, B.S., Göttingen, Ph.D., and private study with some of the best chemical minds in Europe—Bunsen, Dumas, Hofmann, Kirchhoff, Kopp, Wohler and Wurtz.) No recluse, Bolton reveled in the companionship of great minds both in his home and in the workplace. He served as president of the New York Academy of Sciences and was founder of the American Folklore Society. At one time, he was said to have been a member of more scientific societies than any other American. Bolton did not marry until he was fifty (to Henrietta Irving, the great grandniece of author Washington Irving) and moved from New York to Washington in 1896, presumably to be closer to that city's great libraries. Fluent, witty, and charming, Bolton was always in great demand as a speaker. Amiable arid religious (Episcopalian), he was admired for his geniality, refinement, kindly spirit, and encyclopedic knowledge. Following his death at age sixty, close friend and fellow chemist Frank Wigglesworth Clarke eulogized him as being "honest, faithful, straightforward, a delightful friend and companion...."

Herbert T. Pratt
Chief Bibliophile
October 2005

SMITHSONIAN MISCELLANEOUS COLLECTIONS.

———— 850 ————

A SELECT

BIBLIOGRAPHY

OF

CHEMISTRY

1492-1892.

BY

HENRY CARRINGTON BOLTON.

————

CITY OF WASHINGTON:
PUBLISHED BY THE SMITHSONIAN INSTITUTION.
1893.

The Knickerbocker Press
New York

CONTENTS.

PREFACE.

An attempt has been made in the following pages to collect the titles of the principal books on CHEMISTRY published in Europe and America from the rise of the literature to the close of the year 1892. The term Chemistry is taken in its fullest significance, and the Bibliography will be found to contain books in every department of chemical literature, pure and applied; the range of topics will be seen in the Subject-Index and their distribution in the Table at page xiii. The Bibliography is confined, however, to independent works and their translations, and does not, as a rule, include Academic Dissertations (which are so numerous as to require a special catalogue), nor so-called "reprints" or "separates" (Separat-Abdrücke); of the latter only a few score are ordinarily printed and they must be regarded as belonging to periodicals. No attempt has been made to index the voluminous literature of periodicals except in the Section of Biography as noted below. Full bibliographical details have been given whenever possible, but the necessity of transcribing titles from catalogues of many degrees and varieties of imperfection has caused inequality in the fulness of the titles; a considerable number of books have been personally examined, these are indicated by adding the pagination, and for these alone can the compiler be wholly responsible.

To facilitate reference the work is divided into seven Sections:

    I. Bibliography.
    II. Dictionaries.
    III. History.
    IV. Biography.
    V. Chemistry, pure and applied.
    VI. Alchemy.
    VII. Periodicals.

It is believed that these divisions are so clearly marked that few doubts can arise as to the position of a given work; moreover, cross-references are frequently introduced, and it is hoped that this grouping will outweigh the advantages of arrangement in a single alphabet.

Owing to circumstances, partly unavoidable and partly intentional, the seven Sections are unequally developed :

I.   The Section of Bibliography is believed to be quite full and includes to some extent special bibliographies occurring in periodicals ; but bibliographies of individuals are placed in Section IV., Biography, under the name of the chemist.

II.   The Section of Dictionaries is thought to be tolerably full ; in this connection are placed Tables giving a conspectus of chemical bodies (not, however, Tables for Analysis), and works on Nomenclature.

III.   The Section of History is more nearly complete, having been in course of preparation for several years.   It is in fact a second edition of my "Outlines of a Bibliography of the History of Chemistry," published in 1873.   (Ann. N. Y. Lyc. Nat. Hist., Vol. x., New York, 1873.) In this division also periodical literature has to some extent been explored.

IV.   The Section of Biography is less complete, yet pains have been taken to index the necrologies in several periodicals.   The chief of these periodicals are Ber. d. chem. Ges.; Am. J. Sci.; Nature ; Ann. Chem., Crell ; Allg. J. Chem., Scherer and Gehlen ; Chem. News ; Bull. Soc. chim., Paris ; Pop. Sci. Mon., New York ; J. Chem. Soc., London.   In this Section special bibliographies of individuals are placed under their names.

V.   While Section V., Chemistry pure and applied, is more extensive than the other six combined, it is the least satisfactory to the editor.   Besides pure chemistry it comprises works in every department of chemistry applied to the arts, but not works on the arts themselves ; thus agricultural chemistry is given place, but not agriculture : pharmaceutical chemistry, physiological chemistry, etc., find place, but not the sciences of pharmacy and medicine.   It is of course often impossible to draw the line sharply and in doubtful cases the tendency has been to include rather than exclude ; hence many works on technology are included especially when chemistry lies at their foundation.   It is probable, however, that this has not been done uniformly, owing to the difficulty of selection.

Readers should not expect to find in this Section complete lists of the works of a given author ; though some pains have been taken in the case of prominent men to catalogue their writings fully.   (See under Berzelius,

Fresenius, Liebig, Johnston, Orfila, Stöckhardt, etc.) The names of many eminent chemists will be found wanting in Section V. simply because they have published no independent works, although voluminous contributors to periodicals.

VI. The Section of Alchemy is purposely restricted to a comparatively small number of the more important works, as the subject is chiefly of antiquarian interest. This Section could easily be expanded to 5,000 titles, but I do not think profitably; books in this Section are catalogued with special attention to bibliographical details and cross-references are freely inserted; those marked with asterisks are in my private library.

VII. The Section of Periodicals is actually an excerpt from my "Catalogue of Scientific and Technical Periodicals, 1665 to 1882," (Smithsonian Miscellaneous Collections, vol. 29, Washington, 1885), enlarged by the addition of serials published by Societies and brought down to date.

In each Section, excepting those of Biography and Periodicals, the titles are arranged alphabetically by authors, translations of each work following the original in the alphabetical order of the English names of the languages. The order is the same as in the table on page xiii. This table must not, however, be regarded as indicating the relative number of chemical works in the given languages, since the facilities for collecting the information from different countries are very unequal.

In the Section of Biography the titles are placed under the names of the persons described, with cross-references from the authors; in the Section of Periodicals the titles are arranged alphabetically by the first word, articles and " New " excepted, with cross-references from the editors.

Notes and comments, bibliographical and explanatory, have been occasionally introduced, especially in the Sections of History and of Alchemy, to aid students in conceiving the character of a book, or the status of the author.

The sources of the titles are, of course, general and special bibliographies, and libraries (named below). For the principal bibliographies consulted see Section I

In several of the larger libraries the manuscript catalogues of chemistry were transcribed with great care (Königliche Bibliothek, Berlin; Grossh. Bibliothek, Darmstadt); in others I had access to the shelves as well (British Museum, London; Patent Office, London; Universitäts

Bibliothek, Strassburg; and the Libraries in the United States of America).

In the preparation of this volume I have enjoyed the facilities offered by the following Libraries and Institutions, and I hereby express my hearty thanks to the respective Librarians and other Officers for their courtesies and for special privileges : British Museum, London ; Patent Office, London ; University Library, Cambridge ; Free Library, Birmingham ; Bibliothèque Nationale, Paris ; Bibliothèque de la Sorbonne, Paris ; Bibliothèque du Jardin des Plantes, Paris; Bibliothèque du Conservatoire des Arts et Métiers, Paris ; Bibliothèque de la Société chimique de Paris ; Bibliothèque Mazarin, Paris; K. Universitäts- und Landes-Bibliothek, Strassburg; Grossh. Universitäts Bibliothek, Heidelberg ; Grossh. Hof-Bibliothek, Darmstadt ; Königliche Bibliothek, Berlin ; K. Universitäts Bibliothek, Prag; Böhmisch National Museum, Prag; K. K. Hof Bibliothek, Wien; K. K. Universitäts Bibliothek, Wien ; Bibliothek der K. K. technischen Hochschule, Wien ; K. Universitäts Bibliothek, Budapest ; K. Akademie der Wissenschaften, Budapest ; K. K. Oeffentliche Landes- und Studien-Bibliothek, Salzburg ; Bibliothek des Carol Aug. Museum, Salzburg ; K. Hof- und Staats-Bibliothek, München : K. Oeffentliche Bibliothek, Stuttgart ; Library of the Surgeon-General's Office, United States Army, Washington ; Library of Congress, Washington ; Scientific Library of United States Patent Office, Washington ; Public Library, Boston ; Athenæum, Boston ; Massachusetts Institute of Technology, Boston ; Harvard College Library, Cambridge ; Dartmouth College Library, Hanover, N. H ; Amherst College Library, Amherst, Mass.; Library of Columbia College, New York ; Astor Library, New York ; Library of the New York Academy of Sciences.

I am also under obligations to Prof. Charles F. Chandler, Ph.D., M D., LL.D., of Columbia College, New York, and to Prof. Rudolph A. Witthaus, M.D., of the Medical College of the University of the City of New York, for permission to catalogue their valuable private collections.

I am very greatly obliged to Dr. Alfred Tuckerman, of Newport, R. I., for voluntary assistance ; he made a preliminary examination of Poggendorff's Handwörterbuch, as well as of Zuchold's and Ruprecht's Bibliotheca, copying many titles for this Bibliography with accuracy and fidelity.

I am also specially indebted to Prof. Bohuslav Brauner, of Prag, for revision of Bohemian titles of periodicals, and to John Fryer, LL.D., of Shanghai, for a list of treatises in Chinese.

To insure the greatest possible accuracy in the several languages I secured the aid of linguistic experts. The Rev. Prof. Samuel Hart, D.D., of Trinity College, Hartford, Conn., read the Greek and the Latin ; the Rev. Martin Osserwaarde, of New York, read the Dutch ; Mr. Axel Moth, of New York and Copenhagen, read the Danish, Norwegian, and Swedish ; Signor Cesare Poma, Vice-Consul of Italy, of New York City, read the Italian, Spanish, and Portuguese ; Mr. Louis Solyom, of the Library of Congress, Washington City, read the Hungarian, Russian, Polish, Bo- hemian and other Slavonic languages ; Mr. George H. Boehmer, of the Smithsonian Institution, assisted me in the French and the German. The proof-reading continued for twelve months, and matters were so arranged that each expert saw titles only in the language or group of languages which he engaged to read. To these gentlemen I am under great obliga- tions for their assiduity, promptness, and fidelity to my interests.

For the conception and contents of this bibliography I alone am responsible ; its publication has been made possible by the liberality of the Smithsonian Institution, to which I presented the manuscript compiled as a labor of love.

H. CARRINGTON BOLTON.

June 1st, 1893.
UNIVERSITY CLUB,
NEW YORK CITY.

NUMBER OF TITLES IN THE SEVERAL LANGUAGES.

	I. Bibliography.	II. Dictionaries.	III. History.	IV. Biography.	V. Pure and Applied.	VI. Alchemy.	VII. Periodicals.	Totals.
Arabic . . .					5			5
Bohemian . .					11		13	24
Chinese. . .		2			25	2		29
Danish . . .		3	2	3	84		6	98
Dutch . . .	4	3	14	7	277	3	11	319
English. . .	82	57	147	535	1,732	87	125	2,765
Finnish. . .					1			1
French . . .	31	65	146	167	1,563	89	80	2,141
German . .	136	140	308	233	3,072	423	195	4,507
Greek . . .			1		15			16
Gujerathi . .					1			1
Hebrew . .					1			1
Hindi . . .					1			1
Hungarian. .					8		1	9
Italian . . .		19	22	12	450	4	29	536
Japanese . .					4		1	5
Latin . . .	17	16	69	11	403	434		950
Norwegian .			1		24			25
Polish . . .		5	1		31		1	38
Portuguese .		1			8			9
Rumanian . .							1	1
Russian . .	1	6	7		219	2	1	236
Spanish. . .	2	9	5	4	174	2		196
Swedish . .		1	7	7	6		6	117
Welsh . . .					1			1
Totals .	273	327	730	979	8,206	1,046	470	12,031

Titles in the Addenda are included under their respective heads. In Section VI. the numbers indicate individual treatises, not volumes. The larger numbers are only approximate, though determined by actual count.

EXPLANATION OF ABBREVIATIONS AND SIGNS.

For abbreviations of titles of periodicals used in Sections III. and IV., see pages 1159–1164.

* prefixed to a title indicates a work in the private library of the editor.

\+ following a date signifies current at that date.

‖ following a date signifies publication discontinued.

Ill. illustrated.

Pl. plates.

Fol., 4to, 8vo, etc. The sizes given are only approximate ; having been taken largely from catalogues the rules of the American Library Association could not be observed.

Pagination is given only of those works personally examined.

Cross-references in a given Section refer to works in the same Section unless otherwise stated.

A SELECT BIBLIOGRAPHY OF CHEMISTRY.

SECTION I.

BIBLIOGRAPHY.

BIBLIOGRAPHIES OF INDIVIDUAL AUTHORS ARE FOUND UNDER THEIR
NAMES IN SECTION IV., BIOGRAPHY.

ALMANACH FÜR SCHEIDEKÜNSTLER.
See Taschenbuch für Scheidekünstler.

AMERICAN ASSOCIATION FOR THE ADVANCEMENT OF SCIENCE.
Committee on Indexing Chemical Literature, Reports. *See* Reports.

AMERICAN CHEMICAL JOURNAL. General index of volumes I–X (1879–
1888). W. R. Orndorff. Baltimore, 1890. pp. [iv]–87, 8vo.
Comprises an index of authors' names and an index of subjects.

ANNALEN DER CHEMIE UND PHARMACIE. Autoren- und Sach-Register
zu den Bänden I–C (Jahrgang 1832–1856). Bearbeitet von G. C.
Wittstein. Leipzig und Heidelberg, 1861. pp. iv–596, 8vo.
The same, Bände CI–CXVI (1857–1860). Leipzig und Heidelberg, 1861.
pp. 123, 8vo.
The same, Bände CXVII–CLXIV und den Supplementbände I–VIII (1861–
1872). Bearbeitet von Friedrich Carl. Leipzig und Heidelberg,
1874. pp. [iv]–380, 8vo.
General-Register zu den Bänden, CLXV–CCXX (1873–1883) von Liebig's
Annalen der Chemie (früher Annalen der Chemie und Pharmacie).
Bearbeitet von Friedrich Carl. Leipzig, 1885. 8vo.

ANNALEN DER PHYSIK UND DER PHYSIKALISCHEN CHEMIE. Vollständiges
und systematisch geordnetes Sach- und Namen-Register zu den 76
Bänden, 1799–1824. Angefertigt von Heinrich Müller. Leipzig,
1826. pp. x–612–131, 8vo.

ANNALEN DER PHYSIK UND CHEMIE, von J. C. Poggendorff. Namen- und Sach-Register zu den Bänden I bis LX. Bearbeitet von W. Barentin. Leipzig, 1845. pp. 361, 8vo.

Namen- und Sach-Register zu den Bänden LXI bis XC, und den Ergänzungsbänden II bis IV. Bearbeitet von W. Barentin. Leipzig, 1854. pp. 153, 8vo.

The same, Bänden XCI–CXX. Leipzig, 1865. pp. 112, 8vo.

Namen-Register zu Band I bis CL, Ergänzungsband I bis VI nebst Jubelband, und Sach-Register zu Band CXXI bis CL, Ergänzungsband V und VI nebst Jubelband. Bearbeitet von W. Barentin, nebst einem Anhange von J. C. Poggendorff enthaltend Verzeichniss der verstorbenen Auctoren und Zeittafel zu den Bänden. Leipzig, 1875. pp. 334, 8vo.

Sach-Register zu den Bänden 1–160 ; Ergänzungsband 1–8 und Jubelband, 1824–1877. Nach den von W. Barentin herausgegebenen Registern zu je dreissig Bänden, bearbeitet von Fr. Strobel. Leipzig, 1888. pp. viii–719, 8vo.

ANNALES DE CHIMIE. Table générale raisonnée des matières contenues dans les trente premiers volumes, suivie d'une table alphabétique des auteurs qui y sont cités. Paris, An. IX [1801]. pp. viii–430, 8vo.

The same, vols. XXXI–LX. Paris, 1807. pp. 334, 8vo.

The same, vols. LXI–XCVI. Paris, 1821. pp. 420, 8vo.

ANNALES DE CHIMIE ET DE PHYSIQUE, par Gay Lussac et Arago. Table générale raisonnée des matières contenues dans les trente premiers volumes, suivie d'une table alphabétique des auteurs qui y sont cités. Paris, 1831. pp. 454, 8vo.

The same, Tomes XXXI–LX. Paris, 1840. pp. 420, 8vo.

The same, Tomes LXI–LXXV. Paris, 1841. pp. 152, 8vo.

The same, Troisième série. Tomes I–XXX. Paris, 1851. pp. 134, 8vo.

The same, Quatrième série (1864–1873), dressées par U. Gayon. Paris, 1874. pp. 249, 8vo.

The same, Cinquième série (1874–1883), dressées par U. Gayon. Paris, 1885. 8vo.

BARRETT, FRANCIS.
    A Critical Catalogue of Books. *See, in Section IV*, Barrett, F., The Lives of Alchemystical Philosophers ; *also* Waite, A. E.

BATTERSHALL, JESSE PARK.
    Bibliography of Food Adulteration. *See in Section V.*

BAUMER, JOAN WILHELM.
Bibliotheca chemica. Giessen, 1782. 8vo.
Fuchs justly says this is "incomplete."

BECKER, GEORGE FREDERICK.
Atomic Weight Determinations: A Digest of the Investigations published since 1814. The Constants of Nature, part IV. Smithsonian Miscellaneous Collections No. 358. Washington, August, 1880. pp. 150, 8vo.

BECKER, JOHANN HERRMANN.
Versuch einer Literatur und Geschichte der Nahrungsmittelkunde. *See, in Section V*, Becker, J. H.

BERICHTE DER DEUTSCHEN CHEMISCHEN GESELLSCHAFT ZU BERLIN.
General-Register über die ersten zehn Jahrgänge (1868–1877). Bearbeitet von C. Bischoff. Berlin, 1880. pp. 1022, 8vo.
General-Register über die zweiten zehn Jahrgänge (1878–1887). Redacteur: Ferd. Tiemann. Stellvertretender Redacteur: F. v. Dechend. Berlin, 1888. II parts, pp. 1636, 8vo.

BERTHELOT, MARCELLIN.
Bibliography of Explosives. *See, in Section V*, Berthelot, M., Explosive Materials (New York, 1883).

BEUGHEM, CORNELIUS À.
Bibliographia medica et physica novissima perpetuo continuanda : sive Conspectus primus catalogi librorum medicorum, chymicorum, anatomicorum, chyrurgicorum, botanicorum, ut et physicorum, etc., quotquot currente hoc semisæculo, id est ab anno reparatæ salutis 1651 (inclusive), per universam Europam, in quavis lingua orientali, tum Græca, Latina, Gallica, Hispanica, Italica, Anglica, Germanica, et Belgica aut novi aut emendatiores et auctiores typis prodierunt. Amstelædami, 1681. pp. [vi]–503, 18mo.
The alphabetical bibliography is classed by the various languages.

BEYTRAG ZUR GESCHICHTE DER HÖHEREN CHEMIE. Leipzig, 1785.
Contains a Bibliography of Alchemy, with 634 titles. For full title of work, and annotation, *see, in Section III*, Beytrag, *etc.*

BIBLIOTHECA CHEMICA. *See* Zuchold, E. A.

BIBLIOTHECA CHEMICA ET PHARMACEUTICA. *See* Ruprecht, Rudolph.

BIBLIOTHECA CHEMICA, ODER CATALOGUS. *See* Roth-Scholtz, Fr.

BIBLIOTHECA HISTORICO-NATURALIS ET PHYSICO-CHEMICA [*later,* et mathematica] oder systematisch geordnete Uebersicht der in Deutschland und dem Auslande auf dem Gebiete der gesammten Naturwissenschaften [*later,* und der Mathematik] neu erschienenen Bücher. Herausgegeben von E. A. Zuchold [*later, by* Guthe; Metzger; Frenkel; Fricke; Hanstein]. Göttingen, 1851–1892. 42 vols., 8vo.

      Chemistry was dropped in 1887.

BIBLIOTHECA MEDICO-CHIRURGICA PHARMACEUTICO-CHEMICA ET VETERINARIA, oder [vierteljährliche systematisch-] geordnete Uebersicht aller in Deutschland neu erschienenen medicinisch-chirurgisch-geburtshülflichen, pharmaceutisch-chemischen und veterinärwissenschaftlichen Bücher. [*Later,* Herausgegeben von Carl Ruprecht.] Göttingen, 1847–1892. 46 vols., 8vo.

BIBLIOTHECA METALLICA, ODER BERGMÄNNISCHER BÜCHERVORRATH. Dresden, 1730. 4to.

BIBLIOTHECA PHYSICO-MEDICA. Verzeichniss wichtiger älterer sowohl, als sämmtlicher seit 1821 in Deutschland gedruckter Bücher aus den Fächern der Physik, Chemie, Geognosie, Mineralogie, Botanik, Zoologie, Vergleichenden und Menschlichen Anatomie, Physiologie, Pathologie, Therapie, Materia Medica, Chirurgie, Augenheilkunde, Geburtshülfe, Staatsarzneikunde, Pharmacie, Thierarzneikunde, u. s. w., zu finden bei Leopold Voss in Leipzig. Leipzig, 1835. pp. 189, 8vo.

BIDSTRUP, MICHAEL DAV. NIC.

Bibliotheca chymica et cos hermetica, of chymische bibliotheek versiert met de Levensbeschryvinge, zoo veel doenlyk was, over alle Autores, dewelke in chymia als wel ervarene mannen, mitsgaders de waere benaeminge van alle hare schriften, 't oordeel buyten eygen verkiezing uyt de schriften van andere beroemde mannen daerover gevelt hierin beschreven; en tot een welverzeekerde onderwysing trouwhartig meede gedeelt van dien, dewelke alle hare schriften bezit. t'Amsterdam, 1744. pp. [viii]–104, 12mo.

      An alphabetical catalogue of 243 authors, with short biographical notices, and some of their works. A long quotation from the poet Virgil, beginning "*facilis descensus Averni,*" is introduced to show that he was an "adept." This bibliography is more curious than accurate.

BIERENS DE HAAN, DAVID.

Bibliographie néerlandaise historique-scientifique des ouvrages importants, dont les auteurs sont nés aux 16e, 17e et 18e siècles, sur les

BIERENS, DE HAAN, DAVID. [Cont'd.]

sciences mathématiques et physiques avec leurs applications. Extrait du Bulletino di bibliografia et di storia delle scienze matematiche e fisiche, tomes XIV, 1881, XV, 1882, XVI, 1883. Rome, 1883. pp. 424, 4to.

> Valuable. Contains a classified subject-index.

BINZ, JOHANN GEORG.

Verzeichniss von chymischen, alchymischen, physikalischen, theosophischen, kabalistischen und kunstbüchern, welche bei Joh. Georg Binz, Buchhändler in Wein, im Zwettelhof, verkauft werden. 1751. 8vo.

> A trade catalogue of about 2,500 titles.

BLUMHOF, JOHANN GEORG LUDOLPH.

Vollständige systematische Literatur vom Eisen, in mineralogischer, chemischer, technologischer, ökonomischer, camaralistischer und medicinischer Rücksicht. Für Eisenhüttenkundige, Technologen und Litteratoren. Braunschweig, 1803. pp. xviii–271, 8vo.

> The author's claim to completeness is hardly warranted, but this careful bibliography deserves to be better known. It treats of iron from the metallurgical and mining points of view rather than the chemical.

BLYTH, ALEXANDER WINTER.

> Bibliography of the chief works on Toxicology (19th century). *See, in Section V*, Blyth, A. W., Poisons, *etc.*

Бобынинъ, В. В. Русская физико-математическая библіографія. Москва, 1886-91. 2 vols., 8vo.

> Bibliography of the Russian physico-mathematical sciences. Moscow. Contains, however, no chemistry.

BOEHMER, GEORGE RUDOLPH.

Bibliotheca scriptorum historiæ naturalis, œconomiæ, aliarumque artium ac scientiarum ad illam pertinentium realis systematica. Systematisch-literarisches Handbuch der Naturgeschichte, Oeconomie und anderer damit verwandten Wissenschaften und Künste. Leipzig, 1785–1789. 9 parts in 5 vols., 8vo.

> This extensive bibliography deals chiefly with natural history, but has sections on chemistry and on mineralogy. A very full index of names in the last volume occupies over 300 pages.

BOLTON, HENRY CARRINGTON.

Bibliography (A) of Analytical Chemistry for the year 1886 [1887, 1888, 1889, 1890]. J. Anal. Chem., vol. I [II, III, IV, V]. 1887-'91 Easton, Pa. Five parts, 8vo.

BOLTON, HENRY CARRINGTON. [Cont'd.]

Bibliography (A) of Analytical and Applied Chemistry for the year
1891. J. Anal. Appl. Chem., vol. VI. 1892. Easton, Pa. 8vo.

Bibliography (A) of Chemistry for the year 1887. Smithsonian Miscellaneous Collections No. 665. Washington, 1888. pp. 13, 8vo.

Catalogue (A) of Chemical Periodicals. Annals New York Academy
of Sciences, vol. III, pp. 159–216. New York, 1885. 8vo. *Also
in* Chemical News, vols. LI, LII. London, 1885 ; *independently issued*,
London, 1885. pp. 50, 12mo.

　　Supplement. Annals N. Y. Acad. Sci., vol. IV, 1887. pp. 4, 8vo.

Catalogue (A) of Scientific and Technical Periodicals (1665–1882),
together with Chronological Tables and a Library Check-List,
Smithsonian Miscellaneous Collections No. 514. Washington,
1885. pp. x–773, 8vo.

Chemical Literature : An Address delivered before the American Association for the Advancement of Science, at Montreal, August 23,
1882. Proceedings of the A. A. A. S., vol. XXXI. *Also in* Chemical News, XLVI, p. 115 *et seq.* 1883.
　　　Contains bibliographical notes.

Index to the Literature of Manganese (1596–1874). Annals of the
New York Lyceum of Natural History, vol. XI. November, 1875.
New York. pp. 44, 8vo.

Index to the Literature of Uranium. Annals of the New York Lyceum
of Natural History, vol. IX. February, 1870. pp. 15, 8vo.
　　　Second edition, 1789–1885. Smithsonian Report for 1885.
　　　Washington, 1885. pp. 36, 8vo.

List of Elementary Substances announced from 1877 to 1887. Chemical News, LVIII, 188. London, 1888. 4to.
　　　A small contribution to the bibliography of the chemical elements.

Outlines of a Bibliography of the History of Chemistry. Annals New
York Lyceum of Natural History, vol. X, pp. 352–361. New York,
1873. 8vo.
　　　The forerunner of Section III of the present volume.

Problem (A) in Chemical Bibliography : Read before the Chemical
Section of the A. A. A. S., at Buffalo, in 1886. Proceedings Am.
Assoc. Adv. Sci., vol. XXXV, 1886. Salem, 1887. 8vo.

Short Titles of Chemical Periodicals current in 1887. Journal of Analytical Chemistry, vol. I, part I, 1887. pp. 4, 8vo.

BOLTON, HENRY CARRINGTON (AND OTHERS).

Abbreviations of Titles of Chemical Journals. J. Anal. Chem., vol. II,
part I. Jan., 1888.
　　　See also Reports of the Committee on Indexing Chemical Literature.

BOREL, PIERRE [BORELLUS].

Bibliotheca chimica, seu Catalogus librorum philosophicorum hermeticorum, in quo quatuor millia circiter authorum chimicorum, vel
de transmutatione metallorum, re minerali et arcanis tam manuscriptorum quam in lucem editorum, cum eorum editionibus usque
ad annum 1653 continentur. Cum ejusdem Bibliothecæ Appendice
et Corollario. Parisiis, 1654. 12mo.

> *Second edition*, Heidelbergæ, 1656. pp. [xii]-276. *Third
> edition*, 1676. 8vo.
>
> The first extensive catalogue of chemical books, but unsatisfactory from the
> standpoint of modern bibliography. To hundreds of names are
> attached the words "chimicus author," or "chimiæ scriptor," with no
> further data.

BORRICHIUS, OLAUS [OLE BORCH].

Conspectus scriptorum chemicorum illustriorum. Libellus posthumus,
cui prefixa historia vitæ ipsius ab ipso conscripta. Hafniæ
1697. 4to.

> Also reprinted in Mangetus Bibl. chem. curiosa, vol. I, No. 2
> *Compare, in Section III*, Borrichius, O.
> More curious than valuable.

BRITISH ASSOCIATION FOR THE ADVANCEMENT OF SCIENCE.

Index to the Reports and Transactions of the British Association for
the Advancement of Science from 1831 to 1860, inclusive. London, 1864. 8vo.

BRUCKMANN, FRANZ ERNEST.

Catalogus exhibens appellationes et denominationes omnium potus
generum, quæ olim in usu fuerunt et adhuc sunt per totum ter-
rarum orbem quotquot adhuc reperire potuit. Helmstadi, 1722.
4to.

BRÜNNICH, M. TH.

Literatura danica scientiarum naturalium qua comprehenduntur : I Les
progrès de l'histoire naturelle en Dannemarc et en Norvège ; II Bibliotheca patria autorum et scriptorum, scientias naturales tractantium. Hafniæ et Lipsiæ, 1783. pp. 123-242-xiv-[xvi], 12mo.

> *Compare, in Section III*, Brünnich.

BÜCHTING, A.

Bibliotheca pharmaceutica, oder Verzeichniss aller auf dem Gebiete der
Pharmacie in den letzten 20 Jahren 1849-1868 im deutschen
Buchhandel erschienenen Bücher und Zeitschriften. Ein bibliographisches Handbüchlein für Pharmaceutisch-Medicinalbehörden
und Buchhändler. Mit einem ausführlichen Sachregister. Nordhausen, 1869. 8vo.

BÜCHTING, A. [Cont'd.]

Bibliotheca pharmacologia et toxicologica oder Verzeichniss aller auf dem Gebiete der Arzneimittellehre und Giftlehre in den letzten 20 Jahren 1848–1867 im deutschen Buchhandel erschienenen Bücher und Zeitschriften. Mit einem ausführlichen Sachregister. Nordhausen, 1868. 8vo.

BULLETIN DE LA SOCIÉTÉ CHIMIQUE [de Paris]. Table analytique des matières 1re et 2e séries 1858 à 1874, et dans les Répertoires de chimie pure et appliquée, suivie de la table alphabétique des auteurs. Dressées par Ed. Willm. Paris, 1876. pp. viii–511, 8vo.

Tables des années 1875 à 1888 du Bulletin de la Société chimique (Table analytique, Table alphabétique des auteurs). Dressées par Th. Schneider. Paris, 1890. 8vo.

BULLETIN DE PHARMACIE (et des sciences accessoires). Table analytique des auteurs cités et des matières contenus dans les tomes I–XVI (1809–1830). Paris, 1831. 8vo.

> *See* Journal de pharmacie et des sciences accessoires.

CALDWELL, GEORGE CHAPMAN.

Bibliography of works relating to the Adulteration and Testing of Butter, Cheese, Lard, Olive Oil, etc. Second Annual Report of the State Board of Health of New York. Albany, 1882. pp. 544–547, 8vo.

CALLISEN, ADOLPH CARL PETER.

Medicinisches Schriftsteller-Lexicon der jetzt lebenden Aerzte, Wund-ärzte, Geburtshelfer, Apotheker und Naturforscher aller gebildeten Völker. Kopenhagen und Altona, 1830–1845. 33 vols., 12mo.

> A wonderfully laborious and exhaustive compilation, replete with minutest details concerning the literature of medicine and natural science. Chemistry proper, though not included in the title of the book, receives its full quota of attention. The author sacrificed a fortune in compil-ing and publishing these numerous and closely printed volumes.

CATALOGUE DES SCIENCES MÉDICALES PUBLIÉ PAR ORDRE DU GOUVERNE-MENT. Bibliothèque nationale—Département des imprimés. Paris, 1857–1873. 2 vols , 4to.

> Vol. I contains sections on bibliography of medical science, on the history, and on the biography. Vol. II contains sections on spagyric, hermetic, alchemical, and chemical medicine, as well as a section on the history of pharmacy. In all of these much material pertaining to the history and bibliography of chemistry will be found.

CATALOGUE (A) OF MODERN WORKS ON SCIENCE AND TECHNOLOGY, classified under Authors and Subjects. Nineteenth edition. London, 1890. pp. vi–128–32.

> A one-line author list, with a subject-index. Convenient for the trade, as prices are attached.

CATALOGUE OF SCIENTIFIC PAPERS (1800–1883). Compiled and published by the Royal Society of London. London, 1867–1892. 9 vols., 4to.

> A monument of industry and utility, which unfortunately lacks a subject-index. Vol. IX extends only to GIS of the period 1874–1883. Vols. X and XI are expected shortly.

CATALOGUS VON ALTEN UND NEUEN CHYMISCHEN, von Berck-Wercken, Kräutern, Thieren und warmen Bädern und von der Haushaltungs Kunst handlenden Büchern welche bey Johann Daniel Taubers seel. Erben in Nürnberg und in Altdorff verkaufft werden. No. I. Nürnberg und Altdorff, 1722. pp. 24, 12mo.

CATALOGUS VON CHYMISCHEN Büchern welche in der Roth-Scholtzischen Bibliothèque vorhanden seyn. Erstes Stücke. Nürnberg und Altdorff, 1725. pp. 48, 12mo.

> *See* Roth-Scholtz, Fr., Bibliotheca chemica.

CATALOGUS MANUSCRIPTORUM chemico-alchemico-magico-cabalistico medico-physico-curiosorum. n. p. 1786. 8vo.

> A list of the MSS. in the Gräfer collection, Vienna, with an appendix of German authors.

CHANDLER, CHARLES F., AND FERDINAND G. WIECHMANN. Titles of the More Important Books of Reference relating to Chemistry, Prepared for the Use of the Students of the School of Mines. Columbia College. 1881.

> [New York, 1881.] 8vo.
>
> In two parts, Chemical Technology, and Organic Chemistry. Short titles grouped under subjects.

CHEMICAL SOCIETY [London]. Catalogue of the Library of the Chemical Society, arranged according to subjects ; with indexes containing Authors' names and Subjects. London, 1886. pp. viii–328, 8vo.

CHEMISCHES CENTRALBLATT. General-Register, III Folge, Jahrgang I–XII, 1870–1881. Redigirt von Rud. Arendt. Leipzig, 1882–1883. 8vo.

CHEMISCH-TECHNISCHEN MITTHEILUNGEN (DIE) DER NEUESTEN ZEIT.
  Sach-Register zu den ersten acht Heften . . . von L. Elsner. Die
    Jahre 1846–1859 enthaltend. Berlin, 1860. 8vo.
  Sach-Register zu den bisher erschienenen zwanzig Heften. Die Jahre
    1846–1871 enthaltend. Berlin, 1873. 8vo.

CHEMISCH-TECHNISCHES REPERTORIUM herausgegeben von Emil Jacob-
    sen. General-Register zu Jahrgang I–V (1862–66). Berlin, 1867.
    8vo.
  General-Register zu Jahrgang VI–X (1867–71). Berlin, 1873.
  The same, Jahrgang XI–XV (1872–76). Berlin, 1879.
  The same, Jahrgang XVI–XX (1877–81). Berlin, 1884.
  The same, Jahrgang XXI–XXV (1882–86). Berlin, 1889.

CHEVALLIER, A., AND M. P. DE MÈZE.
  Fastes de · la pharmacie française ; exposé des travaux scientifiques
    publiés depuis quarante années par les pharmaciens français avec
    l'indication des ouvrages dans lesquels ces travaux ont été con-
    signés. Suivi d'un dictionnaire des résultats obtenus de l'analyse
    des substances végétales ; précédé d'un annuaire indiquant mois
    par mois l'époque ordinaire de la récolte des plantes indigènes, et
    présentant par ordre alphabétique, le nom des Pharmaciens dont
    les travaux ont enrichi la science. Paris, 1830. pp. xvi-244, 8vo.
      This bibliography covers the period 1792 to 1830, and is confined to France.

CHOULANT, LUDOVICUS.
  Bibliotheca medico-historica sive Catalogus librorum historicorum de re
    medica et scientia naturali systematicus. Lipsiæ, 1842. pp. x–269,
    8vo.
          First Supplement : Additamenta ad Lud Choulanti Biblio-
            thecam Medico-Historicam, edidit Julius Rosenbaum.
            Halis Saxonum, 1842. pp. x–83, 8vo.
          Second Supplement : Specimen secundum. Halis Saxonum,
            1847. pp. xi–166, 8vo.
      Contains, besides works bearing more or less on medical-chemistry, a special
          section on chemistry (pp. 126–129), most of which material has been
          incorporated in the present volume.

CLARKE, FRANK WIGGLESWORTH.
    Constants of Nature. *See, in Section II.*

CLARKE, FRANK WIGGLESWORTH.
  Bibliography of the Preservation of Wood from Decay. Boston City
    Documents, No. 100, 1873 ; and Sci. Proc. Ohio Mechanics Inst.,
    Vol. II, No. 1, March, 1883, p. 20.

COLBY, ALBERT L.

Bibliography on the Adulteration and Examination of Sugars, Syrups, Molasses, Glucose, Confectionery, Honey and Soda-water Syrups. Second Annual Report of the State Board of Health of New York. Albany, 1882. pp. 613–618, 8vo.

Catalogue of the Literature of the Chemistry of Food and Drugs. Second Annual Report of the State Board of Health of New York. Albany, 1882. pp. 705–715, 8vo.

COMPTES RENDUS HEBDOMADAIRES DES SÉANCES DE L'ACADÉMIE DES SCIENCES. Table générale. Tomes I–XXXI, 3 Août, 1835, à 30 Décembre, 1850. Paris, 1853. 4to.

COOPER, WILLIAM.

A Catalogue of Chymicall Books, in 3 parts, collected by William Cooper. London, 1675. 12mo.

> Commonly bound with: The Philosophical Epitaph of W. C. . . . A Brief of the Golden Calf. . . . The Golden Ass well managed. . . . Jehior. London, 1673.

> Parts I and II contain alphabetical catalogue of independent works ; part III is an index to first ten years of Philosophical Transactions. Confined to English literature.

CROASDALE, STUART.

[Bibliography of] the Electrolytic Assay of Copper. *In* J. Anal. Appl. Chem., v, p. 133 (Mch., 1891) and p. 184 (Apr., 1891).

DANTÈS, A.

Tables biographiques et bibliographiques des sciences des lettres et des arts indiquant les œuvres principales des hommes les plus connus en tous pays et à toutes les époques, avec mention des éditions les plus estimées. Paris, 1866. pp. vii–646–[ii].

> This author has adopted the atrocious plan of translating into French the titles of works written in English, German, and other languages, but without stating the fact ; thus his work becomes a snare instead of a help to bibliographers and students of biography. Priestley is said to have made numerous discoveries, but no mention is made of oxygen ; and " his independent character and religious opinions obliged him to retire to America," no mention being made of the terrible Birmingham riots of July, 1791 !

DENIS, FERDINAND, P. PINÇON ET DE MARTONNE.

Nouveau manuel de bibliographie universelle. Manuels-Roret. Paris, 1857. pp. xi–706, 8vo.

> Under the heading " Chimie " this contains a very unsatisfactory bibliography of 51 numbers almost exclusively confined to works written in French or translated into the same language. Under the heading

"Alchimie," a list of 12 titles and not always correct, the authors gravely say Van Helmont was the last of the alchemists! *Compare Section VI.*

DUDLEY, WILLIAM L.

Index to the Literature of Amalgams. In his Vice-Presidential Address to the American Association for the Advancement of Science, at Toronto. Proceedings A. A. A. S. for 1889. Salem, 1890. pp. 161–171, 8vo.

DU FRESNOY, LENGLET.

See Lenglet du Fresnoy.

EDER, JOSEF MARIA.

For a bibliography of the History of Photography, *see, in Section III*, Eder, J. M., Geschichte der Photochemie und Photographie.

EGER, GUSTAV.

Technologisches Wörterbuch in englischer und deutscher Sprache. Durchgesehen und vermehrt von Otto Brandes, Chemiker. Braunschweig, 188*. 2 parts, roy. 8vo.

ELLIOTT, ARTHUR H.

Literature relating to the Testing and Safety of Kerosene Oil. Report on the Methods and Apparatus for Testing Inflammable Oils. Second Annual Report of the State Board of Health of New York. Albany, 1882. pp. 484–491, 8vo.

ENGELHARDT, F. E.

Literature of Wines, Liquors and Malt Liquors. Second Annual Report of the State Board of Health of New York. Albany, 1882. pp. 628–630, 664–667, 679–686, 8vo.

ENGELMANN, WILHELM.

Bibliotheca medico-chirurgica et pharmaceutico-chemica, oder Verzeichniss derjenigen medizinischen, chirurgischen, geburtshülflichen und pharmazeutisch-chemischen Bücher welche vom Jahre 1750 bis zur Mitte des Jahres 1837 in Deutschland erschienen sind. Zuerst herausgegeben von Theod. Christ. Friedr. Enslin, von neuem gänzlich umgearbeitet von W. E. Fünfte durchaus verbesserte und vermehrte Auflage. Leipzig, 1838. pp. 588, 8vo.

Bibliotheca pharmaceutico-chemica, oder Verzeichniss derjenigen pharmazeutisch-chemischen Bücher welche seit Mitte des vorigen Jahrhunderts bis zur Mitte des Jahres 1837 in Deutschland erschienen sind. Mit einem vollständigen Materienregister. (Ein besonderer Abdruck aus der Bibliotheca medico-chirurgica et pharmaceutico-chemica.) Leipzig, 1838. 8vo.

The second part, pp. 495–541, treats of chemistry and pharmacy.

ENGELMANN, WILHELM [Cont'd].

> First edition, 1816; fourth, 1825; sixth, to 1847, Leipzig, 1848, pp. viii–734. Supplement-Heft enthaltend die Literatur bis 1867, Leipzig, 1868. pp. 350, 8vo.
>
> > In the sixth edition chemistry is dropped.

ENSLIN, THEOD. CHRIST. FRIEDR.

> *See* Engelmann, Wilhelm.

ERSCH, J.

> Literatur der Natur- und Gewerbskunde, seit der Mitte des 13ten Jahrhunderts bis auf die neueste Zeit. Amsterdam, 1812.
>
> > Reprinted from Ersch's Handbuch der deutschen Literatur.

ERXLEBEN, JOHANN CHRISTIAN POLYCARP.

> Anfangsgründe der Chemie. Göttingen, 1775. pp. [xxxii]–[lii], 12mo.
>
> > Contains a classified bibliography numbering 135 titles. Throughout the work bibliography is not. neglected. For fuller details see the same in Section V.
>
> Physikalische Bibliothek. Göttingen, 1775–1779. 4 vols., 12mo.
>
> > Contains reviews and notices of books and papers on natural science, including some current chemistry.

FERGUSON, JOHN.

> Bibliographical Notes on Histories of Inventions and Books of Secrets. Three Parts. *Part I also under the title :* Notes on some Books of Technical Receipts, or so-called "Secrets." Read April 20, 1882. From Transactions of the Archæological Society of Glasgow. Glasgow, 1883. pp. 18, 8vo.
>
> *Part II.* Read January 18, 1883. From Transactions of the Glasgow Archæological Society. Glasgow, 1883. pp. 62, 8vo.
>
> *Part III.* Read December 18, 1884. One hundred copies reprinted from Transactions of the Archæological Society of Glasgow. Glasgow, 1885. pp. 42, sm. 4to.
>
> > Part II contains a bibliography of 160 titles, and was followed by a supplement bearing the title :
>
> Account of a copy of the First Edition of the Speculum Majus of Vincent de Beauvais, 1473. Read to the Archæological Society of Glasgow, December 18, 1884. Glasgow, 1885. pp. 25, sm. 4to. One hundred copies reprinted from the Transactions of the Glasgow Archæological Society.
>
> Some Early Treatises of Technological Chemistry. Read to the Philosophical Society of Glasgow, January 6, 1886. [From Transactions of the Philosophical Society of Glasgow.] Glasgow, 1888. pp. 34, 8vo.

FERGUSON, JOHN [Cont'd].

> This may be regarded as supplementary to Professor Ferguson's Biblio-
> graphical Notes on Histories of Inventions and Books of Secrets, and
> exhibits the critical skill, bibliographical accuracy and erudition charac-
> teristic of this author's valuable writings. *Compare, in Section III,*
> Ferguson, John.

FICTULD, HERMANN.

Der längst gewünschte und versprochene chymisch-philosophische
Probier-Stein auf welchem sowohl der wahrhafften hermetischen
Adeptorum als der verführerischen und betriegerischen Sophisten
Schrifften sind probiert und nach deren Werth dargestellt worden,
beschrieben in Zweyen Classen, darvon die Erste bereits herausgegangen, gegenwärtig aber von dem Authore von neuem übersehen,
corrigirt und von seinthero zu Händen gebrachten Authoren vermehret, die zweyte Class aber hinzu gefügt worden ist. Franck-
fort und Leipzig, bey Veraci Orientali Wahrheit und Ernst
Lugenfeind, 1753. pp. 170 and 171, 12mo.

> The first part contains biographical and bibliographical notes in 178 para-
> graphs, relating to *adepts.* The second part contains 182 paragraphs
> relating to *sophisters,* or pretended alchemists, according to Fictuld's
> views. The notes, like those in *this* work, are of unequal value. The
> first edition was published in 1740.

FRANK, OSKAR [Publisher].

Frank's Führer durch die chemische, chemisch-technische und pharmaceutische Literatur vom Jahre 1850-82 nebst einem alphabetisch-
systematisch geordneten Inhaltsverzeichniss dazu. Wien, 1883.
pp. xxx-244, 16mo.

> A useful trade catalogue.

FRIEDLÄNDER, R. [Editor].
> *See :* Naturæ novitates.

FUCHS, GEORG FRIEDRICH CHRISTIAN.

Repertorium der chemischen Litteratur von 494 vor Christi Geburt bis
1806, in chronologischer Ordnung aufgestellt. Jena und Leipzig,
1806-11. 2 vols., 8vo. Vol. I, pp. viii-648 ; vol. II, pp. vi-694.

> This work is highly praised by Petzholdt as exceedingly compendious and
> carefully prepared. It contains not only independent works, but also
> articles from periodical literature to which are added numerous bio-
> graphical and literary notes. Actually it extends only to 1799 inclusive,
> the third volume, 1800 to 1806, never appeared. Only the first volume
> contains an index. Published anonymously.

Versuch einer Uebersicht der chymischen Litteratur und ihrer
Brangschen. Altenburg, 1785. pp. 143, 12mo.

FUCHS, GEORG FRIEDRICH CHRISTIAN [Cont'd].

> Contains a list of journals and a catalogue of independent works and journal-articles arranged under authors' names alphabetically. Biographies of authors.

GALLOUPE, FRANCIS E.

An Index to Engineering Periodicals, 1883 to 1887 inclusive. Boston and New York, 1888. pp. vi–294, 12mo.

> This work, of special value to the engineer, contains a subject-index alphabetically arranged to eighteen periodicals (all in the English language), comprising about 10,000 references, dealing chiefly with Engineering, Railroads, and Manufactures ; Chemical subjects receive attention incidentally, and there is a special heading for Chemistry. The volume is clearly printed, compact and bound.

GATTERER, CHRISTOPH WILHELM JACOB.

Verzeichniss der vornehmsten Schriftsteller über das Bergwerkswesen. Göttingen, 1785. 2 parts, 8vo.

GRACKLAUER, O. [Editor].

Verzeichniss sämmtlicher Schriften über Färberei, Druckerei, Bleicherei, Appretur und Farbenfabrikation welche von 1865–1881 im deutschen Buchhandel erschienen sind.   Gracklauer's Fachkatalog, No. 19. Leipzig, 1881. pp. 12, 8vo.

Verzeichniss sämmtlicher Schriften über Gewerbechemie und chemisch-technische Rezeptbücher welche von 1865–1881 im deutschen Buchhandel erschienen sind. Gracklauer's Fachkatalog, No. 18. Leipzig, 1881. pp. 12, 8vo.

Verzeichniss sämmtlicher Schriften über Beleuchtungswesen, Petroleum, Gasbeleuchtung, Elektrische Beleuchtung welche von 1865–1881 im deutschen Buchhandel erschienen sind. Gracklauer's Fachkatalog No. 15. Leipzig, 1881. pp. 7, 8vo.

Verzeichniss sämmtlicher Schriften über Bierbrauerei, Böttcherei, Branntweinbrennerei, Destillation, Hefenfabrikation und Gährung welche von 1865–1881 im deutschen Buchhandel erschienen sind. Gracklauer's Fachkatalog, No. 16. Leipzig, 1881. pp. 27, 8vo.

Verzeichniss sämmtlicher Schriften über Gewerbe-Literatur und Industrie, Handwerks-, Innungs- und Patentwesen, Kunsthandwerk, Gewerbe-Austellungen, etc., welche von 1865–1881 im deutschen Buchhandel erschienen sind. Gracklauer's Fachkatalog, No. 14. Leipzig, 1881. 4 parts. pp. 264, 8vo.

> Classified minutely.

Verzeichniss sämmtlicher Schriften über Ziegel- und Röhrenfabrikation, Kalk, Asphalt und Cemente welche von 1865–1881 im deutschen

GRACKLAUER, O. [Cont'd].

Buchhandel erschienen sind.   Gracklauer s Fachkatalog, No. 32. Leipzig, 1881. pp. 7, 8vo.

Verzeichniss sämmtlicher Schriften über Zuckerfabrikation welche von 1865–1881 im deutschen Buchhandel erschienen sind.   Gracklauer's Fachkatalog, No. 33.   Leipzig, 1881. pp. 6, 8vo.

HAFERKORN, H. E.

Handy Lists of Technical Literature.   Reference catalogue of books printed in English from 1880 to 1888 inclusive ; to which is added a select list of books printed before 1880 and still kept on publishers' and jobbers' lists.   Part IV, Mines and mining, assaying, metallurgy, analytical chemistry, minerals and mineralogy, geology, palæontology, etc.   Including issues up to May, 1891, and a number of earlier books frequently met with in catalogues.   Together with a list of Periodicals and Annuals in these branches. Milwaukee, Wisconsin, 1891. pp. [vi]–87, 8vo.

HALLOCK, EDWARD J.

Bibliography of Starch-Sugar.   Appendix E to Report on Glucose prepared by the National Academy of Science in response to a request made by the Commissioner of Internal Revenue.   U. S. Internal Revenue.   Washington, D. C., 1884. pp. 44, 8vo.

Index to the Literature of Titanium, 1783–1876.   Extract from Annals of the New York Academy of Sciences.   Vol. I, Nos. 2 and 3. New York, 1877. pp. 22, 8vo.

HERMBSTÄDT, SIGISMUND FRIEDRICH.

Bibliothek der neuesten physisch-chemischen, metallurgischen, technologischen und pharmaceutischen Literatur.   Erster Band [in three parts].   Berlin, 1788–95. 4 vols., 8vo.

> This contains an analysis of the chief chemical publications in the years 1787–92. Vol. I, has a portrait of J. C. F. Meyer.
>
> *Followed by* Vol. v, *under the title :* Annalen der chemischen Literatur herausgegeben von Friedrich Wolff.   Berlin, 1803. 3 parts.

HOFFMANN, CARL AUGUST.

Anzeige aller über [mineral] Wasser erschienen Schriften. *In* Systematischer Uebersicht. *See, in Section II*, Hoffmann, C. A.

HOLTROP, L. S. A.

Bibliotheek voor genees-, heel-, schei-, en artsenijmengkunde, of alphabetische naamlijst van alle boeken, geschriften en stukken, betreffende ontleedkunde, geneeskunde, heelkunde, verloskunde, artsenijmeng-

HOLTROP, L. S. A. [Cont'd].

kunde, scheikunde, kruidkunde, natuurkunde en vee-artsenijkunde, welke in Nederland verschenen zijn, van het jaar 1790 tot 1840 ; zoowel afzonderlijk uitgegeven, als in tijdschriften verspreid, of in de werken der onderscheidene Genootschappen opgenomen ; voorzien van 1° een latijnsch systematisch zaakregister ; 2° eene hollandsch-latijnsche woordenlijst ; 3° eene alphabetische naam-lijst van vertalers en schrijvers van bijvoegselen en aanteckeningen op werken van anderen. Te 's Gravenhage. 1842. 8vo.

Also with the title : Bibliotheca medico-chirurgica et pharmaceutico-chemica, seu catalogus alphabeticus omnium librorum, disserta-tionum, etc., ad anatomiam, artem medicam, chirurgicam, obstetri-ciam, pharmaceuticam, chemicam, botanicam, physico-medicam et veterinariam pertinentium, et in Belgio ab anno 1790 ad annum 1840 editorum, cum separatim tum in diariis criticis et actis societatum. Accedunt : 1° Index systematicus latinus ; 2° Index belgico-latinus ; 3° Index alphabeticus nominum eorum qui ver-siones dederunt, vel aliorum opera annotationibus suis illustraver-unt. Hagæ-Comitis, 1842.

> So long a title needs no remark other than deprecation.

HUGUENY, M. F.

> Bibliographie relative aux eaux potables. *See, in Section V*, Hugueny. Recherches sur la composition chimique . . . qu'on doit exiger des eaux potables. 1865.

INDEX-CATALOGUE OF THE LIBRARY OF THE SURGEON-GENERAL'S OFFICE, UNITED STATES ARMY. Authors and Subjects. Washington, 1881–91. 12 vols. [to *Shut*], 4to. *In progress.*

> The King of Catalogues on special subjects. Invaluable for medical chem-istry. Contains under CHEMISTRY, in Vol. II, (1882), a classified list of works.

INDEX TO FOREIGN SCIENTIFIC PERIODICALS CONTAINED IN THE PATENT OFFICE LIBRARY. *See* Patent Office, London.

JAHRBUCH DER CHEMIE UND PHYSIK. Sachregister zu den drei Jahr-gängen 1823, 1824 und 1825 oder Band VII–XV. Halle, 1826. 8vo.

JACOB, J.

> Notices biographiques et bibliographiques sur les principaux chimistes. . . . *See, in Section V*, Jacob, J., Traité élémentaire de chimie.

JAHRBUCH DER ERFINDUNGEN. Alphabetisches Sach- und Namen-Register zu dem I bis V Jahrgange. Leipzig, 1869. 12mo.

JAHRESBERICHT ÜBER DIE FORTSCHRITTE DER CHEMISCHEN TECHNOLO-
GIE. Herausgegeben von Joh. Rud. Wagner. General-Register
über Band I bis x bearbeitet von Fr. Gottschalk. Leipzig, 1866.
The same, Band XI bis XX. Leipzig, 1876.

JAHRESBERICHT ÜBER DIE FORTSCHRITTE DER REINEN, PHARMACEUTISCHEN
UND TECHNISCHEN CHEMIE, PHYSIK, MINERALOGIE UND GEOLOGIE.
Herausgegeben von Justus Liebig und Hermann Kopp. Register
zu den Berichten für 1847 bis 1856. Giessen, 1858. pp. iv–347.
8vo.

JAHRESBERICHT ÜBER DIE FORTSCHRITTE DER AGRICULTURCHEMIE. General-Register über Jahrg. I–XX (1858–1877). Unter Mitwirkung
von E. von Gerichten, C. Krauch, E. von Raumer, W. Rössler und
O. H. Will, herausgegeben von A. Hilger. Berlin, 1879. 8vo.

JAHRESBERICHT ÜBER DIE FORTSCHRITTE DER CHEMIE UND VERWANDTER
THEILE ANDERER WISSENSCHAFTEN. Herausgegeben von Hermann Kopp und Heinrich Will. Register zu den Berichten für
1857 bis 1866. Giessen, 1868. pp. IV–543.
The same, für 1867 bis 1876. Giessen, 1880. pp. iv–727.

JAHRESBERICHT ÜBER DIE FORTSCHRITTE DER PHYSISCHEN WISSENSCHAF-
TEN, DER CHEMIE UND MINERALOGIE, von Jacob Berzelius. Vollständiges Sach- und Namen-Register. Tübingen, 1847. pp. 180,
8vo.

JAHRESBERICHT ÜBER DIE LEISTUNGEN DER CHEMISCHEN TECHNOLOGIE,
von R. von Wagner. General-Register zu Band XXI–XXX, herausgegeben von F. Fischer. Leipzig, 1889.

JAHRESBERICHTE ÜBER DIE UNTERSUCHUNGEN UND FORTSCHRITTE AUF
DEM GESAMMTGEBIETE DER ZUCKERFABRIKATION. Alphabetisches
Sach-Register zum I–XII Jahrgange, bearbeitet von K. Stammer.
Braunschweig, 1873. 8vo.
The same, Alphabetisches Sach-Register zum XIII bis XXVI Jahrgange
in Band XXVI, 1886.

JAHRESBERICHTE ÜBER DIE FORTSCHRITTE DER THIER-CHEMIE. Sach-
und Autoren-Register über die ersten zehn Jahrgänge von Maty's
Jahresbericht . . . Bearbeitet von Rudolph Andreasch. Wiesbaden, 1881. 8vo.

JOHN, JOHANN FRIEDRICH.
    Chemische Tabellen der Pflanzenanalysen. *See in Section II.*
    Chemische Tabellen des Thierreichs. *See in Section II.*

JOURNAL DE CHIMIE MÉDICALE DE PHARMACIE ET DE TOXICOLOGIE. Table générale des matières et des auteurs de la deuxième Série de 1835–1844. Paris, 1845. 8vo.

JOURNAL DE PHARMACIE ET DES SCIENCES ACCESSOIRES. Table analytique des auteurs cités et des matières contenus dans les tomes XVII-XXVII (1831–1841) du Journal [etc.]. Paris, 1842. 8vo.
    See Bulletin de pharmacie.

JOURNAL FÜR CHEMIE UND PHYSIK. [Edited by J. S. C. Schweigger.] Autoren- und Sach-Register zu sämmtlichen neun und sechzig Bänden. Jahrgänge 1811–1833. Bearbeitet von G. C. Wittstein. München, 1848. pp. 299, 8vo.

JOURNAL FÜR PRACTISCHE CHEMIE. Herausgegeben von Otto Linné Erdmann und Richard Felix Marchand. Sach- und Namen-Register zu Band I–XXX. Leipzig, 1844. pp. 162, 8vo.
The same, Band XXXI–LX. Leipzig, 1854. pp. 113, 8vo.
The same, bearbeitet von Friedrich Gottschalk. Band 61–90. Leipzig, 1865. pp. 276, 8vo.
The same, Band 91–108. Leipzig, 1871. pp. 288, 8vo.

KÄSTNER, ABRAHAM GOTTHELF.
    See Rohr, J. B. von.

KASTNER, CARL WILHELM GOTTLOB.
    Einleitung in die neuere Chemie.
    Contains a bibliography. *See in Section V.*

KEBLER, LYMAN F.
Index to the Literature on the Estimation of Nitrogen by Kjeldahl's Method and its modifications. *In* J. Anal. Appl. Chem., v, 260, May, 1891.
An Index to the Literature on the Estimation of Nitrogen by all other Methods. [See the preceding on Kjeldahl's Methods.] *In* J. Anal. Appl. Chem., v, 264, May, 1891.

KERCHA, A. [*or* KERSCHA, A.].
Pantobiblion, Revue bibliographique internationale des sciences poly-techniques. 12 livraisons par an. St. Pétersbourg. *n. d.* [1891]. Square 8vo.
Also with the German title : Internationale Bibliographie der polytech-nischen Wissenschaften. Monatliche Uebersicht der auf diesen Gebieten neu erschienenen Buch- und Journal-Literatur.
    Of this ambitious work only one part was published.

Kerl, Bruno.

Repertorium der technischen Literatur. Neue Folge die Jahre 1854
bis einschliesslich 1868 umfassend. Im Auftrage des königlich
preussischen Ministeriums für Handel, Gewerbe und öffentliche
Arbeiten. Leipzig, 1871–73. 2 vols., roy. 8vo.

The same, for the years 1869–1873. Leipzig, 1876–78. 2 vols., roy. 8vo.

The same, for 1874, 1875, 1876, 1877, 1878. Leipzig, 1875–1879. 5 vols.,
roy. 8vo. 1 vol. each year.

> An alphabetical index to the technological articles in about 140 journals.
> Contains much chemical matter. Invaluable. Compare Schubarth, of
> which this is a continuation.

Kloss, Georg.

Bibliographie der Freimaurerei und der mit ihr in Verbindung gesetzten
geheimen Gesellschaften. Frankfurt am Main, 1844. pp. xiv–
430–6, 8vo.

> Contains works on Rosenkreutzer, Alchemy, etc.

Korn, Wilhelm Gottlieb [Editor].

Taschenbuch für die Liebhaber der medicinischen Wissenschaften.
Oder : vollständiges Verzeichniss von medicinischen, botanischen,
physikalischen, anatomischen, chymischen und alchymistischen
Büchern so für beygesetzte billige Preise zu haben sind bey W. G.
K., Buchhändler zu Breslau. n. p. 1787. pp. 311, 12mo.

> The section devoted to chemistry and alchemy occupies pages 282–311.
> Useful in determining prices at the date given.

Krebel, R.

Russlands naturhistorische und medicinische Literatur. Schriften und
Abhandlungen in nicht-russischer Sprache. Jena, 1847. pp. vi–
220, 8vo.

> The first division includes Physics, Chemistry and Pharmacy ; pp. 3–21.

Krieger, J. Ch.

Handbuch der Literatur der Gewerbskunde in alphabetischer Ordnung.
Marburg, 1812 and 1829. 2 vols.

Supplement. Marburg, 1822.

> Contains the literature up to 1820.

Krünitz, J. G.

Verzeichniss der vornehmsten Schriften von der Electricität und den
elektrischen Kuren. Leipzig, 1769. 8vo.

Kukula, Richard.

Allgemeiner deutscher Hochschulen-Almanach. Wien, 1888. pp. vi–
1,000.

KUKULA, RICHARD [Cont'd].

> This carefully edited work contains bibliographies of all the living Professors
> and teachers in German and Austrian Universities and High-schools and
> in every branch of knowledge. Though not confined to Science, it
> forms in part a supplement to Poggendorff's Handwörterbuch, so far
> as Germany and Austria are concerned.

LACROIX, EUGÈNE.

Bibliographie des ingénieurs, des architectes, des chefs d'usines industrielles, des élèves des écoles polytechniques et professionelles,
et des agriculteurs. Paris, 1863–67. 3 vols., sm. 4to.

> The volumes are denoted as "première," "deuxième," and "troisième"
> "série." Chemistry is not neglected.

LEEDS, ALBERT R.

Index to the Literature of Peroxide of Hydrogen, 1818–1878. Extract
from Annals of the New York Academy of Sciences, vol. I, No. 13.
New York, 1880. pp. 11, 8vo.

Supplement, 1879–1883. Idem., vol. III, p. 153. New York, 1884. pp.
3, 8vo.

Index to the Literature of Ozone, 1785–1879. Annals of the New
York Academy of Sciences, vol. I, No. 12. New York, 1880. pp.
32, 8vo.

Supplement, 1879–1883. Idem., vol. III, p. 137. New York, 1884. pp.
16, 8vo.

> *Cf., in Section III*, Dachauer ; *also* Moigno.

LEUCHS, JOHANN CARL.

Beschreibung der färbenden und farbigen Körper. Nürnberg, 1825.

> Contains a bibliography of dyeing, etc. For full title, *see in Section V.*

Polytechnische Bücherkunde, oder beurtheilendes Verzeichniss der
vorzüglichsten Bücher über Chemie, Technologie, Fabrikwissenschaft, Mechanik und einzelne Gewerbszweige. Ein Hilfsbuch
für Privatpersonen und Buchhändler zur Kenntniss und Auswahl
zu kaufender Bücher. Nürnberg, 1829. 8vo.

> Second edition, 183–. Third edition, 1841. Fourth edition,
> 1846. 8vo.

Vollständige Essigfabrikation. Nürnberg, 1840.

> Contains a bibliography of vinegar manufacture. For full title, *see in
> Section V.*

LEUPOLD, JACOB.

Prodromus bibliothecæ metallicæ. Leipzig, *n. d.* Second edition,
1726, 8vo. Wolfenbüttel, 1732. 8vo.

LING, ROTH, H.

A Guide to the Literature of Sugar. A book of reference for chemists, botanists, librarians, manufacturers and planters, with comprehensive subject-index. London, 1890. pp. xvi–159, 8vo.

LOVE, EDWARD G.

Bibliography on the Adulteration and Examination of Cereals and the Products and Accessories of Flour and Bread Foods. Second Annual Report of the State Board of Health of New York. Albany, 1882. pp. 584–589, 8vo.

MAFFEI, EUGENIO, Y RAMÓN RUA FIGUEROA.

Apuntes para una biblioteca española de libros, folletos y artículos, impresos y manuscritos, relativos al conocimiento y explotación de la riquezas minerales y á las ciencias auxiliares. Acompañados de reseñas biográficas y de un ligero resumen de la mayor parte de las obras que se citan. Madrid, 1872. 2 vols., 8vo. Vol. I, pp. lxx–529; vol. II, pp. 693.

> This invaluable bibliography is on an admirable plan; the authors are alphabetically arranged, each name being followed by a brief biography. His works are then indicated with full titles, and of each a summary or an analysis is given in smaller type. The work comprises mineralogy, geology, chemistry, metallurgy and kindred topics. Its clear typography and excellent arrangement make it a model. A new edition brought down to date would be welcome. A full subject-index makes it practically useful.

MARTIN, EDWARD W.

Bibliography of Milk. Report on Milk; Second Annual Report of the New York State Dairy Commissioner, 1886. pp. 156–170, 8vo.

MAUMENÉ, E. J.

Traité de la fabrication du sucre. Paris, 1876.

> Contains a bibliography of sugar.

MEULEN, R. VAN DER.

Bibliografie der technische Kunsten en Wetenschappen 1850–1875. Boeken, Plaatwerken en Kaarten in Nederland verschenen over en met betrekking tot : Fabrieks- en Handwerks-nijverheid, de Bouwkundige en Ingenieurs-Wetenschappen, Mechanica, Stoomwegen, Spoorwegen, Telegrafie, enz. Met inhouds opgaaf der voornaamste periodieken, benevens een uitvoerig alfabetisch zaakregister. Amsterdam, 1876. pp. iv–226, 8vo.

> Contains some works on chemical technology.

MOURICK, REINIER VAN.
Naamrol der medicinale en chirurgiale, chemische en natuurkundige
schrijvers. Amsterdam, 1782. 4to.

MUNROE, CHARLES E.
Index to the Literature of Explosives. Part I. Baltimore, 1886. pp.
42, 8vo.
Part II in press (1892).

NASSE, WILHELM.
. . . Uebersicht der physikalisch-chemischen Literatur, . . . 1809.
See, in Section III, Nasse, Wilhelm.

NATURÆ NOVITATES. Bibliographie neuer Erscheinungen aller Länder
auf dem Gebiete der Naturgeschichte und der exacten Wissenschaften. Herausgegeben von R. Friedländer und Sohn. Berlin,
I–XIV Jahrgang, 1879–92. 8vo.

NEUESTE (DAS) UND NÜZLICHSTE DER ERFINDUNGEN, Entdeckungen
und Beobachtungen, besonders der Engländer, Franzosen und
Deutschen in der Chemie, Fabrikwissenschaft, Apothekerkunst,
Oekonomie und Warenkenntniss. Nürnberg, 1794–1828. 24 vols.
Vols. 12–24 also under the title : J. C. Leuchs' neuestes Handbuch für
Fabrikanten, Künstler und Handwerker.
Of vols. 17–24 a special edition bears the title : Technische Encyclo-
pædie. 1012 Abhandlungen über die Fabrikation der vorzüglich-
sten Gegenstände. 1835. 8 vols.
This work is followed by Leuchs' allgemeine polytechnische Zeitung
und Handels-Zeitung. Nürnberg, 1834–47. 14 vols.
See No. 112 of Bolton's Catalogue of Scientific and Technical Periodicals.

NOYES, ARTHUR A.
Index to the Literature of Butines and their Halogen Addition Products
(1863–1888). In Technological Quarterly, Boston, December,
1888. Published at the Massachusetts Institute of Technology.

OUVRAGES QUI ONT ÉTÉ ÉCRITS EN FRANÇOIS AVANT LA FIN DU SEIZIÈME
SIÈCLE, SUR LA CHIMIE. pp. 321–382, volume XXV, Mélanges tirés
d'une grande bibliothèque. [Edited by A. G. Coutant d'Orville
and the Marquis de Paulmy (Marc Antoine René de Voyer
d'Argenson)]. Paris, 1782. 12mo.
A sketch of chemical literature confined to works written in French ; contains some curious items, and includes alchemy.

PANTOBIBLION, REVUE BIBLIOGRAPHIQUE.
See Kercha, A.

PATENT OFFICE, LONDON.
 Index to Foreign Scientific Periodicals contained in the Patent Office
  Library. Printed and published by order of the Commissioners of
  Patents. London, 1867–1876. 7 vols., roy. 8vo.

   This Index covers the period from June, 1866, to Dec., 1872. Each volume
     contains an Index of Authors and one of Subjects.

 List of Scientific and other Periodicals, Transactions of Scientific
  Societies and British Colonial, and Foreign Patents for Inventions,
  Designs and Trade Marks, in the Free Public Library of the Patent
  Office. London, January, 1861. 4to.
 Other editions: April, 1861 ; July, 1861 ; October, 1861 ; January,
  1862 ; April, 1862 ; July, 1862 ; October, 1862 ; January, 1863 ;
  April, 1863 ; July, 1863 ; October, 1863 ; January, 1865 ; July,
  1865 ; January, 1867 ; January, 1870 ; January, 1874 ; July, 1876 ;
  June, 1878 ; May, 1880 ; September, 1882 ; July, 1890.

   Communicated by the Librarian of the Patent Office, Herbert John Allison,
     Esq. Useful checklists of an extensive series of periodicals, including
     pure and applied chemistry.

PAULY, ALPHONSE.
 Bibliographie des sciences médicales—Bibliographie—Biographie—Histoire—Epidémies—Topographies—Endémies. Dédié à l'Association générale des médecins de France. Paris, 1874. pp. xx and
  columns 1758 and Index, roy. 8vo.

PECKHAM, S. F.
 Bibliography of Petroleum. Report on the Production, Technology
  and Uses of Petroleum and its Products. Report of the Tenth
  Census of the United States, vol. x, 1884. pp. 281–301, 4to.

PERRY, NELSON W.
 Bibliography of the Metal Iridium. In W. L. Dudley's paper on
  Iridium, published in Mineral Resources of the United States,
  calendar years 1883 and 1884. Washington, 1885. 8vo.

PFINGSTEN, J. H.
    Bibliothek ausländischer Chemisten.
    See, in Section V, Pfingsten, J. H.

PHARMACEUTICAL JOURNAL. Index to fifteen volumes of the P. J. [vols
  I–xv]. London, 1857. pp. 202, 8vo.
 Index to twelve volumes. Vol. XVI, Old Series (1856), to vol. IX,
  Second Series (1868). London, 1869. pp. 155, 8vo.

PHARMACEUTICAL JOURNAL AND TRANSACTIONS. General Index to Ten Volumes. Second Series vols. X and XI, Third Series vols. I to VIII, July, 1868, to June, 1878. London, 1880. pp. 131, sq. 8vo.

PHARMACEUTICAL SOCIETY OF GREAT BRITAIN. Catalogue of the Library of the P. S. Sixth edition. Compiled by John William Knapman. *Appended ;* Catalogue of the Library of the North British Branch. London, 1889. pp. 580, 8vo.

> This clearly printed, carefully edited dictionary catalogue does its author and the Society much credit.

PICATOSTE Y RODRIGUEZ, F.
Apuntes para una biblioteca científica española del siglo XVI. Estudios biográficos y bibliográficos de ciencias exactas, físicas y naturales y sus inmediatas aplicaciones. Madrid, 1891. 4to.

POGGENDORFF, JOHANN CHRISTIAN.
Biographisch-literarisches Handwörterbuch zur Geschichte der exacten Wissenschaften, enthaltend Nachweisungen über Lebensverhältnisse und Leistungen von Mathematikern, Astronomen, Physikern, Chemikern, Mineralogen, Geologen u. s. w., aller Völker und Zeiten. Leipzig, 1858–63. 2 vols. (vol. I, 1584 col., vol. II, 1468 col.), roy. 8vo.

> Invaluable as a work of reference. Abounds in information concerning chemists of every age and nation. Has been my constant companion while compiling this work.

Lebenslinien zur Geschichte der exacten Wissenschaften seit Wiederherstellung derselben. Berlin, 1853. 14 pp. and 3 tables, 4to.

> The tables are after the plan of Priestley's Chart of Biography.

POISSON, ALBERT.
Bibliographie alchimique du XIXe siècle.
See, in Section VI, Poisson, Albert, Collection [etc.].

POLYTECHNISCHE BIBLIOTHEK. Monatliches Verzeichniss der in Deutschland und dem Auslande neu erschienenen Werke aus den Fächern der Mathematik und Astronomie, der Physik und Chemie, der Mechanik und des Maschinen-Baues, der Baukunst und Ingenieurwissenschaft, des Berg- und Hüttenwesens (der Mineralogie und Geologie), mit Inhaltsangabe der wichtigsten Fachzeitschriften. Jahrgang 1–26, 1866–1891. Leipzig, 1866–91. 26 vols., 8vo. *In progress.*

POLYTECHNISCHE BÜCHER-KUNDE, oder beurtheilendes Verzeichniss der vorzüglichsten Bücher über Chemie, Technologie, Fabrikwissenschaft, Mechanik und einzelne Gewerbszweige. Ein Hülfsbuch

POLYTECHNISCHE BÜCHER-KUNDE, etc. [Cont'd].

für Privatpersonen und Buchhändler zur Kenntniss und Auswahl zu kaufender Bücher. Dritte ganz umgearbeitete Ausgabe. Nürnberg, 1841. pp. viii–224, 12mo.

> In this practical work, issued by C. Leuchs and Company, the titles are arranged under subjects, the latter being alphabetical and minutely detailed. Chemistry proper occupies pages 43 to 52. The preface to the first edition is dated Nürnberg, 1829.

PREYER, W.

Die Blausäure physiologisch untersucht. Bonn, 1870. pp. iv–168, 8vo.

> Contains a bibliography of the subject chronologically arranged from J. Schaub, 1802, to J. G. Wormley, 1869.

PRUGGMAYR, MARTIN MAXIMILIAN.

> For a list of " authors and books that ought to be read by a student of hermetic philosophy," *see, in Section VI*, Chapter III of Pruggmayr's Philoso-phische Untersuchung.

R., G.

Bibliotheca rerum metallicarum. Verzeichniss der bis Mitte 1856 in Deutschland über Bergbau, Hütten- und Salinenwesen und ver-wandten Zweige erschienenen Bücher, Karten und Ansichten. Mit Sachregister. Zweite Auflage. Eisleben, 1857. pp. xxxii–164, 8vo.

> Supplement covering period from July, 1856–64. Eisleben, 1864. pp. xxiv–110, 8vo.

RAMMELSBERG, CARL FRIEDRICH.

> Verzeichniss der wichtigsten Quellen für das Studium der theoretischen Chemie und Stöchiometrie.
>
> *See, in Section V*, Lehrbuch der Stöchiometrie, etc. 1842.

REGNIER, L. R.

> Index bibliographique.
>
> *See, in Section V*, Regnier, L. R., L'intoxication chronique par la morphine.

REINHOLD, JOHANN CHRISTIAN LEOPOLD.

> De Galvanismo.
>
> *See, in Section III*, Reinhold, J. C. L.

REPERTORIUM FÜR DIE PHARMACIE. [Edited by J. A. Buchner.] Auto-ren- und Sach-Register zu der 1. und 2. Reihe des Buchner'schen Repertoriums für die Pharmacie. Jahrgang 1815–48 oder 100 Bände. München, 1848. 3 vols., 12mo.

REPORTS [I to X] OF THE COMMITTEE ON INDEXING CHEMICAL LITERA-
TURE. *In* Proceedings of the American Association for the
Advancement of Science. Vols. XXXII (1883) to XLI (1892).
Salem, 1883–1893. 8vo.

> These Reports contain lists of Indexes to Chemical Literature, and other
> bibliographical data. Edited by H. Carrington Bolton, Chairman.

REUSS, JEREMIAS DAVID.
Repertorium commentationum a societatibus litterariis editarum secundum disciplinarum ordinem digessit J. D. Reuss. Göttingæ, 1803.
4to. (Scientia Naturalis. Chemia, etc., vol. III.)

> An exceedingly useful work, compiled with great diligence. Comprises 16
> vols., of which the third volume of the division of natural science is
> devoted to chemistry and metallurgy (1665–1800). The whole work
> forms a proper introduction to the "Catalogue of Scientific Papers,"
> published by the Royal Society, which covers the years 1800 to 1883.

ROCKWELL, GEORGE JEWETT.
Index to the Literature of Vanadium. Extract from Annals of the New
York Academy of Sciences. Vol. I, No. 5. New York, 1877.
pp. 13, 8vo.

ROHR, JULIUS BERNHARD VON.
Bibliotheca physica curiosa. Leipzig, 1724. 8vo.
Zusätze und Verbesserungen von Abr. Gotthelf Kästner.
Leipzig, 1754. 8vo.

ROSENBAUM, JULIUS.
See Choulant, L., Bibliotheca . . . Additamenta.

ROSENTHAL, G. E.
Literatur der Technologie, d. i. Verzeichniss der Bücher, Schriften und
Abhandlungen, welche von den Künsten, den Manufacturen und
Fabriken, der Handlung, den Handwerkern, und sonstigen Nahr-
ungszweigen, als auch von denen zum wissenschaftlichen Betriebe
derselben erforderlichen Kenntnissen aus dem Naturreiche, der
Mathematik, Physik und Chemie handeln. Nach alphabetischer
Folge des Jacobsonschen Wörterbuchs geordnet. Berlin und
Stettin, 1795. 4to.

ROTH-SCHOLTZ, FRIEDRICH.
Bibliotheca chemica; h. e. collectio auctorum fere omnium qui de
naturæ arcanis, re metallica et minerali, item de melioratione corporum artificiali, etc., hermetice scripserunt. Recensentur etiam

ROTH-SCHOLTZ, FRIEDRICH [Cont'd].

    diversæ librorum editiones aliaque hujus generis manuscripta hactenus inedita. Norimbergæ et Altdorfii. Fasciculus I, 1727. pp. 80, 8vo.

        Second edition, 1735.

        Portrait of Roth-Scholtz.

    Bibliotheca chemica Rothscholtziana darinnen man alle diejenigen Autores findet, die von dem Stein der Weisen, von Verwandlung der schlechten Metalle in bessere, von Bergwercken, von Mineralien, von Kräutern, von Thieren, von Gesund- und Sauer-Brunnen, von Warmen- und anderen Bädern, von der Haushaltungskunst, und was sonsten zu denen drey Reichen der Natur gehöret, geschrieben haben, und in der Roth-Scholtzischen Bibliothèque verhanden seynd. Samt einigen Lebens-Beschreibungen berühmter Philosophorum ans Licht gestellt. Stück II–v. Nürnberg und Altdorff, 1727–29 (1733 completed). pp. 49–328 (imperfect pagination), 8vo.

        Portraits of Flamel, J. F. Helvetius, O. Borrichius.

    [*Another copy.*] Bibliotheca chemica, oder Catalogus von Chymischen-Büchern, darinnen man alle diejenigen Autores findet, die von dem Stein der Weisen, von Verwandlung der schlechten Metalle in bessere, von Bergwercken, von Mineralien, von Kräutern, von Thieren, von Gesund- und Sauer-Brunnen, von Warmen- und anderen Bädern, von der Haushaltungskunst und was sonsten zu denen drey Reichen der Natur gehöret, geschrieben haben, und in der Roth-Scholtzischen Bibliothèque verhanden seyn. Samt einigen Lebens-Beschreibungen berühmter Philosophorum ans Licht gestellt. Nürnberg und Altdorff, 1727. Part I, pp. 14–48.

    Also with a title-page: Catalogus von chymischen Büchern welche in der Roth-Scholtzischen Bibliothèque verhanden seyn.

    Erstes Stück. Nürnberg und Altdorff, 1725. Frontispiece. Zweites Stück, 1727, pp. 49–96. Drittes Stück, 1727, pp. 97–172. Portrait of Nicolas Flammelle.

        Compiled by a Nuremberg bookseller familiar with alchemical literature. It contains the greater part of the bibliography of Borel, which had already become scarce, but it is incomplete, extending only to the letters *Hey*. Poor paper and worse typography, as well as a want of uniformity in style, make this volume wearisome. Roth-Scholtz enumerates elsewhere no less than 79 treatises on alchemy edited and published by himself.

    Veterum philosophorum sigla et imagines magicæ, cum catalogo variorum librorum magico-cabalistico-chemicorum. Herrnstadii, 1732.

ROYAL SOCIETY.

    Scientific Papers, Catalogue of. *See* Catalogue of Scientific Papers

RUPRECHT, RUDOLPH.

Bibliotheca chemica et pharmaceutica. Alphabetisches Verzeichniss der auf dem Gebiete der reinen, pharmaceutischen, physiologischen, und technischen Chemie in den Jahren 1858 bis Ende 1870 in Deutschland und im Auslande erschienenen Schriften. Göttingen, 1872. pp. 125, 8vo.

> A continuation of Zuchold's "Bibliotheca chemica," similarly arranged, but evidently collated with less care and completeness. *Cf.* Zuchold, Ernst A.

SADTLER, SAMUEL P.

> A Handbook of Industrial Organic Chemistry. This contains several special bibliographies ; *see* this title *in Section V.*

SCHEIBLE, J.

Bibliotheca magica, I. Catalog des antiquarischen Bücherlagers von J. S. in Stuttgart. Inhalt : Magie, Alchemie, Astrologie, Zauberei, Teufelsbeschwörung, Hexen- und Gespensterglaube, Schatzgräberei und Wünschelruthe, Geistererscheinungen, Träume und deren Auslegung, Physiognomik, Chiromantie, Wahrsagekunst, Cabbala, Visionen und Offenbarung, Magnetismus, Mesmerismus, Sympathische Heilkunde und andere geheime Künste aller Art. Catalog No. 45. [Stuttgart], 1873. pp. 97, 8vo.

> Contains 1,925 titles.

Bibliotheca magica, II. Catalog des antiquarischen Bücherlagers von J. S. in Stuttgart. Inhalt : Magie, Alchemie, Astrologie, Zauberei, Teufelsbeschwörung, Hexen- und Gespensterglaube, Schatzgräberei und Wünschelruthe, Geistererscheinungen, Träume und deren Auslegung, Physiognomik, Chiromantie, Wahrsagekunst, Cabbala, Visionen und Offenbarung, Magnetismus, Mesmerismus, etc.—Sciences occultes. Cabale. Magie. Apparitions. Demons. Possessions. Exorcismes. Sortilèges. Divinations par les songes, par les signes de la main, par les cartes, etc. Alchimie, médecine spagyrique ou chimique. Astrologie et choses analogues. Catalog No. 47. [Stuttgart], 1874. pp. 34, 8vo.

> This supplement to Bibliotheca magica I. contains titles from 1,926 to 2,661.

Bibliotheca magica et pneumatica. Catalog des antiquarischen Bücherlagers von J. S. in Stuttgart. Inhalt : Handschriften und Werke über Magie, Astrologie, Alchemie, Dämonologie, Zauberei, Teufelsbeschwörung, Hexen- und Gespensterglauben, Schatzgräberei und Wünschelruthe, Geistererscheinungen, Träume und deren Auslegung, Physiognomik, Geomantie, Chiromantie, Metoposcopie, Cabbala, Wahrsagekunst, Visionen und Offenbarung, Magnetismus,

SCHEIBLE, J. [Cont'd].

Mesmerismus, Naturgeheimnisse, Steganographie, egyptische und orientalische Arcanas, Spagyrische Heilkunst und wunderbare Dinge alle Art. [Stuttgart], 1868. No. I. pp. 120, 8vo.

> This trade catalogue of 2,531 titles is a veritable treasury of information.

SCHERER, ALEXANDER NICOLAUS.

Grundzüge der neuern chemischen Theorie. Jena, 1795. pp. xx–400. 8vo. With portrait of Lavoisier. *Contains:* Uebersicht der Literatur der neuern chemischen Theorie. pp. 297–384.

> A very valuable bibliography of the earlier period, systematically arranged, including articles published in journals.

Nachträge zu den Grundzügen der neuern chemischen Theorie. Nebst einigen Nachrichten von Lavoisier's Leben und einer tabellarischen Uebersicht der neuern chemischen Zeichen. Jena, 1796. pp. [xlviii]–574, 8vo.

> The sketch of Lavoisier is translated from that of Jerome Lalande in Magasin encyclopédique, vol. v., No. 18. The bibliography is continued in pp. 547–561.

SCHOTTE, F.

Repertorium der technischen, mathematischen und naturwissenschaftlichen Journal-Literatur. Unter Benutzung amtlicher Materialen mit Genehmigung des Königl. Preuss. Ministeriums für Handel, Gewerbe und öffentliche Arbeiten herausgegeben von F. S. Leipzig, 1869–71. 3 vols., 8vo.

SCHUBARTH, F.

Repertorium der technischen Literatur die Jahre 1823 bis einschliesslich 1853 umfassend. Zum Gebrauche der Königlich technischen Deputation für Gewerbe bearbeitet. Herausgegeben im Auftrage des Königlichen Ministeriums für Handel, Gewerbe und öffentliche Arbeiten. Berlin, 1856. pp. xvi-viii-1049, roy. 8vo.

> An index to the technological articles in over 200 journals for the period named. Arranged alphabetically and containing much chemical literature. Invaluable. Compare Kerl, Bruno, for the continuation.

SCHUBARTH, ERNST LUDWIG.

Literature [und Geschichte] der Chemie. pp. 1–17 of his Lehrbuch der theoretischen Chemie. Berlin, 1829.

> *See, in Section V*, Schubarth, E. L.,

SCOUTETTEN, H.

> A Bibliography of Ozone.
> *See, in Section V*, Scoutetten, L'Ozone. 1856. 8vo.

SCUDDER, SAMUEL H.

Catalogue of Scientific Serials of all Countries, including the Trans-actions of Learned Societies in the Natural, Physical and Mathe-matical Sciences. 1633–1876. Special Publications of Library of Harvard University. Cambridge, 1879. pp. xii–358, 8vo.

An indispensable key to the publications of learned societies.

SPIELMANN, JACOB REINBOLD.

See, in Section V, the author's Institutiones chemiæ.

The author includes in his bibliography Cicero, Homer and Plautus !

STAHL, GEORG ERNST.

G. E. Stahlii scriptorum . . . chymicorum elenchum chronologicum. *See, in Section IV*, Stahl, G. E., Fundamenta, etc. 1721.

STAMMER, K.

Alphabetisches Sach-Register. *See* Jahresbericht . . . Fortschritte . . . Zuckerfabrikation.

STROBEL, FR., UND W. BARENTIN.

See, Annalen der Physik und Chemie. Sach-Register, 1888.

SURGEON-GENERAL's OFFICE, LIBRARY OF THE.

See Index-Catalogue of the Library of the S.-G. O.

SYSTEMATISCHES VERZEICHNISS aller derjenigen Schriften welche die Naturgeschichte betreffen von den ältesten bis auf die neuesten Zeiten. Halle, 1784. pp. viii–446, 8vo.

In the Sections on Mineralogy, etc., chemical works are found.

TABLEAU GÉNÉRAL MÉTHODIQUE ET ALPHABÉTIQUE DES MATIÈRES con-tenues dans les publications de l'académie impériale des sciences de St. Pétersbourg depuis sa fondation. Première partie, pub-lications en langues étrangères. St. Pétersbourg, 1872. pp. xii-488, 8vo.

Contains an alphabetical catalogue of chemical writings, pp. 156–170.

Supplément I, comprenant les publications en langues étrangères depuis 1871 jusqu'au 1. Novembre, 1881. St. Pétersbourg, 1882. pp. vii–56, 8vo.

Chemistry occupies pages 16 to 19.

TASCHENBUCH FÜR SCHEIDEKÜNSTLER UND APOTHEKER. Weimar. Vollständiges Register der Jahre 1780–85 herausgegeben von J. F. A. Göttling. Weimar [1786]. 32mo.

Zweytes vollständiges Register über den Almanach oder Taschenbuch [etc.], 1786–91. Weimar [1792].

Drittes Register, 1792–1797.

Viertes Register, 1798–1803.

Fünftes Register, 1804–1810.

TILDEN, WILLIAM A.

Books on Chemistry. Birmingham Reference Library Lectures. London, Birmingham and Leicester. *n. d.* pp. [ii] 42 to 55–[ii], 8vo.

> A brief sketch with special reference to the books found in the Birmingham Reference Library. Followed by a check-list of 87 titles.

TOLLENS, B.

Kurzes Handbuch der Kohlenhydrate. Breslau, 1888.

> Pages 331 to 360 contain a bibliography of carbohydrates.

TRAPHAGEN, FRANK W.

Index to the Literature of Columþium, 1801–1887. Smithsonian Miscellaneous Collections, No. 663. Washington, 1888. pp. [iv]–27, 8vo.

TROMMSDORFF, JOHANN BARTHOLOMÄ.

Allgemeine chemische Bibliothek des neunzehnten Jahrhundert. Erfurt, 1802–1805. 5 vols., 12mo. Portraits of J. F. Gmelin, J. F. Westrumbs, Lorenz von Crell, A. N. Scherer, Friedrich Hildebrandt.

> Contains summaries and critical reviews of current chemical literature. Vol. v, part 1 (the last published), contains:

Versuch einer systematischen Darstellung der gesammten chemischen Literatur, oder Verzeichniss der in das Gebiet der Chemie gehörenden Schriften welche von Anfang 1800 bis zum Schluss des Jahres 1804 in Deutschland, England, Amerika, Frankreich, Holland, Schweden, Dänemark und Italien wirklich erschienen sind. pp. 93–200.

> This bibliography is of positive utility, exhibiting much technical skill. It is unfortunately limited in its scope to five years.

TUCKERMAN, ALFRED.

Bibliography of the Chemical Influence of Light. Smithsonian Miscellaneous Collections, No. 785. Washington, 1891. pp. 22, 8vo.

Index to the Literature of the Spectroscope. Smithsonian Miscellaneous Collections, No. 658. Washington, 1888. pp. x–423, 8vo.

> Exhaustive, systematic, and accurate.

VALLET DE VIRIVILLE.

Des ouvrages alchimiques attribués à Nicolas Flamel. Extrait du XXIII volume des Mémoires de la Société impériale des Antiquaires de France. pp. 26, 8vo. [Paris].

> A critical study of the extant MSS. and works attributed to Flamel.

VAN, AND VAN DER.
> For Dutch names beginning with Van, or Van der, see next succeeding word.

VAUGHAN, VICTOR C.
A Bibliography of Ptomaines accompanies V. C. V's: Ptomaines and Leucomaines. Philadelphia, 1888. pp. 296–314, 8vo.

VERSUCH EINES SYSTEMATISCHEN VERZEICHNISSES DER SCHRIFTEN UND ABHANDLUNGEN VOM EISEN, als Gegenstand des Naturforschers, Berg- und Hüttenmanns, Künstlers und Handwerkers, Kaufmanns, Staatshaushälters und Gesetzgebers. Berlin, 1782. 8vo.

VERZEICHNISS EINER ALCHYMISTISCHEN BIBLIOTHEK AN SELTENEN MANU-SCRIPTEN UND DRUCKWERKEN AUS ÄLTERER ZEIT. Gotha, 1859. pp. 16, 8vo.
> A trade catalogue commended by Petzholdt.

VERZEICHNISS VON CHYMISCHEN, ALCHYMISCHEN, PHYSIKALISCHEN, THEO-SOPHISCHEN, KABALISTISCHEN UND KUNSTBÜCHERN, welche bei Joh. Georg Binz, Buchhändler in Wien, im Zwettelhof verkauft werden. Wien, 1751. 8vo.

WAITE, ARTHUR EDWARD.
A Bibliography of Alchemy and Hermetic Philosophy. *See, in Section IV,* the author's Lives of Alchemystical Philosophers, pp. 276–306.
> This is carelessly done, being copied for the most part from Barrett, Lenglet du Fresnoy, and other sources, without taking pains to correct errors.

WALLER, ELWYN.
Bibliography of Butter, adulteration, testing, etc., in Report on Butter and its adulterations by E. W., assisted by E. W. Martin, Walter Moeller, and Russell W. Moore. Second Annual Report of the N. Y. State Dairy Commissioners. 1886. pp. 283–290.

WARDER, ROBERT B.
A Bibliography of Geometrical Isomerism. *In* Address [on this subject] to the Chemical Section of the American Association for the Advancement of Science at Indianapolis, 1890. Proceedings A. A. A. S., vol. XXXIX, Salem, 1890. 8vo.

Literature of the Speed of Chemical Reactions. Proceedings American Association for the Advancement of Science, vol. XXXII, Salem, 1883. pp. 3, 8vo.

WARE, EZRA J.

An Index of Researches upon the Production of Ammonia from Atmospheric Nitrogen. *In* Proceedings Michigan State Pharmaceutical Association, 1888. H. J. Brown, Secretary, Ann Arbor, Michigan.

WATT, ROBERT.

Bibliotheca Britannica, or a general index to British and Foreign Literature. Edinburgh, 1824. 4 vols., 4to.

> In vol. III, Subjects, under Chemistry is found a considerable list of very brief titles of chemical books arranged chronologically in groups.

WATTS, HENRY.

Index to the first twenty-five volumes of the Journal of the Chemical Society, 1848-1872 ; and to the Memoirs and Proceedings, 1841-47. London, 1874. 8vo.

WEBB, WALTER W.

Index to the Literature of Electrolysis, 1784-1880. Extract from Annals New York Academy of Sciences, vol. II, No. 10, 1882. New York. pp. 40, 8vo.

> Index des mémoires sur l'électrochimie. Publiés depuis 1784 jusqu'a 1880. Traduit de l'Anglais par Donato Tommasi.
> *See, in Section V,* Tommasi, Donato : Traité théorique et pratique d'électrochimie.

WEIGEL, CHRISTIAN EHRENFRIED.

Grundriss der reinen und angewandten Chemie. Zum Gebrauch academischer Vorlesungen. Greifswald, 1777. 2 vols., 12mo. Folding plates. Vol. I, pp. xl-564 ; vol. II, xxxii-792-[clix].

> This is not a text-book, as might be inferred from the title, but a classified bibliography of much value and interest. The second volume contains an alphabetical index of the authors and their writings that are named in the two volumes. Also a subject index. These two occupy 159 pages.

Einleitung zur allgemeinen Scheidekunst. Leipzig, 1788-1794. 3 vols., 8vo. Vol. III in two parts. Vol. I, [xiv]-556 ; II, [xxiv]-920 ; IIIa, [xxxii]-951 ; IIIb, [xx]-843.

Stück 1. Vorbegriffe. Anfang der allgemeinen Bücherkunde.

Stück 2-3. Fortsetzung der allgemeinen Bücherkunde. Sammlungen und Zeitschriften.

> Petzholdt remarks this contains a quantity of available bibliographical material in the form of notes, but inconveniently arranged. It is practically an extension of the " Grundriss " of the same author.

WIEGLEB, JOHANN CHRISTIAN.

[A short list of works written against alchemy,] *in* Historisch-kritische Untersuchung der Alchemie. Weimar, 1777, pp. 372–377. For full title, *see in Section III*.

Handbuch der allgemeinen Chemie. Zwote neuberichtigte Auflage. Berlin und Stettin, 1786. 2 vols., 8vo.

> Vol. II, pp. 716–740, contains: Verzeichniss der vorzüglichsten Schriften, die zum Behuf der allgemeinen und angewandten Chemie nachgelesen werden können, und als eine auserlesene chemische Bibliothek anzusehen sind.

WILEY, HARVEY W. [Editor].

Bibliography of Beeswax, [and of] Waxes used in Adulterating Beeswax. In Part VI of Foods and Food Adulterants. Bulletin No. 13, Division of Chemistry, U. S. Department of Agriculture. Washington, 1892. pp. 866–871, 8vo.

Bibliography of Honey. Arranged by Years. In Part VI of Foods and Food Adulterants. Bulletin No. 13, Division of Chemistry, U. S. Department of Agriculture. Washington, 1892. pp. 871–874, 8vo.

WITTSTEIN, G. C.

> *See* Journal für Chemie und Physik. *Also*, Repertorium für die Pharmacie. Literatur der Chemie. *See, in Section II*, Wittstein, G. C., Vollständiges etymologisch-chemisch Handwörterbuch, *etc.* 1846–58.

WOLF'S NATURWISSENSCHAFTLICHES VADEMECUM. Alphabetische und systematische Zusammenstellung der litterarischen Erscheinungen auf dem Gebiete der Chemie, Pharmacie und chemischen Technologie. Die Litteratur bis 1890 excl. enthaltend. Mit Register der Schlagwörter. No. II, Abtheilung II, Band 1. Leipzig, *n. d.* [1890]. pp. 91, 12mo.

> A convenient trade catalogue of 2,096 titles.

WOLFF, EMIL THEODOR.

Quellen-Literatur der theoretisch-organischen Chemie, oder Verzeichniss der vom Anfang des letzten Viertheils des vorigen Jahrhunderts bis zum Schluss des Jahres 1844, ausgeführten chemischen Untersuchungen über die Eigenschaften und die Constitution der organischen Substanzen, ihrer Verbindungen und Zersetzungsproducte. Mit steter Berücksichtigung der Literatur der Chemie in ihrer Anwendung auf Agricultur, Physiologie und Pathologie aus den wichtigeren deutschen und französischen Zeitschriften der Chemie und Pharmacie gesammelt, in systematischer Ordnung zusammengestellt und mit ausführlichen Sach- und Namen-Regis-

WOLFF, EMIL THEODOR [Cont'd].

tern versehen. Halle, 1845. 8vo. Col. xii–404. [Columns are numbered in place of pages.]

> A carefully collated and systematically arranged index to researches in organic chemistry within the period named.

Vollständige Uebersicht der elementar-analytischen Untersuchungen organischer Substanzen, nebst Andeutung der verschiedenen Theorien über deren chemische Constitution. Aus den chemischen Journalen nach den Original-Abhandlungen in systematischer Ordnung entworfen. Halle, 1846. pp. xii, col. 808, 8vo.

WOLFF, FRIEDRICH.

> Annalen der chemischen Literatur. *See* Hermbstädt, Sig. Fr., Bibliothek, vol. v.

YOUNG, THOMAS.

A Course of Lectures on Natural Philosophy and the Mechanical Arts. London, 1807. 2 vols., 4to.

> Vol. II contains a valuable "Catalogue of Works relating to Natural Philosophy and the Mechanical Arts, with References to particular passages and occasional abstracts and remarks." pp. 87–738. An edition published in 1845 contains additional authorities.

ZEITSCHRIFT FÜR ANALYTISCHE CHEMIE herausgegeben von C. Remigius Fresenius. Autoren- und Sach-Register zu den Bänden i–x (1862–1871). Wiesbaden, 1872. pp. 108, 8vo.
The same. Bände xi–xx (1872–1881). Wiesbaden, 1882.

ZEITSCHRIFT FÜR PHYSIOLOGISCHE CHEMIE herausgegeben von F. Hoppe-Seyler. Sach- und Namen-Register zu Band i–iv. Strassburg, 1881. pp. 54, 8vo.
Sach- und Namen-Register zu Band v–viii. Strassburg, 1888. pp. 56, 8vo.

ZEITSCHRIFT (NEUE) FÜR RÜBENZUCKER-INDUSTRIE. Vollständiges Autoren- und Sach-Register zu den Bänden i bis xx vom 1 Juli 1878 bis 30. Juni 1888 der Neuen Zeitschrift für Rübenzucker-Industrie. Wochenblatt für die Gesammtinteressen der Zuckerfabrikation. Herausgegeben von Scheibler. Bearbeitet von Ernst Glanz. Berlin, 1889. 4to.

ZUCHOLD, ERNST AMANDUS.

> *See* Bibliotheca historico-naturalis.

ZUCHOLD, ERNST AMANDUS.

Bibliotheca chemica, Verzeichniss der auf dem Gebiete der reinen, pharmaceutischen, physiologischen und technischen Chemie in den Jahren 1840 bis Mitte 1858, in Deutschland und im Auslande erschienenen Schriften. Göttingen, 1859. pp. viii–342, 8vo.

> A most complete contribution to special bibliography. All the works bearing chemistry in their title or relating to the subject, issued between the years named, in 21 different languages, are here alphabetically arranged. A sequel for the years 1858–1870 was issued by Ruprecht in 1872. *Cf.* Ruprecht.

Bibliotheca photographica. Verzeichniss der auf dem Gebiete der Photographie, sowie der damit verwandten Künste und Wissenschaften seit Erfindung der Daguerreotypie bis zu Anfang des Jahres 1860 erschienenen Schriften. Leipzig, 1860. 8vo.

DORVEAUX, PAUL.

Catalogue des thèses de pharmacie soutenues en France (Paris excepté) de 1803 à 1890 inclusivement. Paris, 1892. 8vo.

> Bibliographie des thèses, No. 3.

DORVEAUX, PAUL, ET G. PLANCHON.

Catalogue des thèses soutenues devant l'école de pharmacie de Paris, 1815–1889. Paris, 1891. 8vo.

> Bibliographie des thèses, No. 1.

MAIRE, ALBERT.

Catalogue des thèses de sciences soutenues en France de 1810 à 1890 inclusivement. Paris, 1892. pp. xi–224, 8vo.

> Bibliographie des thèses, No. 2.
> The theses are arranged chronologically, and are provided with both author- and subject-indexes. Christian names of authors are given in full.

SECTION II.

DICTIONARIES AND TABLES.

INCLUDING NOMENCLATURE.

ACCUM, FREDERICK.

A Dictionary of the Apparatus and Instruments employed in operative and experimental chemistry, exhibiting their construction and the method of using them to the greatest advantage. London, 1821. 12mo.

AIKIN, A. AND C. R.

A Dictionary of Chemistry and Mineralogy with an account of the processes employed in many of the most important chemical manufactures, to which are added a description of chemical apparatus and various useful tables of weights and measures, chemical instruments, etc., etc. Illustrated with fifteen engravings. London, 1807. 2 vols., 4to, *with* APPENDIX *entitled :*

An Account of the most important recent discoveries and improvements in chemistry and mineralogy to the present time. London, 1814. 1 vol., 4to.

ANTHON, ERNST FRIEDRICH.

*Handwörterbuch der chemisch-pharmazeutischen und pharmakognostischen Nomenklaturen, oder Uebersicht aller lateinischen, deutschen und französischen Benennungen der chemisch-pharmazeutischen Präparate, so wie der im Handel vorkommenden rohen Arzneistoffe, für Aerzte, Apotheker und Droguisten. Nürnberg, 1833. pp. viii.–724, 8vo.

> Second edition. Leipzig, 1861.
> Manuel de synonymie chimico-pharmaceutique, chimique, technique et pharmaceutique, ou Nomenclature de toutes les dénominations latines, allemandes et françaises des produits chimiques et matières premières médicinales. Deuxième édition revue et augmentée. Bruxelles, 1862. 8vo.

APHELEN, HANS VON (Translator).
> *See* Macquer, P. J. Chymisk Dictionnaire.

AREJULA, JUAN MANUEL.
> Reflexiones sobre la nueva nomenclatura química propuesta por MM.
> de Morveau, Lavoisier, Berthollet, etc., dirigidas á los químicos
> españoles. En Madrid, 1788. 8vo.
>
> > *Cf.* Guyton de Morveau, Lavoisier, etc.

ARENDS, GEORG.
> Synonymen-Lexicon. Eine Sammlung der gebräuchlichsten gleichbe-
> deutenden Benennungen aus dem Gebiete der technischen und
> pharmaceutischen Chemie, der Pharmakognosie und der pharma-
> ceutischen Praxis. Ein Hand- und Nachschlagebuch für Apotheker,
> Chemiker, Droguisten u. A. Leipzig, 1891, pp. viii–672, 8vo.

ARZT, E. H. G.
> Versuch einer systematischen Anordnung der Gegenstände der reinen
> Chemie. Leipzig, 1795. pp. xiv.–282, 12mo.
>
> > A study of the (then) new nomenclature.

ATTFIELD, JOHN.
> The Chemical Nomenclature of the Pharmacopœia, with suggestions
> for its revision. Including opinions on the proposed system by
> chemical, medical and pharmaceutical authorities and additional
> remarks by the author. *n. p.* *n. d.* [London, 1871.] pp. 16, 4to.

AUGUST, E. F., BARENTIN [etc.].
> *See* Handwörterbuch der Chemie und Physik.

AULAGNIER, A. F.
> Dictionnaire des aliments et des boissons en usage dans les divers
> climats et chez les différents peuples ; cet ouvrage contient l'histoire
> naturelle de chaque substance alimentaire, son origine, ses principes
> constituants, ses propriétés, ses altérations, et les moyens de les
> reconnaître et finalement les règles les plus importantes a suivre
> pour conserver la santé. Troisième édition revue et publiée par
> F. M. Adolphe Aulagnier fils. Paris, 1885. pp. vi–884, 8vo.

AVOGADRO, CONTE AMADEO.
> Proposizione di un nuova sistema di nomenclatura chimica. Modena
> 1844. 4to. [Estratto dalle Mem. mat. fis. in Modena, Vol. xxiii.]

BACHER.
> Nomenklatur. *See* Versuch einer französisch, lat. ital. teutsch. Nomenklatur.

BAEDECKER, F.
Chemische Rechentaɪeɪ nach den neueren Atomgewichtszahlen berech-
net. Nebst Erläuterung der Construction und Anweisung zum
Gebrauch. Elberfeld, 1852. Illustrated, 8vo.

BAIRD, J. W., AND ALBERT B. PRESCOTT.
Dictionary of the Action of Heat upon certain Metallic Salts, including
an Index to the principal Literature upon the Subject. New York,
1884. 70 pp., 8vo.

BAJNOK, ANTONIUS.
Nomenclator pharmaceutico-chimicus, continens nomina præparatorum
chemicorum officinalium, tam nova, pure chemica, quam synonyma
antiqua et trivialia. Medicorum et pharmacopolarum usui. Posi-
doniæ, 1850. 8vo.

BARRESWIL, ET AIMÉ GIRARD.
Dictionnaire de chimie industrielle, avec la collaboration de plusieurs
savants. Paris, 1861–64. 4 vols., roy. 8vo. Ill.

BAUDRIMONT, E.
Dictionnaire des altérations et falsifications des substances alimentaires,
médicamenteuses et commerciales, avec l'indication des moyens de
les reconnaître. Sixième édition revue, corrigée et considérablement
augmentée. Paris, 1883. 8vo. Ill.

BAUMANN, A.
Tabelle zur Berechnung der Salpetersäure aus dem gefundenen Volumen
des Stickoxyds durch Multiplikation. München, 1889. Fol.
Tabelle zur gasvolumetrischen Bestimmung der Kohlensäure.
München, 1888. Fol.
Tabelle zur gasvolumetrischen Bestimmung des Stickstoffs. München,
1889. Fol.
Tafeln zur Gasometrie. München, 1885.

BAYLEY, F.
A Pocket-book for Chemists, Chemical Manufacturers. Third edition.
London, 1883.
Fourth edition. London, 1886. Oblong 32mo.
Also New York, 1887.

BEAUME, GUSTAVE.
Essai de classification et de nomenclature des matières organiques.
Le Puy, 1864. 18mo.

BECKER, JOHANN HERRMANN.
> Darstellung der Nahrungsmittel des Menschen nach alphabetischer Ordnung.
> *See, in Section V,* Becker, J. H.

BEILSTEIN, F.
> Handbuch der organischen Chemie.   Hamburg und Leipzig, 1881–83.
> 2 vols., roy. 8vo.
>> Zweite gänzlich umgearbeitete Auflage.   3 vols.   I, 1886 ;
>> II, 1888 ; III, 1890.   Dritte Auflage.   *In progress.*
>
> A stupendous monument of industrious, intelligent compilation.

BELLAVITIS, G.
> Considerazioni sulle nomenclature chimiche, sugli equivalenti chimici,
> ecc. Venezia, 1862. 4to.

BERGMANN, H.
> Chemisch-technisches Receptbuch für die gesammte Metallindustrie.
> Wien, 1887.

BERLICHIUS, ADAMUS GOTTLOB.
> Tabulæ metallurgico-docimasticæ, etc. *See, in Section V,* Clauderus, G.,
> D. G. Clauderi schediasma, etc., 1736.

BERZELIUS, JÖNS JACOB.
> Tabell, som utvisar vigten, etc. *See in Section V.*

BERZELIUS, JOHANN JACOB.
> Chemische Operationen und Geräthschaften, nebst Erklärung chemi-
> scher Kunstwörter, in alphabetischer Ordnung. Aus dem Schwe-
> dischen übersetzt von F. Wöhler. Dresden, 1831. 8vo.
>> Vierte Auflage. Dresden und Leipzig, 1841.

BERZELIUS, J. J.; S. G. B. y J. B.
> Nomenclatura química del célebre sueco Berzelius, sacada de la última
> edicion alemana y puesta en idioma español por M. de los S. G. B.
> y J. B. Barcelona, 1832. 4to.

BERZELIUS, J. J., UND LAGERHJELM.
> Alphabetisches Verzeichniss der Gehalte sämmtlicher bekannter che-
> mischer Verbindungen. Aus dem Französischen mit Bemerkungen
> über chemische Nomenklatur von Meineke. Nürnberg, 1820. 8vo.

BIBRA, ERNST FREIHERR VON.
> Chemische Hülfstafeln. Erlangen, 1846. 8vo.

BIRNBAUM, KARL.
> Kurzes erläuterndes Wörterbuch zu der siebenten Auflage von Otto-
> Birnbaum's Lehrbuch der landwirthschaftlichen Gewerbe, und zu
> den in Einzelausgaben erschienenen Theilen desselben. Braun-
> schweig, 1881. 8vo. Ill. *Compare in Section V,* Otto-Birnbaum.

BOLLMANN, FRIEDRICH.
Photographisch-chemikalisches Lexikon. Braunschweig, 1863. 8vo.
Illustrated.

BONAPARTE, LUCIEN.
Nuova nomenclatura esprimente il rapporto atomico. 1839. 8vo.

BÖTTGER, RUDOLPH, UND A. GRÄGER.
Handwörterbuch der technischen Chemie für Fabrikanten, Gewerbtreibende, Künstler, Droguisten, etc. Weimar, 1867. pp. viii.–580.

BOOTH, JAMES CURTIS, AND CAMPBELL MORFIT.
The Encyclopædia of Chemistry. Philadelphia, 1850. 974 pp., 8vo.
Illustrated.

BOUANT, ÉMILE.
Nouveau dictionnaire de chimie illustré de figures intercalées dans le
texte, comprenant les applications aux sciences, aux arts, à l'agriculture et à l'industrie, à l'usage des chimistes, des industriels, des
fabricants de produits chimiques, des agriculteurs, des médecins,
des pharmaciens, des laboratoires municipaux, de l'école des mines,
des écoles de chimie, etc. Avec la collaboration de professeurs,
d'ingénieurs et d'industriels. Paris, 1887–89. With 650 wood-
cuts. pp. 1160, 8vo.

BOUASSE-LEBEL. ENCYCLOPÉDIE.
Analyse chimique, inorganique, qualitative (par voie humide). Résultats vérifiés d'après les meilleurs auteurs, triés, classés et augmentés
de réactions nouvelles, par L. Errani, à l'usage des écoles, des pharmaciens chimistes, droguistes, etc. Paris, 1877. 8vo.

BOURGUET, LUDWIG.
Chemisches Handwörterbuch nach den neuesten Entdeckungen entworfen. Mit einer Vorrede von Sigm. Fr. von Hermbstädt. [From
3d volume] fortgesetzt von J. B. Richter. Berlin, 1798–1805.
6 vols., 8vo.

BRANDES, RUDOLPH.
Repertorium für die Chemie als Wissenschaft und Kunst, eine möglichst
vollständige alphabetisch-systematisch-geordnete Darstellung des
Wichtigsten über die bekannten Stoffe der Chemie, über die Bestandtheile der Mineralien, Pflanzen- und Thierkörper, mit besonderer Rücksicht auf die praktische Anwendung für die Pharmacie,
Medicin, Agricultur, Fabriken- und Gewerbskunde, so wie nicht
minder auf die Entwickelung der Grundzüge der Wissenschaft und

BRANDES, RUDOLPH. [Cont'd.]

    der Anwendung ihrer Principien auf die Naturerscheinungen über-
haupt und die Physiologie, Kristallologie, Geognosie und Meteorologie insbesondere, mit Zugrundelegung von "Ure's Dictionary
of Chemistry on the Basis of Nicholson's." Hannover, 1826–33.
4 vols., 4to.

        Extends only to the word Blass, and never completed. *Cf.* Ure, Andrew.

BRARD, CYPRIAN PROSPER.

    Dictionnaire usuel de chimie, de physique et d'histoire naturelle.
Ouvrage destiné aux instituteurs, aux artisans et aux gens du
monde. Paris, 1840. 2 vols., 8vo.

BROGNIART, ANTON LOUIS.

    Tableau analytique des combinaisons et des décompositions de diffé-
rentes substances, ou Procédés de chymie pour servir à l'intelligence de cette science. Paris, 1778. pp. 526, 8vo.

BRUGNATELLI, LUIGI.

    Synonymie des nomenclatures chimiques modernes, par Brugnatelli,
traduite de l'Italien par Jean Baptiste van Mons. Bruxelles, 1802.
8vo.

BUENO, PEDRO GUTIERREZ.

        See Gutierrez Bueno, Pedro.

BUYS-BALLOT, C. H. D.

    Repertorium corporum organicorum quæ secundum atomisticam, pro-
centicam et relativam compositionem, annotatis proprietatibus
physicis et præcipuis, e quibus cognoscantur, fontibus, in ordinem
disposita, addita præfatione G. J. Mulder. Trajecti ad Rhenum,
1846. pp. xxiv–134–94, 4to.

CADET, CHARLES LOUIS.

    * Dictionnaire de chimie contenant la théorie et la pratique de cette
science, son application à l'histoire naturelle et aux arts. Paris,
An. xi–1803. 4 vols., 8vo. With plates.

        КАДЕТЪ ШАРЛЬ ЛУИ,
            Словарь химическій. 4 части. С.-Петербургъ, 1810—1813. 4 vols.

CAPAUN–KARLOWA, C. F.

    Chemisch-technische Specialitäten und Geheimnisse mit Angabe ihrer
Zusammensetzung alphabetisch geordnet. Zweite Auflage. Wien,
1886.

CAPRIA, DOMENICO MAMONE.

Dizionario portatile di chimica organica, nel quale travasi descrizione, storia, carattere, composizione e cifra simbolica dei composti, tanto ipotetici che reali di natura organica, ecc. Napoli, 1844. 16mo.

Idee generali di nomenclatura sistematica chimica coordinate. Napoli, 1849. 12mo.

CARNELLEY, THOMAS.

Physico-chemical Constants. Melting- and Boiling-Point Tables. London, 1885–88. 2 vols., roy. 4to.

CASALI, ADOLFO.

Dizionario delle denominazioni e dei sinonimi della chimica e delle scienze arti ed industrie attinenti alla medesima. Bologna, 1872. pp. xvi–574, 8vo.

> *Contains :* Cenni biografici sui più distinti cultori della chimica. pp. 29.

CAVENTOU, J. B.

Nouvelle nomenclature chimique d'après la classification adoptée par Thénard ; ouvrage spécialement destiné aux personnes qui commencent l'étude de la chimie et à celles qui ne sont pas au courant des nouveaux noms. Seconde édition revue, corrigée et augmentée. Paris, 1825. pp. xxxiv–371, 8vo. Folding table.

CHENEVIX, RICHARD.

Remarks upon Chemical Nomenclature. . . . London, 1802.

CHEVALLIER, (JEAN BAPTISTE) ALPHONSE.

Dictionnaire des altérations et falsifications des substances alimentaires, médicamenteuses et commerciales, avec l'indication des moyens de les reconnaître. Paris, 1850–52. 2 vols., 8vo. Plates.

> Deuxième édition, augmentée. Paris, 1854. 2 vols., 8vo.
> > Troisième édition. 1858. 2 vols. 7 plates.
> Wörterbuch der Verunreinigungen und Verfälschungen der Nahrungsmittel, Arzneikörper und Handelswaaren nebst Angabe der Erkennungs- und Prüfungsmittel. Frei nach dem Französischen in alphabetischer Ordnung bearbeitet und mit Zusätzen versehen von A. H. L. Westrumb. Göttingen, 1856. 2 vols., 8vo.
> Diccionario de las alteraciones y falsificaciones de las substancias alimenticias y comerciales, con la indicacion de los medios de reconocerlas . . . Traducido por Ramon Ruiz Gomez. Madrid, 1854–55. 2 vols., 4to.

CHEVALLIER, ALPHONSE, CH. LAMY ET ED. ROBIQUET.

Dictionnaire raisonné des dénominations chimiques et pharmaceutiques contenant tous les termes employés en chimie et en pharmacie, pour désigner les lois, phénomènes, substances, combinaisons ou préparations connus jusqu' à ce jour. Paris, 1853. 2 parts.

CHYMICALL DICTIONARY (A), explaining Hard Places and Words met withall in the Writings of Paracelsus and other obscure Authours. London, 1650. Sm. 4to.

> This is "Printed by Richard Cotes for Thomas Williams at the Bible in Little brittain." It is commonly bound with: A New Light of Alchymie by Michael Sandivogius and Nine Books of the Nature of Things by Paracelsus. All translated by J. F. London, 1650. Many of the explanations throw little light on the meaning of the hard words; *e. g.:* "Acetum Philosophorum is a Mercuriall Water or Virgin's Milke wherein they say metalls are dissolved."

CLARKE, FRANK WIGGLESWORTH.

Specific Gravities, Boiling- and Melting Points and Chemical Formula. Part I, Constants of Nature. Smithsonian Miscellaneous Collections. Washington, 1873. pp. vii–263.

Specific Gravities, Boiling Points and Melting Points. The Constants of Nature, First Supplement to Part I. Smithsonian Miscellaneous Collections. Washington, 1876. pp. 61.

> A Table of Specific Gravity for Solids and Liquids. [New Edition, Revised and Enlarged.] The Constants of Nature, Part I. Smithsonian Miscellaneous Collections, 659. Washington, 1888. pp. xi–409, 8vo.

Table (A) of Specific Heats for Solids and Liquids. The Constants of Nature, Part II. Smithsonian Miscellaneous Collections. Washington, 1876. pp. 58.

Tables of Expansion by Heat for Solids and Liquids. The Constants of Nature, Part III. Smithsonian Miscellaneous Collections. Washington, 1876. pp. 57.

Recalculation (A) of the Atomic Weights. The Constants of Nature, Part V. Smithsonian Miscellaneous Collections. Washington, 1882. pp. 280.

> *Compare* Meyer, Lothar und Karl Seubert.

COTTING, J. R.

A Dictionary of Terms.

> *See, in Section V*, Cotting, J. R.: An Introduction to Chemistry.

COZZI, ANDREA.

Fasi della nomenclatura chimica e applicazione ad essa della teoria atomistica. Firenze, 1842. 8vo.

CUNDILL, J. P.
A Dictionary of Explosives.   Chatham and London, 1889.   pp. xv–109,
8vo.

CZELECHOWSKY, J. R.
Chemisches Wörterbuch zum Gebrauche für Aerzte, Pharmaceuten,
Techniker und Gebildete jeden Standes.   Wien, 1841.   pp. x–691,
roy. 8vo.

DAMMER, O.
Chemisches Handwörterbuch.   Zweite Auflage.   Stuttgart, 1885.
Illustrirtes Lexikon der Verfälschungen und Verunreinigungen der
Nahrungs- und Genusmittel, der Kolonialwaaren, Droguen, gewerb-
lichen Produkte, Dokumente, etc.   Leipzig, 1885.   8vo.   3. Auflage,
1886.   2 vols., 8vo.   Ill.
Kurzes chemisches Handwörterbuch.   Berlin, 1876.
Lexikon der angewandten Chemie.   Die chemischen Elemente und
Verbindungen im Haushalt der Natur und im täglichen Leben, in
der Medizin und Technik, Zusammensetzung der Nahrungsmittel,
Industrieproducte, etc.   Leipzig, 1881.   pp. 527, 8vo.

DECREMPS.
Diagrammes chimiques, ou Recueil de 360 figures (sur 112 planches)
qui expliquent succintement les expériences par l'indication des
agens et des produits a coté de l'appareil, et qui rendent sensible
la théorie des phénomènes en representant le jeu des attractions
par la convergence des lignes.   Ouvrage élémentaire auquel on a
ajouté . . . un Essai de nomenclature chimique.   Paris, 1823.   4to.

DELAURIER, E.
Une nouvelle nomenclature binaire notative pour la chimie.   *See, in Section V*, Delaurier, E.   Essai d'une théorie générale supérieure de
philosophie naturelle [etc.].

DE MORVEAU.
See Guyton de Morveau.

DÉRIARD, AUGUSTE.
Synonymie chimique et pharmaceutique, ou nomenclature de toutes les
dénominations des principaux corps simples et de leurs composés.
Deuxième édition.   Lyon et Paris, 1867.   8vo.

DICCIONARIO TECNOLÓGICO, ò nuevo diccionario universal de artes y
oficios, y de economía industrial y comercial.   Escrito en francés
. . . traducido al castellano por F. S. y C.   Barcelona, 1833–35.
8 vols., 4to, and Atlas in folio.
Contains many chemical articles.

DICKSON, STEPHEN.

An Essay on Chemical Nomenclature, in which are comprised observations on the same subject by Richard Kirwan. London, 1796. pp. xvi–294.

> The author, who was Professor of the Practice of Medicine in Trinity College, Dublin, adopts the French system.

DICTIONNAIRE ABRÉGÉ des termes de l'art et des anciens mots qui ont rapport au Traité de Philalethe, et aux autres Philosophes contenus dans la Bibliothèque alchymique. pp. 570–590 of Richebourg's Bibliothèque. Vol. IV.

> *See, in Section VI*, Richebourg, J. M. D.

DICTIONNAIRE DE CHYMIE.

> By P. J. Macquer. *See* Macquer.

DIETRICH, CHRISTIAN JOHANN.

> *See* Grothuss, Theodor.

DI-GIORGI, S.

Sinonimia chimico-farmaceutica. Mazzara di Vallo, 1889. 16mo.

DÖBEREINER, FRANZ.

Supplement zu J. W. Döbereiner's Grundriss der Chemie. Tabellarische Darstellung der organischen Stoffe in alphabetischer Ordnung für Aerzte, Chemiker und Pharmaceuten. Stuttgart, 1837 pp. (viii)–157, 4to.

> These Tables give the name of the organic body, of its discoverer, references to literature, quantitative composition, occurrence and principal properties. The bodies described are about 1,085 in number. The Tables give a concise summary of the knowledge of organic chemistry half a century ago. For an analogous work showing the present state of the subject, see Beilstein, F.

DÖBEREINER, JOHANN WOLFGANG.

Darstellung der Verhältnisszahlen der irdischen Elemente zu chemischen Verbindungen. Jena, 1816. Folio. Pages not numbered.

Darstellung der Zeichen und Verhältnisszahlen der irdischen Elemente zu chemischen Verbindungen. Erster Theil, die Zeichen und Zahlen der Elementarstoffe und der wichtigsten unorganischen Verbindungen derselben darstellend. Zweyte vermehrte und verbesserte Auflage. Jena, 1823. pp. 27, folio. Twelve Tables.

DORNEUS, GERHARDUS.

Dictionarium chimicum Theophrasti. Francofurti, 1563.

> The author was an alchemist and pupil of Paracelsus who lived in the sixteenth century.

DRAPIEZ, A.

Dictionnaire portatif de chimie et de minéralogie.   Deuxième édition
revue, corrigée et augmentée de plus de 1500 articles, avec 4
planches gravées et 5 tableaux.   Bruxelles, 1825.   8vo.

DUFOUR, ED.

Essai d'une nomenclature chimique et minéralogique.   Nantes, 1871.   8vo.

DUFOUR, L.

Petit dictionnaire des falsifications des substances alimentaires et des
médicaments, avec l'indication des moyens faciles pour les reconnaître.   Paris, 1877.   8vo.
> Deuxième édition, revue et augmentée.   Suivie d'une notice
> sur le Laboratoire municipal de Paris.   Paris, 1882.   16mo.

DULK, FRIEDRICH PHILIPP.

Synoptische Tabellen über die Atomgewichte der einfachen und
mehrerer zusammengesetzter Körper und über das Verhältniss
der Bestandtheile der Letzteren.   Leipzig, 1848.   8vo.

DUNIECKI, J. P. M.

O nomenklaturze chemicznéj.   Wyjątek z najnowszego dzieła Reg-
naulta " Cours élémentaire de chimie " przetłómaczył i do nomenklatury polskiéj zastósował.   Lwów, 1852.   8vo.

EGER, GUSTAV.

Technologisches Wörterbuch in englischer und deutscher Sprache.
Die Wörter und Ausdrucksweisen in Civil- und Militär-Baukunst ;
Schiffsbau ; Eisenbahnbau ; Strassen- Brücken- und Wasserbau ;
Mechanik und Maschinenbau ; Technologie ; Künste ; Gewerbe
und Fabrikindustrie ; Landwirthschaft ; Handel und Schiffahrt ;
Bergbau und Hüttenkunde ; Geschützwesen ; Physik ; Chemie ;
Mathematik ; Astronomie ; Mineralogie ; Botanik, etc., umfassend.
In Verbindung mit P. R. Bedson, O. Brandes, M. Brütt, Ch. A.
Burghardt, Th. Carnelly, J. J. Hummel, G. Lunge, J. Lüroth, G.
Schäffer, W. H. M. Ward, W. Carleton Williams bearbeitet und
herausgegeben.   Technisch durchgesehen und vermehrt von Otto
Brandes.   In zwei Theilen.   Braunschweig.   Erster Theil : Eng-
lisch-Deutsch.   1882.   Zweiter Theil : Deutsch-Englisch.   1884.

EIMBKE, GEORGE.

Versuch einer systematischen Nomenklatur für die phlogistische und
anti-phlogistische Chemie.   Halle, 1793.   pp. vi–234, 12mo.
> A systematic synonymicon attempting to reconcile old and new chemical
> theories.

EISENACH, CARL FRANZ.

Versuch einer tabellarischen Uebersicht der Elementar-Stoffe, zum Theil nach ihren Analogien geordnet, mit Angabe ihrer hauptsächlichen physikalischen und chemischen Eigenschaften, zum Elementar-Unterricht in der Stöchiologie. Entworfen auf J. W. Doebereiner's Veranlassung. Jena, 1838. Large folio.

ELOY, N. F. J.

Dictionnaire historique de la médecine ancienne et moderne, ou mémoires disposés en ordre alphabétique pour servir à l'histoire de cette science et a celle des Médecins, Anatomistes, Botanistes, Chirurgiens et Chymistes de toutes Nations. Mons, 1778. 4 vols., 4to.

ENCICLOPEDIA DI CHIMICA SCIENTIFICA E INDUSTRIALE, ossia dizionario generale di chimica colle applicazioni alla agricoltura e industrie agronomiche, alla farmacia e materia medica, alla fisiologia animale e vegetale, alla patologia, anatomia e tossicologia, all'igiene pubblica e privata, alla merciologia o scienza delle materie prime, alla mineralogia, metallurgia, ecc. Opera originale diretta da Francesco Selmi e compilata da una eletta di chimici italiani. Torino e Napoli, 1868–1881. Sm. 4to. 11 vols., with a Supplement in 3 vols.

> Vol. XI, pp. 503–726, contains an extensive: Compendio storico della chimica.

ENCYCLOPÉDIE CHIMIQUE.
> *See* Fremy, Edmonde.

ENCYCLOPÉDIE MÉTHODIQUE. CHYMIE, PHARMACIE ET MÉTALLURGIE. La chymie par de Morveau ; la pharmacie par Maret ; la métallurgie par Duhamel. Paris, 1786. 6 vols., 4to, and 1 vol. of plates, 1813, 4to.
> Fourcroy and Vauquelin aided de Morveau in this extensive dictionary.

ENCYCLOPÄDIE DER NATURWISSENSCHAFTEN. Herausgegeben von G. Jäger, etc. II Abth., III Theil. Handwörterbuch der Chemie.
> *See* Ladenburg, A.

ERSTINGIUS, ARTHUR CONRAD.

Nucleus totius medicinæ quinque partitus. Pars I continet Lexicon et Dispensatorium pharmaceuticum. Oder : Der Vollkommene und allezeit Fertige Apotheker. Darinnen alle und jede Stücke so würcklich in den Apothecken zu finden, ihre Gestalt und Gehalt Herkunfft, etc., auch was daraus zu machen ist, und wie die Composita auf das beste daraus zu bereiten, erkläret worden sind.

ERSTINGIUS, ARTHUR CONRAD.   [Cont'd.]

Dabey auch nach dem Alphabeth die Kunst-Wörter und vielen anderen Beynahmen mehr folgen, etc.   Pars II continet Lexicon Practico-Chymicum, oder der richtige führend Chymiste.   Cum Appendice brevi loquus chymicus, oder der kurze chymische Redner genannt darinnen alles so zur Chymie gehöret zu finden ist. Pars III continet Lexicon Theoretico-Medicum oder der zur Heilungs Gelahrheit leitende Artzt, etc., welche die medicinischen Kunst-Wörter und was sonsten dazu gehöret erkläret.   Pars IV est Lexicon chyrurgicum ein chyrurgisches Wörter-Buch.   Pars V sistit Lexicon Theoretico-Anatomicum oder der sich selbst erkennende Mensche. Alle diese Theile so wohl aus denen alten als neuem Autoribus (auch Theils aus selbst eigener Erfahrung) mit Mühe und Fleiss colligiret und in solche Ordnung gebracht.   Auch jedweden Theil mit einem deutschen Register vorsehen, damit jedweder die Nahmens, auch was sonsten in der gantzen Medicin vorfält, gleich darinnen finden und nützlich gebrauchen kan. Helmstädt, 1741. 2 vols., sm. 4to.   Portrait of the author by " E. Beckly, pinxit, and Ant. Aug. Beck, sculpsit, 1740," as frontispiece to vol. I.

> The Lexicon Chymicum occupies pages 1-314 of vol. II.   Had the author been able he would doubtless have printed the entire work on the title page, which we have copied in full with '' Mühe und Fleiss.''

EXPLANATORY (AN) DICTIONARY OF THE APPARATUS AND INSTRUMENTS employed in the various operations of philosophical and experimental chemistry.   With seventeen quarto copper plates.   By A Practical Chemist.   London, 1824. pp. vii-295, 8vo.

EXPLICATIO MAXIME VULGARIUM CHARACTERUM QUI APUD MEDICOS ET CHEMICOS OCCURRUNT in Pharmacopœa Harlemensis.   Harlemi, 1741. 8vo.

FAUST, JOHANN MICHAEL.

Lexicon alchymisticum novum.   Frankfurt, 1706.   *In* Compendium alchymisticum novum.   *See in Section VI.*

FECHNER, GUSTAV THEODOR.

Repertorium der neuen Entdeckungen in der unorganischen Chemie. Leipzig, 1830-33. 3 vols., 8vo.
See in Section V, Thénard, L. J.

FIEDLER, CARL WILHELM.

Allgemeines pharmaceutisches chymisches und mineralogisches Wörterbuch.   Mannheim. 1787-90. 2 vols.

FILOPANTI, QUIRICO.

Alcuni misteri di chimica popolarmente spiegati e nuova nomenclatura proposta da esso. Bologna, 1871. 8vo.

FISCHER, N. W.

Systematischer Lehrbegriff der Chemie in Tabellen dargestellt. Berlin, 1836. pp. [vi]–249, 4to.

FONTANELLE, JULIA.

Synoptisk Tabel over physiske, chemiske, medicinske og giftige Egen-skaber hos umiddelbare vegetabilske Grundstoffer og hos nyligen opdagede Alcalier. Oversat fra det Engelske [ved Jo. Chr. Wilh. Wendt]. [1826.]

FOURCROY, A. F.

Dictionnaire de chimie et de métallurgie. *See* Encyclopédie méthodique.

FOURCROY, A. F.

Tableaux synoptiques de chimie pour servir de résumé aux leçons données sur cette science dans les écoles de Paris. Paris. An VIII [1800], 11 pp. folio. 12 folding tables.

> Synoptic Tables of Chemistry intended to serve as a sum-mary of the Lectures delivered in the Public Schools at Paris. Translated by Wm. Nicholson. London, 1801. Fol.

> Uebersicht der Chymie in zwölf Tabellen. Französisch und Deutsch. Herausgegeben von Johann Anton Heid-mann. Wien, 1801. 4to.

> The first part was issued in 1800 with the title System der Chemie.

> System der theoretischen und praktischen Chemie. In Tabellen entworfen. Herausgegeben von Christian Gotthold Eschenbach. Leipzig, 1801. Fol. 12 tables.

> Synoptische Tabellen über den ganzen Umfang der Chemie. . . . Uebersetzt [by J. J. Görres]. Koblenz. Jahr IX [1801]. Fol.

> Oversigt af den chemiske Videnskab. En Oversættelse af Fourcroy's Tableaux synoptiques de chimie. Betydelig forøget ved Carst. Ludv. Schiødt. Kjøbenhavn, 1805.

FOURGERON, J. B.

Nouvelle synonymie chimique contenant tous les changemens produits par les derniers découvertes dans la nomenclature. Paris, 1815. 8vo.

FREDERKING, CARL.

Tabellen über die Zusammensetzung anorganischer, pharmaceutisch und technisch wichtiger chemischer Präparate nebst kurzer . . . Einleitung. Berlin, 1859. Sm. 4to.

FREMV, EDMONDE.

Encyclopédie chimique publiée sous la direction de Fremy par une réunion d' anciens élèves de l'école polytechnique, de professeurs et d'industriels. . . . Paris, *n. d.* [incomplete in 1892]. 10 Tomes in 89 parts, 8vo. Ill.

ANALYSIS OF CONTENTS.
VOL. I.

INTRODUCTION. CONNAISSANCES PHYSIQUES APPLICABLES À LA CHIMIE. (2 PARTS AND AN ATLAS.)

Including in part :

Discours préliminaires sur le développement et les progrès récents de la chimie, par M. Fremy.

Exposé de quelques propriétés générales des corps, par M. Ditte.

Essais sur les équilibres chimiques, par M. G. Lemoine.

Les Laboratoires de chimie, par MM. Fremy, Carnot, Jungfleisch, Terreil, Henrivaux, Girard et Pabst.

VOL. II.

CHIMIE INORGANIQUE. MÉTALLOÏDES.

1re Section.

Nomenclature. — Équivalents. — Atomes. — Oxygène. — Azote. — Air, — Eau. — Composés oxygénés de l'Azote. — Ammoniaque. — Brome. — Iode. — Fluor, par MM. Fremy, Bourgoin, Gaudin. Lemoine, Joly et Urbain. — 1 fort vol., 8vo, avec vignettes.

2me Section.

Soufre. — Sélénium. — Tellure. — Phosphore. — Arsenic. — Carbone. — Cyanogène, par MM. Margottet, Lemoine, Urbain, Ogier et Joannis.

3e Section.

Bore. — Silicium et Silicates, par MM. Joly et Curie.

Complément.

1re Partie : Charbon de bois. — Noir de fumée. — Combustibles minéraux, par MM. Fremy, Urbain, Stanislas Meunier.

2e Partie : Diamant, par M. Boutan.

FREMY, EDMONDE ; Encyclopédie. [Cont'd.]

Appendices.

1ᵉʳ Cahier : Synthèse minéralogique, par M. Bourgeois. 8vo, avec planches.

2ᵉ Cahier : Météorites, par M. Stanislas Meunier. 8vo, avec vignettes.

VOL. III.

CHIMIE INORGANIQUE. MÉTAUX.

1. Généralites sur les Métaux, les Oxydes et les Sels, par M. Rousseau.
2. Potassium, par M. Rousseau.
   Sodium, Cæsium et Rubidium, par MM. Rousseau et de Forcrand.
3. Lithium et Ammonium, par MM. Villiers et de Forcrand.
4. Calcium, Baryum, Strontium, Magnésium et Aluminium, par MM. Nivoit et Margottet.
5. Glucinium, Zirconium, Thorium, Yttrium, Cérium, Lanthane, Didyme, Samarium, Erbium, Holmium, Thulium, Ytterbium, Scandium, Gallium et Indium, par MM. Clève, Lecoq de Boisbaudran et Sabatier.
6. Zinc, Cadmium et Thallium, par M. Sabatier.
7. Niobium, Tantale et Tungstène, par M. Joly.
8. Molybdène, Vanadium et Titane, par M. Parmentier.
9. Fer et Chrome, par MM. Joannis et Moissan.
10. Manganèse, par M. Moissan.
11. Uranium, Étain et Antimoine, par MM. Ditte et Guntz.
12. Cobalt et Nickel, par M. Meunier.
13. Bismuth, par M. Godefroy.
    Plomb, par M. Parmentier.
14. Cuivre et Mercure, par MM. Rousseau et Joannis.
15. Argent, par M. de Forcrand $\begin{cases} 1^{re} \text{ Partie : Étude théorique.} \\ 2^e \quad - \quad \text{Applications.} \end{cases}$
16. Or, par MM. Cumenge et Fuchs.
17. Platine et Métaux qui l'accompagnent, par MM. Debray et Joly.

VOL. IV.

CHIMIE INORGANIQUE. ANALYSE CHIMIQUE.

Résumé d'analyse inorganique, par M. Carnot, Directeur du Laboratoire de l'École des Mines.

Tableaux analytiques, par M. Prunier, professeur à l'École supérieure de Pharmacie.

Analyse des Gaz, par M. Ogier.

Résumé de Chimie analytique appliquée spécialement à l'Industrie et à l'Agriculture, par M. Muntz, Directeur des laboratoires à l'Institut agronomique.

FREMY, EDMONDE ; Encyclopédie. [Cont'd.]

VOL. V.

APPLICATIONS DE CHIMIE INORGANIQUE.

1re Section : Produits Chimiques.

1re Partie : Acide sulfurique et Soude, par M. Sorel.

2e Partie : Généralités, Chlorure de chaux, Phosphates de chaux, Superphosphates, Aluns, Sulfates d'Alumine, Chlorates, par MM. Fremy,
Kolb, Nivoit, Pommier et Péchiney.

2e Section : Industries Chimiques.

1re Partie : Mortiers, et Ciments, par M. Duquesnay.

Matériaux de construction, pierre, brique.

Le Verre et le Cristal, par M. Henrivaux.

Poterie, Faïence.

Porcelaine.

Éclairage électrique, par M. Violle.

Industrie du Gaz, par MM. Marguerite et Camus.

Substances explosives, par MM. Sarreau et Vieille.

Galvanoplastie, par MM. Christophle et Bouillet.

Photographie, par M. Pabst.

2e Partie : Généralités sur la Métallurgie et Cuivre, par MM. Gruner et
Roswag.

Aluminium, par M. Wickersheimer.

Fer, Fonte, par M. Bresson.

Aciers, par M. Bresson.

Étain, par M. Lodin.

Zinc et Plomb, par M. Lodin.

Argent, par M. Roswag.

Désargentation des minerais de plomb, par M. Roswag.

Or, par MM. Cumenge et Fuchs.

Nickel et Cobalt, par M. Villon.

VOL. VI.

CHIMIE ORGANIQUE.

1er Fascicule.

Généralités, Carbures d'hydrogène, par MM. Villiers et Bourgoin.

2e Fascicule.

Alcools et Phénols, par M. Prunier.

VOL. VII.

CHIMIE ORGANIQUE.

3e Fascicule.

Aldéhydes, { 1re Section : Aldéhydes proprement dits et Acétones.
  par M. { 2e Section : Camphres, Aldéhydes à fonction mixte,
Bourgoin. {   Quinons.

FREMY, EDMONDE ; Encyclopédie. [Cont'd.]

4ᵉ Fascicule.

Éthers, par M. Leydié.

5ᵉ Fascicule.

Acides organiques,
   par M. Bourgoin.
{ 1ʳᵉ Section : Acides gras, par MM. Bourgoin et
   Riban.
2ᵉ Section : Acides à fonction simple et à fonction mixte.

VOL. VIII.

CHIMIE ORGANIQUE.

6ᵉ Fascicule.

Alcalis organiques.
{ 1ʳᵉ Section : Alcalis organiques artificiels, par M.
   Bourgoin :
      1ʳᵉ partie : série grasse.
      2ᵉ partie : série aromatique.
2ᵉ Section : Alcaloïdes naturels, par M. Chastaing.

7ᵉ Fascicule.

Amides, Matières albuminoïdes, Composés { 1ʳᵉ partie, série grasse.
   cyaniques, etc., par M. Chastaing.   2ᵉ — série aromatique.

8ᵉ Fascicule.

Radicaux organo-métalliques, par M. Chastaing.
Isomérie de position, par M. Colson.

VOL. IX.

CHIMIE ORGANIQUE.

1ʳᵉ Section : Chimie Biologique.

Microbiologie, par M. Duclaux.

2ᵉ Section : Chimie Physiologique.

1ᵉʳ Fascicule : Structure de la Plante, par M. Fremy.
2ᵉ Fascicule :
   1ʳᵉ Partie : Analyse chimique des liquides et des tissus de l'orga-
      nisme, par MM. Garnier et Schlagdenhauffen.

VOL. X.

APPLICATIONS DE CHIMIE ORGANIQUE.

Contribution à l'étude de la Chimie agricole, par MM. Schlœsing, père
   et fils.
Analyse des végétaux, par MM. Draggendorff et Schlagdenhauffen.
Industries et Sels ammoniacaux par M. Vincent.

FREMY, EDMONDE ; Encyclopédie. [Cont'd.]
 Nutrition de la Plante, par M. Dehérain.
 Fabrication du Papier.
 Fabrication du Sucre.
 Fabrication de la Gélatine.
 Le Bois.
 Substances textiles.
 Matières colorantes, par MM. Girard et Pabst.
 Conservation des aliments.
 Falsifications et moyens de les reconnaître.
 Teinture et apprêts des tissus de coton, par M. Lefèvre.
 Fabrication des couleurs, par M. Guignet.

> The value of a Dictionary of Chemistry, like that of a diamond, is not always proportionate to its size, since in both cases quality is an important consideration.

FRYER, JOHN.
 Hwa hioh tsai liao chung hsi ming mu piao. [Vocabulary of chemical names in Chinese.] Shanghai, 1882.

> *Cf.*, *in Section V*, Fryer, John.

GABBA, LUIGI.
 Manuale del chimico et dell' industriale. Raccolta di tabelle, dati fisici e chimici ad uso dei chimici analitici e tecnici, degli industriali, dei fabricanti di prodotti chimici, degli studenti di chimica, ecc. Milano, 1889. pp. xii–354, 16mo.

GARLANDIUS, JOHANNES.
 Compendium alchimiæ cum dictionario ejusdem artis atque de metallorum tinctura præparatio neque eorundem libello ante annos MDXX eodem authore conscripto. Basiliæ, 1560. pp. xvi–174, 18mo.

> The dictionary or synonymy occupies pages 53 to 70.

GARNIER, JULES.
 Nomenclature chimique française, suédoise, allemande ; et synonymie. Paris, 1844. 12mo.

> Nomenclatura chymica franceza, sueza, allemá e synonymia ; escripta em francés por J. Garnier e vertida em lingua portugueza por J. P. Reis. Lisboa, 1846. 8vo.

GERDING, TH.
 Taschen-Lexikon der Chemie und der damit verbundenen Operationen. Leipzig, 1864. pp. iv–966, 8vo.

> Zakwoordenboek der scheikunde en der scheikundige bewerkingen. Vrij gevolgd naar het Hoogduitsch en

GERDING, TH. [Cont'd.]

> voor Nederlanders bewerkt door R. J. Opwijrda.
> Met voorrede van J. H. van den Broek. Utrecht, 1870.
> Nieuwe goedkoope uitgaaf. Utrecht, 1873, 8vo.

GERGENS, P., AND S. HOCHHEIMER.

> Tabellen über die chemische Verwandtschaft der Körper auf dem nassen
> und trocknen Wege, wie auch die Entstehung der Mittelsalze,
> durch Verbindung der Säuren mit alkalischen Salzen, Erden und
> Metallen, und verschieden zusammengesezte Körper. Mit einem
> Vorbericht worinn in Kurzem verschiedenes Nützliches den An-
> fängern mitgetheilt wird. Frankfurt am Main, 1790. pp. 78, 12mo.
> Three folding tables.

GERSTENHÖFER, MORITZ.

> Hülfsbuch für den gewerblichen Chemiker, oder Sammlung von Formeln,
> Regeln und Tabellen der Mechanik, technischen Wärmelehre und
> Chemie. Zum Gebrauche bei Anlage und rationellen Betriebe von
> chemischen Fabriken, Salinen, metallurgischen Etablissements, Por-
> zellan- und Glasfabriken, Seifensiedereien, Färbereien, Brauereien,
> etc. sowie auch beim Studium der chemischen Technologie. Nach
> den neuesten Forschungen der technologischen Wissenschaften
> bearbeitet. Leipzig, 1851. 8vo.
>
> Общепонятная вспомогательная книга для промышленныхъ химиковъ
> или собраніе формулъ, правилъ и таблицъ механики и химіи для
> руководства при устроеніи и хода чугунныхъ. желѣзныхъ, стальныхъ,
> мѣдныхъ, фарфоровыхъ, стеклянныхъ, мыловаренныхъ и химическихъ
> заводовъ, саловаренъ, красиленъ, пивоваренъ и проч., съ присово-
> купленіемъ новой и выгодной методы устроенія печей разнаго рода.
> Москва, 1852. 8vo.

GIRTANNER, CHRISTOPH.

> Neue chemische Nomenklatur für die deutsche Sprache. Berlin, 1791.
> pp. 22, 8vo.

GÖTTLING, J. F. A.

> Physisch-chemische Encyklopädie oder Physisch-chemischer Haus-
> freund. Jena, 1805-07. 3 vols.

GROTHUSS, THEODOR FREIHERR VON [*really* CHRISTIAN JOHANN
DIETRICH].

> Verbindungsverhältniss oder chemische Aequivalenten-Tafeln u. s. w.
> Nürnberg, 1821. Fol.

GRUNER, J. L. W.

> Tabellarische Uebersicht der Salze und ihrer Bestandtheile nach dem
> phlogistischen und antiphlogistischen System. Hannover, 1795.
> Fol.

GUARESCHI, J.
Enciclopedia di chimica scientifica e industriale colle applicazioni all'
agricoltura ed industrie agronomiche, alla metallurgia, alla merciologia, ecc.   Torino, 1885.
Supplemento annuale alla enciclopedia di chimica scientifica ed
industriale diretto da Icilio Guareschi.   Anno primo.   Torino,
(December) 1884, continued to 1886 [+ ?].

GUTIERREZ BUENO, PEDRO.
Nomenclatura química, para el uso de su escuela pública.   Segunda
edicion . . . Madrid, 1801.

GUYTON DE MORVEAU, LOUIS BERNARD.
La clef de la chimie, ou la nomenclature chimique mis à la portée de
toutes les intelligences, inventée 1782. Publiée par L. M. Pioger.
Mans, 1855.   Imp. fol.

GUYTON DE MORVEAU, LOUIS BERNARD, LAVOISIER, BERTHOLLET ET DE
FOURCROY.
*Méthode de nomenclature chimique proposée par de Morveau.   On y
a joint un nouveau système de caractères chimiques, adaptés à cette
nomenclature par Hassenfratz et Adet.   Paris, 1787.   pp. iii–314,
8vo.   Folding plates.

> This is the corner-stone of modern chemical terminology, and with
> Lavoisier's " Traité élémentaire de chimie" exerted the greatest
> influence on the progress and stability of the science.

Methode der chemischen Nomenclatur für das antiphlogistische System.   Aus dem Französischen.   Wien, 1793.
8vo.
Method of Chymical Nomenclature, proposed by Messrs.
De Morveau, Lavoisier, Berthollet and Fourcroy, to
which is added a new system of chymical characters,
adapted to the nomenclature by Messrs. Hassenfratz
and Adet.   Translated by James St. John.   London,
1788.   8vo.

Cf. Spalding, Lyman.   *Also* Arejula, Juan Manuel.

GYRA, NAUM VON.
Das System der Aequivalente oder folgerechte Herleitung der
Aequivalente, der specifischen Wärme, des chemischen Characters
und sämmtlicher binären, ternären und quarternären Verbindungen aller bekannten und unbekannten Grundstoffe.   Wien, 1852.
8vo.   1 Plate.

HAHNEMANN, SAMUEL.

> Handwörterbuch der Chemie, nach den neuesten Theorien und nach ihrer praktischen Anwendung auf Künste, Gewerbe und Fabriken, sowie auf Pharmazie, Medicin, etc. Mit Hinsicht auf Naturwissenschaften und allgemeine Waarenkunde nach dem Dictionnaire de Chimie par Brismoutier, le Coq et Boisduval bearbeitet und mit den neuesten Entdeckungen ingleichen mit der lateinischen, französischen und englischen Nomenclatur versehen von H. Leng. Ilmenau, 1828. 8vo.
>
>> The author was the founder of Homœopathy, one of the most remarkable delusions of the century.

HAMILTON, GEORGE.

> Suggestions for a New System of Chemical Nomenclature. Liverpool, 1866.

HAMY, R. P. A.

> Tableaux synoptiques de physique et chimie. Lille et Bruges, 1881. pp. 74, long 8vo.
>
>> Notwithstanding the date of these tables, the author (who writes "S. J." after his name) employs the old notation.

HANDWÖRTERBUCH DER CHEMIE, herausgegeben von Ladenburg.

> *See* Ladenburg, A.

HANDWÖRTERBUCH DER CHEMIE UND PHYSIK von E. F. August, F. W. Barentin, W. Beetz, G. Bischof, W. H. Dove, W. Hankel, R. Hoffmann, L. F. Kaemtz, K. F. Kloeden, W. Knop, W. Mahlmann, R. F. Marchand, F. Minding, F. W. G. Radicke, J. A. W. Roeber, L. F. W. A. Seebeck, R. Wagner, E. Zenker, u. a. M. Berlin, 1842–48. 3 vols., 8vo. *For Supplement, see* Wagner, Rudolf.

HANDWÖRTERBUCH DER REINEN UND ANGEWANDTEN CHEMIE. In Verbindung mit mehreren Gelehrten herausgegeben von J. Liebig, J. C. Poggendorff und Fr. Wöhler. Redigirt von Hermann Kolbe. Braunschweig, 1837–64. 9 vols., 8vo. Vol. II in 3 parts.

> A second edition of vols. I and II was published by Hermann von Fehling, 1857–62. 4 vols., 8vo.

HANDWÖRTERBUCH (NEUES) DER CHEMIE. Auf Grundlage des von Liebig, Poggendorff und Wöhler, Kolbe und Fehling herausgegebenen Handwörterbuchs der reinen und angewandten Chemie und unter Mitwirkung von mehreren Gelehrten bearbeitet und redigirt von Hermann von Fehling. [*From* vol. IV, *by* Carl Hell.] Braunschweig, 1871–92. 6 vols. *In progress.*

HANIN, L.

> Tableau de caractères chimiques et pharmaceutiques. Fol. *In his :* Vocabulaire médical. Paris, 1811. 8vo.

HARRIS, J.
Chymical characters. *In his :* Lexicon technicum. London, 1708.

HARRIS, WILLIAM A.
A Technological Dictionary of Insurance Chemistry. Liverpool, 1889.

HARTMANN, C.
Encyclopädisches Wörterbuch der Technologie, der technischen Chemie, der Physik und des Maschinenwesens. Augsburg, 1839–41.

HARTRODT, A.
Die Alkaloide, oder Darstellung der Bereitungsarten, der physischen, chemischen und medicinischen Eigenschaften der bis jetzt bekannten Pflanzenalkalien in alphabetisch tabellarischer Form. Leipzig, 1832. 4to.

HASSENFRATZ ET ADET.
Système de caractères chimiques.
See Guyton de Morveau, Lavoisier : Méthode de nomenclature chimique.

HELLWIG, CHRISTOPH.
Neu eingerichtetes Lexicon medico-chymicum, Oder chymisches Lexicon worinnen nicht allein die Nahmen der nöthigsten Laborum chymicorum sondern auch die gebräuchlichsten Vasa, Oefen, Instrumenta, etc, benennet ; nebst andern nützlichen Dingen. Worbey auch unterschiedliche Stücke, was for Composita daraus præpariret und laboriret werden, und von deren Tugenden, Kräfften, Dosibus, etc. zu finden.   Lateinisch und Teutsch nach dem Alphabeth eingerichtet.   Nebst einem Anhang etlicher Apotheker-Taxe als einer Zugabe.   Franckfurt und Leipzig, 1711.   pp. (vi)–488 12mo.

HEPPE, GUSTAV.
Die chemische Reactionen der wichtigsten anorganischen und organischen Stoffe.   Tabellen in alphabetischer Anordnung zum Gebrauche bei Arbeiten im Laboratorio.   Leipzig, 1875.   pp. xvii–391 roy. 8vo.
Unfortunately the author adopts the old atomic weights, O = 8, etc.

HERBERGER, JOHANN EDUARD.
Systematisch-tabellarische Uebersicht der chemischen Gebilde organischen Ursprungs mit genauer Angabe ihrer Eigenschaften im Zustande der Einfachheit, und in jenem der Verbindung mit andern Körpern.   Für praktische Chemiker, für Aerzte und Apotheker nach den vorzüglichsten Quellen und mit Zuziehung der eigenen Erfahrung bearbeitet.   Erste Lieferung : Die electropositiven organisch-

HERBERGER, JOHANN EDUARD. [Cont'd.]
chemischen Gebilde. Zweite Lieferung: Die electronegativen organisch-chemischen Gebilde. Nürnberg, 1831–1836. Folio. Vol. I : pp. 10 and 35 Tables ; II : pp. viii and 38 Tables + 52 Tables.

HILDEBRANDT, GEORG FRIEDRICH.
Encyclopädie der gesammten Chemie. Erlangen, 1799–1806. 9 vols., 8vo, and Supplement, 1807. Fifteen plates.
Zweite ganz umgearbeitete Auflage. Erlangen, 1809.
Erstes Supplement. Erlangen und Leipzig, 1815.

HOCHHEIMER, CARL FRIEDRICH AUGUST.
Chemisches Farben-Lexicon . . . Leipzig, 1792–94. 2 vols., 8vo.

HOEFER, FERDINAND.
Dictionnaire de chimie et de physique. Paris, 1846. 12mo.
Nomenclature et classifications chimiques suivie d'un lexique historique et synonymique comprenant les noms anciens, les formules, les noms nouveaux, le nom de l'auteur et la date de la découverte des principaux produits de la chimie. Paris, 1845. pp. vii–184, 12mo.

> A convenient handbook with a history of chemical nomenclature. The 62 pages of tables contain the data named of most of the simple and compound bodies known in 1845 ; the size of a volume on the same plan at this date can only be conjectured.

> Nomenclatura e classificazioni chimiche seguite da un lessico storico e sinonímico che comprende i nomi antichi, le formole, i nomi moderni, il nome dell'autore e la data della scoperta dei principali prodotti della chimica. Versione dal Francese, con molte note ed aggiunte di Ferdinando Tonini. Parte I : Nomenclatura e classificazioni dei principali prodotti chimici. Modena, 1847, 8vo.

> Nomenclatura y classificaciones químicas, seguidas de un lexico historico y sinonímico, que comprende los nombres antiguos, las fórmulas, los nombres nuevos, el nombre del autor y fecha del descrubimiento de los principales productos de la química. Madrid, 1853, 8vo.

Alcuni preliminari di chimica generale, nei quali si dà contezza della recente modificazione proposta per la nomenclatura e la classificazione dei corpi ; seguiti da una esposizione breve e chiara della nuova dottrina chimico-organica di Liebig fatta da Ferd. Hoefer. Opusculo pubblicato a vantaggio di tutti quelli che si accingono allo studio della chimica da Francesco Selmi. Modena, 1843. 8vo.

HOFMANN.

Lexicon der chemisch-technischen und pharmaceutischen Präparate
. . . Mit einem Vorworte und Sachregister von E. Winckler.
Dresden, 1861. 8vo.

HOFFMANN, CARL AUGUST.

Systematische Uebersicht und Darstellung der Resultate von zwey
hundert und zwey und vierzig chemischen Untersuchungen mineralischer Wasser von Gesundbrunnen und Bädern in den Ländern
des deutschen Staatenvereins und deren nächsten Begränzungen.
Nebst Anzeige aller über diese Heilwasser erschienenen Schriften.
Berlin, 1815. pp. vi–410, 12mo.

> The chronologically arranged bibliography occupies pages 259 to 408, and
> is important.   The water-analyses are arranged under names of places
> alphabetically.

Tabellarische Bestimmung der Bestandtheile der merkwürdigsten
Neutral und Mittelsalze in hundert Theilen ; nach Bergmann,
Kirwan, Wiegleb, Wenzel und anderen Scheidekünstlern ; nebst
Bezeichnung ihrer Auflösbarkeit sowohl in Wasser als Weingeist
und der Bemerkung der eigenthümlichen Schwere der Körper für
Physiker, Aerzte, Scheidekünstler und Apotheker.   Weimar, 1791.
Fol.   Two tables.

Tabellarische Uebersicht aller zur pharmaceutischen Scheidekunst
gehörigen Werkzeuge und Geräthschaften nebst kurzer Beschreibung der Anwendung derselben.   Weimar, 1791.

HOFFMANN, ROBERT.

Sammlung aller wichtigen Tabellen, Zahlen und Formeln für Chemiker.
Berlin, 1861.   8vo.

> Zweite vermehrte und verbesserte Auflage.   Nach den neu-
> esten Fortschritten der Chemie zusammengestellt von
> Carl Schaedler.   Berlin, 1877.   8vo.

HOPKINS, ALBERT A. [Editor].

Scientific (The) American Cyclopedia of Receipts, Notes and Queries.
New York, 1892.   pp. [iv]-675, 8vo.   Ill.

> Contains in the Appendix, Part III, Chemical Synonyms, pp. 661–675.

HURST, GEORGE H.

A Dictionary of the Coal-Tar Colours.   London, 1892.   pp. iii–106.
8vo.

> Eminently practical.

HUTH, E.

Vorschläge zur Vereinfachung der Zeichensprache und Nomenklatur
in der organischen Chemie.   Breslau, 1888.   8 pp.

> Sammlung naturwissenschaftlicher Vorträge.

JACOBSEN's technologisches Wörterbuch. Berlin, 1796. 8 parts. 4to.

JEHAN, L. F.

Dictionnaire historique des sciences physiques et naturelles depuis l'antiquité la plus reculée jusqu' à nos jours. Origine et progrès de la science chez les différents peuples. Essai d'une explication des prodiges phénomènes singuliers, magie, arts et pratiques diverses, erreurs et préjugés. Histoire naturelle dans l'antiquité et au moyen âge. Notice biographique sur les auteurs qui se sont fait un nom par leurs travaux ou leurs découvertes dans ces branches des connaissances humaines ; examen critique et analyse de leurs ouvrages et de leurs théories. Mouvement philosophique de la science, principes et doctrines à notre époque, etc., etc. Publié par M. l'abbé Migne. Paris, 1857. pp. 1, 107, sm. folio. Also as volume xxx of Migne's Encyclopédie théologique. Paris, 1857.

> Written for a certain type of theologians, this contains much matter not easily found elsewhere.

JOHN, JOHANN FRIEDRICH.

Chemische Tabellen des Thierreichs, oder systematische Uebersicht der Resultate aller bis jetzt zerlegten Animalien mit Rücksicht auf die wichtigsten medicinischen Thatsachen welche aus der Chemie entlehnt sind ; einige wichtige chemische Erscheinungen der Zoochemie und Eigenschaften der animalischen Körper, und die Literatur. Berlin, 1814. pp. viii–138, folio.

Chemische Tabellen der Pflanzenanalysen, oder Versuch eines systematischen Verzeichnisses der bis jezt zerlegten Vegetabilien nach den vorwaltenden nähreren Bestandtheilen geordnet und mit Anmerkungen versehen. Nürnberg, 1814. pp. x–94, folio.

Handwörterbuch der allgemeinen Chemie. In alphabetischer Ordnung. Leipzig, 1817–19. 4 vols., 8vo.

JOHNSON, GULIELMUS.

*Lexicon chymicum cum obscuriorum verborum et rerum hermeticarum tum phrasium Paracelsicarum in scriptis ejus et aliorum chymicorum passim occurrentium planam explicationem continens. Londini, 1652. pp. [xvi]–259, 16mo.

> Another edition. Londini, 1657. pp. 228.
> Another edition. London, 1660. pp. [xii-]259, 16mo. *Followed by :* Lexicon chymicum continens vocabula chymica in priore libro omissa, multis vocabulorum chymicorum characteribus adjectis, e Basilio Valentino,

JOHNSON, GULIELMUS. [Cont'd.]

> Theophrasto Paracelso, Oswaldo Crollio, aliisque
> authoribus chymicis collectis. Liber Secundus. Lon-
> dini, 1660. pp. (xxiv)–72–(xii).
>
> Other editions : Frankfurt, 1676 ; Leipzig, 1678. Also re-
> printed in Manget's Bibliotheca chemica curiosa, I, 217.
>
> The appendix contains a life of Paracelsus. The medley of chemical terms
> and definitions, with methods of divination, mystical terms, supersti-
> tions and fables, illustrates a curious phase of the science.

JUNGKEN, JOHANNES HELFRICUS.

* Lexicon chimico-pharmaceuticum in duas partes divisum, quarum
prior continet processus selectos chimicos, sive medicamentorum
chimicorum vulgo, sed usualium potissimum elaborationes, ex recentiorum curiosorum medicorum laboratoriis noviter prodeuntes,
selectas et probatas. Altera pars composita pharmaceutica tam
usualia, quam alia his subordinata et rationi consentanea exhibet ;
adeo ut medicus practicus, nec non pharmacopœus, lexico hoc
instructus, diffusioribus aliis materiam hanc pertractantibus carere
commodè queat. Opus diu desideratum et proxime in commodum
ita dispositum. Norimbergæ, 1699. pp. [xxxii]–265–[xiii]–423–
[xxxvi], 12mo.

> Second edition in 1709.
>
> Third edition : Lexicon chymico-pharmaceuticum in duas
> partes distinctum ; ubi pars prior continet selectos
> processus chymicos potissimum hactenus magis usuales
> et originaliter et medicorum, non vero pharmaco-
> polarum, laboratoriis prodeuntes ; pars altera exhibet
> composita pharmaceutico-Galenica, tam hactenus usu-
> alia, quam alia his subordinata et correctiora dicta.
> Editio tertia, novitatibus nonnullis auctior reddita, et
> quidem cum præfamine de contractioribus pharmaco-
> poliis, juxta modernæ rei medicæ statum, rite instituen-
> dis, atque medicorum non tantum privatam medica-
> mcntorum chymicorum elaborationem, sed et dispensa-
> tionem a pharmacopolis hinc inde acriter impugnatam
> concernente, etc. Norimbergæ, 1729. pp. 535–[ix],
> 12mo. Folding title-page.
>
> Pharmaceutical-chemistry in two parts, as stated. Arranged alphabetically
> by substances.

KÆPPELIN, R.

Tableaux synoptiques de chimie présentant les corps inorganiques
rangés par classes, ordres, familles, tribus, genres, et espèces,

KÆPPELIN, R. [Cont'd.]

d'après leurs analogies naturelles, avec indication des principaux caractères pour les plus importants d'entre eux, et servant avantageusement pour la recherche des formules et des poids atomiques. Paris, 1842. pp. vi, and 9 tables, folio.

KARMARSCH, K. UND HEEREN.

See Ure, Andrew.

KEIR, JAMES.

Dictionary of Chemistry. Birmingham, 1789. London, 1790. 4to.

KELS, HEINRICH WILHELM.

* Onomatologia chymica practica, oder vollständig practisches Handbuch der Chemie in alphabetischer Ordnung zum Nutzen und Gebrauch für Aerzte, Apotheker, Fabrikanten, Künstler und andere Personen ; unter der Aufsicht und mit einer Vorrede von Johann Friedrich Gmelin. Ulm, 1791. pp. [x]–722 columns–[29].

KERNDT, CARL HULDREICH THEODOR.

Encyclopädie der chemisch-technischen Wissenschaften. Leipzig, 1860.

Poggendorff states only one part (Lieferung) was published up to date.

KERR, J. G.

Chemical Terms in Chinese and English. Canton, 1875 [?].

Cf. in Section V, Kerr, J. G.

KICK, FR. UND GINTL.

See Ure, Andrew ; Karmarsch, K. und Heeren.

KIELMANN, KARL ALBRECHT.

Systematische Darstellung aller Erfahrungen über die einzelnen Metalle. *See in Section V*, Meyer, Johann Rudolph, junior.

KLAPROTH, MARTIN HEINRICH.

Chemisches Wörterbuch. Berlin, 1807–10. 5 vols., 8vo. Supplemente. Berlin, 1815–19. 4 vols., 8vo.

Vol. IV of Supplement contains an Index to all the nine volumes. Vol. I contains portrait of Klaproth.

KLAPROTH, MARTIN HEINRICH, AND F. WOLFF.

* Dictionnaire de chimie. Traduit de l'allemand, avec des notes, par E. J. B. Bouillon-Lagrange et par H. A. Vogel. Paris, 1810–11. 8vo. 4 vols. Vol. I, pp. viii–494, 1810 ; vol. II, pp. 542, 1810 ; vol. III, pp. 512, 1811 ; vol. IV, pp. 569, 1811.

Vol. I contains portrait of Klaproth.

KLENCKE, PHILIPP FRIEDRICH HERMANN.

Illustrirtes Lexikon der Verfälschungen der Nahrungsmittel und Getränke. Leipzig, 1878. 8vo.

> Handwoordenboek der vervalschingen, die in voedingsmiddelen, dranken en andere artikelen van dagelijksch gebruik voorkomen, met aanwijzing der zekerste middelen om die zelf te kunnen opsporen en echtheid van vervalsching te onderscheiden. Een handboek voor iedereen. Naar den tweeden Hoogduitschen druk, voor Nederland bewerkt door A. J. de Bruijn. Utrecht, 1880. 8vo. Ill.

KOLLMYER, A. H.

Chemia Coartata ; or, The Key to Modern Chemistry. London, 1876. 16mo, oblong.

> A series of tables presenting the facts and phenomena of chemistry.

KRAMER, ANTON JOHANN VON.
> *See* Laugier, E.

KRAUSE, G.

Tabelle zum Gebrauche für chemische, technische, mineralogische und pharmaceutische Laboratorien, Real- und Gewerbeschulen, enthaltend die Namen, Symbole, Quantivalenzen, Atom- und Aequivalentgewichte, Volumgewichte, specificirte Gewichte, Härten, Schmelzpunkte [Siedepunkte], specificirte Wärmen, Jahre der Entdeckung und die Namen der Entdecker der chemischen Elemente. Zweite verbesserte Auflage. Köthen, 1876. 4to.

LABOULAYE, CHARLES.

Encyclopédie technologique. Dictionnaire des arts et manufactures et de l'agriculture formant un traité complet de technologie. Paris. First edition, 1847 ; sixth edition, 1886, 4 vols., roy. 8vo.

Dictionnaire des arts et manufactures et de l'agriculture. Description des procédés de l'industrie française et étrangère. Cinquième édition. Paris, 1881. 3 vols., roy. 8vo, and a " Complément." Ill.

> Contains many articles on chemical technology, with special reference to France.

LADENBURG, A.

Handwörterbuch der Chemie herausgegeben von Ladenburg unter Mitwirkung von Berend, Biedermann, Drechsel, Emmerling, Engler, Grehm, Heumann, Jacobsen, Pringsheim, von Richter, Rügheimer, Salkowski, Tollens, Weddige, Wiedemann. Breslau, 1882–89. 6 vols., roy. 8vo. Ill.

> Encyclopädie der Naturwissenschaften, II Abtheilung.

LAGRANGE, EDMONDE JEAN BAPTISTE BOUILLON.
Tableau réunissant les propriétés physiques et chimiques des corps.
Paris, 1799.

LAMPADIUS, WILHELM AUGUST.
Handwörterbuch der Hüttenkunde. . . . Göttingen, 1817.

LANDOLT, H., UND RICHARD BÖRNSTEIN.
Physikalisch-chemische Tabellen. Berlin, 1883. pp. xii–249, roy. 8vo.

LASSAIGNE, JEAN LOUIS.
Dictionnaire des réactifs chimiques employés dans toutes les expériences faites dans les cours publiques et particuliers, les recherches
médico-légales, les expertises, etc. Bruxelles, 1840. 8vo. Plate.
  Dizionario pittoresco e cromoscopico dei reagenti chimici
  usati in tutte le esperienze dei corsi pubblici e parti
  colari, nelle ricerche medico-legali, nei processi verbali,
  negli assaggi, nelle visite farmaceutiche, nelle analisi
  qualificative e quantitative dei corpi semplici e de' loro
  composti, adoperati tanto nelle arti che in medicina.
  Prima traduzione Italiana accresciuta delle nuove
  manipolazioni chimiche di H. Violette e di tutte le più
  recenti scoperte relative, per G. B. Sembenini. Mantova, 1841–42. 2 vols., 8vo.
  Manual de reactivos químicos extractado del diccionario
  de reactivos de J. L. L. traducido libremente del francés
  con algunas adiciones por P. L. Aguilon. Pamplona,
  1846. 8vo.

LAUGIER, E., ET A. DE KRAMER.
Tableaux synoptiques, ou abrégé des caractères chimiques des bases
salifiables. Paris, 1828. 8vo.
  Synoptische Tabellen oder gedrängte Darstellung des chemischen Verhaltens salzfähigen Basen. Aus dem
  französischen übersetzt. Nürnberg, 1829. 8vo.

LENG, H.
  See Hahnemann, Samuel.

LEONHARDI, JOHANN GOTTFRIED.
  Cf. Macquer, P. J.

LEUCHS, J. C.
Polytechnisches Wörterbuch, oder Erklärung der in der Chemie, Physik,
Mechanik, Technologie, Fabrikwissenschaft, in den Gewerben, etc.,
gebräuchlichsten Wörter und Ausdrücke. Zweite Auflage. Nürnberg, 1835.

LEXICON DER GENUSS- UND NAHRUNGSMITTEL.   Stendal, 1810.   8vo.

LIEBIG, POGGENDORFF UND WÖHLER.
    Handwörterbuch der reinen und angewandten Chemie.   *See* Handwörterbuch.

LOBSCHEID, WILLIAM.
    Tourists' Guide and Merchants' Manual.   Shanghai [?], 1884.
    This work contains the principle names connected with chemistry in Chinese.

MACQUER, PIERRE JOSEPH.
    Dictionnaire de chymie contenant la théorie et la pratique de cette
        science, son application à la physique, à l'histoire naturelle, à la
        médecine et à l'économie animale ; avec l'explication détaillée de la
        vertu et de la manière d'agir des médicaments chymiques.   Et les
        principes fondamentaux des arts, manufactures et métiers dépen-
        dans de la chymie.   Paris, 1766.   2 vols., 12mo.   Vol. I, pp.
        xxiv–616 ; II, 688.
            * Seconde édition.   Paris, 1778.   4 vols., 12mo.
            Another "seconde édition."   Paris, 1778.   2 vols., 4to.
            A third "nouvelle édition" was issued in 1769–1770, at
                Yverdon [Suisse].   4 vols., 8vo.
            A Dictionary of Chemistry.   Containing the theory and
                practice of that science ; its application to natural
                philosophy, natural history, medicine and animal econ-
                omy : with full explanations of the qualities and modes
                of acting of chemical remedies : and the fundamental
                principles of the arts, trades and manufactures depen-
                dent on chemistry.   Translated from the French.   With
                plates, notes and additions by the translator.   In two
                volumes.   London, 1771.   pp. xii, vi, 888, 4to.
            Allgemeine Begriffe der Chemie in alphabetischer Ordnung
                aus dem französischen übersetzt und mit Anmerkungen
                vermehrt von Carl Wilhelm Pörner.   Leipzig, 1768–9.
                3 vols., 8vo.
            Chymisches Wörterbuch, oder Allgemeine Begriffe der
                Chymie nach alphabetischer Ordnung.   Aus dem
                franzözischen nach der zweyten Ausgabe übersetzt
                und mit Anmerkungen und Zusätzen vermehrt von
                Johann Gottfried Leonhardi.   Leipzig, 1781–83.   6
                vols., 8vo.   Zweyte Ausgabe.   Leipzig, 1788–91.   7
                vols., 8vo.
            Neue Zusätze und Anmerkungen zu Macquer's Chymischem
                Wörterbuche erster Ausgabe.   Leipzig, 1792.   2 vols.,
                8vo.

MACQUER, PIERRE JOSEPH. [Cont'd.]

Chymisches Wörterbuch oder allgemeine Begriffe der Chymie nach alphabetischer Ordnung. Aus dem französischen nach der zweyten Ausgabe übersetzt und mit Anmerkungen und Zusätzen vermehrt von Johann Gottfried Leonhardi. Dritte ganz umgearbeitete Ausgabe mit Hinweglassung der blossen Vermuthungen und mit Ergänzungen durch die neuern Erfahrungen veranstaltet von Jer. Benj. Richter. Leipzig, 1806–1809. 3 vols., 8vo.

Query. Were the remaining volumes of this edition published ?

Dizionario di chimica di Pietro Giuseppe Macquer ; tradotto dal Francese, e corredato di note e di nuovi articoli da Giovanni Antonio Scopoli. Pavia, 1783–84. 7 vols., 8vo. Tables and plates.

Chymisk Dictionnaire. Indeholdende denne Videnskabs Theorie og Praxin, dens Anvendelse paa Physikken, Naturhistorien, Lægekunsten og den dyriske Forfatning, med omstændig Forklaring over de chymiske Lægemiddelers Kraft og Virknings Maade osv. med tilføjede Anmerkninger. Af det Franske oversat af Hans von Aphelen. Kjøbenhavn, 1771–72. 3 vols.

MATECKI, TEODOR TEOFIL.

Słownictwo chemiczne polskie. Poznań, 1855. 8vo.

*MEDICINISCH-CHYMISCH UND ALCHEMISTISCHES ORACULUM darinnen man nicht nur alle Zeichen und Abkürzungen welche so wohl in den Recepten und Büchern der Aerzte und Apotheker als auch in den Schriften der Chemisten und Alchemisten vorkommen findet, sondern deme auch ein sehr rares chymisches Manuscript eines gewissen Reichs . . . beygefüget. Ulm, 1772. 71 pp., 8vo. Neue Auflage. Ulm, 1783. 8vo.

A key to alchemical symbols more than 2000 in number. *Compare* Sommerhoff.

MEYER, LOTHAR, UND KARL SEUBERT.

Die Atomgewichte der Elemente aus den Originalzahlen neu berechnet. Leipzig, 1883. pp. x–245, 8vo.

This work was published shortly after one having a similar aim by Prof. F. W. Clarke, of the U. S. Geological Survey. *See* Clarke, F. W.

MIGNE, ABBÉ.

Encyclopédie théologique.

See Jehan, L. F.

MITCHELL, JAMES.

A Dictionary of Chemistry, Mineralogy and Geology, in accordance with the present state of those sciences. London, 1823. pp. xviii–630, 12mo. Ill.

MITCHILL, SAMUEL L.

Explanation of the Synopsis of Chemical Nomenclature and arrangement : containing several important alterations of the plan originally reported by the French Academicians. New York, 1801. 8vo.

MONS, J. B. VAN.

Conspectus mixtionum chemicarum, quas ad rationes perpetuas ordinavit signisque alphabeticis expressit. Louvain, 1827.

MORVEAU, LOUIS BERNARD GUYTON DE.
See Guyton de Morveau, L. B.

NEUES HANDWÖRTERBUCH DER CHEMIE.
See Handwörterbuch der reinen und angewandten Chemie. Dritte Auflage.

NICHOLSON, WILLIAM.

A Dictionary of Chemistry, exhibiting the present state of the theory and practice of that science, its application to natural philosophy, the processes of manufactures, metallurgy and numerous other arts dependent on the properties and habitudes of bodies in the mineral, vegetable and animal kingdoms ; with a considerable number of tables, expressing the elective attractions, specific gravities, comparative heats, component parts, combinations and other affections of the objects of chemical research. Illustrated with engravings. London, 1795. 2 vols., 4to ; I, pp. viii–576 ; II from 577–1132. Plates.
New edition. London, 1808.

NISBET, WILLIAM.

A General Dictionary of Chemistry, containing the leading principles of the science in regard to facts, experiments and nomenclature for the use of students. London, 1805. pp. viii–415, 12mo.
Contains a Table of Chemical Nomenclature giving synonymes. Published after the author's death.

OERSTED, JOHANNES CHRISTIANUS.

Tentamen nomenclaturæ chemicæ omnibus linguis scandinavico-germanicis communis. Hafniæ, 1814. pp. 35, 4to. *Contains :* Tabulæ comparativæ technicorum in chemia vocabulorum in linguis scandinavico-germanicis et gallica.

O'NEILL, CHARLES.

A Dictionary of Dyeing and Calico Printing, containing a brief account of all the substances and processes in use in the arts of dyeing and printing textile fabrics. Added an Essay on Coal-tar Colors and their application to dyeing and calico printing by A. A. Fesquet. With an Appendix on Dyeing. Philadelphia, 1869. 8vo.

ONOMASTICA DUO.

I. Philosophicum, medicum synonymum ex variis vulgaribusque linguis.

II. Theophrasti Paracelsi, hoc est earum vocum quarum in scriptis eius solet usus esse, explicatio ; nunc primum in commodum omnium philosophiæ ac medicinæ Theophrasticæ studiosorum, cuiuscunque nationis sint, fideliter publicata.

Gründliche Erklärung in allerlei Sprachen der philosophischen medicinischen und chimischen Namen welcher sich die Aerzte, Apotheker auch Theophrastus zu gebrauchen pflegen. *n. p.* 1574.

> A dictionary of medical and chemical synonymes. The Preface is signed Micaelus Toxites.

ORSONI, FRANCESCO.

> Nomenclatura chimico-quantitativa di alcune sostanze.
> *See, in Section V*, Orsoni, F. Note scientifiche.

OTTLEY, WILLIAM CAMPBELL.

A Dictionary of Chemistry and of Mineralogy as connected with it, in which is attempted a complete list of the names of substances according to the present as well as former systems ; with a vocabulary, copious notes and an introduction pointing out the order in which the chief parts of the work may be perused so as to constitute a regular course of chemistry. New edition containing the recent discoveries. London, 1828. 8vo, wanting in pagination.

PELLETAN, P., FILS.

Dictionnaire de chimie générale et médicale. Paris, 1822–24. 2 vols., 8vo.

PERNETY, ANTOINE JOSEPH.

*Dictionnaire mytho-hermétique, dans lequel on trouve les allégories fabuleuses des poètes, les métaphores, les énigmes et les termes barbares des philosophes hermétiques expliquées. Paris, 1787. pp. xxiv–546, 12mo.

> The author opens his preface thus : " Jamais science n'eut plus besoin de dictionnaire que la Philosophie Hermétique."

PERNETY, ANTOINE JOSEPH. [Cont'd.]

*Les Fables Égyptiennes et Grecques dévoilées et reduites au même principe avec une explication des hieroglyphes et de la guerre de Troye. Paris, 1758. 2 vols., 12mo.

> This and the preceding work of Pernety are attempts to explain ancient mythology on hermetical principles, a method of interpretation developed one hundred and forty years earlier by Michael Maier, the physician of Rudolph II. *See in Section VI*, Maier, Michael.

PHILOSOPHY REFORMED AND IMPROVED IN FOUR TRACTATES. [Crollius, Paracelsus, etc.] London, 1657.

> *Contains :* Alphabetical Glossary of unusual and uncouth words.

POISSON, ALBERT.

> Dictionnaire des symboles hermétiques.
> *See in Section VI*, Poisson, Albert, Collection [etc.].

PÖRNER, CARL WILHELM.

> *See* Macquer, P. J.   Allgemeine Begriffe der Chemie.

POSSANNER, B, VON.

Alkoholometrische Reductions-Tabellen.   Wien, 1890.   8vo.

POZZI, GIOVANNI.

Dizionario di fisica e chimica applicata alle arti secondo le dottrine di Libes, Chaptal, Berthollet e Parkes ; e giusta le teorie moderne ed i metodi più semplici introdottisi nei diversi processi chimici. Milano, 1820.   9 vols., 8vo.   With plates and portrait of the author.

PRECHTL, JOHANN JOSEPH.

Technologische Encyclopädie, oder alphabetisches Handbuch der Technologie, der technischen Chemie und des Maschinenwesens. Stuttgart, 1830–55.   20 vols.; text in octavo and tables in sm. folio. Supplement, 1857–69.   5 vols.; text in octavo and tables in sm. folio.

PRESCOTT, A. B., AND BAIRD, J. W.

> *See* Baird, J. W., A. B. Prescott.

QUARIZIUS, C. G.

Populaires chemisch-technisches Wörterbuch derjenigen Natur- und Kunstprodukte, bei deren Gewinnung chemische Agentien thätig und chemische Kenntnisse erforderlich sind.   Für Cameralisten, Techniker [etc.].   Berlin, 1856.   8vo.

RAMMELSBERG, CARL FRIEDRICH.

Handwörterbuch des chemischen Theils der Mineralogie.   Berlin, 1841.   8vo.   Vier Supplemente, Berlin, 1843–49.   8vo.
Handbuch der Mineralchemie.   Berlin, 1860.

RAMMELSBERG, CARL FRIEDRICH. [Cont'd.]
Zweite Auflage. Leipzig, 1875. 2 vols., 8vo. Ergänzungs-
heft zur zweiten Auflage. Leipzig, 1886.

A sequel to Rammelsberg's Handwörterbuch des chemischen Theils der
Mineralogie.

REALE, N.
Dizionario generale chimico-farmaceutico. Seconda edizione riveduta,
accresciuta e rifatta. Napoli, 1889.

REMLER, JOHANN CHRISTIAN WILHELM.
* Neues chemisches Wörterbuch, oder Handlexicon und allgemeine
Uebersicht der in neuern Zeiten entworfenen französisch-lateinisch-
italienisch-deutschen chemischen Nomenclatur nach Bergmann,
Berthollet [u. s. w.] nebst Beyfügung der alter Nomenclatur und
einem vierfachen Register. Erfurt, 1793. pp. [viii]-355-[iii], 12mo.
Tabellarische Versuch einer französisch-teutschen Nomenclatur der
neueren Chemie. Leipzig, 1793. Fol.

RICH, SIDNEY W.
The New and Old Notation of Chemistry ; in a Complete Set of Tables,
giving the Formulæ of all the more important Compounds, according to each system ; and a Table of the Elements with their New
and Old Atomic Weights. London, 1866.

RICHTER, M. M.
Hülfs-Tabellen für das Laboratorium zur Berechnung der Analysen.
Berechnet und zusammengestellt von M. R. Berlin, 1882. 8vo.
Tabellen der Kohlenstoff-Verbindungen nach deren empirischer Zusam-
mensetzung. Berlin, 1884. pp. viii-517, roy. 8vo.

RIVET, J. B. FRANÇOIS CLAUDE.
Dictionnaire raisonné de pharmacie chimique, théorique et pratique.
Lyon, 1803. 2 vols., 8vo.

ROSS, WILLIAM ALEXANDER.
Alphabetical Manual of Blowpipe Analysis, showing all known methods
old and new. London, 1880. 16mo.
Compare, in Section V, Ross, W. A., The Blowpipe in Chemistry.

RUCHTE, S.
Taschen-Wörterbuch der in der Chemie am häufigsten auftretenden
Körper unter Angabe ihrer chemischen Formeln nach neuester
Schreibweise. Leipzig, 1876. 8vo.

RULAND, MARTIN.
Lexicon alchemiæ [etc.]. See in Section VI.

SADLER, JOHN.

*An Explanation of Terms used in Chemistry. London, 1804. pp. 22, 12mo.

> Accompanies: Outlines of a Course of Lectures on Chemical Philosophy, by Humphry Davy. London, 1804.

SALMON, WILLIAM.

Dictionnaire hermétique, contenant l'explication des termes, fables énigmes, emblèmes et manières de parler des vrais philosophes. Accompagné de deux traités singuliers et utiles aux curieux de l'art par un amateur de la science [Gaston Le Doux]. Paris, 1695. pp. 216–119, 16mo.

> Mystical. *Compare* Pernety.

ШЕРЕРЪ АЛЕКСАНДРЪ.

Опытъ методическаго опредѣленія химическихъ наименованій для россійскаго языка. С.-Петербургъ, 1808.

> SCHERER, ALEXANDER. An attempt at a methodical chemical nomenclature for the Russian language.

SCHERER, ALEXANDER NICOLAUS VON.

Uebersicht der Zeichen für die neuere Chemie. Jena, 1796. Folio. Another edition, 1811.

SCHERER, JOHANN ANDREAS.

Versuch einer neuen Nomenclatur für deutsche Chymisten. Wien, 1792. pp. [xx]–208–[xvi], 8vo, folding tables.

> The author attempts to Germanize De Morveau's new nomenclature. A few examples will show that his scheme was happy in its failure : *Preussischblausäure Bittererde* for prussiate of magnesia ; *Blasenstein-säures Wasserbley* for lithiate of molydenum ; *Sauereisen* for oxide of iron ; *Brandicht schleimsäurichter Spiessglanz* for " pyromucite d'antimoine," etc.

SCHLEIDEN, MATHIAS JULIUS UND E. SCHMID.

Encyclopädie der gesammten theoretischen Naturwissenschaften in ihrer Anwendung auf die Landwirthschaft, umfassend Physik, anorganische Chemie, organische Chemie, Meteorologie, Mineralogie, Geognosie, Bodenkunde, Düngerlehre, Pflanzenphysiologie, Thierphysiologie und Theorie des rationellen Ackerbaues. Braunschweig, 1851. 3 vols., 8vo. Ill.

> Vol. 1 : Physik, anorganische Chemie und Mineralogie. Für Landwirthe bearbeitet von E. Schmid.
>
> Vol. II : Organische Chemie, Meteorologie, Geognosie, Bodenkunde und Düngerlehre. Für Landwirthe bearbeitet von E. Schmid.

SCHLEIDEN, MATHIAS JULIUS UND E. SCHMID. [Cont'd.]
> Vol. III : Die Physiologie der Pflanzen und Thiere und Theorie
> der Pflanzencultur. Für Landwirthe bearbeitet von M. J.
> Schleiden.

SCHMIDT, DAVID PETER HERRMANN.
> Etymologischer chemischer Nomenclator der neuesten einfachen und
> daraus zusammengesetzten Stoffe, nebst Erklärung einiger andern
> chemisch-physikalischen Benennungen. Lemgo, 1839. 8vo.
> Fortsetzung und Nachträge des etymologischen chemischen Nomencla-
> tors der neuesten einfachen und daraus zusammengesetzten Stoffe.
> Lemgo, 1841–47. 8vo.

SCHNACKE, G. E. ALEXANDER.
> Wörterbuch der Prüfungen verfälschter, verunreinigter und imitirter
> Waaren ; mit Angabe des Wesens und der Erkennung der Aecht-
> heit der Waaren. Für Aerzte, Apotheker, Chemiker etc. Gera,
> 1877. 8vo. Ill.

SCHÖPFFER, C.
> Handwörterbuch der Fortschritte der gesammten Technologie. Leip-
> zig, 1862. 8vo.

SCHRÖDER, G. UND J. VON.
> Wandtafeln für die Unterricht in der allgemeinen Chemie und che-
> mischen Technologie. 3 parts. Kassel, 1887. Fol.

SCHULTZ, GUSTAV UND PAUL JULIUS.
> Tabellarische Uebersicht der künstlichen organischen Farbstoffe.
> Berlin, 1888. 4to.
>> Zweite verbesserte und vermehrte Auflage herausgegeben
>> von Gustav Schultz. Berlin, 1891. 4to.

SCHULTZE, C. F.
> Pharmaceutische Synonyma nebst ihren deutschen Bezeichnungen und
> ihren volksthümlichen Benennungen. Berlin, 1889.

SCHUSTER, GOTT.
> Medicinisches-chymisches Lexicon. Chemnitz, 1756. 8vo.

SCOPOLI, GIOVANNI ANTONIO.
> *See* Macquer, P. J., Dictionnaire.

SEIDEL, L.
> Nomenclator synonymorum pharmaceutico-chemicorum, oder chemisch-
> pharmaceutisches Handbuch, enthaltend die Vergleichung der in
> der Pharmazie und Pharmakochemie üblichen älteren und neueren
> Namen mit den gebräuchlichen. Rathenow, 1824. 4to. Plate.

SELMI, FRANCESCO.
    See Enciclopedia di chimica scientifica ed industriale.

СЕВЕРГИНЪ, ВАСИЛІЙ.
    Руководство къ удобнѣйшему разумѣнію химическихъ книгъ иностран-
    ныхъ. С.-Петербургъ, 1815. pp. vi–291–5. 12mo.

        SEVERGHIN, VASILY. A Latin-Russian, French-Russian, and German-
        Russian chemical dictionary.

SEVRIN, L. J.
    Dictionnaire des nomenclatures chimique et minéralogique anciennes,
    comparées aux nomenclatures chimique et minéralogique modernes.
    D'après les ouvrages des chimistes et le traité de minéralogie de
    Hauy. Auquel on a joint trois tableaux synoptiques destinés à
    offrir les principaux caractères des corps simples, et un quatrième
    tableau qui présente les caractères des acides. Avec trois planches
    pour les signes chimiques. Paris, 1807. pp. xxx–31–232, 8vo.
    Eight folding tables and plates.

        The chemical and mineralogical synonymes are separate. The former are
        in two parts: I, Nomenclature moderne et ancienne; II, Nomenclature
        française et latine. The mineralogical synonymes are also in two
        parts: I, ancienne et moderne; II, moderne et ancienne. The tables
        of signs of chemical substances are an expansion of those by Hassen-
        fratz and Adet.

SHARPLES, STEPHEN B.
    Chemical Tables. Cambridge [Massachusetts], 1866. pp. viii–192,
    12mo.

SOMMERHOFF, JOHANNES CHRISTOPHORUS.
    * Lexicon pharmaceutico-chymicum latino-germanicum et germanico-
    latinum, continens terminorum pharmaceuticorum et chymicorum,
    tam usualium quam minus usualium, succinctam ac genuinam
    explicationem; cum versione germanica, et additione signorum,
    quotquot hactenus innotuere characteristica. Cui accessit vocabu-
    larium germanico-latinum locupletissimum vegetabilium, animalium
    et mineralium, in officinis pharmaceuticis et alias usitatorum.
    Adjuncti sunt sub finem characteres metallorum, mineralium,
    planetarum, ponderum aliarumque rerum chymicarum. Opus et
    medicis et pharmacopœis et aliis de notitia harum rerum sollicitis
    necessarium et perutile. Norimbergæ, 1701. pp. xiv–411, folio.
    Portrait of Sommerhoff. Engraved title-page.

        A voluminous compendium of the synonymes used in chemical and pharma-
        ceutical authors. Arranged alphabetically under the Latin name.
        The alchemical symbols are given in connection with each word, so
        far as extant. Invaluable to the student of early chemical literature.
        Compare: Medicinisch-chymisch und alchymistisches Oraculum.

SOUBEIRAN, J. L.

Nouveau dictionnaire des falsifications et des altérations des aliments et des médicaments. Paris, 1874. 8vo.

> Nuevo diccionario de falsificaciones y alteraciones de los alimentos, de los medicamentos y de algunos productos empleados en las artes, en la industria y en la economía doméstica. Exposicion de los medios scientíficos y prácticos para reconocer el grado de pureza, el estado de conservacion y demostrar los fraudes de que son objeto. Traducido, aumentado y anotado por R. Gomez Pamo. Madrid, 1876. 4to. Ill.

SPALDING, LYMAN.

*A New Nomenclature of Chemistry proposed by De Morveau, Lavoisier, Berthollet and Fourcroy, with additions and improvements. Hanover (N. H.), 1799. 12 pp., sm. fol. oblong.

> The author, who was lecturer on chemistry in Dartmouth College, adopts the name Septon for nitrogen, as proposed by Mitchill; nitric acid being septic acid, etc. In other respects De Morveau is generally followed. *Cf.* Guyton de Morveau.

SPINA, DAVID DE.

Manuale sive lexicon pharmaceutico-chymicum, instar compendii medicis practicis et pharmacopœis maxime commodum, continens medicamenta composita polychresta tum usualia tum minus usualia, ex notissimis pharmacopœis et celeberrimus authoribus practicis desumta et par tim ex corpore pharmaceutico domini physici Jüngken huc translata ; editio completa cum indice locupletissimo. Francofurti ad Mœnum, 1700. pp. 1054–25, 8vo.

STORER, FRANK H.

Cyclopædia of Quantitative Chemical Analysis. Boston and Cambridge, 1870. 2 parts, 8vo.

First Outlines of a Dictionary of Solubilities of Chemical Substances. Cambridge [Massachusetts], 1864. 8vo.

STROMEYER, FRIEDRICH.

Tabellarische Uebersicht der chemisch einfachen und zusammengesetzten Stoffe. Mit Rücksicht auf die Synonymie nach den neuesten Entdeckungen entworfen. Göttingen, 1806. Folio.

> Contains 25 tables, each of two pages.

SUAREZ Y NUÑEZ, MIGUEL GERÓNIMO.

Memorias instructivas y curiosas sobre agricultura, comercio, industria,

Suarez y Nuñez, Miguel Gerónimo.   [Cont'd.]
economía, química, botánica, historia natural, etc. . . . Madrid,
1778–91. 12 vols., 4to.

> An encyclopedia containing many articles on chemistry and chemical
> industries with special reference to Spain.

Тамачкинъ, Д.
Энциклопедическій словарь основныхъ химич. свѣдѣній о составѣ орга-
ническихъ и неорганическихъ веществъ въ природѣ съ объясненіемъ
свойствъ тѣлъ и способа добыванія.

> Tamachkine, D.   Encyclopedical Dictionary of elementary chemical in-
> formation about the composition of organic and inorganic substances
> in nature, with an explanation of the properties of the bodies and of
> the mode of procuring them.

Thomson, Robert Dundas.
Cyclopædia of Chemistry with its applications to mineralogy, physiology
and the arts.   London and Glasgow, 1854.   pp. vii–540, 8vo.   Ill.
Dictionary of Chemistry with its applications to mineralogy, physiology
and the arts.   London and Glasgow.   New edition [1856].   *n. d.*

> First edition with title : Encyclopedia, etc.

Thorpe, T. E.
A Dictionary of Applied Chemistry.   London, 1890–92.   3 vols., roy.
8vo.   Vol. I, pp. viii–715 ; II, viii–714 ; III, *in progress.*

Thurneisser, Leonhart.
*Onomasticum und interpretatio, oder auszfürliche Erklerung über
etliche frembde und (bey vielen hochgelarten, die der Lateinischen
und Griechischen Sprach erfahren) unbekannte Nomina, Verba,
Proverbia, Dicta, Sylben, Caracter und sonst Reden.   Deren nicht
allein in des theuren Philosophi und Medici Aurelÿ Theophrasti
Paracelsi von Hohenheim, sondern auch in anderer Authorum
Schrifften, hin und wider weitleufftig gedacht, welche hie zusammen, nach dem Alphabet verzeichnet.   Das Ander theil inwelchem
fast jedes Wort, mit seiner eigenen Schrift, nach der Völcker
Etymologia oder eigenen Art und Weis zureden, beschrieben worden
ist.   Gedruckt zu Berlin durch Nicolaum Voltzen, 1583.   pp. [xii]–
188, folio.   8 Plates.

Tricht, J. P. C. van ; en J. J. Wolterson.
Woordenboek der zuivere en toegepaste scheikunde, (naar het Hoogduitsch).   Rotterdam en Leijden, 1854–70.   12 vols., roy. 8vo.   Ill.

TROMMSDORFF, JOHANN BARTHOLOMÄ.

Allgemeine Uebersicht der einfachen und zusammengesetzten Salze, in vier Tabellen. Gotha, 1789. 4to.

Allgemeines pharmazeutisch-chemisches Wörterbuch,oder Entwickelung aller in der Pharmazie und Chemie vorkommenden Lehren, Begriffe, Beschreibung der Geräthschaften, etc, für Aertze, Apotheker und Chemiker. Erfurt, 1806–22. 4 vols., 8vo. With folding plates. Two supplement vols.

Darstellung der Säuren, Alkalien, Erden und Metalle, ihre Verbindung zu Salzen und ihrer Wahlverwandtschaften, sowohl nach der Berthollet'schen, als Bergmann'schen Affinitätslehre. In 13 Tafeln. Gotha, 1800. Folio. Zweite Auflage, 1806.

Pharmazeutische Nomenclaturtafel . . . Erfurt, 1803. Folio.

Tabelle über alle bis jetzt bekannte Gasarten, ihre Kenntzeichen und Eigenschaften wie und woraus sie erhalten werden und ihre Bestandtheile. Dritte verbesserte Auflage. Weimar, 1804. Royal folio.

TYCHSEN, NICOLAI.

Fransk chemisk nomenklatur, paa Dansk udgivet med anmærkninger af N. T. Kjøbenhavn, 1794. 8vo.

URE, ANDREW.

A Dictionary of Chemistry on the basis of Mr. Nicholson's. London, 1821. 8vo.

> Second edition, 1824. Third edition, 1828. Fourth edition, 1835.

> A Dictionary of Chemistry on the basis of Mr. Nicholson's; with an introductory dissertation containing instructions for converting the alphabetical arrangement into a systematic order of study. American edition, with some additions, notes and corrections by R. Hare, assisted by F. Bache. Philadelphia, 1821. 2 vols., 8vo.

> Nouveau dictionnaire de chimie sur le plan de celui de Nicholson présentant les principes de cette science dans son état actuel, et les applications aux phénomènes de la nature, à la médecine, à la minéralogie, à l'agriculture et aux manufactures, traduit de l'anglais sur l'édition de 1821, par J. Riffault. Paris, 1822–24. 4 vols., 8vo.

> Handwörterbuch der practischen Chemie angewendet auf die anderen Zweige der Naturkunde wie auf Künste und Gewerbe. Nach der neuesten Ausgabe des Origi-

URE, ANDREW.   [Cont'd.]

nals (mit Berücksichtigung der französischen Bearbeitung von Riffault) aus dem Englischen übersetzt.   Nebst
Zusätzen und Anmerkungen.   Mit Vorrede und Anmerkungen von Döbereiner.   Weimar, 1825.   8vo.   14
plates.

For another German extension, *see* Brandes, Rudolph.

A Dictionary of Arts, Manufactures, and Mines ; containing a clear
exposition of their principles and practice.   London, 1839.   8vo.
Supplement, 1844.   8vo.

Fourth edition.   London, 1853.   2 vols.   Fifth edition,
edited by R. Hunt.   London, 1860.   3 vols., 8vo.
Seventh edition, edited by Robert Hunt, assisted by F.
Rudler.   London, 1875.   3 vols., 8vo.

Karmarsch, K. und Heeren.   Technisches Wörterbuch oder
Handbuch der Gewerbskunde.   In alphabetischer Ordnung.   Bearbeitet nach Andrew Ure's " Dictionary of
Arts, etc."   Prag, 1843–44.   3 vols.   Zweite Auflage,
1854–57.   3 vols.   Dritte Auflage ergänzt von Kick
und Gintl.   Prag, 1874–84.   7 vols., 8vo.

VAN, or VAN DER.

For Dutch names beginning with Van, or Van der, *see* next succeeding word.

VARNHAGEN, TH. G. FR.

Versuch einer tabellarischen Uebersicht, sowohl der älteren als neueren chemisch-pharmaceutischen Nomenclaturen.   6 Tables.
Schmalkalden, 1821.   Fol.

Lexicon chemisch-pharmaceutischer Nomenclaturen nebst Ver-
gleichungen der abweichenden Bereitungs-Vorschriften nach den
vorzüglichsten Pharmacopöen.   Durchgesehen und mit Anmerkungen begleitet von E. Witting.   Schmalkalden, 1822.   Zweite Auflage,
1827.   8vo.

VAUQUELIN, LOUIS NICOLAS.

Dictionnaire de chimie et de métallurgie.   Tours, 1815.   6 vols., 4to.

Dictionary of Chemistry containing the principles and
modern theories of the science, with its application to
the arts, manufactures and medicine.   For the use of
seminaries of learning and private students.   Translated from " Le Dictionnaire de chimie, approuvé par
Vauquelin," including the most recent discoveries and
doctrines of the science.   With additions and notes

VAUQUELIN, LOUIS NICOLAS. [Cont'd.]
by Mrs. Almira H. Lincoln [afterwards Mrs. Phelps.]
New York, 1830. pp. 531, 12mo.
Pages xi–xxiv contain a brief history of chemistry.

VERGNIAUD, A. D.
Dictionnaire de chimie. *See in Section V*, Nouveau manuel complet de
chimie.

VERSUCH EINER FRANZÖSISCH LATEINISCH ITALIENISCH TEUTSCHEN NO-
MENKLATUR DER NEUERN CHEMIE. Nach Bacher frei bearbeitet
und vermehrt vom teutschem Herausgeber. Leipzig, 1792.
pp. [iv]–114, 18mo.

VIOLETTE, J. M. HENRI, ET P. J. ARCHAMBAULT.
Dictionnaire des analyses chimiques, ou Répertoire alphabétique des
analyses de tous les corps naturels et artificiels depuis la fondation
de la chimie jusqu' à nos jours, avec l'indication du nom des
auteurs et des recueils où elles ont été insérées : Paris, 1851.
2 vols., 8vo. Vol. I, pp. vii–428 : II, 446.
Second tirage, augmenté de 400 analyses nouvelles. Paris,
1859. 2 vols., 8vo.

WACKENRODER, HEINRICH WILHELM FERDINAND.
Chemische Classification der . . . Körper. *See in Section V.*

WACKENRODER, HEINRICH WILHELM FERDINAND.
Synoptische Tabellen über die chemischen Verbindungen der ersten
Ordnungen. Jena, 1830. Fol.

WAGNER, RUDOLPH.
Ergänzungen zu dem Handwörterbuch der Chemie und Physik. In
Verbindung mit Mehreren herausgegeben. Berlin, 1850. pp.
[vi]–215, 8vo.
Also under the title : Bericht über die neuesten Fortschritte in der
Chemie, Physik und Mineralogie. *See* Handwörterbuch der
Chemie und Physik.

WALTER, FILIP NER.
Wykład nomenklatury chemicznéj Polskiéj i porownanie jéj z nomen-
klaturami : Łacińską francuską, angielską i niemiecką. Kraków,
1844. 8vo.

* WALTIRE, JOHN.
Tables of the various combinations and specific attraction of the substances employed in chemistry. Being a compendium of that

WALTIRE, JOHN. [Cont'd.]

    science : intended chiefly for the use of those gentlemen and ladies who attend the Author's lectures. London, 1769. pp. 24, 12mo.

        The author was a friend of Priestley, and was one of the first chemists to note the deposit of moisture on the inside of a tube after exploding a mixture of air and inflammable gas. *See* Wilson's Life of Cavendish, pp. 58–59.

WANDTAFEL DER PERIODISCHEN GESETZMÄSSIGKEIT DER ELEMENTE NACH MENDELEJEFF. Wien, 1885.

WANDTAFELN DER ATOMGEWICHTE DER CHEMISCHEN ELEMENTE H = 1. Wien, 1885. 2 sheets.

WATTS, HENRY.

    A Dictionary of Chemistry and the allied branches of other sciences. London, 1866–81. 5 vols. and 3 supplements ; Supplement III in 2 parts. Roy. 8vo.

        A new edition of this indispensable work, edited by H. Forster Morley and M. M. Pattison Muir, is in progress. Vols. I and III (A to Phenyl.) in 1892.

WEDEL, GEORG WOLFGANG.

    Tabulæ chymicæ xv in synopsi universam chymiam exhibentes. Jena, 1692. 4to.

WEIDINGER, G.

    Waarenlexicon der chemischen Industrie und der Pharmazie, mit Berücksichtigung der wichtigsten Nahrungs- und Genussmittel. Leipzig, 1868–69. 8vo.

        Waarenlexicon der chemischen Industrie und der Pharmacie. Mit Berücksichtigung der wichtigsten Nahrungs- und Genussmittel. Unter Mitwirkung der Herren Joseph Moeller, Hermann Thoms, K. Thümmel, herausgegeben von T. F. Hanausek. Zweite gänzlich verarbeitete Auflage. Leipzig, 1892. pp. iv–1,000, 8vo.

WELTZIEN, C.

    Systematische Zusammenstellung der organischen Verbindungen. Braunschweig, 1860. pp. lii–720, 8vo.

        Aperçu systématique des combinaisons dites inorganiques. Édition française publiée avec le concours de Ed. Willm. Paris, 1867. pp. vii–105. 4to.

WESTRUMB, A. H. L.

    Wörterbuch, *see* Chevallier, J. B. Alphonse.

WILLIAMSON, DEWAR, FRANKLAND [and others].

Second Report of the Committee consisting of Professors Williamson, Dewar, Frankland, Roscoe, Crum Brown, Odling and Armstrong, Messrs. A. G. Vernon Harcourt, J. Millar Thomson, H. B. Dixon (Secretary), and V. H. Veley, and Drs. F. Japp and H. Forster Morley, reappointed for the purpose of drawing up a statement of the varieties of Chemical Names which have come into use, for indicating the causes which have led to their adoption, and for considering what can be done to bring about some convergence of the Views on Chemical Nomenclature obtaining among English and Foreign chemists. Report of the British Association for the Advancement of Science. 1884.

> Contains : Historical Notes on Chemical Nomenclature, and Tables summarizing the history of the nomenclature of certain typical compounds.

WINCKLER, EMIL.

Technisch-chemisches Recept-Taschenbuch. Enthaltend 1500 Vorschriften und Mittheilungen aus dem Gebiete der technischen Chemie und Gewerbskunde . . . Leipzig, 1860–65. 6 vols., 8vo.

ВИНКЛЕРЪ, Е.

Книга технико-химич. рецептовъ, содерж. въ себѣ 3,727 новѣйш. и полез нѣйш. открытій, свѣдѣній и рецептовъ изъ области техн. химіи и промышленности. Москва 1865.

> WINKLER, E. The book of technological chemical receipts.

WINKLER, EDUARD.

Vollständiges Real-Lexikon der medicinisch-pharmaceutischen Naturgeschichte und Rohwaarenkunde. Leipzig, 1840–42. 2 vols., 8vo.

WITTSTEIN, G. C.

Die chemische Nomenklatur von dem gegenwärtigen Standpunkte der Wissenschaft aus beurtheilt, nebst Vorschlägen zu einer möglichst einfachen und consequenten Durchführung derselben. München, 1849. 12mo. Second edition, 1852.

Vollständiges etymologisch-chemisches Handwörterbuch, mit Berücksichtigung der Geschichte und Literatur der Chemie. Zugleich als synoptische Encyclopädie der gesammten Chemie. München, 1846–58. 3 vols. and 3 Suppl., 8vo. Zweite [Titel-] Auflage, 2 vols. in 3 parts, and 3 Supplements.

WURTZ, ADOLPHE.

Dictionnaire de chimie pure et appliquée, comprenant : la chimie organique et inorganique, la chimie appliquée à l'industrie, à l'agriculture et aux arts, la chimie analytique, la chimie physique et la minéralogie. Paris, 1867–70. 5 vols., 8vo.

WURTZ, ADOLPHE.   [Cont'd.]

Supplément.   Paris.   *n. d.*   2 vols.

Deuxième supplément publié sous la direction de Ch. Friedel.   Paris.   *n. d.*   [In progress.]

Dizionario di chimica pura ed applicata.   Traduzione italiana.   Milano, 1888.   8vo.   [In progress, fascicolo 50 in 1892.]

ZAKWOORDENBOEK DER SCHEIKUNDE.
See Gerding, Th.

ZDZITOWIECKI, JÓZEF SEWERYN.

Niektóre uwagi nad nomenklaturą chemiczną polską przez ... P. J. P. W. [Warszawa].   1832.

Projekt nomenklatury chemicznej.   Warszawa, 1852.   8vo.

ZENNECK, LUDWIG HEINRICH.

Physikalisch-chemische Hülfs-Tabellen.   Zum nöthigen Gebrauche bei öffentliche Vorlesungen sowohl als bei dem Privatstudium der theoretischen und praktischen Chemie.   Leipzig, 1831.   pp. viii-255.   Many folding tables.

ZIUREK, O. A.

Technologische Tabellen und Notizen.   Zum Gebrauche im Fabriken-, Handels-, Gewerbe- und Landwirthschaftlichen Verkehre.   Braunschweig, 1863.   xvi-450, 8vo.

SECTION III.

HISTORY OF CHEMISTRY.

INCLUDING THE HISTORY OF ALCHEMY, PHARMACY, PHYSICS, PHO-
TOGRAPHY, TECHNOLOGY, TOXICOLOGY ; DESCRIPTIONS OF LABORA-
TORIES AND REPORTS OF INTERNATIONAL EXPOSITIONS.

AACHEN.

Die chemischen Laboratorien der königlichen rheinisch westfälischen
technischen Hochschule zu Aachen.   Mit zwei Blatt Zeichnungen.
Aachen, 1879.   4to.

ABILDGAARD, PETER CHRISTIAN.

Dissertatio critico-chymica de utilitate chymiæ in œconomia reipublicæ.
Hafniæ, 1762.

ACKERMANN, EUGÈNE.

Tableau historique de la découverte des éléments chimiques.   Paris,
Juillet, 1888.   1 page folio.

> Tableaux scientifiques, No. 6.

AIKIN, ARTHUR.

An Account of the most recent Discoveries in Chemistry and Mineralogy.
London, 1814.

ALMQVIST, C. J. L.

* Anekdoter såsom bidrag till Guldmakariets Historia.   Manuskriptet
författadt i St. Louis, Missouri i Norra Amerika, men sedermera
aflemnadt till Törnrosens Bok.   Stockholm.   *n. d.*   pp. 84, 8vo.

> The seventh (and last) chapter treats of the adventures of Don Guatimozin,
> an alchemist of Missouri and Mexico.

AMBÜHL, G.

Das neue Kantons Laboratorium in St. Gallen.   (St. Gallen, Ber. nat.
Ges.)   1886.

AMELUNG, PETER.

Tractatus de alchimiæ sive chemicæ artis inventione et progressione,
obscuratione et instauratione, dignitate, necessitate et utilitate,
Lipsiæ, 1607.   8vo.

AMELUNG, PETER. [Cont'd.]

> The author ascribes the origin of chemistry to Adam, and relates many other doubtful tales. His work was attacked by William Bokel, a physician of Stendalia, and Amelung replied with a tract entitled: Tractatus nobilis secundus, continens apologiam, etc. Lipsiæ, 1608. 8vo.

ANDREE, R.

Die Metalle bei den Naturvölkern. Leipzig, 1884.

ARPPE, A. E.

* Anteckningar om Finska Alkemister. Finska Vetenskapssocieteten meddelade den 15 April, 1867. *n. p.* *n. d.* pp. 110, 8vo.

> Published in Helsingfors in 1870 (?).

AUGUSTIN, FRIEDRICH LUDWIG.

Versuch einer vollständigen systematischen Geschichte der galvanischen Electricität und ihrer medicinischen Anwendung. Berlin, 1803. 8vo.

AUWERS, K.

Die Entwickelung der Stereochemie. Theoretische und experimentelle Studien. Habilitationsschrift. Heidelberg, 1890. 8vo.

BACHE, FRANKLIN.

Introductory Lecture to the Course of Chemistry delivered in Jefferson Medical College, October 18th, 1848. Published by the class. Philadelphia, 1848. pp. 20, 8vo.

Introductory Lecture to the Course of Chemistry delivered in Jefferson Medical College, October 13th, 1852. (Published by the class.) Philadelphia, 1852. pp. 16, 8vo.

BAEYER, ADOLF.

Ueber die chemische Synthese. Festrede zur Vorfeier des Allerhöchsten Geburts- und Namensfestes Seiner Majestät Ludwig II, Königs von Bayern, gehalten in der öffentlichen Sitzung der k. b. Akademie der Wissenschaften zu München am 25 Juli, 1878. München, 1878. pp. 22, 4to.

BAEYER, ADOLF UND ALBERT GEUL.

Das neue chemische Laboratorium der Akademie der Wissenschaften zu München. Separatabdruck aus der Zeitschrift für Baukunde, Bd. III. Mit 5 Tafeln. München, 1880. Fol.

BALLING, C. A. M.

Fortschritte im Probirwesen in den Jahren 1879–86. Berlin, 1887. 8vo. Ill.

BAPST, GERMAIN.
Les métaux dans l'antiquité et au moyen âge. L'étain. Paris, 1884. 8vo.

BARRAL ET DUMAS.
> Pièces historiques concernant Lavoisier et N. Leblanc.
> *See in Section V*, Société chimique de Paris (1860).

BARRAL, GEORGES.
* Histoire des sciences sous Napoléon Bonaparte. Paris, 1889. pp. ix–290, 12mo.

> Chapter V contains in alphabetical order biographies of the chief scientists (including chemists) who flourished under the empire.

BATHURST, RATH AND NATH. HENSHAW.
Aerochalinos, or a Register for the Air. London, 1677. 12mo.

> Said by Schubarth to contain the discovery of a vital principle common to saltpeter and to the atmosphere, *i.e.*, oxygen.

BAUER, A.
* Chemie und Alchymie in Oesterreich bis zum Beginnen der XIX Jahrhundert. Eine Skizze. Wien, 1883. pp. [iv]–85, 8vo. Ill.

BAUKE, F.
Die Raoultsche Gefriermethode für die Molekulargewichtsbestimmung und ihr Nutzen für die chemische Forschung. Ein Kapitel aus der Entwicklungsgeschichte der modernen Chemie. Berlin, 1890. 8vo. Ill.

BAUMÉ, ANTOINE.
> Discours historique sur l'æther. *In* Dissertation sur l'æther. *See in Section V*, Baumé, Antoine.

BÉCHAMP, P. J. A.
Essai sur les progrès de la chimie organique depuis Lavoisier. Montpellier, 1857. 8vo.
* Lettres historiques sur la chimie addressées à M. le Professeur Courty. Paris, 1876. pp. cxlix–289, and 332, 8vo.

> This work is dedicated to : " L'illustre et vénérée mémoire de Lavoisier indignement outragé par les chimistes Allemands Kolbe, Liebig et Volhard." This indicates its remarkable partisan character. The letters are in two series : the first, written in 1862–64, are devoted almost wholly to a defence of Lavoisier and his work ; the second series, written in 1871–74, defends the chemists of France and bitterly attacks Prussians. A long introduction sustains the controversial spirit of the author.

BECK, L.

Die Geschichte des Eisens in technischer und kulturgeschichtlicher Beziehung. 2 Abtheilungen. Braunschweig, 1884. pp. 315, 8vo. Ill.

Abtheilung 1, Von der ältesten Zeit bis um das Jahr 1500 nach Christus. Zweite Auflage, 1890.

BECKER, JOHANN HERRMANN.

Versuch einer Literatur und Geschichte der Nahrungsmittelkunde. *See in Section V*, Becker, J. H., Versuch (etc.).

BECKMANN, JOHANN.

Beiträge zur Geschichte der Erfindungen. Leipzig, 1780–1805. 5 vols., 8vo.

* A History of Inventions, Discoveries and Origins. Translated by William Johnston and revised by William Francis and J. W. Griffith. London. 2 vols., 8vo. Bohn's Standard Library ; with engraved portraits of Beckmann and James Watt.

This learned work treats of many subjects related to chemistry. In vol. 1 (of Bohn) : refining gold and silver ore by quicksilver ; archil ; secret poison ; sympathetic ink ; colored glass and artificial gems ; verdigris ; alum ; adulteration of wine ; kermes and cochineal ; ultramarine ; cobalt, zaffre, smalt ; aurum fulminans. In vol. 11 (of Bohn) : chemical names of metals ; zinc ; etching glass ; soap ; madder ; tin ; manganese ; indigo ; Prince Rupert's drops ; steel ; black lead ; sal ammoniac ; Bologna stone ; saltpeter, gunpowder and aquafortis.

BECQUEREL ; et EDMOND BECQUEREL.

Résumé de l'histoire de l'électricité et du magnétisme et des applications de ces sciences à la chimie, aux sciences naturelles et aux arts. Paris, 1858. 8vo.

Appendix to the Authors' Traité de l'électricité et du magnétisme. 3 vols., 8vo.

BÉGIN, ÉMILE AUGUSTE.

Chimie et alchimie. *In* Lacroix (Paul) and Seré (Ferdinand) : Le moyen âge et la renaissance. Paris, 1849. 4to, vol. 2.

BEILSTEIN, F.

Die chemische Grossindustrie auf der Weltaustellung zu Wien im Jahre 1873. Leipzig, 1873. 8vo.

BELL, JACOB.

A Concise Historical Sketch of the Progress of Pharmacy in Great Britain, from the time of its partial separation from the practice of

BELL, JACOB. [Cont'd.]
  medicine until the establishment of the Pharmaceutical Society,
  intended as an introduction to the Pharmaceutical Journal. London, 1843. 8vo.
      [New edition, with continuation] by T. Redwood. London,
      1880. 8vo.

BENZENBERG, J. F.
  Ueber die Dalton'sche Theorie. Düsseldorf, 1830. pp. xvi-192, 8vo.

BERAUD, R. P.
  Dissertation sur la cause de l'augmentation de poids que certaines
      matières acquièrent dans leur calcination. Qui a remporté le prix
      au jugement de l'Académie Royale des Belles Lettres, Sciences
      et Arts de Bordeaux. À la Haye, 1748. pp. 98, 16mo.
          In this interesting booklet the author, a Jesuit Professor of Mathematics at
          the College in Lyons, traces the history of experimentation on the
          increase in weight of metals by calcination, and shows that (1) they lose
          in *sp. gr.;* (2) the increase in weight cannot be caused by the addition
          of corpuscles of fire ; (3) the weight of the air does not contribute to
          the increase in weight of calcined bodies [!] ; and (4) finds the real cause
          of the phenomenon in the foreign bodies contained in the air. The
          author draws his conclusions from the labors of others, with the litera-
          ture of which he is quite familiar.

BERENDES, J.
  Die Pharmacie bei den alten Culturvölkern. Historisch-kritische
      Studien. Mit einem Vorworte von H. Beckurts. Halle a/S., 1891.
      2 vols. Ill.

BERLIN, (LABORATORIUM).
      See Storck, A. ; *also*, Cremer, A. und F. Esser ; *also*, Hofmann, A. W.

BERNARDUS TREVIRENSIS.
  Περὶ χημείας, opus historicum et dogmaticum, ex Gallico in Latinum
      simpliciter versum et nunc primum in lucem editum. Argentorati,
      1567. 8vo.

BERTHELOT, MARCELLIN.
      Des origines de la poudre et des matières explosives. *In* Sur la force des
          matières explosives, *etc.*, vol. 2. 1883. *See in Section V.*

BERTHELOT, MARCELLIN.
  Chimie (La) des Égyptiens d'après les papyrus de Leide. Ann. de chim.
      et phys. [VI], vol. IX, pp. 5-65. Paris, 1886.
  Introduction à l'étude de la chimie des anciens et du moyen âge.
      Paris, 1889. 8vo.

BERTHELOT, MARCELLIN. [Cont'd.]
> With 45 illustrations. Reprinted from "Collection des alchimistes grecs," par Berthelot et Ruelle. *See in Section VI.*

* Origines (Les) de l'alchimie. Paris, 1885. pp. xx–445, 8vo.
> Illustrated with facsimiles of MS. and a portrait of Berthelot.

Quelques figures d'appareils chimiques syriaques et latines au moyen âge. Ann. chim. phys., [VI], vol. XXIII, pp. 433–468. 1891.

* Révolution (La) chimique. Lavoisier. Ouvrage suivi de notices et extraits des registres inédits du laboratoire de Lavoisier. Paris, 1890. pp. xii–334, 8vo.
> Bibliothèque scientifique internationale, vol. LXIX.

BERTHOLLET, CHAPTAL, etc.
> Œuvres diverses. *See in Section V*, Drohojowska.

BERTRAND, JOSEPH.
L'Académie des sciences et les académiciens de 1666 à 1793. Paris, 1868. 8vo.

BERZELIUS, JOHANN JACOB.
> Öfversigt af djur-kemiens . . . tillstånd. *See in Section V.*
> Geschichte des Löthrohrs. *See in Section V*, Berzelius, J. J., Anwendung des Löthrohrs.

BERZELIUS, JOHANN JACOB.
Fortschritte und gegenwärtiger Zustand der thierischen Chemie. Aus dem Englischen übersetzt. Nürnberg, 1815.

Théorie des proportions chimiques et table synoptique des poids atomiques des corps simples, et de leurs combinaisons les plus importantes. Deuxième édition, Paris, 1835. pp. 477, 8vo.
> Contains: "Exposé historique du développement de la théorie des proportions chimiques."

BEYTRAG ZUR GESCHICHTE DER HÖHEREN CHEMIE ODER GOLDMACHER-KUNDE IN IHREM GANZEN UMFANGE. Ein Lesebuch für Alchemisten, Theosophen und Weisensteinsforscher, auch für alle, die wie sie, die Wahrheit suchen und lieben. Leipzig, 1785. pp. [xv]–695, 8vo.
> The author, who conceals his name under the pseudonym "Carbonarius," treats of the history of alchemy from the earliest times in 60 chapters. He gives, in closing, a chronological list of "adepts"; another of fraudulent alchemists; and a bibliography, comprising 634 titles, with annotations. The latter is chronologically arranged under three heads, and is quite good in its details.

BIBRA, ERNST VON.
Historische Skizze des Getreidebaues. *In* Die Getreidearten und das Brod. Nürnberg, 1860. pp. 1–98, 8vo. *Compare in Section V.*

BIERENS DE HAAN.

Bouwstoffen voor de geschiedenis der wis- en natuurkundige wetenschappen in de Nederlanden. [First article in] Verslagen en
mededeelingen der Koninklijke Akademie van Wetenschappen,
Afdeeling Natuurkunde. Tweede reeks, achtste deel. Amsterdam,
1874. Article No. XXXI in the same journal ; derde reeks, zesde
deel. Amsterdam, 1889.

BINKO, H. BOCK.

History of the Colours produced from Coal Tar. Chem. News, vol.
XXIII, p. 53 (1871).

> Setting forth the manufacture of coal-tar colors by Johann Nepomuk
> Jassnüger, of Vienna, as early as 1816.

BIRCH, TH.

History of the Royal Society from 1660 to 1687. London, 1756–57.
4 vols., 4to.

BISCHOFF, JOHANN NICOLAUS.

Versuch einer Geschichte der Färberkunst von ihrer Enstehung an bis
auf unsere Zeiten entworfen und mit einer Vorrede von Johann
Beckmann begleitet. Stendal, 1780. 8vo.

BIZIO, BARTOLOMEO.

Quanto spetta agli Italiani nella chimica scienza, rispetto agli stranieri.
Saggio letto all' Ateneo di Venezia nella tornata ordinaria del 16.
Maggio, 1850. Venezia, 1850. 8vo.

BLEY, LUDWIG FRANZ.

Versuch einer wissenschaftlichen Würdigung der Chemie und Pharmacie auf ihren jetzigen Standpuncten, oder Beleuchtung der Frage :
Was haben diese Wissenschaften seit Ende des 18. Jahrhunderts
geleistet ? Zugleich als Beitrag zur Geschichte dieser Wissenschaften. Halle, 1834–38. 2 vols., 8vo. Vol. I, [xviii]–722.
Vol. II, pp. 542.

Also under the title : Fortschritte und neue Entdeckungen im Gebiete
der Chemie und Pharmacie und der damit verbundenen Hülfswis-
senschaften. Für Chemiker, Apotheker, Techniker, Fabrikanten
und Alle welche sich für diese Wissenschaften interessiren, und
aus ihren Fortschritten Nutzen ziehen wollen. Erste Abtheilung.
Halle, 1834.

> A remarkable work, presenting in a systematic order the results of the
> labors of chemists during the period named, and indicating the sources
> of information. The authorities and literature are given as marginal
> notes, thus preventing interruption of the text.

BLEY, LUDWIG FRANZ. [Cont'd.]

> Vol. I contains a chapter on the origin and earliest state of Chemistry and Pharmacy (pp. 1–22), and a review of the labors of chemists in the first period (1800–1806), pp. 695–722. The latter review is continued in vol. II, pp. 515–542. The author, an apothecary in Bernburg and member of learned societies, designed his work as a sequel to Gmelin's Geschichte der Chemie, and dedicated it to J. B. Trommsdorff on the occasion of his fiftieth anniversary of chemical work. Vol. II contains special chapters on stoichiometry, and on various branches of applied chemistry. Bley's method combines the historical and the useful.

BLOCHMANN, G. M. S.

Beiträge zur Geschichte der Gasbeleuchtung gesammelt und zusammengestellt von Blochmann. Dresden, 1871. pp. viii–124, 8vo.

BLOMSTRAND, C. W.

*Die Chemie der Jetztzeit vom Standpunkte der electro-chemischen Auffassung aus Berzelius Lehre entwickelt. Heidelberg, 1869. pp. xx–417, 8vo.

> The preface is dated at Lund, 1868. A review of the status of chemical theory at the period of writing.

Om de organiska kropparnes konstitution. Historisk-kritisk framställning af de nyare kemiska theorierna. Lund, 1864. 8vo.

BLÜMNER, HUGO.

Technologie und Terminologie der Gewerbe und Künste bei Griechen und Römern. Leipzig, 1875–87. 4 vols., 8vo.

BLUNTSCHLI, F., G. LASIUS UND G. LUNGE.

Die chemischen Laboratorien des Eidgenössischen Polytechnikums in Zürich. Zürich, 1889. Fol. Ill.

BLYTH, ALEXANDER WINTER.

> Essay on the growth of Modern Toxicology. *See in Section V*, Blyth, A. W., Poisons, their effects, *etc.*

BODDY, EVAN MARLETT.

The History of Salt, with Observations on its Geographical Distribution Geological Formation and Medicinal and Dietetic Properties. London, 1881.

BODENSTEIN, ADAM.

De veritate alchymiæ : epistola ad Fuggeros, in qua argumenta alchymiæ infirmantia et confirmantia adducuntur. Basiliæ, 1581. Fol.

BÖCKMANN, F.

> Geschichte der explosiven Stoffe.
> *See in Section V*, Böckmann, F.

BOERHAAVE, HERMANN.
>   A sketch of the History of Chemistry. 1732.
>   *See in Section V*, Boerhaave, H., Elementa chemiæ.

BOISSELET DE SAUCLIÈRES.
Coup d'œil sur l'histoire du galvanisme en France. Paris, 1844. 8vo.

BOLLEY, POMPEJUS ALEXANDER.
Altes und Neues aus Farbenchemie und Färberei. Ueberblick der Geschichte und Rolle der sogenannten Anilinfarben. Sammlung gemeinverständlicher wissenschaftlicher Vorträge. II. Serie. Heft 45. Berlin, *n. d.* [1867]. 8vo.

BOLTON, HENRY CARRINGTON.
An Account of the Progress of Chemistry in the year 1882. Smithsonian Report for 1882, p. 509. Washington, 1883. 8vo.
The same for 1883, 1884, 1885, 1886, in the Reports of the Smithsonian Institution for the years named. Washington. 8vo.

>   The accounts for the years 1883, 1884, 1885 and 1886 are accompanied by necrologies and bibliographies.

Contributions of Alchemy to Numismatics. Read before the New York Numismatic and Archæological Society, Dec. 5, 1889. Author's Edition. New York, 1890. pp. 44, sm. 4to. Three plates. 175 copies reprinted from Am. J. Numismatics.

>   An historical sketch of the circumstances attending the issue of medals and coins struck to commemorate pretended transmutations, with detailed descriptions of 43 pieces.

Early Practice of Medicine by Women. Popular Science Monthly, December, 1880. *Also*, Journal of Science, London, January, 1881.
Historical Notes on the Defunct Elements. Read to the New York Lyceum of Natural History, May 9, 1870. Part I in Amer. Supplement to Amer. Edition of Chem. News, vol. VI, No. 6. Part II in American Chemist, vol. I, p. 1. *Abstract in* Sci. Amer., June 4, 1870 ; Chem. News, XXII, 208 ; College Courant, VI, 411 ; Proc. N. Y. Lyc. Nat. Hist., I, p. 21. (1870.)
Historical Notes on the Gold-cure. Pop. Sci. Monthly, vol. XLI, p. 469. August, 1892.
History of Chemical Notation. Part I. Metallurgic Astronomy and its Symbols. Read before N. Y. Academy of Sciences, Dec. 11, 1882. Part II. Mch. 12, 1883. *Abstracts in* Transactions of N. Y. Acad. Sci., vol. II, Nos. 5 and 6. 1883. 8vo.
Notes on the Early Literature of Chemistry. [Nine papers, as follows :]
No. 1. Were the Alchemists acquainted with Oxygen ? American Chemist, IV, p. 170. (1873.)

BOLTON, HENRY CARRINGTON.   [Cont'd.]

No. II.   Contribution to the History of Sulphur Matches.   Am. Chem.,
IV, 201.   (1873.)

No. III.   Outlines of a Bibliography of the History of Chemistry.   Am.
Chem., IV, 241.   (1873.)

No. IV.   First mentions.   Am. Chem., IV, 288.   (1873.)

No. V.   Definitions of Chemistry and Alchemy.   Am. Chem., V, 215.
(1874.)

No. VI.   Papyrus Ebers, the Earliest Medical Work extant.   Quart.
J. Science, London, Jan., 1876; *also,* Am. Chem., VI, 165;
Weekly Drug News, New York, IX, No. 15 (1884).

No. VII.   The Book of the Balance of Wisdom.   Quart. J. Sci., Oct.,
1876.

No. VIII.   Legends of Sepulchral and Perpetual Lamps.   Monthly J.
Sci., London, Nov., 1879.

No. IX.   Ancient Methods of Filtration.   Popular Science Monthly,
Feb., 1880.

Lunar Society (The), or the festive Philosophers of Birmingham one
hundred years ago.   Trans. N. Y. Acad. Sci., vol. VII, No. 8.
New York, 1888.   8vo.
    Cf. in Section IV, Priestley, Joseph.

Notes on the History of the Magic Lantern.   Philadelphia Photographer, Nov. 17, 1888, p. 688.   Ill.

Recent Progress in Chemistry.   An Address prepared at the request of
the N. Y. Academy of Sciences, and read March 15, 1886.   Transactions N. Y. Acad. Sci., V, No. 6.   New York, 1886.   pp. 26, 8vo.
Also, J. Franklin Inst., CXXII, 199 (1886); Sci. Am. Suppl., XXI,
546 (1886).   *Abstract in* Pop. Sci. Mon., August, 1886.

Views of the Founders of the Atomic Philosophy.   American Chemist,
III, 326.   (1872.)
    See also in Section I, Bolton. H. C.

BONET, M.

Sobre las relaciones que existen entre la química analitica y las demás
ciencias y sobre los servicios que presta á las llamadas naturales en
particular y á las que tienen un carácter técnico ó de aplicación.
Discorso.   Madrid, 1885.   4to.

BONN (LABORATORY).

See Hofmann, A. W.; *also,* Storck, A.

BONN, R.

    Geschichte der Structurformeln.   *See in Section V*, Bonn, R.   Die Structur-
    formeln.

BOOTH, JAMES CURTIS AND CAMPBELL MORFIT.
On Recent Improvements in the Chemical Arts. Smithsonian Report. Washington City, 1852. pp. 216, 8vo.

BORCH, OLE.
See Borrichius, Olaus.

BOREL, PIERRE.
Trésor de recherches et antiquités gauloises et françoises. Paris, 1655. 4to.

> Contains much material pertaining to the history of alchemy, but badly arranged.

BORRICHIUS, OLAUS (OLE BORCH).
*De ortu et progressu chemiæ dissertatio. Hafniæ, typis Matthiæ Godicchenii, sumptibus Petri Haubold, 1668. pp. [xii]–150, 4to. (Reprinted in the Bibliotheca chem. curiosa of Mangetus, vol. 1, No. 1.)

> The author of this celebrated treatise, the most frequently quoted by early historians, was born at Borchen (whence his name), Jütland, in 1626. He was Professor of Philology, Poetry, Chemistry and Botany at the University of Copenhagen, a fact which causes Rodwell to remark that " either professors were difficult to procure in the kingdom of Denmark, or else Olaus Borrichius was an astounding genius." However this may be, he was certainly a man of amazing credulity, and, allowing " the imaginative faculty due to his poetical temperament to exert an undue influence over his sober judgment," he refers the origin of alchemy to the antediluvians, endeavors to prove that Hermes Trismegistus was a real personage, the inventor of all arts, and the father of alchemy, and that the Smaragdine Table was really found by the wife of Abraham, besides accepting the preposterous theories of his contemporaries concerning the elixir of life and the philosophers' stone. This dissertation was highly prized by the alchemists of his day on account of its earnest defence of their principles. Its present value is solely that of a curious example of the extravagant credulity of a learned man.
> For biography of O. B. *see in Section IV.*

*Hermetis, Ægyptiorum et chemicorum sapientia ab Hermanni Conringii animadversionibus vindicata. Deuteronom. XXIII, v. VII. Noli abominari Ægyptium. Hafniæ, sumptibus Petri Hauboldi, 1674. pp. [vi]–448–[viii], 4to.

> This is a reply to Hermann Conringius' " De hermetica medicina." *See* Conring, H.

BOSTOCK, JOHN.
Account of the History and the present state of Galvanism. London, 1818. 8vo.

BOUANT, E.
Histoire de l'eau. Paris, 1882. 32mo. Ill.

> We have not seen this " History of Water," a subject that we should hardly
> expect could be treated in a 32mo volume. We suppose cosmogony and
> the Universal Deluge occupy a portion of the work having so ambitious
> a title.

BOUDET, FÉLIX.
See Leblanc, F. et Dizé.

BOULLET, A. F.
* De l'état des connaissances relatives à l'électricité chez les anciens
peuples d'Italie. Saint-Étienne, 1862. pp. 31, 8vo.

> The author claims extensive knowledge of electricity by the ancient Romans,
> and quotes a passage from Ctesias (Indica, iv.) to prove that lightning-
> rods were in use 416 B.C. In a note by Dr. Michalowski he shows the
> resemblance between the Greek electron (amber) and two Hungarian
> words, Elleuk and tron, which together signify : " He darts forth
> lightning." The whole treatise is interesting, but imaginative.

BRANDE, WILLIAM THOMAS.
Dissertation Third : exhibiting a general view of the progress of chemical philosophy, from the early ages to the end of the eighteenth
century. [London, 1816.] pp. 79, 4to.

> Prefixed to the Supplement to the Encyclopædia Britannica. An American
> edition was issued in Boston in 1818. pp. 120, 8vo.

> Historia de la química, tomada del Manual de Brande. In-
> troduccion al curso de esta ciencia leido en la Uni-
> versidad de Carácas por José M. Vargas. Publicada
> por uno de sus discipulos. Carácas, 1864. pp. 55, 8vo.

BRANTHOME.
Ueberblick der Chemie nach ihrem gegenwärtigen Zustande. Aus dem
französischen mit Anmerkungen von Joh. Bartholomæus Trommsdorff. Strassburg, 1818. 8vo.

BRENDEL, J. G.
De instrumentis quibusdam chemicis Boerhaaveanis. Gottingæ, 1747.
Sm. 4to.

BRENEMAN, A. A.
Historical Summary of the Fixation of Atmospheric Nitrogen. J. Am.
Chem. Soc., XI (1889).

> Contains bibliographic data.

BREYMANN UND KIRSTEIN.
Das chemische Laboratorium der Universität Göttingen. Hannover,
1890. 4to. Ill.

BRIANCHON, CHARLES JULIEN.
Description du laboratoire de chimie de l'école d'artillerie de la garde
royale. Paris, 1822.

BROMEIS, CARL.
Ueber die Entdeckung des Sauerstoffs. [Stuttgart, 1862.] 8vo.

BROUSSE, JULES.
Aperçu historique sur la pharmacie. Paris, 1862. 8vo.

BROWN, SAMUEL.
* Lectures on the Atomic Theory and Essays Scientific and Literary.
Edinburgh and London, 1858. 2 vols, 8vo. Vol. I, pp. x–357;
II, 384.

> These volumes contain four Critical Lectures on the Atomic Theory, and the
> following essays : The Atomic Theory before Christ and since ; Alchemy
> and the Alchemists ; Phlogiston and Lavoisier ; Sir Humphry Davy ;
> The History of Science ; The Humanities of Science. The author
> wrote a pleasing style, and these essays deserve a wider acquaintance.

BRÜNNICH, M. TH.
Les progrès de l'histoire naturelle et des sciences analogues en Dannemarc et en Norvège depuis la fondation de l'Université de Copenhague. Traduit du Danois par N. J. A. Yanssens des Campeaux.
Copenhague, 1783. pp. 123, 12mo.

> This forms the first part of Brünnich's Literatura danica. *See in Section I.*

BÜCHEL, KARL.
Die Lehre von der elementaren Zusammensetzung der Körper in ihrer
Entwicklung von Empedocles bis auf Lavoisier. Programm. Düren,
1876–77. 2 parts. 4to.

BUCHNER, FR. XAVER.
Toxikologische Geschichte des Phosphors. Inauguraldissertation.
Nürnberg, 1846. 8vo.

BUCHNER, LUDWIG ANDREAS.
Ueber den Antheil der Pharmacie an der Entwickelung der Chemie.
Festrede zur Vorfeier des Geburtsfestes Sr. Majestät Max II.,
Königs von Bayern, gehalten in der öffentlichen Sitzung der k.
Akademie der Wissenschaften zu München, am 27. November,
1849. München, 1849. 4to.
Ueber die Beziehungen der Chemie zur Rechtspflege. Festrede.
München, 1875. 4to.

BUCKLEY, ARABELLA B.

A Short History of Natural Science and of the progress of discovery from the time of the Greeks to the present day ; for the use of schools and young persons. London and New York, 1876. pp. xxiii–467, 8vo.

This simple and admirable little work gives to chemistry its share of space.

BUDDEUS, JOANNES FRANCISCUS.

Untersuchung von der Alchemie. *See in Section VI*, Roth-Scholtz Fr. ; Deutsches Theatrum Chemicum, vol. I.

An historical sketch of 146 pages, dated 1727.

BUFF, HEINRICH LUDWIG.

Ein Blick auf die Geschichte der Chemie. Erlangen, 1866. 20 pp., roy. 8vo.

A condensed statement of leading facts in theoretical chemistry from Boyle to Buff.

BURCKHARDT, FRITZ.

* Die Erfindungen des Thermometers und seine Gestaltung im XVII. Jahrhundert. Basel, 1867. 48 pp., 4to. Ill.

BUSCH, G.

Versuch eines Handbuchs der Erfindungen. Eisenach, 1790. 8 vols. Vierte Auflage, 1802–16. *See* Minola, Beiträge.

The earliest complete work of its kind, but the author is careless in verifying statements and sources.

CAHOURS, AUGUSTE.

Histoire des radicaux organiques. Paris, 1860. 8vo.

History of the Organic Radicals. Lecture before the Chemical Society of Paris, March 30, 1860. Translated by W. S. W. Ruschenberger. *n. p. n. d.* 8vo.

CAILLOT, AMÉDÉE.

* Histoire et appréciation des progrès de la chimie au dixneuvième siècle ; thèse présentée et soutenue devant le Jury nommé par arrêté du 9 Juillet, 1838, le jeudi 16 Août, 1838, à six heures du soir. Strasbourg, 1838. pp. 53, 4to.

" *Nullius momenti.*"

CARLSRUHE, (LABORATORIUM).

See Weltzien, Carl.

CARYOPHILUS, BLASIUS.

De antiquis auri, argenti, stanni, aëris, ferri, plumbique fodinis opusculum. Viennæ, 1757. 4to.

CASALI, ADOLFO.
Il passato, il presente e l'avvenire della chimica   Discorso inaugurale alla solenne riapertura degli studi nella libera Università di Ferrara, il giorno 15 Novembre, 1885. Ferrara, 1886. pp. 57, 8vo.

CASTRELL, JOSEPH.
Kritische Uebersicht der herrschenden Theorien über die Constitution der organischen Verbindungen. Zürich, 1847. 8vo.

CAVERNI, R.
Storia del metodo sperimentale in Italia. Tomo I : Dell' origine e de' progressi del metodo sperimentale in Italia. De' principali strumenti del metodo sperimentale (termometro, orologio a pendolo, cannocchiale, barometro, ecc). Firenze, 1891. 8vo.

CECH, C. O.
Die internationale Ausstellung wissenschaftlicher Apparate zu London. Mit besonderer Berücksichtigung der chemischen Gruppe. Heidelberg, 1878. 8vo.

CENTENNIAL OF CHEMISTRY. *See* Proceedings of the C. of C.

CHAPTAL, JEAN ANTOINE CLAUDE.
Essai sur le perfectionnement des arts chimiques en France. Paris, an VIII [1800]. 8vo.

CHÂTEAU, T.
Étude historique et chimique pour servir à l'histoire de la fabrication du rouge turc ou d'Adrianople, et à la théorie de cette teinture. Paris, 1876. 4to.

CHAUVEL.
Histoire de la pharmacie en France. *See in Section V*, Chauvel.

CHEMISCHE (DIE) INDUSTRIE DEUTSCHLANDS auf der Weltausstellung in Philadelphia im Jahre 1876. Berlin, 1876. 8vo.
Cf. Goldschmidt, G.

CHEMNITZ (LABORATORIUM). *See in Section V*, Wunder, Gustav. *See* Gottschaldt, A.

CHENEVIX, RICHARD.
Remarks upon Chemical Nomenclature according to the Principles of the French Neologists. London, 1802. pp. 246, 12mo.
A critical examination of the merits and weaknesses of the French and other novel nomenclatures. He quotes Sage's remark that " oxygen " signifies the " son of a vinegar merchant."

CHEVREUL, MICHEL EUGÈNE.

* Histoire des connaissances chimiques. Paris, 1866. pp. 479, 8vo. Plates.

> Only the first volume was printed.

Histoire des principales opinions que l'on a eues de la nature chimique des corps, de l'espèce chimique et de l'espèce vivante. Paris, 1869. Atlas. 4to.

Reviews of Hoefer's "Histoire de la chimie." A series of 14 articles in Journal des Savants, 1843–1851. 1843, p. 65 ; 1844, 101; 1845, 321 ; 1849, 531, 594, 663, 720 ; 1850, 71, 136, 284, 734 ; 1851, 97, 160, 217.

> For analysis of contents see Malloizel, Œuvres scientifiques de Chevreul. *Cf.* Hoefer, Ferdinand.

Reviews of Cambriel's " Cours de philosophie hermétique," Paris, 1843. A series of four articles in Journal des Savants, 1851, pp. 284, 337, 492, 752.

> *Cf. in Section VI*, Cambriel.

CHIARLONE, D. QUINTIN, Y C. C. MALLAINA.
Ensayo sobre la historia de la farmacia. Madrid, 1847. 4to.

Ходневъ, А.
Историческое развитіе о понятіе о хлорѣ.
Харьковъ. 1847. 8-го

> CHODNEF, ALEXIUS. Historical Sketch of Chlorine. Charkow, 1847.

CHRISTMAS, HENRY.
*The Cradle of the Twin Giants Science and History. London, 1849. 2 vols., 8vo.

> Book v treats historically of alchemy.

CHRISTOFLE, CHARLES.
Histoire de la dorure et de l'argenture électro-chimiques. Avec appendice. Paris, 1851. 8vo.

CHRONOLOGISCHE TABELLE über wichtige Ereignisse aus der Geschichte der Wissenschaften, besonders der Chemie. *In* Chemiker-Kalender für 1889. Berlin, 1889. 16mo.

CIBOT, PIERRE MARTIAL.
Mémoires concernant l'histoire, les sciences, les arts, les mœurs, les usages, etc. des Chinois, par les missionaires de Pekin. Paris, 1776–1814. 16 vols., 4to.

> Contains several articles on the early chemical knowledge of the Chinese, notably on cinnabar and mercury, on borax, on certain minerals, on pottery, etc., by P. M. C., a Jesuit Missionary in China.

CLAASEN, H.
Kurzer Ueberblick über die Zuckerindustrie Deutschlands. Bernburg, 1888.

CLARUS, JOHANN CHRISTIAN AUGUST, UND JOHANN CHRISTIAN LEOPOLD REINHOLD.
    See Sue, Pierre (le jeune).

CLOSMADEUC, G. DE.
La pharmacie à Vannes avant la révolution. Vannes, 1862. 8vo.

COLLOT, T.
Chimie industrielle. La potasse et les sels neutres de potasse, iode, salpêtre et nitrate de soude, à l'Exposition universelle de 1878. Paris, 1879. 8vo.

CONE, ANDREW, AND WALTER R. JOHNS.
Petrolia. A brief History of the Pennsylvania Petroleum Region, its development, growth, resources, etc. From 1859 to 1869. New York, 1870. pp. iv–652, 8vo.

CONRING, HERMANN.
De hermetica Ægyptiorum vetere et Paracelsica nova medicina liber unus ; quo simul in Hermetis Trismegisti omnem, ac universam cum Ægyptiorum tum chemicorum doctrinam animadvertitur. Helmstadii, 1648. 4to. New edition in 1669.

    Conring attacks the extreme antiquity assigned to Alchemy, and provokes a reply by Borrichius, *q. v.*

COOPER, THOMAS.
Introductory Lecture [on Chemistry] at Carlisle College, Pennsylvania. Carlisle, 1812. pp. viii–236, 8vo.

    Reviews the history of chemistry. The notes and references (pp. 101–236) are very full. The author was Professor of Chemistry at Carlisle and afterwards at Columbia, S. C., where he became President of the University. Cooper was a friend of Priestley, whom he followed to America.

CORTENOVIS, ANGELO MARIA.
Che la platina Americana era un metallo conosciuto dagli antichi . . . Bassano, 1790. *Extract in* Ann. chim., XII, 1792.

COSSART, LUDOVICUS.
De eximiis in vita civili chemiæ usibus, præsertim respectu Livoniæ. Regiomonti, 1783. 4to.

COZÉ, R.
Considérations sur l'étude des sciences tirées de l'histoire de la chimie. Strasbourg, 1832. 4to.

Cozzi, Andrea.
Storia dei più grandi progressi della scienza elettrica.   Firenze, 1838.

Cremer, A., und F. Esser.
Das neue chemische Laboratorium zu Berlin.   Berlin, 1868.   Fol.

Cron, Johann Christoph.
De præstantia et utilitate studii chymici.   Göttingæ, 1735.   4to.

Cronsaz, Johannes Petrus.
De physicæ origine et progressionibus.   Groningen, 1724.   4to.

Crum-Brown, Alexander.
The Development of the Idea of Chemical Composition.   Inaugural
Lecture.   Edinburgh, 1869.

Cuvier, Georges.
*Histoire des sciences naturelles pendant la deuxième moitié du XVIIIe
siècle et une partie du XIXe chez tous les peuples connus.   Com-
mencée au Collège de France par G. C., complétée par Magdeleine
de Saint-Agy.   Paris, 1843.   2 vols., 8vo.

> Vol. II contains a section on the history of chemistry.   With characteristic
> national jealousy, Cuvier attributes Priestley's discovery of oxygen to
> his hearing a lecture by the French chemist Bayen, who reduced pre-
> cipitate *per se*.   Bayen, however, noted only the metal and entirely dis-
> regarded the escaping gas.

Rapport historique sur les progrès des sciences physiques et naturelles
depuis 1789 et sur leur état actuel, présenté au gouvernement le
6. Février, 1808, par l'Institut royal de France.   Paris, 1810.   8vo.
Nouvelle édition, 1827.   8vo.

Dachauer, G.
Ozon.   Eine gedrängte Zusammenstellung bisher gewonnener Resultate.
München, 1864.   pp. 204, 8vo.

> *Compare* Moigno, Abbé, L'ozone, etc.   *Also, in Section I*, Leeds, A. R.

Daguerre, Louis Jacques Mandé.
*Historique et description des procédés du Daguerréotype et du
Diorama.   Nouvelle édition corrigée, et augmentée du portrait de
l'auteur.   Paris, 1839.   pp. 76, 8vo.   Portrait of Daguerre and six
plates.

> Das Daguerreotype und das Diorama, oder genaue und
> authentische Beschreibung meines Verfahrens und
> meiner Apparate zu Fixiring der Bilder der Camera ob-
> scura und der von mir bei dem Diorama angewendeten

DAGUERRE, LOUIS JACQUES MANDÉ. [Cont'd.]
>Art und Weise der Malerei und der Beleuchtung. Mit zwei Tafeln Abbildungen. Stuttgart, 1839. pp. 67, 12mo.

DAUBENY, CHARLES.
>*An Introduction to the Atomic Theory, comprising a sketch of the opinions entertained by the most distinguished ancient and modern philosophers with respect to the constitution of matter. Oxford, 1831. pp. xvi–148, 8vo.

DAVY, SIR HUMPHRY.
>Historical View of the Progress of Chemistry ; Davy's Collected Works, vol. IV. London, 1829.

>>A brief sketch, which is found also in Davy's Elements of Chemical Philosophy, part I, vol. I. London, 1812. 8vo.

DEBEAUX, J. O.
>Essai sur la pharmacie et la matière médicale des Chinois. Paris, 1865. 8vo.

DÉHÉRAIN, P. P.
>Découverte de l'oxygène. *In* Revue de l'instruction publique, 1858 and 1859.
>Études pour servir à l'histoire de la chimie. La découverte de la composition de l'eau. Extrait des Annales du Conservatoire des Arts et Métiers, No. 2, Octobre, 1860. Paris, 1860. pp. 56, 8vo.

>>The author analyzes the parts played by Cavendish, Watt, Priestley, Lavoisier and others in the discovery of the chemical nature of water, and summarizes the experimental demonstration of its quantitative composition by Humboldt and Gay Lussac, and by Berzelius, Dulong and Dumas.

DEJEAN, FERDINAND.
>Historia . . . sodæ Hispanicæ, 1773. *See in Section V*, Dejean, F.

DE KONINCK.
>*See* Koninck, de.

DEL MAR, ALEXANDER.
>History of the Precious Metals from the earliest times. London, 1880. 8vo.

DENFFER, JOHANN HEINRICH [*called* JANSEN].
>Alchemistenlogic, oder Vernunftlehre der Scheidekünstler . . . Königsberg, 1762.

DESCRIPTION DE DIVERS PROCÉDÉS POUR EXTRAIRE LA SOUDE DU SEL
MARIN. Faite en exécution d'un arrêté du comité de salut public
du 8 Pluviose, an II de la République Française. Imprimés par
ordre du comité de salut public. Paris, an III. [1795]. pp. 80,
4to. Eleven folding plates.

> The influence of this report on the progress of chemistry is unparalleled. The
> committee in their Résumé say: "Le procédé du C. Leblanc par
> l'intermède de la craie nous paroît celui qui peut être le plus générale-
> ment adopté."

DESCRIPTION DES ATOMES. Paris, 1813. pp. viii–385, 8vo. Nine folding plates.

DEUSING, ANTONIUS.
De manna et saccharo. Groningen, 1659.

DEVENTER, CHARLES MARIUS VAN.
*Schetsen uit de geschiedenis van de scheikunde. Academisch proef-
schrift ter verkrijging van den graad van Doctor in de scheikunde
aan de stedelijke universiteit van Amsterdam . . . 13 en December,
1884. Dordrecht, 1884. pp. [viii]–148, 8vo.

DEWAR, J.
Selected Extracts from different authors on Alchemy in relation to
modern science. (*From* Proc. Roy. Inst., 1884.) London, 1884. 8vo.

DIGBY, SIR KENELM.
> De voornaamste chymicale verborgentheden, etc. *See in Section VI*, Thea-
> trum chemicum, ofte geopende deure der chymische verborgentheden.

DISCOURS HISTORIQUE SUR L'ORIGINE ET LES PROGRÈS DE LA CHYMIE. (1723,)
> *See in Section V*, Senac, J. B.

DONNDORFF, J. A.
Geschichte der Erfindungen in allen Theilen der Wissenschaften und
Künste von der ältesten bis auf die gegenwärtige Zeit, in alphabetischer Ordnung. Quedlinburg und Leipzig, 1817–21. 6 vols.,
8vo.

DRAPER, JOHN WILLIAM.
* History of the Intellectual Development of Europe. New York, 1862.
Fifth edition, 1866. pp. xii–631, roy. 8vo.

> Chapter XIII treats of Arabian contributions to chemistry; and the progress
> of chemistry in its relation to other branches of learning is discussed
> throughout the work.

An Introductory Lecture on the History of Chemistry delivered in the

DRAPER, JOHN WILLIAM. [Cont'd.]
 University of New York, session 1846–47. New York. Printed
 for the Medical Class of the University. 1846–7. pp. 15.
 Priestley's Discovery of Oxygen Gas. Pop. Sci. Monthly, v, 385, Aug.,
 1874.

DRECHSLER, JOHANN GOTTHELF.
 Dissertatio sive præcipuas fermentationis theorias ab antecessoribus
 Lavoisieri excogitatas. Viteberg, 1810. 4to.

DRIESSEN, PETRUS.
 Over het voordeel dat de beoefening der scheikunde voor andere
 kunsten en wetenschappen aanbrengt. Groningen, 1778.

DROHOJOWSKA.
 Les savants modernes et leurs œuvres : Parmentier, Rumford, Liebig.
 Paris, 1886. 8vo. Ill.

DUFRENÉ, HECTOR.
 Étude sur l'histoire de la production et du commerce de l'étain. Paris,
 1881. 8vo.

DUMAS, JEAN BAPTISTE.
  Leçons sur la philosophie chimique. *See in Section V.*

DUMÊNIL, AUGUSTE.
 Geschichtlich-wissenschaftliche Darstellung der Stöchiometrie und
 Electrochemie den Pharmaceuten gewidmet. Hannover, 1824. 8vo.

DUPONT, B.
 Étude historique sur la chimie et la théorie atomique. Limoges, 1869.

DUPONT, M.
 Scheele et la chimie au XVIIIe siècle. Discours prononcé le 24 No-
 vembre, 1881, à la rentrée de l'École de Médecine de Tours.
 Tours, 1881. pp. 30, 8vo.

DUPREZ ET DE KONINCK.
 Les travaux de l'Académie royale de Belgique, 1772–1872, physique,
 météorologie et chimie. Bruxelles, 1872. 2 parts, 8vo.

DÜSSELDORF (Ausstellung).
  See Horadam, J.

DUVAL, ROBERT (ROBERTUS VALLENSIS).
 De veritate et antiquitate artis chemicæ, et pulveris, sive medicinæ
 philosophorum, sive auri potabilis materia et compositione, illiusque

DUVAL, ROBERT (ROBERTUS VALLENSIS). [Cont'd.]
mira vi in tria rerum genera animale, vegetale et minerale testi-
monia et theoremata : ex variis authoribus sacris, theologis, juris-,
peritis, medicis, philosophis et poetis, per R. V. selecta. Parisiis,
1561. 48 leaves, 16mo. Also : *Lugduni-Batavorum, 1593. pp.
46, 18mo.

> A full description of the various editions of this *first history of chemistry* is
> given by Professor Ferguson. *See in Section IV*, Duval, Robert. *Cf.*
> *also in Section I*, Ferguson, John.

EATON, T. J.
History of Alchemy. A paper read before the Kansas City Academy
of Science, Sept. 25, 1877. *In* Western Review of Science and
Industry, Oct., 1877.

ECKHARTSHAUSEN, KARL VON.
*Entwurf zu einer ganz neuen Chimie durch die Entdeckung eines
allgemeinen Naturprinzips wodurch sich das phlogistische System
der alten und das antiphlogistische der neuen Chimisten als zwey
Extreme in ein Mittelsystem vereinigen lassen, worinn allein die
Wahrheit liegt, und die höhere Chimie der ältesten Vorzeit, mit
der gemeinen Schulchimie der jetzigen Zeit vereiniget wird.
Regensburg, 1800. pp. 300, 12mo.

> Considering the period at which this was written, the author exhibits a
> singular theory, chemical, mystical and fanciful. *See* Kopp, Die
> Alchemie, II, 166.

ÉCOLE D'ARTILLERIE DE LA GARDE ROYALE [PARIS]. Description du laboratoire.
> *See* Brianchon, C. J.

* EDELGEBORNE (DIE) JUNGFER ALCHYMIA, oder eine durch Rationes
viele Exempla und Experimenta abgehandelte Untersuchung, was
von der Alchymia zu halten und vor Nutzen daraus zu schöpffen
seye, nebst einem Zusatz von der Medicina Universali, Universal-
Process und einigen Kunst-Stücken aus der Alchymie. Tübingen,
1730. pp. [xxiv]-424, 12m
Also issued under the title :
* Ehren-Rettung der Alchymie, oder Vernünfftige Untersuchung, was
von der herrlichen Gabe, welche die himmlische Weissheit denen
Menschen geschencket, und insgemein mit dem verächtlichen
Nahmen der Alchymie beleget wird, zu halten seye. Durch Ratio-
nes, auch viele curiose Exempla und Experimenta abgehandelt.
Wobey noch von der Medicina Universali Meldung geschiehet.
Sammt einem Anhang des Universal-Processes Zweyer alten
wahrhafften Philosophorum, und kurtzer Consignation etlicher

EDELGEBORNE (DIE) JUNGFER ALCHYMIA, etc. [Cont'd.]

Kunst-Stücke aus der Alchymie. Nicht nur denen Liebhabern dieser edlen Wissenschaften, sondern auch allen andern, was Standes sie seyn mögen, nutzlich und vergnüglich zu lesen. Ehedessen geschrieben von einem bekannten Philosopho. Nun aber auf vieler Verlangen an Tag gegeben von einem Liebhaber der Chymie. Herrenstadt, 1730. pp. [xxiv]–424, 12mo.

> Though bearing different imprints and titles, these two books are otherwise identical. The author conceals his name under the expression : VICTRIX FORTUNÆ SAPIENTIA. My copy of the Ehren-Rettung bears the MS. note : " Hujus opus Author est Joannes Conradus Creiling." Schmieder regards this authorship as probable. The work is divided into five chapters, treating of (1) Whether the transmutation of metals be possible ? (2) Whether and where this has actually occurred ? (3) Experimental evidences of this transmutation. (4) Of the Universal Medicine, Potable Gold, etc. (5) Whether Alchemy is to be recommended to any persons, especially noble Lords. The second chapter contains much material for a history of alchemy of anecdotal character. This work is often quoted by Schmieder in his Geschichte der Chemie.

EDER, JOSEPH MARIA.

Geschichte der Photochemie und Photographie vom Alterthüme bis in die Gegenwart. (Ausführliches Handbuch der Photographie. Zweite Auflage, erstes Heft.) Halle, 1891. pp. 147, 8vo. Ill.

> Contains full-page portraits of Niepce, Daguerre, Fox Talbot, and J. H. Schultze.

ENGLER, C.

Der Stein der Weisen. Festrede bei dem feierlichen Acte des Directorats-Wechsels an der grossherzoglichen badischen technischen Hochschule zu Karlsruhe am 9 November, 1889. Anhang ; Bermerkungen zu Kant's Ansichten über die Chemie als Wissenschaft. Karlsruhe, 1889. pp. 26, roy. 8vo.

Historisch-kritische Studien über das Ozon. [Separat-Abdruck aus der Leopoldina. Heft xv.] Halle, 1879. 4to. Plate.

Эрдманъ, 0.

Объ изученіи химіи. Переводъ съ нѣмецкаго.

С.-Петербургъ 1861.

> ERDMANN, O. On the study of chemistry. St. Petersburg, 1861.

ERNI, HENRY.

Constitution of Organic Compounds ; being a brief account of the different theories advanced on this subject. Boston, 1853. 8vo.

ESPOSIZIONE INDUSTRIALE DEL 1881 A MILANO. Relazioni dei Giurati. Le industrie chimiche : Prodotti chimici, A. Pavesi. Tintoria, L. Gabba. Conceria, A. Cataneo. Milano, 1883. 8vo.

EYSSENHARDT, FRANZ.
* Arzneikunst und Alchemie im siebzehnten Jahrhundert.   Hamburg,
    1890.   pp. 32, 8vo.

> Sammlung gemeinverständlicher Vorträge herausgegeben von Rud. Virchow
> und Wilh. Wattenbach, No. 96.   Treats chiefly of Francesco Giuseppi
> Borri and his adventures.   *Cf*. Borri, F. G., *in Section IV*.

FABBRONI, GIOVANNI VALENTINO MATTIA.
Storia delle opinioni chimiche relativamente alla formazione degli eteri.
    Firenze, 1795.

FABINYI, R.
Das neue chemische Institut der königlichen ungarischen Franz Joseph
    Universität zu Klausenburg.   Budapest, 1882.

FABRE, J. HENRI.
Les inventeurs et leurs inventions.   Histoire élémentaire des principales
    découvertes dans l'ordre des sciences physiques.   Paris, 18—.   8vo.
    Vignettes.

FAIRCHILD, HERMAN LEROY.
*A History of the New York Academy of Sciences, formerly the Lyceum
    of Natural History.   Read in abstract, before the Society, May 10,
    1886.   New York, 1887.   pp. xii–190, 8vo.   Plates and 27 por-
    traits.

> A local history chiefly bearing on the naturalists of the period.

FALCONER, W.
Sketch of the History of Sugar in the early times and through the
    Middle Ages.   Memoirs of the Society of Manchester.   Vol. IV.
    part 2.   pp. 191 *et seq*.

FAULSTICH.
Der Stein der Weisen.   Program.   Berlin, 1814.   12mo.

FAULWETTER, C. ALEX.
Kurze Geschichte der Elektricitätslehre.   Nürnberg, 1793.   5 parts.   8vo.

FAURE, H.
Histoire de la céruse . . . avec histoire de la plomb.   Lille, 1889.   8vo.
    Cf. Pulsifer, W. H.

FERBER, J. J.
Untersuchung der Hypothese von der Verwandlung der mineralischen
    Körper in einander.   Aus den Akten der kaiserl. Akademie der
    Wissenschaften zu St. Petersburg übersetzt mit einigen Anmer-

FERBER, J. J. [Cont'd.]
kungen vermehrt und herausgegeben von der Gesellschaft naturforschender Freunde zu Berlin. Berlin, 1788. pp. 72, 12mo.
A refutation of the claims of the alchemists.

FERGUSON, JOHN.
Eleven Centuries of Chemistry. Address on resigning the Presidency
of the Chemical Section of the Philosophical Society [of Glasgow].
[From the Proceedings of the Philosophical Society of Glasgow,
vol. x, p. 27 and p. 368.] Read before the Chemical Section, 11th
November, 1878. pp. 22, 8vo.
See also in Section I, Ferguson, John.

FERSTEL, R. VON.
Der Bau des chemischen Instituts der Wiener Universität. Wien, 1874.
Fol.

FIGUIER, LOUIS GUILLAUME.
L'Alchimie et les alchimistes. Essai historique et critique sur la philosophie hermétique. Paris, 1855. *Troisième édition, 1860. pp.
iv–421, 12mo.
A readable work, founded mainly on Schmieder's Geschichte der Alchemie,
giving, however, but little credit to the industrious German. Bibliographically rather weak. Entertaining and popular.

Exposition et histoire des principales découvertes scientifiques
modernes. Paris, 1851. 3 vols., 12mo. 4. édition, 1855.

FISCHER, J. C.
Geschichte der Physik seit dem Wiederaufleben der Wissenschaften bis
an das Ende des XVIII Jahrhunderts. Göttingen, 1801–1808.
8 vols., 8vo. Ill.

FITTICA, F.
Ueber die Aufgaben der wissenschaftlichen Chemie. Giessen, 1877. 8vo.

FLORENCOURT, CARL C.
Ueber die Bergwerke der Alten, Göttingen, 1785. 8vo.

FLÜCKIGER, FRIEDRICH AUGUST.
Das Nördlinger Register, Beitrag zur Geschichte der deutschen
Pharmacie im 15 Jahrhundert. [Halle], [1877]. n. t. p.
Documente zur Geschichte der Pharmacie. Halle, 1876.

FOURCROY, ANTOINE FRANÇOIS DE.
See in Section V, Leçons d'histoire naturelle et de chimie. The later edi-
tions contain : Chapter II, Histoire de la chimie. See also, Fourcroy,
A. F. Système des connaissances chimiques.

FOUQUE, VICTOR.
> La vérité sur l'invention de la photographie.
> *See in Section IV*, Niepce, Nicéphore.

FOWNES, GEORGE.
On the Chemistry of Ancient and Modern Times ; an introductory
lecture. *n. p.   n. d.* 8vo.

FRAAS, CARL.
Geschichte der Landwirthschaft, oder geschichtliche Uebersicht der
Fortschritte landwirthschaftlicher Erkenntnisse in den letzten 100
Jahren. Gekrönte Preisschrift. Prag, 1852. 8vo.

FRANCHIMONT, A. P. N.
De verschillende richtingen der chemie, blikken in het verleden, het
heden en de toekomst dier wetenschap. Leiden, 1874. 8vo. ·

FRANCUS, JOHANNES.
De arte chymica ejusque cultoribus epistolæ quatuor. Budissin, 1610. 4to.

FRAZER, PERSIFOR.
The Progress of Chemical Theory, its Helps and Hindrances. Introduction to the Chemical Lecture Course at the Franklin Institute,
November 10th, 1890. J. Frankl. Inst., 3d S., vol. CI, pp. 241,
321, and 409. (Apr., May, June, 1891.) *Also*, Chem. News, vol.
64, p. 6 (July, 1891).

FREDERKING, CARL.
*Grundzüge der Geschichte der Pharmacie und derjenigen Zweige der
Naturwissenschaft auf welchen sie basirt. In zwei Abtheilungen.
Erste Abtheilung, die Perioden der Geschichte der Pharmacie.
Zweite Abtheilung, Lebensbeschreibung der Förderer derselben.
Göttingen, 1874.   pp. viii-303, 8vo.

> Conveniently arranged, but in some cases so brief as to be useless. Contains
> very little bibliography. The author ardently decrys the "homoö-
> pathische Schwindel." The typographical errors are innumerable.
> The author says M. E. Chevreul died in 1844 !

FREMERY, NIC. CORN.
Oratio de chemia et arte pharmaceutica ad majorem perfectionem in
Belgio evehendas. Traject. ad Rhenum, 1822. 8vo. *In* Annales
acad. Rhen. Traject., 1821–22.

FREMY, CARNOT, JUNGFLEISH, TERREIL [*et al.*].
> Les laboratoires de chimie. *See in Section II*, Fremy, E., Encyclopédie
> chimique.

FREMY, EDMONDE.

Discours préliminaire sur le développement et les progrès récents de la chimie. (Extrait de l'Encyclopédie chimique.) Paris, 1881. pp. iii–384, 8vo.

> Covers the latest period, for which reason chiefly it is of value. Contains lists of papers by individual authors, but with no dates nor references !

FRESENIUS, C. REMIGIUS.

* Geschichte des chemischen Laboratoriums zu Wiesbaden. Zur Feier des 25 jährigen Bestehens der Anstalt. Wiesbaden, 1873. pp. 106, 8vo.

> Contains portrait of Fresenius and two plans ; also a Chronologisch-geordnetes Verzeichniss aller Bücher und Abhandlungen welche aus dem chemischen Laboratorium zu Wiesbaden während der ersten 25 Jahre seines Bestehens hervorgegangen sind, mit Vorausschickung der Arbeiten von R. Fresenius, welche vor Errichtung des Laboratoriums veröffentlicht wurden.

FREUND, A.

Historyja chemii, czasopismo. Lwów, 1882.

FRIEDLÄNDER, P.

Fortschritte der Theerfarbenfabrikation und verwandter Industrie-zweige 1877–1887. An der Hand der systematisch-geordneten und mit kritischen Anmerkungen versehenen Deutschen Reichs-patents dargestellt. Berlin, 1888. Sm. 4to. Zweiter Theil, 1887–90. Berlin, 1891.

FUCHS, GEORG FRIEDRICH CHRISTIAN.

Geschichte des Braunsteins, seine Verhältnisse gegen andere Körper und seine Anwendung in Künsten. Jena, 1791.

Geschichte des Zinks in Absicht seines Verhaltens gegen andere Körper und seiner Anwendung auf Arzneiwissenschaft und Künste. Erfurt, 1788. pp. [xvi]–398, 12mo.

FÜRSTENAU, HARTMANN GODFRIED.

Dissertatio de incrementis recentiori ævo in scientia chemica factis. Rintel, 1792. 8vo.

GABBA, C. L.

Contributo alla storia delle falsificazioni e delle adulterazioni degli alimenti. Torino, 18—.

GADD, PETER ADRIAN.

De sacerdote chemico. Aboæ, 1769. 2 parts, 4to.

De fatis scientiæ chemicæ sub epocha patrum. Aboæ, 1763. 4to.

GADD, PETER ADRIAN. [Cont'd.]

Dissertatio chemico-historica, inventa quædam chemica recentiora leviter adumbrans. Aboæ, 1763. 4to.

Dissertatio gradualis, incrementorum scientiæ chemicæ remoras leviter adtingens. Aboæ, 1763. 4to.

GAUTIER, HENRI.

*Essai sur l'histoire de la chimie. Thèse présentée et soutenue à la Faculté de Médecine de Paris le 11 Mars, 1837, pour obtenir le grade de Docteur en Médecine. Paris, 1837. pp. 48, 4to.

> Writing of Priestley, the author says he fled for refuge to North America, and not finding the repose desired even in the remote sources of the Usquehannah, he sought an asylum among the Red-skins, and finally he and his entire family died by poison ! The author treats of the history of chemistry, in three epochs, but apparently has examined but few of the works referred to.

GEHEIMNISSE DER ALTEN bei der eingebrannten durchsichtigen Glas-malerei. Leipzig, 1831. Ill.

GEISSLER, J. G.

Beschreibung und Geschichte der neuesten und vorzüglichsten Instru-mente. Zittau, 1792–1802. 12 parts.

GERDING, TH.

*Geschichte der Chemie. Leipzig, 1867. pp. xxiv–598, 8vo.

> A hastily compiled though compendious history, including notices of living chemists and modern researches.

GERHARD, CARL ABRAHAM.

Versuch einer Geschichte des Mineralreichs. Berlin, 1781–82. 2 vols., 8vo.

GERICKE, PETER.

De lapide philosophorum seu medicina universali vero an falso. Helmstadii, 1742.

GESSERT, M. A.

Geschichte der Glasmalerei bis auf die neueste Zeit. Stuttgart, 1840.

GEUNS, W. A. J. VAN.

Proeve eener geschiedenis van de äquivalentgetallen der scheikundige grondstoffen en van hare soortelijke gewigten in gasvorm, voor-namelijk in betrekking tot de vier grondstoffen der bewerktuigde natuur. Amsterdam, 1853.

GIBBS, WOLCOTT.
Report on the Recent Progress of Organic Chemistry. *In* Proceedings
American Association for the Advancement of Science, vol. IX, 1856.

GIESSEN (LABORATORIUM).
See Hofmann, J. P.

GIGLI, TORQUATO.
*Storia della chimica da Lavoisier fino al 1840. Milano, 1886. pp. 63,
16mo.
Biblioteca del popolo, vol. 201.

GILBERT.
Coup d'œil sur les poisons et les sciences occultes, depuis l'antiquité
jusqu'au XVIIIe siècle. Moulins, 1877. 8vo.

GILBERT, LUDOVICUS GUILIELMUS.
Dissertatio historico-critica de mistionum chemicarum simplicibus et
perpetuis rationibus, etc. Sect. I. Lipsiæ, 1811. 8vo.

GILDEMEISTER, J.
Alchymie. *In* Zeitschrift der deutschen morgenländischen Gesellschaft.
Leipzig, 1876. Bd. xxx. pp. 534–538.
The learned author discusses the etymology of the word *elixir*, and shows
that it signifies the substance capable of effecting transmutations, or the
so-called Philosophers Stone. The citations are in Arabic text. He
identifies the celebrated alchemist Artephius with the poet Al-Toghrâî.

GISH, J. B.
Recent Advances in Chemistry. *In* Trans. M. Soc. Kansas. Lawrence,
1879. Vol. XIII, pp. 78–82.

GLADSTONE, J. H.
The Birth of Alchemy. *In* The Argonaut, January, 1876, p. 1.
An interesting sketch of Chinese alchemical knowledge in remote times.

GMELIN, JOHANN FRIEDRICH.
*Allgemeine Geschichte der Gifte. Drei Theile. Leipzig, 1776–7. pp.
350 ; 8–316 ; [xvi]–525 + sixteen pages of errata ! 12mo.
Part II treats of mineral and part III of vegetable poisons.

Beyträge zur Geschichte des teutschen Bergbaus vornehmlich aus den
mittlern und spätern Jahrhunderten unserer Zeitrechnung. Halle,
1783. 8vo.

De primis chemiæ pneumaticæ originibus. Gottingæ, 1797. pp. 32,
sm. 4to.
This contains also a life of the author.

GMELIN, JOHANN FRIEDRICH. [Cont'd.]
*Geschichte der Chemie seit dem Wiederaufleben der Wissenschaften
    bis an das Ende des achtzehnten Jahrhunderts. Göttingen,
    1797–99. 3 vols., 8vo.
Also under the title: Geschichte der Künste und Wissenschaften seit
    der Wiederherstellung derselben bis an das Ende des achtzehnten
    Jahrhunderts. Von einer Gesellschaft gelehrter Männer ausge-
    arbeitet. Achte Abtheilung. Geschichte der Naturwissenschaften.
    II, Geschichte der Chemie. Göttingen, 1797.

> An unwieldy work with a stupendous amount of detail, badly arranged. It
> excels in bibliographical references.

GOBET, NICOLAS.
Les anciens minéralogistes du royaume de France. Paris, 1779. 2 vols.,
    8vo.

GODIN, LOUIS.
L'Histoire de l'Académie des Sciences de Paris depuis 1680–99. Paris.
    11 vols., 4to.
Table alphabétique des matières contenues dans l'Histoire de l'Académie
    depuis son établissement jusqu'en 1730. Paris. 4 vols., 4to.

GOEBEL, C. C. T. F.
Ueber den Einfluss der Chemie auf die Ermittelung der Völker der
    Vorzeit, oder Resultate der chemischen Untersuchung metallischer
    Alterthümer, insbesondere der in den Ostseegouvernements vor-
    kommenden, behufs der Ermittelung der Völker, von welchen sie
    abstammen. Erlangen, 1842. 8vo.

GOGUET, ANTOINE YVES.
De l'origine des loix, des arts, et des sciences et de leurs progrès chez
    les anciens peuples. Paris, 1758. 3 vols., 4to.

> Dell' origine delle leggi, delle arti, delle scienze, tradotto
>   dall' inglese [ecc.]. Parma, 1802. 6 vols. in 3, 8vo.
> *The Origin of Laws, Arts and Sciences and their Progress
>   among the most ancient nations. Translated from the
>   French. Adorned with cuts. Edinburgh, 1775. 3 vols.,
>   8vo.

> Contains much on early history of technology. The author's erudition is
> aided by his imagination.

GOLDSCHMIDT, G.
Die chemische Industrie. Bericht über die Weltausstellung Philadel-
    phia. Wien, 1877.

GOODE, GEORGE BROWN.
Beginnings (The) of American Science. The third century. An address delivered at the eighth anniversary meeting of the Biological Society of Washington. From the Proceedings of the Biological Society of Washington, vol. IV, 1886–88. Washington, 1888. pp. from 10 to 94, 8vo.

GÖTTINGEN, (LABORATORIUM).
See Breymann und Kirstein.

GOTTSCHALDT, A.
Der Neubau der königlichen technischen Lehranstalten zu Chemnitz. Chemnitz, 1878. 4to.

GOUGE, AUG.
Étude sur l'industrie sucrière à l'Exposition de Philadelphie et sur le raffinage aux États-Unis. Paris, 1877.

GRÄSSE, J. G. TH.
Bierstudien. Ernst und Scherz. Geschichte des Bieres und seiner Verbreitung über den Erdball. Bierstatistik. Bieraberglauben. . . . Brauergeheimnisse. Dresden, 1872. 8vo. Ill.

GRAZ (LABORATORIUM).
See Pebal, L. von.

GREN, FR. ALBR. CARL.
Geschichte der Naturwissenschaft als akademische Vorlesungen vorgetragen ; Ein Fragment aus dessen nachgelassenen Papieren. Ann. Phys., Gilbert, vol. I, pp. 167–204.

GREIFSWALD (LABORATORIUM).
See Müller, G.

GÜLDENFALK, SIEGMUND HEINRICH.
* Sammlung von mehr als hundert wahrhaften Transmutationsgeschichten, oder ganz ausserordentlich merkwürdige Beyspiele von Verwandlung der Metallen in Gold oder Silber nebst der Art und Weise wie damit verfahren worden. Gesammelt und herausgegeben von S. H. G. Frankfurt und Leipzig, 1784. pp. xxxvi–443, 12mo.

> The author gives detailed accounts of 112 transmutations in as many chapters or sections. The work is a veritable encyclopædia of alchemical anecdotes without regard to their authenticity or the probabilities. Schmieder admits his indebtedness to Güldenfalk.

GÜNTHER, SIEGMUND.
Vermischte Untersuchungen zur Geschichte der mathematischen Wissenschaften. Leipzig, 1876. 8vo. 4 tables.

GUETTARD, JEAN ÉTIENNE.

Histoire de la découverte faite en France de matières semblables à celles dont la porcelaine de la Chine est composée. Paris, 1765. 4to. 1766, 12mo.

GUINON, MARNAS ET BONNET.

Exposé historique des travaux relatifs aux matières colorantes des lichens. Paris. 4to.

GURNEY, GOLDSWORTHY.

A course of lectures on Chemical Science as delivered at the Surrey Institution. London, 1823. pp. v–[ii]–310, 8vo.

> The introductory lecture deals with the history of chemistry.

GUTTSTADT, ALBERT.

Die naturwissenschaftlichen und medicinischen Staatsanstalten Berlins. Berlin, 1886.

> The National Scientific Institutions of Berlin. Translated and condensed by George H. Boehmer. Report Regents Smithsonian Institution, July, 1889, p. 89.
>
> An historical sketch of each of the Scientific Institutions in Berlin, including the Royal Academy of Sciences, the Chemical Institute, etc. Valuable for its statistics.

GUYTON DE MORVEAU, LOUIS BERNARD.

> Précis historique et critique des differens systèmes proposés pour l'explication de ce phénomène [Combustion]. *In :* Digressions académiques. *See in Section V*, Guyton de Morveau.

HAESER, HEINRICH.

Lehrbuch der Geschichte der Medicin und der epidemischen Krankheiten. Dritte Bearbeitung. Jena, 1875–82. 3 vols., 8vo.

> First edition, 1845. Second edition, 1865. 2 vols.
>
> A valuable compendium of information concerning the physicians who were also chemists.

HAGEMANN, H. H. (JUN.).

Brevis eudiometriæ historia. Groningen, 1830. 4to.

HAHN, L.

Documenta ad historiam rei pharmaceuticæ Silesiæ. Vratislaviæ, 1848. 8vo.

HALLIWELL, JAMES ORCHARD [Editor].

A Collection of Letters illustrative of the Progress of Science in England from the reign of Queen Elizabeth to that of Charles the

HALLIWELL, JAMES ORCHARD. [Cont'd.]
Second. Historical Society of Science. London, 1841. pp. xxii–124.
A Few Notes on the History of the Discovery of the Composition of
Water. London, 1840. 8vo.

HARRISON, W. JEROME.
A History of Photography written as a practical guide and an introduc-
tion to its latest developments, with a biographical sketch of the
author, and an appendix by Dr. Maddox on the discovery of the
gelatino-bromide process. New York, 1887. pp. 136, 8vo. Ill.
Was Photography discovered a Century ago ? Two articles in Oct. and
Jan. Nos. of The Photographic Quarterly, edited by C. W. Hastings.
London, 1889.

HEBENSTREIT, JOHANN ERNST.
Progressus de notionibus chymicis apud veteres. Lipsiæ, 1756. 4to.

HEIDELBERG (LABORATORIUM).
See Lang, H.

HEINECKEN, JOHANN.
Ueber die wichtigsten Fortschritte in der Physik und Chemie in den
letzten 30 Jahren. Bremen, 1809. 8vo.

HELLER, AUGUST.
Geschichte der Physik von Aristoteles bis auf die neueste Zeit. Stuttgart, 1882–84. 2 vols., roy. 8vo. Vol. I: pp. xii–412, Von Aristoteles
bis Galilei. Vol. II: pp. xvi–744, Von Descartes bis Robert Mayer.
> A comprehensive work. Contains notices of chemists and their works.

HENDERSON, A.
The History of Ancient and Modern Wines. London, 1824. pp. xi–
408. 4to.

HENKEL, JOHANN FRIEDRICH.
*Mineralogische, chemische und alchymistische Briefe von reisenden
und andern Gelehrten an den ehemaligen chursächsischen Berg-
rath J. F. Henkel. Drei Theile. Dresden, 1794. 8vo. Part I, pp.
397–[vi]. Part II, pp. 384–[vii]. Part III, pp. 291–[ix].
> The letters are written to Henkel by a great number of persons, and
> chiefly in reply to questions addressed to them. The lack of an index
> makes the collection less important than otherwise.

HERAPATH, WILLIAM.
On the Early Egyptian Chemistry. Phil. Mag. [4], III (1852).

HERGT, OTTO.

Die Valenztheorie in ihrer geschichtlichen Entwicklung und jetzigen Form. Programm. Herausgegeben vom naturwissenschaftlichen Vereine zu Bremen. Bremen, 1878. 4to.

HERMBSTÄDT, SIGISMUND FRIEDRICH.

Bibliothek der neuesten physisch-chemischen, metallurgischen, technologischen und pharmaceutischen Literatur. Berlin, 1788–95. 4 vols.

Rede über den Zweck der Chemie, über die Methode sie zu studiren und über den Einfluss derselben auf die Arzneywissenschaft. Berlin, 1792. 8vo.

HERVERDI, JOSEPH FERDINAND.

Erklärung des mineralischen Reichs. Ein Beytrag zur Geschichte der Alchemie. Berlin, 1783. pp. 124, 12mo.

HIGGINS, WILLIAM.

On the Origin of the Atomic Theory. *In* Phil. Mag., vol. 48, 1816.

HILDEBRANDT, GEORG FRIEDRICH.

Chemische und mineralogische Geschichte des Quecksilbers. Braunschweig, 1793. pp. x–476, 4to.

HILGER, A.

Blicke in die Vergangenheit und Gegenwart der Chemie. Erlangen, 1887. 4to.

HINTERBERGER, FRIEDRICH UND LUDWIG SCHMUED.

Geschichte des wissenschaftliches Theiles der Chemie in Verbindung mit allgemeiner Geschichte. *In* Jahresbericht der k. k. Ober-Realschule am Schottenfelde in Wien für das Studien-Jahr 1865–1866. Wien, 1866. pp. 1–10, 4to.

General history and chemical discoveries are brought into juxtaposition.

HIORTZBERG, L.

De nexu chemiæ cum utilitate reipublicæ. Upsaliæ, 1752. 4to.

HIRSCHING, WILHELM SIMON CHRISTIAN.

Versuch physicalisch-chemischer Lehrbegriffe zu möglichen Prüfung des Wesens, des Beständnisses, und der Wirkungsart des so berüchtigten Metallverwandelnden Meisterstückes und dessen vorgeblicher Nutzanwendung zu einem allgemeinen Genessmittel, in Absicht einiger Vergnügung einer Natur- und Grundforschenden Wissbegierde entworfen von W. S. C. H. Leipzig, 1754. pp. 488 and Register. 8vo.

Contains the stock narratives of transmutation.

HISTORICAL SOCIETY OF SCIENCE.
> *See* Wright, Thomas ; *also*, Halliwell, J. O.

HOCHGESANG.
> Historische Nachricht von Verfertigung des Glases, mit dazu gehörigen Rissen. Gotha, 1780. 8vo.

HOEFER, FERDINAND.
> Abrégé de l'histoire de la chimie. *In* Éléments de chimie minérale. *See in Section V*.

HOEFER, FERDINAND.
> Histoire de la chimie depuis les temps les plus reculés jusqu'à notre époque ; comprenant une analyse détaillée des manuscrits alchimiques de la bibliothèque royale de Paris ; un exposé des doctrines cabalistique sur la pierre philosophale ; l'histoire de la pharmacologie, de la métallurgie et en général des sciences et des arts qui se rattachent à la chimie, etc. Paris, 1842–43. 2 vols., 8vo. I, x–510 ; II, viii–518.
>> *Histoire de la chimie. Deuxième édition revue et augmentée. Paris, 1866–69. 2 vols., 8vo.
>>> A work of great research, especially in regard to earliest authentic records as derived from ancient manuscripts, though in this respect now supplanted by the magnificent "Collection des Alchimistes Grecques" of Berthelot. The whole work is a valuable compendium, especially for those unfamiliar with the language of Kopp's *Geschichte*.
> Histoire de la physique et de la chimie depuis les temps les plus reculés jusqu'à nos jours. Paris, 1872. pp. 561, 12mo. Ill.
>> The latter part, relating to chemistry, is mainly a condensation of Hoefer's larger work noticed above.

HOFFMANN, G.
> Article : *Chemie, Name,* in Ladenburg's Handwörterbuch der Chemie. Breslau, 1884. Vol. II, pp. 516–530.
>> An erudite essay by an Orientalist on the word chemistry, embodying latest researches in philosophy, and rich in bibliographic references.

HOFMANN, AUGUST WILHELM.
> *Bericht über die Entwickelung der chemischen Industrie während des letzten Jahrzehndes im Verein mit Freunden und Fachgenossen erstattet von A. W. H. Autorisirter Abdruck aus dem "Amtlichen Berichte über die Wiener Weltausstellung im Jahre 1873." Braunschweig, 1875–77. III parts, 8vo.
> The Chemical Laboratories in course of erection in the Universities of Bonn and Berlin. Report addressed to the Right Honourable the Lords of the Committee of her Majesty's most Honourable Privy

HOFMANN, AUGUST WILHELM. [Cont'd.]
    Council on Education. Reprinted from thirteenth report of the
    Science and Art Department of the Committee of Council on Edu-
    cation. Presented to both Houses of Parliament by command of
    her Majesty. London, 1866. pp. viii–72, sm. folio.
        Contains plans of the two Laboratories.
    *Berliner Alchemisten und Chemiker. Rückblick auf die Entwickelung
    der chemischen Wissenschaften in der Mark. Rede gehalten zur
    Feier des Stiftungstages der militärärztlichen Bildungsanstalten am
    2. August, 1882. Berlin, 1882. pp. 79–158, 8vo.
    *Ein Jahrhundert chemischer Forschung unter dem Schirme der Hohen-
    zollern. Rede zur Gedächtnissfeier des Stifters der kgl. Friedrich-
    Wilhelms-Universität zu Berlin am 3. August, 1881, in der Aula der
    Universität gehalten von dem zeitigen Rector. Berlin, 1881. pp.
    74, 4to ; and Berlin, 1882, pp. 77, 8vo.

HOFMANN, J. P.
    Das chemische Laboratorium der Ludwig's Universität zu Giessen.
    Nebst einem Vorwort von Justus Liebig. Heidelberg, 1842. 8vo.
    Ill.

HOFMANN, K. B.
    Das Blei bei den Völkern des Alterthums. Berlin, 1885. 8vo.
        Sammlung gemeinverständlicher wissenschaftliche Vorträge. xx, 472.

HOGHELANDE, THEOBALD VON.
    De alchimiæ difficultatibus liber, in quo docetur quid scire quidque
    vitare debeat chemiæ studiosus ad perfectionem aspirans. Coloniæ,
    1594. 8vo.
        Also in Manget's Bibliotheca chemica curiosa, vol. 1, 336. Cf. in Section
        VI, Manget, J. J.

        *Abhandlung von denen Hindernissen bey der Alchymie.
        Darin gezeiget wird was ein Liebhaber dieser Kunst
        zu wissen, und zu meiden hat, wenn er zur Vollkom-
        menheit gelangen will. Aus dem Lateinischen in das
        Deutsche übersetzet. Gotha, verlegts Christian Mevius,
        1749. pp. [xxx]–176, 12mo.
    Historiæ aliquot transmutationis metallicæ pro defensione alchymiæ
    contra hostium rabiem. Adjecta est Lullii vita, et alia quædam.
    Coloniæ, 1604.
        Beweis dass die Alchymey oder Goldmacherkunst ein sonder-
        bares Geschenk Gottes sei." Leipzig, 1804. 8vo.

HOGHELANDE, THEOBALD VON. [Cont'd.]

> Schmieder remarks that Hoghelande was an important personage in the history of alchemy, for after experiencing serious doubts of the transmutation of metals he became a vigorous defender of the faith and freely made his convictions known. The above work is a collection of marvellous tales concerning veritable (?) transmutations.

HOLLAND, H.

Modern Chemistry. From Quart. Review. *In his* Essays etc. London, 1862. pp. 425–464, 8vo.

HOLLSTEIN, ROBERT.

Kurze Geschichte der Lehre vom Isomorphismus und Polymorphismus. Realprogymnasium zu Lüdenscheid. Wissenschaftliche Abhandlung zum Jahres-Bericht 1884–85. Lüdenscheid. 1885. pp. 20, 4to.

HOPPE, O.

Beiträge zur Geschichte der Erfindungen. Erste Lieferung. Wann, wo und von wem ist die bergmännische Schiessarbeit erfunden und vervollkommnet, und wie steht der Harzer Bergbau zu diesen Fragen? Dazu einige Bemerkungen über das Alter des Feuersetzens und des Schiesspulvers. Clausthal, 1880. pp. 68, 8vo.

> A critical study.

HOPPE-SEYLER, FELIX.

Ueber die Entwickelung der physiologischen Chemie und ihre Bedeutung für die Medicin. Rede zur Feier der Eröffnung des neuen physiologisch-chemischen Instituts der Kaiser-Wilhelms-Universität Strassburg gehalten am 18 Februar, 1884. Strassburg pp. 32, roy. 8vo.

HORADAM, J.

Die chemische Industrie auf der Düsseldorfer Ausstellung, 1880. Berlin, 1880. 8vo.

HORSIN-DÉON, P.

Chimie industrielle à l'Exposition universelle de 1878. No. 3. La fécule, l'amidon et leurs dérivés. Fabrication des glucoses. Paris, 1879. 8vo.

HUE, F.

Le pétrole, son histoire, ses origines, son exploitation dans tous les pays du monde. Paris, 1885. 12mo.

HÜFNER, GUSTAV.

Ueber die Entwickelung des Begriffs Lebenskraft und seine Stellung zur heutigen Chemie. Akademische Antrittsrede gehalten am 10. Juli, 1873 von G. H. Tübingen, 1873. pp. 38, 8vo.

HUME, GILBERT LANGDON.
> History. *See in Section V*, Hume, G. L.

HUMMELIUS, J.
> Tapiarii hermetico-chemici partem alteram præstat Olympus Astralis; ex quo, ceu quarto chemicorum regno, elementi primi excipiendi vires, modus atque dexteritas solis, quibus prospera de sophorum enucleando diademate ridet fortuna, chemicis duobus maxime et curiosis gratis rursum exhibetur laboribus. Francofurti ad Mœnum, 1739. 8vo.

HUREAUX.
> Histoire des falsifications [etc.]. *See in Section V.*

HUXLEY. T. H.
> The Advance of Science in the last Half-Century. [Reprinted from "The Reign of Queen Victoria, a Survey of Fifty Years of Progress," edited by Thomas Humphrey Wood, London.] New York, 1888. pp. 139, 12mo.
>
> Refers incidentally to the advance in chemistry.

INGENKAMP, COSMAS.
> Die geschichtliche Entwickelung unserer Kenntniss von Fäulniss und Gährung. Inaug. Diss. Bonn, 1885, 8vo. Plate.

ISAMBERT.
> La chimie avant Lavoisier. *In* Revue scientifique, deuxième série, vol. vi, pp. 603–607. Paris, 1877.

ITTNER, FRANZ VON.
> Uebersicht der Hauptmomente des gegenwärtigen Zustandes der Chemie. Eleutheria, oder Freiburg. literar. Blätter von Erhardt besonders abgedruckt., 1820, I, 3. Freiburg, 1823. 8vo.

JACKSON, CHARLES T.
> On the History and Uses of Chemistry. In lectures before the American Institute of Instruction. Boston, 1834. 8vo.

JACOB, J.
> Traité élémentaire de chimie expérimentale et appliquée, suivi d'une méthode d'analyse pour reconnaître les métaux, les bases, les acides, les sels et les principaux corps que l'on rencontre dans l'industrie et dans la nature. Illustré de 200 belles gravures sur bois intercalées dans la texte, accompagné d'une notice biographique et bibliographique sur les principaux chimistes auxquels la science doit ses progrès. Paris, 1867. 8vo.

JACQUIN, NICOLAUS JOSEPHUS.
Examen chemicum doctrinæ Meyerianæ de acido pingui, et Blackianæ de aëre fixo, respectu calcis. Vindobonæ, 1769. pp. 96, 12mo.

JAGNAUX, RAOUL.
* Histoire de la chimie. Paris, 1891. 2 vols., roy. 8vo. I, pp. iii–728 ; II, 821.

> This extensive work is divided into four parts : 1. treats of the grand theories of chemistry ; 2, of the metalloids and their chief compounds ; 3, of the metals ; 4, of organic chemistry. The author omits albuminoid substances as belonging to biology rather than chemistry. In the preface the author exhibits characteristic national bias and says : " La chimie scientifique est donc dans ses grandes lignes, une science française. C'est pour le démontrer que le présent ouvrage a été écrit."

JAUSSIN, LOUIS ARMAND.
Ouvrage historique et chimique où l'on examine s'il est certain que Cléopatre ait dissous sur le champ la perle qu'on dit qu'elle avala dans un festin. Paris, 1749.

JOHNSON, W. B.
History of the Progress and Present State of Animal Chemistry. London, 1803. 3 vols., 8vo. I, pp. vi–[ii]–411 ; II, pp. [iv]–477 ; III, pp. [iv]–407.

> A remarkable work, considering the period at which written—exhaustive and historical.

JOHNSTON, JAMES FINLAY WEIR.
Report on the Recent Progress and Present State of Chemical Science. Report of Oxford Meeting of B. A. A. S. London, 1833. pp. 414–529.

> Abriss einer Geschichte der neueren Fortschritte und des gegenwärtigen Zustandes der Chemie. Nach dem Englischen des J. F. W. J. im Report of the first and second meetings of the British Association for the Advancement of Science, London, 1833. Bearbeitet und ergänzt von C. Rammelsberg. Berlin, 1837. 8vo.

JULIA–FONTENELLE, ÉTIENNE.
Recherches historiques, chimiques et médicales sur l'air marécageux. Paris, 1823. 8vo.

JUNGIUS, JOACHIM.
See Wohlwill, Emil.

K., H.
The Alchemists. The Mirror of Literature, Amusement and Instruction. London, 1840. 8vo, page 304.

KAPP, GEORG CHRISTIAN FRIEDRICH.

Systematische Darstellung der durch die neuere Chemie in der Arzneikunst bewirkten Veränderungen und Verbesserungen.   Hof, 1805.
8vo.

KARMARSCH, KARL.

* Geschichte der Technologie seit der Mitte des achtzehnten Jahrhunderts.   Auf veranlassung und mit unterstützung seiner Majestät
des Königs von Bayern Maximilian II., herausgegeben durch die
historische Commission bei der königl. Academie der Wissenschaften.   München, 1872.   pp. vii–932, 8vo.

Also under the title : Geschichte der Wissenschaften in Deutschland.
Neuere Zeit.   Elfter Band.

KARSTEN, CARL JOHANN BERNHARD.

Philosophie der Chemie.   Berlin, 1843.   pp. viii–327, 8vo.
Revision der chemischen Affinitätslehre mit beständiger Rücksicht auf
Berthollet's neue Theorie.   Leipzig, 1803.   pp. 273, 8vo.

KASTNER, CARL WILHELM GOTTLOB.

Beyträge zur Geschichte chemischer Entdeckungen.   *In* J. der Pharm.
XVI, 1808.   pp. 61–74.

> The author discusses Kunkel's remarks on several minor discoveries—viz.,
> fulminating quicksilver, fulminating silver, mercuric oxide, etc.

Chronologische Uebersicht der Geschichte der Chemie.   *In* Einleitung
in die neuere Chemie.   Halle, 1814.   *See in Section V.*

> In six parallel columns are given (a) Dates ; (b) Contemporary events in the
> intellectual progress of the human race ; (c) Epochs of Chemistry ; (d)
> Discoveries and improvements ; (e) Inventions and their applications ;
> (f) Names of the discoverers and inventors with titles of their principal
> works.   These tables occupy pages 523 to 696 of the volume.

Geschichte der Physik und Chemie.   Leipzig, 1833.

KEKULÉ, AUGUST.

* Die wissenschaftlichen Ziele und Leistungen der Chemie.   Rede gehalten beim Antritt der Rectorats der rheinischen Friedrich-
Wilhelms-Universität am 18 October, 1877.   Bonn, 1878.   pp. 29,
8vo.

KELL, R.

Sebalt Schwertzer als kursächsischer Faktor und kaiserlicher Berghauptmann.   Zur Geschichte der Alchemie im 16. Jahrhundert.   Leipzig,
1881.

KERNOT, F.
Storia delle farmacia e dei farmacisti appo i principali popoli del mondo.
Napoli, 1871. 8vo.

KIELMEYER.
Die Entwickelung der Färberei, Bleicherei, Druckerei. *In* Dingler's
polytechn. Journal, vol. 234, p. 63.

KINGZETT, CH. TH.
History of the Alkali Trade. *See in Section V*, Kingzett, Ch. Th.

KIRCHER, ATHANASIUS.
De origine alchymiæ ; *also* De lapide philosophorum. *In* Mundus
subterraneus, vol. II, liber XI, sectio 1 and 2. Amsterodami, 1665.
Folio.

> Athanasius Kircher, a celebrated historian, philosopher, mathematician and
> physical philosopher, was born at Fulda in 1601, and died at Rome in
> 1680. He filled the chairs of philosophy and oriental languages in the
> College of Würtzburg and in the Jesuits' College at Avignon. He was
> afterwards professor of mathematics in the Jesuits' College at Rome.
> Kircher was a man of " wide and varied, but ill-digested erudition, and
> a most voluminous writer." Although credulous to an absurd de-
> gree, in the dissertation " De origine alchymiæ " he violently attacks
> the alchemists and their pretended transmutations of the baser metals
> into gold. It is reprinted in Mangetus' Bibliotheca chemica curiosa,
> together with replies to his attacks by Clauder and by Blauenstein.
> Lenglet du Fresnoy says Kircher wrote against alchemy because he was un-
> successful in his attempts at transmutation.

KIRCHMAIER, GEORG CASPAR.
Dissertatione de metallorum metamorphosi. Wittenberg, 1693. 4to.

> The author thinks Adam was the first to use iron and other metals for in-
> dustrial and economic purposes, to which opinion we have no objections
> based on evidence.

KITAIBEL, PAUL.
Ueber seinen Antheil an der Entdeckung des Tellurs. *In* Gehlen's J.
Vol. 1 (1803).

KLAPROTH. J.
Sur les connoissances chimiques des Chinois dans le 8e siècle. *In* St.
Pétersburg Acad. Sci. Mémoires. Vol. II, 1810.

KLAUSENBURG (LABORATORIUM).
See Fabinyi, R.

KLEINERT, E.
Dissertatio chemica de J. B. Richteri doctrina. Vratislaviæ, 1851. 8vo.

Князевъ, А.
>   Очеркъ исторіи химіи.
>>   Вильна 1875.
>>>   KNYAZEFF, A.   A sketch of the history of chemistry.   Wilna, 1875.

KNAPP, F.
>   Geschichte der Gasbeleuchtung.   *See in Section V*, Schilling, N. H.,
>   Handbuch.

KOBELL, FRANZ VON.
Ueber den Einfluss der Naturwissenschaften, insbesondere der Chemie
    auf die Technik.   Landshut, 1841.   8vo.

Кочубей, П.
>   Описаніе замѣчат. лабораторій Германіи и Бельгіи.
>>   С.-Петербургъ 1854.
>>>   KOCHUBEЇ, P.   Description of the principal laboratories of Germany and
>>>   Belgium.   St. Petersburg, 1854.

KOLB, J.
Sur l'évolution actuelle de la grande industrie chimique.   Société
    industrielle du nord de la France.   Lille, 1883.   8vo.

KOLBE, HERMANN.
Chemische (Das) Laboratorium der Universität Leipzig und die seit
    1866 darin ausgeführten chemischen Untersuchungen.   Braun-
    schweig, 1872.   pp. xlviii–664, 8vo.   2 plates.
Accompanied by : Photographische Ansichten vom Laboratorium der
    Universität Leipzig.   Braunschweig, 1872.   Three folio sheets in
    cover.
Chemische (Das) Laboratorium der Universität Marburg und die seit
    1859 darin ausgeführten chemischen Untersuchungen nebst Ansich-
    ten und Erfahrungen über die Methode des chemischen Unter-
    richts.   Braunschweig, 1865.   pp. x–524, roy. 8vo.   Three folding
    plates.
Chemischer Rückblick auf die Jahre 1872 und 1873.   Leipzig, 1873–74.
Zur Entwickelungsgeschichte der theoretischen Chemie.   (Separat-
    Abdruck aus dem Journal für praktische Chemie.)   Leipzig, 1881.
    8vo.

KONINCK, LE G. DE.
Rapport sur les travaux de chimie présenté à l'académie royale des
    sciences, des lettres et des beaux-arts de Belgique pendant la
    période séculaire 1772–1872.   8vo.

KOPENHAGEN (WELTAUSSTELLUNG.)
>   *See* Kronberg, M.

KOPP, E.
  Wiener Weltausstellung. Schweiz. Chemische Industrie. Schaffhausen,
  1874.

KOPP, HERMANN.
  *Alchemie (Die) in älterer und neuerer Zeit. Ein Beitrag zur Cultur-
  geschichte. *Erster Theil :* Die Alchemie bis zum letzten Viertel
  des 18. Jahrhunderts. *Zweiter Theil :* Die Alchemie vom letzten
  Viertel des 18. Jahrhunderts an. Heidelberg, 1886. 2 vols., I,
  xiv–260 ; II, vi–425, 8vo.

> The second volume contains as Anhang, "Beitrag zur Bibliographie der
> Alchemie." This bibliography, while very rich in material, is un-
> fortunately arranged. The author himself says the titles are grouped
> "nach einem gemischten System." This defect is only partially com-
> pensated by an index.

  *Aurea catena Homeri. Braunschweig, 1880. pp. 50, 8vo.

> Dedicated to Friedrich Wöhler on the eightieth anniversary of his birth.

  *Aus der Molecular-Welt. Eine Gratulations-Schrift an R. Bunsen.
  Heidelberg, 1885. 8vo.
  *Beiträge zur Geschichte der Chemie. Braunschweig, 1869–1875. Three
  parts. Parts I and II, pp. xi–530, 8vo.
  Part III also under the title : Ansichten über die Aufgabe der Chemie
  und über die Grundbestandtheile der Körper bei den bedeutenderen
  Chemikern von Geber bis G. E. Stahl. Die Entdeckung der Zu-
  sammensetzung des Wassers. 1875. pp. ix–310, 8vo.

> Erudite, comprehensive, important and valuable.

  *Entwickelung (Die) der Chemie in der neueren Zeit. München, 1873.
  pp. xxii–854, 8vo.
  Also under the title : Geschichte der Wissenschaften in Deutschland.
  Neuere Zeit. Auf Veranlassung und mit Unterstützung seiner
  Majestät des Königs von Bayern, Maximilian II, herausgegeben
  durch die Historische Commission bei der königl. Academie der
  Wissenschaften. Vol. x.
  *Geschichte der Chemie. Braunschweig, 1843–47. 4 vols., 8vo. I, xix–
  455, II, x–426, III, xii–372, IV, xvi–448. With portraits of La-
  voisier, Berzelius, Davy and Liebig.

> A classical work, above praise. Is rather scarce ; a new edition brought
> down to date and printed in Roman type is desirable.

  *Remarques concernant "Les Origines de l'Alchimie" de M. Berthelot et
  les "Beiträge zur Geschichte der Chemie" de H. Kopp. Mémoire
  sur les volumes moléculaires des liquides. Paris et Heidelberg,
  1886. pp. xvi–32, 8vo.

KOPP, HERMANN. [Cont'd.]

>The author complains of the neglect by Berthelot to recognize his publications.

*Sonst und Jetzt in der Chemie. Ein populär-wissenschaftlicher Vortrag. Braunschweig, 1867. pp. 34, 8vo.

Коппъ Германъ.
Былое и современное химіи.
Москва 1870.

Ueber den Verfall der Alchemie und die hermetische Gesellschaft. Denkschriften der Gesellschaft für Wissenschaft und Kunst in Giessen. Erster Band, erstes Heft. Giessen, 1847. pp. 1–34, 8vo.

>A lecture delivered January 31, 1845.

Ueber den Zustand der Naturwissenschaften im Mittelalter. Heidelberg, 1869. 4to.

>*L'État des sciences au moyen âge. Discours rectoral à la séance solennelle de l'université de Heidelberg. Traduit de l'allemand par L. Koch. Revue des Cours Scientifiques de la France et de l'Étranger. Septième année, Nro. 26. Mai 28, 1870. pp. 402–409, 4to.

KORTUM, KARL ARNOLD.

*Verteidiget die Alchemie gegen die Einwürfe einiger neuen Schriftsteller besonders des Herrn Wieglebs. Duisburg, 1789. pp. 360, 8vo.

*Noch ein Paar Worte über Alchimie und Wiegleb, oder erster Anhang der Vertheidigung der Alchimie wider die Einwürfe der neuesten Gegner. Duisburg, 1791. pp. 80, 8vo.

>An earnest and honest defense of alchemy by a credulous physician, in reply to the attacks of Wiegleb, J. C., q. v.

KRAFFT, FRIEDRICH.

*Ueber die Entwickelung der theoretischen Chemie. Vortrag gehalten in der Aula der Universität Basel am 16. November, 1875. Basel, 1875. pp. 56, 8vo.

KRAUSE, EMIL.

Beiträge zur Geschichte der Gährungstheorie. Lübben, 1873, 4to. Schulprogramm.

KRONBERG, M.

Die heutige nordische chemische Industrie und ihre Vertretung auf der Kopenhagener Weltausstellung. Berlin, 1889.

Крупскій, А.
  Русская химическая промышленность во время международныхъ выставокъ: Московской 1872 и Вѣнской 1873.
      С.-Петербургъ 1874
        Krupskiĭ, A. Russian chemical industry at the international exhibitions of Moscow in 1872 and that of Wilna in 1873. St. Petersburg, 1874.

Kühn, Otto Bernhard.
  De utilitate quæ ex arte experimenta chemica recte instituenda profluit, annexa de nonnullis suis laboribus chemicis narratione parva ; præfatur ad audiendam orationem qua . . . D. 15 Mens. Augusti A. 1829, invitat O. B. K. Lipsiæ, *n. d.* [1829]. 4to.

Kurrer, Wilhelm Heinrich von.
  Geschichte der Zeugdruckerei, der dazu gehörigen Maschinen und Hülfswerkzeuge und der Erfindungen im Gebiet des Colorits für den Baumwollen-, Leinen-, Seiden- und Schaf-wollendruck, bis auf die neueste Zeit. Nebst einer ausführlichen Uebersicht des gegenwärtigen Standes dieser Kunst, in technischer, commerzieller und statistischer Hinsicht. Nürnberg, 1840. Ill.

Lacroix, Paul.
  *Chemistry and Alchemy [in the Middle Ages]. Science and Literature in the Middle Ages and at the period of the Renaissance. New York, 1878. pp. 174–199, sm. 4to. Illustrated.
        Not free from errors, being by a littérateur rather than a scientist. The illustrations are pleasing.

Ladenburg, A.
  *Vorträge über die Entwickelungsgeschichte der Chemie in den letzten 100 Jahren. Braunschweig, 1869. pp. xi–318, 8vo.
        Zweite verbesserte und vermehrte Auflage. 1887.

Lamarck, Jean Baptiste Pierre Antoine de Monnet, Chevalier de.
  Réfutation de la théorie pneumatique ou de la nouvelle doctrine des chimistes moderne présentée article par article, dans une suite de réponses aux principes rassemblés et publiés par le citoyen Fourcroy dans sa Philosophie Chimique ; précédée d'un supplément complémentaire de la théorie exposée dans l'ouvrage intitulé : Recherches sur les causes principaux faits physiques, auquel celui-ci fait suite et devient nécessaire. Paris. An IV [1796]. pp. 484, 8vo.
        An attack on the Lavoisierian theory of combustion. Lamarck refers to the " pretended existence of a material called oxygen which the pneumatic chemists have never seen nor studied, and the existence of which they

LAMARCK, JEAN BAPTISTE PIERRE ANTOINE DE MONNET, CHEVALIER DE. [Cont'd.]

> imagine to explain the effects of fixed acidific fire." Lamarck reprints Fourcroy's treatise on the even-numbered pages and his refutations on the opposite odd-numbered ones. *See* Fourcroy, Philosophie chimique.

LAMBL, J., F. WOLFF, W. WEISS UND E. JONÁK.

Bericht über chemische Producte, Färberei und Druckerei, Kautschuk, Leder, Papier, Tobakfabrikate auf der Pariser Agricultur- und Industrie-Ausstellung im Jahre 1855. Leipzig, 1858.

LANDOLT, H.

Das chemische Laboratorium der Universität Aachen. 1868. 4to.

LANG, H.

Das chemische Laboratorium in Heidelberg. Carlsruhe, 1858. Fol.

LANGE, WILLEM DE.

*Boerhaave's " Elementa chemiæ." Academisch Proefschrift. Rotterdam, 1884. pp. 79, 8vo.

LARSSON, A.

Något om den kemiska industrien i Sverige. Helsingborg, 1886. 8vo.

LASSWITZ, KURD.

Geschichte der Atomistik vom Mittelalter bis Newton. Hamburg, 1889–90. 2 vols., 8vo.

> The author treats of the borderline between physics and philosophy in an historical manner, and does not neglect its bearings on chemical philosophy.

LAUGEL, A.

Les découvertes de la chimie physiologique. Travaux de Pasteur. Paris, 8vo.

LAUTH, C.

Les produits chimiques et pharmaceutiques à l'Exposition universelle internationale de 1878 à Paris. Paris, 1881. 8vo.

LEBLANC, F. ET DIZÉ.

Notice historique sur la découverte de la soude artificielle, lue à la Société de pharmacie par Félix Boudet. Paris, 1852. 8vo.

LECLERC, LUCIEN.

Histoire de la médecine arabe. Paris, 18—. 2 vols., 8vo.

LEEDS, ALBERT R.

The Lines of Discovery in the History of Ozone. *In* Chem. News.,
xli, 138 (1880).

> *Cf. in Section I*, Leeds, A. R.

LEEMANS, C.

*Papyri græci musei antiquarii publici Lugduni-Batavi. Regis augus-
tissimi jussu edidit, interpretationem Latinam, adnotationem,
indices et tabulas addidit. Tomus II. Lugduni-Batavorum, 1885.
pp. viii–310, 4to. Tab. IV.

> Contains the text of several earliest alchemical treatises. *See* Berthelot and
> Ruelle.

LEIPZIG (LABORATORIUM).

> *See* Kolbe, Hermann.

LEMOINE, GEORGES.

Les progrès de la chimie dans les dix dernières années (1868–1878).
Congrès bibliographique international tenu à Paris du 1ᵉʳ au 4
Juillet, 1878, sous les auspices de la Société bibliographique.
Extrait du compte-rendu des travaux. Paris, 1879. pp. 22, 8vo.

LENGLET DU FRESNOY, NICOLE.

*Histoire de la philosophie hermétique. Accompagnée d'un catalogue
raisonné des écrivains de cette science. Avec le véritable Philalèthe, revû sur les originaux. Paris, 1742. 3 vols., 12mo. Vol.
I, pp. xxiii–506 ; vol. II, pp. xxxii–120–360 ; vol. III, pp. xxii–432.

> The author of this exceedingly curious work was an Abbé of some distinction
> as a littérateur. He was born in 1674, and died in 1755. While appar-
> ently accepting the truth of the legends relating to the great antiquity of
> alchemy, and narrating accounts of veritable transmutations at consid-
> erable length, he at the same time exposes the frauds practised by the
> adepts, and quotes entire the celebrated essay of Geoffroy : " Des
> Supercheries concernant la Pierre Philosophale," which rang the death-
> knell of the Hermetic Art. The first volume of Du Fresnoy's work
> contains only historical matter, concluding with a "Chronologie des
> plus célèbres auteurs de la philosophie hermétique." In this chronology,
> which begins with " Hermes, 1996 B.C.," he includes Moses, Cleopatra,
> and Caligula, *Adepts* being marked by an asterisk. The second volume
> continues the history, and includes the " Introitus apertus ad occlusum
> regis palatium " of Philalethes, entire, both in French and in Latin.
> The third volume is wholly devoted to bibliography ; the classification
> is convenient and supplemented by an alphabetical index of authors.
> Lenglet's comments are interesting and sometimes amusing. This
> bibliography is the best published to that date. For a review of the
> work see Michault, J. B.

LENSE, JOSUA.

Disputatio medico-chymica prima de ortu et progressu chymiæ.   Lugduni-Batavorum, 1702.   pp. 12–[ii], 4to.

LENZ, JOHANN CHRISTOPHER.

Sammlung merkwürdiger Begebenheiten unterschiedlicher Adepten
und ihrer philosophischen Tinctur ; nebst der Geschichte Nicolaus
Flamelli. Hildesheim, 1780. 8vo.

LEQUIN, ROUX, MICHAUD, RICHE ET D'A.

Exposition universelle internationale de 1889 à Paris.   Rapport sur
les produits chimiques et pharmaceutiques. Paris, 1891. 8vo.

LEWINSTEIN, GUSTAV.

*Die Alchemie und die Alchemisten.   Sammlung gemeinverständlicher
wissenschaftlicher Vorträge herausgegeben von Rud. Virchow und
Fr. v. Holtzendorff. V. Serie, Heft 113. Berlin, 1870. pp. 36.

LEWIS, WILLIAM.

Historie der Farben. Aus dem Englischen übersetzt von Johann
Heinrich Ziegler. Zürich, 1766. 8vo.

LEWIS, WILLIAM.

Historie des Goldes [etc.].   *See in Section V.*

LIBAVIUS, ANDREAS.

Περὶ χρυσοποιήσεως dissertatio, in qua conferuntur inter se argumenta
eorum, qui de transmutatione metallorum, hydrargyri in argentum
et aurum, aut omnino in aurum, inter se contendunt. Jenæ, 1591.
4to.

LIBES, A.

*Histoire philosophique des progrès de la physique. Paris, 1810. 3
vols., 8vo.

A philosophical work.   Vol. I treats of the period ending with Gilbert ; II
of Boyle, Huyghens and Cassini ; III of Franklin, Saussure and
Coulomb.   Each volume is well indexed.   Bibliography and biography
are subordinate.

LIBRI, GUILLAUME.

* Histoire des sciences mathématiques en Italie depuis la renaissance des
lettres jusqu' à la fin du dix-septième siècle. Paris, 1838–41. 4
vols., 8vo.

A learned work based on original documents, extracts from which are given
in the Notes in each volume.

LIEBIG, JUSTUS VON.
Ueber das Studium der Naturwissenschaften und über den Zustand der Chemie in Preussen. Braunschweig, 1840. 8vo.
> Om naturvidenskabernes studium, oversat af Harald Thaulow. Christiania, 1841. 8vo.
> Over de beoefening der natuurkundige wetenschappen en over den toestand der scheikunde in Pruissen. Uit het Hoogduitsch. Utrecht, 1841. 8vo.

Ueber das Studium der Naturwissenschaften. Eröffnungsrede zu seinen Vorlesungen über Experimental-Chemie im Wintersemester 1852–'53. München, 1852. 8vo.
> Sopra lo studio delle scienze naturali. Discorso. Tradotto dal Tedesco da Giandomenico Bruno. Torino, 1853. 8vo.

LIEPMANN, H. K.
Die Mechanik der Leucipp-Democritischen Atome unter besonderer Berücksichtigung der Frage nach dem Ursprung der Bewegung derselben. Leipzig, 1886. 8vo.

LIGHTFOOT, JOHN.
The Chemical History and Progress of Aniline Black. Burnley, 1871.

LIMOUSIN, F., LE BLANC ET SCHMITZ.
Le matérial des arts chimiques de la pharmacie et de la tannerie à l'Exposition universelle internationale de 1878 à Paris. Paris, 1884. 8vo.

LINCK, JOHANN WILHELM.
Epistola gratulationis ad Carolum Glo. Kühn de historia nonnullisque instrumentis chemiæ. Lipsiæ, 1783. 4to.
Grundsätze der Pharmazie nebst Geschichte und Litteratur derselben, zur Erklärung der neuen verbesserten österreichischen Provinzial-Pharmacopöe. Wien, 1800. 2 vols, 8vo. Vol. I in two parts; I, pp. viii–304; II, pp. 551; III, pp. 510. Seven folding plates.
> Contains a history of the literature of pharmacy in four periods, with notes on the bibliography. In the plates the symbols of chemical substances are very fully delineated on the phlogistic theory, and on the antiphlogistic theory, the latter after Hassenfratz and Adet.

LIPPMANN, EDMUND O. VON.
*Geschichte des Zuckers, seiner Darstellung und Verwendung, seit den ältesten Zeiten bis zum Beginne der Rübenzuckerfabrikation. Ein Beitrag zur Kulturgeschichte. Leipzig, 1890. pp. xvi–474, 8vo. Map. Ill.

LONDON, ROYAL INSTITUTION, LABORATORIES OF.
> *See* Spottiswoode, William.

LONGCHAMP.
Bibliothèque du chimiste.   Paris, 1834.   Tome septième.   pp. 628, 8vo.
Folding table.

> This collection was intended to embrace 15 volumes in three series, and the author began with vol. VII, the only one published! Contains biographies of Bayen and Lavoisier and portions of their works.

LOSSAU, CHRISTIAN JOACHIM.
Dissertatio inauguralis medicina de valore chemiæ hodiernæ.   Kiliæ,
1725.   4to.

LOWIG, KARL J.
J. B. Richter der Entdecker der chemischen Proportionen.   Breslau,
1874.   4to.

LOYS, DE.
Abrégé chronologique pour servir à l'histoire de la physique jusqu'à
nos jours.   Strasbourg, 1786, 1787, 1789.   3 vols., 8vo.

> A German translation by Carl Glo. Kühn was published at Leipzig, 1798–99, in 8vo.

LUDWIG, J. F. H.
> Geschichte der Apotheker, *see* Philippe, A.

LUNGE, GEORG.
Bericht über Gruppe 15 der Schweizerischen Landesausstellung Zürich,
1883.   Chemische Industrie.   Zürich, 1884.   8vo.

LUNN, FRANCIS.
Historical Introduction to Article "Chemistry" in the Encyclopædia
Metropolitana, 1845.   pp. 587–607, 4to.

LUSSON, F.
Les origines de la chimie.   La Rochelle, 1880.   8vo.

LUXARDO, O.
Il laboratorio di chimica del r. istituto tecnico di Mantova.   Mantova,
1882.   4to.

MACADAM, S.
Address on the Recent Progress and State of· Chemical Philosophy.
British and Foreign Med. Chir. Review.   London, 1848.   Vol. I,
pp. 476–503.

MACHER, MATHIAS.
Das Apothekerwesen in den k. k. österreichischen Staaten ; eine
Darstellung der Geschichte des Apothekerwesens . . Wien,
1840. 8vo.

MACKAY, CHARLES.
The Alchymists. *In his :* Memoirs of Extraordinary Popular Delusions
and the Madness of Crowds. London, 1841. 3 vols., 8vo. Ill.
Also, London, 1852, 2 vols., 12mo. Philadelphia, 1850.

> An entertaining, popular essay. Editions are numerous.

MACQUER, PIERRE JOSEPH ET ANTOINE BAUMÉ.
Plan d'un cours de chimie expérimental et raisonnée, avec un Discours
historique sur la chimie. Paris, 1757. 12mo.

MAINDRON, ERNEST.
*L'Académie des Sciences. Histoire de l'Académie—Fondation de
l'Institut National—Bonaparte membre de l'Institut National.
Paris, 1888. pp. iv–344, roy. 8vo. 8 plates, 53 engravings in the
text, plans and autographs reproduced from original documents.

> Contains portraits of all the perpetual secretaries, including the following :
> Cuvier, Arago, Flourens, Fourier, Dulong, J. B. Dumas, de Beaumont,
> Jamin, Vulpian. A full bibliography of the Academy completes this
> work, important in the history of every science.

MAIR, WILHELM.
Vergleichende Würdigung der Verdienste J. Th. Desagulier, W. Grawe-
sands und Pet. van Mossenbroek's um die Experimentalphysik.
Ein Beitrag zur Geschichte der Physik. Gekrönte Preisschrift.
München, 1834. 8vo.

MANCHESTER (LABORATORY).
> *See* Roscoe, H. E., and Waterhouse.

MANTOVA (LABORATORIO).
> *See* Luxardo, O.

MARBURG (LABORATORIUM).
> *See* Kolbe, Hermann.

MARCHAND, R. F.
Ueber die Alchemie. Ein Vortrag im wissenschaftlichen Vereine zu
Berlin am 20. Februar 1847. Halle, 1847. pp. 45, 12mo.

MARIE, MAXIMILIEN.
Histoire des sciences mathématiques et physiques. Paris, 1883–88.
12 vols., 8vo.

MARNE, N. H.

Ueber die Anzahl der Elemente. Ein Beytrag zur allgemeinen Naturlehre. Berlin und Leipzig, 1786. 8vo.

MARTIN, A.

*Repertorium der Photographie. I. Vollständige Anleitung zur Pho-
tographie auf Papier. II. Literatur der Photographie auf Metall.
Wien, 1846. pp. viii–134, 12mo.

> Contains a brief history and a valuable bibliography of photography up to
> the date of publication.

MARTIN, W. A. P.

*Alchemy in China. Read before the American Oriental Society,
October, 1868. Published in the China Review, January, 1879,
and in : The Chinese, their Education, Philosophy and Letters.
New York, 1881. pp. 167–193, 8vo.

MARTIUS, C. A.

Verzeichniss der chemischen Fabriken Deutschlands. Berlin, 1880. 4to.

MARX, K. F. H.

Geschichtliche Darstellung der Giftlehre. Göttingen, 1827–29. 2 vols.,
8vo.

> Valuable for its reliable bibliography of toxicology prior to its date. It
> forms an introduction to an incompleted work entitled : " Die Lehre
> von den Giften in medizinischer, gerichtlicher und polizeylicher Hin-
> sicht," by the same author.

MASSON, JOHN.

Atomic Theory of Lucretius and Modern Doctrines. London, 1884. 8vo.

MATTHEWS, WILLIAM.

An Historical Sketch of the Origin, Progress, and Present State of Gas-
Lighting. London, 1827. pp. xxxii–434, 12mo.
Second edition. London, 1832. pp. x–440.

MEADE, WILLIAM.

Outlines of the Origin and Progress of Galvanism, with its Application
to Medicine. In a letter to a friend. Dublin, 1805. pp. 74, 8vo.
2 plates.

MEIDINGER, HERR VON.

*Die Richtigkeit der Verwandlung derer Metalle aus der wahrhaften
Begebenheit welche sich im Jahr 1761 auf der kurfürstlichen trier-
ischen Münzstatt zu Koblenz mit einem Adepten Namens Georg
Stahl zugetragen hat. Beschrieben von dem damaligen kurtrier-
ischen Münzdirektor, k. k. wirkliche Hofrathe Herrn von M. Leipzig, 1783. pp. 60, 12mo.

MELSENS.

Note historique sur J. B. van Helmont à propos de la définition et de la théorie de la flamme. Opinions des anciens chimistes et physiciens sur la chaleur, le feu, la lumière et la flamme dans leurs rapports avec les idées et les travaux de van Helmont. Présentée à la classe des sciences le 10 octobre, 1874.

MENN, JOHANN GEORG.

Rede von der Nothwendigkeit der Chemie. Köln, 1777. 4to.

MÉTHERIE, DE LA.

Journal de physique, 1789. Discours préliminaire. Paris, 1789.
>   Review of progress of investigations in gas chemistry.

MEYER, ERNST VON.

* Geschichte der Chemie von den ältesten Zeiten bis zur Gegenwart. Zugleich Einführung in das Studium der Chemie. Leipzig, 1889. pp. xi–466, 8vo.
>   An ably written condensed history covering the entire period of chemistry, and amplifying specially the progress since Lavoisier.

>   * A History of Chemistry from earliest times to the present day, being also an introduction to the study of the science. Translated with the author's sanction by George M'Gowan. London and New York, 1891. pp. xxii–556, 8vo.

MEYER, JOHANN RUDOLPH, JR.

Systematische Darstellung aller Erfahrungen in der Naturlehre entworfen.
Vol. I and II : Systematische Darstellung aller Erfahrungen über allgemeine verbreitete Potenzen von Ludwig von Schmidt, genannt Phiseldeck. 2 vols.
Vol. III : Systematische Darstellung aller Erfahrungen über die einzelnen Metalle in zwei Bänden, von Karl Albrecht Kielmann. Aarau, 1807. 3 vols., 4to. Vol. I: pp. xxxii–543 and eleven plates. II: pp. iv–420 and plates 12–17. III: pp. xxxiv–498 and one plate.
>   This is an ambitious attempt to present a summary of all facts based on experiments with the known elementary bodies. Vol. II treats of light, heat, electricity, galvanism, magnetism, oxygen, hydrogen, water, nitrogen, carbon and atmospheric air ; vol. III, of the metals and their known combinations with the preceding and with each other. The whole is very systematically arranged, and certainly gives an excellent view of chemical knowledge at the time. The work abounds in citations and bibliographical data. Historical notes accompany each metal.

MEYER, RICHARD.

Ueber Bestrebungen und Ziele der wissenschaftlichen Chemie. Vortrag
gehalten am 30 Januar, 1879. Berlin, 1880. pp. 52, 12mo.

> Sammlung gemeinverständlicher wissenschaftlicher Vorträge, herausgegeben
> von Rud. Virchow und Fr. von Holtzendorff. xv Serie, Heft 342.

MEYER, VICTOR.

Chemische Probleme der Gegenwart, Vortrag gehalten in der ersten
allgemeinen Sitzung der 62. Versammlung Deutscher Naturforscher
und Aerzte am 18 September, 1889, zu Heidelberg. Zweite Auflage.
Heidelberg, 1890. pp. 46, 8vo.

Ergebnisse und Ziele der stereochemischen Forschung. Vortrag
gehalten in der Sitzung der deutschen chemischen Gesellschaft zu
Berlin am 23. Januar 1890. Heidelberg, 1890. 8vo.

MICÉ, L.

Rapport méthodique sur les progrès de la chimie organique pure en
1868, avec quelques détails sur la marche de la chimie physique.
Paris, 1869. pp. 446, 8vo.

Rapport méthodique sur les progrès de la chimie biologique en 1882.
Extrait du Journal de médecine de Bordeaux. Bordeaux, 1883.
pp. 86, 8vo.

> Limited to progress in France, though not so stated in the text or title.

MIKOWEC, FERDINAND B.

Die Alchimisten in Böhmen unter Rudolf II. Oesterreichische Blätter
für Literatur und Kunst. No. 42–44. Oktober, 1854. Sm. fol.

MILANO (ESPOSIZIONE).
> *See* Esposizione industriale.

MILLER, SAMUEL.

Brief Retrospect of the 18th Century. Vol. 1, 77. London, 1803.
> Contains a sketch of the history of chemical philosophy.

MINOLA.

Beiträge zu Busch : Handbuch der Erfindungen. Erster Band. Leipzig [?], 1806. *See* Busch, G.

MITCHELL, T. D.

A Cursory View of the History of Chemical Science, and some of its
more important uses to the physician, being an introduction to the
course of lectures for the session 1837–38. Lexington, 1837.

MOIGNO, ABBÉ.

L'ozone, ce qu'il est, ses propriétés physiques et chimiques. Analyse des recherches dont il a été l'objet. Paris, 1878. 12mo.

> Compare Dachauer, also Odling, *in Section III ;* and Leeds, A. R., *in Sections I and III.*

MOORE, NATHANIEL F.

Ancient Mineralogy ; or, An Inquiry respecting mineral substances mentioned by the Ancients : with occasional remarks on the uses to which they were applied. New York, 1836.

> * Second edition, New York, 1859. pp. 250, 12mo.

MOORE, WILLIAM D.

An Outline of the History of Pharmacy in Ireland. Extracted from the Dublin Quarterly Journal of Medical Science, No. XI, August, 1848. Dublin, 1848. 8vo.

MORHOF, DANIEL GEORG.

> De metallorum transmutation [etc.]. *See in Section VI.*

MORIDE, EDOUARD.

Histoire de la savonnerie et de ses matières premières. Paris, 1887. pp. 110, 12mo.

MOSCATI, PIETRO.

Discorso accademico dei vantaggi della educazione filosofica nello studio della chimica. Recitato nell' aprimento della nuova Scuola chimico-farmaceutica dello Spedal Maggiore, dal regio . . . il giorno 4 Febbrajo, 1784. Milano, *n. d.* [1784]. pp. [viii]–90–[viii], 8vo.

> An interesting volume, considering the date and the country of its issue. The history of chemistry is followed by chronological tables of history up to 1695, and by a bibliography bearing the special title : Collezione scelta di libri d'argomento chimico ad uso degli amatori. This begins with Borrichius and other historians, and treats in separate sections of general treatises, physical, mineralogical, and pharmaceutical chemistry. The work contains a supplement giving heads of a course of twenty lectures by Moscati delivered in the Laboratory of the "Spedal Maggiore."

MOSCH, C. F.

Geschichte des Bergbaues in Deutschland. Liegnitz, 1829.

MOSCOW INTERNATIONAL EXHIBITION.

> *See* Крупскій, А.

MOYRET.

> L'histoire chimique de la teinture de la soie. *See in Section V*, Moyret : Traité de la teinture.

MUCK, F.
Die Entwickelung der Steinkohlenchemie in den letzten 15 bis 20 Jahren
und die ehemaligen Ziele der Steinkohlenforschung überhaupt.
N. p., n. d.

MÜLLER, C. C. H.
* Geschichte der Hamburger Apotheken. Historisch-biographische
Skizze vom Jahre 1265 bis auf die neueste Zeit. Hamburg, 1888.
pp. 112, 8vo. Three facsimiles.

MÜLLER, G.
Das chemische Laboratorium der Universität Greifswald. Berlin,
1864. Fol.

MÜLLER, THEOPHILUS.
Commentationi duo de oleis variisque illa extrahendi modis et de quibusdam ortum et progressum chimiæ illustrantibus. Hamburg, 1688.
12mo.

MÜNCHEN (LABORATORIUM).
      See Baeyer, A. und Geul. *Also* Voit und Liebig.

MULDER, E.
Historisch-kritisch overzigt van de bepalingen der æquivalent-
gewigten van 13 eenvoudige ligchamen. Utrecht, 1853.

MULDER, L.
Historisch-kritisch overzigt van de bepalingen der æquivalent-
gewigten van 24 metalen. Utrecht, 1853.

MUÑOZ DE LUNA, RAMON TORRES.
Los 4 elementos de Aristoteles en el siglo 19. Madrid, 1858. 8vo.

MURHARD, FRIEDRICH WILHELM AUGUST.
Geschichte der Physik seit dem Wiederaufleben der Wissenschaften bis
auf Ende des 18 Jahrhunderts. Göttingen, 1798–99. 8vo.
      Deals with aëronautics, barometers and hygrometers.

MURR, CHRISTOPH GOTTLIEB VON.
* Litterarische Nachrichten zu der Geschichte des sogenannten Gold-
machens. Leipzig, 1805. pp. vi–154, 12mo.
      Chronologically arranged notes, chiefly from original sources, on the history
      of transmutations. With bibliographical references.

NAPIER, JAMES.

The Ancient Workers and Artificers in Metal. London, 1856. 12mo.

* Manufacturing Arts in Ancient Times, with special reference to Bible history. Paisley, 1879. pp. vii–367, 8vo.

> A readable treatise by a practical man, treating chiefly of ancient metallurgy and technology. The chemical knowledge of the Israelites is fully developed.

NAQUET, A.

> Histoire de la synthèse en chimie organique. *In* Principes de chimie. *See in Section V*, Naquet, A.

NASSE, WILHELM.

Ueber Naturphilosophie in Bezug auf Physik und Chemie. Ein Beitrag zur kritischen Uebersicht der physikalisch-chemischen Literatur. Freiberg, 1809. 8vo.

NAU, BERNHARD SEBASTIAN.

Kurzer historischer Abriss des Ursprungs und der weitern Fortschritte in der Naturgeschichte, Chemie, Mathematik und Physik. Frankfurt am Main, 1792. 8vo.

NAUDIN, L.

Aperçu historique de l'industrie de l'alcool ; progrès depuis la fin du xviiie siècle. Paris, 1881. 8vo.

NENTER, GEORG PHILIPP.

* Bericht von der Alchemie. Darinnen von derselben Ursprung, Fortgang und besten scriptoribus gehandelt, auf alle Einwürffe der adversariorum geantwortet und klar bewiesen wird, dass wahrhafftig durch die Alchemie der rechte Lapis Philosophorum als eine Universal Medicin könne bereitet werden. Zum Druck befördert durch Friederich Roth-Scholtzen. Nürnberg, 1727. pp. 147–218 [following Buddeus' Untersuchung]. 12mo.

> The first chapter discusses the origin of alchemy, the second contains a list of 21 authors and some of their works.
> *See in Section VI*, Roth-Scholtz : Deutsches Theatrum Chemicum.

NENDTWICH, C. M.

Grundriss der Stöchiometrie nebst einem geschichtlichen Ueberblick derselben für angehende Chemiker und Pharmaceuten. Ofen, 1839. pp. iv–64, 8vo.

> The historical introduction of 12 pages is not without interest.

NETTER, WILLIAM (Translator).

> *See* Peters, Hermann.

NEUBAUER, C.

Ueber die Fortschritte der Chemie in den letzten Decennien. Rede
zur Feier des 25 jährigen Bestehens des chemischen Laboratoriums
zu Wiesbaden am 3. Mai, 1873. Wiesbaden, 1873. pp. 18, 8vo.

NEUMANN, K. C., UND J. V. DIVIS.

Entwurf einer Geschichte der Zuckerindustrie in Böhmen. Erste
Periode (1787–1830), und Beiträge zur Geschichte der Zucker-
industrie in Böhmen. Zweite Periode (1830–1860). Aus dem
Böhmischen übersetzt. Zwei Theile. Prag, 1891. 8vo. Ill.

NEWLANDS, JOHN A. R.

On the Discovery of the Periodic Law and on relations among the
Atomic Weights. London, 1884. pp. viii–39, 12mo. Folding
plates.

NICOLICH, EMANUELE.

La pietra filosofale. Programma dell' I. R. scuola reale superiore in
Pirano pubblicato dalla direzione alla fine dell' anno 1878–79.
Trieste, 1879. pp. 47, roy. 8vo.

NICOT, A.

La chimie et la pharmacie à l'Exposition universelle de 1889. Extrait
du "Bulletin général de thérapeutique," Nos. des 15, 30 novembre
et 15 décembre, 1889. Paris, 1890. 8vo.

NIEPCE, ISIDORE.

Post tenebras lux ; Historique de la découverte improprement nommée
Daguerréotype, précédée d'une notice sur son véritable inventeur
feu Joseph Nicéphore Niepce de Chalons-sur-Saône, par son fils
Isidore Niepce. Paris, 1841. 8vo. *Cf.* Fouque, Victor.

NOAD, HENRY M.

Lectures on Chemistry, including its Applications in the Arts and the
Analysis of organic and inorganic compounds. London, 1855.
pp. 505, 8vo.

> An historical sketch (pp. 1–34) forms an introductory lecture.

NOELTING, E.

Histoire scientifique et industrielle du noir d'aniline. Mulhausen i. E.,
1889. Roy. 8vo.

NÜRNBERGER, JOSEPH EMIL.

Natur- und gewerbswissenschaftliche Berichte, oder Darstellung der
neuesten Physik und Technologie. Kempten, 1837. 8vo.

ODLING, WILLIAM.

On the History of Ozone. Read before the Royal Institute of Great Britain. Chem. News., vol. XXVI, p. 281. (1872.)

> *Cf.* Dachauer, Leeds and Moigno.

ODOMARUS, MAGISTER.

> Historiola antiqua de argento in aurum verso. *See in Section VI*, Theatrum Chemicum, vol. III, p. 170.

OERSTED, HANS CHRISTIAN.

Betrachtungen über die Geschichte der Chemie. Eine Vorlesung. J. Chem. Phys., Gehlen., vol. III, p. 194–231 (1807). *Also in* vol. II, of Der Geist in der Natur. München, 1850. pp. 371–428, 12mo.

OHSSON, CONSTANTIN D'.

Om kemiens framsteg från dess ursprung till närvarande tid. Stockholm, 1817.

Kort øfversigt af kemiens historia ifrån de äldsta tider. Tal, hållit . . . uti kongl. Vetenskaps-Akademien. Stockholm, 1826. 8vo.

OLLIFFE, CHARLES.

* Les alchimistes d'autrefois. Paris, 1842. pp. xvi–291, 32mo.

> Distinguished chiefly by its minute size, the text measuring 65 *mm.* x 38 *mm.*

OLSCHANETZKY, M. A.

Entdeckung des Sauerstoffs Hamburg, 1890. pp. 47, 8vo.

> Sammlung gemeinverständlicher wissenschaftlicher Vorträge, herausgegeben von Rud. Virchow und Wilh. Wattenbach, No. 105.

OSTEN, HANS VON.

* Eine grosse Herzstärkung für die Chymisten, nebst einer Dose voll gutes Niesepulver für die unkundigen Widersprecher der Verwandlungskunst der Metalle im Kloster zu Odenburg seit Anno 1426 aufbehalten durch H. von O. welche von wenigen Monaten von einem Maurergesellen daselbst gefunden worden. Begleitet mit einer Zuschrift an die Chymisten, und einer wahrhaften Nachricht dieser Geschichte nebst dem dazu gehörigen Kupfer. Berlin, 1771. pp. 108, 18mo. Folding plate.

> An exposure of alchemical knavery ; the author mentions categorically no less than 45 tricks and deceptions practised.

OSTWALD'S KLASSIKER der exacten Wissenschaften. *See in Section V.*

OUDEMANS, A. C., JR.

Historisch-kritisch overzigt van de bepaling der æquivalent-gewigten van twee en twintig metalen. Leiden, 1853. 8vo.

> *Cf.* Mulder, L.

Paets van Troostwijk, A.

Korte schets van de geschiedenis der scheikunde. Amsterdam, 1796. 8vo.

Panciroli, Guido.

* Rerum memorabilium sive deperditarum pars prior, commentariis illustrata et locis prope innumeris postremum aucta ab Henrico Salmuth, Ambergen. syndico emerito. Francofurti, sumptibus Godefridi Tampachi. *n. d.* pp. [xii]–350–[xxii], 4to.

[Part II :] Nova reperta, sive rerumm emorabilium recens inventarum et veteribus incognitarum pars posterior, ex Italico Latine reddita, nec non commentariis illustrata et locis prope innumeris postremum aucta ab Henrico Salmuth, cum indice rerum et verborum copiosissimo. Francofurti, sumptibus Godefridi Tampachi, typis Casparis Rötetii, 1631. pp. [viii]–313–[xvii], 4to.

* The history of many memorable things lost, which were in use among the ancients ; and an account of many excellent things found now in use among the moderns, both natural and artificial. Done into English and illustrated with a new commentary of choice remarks, pleasant relations, and useful discourses from· Salmuth's large annotations with several additions throughout. London, 1715. 2 vols., 18mo.

Contains in vol. II, chap. VII, a quaint section on chymistry.

Goguet says of Panciroli : " This is in general a very crude indigested compilation, in which the author sticks at nothing. The falsest facts and most ridiculous tales are adopted for truths. The work is an example of the greatest negligence joined to the strongest itch for making a book."

Paris (Exposition of 1855).

See Lambl, J., F. Wolff (etc.).

Paris (Exposition of 1878).

See Limousin, F., Le Blanc et Schmitz. *Also* Collot, T. *Also* Lauth, C. *Also* Poirier.

Paris (Exposition of 1889.)

See Persoz, J.

Paris. Bruhat, J.

Le laboratoire municipal de Paris, sa direction, ses documents sur les falsifications des produits alimentaires et les travaux du laboratoire. Paris, 1884. 8vo.

Pariset, E.

Histoire de la soie. Paris, 1862. 2 vols.

PARKES, SAMUEL.
> Chemical Essays. *See in Section V.*
> The work has valuable historical notes.

PAULIN, JACOB.
Dissertatio gradualis de primordiis chemiæ. Upsaliæ [1779]. Sm. 4to.

PEBAL, LEOPOLD VON.
Das chemische Laboratorium der Universität Graz. Wien, 1880. pp. 28, 4to. 8 plates.

PECKSTON, T. S.
> Theory and Practice of Gas-Lighting. *See in Section V.*
> Contains a sketch of the early history of Gas-lighting.

PELLEGRINO, LUIGI.
La chimica e le arti in Europa, ossia esame filosofico della vita e delle arti, tratto dalle vicende della chimica, da servire come elemento alla storia filosofica delle arti in Europa. Messina, 1844. 8vo.

PENNÈS, J. A.
Notice pour servir à l'histoire générale de la pharmacie presentée devant l'assemblée générale des pharmaciens de la Seine le 14 avril 1869. Paris, 1869. 8vo.

PERNETY, ANTOINE JOSEPH.
> Les Fables Égyptiennes et Grecques dévoilées. *See in Section II ; and compare in Section III*, Tollius, Jacob.

PERSOZ, J.
Exposition universelle internationale de 1889 à Paris. Rapport sur les procédés chimiques de blanchiment, de teinture, de l'impression et d'apprêt. Paris, 1891. 8vo.

PEST (LABORATORIUM).
> *See* Than, C. von.

PETERS, HERMANN.
* Aus pharmazeutischer Vorzeit in Bild und Wort. Berlin, 1886. pp. x–224. Neue Folge. Berlin, 1889. pp. xi–287. 2 vols., 8vo.

> Handsomely illustrated with facsimiles of early cuts, title-pages and portraits, 83 in number. One section deals with : " Goldmacherkunst." Vol. II has 92 illustrations. The chapter on mineralische Arzneistoffe is an interesting contribution to medical chemistry. For an American translation of vol. I see the following :

> * Pictorial History of Ancient Pharmacy with sketches of early medical practice. Translated from the German

PETERS, HERMANN.   [Cont'd.]
and revised with numerous additions by William Netter.
Chicago, 1889. pp. xiv–184, 8vo. Ill.

    The additions by the American editor are numerous and valuable.

PETERSEN, J.
Fotografiens Forhistorie, biografisk Foredrag. Ved Fotografiens fem-
tiaarige Jubilæumsfest i Christiania 18–22 Juni, 1889. Kjøbenhavn,
1889. pp. 61, 8vo. Ill.

PETTIGREW, THOMAS JOSEPH.
* On Superstitions connected with the History and Practice of Medicine
and Surgery. London, 1844. pp. viii–167, 8vo. Plates.

    Contains a section on alchemy. A work packed full of information on the
    subjects named.

PETZOLD, WILHELM.
Die Grundzüge der chemisch-physikalischen Atomtheorie. Wissenschaftliche Abhandlung zum Programm des Gymnasiums zu
Neubrandenburg, Ostern, 1877. 4to. Neubrandenburg, *n. d.*
pp. 40, 4to.

PFAFF, CHRISTIAN HEINRICH.
Der Elektro-Magnetismus. Eine historisch-kritische Darstellung der
bisherigen Forschungen auf dem Gebiete derselben, nebst eigenthümlichen Versuchen. Hamburg, 1824. 8vo. 8 plates.

PFEIFER, XAVER.
Die Controverse über das Beharren der Elemente in den Verbindungen
von Aristoteles bis zur Gegenwart. Historisch und kritisch
dargestellt. Dillingen, 1879. 8vo.

PHILADELPHIA (WELTAUSSTELLUNG).
    Die chemische Industrie. *See* Goldschmidt, G. *Also* Chemische (Die)
    Industrie Deutschlands. *Also* Wilhelm.

PHILIPPE, A.
Histoire des apothicaires chez les principaux peuples du monde depuis
les temps les plus reculés jusqu'à nos jours ; suivie du tableau de
l'état actuel de la pharmacie en Europe, en Asie, en Afrique et en
Amérique. Paris, 1853. 8vo.

    Geschichte der Apotheken bei den wichtigsten Völkern der
    Erde aus dem französischen übersetzt von H. Ludwig.
    Jena, 1854. 8vo. Zweite Auflage, Jena, 1858. 8vo.

PICKEL.

Rede von dem Nutzen und Einflusse der Chemie auf das Wohl eines Staates und auf verschiedene Künste und Wissenschaften. Würzburg, 1785. 4to.

PICTON, HAROLD W.

*The Story of Chemistry. With a preface by H. Roscoe. London, 1889. pp. viii–382, 8vo. Illustrated. Portrait of Dalton.

A readable general history, not strong in bibliography.

PIESSE, CHARLES H.

History of the Art of Perfumery. *See in Section V*, Piesse, C. H.

PLEISCHL, ADOLPH MARTIN.

Das chemische Laboratorium an der Universität zu Prag ; Entstehung und gegenwärtiger Zustand derselben, sammt Nachrichten über einige darin vorgenommene Arbeiten, nebst einigen Abhandlungen chemisch-medizinischen Inhalts. Prag, 1820. 8vo. Plate.

PLUGGE, P. C.

De invloed der scheikunde op de ontwikkeling der pharmacotherapie. Groningen, 1890. 8vo.

POGGENDORFF, JOHANN CHRISTIAN.

*Geschichte der Physik. Vorlesungen gehalten an der Universität zu Berlin. Leipzig, 1879. pp. [viii]–937. 8vo. Ill.

POIRIER.

Les matières colorantes et produits chimiques à l'Exposition de 1878. 8vo.

Поповъ, А.

Сборникъ работъ химической лабораторіи Императорскаго Варшавскаго университета, 1870—1876 г.

Варшава 1876. pp. x–219, 8vo.

POPPE, ADOLPH.

*Chronologische Uebersicht der Erfindungen und Entdeckungen auf dem Gebiete der Physik, Chemie, Astronomie, Mechanik und industriellen Technik von den ältesten Zeiten bis auf unsere Tage. Frankfurt, 1856. 8vo.

Zweite Auflage, 1857. pp. 74. Dritte Auflage, Frankfurt am Main, 1881.

In the third edition the word Alphabetisch is prefixed to the above title.

POPPE, JOHANN HEINRICH MORITZ VON.

Geschichte aller Erfindungen und Entdeckungen im Bereiche der Gewerbe, Künste und Wissenschaften von der frühesten Zeit bis auf unsere Tage. Zweite Auflage. Frankfurt-a-M., 1847. 8vo.

Geschichte der Technologie seit der Wiederherstellung der Wissenschaften bis an das Ende des achzehnten Jahrhunderts. Göttingen, 1807–11. 3 vols.

POST, JULIUS, UND JOS. LANDGRAF.

Rückblick auf die Fortschritte der chemischen Grossindustrie im Jahre 1876. Berlin, 1877. 8vo.

POTT, A. F.

Chemie oder Chymie? *In* Zeitschrift der deutschen morgenländischen Gesellschaft. Bd. xxx. pp. 6-20. Leipzig, 1876.

> An erudite and critical study of the etymology of chemistry leads the author to decide against "chymistry" and to favor the Egyptian root-source. At the close of his paper he directs attention to the analogy between early Greek symbols for chemical bodies and the signs used by the Egyptians. *Cf.* Gildemeister ; *also* Smith, R. A.

POUCHET, F.

Histoire des sciences naturelles au moyen âge. Albertus Magnus et son époque. Paris, 1853. 8vo.

POWELL, BADEN.

*History of Natural Philosophy from the earliest periods to the present time. London, 1842. pp. xvi-396, 12mo.

> A volume in the series known as Dr. Lardner's Cabinet Cyclopædia.

PRAG (LABORATORIUM).

> *See* Pleischl, A. M.

*PROCEEDINGS AT THE CENTENNIAL OF CHEMISTRY, held August 1, 1874, at Northumberland, Pa. Reprinted from the August, September and December numbers of the American Chemist. Philadelphia, 1875. pp. iv, 35–114, 195–209, 4to.

> A celebration of the Centenary of Priestley's Discovery of Oxygen.

PUECH, A.

Les pharmaciens d'autrefois à Nîmes, étude historique d'après les documents inédits. Paris, 1881. 8vo.

PULSIFER, WILLIAM H.

*Notes for a History of Lead, and an inquiry into the development of the manufacture of white lead and lead oxides. New York, 1888. pp. vii-389, roy. 8vo.

PULSIFER, WILLIAM H. [Cont'd.]
>A compendious treatise on the history of the metal in all ages, its mining, metallurgy and uses. *Cf.* Hofmann, K. B.

QUETELET, ADOLPH.
Histoire des sciences mathématiques et physiques chez les Belges. Bruxelles, 1864. pp. 479, roy. 8vo.
Sciences mathématiques et physiques chez les Belges au commencement du XIX siècle. Bruxelles, 1866. pp. iv–754, roy. 8vo.

RAMMELSBERG, C. [Translator].
>*See* Johnston, James F. W.

RATTON, J. J. L.
A Handbook of Common Salt. Madras, 1877. 8vo.
>Contains brief historical sketch. *Cf.* Schleiden, M. J.

RATTRAY, JACOB.
Dissertatio exhibens vitrei thermometri historiam simul et acroamaticam delineationem. Lugduni Batavorum, 1679. 4to.

RAU, ALBRECHT.
*Die Grundlage der modernen Chemie ; eine historisch-philosophische Analyse. Braunschweig, 1877. pp. 114, 8vo.
Die Lehre von der chemischen Valenz und ihr Verhältniss zur electrochemischen Theorie. Eine historisch-kritische Studie. Leipzig, 1879. 8vo.

REED, WILLIAM.
The History of Sugar and Sugar-yielding Plants. Together with an epitome of every notable process of sugar extraction and manufacture, from the earliest times to the present. London, 1866. pp. viii-206, 12mo.

REGNAULT.
L'origine ancienne de la physique nouvelle. Amsterdam, 1735. 3 vols., 8vo.

REIBER, FERDINAND.
Études Gambrinales. Histoire et archéologie de la bière et principalement de la bière de Strasbourg. Paris, 1882. 8vo.

REINHOLD, JOHANN CHRISTIAN LEOPOLD.
De galvanismo specimen I, II. Lipsiæ, 1798. 4to.
>Contains the history and literature to date of issue.
>*See* Sue, Pierre (le jeune).

REITEMEIER, JOH. FR.

Geschichte des Bergbaues und Hüttenwesens bey den alten Völkern. Göttingen, 1785. 8vo.

REMSEN, IRA.

On the Relations of Organic Chemistry to Chemistry. Address before the Sub-Section of Chemistry, American Association for the Advancement of Science, August, 1879. Proceedings Am. Assoc. Adv. Science, vol. XXIX. Salem, 1880. 8vo.

REUVENS, C. J. C.

*Lettres à M. Letronne sur les Papyrus bilingues et grecs et sur quelques autres monumens gréco-égyptiens du Musée d'Antiquités de l'Université de Leide. Avec atlas [in folio]. A Leide, 1830. pp. 1–89 ; 1–57 ; 1–164 ; 4to.

> Contains brief notice of papyrus No. 66. *See* Leemans ; *also*, Berthelot and Ruelle.

REYER, E.

Zinn, eine geologisch-montanistisch historische Monografie. Berlin, 1881. pp. iv–248, 8vo.

> Contains a history of tin, also of the chief tin-mines, and a bibliography.

REYHER, SAMUEL.

* Dissertatio de nummis quibusdam ex chymico metallo factis. Kiliæ Holsatorum, 1692. pp. 144, sm. 4to.

> Contains engravings of nine coins of hermetic origin, struck to commemorate transmutations. *Cf.* Bolton, H. C., Alchemy and Numismatics.

RHAMM, A.

*Die betrüglichen Goldmacher am Hofe des Herzogs Julius von Braunschweig. Nach den Processakten dargestellt. Wolfenbüttel, 1883. pp. 128, 8vo.

> A sketch of Philipp Sommering, or Therocyclus, and his alchemical adventures at the court of Wolfenbüttel, drawn from the original documents.

RHYNE, GUILIELMUS TEN

De chemiæ et botanicæ antiquitate et dignitate. Londini, 1683. 8vo.

RICHARDS, J. W.

> Aluminium, its History, etc. *See in Section V.*

RINMANN, SVEN.

Försök till järnets historia. Stockholm, 1782. 4to.

> *The same in German.* Berlin, 1785. 2 vols., 8vo.
> Geschichte des Eisens. Aus dem schwedischen von K. J. B. Karsten. Liegnitz, 1815. 2 vols., 8vo.

Rioz y Pedraja, Manuel.
Discurso sobre la importancia filosófica de la química, considerada en su historia y en sus relaciones con otras ciencias que, como ella, tienen por objeto el estudio de la naturaleza. Madrid, 1853. 4to.

Ritter, Johann Wilhelm.
Versuch einer Geschichte der Schicksale der chemischen Theorie in den letzten Jahrhunderten. J. für Chem., Gehlen, vol. vii, 1808. pp. 1–66.
    A study of the phlogistic and antiphlogistic theories.

Robertus Vallensis.
    See Duval, Robert.

Rochas, Albert de.
* La Science dans l'antiquité. Les origines de la science et ses premières applications. Les peuples préhistoriques, la civilisation Égyptienne, la science Grecque. L'origine du feu ; la statue de Memnon ; les prestiges des temples ; les automates d'Homère et de Héron ; les miroirs ardents ; les premiers appareils de physique ; les machines de guerre, etc. Avec 117 figures dont cinq planches hors texte. Paris, n. d. [1888]. pp. 288, 8vo.
    A popular work only remotely touching the history of chemistry.

Les doctrines chimiques au xviie siècle. Extrait du "Cosmos," Paris, 1888. 4to.

Rodwell, George F.
On the Supposed Nature of Air prior to the Discovery of Oxygen. Chem. News, vol. viii, p. 113 (Sept. 5, 1863), and succeeding numbers.
* The Birth of Chemistry. "Nature Series." London, 1874. pp. xii–136, 12mo. With a portrait of Mayow, and Illustrations. Reprinted from Nature, vols. vi and vii, 1872–73.

Rössing, Adelbert.
    Geschichtliche Entwickelung der chemischen Theorien. Part i of Einführung in das Studium der theoretischen Chemie, q. v. in Section V.

Rolfinck, G.
    Chimia in artis formam redacta. See in Section V.
    Contains a treatise on the origin and antiquity of chemistry.

Roloff, Johann Christoph Heinrich.
Zur Geschichte der [durch ihn veranlassten] Entdeckung des Kadmiums. Gilb. Ann. 60. (1818.)

ROMEGIALLI, A.
Progressi di chimica industriale dal 1883 al 1889. Torino, 1889. 8vo.
     An appendix to the Italian edition of Rudolph Wagner's Handbuch der
     chemischen Technologie. *See* Wagner, Cossa e Romegialli.

ROSA, GABRIELE.
L'alchimia dalla sua origine sino al secolo XIV, e "la Compostella,"
     opere di Frate Bonaventura d'Iseo. Dissertazione. Brescia, 1846.
     8vo.

ROSCOE, HENRY E.
The History of the Chemical Elements : a lecture delivered in the
     City Hall, Glasgow, under the auspices of the Glasgow Science
     Lectures Association, on 8th February, 1875. London, 1875. 8vo.
Record of Work done in the Chemical Department of the Owen's
     College. 1857–1887. London and New York, 1887. 52 pp., 8vo.
     Printed for private circulation.

ROSCOE, H. E., AND WATERHOUSE, A.
Description of the Chemical Laboratories at the Owen's College.
     Manchester. 4to.

ROSENBERGER, F.
*Die Geschichte der Physik in Grundzügen mit synchronistischen
     Tabellen der Mathematik, der Chemie und beschreibenden Natur-
     wissenschaften. III Parts. Braunschweig, 1882–90. 8vo.
Part I, Geschichte der Physik im Alterthum und im Mittelalter, 1882.
     Part II, Geschichte der Physik in der neueren Zeit. 1884. Part
     III, Geschichte der Physik in den letzten hundert Jahren. Zwei
     Abtheilungen. 1887–1890.

ROSSIGNOL, J. P.
Les métaux dans l'antiquité. Origines religieuses de la métallurgie ou
     les dieux de la Samothrace representés commes métallurges d'après
     l'histoire et la géographie de l'orichalque. Paris, 1863. 8vo.

ROULAND, N.
Tableau historique des propriétés et des phénomènes de l'air, con-
     sidéré dans ses différens états et sous ses divers rapports. Paris,
     1784. pp. xvi–636, 12mo.

ROUSSEAU, GEORG LUDWIG CLAUDIUS.
Vertheidigungsrede der Chemie wider die Vorurtheile der Zeit.   .
     Ingolstadt, 1774.

ROUTLEDGE, ROBERT.
> * A Popular History of Science.   London, 1881.   pp. xviii–673.   8vo.
> Portraits and illustrations (331).
>> Chemistry receives its share of attention in this popular survey of the progress
>> of science from Thales to Sir William Thomson.

ROUX, AUGUSTIN, ET LE BARON D'HOLBACH.
> Recueil des mémoires les plus intéressants de chymie et d'histoire
> naturelle contenus dans les actes de l'Académie d'Upsal et dans
> les mémoires de l'Académie royale des sciences de Stockholm,
> publiés depuis 1720 jusqu'en 1760.   Traduit du Latin et de
> l'Allemand.   Paris, 1764.   2 vols., 12mo.   I, xii–336 ; II, iv–687.

RÜDORFF, FRIEDRICH.
> * Die Fortschritte der Chemie in den letzten fünfundzwanzig Jahren.
> Rede zum Geburtsfeste Seiner Majestät des Kaisers und Königs
> in der Aula der königlichen technischen Hochschule zu Berlin am
> 21. März, 1887.   Berlin, 1887.   pp. 16, 4to.
>> A brief sketch for a popular audience.

SACHS, PHILIPP JACOB, VON LEWENHEIM.
> Aurum Chymicum.   *In* Manget's Bibliotheca chemica curiosa, vol. I, p.
> 192.   *See in Section VI*, Manget, J. J.
>> Historical testimony to the verity of transmutations.

SAINT HILAIRE, AUGUSTE DE.
> Histoire de l'indigo depuis l'origine des temps historiques jusqu'à
> l'année 1833.   Extrait des Nouvelles annales des voyages.   Paris,
> *n. d.* [1837].   8vo.

SALMASIUS, CLAUDIUS.
> De saccharo et manna commentarius.   Paris, 1663.   8vo.

SANTAGATA, ANTONIO.
> Discorso sulla importanza della chimica.   Bologna, 1823.   4to.

SAVÉRIEN.
> Histoire des progrès de l'esprit humain dans les sciences naturelles et
> dans les arts qui en dépendent ; savoir l'espace, le vide, le temps,
> le monument et le lieu ; la matière ou les corps, la terre, l'eau, l'air,
> le son, le feu, la lumière et les couleurs, l'électricité, l'astronomie
> physique, le globe terrestre, l'économie animale, la chimie, la
> verrerie, la teinture.   Avec un abrégé de la vie des plus célèbres
> auteurs dans ces sciences.   Paris, 1775.   pp. xvi–402, 8vo.
>> The history of chemistry occupies pages 296–308 ; the history of fire, pages
>> 160–191.   *See in Section IV*, Savérien.

SCHAEFER, HEINRICH WILHELM.

*Die Alchemie. Ihr ägyptisch-griechischer Ursprung und ihre weitere historische Entwicklung. Jahresbericht über das Schuljahr 1886–1887. Königliches Gymnasium und Realgymnasium zu Flensburg. Flensburg, 1887. pp. 52. 4to.

SCHAEFER, TH.

Ueber die Bedeutung der Alchemie. Wissenschaftliche Abhandlung zu dem Programm der Hauptschule zu Bremen. Bremen, 1885. pp. 32, 4to.

SCHAER, EDUARD.

Aus der Geschichte der Gifte. Vortrag. Basel, 1883. 8vo.
Oeffentliche Vorträge gehalten in der Schweiz, VII, 7.

SCHARLING, E. A.

Bidrag til at oplyse de forhold under hvilke chemien har været dyrket i Danmark. Kjøbenhavn, 1857. 4to.

SCHEERER, THEODOR.

Ueber die Fortschritte der Chemie in der Gebiete der Metallurgie, Mineralogie und Geologie während des letzten Jahrhunderts (1766–1866). Separatabdruck aus dem zweiten Theile der Festschrift zum hundertjährigen Jubiläum der königlichen sächsischen Bergakademie zu Freiberg. Dresden, *n. d.* [1866]. 4to.

SCHEIBLER, C.

Actenstücke zur Geschichte der Rübenzuckerfabrikation in Deutschland während ihrer ersten Entwickelung. Berlin, 1875. 8vo.

SCHERER, JOHANN BAPTISTE ANDREAS VON.

Beweis dass Johann Mayow vor hundert Jahren den Grund zur antiphlogistischen Chemie und Physiologie gelegt hat. Wien, 1793. pp. xxiv–188, 8vo. With portrait of Mayow on title-page.
Johann Mayow died in 1688. He certainly was acquainted with the action of oxygen in respiration of animals and plants, anticipating Priestley and Lavoisier about a century. *See in Section V*, Mayow, Johannes.

Beweis dass Mayow und Pechlin den Grund zu den neueren Theorien des lebenden Organismus gelegt haben. Wien, 1802. 8vo.
Joh. Nicolaus Pechlin died in 1706.

Geschichte der Luftgüte-Prüfungslehre für Aerzte und Naturfreunde kritisch bearbeitet. Wien, 1785. 2 parts. 8vo. Folding plate. Part I, pp. xvi–214; Part II, pp. 228.

SCHERER, JOHANN BAPTISTE ANDREAS VON.   [Cont'd.]
  Versuch einer neuen Nomenclatur für deutsche Chymisten.   Wien,
    1792. pp. [xx]-208-[xvi], 8vo.   Two folding tables.

> A modification of Lavoisier's nomenclature adapted to the German, and sub-
> stantially preserved till the present day. Alphabetized under the French
> names.

SCHEURER–KESTNER.
  Nicolas Leblanc et la soude artificielle (Conférence de la Société
    Chimique de Paris). Paris, 1885. 8vo.

SCHIENDL, C.
  * Geschichte der Photographie.   Mit den Bildnissen der Erfinder und
    Gründer der Photographie und einer Abbildung der ersten Photo-
    graphie. Wien, Pest, Leipzig, 1891. pp. vii–380. 8vo.

SCHLEIDEN, M. J.
  Das Salz. Seine Geschichte, Symbolik und Bedeutung im Menschen-
    leben. Leipzig, 1875. 8vo.
        Cf. Ratton, J. J. L.

SCHLOTTMANN.
  Kritische Geschichte der Theorien des Galvanismus.   Breslau, 1856.

SCHMIDT, D. P. H.
  Etymologischer chemischer Nomenclator der neuesten einfachen und
    daraus zusammengesetzten Stoffe, nebst Erklärung einiger andern
    chemisch-physicalischen Benennungen. Lemgo. 1839. 8vo.

SCHMIDT, LUDWIG VON.
        Systematische Darstellung (etc.).   See Meyer, Johann Rudolph, junior.

SCHMIEDER, KARL CHRISTOPH.
  * Geschichte der Alchemie.   Halle, 1832. pp. x–613, 8vo.

> In this History of Alchemy the author, Director of a High School in Cassel,
> endeavors to establish by historic proofs the reality of the transmutation
> of metals. He recognizes two distinct sciences, chemistry and alchemy,
> and claims they existed independently of each other from the earliest
> ages. Alchemy, he states, has a threefold dogma : I. It is possible to
> prepare by true art perfect gold from substances which contain no gold.
> II. The same is true of silver.   III. This artificial preparation is a
> wonderful medicine, panacea of life.
> Starting with this statement, he investigates the authenticity of the his-
> toric records of transmutation, and, sparing no pains in deciphering
> musty manuscripts of a former age, he concludes that we must acknowl-
> edge the reality of the transmutation of metals. He confesses that
> impostors abounded, but he claims five persons as true adepts and

SCHMIEDER, KARL CHRISTOPH.   [Cont'd.]

> gives their personal history with narratives of their wonderful accomplishments.   He calls attention to the fact that the five persons named lived at succeeding periods, and concludes that the Philosopher's Stone was secretly handed down from one to the other.
>
> The whole aim and scope of this strange work, and especially the conclusions drawn, seem more appropriate to the times of Borrichius than to the second quarter of the enlightened 19th century.

SCHNEIDER.

Zur Geschichte der Physik im 17. Jahrhundert.   Programm des königlichen Gymnasiums in Ellwangen am Schlusse des Schul-Jahres 1884–85.   Ellwangen, *n. d.*   42 pp., 4to.

SCHORLEMMER, CARL.

On the Origin of the Word Chemistry.   Read before the Manchester Literary and Philosophical Society, Nov. 18, 1879.   Chem. News, XL, 309 (1879).

* The Rise and Development of Organic Chemistry.   London, 1879. pp. 124, 12mo.

> Origine et développement de la chimie organique.   Ouvrage traduit de l'anglais avec autorisation de l'auteur par Alexandre Claparède.   Paris, 1885.   pp. ix–171, 12mo.

*Der Ursprung und die Entwickelung der organischen Chemie. Braunschweig, 1889.   pp. vi–199, 8vo.

> An extension of the English essay published in 1879.

SCHREGER, CHRISTIAN HEINRICH THEODOR.

Chemie (geschichtliche Uebersicht) *in* Allg. Encyclopaedie der Wissenschaft und Kunst.   Ersch und Gruber.   Theil 16.   Leipzig, 1827.   4to.

Kurze Beschreibung der chemischen Geräthschaften älterer und neuerer Zeit als Beitrag zur Geschichte der Erfindungen in der Chemie.   Nebst einer Vorrede des Hofrath [Friedrich] Hildebrandt.   Fürth, 1802.   3 vols., 8vo.   Plates.   I, pp. x–333 ; II, pp. 393 ; III, pp. 266.

> The author treats in vol. I of pharmaceutical and technical apparatus ; in vol. II, of pneumatic apparatus ; in vol. III of chemico-physical apparatus.   The arrangement is a systematic one, and there is no index.   The bibliography of the subject has not been neglected.

SCHRÖDER, FRIEDRICH JOSEPH WILHELM.

Geschichte der ältesten Philosophie und Chemie oder sogenannten hermetischen Philosophie der Ægyptier.   Marburg, 1775.   8vo.

> Schröder was Professor of Chemistry and Medicine at the University of Marburg.   This work is written in defence of the " Higher Chemistry," a term applied to alchemy by Wenzel shortly before.
>
> Also in Schröder's Neue Sammlung der Bibliothek für die . . . Chemie. *See in Section VI.*

SCHRÖDER, WILHELM FREIHERR VON.

*Fürstliche Schatz- und Rent-Kammer nebst seinem nothwendigen Unterricht vom Goldmachen. Leipzig, 1704. pp. [xxviii]–474–[LX]–[v]–70, 12mo.

> The second part contains a chapter treating historically of transmutation.

SCHUBARTH, ERNST LUDWIG.

Geschichte und Literatur der Chemie. pp. 1–17 of Lehrbuch der theoretischen Chemie. Berlin, 1829.

> *See in Section V*, Schubarth, E. L.

SCHÜTZENBERGER.

Histoire de la chimie (Extrait). Paris, *n. d.* 8vo.

SCHWEIGGER, JOHANN SALOMO CHRISTOPH.

Ueber naturwissenschaftliche Mysterien in ihrem Verhältnisse zur Literatur des Alterthums. Halle, 1843. 4to.

SEBELIEN, JOHN.

> Beiträge zur Geschichte der Atomgewichte. *See in Section V.*

SELMI, FRANCESCO.

> Compendio storico della chimica. [Napoli. 1878.] *In* vol. II, pp. 503-726, of Selmi's Enciclopedia di chimica. *See in Section II*, Enciclopedia di chimica.

SENAC, J. B.

> Discours historique sur l'origine . . . de la chymie. *In* Nouveau cours de chymie. *See in Section V*, Senac, J. B.

SENNERT, DANIEL.

De chymicorum cum Aristotelicis et Galenicis consensu ac dissensu liber I, controversias plurimas tam philosophis quam medicis cognitu utiles continens. Wittenbergæ, 1619. pp. [xxxii]–709 (and Index), 12mo.

> Third edition. Frankfurt and Wittenberg, 1655. pp. [xx]–434 (and Index), sm. 4to.

> In chapters II and III the author discusses the utility, necessity, antiquity, origin and progress of chemistry.

Tractatus de alchimiæ transmutatoriæ certitudine ; additæ sunt Questiones physico-medico-controversæ. Wittenbergæ, 1624. 4to.

SEYFFER, O. E. J.

Geschichtliche Darstellung des Galvanismus. Stuttgart, 1848.

SIEBERT, G.

*Kurzer Abriss der Geschichte der Chemie. Wien und Leipzig, 1886. pp. iv–124, 8vo.

SIGAUD DE LA FOND.

Précis historique et expérimental des phénomènes électriques depuis l'origine de cette découverte jusqu' à ce jour. Paris, 1781. 8vo. Second edition. Paris, 1785. 8vo.

    See also in Section V.

SILLIMAN, BENJAMIN [Jr.].

*American Contributions to Chemistry. An address delivered on the occasion of the Celebration of the Centennial of Chemistry at Northumberland, Pa., August 1, 1874. (Reprinted from the American Chemist for August, September and December, 1874.) Philadelphia, 1874. pp. 176, 12mo.

    Biographical and bibliographical; an invaluable source of information for the part played by the United States in the progress of chemistry.

Century (A) of Medicine and Chemistry. A Lecture delivered Sept. 14, 1871. New Haven, Connecticut, 1871. 12mo.

SILVESTRI, ORAZIO.

Il presente ed il passato della chimica considerata nei suoi rapporti con le altre scienze naturali. Catania, 1864. 16mo.

SMITH, J. LAWRENCE.

The Century's Progress in Industrial Chemistry. Am. Chem. Vol. v, p. 61–70. New York, 1875.

The Progress and Condition of several Departments of Industrial Chemistry. *In* U. S. Commissioners for the Paris Exhibition, 1867, Reports, vol. II.

SMITH, R. ANGUS.

The Word "Chemia" or "Chemistry." (From the seventh volume of the third series of Memoirs of the Manchester Literary and Philosophical Society, Session 1879–80.) London, 1880. pp. 101–126, 8vo.

    Also in Chem. News, XLII, p. 68 and p. 244 (1880).

*A Century of Science in Manchester. (In a series of Notes.) For the Hundredth Year of the Literary and Philosophical Society of Manchester (1881). London, 1883. xii–476, 8vo.

SMITH, T. P.

Sketch of the Revolutions in Chemistry. Philadelphia, 1798. pp. 40, 8vo.

SMYTH, GEORGE A.

Entwickelung der theoretischen Ansichten über die gepaarten Schwefel-Verbindungen. Berlin, 1877. 8vo.

SPICA, P.
> Sguardo storico sulla chimica : prelezione al corso di chimica generale
> nella Regia Universitá di Padova. Padova, 1883. 8vo.

SPIELMANN, JACOB REINBOLD.
> Dissertatio sistens historiam aëris factitii. Argentorati, 1776. 4to.
>> For a German version *see in Section V*, Mancherlei.

SPOTTISWOODE, WILLIAM.
> On the Old and New Laboratories at the Royal Institution [London].
> Extract from J. Roy. Inst., Jan. 17, 1873. Vol. VII, pp. 1–10, 8vo.

SPRAT, TH.
> History of the Royal Society of London. Second edition. London,
> 1702. 4to. ⟨ Ill.

SPRENGEL, CURTIUS.
> De artis chemicæ primordiis commentariolum . . . disputationem in-
> auguralem, die 17 Sept., 1823. . . . Halæ, *n. d.* [1823]. Two parts.
> Pt. I, pp. 12 ; pt. II, pp. 15 ; 12mo.

STÖCKHARDT, JULIUS ADOLPH.
> Die Fortschritte der Chemie in Deutschland. *In* Germania, die Ver-
> gangenheit, Gegenwart und Zukunft der deutschen Nation. Leip-
> zig, 1857. Vol. I. 8vo.

STÖLZEL, CARL.
> Die Entstehung und Fortentwicklung der Rübenzucker-Fabrikation.
> Braunschweig, 1851. 8vo.

STORCK, A.
> Les laboratoires de chimie en Allemagne. Laboratoires de Bonn et de
> Berlin. Lyon, 1876. 8vo.

STORIA ED ORIGINE DELLA FARMACIA IN MILANO. Milano, 1812. 8vo.

STRASSBURG, UNIVERSITÄT.
> Das chemische Institut. Festschrift für die 58 Versammlung deutscher
> Naturforscher und Aerzte. Die naturwissenschaftlichen und
> medicinischen Institute der Universität und die naturhistorische
> Sammlung der Stadt Strassburg. [1885.] 4to.
> Das chemische Institut. Festschrift zum Einweihung der Neubauten
> der Kaiser-Wilhelms Universität. Strassburg, 1884. pp. viii–150.
> 4to.
>> The chemical laboratory is described in pages 55–59, with plans.

STRATINGH, SEBALDUS.

Oratio de chemiæ recentioris incrementis atque præstantia. *In* Annales academicae. Groningen, 1823–24. 8vo.

STRECKER, ADOLPH.

Das chemische Laboratorium der Universität Christiania und die darin ausgeführten chemischen Untersuchungen. Auf Veranlassung des Academischen Collegiums herausgegeben. Universitäts-Programm für das zweite Halbjahr 1854. Christiania, 1854. pp. [vi]–104, 4to. Two folding plates.

STRUMPF, L. F.

Die neuesten Entdeckungen der angewandten Chemie. Berlin, 1845–51. 2 vols., 8vo.

STURMIUS, JOHANNES CHRISTOPHORUS.

Collegium experimentale sive curiosum, in quo primaria hujus seculi inventa et experimenta physico-mathematica, speciatim campanæ urinatoriæ, camera obscuræ, tubi Torricelliani seu baroscopii, antliæ pneumaticæ, thermometrorum, hygroscopiorum, telescopiorum, microscopiorum, etc., phænomena et effecta, partim ab aliis jam pridem exhibita, partim noviter istis superaddita, per ultimum quadrimestre anni 1672 viginti naturæ scrutatoribus ex parte illustri nobilique prosapia oriundis, et spectanda oculis subjecit, et ad causas suas naturales demonstrativa methodo reduxit, quodque nunc accessione multa, demonstrationum ac hypothesium veritatem porro illustrante, confirmante et a nunnullorum cavillationibus vindicante, locupletatum amicorum quorundam suasu et consilio publicum adspicere voluit. Norimbergæ, 1676. pp. [xxii]–168–122–[xii], sm. 4to. Illustrations and folding plates.

> Though pertaining to physics rather than chemistry, this early summary of progress deserves notice. Its scope is fully indicated in the long title.

SUE, PIERRE (LE JEUNE).

Histoire du galvanisme et analyse des différens ouvrages publiées sur cette découverte, depuis son origine jusqu'à ce jour. Paris, 1802. 2 vols.

> A German edition was published by Jo. Chr. Aug. Clarus and Jo. Chr. Leop. Reinhold at Leipzig in 1803. 2 vols., 8vo. Ill.

SVÁTEK, JOSEF.

* Die Alchemie in Böhmen. *In* Culturhistorische Bilder aus Böhmen. Wien, 1879. pp. 43–94, 8vo.

T. BEYTRÄGE ZUR GESCHICHTE DER CHEMIE UND ALCHEMIE in den ältesten und mittleren Zeiten. Litteratur und Völkerkunde. Dessau, 1783. Vol. II, pp. 781–807.

TALBOT, WILLIAM HENRY FOX.
On the colored rings produced by iodine on silver, with remarks on the History of Photography. Phil. Mag. [3]. Vol. XXII, pp. 94–97 (1843).

TEIXEIRA MENDES, R.
La philosophie chimique d'après Auguste Comte, indications générales sur la théorie positive des phénomènes de composition et de décomposition suivies d'une appréciation sommaire de l'état actuel de la chimie. Distribution gratuite. Centre Positiviste du Brésil. Ordre et Progrès. Rio de Janeiro, 1887. pp. xx–251, 8vo.

> The author cites Adolphe Wurtz as an example of "irrationalité académique," accuses Marcellin Berthelot of false reasoning, and endeavors to show that August Comte anticipated Gerhardt in his views of theoretical chemistry.

TERQUEM, A.
La science romaine à l'époque d'Auguste. Paris, 1885. Ill.

TETZLAFF.
Darstellung der atomistischen Hypothese nach den modernen Theorien der Chemie. Programme des Gymnasiums zu Stralsund, Ostern, 1886. Stralsund, 1886. pp. 32, 4to.

THAN, C. VON.
Das chemische Laboratorium der königlichen Universität in Pest. Wien, 1872.

THEOPHRASTUS.
Περὶ λίθων.

> Theophrastus died 285 B.C.; of his work on minerals many editions and translations have been published: in Latin by De Laet, Leyden, 1647; in English by John Hill, London, 1746, 8vo; in French, Paris, 1754; in German by Baumgärtner, Nürnberg, 1770.

THOMSON, J.
> See Tissandier, Gaston.

THOMSON, THOMAS.
*History of the Royal Society from its institution to the end of the Eighteenth Century. London, 1812. pp. viii–552–xci, sm. folio.

> Book IV, pp. 464–521, deals in three chapters with the history of chemistry as set forth in the Philosophical Transactions.

*The History of Chemistry. London, 1830. 2 vols., 12mo. Vol. I, pp. x–349, portrait of Jos. Black. Vol. II, pp. [iv]–325. Vol. II also forms No. X of the "National Library," conducted by G. R. Gleig.

THOMSON, THOMAS.   [Cont'd.]

> This entertaining work was long the only History of Chemistry in the English language. It is rather uneven in its treatment, but the progress of analytical chemistry is reviewed with critical skill. The author was Professor of Chemistry in the University of Glasgow. For a biography of the author, *see Section IV*, under his name.

TIEBOEL, BOUDEWYN.

Welke zyn de eigenlyke oorzaken waarom de scheikunde by onze nabuuren . . . in meer aanzien en algemeene beoefening is, dan in ons Vaderland? [Middelburg (?), 1766]. 8vo.

TIMBS, JOHN.

Things not generally known familiarly explained. Curiosities of Science, Second Series. A book for old and young. London, 1860. 12mo.

> Pages 1–47 on Alchemy and Chemistry.

TISSANDIER, GASTON.

A History and Handbook of Photography translated from the French. Edited by J. Thomson. With upwards of seventy illustrations. London, 1876. pp. xvi–326, 8vo. Ill.

> Second and Revised Edition. With some Specimens of Permanent Processes and an Appendix by the late Henry Fox Talbot. London, 1878. pp. 400. Ill.

> The history occupies pages 1–91.

TOLLIUS, JACOB.

Fortuita, in quibus, præter critica nonnulla, tota fabularis historia Græca, Phœnicia, Ægyptiaca ad chemiam pertinere asseritur. Amstelodami, 1687. 8vo. Ill.

> An erudite work attempting to give an alchemical interpretation to Grecian and Egyptian fables. The author was an admirer of Basil Valentine, an "inquietus homo, in summa paupertate mortuus" (Choulant). *Cf. in Section II*, Pernety, Antoine Joseph.

TONNI-BAZZA, LORENZO.

Dell' alchimia e degli alchimisti. Dissertazione. Pavia, 1858. 8vo.

TRÉSOR DE RECHERCHES ET ANTIQUITÉS GAULOISES ET FRANÇOISES. Paris, 1655. 4to.

> According to Schmieder this contains much material on the history of alchemy, but badly arranged.

TROMMSDORFF, JOHANN BARTHOLOMÄ.

Geschichte des Galvanismus oder der galvanischen Electricität, vor-

TROMMSDORFF, JOHANN BARTHOLOMÄ. [Cont'd.]
züglich in chemischer Hinsicht. (Besonders abgedruckt aus J. B.
Trommsdorff's Chemie im Felde der Erfahrung.) Erfurt, 1803.
pp. [vi]–260, 8vo.

> *See in Section V*, Trommsdorff's Systematisches Handbuch der gesammten
> Chemie.

Historisches Taschenbuch für Aerzte, Chemiker und Apotheker auf
das Jahr 1803 [–05]. Erfurt, 1803–05. 3 vols.

> Vol. I contains 6 portraits of chemists. Vol. II contains portrait of Stahl,
> and vol. III that of Gren. These volumes contain : Versuch einer all-
> gemeinen Geschichte der Chemie. *Cf. in Section I*, Trommsdorff.

Uebersicht der wichtigsten Entdeckungen in der Chemie. Weimar,
1792. Folio.

TROOST, L.
Un laboratoire de chimie au dix-huitième siècle—Scheele.—Conférence
faite à la Sorbonne dans la soirée scientifique du 19 Janvier, 1866.
Suivie d'un résumé des principaux travaux de Scheele. Paris,
1866. pp. 66, 12mo. Actualités scientifiques. Illustrated.

TROOSTWIJK, A. PAETS VAN.
> *See* Paets van Troostwijk, A.

UPMANN, J.
> Geschichte des Schiesspulvers. *See in Section V*, Bolley's Handbuch [etc.].

URECH, FRIEDRICH.
Itinerarium durch die theoretische Entwickelungsgeschichte der Lehre
von der chemischen Reactionsgeschwindigkeit. Berlin, 1885.

VALENTINI, M. BERNHARDUS.
Historia physices experimentalis, qua antiquitatem ejus, ortum et progressum adstruebat. Giessen, 1688. 4to.

VAN, VAN DER.
> For Dutch names beginning with Van, or Van der, see next succeeding word.

VARGAS, JOSÉ M.
> *See* Brande, W. T. Historia de la química, *etc.* 1867–80.

VEITCH, JOHN.
Lucretius and the Atomic Theory. Glasgow, 1875. 12mo.

VENTURI, ALB.
De mellis origine et usu : dissertatio historico-medica. Coloniæ
Agrippinæ, 1765. 8vo.

VIREY, J. J.

Discours sur l'histoire et les progrès des sciences pharmaceutiques ou naturelles et chimiques jusqu'aux temps actuels. *In* Mémoires de l'Académie royale de Médecine. 1828. Vol. I, part I, pp. 323–339.

VLASTO, E.

Les origines de l'alchimie par M. Berthelot ; analyse. Paris, 1886. 8vo. pp. 24.

> *Cf.* Berthelot, Marcellin.

VOGEL, A.

Ueber die Entwickelung der Agriculturchemie. München, 1869. 4to.

VOGEL, H. W.

Progress of Photography since 1879. Philadelphia, 1884. 8vo.

VOIT UND LIEBIG.

Das chemische Laboratorium der königlichen Universität in München. Braunschweig, 1859.

VOLTELEN, FLORENTIUS JACOBUS.

De chemiæ hoderniæ pretio rite constituendo. Lugduni Batavorum, 1784. 4to.

VULPIUS, G.

Ueber die Alchemisten. Ein im historisch-philosophischen Verein in Heidelberg gehaltener Vortrag. Heidelberg, 1874. 8vo.

WAGNER, RUDOLPH.

> Bericht über die neuesten Fortschritte in der Chemie.
> *See in Section II,* Handwörterbuch der Chemie und Physik.

WAGNER, RUDOLPH.

Die Geschichte der Chemie von der Kindheit des Menschengeschlechts bis auf unsere Tage. Leipzig, 1853. 8vo.

> *Zweite vermehrte Auflage. Leipzig, 1855. pp. viii–108, 8vo.

WALLERIUS, JOHAN GOTSCHALK.

> Om historien af chemien och dess namn.
> *In his* Chemia physica. *See in Section V.*

WARE, L. S.

The Sugar Beet. History of Beet Sugar Industry in Europe [etc.]. London [?].

WARTMANN, ÉLIE FRANÇOIS.

Essai historique sur les doctrines et les phénomènes de l'électrochimie. Genève, 1838. 8vo.

WEDEL, GEORG WOLFFGANG.
  * Introductio in alchimiam. Jenæ, 1706. pp. [iv]–58–[ii], 4to.
            * Vernünfftige Gedancken vom Gold-Machen. Nebst einer
                  Vorrede Christian Gottfried Stentzels. Wittenberg,
                  Zweyte Auflage, 1734. pp. [xxvi]–100, 12mo.

> Has chapters on the antiquity and writers on alchemy. The preface is
> dated 1707. Not mentioned by Lenglet Du Fresnoy, by Kopp nor by
> Fuchs.

WEECH, VON
  Verfolgte Alchymisten. Zeitschrift für die Geschichte des Oberrheins,
      herausgegeben von dem Grossherzoglichen General-Landesarchive
      zu Karlsruhe. Vol. xxv, pp. 468–470. 1873.

> Two original letters, dated 1605 and 1607, showing that Alexander Seton
> (or Sydon) and George Honauer were hung for their deceptions by order
> of Duke Frederick I. of Württemberg (1593–1608). An interesting
> contribution to the history of alchemy.

WEHRLE, ALOYS.
  Dissertatio inauguralis chemica sistens historiam acidi muriatici.
      Viennæ, 1819. 8vo.
            Geschichte der Salzsäure, oder zusammenhängende Ueber-
                  sicht aller Verbindungen derselben. . . . Wien, 1819. 8vo.

WEIDNER, JOHANN.
  De arte chymica ejusque cultoribus. Basiliæ, 1576. 4to.

WEIGEL, CHRISTIAN EHRENFRIED.
  Beiträge zur Geschichte der Luftarten in Auszügen, als ein Nachtrag zu
      dem historischen kurzen Begriffe elastischer Ausflüsse in Herrn
      Lavoisier physikalisch-chemischen Schriften, Bd. 1, Th. 1. Greifs-
      wald, 1784. pp. [xvi]–518, 12mo. Folding plate.

> The extracts are from the works of Rey, Mayow, Papins, Leeuwenhoeck,
> Hauksbee, Lowther, Desaguliers, Maud, d'Arguier, Priestley, and
> many others.

  Einfluss (Der) chemischer Kenntnisse auf die Oekonomie besonders des
      Schwedischen Pommerns. Greifswald, 1775. 4to.
  Einladungsschrift von der Nothwendigkeit der Versuche beim Vortrage
      der Scheidekunst. Nebst einer Anzeige seiner Vorlesungen.
      Greifswald, 1796. 4to.
  Vom Nutzen der Chemie insbesondere in Absicht auf Pommern be-
      trachtet. Eine Antritts-Rede. Greifswald, 1774. 4to.

WEIHRICH, G.
  Beiträge zur Geschichte des chemischen Unterrichts an der Universität
      Giessen. Giessen, 1891. 4to.

WEIHRICH, G. [Cont'd.]

Ansichten (Die) der neueren Chemie. Programm des grossherzoglichen Gymnasiums zu Mainz, Schuljahr 1871–72. Mainz, 1872. pp. 42, 4to.

WEITZ, MAX.

*Geschichte der Chemie in synchronistischer Darstellung. Ein kurzgefasster Leitfaden für Fachmänner und Laien, Studirende und Praktiker, für Schüler und zum Selbstunterricht. Berlin, 1889. pp. 36, long 8vo.

WELD, CHARLES RICHARD.

A History of the Royal Society with Memoirs of the Presidents. Compiled from authentic documents. London, 1848. 2 vols., 8vo. Vol. I, pp. xx–527 ; vol. II, viii–611. Illustrated.

WELTZIEN, CARL, AND H. LANG.

Das chemische Laboratorium an der grossherzoglichen polytechnischen Schule zu Carlsruhe. Carlsruhe, 1853. Fol.

WENGHÖFFER, L.

Die wichtigsten Forschungsresultate auf dem Gebiete der Chemie der Kohlenstoffverbindungen im Jahre 1882. Stuttgart, 1883.

WERGE, J.

The Evolution of Photography with a Chronological Record of Discoveries, Inventions, etc. Contributions to photographic literature and reminiscences extending over forty years. [London] 18—. Illustrated.

WESTRUMB, JOHANN FRIEDRICH.

Geschichte der neu entdeckten Metallisirung der einfachen Erden. Nebst Versuchen und Beobachtungen. Hannover, 1791. pp. x–144, 16mo.

> The author shows that the alleged discovery of metallic bases of the earths by Ruprecht and Tondy is fallacious.

WHEWELL, WILLIAM.

*History of the Inductive Sciences from the Earliest to the Present Times. London, 1837. 3 vols., 8vo.

> Vol. III has a chapter on the history of chemistry, but this section is meagre; Joseph Priestley and his discoveries are dismissed in nine lines, and James Watt is not even mentioned.

Geschichte der inductiven Wissenschaften, der Astronomie, Physik, Mechanik, Chemie, Geologie, etc., von der

WHEWELL, WILLIAM. [Cont'd.]
frühesten bis zu unserer Zeit. Nach dem englischen mit Anmerkungen von J. J. von Littrow. Stuttgart, 1840–41. 3 vols., 8vo.

WHITE, ANDREW DICKSON.
* The Warfare of Science. New York, 1876. pp. 151, 12mo.

In this history of the interference of religion with science, chemistry and physics receive attention. Additional chapters are now [1892] appearing serially in the Pop. Sci. Monthly, New York.

WIEBEKE.
Geschichtliche Entwickelung unserer Kenntniss der Ptomaïne und verwandter Körper. Frankfurt a. Oder, 1886. pp. 22, 8vo.

WIEDEMANN, EILHARD.
Zur Chemie der Araber. Zeitschrift der deutschen morgenländischen Gesellschaft. Leipzig, 1878. Vol. XXXII, pp. 575–580.

The author treats of the etymology of the word Alembic, and refers to an Arabic MS. of Geber. Citations are in Arabic text.

WIEGLEB, JOHANN CHRISTIAN.
* Geschichte des Wachsthums und der Erfindungen in der Chemie in der ältesten und mittlern Zeit. Aus dem lateinischen übersetzt, mit Anmerkungen und Zusätzen. Berlin und Stettin, 1792. pp. [xii]–256. 8vo.
* Geschichte des Wachsthums und der Erfindungen in der Chemie in der neuern Zeit. Berlin und Stettin, 1790–91. 2 vols., 8vo.
* Historisch-kritische Untersuchung der Alchemie, oder der eingebildeten Goldmacherkunst; von ihrem Ursprunge sowohl als Fortgange, und was nun von ihr zu halten sey. Weimar, 1777. pp. [xxii]–437.

Also a new title-page, Weimar, 1793. *See in Section VI*, Wiegleb, J. C.

WIEN (LABORATORIUM).
See Ferstel, R. von.

WIEN (WELTAUSSTELLUNG).
See Beilstein.

WIESBADEN (LABORATORIUM).
See Fresenius.

WILD, JOHANN RUDOLPH, D. J.
* Versuch einer Charakteristik des Verhältnisses der Alchemie zur Magie, Astrologie und verwandten ähnlichen Wissenschaften, mit besonderer Berücksichtigung der alchemistischen Zeichen. Cassel, 1841. pp. 68, 8vo. Eight plates.

WILHELM.

Droguen und Chemikalien auf der Weltausstellung in Philadelphia. Wien, 1877.

WILKINSON, C. H.

* Elements of Galvanism in theory and practice ; with a comprehensive view of its History, from the first experiments of Galvani to the present time. Containing also practical directions for constructing the galvanic apparatus and plain systematic instructions for performing all the various experiments. Illustrated with a great number of copper plates. London, 1804. 2 vols., 8vo. Vol. I, pp. xv–468, three plates. Vol. II, pp. xi–472–[40], nine plates.

WILLBRAND, J. B.

Bedenken und Zweifel betreffend das Verhältniss der chemischen Theorien zu den Erfordernissen des Wissens überhaupt, und zur Physiologie so wie zur ärztlichen Praxis insbesondere. Mainz, 1842. 8vo.

WINTERL, JACOB JOSEPH.

Prolusio ad chemiam sæculi XIX. Ofen, 1800. 8vo.

WISSER, JOHN P.

* History of Chemistry. Course of sciences applied to military art. U. S. Artillery School. Fort Monroe, Va., 1885. Square 8vo.

WITT, OTTO N.

Geschichtlicher Ueberblick über die Entwickelung der Textilgewerbe. *In* Chemische Technologie der Gespinnstfasern. *See in Section V.*

WITTSTEIN, G. C.

Vollständiges etymologisch-chemisches Handwörterbuch, mit Berücksichtigung der Geschichte . . . der Chemie, *etc.* 1846–58.
See in Section II, Wittstein, G. C.

WOHLWILL, EMIL.

* Joachim Jungius und die Erneuerung atomistischer Lehren im 17. Jahrhundert. Ein Beitrag zur Geschichte der Naturwissenschaft in Hamburg. Sonder-Abdruck aus Band X der Abhandlungen aus dem Gebiete der Naturwissenschaften. Festschrift zur Feier des 50 jährigen Bestehens des Naturwissenschaftlichen Vereins in Hamburg, 1887. pp. 66, 4to.

WOLTTER, JOHANN ANTONIUS DE.

Utilitatem artis chemiæ ad rem publicam ipsumque principem redeuntem . . . Monachii, *n. d.* [1764]. pp. [28], sm. 4to.

WRIGHT, THOMAS [Editor].

* Popular Treatises on Science written during the Middle Ages in Anglo-Saxon, Anglo-Norman and English. Historical Society of Science. London, 1841. pp. xvi–140, 8vo.

WURTZ, ADOLPHE.

Aperçu historique sur le développement de la chimie. *See in Section V*, Introduction à l'étude de la chimie.

WURTZ, ADOLPHE.

* Histoire des doctrines chimiques depuis Lavoisier jusqu' à nos jours. Paris, 1869. pp. iii–280, 12mo.

> Valuable ; well known for its much criticised opening sentence : " La chimie est une science française."

Geschiedenis der chemische theoriën van Lavoisier tot op onzen tijd. Naar de hoogduitsche bewerking vertaald door W. Wolthuis. Breda, 1874. 8vo.

Geschichte der chemischen Theorien seit Lavoisier bis auf unsere Zeit. Uebersetzt von Alphonse Oppenheim. Berlin, 1870. 8vo.

Исторія химич. доктринъ отъ Лавуазье и до настоящаго времени. Перев. Бутлерова. С.-Петербургъ 1869.

History of chemical doctrines from Lavoisier up to the present time.

* La théorie atomique. Paris, 1880. pp. iv–246, 8vo. One folding plate.

> Bibliothèque scientifique internationale.

La théorie atomique, quatrième édition, précédé d'une introduction sur la vie et les travaux de l'auteur par Ch. Friedel. Paris, 1886.

Die atomische Theorie. Autorisirte Ausgabe. Leipzig, 1879. 8vo.

Internationale wissenschaftliche Bibliothek, vol. XXXVII.

Progrès de l'industrie des matières colorantes artificielles. [Extract from Rapports du Jury de l'Exposition de Vienne, 1873.] Paris, 1874. 8vo. Five plates and 29 specimens.

* Sur quelques points de philosophie chimique. Leçons professées les 6 et 20 Mars, 1863, devant la Société Chimique [de Paris], par M. Adolphe Wurtz, Président de la Société. Paris, 1864. 8vo.

> An admirable discussion of the development and principles of modern chemical philosophy.

Wurzer, Ferdinand.

Rede über die vornehmsten Schicksale der Chemie, ihren Einfluss auf die gesammte Naturkunde und über die durch sie dem Staate erwachsenden Vortheile. Bonn, 1793. 4to.

Brevis narratio de donis nonnullis quæ nuperis temporibus chemiæ debemus. Marburgi, 1821. 4to.

Wüstenfeld, Ferdinand.

*Geschichte der arabischen Aerzte und Naturforscher. Göttingen, 1840. pp. xvi–167–xiv, 8vo.

From original sources by an Arabic scholar.

Yearbook of the Scientific and Learned Societies of Great Britain and Ireland ; comprising lists of the papers read during 1883 [–1887] before societies engaged in all departments of research, with the names of their authors. Compiled from official sources. London, 1884–88. 5 vols., 8vo.

Valuable works of reference.

Yeats, G. D.

*Observations on the Claims of Moderns to some Discoveries in Chemistry and Physiology. London, 1798. pp. xxxvi–403, 8vo.

A comprehensive study of the contributions of Mayow to the doctrines of combustion and respiration. Portrait of Mayow. *Cf. in Section V*, Mayow, Johannes.

Zier, E.

Beitrag zur Geschichte der Verbreitung und Vervollkommnung der Rübenzuckerfabrikation in Deutschland ; oder was geschah dafür in den Jahren 1832 bis 1836 ? Zerbst, 1836. 4to.

Zippe, F. X. M.

Geschichte der Metalle. Wien, 1857.

Zürich (Laboratorium).

See Bluntschli, Lasius und Lunge.

Zürich (Landausstellung).

See Lunge, G.

SECTION IV.

BIOGRAPHY.

INCLUDING BIBLIOGRAPHIES OF INDIVIDUALS.

Biographies and bibliographies are placed under the names of the chemists described, with cross-references from the authors. For abbreviations used in this section, see preface.

ABEL, JOHN SANGSTER.
    Obituary. J. Chem. Soc., London, xxv, 342 (1872).

ABERLE, KARL. Schädel und Abbildungen des Paracelsus, *see* Paracelsus.

ACHARD, FRANZ CARL.
    Portrait in All. J. Chem., vol. IV, frontispiece

ADELUNG, J. F.
    * Geschichte der menschlichen Narrheit oder Lebensbeschreibungen
      berühmter Schwarzkünstler, Goldmacher, Teufelsbanner, Zeichen-
      und Liniendeuter, Schwärmer, Wahrsager und anderer philosophi-
      scher Unholden. Leipzig, 1785–89. 7 vols., 12mo.

        Contains biographies of: Vol. I, Nicol. Barnaud, J. F. Borro, J. A.
        Augurelli, J. J. Becher. Vol. II, C. von Drebbel. Vol. III, Nicolas
        Flamel. Vol. IV, J. R. Glauber, J. B. van Helmont, David Beuther.
        Vol. V, Heinrich Khunrath. Vol. VI, Delisle and Aluys, Michael
        Sendivog. To each biography is appended a full bibliography. Pub-
        lished anonymously.

AFZELIUS, ADAM.
    Obituary. Am. J. Sci., xxxiii, 211 (January, 1838).

AGRICOLA, GEORG.
    BECHER, FR. L. Die Mineralogen Georg Agricola zu Chemnitz im
        sechszehnten und A. G. Werner zu Freiberg im neunzehnten Jahr-
        hunderte. Winke zu einer biographischen Zusammenstellung aus
        Sachsens Culturgeschichte. Freiberg, 1819. pp. 67, 12mo.

            See Figuier, Vie des savants.

    JACOBI, G. H. Der Mineralog Georgius Agricola und sein Verhältniss
        zur Wissenschaft seiner Zeit. Werdau, 1889.

AGRIPPA, HENRY CORNELIUS. The Life of Henry Cornelius Agrippa. *See, in Section VI*, Agrippa, H. C., The Vanity of Arts.

ALBERTONI, P. Life of Galvani. *See* Galvani, Luigi.

ALBERTUS MAGNUS (OR ALBRECHT GRAF VON BOLLSTADT). For a list of more than 35 biographies and essays on A.M., *see, in* Chevalier, Ulysse : Répertoire.

ALBERTUS MAGNUS.

FERGUSON, JOHN. On a copy of Albertus Magnus' De secretis Mulierum printed by Machlinia. Communicated to the Society of Antiquaries [of London]. (Extract from The Archæologia, vol. XLIX.) Westminster, 1886. 9 pp., 4to. With a facsimile.

A critical bibliographical study.

*LIECHTY, REINHARD DE. Albert le Grand et Saint Thomas d'Aquin, ou la science au moyen âge. Paris, 1880. pp. ii–252, 12mo.

By a Roman Catholic priest, and written with characteristic religious bias.

ALBERTUS MAGNUS. Poisson, Albert : Notice biographique sur Albert le Grand. *See, in Section VI*, Poisson, A.

ALETHOPHILUS, Life of Hermes. *See* Hermes Trismegistus.

ALEXANDER, ROBERT.
Obituary. Chem. News, LIV, 259 (1886).

ALLAN, JAMES.
Obituary. Chem. News, XIII, 167. Apr. 6, 1866.
Obituary. J. Chem. Soc., London, xx, 386 (1867).

ALLEN, FREDERICK.
Obituary. J. Chem. Soc., London, XLIX, 342 (1886).

ALLEN, WILLIAM. Life of, *see* Walker, William, Jr.

ALLIBONE, S. AUSTIN.
A Critical Dictionary of English Literature and British and American Authors living and deceased from the earliest accounts to the latter half of the nineteenth century ; containing over forty-six thousand articles (authors). With forty indexes of subjects. Philadelphia, 1878. 3 vols., roy. 8vo. Supplement, 1892. 2 vols., roy. 8vo.

Contains biographical and bibliographical notices of 502 chemists.

AMBERG, FRIEDRICH. Zur Geschichte des Paracelsus, *see* Paracelsus.

AMPÈRE, ANDRÉ MARIE. Life of, *see* Arago, François ; *also* Mangin, Arthur.

ANASTASI. Life of Nicolas Leblanc. *See* Leblanc, Nicolas.

ANDERSON, THOMAS.
Obituary. J. Chem. Soc., London, XXVIII, 1309 (1875).

ANDREWS, THOMAS.
Obituary. J. Chem. Soc., London, XLIX, 342 (1886).

APJOHN, JAMES.
Obituary. J. Chem. Soc., London, LI, 469 (1887).

APJOHN, RICHARD.
Obituary. J. Chem. Soc., London, XXXIII, 227 (1878).
Obituary. Chem. News, XXXVI, 150 (1877).

ARAGO, FRANÇOIS.
*Oeuvres de F. A. Notices biographiques. Paris et Leipzig, 1854. 3 vols., 8vo.

> Contains following biographies : Vol. I, Arago, Fresnel, Volta, Young, Fourier, Watt, Carnot. Vol. III, Gay Lussac, Malus [and Astronomers]. Vol. II, Ampère, Condorcet, Bailly, Monge, Poisson.

Biographies of distinguished scientific men, translated by W. H. Smyth. London, 1857. 8vo. *Also :* Boston, 1858. 2 vols.

> The life of Condorcet was also translated by M. A. Henry, and published in Annual Report of the Smithsonian Institution for 1878. Washington, D. C., 1879. pp. 180–235, 8vo.

ARNALDUS DE VILLA NOVA.
HAURÉAU, JEAN-BARTHÉLEMY. Arnauld de Villeneuve, médecin et chimiste. Histoire littéraire de la France. Tome XXVIII, pp. 26–126 (1881).

> An important critical study by a master hand. The titles of works described number 123.

MENENDEZ PELAYO, M. Arnaldo de Vilanova, médico catalan del siglo XIII. Ensayo histórico, seguido de tres opusculos ineditos de Arnaldo, y de una coleccion de documentos relativos a su persona. Madrid, 1879. 16mo.

Life of Arnauld de Villeneuve. For a list of more than thirty biographies of A. de V., *see* Chevalier (Ulysse), Répertoire des sources historiques du moyen âge. Tome I. Paris, 1877–83, and Supplément, Paris, 1888. Vol. I, 2,370 col., 8vo. *See also* Figuier, Vie des savants.

> POISSON, ALBERT. Notice biographique sur Arnauld de Villeneuve. *See, in Section VI*, Poisson, Albert.

ARNOLD, EDWARD.
    Obituary.   J. Chem. Soc., London, xxv, 342 (1872).

ARNOT, WILLIAM.
    Obituary.   J. Chem. Soc., London, xxxix, 190 (1881).

ATKINSON, G. T.
    Obituary.   J. Chem. Soc., London, xlvii, 329 (1885).

AUGURELLI, J. A.   Life and bibliography, *see* Adelung, J. F.

AULAGNIER, ALEXIS-FRANÇOIS.   Notice historique sur la vie et les travaux de A. F.
        Aulagnier.   *See*, *in Section II*, Aulagnier : Dictionnaire des aliments et des
        boissons.

AURIVILLIUS, P. F.   Life of Torbern Bergman.   *See* Bergman, Torbern.

AVERROËS (IBN-ROSHD).
    Life and portrait.   Pop. Sci. Mon., xxv, 405, July, 1884.

AVOGADRO, AMADEO.
    BOTTO, G. D.   Cenni biografici sulla vita e sulle opere del Conte
        Amadeo Avogadro.   Torino, 1858.   4to. [Extract from : Memorie
        della reale accademia delle scienze di Torino.   Serie ii, Tomo
        xvii.]

BACHHOFFNER, GEORGE HENRY.
    Obituary.   J. Chem. Soc., London, xxxvii, 255 (1880).

BACON, FRANCIS.
    JANET, P.   Baco Verulamius alchemicis philosophis quid debuerit.
        Paris, 1889.

BACON, ROGER.
    Life of.   For a list of more than 30 biographies and essays on R. B.,
        see Chevalier (Ulysse), Répertoire des sources historiques du moyen
        âge.   Tome i.   Paris, 1877–83.
    CHARLES, ÉMILE.   R. B. sa vie, ses ouvrages, ses doctrines d'après des
        textes inédit.   Paris et Bordeaux, 1861.   8vo.

BACON, ROGER.   Poisson, Albert : Notice biographique sur Roger Bacon.   *See*, *in Section
        VI*, Poisson, Albert.   Life of, *see* Figuier, Vie des savants ; *also* Savérien.

BAILEY, JACOB W.
    Obituary.   Am. J. Sci. [2], vol. xxiii, 447, May, 1857. *Also*, vol.
        xxv, p. 133, 1858.

BAILLY, FRANCIS.   Life of, *see* Arago, François.

BAKER, WILLIAM.
Obituary. J. Chem. Soc., London, xxxv, 265 (1879).

BALARD, ANTOINE JÉRÔME.
Obituary. J. Chem. Soc., London, xxxi, 512 (1877).
DUMAS, JEAN BAPTISTE. Éloge sur A. J. B. Discours et éloges
académiques, Paris. vol. ii, p. 85.
Éloge de A. J. Balard. [Paris], 1879, 4to.
HENNINGER, ARTHUR. Nekrolog auf Balard. Ber. d. chem. Ges.,
ix, p. 639 (1876).

BALMAIN, WILLIAM HENRY.
Obituary. J. Chem. Soc., London, xxxvii, 256 (1880).

BARBAGLIA, GIOVANNI ARCANGELO.
Necrolog. Ber. d. chem. Ges., xxv, p. 137 (1892).

BARBIER, JEAN BAPTISTE GRÉGOIRE.
TAVERNIER, M. Éloge de J. B. G. Barbier décédé directeur honoraire
de l'École préparatoire de médecine et de pharmacie d'Amiens,
etc. Prononcé à la séance publique de l'Académie d'Amiens le
31 août, 1856. 8vo.

BARFF, FREDERICK SETTLE.
Obituary. J. Chem. Soc., London, li, 471 (1887).

BARKER, GEORGE F.
Life and portrait. Pop. Sci. Mon., xv, 693, Sept., 1879.

BARNAUD, NICOLAUS. Life of, see Adelung, J. F.

BARRAL, J. A.
Note sur les titres et travaux scientifiques de M. Barral. Paris, 1850. 4to.

BARRETT, FRANCIS.
* The Lives of Alchemystical Philosophers ; with a critical catalogue of
books in occult chemistry and a selection of the most celebrated
treatises on the theory and practice of the Hermetic Art. London,
1815. pp. 386, 8vo.
> Contains superficial biographies of 45 so-called adepts, a list of 750 alchemi-
> cal books, and selections from the most incredible treatises on the
> hermetic art. Bibliographically it is very inexact, giving one-line titles
> only. The work is quite scarce and commands a price disproportionate
> to its value for students who read other languages than the English.
> Published anonymously. For a revised edition of a portion, see
> Waite, A. E.

BARTH, L.   *See* Hlasiwetz, Heinrich,

BARTHENAU, LUDWIG BARTH VON.
   C. Senhofer und G. Goldschmiedt, Nekrolog auf B.   With portrait and
      bibliography of Barthenau.   Ber. d. chem. Ges., XXIV, 1089–1115
      (1891).

BARTHOLDY, GEORGES CHARLES.
   K. J. B.   Notice biographique sur Georges Charles Bartholdy, membre
      de la Société Littéraire de Colmar, ancien professeur de chimie,
      ancien Maire de la ville de Münster.   Colmar, 1851, 8vo.
         From Biographies alsaciennes.

BASILIUS VALENTINUS.   *See* Valentinus, Basilius.

BAUERNFEIND, C. M. VON.   Count Rumford, *see* Thompson, Benjamin.

BAUMÉ, ANTOINE.
   DEYEUX, NICOLAS.   Notice sur Antoine Baumé.   Ann. de chim., LV,
      105, (1805).
   CADET, C. L.   Éloge d'Antoine Baumé, Apothicaire.   Bruxelles, An.
      XIV (1805).   pp. 24, 8vo.

BAUMGARTEN–CRUSIUS, LUDWIG.   *See* Trismegistus, Hermes.

BAUMHAUER, EDOUARD HENRI VON.
   Obituary.   Am. J. Sci. [3], XXX, 408 (1885).
   Obituary.   Chem. News, LI, 69 (1885).

BAYEN, PIERRE.
   Biographical Notices of P. B. : Lassus, Annales de chimie, vol. XXVI,
      p. 278.   Parmentier, Recueil périodique de la société de médecine
      de Paris, No. 21.   Allg. J. Chem., Scherer., I, 331, 1798.

BAYEN, PIERRE.   Life of.   *See, in Section V*, Bayen, Pierre, Opuscules chimiques.   *Also,
      in Section III*, Longchamp.

BAYLE, PIERRE.
   Dictionnaire historique et critique.   3e édition.   Rotterdam, 1720.
      4 vols., fol.
         Contains biographies of Agricola (Geo.), Agricola (H. C.), Albertus Mag-
         nus, Averroës, Bacon (Roger), Borrichius (Olaus), Dioscorides, Leucip-
         pus, and Lucretius.

BECHER, F. L.   Life of George Agricola, *see* Agricola, Georg.

BECHER, JOHANN JOACHIM.   Life and bibliography, *see* Adelung, J. F.   Verzeichniss
      derer Schriften J. J. Becheri, *see, in Section V*, Becher, J. J. : Opuscula
      chymica rariora.

BECHER, JOHANN JOACHIM.

> BUCHER, URBAN GOTTFRIED. Das Muster eines nützlich-Belehrten in der Person Herrn Doctor Johann Joachim Bechers, nach seinen philologischen, mathematischen, physicalischen, politischen und moralischen Schrifften beurtheilet und nebst seinem Lebens-Lauff vorgestellet. Nürnberg und Altdorff, 1722. pp. [xiv]-160, 12mo. Two portraits of Becher, and folding plates.

> * ROTH-SCHOLTZ, FRIEDRICH. Joh. Joachim Bechers chymischer Rosen-Garten, samt einer Vorrede und kurtz gefassten Lebens-Beschreibung. Nürnberg, 1717. pp. 96, 12mo.
>
>> Contains, besides the life of Becher, a list of his published and his promised (!) works, 37 in number. Contains also a bibliography of the many biographies of Becher. *Cf., in Section V*, Becher, J. J.

> R., J. F. Närrische Weissheit und weise Narrheit; oder ein hundert so politische als physicalische, mechanische und mercantilische Concepten und Propositionen deren etliche gut gethan, etliche zu nichts worden. Anjetzo von neuem herausgegeben mit einem Vorbericht an den Leser darinnen erstlich von des Herrn Bechers Person nach ihren Tugenden und Lastern, und dem daraus entstandenen Glück und Unglück; Hernach von seinen Schrifften so wol insgemein als auch von gegenwärtigen Tractat insonderheit gehandelt wird A., 1707. 12mo.

BECQUEREL, EDMOND.

> Analyse succinct des travaux scientifiques de Edmond Becquerel, professeur à l'Institute national agronomique, etc. Paris, 1850. 4to.

BEER, ALEXANDER. *See* Müller, Albrecht.

BEILSTEIN, F. *See* Hübner, Hans.

BELL, JACOB.

> Obituary. Q. J. Chem. Soc. London, XIII, 167 (1861).

BELL, JAMES WILSON.

> Obituary. J. Chem. Soc. London, XXXVII, 257 (1880).

BÉRARD, A. Éloge de M. Orfila. *See* Orfila, M. J. B.

BÉRARD, E. P. Notice biographique sur Félix Leblanc. *See* Leblanc, F.

BERGMAN, TORBERN.

> AURIVILLIUS, PETRUS FABIANUS. Oratio parentalis quam in memoriam viri amplissimi et celeberrimi Torbern Bergman. . . . in publico die xv Junii 1785 habuit. Ex Suevico in Latinum sermonem translata. Lipsiæ, 1787. pp. 46, 4to.

BERGMAN, TORBERN.   [Cont'd.]
  Biography of T. B.   Chem. Annalen, Crell, 1787, I, 74–96.
  Epitaph on his tomb.   Chem. Ann., Crell, 1784, II, p. 378.
  CRELL, L.   Nachricht von dem Tode des Torbern Bergmann.   Chem.
      Ann., Crell, 1784, II, p. 379.
  HJELM, PETER JACOB.   Gedächtnissrede auf dem Herrn Torbern Berg-
      man gehalten in der königlichen Wissenschafts-Akademie zu Stock-
      holm den 3. May, 1786.   Aus dem schwedischen übersetzt.   Greifs-
      wald, 1790.   pp. 92, sm. 4to.   Medallion-portrait.
          A biography and bibliography of 106 titles.

BERTHELOT, MARCELLIN.
  Notice sur les travaux scientifiques de Berthelot, préparateur de chimie
      au Collège de France.   Paris, 1857.   4to.
  Life and portrait.   Pop. Sci. Mon., XXVII, 113, May, 1885.

BERTHIER, PIERRE.
  Biography.   Ann. des Mines, XV, 1869.
  Obituary.   Am. J. Sci. [2], XXXIII, 108, May, 1862.

BERTHOLLET, CLAUDE LOUIS.
  CUVIER, GEORGES.   Éloge de Claude Louis Berthollet.   [Paris], 1826.
      4to.
  MÜLLER.   Berthollet's Leben nach der Beschreibung von Hugh Col-
      quhoun.   [Erlangen], 1828.

BERTHOLLET, CLAUDE LOUIS.   Life of, *see* Mangin ; *also* Montyon et Franklin.

BERZELIUS, JÖNS JACOB.
  Obituary.   Am. J. Sci. [2], VI, 448, Nov., 1848.
  CANNOBIO, GIAMBATTISTA.   Elogio di Berzelius.   Prolusione al corso
      di chimica applicata alla farmacia.   Genova, 1849.   8vo.
  LOUYET, P.   Notices sur la vie et les travaux de J. J. Berzelius.
      Bruxelles, 1849.   18mo.
              Levensschets van J. J. Berzelius.   In het Nederduitsch
              overgebracht door L. Mulder.   Rotterdam, 1849.   8vo.
  MARTIUS, CARL FRIEDRICH PHILIPP VON.   Denkrede auf J. J. Berzelius.
      München, 1848.   4to.
  Minnesfest öfver J. J. Berzelius förad af Litteratur-Sällskapet i Stock-
      holm.   (Minnestal af P. A. Siljeström.)   Stockholm, 1849.   8vo.
  MULDER, GERARDUS JOHANNES.   Berzelius herdacht.   Uitgesproken
      bij de oefening der scheikundige lessen aan de Hoogeschool te
      Utrecht.   Rotterdam, 1848.   8vo.

BERZELIUS, JÖNS JACOB. [Cont'd.]

* ROSE, HEINRICH. Gedächtnissrede auf Berzelius, gehalten in der öffentlichen Sitzung der Akademie der Wissenschaften in Berlin am 3 Juli, 1851. Berlin, 1852. pp. 61, 4to.

ROSE, H. Biography of Berzelius. Am. J. Sci. [2], XVI, 1, 173, 305, Nov., 1853; XVII, 103, May, 1854.

SVANBERG, LARS FREDRIK. Biografi öfver J. J. Berzelius. [Stockholm], 1849. 8vo.

BERZELIUS, J. J., UND FRIEDRICH WÖHLER. Briefe, *see* Hjelt, E.

BESANEZ, EUGEN GORUP VON. *See* Gorup-Besanez, E. von.

BEUTHER, DAVID. SPRÖGEL, CHRISTOPH. Beuther's Person und Schrifften, *see, in Section VI*, Beuther, D., Universal und Particularia. Bibliography and life, *See* Adelung, J. F.

BEYER, CARL.

L. CLAISEN. Nekrolog und Bibliographie auf C. B Ber. d. chem. Ges., XXIV, 1117–1121 (1891).

BEYRICH, FERDINAND.

Nekrolog. Ber. d. chem. Ges., II, 781 (1869).

BIOT ET GARDEUR LE BRUN. Biography of Gay Lussac, *see* Gay Lussac.

BIRD, ALFRED.

Obituary. J. Chem. Soc., London, XXXV, 266. (1879.)

BIRNBAUM, KARL.

HOFMANN, A. W. Nekrolog auf K. Birnbaum. Ber. d. chem. Ges., XX, p. 473 (1887).

BIRNBAUM, KARL. *See* Weltzien, Karl.

BIZIO, BARTOLOMEO. Elogio del Prof. Luigi Brugnatelli. *See* Brugnatelli, Luigi.

BLACK, JOSEPH. Life of. *See* Brougham, Lord, *also* Young, Thomas.

BLACK, JOSEPH.

Nekrolog. All. J. Chem., Scherer., vol. VI, p. 98 (1801). Portrait.

THOMSON, THOMAS. Biographical Account of Joseph Black. Ann. Phil., IV, 321 (1814).

BLAIKIE, ADRIAN.

Obituary. J. Chem. Soc., London, XLVII, 330 (1885).

BLEY, LUDWIG FRANZ. Das Leben von Rudolph Brandes. *See* Brandes, Rudolph.

BLOMSTRAND, C. W.   *See* Palmstedt, Carl.   Minnesteckning öfver Carl Wilhelm Scheele.
   See Scheele, C. W.

BLOXAM, CHARLES LOUDON.
   Obituary.   J. Chem. Soc., London, LIII, 508 (1888).
   Obituary.   Chem. News, LVI, 238 and 248 (1887).

BLOXAM, THOMAS.
   Obituary.   J. Chem. Soc., London, XXVI, 773 (1873).

BLYTH, JOHN.
   Obituary.   J. Chem. Soc., London, XXV, 343 (1872).

BOË, DE LA, FRANCISCUS.   *See* Sylvius.

BOERHAAVE, HERMANN.
   * BURTON, WILLIAM.   An Account of the Life and Writings of Hermann
      Boerhaave.   In Two Parts, with an Appendix.   Second Edition.
      London, 1746.   pp. vii–[iii]–226, 12mo.   Portrait.
         First edition, 1743.   Contains bibliography of H. B.   Bound with the second
         edition : Books printed for Bernard Lintot at the Cross-Keys between
         the Temple-Gates in Fleet Street.   pp. [xxviii].
   JOHNSON, S.   Life of Hermann Boerhaave.   London, 1834.   8vo.   In
      Dutch : Amsterdam, 1836.   12mo.
   MATY, MATTHEW.   Essai sur le caractère du grand médecin, ou Éloge
      historique et critique de Hermann Boerhaave.   Cologne, 1747.   8vo.
         A German translation was published at Leipzig and Freyburg, 1748, in 8vo.
   SCHULTENS, A.   Oratio academica in memoriam Hermanni Boerhaave.
      Lugduni-Batavorum, 1738.   4to.
         Academische redevoering . . . ter gedachtenisse uit het
            latyn vertaald door Jan Jacob Schultens.   Leyden,
            1739.   4to.
         See Savérien, A.: Histoire des philosophes modernes.   Vol. VII, p. 259.
         Portrait.

BOERHAAVE, HERMANN.   *See, in Section III*, Lange, Willem de.   *See* Fontenelle,
      Éloges ; Figuier, Vie des savants ; Robertson, A.

BÖTTICHER, JOHANN FRIEDRICH.   *See* Böttger, J. F.

BÖTTGER, JOHANN FRIEDRICH.
   ENGELHARDT, CARL AUGUST.   J. F. Böttger, Erfinder des sächsischen
      Porzellans.   Biographie aus authentischen Quellen.   Nach dem
      Tode des Verfassers vollendet und herausgegeben von August
      Moritz Engelhardt.   Nebst einer kurzen Darstellung der Staats-
      Gefängnisse und merkwürdigen Staatsgefangenen in Sachsen seit
      dem sechszehnten Jahrhundert.   Leipzig.   pp. x–659, 8vo.   Por-
      trait of Böttger.

BOETTINGER, C.   *See* Kekulé, August.

BOGUSKI, J. G.   *See* Fudakowski, Hermann von B.

BOIS-REYMOND, EMIL DU.   *See* Jones, Henry Bence.   Voltaire, *see* Voltaire.

BOISSARD, JEAN JACQUES.
Icones [quinquaginta] virorum illustrium doctrina et eruditione præ-
stantium ad vivum effectæ, cum eorum vitis descriptis ; omnia
recens in æs artificiose incisa per Theod. de Bry.   4 parts in 1 vol.
Francofurti, 1597–99.   4to.
> Contains 198 finely engraved portraits.

BOLLEY, POMPEJUS ALEXANDER.
KOPP, EMIL UND SCHERR, JOHANNES.   Nekrolog auf A. P. Bolley.   Ber.
d. chem. Ges., III, p. 813 (1870).
MÜHLBERG, F.   Zur Erinnerung an P. A. Bolley.   Aarau, 1871.   4to.
Obituary.   Chem. News, XXII, 104 (1870).
WISLICENUS, JOHANNES.   Gedächtnissrede auf P. A. Bolley, am 3.
August, 1871, dem ersten Jahrestage seines Todes zur Einweihung
seines Denkmals gehalten in der Aula des schweizerischen Poly-
technikums.   Zürich, 1871.   pp. 19, 8vo.   Bust portrait.

BOLTON, HENRY CARRINGTON.   *See* Priestley, Joseph, Scientific Correspondence of.

BOLTZMAN, LUDWIG.   Skizze des Gustav Robert Kirchhoff, *see* Kirchhoff, G. R.

BONJOUR, J. F.
JACQUES BONJOUR.   Notice biographique sur J. F. B.   Chimiste, né à
Onglières, près Nozeroy (Jura), le 12 décembre, 1754, mort com-
missaire des salines de l'Est, à Dieuze, en Lorraine, le 24 février,
1811.   Lons-le-Saunier, 1853.   8vo.

BOON-MESCH, A. H. VAN DER.   Redevoering over Humphry Davy.   *See* Davy, Humphry.

BOOTH, JAMES CURTIS.
DU BOIS, PATTERSON.   In Memoriam, James Curtis Booth.   Read
before the American Philosophical Society, October 5, 1888.
Life of.   Pop. Sci. Mon., vol. XL, p. 116 (November, 1891).   Portrait.
Life and portrait.   Scientific American, June 9, 1888.

BORCH, OLAF.   *See* Borrichius, Olaus.

BORNEMAN.   Ligprædiken over Olaf Borch.   *See* Borrichius, O.

BORRI, FRANCESCO GIUSEPPI.   [*Or* Burrhi, Franciscus Joseph.]   Biographical sketch of,
in German, *see, in Section VI*, Borri, F. G.   For biography, *see, in Section
III*, Eyssenhardt, F.   For life and bibliography, *see* Adelung, J. F.
Relatio fidei actionum ac vitæ Burrhianæ.   *See, in Section VI*, Borri, F. G.

BORRICHIUS, OLAUS.

BORNEMAN. Ligprædiken over Olaf Borch. Kjøbenhavn, 1690. Fol.
KOCH, E. F. Oluf Borck en literær-historisk-biografisk skildring.
Kjøbenhavn, 1866. pp. 143, 8vo.

BORRICHIUS, OLAUS. Life of, *see the same in Section I under* "Conspectus scriptorum
chemicorum," etc. *Compare also :* Borrichius, Olaus, Dissertatio de ortu et
progressu chemiæ, *in Section III.*

BORRO, F. J. *See* Borri, Francesco Guiseppi.

BOTTO, G. D. Life of Amadeo Avogadro, *see* Avogadro, Amadeo.

BOUDET, F. Éloge de Louis Antoine Planche. *See* Planche, L. A.

BOUIS, JULES.

Notice biographique. Bull. soc. chim., vol. XLVII, xiii (1887).

BOULLAY, PIERRE FRANÇOIS GUILLAUME. Life of C. L. Cadet-Gassicourt. *See* Cadet-
Gassicourt.

BOULTON, MATTHEW. Life of, *see* Young, Thomas.

BOUSSINGAULT, JEAN BAPTISTE JOSEPH DIEUDONNÉ.

Obituary. J. Chem. Soc., London, LIII, p. 509 (1888).
Life and portrait. Pop. Sci. Mon., XXXIII, 836, Oct., 1888.

BOUVERIE, C. J. Life of T. L. Phipson, *see* Phipson, T. L.

BOWMAN, JOHN EDDOWES.

Obituary. Q. J. Chem. Soc., London, IX, 159 (1857).

BOYLE, ROBERT.

THOMPSON, CHARLES O. Robert Boyle, a study in biography. Paper
read at the Semi-Annual Meeting of the American Antiquarian
Society at Boston, April 26, 1882. Reprinted from Proceedings of
the American Antiquarian Society. Worcester, Mass., 1882. pp.
28, 8vo.

> Life of, *see* Cap, P. A. ; *also* Figuier, Vie des savants ; *also in* his Works by
> Th. Birch (London, 1744) ; *also* Robertson, A. ; *also* Hoefer, Ferdi-
> nand. *Also* Savérien : Histoire des philosophes modernes, vol. VI, p.
> 63. Portrait.

BRACONNOT, HENRI.

NICKLÈS, J. Sa vie et ses travaux. Paris [1856 ?].
> Contains a bibliography of H. B.

BRADY, HENRY BOWMAN.

Obituary. J. Chem. Soc., London, LX, 452 (1891).

BRANDE, WILLIAM THOMAS.
  Obituary.   Chem. News, vol. XIII, p. 107, March 2, 1866.
  Obituary.   Am. J. Sci. [2], XLI, 428 (1866).
  Obituary.   J. Chem. Soc., London, XIX, 509 (1866).

BRANDES, RUDOLPH.
  BLEY, LUDWIG FRANZ.   Leben und Wirken von R. B. in besonderer
    Beziehung auf seine Verdienste um die Pharmacie und dem
    Apotheker-Verein in Norddeutschland geschildert.   Hannover,
    1844. 8vo.
  DRESEL, A.   Eine Beschreibung des am 18. October, 1848, gefeierten
    Festes der Einweihung des ihm errichteten Denkmals, nebst den
    bei dieser Gelegenheit gehaltenen Reden.   Lemgo, 1849. 8vo.

BRAYLEY, EDWARD WEDLAKE.
  Obituary.   J. Chem. Soc., London, XXIII, 292 (1870).

BRAZIER, JAMES SMITH.
  Obituary.   J. Chem. Soc., London, LV, 289 (1889).
  Obituary.   Chem. News, LIX, 46 (1889).

BREWSTER, SIR DAVID.
  Life and portrait.   Pop. Sci. Mon., XXVI, 546, Feb., 1885.
  Obituary.   Chem. News, XVII, 84, Feb. 14, 1868.

BROECKX, C.   Notice sur J. I. H. Pypers, *see* Pypers, J. I. H.   Notice sur A. D. Sas-
    senius, *see* Sassenius, A. D.

BRODIE, BENJAMIN COLLINS.
  Obituary and bibliography (in part).   J. Chem. Soc., London, XXXIX,
    182 (1881).
  Obituary.   Am. J. Sci. [3], XXI, 86 (1881).

BROUGH, JOHN CARGILL.
  Obituary.   J. Chem. Soc., London, XXVI, 774 (1873).

BROUGHAM, HENRY, LORD.
  Lives of Philosophers of the Time of George III.   London and Glas-
    gow, 1855. pp. XIII–492, 8vo.
        Contains biographies of Black, Watt, Priestley, Cavendish, Davy, Lavoisier,
        Sir Joseph Banks, and D'Alembert.

BROWN, EDWIN ORMOND.
  Obituary.   J. Chem. Soc., London, XLIX, 344 (1886).

BRUCE, A. CAMERON.
  Obituary.   J. Chem. Soc., London, XXXIX, 189 (1881).

BRÜNING, ADOLF VON.
  HOFMANN, A. W. Nachricht von dem Tode des Adolf von Brüning.
    Ber. d. chem. Ges., XVII, p. 949 (1884).

BRUGNATELLI, LUIGI VINCENZO.
  BIZIO, BARTOLOMEO. Elogio del professore Luigi Brugnatelli, letto
    all' ateneo di Venezia il giorno 19 di Luglio, 1827. Venezia, 1832.
    pp. 33, 8vo.
  CATTANEO, ANTONIO. Cenni sulla vita di Luigi Vincenzo Brugnatelli.
    Milano, 1836, 4to. Portrait.
  COSSA, A. Cenni sulla vita e sugli scritti di L. V. B. Pavia, 1857. 8vo.
        Contains a complete bibliography of the writings of Brugnatelli.

BRUSH, GEORGE JARVIS.
  Life and portrait. Pop. Sci. Mon., XX, 117, Nov., 1881.

BUCHER, URBAN GOTTFRIED. Life of J. J. Becher, see Becher, J. J.

BUCHKA, K. See Leuckart, Rudolph.

BUCHNER, JOHANN ANDREAS.
  Life of, Repert. f. Pharm., vol. V (1818).

BUCHHOLZ, WILHELM HEINRICH SEBASTIAN.
  Nekrolog. All. J. Chem., Scherer, vol. II, p. 591 (1799).
  TROMMSDORFF, J. B. Nekrolog und Bibliographie. J. der Pharm., XVI,
    2 (1799).

BUFF, HEINRICH LUDWIG.
  KRAUT. Nekrolog auf H. L. Buff. Ber. d. chem. Ges., VI, p. 688
    (1873). Portrait of Buff.

BUFF, HEINRICH.
  HOFMANN, A. W. Nekrolog auf H. Buff. Ber. d. chem. Ges., XII, p. 1
    (1879).

BULK, CARL.
  LIEBERMANN, C. Nachricht von dem Tode des C. Bulk. Ber. d.
    chem. Ges., p. 2,099 (1886).

BUNSEN, ROBERT WILHELM.
  Life and portrait of, Pop. Sci. Mon., XIX, 550, Aug., 1881.
  ROSCOE, SIR HENRY E. Sketch of R. W. B. With portrait. Nature,
    XXIII, 597 (1881).

BURGESS, WILLIAM ROSCOE. Life of Faraday, see Faraday, Michael.

BURR, THOMAS WILLIAM.
Obituary. J. Chem. Soc., London, XXVIII, 1,313 (1875).

BURRHI, FRANCISCUS JOSEPH. *See* Borri, F. G.

BURTON, COSMO INNES.
Obituary. J. Chem. Soc., London, LX, 453 (1891).

BURTON, W. Life of Hermann Boerhaave, *see* Boerhaave, H.

BUSSY, ANTOINE ALEXANDER BRUTUS.
Notice sur les principaux travaux chimiques de M. Bussy, ancien élève de l'école polytechnique, professeur de chimie, directeur de l'école de pharmacie de Paris. Paris, 1850. 4to.

BUTLEROW, ALEXANDER MIKHAÏLOVITCH.
Obituary. J. Chem. Soc., London, LI, 472 (1887).

BUTLEROW, A. M. *See* Fritzsche, Carl Julius.

BUZAREINQUES, LOUIS CHARLES FRANÇOIS GIROU DE.
DUVAL, J. Notice biographique sur de B. avec tableau chronologique des écrits. Batignolles, 1858. 8vo. Biographies Aveyronnaises.

CABANY, SAINT MAURICE. Notice sur J. G. Dizé. *See* Dizé, J. G.

CADET, C. L. Life of Antoine Baumé, *see* Baumé, Antoine.

CADET-GASSICOURT, CHARLES LOUIS.
PARISET, E. Éloge de Charles-Louis Cadet-Gassicourt. Histoire des membres de l'Académie de médecine. Paris, 1850. Vol. I, p. 130.
SALVERTE, A. J. E. BACONNIÈRE. Notice sur la vie et les ouvrages de Charles Louis Cadet-Gassicourt. Paris, 1822. 8vo.
VIREY, J. J. Notice sur la vie et les travaux de Charles Louis Cadet-Gassicourt. Paris, *n. d.* [1823?]. 8vo.

CADET-GASSICOURT, LOUIS-CLAUDE.
Nekrolog. All. J. Chem., Scherer. Vol. VII, p. 488. 1801. With bibliography.
BOULLAY, PIERRE FRANÇOIS GUILLAUME. Notice historique sur la vie et les travaux de Louis Claude Cadet-Gassicourt. Paris. An XIV [1806], 8vo.

CADET-GASSICOURT, F. Éloge de P. A. Reymond, *see* Reymond, P. A.

CAHOURS, AUGUSTE.
Necrology. Revue générale des sciences pures et appliquées. II, No. 7. 1891.

CAILLAU, J. M.   Mémoire sur J. B. Van Helmont.   *See* Van Helmont, J. B.

CALLISEN, A. C. P.   Life and bibliography of J. B. Trommsdorff.   *See* Trommsdorff,
J. B.

CALVERT, FREDERICK CRACE.
Bibliography of F. C. C.   Chem. News, XXXI, 56 (1875).
Obituary..   Chem. News, vol. XXVIII, p. 224 (1873).
Obituary.   J. Chem. Soc., London, XXVII, 1198 (1874).
SELL, EUG.   Nekrolog auf C. Calvert.   Ber. d. chem. Ges., VI. p.
1587 (1873).

CAMPBELL, DUGALD.
Obituary.   Chem. News, XLV, 232 (1882).
Obituary.   J. Chem. Soc., London, XLIII, 252 (1883).

CAMPBELL, W. R.   Life of William Ripley Nichols.   *See* Nichols, William Ripley.

CANNIZZARO, STANISLAO.   Vita e opere di Raffaele Piria.   *See* Piria, Raffaele.

CANNOBIO, GIAMBATTISTA.   Elogio di Berzelius, *see* Berzelius, J. J.

CAP, PAUL ANTOINE.
Études biographiques pour servir a l'histoire des sciences.   Première
série; chimistes, naturalistes.   Paris, 1857.   pp. vi–408, 12mo.
> Contains biographies of Paracelse, Bernard Palissy, Pierre Belon, Nic. Houël,
> Van Helmont, Moïse Charas, Robert Boyle, Nic. Lémery, Rouelle ainé,
> van Mons, Labarraque, Bernard Courtois, Al. Dupasquier, Benj. Deles-
> sert, Bonafous, and Séba.   Only two or three memoirs are followed by
> bibliographies.

CAP, PAUL ANTOINE.   Scheele, chimiste Suédois, *see* Scheele, Carl Wilhelm.   Vie de
Rouelle ainé, *see* Rouelle, Guillaume François.

CARIUS, LUDWIG.
LADENBURG.   Nekrolog auf L. Carius.   Ber. d. chem. Ges., IX, p.
1,996 (1876).   Portrait of Carius in X (1877).

CARNELLEY, THOMAS,
Obituary and bibliography.   J. Chem. Soc., London, LX, 455 (1891).
Also, Nature, Sept. 25, 1890.

CARNOT, LAZARE NICOLAS MARGUERITE.   Life of, *see* Arago, François.

CARO, H.   *See* Griess, Peter.

CARPENTER, WILLIAM LANT.
Obituary.   J. Chem. Soc., London, LX, 461 (1891).

CARPENTIER, PAUL. Life of Daguerre, *see* Daguerre, L. J. M.

CARSON, JOS. Memoirs of James B. Rogers, *see* Rogers, J. B.

CARSTANJEN. *See* Mey, Max.

CASALI, ADOLFO. Cenni biografici sui più distinti cultori della chimica. pp. 29. *See, in Section II,* Casali, Dizionario, etc.

CATTANEO, ANTONIO. Vita di L. V. Brugnatelli, *see* Brugnatelli. Vita di A. F. Fourcroy, *see* Fourcroy, A. F.

CAVENDISH, HENRY.
  CUVIER, GEORGES. Éloge de Cavendish. Mém. Acad. Sciences, 1 s., XII, cxxvi–cxliv (1811).
  * WILSON, GEORGE. The Life of the Honourable Henry Cavendish, including abstracts of his more important scientific papers, and a critical inquiry into the claims of all the alleged discoverers of the composition of water. London. For the Cavendish Society, 1851. 478 pp., 8vo. Portrait of H. C.

CAVENDISH, HENRY. *See* Brougham, Lord ; Young, Thomas, Biographies ; Walker, Wm.

CHABANEAU.
  Notice sur C., chimiste perigourdin. Par Jules Delanou. Perigueux, 1857. 8vo.

CHABOT, PHILIP JAMES.
  Obituary. J. Chem. Soc., London, XXI, p. xxxiv (1868).

CHANCEL, G.
  FORCRAND, R. DE. Notice sur la vie et les travaux de G. C. par R. de F. Bull. Soc. chim., [3] v, 1891. Bibliography and portrait.

CHANDLER, CHARLES F.
  Life and Portrait. Pop. Sci. Mon., XVI, 833, April, 1880.
  President Chandler and the New York City Health Department, 1866–1883. Reprint from The Sanitary Engineer of May 17, 1883. 15 pp., sm. 12mo.

CHAPMAN, ERNEST THEOPHRON.
  Obituary. J. Chem. Soc., London, XXVI, 775 (1873).
  Obituary. Chem. News, XXVI, p. 10 (1872).

CHAPTAL, JEAN ANTOINE (DE CHANTELOUP).
  FEUCHTWANGER, LOUIS (transl.) Necrology of J. A. C. Am. J. Sci., XXVI, 127, July, 1834.

CHAPTAL, JEAN ANTOINE (DE CHANTELOUP).   [Cont'd.]
FLOURENS, MARIE J. P.   Éloge historique de Jean Antoine Chaptal.
Liste des ouvrages de Chaptal.   Mém. Acad. Sciences.   2 S. xv,
i–xliii (1838).

CHAPTAL, JEAN ANTOINE.   Life and Portrait, *see* Montyon et Franklin.

CHARAS, MOÏSE.   Life of.   *See* Cap, P. A.

CHARLES, ÉMILE.   Roger Bacon, life and works, *see* Bacon, Roger.

CHENEVIX.
Obituary.   Am. J. Sci., xx, 305, July, 1831.

CHEVALIER, ULYSSE.
Répertoire des sources historiques du moyen âge.   Tome premier.   Bio-
bibliographie.   Paris, 1877–83.   Roy. 8vo, pp. xx–columns 2,369.
Complément–Supplément.   Paris, 1888.   Columns 2,372–2,846.
Contains extensive and valuable lists of biographies.

CHEVALLIER, A.
Énumération des titres et travaux scientifiques de A. C.   Paris, 1862.
8vo.

CHEVALIIER, JEAN BAPTISTE ALPHONSE.   Notice de Louis Nicolas Vauquelin, *see*
Vauquelin, L. N.

CHEVREUL, M. E.   Biography of J. J. Ebelman.   *See* Ebelman, J. J.

CHEVREUL, MICHEL EUGÈNE.
Life and Portrait.   Pop. Sci. Mon., xxvii, 548, Aug., 1885.
HOFMANN, A. W.   Nekrolog auf M. E. Chevreul.   Ber. d. chem. Ges.,
xxii, p. 1163 (1889).
Obituary.   Chem. News, lix, 179 (1889).
Obituary.   J. Chem. Soc., London, lvii, 445 (1890).
*MALLOIZEL, GODEFROY.   Œuvres scientifiques de Michel Eugène
Chevreul, doyen des étudiants de France, 1806–1886.   Avec une
introduction de J. Desnoyers et une préface de Charles Brongniart.
Paris, 1886.   pp. 298, 8vo.
With a portrait of Chevreul and a full bibliography of his writings, etc.
Célébration du centenaire de M. Chevreul.   31 Août 1786, 31 Août,
1886.   Rouen, 1886.   56 pp.   With heliotype of medal.
FAVERI, S. DE.   Chevreul e la chimica del suo tempo.   Terza edizione.
Roma, 1891.   16mo.
Hommage à Chevreul le 31 août (Mémoires de Berthelot, Demarçay,
Gautier, etc.).   Paris, 1886.   4to.

CHILTON, JAMES RENWICK.
Obituary. Am. J. Sci. [2], XXXVI, 314, Nov., 1863.

CHRISTIANI, ARTHUR.
MARTIUS, C. A. Nekrolog auf A. Christiani. Ber. d. chem. Ges., XX,
p. 3271 (1887).

CLAISEN, L. *See* Beyer, Carl.

CLAPHAM, ROBERT CALVERT.
Obituary. Chem. News, XLV, 21 (1882).
Obituary. J. Chem. Soc., London, XLI, 236 (1882).

CLARK, THOMAS.
Obituary. Chem. News, XVI, 292, Dec. 6, 1867.
Obituary. J. Chem. Soc., London, XXI, viii (1868).

CLAUDERUS, GABRIEL. Life, by his nephew, G. F. Clauderus, *see, in Section V,* Clauderus, D. G. Clauderi, schediasma, *etc.* 1736.

CLEMENTS, GEORGE WILLIAM HOLLIDAY.
Obituary. J. Chem. Soc., London, XLIX, 345 (1886).

CLEVE, P. T. Minnesblad öfver Carl Wilhelm Scheele, *see* Scheele, C. W.

COLEMAN, JOSEPH JAMES.
Obituary. J. Chem. Soc., London, LV, 290 (1889).

CONDORCET, MARIE J. A. N. C. Life of, *see* Arago, François.

CONINGTON, FRANCIS THIRKILL.
Obituary. J. Chem. Soc., XVII, 435 (1864).

COOKE, JOSIAH PARSONS, JR. *See* Rogers, William Barton.

COOKE, JOSIAH PARSONS, JR.
Life and portrait. Pop. Sci. Mon., X, 491, Feb., 1877.

COOPER, JOHN THOMAS.
Obituary. Q. J. Chem. Soc., London, VIII, 109 (1856).

CORRY, JOHN. Life of Joseph Priestley, *see* Priestley, Joseph.

COURTOIS, BERNARD. Life of, *see* Cap, P. A.

COWPER, CHARLES.
Obituary. Q. J. Chem. Soc., London, XIV, 350 (1862).

COWPER, RICHARD.
   Obituary.   J. Chem. Soc., London, LI, 473 (1887).

CRELL, LORENZ.   *See* Marggraf, Andreas Sigismund.

CROOKES, WILLIAM.
   Life and portrait.   Pop. Sci. Mon., X, 739, Apr., 1877.
   Sketch of his life, with portrait.   The Electrician, XXVI, p. **322**, Jan.
      16, 1891.

CRUM, WALTER.
   Obituary.   Chem. News, XV, 242, May 10, 1867.
   Obituary.   J. Chem. Soc., London, XXI, xvii (1868).

CUVIER, GEORGES.   Éloge de Fourcroy, *see* Fourcroy, A. F.   Éloge de Cavendish, *see*
      Cavendish, Henry.   Éloge de Davy, *see* Davy, H. ; Notice d'Arcet, *see*
      Darcet, Jean ; Éloge de Duhamel, *see* Duhamel, J. P. F. G.

CUVIER, GEORGES.
   Recueil des éloges historiques lus dans les séances publiques de
      l'Institut de France.   Paris, 1861 (Nouvelle edition).   3 vols., 8vo.
         Vol. I contains biographies of Priestley, Fourcroy, Cavendish ; II, Rumford,
            Duhamel, Berthollet ; III, Davy, Vauquelin.

CZÓGLER, ALAJOS.
   A fizika története életrajzokban.   Budapest, 1882.   2 vols., with 23
      portraits.
         Contains biographies of : *Leonardo, Kopernicus*, Porta, *Galilei, Kepler*, Gil-
            bert, Stevin, Snell, Gassendi, *Descartes*, Torricelli, Borelli, Grimaldi,
            Pascal, *Guericke*, Boyle, Mariotte, Amontons, Hooke, *Huyghens,
            Newton*, Halley, Bradley, Bernouilli, D., *Watt, Franklin*, Coulomb,
            Volta, Chladni, Lavoisier, Laplace, Rumford, *Gay Lussac*, Davy,
            Wollaston, Dulong, *Young*, Malus, *Biot, Gauss, Arago, Fresnel*,
            Brewster, Oersted, *Ampère*, Seebeck, Melloni, Ohm, *Faraday, Robert
            Mayer*.   Portraits accompanying the biographies of those whose names
            are italicized.

DAGUERRE, L. J. M.   *See, in Section III*, Schiendl, C., Geschichte der Photographie.

DAGUERRE, LOUIS JACQUES MANDÉ.
   CARPENTIER, PAUL.   Notice sur Daguerre, peintre, inventeur du
      diorama, [etc.].   Paris, 1855.   pp. 24, 8vo.   With bust-portrait.
      Extract from : Annales de la Société libre des Beaux Arts, vol.
      XVIII.

DALE, JOHN.
   Obituary.   J. Chem. Soc., London, XXV, 344 (1872).
   Obituary.   J. Chem. Soc., London, LVII, 446 (1890).

DALTON, JOHN. *See, in Section V,* Ostwald's Klassiker. Life of, *see* Walker, Wm. Jr.

DALTON, JOHN.

> CLAY, CHARLES. A Reminiscence of Dr. Dalton. A Paper read before the Manchester Literary and Philosophical Society. April 15, 1884. Chem. News, L, 59 (1884).

> * HENRY, WILLIAM CHARLES. Memoirs of the Life and Scientific Researches of John Dalton. London, 1854. (Printed for the Cavendish Society.) pp. xii–249, 8vo.

>> With a bust-portrait, fac-similes, and a bibliography.

> ROSCOE, HENRY E. John Dalton and his Atomic Theory. A Lecture delivered in the Hulme Town Hall, Manchester, on Wednesday, November 4, 1874. Science Lectures for the People. Manchester, Sixth Series, 1874. *n. d.*

> * SMITH, ROBERT ANGUS. Memoir of John Dalton and History of the Atomic Theory up to his Time. London, 1856. pp. ix–316, 8vo. Portrait.

> The Worthies of Cumberland. John Dalton ; by Henry Lonsdale. London, 1874

DALTON, JOHN C. Galen and Paracelsus, *see* Paracelsus.

DANIELL, J. F.

> Obituary. Am. J. Sci. [2.], II, 145, Nov., 1846.

DARCET, JEAN.

> DIZÉ, J. G. Précis historique sur la vie et les travaux de Jean D'Arcet. Paris, An x [1802], 8vo.

> CUVIER, GEORGES. Notice historique sur Jean d'Arcet. Mémoires de l'Acad. des Sciences, IV, 1 S. Hist. (74–88). An XI [1803].

DAUBENY, CHARLES GILES BRIDLE.

> Obituary. J. Chem. Soc., London, XXI, xviii (1868).
> Obituary. Chem. News, XVI, 316, Dec. 20, 1867.
> Obituary. Am. J. Sci. [2], XLV, 124, 272 (1868).

DAUBRÉE, A.

> Notices des travaux de M. Daubrée, doyen de la Faculté des sciences de Strasbourg, etc. Paris, 1857. 4to.

DAVANNE, A. Nicéphore Niepce, *see* Niepce, Nicéphore.

DAVY, EDMUND.

> Obituary. Q. J. Chem. Soc., London, XI, 184 (1859)

DAVY, SIR HUMPHRY.

Necrology of Sir H. D.   Am. J. Sci., XVII, 157, January, 1830.   Portrait.

BOON MÉSCH, A. H. VAN DER.   Redevoering over Humphry Davy, den gelukkigen toepasser zijner wetenschap op de belangen der maatschappij: en eenige aanteekeningen.   Leiden, 1837.   8vo.

CRUSE, W.   Sir Humphry Davy.   Königsberg, 1841.

CUVIER, GEORGES.   Éloge de Sir Humphry Davy.   Mém. Acad. Sciences. 2 S. XII, i–xxxviii (1833).

Éloge de Sir Humphry Davy.   [Paris], 1830.   4to.

*DAVY, JOHN.   Memoirs of the Life of Sir H. D.   London, 1836.   2 vols., 8vo, I, pp. xii–507 ; II, vii–419.   Portrait.

> *Also in the* Collected Works of Sir H. Davy, *see, in Section V.*

> NEUBERT, CARL.   Denkwürdigkeiten aus dem Leben Sir Humphry Davy herausgegeben von seinem Bruder John Davy, Deutsch bearbeitet von C. N.   Eingeleitet von Rudolph Wagner.   Leipzig, 1840.   2 vols., 12mo. Vol. I, pp. xxvi–368.   Vol. II, pp. 266.

*DAVY, JOHN.   Fragmentary Remains, literary and scientific, of Sir Humphry Davy, with a Sketch of his Life and selections from his correspondence.   London, 1858.   pp. viii–330, 8vo.

> Contains a bibliography of Davy.

FERGUSON, JOHN.   Sir Humphry Davy.   Good Words for 1879. London, 1879.   pp. 112, 185, and 304, roy. 8vo.

> A biographical sketch illustrated with portrait and views.

*PARIS, JOHN AYRTON.   The Life of Sir Humphry Davy.   London, 1831.   pp. xv–547, 4to.   Portrait.

> This contains also " A Sketch of the History of Chemical Science, with a view to exhibit the revolutions produced in its doctrines by the discoveries of Sir H. D.   Also a bibliography of Davy's publications.

The Centenary of the Birth of Sir H. D.   December 17, 1878.   Chem. News, XXXVIII, 289 (1878).

Life and portrait, Pop. Sci. Mon., XIV, 813, Apr., 1879.

DAVY, SIR HUMPHRY.   Life, *see* Brougham, Lord ; Walker, Wm., Jr.; Brown, Samuel, *in Section III ;* Hoefer, Ferdinand ; Montyon et Franklin.

DEACON, HENRY.

Obituary.   J. Chem. Soc.   London, XXXI, 494 (1877).

DEBRAY, JULES HENRI.

Obituary.   Chem. News, LVIII, 71 (1888).

Obituary.   J. Chem. Soc.   London, LV, 291 (1889).

DEE, JOHN.

BAILEY, JOHN EGLINGTON. Diary for the years 1595 to 1601 of Dr. John Dee, Warden of Manchester from 1595 to 1608. Edited from the Original MSS. in the Bodleian Library. London, 1880. 4to. Portrait.

> Only 20 copies printed. Not published.

DISRAELI, I. The occult Philosopher, Dr. Dee. *In* Amenities of Literature, vol. II.

ROBY, J. Biography and legends of J. D. *In* J. Roby's Traditions of Lancashire, vol. I. London, 1829. 12mo.

* The Private Diary of Dr. John Dee and the Catalogue of his Library of Manuscripts, from the original manuscripts in the Ashmolean Museum at Oxford, and Trinity College Library, Cambridge. Edited by James Orchard Halliwell. London, 1842. pp. viii–102–35, square 8vo. Printed for the Camden Society.

> Not properly chemical, but very curious, and illustrating the magical and alchemical phase of chemistry. The catalogue of MSS. contains 224 titles.

DEIMAN, J. R.

Portrait in All. J. Chem., Scherer., vol. VII, frontispiece, 1801.

DELANOU, JULES. Notice sur Chabaneau, *see* Chabaneau.

DE LA RIVE, ARTHUR-AUGUSTE.

DUMAS, J. B. Éloge sur A. de la Rive. Discours et éloges académiques. Paris, 1885. Vol. I, p. 252.

Obituary. Am. J. Sci. [3], xv, 160 (1878).

SORET, J. L. Auguste de la Rive. Genève, 1877. Portrait.

DE LA RIVE, A. Life of Michael Faraday, *see* Faraday, Michael.

DE LA RUE, THOMAS.

Obituary. J. Chem. Soc., London, xx, 387 (1867).

DE LA RUE, WARREN.

Obituary. J. Chem, Soc., London, LVII, 441 (1890).

HOFMANN, A. W. Nekrolog auf Warren de la Rue. Ber. d. chem. Ges., XXII, p. 1169 (1889).

DELISLE AND ALUYS. Life and bibliography, *see* Adelung, J. F.

DEMOCRITUS AND SYNESIUS.

FERGUSON, JOHN. On the First Editions of the Chemical Writings of Democritus and Synesius. Read before the Society 19th Novem-

DEMOCRITUS AND SYNESIUS. [Cont'd.]
ber, 1884. [Reprinted from the Transactions of the Philosophical Society of Glasgow.] [Glasgow, 1884.] pp. 11, 8vo.
POSTSCRIPT. Glasgow, May 8, 1885. pp. 2, 8vo.

DESCLOSIÈRES, GABRIEL.
Biographie des grands inventeurs dans les sciences et l'industrie, donnant un aperçu de l'histoire de leurs développements par le récit de la vie des hommes illustres qui en ont assuré le progrès. Ouvrage indiqué par le Ministre de l'Instruction publique pour les Bibliothèques libres et communales. Adopté par le Ministre de l'Intérieur pour les Établissements pénitentiaires. Par la ville de Paris pour être donné en prix dans les Écoles ayant obtenu en 1882 une mention honorable de l'Académie française. Sixième édition. Paris, *n. d.* 2 vols., 12mo. Ill.

> Signature 9 of vol. 1 treats of chemistry. The author says of Priestley: It was in the sumptuous residence of the noble lord [Shelburne] that the celebrated chemist died in 1804!

DESSAIGNES, A. R.
DESSAIGNES, V. Travaux de chimie organique. Précédés d'une notice biographique d'A. R. Dessaignes. Vendôme, 1886. 8vo.

DESSAIGNES, VICTOR.
Obituary. J. Chem. Soc., London, XLI, 236 (1882). *Also*, XLVII, 309 (1885).

DEYMANN. Éloge de Lavoisier, *see* Lavoisier, A. L.

DEYEUX, NICOLAS. Notice sur Antoine Baumé, *see* Baumé, Antoine.

DIRCKS, HENRY.
Obituary. J. Chem. Soc., London, XXVIII, 1314 (1875).

DISRAELI, I. Life of Dr. Dee, *see* Dee, John.

DITTMAR, WILLIAM.
Necrology. Nature, XLV, 493, Mch., 24, 1892.

DIZÉ, J. G. Vie de Jean D'Arcet, *see* Darcet, J.

DIZÉ, J. G.
SAINT MAURICE CABANY. Notice biograpique sur J. G. Dizé. Deuxième édition. Paris, 1845.

DÖBEREINER, JOHANN WOLFGANG.
VOGEL, AUGUST. Denkrede auf J. W. D., gehalten in der öffentlichen

DÖBEREINER, JOHANN WOLFGANG. [Cont'd.]
Sitzung der königlichen bayerischen Akademie der Wissenschaften am 27 November, 1849. München, 1849. 4to.
Obituary. Am. J. Sci. [2], VIII, 450, Nov., 1849.

DOLLFUS-AUSSET.
Matériaux pour la coloration des étoffes. Paris, 1865. 2 vols., roy. 8vo.
Vol. I is essentially an alphabetical dictionary of the biographies and writings of authors on dyeing.

DONKIN, W. F.
Obituary. J. Chem. Soc., London, LV, 292 (1889).

DORTOUS DE MAIRAN. Éloge de Nicolas Lémery, see Lémery, Nicolas.

DOWSON, EDWARD.
Obituary. J. Chem. Soc., London, XXXVII, 258 (1880).

DRAPER, HENRY.
Obituary. Am. J. Sci., XXIV, 482 (1882).

DRAPER, JOHN WILLIAM.
Life. Biographical Notices of the National Academy of Sciences. Washington City, 1886. Vol. II, p. 349.
Life and portrait. Pop. Sci. Mon., IV, 361, Jan., 1874.
Obituary. Am. J. Sci. [3], XXIII, 163, (1882).
LOVERING, JOSEPH. Necrology of John William Draper. Proceedings Am. Acad. Arts and Sci. Boston, 1882.
MARTIN, BENJAMIN N. A Sketch of John William Draper. Magazine of American History, April, 1882.

DREBBEL VAN ALKMAR, CORNELIS. Kort verhaal van het leven des beroemden natuurkenners Cornelis Drebbel. See, in Section VI, Drebbel v. A., C. Life and bibliography, see Adelung, J. F.

DRESEL, A. [Erinnerungen an] Rudolph Brandes, see Brandes, R.

DUBOIS, E. F. Éloge de M. Orfila. see Orfila, M. J. B.

DUBOIS, FRANÇOIS. See Sylvius.

DUCLAUX, E.
Notice sur les travaux scientifiques de E. Duclaux. Sceaux, 1884. 4to.

DUFFY, PATRICK.
J. Chem. Soc., London, LIII, 513 (1888).

DUHAMEL, J. P. F. GUILLOT.
CUVIER, GEORGES. Éloge historique de Duhamel. Mém. Acad. Sciences. 2 S., VI, clx–clxxvi (1823).

DUMAS, JEAN BAPTISTE ANDRÉ.

Discours et éloges académiques.   Paris, 1884.   2 vols., 8vo.

Notice of the éloges of Dumas.   Nature, XXX, 15 (1884).

Obituary.   Chem. News, XLIX, 193 (1884).

Obituary.   J. Chem. Soc., London, XLVII, 310 (1885).

Life and portrait.   Pop. Sci. Mon., XVIII, 257, Dec., 1880.

COOKE, J. P.   J. B. A. Dumas.   Am. J. Sci. [3], XXVIII, 289 (1884).

HOFMANN, A. W.   Zur Erinnerung an Jean Baptiste André Dumas.
Berlin, 1885.   8vo.   Portrait.

MAINDRON, ERNST.   L'œuvre de J. B. Dumas, avec une introduction
par Schützenberger, accompagné d'un portrait de Dumas.   Paris,
1886.   8vo.

PASTEUR, LOUIS.   M. Darboux.   Inauguration de la statue de Jean
Baptiste Dumas à Alais le 21 octobre, 1889.   Paris, 1889.   pp. 73,
4to.

DUMAS, JEAN BAPTISTE.   Éloge de Michel Faraday, *see* Faraday, Michael.   Éloge de
Balard, *see* Balard, A. J.   Éloge de Pelouze, *see* Pelouze, Th. J.   *See* Piria,
Raffaele.   Obituary, *see* Voit, C. von.

DUPASQUIER, ALPHONSE.

C. E.   Notice de la vie et les travaux de Dupasquier.   Lyon, 1849.   8vo.

BONNOT, AMÉDÉE.   Éloge de A. D.   Lyon, 1849.   8vo.

DUPPA, BALDWIN FRANCIS.

Obituary.   J. Chem. Soc., London, XXVII, 1199 (1874).

SELL, EUGEN.   Nekrolog auf B. F. Duppa.   Ber. d. chem. Ges., VII,
p. 1588 (1873).

DUVAL, J.   Notice de Buzareingues, G. de, *see* Buzareingues, G. de.

DUVAL, ROBERT [ROBERTUS VALLENSIS].

*FERGUSON, JOHN.   The First History of Chemistry.   [Read before
the (Philosophical) Society (of Glasgow), 6th January, 1886.]   pp.
17, 8vo.

Discusses the little treatise of Robert Duval from the bibliographic point of
view.   *See, in Section III*, Duval, Robert.

EBELMEN, JACQUES JOSEPH.

Notice sur les travaux scientifiques de M. E.   [Paris, 1851.]   pp. 24, 4to.

CHEVREUL, M. E.   Biography and bibliography of J. J. E.   *In* Ebelmen's Chimie céramique [etc.], *q. v. in Section V*.

SALVÉTAT.   Recueil des travaux scientifiques de Ebelmen, revu et
corrigé et suivi d'une notice sur E. et sur ses travaux par M. E.
Chevreul.   Paris, 1861.   3 vols., 8vo.   Folding plates.

Contains a biography and bibliography of Ebelman.

EGGERTZ, VICTOR.
   Obituary. Chem. News, LX., 119 (1889).

ELLIS, GEORGE E. Life of Count Rumford, *see* Thompson, Benjamin.

ELTOFT, THOMAS.
   Obituary. J. Chem. Soc., London, XXXIX, 189 (1881).

EMICH, F. *See* Maly, Richard.

ENGELBACH, THEOPHIL.
   OPPENHEIM, A. Nekrolog auf Th. Engelbach. Ber. d. chem. Ges.,
      x, p. 917 (1877).

ENGELHARDT, C. A. Life of J. F. Böttger. *See* Böttger, J. F.

ERDMANN, OTTO LINNÉ.
   Obituary. Am. J. Sci. [2], XLIX, 144 (1870).
   Obituary. J. Chem. Soc., London, XXIII, 306 (1870).
   KOLBE, H. Nekrolog auf O. L. Erdmann. Ber. d. chem. Ges., III,
      p. 374 (1870).

ERLENMEYER, EMIL. Liebig und die reine Chemie, *see* Liebig, Justus von.

ETTI, CARL.
   WEGSCHEIDER, R. Nekrolog auf Carl Etti. Ber. d. chem. Ges., XXIII,
      p. 910 (1890).

EUSÈBE-GRIS.
   Notice biographique et nécrologique sur Eusèbe-Gris, ancien professeur
      de chimie, membre de plusieurs sociétés savantes. Châtillon-sur-
      Seine, 1849. 8vo.
         From Annales de la Société d'Horticulture de France.

EVANS, HENRY SUGDEN.
   Obituary. J. Chem. Soc., London, LI, 475 (1887).

FABRI-SCARPELLINI, ERASMO. Sopra i lavori chimiche de Pietro Peretti. *see* Peretti, Pietro.

FARADAY, MICHAEL.
   Obituary. J. Chem. Soc., London, XXI, p. xxi (1868).
   Obituary. Chem. News, XVI, 110, August 30, 1867
   Obituary. Am. J. Sci. [2], XLIV, 293 (1867).
   Sketch of M. F. ; Scientific Worthies. Nature, VIII, 397 (1873). With
      portrait.
   BURGESS, WILLIAM ROSCOE. Michael Faraday [A biographical sketch],
      London [1877]. 12mo.

FARADAY, MICHAEL.   [Cont'd.]

DE LA RIVE, A.   The life and works of M. Faraday.   Am. J. Sci. [2],
    XLV, 145 (1868).

His Life and Works.   Translated from Bibliothèque Universelle,
    October 25, 1867, Archives des Sciences, pp. 131–176.   Annual
    Report of the Smithsonian Institution for 1867.   Washington, 1868.
    pp. 227–245, 8vo.

DUMAS, JEAN BAPTISTE.   Éloge sur M. F.   Discours et éloges aca-
    démiques, vol. I, p. 49.

Discours prononcé à la mémoire de Faraday, devant la Société chimique
    de Londres, le jeudi 17 juin, 1869.   Bull. soc. chim., XII, p. 172.

Éloge historique de Michel Faraday.   Mém. Acad. Sciences.   2 S.,
    XXXVI, vii–lxiv (1870).

GEIKIE, JOHN CUNNINGHAM.   Michael Faraday and Sir David Brewster,
    philosophers and Christians : Lessons from their lives.   London,
    [1868].   8vo.

GLADSTONE, JOHN HALL.   Michael Faraday [a biography].   London,
    1872.   12mo.

    For a review of this work, by W. F. Barrett, see Nature, VI, 411 (1872).

The Life of Faraday.   A lecture. . . . Science Lectures.   Series 4.
    1866.   8vo

JONES, HENRY BENCE.   The Life and Letters of Michael Faraday.
    London, 1870.   2 vols., 8vo.

    For a review of this work, by J. H. Gladstone, see Nature, I, 401 (1870).

MARTIN, SAMUEL.   Michael Faraday : philosopher and Christian.   A
    Lecture.   London, 1867.   8vo.

Memorial to Faraday.   Proceedings at a public meeting at the Royal
    Institution. . . . June 21, 1869.   [London, 1869.]   8vo.

* TYNDALL, JOHN.   Faraday as a Discoverer.   London and New York,
    1868.   pp. viii–171, 12mo.   Two portraits of Faraday.

    The same in French, transl. by Abbé Moigno.   Paris, 1868.   12mo.

On Faraday as a discoverer.   Am. J. Sci. [2], XLVI, 34, 180.   1868.

FAVRE, P. A.

Travaux scientifiques de P. A. Favre, professeur de chimie à la faculté
    des sciences de Marseille.   Paris, 1857.   4to.

FEHLING, HERMANN VON.

Obituary.   J. Chem. Soc., London, XLIX, 346 (1886).

HÖFMANN, A. W.   Nekrolog auf H. von Fehling.   Ber. d. chem. Ges.,
    XVIII, p. 1811 (1885).   Portrait of H. v. F.

FERGUSON, JOHN.   *See* Paracelsus ; Democritus and Synesius ; Albertus Magnus ; Duval,
    Robert ; Davy, Sir H.

FIELD, FREDERIC.
 Obituary. Chem. News, LI, 190 (1885).
 Obituary. J. Chem. Soc., London, XLIX, 347 (1886).

FIELD, HENRY W.
 Obituary. J. Chem. Soc., London, LV, 293 (1889).

FIGUIER, LOUIS.
 * Vie des savants illustres, depuis l'antiquité jusqu' au dix-neuvième
 siècle. Paris, *n. d.* 5 vols., roy. 8vo. Illustrated with portraits
 and engravings.

> A handsome work in popular style. Vol. I, Savants de l'antiquité ; II,
> Savants du moyen âge ; III, Savants de la Renaissance ; IV, Savants du
> XVII siècle ; V, Savants du XVIII siècle. Besides philosophers, mathe-
> maticians, physicists, naturalists, and astronomers, contains biographies
> of following chemists : Geber, *Roger Bacon, Arnaldo Villanova,* Ray-
> mund Lully, *Paracelsus, Agricola, Van Helmont, Robert Boyle, Nicolas
> Lémery, Boerhaave,* Réaumur, *Rouelle, Lavoisier.* Portraits are given
> of those whose names are in italics.

FIKENSCHER, GEORG W. A. Geschichte von Baron Krohneman. *See* Krohneman, C. W.

FILIPPUZZI, FRANCESCO.
 HOFMANN, A. W. Nekrolog auf F. Filippuzzi. Ber. d. chem. Ges.,
 XIX, p. 2941 (1886).

FISCHER, EMIL. *See* Griess, Peter.

FISCHER, GEORGE P. Life of Benj. Silliman [Sr.], *see* Silliman, B.

FITZ, ALBERT.
 HOFMANN, A. W. Nachricht von dem Tode des A. Fitz. Ber. d.
 chem. Ges., XVIII, p. 1505 (1885).

FLAMEL, NICOLAS.
 VALLET DE VIRIVILLE. Des ouvrages alchimiques attribués à Nicolas
 Flamel. Paris, 1856. 8vo.
 * VILLAIN, ABBÉ. Histoire critique de Nicolas Flamel et de Pernelle
 sa femme ; recueillie d'actes anciens qui justifient l'origine et la
 médiocrité de leur fortune contre les imputations des Alchimistes.
 On y a joint le Testament de Pernelle et plusieurs autres pièces
 intéressantes. Paris, 1761. pp. [xii]-403-[iv], 12mo. Portrait
 and folding plate.

> After a critical examination of original documents the author shows the base-
> lessness of the claims of alchemists with respect to Flamel. Published
> anonymously.

FLAMEL, NICOLAS. Life and Writings. For a list of more than 18 biographies and essays on N. F. see Chevalier, Ulysse, Répertoire. For bibliography *see* Adelung, J. F. Life, *see*, *in Section III*, Lenz, Jo. Christopher.

FLEISCHER, ANTON.
STEINER, A. Nekrolog auf Anton Fleischer. Ber. d. chem. Ges., XI, p. 2308 (1878).

FLOURENS, MARIE J. P. Éloge de J. A. Chaptal, *see* Chaptal, J. A. Éloge de L. J. Thénard, *see* Thénard, Louis Jacques.

FLÜCKIGER, F. A. Zur Erinnerung an Scheele, *see* Scheele, Carl Wilhelm.

FONTENELLE, BERNARD LE BOUYER DE.
Éloges des académiciens. Paris, 1719. 3 vols., 12mo. And in his Œuvres, Paris, 1790–92. Vols. 6 and 7.
Contains among others notices of Lémery, Homberg, and Boerhaave.

FONVIELLE, W. DE. Vie de Joseph Priestley, *see* Priestley, Joseph.

FORBES, DAVID.
Obituary. J. Chem. Soc., London, XXXI, 496 (1877).
Obituary. Chem. News, XXXIV, 260 (1876).

FORCRAND, R. DE. La vie de G. Chancel, *see* Chancel, G.

FORSTER, JOHANN REINHOLD.
Biographical Notices of J. R. F.: Sprengel, Kurt; Neu Teutsch. Merkur. 1799. p. 33 and p. 234. *Also:* Monthly Magazine, June, 1798, p. 403 and p. 241. All. J. Chem., vol. III, p. 617, with bibliography.

FOUCHY, GRANDJEAN DE. Éloge de Claude Joseph Geoffroy. *See* Geoffroy, C. J.

FOUQUE, VICTOR. Vie et travaux de Nicéphore Niepce, *see* Niepce Nicéphore.

FOURCROY, ANTOINE FRANÇOIS DE.
Portrait in All. J. Chem., Scherer, vol. VIII. Frontispiece. 1802.
CATTANEO, ANTONIO. Cenni su la vita di Anton Franc. Fourcroy. Milano, 1839. pp. 50, 4to. Portrait of F.
CUVIER, GEORGES. Éloge historique du comte de Fourcroy. Mém. Acad. Sciences, XI, 1 S., Hist. xcvii–cxxviii (1810).

FOURCROY, ANTOINE FRANÇOIS DE. *See* Young, Thomas, Biographies. Vie de Lavoisier, *see* Lavoiser, A. L.

FOURIER, JEAN BAPTISTE JOSEPH. Life of, *see* Arago, François.

FOWNES, GEORGE.
Obituary. Q. J. Chem. Soc., London, II, 184–187 (1850).
Obituary. Am. J. Sci. [2], VII, 452. May, 1849.

Fox, Talbot.   *See* Talbot, Wm. Henry Fox.

Fraunhofer, Joseph von.
   [Biography] redigirt von E. Lommel.   Herausgegeben im Auftrage der
      mathematisch-physikalischen Classe der königlichen bayerischen
      Akademie der Wissenschaften. (1888.)   München, 1889.   8vo.
      Portrait.

Freire-Marreco, A.
   Obituary.   J. Chem. Soc., London, XLI, 238 (1882).

Fremy, Edmonde.
   Analyse succincte des mémoires de chimie publiés par M. F.   Paris,
      1844.   pp. 25, 4to.
   Notice sur les mémoires de chimie publiés par E. Fremy, professeur
      de chimie à l'École Impériale polytechnique et au Muséum d'histoire
      naturelle.   Paris, 1857.   4to.

Fresenius, Carl Remigius.   Bibliography and Portrait.   *See, in Section III*, Fresenius,
      C. R. : Geschichte des chemischen Laboratoriums zu Wiesbaden.

Fresnel, Augustin Jean.   Life of.   *See* Arago, François.

Friedel, Charles.
   Notice sur les travaux scientifiques de C. F.   Paris, 1876.   pp. 68, 4to.
      Contains a bibliography of C. F.

Friedel, Charles.   Notice sur la vie de Charles Adolphe Wurtz, *See* Wurtz, C. A.

Fritzsche, Carl Julius.
   Obituary.   J. Chem. Soc., London, XXV, 345 (1872).
   Butlerow, A.   Nekrolog auf C. J. Fritzsche.   Ber. d. chem. Ges., v,
      p. 132 (1872).

Fuchs, Johann Nepomuk von.   Nekrolog und Bibliographie, *see, in Section V*, Fuchs,
      J. N. von.

Fudakowski, Hermann Boleslaw.
   Boguski, J. G.   Nekrolog auf H. B. Fudakowski.   Ber. d. chem. Ges.,
      XIII, p. 1038 (1879).

Gadolin, Johan.
   Tigerstedt, Robert.   Johan Gadolin ; Ett bidrag till de induktiva
      vetenskapernas historia i Finland.   *In* Finska vetenskaps Societet
      Bidrag till kännedom [etc.], 1877.   8vo.

Gahn, Johann Gottlieb.
   Nekrolog.   All. nord. Ann. Chem., vol. VI, p. 200.   (1821).

GAILLON, M.   Nécrologie de, *see* Girardin, J.

GALVANI, LUIGI.
  ALBERTONI, P.   Galvani e le sue opere.   Bologna, 1888.
  ALIBERT, JEAN LOUIS.   Elogio storico di L. G.   Traduzione dal fran-
    cese.   Bologna, 1802.   8vo.
  * MEDICI, MICHELE.   Elogio di Luigi Galvani, nel teatro anatomico dell'
    antico archiginnasio di Bologna in occasione d'una radunanza
    semipubblica tenutavi dalla Societá medico-chirurgica il 6 novem-
    bre, 1844.   Bologna, 1845.   pp. 29, 4to.
        With an engraved portrait by A. Marchi.

GARTHSHORE.   Life of Jan Ingenhousz, *see* Ingenhousz, Jan.

GASSIOT, JOHN PETER
  Obituary.   J. Chem. Soc., London, XXXIII, 227 (1878).

GASTINEAU, BENJAMIN.
  * Les génies de la science et de l'industrie.   Paris [1870].   pp. 181, 18mo.
        In this carelessly written work the author says that Joseph Priestley, exiled
        to America, became the friend of Jefferson Davis !   No German chemist
        is mentioned !

GAUTIER, A.
  Notices sur les travaux scientifiques de A. G.   Paris, 1888.   4to.

GAY LUSSAC, LOUIS JOSEPH.
  Obituary.   Am. J. Sci. [2], x, 137, Nov., 1850.
  BIOT ET GARDEUR LE BRUN.   Notices biographiques sur Gay-Lussac.
    Châlons, 1850.   8vo et 4to.
  [A biography in connection with the inauguration of a statue to his
    memory at Limoges, France, August 11, 1890.]   Revue générale
    des sciences pures et appliquées.   Vol. 1, No. 13, Aug. 15, 1890.

GAY-LUSSAC.   Untersuchung über das Jod, *see* Ostwald's Klassiker *in Section V*.
    Life of G. L., *see* Arago, François, *also* Mangin.

GEBER.   Life of, *see* Figuier, Vie des savants.

GEIKE, JOHN CUNNINGHAM.   Michael Faraday and Sir David Brewster, *see* Faraday,
    Michael.

GEISSLER, HEINRICH.
  HOFMANN, A. W.   Nachricht von dem Tode des H. Geissler.   Ber.
    d. chem. Ges., XIII, p. 147 (1879).
  Gedenkblatt zur Erinnerung an H. G., Glastechniker, geboren zu

GEISSLER, HEINRICH. [Cont'd.]

Igelshieb in Thüringen 1816, gestorben zú Bonn am 27. Januar 1879. Zur Feier des 50 jährigen Bestehens der Firma seinen Freunden mit beigeheftetem Bildniss gewidmet. Bonn am Rhein, 1890. pp. [iv], 8vo. Two plates.

GEOFFROY, CLAUDE JOSEPH.

FOUCHY, GRANDJEAN DE. Éloge de Claude Joseph Geoffroy. Éloges des académiciens de l'Académie des sciences, 1761. .Vol. I, p. 204.

GERHARD, KARL ABRAHAM.

Nekrolog. All. nord. Ann. Chem., vol. VI, p. 200 (1821).

GERHARDT, CHARLES FRÉDÉRIC.

Obituary. Q. J. Chem. Soc., London, x, 187 (1858).

* PAPILLON, I. H. FERNAND. La vie et l'œuvre de Charles Frédéric Gerhardt suivi de notes et de développements scientifiques. Paris, 1863. pp. 72, 8vo.

> Contains a bibliography of Gerhardt and short biographies of Lavoisier, Thénard, Berzelius, Dumas, Liebig, Laurent, Wurtz, Berthelot.

Notice analytique sur les travaux de M. Ch. Gerhardt, professeur à la faculté des sciences de Montpellier. Paris, 1850. 4to.

For biography and portrait, *see, in Section V*, Жераръ, К. Ф. Введеніе.

GERICHTEN, E. VON. *See* Gorup-Besanez, E. von.

GERSTL, RUDOLF.

Obituary. J. Chem. Soc., London, XLI, 237 (1882).

GEUTHER, ANTON. *See* Ludwig, Johann F. H.

GEUTHER, ANTON.

Obituary. J. Chem. Soc., London, LVII, 448 (1890).

LIEBERMANN, C. Nachricht von dem Tode des A. Geuther. Ber. d. chem. Ges , XXII, p. 2388 (1889.)

GEYER, BENGT REINHOLD.

Nekrolog. All. nord. Ann. Chem., vol. II, p. 145 (1819).

GEYGER, ADOLPH.

HOFFMANN, A. W. Nekrolog auf A. Geyger. Ber. d. chem. Ges., XX, p. 3025 (1887). Portrait of A. G.

GIRARDIN, JEAN.

Notices sur diverses questions de chimie agricole et industrielle, suivies

GIRARDIN, JEAN. [Cont'd.]
    de plusieurs notices nécrologiques (sur MM. Robiquet, Planche et
    Gaillon). Rouen, 1841. 8vo.
Notice sur les travaux de J. G. [Rouen, c. 1842.] pp. 10, 4to.

GIRTANNER, CHRISTOPH.
    Nekrolog. All. J. Chem., Scherer., vol. VI, p. 79 (1801).

GLADSTONE, JOHN HALL. Life of Michael Faraday, see Faraday, M.

GLASS, WILLIAM.
    Obituary. J. Chem. Soc., London, XXXVII, 258 (1880).

GLAUBER, J. R. Life and bibliography, see Adelung, J. F.

GLEICHMANN. Nachrichten von Theoph. Paracelsus, see Paracelsus.

GMELIN.
    Stammbaum der Familie Gmelin. Karlsruhe, 1877.

GMELIN, JOHANN FRIEDRICH. For a life of J. F. G., see, in Section III, Gmelin, J. F.,
    De primis chemiæ pneumaticæ originibus.

GMELIN, LEOPOLD.
    Obituary. Q. J. Chem Soc., London, VII, 144 (1855).

GNEHM, R. See Kopp, Emil.

GOETZIUS, JOH. CHRISTOPH. See Stahl, Georg Ernest.

GÖLDI, EMIL A. See Michler, Wilhelm.

GOLDSCHMIDT, CARL THEODOR.
    LIEBERMANN, C. Nekrolog auf C. T. Goldschmidt. Ber. d. chem.
        Ges., IX, p. 108 (1876).

GOLL, OSCAR FRIEDRICH.
    HOFMANN, REINHARDT. Nekrolog auf O. F. Goll. Ber. d. chem. Ges.,
        IV, p. 213 (1871).

GORUP-BESANEZ, EUGEN VON.
    Hilger, A. Nekrolog auf E. v. Gorup-Besanez. Ber. d. chem. Ges.,
        XII, p. 1029 (1879). Portrait of Gorup-Besanez in XIII (1880).
    GERICHTEN, E. VON. Nekrolog auf Gorup-Besanez. Ber. d. chem.
        Ges., XI, p. 2163 (1878).
    LOMMEL, EUGEN. Rede am Grabe des Eugen Freiherr Gorup von Besa-
        nez gehalten am 26 November, 1878. Erlangen, 1879. pp. 14, 4to.
        Contains a bibliography of Gorup-Besanez.

GOSSAGE, WILLIAM.
Obituary. J. Chem. Soc., London, XXXIII, 229 (1878).
Obituary. Chem. News, XXXV, 186 (1877).

GOTTLIEB, JOHANN.
MALY, RICHARDT. Nekrolog auf J. Gottlieb, Ber. d. chem. Ges., VIII,
p. 448 (1875).

GRAHAM, JOHN.
Obituary. J. Chem. Soc., London, XXII, v (1869).
GRAHAM, THOMAS.
Obituary. Am. J. Sci. [2], XLIX, 144 (1870).
Obituary. Chem. News, XX, 152, Sept. 24, 1869. Also the same vol.,
p. 186.
Obituary. J. Chem. Soc., London, XXIII, 293 (1870).
* COOKE, JOSIAH PARSONS. Memoir of Thomas Graham. Reprinted
from the Proceedings of the American Academy of Arts and
Sciences (vol. VIII, May 24, 1870), in Scientific Culture, New
York, 1891.
HOFMANN, A. W. Nekrolog. Ber. d. chem. Ges., II, 753 (1869). Portrait.
* SMITH, R. ANGUS. The Life and Works of Thomas Graham, illustrated
by 64 unpublished letters ; prepared for the Graham Lecture Com-
mittee of the Glasgow Philosophical Society. Edited by J. J.
Coleman. Glasgow, 1884. pp. vi–114, 8vo. Portrait.
Contains a bibliography, each title accompanied by an abstract.

GREGORY, WILLIAM.
Obituary. Q. J. Chem. Soc., London, XII, 172 (1860).

GREN, FRIEDRICH ALBRECHT CARL.
Nekrolog. All. J. Chem., Scherer, vol. II, p. 357 (1799).
TROMMSDORFF, JOH. B. Nekrolog auf F. A. C. Gren. J. der Pharm.,
VI, 2 (1799).
Contains also a bibliography of Gren.

GRIESS, PETER.
SCHEIBLER, C. Nachricht von dem Tode des P. Griess. Ber. d. chem.
Ges., XXI, p. 2799 (1888).
Nekrolog auf P. G. In three Parts. Part I by A. W. von Hofmann ;
Part II by Emil Fischer ; Part III by H. Caro. Portrait. Ber.
d. chem. Ges., XXIV, pp. 1007–1078 (1891).

GRIFFIN, JOHN JOSEPH.
Obituary. Chem. News, XXXV, 264 (1877).
Obituary. J. Chem. Soc., London, XXXIII, 229 (1878).

GRIMAUX, EDOUARD.
Notice sur les travaux scientifiques de E. G.   Paris, 1881.   pp. 33, 4to.

GRIMAUX, EDOUARD.   La vie de Lavoisier.   *See* Lavoisier, Ant. Laurent.

GROSJEAN, JEAN JOSEPH BEAUMONT JEANNERET.
Obituary.   J. Chem. Soc., London, XLIII, 253 (1883).

GRUBER, MAX.   *See* Kretschy, Michael.

GUTHRIE, SAMUEL.
Memoirs of Samuel Guthrie and the History of the Discovery of Chloroform, by O. Guthrie.   Chicago, 1887.

GUYTON DE MORVEAU, LOUIS BERNARD.   Life of, *see* Young, Thomas.

HADOW, EDWARD ASH.
Obituary.   Chem. News, XIV, 82, Aug. 17, 1866.
Obituary.   J. Chem. Soc., London, XX, 388 (1867).

HAGENBACH, EDWARD.   Life and bibliography of Christian Friedrich Schönbein, *see* Schönbein, C. F.

HALES.   *See* Savérien, A., Histoire des philosophes modernes, vol. VIII, p. 179.   Portrait.

HALL, THOMAS.
Obituary.   J. Chem. Soc., London, XXXIII, 230 (1878).

HANBURY, DANIEL.
Obituary.   J. Chem. Soc., London, XXVIII, 1314 (1875).

HANCKWITZ, AMBROSE GODFREY.
Sketch of his life.   Pharmaceutical Journal, London, 1858–59, pp. 126, 157, and 215.   *Also* Chem. News, XXXVII, 130 (1878).

HARCOURT, WILLIAM VENABLES VERNON.
Obituary.   J. Chem. Soc., London, XXV, 348 (1872).

HARE, ROBERT.
Biographical sketch.   Am. J. Sci. [2], vol. XXV (1858).

HARRIS, WILLIAM HARRY.
Obituary.   J. Chem. Soc., London, XXVII, 1200 (1874).

HARTMANN, FRANZ.   Life and teachings of Paracelsus, *see* Paracelsus.

HARVEY, ALEXANDE
Obituary.   J. Chem. Soc., London, XXXI, 499 (1877).

HARVEY, ROBERT.
   Obituary.   J. Chem. Soc., London, XLVII, 331 (1885).

HASENCLEVER, FRIEDRICH WILHELM.
   LANDOLT, H.   Nekrolog auf F. W. Hasenclever.   Ber. d. chem. Ges.,
      VIII, p. 703 (1875).

HATTON, FRANK.
   Obituary.   J. Chem. Soc., London, XLIII, 257 (1883).

HAURÉAU, J. B.   Life and bibliography of Arnaldus de Villa Nova, *see* Arnaldus de Villa
      Nova.

HAYS, B. FRANK.   Lifework of Carl Wilhelm Scheele, *see* Scheele, C. W.

HEARDER, JONATHAN.
   Obituary.   J. Chem. Soc., London, XXXI, 500 (1877).

HEINTZ, WILHELM.
   WISLICENUS, J.   Nekrolog auf W. Heintz.   Ber. d. chem. Ges., XVI,
      p. 3121 (1883).   Portrait of W. Heintz.

   Obituary.   J. Chem. Soc., London, XXXIX, 181 (1881).

HELLOT, JEAN.   Biographies of this chemist will be found in Biographie universelle,
      Hoefer's Nouvelle biographie générale, etc.

HELMONT, JOHANN BAPTIST VAN.   Biography, *see* Cap, P. A.; Figuier, Vie des savants;
      Rixner, Th. A.; Adelung, J. F. (Bibliography).   Life of, by his son Fran-
      ciscus Mercurius van Helmont, *see*, *in Section V*, Helmont, J. B. van,
      Opera omnia [etc.].

HELMONT, JOHANN BAPTIST VAN.
   Bibliography of essays on the life and works of Van Helmont, Gazette
      médicale de Paris, 1868.   p. 457.
   CAILLAU, J. M.   Mémoire sur Jean Baptiste van Helmont et ses écrits.
      Bordeaux, 1819.   8vo.
   LOOS.   J. B. van Helmont.   Heidelberg, 1807.   12mo.
   MANDON, DR.   J. B. van Helmont, sa biographie, histoire critique de
      ses œuvres.   (Extrait des mémoires de l'Académie royale de
      Belgique.)   Bruxelles, 1868.   4to.
   MASSON, H.   Essai sur la vie et les ouvrages de Van Helmont.   Brux-
      elles, 1857.   18mo.
   POULTIER D' ELMOTTE.   Essai philosophique et critique sur la vie et
      les ouvrages de Jean Baptiste van Helmont.   Bruxelles, 1817.   8vo.
   ROMMELAERE, DR.   Études sur J. B. van Helmont.   Bruxelles, 1868.
      4to.   Extrait des mémoires de l'Académie de Belgique.

HENNINGER, ARTHUR.   *See* Balard, A. J.

HENNINGER, ARTHUR.
  HOFMANN, A. W.   Nachricht von dem Tode des A. Henninger.   Ber.
    d. chem. Ges., XVII, p. 2812 (1884).

HENRY, JOSEPH.
  TAYLOR, WILLIAM B.   An Historical Sketch of Henry's Contribution
    to the Electro-Magnetic Telegraph.   (From the Smithsonian Re-
    port for 1878.)   Washington, 1879.   pp. 103, 8vo.
  GRAY, ASA.   Biographical Memoir of J. H.   Annual Report of the
    Smithsonian Institution, 1878.   Washington, 1879.   8vo.
  Life and portrait.   Pop. Sci. Mon., II, 741, April, 1873.
  * A Memorial of Joseph Henry.   Published by order of Congress.   Wash-
    ington, 1880.   pp. iv–528, sm. 4to.   Portrait.
        Contains a bibliography of his works.   Full and valuable.

HENRY, WILLIAM.   Life of, *see* Walker, Wm., Jr.

HENRY, WILLIAM CHARLES.   Memoirs of John Dalton, *see* Dalton, John.

HERAPATH, THORNTON JOHN.
  Obituary.   Q. J. Chem. Soc., London, XII, 171 (1860).

HERAPATH, WILLIAM [senior].
  Obituary.   Chem. News, XVII, 97, Feb. 21, 1868.
  Obituary.   J. Chem. Soc., London, XXI, xxiv (1868).

HERAPATH, WILLIAM BIRD.
  Obituary.   J. Chem. Soc., London, XXII, vi (1869).

HERMES TRISMEGISTUS.   *See* Trismegistus, Hermes.

HESS, G. H.   Arbeiten von Richter, *see* Richter, Jeremias Benjamin.

HIGGIN, JAMES.
  Obituary.   J. Chem. Soc., London, XLIX, 351 (1886).

HILGER, A.   *See* Gorup-Besanez, E. v.

HILDEBRAND, H.   Der Alchemist Basilius Valentinus, *see* Valentinus, Basilius.

HINDMARCH, WILLIAM MATTHEWSON.
  Obituary.   J. Chem. Soc., London, XX, 391 (1867).

HIRST, GEORGE.
  Obituary.   J. Chem. Soc., London, LI, 475 (1887).

HJELM, PETER JACOB.   Biography and bibliography of Torbern Bergman, *see* Bergman,
      Torbern.

HJELT, E. Briefe von Fried. Wöhler and J. J. Berzelius, *see* Wöhler, Friedrich.

HLASIWETZ, HEINRICH. *See* Rochleder, Friedrich.

HLASIWETZ, HEINRICH.
> BARTH, L. Nekrolog auf H. Hlasiwetz. Ber. d. chem. Ges., IX, p. 1961 (1876). Portrait of Hlasiwetz in vol. x (1877).

HOBLER, FRANCIS HELVETIUS.
> Obituary. J. Chem. Soc., London, XXXI, 501 (1877).

HOEFER, FERDINAND.
> * La chimie enseignée par la biographie de ses fondateurs, R. Boyle, Lavoisier, Priestley, Scheele, Davy, etc. Paris, 1865. pp. iv–305, 12mo.
>> A popular compilation of comparatively little value to a student.
>> Kemiens grundsanningar framställda i lefnadsteckningar af dess heroer. Fri bearbetning efter Hoefers "la chimie enseignée par la biographie de ses fondateurs." Med ändringar och tillägg af H. Santesson. Med förord af C. W. Blomstrand. Stockholm, 1870.

HOFFMANN, FRIEDRICH. *See* Hofmann, A. W. von.

HOFMANN, AUGUST WILHELM VON.
> Life and portrait. Pop. Sci. Mon., XXIV, 831, April, 1884.
> HOFFMANN, FRIEDRICH. Nekrolog auf A. W. von. Pharmaceutische Rundschau, x, 125 (Juni, 1892).

HOFMANN, A. W. *See* Birnbaum, Karl ; Brüning, Adolf von ; Buff, Heinrich ; Chevreul, Michel-Eugène ; Dumas, J. B.; Fehling, Hermann von ; Filipuzzi, Francesco ; Fitz, Albert ; Geissler, Heinrich ; Geyger, Adolph ; Griess, Peter ; Henniger, Arthur ; Kirchhoff, Gustav ; Kolbe, Hermann ; La Coste, Wilhelm ; De La Rue, Warren ; Liebig, Justus von ; Linnemann, Eduard ; Magnus, Gustav ; Mendelssohn-Bartholdy, P. ; Mendius, Otto ; Merck, Georg ; Oppenheim, Alphons ; Pebal, Leopold von ; Quesneville, G. A. ; Reimer, Karl Ludwig ; Römer, Hermann ; Schlieper, Adolph ; Skalweit, Joh. ; Smith, Robert Angus ; Sonnenschein, Franz L.; Valentin, W. G. ; Websky, Martin ; Will, Heinrich ; Wöhler, Friedrich ; Wurtz, Adolphe ; Zinin, Nicolaus.

HOFMANN, AUGUST WILHELM VON.
> * Zur Erinnerung an vorangegangene Freunde. Gesammelte Gedächtnissreden. Mit Portraitzeichnungen von Julius Ehrentraut. Braunschweig, 1889. 3 vols., 8vo.
>> Vol. I, pp. xii–401, with portraits of Thomas Graham, Gustav Magnus, Justus Liebig, Alphons Oppenheim, Heinrich Buff, Paul Mendelssohn-

HOFMANN, AUGUST WILHELM VON.   [Cont'd.]

 Bartholdy, and Hermann von Fehling. Vol. II, pp. 408, with portraits of Friedrich Wöhler, J. B. A. Dumas, Leopold von Pebal. Vol. III, pp. 432, with portraits of Quintino Sella, Gustav Kirchhoff, Adolph Geyger, Adolphe Wurtz. Also fac-similes of letters by Graham, Magnus and Wöhler, and plate of the Liebig monument. The work of a master who had the advantages of intimate acquaintance with the persons and an appreciation of their scientific labors.

HOGARTH, JAMES.

 Obituary. J. Chem. Soc., London, XLV, 615 (1884).

HOHENHEIM, PHILIPPUS AUREOLUS THEOPHRASTUS PARACELSUS BOMBAST VON. *See* Paracelsus.

HOLLAND, J. W.   *See* Rogers, Robert E.

HOLTZ, J. F.   *See* Schering, Ernst. F. C.

HOMBERG, WILHELM. *See* Savérien, A., Histoire des philosophes modernes, vol. VII, p. 173. Portrait. Éloge de W. H. *See* Fontenelle.

HOWARD, ROBERT.

 Obituary. J. Chem. Soc., London, XXV, 349 (1872).

HÜBNER, HANS JULIUS ANTON EDWARD.

 BEILSTEIN, F. Nekrolog auf H. Hübner. Ber. d. chem. Ges., XVII, Ref. p. 763 (1884).

 LANDOLT, H. Nekrolog auf H. Hübner. Ber. d. chem. Ges., XVII, p. 1581 (1884).

HUDSON, FEARNSIDE.

 Obituary. Chem. News, XIII, 167. April 6, 1866.

 Obituary. J. Chem. Soc., London, XX, 382 (1867).

HUDSON, JAMES WILLIAM.

 Obituary. J. Chem. Soc., London, XLVII, 331 (1885).

HUGGON, MR.

 Obituary. J. Chem. Soc., London, XXXVII, 259 (1880).

HUMBERT, G. Criticism of J. von Liebig, *see* Liebig, Justus von.

HUMBOLDT, ALEXANDER VON.

 AGASSIZ, LOUIS. Necrology of Alexander von Humboldt. Am. J. Sci. [2], XXVIII, 96, Nov., 1859.

 Obituary Notice of A. v. Humboldt. Am. J. Sci. [2], XXVIII, 164, Nov., 1859.

HUMPIDGE, THOMAS SAMUEL.
  Obituary. Chem. News, LVI, 248 (1887).
  Obituary. J. Chem. Soc., London, LIII, 513 (1888).

HUNT, EDWARD.
  Obituary. J. Chem. Soc., London, XLV, 616 (1884).

HUNT, T. STERRY.
  BENJAMIN, MARCUS. Life and works of T. S. H. Scientific American,
    LXVI, 182, March 19, 1892. Portrait.
  Life and portrait. Pop. Sci. Mon., VIII, 486, Feb., 1876.
  Obituary. Nature, XLV, 400, February 25, 1892.
  Obituary. Am. J. Sci. [3], XLIII, 246, March, 1892.

HUNTER, JOHN.
  Obituary. J. Chem. Soc., London, XXVI, 777 (1873).

HUSEMANN, AUGUST.
  HUSEMANN, THEODOR. Nekrolog auf August Husemann. Ber. d.
    chem. Ges., X, p. 2297 (1877).

HUSEMANN, THEODOR. *See* Husemann, August.

HUSTLER, WILLIAM.
  Obituary. J. Chem. Soc., London, XXV, 350 (1872).

HUTCHESON, JOHN B.
  Obituary. J. Chem. Soc., London, LX, 463 (1891).

HUTCHINSON, CHARLES HERBERT.
  Obituary. Chem. News, XLVII, 212 (1883).

INGENHOUSZ, JAN.
  GARTHSHORE. Biographical Account of Jan Ingenhousz. Ann. Phil.
    Thomson, X, 161, 1817.
  Portrait, *in* All. J. Chem., Scherer, vol. X, frontispiece, 1803.

INGENHOUSZ, JAN. Life of, *see* Young, Thomas.

JABIR IBN HAIYĀN AL TARSUSI. *See* Geber.

JACKSON, CHARLES THOMAS.
  Obituary. Am. J. Sci. [3], XX, 351 (1880).
  Life and portrait. Pop. Sci. Mon., XIX, 404, July, 1881.

JACOB, J. Notice biographique et bibliographique sur les principaux chimistes. *See in*
    Section V, Jacob, J., Traité élémentaire [etc.].

JACOBSEN, OSCAR.
  LIEBERMANN, C.   Nachricht von den Tode des O. Jacobsen.   Ber. d.
    chem. Ges., XXII, p. 2387 (1889).

JACQUIN, JOSEPH FRANZ EDLER VON.
  Portrait.   All. J. Chem., Scherer, vol. IX, frontispiece, 1802.

JACQUIN, NICOLAUS JOSEPH VON.
  Nekrolog.   All. nord. Ann. Chemie, vol. V, p. 230, 1820.

JANET, P.   Life of Francis Bacon, *see* Bacon, Francis.

JENNINGS, FRANCIS MONTGOMERY.
  Obituary.   J. Chem. Soc., London, XLVII, 332 (1885).

JOHNSON, S.   Life of Hermann Boerhaave, *see* Boerhaave, H.

JOHNSON, MATTHEW WARTON.
  TOOKEY, CHARLES.   Obituary.   J. Chem. Soc., London, XVI, 435 (1863).

JOHNSON, PERCIVAL NORTON.
  MATTHEY, G.   Obituary of P. N. J.   J. Chem. Soc., London, XX, 392,
    (1867).

JOHNSON, RICHARD.
  Obituary and bibliography.   J. Chem. Soc., London, XXXIX, 188 (1881).

JOHNSTON, JAMES F. W.
  Obituary.   Q. J. Chem. Soc., London, IX, 157 (1857).

JOHNSTON, JOHN.   Life of Count Rumford, *see* Thomson, Benjamin.

JOLLY, J. G. VON, J. B. DUMAS, A. WURTZ, H. KOLBE.
  VOIT, C. VON.   Nekrologe auf Jolly, Dumas, Wurtz, Kolbe, R. A.
    Smith.   München, 1885.   8vo.

JONES, HENRY BENCE.
  BOIS-REYMOND, E. DU.   Nekrolog auf H. B. Jones.   Ber. d. chem.
    Ges., VI, p. 1,585 (1873).
  Obituary.   Chem. News, vol. XXVII, p. 206 (1873).
  Obituary.   J. Chem. Soc., London, XXVII, 1201 (1874).
      Life of Faraday, *see* Faraday, Michael.

JÖRG, LEONHARD.   Die Naturwissenchaft des Paracelsus, *see* Paracelsus.

JOULE, JAMES PRESCOTT.
  Obituary.   J. Chem. Soc., London, LVII, 449 (1890).

JOY, CHARLES ARAD.
Obituary. Am. J. Sci. [3], XLII, 78, July, 1891.

JOY, LAURA R. (Translator). Recollections of F. Wöhler, *see* Wöhler, Friedrich.

JUNGFLEISCH, E. La vie de E. M. Péligot, *see* Péligot, E. M.

KAHLBAUM, AUGUST WILHELM.
LANDOLT, H. Nekrolog auf A. W. Kahlbaum. Ber. d. chem. Ges., XVII, p. 1582 (1884).

KAISER, CAJETAN GEORG. Nekrolog von Johann. Nepomuk von Fuchs, *see, in Section V*, Fuchs, J. N. von.

KAISER, GEORG CAJETAN VON.
V. A. Nekrolog auf G. C. von Kaiser. Ber. d. chem. Ges., IV, p. 894 (1871).

KARSTEN, CARL JOHANN BERNHARD.
KARSTEN, GUSTAV. Umrisse zu Carl Johann Bernhard Karsten's Leben und Wirken. Berlin, 1854. pp. 184, 8vo.
> With portrait of C. J. B. K. Contains bibliographies of four members of the Karsten family. Published anonymously.

KEIR, JAMES.
* MOILLIET, JAMES KEIR. Sketch of the Life of James Keir, with a selection from his correspondence. London, *n. d.* [1868]. pp. 164, 8vo.
> Printed anonymously for private circulation.

KEKULÉ, AUGUST.
BOETTINGER, CARL. Ueber Kekulé und seine Bedeutung in der Chemie. Ein Vortrag hervorgegangen aus einem Briefe des Kekulé. Darmstadt, 1892. 8vo.

KELL, RICHARD. Sebalt Schwertzer [alchemist]. *See* Schwertzer, Sebalt.

KELLEY, EDWARD. Biography and Bibliography. *See, in Section VI*, Roth-Scholtz, F. Deutsches Theatrum Chemicum. Vol. III, p. 734.

KHUNRATH, HEINRICH. Bibliography of Khunrath, H. *See, in Section VI*, Khunrath, H., Wahrhafter Bericht. *See*, for life and bibliography, Adelung, J. F.

KIDDER, JEROME H.
Necrology of J. H. K. Annual Report of the Board of Regents of the Smithsonian Institution . . . to July, 1889. Washington, 1890, p. 66.

KIRCHHOFF, GUSTAV ROBERT.
  Life and portrait.   Pop. Sci. Mon., XXXIII, 120, May, 1888.
HOFMANN, A. W.   Nekrolog auf G. Kirchhoff.   Ber. d. chem. Ges.,
    XX, p. 2771 (1887).   Portrait of G. Kirchhoff.
BOLTZMAN, LUDWIG.   Festrede zur Feier des 301 Gründungstages der
    Karl-Franzens-Universität zu Graz gehalten am 15 November,
    1887, von Ludwig Boltzmann.   Leipzig, 1888.   pp. viii–32, 8vo.
    Portrait of Kirchhoff.

KIRÉEVSKY.
  * Histoire des législateurs chimistes : Lavoisier, Berthollet, Humphry
    Davy.   Frankfurt a. M., 1845.   pp. 157, 8vo.

KIRWAN, RICHARD.
  Portrait, Neues all. J. Chem., Gehlen, vol. 1, frontispiece, 1803.

KNAPP, FRIEDRICH.   See Varrentrap, Franz.

KNOP, CONRAD ALEXANDER TRAUGOTT.
  PAALZOW, A.   Nekrolog auf C. A. T. Knop.   Ber. d. chem. Ges., VI,
    p. 1581 (1873).

KOCH, E. F.   Biografisk skildring af Oluf Borck, see Borrichius, O.

KÖNIG, CARL.
  LANDOLT, H. : Nekrolog auf C. König.   Ber. d. chem. Ges., XVIII,
    p. 773 (1885).

KOLBE, H.   See Erdmann, Otto Linné.

KOLBE, ADOLPH WILHELM HERMANN.
  HOFFMANN, A. W.: Nekrolog auf H. Kolbe.   Ber. d. chem. Ges., XVII,
    p. 2809 (1884).
  Obituary.   Am. J. Sci. [3], XXIX, 84, January, 1885.
  Obituary.   Chem. News, L, 282 (1884).
  Obituary.   J. Chem. Soc., London, XLVII, 323 (1885).

KOLBE, HERMANN.   Obituary of, see Voit, C. von.

KONINCK, LAURENT GUILLAUME DE.
  Obituary.   Chem. News, LVI, 66 (1887).

KOPP, EMIL.
  Obituary.   Am. J. Sci. [3], XI, 80, 1876.
  GNEHM, R.   Nekrolog auf E. Kopp.   Ber. d. chem. Ges., IX, p. 1950
    (1876).   Portrait of Kopp in vol. X (1877).

KOPP, EMIL UND SCHERR.　*See* Bolley, P. A.

KOPP, HERMANN FRANZ MORITZ.
　THORPE, T. E.　Obituary.　Nature, XLV, 441, March 10, 1892.
　Necrology.　Ber. d. chem. Ges., XXV, 505 (1892).

KRAUT.　*See* Städeler, Georg, *and* Buff, Heinrich.

KRETSCHY, MICHAEL.
　GRUBER, MAX.: Nekrolog auf Michael Kretschy.　Ber. d. chem. Ges.,
　　XVII.　Ref. p. 761 (1884).

KROHNEMAN, CHRISTIAN WILHELM, BARON VON.
　* FIKENSCHER, GEORG WOLFGANG AUGUSTIN.　Christian Wilhelm, Baron
　　von Krohneman.　Geschichte dieses angeblichen Goldmachers,
　　eines der grössesten merkwürdigsten Betrügers des siebenzehnten
　　Jahrhunderts.　Aus archivalischen Quellen bearbeitet.　Nürnberg,
　　1800.　pp. 223, 8vo.
　　　　Contains engravings of alchemistic coins.

KUNCKEL VON LÖWENSTERN.　*See* Savérien, A.: Histoire des philosophes modernes, vol.
　　VII, p. 71.　Portrait.

LA COSTE, WILHELM.
　HOFMANN, A. W.　Nachricht von dem Tode des W. La Coste.　Ber. d.
　　chem. Ges., XIX, p. 1 (1886).
　MICHAELIS, A.　Nekrolog auf W. La Coste.　Ber. d. chem. Ges., XIX,
　　Ref. p. 903 (1886).

LADENBURG, A.　*See* Carius, Ludwig.

LALANDE, JÉRÔME.　La vie de Lavoisier, *see* Lavoisier, A. L.

LAMBERT, CHARLES.
　Obituary.　J. Chem. Soc., London, XXXI, 504 (1877).

LANDOLT, H.　*See* Hasenclever, F. W. ; Hübner, Hans ; Kahlbaum, August W. ;
　　König, Carl ; Lehmann, Arthur ; Löwig, Carl.

LASÈGUE, E. C.　Stahl et sa doctrine, *see* Stahl, G. E.

LAUGIER, ANDRÉ.
　Nécrologie.　Ann. chim. phys. [2], LII, 190 (1833).

LAURENT, AUGUSTE.
　Obituary.　Q. J. Chem. Soc., London, VII, 149 (1855).

LAVOISIER, ANTOINE LAURENT, Life of. *See* Brougham, Lord ; Figuier, Vie des
    savants ; Mangin ; Kiréewsky ; Brown, Samuel (*in Section III*) ; Long-
    champ (*in Section III*) ; Hoefer, Ferdinand ; Montyon et Franklin.

    Scherer, Alex. Nic., Nachrichten von Lavoisier's Leben, *see*, *in Section V*,
    Scherer's Grundzüge der neuern chemischen Theorie.

LAVOISIER, ANTOINE LAURENT.

BERTHELOT, MARCELLIN.   Notice historique sur Lavoisier.   Lue dans
    la séance publique annuelle de l'Académie des sciences du 30
    décembre, 1889.   Paris, 1889.   pp. 56, 4to.

DEYMANN.   Éloge de Lavoisier.   Nieuwe scheikundige bibliotheek.   1
    Deel.   Amsterdam, 1799.   8vo.   Addressed to the Society *Con-
    cordia et Libertate.*

[FOURCROY, A. F.]   Notice sur la vie et les travaux de Lavoisier pré-
    cédée d'un discours [by Mulot] sur les funérailles et suivi d'une
    ode sur l'immortalité de l'âme.   Paris. An. IV [1796].   [pp. 60], 8vo.

* GRIMAUX, EDOUARD.   Lavoisier, 1743–1794, d'après sa correspondance,
    ses manuscrits, ses papiers de famille et d'autres documents inédits.
    Avec dix gravures hors texte.   Paris, 1888.   pp. vii–399, roy. 8vo.

    This handsome volume contains portraits of Lavoisier, a facsimile letter, a
    bibliography, etc.

HJORTDAHL, TH.   Om Lavoisier og den franske chemi.   Særskilt
    Aftryk af Forhandlinger i Christiania Videnskabs-Selskab., 1871.
    [Christiania, 1872.]   8vo.

LALANDE, JÉRÔME.   Notice sur la vie et les ouvrages de Lavoisier.
    Magasin encyclopédique, v, p. 174, 1795.   8vo.

* MARVIN, DAN, Jr.   Lavoisier.   An essay read before the Chemical
    Society of Columbia College [New York], April 8, 1863.   Brooklyn.
    n. d.   pp. 11, 8vo.

MINE, EDWARD.   Notice sur les travaux physiologiques de Lavoisier.
    Paris, 1885.   8vo.

NIEUWLAND, P.   Schets van het scheikundig leerstelsel van Lavoisier.
    Amsterdam, 1791.   8vo.

LAVOISIER, Scientific work of.   *See*, *in Section III*, Berthelot, La révolution chimique.

LAVOISIER AND PHLOGISTON.   *See, in Section III*, Brown, Samuel, Lectures.

LEBLANC, FÉLIX.

BÉRARD, E. P.   Notice biographique sur Félix Leblanc.   Bull. soc.
    chim., vol. 47, i (1887).

    Contains a bibliography of F. L.

LEBLANC, NICOLAS.

*ANASTASI, AUG.   Nicolas Leblanc, sa vie, ses travaux et l'histoire de
    la soude artificielle.   Paris, 1884.   pp. 231, 16mo.

LEBLANC, NICOLAS. [Cont'd.]
This life of Leblanc, by his grandson, contains : Rapport relatif à la découverte de la soude artificielle (1856) ; Mémoire sur le minéral connu sous le nom de nickel (1798) ; and : Inventaire du laboratoire de Lavoisier.

LE CANU, L. R. Souvenirs de M. Thénard, *see* Thénard, Louis Jacques.

LE CAT, CLAUDE NICOLAS.
Précis de la vie et des travaux de C. N. Le C. J. des beaux arts et sciences, Nov., 1768.

LEEDS, ALBERT R. *See* Morton, Henry,

LEEDS, ALBERT RIPLEY.
Biography and Portrait. The Stevens Link. Published by the students of the Stevens Institute of Technology, Hoboken. New York, 1892. 8vo.

LEESON, HENRY BEAUMONT. Obituary. J. Chem. Soc., London, XXVI, 778 (1873).

LEFÈVRE, A. *See* Savérien, A., Histoire des philosophes modernes, vol. VII, p. 37.
Portrait. Éloge de R. P. Lesson, *see* Lesson, R. P.

LEHMANN, ARTHUR.
LANDOLT, H. Nachricht von dem Tode des A. Lehmann. Ber. d. chem. Ges., XXI, p. 139 (1888).
SACHSE, ULRICH. Nekrolog auf A. Lehmann. Ber. d. chem. Ges., XXII. Ref. p. 867 (1889).

LEHMANN, CARL GOTTHELF.
Obituary. J. Chem. Soc., London, XVI, 433 (1863).

LÉMERY, NICOLAS. *See* Savérien, A.: Histoire des philosophes modernes, vol. VII, p. 129. Portrait. Éloge de, *see* Fontenelle ; Life of, *see* Cap, P.-A. ; *also*, Figuier, Vie des savants.

LÉMERY, NICOLAS.
DORTOUS DE MAIRAN. Éloge de Nicolas Lémery. Éloges des académiciens de l'Académie des sciences. Paris, 1747.
TONNET, JOSEPH. Notice sur Nicolas Lémery, chimiste. Niort, 1844. 8vo.

LENGLET DU FRESNOY.
*[MICHAULT, JEAN BERNARD.] Mémoires pour servir à l'histoire de la vie et des ouvrages de M. l'abbé Lenglet du Fresnoy. Londres et Paris, 1761. pp. 225, 12mo.
Contains a life of this bibliographer of chemistry and critical reviews of his works.

LESSING, MICHAEL BENEDICT.   Leben und Denken des Paracelsus, *see* Paracelsus.

LESSON, RÉNÉ PRIMEVÈRE.

A. LEFÈVRE.   Éloge historique de Réné-Primevère Lesson, né à Rochefort le 20 mars, 1794, est mort dans la même ville à l'age de 55 ans
et un mois   Premier pharmacien en chef de la marine, professeur
de chimie et de physique médicale.   Rochefort, 1850.   8vo.

LETHEBY, HENRY.

Obituary.   J. Chem. Soc., London, xxix, 618 (1876).
Obituary.   Chem. News, xxxiii, 146 (1876).

LEUCKART, RUDOLF.

BUCHKA, K.: Nekrolog auf R. Leuckart.   Ber. d. chem. Ges., xxii.
Ref. p. 861 (1889).

LICHTENBERG, GEORG CHRISTOPH.

Nekrolog.   All. J. Chem., Scherer, vol. iii, p. 609 (1799).

LIEBEN, ADOLPH.   *See* Schrötter, Anthon.

LIEBERMANN, C.   *See* Goldschmidt, Carl Theodor ; Neubauer, Carl Th. L.; Mohr, Karl
Friedrich ; Zimmermann, Clemens ; Bulk, Carl ; Jacobson, Oscar ; Geuther,
Anthon.

LIEBIG, GEORG FREIHERR VON.   *See* Liebig, Justus von.

LIEBIG, JUSTUS VON.

Obituary.   J. Chem. Soc., London, xxvii, 1204 (1874).
Obituary.   Chem. News, vol. xxvii, p. 206 (1873).
Life and portrait.   Pop. Sci. Mon., iii, 232, June, 1873.
An autobiographical sketch.   Read at a joint meeting of Societies in
the Chemical Laboratory, University College, Liverpool, on Wednes-
day evening, March 18, 1891, by J. Campbell Brown.   Chem. News,
vol. 63, p. 265 (June, 1891).   Pop. Sci. Mon., xl, 655 (1892).
BENEKE, F. W.   J. von L., his merits in the Promotion of Practical
Medicine.   A Memorial Address delivered at the Annual Meeting
of the Society for the Advancement of the Natural Sciences in
Marburg, June 11, 1874.   Glasgow, 1881.
* ERLENMEYER, EMIL.   Ueber den Einfluss des Freiherrn Justus von
Liebig auf die Entwickelung der reinen Chemie.   Eine Denkschrift.
München, 1874.   pp. 65, 4to.
Denkreden auf Justus von Liebig, von Pettenkofer, Bischoff, Erlen-
meyer und Vogel.   München, 1874.   4 parts, 4to.
HOFMANN, A. W.   The Life-Work of Liebig in Experimental and
Philosophic Chemistry, with allusions to his Influence on the De-

LIEBIG, JUSTUS VON. [Cont'd.]
velopment of the Collateral Sciences and of the Useful Arts. J.
Chem. Soc., London, XXVIII, 1065 (1875). Portrait.

* The Life-Work of Liebig in Experimental and Philosophic Chemistry, with Allusions to his Influence on the Development of the
Collateral Sciences and of the Useful Arts. A Discourse delivered
to the Fellows of the Chemical Society of London, in the theatre
of the Royal Institution of Great Britain on March the 18th, 1875.
London, 1876. pp. 146, 8vo. Portrait.

Festrede gehalten bei der Enthüllung des Liebig-Denkmals. Ber. d.
chem. Ges., XXIII, p. 792 (1890). Engraving of the monument.

Nekrolog auf J. v. Liebig. Ber. d. chem. Ges., VI, p. 465 (1873).
Portrait of Liebig in vol. VIII, p. 1701.

LIEBIG, GEORG FREIHERR VON. Eigenhändige biographische Auf-
zeichnungen. Ber. d. chem. Ges., XXIII, p. 817 (1890). Portrait of
Liebig.

HUMBERT G. Herr v. Liebig und die Stickstoff-Theoretiker. Berlin,
1858. 8vo.

MEISSNER, P. T. Justus Liebig, Dr. der Medicin . . . analysirt
von P. T. M. Frankfurt am Main, 1844. 8vo.

PETTENKOFER, MAX VON. Zum Gedächtniss des Justus Freiherrn von
Liebig. Rede gehalten im Auftrage der mathematisch-physikali-
schen Klasse der königl. bayerischen Akademie der Wissenschaften zu München in der öffentlichen Sitzung am 28 März,
1874. München, 1874. 8vo.

SCHRÖTTER, A. R. VON. Justus von Liebig. Eine Denkrede. Wien,
1873. 8vo.

THUDICHUM, J. L. W. The Discoveries and Philosophy of Liebig.
Five [Cantor] lectures. . . . London, 1869. Reprinted from
the Journal of the Society of Arts.

VOGEL, AUGUST. Justus Freiherr von Liebig als Begründer der Agrikultur-Chemie. Eine Denkschrift. [Königliche bayerischen Akade-
emie der Wissenschaften], München, 1874. pp. 61, 4to.

* WÖHLER, EMILIE, UND A. W. HOFMANN. Aus Justus Liebig's und
Friedrich Wöhler's Briefwechsel in den Jahren 1829–1873. Braun-
schweig, 1888. 2 vols., 8vo. Vol. I, pp. x–385 ; vol. II, pp. 1–362.
Illustrated.

LIEBIG, JUSTUS VON, UND FRIEDRICH WÖHLER.
HOFMANN, A. W. VON. J. von Liebig und F. Wöhler. Zwei Gedächt-
nissreden. Mit dem Bruchstück einer Autobiographie Liebig's als
Anhang mit den Porträts von Liebig und Wöhler, sowie den Ab-
bildungen der Denkmäler in München, Giessen und Göttingen.
Leipzig, 1891. pp. iv–80, 8vo. Ill.

LIEBREICH, O.   *See* Werther, August F. G.

LIECHTY, REINHARD DE.   *See* Albertus Magnus.

LINNEMANN, EDUARD.
  HOFMANN, A. W.   Nachricht von dem Tode E. Linnemann.   Ber. d.
  chem. Ges., XIX, p. 1149 (1886).

LIVERSIDGE, ARCHIBALD.
  List of Scientific Papers and Reports by A. L.   Sydney [New South
  Wales].   *n. d.* [1881].   pp. 8, 8vo.

LOCHER, HANS.   Paracelsus der Luther der Medicin, *see* Paracelsus.

LÖWIG, CARL.
  LANDOLT, H.: Nekrolog auf Carl Löwig.   Ber. d. chem. Ges., XXIII,
  p. 905 (1890).

LÖWIG, KARL J.   Denkschrift über J. B. Richter, *see* Richter, J. B.

LOMMEL, EUGEN.   Necrology of von Gorup Besanez.   *See* von Gorup Besanez.   Life of
  Joseph von Fraunhofer, *see* Fraunhofer, Joseph von.

LOOS.   Life of J. B. Van Helmont, *see* Van Helmont, J. B.

LOUYET, P.   Life of Berzelius, *see* Berzelius, J. J.

LOVERING, JOSEPH.   Necrology of John William Draper, *see* Draper, J. W.

LUC, J. A. DE.
  Portrait *in* Neues all. J. Chem., Gehlen, vol. v.   Frontispiece.   1805.

LUDWIG, JOHANN FRIEDRICH HERMANN.
  GEUTHER, A.   Nekrolog auf J. F. H. Ludwig.   Ber. d. chem. Ges.,
  VI, p. 1578 (1873).

LULL, RAMÓN.   *See* Lullius, Raymundus.

LULLIUS, RAYMUNDUS.   Life of, *see* Figuier : Vie des savants.   Colletet : La vie de
  Raymond-Lulle, *see*, *in Section VI*, Lullius, Raymundus : La clavicule ou
  la science de R. L.   Notice biographique sur Raymond Lulle, *see*, *in
  Section VI*, Poisson, Albert.

LULLIUS, RAYMUNDUS.
  PERROQUET.   La vie et le martyre du Docteur Illumine le bienheureux
  Raymond Lulle, avec une apologie de sa sainteté et de ses œuvres
  contre le mensonge, l'envie et la médisance.   Vendosme, 1667.   pp.
  xvi-390 and index, 12mo.
    Contains a one-line bibliography.

LULLIUS, RAYMUNDUS. [Cont'd.]

SEGNI, JUAN. Vida y hechos del admirable doctor y martyr Ramón Lull, vezino de Mallorca. Mallorca, 1606.

VERNON, JEAN MARIE DE. L'histoire véritable du bien heureux Raymond Lulle, martyr du tiers ordre S. François, et la réparation de son honneur. Paris, 1668. pp. [xxviii]–388, 12mo.

> Contains a bibliography of Lully, but giving no details.

WEYLER Y LAVIÑA, D. F. Raimundo Lulio, juzgado por si mismo . . . Palma, 1866.

Vita Lulii. *In* Theobald von Hoghelande's Historiæ aliquot transmutationis metallicæ. *See, in Section III*, Hoghelande, Th. von.

LUTZ VON LAUFELFINGEN. *See in Section VI.*

MACAIRE, J. F. Notice de Théodore de Saussure, *see* Saussure, Th. de.

MACLEAN, JOHN.

* MACLEAN, JOHN [Jr.]. A memoir of John Maclean by his son. For private distribution only. Princeton, 1876. pp. 64, 12mo.

> Dr. Maclean was the first Professor of Chemistry in the College of New Jersey (1795).

MAGNUS, HEINRICH GUSTAV.

Obituary. J. Chem. Soc., London, XXIV, 610 (1871).

HOFMANN, AUG. WILH. Zur Erinnerung an Gustav Magnus. Nach einem am 14 December, 1870, in der General-Versammlung der deutschen chemischen Gesellschaft zu Berlin gehaltenen Vortrage. Mit Portrait und Facsimile. Berlin, 1871. pp. 111, 8vo.

MALAGUTI, FAUSTINO GIOVITA MARIANO.

Obituary. J. Chem. Soc., London, XXXV, 266 (1879).

MALLOIZEL, GODEFROY. Life and bibliography of Michel Eugène Chevreul, *see* Chevreul, M. E.

MALUS, ÉTIENNE LOUIS. Life of, *see* Arago, François.

MALY, RICHARD. *See* Gottlieb, Johann.

MALY, RICHARD.

F. EMICH. Nekrolog auf R. M. Ber. d. chem. Ges., XXIV, 1079–1088 (1891). Portrait and bibliography of Maly.

MANDON, DR. Biographie de J. B. Van Helmont. *See* Van Helmont, J. B.

MANGIN, ARTHUR.
   *Les savants illustres de la France. Paris, *n. d.* pp. vii–524. 8vo.
   Illustrated with sixteen portraits.

>   Contains biographies of *Lavoisier*, Fourcroy, Berthollet, *Gay Lussac*,
>   Thénard, Ampère, and other physicists and naturalists, with portraits of
>   those in italics.

MANSFIELD, CHARLES BLACHFORD.
   Obituary. Q. J. Chem. Soc., London, VIII, 111 (1856).

MAQUENNE, L.
   Notices sur les travaux scientifiques de Maquenne. Paris, 1889. 4to.

MARGGRAF, ANDREAS SIGISMUND.
   Portrait *in* Neues all. J. Chem., Gehlen, vol. II, 1804, frontispiece.
   CRELL, LORENZ. Lebensgeschichte Andreas Sigismund Marggrafs.
   Chem. Ann., Crell, 1786, I, p. 181.

MARRECO, A. FREIRE.
   Obituary. Chem. News, XLV, 108 (1882).

MARSH, J. F. On some Unpublished Correspondence of Joseph Priestley, *see* Priestley,
   Joseph.

MARTIN, BENJAMIN N. Sketch of John William Draper. *See* Draper, J. W.

MARTIN, SAMUEL. Life of Faraday, *see* Faraday, Michael.

MARTIUS, C. A. *See* Christiani, Arthur.

MARTIUS, CARL FRIEDRICH PHILIPP VON. Denkrede auf Berzelius, *see* Berzelius, J. J.
   Denkrede auf Schweigger, *see* Schweigger, J. S. C.

MARVIN, DAN, JR. Life of Lavoisier, *see* Lavoisier, A. L.

MARX, KARL FRIEDRICH HEINRICH. Zur Würdigung des Theophrastus, *see* Paracelsus.

MASSON, H. Vie de J. B. Van Helmont. *See* Van Helmont, J. B.

MATTEUCCI, CH. *See* Piria, Raffaele.

MATTHEWS, FRANCIS COOK.
   Obituary. J. Chem. Soc., London, XXVIII, 1317 (1875).

MATTHIESSEN, AUGUSTUS.
   Obituary. Am. J. Sci. [2], L, 437 (1870).
   Obituary. Chem. News, XXII, 189 (1870).
   Obituary. J. Chem. Soc., London, XXIV, 615 (1871).

MATY, MATTHEW. Éloge de Hermann Boerhaave. *See* Boerhaave, H.

MAUMENÉ, E. J.
Notice sur les travaux scientifiques de E. J. Maumené. Paris, 1884. 4to.

MAY, JOSEPH. *See* Priestley, Joseph.

MAYOW, JOHANN.
Portrait, *in* Neues all. J. Chem., Gehlen, vol. III, 1804, frontispiece.

MAYOW, JOHN. Life and work of, *see, in Section III*, Scherer, J. A.

MEDICI, MICHELE. *See* Galvani, Luigi.

MEDLOCK, HENRY.
Obituary. J. Chem. Soc., London, XXVIII, 1317 (1875).

MELDRUM, EDWARD.
Obituary. J. Chem. Soc., London, XXIX, 619 (1876).

MENDELEEFF, DMITRI IVANOWITSH.
THORPE, T. E. [Life and works of D. I. M.] Nature, XL, 193 (1889).
Portrait.

MENDELSSOHN-BARTHOLDY, PAUL.
HOFMANN, A. W. Nekrolog auf P. Mendelssohn-Bartholdy. Ber. d.
chem. Ges., XIII, p. 299 (1880). Portrait.

MENDIUS, OTTO.
HOFMANN, A. W. Nachricht von dem Tode des O. Mendius. Ber. d.
chem. Ges., XVIII, p. 1607 (1885).

MENENDEZ, PELAYO, M. *See* Arnaldus de Villa Nova.

MENSING, J. G. W. Leben des J. B. Trommsdorff, *see* Trommsdorff, J. B.

MERCER, JOHN.
Obituary. J. Chem. Soc., London, XX, 395 (1867).
Obituary. J. Chem. Soc., London, XXXVII, 260 (1880).
PARNELL, E. A. The life and labors of John Mercer, the self-taught
Chemical Philosopher, including numerous recipes used at the
Oakenshaw Calico Print Works. London, 1886. 8vo. Portrait.

MERCK, GEORG.
HOFMAN, A. W. Nekrolog auf G. Merck. Ber. d. chem. Ges., VI, p.
1582 (1873).

MERRICK, J. M.
Obituary. Chem. News, XXXIX, 162 (1879).

MEY, MAX.
   CARSTANJEN.   Nekrolog auf M. Mey.   Ber. d. chem. Ges., IX, p. 212
      (1876).

MEYER, JOHANN FRIEDRICH,   Life and labors, by E. G. Baldinger.   *See, in Section V,*
      Wiegleb : Kleine chymische Abhandlungen, 1767.

MEYER, LOTHAR.   Nekrolog auf Leopold von Pebal, *see* Pebal, L. von.

MICHAELIS, A.   *See* La Coste, W.

MICHAULT, JEAN BERNARD.   La vie de Lenglet du Fresnoy, *see* Lenglet du Fresnoy.

MICHEL, MIDDLETON.   *See* Smith, J. Lawrence.

MICHLER, WILHELM.
   GÖLDI, EMIL A. Nekrolog auf W. Michler.   Ber. d. chem. Ges., XXII,
      Ref. p. 873 (1889).

MIDDLETON, JAMES.
   Obituary.   J. Chem. Soc , London, XXIX, 619 (1876).

MILBURN, WM. HENRY.   Biography of Thomas Young, *see* Young, Thomas.

MILLER, WILLIAM ALLEN.
   Obituary.   Chem. News, XXII, 177 (1870).
   Obituary.   J. Chem. Soc., London, XXIV, 617 (1871).

MILLER, WILLIAM HALLOWES.
   Obituary.   J. Chem. Soc., London, XXXIX, 188 (1881).
   COOKE, JOSIAH PARSONS.   Memoir of W. H. Miller.   Reprinted from
      the Proceedings of the American Academy of Arts and Sciences
      (vol. XVI, May 24, 1881), *in* " Scientific Culture."   New York,
      1891.

MILLON, NICOLAS AUGUSTE EUGÈNE.
   REISET, J.   Millon, sa vie, ses travaux de chimie, et ses études écono-
      miques et agricoles sur l'Algérie.   Paris, 1870.   8vo.   Portrait of
      Millon.

MITSCHERLICH, EILHARDT.
   Obituary.   Am. J. Sci. [2], XXXVI, 451, Nov., 1863.
   Obituary.   J. Chem. Soc., London, XVII, 440 (1864).
   *ROSE, GUSTAV.   Eilhardt Mitscherlich.   Gedächtnissrede gehalten in
      der deutschen geologischen Gesellschaft.   Berlin, 1864.   pp. 54,
      8vo.

MOEHSEN, J. C. W.   Leben Thurneissers zum Thurn, *see* Thurneisser, Leonhard von.

MOHR, KARL FRIEDRICH.

Life and portrait. Pop. Sci. Mon., XVII, 402, July, 1880.

LIEBERMANN, C. Nachricht von dem Tode des K. F. Mohr. Ber. d. chem. Ges., XII, p. 1932 (1879).

MOILLET, JAMES KEIR. Life of James Keir, *see* Keir, James.

MOISSAN, H.

Notices sur les travaux scientifiques de M. Paris, 1889. 4to.

MONS, VAN, J. B. Life of van Mons, *see* Cap, P. A.

MONTGOMERY, JAMES WALKER.

Obituary. J. Chem. Soc., London, LVII, 453 (1890).

MONTYON ET FRANKLIN, SOCIÉTÉ.

Portraits et histoire des hommes utiles, hommes et femmes de tous pays et de toutes conditions qui ont acquis des droits à la reconnaissance des hommes par des traits de dévoument, de charité ; par des fondations philanthropiques ; par des travaux, des tentatives, des perfectionnemens, des découvertes utiles à l'humanité, etc. Publiés et propagés pour et par la Société Montyon et Franklin. Paris, 1833–40. 4 vols., 8vo.

> This collection contains biographies, with portraits, of Franklin, Watt, Lavoisier, Berthollet, Davy, d'Arcet, Chaptal, Abbé Rozier, Sir James Banks, Kopernicus, Galileo, Descartes, Hauy brothers, Palissy, Newton, Cuvier, Legendre, les Jussieu and many others not falling within the scope of science.

MORTON, HENRY.

SELLERS, COLEMAN, AND ALBERT R. LEEDS. Biographical Notice of President H. M. Prepared by C. S. and A. R. L. on the occasion of the presentation to the Trustees and Faculty [of the Stevens Institute of Technology] by the Alumni Association of a Portrait of H. M., February 15, 1892. New York, 1892. *See* notice in Chem. News, LXVI, p. 49 (July 22, 1892).

MÜLLER, ALBRECHT.

BEER, ALEXANDER. Nekrolog auf A. Müller. Ber. d. chem. Ges., XXII, Ref. p. 865 (1889).

MUHLBERG, F. Leben des P. A. Bolley, *see* Bolley, P. A.

MUIR, MATTHEW MONCRIEFF PATTISON.

* Heroes of Science—Chemists. London, 1883. pp. vii–332, 12mo.

> Contains an historical sketch of chemical progress from the time of Black to the present.

MUIRHEAD, JAMES PATRICK. Life and correspondence of James Watt, *see* Watt, James.

MULDER, GERARDUS JOHANNES.   Sketch of Berzelius, *see* Berzelius, J. J.

MULDER, GERARDUS JOHANNES.
  Obituary.   J. Chem. Soc., London, XXXIX, 181 (1881).

MURPHY, MICHAEL.
  Obituary.   J. Chem. Soc., London, XXXIII, 232 (1878).

MUSPRATT, FREDERICK.
  Obituary.   J. Chem. Soc., London, XXVI, 780 (1873).

MUSPRATT, JAMES SHERIDAN.
  Obituary.   Chem. News, vol. XXIII, 82 (1871).
  Obituary.   J. Chem. Soc., London, XXIV, 620 (1871).
          Contains a bibliography of his papers.
  Biography of S. M., by a London Barrister-at-Law, and a third
      edition of the Influence of Chemistry in the Animal, Vegetal and
      Mineral Kingdoms, by S. M.   London, 1852.   8vo.

NAPIER, JAMES.
  Obituary.   J. Chem. Soc., London, XLVII, 333 (1885).

NAQUET, ALFRED.
  PROTH, MARIO.   Alfred Naquet [Biography of].   Paris, 1883.   16mo.
      Portrait.

NASON, HENRY BRADFORD.
  Life and Portrait.   Pop. Sci. Mon., XXXII, 694, Mch., 1888.

NEUBAUER, CARL THEODOR LUDWIG.
  LIEBERMANN, C.   Nachricht von dem Tode des C. T. L. Neubauer.
      Ber. d. chem. Ges., XII, p. 1931 (1879).

NEUBERT, CARL.   *See* Davy, Sir Humphry ; *also* Davy, John.

NEUMANN, CASPAR.   For a critical bibliography of Neumann, *see*, *in Section V*, Neu-
      mann, Caspar, Chymia medica, etc.

NICHOLS, WILLIAM RIPLEY.
  * CAMPBELL, W. R.   William Ripley Nichols.   A Memorial.   Funeral
      Address [delivered] in the Highland Congregational Church,
      August 5, 1886.   Boston.   Printed for Private Circulation.   1887.
      pp. 24, 12mo.   With heliotype portrait.
  STORER, FRANK H.   Necrology of William Ripley Nichols.   *In* Proc.
      Am. Acad. Arts and Sci., Boston, 1886.
          Contains a bibliography of the writings of Nichols.

NICHOLSON, EDWARD CHAMBERS.
Obituary. J. Chem. Soc., London, LX, 464 (1891).

NICKELS, BENJAMIN.
Obituary. J. Chem. Soc., London, LVII, 452 (1890).

NICKLÈS, J. Vie et bibliographie de Henri Braconnot, *see* Braconnot, H.

NIEPCE, NICÉPHORE. *See, in Section III*, Eder, J. M., Geschichte der Photochemie; *also*, Niepce, Isidore.

NIEPCE, NICÉPHORE.
*DAVANNE, A. Nicéphore Niepce, Inventeur de la photographie. Conférence faite a Chalon-sur-Saône pour l'inauguration de la statue de Nicéphore Niepce le 22 Juin, 1885. Paris, 1885. pp. 33, 8vo. Heliotype of statue.
*FOUQUE, VICTOR. La vérité sur l'invention de la photographie. Nicéphore Niepce, sa vie, ses essais, ses travaux, d'après sa correspondance et autres documents inédits. Paris, 1867. pp. 282, 8vo.
   With portrait-bust and facsimile. The author claims the invention of photography for Niepce and strongly attacks the claims of the friends of Daguerre for that honor. The author bases his statements on autographic letters, 101 in number, not before published. The volume contains a biography of Niepce.

NOAD, HENRY MINCHIN.
Obituary. J. Chem. Soc., London, XXXIII, 233 (1878).

NORTHCOTE, AUGUSTUS BEAUCHAMP.
Obituary. J. Chem. Soc., London, XXIII, 299 (1870).

NORTON, JOHN PITKIN.
Obituary. Am. J. Sci. [2], XIV, 448, Nov., 1852.
Memorials of J. P. N., late Professor of Analytical and Agricultural Chemistry in Yale College, New Haven, Conn. Published for private distribution. Albany, 1853. pp. 85, 4to. Portrait.

OERSTED, HANS CHRISTIAN.
Obituary. Am. J. Sci. [2], XII, 147, Nov., 1851.

OPPENHEIM, ALPHONS. *See* Engelbach, Theophil.

OPPENHEIM, ALPHONS.
HOFMANN, A. W. Nekrolog auf Oppenheim. Ber. d. chem. Ges., x, p. 2262 (1877).
Obituary. Chem. News, XXXVI, 151 (1877).

ORFILA, MATHÉO JOSÉ BONAVENTURE.

BÉRARD, A.   Éloge de M. Orfila prononcé dans la séance de rentrée de
la Faculté, le 15 Novembre, 1854.   Paris, 1854.   8vo.

DUBOIS, E. F. (d'Amiens).   Éloge de Matthieu Joseph Bonaventure
Orfila.   Éloges lus dans les séances publiques de l'Académie de
médecine.   Paris, 1864.   Vol. I, p. 339.

PAALZOW, A.   *See* Knop, Conrad A. T.

PAGE, DAVID.

Obituary.   J. Chem. Soc., London, LVII, 453 (1890).

PALMSTEDT, CARL.

BLOMSTRAND, C. W.   Nekrolog auf C. Palmstedt.   **Ber. d. chem. Ges.,**
III, p. 579 (1870).

PAPILLON, J. H. FERNAND.   La vie de Charles Frédéric Gerhardt, *see* Gerhardt, C. F.

PARACELSUS.   *See* Savérien, A.: Histoire des philosophes modernes, vol. vii, p. i, portrait.
Life of, *see* Cap, P. A.; *also*, Figuier : Vie des savants ; Rixner, Th. A.,
and Th. Siber.   *Compare Section I.*

PARACELSUS.   [PHILIPPUS AUREOLUS THEOPHRASTUS PARACELSUS BOM-
BASTUS VON HOHENHEIM.]

Gewürdigt in der zur Feyer des Geburtsfestes seiner Majestät des
Kaisers Alexander des Ersten, Selbstherrschers aller Reussen, etc.,
am 12 December, 1820, gehaltenen Hauptversammlung der pharmaceutischen Gesellschaft zu St. Petersburg von dem Director
derselben.   All. nord. Ann. Chem., vol. VI, pp. 243–296, 1821.

*ABERLE, KARL.   Theophrastus Paracelsus und dessen Ueberreste in
Salzburg.   Salzburg, 1878.   8vo. (Sonder-Abdruck aus den im
Selbstverlage der Gesellschaft für Salzburger Landeskunde erschienenen Mittheilungen.   XVIII Band, II Heft.   pp. 186–247.)   With
an engraving of the skull of Paracelsus.

* Grab-Denkmal, Schädel und Abbildungen des Theophrastus Paracelsus.   Beiträge zur genaueren Kenntniss derselben.   Mittheilungen der Gesellschaft für Salzburger Landeskunde, XXVII Ver-
einsjahr, 1887.   Heft I.   Salzburg, *n. d.*   pp. 74, 8vo.   With eight
portraits and engraving of monument.

> The author shows that the very numerous extant portraits (engravings) of
> Paracelsus may be referred to four types.   The portrait attached to the
> monument erected to Paracelsus in the church of St. Sebastian is of
> Paracelsus' *Father ;* this is established by oil paintings of both preserved
> in Salzburg.

PARACELSUS. [Cont'd.]

AMBERG, FRIEDRICH. Zur Geschichte des Theophrastus Paracelsus. Historische Darstellung der historischen Gesellschaft in Jena, v, pp. 139–200.

*DALTON, JOHN C. Galen and Paracelsus. [Reprinted from the N. Y. Medical Journal, May, 1873.] New York, 1873. pp. 29, 8vo.

*FERGUSON, JOHN. Bibliographia Paracelsica, an examination of Dr. Friedrich Mook's "Theophrastus Paracelsus. Eine kritische Studie." Privately printed. Glasgow. Two parts, 1877 and 1885. Part I, pp. 40 ; Part II, pp. 54, 8vo.

> Criticises Mook's Essay, and adds to his bibliography.

GLEICHMANN. Historische Nachrichten von Theoph. Paracelsus ab Hohenheim. Jena und Leipzig, 1732. 8vo.

*HARTMANN, FRANZ. The Life of Philippus Theophrastus Bombast of Hohenheim, known by the name of Paracelsus, and the substance of his teachings concerning cosmology, anthropology, pneumatology, magic and sorcery, medicine, alchemy and astrology, philosophy and theosophy, extracted and translated from his rare and extensive works and from some unpublished manuscripts. London, 1887. pp. xiii–220, 8vo.

> The scope of this work is indicated in its title. It contains a bibliography of Paracelsus and an Explanation of Terms occurring in his writings. The author affects a mystical treatment of the subject.

*HERING, CONSTANTINE. Catalogue of a very rare and curious collection of the different editions of the works of Theophrastus Bombastus Paracelsus, together with several hundred commentaries and translations collected during fifty years by Constantine Hering. Philadelphia, 1881. pp. 20, 8vo.

> This catalogue, which is quite imperfect bibliographically, contains 189 titles, some of which however have little to do with Paracelsus and his doctrines. The collection is now preserved in the Homœopathic Medical College, Philadelphia, Pennsylvania.

JÖRG, LEONHARD. Die Naturwissenschaft des Paracelsus. Programm der K. Studienanstalt zu Landau am Schlusse des Studienjahres 1881–82. Landau, 1882. pp. 30, 8vo.

*LESSING, MICHAEL BENEDICT. Paracelsus, sein Leben und Denken. Drei Bücher. Berlin, 1839. pp. xvi–250, 8vo. Portrait.

LOCHER, HANS. Theophrastus Bombastus von Hohenheim der Luther der Medicin und unser grösster Schweizerarzt. Eine Denkschrift auf die Feier des Züricher Jubilarfestes vom 1. Mai, 1851, und ein Beitrag zur Würdigung vaterländischer Verdienste in jedem gebildeten Kreise. Zürich, 1851. pp. vi–68, 8vo. Portrait.

PARACELSUS. [Cont'd.]

MARX, KARL FRIEDRICH HEINRICH.   Zur Würdigung des Theophrastus
von Hohenheim. Göttingen, 1842. pp. 140, 4to.

* MOOK, FRIEDRICH.   Theophrastus Paracelsus.   Eine kritische Studie.
Würzburg, 1876. pp. [vi]–136, 4to.

> Contains a bibliography of 276 titles. *See, under Paracelsus*, Ferguson,
> John, *also* Rohlfs, H., *and* Schubert, E.

POISSON, ALBERT.   Notice biographique sur Paracelse. *See in Section
VI.*

RITTMANN, A.   Das reformirte Deutschland und sein Paracelsus.
Als viertes Heft der culturgeschichtlichen Abhandlungen über die
Reformation der Heilkunst. Wien, 1875. pp. 57, 8vo.

> A study of Paracelsus' contributions to medical art.

ROHLFS, HEINRICH.   Mooks' Theophrastus Paracelsus, eine kritische
Studie. *In* Deutsches Archiv für Geschichte der Medicin und
Medicinischen Geographie, 5. Jahrgang. Leipzig, 1882. p. 213
et seq, 8vo.

> *See, under Paracelsus*, Mook, F. ; *also*, Schubert, Ed., *and* Ferguson, John.

SCHERER, ALEXANDER NICOLAUS VON.   Theophrastus Paracelsus gewürdigt . . . St. Petersburg, 1822. 8vo.

SCHUBERT, EDUARD, AND KARL SUDHOFF.   Paracelsus-Forschungen.
Erstes Heft.   Inwiefern ist unser Wissen über Theophrastus von
Hohenheim durch Friedrich Mook und seinen Kritiker Heinrich
Rohlfs gefördert worden ?   Eine historisch-kritische Untersuchung.
Frankfurt a. M., 1887. pp. vi–89, 8vo.

SOANE, GEORGE.   The Life and Doctrines of Paracelsus. *In* New
Curiosities of Literature, vol. I, p. 194.

* STEINER, LEWIS H.   Paracelsus and his influence on Chemistry and
Medicine. Chambersburg, Pa., 1853. pp. 18, 8vo.

> A lecture before the Baltimore Medical Institute delivered April 5, 1853.

SUAVIUS.   Vita et catalogus operum et librorum Phil. Aur. Theophr.
Bombast. Paracelsi de Hohenheim.   T. Paracelsi philosophiæ et
medicinæ utriusque universæ compendium. Basileæ, 1568.

WERNECK, DR.   Zur Geschichte des Paracelsus. Clarus und Radius.
Beiträge zur Medicin. III Band. Leipzig, 1836. pp. 209–238, 8vo.

PARACELSUS.   Essays on the Life and Works of Paracelsus have been published by
Bordes-Pagès in Revue indépendante, 10 April, 1847 ; Michéa in Gazette
médicale, 7 and 14 May, 1842 ; Cruveilhier in Philosophie des sciences
médicales, Œuvres choisies, Paris, 1861, 18mo ; Franck, Notice sur Paracelse
et l'alchimie au seizième siècle lue à la séance publique des cinq Académies
le 25 décembre, 1853.   I have made no attempt at completeness in this brief
list of the biographies of Paracelsus.

PARIS, JOHN AYRTON. Life of Sir H. Davy, *see* Davy, H.

PARISET, E. Éloge de C. L. Cadet-Gassicourt. *See* Cadet-Gassicourt, C. L.

PARKMAN, THEODORE.
　Obituary. Am. J. Sci. [2], xxxv, 155, May, 1863.

PARMENTIER. Éloge de Pierre Bayen. *See, in Section V*, Malatret, P.

PARRISH, EDWARD.
　PROCTER, WILLIAM, JR. A Memorial of E. P. Read at the Annual
　　Meeting of the Philadelphia College of Pharmacy, March 31, 1873.
　　Philadelphia, 1873. pp. 7, 8vo.

PARRY, GEORGE.
　Obituary. J. Chem. Soc., London, xxxi, 506 (1877).

PARSONS, HENRY BETTS.
　* PRESCOTT, ALBERT B. Henry Betts Parsons. A Memorial Address
　　delivered before the Alumni Association of the School of Pharmacy,
　　University of Michigan, with a Bibliography of the published
　　writings of Prof. Parsons. [Detroit, 1886.] pp. 16, 8vo.
　　　　For portrait of Parsons, *see* Proc. Am. Pharm. Assoc., 1885.

PASTEUR, LOUIS.
　Life and portrait. Pop. Sci. Mon., xx, 823, Apr., 1882.
　Notice des travaux de L. Pasteur. [Paris, *n. d.*] pp. 16, 4to.
　Histoire d'un savant par un ignorant. Paris, 1884. pp. 392, 18mo.
　　Portrait.
　ROSCOE, SIR HENRY E. The Life Work of a Chemist. An Address
　　delivered to the members of the Birmingham and Midland Insti-
　　tute in the Town Hall, Birmingham, on October 7, 1889. *In*
　　Nature, October 10, 1889, vol. XL, pp. 578–583. *Also in* Report
　　Regents Smithsonian Institution, July, 1889. pp. 491–506. Wash-
　　ington, 1890.

PATTINSON, HUGH LEE.
　Obituary. Q. J. Chem. Soc., London, XII, 169 (1860).

PATTISON-MUIR, M. M. *See* Muir, M. M. Pattison.

PEAKE, WILLIAM HENRY ASTON.
　Obituary. J. Chem. Soc., London, XLV, 617 (1884).

PEBAL, LEOPOLD VON.
　HOFMANN, A. W. Nekrolog auf L. Pebal. Ber. d. chem. Ges., xx,
　　p. 467 (1887).
　MEYER, LOTHAR Nekrolog auf L. von Pebal. Ber. d. chem. Ges.,
　　xx. Ref. p. 997 (1887). Portrait.

PELIGOT, EUGÈNE MELCHIOR.

JUNGFLEISCH, E. Notice sur la vie et les travaux de E. M. P. Bull. Soc. chim. [3], v, xxi, 1891. Portrait.

Notice des travaux de E. P. [Paris], *n. d.* pp. 34, 4to.

PELLETIER, BERTRAND.

Biographical Notices. *Lassus:* Decade philosophique, An. vi, No. 6, p. 330 ; *also* Mag. encycl., 3e an. vol. iv, pp. 187–192. *Darcet:* Lycée des arts, An. vi. *Sedillot:* Recueil périodique de la Société de Médecine de Paris, No. 15 (1797). *Also,* Moll's Jahr. d Berg u. H., vol. ii, p. 405. *Bouillion Legrange:* Trommsdorff's Journal der Pharmacie, vol. v, p. 345. *Guyton:* Journal de l'école polyt. Cah. 5, p. 185. All. J. Chem., Scherer, i, 330, 1798.

PELOUZE, JULES THÉOPHILE.

Obituary. Am. J. Sci. [2], xliv, 137, 1867.

Obituary. Chem. News, xv, 294, June 7, 1867.

Obituary. J. Chem. Soc., London, xxi, xxv (1868).

DUMAS, J. B. Éloge sur J. P. Discours et éloges académiques, vol. i, p. 125.

PENNY, FREDERICK.

Obituary. J. Chem. Soc., London, xxiii, 301 (1870).

PERCY, JOHN.

Obituary. Chem. News, lx, 11 (1889).

PERETTI, PIETRO.

FABRI-SCARPELLINI, ERASMO. Sopra i lavori chimico-farmaceutici del Pietro Peretti ; breve commentario. Roma, 1850. 8vo.

PERROQUET. La vie de Raymond Lulle, *see* Lullius, Raymundus.

PETTENKOFER, MAX VON.

Life and portrait. Pop. Sci. Mon., xxiii, 841, Oct., 1883.

PETTENKOFER, MAX VON. Zum Gedächtniss des Liebig, *see* Liebig, Justus von.

PETZHOLDT, GEORG PAUL ALEXANDER.

PETZHOLDT, JULIUS. Biographisch-litterarische Skizze, von G. P. A. P. Abdruck aus dem Neuen Anzeiger für Bibliographie und Bibliothekswissenschaft, Oktober-Heft, 1857. Dresden, 1857. 12mo.

PFINGSTEN, JOHANN HERMANN.

Bibliothek ausländischer Chymisten, Mineralogen u. s. w., nebst derley biographische Nachrichten. Nürnberg, 1781–84. 4 vols., 8vo.

PFINGSTEN, JOHANN HERMANN. [Cont'd.]

> For fuller title and details, *see the same, in Section V.* The volumes contain
> biographies and portraits of Walch, Gesner, Brückmann and Hundert-
> mark.

PHIPSON, THOMAS LAMB.

BOUVERIE, C. J. The Scientific and Literary Works of Thomas Lamb
Phipson, with a short biographical notice. London, 1884. pp.
32, 8vo.

PIESSE, GEORGE WILLIAM SEPTIMUS.

Obituary. J. Chem. Soc., London, XLIII, 255 (1883).

PIRIA, RAFFAELE.

Obituary. J. Chem. Soc., London, XIX, 512 (1866).

* CANNIZZARO, STANISLAO. Discorso sulla vita e sulle opere di Raffaele
Piria il giorno 14 marzo 1883, nella regia università di Torino
inaugurandosi un busto del Piria. Torino, 1883. pp. 88, 8vo.

> With portrait (photograph) and facsimile of a letter.

DUMAS, J. B. Nécrologie de R. P. Bull. soc. chim., vol. 4, p. 182
(1865).

MATTEUCCI, CH. Nécrologie de R. P. Bull. soc. chim., vol. 4, p. 184
(1865).

PLANCHE, LOUIS ANTOINE.

BOUDET, F. Éloge de Louis Antoine Planche, pharmacien ; membre
de l'Académie royale de médecine, etc. Paris, 1841. 8vo. [Né
en 1776, mort le 7 mai, 1840.]

PLANCHE, LOUIS ANTOINE. Nécrologie de, *see* Girardin, Jean.

PLAYFAIR, SIR LYON.

Life and portrait. Pop. Sci. Mon., XXVIII, 117, Nov., 1885.

PLUNKETT, WILLIAM.

Obituary. J. Chem. Soc., London, XLV, 618 (1884).

POGGENDORFF, J. C.

Obituary. Am. J. Sci. [3], XIII, 246 (1877).

POISSON, SIMÉON DENIS. Life of, *see* Arago, François.

PORRETT, ROBERT.

Obituary. J. Chem. Soc., London, XXII, vii (1869).

POTT, JOHANN HEINRICH.

POTT, ROBERT. J. H. P., ein Beitrag zur Geschichte des Zeitalters
der Phlogistontheorie. Jena, 1876. pp. 23, 4to.

POTTS, LAWRENCE HOLKER.
Obituary.　Q. J. Chem. Soc., London, IV, 346 (1852).

POULTIER D'ELMOTTE.　Vie de J. B. Van Helmont.　*See* Helmont, J. B. van.

PRAUSNITZ, G.　*See* Richter, Victor von.

PRENTICE, EDWARD HENRY.
Obituary.　J. Chem. Soc., London, XXV, 352 (1872).

PRESCOTT, ALBERT B.　Life of H. B. Parsons, *see* Parsons, H. B.

PRESCOTT, ALBERT BENJAMIN.
BENJAMIN, MARCUS.　Biographical sketch of A. B. P.　Scientific Am.,
　　vol. LXV, p. 119, Aug. 22, 1891.
Index to the American and European Publications of Original Articles
　　on Chemistry and Pharmacy, and works on Analytical Chemistry
　　by A. B. P.　*n. t. p.*　[Ann Arbor] [1880] 8vo.
　　　　This bibliography of Prescott embraces 46 numbers.

PRIESTLEY, JOSEPH.
Portrait *in* Neues all. J. Chem., Gehlen, vol. VI.　Frontispiece.　1806.
Nekrolog.　Neues all. J. Chem., Gehlen, vol. II, p. 706, 1804.
* Memoirs of Joseph Priestley to the year 1795, written by himself, with
　　a continuation to the time of his decease, by his son Joseph Priestley ;
　　and observations on his writings by Thomas Cooper and the Rev.
　　William Christie.　London, 1803.　2 vols., 8vo.　Portrait by Part-
　　ridge after Stuart.
* BOLTON, HENRY CARRINGTON.　Scientific Correspondence of Joseph
　　Priestley.　Ninety-seven letters addressed to Josiah Wedgwood,
　　Sir Joseph Banks, Capt. James Keir, James Watt, Dr. William
　　Withering, Dr. Benjamin Rush, and others.　Together with an
　　Appendix : I. The Likenesses of Priestley in Oil, Ink, Marble, and
　　Metal.　II. The Lunar Society of Birmingham.　III. Inventory
　　of Priestley's Laboratory in 1791.　Edited with copious biographi-
　　cal, bibliographical and explanatory notes.　New York, privately
　　printed, 1892.　pp. vii–240, 8vo.　Portraits of Priestley and Wedg-
　　wood.
* CORRY, JOHN.　The Life of Joseph Priestley, etc., etc., with critical
　　observations on his works.　Birmingham, 1804.　pp. 112, 18mo.
　　Portrait.
* FONVIELLE, W. DE.　Célébration du premier centenaire de la décou-
　　verte de l'oxygène 1er août, 1774.　La vie et les travaux du Docteur
　　Priestley.　Paris, 1875.　36 pp., 18mo.
　　　　A lecture delivered August 2d in the Salle des Écoles, Paris.　Contains an
　　engraving of the Birmingham statue.

PRIESTLEY, JOSEPH. [Cont'd.]

HENRY, WILLIAM. An Estimate of the Philosophical Character of Dr. Priestley. Am. J. Sci., XXIV, 28, 1833.

MARSH, JOHN FICKETT. On some Correspondence of Dr. Priestley preserved in the Warrington Museum and Library. (Read 19th April, 1855.) Transactions of the Historic Society of Lancashire and Cheshire, vol. VII, pp. 65–81. London, 1855. 8vo.

> Extracts from letters of a domestic character.

* MAY, JOSEPH. Joseph Priestley. A Discourse delivered in the First Unitarian Church of Philadelphia on Sunday, March 18, 1888. Printed by request. Philadelphia, n. d. pp. 24, 8vo.

* RUTT, JOHN TOWILL. Life and Correspondence of Joseph Priestley. London, 1831. 2 vols., 8vo.

> An expansion of Priestley's "Memoirs," by one interested chiefly in the theological and literary labors of the Rev. Dr., whose chemical work is lightly touched upon. Contains hundreds of Priestley's letters.

[Seven letters of Joseph Priestley, with annotations; an unsigned paper written apparently by the Editor.] Christian Reformer, New Series, vol. VII, 1851. London, 1851. pp. 100–108.

> These letters chiefly on domestic matters are selected from a series preserved in the Museum and Library of Warrington. The series was collected by John Fickett Marsh, and extend from 1790 to 1802. They were nearly all addressed to Priestley's brother-in-law, John Wilkinson. *See under* Priestley, J., Marsh, J. F., *and* Bolton, H. C.

* THORPE, T. E. Joseph Priestley : his Life and Chemical Work. A Lecture delivered in the Hulme Town Hall, Manchester, on Wednesday, November 18, 1874. Science Lectures for the People, Sixth Series, 1874. Manchester, n. d.

* TIMMINS, SAMUEL. Dr. Priestley in Birmingham. Read to the Archæological Section of the Birmingham and Midland Institute, January 29, 1875. Transactions, 1875. Birmingham, 1877. 4to.

* WARE, HENRY, JR. A Memoir of the life of Joseph Priestley, by Henry Ware, Jr. Introductory to Views of Christian Truth, Piety, and Morality selected from the writings of Jos. Priestley. Cambridge [Mass.], 1834. pp. lxxx–207, 12mo.

YATES, JAMES. Memorials of Joseph Priestley. Christian Reformer, 1860. 13 pp.

PRIESTLEY, JOSEPH. Life, *see* Brougham, Lord ; Hoefer, Ferdinand. *Also,* Pop. Sci. Mon., v, 480, Aug., 1874. *See also in Section III*, Proceedings at the Centennial of Chemistry held in 1874.

PRIESTLEY (THE) MEMORIAL AT BIRMINGHAM, August, 1874. London, 1875. pp. 164. Portrait.

PRIESTLEY (THE) MEMORIAL. [Cont'd.]
> Contains reports of the centenary celebrations at Birmingham, Leeds, Paris, and Northumberland. Published by the British and Foreign Unitarian Association.

PROTH, MARIO. Life of Alfred Naquet, *see* Naquet, Alfred.

PROUST, LUIS JOSÉ.
> Life and bibliography. *See in Section I*, Maffei, Eugenio, y Ramón Rua Figueroa, Apuntes para una biblioteca española. Madrid, 1873. Vol. II, p. 68, 8vo.

PROUT.
> Obituary. Am. J. Sci. [2], x, 138, Nov., 1850.

PRUGGMAYR, MARTIN MAXIMILIAN.
> Verzeichniss der Verfasser und Namen der philosophischen Adepten . . . *See in Section VI, the author's* Philosophische Untersuchung des wahrhaften Lebens-Elixieres.

PUGH, EVAN.
> Obituary. J. Chem. Soc., London, XVIII, 346 (1865).

PYPERS, JOSEPH IGNACE HUBERT.
> BROECKX, C. Notice sur Joseph Ignace Hubert Pypers, secrétaire de la Société de pharmacie d'Anvers. Avec portrait. Anvers, 1850. 8vo.

QUESNEVILLE, GUSTAV AUGUSTIN.
> Obituary. Chem. News, LX, 269 (1889).
> HOFMANN, A. W. Nekrolog auf G. A. Quesneville. Ber. d. chem. Ges., XXII, p. 3169 (1889).

RADZIEJEWSKI, SIEGMUND.
> SALKOWSKI. Nekrolog auf S. Radziejewski. Ber. d. chem. Ges., VII, p. 1801 (1874).

RAMMELSBERG, C. *See* Rose, Gustav.

RAYMUND LULLY. *See* Lullius, Raymundus.

RÉAUMUR, RÉNÉ ANTOINE FERCHAULT DE. Life of, *see* Savérien, A.: Histoire des philosophes modernes, vol. VIII, p. 205. Portrait. *Also*, Figuier: Vie des savants.

REDTENBACHER, JOSEPH.
> Obituary. J. Chem. Soc., London, XXIII, 311 (1870).

REGNAULT, HENRY VICTOR.
  Obituary. J. Chem. Soc., London, XXXIII, 235 (1878).

REIMER, KARL LUDWIG.
  HOFMANN, A. W. Nekrolog auf K. L. Reimer. Ber. d. chem. Ges.,
    XVI, p. 99 (1883).

REISET, J. La vie de Millon, *see* Millon, N. A. E.

RENWICK, JAMES.
  Obituary. Am. J. Sci. [2], XXXV, 306, May, 1863.

RENWICK, JAMES. Life of Count Rumford, *see* Thompson, Benjamin.

RÉVEIL, PROFESSOR.
  Obituary. Chem. News, vol. 11, p. 305 (1865).

REYMOND, P. A.
  CADET-GASSICOURT, F. Éloge de P. A. Reymond, prononcé le 15
    novembre, 1854, dans la séance annuelle et publique de rentrée de
    l'École de pharmacie de Paris. Paris, 1855. 8vo. Extrait du
    Répertoire de pharmacie, janvier, 1855.

REYNOLDS, JOHN WILLIAM.
  Obituary. J. Chem. Soc., London, XXIX, 620 (1876).

RHEES, WM. J. Life and writings of James Smithson, *see* Smithson, James.

RICHARDSON, THOMAS.
  Obituary. Chem. News, XVI, 40, July 19, 1867. Condensed biography,
    Chem. News, XIX, 140, March 19, 1869. With portrait.

RICHTER, JEREMIAS BENJAMIN.
  HESS, G. H. Ueber Richter's Arbeiten. J. prakt. Chem., XXIV, 1841.
  LÖWIG, K. J. Der Entdecker der chemischen Proportionen. Eine
    Denkschrift. Breslau, 1874. pp. 56, 4to.

RICHTER, VICTOR VON.
  G. PRAUSNITZ. Nekrolog auf V. v. R. Ber. d. chem. Ges., XXIV,
    1123–1130 (1891).

RITTMAN, A. Das reformirte Deutschland und sein Paracelsus. *See* Paracelsus.

RIXNER, THADDÄ ANSELM, UND THADDÄ SIBER.
  * Leben und Lehrmeinungen berühmter Physiker am Ende des XVI und
    am Anfange des XVII Jahrhunderts, als Beyträge zur Geschichte
    der Physiologie in engerer und weiterer Bedeutung. Sulzbach,
    1819–26. 7 vols., 8vo.

RIXNER, THADDÄ ANSELM, UND THADDÄ SIBER. [Cont'd.]
> Vol. I, Paracelsus; II, Cardanus; III, Patritius; IV, Telesius; V, Brunus; VI, Campanella; VII, Van Helmont. Each biography is accompanied by a portrait, and is based on the author's writings, from which systematically arranged extracts are given.

ROBERTSON, ADRIAAN.
Boyle en Boerhaave beschouwd als Scheikundigen. Academisch Proefschrift ter verkrijging van den graad van Doctor in de artsenijbereidkunde aan de Universiteit van Amsterdam. Rotterdam, *n. d.* pp. 129, 8vo.

ROBIN, EDOUARD.
LACOSTE, ALEXIS DE. Essai biographique sur les travaux en chimie de M. Edouard Robin. Paris, 1853. 8vo.

ROBIQUET, M. Nécrologie de. *See* Girardin J.

ROBISON, JOHN.
Nekrolog. All. nord. Ann. Chem., vol. V, p. 235 (1820).

ROCHLEDER, FRIEDRICH.
HLASIWETZ, H. Nekrolog auf F. Rochleder. Ber. d. chem. Ges., VIII, p. 1702 (1875). Portrait of F. Rochleder.

RÖMER, HERMANN.
HOFMANN, A. W. Nekrolog auf H. Römer. Ber. d. chem. Ges., XVIII, p. 285 (1885).

ROGERS, JAMES B.
Obituary. Am. J. Sci. [2], XIV, 290. Nov., 1852.
CARSON, JOSEPH. A Memoir of the Life and Character of J. B. R. Delivered by request of the Faculty, Oct. 11, 1852, and published by the class. Philadelphia, 1852, pp. 22, 8vo.

ROGERS, ROBERT EMPIE.
HOLLAND, J. W. A Eulogy on the life and character of R. E. R. Introductory to the Course of 1885–86 at Jefferson Medical College. Delivered Sept. 30, 1885. Philadelphia, 1885. pp. 26, 8vo.
RUSCHENBERGER, W. S. W. A sketch of the life of R. E. R., with biographical notices of his father and brothers. Read before the American Philosophical Society November 6, 1885. Philadelphia, 1885. pp. 45, 8vo.
> Contains a bibliography.

ROGERS, THOMAS KING.
Obituary. J. Chem. Soc., London, XLVII, 335 (1885).

Rogers, William Barton.

> Cooke, Josiah Parsons. Notice of W. B. R., Founder of the Massa-chusetts Institute of Technology. From Proceedings of the American Academy of Arts and Sciences, vol. xviii, p. 426. Cambridge, 1883. pp. 428–438, 8vo.
>
> Walker, Francis A. Memoir of W. B. R., read before the National Academy [of Sciences], April, 1887. Washington, D. C. *n. d.* pp. 13, 8vo.

Romanis, Robert.

> Obituary. J. Chem. Soc., London, lvii, 455 (1890).

Rommelaere, Dr. Études sur J. B. Van Helmont. *See* Van Helmont, J. B.

Ronalds, Edmund.

> Obituary. J. Chem. Soc., London, lvii, 456 (1890).

Roscoe, Sir Henry E.

> Life and portrait. Pop. Sci. Mon., xxvi, 402, Jan., 1885.

Roscoe, Sir Henry E. Life of Dalton. *See* Dalton, John.

Rose, Gustav.

> Rammelsberg, C. Nekrolog auf G. Rose. Ber. d. chem. Ges., vi, p. 1573 (1873).

Rose, Gustav. Gedächtnissrede auf Mitscherlich. *See* Mitscherlich, Eilhardt.

Rose, Heinrich.

> Obituary. J. Chem. Soc., London, xvii, 437 (1864).
>
> Obituary. Am. J. Sci. [2], xxxvii, 304, May, 1864 ; *also*, xxxviii, 305, Nov., 1864.

Rose, Heinrich. Gedächtnissrede auf Berzelius. *See* Berzelius, J. J.

Roth-Scholtz, Friedrich. Lebensbeschreibung J. J. Bechers. *See* Becher, J. J.

Rouelle, Guillaume François (Ainé).

> Cap, Paul Antoine. Guillaume François Rouelle. Biographie chi-mique. Paris, 1842. 8vo. Extract from J. de pharm. et de chimie.
>> *See* Figuier : Vie des savants.

Rumford, Count. *See* Thompson, Benjamin.

Rumney, Robert.

> Obituary. J. Chem. Soc., London, xxvi, 780 (1873).

Rupstein, Friedrich.

> Tiemann, Ferd. Nekrolog auf F. Rupstein. Ber. d. chem. Ges., viii, p. 1712 (1875).

RUTHERFORD, DANIEL, Life of.   *See* Walker, Wm., Jr.

RUTT, JOHN TOWILL.   Life of Joseph Priestley.   *See* Priestley, Joseph.

SACHSE, ULRICH.   *See* Lehmann, Arthur.

SAGE, BALTHAZAR GEORGES.
Analyse du lait de vache suivie de la liste chronologique des ouvrages
publiés [par l'auteur] dans l'espace de 51 ans   Paris, 1820.   8vo.
Exposé sommaire des principales découvertes faites [par l'auteur] dans
l'espace de 54 années.   Paris, 1813.   8vo.

SAINT MAURICE CABANY.   Notice de J. G. Dizé.   *See* Dizé, J. G.

SAINTE-CLAIRE-DEVILLE, ÉTIENNE HENRI.
Obituary.   Chem. News, XLIV, 34 (1881).
Obituary.   J. Chem. Soc., London, XLI, 235 (1882).
DUMAS, J. B.   Éloge sur S.-C.-D.   Discours et éloges académiques.
Vol. II, p. 281.

SALKOWSKI.   *See* Radziejewski, Siegmund.

SALVERTE, A. J. E. BACONNIÈRE.   Vie de C. L. Cadet-Gassicourt.   *See* Cadet-Gassicourt,
C. L.

SALVÉTAT.   Travaux de Ebelmen.   *See* Ebelmen, J. J.

SASSENIUS, A. D.
BROECKX, C.   Notice sur A. D. Sassenius pharmacien, professeur de
chimie et de botanique à l' Université de Louvain.   Anvers, 1850.
8vo.

SAUSSURE, TH. DE.
MACAIRE, ISAAC FRANÇOIS.   Notice sur la vie, etc., de Th. de S.
Genève, 1845.

SAVÉRIEN, A.
* Histoire des philosophes modernes avec leur portrait gravé dans le
goût du crayon d'après les planches in 4to dessinées par les plus
grands peintres.   Histoires des métaphysiciens.   À Paris, de l'imprimerie de Brunet, imprimeur de l'Académie Françoise.   1762.
8 vols., 16mo.   Portrait of Savérien.
Third edition, 1767.
> Contains 67 biographies of men of letters, philosophers, and scientists, accom-
> panied by 53 portraits.   Among them the following : Fr. Bacon, Gas-
> sendi, Descartes, Pascal, Newton, Leibnitz, Halley, Bernouilli, Coper-
> nicus, Tycho Brahé, Galileo, Kepler, Huyghens, Boyle, Paracelsus,
> Lémery, Boerhaave, Hales and Réaumur, with portraits of each.
> *Also* Homberg, Kunckel, Lefèvre.   *See also, in Section III*, Savérien.

SCHACHT, L.   Necrology of Franciscus de la Boe Sylvius.   *See* Sylvius, F. d. l. B.

SCHAEDLER, CARL.
* Biographisch-litterarisches Handwörterbuch der wissenschaftlich-bedeutenden Chemiker.   Berlin, 1891.   pp. vi-162, 12mo.

> A convenient handbook, though not very full.   The author repeats the fable of the death of Priestley by poison, and asserts that he made his memorable investigations while travelling through Holland, France and Germany.   The compiler omits many prominent Frenchmen and Englishmen, and mentions but four Americans.

SCHEELE, CARL WILHELM.
Life and portrait.   Pop. Sci. Mon., XXXI, 839, Oct., 1887.

> Life, *see* Hoefer, Ferdinand.

Portrait, *in* All. J. Chem., Scherer, vol. v.   Frontispiece.

BLOMSTRAND, C. W.   Minnesteckning öfver Carl Wilhelm Scheele.   Stockholm, 1886.   8vo.

CAP, PAUL ANTOINE.   Scheele, chimiste Suédois.   Étude biographique.   Paris, 1863.   pp. 40, 8vo.   (Extrait du Journal de pharmacie et de chimie, Avril et Mai, 1863.)

* CLEVE, P. T.   Carl Wilhelm Scheele.   Ett Minnesblad på hundrade årsdagen af hans död.   Köping, *n. d.* [1886].   pp. 54.   With portrait and facsimile.

FLÜCKIGER, F. A.   Zur Erinnerung an Scheele ein Jahrhundert nach seinem Ableben.   Separat-Abdruck aus dem Archiv der Pharmacie, Bd. 24, Heft 9, 1886.   pp. 50, 8vo.

> With a portrait of Scheele.

HAYS, B. FRANK.   The Life Work of Carl Wilhelm Scheele.   A Paper read before the Alumni Association of the College of Pharmacy of the City of New York, Nov. 21, 1884,   New York, 1884.   pp. 14, 16mo.   From the Pharmaceutical Record, December 1, 1884.

LECONTE, A.   Notice biographique suivie de : Le laboratoire de Scheele (1742–1786).   Lyon, 1885.   8vo.

SCHEERER, JOHANN AUGUST THEODOR.
Obituary.   Am. J. Sci. [3], x, 320, 1875.

SCHEIBLER, C.   *See* Griess, Peter.   *See also*, Stöckhardt, Adolf J.

SCHENCK, ROBERT.
Obituary.   J. Chem. Soc., London, XXIX, 621 (1876).

SCHERER, JOHANN JOSEPH.
WAGNER, R.   Nekrolog auf J. J. Scherer.   Ber. d. chem. Ges., II, 108 (1869).

SCHERING, ERNST FRIEDRICH CHRISTIAN.
HOLTZ, J. F. Nekrolog auf E. F. C. Schering. Ber. d. chem. Ges.,
XXIII, p. 900 (1890). Portrait of Schering.

SCHLIEPER, ADOLPH.
HOFMANN, A. W. Nekrolog auf A. Schlieper. Ber. d. chem. Ges.,
XX, p. 3167 (1887).

SCHÖNBEIN, CHRISTIAN FRIEDRICH.
Obituary. J. Chem. Soc., London, XXII, x (1869).
* HAGENBACH, EDUARD. C. F. Schönbein, Programm für die Rectorats-
feier der Universität [Basel]. Basel, 1868. pp. 64–xxii, 4to.
> The author writes in separate sections of Schönbein's life and scientific
> labors. Annexed is a bibliography of Schönbein's writings.

SCHRÖTTER, ANTON, RITTER VON KRISTELLI.
LIEBEN, AD. Nekrolog auf A. Schrötter. Ber. d. chem. Ges., IX, p.
90 (1876). Portrait of Schrötter.
Obituary. J. Chem. Soc., London, XXIX, 622 (1876).

SCHULTENS, A. In memoriam Hermanni Boerhaave, *see* Boerhaave, H.

SCHULZE, J. H. *See, in Section III*, Eder, J. M., Geschichte der Photochemie.

SCHULZE, FRANZ.
Nekrolog. Ber. d. chem. Ges., VI, p. 775 (1873).

SCHUTZENBERGER, PAUL.
Exposé des titres et des travaux scientifiques de P. S. Paris, 1884.
pp. 63, 4to.

SCHWEIGGER, JOHANN SALOMO CHRISTOPH.
MARTIUS, CARL FRIEDRICH PHILIPP VON. Denkrede auf Johann Salomo Christoph Schweigger, gehalten in der öffentlichen Sitzung
der königl. bayr. Akademie der Wissenschaften am 28 November,
1857. München, 1858. pp. 11, 4to.

SCHWERTZER, SEBALT.
* KELL, RICHARD. S. S. als Kursächsischer Faktor und Kaiserlicher
Berghauptmann. Inaugural-Dissertation zur Erlangung der Dok-
torwürde in der philosophischen Fakultät der Universität Leipzig,
verfasst von Richard Kell aus Kirchberg. Leipzig, 1881. pp. 80,
8vo.
> An historical study of this noted alchemist at the court of Elector Augustus
> of Saxony, based on manuscripts preserved in the Royal Saxon Archives
> of State.

SCOPÓLI, J. A.
  CRELL, L. Zum Andenken von J. A. Scopoli. Chem. Ann., Crell., 1788, II, p. 534.
  PANORMITA, J. T. Monumentum Joannis Antoni Scopuli. Chem. Ann., Crell, 1788, II, p. 534.
  Ad. B. Epiphani in ejusdem Scopuli funere indictivo. Chem. Ann., Crell., 1788, II, p. 536.

SCOTT, HENRY YOUNG DARRACOTT.
  Obituary. J. Chem. Soc., London, XLV, 619 (1884).

SEAR, FREDERICK.
  Obituary. J. Chem. Soc., London, LI, 477 (1887).

SEEBECK, L. F. W. A. Gedächtnissrede auf A. Volta, see Volta, A.

SELL, EUGEN. See Duppa, Baldwin. See also Calvert, Crace.

SELLERS, COLEMAN, see Morton, Henry.

SENDIVOGIUS, MICHAEL. Life and bibliography, see Adelung, J. F.

SENHOFER, C., UND G. GOLDSCHMIEDT. See Barthenau, Ludwig Barth von.

SIEMENS, CHARLES WILLIAM.
  Obituary. J. Chem. Soc., London, XLV, 624 (1884).

SILLIMAN, BENJAMIN (JR.). Sketch of John Lawrence Smith, see Smith, J. L.

SILLIMAN, BENJAMIN (JR.).
  Life and portrait. Pop. Sci. Mon., XVI, 550, Feb., 1880.
  Obituary. Am. J. Sci. [3], XXIX, 85, February, 1885.
  Obituary. Chem. News, LI, 69 (188

SILLIMAN, BENJAMIN (SR.).
  Biographical notice in Biographical Memoirs of the National Academy of Sciences, vol. I, p. 99. Washington City, 1877. 8vo.
  Life and portrait. Pop. Sci. Mon., XXIII, 259, June, 1883.
  Obituary. Am. J. Sci. [2], XXXIX, 422.
  FISHER, GEORGE P. Life of Benjamin Silliman, chiefly from his manuscript reminiscences, diaries and correspondence. New York, 1866. 2 vols., 8vo. Portrait.

SKALWEIT, JOH.
  HOFMANN, A. W. Nekrolog auf Joh. Skalweit. Ber. d. chem. Ges., XX, p. 2771 (1887).

SMEE, ALFRED.
Obituary. J. Chem. Soc., London, XXXI, 509 (1877).

SMITH, FRANK.
Obituary. J. Chem. Soc., London, XXXI, 511 (1877).

SMITH, HENRY.
Obituary. J. Chem. Soc., London, LX, 465 (1891).

SMITH, HENRY JOHN STEPHEN.
Obituary. J. Chem. Soc., London, XLIII, 255 (1883).

SMITH, JAMES HILL.
Obituary. J. Chem. Soc., London, XLV, 627 (1884).

SMITH, JOHN LAWRENCE.
Life and portrait. Pop. Sci. Mon., VI, 233, Dec., 1874.
Obituary. Am. J. Sci. [3], XXVI, 414, 1883.
MICHEL, MIDDLETON. In Memoriam, J. L. S. From Year-book City of Charleston, South Carolina, *n. p.* 1884. pp. 11, 8vo.
SILLIMAN, BENJAMIN [JR.]. Sketch of the Life and Scientific Work of John Lawrence Smith. With a complete list of his published memoirs, etc. Washington, D. C., 1884. pp. 32, 8vo. *In* Bulletin of the National Academy of Sciences.
Biographical notice in Biographical Memoirs of the National Academy of Sciences, vol. II, p. 217. Washington City, 1886. 8vo.

SMITH, JOSEPH DENHAM.
Obituary. J. Chem. Soc., London, LV, 294 (1889).

SMITH, ROBERT ANGUS. Obituary of, *see* Voit, C. von.

SMITH, ROBERT ANGUS.
Obituary. Am. J. Sci. [3], XXVIII, 79, 1884.
Obituary. J. Chem. Soc., London, XLVII, 335 (1885).
Obituary. Chem. News, XLIX, 222 (1884).
HOFMANN, A. W. Nekrolog auf R. A. Smith. Ber. d. chem. Ges., XVII, p. 1211 (1884).

SMITH, ROBERT ANGUS. Memoir of John Dalton, *see* Dalton, John. Life of Thomas Graham, *see* Graham, Thomas.

SMITHSON, JAMES.
* RHEES, WILLIAM J. The Scientific Writings of James Smithson. (Smithsonian Miscellaneous Collections, No. 327.) Washington, 1879. pp. vii–159, 8vo.

SMITHSON, JAMES. [Cont'd.]
* James Smithson and his Bequest. (Smithsonian Miscellaneous Contributions, No. 330.) Washington, 1880. pp. viii–68, 8vo.

> Contains three portraits of Smithson, facsimiles, and engraving of the Smithsonian Institution.

SÖMMERING, PHILIPP. For the adventures of this alchemist, *see, in Section III*, Rhamm, A. : Die betrüglichen Goldmacher [etc.].

SONNENSCHEIN, FRANZ LEOPOLD.
HOFMANN, A. W. Nachricht von dem Tode des F. L. Sonnenschein. Ber. d. chem. Ges., XII, p. 401 (1879).

SOUTHBY, E. R.
Obituary. J. Chem. Soc., London, LI, 476 (1887).

SPENCE, PETER.
Obituary. J. Chem. Soc., London, XLV, 622 (1884).
Obituary. Chem. News, XLVIII, 19 (1883).

SPOTTISWOODE, WILLIAM.
Obituary. J. Chem. Soc., London, XLV, 628 (1884).

SPRÖGEL, J. CHR. Life of David Beuther, *see* Beuther, David.

STÄDELER, GEORG.
KRAUT. Nekrolog auf G. Städeler. Ber. d. chem. Ges., IV, p. 425 (1871).

STAHL, GEORG ERNEST.
GOETZIUS, JOH. CHRISTOPH. Georgi Ernesti Stahlii aliorumque ad ejus mentem disserentium scripta serie chronologica. Norimbergæ, 1726. pp. ii–184 and index, 4to.

> This bibliography of the writings of Stahl embraces 276 titles, with annotations.

LASÈGUE, E. C. De Stahl et sa doctrine médicale. (Thèse.) Paris, 1846. 4to.

STAHL, G. E. For bibliography, *see, in Section V*, Stahl's Fundamenta chymico-pharmaceutica.

STAPLES, BENJAMIN CHARLES.
Obituary. J. Chem. Soc., London, XXVI, 782 (1873).

STARK, MICHAEL JOHN.
Obituary. J. Chem. Soc., London, XXVIII, 1318 (1875).

STARK, WILLIAM.
Obituary.   J. Chem. Soc., London, XVII, 436 (1864).

STAS, JEAN SERVAIS.
Necrology.   Ber. d. chem. Ges., vol. XXV, p. 1 (1892).

STEINER, A.   *See* Fleischer, Anton.

STEINER, LEWIS H.   Paracelsus and his influence.   *See* Paracelsus.

STENHOUSE, JOHN.   Obituary.   J. Chem. Soc., London, XXXIX, 185 (1881).

STEWART, WILLIAM AULD.
Obituary.   J. Chem. Soc., London, XXXV, 266 (1879).

STÖCKHARDT, JULIUS ADOLF.
SCHEIBLER, C.   Nachricht von dem Tode des J. A. Stöckhardt.   Ber. d. chem. Ges., XIX, p. 1471 (1886).
Life and portrait.   Pop. Sci. Mon., XIX, 261, June, 1881.

STODDART, WILLIAM WALTER.
Obituary.   J. Chem. Soc., London, XXXIX, 190 (1881).

STONE, DANIEL.
Obituary.   J. Chem. Soc., London, XXVII, 1202 (1874).

STORER, FRANK H.   Bibliography and Life of Wm. Ripley Nichols, *see* Nichols, Wm. Ripley.

STOREY, JOSEPH.
Obituary.   J. Chem. Soc., London, LV, 296 (1889).

STRECKER, ADOLPH.
Obituary.   J. Chem. Soc., London, XXV, 353 (1872).
WAGNER, RUDOLF.   Neckrolog auf A. Strecker.   Ber. d. chem. Ges., V, p. 125 (1872).   Portrait of A. Strecker.

STROMEYER, AUGUST.
Obituary.   Chem. News, LVI, 258 (1887).

STURGEON, WILLIAM.
Obituary.   Am. J. Sci. [2], XI, 444, May, 1851.

SUAVIUS.   Vita Paracelsi, *see* Paracelsus.

SUCHTEN, ALEXANDER VON.
Biography.   I. N. v. E. I.   Alchymia denudata, II Theil, drittes Kapitel.   pp. 100–111.
> Suchten is defended from the charges made by Kunckel, and extolled as a worthy alchemical adept.

SVANBERG, JÖNS.
Biografi öfver J. S. [Stockholm, 1852.] pp. 15, 8vo.

SVANBERG, LARS FREDRIK. Biografi öfver Berzelius, *see* Berzelius, J. J.

SYLVIUS, FRANCISCUS DE LA BOE.
SCHACHT, L. Oratio funebris in obitum Francisci de Boë Sylvii.
Lugduni-Batavorum, 1673. 4to.

SYNESIUS. Bibliography of. *See* Democritus and Synesius, Ferguson, John.

TALBOT, WM. HY. FOX. *See, in Section III*, Eder, J. M.: Geschichte der Photochemie.

TAVERNIER, M. Éloge de J. B. G. Barbier. *See* Barbier, J. B. G.

TAYLOR, WILLIAM B. Henry's Contributions to the Electro-Magnetic Telegraph, *see* Henry, Joseph.

TENNANT, CHARLES, Life of, *see* Walker, Wm., Jr.

TENNANT, JAMES.
Obituary. J. Chem. Soc., London, XXXIX, 190 (1881).

TENNANT, SMITHSON, Life of, *see* Young, Thomas.

TENNENT, JOHN.
Obituary. J. Chem. Soc., London, XXI, xxix (1868).

TESCHEMACHER, EDWARD FREDERICK,
Obituary. J. Chem. Soc., London, XXXIII, 234 (1878).
SMITH, DENHAM. Obituary. J. Chem. Soc., London, XVI, 434 (1863).

THÉNARD, LOUIS JACQUES. Life of, *see* Mangin.

THÉNARD, LOUIS JACQUES.
Biographical notice of Thénard. Am. J. Sci. [2], XXV, 430, May, 1858.
Obituary. Am. J. Sci. [2], XXIV, 408, Nov., 1857.
Obituary. Q. J. Chem. Soc., London, XI, 182 (1859).
LE CANU, L. R. Souvenirs de M. Thénard. Lus en séance de rentrée
de l'École de pharmacie, 11 Novembre, 1857. Paris, 1857. pp.
63, 8vo.
FLOURENS, M. Éloge historique de Louis-Jacques Thénard. Institut
Impérial de France. Lu dans la séance publique du 30 Janvier,
1860. 2 S., vol. XXXII, Mém. Acad. Sci. pp. I–xxxv (1864).
With a bibliography of 35 titles.

THOMAS, SIDNEY GILCHRIST.
Obituary. J. Chem. Soc., London, XLVII, 337 (1885).

THOMAS, THOMAS.
Obituary. J. Chem. Soc., London, XXVI, 782 (1873).

THOMPSON, BENJAMIN, COUNT RUMFORD.
BAUERNFEIND, C. M. von. Benjamin Thompson, Graf von Rumford. München, 1889. 4to.
CUVIER, GEORGES. Biographical Memoir of Sir B. T. Rumford. Am. J. Sci., XIX, 28, January, 1831.
* ELLIS, GEORGE E. Memoir of Sir Benjamin Thompson, Count Rumford, with notices of his daughter. Published in connection with an edition of Rumford's complete works by the American Academy of Arts and Sciences, Boston. Boston, *n. d.* pp. xvi–680, 8vo.
    With portraits, facsimiles and plates.
JOHNSTON, JOHN. Sketch of the early history of Count Rumford, in which some of the mistakes of Cuvier and others of his biographers are corrected. Read before the Natural History Society of the Wesleyan University, June 30, 1837. Am. J. Sci., XXXIII, p. 21.
RENWICK, JAMES. Life of Benjamin Thompson, Count of Rumford. Library of American Biography conducted by Jared Sparks. Second Series, vol. V. Boston, 1855. pp. 4–216, 12mo. Portrait.
Life and portrait, Pop. Sci. Mon., IX, 231, June, 1876.

THOMPSON, BENJAMIN, COUNT RUMFORD. Life, *see* Young, Thomas; Cuvier, G.; Walker, Wm.

THOMPSON, CHARLES O. *See* Boyle, Robert.

THOMSON, ALEXANDER MORRISON.
Obituary. J. Chem. Soc., London, XXV, 357 (1872).

THOMSON, JAMES.
Obituary. Q. J. Chem. Soc., London, IV, 347 (1852).

THOMSON, ROBERT DUNDAS.
Obituary. J. Chem. Soc., London, XVIII, 344 (1865).

THOMSON, THOMAS.
Obituary. Q. J. Chem. Soc., London, VI, 152–155 (1854).

THOMSON, THOMAS. Life of Joseph Black, *see* Black, Joseph.

THORPE, T. E. Life of Joseph Priestley, *see* Priestley, Joseph.

THURNEISSER ZUM THURN, LEONHARD.
FRANZ, R. Ueber den Alchemisten L. T. Berlin, 1875. 4to.
* MOEHSEN, J. C. W. Leben Leonhard Thurneissers zum Thurn. Ein Beitrag zur Geschichte der Alchemie wie auch der Wissenschaften

THURNEISSER ZUM THURN, LEONHARD. [Cont'd.]
und Künste in der Mark Brandenburg gegen Ende des sechszehnten Jahrhunderts. Beiträge zur Geschichte der Wissenschaften in der Mark Brandenburg von den ältesten Zeiten an bis zu Ende des sechszehnten Jahrhunderts. Berlin und Leipzig, 1783. * Sm. 4to, pp. 198. With one plate of coins and medals.

> An entertaining and thorough treatise. Contains a bibliography of Thurneisser's works.

TIDY, C. MEYMOTT.
Obituary. Chem. News, LXV, 143, Mch. 18, 1892.

TIEMANN, FERD. *See* Rupstein, Friedrich.

TIGERSTEDT. Life of Joh. Gadolin, *see* Gadolin, Joh.

TILLEY, THOMAS GEORGE.
Obituary. Q. J. Chem. Soc., London, II, 188 (1850).

TILLMAN, SAMUEL D.
Obituary. Am. J. Sci. [3], x, 402, 1875.

TONNET, JOSEPH. Notice sur Nicolas Lémery, *see* Lémery, Nicolas.

TORREY, JOHN.
Obituary. Am. J. Sci. [3], v, 324, 1873.
GRAY, A. A Biographical Notice of John Torrey. Am. J. Sci. [3] v, 411, 1873.

TRIBE, ALFRED.
Obituary. J. Chem. Soc., London, XLIX, 352 (1886).

TRISMEGISTUS, HERMES.
*ALETHOPHILUS. Hermetis Trismegisti Einleitung in's höchste Wissen: von Erkenntniss der Natur und des darin sich offenbarenden grossen Gottes. Stuttgart, 1855. pp. 256, 32mo. (Wunder-Schauplatz, I.)

> Contains life and genealogy of Hermes, based on Borrichius.

BAUMGARTEN-CRUSIUS, LUDWIG FRIEDRICH OTTO. Programmata de librorum hermeticorum origine atque indole. Jenæ, 1827. 4to.

> A study of the works ascribed to Hermes Trismegistus, by a theologian.

* PIETSCHMANN, RICHARD. Hermes Trismegistos nach ägyptischen, griechischen und orientalischen Ueberlieferungen dargestellt. Leipzig, 1875. pp. 58, 4to.

TROMMSDORFF, JOHANN BARTHOLOMAEUS.

CALLISEN, AD. CARL PETER. Dem Andenken des verdienten Chemikers Joh. Barthol. Trommsdorff. Eine Skizze. Zweite durchgesehene und vermehrte Auflage mit eigenen Berichtigungen des Verstorbenen und mit einem Anhang. Kopenhagen, 1887. 8vo.

MENSING, J. G. W. Des Geheimen Hofraths und Professors Joh. B. Trommsdorff Lebensbeschreibung. Erfurt, 1839. pp. 88, 8vo. With a medallion-portrait.

> Contains a systematic bibliography of Trommsdorff's publications.
> *See* Gren, Friedrich Albrecht Carl.

TROOST, L.

Notice sur les travaux scientifiques de L. Troost. Paris, 1884. 4to.

TSCHIRCH, A. *See* Ziurek, Otto.

TUSON, RICHARD VINE.

Obituary. Chem. News, LVIII, 230 (1888).

TYNDALL, JOHN. Faraday as a Discoverer, *see* Faraday, Michael.

VALENTIN, WILLIAM GEORGE.

Obituary. Chem. News, XXXIX, 204 (1879).

Obituary. J. Chem. Soc., London, XXXVII, 260 (1880).

HOFMANN, A. W. Nachricht von dem Tode des W. G. Valentin. Ber. d. chem. Ges., XII, p. 1041 (1879).

VALENTINUS, BASILIUS. Life of, *See, in Section VI*, Valentinus, Basilius : Chymische Schriften.

VALENTINUS, BASILIUS.

* HILDEBRAND, H. Der Alchemist Basilius Valentinus. Einladungschrift des Herzogl. Francisceums in Zerbst zu den am 5. und 6. April abzuhaltenden öffentlichen Prüfungen der Gymnasial-, Real- und Vorklassen. Zerbst, 1876. pp. 38, 4to.

> The essay occupies pp. 1–18.

VALLET DE VIRIVILLE. *See* Flamel, Nicolas.

VAN, *and* VAN DER. For Dutch names beginning with Van, or Van der, see next succeeding word.

VARRENTRAP, FRANZ.

KNAPP, FRIEDRICH. Nekrolog auf F. Varrentrap. Ber. d. chem. Ges., X, p. 2291 (1877).

VAUQUELIN, LOUIS NICOLAS.
> CHEVALLIER, (JEAN BAPTISTE) ALPHONSE. Inauguration d'un monument à la mémoire de Louis Nicolas Vauquelin. Notice biographique de ce chimiste. Paris, 1850. 8vo. From Journal de chimie médicale, Avril, 1851.
> CHEVALLIER ET ROBINET. Éloge de L. N. Vauquelin. Paris, [1830]. 8vo. Extract from Journal de la Société de pharmacie.
> CUVIER, GEORGES. Éloge historique de L. N. Vauquelin. Mém. Acad. Sciences, 2 S., XII, xxxix–lvi (1833).
> VIREY, M. Séance publique tenue le 21 Avril 1830 en commémoration de Vauquelin. Discours prononcé par M. Virey. Société de pharmacie. pp. 62, 8vo. [*n. d., n. p.*]

VERNON, JEAN MARIE DE. Life and bibliography of Raymond Lully, *see* Lullius, Raymundus.

VILLAIN, ABBÉ. Histoire de Nicolas Flamel, *see* Flamel, Nicolas.

VILLANOVA, ARNALDO DE. *See* Arnaldus de Villanova.

VIREY, J. J. Vie de C. L. Cadet-Gassicourt, *see* Cadet-Gassicourt, C. L.

VOELCKER, JOHN CHRISTOPHER AUGUSTUS.
> Obituary. J. Chem. Soc., London, XLVII, 339 (1885).

VOGEL, AUGUST. Liebig als Begründer der Agricultur-Chemie, *see* Liebig, Justus von. Denkrede auf Johann Wolfgang Döbereiner, *see* Döbereiner, J. W.

VOGEL, HEINRICH AUGUST VON.
> VOGEL, AUGUST [the Younger]. Denkrede auf H. A. von Vogel. . . . München, 1868. 8vo.

VOIT, C. VON.
> Nekrologe auf J. G. von Jolly, J. B. Dumas, A. Wurtz, H. Kolbe, R. A. Smith, und andere. München, 1885. 8vo.

VOLTA, ALESSANDRO.
> ARAGO, FRANÇOIS. Éloge historique de A. Volta. Mém. Acad. Sciences. 2 S., XII, lvii–civ (1833).
> SEEBECK, L. F. W. A. Gedächtnissrede auf A. Volta. Dresden und Leipzig, 1846. 8vo.

VOLTAIRE, FRANÇOIS MARIE AROUET DE.
> BOIS-REYMOND, EMIL DU. Voltaire in seiner Beziehung zur Naturwissenschaft. Festrede in der öffentlichen Sitzung der königl. Preuss. Akademie der Wissenschaften zur Gedächtnissfeier Friedrich's II., am 30 Januar, 1868. Berlin, 1868. pp. 30, 8vo.

WAGNER, RUDOLF. *See* Strecker, Adolph, *also* Scherer, Johann Joseph von.

WAITE, ARTHUR EDWARD.
* Lives of Alchemystical Philosophers based on materials collected in 1815 and supplemented by recent researches. With a philosophical demonstration of the true principles of the magnum opus or great work of alchemical reconstruction and some account of the spiritual chemistry. To which is added a bibliography of alchemy and hermetic philosophy. London, 1888. pp. 1–315, 8vo.

> The author uses the material in an anonymous work of the same title usually attributed to Francis Barrett, re-writes the sections, and adds considerable matter of his own. He omits, however, the excerpts from alchemical authors. The bibliography is disappointing, being meagre in details and little more than a reprint of Barrett's one-line titles augmented by titles from Lenglet du Fresnoy, without verification of the data.

WALDIE, DAVID.
Obituary. J. Chem. Soc., London, LVII, 456 (1890).

WALKER, WILLIAM, JUNIOR.
* Memoirs of the Distinguished Men of Science of Great Britain living in the Years 1807–08. With an Introduction by Robert Hunt. London, 1862. pp. xii–228, 8vo. With folding plate containing portraits of 51 scientists grouped in the Royal Institution.

> This contains biographical sketches of 51 British scientists, among whom are the following chemists : Henry Cavendish ; John Dalton ; Sir H. Davy ; Wm. Henry ; Count Rumford ; Charles Tennant ; Thomas Thomson ; Richard Watson, Bishop of Llandaff ; James Watt ; Wm. H. Wollaston ; Daniel Rutherford ; Wm. Allen.

WALLACE, WILLIAM.
Obituary. J. Chem. Soc., London, LXI, 296 (1889).
Obituary. Chem. News, LVIII, 266 (1888).

WARD, W. SYKES.
Obituary. J. Chem. Soc., London, LIII, 518 (1888).

WARINGTON, GEORGE.
Obituary. J. Chem. Soc., London, XXVII, 1203 (1874).

WARRINGTON, ROBERT.
Obituary. Chem. News, XVI, 316, Dec. 20, 1867.
Obituary. J. Chem. Soc., London, XXI, xxxi (1868).

WATSON, HENRY HOUGH.
Obituary and bibliography. J. Chem. Soc., London, LI, 477 (1887).

WATSON, RICHARD, BISHOP OF LLANDAFF. Life of, *see* Young, Thomas; *also* Walker, Wm., Jr.

WATT, JAMES.
ARAGO, FRANÇOIS. Éloge historique de James Watt. Paris. Mémoires de l'Académie des Sciences pour l'an 1839. Paris, 1839. 2 S., XVII, xli–clxxxviii.

> Historical Éloge of James Watt by M. Arago. Translated from the French with additional Notes and an Appendix by James Patrick Muirhead. London, 1839. pp. ix–261, 8vo. Portrait.

> Life of James Watt, by M. Arago. Reprinted from the Edinburgh New Philosophical Journal for October, 1839. Edinburgh, 1839.
> Of this the third edition bears the following title :

> Life of James Watt with memoir on machinery considered in relation to the prosperity of the working classes. By M. Arago. To which are subjoined Historical Account of the Discovery of the Composition of Water by Lord Brougham, and Eulogium of James Watt by Lord Jeffrey. Edinburgh, 1839. pp. 222, 18mo.

* MUIRHEAD, JAMES PATRICK. The Life of James Watt, with selections from his Correspondence. With portraits and wood-cuts. London, 1858. pp. xvi–580.

* MUIRHEAD, JAMES PATRICK. Correspondence of the late James Watt on his Discovery of the Theory of the Composition of Water, with a letter from his Son. Edited with introductory remarks and an appendix. London, 1846. pp. cxxvii–264. Portrait.
> *See also* Walker, Wm., Jr.

WATTS, HENRY.
Autobiography. J. Chem. Soc., London, XLVII, 343 (1885).
Obituary. Am. J. Sci. [3], XXIX, 172, February, 1885. *Also*, XXIX, 268, March, 1885.
Obituary. Chem. News, L, 10 (1884).

WAY, JAMES THOMAS.
Obituary. J. Chem. Soc., London, XLV, 629 (1884).

WEBSKY, MARTIN.
HOFMANN, A. W. Nekrolog auf M. Websky. Ber. d. chem. Ges., XIX, p. 3077 (1886).

WEGSCHEIDER, R. *See* Etti, Carl.

WELDON, WALTER.
Obituary.   J. Chem. Soc., London, LI, 480 (1887).

WELTZIEN, KARL.
Obituary.   J. Chem. Soc., London, XXIV, 622 (1871).
BIRNBAUM, K.   Nekrolog auf K. Weltzien.   Ber. d. chem. Ges., VII,
p. 1698 (1875).

WERNECK.   Paracelsus und Medicin.   *See* Paracelsus.

WERNER, ABRAHAM GOTTLOB.
Portrait *in* Neues all. J. Chem., Gehlen, vol. IV.   Frontispiece, 1805.

WERNER, A. G.   Biography of.   *See* Agricola, G. ; Becher, F. L.

WERTHER, AUGUST FRIEDRICH GUSTAV.
LIEBREICH, O.   Nekrolog auf A. F. G. Werther.   Ber. d. chem. Ges.,
III, p. 372 (1870).

WETHERILL, C. M.
Obituary.   Am. J. Sci. [3], I, 478, 1871.

WEYLER Y LAVIÑA, D. F.   *See* Lullius, Raymundus.

WIEGLEB, JOHANN CHRISTIAN.
Nekrolog.   All. J. Chem., Scherer, vol. IV, p. 684.

WIGNER, GEORGE WILLIAM.
Obituary.   J. Chem. Soc., London, XLVII, 345 (1885).

WILCOCK, EDGAR.
Obituary.   J. Chem. Soc., London, XLIII, 255 (1883).

WILL, HEINRICH.
Obituary and bibliography.   J. Chem. Soc , London, LX, 466 (1891).
HOFMANN, A. W.   Nekrolog auf Will.   Ber. d. chem. Ges., XXIII, p. 852
(1890).   Portrait of Will.

WILLIAMS, JOHN.
Obituary.   J. Chem. Soc., London, LV, 298 (1889).

WILLS, THOMAS.
Obituary.   J. Chem. Soc., London, XXXVII, 261 (1880).
Obituary.   Chem. News, XXXIX, 217 (1879).
The Life of T. W. by his Mother, Mary Wills, and her Friend, J. Luke.
London, 1880.

WILSON, GEORGE.
> GLADSTONE, J. H. Obituary. Q. J. Chem. Soc., London, XIII, 169 (1861).

WILSON, GEORGE. Life of Henry Cavendish. *See* Cavendish, Henry.

WILSON, J. CHAPMAN.
> Obituary. J. Chem. Soc., London, XXVI, 783 (1873).

WINSOR, WILLIAM.
> Obituary. J. Chem. Soc., London, XXI, xxxiv (1868).

WISLICENUS, J. *See* Heintz, Wilhelm. Gedächtnissrede auf P. A. Bolley. *See* Bolley, P. A.

WITHERING, WILLIAM.
> * The Miscellaneous Tracts of the late William Withering, to which is prefixed a memoir of his life, character and writings in two volumes. London, 1822. 2 vols., 8vo. Vol. I, pp. vi–496 ; II, iv–503.

WÖHLER, EMILIE, UND A. W. HOFMANN. Aus Justus Liebig's und Friedrich Wöhler's Briefwechsel. *See* Liebig, Justus von.

WÖHLER, FRIEDRICH.
> Obituary. Chem. News, XLVI, 183 (1882).
> Obituary. J. Chem. Soc., London, XLIII, 258 (1883).
> Life and portrait. Pop. Sci. Mon., XVII, 539, Aug., 1880.
> * HOFMANN, A. W. Zur Erinnerung an Friedrich Wöhler. Berlin, 1883. 8vo.
> > Histoire de la vie et des travaux de Fr. Wöhler. Traduit de l'allemand. Paris, 1883. 4to.
> HOFMANN, A. W. Festrede gehalten bei der Enthüllung des Wöhler-Denkmals in Göttingen. Ber. d. chem. Ges., 1890, p. 833. Portrait of Wöhler and engraving of the monument.
> * JOY, LAURA R. Early Recollections of a Chemist. Translated by Laura R. Joy. (Reprinted from the American Chemist for October, 1875.) Philadelphia, 1875. pp. 16, 16mo.
> THORPE, T. E. The Chemical Work of Wöhler. A Lecture delivered in the Royal Institution, Feb. 15, 1884. Chem. News, XLIX, 90 (1884).
> * Zur Feier der achtzigsten Wiederkehr von Friedrich Wöhler's Geburtstag, am 31sten Juli, 1880. pp. 34, 8vo. *n. d., n. p.* Portrait and facsimile.

WÖHLER, FRIEDRICH, UND J. J. BERZELIUS.
> * HJELT, EDV. Bruchstücke aus den Briefen F. Wöhlers an J. J. Berzelius. Berlin, 1884. pp. 56, 16mo.

WÖHLER, FRIEDRICH, UND LIEBIG, JUSTUS.   Briefwechsel.   *See* Liebig, Justus von.

WOLLASTON, WILLIAM HYDE.
 Necrology.   Am. J. Sci., XVI, 216, 1829.   *Also,* XVII, 159, 1830.

WOLLASTON, WM. H.   Life of.   *See* Walker, Wm., Jr.

WOLLIN, CHRISTIAN.
 Nekrolog.   All. J. Chem., Scherer, vol. III, p. 392.

WROBLEWSKI, SIGMUND.
 Obituary.   Chem. News, LVII, 201 (1888).

WURTZ, CHARLES ADOLPHE.
 FRIEDEL, C.   Notice sur la vie et les travaux de Charles Adolphe
  Wurtz.   Paris, 1885.   8vo.   Portrait.
   Cf. in Section III, Wurtz, Adolphe, La théorie atomique.
 HOFMANN, A. W.   Nekrolog auf A. Wurtz.   Ber. d. chem. Ges., XVII,
  p. 1207 (1884) ; also XX, Ref., p. 815 (1887).   Portrait.
 Obituary.   J. Chem. Soc., London, XLVII, 328 (1885).
 Obituary.   Chem. News, XLIX, 222 (1884).

WURTZ, CHARLES ADOLPHE.   Life of, *see* Voit, C. von ; Life and portrait, Pop. Sci.
  Mon., XXII, 114, Nov., 1882.

YORKE, PHILLIP.
 Obituary.   J. Chem. Soc., London, XXVIII, 1319 (1875).

YOUNG, JAMES.
 Obituary.   Chem. News, XLVII, 245 (1883).
 Obituary.   J. Chem. Soc., London. XLV, 630 (1884).

YOUNG, THOMAS.
 ARAGO, FRANÇOIS.   Éloge historique du Thomas Young.   Mém. Acad.
  Sciences, 2 S., XIII, lvii–cv (1835).
 MILBURN, WILLIAM HENRY.   Biographical notice of T. Y., with portrait,
  Harper's Monthly, LXXX, 670 (1890).
 PEACOCK, GEORGE.   Life of T. Y.   London, 1855.   pp. xi–[iii]–514,
  8vo.   With a portrait.
 Memoir of the Life of Th. Young.   Am. J. Sci., XXII, 232, July, 1832.
 Biographies of Men of Science [comprising Cavendish, Smithson, Ten-
  nant, Count Rumford ; Richard Watson, Bishop of Llandaff ; Four-
  croy, Ingenhousz, Robison, Dolomieu, Coulomb, Borda, de la
  Condamine, Lagrange, Fermat, Malus, Lalande, Lambert, Maske-
  lyne, Atwood]   Miscellaneous Works, edited by George Peacock,
  vol. II.   London, 1855.   8vo.

YOUNG, THOMAS. [Cont'd.]

Young also wrote biographies of the following men of science for the Supplement to the Encyclopedia Britannica : Ed. Beccaria, Black, Boulton, Bramah, Brisson, Camus, Cavallo, Duhamel, F. Fontana, G. Fontana, J. R. Forster, J. G. A. Forster, Frisi, Guyton de Morveau, Lemonnier, De Luc, Méchain, Messier, Pallas, and Rush.

ZIMMERMANN, CLEMENS.

LIEBERMANN, C. Nachricht von dem Tode des C. Zimmermann. Ber. d. chem. Ges., XVIII, p. 1021 (1885).

ZININ, NICOLAUS.

HOFMANN, A. W. Nachricht von dem Tode des Nicolaus Zinin. Ber. d. chem. Ges., XIII, p. 449 (1880).

ZIUREK, OTTO OSKAR ALBERT.

TSCHIRCH, A. Nekrolog auf O. Ziurek. Ber. d. chem. Ges., XIX, Ref., p. 893 (1886).

SECTION V.

CHEMISTRY, PURE AND APPLIED.

AARON, C. H.

Assaying. In three Parts. Part I : Gold and Silver Ores. Part II : Gold and Silver Bullion. Part III : Lead, Copper, Tin, Mercury. San Francisco, 1884. 8vo.

АБАШЕВЪ, Д.

Изслѣдованія о явленіяхъ взаимнаго растворенія жидкостей. Москва, 1858.

> ABASHEF, D. Investigation of the phenomena of the mutual action of solutions. Moscow, 1858.

ABBOTT, HELEN C. DE S.

Plant Analysis as an Applied Science. A lecture delivered before the Franklin Institute Jan. 17, 1887. Philadelphia, 1887. 8vo.

Plant Chemistry as illustrated in the production of Sugar from Sorghum. A lecture delivered before the Alumni Association of the Philadelphia College of Pharmacy Feb. 8, 1887. Philadelphia, 1887.

ABEL, SIR FREDERICK AUGUSTUS.

Explosive Agents applied to Industrial Purposes. London, 1880.

> See Noble, Captain, and F. A. Abel.

> Les agents explosifs appliqués dans l'industrie. Mémoire lu le 23 mars, 1880, devant la société des ingénieurs civils de Londres. Traduit avec l'autorisation de l'auteur par Gustave Richard. Paris, 1881. pp. 92, 12mo.

On Recent Investigations and Applications of Explosive Agents. A lecture . . . Edinburgh, 1871. 8vo.

ABEL, F. A., AND CHARLES LOUDON BLOXAM.

Hand-book of Chemistry, theoretical, practical and technical . . . with a preface by [A. W.] Hofmann. London (also Philadelphia), 1854. 8vo.

> Second edition, 1858.

ABENIUS, P. V.
Undersökningar inom diazinserien.   Upsala, 1891.   8vo.

ACADEMIA DEL CIMENTO.
  * Essays of Natural Experiments made in A. del C.   Written in Italian
    by the Secretary of that Academy.   Englished by Richard Waller.
    London, printed for Benjamin Alsop at the Angel and Bible in the
    Poultrey, over-against the Church, 1684.   pp. [xxii]–160–[x], 8vo.
    19 Plates.

ACCUM, FREDERICK (CHRISTIAN).
  Chemical Amusements, a series of curious and instructive experiments
    in chemistry which are easily performed and unattended by danger.
    London, 1817.   12mo.
          Third edition, with plates, and enlarged.   London, 1818.
            12mo.   Ill.
          Manuel de chimie amusante, ou nouvelles récréations chi-
            miques, contenant une suite d'expériences d'une exécu-
            tion facile et sans danger, ainsi qu'un grand nombre de
            faits curieux et instructifs.   Traduit de l'anglais de
            Frédérick Accum, Samuel Parkes et Joseph Griffin,
            par A. D. Vergnaud.   Quatrième édition, revue, corrigée
            et considérablement augmentée.   Paris, 1835.   pp. 331,
            32mo.   Folding plate.
          Chemische Unterhaltungen.   Eine Sammlung merkwürdiger
            . . . Erzeugnisse der Erfahrungschemie.   Aus dem
            englischen.   Kopenhagen, 1819.   8vo.   Ill.
          Chemische Belustigung.   Nach der dritten englischen Aus-
            gabe vom Verfasser deutsch bearbeitet.   Nürnberg,
            1824.   8vo.
          Divertimento chimico contenente esperienze curiose.   Mi-
            lano, 1820.   2 vols.   Plates.
          ACCUM, F. A., E POZZI.   La chimica dilettevole o serie di
            sperienze curiose e instruttive di chimica che si esegui-
            scono con facilità e sicurezza.   Nuova edizione con
            tavole in rame.   Milano, 1854.   8vo.
  * Culinary Chemistry, exhibiting the scientific principles of cookery, with
    concise instructions for preparing good and wholesome pickles,
    vinegar conserves, fruit jellies, marmalades and various other ali-
    mentary substances employed in domestic economy.   With obser-
    vations on the chemical constitution and nutritive qualities of
    different kinds of food.   With copper plates.   London, 1821.   pp.
    xxiv–356, 12mo.

ACCUM, FREDERICK (CHRISTIAN).   [Cont'd.]

Description of the Process of manufacturing Coal Gas, etc., with eleva-
tions, sections and plans of the most improved sorts of apparatus
now employed at the Gas Works in London.   London, 1819.   8vo.
Ill.

> Second edition, 1820.   8vo.   7 plates.

Manual of a Course of Lectures on Experimental Chemistry and on
Mineralogy.   With an account of the action of chemical tests.
London, 1810.   12mo.

Practical (A) Essay on the Analysis of Minerals.   London, 1804.   8vo.

Practical (A) Essay on Chemical Reagents or Tests.   Illustrated by a
series of experiments.   London, 1816.   8vo.   *Also* Philadelphia,
1817.

>> Practical Treatise on the use and application of chemical
tests, and concise directions for analyzing, illustrated by
experiments.   Second edition, with plates, enlarged.
London, 1818.   12mo.   Ill.

>> Another edition : Improved and brought down to the present
state of chemical science by J. Maugham.   London,
1828.   12mo.

>> Traité pratique sur l'usage et le mode d'application des ré-
actifs chimiques, traduit par Riffault.   Paris, 1819.
Plates.

>> Trattato pratico per l'uso ed applicazione de' reagenti
chimici, traduzione fatta sulla seconda edizione Inglese,
con aumentazioni di G. Pozzi.   Milano, 1819.   2 vols.
in 1.   Ill.

System (A) of Theoretical and Practical Chemistry.   London, 1803.
2 vols, 8vo.

> Second edition, London, 1807.   *Also* * Philadelphia, 1808.
2 vols., 8vo.   •

* Treatise (A) on Adulterations of Food and Culinary Poisons.   Exhibiting the fraudulent sophistications of bread, beer, wine, spirituous
liquors, tea, coffee, cream, confectionery, vinegar, mustard, pepper,
cheese, olive oil, pickles and other articles employed in domestic
economy, and methods of detecting them.   London, 1820.   pp.
xvi–372, 12mo.   *Also* Philadelphia, 1820.   12mo.

>> This early work on adulteration of food attracted much attention when
issued.   From the representation of a skull and cross bones on the
cover, and the legend, "There is death in the pot, II Kings, C. 4, v.
40," the book was popularly called " Death in the Pot."

>> Von der Verfälschung der Nahrungsmittel und von den
Küchengiften, oder von den betrügerischen Verfäl-

ACCUM, FREDERICK (CHRISTIAN). [Cont'd.]

> schungen des Brodes, Bieres, Weins, des Liqueurs, des
> Thees, Kaffees, Milchrahms, Confects, Essigs, Senfs,
> Pfeffers, Käse, Olivenöls, der eingelegten Gemüse und
> Früchte und anderer in der Haushaltung gebräuchlichen
> Artikel, und von den Mitteln, dieselben zu entdecken.
> Nach der zweiten Ausgabe aus dem englischen über-
> setzt von L. Cerutti, und mit einer Einleitung versehen
> von C. G. Kühn. Zweite Ausgabe. Leipzig, 1841. 8vo.

> First edition, Leipzig, 1822. 8vo.

Treatise (A) on the Art of Brewing. Second edition. London, 1821.
8vo. Ill.

> Manuel théorique et pratique du brasseur, ou l'art de faire
> toutes sortes de bière, contenant tous le procédés de
> cet art tels qu'ils sont usités à Londres, suivi d'un ex-
> posé des altérations frauduleuses de la bière et des
> moyens de les découvrir. Traduit de l'anglais par
> Riffault. Paris, 1825. 8vo.

> Abhandlung über die Kunst zu brauen. Aus dem eng-
> lischen. Hamm., 1821. 8vo. Ill.

Practical Treatise on Gas-light, with a summary description of the
apparatus and machinery best calculated for illuminating streets,
houses and manufactories with carburetted hydrogen, or coal
gas. London, 1818. 8vo. Folding plates.

> Praktische Abhandlung über das Gaslicht, eine vollständige
> Beschreibung des Apparates und der Maschinen, um
> Strassen, Häuser und Manufacturen damit zu beleuch-
> ten. Aus dem englischen übersetzt von Lampadius.
> Weimar, 1819. 8vo.

> Accum was born in Germany in 1769. He long resided in London, and as
> Professor of Chemistry at the Surrey Institution, London, conducted
> a laboratory for instruction. There Benjamin Silliman, Senior, in
> 1804–5, and later James Freeman Dana (both of the United States of
> America), pursued their studies.
> For other works by Accum, *see in Section II. Cf.* Gordon, D.

ACHARD, FRANZ CARL.

Chymisch-physische Schriften. Berlin, 1780. pp. 367, 8vo. 10 folding tables.

Europäische (Die) Zuckerfabrikation aus Runkelrüben, in Verbindung
mit der Bereitung des Brandweins, des Rums, des Essigs und eines
Coffee-Surrogats aus ihren Abfällen, beschrieben und mit Kupfern
erläutert durch ihren Urheber, F. C. A. Drey Theile. Leipzig,
1809. 4to. Ill.

ACHARD, FRANZ CARL. [Cont'd.]

        Zweite Auflage. Breslau und Leipzig, 1813. 8vo. Ill.

        Traité complet sur le sucre européen de betteraves, sur la manière d'en extraire économiquement le sucre et le sirop. Paris, 1811. 8vo.

Recherches sur les propriétés des alliages métalliques. Berlin, 1788. 4to.

Sammlung physikalischer und chemischer Abhandlungen. Berlin, 1784. 8vo. Ill.

ACLAND, T. D.

Introduction to the Chemistry of Farming, specially prepared for practical farmers, with records of field experiments. London, 1891. 8vo.

ADAM, A. F.

Étude sur les principales méthodes d'essai et d'analyse du lait, suivie de la description d'un nouveau procédé pour l'analyse complète de ce liquide. Paris, 1879. 8vo.

ADAMS, GEORGE.

*Lectures on Natural and Experimental Philosophy, considered in its present state of improvement, describing, in a familiar and easy manner, the principal phenomena of nature; and shewing, that they all co-operate in displaying the Goodness, Wisdom and Power of God. London, 1794. 5 vols., 8vo.

        The fifth volume consisting exclusively of the Plates and Index. In the first volume 1 illustration.

ADELUNG, JOHANN CHRISTOPH.

Mineralogische Belustigungen zum Behufe der Chemie, und Naturgeschichte des Mineralreichs. Leipzig und Kopenhagen, 1768–71. 6 vols.

ADERHOLDT, A. E.

Unorganische Chemie. Ein Leitfaden für den Unterricht in Gymnasien, Realschulen, höhere Bürgerschulen, etc., und Taschenbuch für Repertoria und Examinatoria. Dritte mit Berücksichtigung der neueren Entdeckungen und Ansichten bearbeitete Auflage. Weimar, 1868. 16mo.

        Zweite Auflage, 1859.

ADET, PIERRE AUGUST.

Leçons élémentaires de chimie à l'usage des Lycées. Ouvrage rédigé par ordre du gouvernement. Paris, 1804. 8vo.

ADET, PIERRE AUGUST. [Cont'd]
> Grundzüge der Chemie auf Befehl der französischen Regie-
> rung entworfen. Aus dem französischen übersetzt und
> mit einer Vorrede und mit Zusätzen begleitet von
> Huber. Basel und Aarau, 1805. 8vo.

ADLER, JOSEPH.
> La diffusion de Jules Robert, comptes rendus, rapports, communica-
> tions, jugements relatifs à ce procédé d'extraction des jus sucrés.
> Paris, 1868.

ADRIAN, H.
> Scalella Chemica. London, 1887. 8vo.

ADRIANCE, J. S.
> Laboratory Calculations and Specific Gravity Tables. New York,
> 1886. 12mo.

ÆPLINIUS, CHRISTIAN LUDWIG.
> Dissertatio inauguralis physico-chemica de solutionis corporum chemicæ
> fundamento. Halæ Magdeburgicæ, *n. d.* [1736]. pp. 23-[viii],
> 4to.

AFZELIUS, JOHANN [*called* ARVIDSON].
> De acido sacchari. Upsaliæ, 1776.
> De niccolo. Upsaliæ, 1775.

AFZELIUS, JOHANN, AND PETER OEHRN.
> De acido formicarum. Upsaliæ, 1777.

AGGIUNTI, NICCOLÓ.
> Diverse conclusioni di fisica. Romæ, 1687.
>> Contains observations on capillarity which cause Nelli to regard him as the
>> discoverer.

AGNOLESI, POMPILIO.
> Lettere sull' apparecchio per discoprire il fosforo. Firenze, 1864,
> 16mo.
> Vade mecum di tossicologia clinica. Napoli, 1880. 16mo.

AGRESTINI, A.
> Dosamento del biossido di zirconio in presenza di acido titanico.
> Urbino, 1886. 8vo.

AGRICOLA, GEORG.
> De mensuris et ponderibus, ... de precio metallorum et monetis.
> Basileæ, 1580. Fol.

AGRICOLA, GEORG. [Cont'd.]

De ortu et causis subterraneorum—de natura eorum quæ effluunt ex terra—de natura fossilium—de veteribus et novis metallis—Bermannus, sive de re metallica dialogus. Basiliæ, 1558.

Another edition, 1546.

De re metallica libri XII, quibus officia, instrumenta, machinæ, ac omnia denique ad metallicam spectantia non modo luculentissime describuntur, sed et per effigies suis locis insertas, adjunctis Latinis Germanicisque appellationibus, ob oculos ponuntur, ut clarius tradi non possint. Eiusdem de animantibus subterraneis liber, ab autore recognitus, cum indicibus diversis, quicquid in opere tractatum est pulchre demonstrantibus. Basiliæ, 1546. 4to. Ill.

*Second edition, Basiliæ, 1556, pp. 502 + indexes. 4to. Ill.

Other editions in 1558, '61, '71, 1621, 1657, etc.

Well known and deservedly prized for its illustrations of mining operations.

Dell' arte de' metalli, aggiuntovi il libro che tratta degli animali di sotterra, tradotto da Michelangelo Florio. Basiliæ, 1563. Folio.

Vom Bergwerck XIII Bücher; Darin alle Aemter, Instrument, Gezeuge . . . mit schönen figuren verbildet und klärlich geschrieben seindt. Gedruckt zu Basel durch J. Froben, 1557. Fol. Ill.

Bergwerck Buch; darinnen nicht allein alle Empter, Instrument, Gezeug . . . beschrieben : sondern auch, wie ein recht verständiger Bergmann seyen soll . . . Item wie das Gold vom Silber . . . zu scheiden sey. Basel, 1621. Fol. Ill.

Mineralogische Schriften. Uebersetzt von E. J. T. Lehmann. Freiburg, 1806–'13. 4 vols.

AGRICULTURCHEMISCHE UNTERSUCHUNGEN UND FÜTTERUNGS-, DÜNG-UNGS- UND CULTUR-VERSUCHE. Angestellt und gesammelt bei der landwirthschaftlichen Versuchsstation in Möckern auf dem Gute der Leipziger ökonomischen Societät in dem Jahren 1851, 1852, 1853, 1854. Leipzig, 1855. pp. xii–172, viii–136, v–54, vi–80, roy. 8vo.

The four reports have separate title-pages and independent pagination. The prefaces are signed Wilhelm Crusius.

AITKEN, SIR WILLIAM.

On the Animal Alkaloids, the Ptomaïnes, Leucomaïnes and Extractives in their pathological relations. Second edition. London, 1889. pp. xi–112, 12mo.

AJASSON DE GRANDSAGNE, J. B. F., AND J. M. L. FOUCHÉ.
    Nouveau manuel complet de chimie . . . Paris, 1828.   18mo.
            Nuevo manual completo de química general, aplicada á la
            medicina escrito en frances y traducido al español por
            Rafael Fernandez y Francisco de Galvez Padilla.
            Sevilla, 1841.

ÅKERMAN, JOACHIM.
    Elementarkurs i Kemien.   Stockholm, 1831.
    Föreläsningar uti kemiskt Technologi.   Stockholm, 1832.

ALBERTI, WILHELM CHRISTOPH.
    Anleitung zur Salmiakfabrikation.   Berlin und Leipzig, 1780.

ALBINEUS, NATHAN.
            See Aubigné de la Fosse, Nathan.

ALBRECHT, CARL.
    Ueber einige Pyrogallussäure und Phloroglucinderivate, und die
        Beziehungen derselben zu Daphnetin und Aesculetin.   Berlin,
        1884.

ALBRECHT, M.
    Das Paraffin und die Mineralöle.   Stuttgart, 1875.   Ill.

ALBUQUERQUE, L. D. S. M. DE.
    Curso elementar de physica e de chymica.   Lisboa, 1824.   3 vols., 8vo.
        Tables and plates.

ALCALÁ, FRANCISCO ALVAREZ.
    Formulario universal, ó guia del medico, del cirujano y del farmacéu-
        tico.   Segunda edicion refundida y considerablemente aumentada.
        Madrid.   4 tomos, 8vo.
    Nuevos elementos de química, aplicada á la medicina y á las artes,
        redactados con arreglo á las ultimas ediciones de los tratados de
        Orfila, Thénard, Dumas. etc.   Madrid, 1839.   2 vols., 4to.

ALDINI, JOHN.
    An Account of the Late Improvements in Galvanism, with a series of
        curious and interesting experiments performed before the Commis-
        sioners of the French National Institute and repeated lately in the
        Anatomical Theatres of London.   To which is added an Appendix
        containing the author's experiments on the body of a malefactor
        executed at Newgate.   Illustrated with engravings.   London,
        1803.   pp. xi–221, 4to.

ALESSANDRI, P. E.

Cereali, farine, sostanze feculacee, pane e paste alimentari. Milano, 1885. 8vo.

Metodo sistematico per l'analisi chimica qualitativa delle sostanze inorganiche e per le ricerche tossicologiche. Milano, 1880. 8vo. Ill. Second edition, 1885.

Studio fisico-chimico delle principali materie coloranti derivate dal catrame. Milano, 1889.

ALESSANDRI, P. E., AND L. MAGGI.

Acque potabili considerate come bevanda dell' uomo e dei bruti. Milano, 1887. pp. xiv–411, 12mo. Ill.

> A volume of the series entitled:

Alterazione e falsificazione delle sostanze alimentari e di altre importanti materie di uso comune. Manuali scritti da un gruppo di persone competenti e appartenenti alle Università e ad altri Istituti scientifici del Regno sotto la direzione del Egidio Pollacci.

ALESSI, A.

La chimica del carbonio esposta nei suoi principii più elementari agli studenti di Istituto Tecnico. Pisa, 1881. 8vo.

АЛЕКСѢЕВЪ. П.

Лекціи органич. химіи. С.-Петербургъ, 1868.

> ALEXEEF, P. Lectures on organic chemistry. St. Petersburg, 1868.

ALÉXÉYEFF, P.

Méthodes de transformation des combinaisons organiques traduit du russe par Georges Darzens et Léon Lefèvre. Introduction par Ed. Grimaux. Paris, 1891. pp. xiii–215. 8vo.

> A clear, concise work, filling an important gap in chemical literature.

АЛИБЕРТЪ, П. П.

Разысканіе и открытіе самороднаго графита. С.-Петербургъ, 1856. 8o

> ALIBERT, I. P. Discovery and exploitation of graphite. St. Petersburg, 1856. 8vo.

ALKER, CH.

Electro-métallurgie. Ses diverses applications dans les arts et dans l'industrie. Bruxelles, 1882. 12mo.

ALLAIN, P. A.

Traité de chimie élémentaire d'après les équivalents ayant l'hydrogène pour unité. Paris, 1847. 8vo.

ALLEN, ALFRED HENRY.
Commercial Organic Analysis. A treatise on the properties, proximate analytical examination and modes of assaying the various organic chemicals and products employed in the arts, manufactures and medicine. With concise methods for the detection and determination of their impurities, adulterations and products of decomposition. Second edition, revised and enlarged. London, 1889–92. 3 vols., 8vo.
Vol. I : Alcohols, Ethers, Vegetable Acids and Fibres, Starch and its Isomers, etc.
Vol. II : Fixed Oils and Fats, Hydrocarbons and Mineral Oils, Phenols and their Derivatives, Coloring Matters, etc.
Vol. III, Part 1 : Acid Derivatives of Phenols, Aromatic Acids, Tannins, Dyes and Coloring Matters.
Vol. III, Part 2. (In press.)
Introduction (An) to the Practice of Commercial Organic Analysis. London, 1879–82. 2 vols., 8vo. Ill.

ALSTON, CHARLES.
Dissertation (A) on Quicklime and Lime Water. Edinburgh, 1753.
Second Dissertation (A) on Quicklime and Lime Water. Edinburgh, 1755.

ALTMANN, E.
Grundriss der Chemie. Ein Leitfaden für den Unterricht an landwirthschaftlichen Lehranstalten und zum Selbstunterricht. Teil I : Unorganische Chemie. Leipzig, 1881. pp. 7–122, 8vo. Ill.

ALVAREZ ALCALÁ, FRANCISCO.
See Alcalá, Francisco Alvarez.

ALWENS, FRIEDRICH.
Stöchiometrische Schemata zur "Anleitung zur qualitativen chemischen Analyse," von R. Fresenius. Würzburg, 1854. 8vo.

AMATO, DOMENICO.
Del carbonio quale base del mondo organico. Prelezione. Catania, 1885.

AMBÜHL, G.
Die Lebensmittelpolizei. Anleitung zur Prüfung und Beurtheilung von Nahrungs- und Genussmitteln. Ein Buch aus der Praxis. Für die Gesundheitskommissionen des Kantons St. Gallen bearbeitet. St. Gallen, 1883. pp. iv–254, 12mo. Folding tables.

AMOR Y DELGRADO, A. M.

Elementos de química, obra útil para el repaso de esta asignatura á los alumnos de las facultades de ciencias medicina, farmacia, y á los de los estudios de ingenieros, etc. Madrid, 1881. 4to.

AMSEL, HUGO.

Grundzüge der anorganischen und organischen Chemie für Mediciner und Pharmazeuten, Chemiker, etc. Berlin, 1888.

Leitfaden für die Darstellung chemischer Präparate. Zum Gebrauche für Studierende. Stuttgart, 1891. pp. viii–102, 16mo.

ANDERSON, J. H.

The Public School Chemistry : being at once a Syllabus for the Master and an abstract for the Boy. Second edition, revised, corrected, and enlarged. London, 1886. 8vo.

ANDERSON, THOMAS.

Elements of Agricultural Chemistry. Edinburgh, 1860. 8vo.

On Certain Products of the Composition of the Fixed Oils in Contact with Sulphur. Edinburgh, 1847. 4to.

On the Constitution and Properties of Picoline, a new Organic Base from Coal Tar. Edinburgh, 1846. 8vo. [From the Trans. Roy. Soc., Edinburgh, vol. xvi.]

Products of Destructive Distillation of Animal Substances. [From the Transactions of the Royal Society of Edinburgh, vol. xvi, et seq.] Edinburgh, 1848–68. 5 pts., 4to.

ANDÈS, L. E.

Die trocknenden Oele [etc.]. See Bolley's Handbuch [etc.], Neue Folge.

ANDRAL ET GAVARRET.

Recherches sur la quantité d'acide carbonique exhalé par le poumon dans l'espèce humaine. Paris, 1843. pp. 30, 8vo. Folding plate.

ANDRÉ, G.

Étude chimique et thermique de quelques oxychlorures métalliques. Paris, 1884. 4to.

ANDREÆ, B.

Die neuesten Erfolge des Wassergases in der Leuchtgasindustrie, deren technische, finanzielle und volkswirthschaftliche Bedeutung. Wien, 1884. 8vo.

ANDREWS, THOMAS.

Scientific Papers ; with a Memoir by P. G. Tait and A. Crum Brown. London, New York, 1889. pp. lxii–514, 8vo. Portrait and illustrations.

ANDREWS, THOMAS, AND PETER G. TAIT.
On the volumetric relations of ozone and the action of the electrical discharge on oxygen and other gases. London, 1860. 4to. Plate.

ANDRIA, NICCOLA.
Elementa chemicæ philosophicæ. Napoli, 1786.
Instituzioni di chimica filosofica. Napoli, 1813.
Trattato delle acque minerali. Napoli, 1775. 8vo.

ANDRIESSEN, ADOLPH.
Lehrbuch der unorganischen Chemie für Schulen. Braunschweig, 1860. 8vo. Ill.

ANGARYD, E. H.
Fortschritte und Verbesserungen der Wollenstückfärberei seit 1877. Leipzig, 1885. 5 parts.

ANGELL, ARTHUR, AND OTTO HELMER.
Butter, its Analysis and Adulterations. Specially treating on the detection and estimation of foreign fats. London, 1874. pp. iv–52.

ANGENOT, CHARLES.
Leçons sur le pétrole et ses dérivés données à l'Institut supérieur de commerce. Anvers, 1885. 8vo.

ANKERSMIT, PIETER.
Scheikundig overzicht der suikers. Akademisch proefschrift ter verkrijging van den graad van Doctor in de Wis- en Natuurkundige Wetenschappen aan de Hoogeschool te Utrecht . . . Amsterdam, 1859, pp. [ix]–112, 8vo.

ANKUM, C. H. VAN.
Scheikundig onderzoek van Nederlandsche wateren. Verhandeling uitgegeven door de Hollandsche Maatschappij der Wetenschappen te Haarlem. Met de gouden Medaille bekroond, in den Jare 1852. Haarlem, 1853. 4to. 2 plates in fol.

ANLEITUNG ZUR BEREITUNG DES ZUCKERS AUS AHORN UND RUNKELRÜBEN, deren Anbau und übrigen Benützungen. Wien, 1812.

ANTICO, G.
Lezioni di chimica. Ostuni, 1891. 8vo.

ANTISELL, THOMAS.
The Manufacture of Photogenic or Hydro-Carbon Oils. New York, 1859. 8vo.
Another edition. New York, 1866. 8vo.

ANTONI, ALESSANDRO VITTORIO PAPACINO D'

Examen de la poudre.   Traduit de l'italien par Gratien de Flavigny.
    Amsterdam, Genève et Paris, 1773.   pp. [vi]–240, 8vo.   Ill.

APJOHN, JAMES.

Manual of the Metalloids.   Galbraith and Haughton's scientific manuals.
    Experimental and Natural Science Series.   Second edition.
    London, 1865.   pp. xii–600, 12mo.

APPLETON, JOHN HOWARD.

Beginners' (The) Handbook of Chemistry.   The non-metals.   New
    York, 1884.   Second edition, 1888.

Chapters on the Carbon Compounds.   Providence, 1892.

Chemistry developed by the facts and principles drawn chiefly from the
    non-metals.   New York, 1884.   16mo.

First Report Book ; for chemical experiments.   Boston, 1891.

Introduction (An) to Qualitative Chemical Analysis arranged for the
    use of the students in Brown University Laboratory.   Providence,
    1869.   pp. 59, 16mo.

Lessons in Chemical Philosophy.   Second edition.   New York, Boston,
    Chicago, 1890.   pp. vii–256, 8vo.   Ill.

Metals (The) of the Chemist.   Providence, 1891.

Second Report Book ; for Qualitative Analysis.   Providence, 1891.

Short (A) Course in Qualitative Chemical Analysis.   Boston, 1878.   12mo.
        Third edition.   Philadelphia, 1878.   pp. 112, 12mo.

Short (A) Course in Quantitative Chemical Analysis.   Boston, 1881.

Third Report Book ; for Quantitative Analysis.   Providence, 1891.

Young (The) Chemist.   A Book of Laboratory Work for Beginners.
    Second edition.   Philadelphia, 1878.   8vo.

ARATA, PEDRO N.

Relacion de los trabajos practicados por la officina química municipal
    de la ciudad de Buenos Aires, 1884.   Buenos Aires, 1885.

ARBO, NICOLAUS.

Dissertatio inauguralis sistens analysin nitri physico-chymicam.
    Hafniæ, 1752.   4to.

ARCET, D'.   *See* D'Arcet.

ARCHER, CLEMENT.

Miscellaneous Observations on the effects of Oxygen on the Animal
    and Vegetable Systems.   Interspersed with chemical, physiological
    . . . remarks ; and an attempt to prove why some plants are ever-
    green and others deciduous.   Bath, 1798.   8vo.

ARENA, F.

Trattato pratico di chimica clinica sull' analisi dell' urina con appendice sull' analisi delle feci, per uso dei medici, studenti di medicina, di chimica e farmacisti.   Napoli, 1884.   8vo.

ARENDS, L. A. F.

Chemie des menschlichen Körpers und der Nahrungsmittel .   .   . Berlin, 1857.   2 vols., 8vo.

ARENDT, RUDOLPH.

Grundzüge der Chemie.   Methodisch bearbeitet.   Hamburg, 1884. 8vo.   Ill.

> Zweite Auflage.   Hamburg, 1887.   8vo.

Lehrbuch der anorganischen Chemie.   Nach den neuesten Ansichten der Wissenschaft, auf rein experimenteller Grundlage für höhere Lehranstalten und zum Selbstunterricht methodisch bearbeitet Dritte durchgesehene und vermehrte Auflage.   Leipzig, 1875.   pp. xiv–578.   8vo.   Ill.

> Учебникъ неорганической химіи.   Москва, 1871.

Leitfaden für den Unterricht in der Chemie.   Hamburg, 1884.   8vo.

> Vierte verbesserte und vermehrte Auflage.   Hamburg, 1892. 8vo.   Ill.

Organisation, Technik und Apparat des Unterrichts in der Chemie an niederen und höheren Lehranstalten.   Eine Ergänzungsschrift zu des Verfassers Lehrbuch der anorganischen Chemie   Leipzig, 1868.   pp. [ii]–135.   8vo.

Technik der Experimental-Chemie.   Anleitung zur Ausführung chemischer Experimente für Lehrer und Studirende sowie zum Selbstunterricht.   Leipzig, 1881.   2 parts.   8vo.   Ill.

> Zweite Auflage.   Hamburg, 1891–92.

ARMSTRONG, HENRY EDWARD.

Introduction to the Study of Organic Chemistry ; Chemistry of Carbon and its compounds.   London, 1874.   pp. x–349.   12mo.   Ill. Text-books of Science.

> Third edition.   London, 1882.   12mo.   Ill.

ARMSTRONG, JAMES E., AND JAMES H. NORTON.

Laboratory Manual of Chemistry.   New York, Cincinnati and Chicago, n. d.   [1892].   pp. 144, 12mo.

ARNOLD, C.

Kurze Anleitung zur qualitativen chemischen Analyse und medicinisch-chemischen Analyse.   Namentlich zum Gebrauche für Mediciner und Pharmaceuten bearbeitet.   Zweite Auflage.   Hannover, 1887.

ARNOLD, C. [Cont'd.]

> Dritte Auflage. Hannover, 1890. 8vo.

Repetitorium der Chemie. Mit besonderer Berücksichtigung der für
die Medizin wichtigen Verbindungen sowie der " Pharmacopœa
Germanica." Hamburg, 1884. 8vo.

> Zweite Auflage, 1887.
> Dritte Auflage, 1890.

ARNOLD, R.

Ammoniak und Ammoniak-Präparate. Ein praktisches Handbuch für
Fabrikanten, Chemiker [etc.]. Berlin, 1888. Ill.

> Ammonia and Ammonium Compounds. Comprising their
> manufacture from gas-liquor, and from spent oxide
> (with the recovery from the latter of the bye-products,
> sulphur, sulphocyanides, prussian-blue, etc.) ; special
> attention being given to the analysis, properties, and
> treatment of the raw materials and final products. A
> practical manual for manufacturers, chemists, gas-
> engineers and drysalters, from personal experience,
> and including the most recent discoveries and improve-
> ments. Translated from the German by H. G. Colman.
> Illustrated by numerous woodcuts. London, 1889.
> pp. vii–130, 8vo.

ARPHE Y VILLAPHAÑE, JUAN DE.

Qvilatador de la plata, oro, y piedras. Valladolid, por Alonso y Diego
Fernãdez de Cordova, 1572. Sq. 8vo. Ill.

ARRATA.

Coleccion de procederes químicos aplicables á la economía doméstica,
á la medicina y artes. Vitoria, 1842. 4to.

ARRONET, H.

Quantitative Analyse des Menschenblutes, nebst Untersuchungen zur
Controlle und Vervollständigung der Methode. Dorpat, 1887.
pp. 71, 8vo.

ARTUS, WILIBALD.

Grundzüge der Chemie in ihrer Anwendung auf das praktische Leben.
Für Gewerbetreibende und Industrielle im Allgemeinen, sowie für
jeden Gebildeten. Wien, 1880. Ill.

Lehrbuch der Chemie zum Gebrauche bei Vorträgen, sowie auch zum
Selbststudium für Mediciner, Pharmaceuten, Landwirthe und
Techniker. Leipzig, 1846. 8vo.

ARTUS, WILIBALD. [Cont'd.]
Zweite verbesserte und vermehrte Auflage.   Leipzig, 1851.
8vo.
Leichtfassliche Anleitung zur Auffindung der Mineralgifte.   Ein
Leitfaden bey gerichtlich-chemischen Untersuchungen zum Ge-
brauche für Aerzte und Apotheker, nebst einem Anhange über
Prüfung des Weines, Essigs und Bieres.   Leipzig, 1843   pp. xii–247.

ARVIDSON. *See* Afzelius, Johann.

ARZT, PHILIPP EDMUND GOTTLOB.
Versuch einer systematischer Anordnung der Gegenstände der reinen
Chemie.   Leipzig, 1795.

ASCHOFF, LUDWIG PHILIPP.
Anweisung zur Prüfung der Arzeneimittel auf ihre Güte, Aechtheit und
Verfälschung . . . Lemgo, 1829.

ASH, JOHN.
Experiments and Observations to Investigate by Chemical Analysis
the Medicinal Properties of the Mineral Waters of Spaa and Aix-
la-Chapelle.   London, 1788.

ASKINSON, GEORGE WILLIAM.
Fabrikation (Die) der ätherischen Oele.   Anleitung zur Darstellung
derselben nach den Methoden der Pressung, Destillation, Extraction, Deplacirung, Maceration und Absorption, nebst einer ausführlichen Beschreibung aller bekannten ätherischen Oele in Bezug
auf ihre chemischen und physikalischen Eigenschaften und
technische Verwendung, sowie der besten Verfahrungsarten zur
Prüfung der ätherischen Oele auf ihre Reinheit.   Zweite verbesserte
und vermehrte Auflage.   Wien, Pest, Leipzig, 1876.   Ill.
Parfümerie-Fabrikation (Die).   Vollständige Anleitung zur Darstellung
aller Taschentuch-Parfums, Riechsalze, Riechpulver, Räucher-
werke, aller Mittel zur Pflege der Haut, des Mundes und der
Haare, der Schminken, Haarfärbemittel und aller in der Toilette-
kunst verwendeten Präparate, nebst einer ausführlichen Schilderung
der Riechstoffe.   Dritte sehr vermehrte und verbessere Auflage.
Leipzig, Wien und Pest, 1876.   Ill.
Zweite Auflage, 1883
Guide du parfumeur.   Odeurs—Essences—Extraits et vinaigres de toilette—Poudres—Sachets—Pastilles—Émulsions—Pommades—Dentifrices.   Édition française par
G. Calmels.   Paris, *n. d.*   12mo.   Ill.

ASSCHE, FR. VAN.

Des sucres, théorie moléculaire de leurs fonctions.   Paris, 1378. 8vo.

ASSMUSS, EDUARD PHILIBERT.

Die trockene Destillation des Holzes und Verarbeitung der durch
dieselbe erhaltenen Rohproducte auf feinen wie auf Essigsäure,
essigsaure Salze, Terpentinöl, Wagenschmiere, Kienruss, etc. Ein
Handbuch für Techniker, Chemiker und Fabrikanten.   Nach
eigenen mehrjährigen Erfahrungen. Berlin, 1867. pp. viii–144.
8vo.   Ill.

ATCHERLEY, ROWLAND J.

Adulterations of Food.   With short processes for their detection.
London, 1874.   pp. viii–112, 16mo.   Plates.

ATKINSON, J. J.

A Practical Treatise on the Gases met with in Coal Mines.   New
York, 1875.   pp. 53, 24mo.

ATLEE, W. L.

The Chemical Relations of the Human Body with surrounding agents.
Philadelphia, 1845.   8vo.

ATTFIELD, JOHN.

Introduction to Pharmaceutical Chemistry.   London, 1867.   Ill.
Chemistry ; general, medical and pharmaceutical, including
the chemistry of the British Pharmacopœia.   [Second
edition of Introduction to Pharmaceutical Chemistry.]
London, 1869.   Ill.
[Third edition] from the second English edition.   Including
the chemistry of the U. S. Pharmacopœia.   Philadelphia,
1871.   Ill.
Fourth [English] edition.   London, 1872.
Fifth [U. S.] edition.   Philadelphia, 1873.
Sixth [English] edition.   London, 1875.
Seventh [U. S.] edition.   Philadelphia, 1876.
Eighth [U. S.] edition.   Philadelphia, 1879.
Ninth [English] edition.   London, 1881.
Tenth [U. S.] edition, Philadelphia, 1883.
Eleventh [English] edition.   London, 1885.
Thirteenth [English] edition.   London, 1889.

AUBERT, A.

Innervation der Kreislaufsorgane.   *See* Handbuch der Physiologie.

AUBIGNÉ DE LA FOSSE, NATHAN. [ALBINEUS.]
Bibliotheca chymica contracta. Genf, 1653.
Lumen novum chymicum. Genf, 1654.
Cf., in Section VI, Albineus, Nathan.

AUERBACH, G.
Das Anthracen und seine derivate für Technik und Wissenschaft.
Zweite vermehrte Auflage. Braunschweig, 1880. pp. xi-280, 8vo.
First edition, 1873.
Anthracen its Constitution, Properties, Manufactures and
Derivatives, including Alizarin, Anthrapurpurin, etc.,
with their applications in Dyeing and Printing. Trans-
lated and revised by W. Crookes. London, 1877. 8vo.

AUGUSTIN, FRIEDRICH LUDWIG.
Vom Galvanismus und dessen medicinische Anwendung. Berlin, 1801.

AUSFÜHRUNG (DIE) DER MALZANALYSE nach den Vereinbarungen auf
dem land- und forstwirthschaftlichen Congresse in Wien, 1890.
Wien, 1892. Roy. 8vo.

AUSTEN, PETER TOWNSEND.
Chemical Lecture Notes. New York, 1888. 12mo.
Kurze allgemeine Einleitung zu den aromatischen Nitroverbindungen.
Leipzig und Heidelberg, 1876. 8vo.

AUSTEN, WILLIAM CHANDLER ROBERTS.
An Introduction to the Study of Metallurgy. London, 1891. 8vo.
Griffin's Scientific Text-books.

AUSTIN, G. L.
Water-Analysis. A Handbook for Water-Drinkers. Boston and New
York, 1883. pp. 48, 18mo.

AUWERS, CARL.
Die Entwicklung der Stereo-Chemie ; theoretische und experimentelle
Studien. Heidelberg, 1890. 8vo. Ill.

AVELING, EDWARD BIBBINS.
Chemistry of the Non-Metallics. London, 1886. pp. 215, 8vo. Ill.
Hughes's Matriculation Manuals.
Mechanics and Experimental Science as required for the matriculation
examination of the University of London. Chemistry. London,
1888. 8vo. Ill.
The same. Key to Chemistry. London, 1888. 8vo.

AVERY, ELROY M.
Avery's Complete Chemistry.   New York [1883].   12mo.

AVOGADRO, AMADEO.
Fisica dei corpi ponderabili ; ossia trattato della costituzione generale
    dei corpi.   Torino, 1837–41.   4 vols., 8vo.   Plates.
Mémoire sur les volumes atomiques et sur leur relation avec le rang
    que les corps occupent dans la série électro-chimique.   Turin,
    1844.   4to.   [From Mém. de Turin, Sér. 2, T. 8.]
(2e) Mémoire sur les volumes atomiques des corps composés.   Turin,
    1845.   4to.   [From Mém. de Turin, Sér. 2, T. 8.]
(3e) Mémoire sur les volumes atomiques.   Détermination des nombres
    affinitaires des différents corps élémentaires par la seule considéra-
    tion de leur volume atomique et de celui de leurs composées.
    Turin, 1849.   4to.   [From Mém. de Turin, Sér. 2, T. 11.]
(4e) Mémoire sur les volumes atomiques.   Détermination des volumes
    atomiques des corps liquides à leur température d'ébullition ;
    nombres affinitaires qui s'en déduisent pour quelques uns des
    corps élémentaires.   Turin, 1850.   4to.   [From Mém. de Turin,
    Sér. 2, T. 12.]
Mémoire sur les conséquences qu'on peut déduire des expériences de
    M. Regnault sur la loi de compressibilité des gaz.   Turin, 1851.
    4to.   [From Mém. de Turin, Sér. 2, T. 13.]
Nouvelles considérations sur la théorie des proportions déterminées
    dans les combinaisons et sur la détermination des masses des
    molécules des corps.   [Torino], 1821.   4to.
Nouvelles recherches sur le pourvoir neutralisant de quelques corps
    simples.   [Torino], 1836.   4to.

AVOGADRO, AMADEO, UND AMPÈRE.
Die Grundlagen der Molekulartheorie.   Abhandlungen von A. Avogadro
    und Ampère.   Leipzig [1889].   8vo.   Plates.
        Ostwald's Klassiker der exakten Wissenschaften.

BAADER, FRANZ XAVER VON.
Vom Wärmestoff, seiner Vertheilung, Bindung und Entbindung.   Wien
    und Leipzig, 1786

BABCOCK, S. M.
Report of the Chemist to the New York Agricultural Experiment
    Station, Geneva, N. Y.   (Extract from the Fifth Annual Report of
    the New York Agricultural Experiment Station for 1887.)   Dis-
    tributed January 30, 1887.   Elmira, N. Y., 1887.

BABO, LAMBERT VON.

Anleitung zur chemischen Untersuchung des Bodens für Landwirthe. Frankfurt am Main, 1843. 8vo.

Of this a popular edition was published under the following title :

Ackerbauchemie, oder kurze Darstellung dessen was der Landmann von chemischen Kenntnissen bedarf, um seinen Acker zweckmässig zu behandeln. Frankfurt am Main, 1845. 12mo.

Neue mit einem Anhang vermehrte Auflage. In 17 Abendunterhaltungen. Frankfurt am Main, 1851. 12mo. Tables in 4to and folio.

Akkerbouw-scheikunde, of kort begrip der scheikundige grondwaarheden welke de landbouwer behoort te kennen ; ten einde zijnen akker behoorlijk te behandelen. Onder toezicht van eenen landbouwer voor Nederlanders bewerkt. Deventer, 1847. 8vo.

Les veillées du cultivateur, ou catéchisme de chimie agricole. Traduit de l'allemand par Mme. Parisot de Cassel. Moulins, 1849. 8vo.

Земледѣльческая химія, или краткое изложеніе химическихъ свѣдѣній, необходимыхъ каждому земледѣльцу. С.-Петербургъ, 1847. 8⁰

Erzeugung (Die) und Behandlung des Traubenweins. Frankfurt a. M., 1846.

Hauptgrundsätze (Die) des Ackerbaus. Frankfurt a. M., 1851.

Ueber die Anwendung eines logarithmischen Proportionalkreises zur Ausführung und Controle chemischer Berechnungen. Heidelberg, 1855. 8vo.

BABU, L.

Précis d'analyse qualitative ; recherches des métalloides et des métaux usuels, dans les mélanges de sels, les produits d'art, et les substances minérales. Paris, 1888. 8vo

BACHE, FRANKLIN.

A System of Chemistry for the Use of Students of Medicine. Philadelphia, 1819. pp. xvi–624, 8vo.

BACHHOFFNER, GEORGE H.

Chemistry as Applied to the Fine Arts. London, 1837. Ill. Second edition. London, 1884.

BACHMANN, JOSEPH.

Chemische Abhandlungen über das Mangan. Wien, 1829.

BACON, ROGER.
De secretis operibus artis et naturæ et de nullitate magiæ.   Paris, 1542.
Opus majus.   London, 1733.
>   *Cf. also, in Section VI*, Bacon, Roger.

BAENITZ, C.
Lehrbuch der organischen Chemie in populärer Darstellung.   Nach
methodischen Grundsätzen für gehobene Lehranstalten, sowie zum
Selbstunterrichte bearbeitet.   Berlin, 1873.   8vo.   Ill.

BAER, W.
Chemie für Schule und Haus.   Populäre Darstellung der Lehren der
Chemie erläutert an einfachen Experimenten, die von Jedermann
leicht angestellt werden können.   Leipzig, 1859.   8vo.   Ill.
Chemie (Die) des praktischen Lebens.   Populäre Darstellung der Lehren
der Chemie in ihrer Anwendung auf die Gewerbe, die Land- und
Hauswirthschaft, sowie auf die Vorgänge im menschlichen Körper
nebst einer Anleitung zur Ausstellung der einfachsten chemischen
Versuche.   Zweite Auflage.   Leipzig, 1861.   2 vols., 8vo.   Ill.

BAHNSON, J. J.
Forelæsninger over organisk Chemie ved den kongelige militaire
Höjskole.   Kjøbenhavn, 1867.   4to.

BAILLY, VICTOR.
Eaux thermales de Lamotte-les-Bains, arrondissement de Grenoble.
Paris, 1844.   18mo.

BAIN, A.
Lógica de la química.   Version española por A. Ordax.   Madrid, 1881.
8vo.

BAKER, THOMAS R.
A Short Course in Chemistry based on the experimental method.
Lancaster, 1883.   pp. 152, 8vo.   Ill.

BAKER, J.
Epitome of Chemistry, in lectures illustrated by 500 experiments,
[with] Appendix.   Portsmouth, 1838.

BALAGUER, F.
Manual de industrias químicas inorgánicas.   Bibliotheca enciclopédica
popular ilustrada.   Madrid, 1879.   12mo.

BALAGUER Y PRIMO, FRANCISCO.
Manual práctico de análisis de los vinos.   Segunda edicion corregida y
considerablemente aumentada.   Madrid, 1873.   8vo.   Ill

BALARD, ANTOINE JÉROME.
> This French chemist, distinguished by his discovery of Bromine, seems not
> to have published any independent works.   His announcement of
> Bromine is found in *Ann. chim. phys.*, XXXII, 1826.

BALDUINUS, CHRISTIANUS ADOLPHUS.
*Phosphorus hermeticus, sive Magnes luminaris.   Francofurti et Lipsiæ,
1675.   pp. [20], 24mo.   Folding plate.
> " Baldwin's phosphorus," so-called, described in this work, is thought to be
> anhydrous calcium nitrate.   " Homberg's phosphorus," discovered in
> 1693, was an oxychloride of calcium.   " Canton's phosphorus," dis-
> covered in 1768, was calcium sulphide.   " Brand's phosphorus," the
> elementary substance, was discovered in 1669.   The Bologna Stone,
> luminous in the dark, was discovered about 1603.   *See also, in Section
> VI*, Balduinus, C. A.

BALLING, CARL A. M.
Compendium der metallurgischen Chemie.   Propädeutik für das
Studium der Hüttenkunde.   Bonn, 1882.   8vo.
Probirkunde (Die).   Anleitung zur Vornahme docimastischer Unter-
suchungen der Berg-, und Hüttenprodukte.   Braunschweig, 1879.
8vo.   Ill.
> Manuel pratique de l'art de l'essayeur.   Guide pour l'essai
> des minérais, etc.   Traduit par L. Gautier.   Paris, 1881.
> 8vo.

BALLING, CARL JOSEPH NAPOLEON.
Anleitung zum Gebrauch des Saccharometers.   Prag, 1855.
Branntweinbrennerei (Die) wissenschaftlich begründet und praktisch
dargestellt.   Dritte vermehrte und verbesserte Auflage.   Prag,
1865.   2 vols , 8vo.
Gährungschemie (Die) wissenschaftlich begründet und in ihrer Anwendung auf die Weinbereitung, Bierbrauerei, Branntweinbrennerei
und Hefenerzeugung praktisch dargestellt.   Prag, 1845–47.   4 vols.,
8vo.   Ill.   Vol. I, pp. xvi–327, 1845 ; vol. II, pp. xviii–500, 1845 ;
vol. III, pp. xvi–411.
> Zweite Auflage, 2 vols. in 4 parts.   Prag, 1854–55.
> Dritte Auflage, 4 vols., 1865.
Saccharometrische (Die) Bier- und Branteweinprobe.   Prag, 1846.
Ueber einige des wichtigsten Gegenstände des Eisenhüttenwesens.
Leipzig, 1829.

BALLISTERO, F. P.
Estudios contemporanéos de química legal.   Parte I : Ptomaïnas y
Leucomaïnas.   Madrid, 1891.   4to.

BALLÓ, M.

Das Naphtalin und seine Derivate in Beziehung auf Technik und Wissenschaft dargestellt. Braunschweig, 1870. 8vo.

BALMAIN, WILLIAM H.

Lessons on Chemistry, for the use of pupils in schools, junior students in universities, and readers who wish to learn the fundamental principles and leading facts ; with questions for examination, a glossary of chemical terms and chemical symbols, and an index. With numerous wood-cuts, illustrative of the decompositions. London, 1844. 8vo.

BALTHASAR, THEODOR.

Dissertatio chymico-medica, pro licentia, de sale communi. Altorf, 1701.

     Kurze Beschreibung der vortrefflichen Eigenschaften des edlen gemeinen Salzes. . . . Erlangen, 1708.

BALTZER, L.

Nahrungs- und Genussmittel der Menschen in ihrer chemischen Zusammensetzung und physiologischer Bedeutung. Zweite Auflage. Leipzig, 1874.

BAMJI, M. C.

The Student's Chemistry, including Qualitative Analysis. Poona, 1890. 2 vols., 8vo. (Published anonymously.)

BANCROFT, EDWARD.

Experimental Researches concerning the Philosophy of Permanent Colors and the best means of producing them, by Dyeing, Calico Printing, etc. Philadelphia, 1814. 2 vols., 8vo.

BANDLIN, O.

Die Gifte und ihre Gegengifte. Basel, 1869-73. 3 vols., 12mo.

BANISTER, H.

Gas Manipulation, with an account of the apparatus employed in the analysis of coal and coal gas. London, 1863. 8vo. Ill.

BANISTER, HENRY, AND SUGG, W. T.

Gas Manipulation with a description of the various instruments and apparatus employed in the analysis of coal and coal gas. London, 1867. pp. viii-135. 8vo. Ill.

BARBA, A. A.

Arte de los metales ; en que se enseña el verdadero beneficio de los de
oro y plata por azogue. . . . Nuevamente añadido con el
tratado de las antiguas minas de España, que escribió Don Alonso
Carrillo y Laso. Lima, 1817. 4to.

Métallurgie, on l'art de tirer et de purifier les métaux, traduite de l'espagnol avec les dissertations les plus rares
sur les mines et les opérations métalliques. Paris, 1851.
2 vols., 12mo.

BARBAGLIA, G. A.

Alcaloidi e ptomaine. Pisa, 1888. 8vo.

BARBE, ——.

L'emploi des engrais chimiques. Paris, 1885.

BARBET, ÉMILE.

Analyse des liquides sucrés. Méthodes exactes et méthodes rapides.
Paris, 18—.

Les appareils de distillation et de rectification ; Étude comparative de
leur consommation de vapeur et des résultats obtenus comme purification de l'alcool. Théorie de la rectification, épuration des
eaux-de-vie, tafias, genevièvres, etc. Paris, 1890. Ill.

Rôle du noir animal en sucrerie. Paris, 18—.

BARBIÉ, G. A.

Elementi di chimica generale ad uso delle scuole elementari. Torino,
1869. 8vo.

BARCHUSEN, JOHANN CONRAD.

Acroamata in quibus complura, ad iatrochemiam atque physicam spectantia, jucunda rerum varietate explicantur. Trajecta Batavorum,
1703.

Compendium ratiocinii chemici, more geometrarum concinnatum.
Leyden, 1712.

Elementa chemiæ, quibus subjuncta est confectura lapidis philosophicis
imaginibus repræsentata. Ultrajecti, 1703. 8vo.

　　　*Editio secunda. Apud Theodorum Haak. Lugduni-
Batavorum, 1718. pp. [xii]–532–[xx], 4to. Plates.

　　The 78 emblems treating of hermetic mysteries are in no wise explained by
　　the author, which leads Lenglet du Fresnoy to remark of him : " Il n'a
　　pas sçû, non plus que beaucoup d'autres, le secret des vrais chimistes."

*Pyrosophia, succincte atque breviter iatro chemiam, rem metallicam
et chrysopœiam pervestigans ; opus medicis, chemicis, pharma-

BARCHUSEN, JOHANN CONRAD.   [Cont'd.]
copœis, et metallicis non inutile. Lugduni-Batavorum, impensis
Cornelii Boutestein, 1698. pp. [xvi]–470, sm. 4to. Plates.

> Barchusen, b. 1666, d. 1732, was professor of medicine and chemistry at
> Utrecht.
>
> "Pyrosophia" has been supposed to be the first work in which the word
> *affinity* occurs, but Kopp shows it was used by Albertus Magnus.
> Lenglet du Fresnoy says of this work : " La seconde et la troisième
> partie de cet ouvrage . . . sont très curieuses et méritent d'être
> lues. L'auteur qui étoit habile ne disconvient pas de la transmutation
> des métaux."

BARCKHAUSEN, J. C.   *See* Barchusen, J. C.

BARENTIN, FRIEDRICH WILHELM.
De boratibus dissertatio chemica. Berolini, *n. d.* [1834]. 8vo.

BARFF, FREDERICK SETTLE.
An Introduction to Scientific Chemistry, designed for the use of schools
and candidates for university matriculation examinations. London,
1869. 8vo.
> Second edition, London, 1870.
> Third edition, London, 1871.
> Another edition, London, 1884. pp. xv–317–xi, 8vo.

BARFOED, C. T.
Lærebog i den analytiske Chemie. Prövemidlerne og den uorganiske
qualitative Analyse. Anden Udgave. Kjøbenhavn, 1880. 8vo.
> Lehrbuch der organischen qualitativen Analyse. Kopen-
> hagen, 1881. 8vo. Ill.
Organiske (De) stoffers qualitative analyse. Kjøbenhavn, 1878.
3 parts.

BARILLOT, ERNEST.
> *See* Chastaing, P., et E. Barillot.

BARILLOT, ERNEST.
Manuel de l'analyse des vins. Dosage des éléments, naturels, recherche
analytique des falsifications. Paris, 1889. pp. xii–131, 12mo. Ill.

BARKER, A.
Introduction to Chemical Analysis (Inorganic-Qualitative) for the use
of schools and science classes. Leeds and London, 1884. 12mo.

BARKER, GEORGE F.
A Text-book of Elementary Chemistry, theoretical and inorganic. New
Haven, 1870. 12mo. Ill.

BARKER, GEORGE F. [Cont'd.]
>   Eighth edition, New Haven, 1872.
>   Twelfth edition, Louisville, Ky., *n. d.* [1875]. pp. vi–342,
>   12mo. Ill.
>   Second edition revised, Louisville, Ky., 1891.

BARNER, JACOB.
Chymia philosophica cum doctrina salium. . . . Norimbergæ, 1689.
Exercitium chymicum delineatum. Pataviæ, 1670.
Spiritus vini sine acido. . . . Lipsiæ, 1675.

BARNES, J.
New (The) London Pocket Book or Memoranda Chemica, adapted to
    the daily use of the student. London, 1844. 12mo.
Tables for the Qualitative Analysis of simple Salts and Easy Mixtures,
    for the use of students preparing for the government science
    Oxford and Cambridge local examinations. Manchester, 1882.
    12mo.

BARNEVELD, WILLEM VAN.
De zamenstelling van het water op Lavoisiaansche gronden proefonder-
    vindelÿk verklaard. Amsterdam, 1791. 8vo.
Verhandeling over het regenwater, hetwelk met loodstof bezwangerd
    is. Amsterdam, 1807.

BARRAL. J. A.
>   Leçon (etc.). *See* Société chimique de Paris.

BARRAL, J. A.
Statique chimique des animaux. Emploi agricole du sel. Paris, 1850.
    12mo.

BARRESWIL ET GIRARD.
>   *See in Section II.*

BARRESWIL, LOUIS CHARLES, ET DAVANNE.
Chimie photographique, contenant les éléments de chimie expliquées
    par les manipulations photographiques, les procédés, etc., les
    recettes les plus nouvelles et les derniers perfectionnements ; la
    gravure et la lithophotographie. Paris, 1854. 8vo. Ill.
>   Deuxième édition, Paris, 1858.
>   Quatrième édition, Paris, 1864.
>   *In German* by Christian Heinrich Schmidt. Weimar, 1854. 8vo.

BARRESWIL, LOUIS CHARLES, ET ASCANIO SOBRERO.

Appendice à tous les traités d'analyse chimique, recueil des observations publiées depuis dix ans sur l'analyse qualitative et quantitative. Avec planches et figures dans le texte. Paris, 1843. 8vo.

Analytische Chemie. Deutsch bearbeitet von Fr. Ant. Kussin. Wien, 1844. 8vo. Ill.

BARRUEL, ÉTIENNE.

Sur l'extraction en grand du sucre de betterave. Paris, 1811.

BARRUEL, G.

Traité de chimie technique appliquée aux arts et à l'industrie, à la pharmacie et à l'agriculture. Paris, 1856–63. 7 vols., 8vo. Ill. Vol. I, pp. viii–588, 1856; vol. II, pp. 533 and 1 plate, 1856; vol. III, pp. 613, 1857; vol. IV, pp. 470, 1858; vol. V, pp. 512, 1860; vol. VI, pp. 549, 1861; vol. VII, pp. 543, 1863.

BARRY, SIR EDWARD.

Observations, Historical, Critical, and Medical, on the Wines of the Ancients and the analogy between them and modern wines. With general observations on the principles and qualities of water, and in particular on those of Bath. London, 1775. pp. xii–479, 4to. Frontispiece.

BARTELS, ERNST DANIEL AUGUST.

Grundlinien einer neuen Theorie der Chemie und Physik. Hannover, 1804. 8vo.

BARTELS, R.

Ueber die Einwirkung des Antimonwasserstoffs auf Metallsalzlösungen. Berlin, 1889. 8vo.

BARTH, MAX.

Die Weinanalyse. Kommentar der im kaiserlichen Gesundheitsamte 1884 zusammengestellten Beschlüsse der Kommission zur Beratung einheitlicher Methoden für die Analyse des Weines. Zugleich ein Leitfaden zur Untersuchung und Beurtheilung von Weinen, für Chemiker und Juristen bearbeitet. Mit einem Vorwort von J. Nessler. Hamburg und Leipzig, 1884. vii–[ii]–71, 24mo. Ill.

BARTHE, L.

Synthèses au moyen des éthers cyanacétiques et cyanosucciniques. Bordeaux, 1891. 4to. Ill.

BARTHENAU, L. BARTH VON.

Die nächsten Aufgaben der chemischen Forschung. Wien, 1880. 8vo.

BARTHOLINUS, CASPAR.
  De mixtione eamque consequentibus temperamento, coctione, putredine,
    petrificatione, etc. Hafniæ, 1617. Rostochii, 1618.
            Another edition, 1621.
  De terra, aëre et igne institutio physica succincta ; cum præmissa ele-
    mentorum theoria generali. Rostochii, 1619. 8vo.

BARTLEY, ELIAS H.
  Text-Book (A) of Medical Chemistry for medical and pharmaceutical
    students and practitioners. Philadelphia, 1885. pp. viii–376. 8vo.
    Ill. Second edition, Philadelphia, 1889. 8vo. Ill.

BASILIUS VALENTINUS. *See in Section VI*, Valentinus, Basilius.

BASSET, N.
  Chimie de la ferme, leçons familières sur les notions de chimie élé-
    mentaire utiles au cultivateur et sur les opérations chimiques les
    plus nécessaires à la pratique agricole. Paris, 1858. 18mo. Ill.
            Guide pratique de chimie agricole, leçons familières [etc.,
              as above]. Paris, 1863. 18mo.
  Guide pratique du fabricant de sucre, contenant l'étude théorique
    et technique des sucres de toute provenance, la saccharimétrie
    chimique et optique, la description et l'étude culturale des plantes
    saccharifères, les procédés usuels et manufacturiers de l'industrie
    sucrière, et les moyens d'améliorer les diverses parties de la fabri-
    cation ; avec de nombreuses figures intercalées dans le texte.
    Paris, 1861. 8vo.
            Nouvelle édition entièrement refondue et considérablement
              augmentée. Paris, 1872–75. 3 vols, 8vo. Ill.
  Précis de chimie pratique, ou Éléments de chimie vulgarisée. Paris,
    1860. 18mo. Ill.
  Traité complet d'alcoolisation générale, guide du fabricant d'alcools,
    renfermant la marche à suivre pour obtenir l'alcool de toutes les
    substances alcoolisables, les moyens de débarrasser l'alcool des
    odeurs propres et de celles d'empyreume, ainsi que l'indication
    des rendements au point de vue de la fabrication par les méthodes
    les plus économiques, et toutes les règles, formules et tables de ré-
    duction qui peuvent être utiles au distillateur. Paris, 1854. pp.
    500, 12mo. Plates.

BASTIDE, ÉTIENNE.
  Les vins sophistiques, procédés simples pour reconnaitre les sophistica-
    tions les plus usuelles, coloration artificielle, plâtrage, salicylage,
    vinage, mouillage, etc. Paris, 1889. pp. 154, 12mo. Ill.

BATSCH, AUGUST JOHANN GEORG CARL.
  Die ersten Gründe der systematischen Chemie.   Jena, 1788, 8vo.
  Versuch einer historischen Naturlehre oder einer allgemeinen und
    besonderen Geschichte der körperlichen Grundstoffe.   Halle,
    1789-91.   2 parts, 8vo.

BATTERSHALL, JESSE PARK.
  Food Adulteration and its Detection.   New York, 1887, 8vo.
    Contains : Bibliography, including Periodicals, Reports and General Works
      chronologically arranged.

BAUDRIMONT, A.
  Du sucre et de la fabrication.   Suivi d'un précis de la législation qui
    régit cette industrie par A. Trébuchet.   Paris, 1841.   8vo.   Ill.
  Introduction à l'étude de la chimie par la théorie atomique.   Paris,
    1833.   pp. 208, 8vo.
  Traité de chimie générale et expérimentale, avec les applications aux
    arts, à la médecine et à la pharmacie.   Avec 260 figures intercalées
    dans le texte.   Paris, 1844–46.   2 vols., 8vo.

BAUER, ALEXANDER, UND FR. HINTERBERGER.
  Lehrbuch der chemischen Technik.   Wien, 1859.   pp. 574, 8vo.   Ill.

BAUER, E.
  Gährungstechnische Untersuchungsmethoden für die Praxis der Spi-
    ritus- und Presshefe-Industrie, mit besonderer Berücksichtigung
    der Bestimmung stickstoffhaltiger organischer Substanzen und
    der Kohlehydrate.   Braunschweig, 1891.   8vo.

BAUERMANN, H.
  Treatise on the Metallurgy of Iron.   First American Edition.   New
    York, *n. d.* [1868].   12mo.

BAUMANN, A.
  Tabelle zur Berechnung der Salpetersäure aus dem gefundenen Volu-
    men des Stickoxyds durch Multiplication.   München, 1889.   Fol.
  Tabelle zur gasvolumetrischen Bestimmung der Kohlensäure.   Mün-
    chen, 1889.   Fol.
  Tabelle zur gasvolumetrischen Bestimmung der Stickstoffs.   München,
    1889.   Fol.
  Tafeln zur Gasometrie.   Zum Gebrauche in chemischen und physika-
    lischen Laboratorien, sowie an hygienischen Instituten.   München,
    1885.   8vo.

BAUMBAUER, H.
  Leitfaden der Chemie.   Theil 1, Anorganische Chemie.   Freiburg,
    1884.   8vo.

BAUMÉ, ANTOINE.
  * Chymie expérimentale et raisonnée. Paris, 1773. 3 vols , 8vo. With
    portrait of author and many folding plates.

> The author was demonstrator of chemistry with Macquer for twenty-five
> years, during which time they gave sixteen courses in chemistry, each
> course comprising more than 2,000 experiments. His works are written
> in a clear, simple style ; he adopts the prevailing theory of phlogiston,
> but in no partisan spirit. He rejects firmly and briefly the baseless
> pretensions of the alchemists.

  Manual of Chemistry, translated from the French of Beaumé
    [*sic*]. Second edition. London, 1786
> The translator is believed to be J. Aikin.

  Erläuterte Experimental-Chemie aus dem französischen
    übersetzt von Johann Carl Gehlen. Leipzig, 1775. 3
    vols., 8vo.

  Handbuch der Scheidekunst, oder Beschreibungen der
    chemischen Behandlungen und ihrer Erzeugnisse.
    Aus dem französischen des Baumé ins deutsche über-
    setzt und mit Anmerkungen vermehrt von Franciscus
    Xavier von Wasserberg. Wien, 1774. 8vo.

Dissertation sur l'æther dans laquelle on examine les différens produits
    du mélange de l'esprit de vin avec les acides minéraux. Paris,
    1757. pp. xii–332, 12mo.
> *Contains :* Discours historique sur l'æther vitriolique, pp. 1–26.

Opuscules chimiques faisant suite à la chimie expérimentale et raison-
    née. Paris, An VI [1798]. 8vo.

  Kleine chemische Schriften aus dem französischen über-
    setzt. Frankfurt a. M., 1800. 8vo.

BAUMÉ, F. B. T.
  Versuch eines chemischen Systems der Kenntnisse von den Bestand-
    theilen des menschlichen Körpers. Aus dem französischen über-
    setzt von C. F. B. Karsten. Mit einigen Anmerkungen und einer
    Vorrede begleitet von Sigismund Friedrich Hermbstädt. Berlin,
    1802. 8vo.
> The original was published in 1798.

BAUMER, JOHANN WILHELM.
  Fundamenta chemiæ theoretico-practicæ. Giessæ, 1783. 8vo.
    Another edition. Marburg, 1787.

BAUMERT, GEORG.
  Lehrbuch der gerichtlichen Chemie, mit Berücksichtigung sanitäts-
    polizeilicher und medicinisch-chemischer Untersuchungen. Zum

BAUMERT, GEORG. [Cont'd.]
  Gebrauche bei Vorlesungen und im Laboratorium bearbeitet.
  Braunschweig, 1889–90. 8vo. Ill.

BAUMGARTEN, ANDREAS VON.
  Aräometrie oder Anleitung zur Bestimmung des spezifischen Gewichts
  und zur Verfertigung genauer Aräometer. Wien, 1820.

BAUMGARTEN, A.
  Ueber das Vorkommen des Vanadiums in dem Aetznatron des Handels.
  Göttingen, 1865. 8vo.

BAUMHAUER, EDUARD HEINRICH VON.
  Beknopt leerboek der onbewerktuigde scheikunde. Derde uitgave.
  Utrecht en Amsterdam, 1864. 8vo.
  Beziehungen (Die) zwischen dem Atomgewichte und der Natur der
  chemischen Elemente ; mit einer Tafel. Braunschweig, 1870. 8vo.
  Specimen meteorologico-chemicum de ortu lapidum meteoricorum,
  annexis duorum lapidum analysibus chemicis.   Trajecta ad
  Rhenum, 1844.

BAUP, SAMUEL.
  Sur la fixation du chiffre des équivalents chimiques. Paris, 1841. 8vo.

BAYEN, PIERRE.
  Analyses des eaux de Bagnères de Luchon.   Paris, 1765.
  Moyen d'analyser les serpentins, porphyres, ophites, granets, jaspes,
  schistes, jades et feldspaths. Paris, 1778.
  *Opuscles chimiques. Paris. An VI [1797], 2 vols., 8vo. Vol. I, pp.
  lxxiv–395 ; II, pp. 468.
         Published after Bayen's death by Malatret and Parmentier. Includes a bio-
         graphical notice of Bayen by the latter. In vol. I, p. 299, is shown how
         near Bayen came to discovering oxygen before Priestley, and how he
         failed.

BAYEN, PIERRE, ET LOUIS MARTIN CHARLARD.
  Recherches chimiques sur l'étain faites et publiées par ordre du gouvernement, ou réponse à cette question : Peut-on sans aucun danger
  employer les vaisseaux d'étain dans l'usage économique. Paris,
  1781. pp. viii–285, 8vo.
         Contains an historical sketch of interest.

BAYLEY, THOMAS.
  The Assay and Analysis of Iron and Steel, Iron Ores and Fuel.
  Reprinted from "The Mechanical World" with Additions.
  London, Manchester, New York, 1884. pp. x–91, 12mo. Ill.

BEALE, LIONEL.
Tables for the Chemical and Microscopical Examination of Urine in
    Health and Disease.   London, 1857.   8vo.

BEAUVISAGE, GEORGES.
Les matières grasses ; caractères, essai et falsification des beurres, huiles,
    graisses, suifs, etc.   Paris, 1891.   16mo.   Ill.
    Bibliothèque des connaissances utiles.

BECCARI, G.
Chimica agraria per le Scuole d'agricoltura del Regno.   Milano, 1891.

BÉCHAMP, A.
Mémoire sur les matières albuminoïdes.   Paris, 1884.   4to.
    Conférence (etc.), see Société Chimique de Paris.

BECHER, JOHANN JOACHIM.
Actorum laboratorii chymici Monacensis, seu physicæ subterraneæ
    libri duo, quorum prior profundam subterraneorum genesin, nec
    non admirandam globi terr-aque-aërei super et subterranei fabri-
    cam, posterior specialem subterraneorum naturam, resolutionem
    in partes partiumque proprietates exponit ; accesserunt sub finem
    mille hypotheses seu mixtiones chymicæ, ante hac nunquam visæ,
    omnia, plusquam mille experimentis stabilita, sumptibus et per-
    missu Electoris Bavariæ   .   .   .   Francofurti, 1669.   pp. [xxxiv]–
    633–[vii], 12mo.   Frontispiece.
        Another edition.   Francofurti, 1681.
* Chymischer Glücks-Hafen ; oder grosse chymische Concordantz und
    Collection von 1500 chymischen Processen : Durch viel Mühe und
    Kosten auss den besten Manuscriptis und Laboratoriis in diese
    Ordnung, wie hier folgendes Register aussweiset, zusammenge-
    tragen.   Franckfurt, in Verlegung Johann Georg Schiele, Buch-
    händlers, 1682.   pp. [viii]–810–[xxxv], sm. 4to.
        A large collection of chemical recipes by a noted German physician and
        alchemist.   Becher and Stahl introduced the theory of phlogiston into
        chemical science.
Chymischer Rosen-Garten, samt einer Vorrede und kurtz gefassten
    Lebens-Beschreibung Herrn J. J. Bechers, zum Druck befördert
    von Friedrich Roth-Scholtze.   Nürnberg, bey Johann Daniel Tau-
    bers seel. Erben.   Anno 1717.   pp. 96, 16mo.
* Chymisches Laboratorium oder unter-erdische Naturkündigung.   Da-
    rinnen enthalten wird : 1. Die tieffe Zeugung derer unter-erdischen
    Dinge : Wie auch der wunderbare Bau der ober- und unter-erdi-
    schen Erd-, Wasser- und Lufft-Kugel : Und dann die absonderliche
    Natur der unter-erdischen Dinge Aufflöss- und Zerlegung in ihre

BECHER, JOHANN JOACHIM. [Cont'd.]

Theile, und derselben Eigenschafft. II. Neue chymische Proben,
einiger künstlichen gleich darstelligen Verwandelung derer Metallen nach Anleitung der in vorigen Jahren in Druck gegebenen
Physicæ subterraneæ. III. Ein nochmaliger Zusatz und philosophischer Beweissthum derer chymischen, die Wahr- und Möglichkeit
derer Metallen Verwandelung in Gold, bestreitenden Lehr-Sätze.
IV. Ein chymischer Rätseldeuter, derer verdunckelten Wort-Sätze
Urhebung und Geheimnisse offenbahrend und aufflösend. Franckfurt, zu finden bey Philipp Fievet, Buchhändlung, 1680–90. 12mo.
[Pagination as below.]

[*Contents :*]

I. Physicæ subterraneæ : oder der unter-erdischen Naturkündigung.
1690. pp. [xxviii]-732.

II. [*On the title-page numbered 3.*] Nochmaliger Zusatz über die
unter-erdische Naturkündigung. Das ist : Philosophisches Be-
weisthum oder chymische, die Wahr- und Mögligkeit derer
Metallen Verwandelung in Gold bestreitende Lehr-Sätze an den
unüberwindlichsten Kayser Leopoldum. 1680. pp. 175-[xv].

III. [*On the title-page numbered 2.*] Experimentum chymicum novum :
oder Neue chymische Prob, worinnen die künstliche gleich-
darstellige Transmutation, oder Verwandelung, derer Metallen
augenscheinlich dargethan : An statt einer Zugabe, in die Physicam subterraneam : und Antwort auff D. Rollfinken Schriften
von der Nicht-Wesenheit des Mercurii derer Cörper. Ein Werck
voller üblichen Proben, wie auch derer Philosophen erklärter
vornehmer Sprüche ; dem chymie-liebenden Leser nicht unbe-
liebig fallend. 1680. pp. 192.

IV. Oedipus Chymicus, oder chymischer Rätseldeuter, worinnen derer
verdunckelten chymischen Wortsätze Urhebungen und Geheimnissen offenbahret und aufgelöset werden. Allen der Artzney-
und Chymiæ-Kunst beflissenen gar nützlich und nothwendig zu
lesen. Auf Begehren und mit sonderbarem Fleiss aus dem lateinischen in teutsch übersetzet in Druck gegeben. 1680. pp. 156-[iv].

* Institutiones chimicæ prodromæ, id est Oedipus Chimicus, obscuriorum terminorum et principiorum chimicorum mysteria aperiens et
resolvens. Amstelodami, 1664. pp. 202-[viii], 32mo.

> With engraving of coin struck by Ferdinand III in 1648 at Prague. Useful
> in defining symbolic expressions. *Also in* Manget's Bibliotheca chemica
> curiosa, vol. I, p. 306. *See, in Section VI,* Manget, J. J.

Magnalia Naturæ ; or, the Philosophers-Stone lately expos'd to publick
sight and sale. Being a true and exact account of the manner
how Wenceslaus Seilerus, the late famous projection-maker, at the

BECHER, JOHANN JOACHIM. [Cont'd.]

Emperour's Court, at Vienna, came by, and made away with a very great quantity of pouder of projection, by projecting with it before the Emperor, and a great many witnesses, selling it, &c., for some years past. Published at the request, and for the satisfaction of several curious and ingenious, especially of Mr. Boyl, &c. London, 1680. pp. [vi]–31, 4to.

Naturkündigung der Metallen, mit vielen curiosen Beweissthumen, natürlichen Gründen, Gleichnüssen, Erfahrenheiten, und bishero ohngemeinen Aufmerckungen vor Augen gestellet. Zur Erhaltung der Warheit, Erläuterung der spagierischen Philosophi und Gefallen der Liebhaber. Franckfurt, 1661.

* Opuscula chymica rariora, addita nova præfatione ac indice locupletissimo, multisque figuris æneis illustrata a Friderico Roth-Scholtzio. Norimbergæ et Altorfii, apud hæredes J. D. Tauberi, 1719. pp. [xii]–50–310, 12mo. Ill.

[*Contents :*]

Verzeichniss derer Schriften Joannis Joachimi Becheri. p. 13.

Tripus Hermeticus fatidicus pandens oracula chymica, seu

   I. Laboratorium portatile cum methodo vere spagyrice, sc. juxta exigentiam naturæ, laborandi ; accessit pro praxi et exemplo,

   II. Magnorum duorum productorum nitri et salis textura et anatomia; atque in omnium præcedentium confirmationem adjunctum est,

   III. Alphabetum minerale seu viginti quatuor theses de subterraneorum et mineralium genesi, textura et analysi.

His accessit concordantia mercurii lunæ. Omnia juxta authoris doctrinam et principia in physica sua subterranea ejusque supplementis conscripta, adeo ut hic Tripus Hermeticus commentarius practicus super præfatam physicam subterraneam vere dici queat, utpote scriptum raris experimentis, multis figuris et profundis speculationibus innixum, ut lectori per se patebit.

Tractatus primus.

Scyphus Becherianus, sive Laboratorium portatile, quo omnes labores chymici excogitabiles et practicabiles levi sumptu, brevi tempore, jucunda operatione elaborari possunt ; accessit ejusdem methodus vera et genuina philosophice et spagyrice laborandi. p. 31.

Tractatus secundus.

Centrum mundi concatenatum seu duumviratus Hermeticus, sive magnorum mundi duorum productorum nitri et salis textura et anatomia, aëris nempe et maris consideratio ; pro commentario in posteriora duo capita supplementi primi physicæ suæ subterraneæ. p. 63.

BECHER, JOHANN JOACHIM. [Cont'd.]

Tractatus tertius.

Alphabetum minerale seu viginti quatuor theses chymicæ de mineralium, metallorum cæterorumque su'bterraneorum genesi, principiis,
differentiis, mixtione et solutione ; cornubiæ in Anglia inter ipsas
multifarias mineras earumque examinationes autopsia et praxi
congestæ et demonstratæ et juxta principia philosophica physicæ
subterraneæ J. J. Becheri. Truro. Anno 1682.

Tractatus quartus.

Concordantia mercuriorum lunæ. p. 150.

Tractatus quintus.

Concordantia menstruorum. p. 183.

Tractatus sextus.

Bericht von Erfind- und Zubereitung eines compendieusen Ofens, den
man auch auf Reisen bequem mit sich führen kan. In teutscher
Sprache hier beigefüget durch Friederich Roth-Scholtzen. p. 195.

Tractatus septimus.

Zu Erfüllung des ohnedem leeren Platzes hat man hier die Beschreibung des Herrn Obristen von Schellenbergs Universal Ofens mit
beyfügen sollen. p. 202.

Tractatus octavus.

Chymischer Rosengarten samt einer Vorrede hier beygefüget durch
F. Roth-Scholtzen. p. 207.

[By a blunder of the printers, the ninth treatise is placed after the tenth.]

Tractatus decimus.

Bericht von dem Sande als einem ewig-währendem Metall oder Berg-
Wercke diesen opusculis chymicis mit beygefüget durch F. Roth-
Scholtzen. p. 256.

Tractatus nonus.

Pantaleon delarvatus. p. 295.

Physica subterranea, profundam subterraneorum genesin e principiis
hucusque ignotis ostendens. (Experimentum chymicum novum,
quo artificialis et instantanea metallorum generatio et transmutatio ad oculum demonstratur ; loco supplementi et responsi ad D.
Rolfincii schedas de non entitate mercurii corporum. Supplementum secundum : Demonstratio philosophica seu theses chymicæ
possibilitatem transmutationis . . . evincentes. Experimentum novum de minera arenaria perpetua . . . loco supplementi

BECHER, JOHANN JOACHIM. [Cont'd.]

tertii.) Editio novissima, præfatione utili præmissa ; libro tersius et
curatius edendo operam navavit et specimen Beccherianum funda-
mentorum, documentorum, experimentorum subjunxit G. E. Stahl.
Lipsiæ, 1703. 2 vols., 8vo.

Editio Novissima. Lipsiæ, 1738. Sm. 4to. Frontispiece.

* Trifolium Becherianum Hollandicum oder drey neue Erfindungen,
bestehende in einer Seiden- Wasser- Mühle- und Schmeltz-Wercke.
Zum ersten Mahl in Holland vorgeschlagen und Werckstellig
gemacht : Mit gründlicher Anweisung wie es mit denselbigen
Sachen beschaffen ist. Auss der niederländischen in die hochteutsche Sprach übersezzet. Franckfurt, in Verlegung Johann
David Zunners, 1679. 12mo.

BECK, LEWIS C.

Adulterations of Various Substances used in Medicine and the Arts, with
the means of detecting them. New York, 1846. 12mo.

Manual (A) of Chemistry . . . intended as a Text-Book for Medical Schools. Third edition. New York, 1838. 8vo.

First edition, 1831.

BECKE, W. VON DER.

Die Milchprüfungs-Methoden nach vergleichenden Untersuchungen
bearbeitet und zusammengestellt. Mit einer Vorrede von J.
König. Bremen, 1882. pp. vi–105, 8vo. Ill.

BECKER, CARL FERDINAND.

Theoretisch-praktische Anleitung zur künstlichen Erzeugung und
Gewinnung des Salpeters nach eigenen und der in Frankreich
gemachten Erfahrungen. Braunschweig, 1814.

BECKER, FRANZ JOSEPH VON.

Om Kolhydraternes Förändring inom den lefvande Djurkroppen.
Helsingfors, 1853. 8vo.

BECKER, JOHANN HERRMANN.

Versuch einer allgemeinen und besondern Nahrungsmittelkunde. Mit
einer Vorrede von S. G. Vogel. Stendal, 1810–22. 2 vols. in 5
parts, 8vo.

Part I also under the title : Versuch einer Literatur und Geschichte der
Nahrungsmittelkunde. Vol. II also under the title : Darstellung der
Nahrungsmittel des Menschen nach alphabetischer Ordnung. 2 Parts.
1818–22.

BECKER, JOHANN PHILIPP.

Chemische Anekdoten, oder Versuche über einige zweifelhafte und noch keine authentike Gültigkeit erlangte Sätze. Leipzig, 1788. pp. [xii]–253, 8vo.

Chemische Untersuchung der Pflanzen und deren Salze nebst andern dahin gehörigen Materien. Leipzig, 1786. pp. xxxii–286, 8vo.

BECKERHINN, CARL.

Kurzes Handbuch der theoretischen Chemie. Wien, 1877. 8vo. Ill.

Lehrbuch der Chemie für die kaiserliche königliche Infanterie- und Kavallerie Kadeten-Schulen. Wien, 1878. 8vo. Ill.

BECKERS, A.

Das Gesammte der Färberei und Druckerei mit Anilin-Farbstoffen auf Wolle, Baumwolle und Seide. Zweite Auflage. Berlin, 1865. 8vo.

    Anilin-Färberei. Das Gesammte der Färberei (etc.). Dritte vermehrte Auflage. Berlin, 1867. pp. 268, 8vo.

    Fünfte Auflage bearbeitet von M. Reimann. Berlin, 1874.

BECQUEREL, ALEXANDRE EDMOND.

Des forces physico-chimiques et de leur intervention dans la production des phénomènes naturels. Paris, 1875. 8vo, and atlas in 4to.

* La lumière, ses causes et ses effets. Paris, 1867–'68. 2 vols., 8vo. Ill.

    Vol. II deals with the chemical and physiological effects of light.

BECQUEREL, ANTOINE CÉSAR.

Éléments d'électro-chimie appliquée aux sciences naturelles et aux arts. Avec 3 planches. Paris, 1843. 8vo.

    Deuxième édition. Paris, 1864. 8vo. Ill.

    Elemente der Electro-Chemie in ihrer Anwendung auf die Naturwissenschaften und die Künste. Aus dem französischen. Erfurt, 1845.

    Zweite wohlfeilere Ausgabe. Erfurt, 1848. 8vo. Dritte Ausgabe. 1857.

    See also, in Section III, Becquerel.

BECQUEREL, EDMOND ET PASTEUR.

    Leçon (etc.). *See* Société chimique de Paris.

BECQUEREL, LOUIS ALFRED, ET A. RODIER.

Traité de chimie pathologique appliquée à la médecine pratique. Paris, 1854. pp. x–608. 8vo.

    Pathological Chemistry, in its application to the practice of

BECQUEREL, LOUIS ALFRED, ET A. RODIER.   [Cont'd.]
> medicine. Translated by S. T. Speer. London, 1857.
> 8vo.
> Tratado de química patológica aplicada à la medicina
> · practica. Traducido por Teodoro Yanez y Font.
> Madrid, 1862. 4to.

Recherches sur la composition du sang dans l'état de santé et dans
l'état de maladie. Paris, 1844. 8vo. *In German by* Eisenmann
1845, 8vo, and 1847, 8vo.

BEDDOES, THOMAS.
Contributions to physical and medical knowledge, principally of the
West of England. London, 1799.

> Including among others: Humphry Davy, Researches on Heat, Light and
> Respiration.

BEDDOES, THOMAS, AND JAMES WATT.
Considerations of the Medicinal Use of Factitious Airs, and on the
manner of obtaining them in large quantities. Bristol, 1794–96.
Five parts. 8vo. Part I by Th. B., Part II by J. W.

> *A second edition bears the title :*
>> * Considerations on the Medicinal Use and on the Produc-
>> tion of Factitious Airs. Bristol, 1795–96. 3 parts, pp.
>> 172–40 ; 5 plates ; –xii–122 ; folding tables. 8vo.

> Part II has separate pagination and bears the title : Description of a Pneu-
> matic Apparatus with directions for procuring the factitious airs. By
> James Watt, 1795. *For a Supplement to Part II see :* WATT, JAMES,
> Supplement. Part III contains letters from Carmichael, Darwin, Ewart,
> Ferriar, Garnet, Johnstone, Pearson, Thornton and Trotter ; also from
> Atwood, Barr, W. W. Capper, Gimbernat, Sandford and others.

BEEN, JOHANNES NICOLAUS
Pertractatio trium quæstionum physicarum de auro, sole chymicorum.
Hafniæ, 1706. 4to.

BEETS, MARTINUS NICOLAAS.
Grondbeginselen der artsenijmengkundige scheikunde. Haarlem,
1812, 8vo.
Volks-Scheikunde. Amsterdam, 1815. Ill.

BEGUIN, JEAN [BEGUINUS, JOHANNES].
Tyrocinium chymicum, e naturæ fonte et manuali experientia de-
promptum. Paris, 1608. 12mo.
> Secunda editione recognitum. Coloniæ, 1612. 12mo.
Postrema editione. Coloniæ, 1615.

BEGUIN, JEAN. [Cont'd.]

> Editio sexta per Christophoro Glückradt. *n. p.* 1625.

> * Tyrocinium chymicum ; antehac a viris clarissimis Christophoro Glückradt et Jeremia Barthio notis elegantibus illustratum, formulisque medicamentorum optimis et secretis locupletatum ; nunc vero a Jo. Georgio Pelshofero notis et medicamentorum formulis in unum systema redactis ; hac novissima editione triplici indice ornatum, perillustri et eximia aromatario Antonio de Sgobbis. Venetiis, apud Baleonium, 1643. pp. [lvi]–480–[xliv].

> Tyrocinium chymicum, a Christophoro Glückradt et Jeremia Barthio notis elegantibus illustratum ; nunc vero a Johanne Georgio Pelshofero . . . in publicum emissum. Wittebergæ, 1650. 8vo.

> Tyrocinium chymicum commentario illustratum a Gerardo Blasio. Editio secunda. Amstelodami, 1669. 8vo.

> Tyrocinium chymicum, notæ Johannis Hartmanni olim editæ a Christoph. Glückradt. Francofurti ad Mœnum, 1684. Fol.

> * Tyrocinium Chymicum, or Chymical Essayes aquired from the Fountaine of Nature and Manuall Experience. London, 1669. pp. [vi]–136–[iv], 16mo. Illustrated title-page.

> Les élémens de chymie de Maistre Jean Beguin. Revueux, expliquez et augmentez par Jean Lucas de Roy. Troisième édition. Paris, *n. d.* [1624 ?].

> Les élémens de chymie. Reveus, expliquez et augmentez par Jean Lucas de Roy, médecin Boleducois. Rouen, chez Jean Baptiste Behourt, ruë aux Juifs prés le palais. 1632. pp. [xvi]–432–[xlviii], 16mo.

> First edition, Paris, 1615. Second, 1620. Dernière édition, Genève, 1624. 8vo. Other editions, Rouen, 1626, 1637, 1660 ; Lyon, 1665.

BÉHAL, AUGUSTE.

Études théoriques sur les composés azoiques et leur emplois industriels. Paris, 1889.

BEHREND, G.

Das Brauen mit ungemälztem Getreide. Halle, 1884.

BEHRENS, E. A.

Ueber Steinkohlentheer und über Steinkohlentheerpech. [Leipzig, 1872.] 8vo.

BEHSE, W. H.

Die Chemie in der Werkstatt. Leichtfassliche Darstellung der chemischen Erscheinungen, wie sie im Berufe des Bauhandwerkers, Metallarbeiters, Landwirths, u. s. w. täglich vorkommen. Eingerichtet sowohl zum Selbststudium, als auch zum Gebrauche beim Unterricht an Gewerb-, Real-, Baugewerk- und Handwerkerschulen. In two parts. (Anorgan. Chemie.—Organ. Chemie.) Weimar, 1872. 8vo. Ill.

BEILSTEIN, F.
Handbuch der organischen Chemie. *See in Section II.*

Бейльштейнъ, Р.
Руководство къ качествен. химич. анализу. С.-Петербургъ, 1867. (Other editions, 1873, etc.)

Inleiding tot de qualitatieve chemische analyse. Uit het Hoogduitsch vertaald onder toezicht van H. Yssel de Schepper. Deventer, 1868. 8vo.

A Manual of Qualitative Chemical Analysis, translated by William Ramsay. London, 1873. *Also* New York, *n. d.* pp. 70, 12mo. Printed in Glasgow.

Manuel d'analyse chimique qualitative. Traduction française publiée avec autorisation de l'auteur, sur la cinquième édition allemande par A. et P. Buisine. Lille, 1883. 8vo.

Anleitung zur qualitativen chemischen Analyse. Vierte verbesserte Auflage. Leipzig, 1877. 8vo.

BEILSTEIN'S (F.) Lessons in Qualitative Chemical Analysis. Arranged on the basis of the fifth German edition. With copious additions, including chapters on Chemical Manipulations, Analysis of Organic Substances, and Lessons in Volumetric Analysis, by Charles O. Curtman. Second edition, revised and greatly enlarged with additional chapters on Analysis of Drinking Water and of Urine. Saint Louis, Mo., 1886. pp. xii–200, 12mo. Ill.

Curtman's additions are so numerous as to make this edition of Beilstein's work practically a new one.

Бейльштейнъ, Р.
Руководство къ количеств. анализу. С.-Петербургъ, 1868.

BEILSTEIN, F. Manual of Quantitative Analysis. St. Petersburg, 1868.

Бекъ, В.

Таблицы качеств. химич. анализа. С.-Петербургъ. 1862.

> Beck, V. Tables of qualitative chemical analysis. St. Petersburg, 1862.

Бекетовъ, Никол.

О хлороформѣ. Казань, 1850. 4º

> Beketof, Nicolas. On Chloroform. Kasan, 1850.

Бекетовъ, Никол.

О нѣкоторыхъ новыхъ случаяхъ химическаго сочетанія. С.-Петербургъ, 1853. 8º

> Beketof, Nicolas. On Some New Chemical Compounds. St. Petersburg, 1853.

Bell, James.

The Chemistry of Foods. With microscopic illustrations. Two parts. London, n. d. [1881]. 12mo. Part i: Tea, Coffee, Cocoa, Sugar, etc. pp. 120. Part ii: Milk, Butter, Cheese, Cereal Foods, Prepared Starches, etc. pp. viii–179. South Kensington Museum Science Handbooks; Branch Museum, Bethnal Green.

> Die Analyse und Verfälschung der Nahrungsmittel. Uebersetzt von C Mirus. Mit einem Vorwort von E. Sell. Vol. i: Thee, Kaffee, Kakao, Zucker, etc. Vol. ii: Milch, Butter, Käse und Cerealien, präparirte Stärkemehl, etc. Uebersetzt von P. Rasenack. Berlin, 1882–85. 2 vols., 8vo.

Bell, Sir I. Lowthian.

Chemical Phenomena of Iron Smelting, an experimental and practical examination of the circumstances determining the capacity of the 'blast furnace, temperature of the air. London, 1872. 8vo. Plates.

Bellani, Angelo.

Ricerche fisico-chimiche sul fosforo particolarmente considerato come mezzo eudiometrico. Pavia, 1813. 4to.

Bellenghi di Forti, Filippo.

Processo sulle tinte che si estraggono dai legni ed altre piante indigene. Ancona, 1811. 8vo.

Bellini, Ranieri.

Manuale di tossicologia, Pisa, 1878. 16mo.

Belloc, A.

Les quatres branches de la photographie. Traité complet théorique et pratique des procédés de Daguerre, Talbot, Niepce de Saint-Victor

BELLOC, A. [Cont'd.]
  et Archer. Précédé des Annales de la photographie et suivi
  d'éléments de chimie et d'optique appliqués à cet art. Avec deux
  portraits. Paris, chez l'auteur, 1855. 8vo.

Compendium des quatre branches de la photographie. Traité complet,
  théorique et pratique des procédés de Daguerre, Talbot, Niepce
  de Saint-Victor et Archer. Applications diverses. Précédé des
  Annales de la photographie et suivi d'éléments de chimie et d'optique
  appliqués à cet art. Paris, 1858. 8vo.

BELLUCCI, GIUSEPPE.
  Sull' ozono, note e riflessioni. Prato, 1869. pp. 456, 12mo.

> Contains an historical sketch of ozone. Closely printed in small type and no
> index !

BELLUOT, ANTONIO.
  Tratado de química práctica y casera, ó colleccion de recetas. . . .
  Valladolid, 1843. 8vo.

BELZA, JÓZEF.
  Zasady technologii chemicznej gospodarskiej, obejmujący naukę o
  wypalaniu wódki, warzeniu piwa i otrzymywaniu octu i cukru z
  buraków i t. p. z. dodaniem pomnożonych wiadomósci Kar. Kurka
  tyczących się gorzelnictwa i piwowarstwa. Wydanie drugie przerob.
  Warszawa, 1851. 12mo.

> Elements of Chemistry with reference to Technology and Domestic Economy.
> Warsaw, 1851.

Krotki rys chemii z dodaniem treściowego zastósowania jéj do rolnictwa.
  Z drzeworytami w tekście. Warszawa, 1852. 12mo.

> A Brief Sketch of Chemistry Applied to Agriculture. Warsaw, 1852.

BEMMELEN, J. M.
  De scheikunde als leer der stofwisseling. Redevoering ter aanvaarding
  van het gewoon hoogleeraarsambt in de faculteit der wis- en natuurkunde aan de Hoogeschool te Leiden, uitgesproken den 25ten April,
  1874. Leiden, 1874. 8vo.

BENDER, ADOLF.
  Das Furfuran und seine Derivate. Berlin, 1889. pp. viii–83, 8vo.

BENDER, CARL.
  Die Bedeutung und Verwerthung der Atomenlehre in der Chemie.
  Vortrag. . . . Nördlingen, 1871. 8vo.

BENEDIKT, RUDOLF.
Analyse der Fette und Wachsarten. Berlin, 1886. 8vo.
   Zweite Auflage. Berlin, 1892. pp. ix–465, 8vo.
Die künstliche Farbstoffe (Theerfarben) ; ihre Darstellung, Eigen-
  schaften, Prüfung, Erkennung und Anwendung. Kassel, 1883. 8vo.
   Chemistry of the Coal-tar Colours. Translated and edited
   with additions by E. Knecht. Second edition. London,
   1889.

BENOIT, ÉMILE.
Du manganèse, étude de chimie analytique au double point de vue de
  la pharmacie et de l'industrie, contenant un procédé de préparation
  du peroxyde de manganèse, à l'état de pureté et une méthode
  nouvelle de dosage du manganèse à l'état métallique dans les
  minérais. Paris, 1885.
Traité des manipulations chimiques et de l'emploi du chalumeau, suivi
  d'un Dictionnaire descriptifs des produits de l'industrie susceptibles
  d'être analysés. Paris, 1854. 8vo.
   The Dictionnaire has distinct pagination.

BENRATH, H. E.
Die Glasfabrikation. Braunschweig, 1875. 8vo. 200 ill.
   See Bolley's Handbuch der chemischen Technologie, Neue Folge.

BENVENUTI, GIUSEPPE.
Riflessioni ed esperienze sulla natura, qualità e scelta dell' acqua. In
  Lucca, 1769. 4to.

BENZENBERG.
Ueber die Dalton'sche Theorie. Düsseldorf, 1830. 8vo.

BEOBACHTUNGEN ÜBER DIE DARSTELLUNG DES ZUCKERS und eines brauch-
  baren Syrups aus einheimischen Gewächsen. Berlin, 1799.

BERGLUND, E.
Handledning vid de första Arbeterna på Laboratorium. Lund, 1879. 8vo.
Lärobok i qualitativ oorganisk Analys. Lund, 1885. 8vo.

BERGMANN, A. L.
Das Kreosot in chemischer, pharmaceutischer und therapeutischer
  Beziehung. Nürnberg, 1835. 8vo.

BERGMANN, HEINRICH.
Chemisch-technisches Receptbuch für die gesammte Metall-Industrie.
  Eine Sammlung ausgewählter Vorschriften für die Bearbeitung

BERGMANN, HEINRICH. [Cont'd.]

aller Metalle, Decoration und Verschönerung daraus gefertigter Arbeiten, sowie deren Conservirung. Ein unentbehrliches Hilfs- und Handbuch für alle Metalle verarbeitenden Gewerbe. Wien, Pest, Leipzig, 188–.

BERGMANN, J. G.

Farmaceutisk-kemiska Analys. Göteborg, 1887.

BERGMAN, TORBERN (OLAF).

Anleitung zu Vorlesungen über die Beschaffenheit und den Nutzen der Chemie, und die allgemeinsten Verschiedenheiten natürlicher Kör- per. Aus dem schwedischen übersetzt. Stockholm und Leipzig, 1770. 8vo.

> * An Essay on the Usefulness of Chemistry and its applica- tion to the various occasions of life. Translated from the original. London, printed for John Murray, 1783. pp. [iv]–163, 8vo.

De primordiis chemiæ. Upsala, 1779. 4to.

> An Essay read before the Academy of Sciences at Upsala June 4, 1779. For a German translation, *see* Wiegleb, J. C.: Geschichte des Wachsthums . . . der Chemie (pp. 1–120). For an English translation, *see* Bergman's Physical and Chemical Essays, translated by Edmund Cullen, vol. III. Bergman, born 1735, died 1784, was Professor of Chemistry at University of Upsala.

Historiæ chemiæ medium seu obscurum ævum a medio sæculi VII ad medium sæculi XVII. Upsala, 1782, 4to.

> An Essay publicly read at Upsala June 11, 1782. For a German translation, *see* Wiegleb, J. C.: Geschichte . . . der Chemie, pp. 121–260. For an English translation, *see below* Bergman's Physical and Chemical Essays, translated by Edmund Cullen, vol. III.

Opuscula physica et chemica, pleraque antea seorsim edita, ab autore collecta revisa et aucta. Holmiæ, Upsalæ et Aboæ, 1779–90. 6 vols, 8vo. Plates.

> Editio nova emendatior. Lipsiæ, 1788–1790. 6 vols, 8vo.
>
> Vols. IV–VI were edited after the author's death by Ernest Benjamin Gottlieb Hebenstreit. Illustrated with plates. Vol. VI contains a full index.

> Opuscules chymiques et physiques. Recueillis, revus et augmentés par lui-même. Traduits par Guyton de Morveau. Dijon, 1780–'85. 2 vols., 8vo.

> * Physical and Chemical Essays, translated from the original Latin by Edmund Cullen. To which are added notes and illustrations by the translator. London, printed for John Murray, No. 32 Fleet-Street ; Balfour and

BERGMAN, TORBERN. [Cont'd.]

Co., W. Gordon and J. Dickson, at Edinburgh ; and L. White, at Dublin, 1784. 3 vols. [vol. III, 1791], 8vo. Plates.

Vol. III *contains :* Of the Origin of Chemistry, read at Upsala June 4, 1779. *Also :* The History of Chemistry during the Obscure or Middle Age, read at Upsala June 11, 1782.

Kleine physische und chymische Wercke. Aus dem lateinischen übersetzt von H. Tabor. Frankfurt am Main, 1782–85. 3 vols., 8vo.

Opusculi chimici et fisici tradotti in Italiano con aggiunte e note. Napoli, 1787–88. 2 vols., 8vo.

* Traité des affinités chymiques ou attractions électives, traduit du Latin. Augmenté d'un supplément et de notes. Paris, 1788. pp. [viii]–444, 8vo. 3 folding tables and 4 folding plates.

A Dissertation on Elective Attractions. Translated from the Latin. London, 1785. 8vo.

Tavole delle affinità e delle combinazioni chimiche trasportate in parole colla nuova nomenclatura chimica da Andrea Silvestri. Milano, 1801. 8vo.

* Grundriss des Mineralreichs, in einer Anordnung nach den nächsten Bestandtheilen der Körper. Aus dem lateinischen mit einigen Zusätzen von Joseph Xav. Lippert. Wien, in der Krauszischen Buchhandlung, 1787. pp. 207, 16mo.

BERGMANN, TORBERN.

See, in Section III, Wiegleb, J. Chr.: Geschichte des Wachsthums . . . der Chemie.

BERING, HUGO,

Kurze Anleitung zur Ausführung maassanalytischer Untersuchungen für Fabrikanten, Berg- Hüttenmänner, Chemiker, etc. Leipzig, 1861. 8vo. Ill.

Берингъ, Гуго.

Химическій анализъ посредствомъ титированія. С.-Петербургъ, 1867.

BERING, HUGO. Chemical analysis by means of standard solutions. St. Petersburg, 1867.

BERINGER, CORNELIUS AND JOHN J.

A Text-book of Assaying, for the use of those connected with mines. London, 1889. 8vo.

Second edition. London, 1890. 8vo.

BERKENHOUT, JOHN.

First Lines of the Theory and Practice of Philosophical Chemistry. London, 1788. 8vo. Ill.

BERLIN, NILS JOHANNES.
Anvisning till de allmännaste Gifters Upptäckande på kemisk väg.
    Stockholm, 1845. 8vo.
Elementar-Lærobok i oorganisk Kemi. Lund, 1857.
            3die Upplagan bearbetad af C. W. Blomstrand. Lund, 1870.
Grunderna för den qvalitativa kemiska Analysen. Lund, 1847. 12mo.
        Andra förbättrade och tillökta Upplagan. Lund, 1856.
Kort Lärobok i Kemi. Till Elementar-Läroverkens Tjenst utgifven.
    Lund, 1860.
        Andra upplagan. Lund, 1865.
Tentamina chemici observationes quasdam circa salia oxydi chromici
    sistens. Upsaliæ, 1833.
        Anmärkningar vid Chrom Oxiden och några dess Salter.
        Upsala, 1835.
Wäxt-Chemien i Sammandrag. Stockholm, 1835.

BERNARD, CLAUDE.
Leçons sur les effets des substances toxiques et médicamenteuses.
    Paris, 1857. pp. vii–488, 8vo.

BERNARD, J.
Repetitorium der Chemie. i. Anorganische Chemie. Dritte Auflage
    bearbeitet von J. Spenrath. Aachen, 1887. ii. Organische Chemie.
    Zweite Auflage. Aachen, 1888.

BERNAYS, ALBERT JAMES.
Household Chemistry, or Rudiments of the Science applied to Every-
    Day Life. London, 1852. pp. xvi–188, 12mo. Ill.
        Third edition. London, 1866.
        The Student's Chemistry. Being the 7th edition of "House-
            hold Chemistry," or the science of home life. London,
            1870. 8vo.
Notes for Students in Chemistry, being a Syllabus of Chemistry and
    Practical Chemistry. Fifth edition. London, 1870. 16mo.
        Sixth edition. London, 1878. pp. xiii–131, 16mo.
Notes on Analytical Chemistry for Students in Medicine. (Extracted
    from the fifth edition of "Notes for Students in Chemistry.")
    Second edition, London, 1886.
        Third edition. London, 1889. 8vo.

BERNSTEIN, A.
Die neuere Chemie. Berlin, 1880. 8vo.
Kurzes Lehrbuch der organischen Chemie. Braunschweig, 1887.

BERNTHSEN, A.

Kurzes Lehrbuch der organischen Chemie. Braunschweig, 1887. 8vo.
>   Zweite Auflage. 1890.
>   A Text-book of Organic Chemistry. Translated from the
>   German by E. McGowan. London, 1889.

BERR, FRANZ.

Anfangsgründe der Chemie, als Lehrbuch für Unter-Realschulen.
    Brünn, 1853. 8vo.
Zweite Auflage. Brünn, 1854.
Dritte verbesserte und vermehrte Auflage. Brünn, 1858.
>   Elementi di chimica, proposti ad uso delle scuole reali
>   inferiori. Edizione italiana per cura di Francesco
>   Businelli. Brünn, 1858. 8vo.

BERSCH, JOSEPH.

Essig-Fabrikation (Die). Eine Darstellung der Essigfabrikation nach
    den ältesten und neueren Verfahrungs-Weisen, der Schnell-Essig-
    fabrikation, der Bereitung von Eisessig und reiner Essigsäure
    aus Holzessig, sowie der Fabrikation des Wein-, Trestern Malz-,
    Bieressigs und der aromatischen Essigsorten nebst der praktischen
    Prüfung des Essigs. Dritte erweiterte und verbesserte Auflage.
    Wien, 1886. Ill.
Fabrikation (Die) der Anilinfarbstoffe und aller anderen aus dem
    Theere darstellbaren Farbstoffe (Phenyl-, Naphtalin-, Anthracen-
    und Resorcin Farbstoffe) und deren Anwendung in der Industrie.
    Wien, 1878. 8vo. Ill.
Fabrikation (Die) der Erdfarben. Enthaltend : Die Beschreibung aller
    natürlich vorkommenden Erdfarben, deren Gewinnung und Zube-
    reitung. Handbuch für Farben-Fabrikanten, Maler, Zimmermaler,
    Anstreicher und Farbwaaren-Händler. Wien, Pest, Leipzig, 188-.
    Ill.
Fabrikation (Die) der Mineral- und Lackfarben. Enthaltend : Die
    Anleitung zur Darstellung aller künstlichen Maler- und An-
    streicher-Farben der Email- und Metall-Farben. Ein Handbuch
    für Fabrikanten, Farbwaarenhändler, Maler und Anstreicher.
    Dem neuesten Stande der Wissenschaft entsprechend dargestellt.
    Wien, Pest, Leipzig, 18—. Ill.
Gährungs-Chemie für Praktiker. Berlin, 1879–86. 5 vols., 8vo.
    I. Hefe und Gährungserscheinungen. II. Malz- und Dextrinfabri-
    kation. III. Bierbrauerei. IV. Spiritusfabrikation und Presshefe-
    bereitung. V. Schnell-Essigfabrikation und Fabrikation von Wein-
    essig.

BERSCH, JOSEPH. [Cont'd.]

Verwerthung (Die) des Holzes auf chemischem Wege. Eine Darstellung der Verfahren zur Gewinnung der Destillationsproducte des Holzes, der Essigsäure, des Holzgeistes, des Theeres und der Theeröle, des Creosotes, des Russes des Röstholzes und der Kohlen. Die Fabrikation von Oxalsäure, Alkohol und Cellulose, der Gerb- und Farbstoff-Extracte aus Rinden und Hölzern, der ätherischen Oele und Harze geschildert für Praktiker. Wien, 1883. 8vo. Ill.

Wein (Der) und sein Wesen. Wien, 1878–79. 2 vols. I. Enstehung des Weines. II. Kellerwirthschaft.

BERTA, ANTONIO.

Saggio di un dizionario dei termini chimici. Padova, 1842. 8vo.

BERTHELOT, MARCELLIN.

Chimie organique fondée sur la synthèse. Paris, 1860. 2 vols., 8vo.

Essai de mécanique chimique fondée sur la thermochimie. Paris, 1879. 2 vols., 8vo. Ill. Vol. 1, pp. xxxi–566. Vol. 2, pp. xi–774.
　　　Vol. 1 has a portrait of the author.

Leçons sur les méthodes générales de synthèse en chimie organique, professées en 1864 au Collège de France. Paris, 1864. pp. xxi–524, 8vo.

　　　Lecciones sobre los métodos generales de síntesis en química orgánica, traducidos por Vicente Martin de Argenta. Madrid, 1874. 4to.

Leçons sur les principes sucrés, professées devant la Société chimique de Paris en 1862. Paris, 1862. 8vo.

Force (La) de la poudre et des matières explosives. Deuxième édition. Paris, 1872. 18mo.

　　　Sur la force des matières explosives d'après la thermochimie. Troisième édition (avec figures), revue et considérablement augmentée. Paris, 1883. 2 vols., 8vo. Vol. 1, pp. xxv–405 ; vol. 2, pp. 445.

　　　Explosive Materials. A series of lectures delivered before the College de France at Paris. Translated by Marcus Benjamin. To which is added a short historical sketch of Gunpowder. Translated from the German of Karl Braun by John P. Wisser. And a bibliography of works on explosives, reprinted from Van Nostrand's Magazine. New York, 1883. pp. 180, 24mo.

Synthèse (La) chimique. Paris, 1876. pp. viii–294, 8vo. Bibliothèque scientifique internationale.

BERTHELOT, MARCELLIN. [Cont'd.]
> Die chemische Synthese Autorisirte Ausgabe. Leipzig,
> 1877. 8vo. Internationale wissenschaftliche Bibliothek,
> vol. xxv.
>
> La sintesi chimica. Milano, 1876. 16mo.

BERTHELOT, MARCELLIN, ET PÉAN DE SAINT GILLES.
> Recherches sur les affinités. De la formation et de la composition des
> éthers. Avec table. Paris, 1862–63. 4 parts. 8vo.

BERTHELOT, MARCELLIN, ET E. JUNGFLEISCH.
> Traité élémentaire de chimie organique. Seconde édition avec de
> nombreuses figures. Revue et considérablement augmentée.
> Paris, 1891. pp. xx–489, 8vo.
>
>> First edition, Paris, 1872. 8vo.

BERTHELT, A.
> Chemie. Für Schulen und zum Selbstunterrichte. Siebente verbes-
> serte Auflage bearbeitet von R. Kell. Leipzig, 1883. 8vo.

BERTHIER, PIERRE.
> Chimie minérale et Analyse des substances minérales; travaux de 1829,
> 1830, et 1831. (Extrait des Annales des mines.) Paris, 1833.
> 8vo.
>
> Traité des essais par la voie sèche, ou des propriétés, de la composition,
> et de l'essai des substances métalliques et des combustibles.
> Paris, 1834. 2 vols., 8vo.
>
>> Another edition, Paris, 1848. 2 vols., 8vo.
>>
>> Handbuch der metallurgisch-analytischen Chemie. Ueber-
>> setzt und mit eigenen Erfahrungen und Zusätzen ver-
>> mehrt von Carl Moritz Kersten. Leipzig, 1835–36.
>> 2 vols., 8vo. Ill.

BERTHOLD, A. A.
> See Bunsen (R. W.) and Berthold: Eisenoxyhydrat, etc. 1837.

BERTHOLLET, CLAUDE LOUIS.
> Élémens de l'art de la teinture. Paris, 1791. 2 vols., 8vo.
>
>> Deuxième édition. Avec une description du blanchiment
>> par l'acide muriatique oxigéné. Par C. L. et Amédée
>> B. Berthollet. Paris, 1804. 2 vols., 8vo. Ill.
>>
>> Elements of the Art of Dyeing; translated from the French
>> by W. Hamilton. London, 1791. 2 vols. 1 plate.
>>
>> Elements of the Art of Dyeing, with a description of the
>> art of bleaching by oxymuriatic acid. Translated

BERTHOLLET, CLAUDE LOUIS. [Cont'd.]
> from the French, with notes and engravings, by
> Andrew Ure. London, 1824. 2 vols., 8vo.
>> Another edition, London, 1841. 8vo.
> Essai de statique chimique. Paris, An XI [1803]. 2 vols., 8vo.
>> Essay on Chemical Statics, with copious Explanatory Notes,
>> and an Appendix on Animal and Vegetable Substances,
>> translated by B. Lambert. London, 1804. 2 vols., 8vo.
>> Versuch einer chemischen Statik, das ist einer Theorie der
>> chemischen Naturkräfte. Aus dem französischen von
>> G. W. Bartholdy und mit Erläuterungen begleitet von
>> Ernst Gottfried Fischer. Berlin, 1810–11. 2 vols., 8vo.
> Recherches sur les lois de l'affinité, par le citoyen Berthollet. Paris,
> An IX [1801]. pp. 105, 8vo.
>> Researches into the Laws of Chemical Affinity. Translated
>> from the French by M. Farrell. Baltimore, 1809. pp.
>> iv–212, 12mo.
>> Ueber die Gesetze der Verwandschaft in der Chemie. Aus
>> dem französischen übersetzt mit Anmerkungen, Zusät-
>> zen und einer synthetischen Darstellung von Berthollets
>> Theorie versehen. Von Ernst Gottfried Fischer. Ber-
>> lin, 1802. pp. xii–332, 8vo.

Cf. Schnaubert, Ludwig ; *also*, Karsten, C. J. B.

BERTHRAND, C. A. H. A.
Manuel médico-légal des poisons introduits dans l'estomac, et des
moyens thérapeutiques qui leur conviennent. Suivi d'un plan
d'organisation médico-judiciaire, d'un tableau de classification
générale des empoisonnements, et du rapport fait à la Société de
médecine de Paris. Paris, 1817. pp. xxxii–384, 8vo.

BERZELIUS, JÖNS JACOB.
> *See* Mueller, Ludwig, Berzelius' Ansichten, *etc.*

BERZELIUS, JÖNS JACOB.
Afhandling om Galvanismen. Stockholm, 1802. 8vo.
Atomgewichts-Tabellen. (Gratiszugabe zu des Verfassers Lehrbuch
der Chemie ; fünfte Auflage.) Leipzig, 1847. 8vo.
Chimie du fer d'après Berzelius. Traduit par le chevalier Hervé.
Paris, 1826. pp. xvi–200, 8vo.
De l'analyse des corps inorganiques. Traduit de l'allemand [by E——r,
i. e., E. Esslinger]. Paris, 1827. pp. iii–232, 8vo. One plate.
> The Analysis of Inorganic Bodies. Translated from the
> French edition by G. O. Rees. London. 1833. 12mo.

BERZELIUS, JÖNS JACOB.   [Cont'd.]

Analisi chimica d'ogni specie di minerali.   Firenze, 1822.

Elemente der Chemie der unorganischen Natur.   Aus dem schwedischen von Blumhof.   Leipzig, 1816.   8vo.

Essai sur la théorie des proportions chimiques et sur l'influence chimique de l'électricité.   Traduit du suédois sous les yeux de l'auteur, et publié par lui-même.   Paris, 1819.   pp. xvi–120–[ii], 8vo.

> Théorie des proportions chimiques, et table synoptique des poids atomiques des corps simples et de leurs combinaisons les plus importantes.   Deuxième édition, revue corrigée et augmentée, Paris, 1835.   pp. 477, 8vo.
>
> *Contains :* Exposé historique du développement de la théorie des proportions chimiques.   pp. 1–11.

> Versuch über die Theorie der chemischen Proportionen und über die chemischen Wirkungen der Electricität; nebst Tabellen über die Atomengewichte der meisten unorganischen Stoffe und deren Zusammensetzungen. Nach den schwedischen und französischen Originalausgaben bearbeitet von K. A. Bloede.   Dresden, 1820. 8vo.   With an alphabetic table in 4to.

Föreläsningar i Djurkemien.   Stockholm, 1806–1808.   2 parts, 8vo.

Försök, att, genom Användandet af den electrokemiska Theorien och de kemiska Proportionerna, grundlägga ett rent vetenskapligt System för Mineralogien.   Stockholm, 1814.   8vo.

> An attempt to establish a pure scientific System of Mineralogy by the application of the Electro-chemical Theory and the Chemical Proportions.   Translated from the Swedish by J. Black.   [Revised by T. Thomson.] London, 1814.   8vo.

> Nouveau système de minéralogie.   Paris, 1819.

> Versuch durch Anwendung der electrisch-chemischen Theorie . . . ein . . . System der Mineralogie zu begründen . . . Aus dem schwedischen übersetzt von A. F. Gehlen.   Nürnberg, 1815.   8vo.

> Neues chemisches Mineralsystem nebst einer Zusammenstellung seiner ältern hierauf bezüglichen Arbeiten.   Herausgegeben von C. F. Rammelsberg.   Als zweite Auflage von Berzelius' neuern System der Mineralogie aus dem schwedischen übersetzt von Christian Gmelin und W. Pfaff.   Nürnberg, 1847.   8vo.

Lärbok i Kemien.   Upsala, 1808–18.   3 vols.

> Another edition.   Stockholm, 1817–30.   6 vols. and 1 vol. Tables.

BERZELIUS, JÖNS JACOB. [Cont'd.]

Lehrbuch der Chemie. Uebersetzt von Blöde [und K. Palmstedt]. Dresden, 1824. 2 vols.

Lehrbuch der Chemie, nach des Verfassers schwedischer Bearbeitung der Blöde-Palmstedt'schen Auflage übersetzt von F. Wöhler. Dresden, 1825–31. 8 vols. in 4, 8vo.

Third edition, Dresden and Leipzig, 1833–41. 10 vols.

Fourth edition, Dresden, 1835.

Fifth edition, Dresden, 1843–48. 5 vols.

A fourth edition of vols. 1–7, 1835–1838. This is completed in vols. VIII, IX, X [1839–'41] of the third edition.

*Lehrbuch der Chemie. Fünfte umgearbeitete Original-Auflage (Zweite wohlfeilere Ausgabe). Leipzig, 1856. 4 vols. in 5.

Preface dated Stockholm, 1842.

Leerboek der scheikunde naar de derde omgewerkte en vermeerderde oorspronkelijke uitgave vertaald, onder medewerking van G. J. Mulder, door A. S. Tischauser, B. Eikma en A. F. van der Vliet. Leyden, 1840–44. 3 parts, 8vo.

Another edition, Rotterdam 1834–41. 6 parts.

Traité de chimie traduit par A. J. L. Jourdan sur les manuscrits inédits de l'auteur et sur la dernière édition allemande. Paris, 1829–33. 8 vols.

After vol. 1, traduit par Esslinger.

Traité de chimie minérale, végétale et animale. Deuxième édition française traduite par Esslinger et Hoefer sur les manuscrits inédits de l'auteur et en partie sur la cinquième et dernière édition allemande. Paris, 1845–50. 8 vols., 8vo. Ill.

Traité de chimie. Nouvelle édition entièrement refondue d'après la quatrième édition allemande publièe en 1838 par B. Valerius. Bruxelles, 1839–46. 5 vols., roy. 8vo.

Trattato elementare di chimica teorica e pratica, con aggiunta di C. Frisiani. Milano, 1820. 4 vols.

Trattato di chimica. Prima edizione napolitana, conforme alla quarta edizione tedesca, di Giovanni Guarini. Napoli, 1840–43. 9 vols., 8vo.

Tratado de química. Nueva edicion, completamente refundida, segun la cuarta edicion alemana publicada en 1838 por B. Valerius, traducida del francés al castellano par Rafael Saez y Palacios y Carlos Ferrari Scardini. Madrid, 1845–52. 5 vols., 8vo.

BERZELIUS, JÖNS JACOB   [Cont'd.]

Lehrbuch der Chemie in vollständigem Auszüge, mit Zusätzen und Nachträgen aller neueren Entdeckungen und Erfindungen zu Vorlesungen und zum Selbsstudium für Aerzte, Apotheker, Fabrikanten, Kameralisten, Landwirthe, Gewerbtreibende, etc., bearbeitet von H. F. Eisenbach und E. A. Hering. Stuttgart, 1832. 3 vols., 8vo.

A condensed edition of the foregoing Lehrbuch.

Lehrbuch der Chemie in gedrängter Form. Bearbeitet und mit den neuesten Entdeckungen bereichert von Fr. Schwarze und Anderen. Quedlinburg, 1838–40. 4 vols., 8vo.

Lärobok i Djur-Chemien. Sammandragen öfversättning af Lehrbuch der Thier-Chemie. Westerås, 1849. 8vo.

Lehrbuch der Thierchemie, aus dem schwedischen übersetzt von F. Wöhler. Dresden, 1831. 8vo.

Lehrbuch der Pflanzen- und Thierchemie. Dritte Auflage. Dresden, 1837–41.

Några Underrättelser om artificiella Mineral-Vatten. Stockholm, 1803. 8vo.

Nova analysis aquarum Medeviensium. Upsal, 1800.

Öfversigt af Djur-Kemiens Framsteg och närvarande Tillstånd. Tal hållet för Kgl. Vetenskaps-Academien, 1810. Stockholm, 1812. 8vo.

A View of the Progress and Present State of Animal Chemistry. Translated from the Swedish by Gustavus Brunnmark. London, 1813. pp. vii–115, 8vo.

Uebersicht der Fortschritte und des gegenwärtigen Zustandes der thierischen Chemie. Aus dem schwedischen ins englische übersetzt von Gustav Brunnmark. Aus dem englischen ins deutsche übersetzt von G. C. L. Sigwart. Tübingen, 1814. Nürnberg, 1815. 8vo.

Om Blåsrörets Användande i Kemien och Mineralogien. Stockholm, 1820. pp. [vi]–302, 8vo. 4 plates.

The Use of the Blowpipe in Chemical Analysis, and in the examination of Minerals. . . . Translated from the French of M. Fresnel by J. G. Children. With a sketch of Berzelius' System of Mineralogy . . . and numerous notes and additions by the translator. London, 1822. 8vo.

BERZELIUS, JÖNS JACOB. [Cont'd.]

The use of the Blowpipe in Chemistry and Mineralogy. Translated from the 4th enlarged and corrected edition by J. D. Whitney. London, 1847. 8vo.

Also London, 1849. 8vo.

De l'emploi du chalumeau dans les analyses chimiques et les déterminations minéralogiques. Traduit de l'allemand par M. Fresnel. Paris, 1821. 8vo. Quatrième édition, Paris, 1842. 8vo. 4 plates.

Die Anwendung des Löthrohrs in der Chemie und Mineralogie. Aus der Handschrift übersetzt von Heinrich Rose. Nürnberg, 1821. pp. xvi–311, 8vo. 4 plates.

Zweite verbesserte Auflage, Nürnberg, 1828.

Dritte verbesserte Auflage, Nürnberg, 1837.

Vierte verbesserte Auflage, Nürnberg, 1844. 8vo. 4 plates.

Объ употребленіи паяльной трубки при химическихъ и минерало гическихъ изслѣдованіяхъ, соч. Я Берцеліуса; пер. съ А. Таскинъ С.-Петербургъ, 1831. 8o

Saidschitzer (Das) Bitterwasser, chemisch untersucht von J. Berzelius, mit Bemerkungen über seine Heilkräfte von A. E. Reuss. Zweite Auflage. Prag, 1843. 12mo.

Tabell, som utvisar Vigten af större Delen vid den oorganiska Kemiens studium märkvärdiga enkla och sammansatta Kroppars Atomer, jemte deras Sammansättning, räknad i procent. Bihang till Tredjedelen af Lärboken i Kemien. Stockholm, 1818. 4to.

Ueberblick über die Zusammensetzung der thierischen Flüssigkeiten. Aus dem englischen übersetzt von J. S. C. Schweigger. Nürnberg, 1814. 8vo.

BESANA, C.

Manuale di chimica applicata al caseificio. Milano, 1876.

BESCHREIBUNG EINIGER ZUM GEBRAUCH DER DEPHLOGISTISIRTEN LUFT bey dem Blaserohr und Schmelzfeuer eingerichteten Maschinen, samt einer Anweisung sich die dephlogistisirte Luft in Menge zu verschaffen. Tübingen, 1785. pp. 45, 16mo. 5 folding plates.

> The apparatus is a modification of Achard's. The plates are interesting to students of the history of Aëronautics.

BESEKE, JOHANN MELCHIOR GOTTLIEB.

Entwurf eines System der transzendentellen Chemie. Leipzig, 1787. 8vo.

* Ueber Elementarfeuer und Phlogiston als Uranfänge der Körperwelt, insbesondere über elektrische Materie. In einem Schreiben an Achard in Berlin. Leipzig, 1786. pp. 52, 8vo.

BEST, R.

Tables of Chemical Equivalents, incompatible substances, and poisons and antidotes. Lexington (Kentucky), 1825. 8vo.

BETRACHTUNGEN DER CHEMISCHEN ELEMENTE, ihrer Qualitäten, Aequivalente und Verbreitung ; weihet den Manen seiner Lehrer . . . aus Anlass der dritten Secularfeier der Universität zu Jena, ein Veteran derselben. Prag, 1858. 8vo.

BEUGHEM, CORNELIUS À.

Syllabus recens exploratorum in re medica, physica et chymica, prout in miscellaneis medico-physicis naturæ curiosorum Germaniæ, Galliæ, Daniæ et Belgii sparsim extant, in ordinem redactus et juxta indicem harmonice adornatus. Amstelodami, 1696. pp. [xliv]–316, 12mo.

BEUGNOT.

Traité de chimie médicale. Paris, 1855. 8vo.

BEUTE, F.

Anleitung zur ersten Ausführung chemischer Arbeiten in landwirthschaftlichen Lehranstalten. Uelzen, 1885. 8vo.

BEUTTLER, JAMES OAKLEY.

Inorganic Chemistry. The Chemistry of the Non-Metals. London, *n. d.* [1890]. 8vo.

BEWLEY, R.

A Treatise on Air, containing new Experiments and Thoughts on Combustion. London, 1791. 8vo.

BEYCKERT, JOHANN PHILIPP.

De pyrophoris. Argentorati, 1731. 4to.

BIBLIOTHEK DES WISSENSWÜRDIGSTEN AUS DER TECHNISCHEN CHEMIE UND GEWERBSKUNDE. Technisch-chemische Notizen, Recepte, Erfahrungen und Abhandlungen zum Gebrauche für Chemiker und Techniker, Apotheker, etc. Nach den neuesten Quellen zusammengestellt, sowie auf Grunde eigener Erfahrungen bearbeitet von Emil Winckler. Leipzig, 1872. 8vo.

BIBRA, ERNST VON.

Chemische Fragmente über Leber und Galle. Braunschweig, 1849.
Chemische Untersuchung verschiedener Eiterarten. Berlin, 1842.
Chemische Untersuchungen über die Knochen und Zähne des Menschen und der Wirbelthiere, mit Rücksichtnahme auf ihre physiologischen und pathologischen Verhältnisse. Schweinfurth, 1844. 8vo. Ill.

BIBRA, ERNST VON. [Cont'd.]
  Getreidearten (Die) und das Brod. Nürnberg, 1860. pp. viii–502, 8vo.
> Contains, pp. 1–98, an extended historical sketch of cereals and their culti-
> vation from the earliest times to date.

  Hülfstabellen zur Erkennung zoöchemischer Substanzen. Erlangen,
    1846. Fol.
> Beknopte handleiding tot de herkenning van zoöchemische
> stoffen. Schoonhoven, 1847. 8vo.

  Narkotischen (Die) Genussmittel der Menschen. Nürnberg, 1855.
  Vergleichende Untersuchung über das Gehirn der Menschen und der
    Wirbelthiere. Mannheim, 1854.
  Wirkung (Die) des Schwefeläthers. Erlangen, 1847.

BIDDER, UND C. SCHMIDT.
  Die Verdauungssäfte und der Stoffwechsel. Eine physiologisch-
    chemische Untersuchung. Mitau und Leipzig, 1852. pp. x–413,
    8vo.

BIECHELE, MAX.
  Chemischen (Die) Gleichungen der wichtigsten anorganischen und or-
    ganischen Stoffe. Mit besonderer Berücksichtigung der deutschen
    und österreichischen Pharmacopoe, sowie der massanalytischen
    Untersuchungen der Arzneistoffe. Nach den neuesten chemischen
    Anschauungen bearbeitet. Eichstädt, 1885. pp. iv–952–vi, 8vo.
  Stöchiometrie, mit besonderer Berücksichtigung der deutschen Pharma-
    copoe, sowie der massanalytischen Untersuchungen der Arznei-
    stoffe. Eichstädt, 1887.

BIEDERT, PH.
  Untersuchungen über die chemischen Unterschiede der Menschen- und
    Kuhmilch. Stuttgart, 1884. pp. viii–69, 8vo.

BILLET, H.
  Chimie : résumé des première leçons. Troisième édition. Boulogne-
    sur-Mer, 1886. 8vo.

BILLICH, ANTON GÜNTHER.
  De tribus chymicorum principiis et quinta essentia exercitatio. Bremæ,
    1621. 8vo.

BILTZ, E.
  Der Schutz des Chloroforms vor Zersetzung am Licht und sein erstes
    Vierteljahrhundert. Zeitgemässe historische und chemische Studien.
    Erfurt, 1892. 8vo.

BINZ UND H. SCHULZ.

Die Arsengiftwirkungen von chemischen Standpunkt. Leipzig und Berlin, 1879–81.

BIRD, F. J.

The American Practical Dyer's Companion ; accompanied by 170 dyed samples of raw materials and fabrics. Philadelphia, 1882. 8vo.

BIRINGUCCI, VANUCCIO. [BIRINGOCCIO].

Pirotecnica, nella quale si tratta non solo della diversità delle miniere, ma anco di quanto si ricerca alla prattica di esse, e che s'appartiene all' arte della fusione o getto de' metalli. Venezia, 1540.

> *La pyrotechnie, ou art du feu, contenant dix livres ausquels est amplement traicté de toutes sortes et diversité de minières, fusions et séparations des métaux ; des formes et moules pour getter artilleries, cloches et toutes autres figures ; des distillations, des mines, contremines, pots, boulets, fusées, lances et autres feux artificiels, concernant l'art militaire et autres choses dépendantes du feu. Composée par Vanoccio [sic] Biringuccio Siennois, et traduite d'italien en françois par feu maistre Jaques Vincent. A Paris, chez Claude Fremy à l'enseigne S. Martin, rue St. Jaques, 1556. pp. viii, ff. 228, 4to. Ill.

BIRNBAUM, CARL.

Kurzes Lehrbuch der landwirthschaftlichen Gewerbe. Chemische Technologie landwirthschaftlicher Producte. Zugleich als achte Auflage von Friedrich Julius Otto's Lehrbuch der rationellen Praxis der landwirthschaftlichen Gewerbe. Braunschweig, 1886. 3 vols., 8vo. Ill.

> Vol. I. Die Fabrikation der Stärke, des Dextrins, des Stärkezuckers, der Zuckercouleur, das Brotbacken und die Rübenzucker-Industrie. Braunschweig, 1886. 8vo.

Leitfaden der chemischen Analyse für Anfänger. Fünfte Auflage. Leipzig, 1886. 8vo.

> Sechste Auflage. Leipzig, 1891. 8vo.
> A Laboratory Guide for Beginners in Chemical Analysis. From the German. Brunswick [Maine], 1870. 12mo.
> Apparently a separate issue of the introductory pages on Notation and Nomenclature translated from the German, with the title belonging to the entire work prefixed.

Löthrohrbuch. Anleitung zur Benutzung des sogenannten trocknen Weges bei chemischen Analysen. Braunschweig, 1872. 8vo. Ill.

Prüfung der Nahrungsmittel und Gebrauchsgegenstände im Grossherzogthum Baden. Karlsruhe, 1883.

BIRNBAUM, CARL.

> Das Brodbacken. See Otto-Birnbaum's Lehrbuch, vol. VIII.
> Die Torf-Industrie. See Otto-Birnbaum's Lehrbuch, vol. XI.

BIRNBAUM, CARL. [Cont'd.]
Kurzes erläuterndes Wörterbuch. *See* Otto-Birnbaum's Lehrbuch, vol. XIV.
Handbuch der chemischen Technologie. *See* Bolley's Handbuch, etc.

BISCHOF, CARL GUSTAV CHRISTOPH.
Lehrbuch der Stöchiometrie. *See* Hildebrandt, G. F.: Lehrbuch der Chemie.

BISCHOF, CARL GUSTAV CHRISTOPH.
Lehrbuch der chemischen und physikalischen Geologie. Bonn, 1846–55. 2 vols., 8vo.
Zweite Auflage. Bonn, 1863–71. 3 vols. and Supplement.
Elements of Physical and Chemical Geology. Translated from the German. Printed for the Cavendish Society. London, 1855–56. 2 vols., 8vo.
Lehrbuch der reinen Chemie. Erster Band. Bonn, 1823.

BISCHOF, H., ET L. R. DE FELLENBERG.
Expertise chimico-légale à l'occasion d'un empoisonnement. Lausanne, 1847. 8vo.

BISCHOFF, CARL.
Die practischen Arbeiten im chemischen Laboratorium. Mit 90 eingedruckten Abbildungen. Berlin, 1862. 8vo.

BIZIO, BARTOLOMEO.
Dinamica chimica. Venezia, 1850–53. 2 vols., 8vo.
La soluzione senza socc. di affinità chimica. Venezia, 1860. 4to.

BIZZOZERO, G.
Manuel de chimie clinique. Paris, 1885. 8vo.

BLACK, G. W. The Formation of Poisons by Micro-Organisms. A biological study of the germ theory of disease. Philadelphia, 1884. pp. vi–178, 8vo.

BLACK, JAMES G.
Chemistry for the Gold Fields, including lectures on the non-metallic elements, metallurgy and the testing and assaying of metals, metallic ores and other minerals by the test-tube, the blow-pipe and the crucible. Dunedin [New Zealand], 1886.

BLACK, JOSEPH.
Lectures on the Elements of Chemistry delivered in the University of Edinburgh. Published from his manuscripts by John Robison. Edinburgh, 1803. 2 vols., 8vo. I, pp. lxvi–[x]–556 ; II, pp. 762.
* First American from the last London edition. Philadelphia, 1807 3 vols., 8vo. Folding plates. Portrait of Black.

BLACK, JOSEPH. [Cont'd.]

Vorlesungen über die Grundlehren der Chemie aus seiner Handschrift herausgegeben von Johann Robison. Aus dem englischen übersetzt und mit Anmerkungen versehen von Lorenz von Crell. Hamburg, 1804. 4 vols., 8vo. Portrait of Black.

BLACK, JOSEPH, AND WILLIAM CULLEN.

Experiments upon Magnesia alba, Quicklime and other alcaline substances by Joseph Black. To which is annexed an Essay on the cold produced by evaporating fluids, and of some other means of producing cold, by W. C Edinburgh, 1777. pp. 133, 16mo.
Another edition, 1782. pp. 135, 16mo.

BLAIR, ANDREW ALEXANDER.

The Chemical Analysis of Iron. A complete account of all the best-known methods for the analysis of iron, steel, pig-iron, iron-ore, limestone, slag, clay, sand, coal, coke and furnace and produce gases. Philadelphia, 1888. 8vo. Ill.

Second edition. Philadelphia, 1891. 8vo. Ill.

Die chemische Untersuchung des Eisens. Eine Zusammenstellung der bekanntesten Untersuchungsmethoden von Eisen, Stahl, Roheisen, Eisenerz, Kalkstein, Schlacke, Thon, Kohle, Kokes, Verbrennungs- und Generatorgasen. Vervollständigte deutsche Ausgabe von L. Rürup. Berlin, 1892. 8vo. Ill.

BLAIR, D.

See Phillips, Sir Richard.

BLAIR, J. A.

The Organic Analysis of Potable Waters. London, 1890. pp. viii–118. 12mo.

BLANCARD, STEPHAN.

* Neue (Die) heutiges Tages gebräuchliche Scheide-Kunst, oder Chimia. Nach den Gründen des fürtreflichen Cartesii und des Alcali und Acidi eingerichtet. Hannover und Wolffenbüttel. Verlegts Gottlieb Heinrich Grentz, Buchhändl. 1690. pp. 179–[xi], 16mo. Ill.

Hedendaagsche (De) chymie. Amsterdam, 1723.

Pharmacie en chymie. Amsterdam, 1686.

* Theatrum chimicum, oder eröffneter Schau-Platz und Thür zu den Heimligkeiten in der Scheide-Kunst von denen berühmtesten Männern, die iemals in der Scheide-Kunst sich selbst bemühet und davon geschrieben, als Schröder, Angelus Sala, Rolfinck, Le Febure, Crollius, Charras, Beguin und andern itzo noch lebenden

BLANCARD, STEPHAN. [Cont'd.]

auffgethan, nun aber von einem Liebhaber der Kunst also ins Gesichte gestellet. Nebenst einer Vermehrung wie die geringen Metallen und gemeinen Steine zu verbessern sind, durch Kenelmus Digby. Mit unterschiedenen Kupffern versehen und aus dem niederländischen ins hochteutsche übersetzt. Leipzig, verlegts Johann Friedrich Gleditsch, 1694. 2 parts. pp. [vi]–658–183, 12mo. Plates.

BLAREZ, C.

Sur quelques appareils à l'usage des laboratoires de chimie. Bordeaux, 1886.

BLAS, CH.

Analyse pyrognostique par la méthode de Bunsen, suivie de la détermination méthodique des minéraux d'après la division dichotomique de Laurent et suivant un plan nouveau. Bruxelles, 1885. 12mo.

Application de l'électrolyse à l'analyse chimique. Avec un essai d'une méthode générale d'analyse électrolytique. Louvain, 1881. 8vo.

Contribution à l'étude et à l'analyse des eaux alimentaires et spécialement des eaux de la ville de Louvain et de quelques autres localités de la Belgique. Bruxelles et Louvain, 1884. pp. 174, 8vo.

Méthode de l'analyse qualitative minérale par la voie humide, avec un appendice relatif à la recherche des acides organiques, ainsi que des alcaloïdes et des principes immédiats les plus importants des plantes. Deuxième édition augmentée d'un essai de méthode d'analyse électrolytique. Louvain, 1882. 12mo.

Traité élémentaire de chimie analytique ; Tome 1er. Analyse qualitative par la voie sèche, ou analyse au chalumeau. Louvain, 1885. 8vo.

Deuxième édition. Louvain, 1886.

Traité de chimie analytique. Troisième édition revue et augmentée. Tome II : Analyse qualitative par la voie humide, y compris la recherche des principaux acides organiques et alcaloïdes, ainsi que l'analyse électro-lytique. Louvain, 1890.

BLAS Y MANADA.

Tratado elementar de análisis químico, mineral y organico cualitativo y cuantitativo. Madrid, 1889. 4to.

BLASIIS, F. DE.

Istruzione teorico-pratica sul modo di fare il vino e conservarlo, e della coltivazione degli ulivi e della vigna bassa. Opera divisa in sei parti. Settima edizione. Firenze, 1880. 16mo.

BLENNARD, A.
Lectures sur la chimie et la physique, mises à la portée de tout le
    monde, avec une introduction par Gaston Tissandier. Paris, 1884.
    8vo.

BLEY, LUDWIG FRANZ.
Taschenbuch für Aerzte, Chemiker und Badereisende die Bestandtheile
    der vorzüglichen Mineralquellen Deutschlands, etc., etc., ent-
    haltend. Leipzig, 1831.
Zuckerbereitung (Die) aus Runkelrüben.   Zweite vermehrte Ausgabe.
    Nebst einem Anhang   .   .   .   von Franz Wilh. Schweigger-Seidel.
    Halle, 1836. 8vo. Ill.

BLEY, LUDWIG FRANZ.
            Versuch einer wissenschaftliche Würdigung der Chemie. *See, in Section III.*

BLICHFELD, PETRUS CHR.
Disputatio de principiis chymicis et qualitatibus falso ab iis derivatis.
    Hafniæ, 1671. 4to.

BLOCHMANN, R.
Erste Anleitung zur qualitativen chemischen Analyse. Königsberg, 1890.

BLOMSTRAND, C. W.
Kort Lärobok i oorganisk Kemi. Lund, 1873.
        3die Uppl. Lund, 1886. 8vo.
Lärobok i organisk Kemi. Lund, 1877. 8vo.
        See also, in Section III, Blomstrand, C. W.

BLONDOT, N.
Nouveaux perfectionnements à la méthode de Marsh, pour la recherche
    chimico-légale de l'arsénic. Paris, 1845. 8vo.
Sur la recherche toxicologique du phosphore par la coloration de la
    flamme. Nancy, 1861. 8vo.
Sur la recherche de l'arsénic par la méthode de Marsh. Mémoire lu à
    l'Académie de médecine le 5 mai, 1857. Nancy, 1857. 8vo. From
    Mém. de l'Acad. de Stanislas.
Traité analytique de la digestion considérée particulièrement dans
    l'homme et dans les animaux vertébrés. Paris, 1843. pp. 471, 8vo.

BLOXAM, CHARLES LOUDON.
Chemistry, Inorganic and Organic, with experiments, and a comparison
    of equivalent and molecular formulæ. London, 1867. 8vo.
        Second edition. London, 1872.
        Fourth edition. London, 1880.

BLOXAM, CHARLES LOUDON.   [Cont'd.]

> Seventh edition, by John Millar Thompson and Arthur G. Bloxam.   London, 1890.   pp. xii–790, 8vo.   Ill.   Also American editions.

Laboratory Teaching, or progressive exercises in practical chemistry. London, 1869.   8vo.

> Second edition revised.   London, 1871.   8vo.
>
> Third edition.   London, 1874.   8vo.
>
> Fourth edition.   London, 1879.   8vo.
>
> Hwa hio fen yuan.   Translated [into Chinese] by John Fryer. Shanghai, 187–.   2 vols.
>
> Enseignement du laboratoire ou exercises progressifs de chimie pratique.   Traduit sur la troisième édition par G. Darin.   Paris, 1875.   8vo.

Metals, their Properties and Treatment.   London, 1870.

> New edition, partially rewritten and augmented by Alfred K. Huntington.   London, 1888.   pp. xii–439, 12mo.   Ill.

BLUM, WILHELM.

Natürliche und künstliche Mineralwasser.   Braunschweig, 1853.   8vo.   Ill.

BLYTH, ALEXANDER WYNTER.

A Manual of Practical Chemistry : the analysis of foods and the detection of poisons.   London, 1879.   pp. xvii–468, 8vo.   Ill.

Foods : their Composition and Analysis.   A Manual for the use of analytical chemists and others.   With an introductory essay on the history of adulterations.   Third edition.   London, 1888.   8vo.

> This forms one volume of the second edition of "A Manual of Practical Chemistry."   First edition, London, 1882.

Poisons : their Effects and Detection.   A manual for the use of analytical chemists and experts.   With an introductory essay on the growth of modern toxicology.   London, 1884.   pp. xxxvi–712, 8vo.

> This forms one volume of the second edition of "A Manual of Practical Chemistry."   Contains an historical sketch and a bibliography (pp. 1–19).

BLYTH, J.

Outlines of Qualitative Chemical Analysis, for the use of agricultural students.   London, 1849.   12mo.

BLYTH, THOMAS ALLEN.

Metallography as Separate Science ; or the Student's Handbook of Metals : designed as an elementary work for the use of schools and science classes, and consisting of notes of fifty-five metals, their

BLYTH, THOMAS ALLEN.   [Cont'd.]

   various properties, their history, the localities in which they are
   found and the principal uses to which they are applied. London,
   1871. 8vo.

BOBIERRE, ADOLPHE.

   Laboratoire de chimie agricole de la Loire Inférieure, 1850–75.   Compte
   rendu des travaux. Paris, 1876. 1 vol., 8vo.
   Traité de manipulations chimiques. Paris, 1844. 8vo.

BODE, FR.

   Beiträge zur Theorie und Praxis der Schwefelsäure-Fabrikation.
   Berlin, 1872. 8vo.

BODEMANN, THEODOR.

   Anleitung zur berg- und hüttenmännischen Probierkunst.   Clausthal,
   1845. 8vo.
             Zweite Auflage bearbeitet von B. Kerl. Clausthal, 1857. 8vo.
             Treatise on the Assaying of Lead, Copper, Silver, Gold
                and Mercury. Translated by W. A. Goodyear. New
                York, 1878. 12mo.

BÖCKER, FRIEDRICH WILHELM.

   Lehrbuch der praktischen medicinischen Chemie für praktische Aerzte
   und Studirende der Medicin, oder Anleitung zur qualitativen und
   quantitativen zoochemischen Analyse. Weimar, 1855. 12mo. Ill.
             Medical Chemistry for Physicians and Students of Medi-
                cine, or practical instructions in zoo-chemical analysis.
                Translated and arranged by A. Zumbrock. Phila-
                delphia, 1855. 12mo.
   Vergiftungen (Die) in forensischer und klinischer Beziehung.   Iserlohn,
   1857. 8vo. Ill.

BÖCKMANN, CARL WILHELM.

   Versuche über das Verhalten des Phosphorus in verschiedenen Gasarten.
   Herausgegeben von Friedrich Hildebrandt. Erlangen, 1800. pp.
   xvi–342–[ii], 8vo. Three folding plates.

BÖCKMANN, FR.

   Chemisch-technische Untersuchungsmethoden der Grossindustrie, der
   Versuchsstationen und Handelslaboratorien. Unter Mitwirkung
   von C. Balling, M. Barth, Th. Beckert, etc., herausgegeben. Zwei
   Bände. Mit 155 in den Text gedruckten Abbildungen. Zweite
   vermehrte und umgearbeitete Auflage. Berlin, 1887. pp. 1222, 8vo.
             First edition. Berlin, 1883. 8vo.

BÖCKMANN, FR. [Cont'd.]

Celluloid (Das), seine Rohmaterialien, Fabrikation, Eigenschaften und technische Verwendung. Für Celluloid- und Celluloidwaaren Fabrikanten, für alle Celluloid verarbeitenden Gewerbe, Zahnärzte und Zahntechniker. Wien, Pest, Leipzig, 1880. Ill.

Explosiven (Die) Stoffe, ihre Geschichte, Fabrikation, Eigenschaften, Prüfung und praktische Anwendung in der Sprengtechnik. Mit einem Anhange, enthaltend: Die Hilfsmittel der submarinen Sprengtechnik (Torpedos und Seeminen). Bearbeitet nach den neuesten wissenschaftlichen Erfahrungen. Wien, Pest, Leipzig, 1880. 8vo. Ill.

Kurzgefasstes Lehrbuch der unorganischen Chemie für den ersten chemischen Unterricht an Baugewerks-, Gewerbe-, Real- und Handelsschulen, landwirthschaftlichen Lehranstalten und Schullehrerseminarien. Nebst einem Anhang enthaltend die Chemie der Baumaterialien mit den Abschnitten: Luftmörtel, Ziegelthon, Wassermörtel, Kitte, Oele, Farbstoffe, Holz. Leipzig, 1878. 8vo.
Zweite Auflage. Leipzig, 1880.

BOEDEKER, CARL HEINRICH DETLEV [Translator].
See Regnault, Henri Victor, Cours élémentaire de chimie.

BOEKE, J. D.

Stoechiometrische vraagstukken ten gebruike bij het onderwijs in de scheikunde. Alkmaar, 1872. 8vo.
Sammlung stoechiometrischer Aufgaben zum Gebrauche beim chemischen Unterrichte, sowie beim Selbststudium. Nach der dritte holländischen Auflage bearbeitet. Berlin, 1882. 8vo.

BOERHAAVE, HERMAN.

Institutiones et experimenta chemiæ. Parisiis, 1724. 2 vols., 8vo.
This is called by Boerhaave himself the "surreptitious edition," being published without his participation.
*A new method of Chemistry including the Theory and Practice of that Art, laid down on mechanical principles, and accommodated to the uses of life. The whole making a clear and rational system of chemical philosophy. To which is prefixed a critical history of chemistry and chemists, from the origin of the art to the present time. Translated from the printed edition, collated with the best manuscript copies, by P. Shaw and E. Chambers. With additional notes and sculp-

Boerhaave, Herman. [Cont'd.]

tures. London, printed for J. Osborn and T. Longman,
at the ship in Pater-noster-Row, 1727. pp. xvi–383–
335–[xliii], 4to. Plates.

<small>This is an English translation of the surreptitious edition, Institutiones et
experimenta chemiæ.</small>

Elementa chemiæ, quæ anniversario labore docuit in publicis priva-
tisque scholis, cum tabulis aëneis. Lugduni Batavorum, 1732. 2
vols., 8vo. Seventeen plates. Vol. i, pp. [lxii]–895 ; ii, [viii]–538–
[xliii].

Other editions : Paris, 1733 ; 2 vols., 4to. Paris, 1753.
Hagæ Comitis, 1746 ; 2 vols., 8vo.

<small>Vol. i, part i, is a condensed history of chemistry, logically arranged, form-
ing an introduction to this remarkable work</small>

* A New Method of Chemistry, including the History,
Theory and Practice of the Art : translated from the
original Latin of Boerhaave's Elementa chemiæ, as
published by himself. To which are added Notes and
an Appendix shewing the necessity and utility of
enlarging the bounds of Chemistry. With Sculptures.
By Peter Shaw. Third Edition. London, 1753. 2
vols., 4to. Vol. i, pp. xxx–593, 17 plates ; vol. ii, pp.
[i]–410–[xxxvii], 8 plates.

<small>A translation of the authentic edition, Elementa chemiæ, 1732.</small>

Elements of Chemistry. Translated from the Latin by
Timothy Dallowe. London, 1735. 2 vols., 4to.

Anfangsgründe der Chymie, oder gründliche Anweisung
auf was Art die natürlichen Cörper können chymisch
aufgeschlossen und daraus heilsame Artzeneyen bereitet
werden. Aus dem lateinischen . . . übersetzt. Nebst
einem . . . Anhange von chymischen Geräthschaften,
von Anwendung sowohl der florentischen als Fahren-
heitischen Thermometers . . . Berlin, 1762. 8vo.
3 parts.

Zweite Auflage. Berlin, 1791.

Elements of Chemistry, faithfully abridged from the German
edition published and signed of himself at Leyden with
all the cuts and experimentations contained in the
original, to which are added curious and useful notes
by a physician. Second edition. London, 1734.

Élémens de chymie par Herman Boerhaave, traduits du
Latin par J. N. S. Allamand. La Haye, 1748. 2 vols.,
8vo.

BOERHAAVE, HERMAN.   [Cont'd.]
>   Other editions : Leide, 1752 ; 2 vols., 8vo.   Paris, 1754 ; 6
>   vols., 12mo.

BÖTCHER, NICOLAUS.
Dissertatio physico-medica de aëre dephlogisticato.   Hafniæ, 1786.

BÖTTGER, RUDOLPH.
Beiträge zur Physik und Chemie.   Frankfurt am Main, 1838–46.
Three parts.   8vo.   Ill.
>   Part II under the title : Neuere Beiträge.

Ueber die Einrichtung und Behandlung der Döbereinerschen Zünd-
maschine.   Sondershausen.   Zweite Auflage, 1838.

BOHN, JOHANN.
Meditationes physico-chymicæ de aëris in sublunaria inflexu.   Lipsiæ,
1685.   4to.
>   *Prefixed to his* Dissertationes chymico-physicæ, etc.

BOILLOT.
De la combustion.   Phénomènes généraux, modifications apportées à
la théorie de Lavoisier.   Paris, 1869.   12mo.

Бокій, П.
Основанія химніи.   Тифлисъ, 1874.
>   BOKĬĬ, I.   First principles of Chemistry.   Tiflis, 1874.

BOLLEY, POMPEJUS ALEXANDER.
Handbuch der technisch-chemischen Untersuchungen.   Eine Anleitung
zur Prüfung und Werthbestimmung der im gesammten Gewerbs-
wesen oder der Hauswirthschaft vorkommenden und zur
chemischen Untersuchung geeigneten Natur- und Kunsterzeug-
nisse.   Frauenfeld, 1853.   8vo.   Ill.
>   Dritte Auflage.   Leipzig, 1865.
>   Bolley's Handbuch der technisch-chemischen Untersu-
>   chungen.   Eine Anleitung zur Prüfung und Werthbe-
>   stimmung der im gesammten Gewerbswesen oder der
>   Hauswirthschaft vorkommenden und zur chemischen
>   Untersuchung geeigneten Natur- und Kunsterzeugnisse.
>   Vierte Auflage ergänzt und bearbeitet von Emil Kopp,
>   unter gefälliger Mitwirkung von Rob. Gnehm und
>   Georg Wyss.   Leipzig, 1874.   pp. viii–856.   8vo.   Ill.
>   Fünfte Auflage nach Emil Kopp's Tode ergänzt und bear-
>   beitet von C. Stahlschmidt.   Leipzig, 1879–80.   8vo.   Ill.
>   Sechste Auflage.   Bearbeitet von C. Stahlschmidt.   Leipzig,
>   1888–89.   pp. i–544.   8vo.   Ill.

BOLLEY, POMPEJUS ALEXANDER. [Cont'd.]

Manual of Technical Analysis, a Guide for the testing and valuation of the various natural and artificial substances employed in the arts and in domestic economy. Founded on the "Handbuch der technisch-chemischen Untersuchungen" of P. A. B. by B. H. Paul. London, 1847. 8vo. Bohn's Scientific Library.

Manuel pratique d'essais et de recherches chimiques appliqués aux arts et à l'industrie. Guide pour l'essai et la détermination de la valeur des substances naturelles ou artificielles employées dans les arts, l'industrie, etc. Traduit de l'allemand sur la troisième édition et augmenté par L. A. Gautier. Paris, 1869. pp. viii–747, 12mo. Ill.

Manuel complet d'essais et de recherches chimiques appliqués aux arts et à l'industrie. Guide pratique pour l'essai et la détermination de la valeur des substances naturelles ou artificielles employées dans les arts et dans l'industrie, pour la recherche des altérations et de falsifications des substances alimentaires, etc. Quatrième édition, revue et augmentée par E. Kopp, avec la collaboration de Rob. Gnehm et G. Wyss. Deuxième édition française, traduite par L. Gautier. Paris, 1875. 12mo. Ill.

Mist (Der) seine chemische Zusammensetzung, seine Wirkung als Düngemittel und seine Zubereitungsweise. Für deutsche Landwirthe bearbeitet nach dem Plane von J. Girardin's Vorlesungen über diesen Gegenstand gehalten an der Landwirthschaftsschule zu Rouen. Braunschweig, 1846. 8vo.

De mest, zijne scheikundige zamenstelling, uitwerkingen en de wijze om die te behandelen. Uit het Hoogduitsch. Onder toezigt van een landbouwer voor Nederlanders bewerkt. Deventer, 1851. 8vo.

BOLLEY'S HANDBUCH DER CHEMISCHEN TECHNOLOGIE. In Verbindung mit mehreren Gelehrten und Technikern bearbeitet und herausgegeben von P. Bolley. [Fortgesetzt von K. Birnbaum. Nach dem Tode des Herausgebers weitergeführt von C. Engler.] Braunschweig, 1862–89. 8vo. 8 vols. in 49 parts, besides 5 parts of a "Neue Folge."

[Contents :]

Vol. I.   1. Die chemische Technologie des Wassers. Von P. B.   1862.
For second edition see Neue Folge below.

BOLLEY'S HANDBUCH DER CHEMISCHEN TECHNOLOGIE. [Cont'd.]

2 (a). Das Beleuchtungswesen, von P. Bolley und G. Wiedemann. 1862. Erste Hälfte. Kerzen-Lampen- und elektrische Beleuchtung. Zweite Hälfte. Die Gasbeleuchtung aus verschiedenen Materialen.

2 (b). Die Industrie der Mineralöle 1. Theil. Die Erdöl-Industrie von Hans Höfer und Ferdinand Fischer. Erste Lieferung, Das Erdöl (Petroleum) und seine Verwandten von Hans Höfer. 1888. Ill.

3 (a). Die chemische Technologie der Brennstoffe, von Ferdinand Fischer. 1880-87.

3 (b). Die Industrie der Steinkohlentheer-Destillation und Ammoniakwasser-Verarbeitung. Von G. Lunge. 1882.

Vol. II.   1 (a). Die Technologie der chemischen Produkte welche durch Grossbetrieb aus unorganischen Materialen gewonnen werden. Handbuch der Soda-Industrie und ihrer Nebenzweige. Von Ph. Schwarzenberg und G. Lunge. 1865–79. 2 vols.

1 (b). Die Technologie der chemischen Produkte welche durch Grossbetrieb aus unorganischen Materialen gewonnen werden. Die Stassfurter Kali-Industrie. Von Emil Pfeiffer. 1887.

2. Die Fabrikation chemischer Producte aus thierischen Abfällen, 1862. Zweite Auflage von H. Fleck. 1878. 8vo. Ill.

Vol. III.   1. Die Fabrikation des Glases. Von W. Stein. 1862. Ill.

Vol. IV.   1. Die Bierbrauerei, Branntweinbrennerei und die Liqueurfabrikation. Von Jul. Fr. Otto. 1865.

2. Die Essig-, Zucker- und Stärkefabrikation, Fabrikation des Stärkegummis, Stärkesyrups und Stärkezuckers, sowie die Butter und Käsebereitung. Von Fr. Julius Otto. 1867.

3. Der Weinbau und die Weinbereitungskunde, sowie die Bereitung des Obstweins und Krauts. Von Friedrich Mohr. 1865.

Vol. V.   1. Chemische Verarbeitung der Pflanzen- und Thierfasern. Die Spinnfasern und die im Pflanzen- und Thierkörper vorkommenden Farbstoffe. Die künstlich erzeugten organischen Farbstoffe. Von P. Bolley, E. Kopp, Richard Meyer. 1867–83. 5 Lieferungen.

2. Chemische Technologie der Gespinnstfasern, ihre Geschichte, Gewinnung, Verarbeitung und Veredlung. Von Otto R. Witt. Erster Lieferung : Seide, Wolle, Seidenhaare, Baumwolle, Flachs, Hanf, Jute und andere Pflanzenfasern. Von Otto R. Witt. 1888. Zweite Lieferung, 1891. [Not completed.]

BOLLEY'S HANDBUCH DER CHEMISCHEN TECHNOLOGIE. [Cont'd.]

 3. Die Fabrikation des Russes und der Schwärze aus Abfällen und Nebenproducten, insbesondere der Theer- und Mineralöl-Destillerien, der Braunkohlenschweelereien, Weinsäurefabriken, etc. Von Hippolyt Köhler. 1889. Ill.

Vol. VI. 1 (a). Die chemische Technologie der Baumaterialen und Wohnungseinrichtungen. 1. Abtheilung. Chemische Technologie des Holzes als Baumaterial. Von Adolph Mayer. 1872.

 1 (b). Die chemische Technologie der Mörtelmaterialen. Von G. Feichtinger. 1885.

 2. Die Darstellung der Seifen, Parfümerien und Kosmetika. Von C. Deite. 1867.

 3 (a). Das Schiesspulver, dessen Geschichte, Fabrikation, etc. Von J. Upmann. 1874.

 3 (b). Explosivkörper und die Feuerwerkerei. Von E. von Meyer. 1874.

 3 (c). Die Zündwaaren-Fabrikation. Von W. Jettel. 1871.

 4 (a). Grundzüge der Lederbereitung, etc. Von Christian Heinzerling. 1882. Ill.

 4 (b). Die Fabrikation der Kautschuk- und Guttapercha-waaren, sowie des Celluloids und der wasserdichten Gewebe. Von Christian Heinzerling. 1883. Ill.

 5 (a). Die Fabrikation des Papiers. Von Egbert Hoyer. 1887.

 6 (b). Handbuch der Sprengarbeit von Oscar Guttmann. Braunschweig, 1892. 8vo. Ill.

Vol. VII. Die Metallurgie. Von C. Stölzel. 2 vols. 1863–86.
   Contains a valuable bibliography of metallurgy. pp. i–lxxxvi.

Vol. VIII. 1. Die Metallverarbeitung. Von A. Ledebur. 1882.
   Contains several bibliographies of different branches of the subject.

 2. Die Erzeugung der Eisen- und Stahlschienen. Von Alph. Petzholdt. 1874.

NEUE FOLGE. 1. Die Fabrikation chemischer Producte aus thierischen Abfällen ; Phosphorfabrikation, Leimfabrikation, Fabrikation des Blutlaugensalzes, des Pariser und Berliner Blau, der Ammoniaksalze des Salmiakgeistes. Zweite Auflage, 1880. 8vo. Ill.

 2. Die Glasfabrikation. Von H. E. Benrath. 1880. 201 ill.

 3. Die chemische Technologie des Wassers, von Ferdinand Fischer. 1880. 271 ill.

 4. Die trocknenden Oele, ihre Eigenschaften, Zusammensetzung und Veränderungen. Von L. E. Andés. 1882. Ill.

 5. Die Industrie des Steinkohlentheers und Ammoniaks. Von Georg Lunge. Dritte vermehrte und verbesserte Auflage, 1888. 195 ill.

Боллей, П.
   Химическая технологія перев. Алексѣева. С.-Петербургъ, 1875.

> BOLLEY, P. Chemical Technology. St. Petersburg, 1875. *Cf.* Habig, H. Ch.

Боллей, П.
   Химическая переработка растительныхъ и животныхъ волоконъ. Перев.
   Гейнемана Л. С.-Петербургъ, 1874.

> BOLLEY, P. Chemical transformation of vegetable and animal fibres. Translated by Heineman. St. Petersburg, 1874.

BOLLEY, P., EMIL KOPP, AND R. MEYER.
> Die künstlich erzeugten organischen Farbstoffe. *See* Bolley's Handbuch [etc.].

BOLLEY, P. A., AND EMIL KOPP.
   Traité des matières colorantes artificielles dérivées du goudron de houille, traduit de l'allemand et augmenté des travaux les plus récents par L. Gautier. Paris, 1874. 8vo. Ill.

BOLTON, HENRY CARRINGTON.
   The Students' Guide in Quantitative Analysis, intended as an aid to the Study of Fresenius' System. New York, 1882. pp. xii, from 14–124-[iv], 125–127. Ill.
> Second edition : New York, 1885. pp. xii, from 14–131. Folding table.
> Third edition, revised and enlarged. New York, 1889.

BOLTON, HENRY CARRINGTON.
> *See, in Section IV*, Priestley, Joseph, Scientific Correspondence of.

BONAME, PH.
   Culture de la canne à sucre à la Guadeloupe ; deuxième édition revue et considérablement augmentée. Paris, 18—.

BONDT, NICOLAS, PAETS VAN TROOSTWYK, DEIMAN, NIEUWLAND, *et al.*
   Recherches physico-chimiques. Four parts. Amsterdam, 1792–94.

BONET Y BONFILL, MAGIN.
   Memoria sobre los adelantos hechos por varias industrias químicas. Madrid, 1861. 4to.

BONILLA MIRAT, S.
   Tratado elementar de química general y descriptiva. Valladolid, 1882. 4to.

BONN, R.
   Die Structurformeln. Geschichte, Wesen und Beurtheilung des Werthes derselben. Frankfurt an der Oder, 1887. 8vo.

BONNAMI, H.
Fabrication et contrôle des chaux hydrauliques et des ciments.
Théorie et pratique, influences réciproques des différentes opérations et de la composition sur la solidification, énergie thermodynamique, thermochimie.   Paris, 1888.

BONNET, V.
Précis d'analyse microscopique des denrées alimentaires.   Caractères.
Procédés d'examen.   Altérations et falsifications.   Avec une préface
par Léon Guignard   Paris, 1890.   pp. viii–200, 12mo.   20 colored
plates and 163 woodcuts.

BONNIER, G.
Leçons de choses combustibles ; métaux ; matériaux de construction ;
eau ; air ; saisons.   Paris, 1882.   18mo.   Ill.

BONSDORFF, PEHR ADOLPH VON.
Dissertatio chemica, nova experimenta naturam pargasitæ illustrantia
proponens.   Pt. I et II.   Åbo, 1817–18.
Tentamen mineralogica-chemica de pargasita.   Åbo, 1816.

BONTEKOF-DECHER, CORNELIS.
Alle de philosophische, medicinale en chymische werken van den Heer
C. B-D.   Amsterdam, 1689.   4to.   2 vols.
Portrait of author in vol. I.
Van het acidum en alkali.   Haag, 1683.

BONZANI, FR.
Elementi di chimica inorganica.   Savona, 1865.   12mo.

BOON MESCH, ANTONY HENDRIK VAN DER.
Leerboek der scheikunde met toepassing op kunsten en fabryken.
Lugduni-Batavorum, 1831.   3 vols.   8vo.
De chymiæ materia ratione et usu.   Lugduni-Batavorum, 1827.   4to.

BOON MESCH, HENDRIK CAREL VAN DER.
De acido muriatico oxygenato.   Ultrajecti, 1819.   8vo.

BORCK, JOHAN BERNHARD.
Försök till en praktisk Elementar-Lärobok i oorganiske Kemien.   Med
öfver 300 i Texten inflätade, samt genom Figurer förtydligade
Öfningsförsök för Nybegynnare.   Lund, 1853.

BORDAS, F.
Étude sur la putréfaction.   Paris, 1892.   8vo.   Ill.

BORGMANN, EUGEN.

Anleitung zur chemischen Analyse des Weines.   Mit Vorwort von C.
Remigius Fresenius. Mit zwei Tafeln in Farbendruck und drei-
undzwanzig Holzschitten im Texte. Wiesbaden, 1884. pp. viii-
168, 12mo.

BORIAS, EDMOND.

Traité théorique et pratique de la fabrication du gaz et de ses divers
emplois à l'usage des ingénieurs, directeurs et constructeurs d'usines
à gaz. Paris, 1890. pp. ii–494. 8vo.

BORIE, L.

Catéchisme toxicologique, ou essai sur l'empoisonnement.   Tulle, 1841.
12mo.

BORK, H.

Die Chemie.   Leitfaden für den chemischen Kursus in der Secunda des
Gymnasiums.   Methodisch bearbeitet.   Paderborn, 1886.   8vo.

BORNEMANN, G.

Die fetten Oele der Pflanzen und Thierreiche.   Fünfte Auflage, 1889.
Atlas of 12 plates.

Die flüchtigen Oele des Pflanzenreiches, ihr Vorkommen, ihre Gewinnung und Eigenschaften, ihre Untersuchung und Verwendung.
Nebst einem Kapitel : Botanische Betrachtungen über das Vorkommen der ätherischen Oele, von R. L. Vetters.   Fünfte vollständig neubearbeitete Auflage von Fontenelle's Handbuch der
Oelfabrikation.   Weimar, 1891.   8vo.

BORNTRAEGER. A.

Determinazione degli zuccheri.   Roma, 1890.

BORRICHIUS, OLAUS [OLE BORCH].

Docimastice metallica clare et compendiario tradita.   Hafniæ, 1677.
pp. 46, sm. 4to.

Metallische Probier-Kunst deutlich und kurtz beschrieben.
Verteutscht durch Georg Kus.   Kopenhagen, 1680.

A Swedish translation by Jac Fischer was published at Stockholm in 1738.
See also, in Sections I, III, and VI, Borrichius, O.

BOTET Y JOMELLÀ, RAMÓN.

Resúmen de química legal.   Apéndice añadido á la análisis química
del Dr. Will.   Lérida, 1857.   4to.   Tables.

Cf. Will, Heinrich.

BOTTLER, M.

Graphische Darstellungen zur Vergleichung der Mineralquellen deutscher und deutsch-österreichischer Kurorte. Kissingen, 1891. 8vo.

BOUANT, E.

Aide mémoire de chimie. Paris, 1885. 12mo.

Cours de chimie à l'usage des élèves de la classe de mathématiques spéciale. Paris, 1885.

Cours de physique et de chimie. (Programme des écoles normales primaires des instituteurs). Deuxième édition. Paris, 1883. 12mo.

BOUCHARDAT, APOLLINAIRE.

Chimie élémentaire avec ses principales applications aux arts et à l'industrie. Rédigée d'après les derniers programmes officiels. Troisième édition corrigée, augmenté et orné de 64 figures intercalées dans le texte. Paris, 1847. 12mo.

> Die Chemie in ihrer Anwendung auf Künste und Gewerbe. Aus dem französischen übersetzt und mit vielen Zusätzen und Nachträgen bereichert von G. Kissling. Ludwigsburg, 1845. 8vo.
>
> Elementos de química aplicada á las artes, á la industria y á la medicina . . . traducidos y considerablemente aumentados por O. Bofill y J. Martí. Barcelona, 1843–44. 2 vols., 8vo. Ill.
>
> Elementos de química con sus principales aplicaciones á la medicina, á las artes y á la industria, adornados con 63 figuras intercaladas en el testo. Traducidos de la segunda edicion y adicionados por D. Gregorio Lesana y D. Juan Chavarri. Madrid, 1845. 8vo.
>
> Tratado completo de química con sus principales aplicaciones á las artes y á la industria, ilustrado con sesenta figuras intercaladas en el testo. Nueva traduccion por D. Antonio Sanchez de Bustamentè. Madrid, 1848. 4 vols., 8vo.

BOUCHARDAT, A., ET TH. A. QUEVENNE.

Du lait. Premier fascicule. Instruction sur l'essai et l'analyse du lait. (Chimie légale du lait.) Deuxième fascicule. Du lait en général. Des laits de femme, d'anesse, d'chèvre, de brebis, de vache en particulier. Paris, 1857. pp. iii–210, 8vo.

BOUDET, FELIX HENRI, ET BOISSENOT.

Essai chimique sur la cire d'abeilles. Paris, 1827.

Essai critique et expérimental sur le sang. Paris, 1833.

BOUDRÉAUX, A.
Traité élémentaire de manipulations chimiques.   Tome I.   Clamart,
1888. pp. 216, 12mo. Ill.

BOUILLET, JEAN.
Mémoires sur l'huile de pétrole et les eaux minérales de Gabian.
Béziers, 1752.

BOUILLET ET CHRISTOPHLE.
Galvanoplastie.   *See, in Section II*, Fremy: Encyclopédie chimique, vol. v.

BOUILLON-LAGRANGE.
See Lagrange, Edmonde Jean Baptiste Bouillon.

BOUILLOT, C. L.
Étude synoptique de la chimie élémentaire.   Sainte Étienne, 1876.   8vo.

BOUIS, JULES.
Empoisonnement par les gaz.   Paris, 1859.   8vo.

BOULIN, P.
Manuel pratique de la fabrication de la bière.   Paris, 1889.
Manuel pratique du fabricant de sucre.   Sucre de betteraves et sucre
de cannes.   Paris, 1889.

BOUQUET, J. P.
Histoire chimique des eaux minérales et thermales de Vichy, Cusset,
etc.   Paris, 1855.

BOURGEOIS.
Synthèse minéralogique.   *See, in Section II*, Fremy: Encyclopédie chimique.

BOURGET.
Manuel de chimie clinique.   Analyse de l'urine, des calculs concrétions
et sédiments, des transsudats et exsudats liquides, des liquides
kystiques et du suc gastrique.   Paris, 1891.   12mo.

BOURGOIN, A. EDME.
Aldéhydes proprement dits et acétones.   Paris, 1884.   8vo.
Carbonyles, quinones, aldéhydes à fonction mixte.   Paris, 1885.   8vo.
Chimie organique.   Des alcalis organiques.   Paris, 1869.   8vo.
Chimie organique.   Principes de la classification des substances or-
ganiques.   Paris, 1876.   8vo.
De l'isomérie.   Paris, 1866.   8vo.
Cf., in Section II, Fremy : Encyclopédie chimique, vol. II, vol. VI, etc.

BOUSSINGAULT, JEAN BAPTISTE JOSEPH DIEUDONNÉ.

Agronomie, chimie agricole et physiologie.   Deuxième édition.   Paris,
    1860–84.   2 vols., 8vo

        Troisième édition, Paris, 1886.   2 vols., 8vo.

Économie rurale, ou chimie appliquée à l'agriculture, considérée dans
    ses rapports avec la chimie, la physique et la météorologie.   Paris,
    1844.   2 vols., 8vo.

            Deuxième édition, revue et corrigée.   Paris, 1851.   2 vols.,
            8vo.

            Rural Economy in its relations with Chemistry, Physics and
            Meteorology ; or, Chemistry applied to Agriculture.
            Translated, with an introduction and notes, by George
            Law.   New York and Philadelphia, 1845.   pp. iv–507.
            12mo.

            Die Landwirthschaft in ihren Beziehungen zur Chemie,
            Physik und Meteorologie, deutsch bearbeitet von N.
            Graeger.   Halle, 1844.   2 vols., 8vo.

            Zweite verbesserte Auflage. Halle, 1851.   2 vols. 8vo. Vol. III :
            Die Zusätze und Verbesserungen der zweiten Auflage
            des Originals enthaltend.   Für die Besitzer der 1. und
            2. Auflage der Uebersetzung.   Halle, 1854.   Vol. IV :
            Supplement-Band, *also under the title :* Beiträge zur
            Agricultur-Chemie und ˙Physiologie.   Halle, 1856.
            8vo.   Ill.

            L'economia rurale considerata ne' suoi rapporti con la
            chimica, la fisica e la meteorologia.   Prima versione
            Italiana con note di Jac. Bologna.   Venezia, 1850.
            2 vols., 8vo.

            Cenni sul valore comparativo dei concimi e delle materie
            nutritive desunti dall' " Economia rurale," e pubblicati
            dal Comizio agrario di Vercelli l'anno 1845.   Vercelli,
            1845.   4to.   Ill.

Mémoires de chimie agricole et de physiologie.   Paris, 1854.   8vo.

BOUSSINGAULT, J. B. J. D., ET DUMAS.
        See Dumas, Jean Baptiste.

BOUTAN.
            Diamant.   *See, in Section II*, Fremy: Encyclopédie chimique, vol. II, com-
            plément.

BOUTET DE MONVEL, B.

Cours de chimie rédigé conformément aux derniers programmes de
    l'enseignement scientifique dans les lycées et à celui du baccalauréat
    des sciences.   Paris, 1856   18mo.   Ill.

BOUTET DE MONVEL, B. [Cont'd.]

       Deuxième édition. Paris, 1857. 18mo.

       Cours de chimie, comprenant les matières indiquées par les programmes officiels arrêtés le 24 mars, 1855, pour l'enseignement de la chimie dans la classe de mathématiques élémentaires, avec de nombreuses figures dans le texte. Septième édition. Paris, 1868. 18mo.

       Elementi di chimica generale per gli Istituti tecnici ed i Licei. Seconda edizione. Milano, 1869. 8vo.

Notions de chimie, rédigées conformément au dernier programme de l'enseignement de la classe de seconde (section des lettres). Quatrième édition. Paris.

       Troisième édition, Paris, 1853.

       Cinquième édition, 1861.

       Sixième édition, 1863.

       Septième édition, 1865.

       Huitième édition, 1866, conformés au programme officiel arrêté le 25 mars, 1865, pour l'enseignement de la chimie dans les classes de philosophie.

       Neuvième édition, Paris, 1867.

       Nociones de química con numerosos grabados en el texto. Traducidas de la 8a edicion, por A. Ramón de la Sagra. Paris, 1866. 18mo.

BOUTLEROW.

       See Butlerof.

BOUTRON ET F. BOUDET.

Hydrotimêtrie. Nouvelle méthode pour déterminer les proportions des matières minérales en dissolution dans les eaux de sources et de rivières. Huitième édition. Paris, 1887. 8vo.

       Sixième édition, Paris, 1877.

       Septième édition, Paris, 1882.

       Hidrotimetría. Nuevo método para determinar las proporciones de las materias minerales disueltas en las aguas de redaccion del semanario farmaceutico. Madrid, 1879. 4to.

BOWDITCH, W. R.

The Analysis, Technical Valuation, Purification and Use of Coal Gas. London, 1867. pp. iv–300, 8vo.

On Coal-Gas. A discourse delivered to some directors and managers of Gas-Works, June 13, 1860, and published at their request. London, 1860.

Bowman, John Eddowes.

Introduction (An) to Practical Chemistry, including Analysis.   London,
1848,   16mo.

Second edition, London, 1854.   8vo.

Third edition, edited by C. L. Bloxam, London, 1858.   8vo.

Fifth edition, London, 1866.   8vo.

Seventh edition, London, 1878.   8vo.

Eighth English edition, revised by C. L. Bloxam.   London,
1885.   8vo.

Practical (A) Handbook of Medical Chemistry.   London, 1850.   8vo.

Second edition, London, 1852.   8vo.

Fourth edition, edited by C. L. Bloxam, London, 1862.   8vo.

For an American edition, see Greene, William H.

Bowring, Juan.

Aplicacion de la química y de la electricidad al beneficio de los metales
de plata.   México, 1858.   8vo.

Boyle, Robert.

Opera varia physico-mechanico-philosophica [etc.].   Genevæ, 1680–96.
4 vols.

* Some Considerations Touching the Usefulnesse of Experimental
Natural Philosophy, propos'd in Familiar Discourses to a Friend,
by way of Invitation to the Study of it.   Oxford.   Printed by Hen:
Hall, Printer to the University, for Ric: Davis, Anno Domini,
1663.   2 parts.   pp. [xiv]–127–[vi]–[ii]–417–[xvii], sq. 8vo.

Sceptical (The) Chymist : or chymico-physical Doubts and Paradoxes
touching the Experiments whereby vulgar Spagirists are wont to
endeavour to evince their Salt, Sulphur and Mercury to be the
true Principles of Things.   Oxford, 1661.   12mo.

* Second edition, Oxford, 1680.   *Contains also :* Divers Experiments and Notes about the producibleness of
chymical Principles.   Oxford.   Printed by Henry Hall
for Ric. Davis and B. Took at the Ship in St. Paul's
Church-Yard.   1680.   pp. [xxi]–440–[xxvii]–268, 12mo.

The learned Irish experimental philosopher was one of the first to abandon
the three imaginary principles of alchemical philosophy, and to pro-
pose the molecular or atomic idea in chemistry.

* Works (The) of the Honourable R. B. epitomiz'd by Richard Boulton
of Brazen Nose College in Oxford.   Illustrated with Copper Plates.
London, 1699.   Four vols., 12mo.   Vol. I, pp. [xxviii]–482–[ix].
Portrait and 8 plates.   1699.   Vol. II, pp. [x]–523–[ix].   7 plates.
1700.   Vol. III, pp. [xiv]–552–[viii].   5 plates.   1700.   Vol. IV, pp.
[xii]–365–[iii]–122.   1700.

BOYLE, ROBERT. [Cont'd.]

> Vol. IV contains : A general idea of the Epitomy of the works of Robert
> Boyle ; To which are added General Heads for the Natural History of
> a Country. By R. Boulton. London, 1700.

The Works of the Honourable R. B. in five volumes. To which is prefixed the Life of the Author. London, 1744. 5 vols., fol. Plates.
I, pp. viii–583. II, pp. 565. III, pp. 803. IV, pp. 556. V. pp.
736–[lxxxvi]. Portrait, and many folding plates.

> Edited by Thomas Birch. The dedication is signed by Andrew Millar.

* The Philosophical Works of the Honourable R. B., abridged, metho-
dized and disposed under the General Heads of Physics, Statistics,
Pneumatics, Natural History, Chymistry and Medicine. The
whole illustrated with Notes, containing the improvements made in
the several parts of natural and experimental Knowledge since his
time. By Peter Shaw. London, 1725. 3 vols., 4to. Ill. Vol. I,
pp. [iv]–xliii–730. 1 plate. Vol. II, pp. xx–726. 19 plates. Vol.
III, pp. [iv]–xv–756. 1 plate.

BOZZI, AUGUSTUS.

> *See* Granville, Augustus Bozzi.

BRANCHE, L.

Le chlorure de sodium et les eaux chlorurées sodiques, eaux minérales
et eaux de mer. Lyon, 1885. pp. 295, 8vo.

BRANCHI, GIUSEPPE.

Sopra alcune proprietà del fosforo ; esperienze ed osservazioni. Pisa,
1813, 8vo.
Trattato sulle falsificazioni delle sostanze specialmente medicinale.
Pisa, 1823–24. 2 vols.

BRANCHI, NICCOLÓ ANTONIO.

Indice d'esperienze chimiche che saranno mostrate nel corrente anno
1752 [*and* 1753] in Firenze nel laboratorio della Specieria del
Cignale in Mercato nuovo. Pisa, 1752 and 1753.

BRANDE, WILLIAM THOMAS.

A Manual of Chemistry, containing the principal facts of the science
arranged in the order in which they are discussed and illustrated in
the Lectures at the Royal Institution of Great Britain. London,
1819. 8vo.

> First American from second London edition   With notes
> and emendations by J. MacNeven. New York, 1821.
> 3 vols in 1, 8vo.

BRANDE, WILLIAM THOMAS.   [Cont'd.]

> Second American from second London edition. New York,
> 1826. 3 vols. in 1.
>
> Third American from the second London edition. . . .
> To which are added notes and emendations by W. J.
> Macneven. New York, 1829. 8vo.
>
> Fourth edition, greatly enlarged, London, 1836.
>
> Sixth edition, London, 1848. 2 vols., 8vo.
>
> Handbuch der Chemie für Liebhaber. Aus dem englischen.
> Leipzig, 1820. 8vo.

Tables in Illustration of the Theory of Definite Proportionals ; shewing
the prime equivalent numbers of the elementary substances, and
the volume and weights in which they combine. London, 1828.
4to.

The Subject-Matter of a Course of Ten Lectures on some of the Arts
connected with Organic Chemistry as applied to manufactures, including dyeing, bleaching, calico-printing, sugar-manufacture, the
preservation of wood, tanning, etc., delivered before the members
of the Royal Institution in the session of 1852. Arranged by permission, from the lecturer's notes, by J. Scoffern. London, 1854.
8vo.

BRANDE, WILLIAM THOMAS AND A. S. TAYLOR.

> Chemistry. Philadelphia, 1863. pp. xii–696, fol.

BRANDES, RUDOLPH.

Mineralquellen (Die) und Schwefelschlammbäder zu Meinberg . . .
Lemgo, 1832.

Mineralquellen (Die) und das Mineralschlammbad zu Tatonhausen
. . . Lemgo, 1830.

Monographie des Ammoniaks. Hannover, 1820. 4to.

Repertorium der Chemie . . . Hannover, 1827–33. 4 vols. [not
completed].

Ueber das Chlor und seine Verbindungen. Lemgo, 1831.

BRANDES, RUDOLPH [Editor].
See in Section VII.

BRANFORD, V. V.

Atlas of Chemistry, Inorganic and Organic. Designed for the use of
Medical and Science Students. Part 1. Edinburgh, 1891. 8vo.

BRANNT, WILLIAM T.

Metallic Alloys, a Practical Guide for the Manufacture of all kinds of
Alloys, Amalgams and Solders. Edited chiefly from the German

BRANNT, WILLIAM T.   [Cont'd.]
of A. Krupp and Wildberger, with many additions.   London, 1889.
8vo. Ill.

Practical (A) Treatise on Animal and Vegetable Fats and Oils.   Edited
chiefly from the German of [C] Schaedler.   London, 1888.

Practical (A) Treatise on the Manufacture of Soap and Candles, based
upon the most recent experiences in the science and the practice.
Comprising the chemistry, the raw materials, the machinery and
utensils, and various processes of manufacture, including a great
variety of formulas.   Philadelphia, 1888. 8vo.   Ill.

> Contains list of Patents relating to soap and candles issued by the Govern-
> ment of the United States of America, 1790-1888.

Practical (A) Treatise on the Manufacture of Vinegar and Acetates,
Cider and Fruit-Wines ; preparation of fruit-butters, jellies, mar-
malades, catchups, pickles, mustards, etc.   Edited from various
sources.   Philadelphia and London, 1890.   pp. xxxii–18 to 479,
8vo.   Ill.

Practical (A) Treatise on the Raw Materials and the Destillation and
Rectification of Alcohol, and the preparation of alcoholic liquors,
liqueurs, cordials and bitters.   Edited chiefly from the German of
K. Stammer, F. Elsner and E. Schubert.   Philadelphia, 1885. 8vo.

BRANNT, WILLIAM T., AND WILLIAM H. WAHL.
Techno-chemical Receipt Book (The).   Philadelphia and London, 1886.

BRANTHOME, YVES MARIE.
Précis des leçons de chimie données á la faculté de Strasbourg   . . .
Paris, 1825.

BRASCHE, O.
Ueber Verwendbarkeit der Spektroskopie zur Unterscheidung der Far-
benreactionen der Gifte im Interesse der forensischen Chemie.
Dorpat, 1891. 8vo.

BRAUMÜLLER, J. G.
Ueber die Veredlung einiger vorzüglicher Landesproducte.   1799.
> Treats of the cultivation of the sugar-beet.

BRENDEL, ZACHARIAS.
Chymia in artis formam redacta.   Jenæ, 1630.

BRESCH, R.
Chemismus, Magnetismus und Diamagnetismus im Lichte mehrdimen-
sionaler Raumanschauung.   Leipzig, 1882.

BRESLAUER, M.

Chemische Untersuchung der Luft für hygienische Zwecke. Berlin, 1885. 8vo.

Die chemische Beschaffenheit der Luft in Brandenburg. Berlin, 1886.

BRESSON.

Fer, Fonte, Aciers. *See, in Section II*, Fremy: Encyclopédie chimique, vol. v.

BREVANS, J. DE.

La margarine, etc. Par Ch. Girard et J. de Brevans. *See* Girard, Ch.

BREVANS, J. DE.

Fabrication (La) des liqueurs et des conserves. Introduction par Ch. Girard. Paris, 188–. 16mo. Ill.

Bibliothèque des connaissances utiles.

BRIANCHON, CHARLES JULIEN.

Essai chimique sur les réactions foudroyantes. Paris, 1825.

BRIANT, L.

Laboratory Text-book for Brewers. London, 1885.

BRIEGER, L.

Untersuchungen über Ptomaine ; [and] Weitere Untersuchungen. Berlin, 1885–86. 3 parts, 8vo.

BRIEM, HERMANN.

Die Rübenbrennerei. Dargestellt nach den praktischen Erfahrungen der Neuzeit. Wien, Pest, Leipzig, 18—. Ill.

BRIGGS, WILLIAM.

Synopsis of Non-Metallic Chemistry. London, 1892.

BRIGGS, WILLIAM, AND R. W. STEWART.

Analysis of a Simple Salt, with a Selection of Model Analyses. London, 1891. 12mo.

BRINKMANN, J. P.

Beyträge zu einer neuen Theorie der Gährungen. Düsseldorf und Leipzig. 1774. 8vo.

BRISSON, MATHURIN JACQUES.

Éléments ou principes physico-chymiques. Paris, 1800. 4 vols.

The Physical Principles of Chemistry, to which is added a short Appendix by the translator. London, 1801. pp. ix–424–xxiv. 8vo. Plates.

BRISSON, MATHURIN JACQUES. [Cont'd.]
Principes élémentaires de l'histoire naturelle et chimique des substances
    minérales. Paris, 1797.
>   Elements of Natural History and Chymical Analysis of
>   Mineral Substances. Translated from the French.
>   London, 1800.
Pésanteur spécifique des corps. Paris, 1787. 4to.
>   Spezifischen (Die) Gewichte der Körper. Aus dem franzö-
>   sischen . . . übersetzt, und mit Anmerkungen be-
>   sonders die Litteratur betreffend vermehrt von Johann
>   Georg Ludolph Blumhof. Mit Zusätzen von Kästner.
>   Leipzig, 1795. pp. xxxii–392. 8vo. Folding plates.

BRIX, ADOLPH FERDINAND WENCESLAUS.
Alkoholometer (Das) . . . Berlin, 1850.
Alkoholometer (Das) und dessen Anwendung . . . Berlin, 1856.
Ueber die Beziehungen welche zwischen den Procentgehalten ver-
    schiedener Zuckerlösungen in Wasser, den zugehörigen Dichten
    und Aräometergraden stattfinden. Berlin, 1854.

BRIX, CHRISTIANUS FREDERICUS.
Dissertatio chemica de salibus alkalino-volatilibus. Hafniæ, 1756–57.
    2 parts, 4to.

BRIZÉ-FRADIN, C. A.
La chimie pneumatique appliquée aux travaux sous l'eau, dans les
    puits, les mines, les fossés. Paris, 1808. Ill.

BRODIE, SIR BENJAMIN C.
Calculus (The) of Chemical Operations. London, 1866–76. 2 parts, 4to.
>   Le calcul des opérations chimiques, soit une méthode pour
>   la recherche par le moyen de symboles des lois de la
>   distribution du poids dans les transformations chimiques.
>   Traduit de l'anglais par A. Naquet. Extrait du Moni-
>   teur scientifique-Quesneville. Paris, 1879. pp. 155, 4to.
Ideal Chemistry. A Lecture. London, 1880. pp. 64, 16mo.
On the Conditions of Chemical Change. London, 1850. 4to.

BROECK, VICTOR VAN DEN.
Traité abrégé de docimasie, ou résumé des leçons données à l'École des
    mines du Hainaut ; accompagné de seize planches de figures, dont
    treize coloriées. Mons, 1841. 8vo.

BROEK, J. H. VAN.

Handleiding der scheikunde ten gebruike bij het onderwijs aans Rijks-
kweekschool voor militaire geneeskundigen.   Scheikunde der onbewerktuigde ligchamen.   Utrecht, 1857.   8vo.

BROMEIS, JOHANN CONRAD.

Die Chemie mit besonderer Rücksicht auf Technologie zum Gebrauche
bei Vorträgen an Universitäten und höheren Lehranstalten beim
Unterricht an Gymnasien, Real- und Gewerbeschulen sowie zum
Selbstunterricht.   Stuttgart, 1854.   8vo.

> Die anorganische Chemie mit besonderer Rücksicht auf
> Technologie zum Gebrauche bei Vorträgen an Univer-
> sitäten und höheren Lehranstalten sowie zum Selbst-
> unterricht.   Zweite stark vermehrte Auflage.   Nach
> dem Tode des Verfassers herausgegeben von Th.
> Bromeis.   Stuttgart, 1866.   8vo.   Ill.

BRONNER, PAUL.

Lehrbuch der Essigfabrikation mit Einschluss der Holzessigfabrikation
und die Darstellung der essigsauren Salze (Acetate).   (Zugleich
als dritte Auflage von Fr. Jul. Otto's Lehrbuch der Essigfabrikation
und als Lieferung 16 von Otto-Birnbaum's landwirthschaftlichem
Gewerbe.)   Braunschweig, 1876.   8vo.

> Cf. Otto-Birnbaum's Lehrbuch.

BROUARDEL, P. ET G. OGIER.

Le laboratoire de toxicologie.   Méthodes d'expertises toxicologiques,
travaux du laboratoire.   Paris, 1891.   8vo.   Ill.

BROWN, ALEXANDER CRUM.

Chemistry.   London and Edinburgh, 188–.

> Chambers' Elementary Science Manuals.

> Chimica: traduzione da L. Pratesi.   Napoli ed Roma,
> 1880.   16mo.

On the Theory of Isomeric Compounds.   Edinburgh, 1864.   4to.

BROWN, A. M.

The Animal Alkaloids, Cadaveric and Vital, or, the Ptomaines and
Leucomaines chemically, physiologically and pathologically considered in relation to scientific medicine.   With an introduction by
Armand Gautier.   Second edition.   London, 1889.   pp. xxv–252.
8vo.

BROWN, J. CAMPBELL.
Practical Chemistry. Analytical tables and exercises for students.
Second edition. London and Liverpool, 1883.
Third edition, London, 1888.

BROWN, JOHN CROUMBIE.
Lectures on Chemistry. Cape Town, 1846. 12mo.

BROWN, WALTER LEE.
Manual of Assaying Gold, Silver, Copper and Lead Ores. Revised,
corrected and considerably enlarged with a chapter on the Assaying of Fuels. Chicago and London, 1890. 8vo.
Fourth edition, Chicago, 1892.

BROWN, WILLIAM SYMINGTON.
Chemistry for Beginners. Second edition. Boston, 1855. 16mo.

BROWNE, G. LATHAM, AND G. G. STEWART.
Reports of Trials for Murder by Poisoning by prussic acid, strychnia,
antimony, arsenic, aconitia, including the trials of Tawell, W.
Palmer, Dove, Madeline Smith, Dr. Pritchard, Smethwist, and Dr.
Lamson. London, 1883.

BROWNING, JOHN.
How to work with the Spectroscope. A Manual of Practical Manipula-
tion with Spectroscopes of all kinds . . . and accessory apparatus.
London, *n. d.* pp. 68, 8vo. Ill.

BROWNRIGG, WILLIAM.
Art of Making Common Salt. London, 1748. 8vo.

BRUCKMÜLLER, A.
Lehrbuch der anorganischen Chemie. Zweite vermehrte Auflage.
Wien, 1882.

BRUGNATELLI, LUIGI VALENTINO.
Elementi di chimica, appoggiati alle più recenti scoperte [etc.]. Pavia,
1795–97. 2 vols., 8vo. *Also,* Venezia, 1800. 3 vols.
Seconda edizione, Pavia, 1803.
Guida allo studio della chimica generale. Pavia, 1819–24. 3 vols.
Sunto delle sue lezioni di chimica organica raccolto per cura di E.
Zenoni e A. Vidari. Pavia, 1874. 4to.

BRUGNATELLI, LUIGI VALENTINO [Editor].
See, in Section VII, Annali di chimica ; *also* Giornale di fisica.

BRUHAT, I.

Le laboratoire municipal de Paris.   *See* Paris.

Брыковъ.

О манганцовыхъ препаратахъ.   1851.   8o

BRUIKOF.   On Preparations of Manganese.   [St. Petersburg ?] 1851.

BRUNET Y TALLEDA, A.

Sinopsis de la clasificacion y principales caracteres de los ácidos y bases, asi orgánicos como inorgánicos, que más comunmente se encuentran en las operaciones de análises química.   Madrid, 1874.   8vo.

BRUNNER, C.

Grundriss der Chemie.   Leipzig, 1891.   8vo.

BRUNNER, H.

Guide pour l'analyse chimique qualitative des substances minérales et des acides organiques et alcaloïdes les plus importants.   Lausanne, 1889.   8vo.

BRUNSCHWICK, IHERONIMUS.

*Liber de arte distillandi de compositis.   Das buch der waren kunst zu distillieren die composita und simplicia und ds Buch thesaurus pauperum, ein Schatz der armen genant Micarium . . . [Colophon :] Und hie dis buch seliglich getruckt und gendigt in der keisserlichen fryen stat Strassburg uff sant Mathis abent in dem Jar 1512.   Ff. 1–18, 9–344–[vi], fol.   Ill.

[CONTENTS :]

Erstes Buch . . . Die ware kunst der distillierung der composita, als quinta essentia, Aurum potabile, Aqua vite simplex und composite und Balsam artificialiter . . . fol. 9.

Zweites Buch . . . Wie man die Simplicia und Composita zusamen vermischen und componieren sol in einer gemein zum allen krankheiten . . . fol. 139.

Drittes Buch : Die Leren der zusammen vermisten wasser als die composita, welcher zugehörig seint dem ganzen leib, von dem haupt an biss zu den füssen, es sey von ussen oder von innen.   Fol. 201.

Viertes Buch : Ein kurzer begriff uss der lere des glosieres uber Johannem Rubicissi.   Fol. 267.

Fünftes Buch : Micarium medicine, vel Thesaurus pauperum.   Fol. 283.

> The author is Hieronymus Saler, generally called Brunschwick, or Brunswieg.   The first edition of "Liber de arte distillandi de compositis" was published at Strassburg in 1500.   Fol.   Later editions are numerous.

BRUNSCHWICK, IHERONIMUS. [Cont'd.]
> An English edition has the title: Hieronimus Bruynswayke's virtuous Book of Distillation of the Waters of all manner of Herbs, with the figures of the Stillatories. Translated by Laurence Andrew. London, 1527. Fol.

BRUNTON, T. L., and J. T. CASH.
Contributions to our Knowledge of the Connection between Chemical Constitution, Physiological Action and Antagonism. London, 1884. 4to.
Part II under the title : Contributions to the Study of the Connection between Chemical Constitution and Physiological Action. London, 1891. 4to. Ill.

BRUYLANTS, G.
Chimie physiologique. Louvain, 1888. 8vo.

BUCCI, ANT.
Osservazioni circa il flogisto, e le differenti specie d'aria secondo le moderne scoperte. Pavia, *n. d.* 8vo.

BUCHHOLZ, J. A.
Grundriss der Chemie. Nürnberg, 1826.

BUCHKA, KARL.
Chemie (Die) des Pyridins und seiner Derivate. Unter Benutzung eines Manuscriptes des Arthur Calm herausgegeben. Braunschweig, 1889-91.
Lehrbuch der analytischen Chemie. Wien, 1891–92. 2 parts, 8vo. Ill.

BUCHNER, AUGUST WILHELM.
Neueste Entdeckungen über die Gerbsäure oder den sogenannten Gerbstoff. Gekrönte Preisschrift. Nebst einem Vorwort von Phil. Lorenz Geiger. Frankfurt am Main, 1833. 8vo.

BUCHNER, JOHANN ANDREAS.
Erster Entwurf eines Systems der chemischen Wissenschaft und Kunst. München, 1815.
Grundriss der Chemie. Nürnberg, 1830–36. 3 vols., 8vo.
> This also forms part of the author's Vollständiger Inbegriff der Pharmacie. Nürnberg, 1821–27. 7 vols.

Grundriss der Physik als Vorbereitung zur Chemie . . . München, 1825.
Toxikologie. Ein Handbuch für Aerzte und Apotheker sowie auch für Polizei und Kriminal–Beamten. Nürnberg, 1822.
> Zweite Auflage, 1827. 8vo.

BUCHNER, JOHANN ANDREAS [Editor].
    See, in Section VII, Repertorium der Pharmacie.

BUCHNER, LUDWIG ANDREAS.
    Betrachtungen über die isomeren Körper . . . Nürnberg, 1836.
    Neue chemische Untersuchung der Angelicawurzel. Nürnberg, 1842.
    Versuche über das Verhalten der Auflösungen chemischer Stoffe zu
      Reagentien bei verschiedenen Graden von Verdünnung, sowie über
      die Gränzen der Wahrnehmung chemischer Reactionen. Eine
      gekrönte Preisschrift. Nürnberg, 1843. 4to.

BUCHNER, G.
    Die Metallfärbung und deren Ausführung mit besonderer Berücksichti-
      gung der chemischen Metallfärbung. Berlin, 1891. 8vo.

BUCHOLZ, CHRISTIAN FRIEDRICH.
    Beiträge zur Erweiterung und Berichtigung der Chemie. Erfurt,
      1799–1802. 3 vols., 8vo.
    Chemische Analyse der Schwefelquellen des Güntherbades bei Sonders-
      hausen . . . Sondershausen, 1816.
    Gemachten (Die) Erfahrungen und Meinungen über die Darstellung
      des Zuckers aus Pflaumen. Leipzig, 1813.
    Theorie und Praxis der pharmaceutisch-chemischen Arbeiten. Leip-
      zig, 1812. 2 vols.
             Zweite Auflage, 1818.
    Versuche zur Bereitung des Zinnobers auf nassem Wege. Erfurt, 1801.

BUCHOLZ, CHRISTIAN FRIEDRICH [Editor].
    See, in Section VII, Almanach für Scheidekünstler.

BUCHOLZ, WILHELM HEINRICH SEBASTIAN.
    Chymische Versuche über das Meyerische Acidum pingue. Weimar,
      1771. 8vo.
        See Meyer, Johann Friedrich, *and cf.* Cranz ; *also,* Jacquin.
    Chymische Versuche über einige der neuesten einheimischen antisep-
      tischen Substanzen. Weimar, 1776.
    De sulphure minerali. Jenæ, 1762.
    Ueber die antiseptischen Kräfte des Wolverley ; Achard's Manier Berg-
      krystalle mittelst fixer Luft zu erzeugen . . . Erfurt, 1785.
    Ueber die vorgebliche giftige Eigenschaft des Witherits, der Schwererde
      und der salzsauren Schwererde. Weimar, 1792.

BUCKERIDGE, H. L.
    Chemical Student's Manual for the Lecture-room and Laboratory,
      embracing non-metals, the analysis of simple salts, historical and
      much other useful memoranda. London, 1886.

BUCKMASTER, J. C.

Elements of Inorganic Chemistry. Part I : Elementary Stage. Tenth
edition, revised by G. Jarmain and C. A. Buckmaster. London,
1873. 12mo.

Part II : Advanced Stage.. Revised by C. A. Buckmaster and G. Jarmain. Twelfth edition. London, 1880. pp. 217, 16mo.

BÜCHNER, PH. TH.

Lehrbuch der anorganischen Chemie nach den neuesten Ansichten der
Wissenschaft. Braunschweig, 1872. pp. xvi–964. 8vo. Ill.

Zweite verbesserte Auflage. Braunschweig, 1878, 8vo. Ill.

BÜLOW, K.

Beiträge zur Trennung des Quecksilbers von den Metallen der sogenannten Arsen- und Kupfergruppe. Göttingen, 1890. 8vo.

BÜRKNER, J.

Populäre Chemie und ihre Anwendung auf Gewerbe. Vorgetragen im
Gewerbe-Verein zu Breslau in die Jahre 1836–38. 9 Parts. Brieg,
1839 [?].

BUFF, HEINRICH L.

Grundlehren der theoretischen Chemie und Beziehungen zwischen den
chemischen und physikalischen Eigenschaften der Körper. Erlangen, 1866. 8vo.

Kurzes Lehrbuch der anorganischen Chemie entsprechend den neueren
Ansichten. Erlangen, 1868. 8vo.

Versuch eines Lehrbuchs der Stöchiometrie. Ein Leitfaden zur Kenntniss und Anwendung der Lehre von den bestimmten chemischen
Proportionen. Nürnberg, 1829, 8vo.

Zweite Auflage, Nürnberg, 1842. 8vo.

In the second edition the prefix " Versuch " is dropped.

BUFF, H., KOPP, H., UND F. ZAMMINER.

Lehrbuch der physikalischen und physischen Chemie. *See* Graham-Otto.

BUISINE, A.

Recherches sur la composition chimique du suint du mouton. Lille,
1887.

БУНГЕ, А.

Очерки неорганической химіи. Опытъ болѣе тѣснаго соединенія неорганиче
ской химіи съ органическою. Кіевъ, 1867.

BUNGE, A.   Outlines of inorganic chemistry.   An essay on a closer connection between organic and inorganic chemistry.   Kiev, 1867.

Бунге, Н. А.

О дѣйствіи солода на крахмалъ. Кіевъ, 1873.

> BUNGE, N. A. On the action of malt upon starch. Kiev, 1873.

BUNGE, G.

Lehrbuch der physiologischen und pathologischen Chemie. In einund-
zwanzig Vorlesungen für Aerzte und Studirende. Leipzig, 1887. 8vo.

> Zweite Auflage, 1889. pp. iv–404, 8vo.
>
> > Cours de chimie biologique et pathologique.[a] Traduit sur
> > la deuxième édition allemande par A. Jaquet. Paris,
> > 1891. 8vo.
> >
> > Text-book of Physiological and Pathological Chemistry. In
> > twenty-one lectures for physicians and surgeons. Trans-
> > lated from the second German edition by the late L. C.
> > Wooldridge. London, 1890. pp. xii–469, 8vo.

BUNSEN, ROBERT WILHELM.

Anleitung zur Analyse der Aschen und Mineralwasser. Heidelberg,
1874. 8vo. Ill.

> Zweite Auflage, Heidelberg, 1887.

Enumeratio ac descriptio hygrometrorum. Göttingæ, 1830.

Flammen-Reactionen. Heidelberg, 1880. 8vo.

> Zweite Auflage, Heidelberg, 1886.
>
> *Cf.* Wartha, V.
>
> > Essais microchimiques par la voie sèche, procédé de Bunsen.
> > Résumé à l'usage des laboratoires d'instruction par L.
> > De Koninck. Liège, 1885. 8vo. Ill.

Gasometrische Methoden. Braunschweig, 1857. 8vo. Ill.

> Zweite umgearbeitete und vermehrte Auflage. Braunschweig,
> 1877. 8vo.
>
> > Gasometry: comprising the leading properties of gases.
> > Translated from the German by Henry E. Roscoe.
> > London, 1857. 8vo.

Ueber eine volumetrische Methode von sehr allgemeiner Anwendbarkeit.
Leipzig, 1854. 8vo.

Untersuchungen über die Kakodylreihe (1837–1843). Herausgegeben
von A. von Baeyer. Leipzig, 1891. 8vo.

> Oswald's Klassiker der exacten Wissenschaften.

BUNSEN, ROBERT WILHELM, UND ARNOLD ADOLPH BERTHOLD.

Das Eisenoxydhydrat, ein Gegengift des weissen Arseniks oder der
arsenigen Säure. Göttingen, 1834. 8vo.

> Zweite vermehrte Auflage. Göttingen, 1837. pp. viii–128, 8vo.

BUNSEN, ROBERT WILHELM, AND HENRY E. ROSCOE.
Photochemical Researches. London, 1858–63. 5 parts, 4to, with 13 plates.

BURCKER, E.
Traité des falsifications et altérations des substances alimentaires et des boissons. Paris, 1892. pp. iii–474, 8vo. Ill. One folding table at page 118.

BURG, E. A. VAN DER.
Handleiding ter beoefening der qualitative chemische analyse. Tiel, 1874. 8vo.

BURGEMEISTER, A.
Das Glycerin, seine Geschichte, Eigenschaften, Darstellung, Zusammensetzung, Anwendung und Prüfung nebst den wichtigsten Zersetzungen und Verbindungen. Eine von dem Verein zur Beförderung des Gewerbefleisses in Preussen gekrönte Denkschrift. Berlin, 1871. 8vo.

BURGER.
Untersuchungen über die Möglichkeit und den Nutzen der Zuckerbereitung aus inländischen Pflanzen. Wien, 1811.

BURGMANN, ARTHUR.
Petroleum und Erdwachs. Darstellung der Gewinnung von Erdöl und Erdwachs (Ceresin), deren Verarbeitung auf Leuchtöle und Paraffin, sowie aller anderen aus denselben zu gewinnenden Producte, mit einem Anhang, betreffend die Fabrikation von Photogen, Solaröl und Paraffin aus Braunkohlentheer. Mit besonderer Rücksichtnahme auf die aus Petroleum dargestellten Leuchtöle, deren Aufbewahrung und technische Prüfung, Wien, 1880. Ill.

    Cf. Strippelmann.

BURMAN-BECKER, JOHANN GOTTFRIED.
Veiledning til at foretage chemiske Analyser. Kjøbenhavn, 1829. Ill.

BURNS, W.
Illuminating and Heating Gas ; a manual of the manufacture of gas from tar, oil and other liquid hydrocarbons, and extracting oil from sewage sludge. London, 1887.

BUSSY, ANTOINE ALEXANDRE BRUTUS, ET A. F. BOURTON-CHALARD.
Traité des moyens de reconnaître les falsifications des drogues simples et composées et d'en constater le degré de pureté. Paris, 1829. 8vo.

Bussy, Antoine Alexander Brutus, et A. F. Bourton-Chalard. [Cont'd.]

> Tratado de los medios de averiguar las fabricaciones de las drogas escrita en francés. Traducida al castellano par José Luis Casaseca. Madrid. 4to.

Бутлеровъ, А. М.

Основныя понятія химіи. С.-Петербургъ 1886.

> Butlerof, A. M. Fundamental notions of chemistry. St. Petersburg, 1886.

Бутлеровъ, А.

Введеніе къ полн. изученію орган. химіи. Казань, 1865—1866. 3 parts.

> Lehrbuch der organischen Chemie zur Einführung in das specielle Studium derselben. Aus dem russischen übersetzte ; Deutsche Ausgabe vom Verfasser revidirt und mit Zusätzen vermehrt. Leipzig, 1868. pp. xii-752, sm. fol.

Буттацъ, Францъ.

О фосфорѣ, его существѣ, его раствореніи и о его способѣ употреблять внутрь какъ лѣкарство ; опытами доказано. Курскъ, 1804.

> Buttatz, Frantz. Phosphorus, its occurrence, solution, and its medical use tested by experience. Kursk, 1804.

Buys-Ballot, Christoph Heinrich Diedrich.

Repertorium corporum organicorum . . . Trajecta ad Rhenum. 1846.

Tabulæ repertoriæ chemicæ. Schets eener physiologie van het onbewerktuigde rÿk der natuur. Utrecht, 1849.

Caccinni, P. Emmanuele.

Chimica teoretica ed industriale. Venezia, 1870. 3 vols., 16mo.

Cadet de Gassicourt, Charles Louis.

Chimie (La) domestique, ou Introduction à l'étude de cette science. Paris, 1801. 8 vols., 12mo.

Cadet de Gassicourt, Charles Louis.

> Dictionnaire de chimie. *See, in Section II.*

Cadet de Gassicourt, Louis Claude.

Analyses des eaux minérales de Passy. Paris, 1755.

Mémoire sur la terre foliée de tartre. Paris, 1764.

Réponse à plusieurs observations de M. Baumé sur l'éther vitriolique. Paris, 1775.

CADET DE VAUX, ANTOINE ALEXIS.

Aperçus économiques et chimiques sur l'extraction du sucre de betterave. Paris, 1812.

Sur la gélatine des os et son application. . . . Paris, 1803.

Traité du blanchissage domestique à la vapeur. Paris, 1805.

CAEN, LABORATOIRE DE CHIMIE DE. Bulletin des travaux de 1875. Caen, 1876. 8vo.

CAESALPIN, ANDREAS.

De metallicis libri tres. Romæ, 1596. *Also* Noribergæ, 1602. Sm. 4to.

CAHOURS, AUGUSTE ANDRÉ THOMAS.

Leçon (etc.). *See* Société chimique de Paris.

CAHOURS, AUGUSTE ANDRÉ THOMAS.

Leçons de chimie générale élémentaire professées à l'École centrale des arts et manufactures. Paris, 1855–56. 2 vols. Ill.

Deuxième édition, Traité de chimie (etc.). Paris, 1858–60. 3 vols., 12mo. Ill.

Quatrième édition. Paris, 1874–78. 6 vols., 12mo. Ill.

Курсъ элементарной общей химіи. Перев. Аверкіева и Ильина. С.-Петербургъ 1862—1863. 2 тома.

Lecciones de química general elementar, explicadas en la Escuela central de Artes y Manufacturas de París, y dedicadas á M. Chevreul. Traducidas por Ramon Ruiz Gomez. Madrid, 1856–57. 2 vols., 4to. Ill.

Segunda edicion. Madrid, 1859.

CAHOURS, AUGUSTE, ET ALFRED RICHE.

Chimie des demoiselles, leçons professées à la Sorbonne. Paris, 1869. 8vo. Ill.

Chimica delle signorine. Milano, 1883. 16mo.

Бесѣды о химіи. Перев. Жученко. С.-Петербургъ, 1874.

CAILLETET, CYRILLE.

Guide pratique de l'essai et du dosage des huiles employées dans le commerce ou servant à l'alimentation des savons et de la farine de blé. Manuel pratique à l'usage des commerçants et des manufacturiers. Paris, *n. d.* pp. 104, 12mo.

CAIRNS, FREDERICK A.

A Manual of Quantitative Chemical Analysis for the use of students. New York, 1879. 8vo.

New edition, revised and edited by Elwyn Waller. New York, 1881. pp. viii–279, 8vo. Ill.

CALDWELL, GEORGE C.

Agricultural Qualitative and Quantitative Chemical Analysis, after E. Wolff, Fresenius, Krocker and others. New York, 1869. 8vo.

CALDWELL, GEORGE C., AND A. BRENEMAN.

Manual of Introductory Chemical Practice for the use of students in Colleges and High Schools. Second edition, revised and corrected. New York, 1878. pp. xlvii–123, 8vo. Ill.

CALM, ARTHUR.

      See Buchka, K.

CALMELS, G., ET G. SAULNIER.

Guide pratique du fabricant de savons. Savons communs, savons de toilette mousseux, transparents, médicinaux, pâtes et émulsions, analyse des savons. Paris, 1887. pp. 244–[iv], 12mo. Ill.

CAMERON, CHARLES ALEXANDER.

Chemistry (The) of Agriculture : the foods of plants, including the composition, properties and adulteration of manures. Dublin, 1857. 8vo.

Chemistry (The) of Food in relation to the breeding and feeding of live stock. London, 1868. 8vo.

CAMERON, JAMES.

Oils, Resins and Varnishes. Their Chemistry, Manufacture and Uses. London, 1886. Ill.

      Also Philadelphia, 1886.

Soaps and Candles. London, 1888. pp. ix–306, 12mo.

      Churchill's Technological Handbooks.

CAMPANA, ANTONIO FRANCESCO.

Lezioni di fisica e chimica per gli anni 1796–1802. Ferrara [?].

CAMPANI, R.

Sommario di chimica generale : parte speciale. Pisa, 1880. 8vo.

CAMPANO, L.

Manual del cervecero y fabricante de bebidas gaseosas y fermentadas, obra extractada de los mejores métodos modernos. Segunda edicion. Paris, 1881. 12mo.

CAMPARI, G.

Analisi chimica dell' urina, sedimenti e calcoli, colla descrizione degli utensili e dei reattivi necessarii per eseguirla. Parma, 1885.

Studii sperimentali sulla distillazione secca dei legni resinosi per la produzione del catrame, della fabbricazione di olii di resina illuminanti, e sopra una nuova lampada per la combustione degli olii di resina molto carburati. Parma, 1884. 8vo.

CAMPBELL, DUGALD.
A practical Text-Book of Inorganic Chemistry, with qualitative and quantitative analysis. London, 1849. 8vo.

CAMUS (ET MARGUERITTE).
Industrie du gaz. *See, in Section II,* Fremy : Encyclopédie chimique, vol. v.

CANDLOT, E.
Étude pratique sur le ciment de Portland, fabrication, propriétés, emploi. Paris, 1886. 8vo. Ill.

CANEFRI, CESARE NICCOLO.
Lezioni di chimica applicata alla farmacia. Milano, 1793.

CANEPARIUS, PETRUS MARIA.
* De atramentis cujuscunque generis, opus sane novum, hactenus à nemine promulgatum ; in sex descriptiones digestum. Londini, excudebat J. M. impensis Jo. Martin, Ja. Alestry, Tho. Dicas, apud quos veneunt ad insigne Campanæ, in cœmeterio Paulino, 1660. pp. [xvi]-568, sm. 4to.

CANESTRINI, E.
Sedici lezioni di chimica per i Licei. Verona, 1886. 8vo.

CANNIZZARO, STANISLAO.
Lezioni sulla teoria atomica. Genova, 1858.
Обзоръ развитія понятій объ атомѣ, частицѣ и эквивалентѣ и различныхъ системъ формулъ. Перев. Алексѣева. Кіевъ, 1873.
Sunto di un corso di filosofia chimica fatto nella regia Università di Genova ; e nota sulle condensazioni di vapore. Roma, 1880. pp. 80, 8vo.
Abriss eines Lehrganges der theoretischen Chemie, vorge-tragen an der Universität Genua (1858). Herausge-geben von L. Meyer. Leipzig, 1891. 8vo.
Ostwald's Klassiker der exakten Wissenschaften.
О предѣлахъ и о формѣ теоретическаго преподаванія химіи. Перев. Алексѣева. Кіевъ, 1873.

CANOBBIO, GIANBATTISTA.
Manuale di chimica. . . . Genova, 1835-37. 2 vols.

CANTALUPI, A.
Dell' acqua potabile. Trattato pratico elementare sulla ricerca, deriva-zione, condotta e distribuzione dell' acqua nelle città e nelle borgate pel servizio pubblico e privato. Milano, 1891. 8vo.

CAPAUN–KARLOWA, C. F.

Chemisch-technische Specialitäten und Geheimnisse mit Angabe ihrer Zusammensetzung nach den bewährtesten Chemikern. Alphabetisch zusammengestellt. Wien, 1878. 8vo.

Unsere Lebensmittel, eine Anleitung zur Kenntniss der vorzüglichsten Nahrungs- und Genussmittel, deren Vorkommen und Beschaffenheit in gutem und schlechtem Zustande, sowie ihre Verfälschungen und deren Erkennung. Wien, 1879.

CAPDEVILA, RAMÓN.

Lecciones de los principios de química que se deben explicar á los alumnos del real Colegio de Medicina y Cirugia de San Carlos. Madrid, 1831. 8vo.

CAPEZZUOLI, SERAFINA.

Trattato di chimica organica applicata alla medicina e più specialmente alla fisiologia, alla patologia, alla pratica medico-chirurgica, all' igiene, alla medicina forense. Firenze, 1855–60. 2 vols., 8vo. I, pp. 789; II, pp. 349.

CAPPANERA, R.

Elementi di fisica e chimica. Napoli, 1887. 16mo.

CAPRIA, DOMENICO MAMONE.

     See, in Section II, for Nomenclature.

CAPRIA, DOMENICO MAMONE.

Elementi di chimica filosofico-sperimentale. Napoli, 1841. 8vo.
     Quarta edizione, Napoli, 1844. 8vo.

CAPRON, J. RAND.

Photographed Spectra. One hundred and thirty-six photographs of metallic, gaseous and other spectra printed by the permanent autotype process, with introduction, description of plates and index, and with an extra plate of the solar spectrum (showing bright lines) compared with the air spectrum. London, 1877. pp. 84, 8vo. 37 plates.

CARBONELL Y BRAVO, FRANCISCO.

Pharmaciæ elementa chemiæ recentioris fundamentis innixa. Barcelonæ, 1796.

CARDAN, JÉROME.

* Les livres de Hierome Cardanus de la subtilité, et subtiles inventions, ensemble les causes occultes et raisons d'icelles, traduis de

CARDAN, JÉROME. [Cont'd.]

Latin en Françoys par Richard le Blanc. Paris, par Claude Micard, rue S. Jaques, à l'enseigne de la coupe d'Or, 1566. Ff. [lxxii]–478, 16mo. Ill.

> This edition is based on the text of 1554 and contains the paragraphs that had been condemned. The complete works of Cardanus were published in 10 folio volumes at Leyden, 1663, and embrace 408 treatises, mathematical, philosophical, medical, etc.

CARDILUCIUS, JOHANNES HISKIAS.

* Königlicher chymischer und artzneyischer Palast, worin über das weltberühmte Buch genannt Basilica chymica : Eine durch alle Capitel des gantzen Werckes vollständige Vermehr- und Erläuterung gestellet, und diejenige hohe Secreta als Laudanum Mercuriale, und andere, welche bisher in allen Exemplarien gedachter Basilicæ Crolliano Hartmannianæ ausgelassen worden, aus des Authoris Manuscript treulich ersetzt werden nebenst offenhertziger Communication vieler spagyrischer und artzneyischer Secreten. Alles dem deutschen Vaterlande zu Dienst und Gefallen deutsch und deutlich publiciret. Nürnberg, 1684. pp. 1010–xxii. 3 parts. Frontispiece and ill.

CAREY, G. G.

Five hundred useful and amusing Experiments in Chemistry, and in the Arts and Manufactures ; with observations on the properties of the substances employed and their application to useful purposes. New edition. London, 1825. 18mo. Plates.

CARL, ANTON JOSEPH.

De igne et gravitate calcis metallicæ. Ingolstadt, 1772.

De oleis. Ingolstadt, 1760.

Zymotechnia vindicata et applicata. Ingolstadt, 1759.

CARL, JOHANN SAMUEL.

De analysi chemico-medica reguli antimonii medicinalis. Halæ, 1698.

Ichnographia chymiæ fundamentalis, ex specimine Stahliano doctrinæ Beccherianæ in compendium redacto, collecta cura et usu auditorii privati. Budinguæ, 1722. 16mo.

Lapis Lydius philosophico-pyrotechnicus ad ossium fossilium docimasiam analytice demonstrandam adhibitus, et per multa experimenta chymico-physica in lucem publicam missus. Francofurti ad Mœnum, 1704.

CARLES, P.

Influence exercée sur les réactions chimiques par les agents physiques autres que la chaleur. Paris, 1880. 4to.

CARLEVARIS, PROSPERO.

Corso elementare di chimica moderna, adorno di molte tavole illustrative. Torino, 1871–73. 3 vols, 8vo, and 3 vols. of plates.

Lezioni di chimica applicata all' agricoltura, dette nel Regio Istituto Tecnico di Torino, 1853. 8vo.

CARLSSON, C. R.

Enkla kemiska och fysikaliska försök för folkskolor. Upsala, 1874. 8vo.

CARNOT.

Les laboratoires de chimie [and other articles]. *See, in Section II*, Fremy : Encyclopédie chimique.

CARPENTER, W. L.

A Treatise on the Manufacture of Soap and Candles, Lubricants and Glycerin. London, 1885.

CARRACIDO, JOSÉ R.

La nueva química. Introduccion al estudio de la química, según el consepto mecánico. Madrid, 1887.

Tratado de química orgánica teórico y práctico aplicado especialmente á las ciencias médicas. Madrid. *n. d.* [1891]. pp. 924. Ill.

CARTHEUSER, FRIEDRICH AUGUST.

Dissertationes physico-chemico-medicæ. Francofurti, 1774.

Elementa chemiæ-medicæ dogmatico-experimentalis ; una cum synopsi materiæ medicæ selectioris. Halæ Magdiburgicæ, 1736. 8vo.

> Editio secunda. Francofurti, 1753.
> Other editions. Halæ, 1763 ; 1766.

Rudimenta hydrologiæ systematicæ. Francofurti i. O., 1758.

Ueber die Verfälschung der Weine. Giessen, 1779.

Vermischte Schriften aus der Naturwissenschaft, Chymie und Arzneygelahrtheit. Leipzig und Magdeburg, 1759.

CARTUYVELS, JULES, F. RÉNOTTE ET E. RÉBOUX.

La diffusion et les procédés récents de fabrication du sucre au moyen de l'osmose ou de la séparation. Paris, 1884. 8vo.

CASALI, ADOLFO.

Corpi (I) indecomposti e le loro principali proprietà fisiche e chimiche. Quadro sinottico redatto per uso degli studenti dell' istituto tecnico di Bologna. Bologna, 1870. 8vo. Ill.

Elementi di chimica ad uso degli studenti. Bologna, 1888. 16mo.

Elementi di chimica generale. Seconda edizione, riveduta ed ampliata, con un sommario storico della scienza. Bologna, 1888. pp. 241, 16mo.

CASALI, ADOLFO. [Cont'd.]

Principi fondamentali di chimica inorganica, chimica organica e analisi minerale qualitativa. Bologna, 1876. Ill.

Ragguagli sui lavori eseguiti nel laboratorio chimico agrario di Bologna. Bologna, 1885. 16mo.

CASARES, ANTONIO.

Manual de química general con aplicacion á la industria, y con especialidad á la agricultura. Madrid, 1857. 2 vols., 8vo.

Tercera edicion, Madrid, 1873. 2 vols., 4to.

Tratado elementar de química general. Madrid, 1848. 2 vols., 8vo. Ill.

Tratado práctico de análisis química de las aguas minerales y potables, con indicacion de las fuentes de aguas minerales más notables de España, su composicion, enfermedades á cuya curacion se aplican y número de enfermos que á ellas acuden. Madrid, 1867. 4to.

CASARTELLI, LUDWIG.

Ausführliches Handbuch der Fabrikation des Salmiaks, und der in Verbindung damit zu erzielenden Nebenproducte. Enthaltend sämmtliche älterer und neuere Darstellungs–Methoden nebst Anleitung, den Salmiak auf die wohlfeilste Weise ohne Anwendung der bisher verwendeten stickstoffhaltigen Substanzen darzustellen. Quedlinburg, 1850. 8vo. Ill.

Blausalz-Fabrikation (Die). Ausführliche Anleitung zur Darstellung des eisenblausauren Kalis im Grossen nach allen dazu angegebenen Methoden [etc.]. Quedlinburg, 1849. 8vo.

CASSAL, CHARLES E.

The Extension of Public Analysis. London, 1890. 8vo.

CASSELMANN, ARTHUR.

Die Analyse des Harns in Fragen und Antworten für Mediciner und Pharmaceuten zusammengestellt. Zweite vermehrte Auflage, Giessen, 1874. 8vo.

Guide pour l'analyse de l'urine, des sédiments et des concrétions urinaires. Paris, 1873. 8vo.

КАССЕЛЬМАНЪ, АРТУРЪ.

Краткое руководство судебной химіи для фармацевтовъ, врачей и судебныхъ слѣдователей. С.-Петербургъ, 1871.

CASSELMANN, ARTHUR. Short manual of legal chemistry for the use of chemists and druggists, doctors and coroners. St. Petersburg, 1871.

CASSELMANN, ARTHUR, AND CARL FREDERKING.

Lehrbuch der gesammten Pharmacie und ihrer Hilfswissenschaften für Apotheker und Aerzte. Riga, 1869–70. 8vo. 2 parts.

CASSELMANN, WILHELM THEODOR OSCAR.

Leitfaden für den wissenschaftlichen Unterricht in der Chemie. Für
Realgymnasien, Gymnasien, Realschulen, und zum Selbstunterrichte. Fünfte Auflage bearbeitet von G. Krebs. Wiesbaden,
1886. 8vo.

First edition, Wiesbaden, 1847–50.

CASSIUS, ANDREAS.

De extremo illo et perfectissimo naturæ opificio ac principe terrenorum
sidere, auro, et admiranda ejus naturæ generatione, affectionibus,
effectis atque ad operationes artis habitudine, cogitata, experimentis
illustrata. Hamburgi, 1685.

> Contains the discovery of the so-called Purple of Cassius (a mixture of the
> oxides of gold and tin), first prepared by the Father of the author, also
> named Andreas.

CASTELEIN, PETRUS JOHANNES.
See Kasteleyn, P. J.

CASTELL-EVANS, JOHN.

A New Course of Experimental Chemistry, including the principles of
qualitative and quantitative analyses. Being a systematic series of
experiments and problems for the laboratory and class-room.
London, 1888.

CASTELLI, PIETRO.

Chalcantinum dodecaporion sive duodecim dubitationes in usum olei
vitrioli. . . . Romæ, 1619.

> Breve ricordo dell'elettione, qualità e virtù dello spirito e
> dell' olio acido del vitriolo. Romæ, 1621.

Responsio chymiæ de effervescentia et mutatione colorum in mixtione
liquorum chymicorum. Messinæ, 1654.

CASTILLON, A.

Récréations chimiques. Ouvrage illustré de 34 vignettes par H. Castelli
et faisant suite aux récréations physiques du même auteur. Paris,
1866. 16mo.

> Troisième édition, Paris, 1874. 16mo.

CASTRO Y DIAZ, LUIS DE.
Prácticas de química. Madrid, 1858. 4to. Ill.

CATECISMO DE QUÍMICA con una estampa que representa los utensilios
químicos. Londres, 1824. pp. iv–116, 24mo.

CATINELLI, CARL.
   Kritische Bemerkungen über F. X. Hlubek's Beleuchtung der organischen
      Chemie des J. Liebig.   Wien, 1843.   8vo.

CATTANEO, ANTONIO.
   Il latte e suoi prodotti.   Milano [?], 1839.
   Del tabacco.   Milano [?], 1843.

CATTANEO, ANTONIO [Editor].
      See, in Section VII, Giornale di farmacia, chimica (etc.).

CATULLO, TOMMASO.
   Quesiti di chimica e di storia naturale scelti dal corso di lezioni date
      dall' autore l'anno 1816, nell'i. r. liceo convitto di Verona, 1816.
   The same for 1817.

CAUCHY, PHILIPPE FRANÇOIS.
   Principes généraux de chimie inorganique, avec un Tableau synoptique
      des corps inorganisés d'origine inorganique.   Bruxelles, 1838.   8vo.

CAUDA, VALERIO.
   Laboratorio di chimica docimastica presso la scuola d'applicazione per
      gl'ingegneri di Torino.   Riassunto delle analisi dei minerali eseguite
      negli anni 1861-62-63-64-65-66-67-68 d'ordine del governo ed a
      richiesta dei privati.   Torino, 1869.   8vo.

CAUVET, D.
   Procédés pratiques pour l'essai des farines.   Caractères, altérations,
      falsifications, moyens de découvrir les fraudes.   Paris, 1886.   pp.
      97, 12mo.   Ill.

CAVALLO, TIBERIUS.
   * A Treatise on the Nature and Properties of Air and other permanently
      elastic fluids, to which is prefixed an introduction to Chymistry.
      London, 1781.   pp. viii–835, 4to.   Ill.
         Dedicated to Sir Joseph Banks.

            Abhandlung über die Natur und Eigenschaften der Luft,
               und der übrigen beständig elastischen Materien nebst
               einer Einleitung in die Chymie.   Aus dem englischen
               übersetzt mit drey Kupfertafeln.   Leipzig, 1783.   pp.
               [viii]–758–[xvi], 12mo.   3 folding plates.

CAVANNA VIANI, MARIA.
   Lezioni di scienze naturali per uso delle scuole ginnasiali, tecniche e
      normali secondo gli ultimi programmi.   Parte I. Elementi di
      fisica.   Parte II.   Elementi di chimica, mineralogia e geologia.
      Milano, 1891.   8vo.   Ill.

CAVENDISH, HENRY.
Expériences sur l'air. Mémoire lu à la Société Royale le 15 Janvier, 1784. Londres, 1785. pp. viii–68, 12mo.
> Translated from the Phil. Trans. for 1784.

CAVENDISH SOCIETY, PUBLICATIONS OF. London, 1848–72. 29 vols., 8vo., and 4to Atlas. [Containing :]
(1) Gmelin's Handbook of Chemistry. *See* Gmelin, Leopold.
(2) Lehmann's Physiological Chemistry. *See* Lehmann, Carl Gotthelf.
(3) Bischof's Elements of Chemical and Physical Geology. *See* Bischof, Carl G. C.
(4) Laurent's Chemical Method. *See* Laurent, Auguste.
(5) Life and Works of Henry Cavendish. *See, in Section IV*, Cavendish, H.; Wilson, George.
(6) Graham's Chemical Reports and Memoirs. *See* Graham, Thomas.
(7) Life and Scientific Researches of John Dalton. *See, in Section IV*, Dalton, John ; Henry, W. C.
(8) Funk's Atlas of Physiology. *See* Funk, Otto.

CAVENTOU, J. B.
> Nouvelle nomenclature chimique. *See, in Section II*, Caventou, J. B.

CAZENEUVE, P.
La coloration des vins par les couleurs de la houille. Méthodes analytiques et marche systématique pour reconnaître la nature de la coloration. Paris, 1886. 12mo.

CAZENEUVE, P.
> Conférence (etc.). *See* Société chimique de Paris.

CECH, KAREL.
Puvôd k chmelařstvi a pivovarstvi. v Praha, 1884.

CECCHI, F.
Nozioni elementari di chimica ad uso dei Licei, ovvero breve introduzione allo studio di questa scienza secondo le teorie moderne. Firenze, 1869. 12mo.
> Another edition, Firenze, 1875.
> Quarta edizione, Firenze, 1880.
> Quinta edizione con appendice di G. Giovannozzi. Firenze, 1888. 16mo.

CELLIO, MARCO ANTONIO.
Il fosforo, ovvero la pietra Bolognese preparata per rilucere frà l'ombre. Roma, 1680. pp. 102, 18mo. Illuminated title-page.

CENTNER, B.

Die Praxis der Bleicherei und Appretur, Baumwolle, Leinen, Jute im Rohzustande, sowie als Garn und Gewebe. Leipzig, 1886. 8vo.

CENTROPHILUS, MARTIN.

Die neuesten und wichtigsten mechanischen und chemischen Erfindungen für Ganzgelehrte, Halbgelehrte und Ungelehrte, für Physiker, Chemiker, Mechaniker, Technologen, Oekonomen Cameralisten, Polizeibeamte, Künstler, Fabrikanten, Draisinenlaufer, Wassertreter, Luftschiffer und Ziegelstreicher. In einer Reihe von Briefen aus Japan von C. M. Herausgegeben von dessen Freunde Peter Punktophilus. Erste Hälfte. Frankfurt a. d. Oder, 1818. pp. xvi–80, 16mo.

CERESOLI, FEDERICO.

Principi elementari di chimica. Milano, 1855. 8vo.

CERTES, A.

Analyse micrographique des eaux. Paris, 1883. 8vo. Ill.

CESALPINO, ANDREAS.

See Caesalpin, Andreas.

CESARI, G.

Manuale di farmacologia pratica ed analitica ovvero delle alterazioni ed adulterazioni dei medicamenti. Napoli, 1883. 8vo.

CEULEN, J. A.

De filter met meervoudige uitwerking ter vergadering van het beenzwart. Gevolgd door eene praktische handleiding voor de toepassing der diffusie en eenige praktische wenken met betrekking tot de suikerfabricatie. Amsterdam, 1885.

CHALLENGER EXPEDITION, CHEMISTRY OF.

See Report on the Scientific Results [etc.].

CHALON, J.

Quelques expériences de chimie. Verviers, 1884. 12mo.

CHALON, PAUL F.

Traité théorique et pratique des explosifs modernes et dictionnaire des poudres et explosives. Seconde édition revue et augmentée. Paris, 1889. pp. xi–507.

Première édition Paris, 1886.

CHAMBERT, J. F. R.

Chimie usuelle répondant aux nouveaux programmes officiels. Deuxième
édition, entièrement refondue et augmentée. Ouvrage destiné à
tous les établissements d'enseignement primaire supérieur et
secondaire. Paris, 1884. 18mo. Ill.

Notions de chimie usuelle, développées par 700 problèmes nouveaux et
317 expériences faciles nécessitant peu d'appareils. Lille, 1875.
12mo.

Solutions des 700 problèmes de chimie usuelle. Ouvrage utile aux
instituteurs d'enseignement secondaire, lycées pensionnats, etc.
Paris, 1876. 12mo.

CHAMPION, P.

La dynamite et la nitroglycérine, historique, préparation, propriétés,
emploi, modes d'explosion, appareils électriques, applications à la
guerre et à l'industrie. Torpilles. Paris, 1872. pp. viii–249.
16mo. Ill.

CHANCEL, G.

Cours élémentaire d'analyse chimique, à l'usage des médecins, des
pharmaciens et des aspirants aux grades universitaires. Paris, 1851.
18mo.

> *Cf.* Gerhardt, C. F.

CHANDELET.

Art du raffineur ou traité théorique et pratique du sucre de cannes.
Paris, 1828. 12mo.

CHANDELON, T.

Traité de toxicologie et de chimie légale appliquée aux empoisonnements. Liège, 1888. pp. 462, 8vo.

CHANDLER, CHARLES F.

Miscellaneous Chemical Researches. Inaugural dissertation. Göttingen,
1856. 8vo.

CHANSAREL.

Nouvelle doctrine chimique, suivie d'une dissertation sur les poisons et
contre-poisons, et proposition de nouveaux moyens de traiter de
l'empoisonnement, avec des observations sur la toxicologie générale
de M. Orfila. Paris, 1824. pp. 200, 8vo.

> In this work the author, a pharmaceutist of Bordeaux, regards water as an
> elementary body and the metals as compounds. He claims that the
> hydrogen evolved upon dissolving a metal in water comes from the
> metal, although he admits acquaintance with the writings of Cavendish,

CHANSAREL. [Cont'd.]
> Lavoisier, Fourcroy, Thénard, Vauquelin, Gay Lussac, Berzelius, Berthollet and others ! His toxicological memoir is controversial.

Nueva doctrina química . . . traducida por Juan Aledo. Barcelona, 1826. 8vo.

CHAPPUIS, J.
Étude spectroscopique sur l'ozone. Paris, 1882. 4to.

CHAPTAL DE CHANTELOUP, JEAN ANTOINE CLAUDE.
Art (L') de la teinture du coton en rouge. Paris, 1807.
Art (L') de faire, de gouverner et de perfectionner les vins. Paris, 1801. 8vo.
> Deuxième édition, Paris, 1819.
> Troisième édition, Paris, 1839.

Art (L') des principes chimiques du teinturier dégraisseur. Paris, 1808.
Chimie appliquée à l'agriculture. Paris, 1823. 2 vols., 8vo.
> Deuxième édition, Paris, 1829. 2 vols., 8vo.
>> Chymistry applied to Agriculture. First American, translated from the second French edition. Boston, 1835. pp. xl–365, 12mo.
>> Chymistry applied to Agriculture. With a preliminary chapter on the organization, structure, etc., of plants by Sir Humphry Davy, and an Essay on the use of lime as a manure by M. Puvis, with introductory observations to the same by James Renwick. Translated and edited by William P. Page. New York, 1840. pp. 359, 18mo.

Chimie appliquée aux arts. Paris, 1807. 4 vols., 8vo. Folding plates.
Vol. I, pp. lxxix–302, 10 plates; vol. II, pp. viii–544, 1 plate ; vol. III, pp. viii–534, 1 plate ; vol. IV, pp. viii–554.
> Chemien anvendt paa Kunster og Næringsdrift. En Oversættelse gjennemseet og forsynet med Anmærkninger ved Hans Christian Ørsted. Kjøbenhavn, 1820.
> Chemistry applied to Arts and Manufactures. London, 1807. 4 vols., 8vo. I, pp. lxviii–259, folding plates ; II, xii–448, plates ; III, xvi–512, plates ; IV, xvi–520–xx.
> Die Chemie in ihrer Anwendung auf Künste und Handwerke. Aus dem französischen übersetzt mit Anmerkungen begleitet von Sigismund Friedrich Hermbstädt. Berlin, 1808. 4 parts in 2 vols. 8vo.
> Química aplicada á las artes. Traducida del francés por Francisco Carbonell y Bravo. Barcelona, 1816. 4 vols., 4to. Ill.

CHAPTAL DE CHANTELOUP, JEAN ANTOINE CLAUDE.   [Cont'd.]

Élémens de chymie.   Paris, 1790.   3 vols., 8vo.

> Deuxième édition, Paris, An III [1794].   3 vols., 8vo.
>
> Troisième édition, Paris, 1796.   3 vols., 8vo.   I, pp. xcii–361 ; II, 448 ; III, 495.
>
> Quatrième édition, Paris, 1803.
>
> Elements of Chemistry, translated from the French. London, 1791.
>
> Second American edition, Philadelphia, 1801.   3 vols., 8vo.
>
> Third American edition, Boston, 1806.   3 vols., 8vo.
>
> Fourth American edition by James Woodhouse.   Philadelphia, 1807.
>
> Anfangsgründe der Chemie aus dem französischen übersetzt von Friedrich Wolff, nebst einer Vorrede von Sigismund Friedrich Hermbstädt.   Königsberg, 1791–1805.   4 vols., 8vo.
>
> Elementi di chimica tradotti da N. Dallaporta.   Venezia, 1792.   5 vols.
>
> Elementos de química, traducidos por Higinio Antonio Lorente.   Madrid, 1793–94.   3 vols., 4to.
>
>> Supplemento, por Juan Manuel Munarriz.   Madrid, 1801.   4to.

Essai sur le blanchiment.   Paris, 1801.

Tableau analytique du cours de chymie, fait à Montpellier.   Montpellier, 1783.   pp. [iv]–209, 12mo.

Tableau des principaux sels terreux . . .   Paris, 1798.

Traité des salpêtres et goudrons.   Montpellier, 1796.   8vo.

CHARAS, MOISE.

Pharmacopée royale galénique et chimique.   Paris, 1676.

> Later editions were edited by Nicolas Lémery, Paris, 1697, 4to, and 1706.
>
> Royal pharmacopœia according to the practice of the physicians of France, by Moses Charras.   Englished. London, 1678.   6 plates.

CHARLAND, LOUIS MARTIN.

> See Bayen, Pierre : Recherches chimiques.

CHARLES, T. CRANSTOUN.

Elements of Physiological and Pathological Chemistry.   London, 1884. Ill.   Another edition, 1885.

> The Elements of Physiological and Pathological Chemistry. A Handbook for medical students and practitioners, containing a general account of nutrition, foods and

CHARLES, T. CRANSTOUN. [Cont'd]
> digestion, and the chemistry of the tissues, organs, secretions and excretions of the body in health and disease. Together with the methods for preparing or separating their chief constituents, as also for their examination in detail, and an outline syllabus of a practical course of instruction for students. Philadelphia, 1884. pp. 463, roy. 8vo. Ill.

CHASTAING, P.
> Alcaloïdes naturels ; Matières albuminoïdes ; Radicaux organo-métalliques.
> *See, in Section II*, Fremy : Encyclopédie chimique, vol. VIII.

CHASTAING, P.
Action de l'air et de la lumière sur les médicaments chimiques. Paris, 1879.
Étude sur la part de la lumière dans les actions chimiques. Paris, 1877. 4to.

CHASTAING, P. ET E. BARILLOT.
Chimie organique. Essai analytique sur la détermination des fonctions. Méthodes générales pour la séparation, la caractérisation et l'analyse des composés organiques. Paris, 1888. 16mo.

CHATEAU, THÉODORE.
Traité complet des corps gras industriels, contenant l'histoire des provenances, du mode d'extraction, des propriétés physiques et chimiques, du commerce des corps gras [etc.]. A l'usage des chimistes, des pharmaciens, des parfumeurs, des fabricants d'huile [etc.]. Paris, 1862. 18mo.
> Guide pratique de la connaissance et de l'exploitation des corps gras industriels, contenant l'histoire des provenances, des modes d'extraction, des propriétés physiques et chimiques, du commerce des corps gras ; des altérations et des falsifications dont ils sont l'objet, et des moyens anciens et nouveaux de reconnaître ces sophistications. Deuxième édition. Paris, 1864. pp. xiv-27.
Die Fette. [Translated by Hartmann.] Leipzig, 1864. 8vo.

CHATIN, AD.
Existence de l'iode dans toutes les plantes d'eau douce. Conséquences de ce fait pour la géognosie, la physiologie végétale, la thérapeutique, et peut-être pour l'industrie. Paris, 1851. 12mo.

CHAUDRON-JUNOT, J.

Argyrolithe (argent de pierre). Réduction, par voie électro-chimique et par les procédés brevetés en France et dans les pays étrangers, des métaux : le silicium, le tungstène, le molybdène, le chrome, le titane, l'aluminium, le magnesium, l'urane, considérés jusqu'ici comme irréducibles et sur leur application dans les arts et l'industrie. Paris, 1855. 8vo.

Notice sur la réduction, par voie électro-chimique, du silicium, du tungstène, d'aluminium et des autres métaux considérés jusqu'ici comme irréducibles, et sur leur application aux arts et à l'industrie. Paris, 1855. 8vo.

CHAULNES, MARIE JOSEPH LOUIS D'ALBERT D'AILLY, DUC DE.

Nouvelle méthode de saturer d'air fixe à la fois et en moins d'une minute 30 pintes d'eau . . . Paris, 1778.

CHAUVEL.

Essai de déontologie pharmaceutique, ou traité de pharmacie professionelle, précédé d'un histoire de la pharmacie en France, et suivi de quelques réflexions sur les principes généraux qui doivent servir de base à sa réorganisation. Saint-Brieuc, 1854. 8vo.

CHEEVER, BYRON W.

Laboratory Notes on Quantitative Analysis. Part I. Ann Arbor, 1885. 12mo.

Laboratory Notes on some Select Methods in Quantitative Analysis. Part II. Ann Arbor, 1885.

Select Methods in Quantitative Analysis. Arranged from Professor Cheever's manuscript by Frank Clemes Smith. Parts I and II. Second edition. Ann Arbor, 1888. pp. 100, 12mo.

CHEMIA POLICYJNO-PRAWNA, wydana przez Radę lekarską Królestwa Polskiego (redakcyi Członka Rady J. Bełzy). Warszawa, 1844. 8vo.

CHEMIA ROLNICZA. Z przedmową K. G(arbińskiego), pod tegoż kierunkiem, sposobem popularnym wyłożona przez Wład. G(arbinskiego). Warszawa, 1846. 8vo. (Nakład redakcyi Roczników gospodarstwa krajowego.)

CHEMISTRY (THE) OF ARTIFICIAL LIGHT ; including the history of wax, tallow and sperm candles and the manufacture of gas, their various illuminating powers compared with animal and vegetable oils and a descriptive sketch of lamps and other apparatus. London, 1856. 8vo.

CHEMISTRY. A Manual for Beginners. " Forti et fideli nihil difficile."
London, 1886. 24mo.

Химическіе аппараты, инструменты и реактивы для высшихъ и среднихъ
учебныхъ заведеній, фабрикъ и проч. изготовляемые при мастерской
Гедвилло. Москва 1869.
> Chemical apparatus, instruments and reagents for high and middle schools,
> etc., prepared in the workshop of Hedvillo. Moscow, 1869.

CHEUVREUSSE, N.
Cours de chimie appliquée aux arts militaires. Metz, 1823.

CHEVALLIER, JEAN BAPTISTE ALPHONSE, ET O. REVEIL.
Notice sur le lait : les falsifications qu'on fait subir ; instructions sur
les moyens à employer pour les reconnaître. Avec des tables de
corrections pour le lait écrèmé et pour le lait avec sa crème.
Paris, 1856. 8vo.

CHEVALLIER, JEAN BAPTISTE ALPHONSE, ET J. BARSE.
Manuel pratique de l'appareil de Marsh, ou guide de l'expert toxicolo-
giste dans la recherche de l'antimoine et de l'arsenic, contenant un
exposé de la nouvelle méthode Reinsch applicable à la recherche
médico-légale de ces poisons. Paris, 1843. 8vo.

CHEVALLIER, JEAN BAPTISTE ALPHONSE, ET A. CHEVALLIER FILS.
Recherches chronologiques sur les moyens appliqués à la conservation
des substances alimentaires de nature animale et de nature végétale.
Paris, 1858. 8vo.

CHEVALLIER, JEAN BAPTISTE ALPHONSE, ET ANSELME PAYEN.
> Traité élémentaire des réactifs. *See* Payen, Anselme, et A. Chevallier.

CHEVREUL, MICHEL EUGÈNE.
Considérations générales sur l'analyse organique et sur ses applications.
Paris, 1824. 8vo.
Leçons de chimie appliquée à la teinture. Paris, 1829–30. 2 vols., 8vo.
> Each vol. contains 15 " leçons " separately paginated.

Recherches chimiques sur les corps gras d'origine animale. Paris,
1823. pp. xvi–484, 8vo.
> This classical work was republished in 1889 as follows :
>> Recherches chimiques sur les corps gras d'origine animale,
>> précédées d'un avant propos de M. Arnaud. Nouvelle
>> édition. Paris, 1889. 4to. Ill.

CHEYNE, W. WATSON, W. H. CORFIELD AND CHARLES E. CASSAL.
Public Health Laboratory Work. Printed and published for the Ex-
ecutive Council of the International Health Exhibition, and for the
Council of the Society of Arts. London, 1884. pp. 92, 8vo.

CHIAVELLO, BENEDETTO [CHARELLUS].

Chimica filosofica ovvero Problemi naturali sciolti in uso morale ; libri tre. Messina, 1696, 1701 e 1702.

CHILDREN, JOHN GEORGE.

An Essay on Chemical Analysis. London, 1819. 8vo.

This is actually a translation of a work by Thénard, with additions.

CHITTENDEN, R. H.

Studies from the Laboratory of Physiological Chemistry, Sheffield Scientific School of Yale College. For the year 1884–'85. New Haven, 1885. pp. iv–198, 8vo.

The same, vol. II, for the years 1885–86. New Haven, March, 1887. pp. iv–236. 8vo.

CHODKIEWICZ, ALEXANDER.

Chemiia. w Warszawie, 1816–20. 7 vols., 12mo.

This extensive Polish treatise seems to be based on Fourcroy's work.

CHODNEFF, A.

See Khodnef, A.

CHRESTIEN, A. T.

Cours de chimie médicale et de pharmacie. Discours d'ouverture, Montpellier, 1856. 8vo.

Χρηστομάννος, Ἀναστ. Κ.

Ἀναλυτικοὶ πίνακες ἤτοι μέθοδος τῆς ποιοτικῆς ἀναλύσεως διὰ τῆς ὑγρᾶς ὁδοῦ εἰς χρῆσιν τῶν ἰατρῶν, φαρμακοποιῶν καὶ χημικῶν. Ἐν Ἀθήναις, 1865. 8vo.

Ἐγχειρίδιον χημείας κατὰ τὰς νεωτέρας τῆς ἐπιστήμης θεωρίας, πρὸς χρῆσιν τῶν φσιτητῶν τοῦ ἐθνικοῦ πανεπιστημίου. Μέρος πρῶτον. Ἐν Ἀθήναις, 1887.

Μέρος II; Μέταλλα. Μέρος III; ὀργανικὴ χημεία. Ἐν Ἀθήναις, 1887.

Στοιχεῖα χημείας πρὸς χρῆσιν τῶν ἑλληνικῶν γυμνασίων καὶ ἀνωτέρων ἐκπαιδευτηρίων. Ἔκδοσις δευτέρα ἐπεξεργασθεῖσα καὶ αὐξηθεῖσα. Ἐν Ἀθήναις, 1883. 8vo.

CHRISTENSEN, CHR.

Ledetraad i den analytiske Kemi, udarbejdet til Brug ved Landbrugs-undervisningen i Tune. Anden Udgave. Trykt som Manuskript. Kjøbenhavn, 1879. 8vo.

Organisk Kemi, til Brug ved Folkehøjskoler og mindre Landboskoler. Kjøbenhavn, 1881. 8vo.

CHRISTENSEN, CHR. [Cont'd.]
Uorganisk Kemi, udarbejdet til Brug ved Folkehøjskoler og mindre
Landboskoler. Kjøbenhavn, 1879.
>
> Tredie Udgave, 1884.
> Fjerde Udgave, 1887.

CHRISTENSEN, M.
Huslige (Den) Chemiker, en populair Anvisning til at undersöge for-
skellige Aritikler som bruges i Husholdningen, Lægekunsten og
Kunstfagene. Oversat efter det Engelske og forsynet med adskil-
lige Anmærkninger og Tillæg. Kjøbenhavn, 1835. 8vo. Ill.

CHRISTISON, ROBERT.
A Treatise on Poisons in Relation to Medical Jurisprudence, Physiology
and the Practice of Physic. Edinburgh, 1829. pp. xx–698, 8vo.
Ill.
>
> Second edition, 1832. 8vo.
> Third edition, 1836. 8vo.
> First American from the Fourth Edinburgh edition, Phila-
> delphia, 1845. 2 vols., 8vo.
>
> For an abridgment, *see* Ducatel, Julius Timoléon.

CHRISTOBAL, JOSEF MARIA DE S., Y JOSEF GARRIGA Y BUACH.
Curso de química general aplicada á las artes. Paris, 1804–05. 2 vols.,
8vo. 24 Plates.

CHRISTOPHER, W. S.
Chemical Experiments for Medical Students. Arranged after Beilstein.
Cincinnati, 1888. 8vo.

CHRISTOPHLE ET BOUILLET.
> Galvanoplastie. *See, in Section II,* Fremy : Encyclopédie chimique, vol. v.

CHRISTY, DAVID.
The Chemistry of Agriculture ; or, The Earth and Atmosphere, as related
to vegetable and animal life. With new and extensive analytical
tables. Cincinnati, 1853. 12mo.

CHURCH, ARTHUR HERBERT.
Chemistry (The) of Paints and Painting. London, 1890. 8vo.
Food ; some account of its sources, constituents and uses. New York
and London, 1877. 16mo.
> South Kensington Museum Science Handbook.

Laboratory Guide for Students of Agricultural Chemistry. London,
1864. 8vo.

CHURCH, ARTHUR HERBERT.   [Cont'd.]
>   Third edition, London, 1874.
>   Fifth edition, London, 1882.
>   Sixth edition, London, 1888.   pp. 286, 8vo.

CHURCH, HENRY JAMES.

Chemical Processes of the British Pharmacopœia, and the behavior, with reagents, of their products.   London, 1864.

CIOTTO, FRANCESCO.

Del iodio, delle sue chimiche combinazioni e dei suoi preparati farmaceutici.   Edizione arricchita di appendice per la parte scientifica e corredata da esteso indice bibliografico dei lavori risguardanti il jodio in se stesso, nei suoi composti, nelle sue applicazioni in medicina, in chirurgia, in veterinaria, in farmacia, nell' industria e nelle arti, e nei suoi rapporti coll' igiene, colla fisiologia animale e vegetale, colla fisica, ecc.   Venezia, 1858.   8vo.

CLAASSEN, H.

Kurzer Ueberblick über die Zuckerindustrie Deutschlands.   Nienburg a. S., 1888.   8vo.   Ill.

CLARCKE, WILLIAM.

The Natural History of Nitre, or a Philosophical Discourse of the Nature, Generation, Place and Artificial Extractions of Nitre, with its virtues and uses.   London, 1670.   12mo.
>   Naturalis historia nitri, sive discursus philosophicus de natura, generatione, loco, etc., nitri, ejusque virtutibus et usibus.   Francofurti et Hamburgi, 1675.   8vo.

CLARK AND SMITH.

Chemical Report on Various Specimens of Water from Chalk Springs near Waterford : Specification of the Patent granted to Prof. Clark for rendering Water less impure and hard, and Homer's Analysis of a Medicinal Mineral Water at Helivan near Cairo.   London, 1842–43.   8vo.

CLARKE, EDWARD DANIEL.

The Gas Blow-Pipe, or art of fusion by burning the gaseous constituents of water : giving the history of the philosophical apparatus so denominated ; The proofs of analogy in its operations to the nature of volcanoes.   Together with an Appendix containing an account of experiments with this blow-pipe.   London, 1819.   pp. iii–109, 8vo.   Ill.

CLARKE, FRANK WIGGLESWORTH.

A Report of Work done in the Washington Laboratory during the fiscal year 1883–'84.   F. W. C., Chief Chemist ; T. M. Chatard, Assistant Chemist. Washington, 1884. 8vo. Bulletin No. 9 of the U. S. Geological Survey.

Report of Work done in the Division of Chemistry and Physics, mainly during the fiscal year 1884–'85. Washington, 1886. Bulletin No. 27 of the U. S. Geological Survey.

The same for 1885–'86. Washington, 1887. Bulletin No. 42.

The same for 1886–'87. Washington, 1889. Bulletin No. 55.

The same for 1887–'88. Washington, 1890. Bulletin No. 60.

The same for 1888–'89. Washington, 1890. Bulletin No. 64.

The same for 1889–'90. Washington, 1891. Bulletin No. 78 of the U. S. Geological Survey.   Department of the Interior.

CLASSEN, ALEXANDER [Editor].
      See Sonnenschein, F. L.; *also* Mohr, Friedrich.

CLASSEN, ALEXANDER.

Grundriss der analytischen Chemie. Bonn, 1873. 2 vols., 8vo.
         Zweite gänzlich umgearbeitete Auflage. Stuttgart, 1879. 2 vols., 8vo.

Handbuch der analytischen Chemie.   Vierte vermehrte und verbesserte Auflage. Theil I : Qualitative Analyse. Stuttgart, 1889. 8vo.
         Précis de chimie analytique. Première partie : analyse qualitative inorganique et organique. Édition française publiée avec l'autorisation de l'auteur, par V. Francken et L. Lebrun. Liège, 1875. 8vo.

Handbuch der analytischen Chemie. Theil II : Qualitative Analyse. Dritte Auflage, Stuttgart, 1885.

Vierte Auflage, Stuttgart, 1891.
         Précis de chimie analytique. Seconde partie : Analyse quantitative. Édition française publiée avec un appendice sur l'examen des matières sucrées par V. Francken et L. Lebrun. Liège, 1876. 8vo. Ill.
         Précis d'analyse chimique quantitative. Traduit sur la troisième édition par L. Gautier. Paris, 1888. pp. 470, 12mo. Ill.

Quantitative Analyse auf elektrolytischem Wege. Aachen, 1882.
         Quantitative chemische Analyse durch Electrolyse. Nach eigenen Methoden. Zweite gänzlich umgearbeitete und vermehrte Auflage. Berlin, 1886. Ill.
         Quantitative Chemical Analysis by Electrolysis, according to original methods. Authorized Translation from

CLASSEN, ALEXANDER. [Cont'd.]
> the second revised and enlarged German edition, by
> William Hale Herrick. New York, 1887. 8vo.
> Analyse électrolytique quantitative. D'après la deuxième
> édition allemande par C. Blas. Louvain, 1886. 8vo.

Tabellen zur qualitativen Analyse. Im Anschlusse an den Grundriss
der analytischen Chemie. Stuttgart, 1876. 8vo.
> Theil 1 : Zweite verbesserte Auflage. Stuttgart, 1888. 7 Tabellen
> in 4to.

CLAUDIUS.
Anbau der Runkelrübe. 1836.

CLAUS, ADOLPH.
Die Grundzüge der modernen Theorie in der organischen Chemie.
Freiburg, 1871. 8vo.

CLAUS, CARL ERNST.
Grundzüge der analytischen Phyto-Chemie. Dorpat, 1837. 8vo.
Fragment einer Monographie des Platins und der Platinmetalle 1865–
1883. St. Petersburg, Riga, Leipzig, 1883. 8vo.
Beiträge zur Chemie der Platinmetalle. Festschrift zur Jubelfeier des
50 jährigen Bestehens der Universität Kasan. Dorpat, 1854. 8vo.
Methodische Reactions-Tabellen behufs chemischer qualitativ-analy-
tischer Untersuchungen zum Gebrauch für Mediciner und Pharma-
ceuten. Dorpat, 1862. Fol.
On Platinum Residues and on Ruthenium. Kasan, 1844. 8vo.
> Published in the Russian language ; the original I have not seen. The
> treatise was awarded the Demidoff Prize.

CLAUSIER, JEAN LOUIS.
Introduction à la chymie . . . Paris, 1741.

CLEEFF, G. DOYER VAN.
Handleiding bij het qualitatief scheikundig onderzoek voor eerstbegin-
nenden. Onderzoek van zouten. Utrecht, 1879. 8vo.
Leerboek der scheikunde. Haarlem, 1886. Roy, 8vo.

CLEGG, S.
A Practical Treatise on the Manufacture and Distribution of Coal-gas,
its introduction and progressive improvement. Third edition.
London, 1859. Ill. Plates.
> Another edition, 1841.

CLEMANDOT.
Fabrikation des Zuckers aus Runkelrüben. Wien, 1831.

CLEMENTS, HUGH.

Manual of Organic Chemistry, practical and theoretical, for Colleges and Schools, Medical and Civil Service examinations, and especially for elementary, advanced and honours students at the classes of the Science and Art Department, South Kensington. London, 1879. 12mo.

CLERC, A.

Chimica popolare illustrata. Milano, 1886. 8vo.

CLERGET.

Analyse des sucres et des substances sacchariferes au moyen des propriétés optiques de leurs dissolutions. 1846. 4to.

CLEVE, P. T.

See, in Section II, Fremy : Encyclopédie chimique, vol. III.

CLEVE, P. T.

Lärobok i kemiens grunder. 3die Hefte. Lärobok i oorganisk och organisk Kemi för Begynnare. Stockholm, 1886. 8vo.

Lärobok i oorganisk Kemi. Stockholm, 1873. 8vo.

CLOËTZ.

Leçon (etc.). See Société chimique de Paris.

CLOWES, FRANK.

Practical Chemistry and Qualitative Inorganic Analysis, specially adapted for Colleges and Schools. London, 1874. 8vo.

Second edition. London, 1877. 8vo. Ill.

Third edition, London, 1880.

Fourth edition, London, 1885.

A Treatise on Practical Chemistry and Qualitative Analysis, specially adapted for Colleges and Schools. Fifth edition. London, 1890.

An Elementary Treatise on Practical Chemistry and Qualitative Inorganic Analysis. Specially adapted for use in the laboratories of Schools and Colleges and by beginners. Second American from the third and revised English edition. Philadelphia, 1883. 12mo. Ill.

A Treatise on Practical Chemistry and Qualitative Inorganic Analysis, adapted for use in the laboratories of Colleges and Schools. From the fourth English edition. Philadelphia, 1885. pp. xiii-376, 8vo. Ill.

COCHRANE, ARCHIBALD, EARL OF DUNDONALD.

Present (The) State of the Manufacture of Salt explained. London, 1785.

Principles (The) of Chemistry applied to the Improvement of the Practice of Agriculture. London, 1799. 8vo.

Treatise (A) showing the intimate connexion that subsists between Agriculture and Chemistry. London, 1795.

COCK, G., W. WIGNER AND H. HARLAND.

Sugar Growing and Refining. London, 1885.

COELHO DE SEABRA, VICENTE.

Elementos de chimica offerecidos á Sociedade litteraria do Rio de Janeiro para o uso do seu curso de chimica. Coimbra, 1788–90. 8vo. Part I, 1788, pp. xii–54 ; Part II, 1780, pp. 55–485.

> A "Dissertaçao sobre o calor," by the same author, is appended, as announced in a foot-note on the title-page.

COHAUSEN, JOHANN HEINRICH.

*Lumen novum phosphoris accensum ; sive, Exercitatio physico-chymica de causa lucis in phosphoris tam naturalibus quam artificialibus, exarata ad provocationem celeberrimæ Regiæ in Galliis Burdegalensium Academiæ. Amstelodami, apud Ioannem Oosterwyk, 1717. pp. [xxvi]–306–[xviii], 24mo. Illustrated title-page. Plates.

COHEN, JULIUS B.

The Owens College Course of Practical Organic Chemistry. London, 1887. pp. xvi–200.

COHN, WILHELM.

Die künstlichen Dungmittel, ihre Darstellung und Verwendung. Eine technisch-landwirthschaftliche Studie. Braunschweig, 1883. 8vo.

COIT, J. MILNOR.

The Elements of Chemical Arithmetic, with a short system of elementary qualitative analysis. Boston, 1887. pp. iv–92, 12mo.

COLDING, JOHANN PETER.

Om Gasbelysningen eller : om Gaslysets physiske og øconomiske Fortrin fremfor ethvert andet Belysnings- Middel, til dermed at oplyse Gader og Huse, m. m. Kjøbenhavn, 1819.

COLEY, HENRY.

A Treatise on Medical Jurisprudence. Part I : Comprising the consideration of poisons and asphyxia. New York, 1832. 8vo.

COLIZZI, GIUSEPPE (ABBÉ).

Osservazioni sullo stato attuale della chimica ; e se le recenti scoperte, delle quali vuolsi arricchita, sien giunte a tal grado di evidenza e certezza da meritare la piena fiducia di coloro che la coltivano. Perugia, 1839. pp. 260, 8vo.

Trattato fisico-chimico dell' arte di analizzare le acque minerali, e d'imitarle. Macerata, 1803. pp. vii–180, sm. 4to. One folding plate and five folding tables. Also folding table at page 14.

COLLECTANEA CHYMICA Leydensia Maëtsiana, Margraviana, le Mortiana, olim trium in Academia Lugduno Batava facultatis chymicæ professorum . . . ante hoc collecta, digesta, edita a Christoph. Love Morley, nunc autem . . . aucta . . . correcta . . . per Theodorum Muykens. Lugduni Batavorum, 1693.

> * Collectania chymica Leidensia, oder auserlesene mehr als 700 chymische Processe welche von Herrn Maëthio, Margravio und le Mortio ehedessen dreyen berühmten Professoribus der Chymie zu Leyden, denen damahls aus allen Theilen Europæ gegenwärtigen Auditoribus so wohl publice, als privatim nicht nur gewiesen, sondern auch mündlich dictirt worden vor diesem von Christoph Ludwig Morley in Ordnung zusammengetragen und ans Licht gebracht, nachmahls durch Theodorum Muykens mit vielen neuen, schönen und accuraten Experimenten vermehret, in richtigere Ordnung gestellet, allent halben verbessert, und von überflüssigen Processen gesaubert ; nun aber auf Ersuchen guter Freunde ins Teutsche übersetzt und mit doppelten Registern versehen. Ein Werck, so allen Medicis, Chymicis, Physicis, Apothekern, und jeden seine Gesundheit liebenden höchst nöthig und nützlich. Editio secunda. Verlegts Henrich Christoph Cröker, Buchhändler, Anno 1700. pp. [vii]–785–[xliii], 16mo. Illustrated title-page.
>
> Another edition, Jena, 1726.
>
> Morley's Collection first appeared in Latin, Jena, 1684. The first German edition by Muykens is dated 1696.

COLLIER, PETER.

Investigation of the juices of the stalks of various varieties of sorghum and corn—average contents of sucrose, glucose and solids. Washington, 1881.

COLSON.

Isomérie de position. *See, in Section II*, Fremy : Encyclopédie chimique, vol. VIII.

COLVIN, VERPLANCK.

On Certain New Phenomena in Chemistry. Read before the Albany Institute January 2, 1872. 8vo. [*n. p.* *n. d.*]

COMBE, ANDREW.

The Physiology of Digestion, considered with relation to the principles of dietetics. Ninth edition, edited and adapted to the present state of physiological and chemical science by James Coxe. London, 1849. Roy. 12mo.

COMBES, A.

Conférence (etc.). *See* Société chimique de Paris.

COMBES, JULES LUDOMIR.

De l'atmosphère et de l'air atmosphérique. Leur importance con- sidérée sous le point de vue physique, chimique, physiologique et géologique. Agen, 1851. 8vo.

COMBRUNE.

L'art de brasser, renfermant les principes de la théorie et ceux de la pratique. Paris, 1802. 8vo.

COMMERSON, E., ET E. LAUGIER.

Guide pour l'analyse des matières sucrées, sucres bruts, mélasses et liquides sucrés, produits et résidus de fabrication betterave et canne, essais relatifs à la distillation et à la fabrication du glucose, essais des noirs, calcaires, etc. Deuxième édition. Suivi d'analyse des cendres, recherche de matières organiques par E. Laugier. Paris, 1878. pp. 338, 8vo. Plates.

Troisième édition entièrement modifiée. Paris, 1884. pp. 248, 8vo.

COMSTOCK, JOHN LEE.

Conversations on Chemistry, illustrated by Experiments. Eighth American from the sixth London edition, revised and enlarged. Hartford, 1822. 12mo.

Tenth American from the eighth London edition, revised and enlarged. With a new series of questions by J. L. Blake. Hartford, 1826. 8vo.

Eleventh American edition, Hartford, 1829.

Twelfth American edition from the last London edition, with additions and corrections. Hartford, 1830. 12mo.

COMSTOCK, JOHN LEE. [Cont'd.]
    Another edition. Hartford, 1855. 12mo.
        Cf. Marcet, Jane.
    Elements of Chemistry ; in which the recent discoveries in the science
        are included, and its doctrines familiarly explained ; designed for
        the use of schools and academies. New York, 1850. 8vo. Ill.
            Earlier edition, Hartford, 1831. Twenty-sixth edition, New York, 1838.
    Young (The) Chemist. New York, 1835. 24mo.

CONFÉRENCES FAITES À LA SOCIÉTÉ CHIMIQUE DE PARIS.
        See Société chimique de Paris.

CONFÉRENCES FAITES AU LABORATOIRE DE M. FRIEDEL.
        See Cours de la Faculté des Sciences de Paris.

CONFIGLIACHI, PIETRO.
    Sull' analisi dell' aria contenuta nella vescica natatoria dei pesci. Me-
        moria . . . In Pavia, 1809. 4to

CONGRÈS BETTERAVIER . . . PARIS.
    Compte rendu des travaux du congrès betteravier tenu à Paris les 6 et
        7 février, 1882, publié au nom du bureau par A. Ladureau. Lille,
        1882. 8vo.

CONGRÈS INTERNATIONAL DE CHIMIE, tenu au Conservatoire des Arts et
        Métiers du 30 juillet au 3 août, 1889. Procès-verbaux des séances.
        Ministère du commerce, de l'industrie et des colonies. Exposition
        universelle internationale de 1889 ; Direction générale de l'exploita-
        tion. Paris, 1889. 8vo.

CONGRÈS SUCRIER DE SAINT-QUENTIN.
    Compte-rendu sténographique. Paris, 1882. 8vo.

CONINCK, OECHSNER DE.
    Nouvelles recherches sur les bases de la série pyridique et sur les bases
        de la série quinoléique. Paris, 1890. pp. xiv–15–128, 8vo.

CONINGTON, E. T.
    Tables for Qualitative Analysis, to accompany Conington's Handbook
        of Analysis. Second edition. London, 1859. 12mo.

CONINGTON, F. P.
    Handbook of Chemical Analysis. London, 1858.

CONRAD, J.
    Leitfaden für den Unterricht in der Chemie. Zum Gebrauche in der
        Gewerbeschule unter ergänzender Verwendung der einschlägigen
        Objecte, Apparate und Experimente von Seiten des Lehrers. Her-
        mannstadt, 1883. 8vo.

CONRADI, JOHANN LUDWIG.
Dissertatio inauguralis chymico-medica, sistens historiam solutionis corporum particularis.   Marburgi, *n. d.* [1729].   pp. 34, 4to.

CONRING, HERRMANN.
De aquis.   Helmstadtii, 1638.
De fermentatione.   Francofurti, 1643.
De terris earumque ortu et differentiis.   Helmstadtii, 1638.
De sale, nitro et alumine.   Helmstadtii, 1672.
     See also, in Section III.

CONSANI, J.
Manuale delle droghe e prodotti chimici per uso del commercio.   Livorno, 1874.   8vo.

CONTARINI, GASPARO.
De elementis et eorum mixtionibus.   Paris, 1548.

CONVERSATIONS ON CHEMISTRY.   Calcutta, 1847.   8vo.
     Published in the Hindi language.   *Cf.* Comstock, J. L. ; *also* Marcet, Jane.

COOK, ERNEST H.
Introductory Inorganic Analysis.   A first course of chemical testing.   London, 1888.   8vo.

COOKE, JOSIAH PARSONS, JR.
Chemical Problems and reactions to accompany Stöckhardt's "Elements of Chemistry."   Cambridge [Mass.], 1857.
          Another edition: With tables, logarithms and antilogarithms.   Philadelphia, 1871.   12mo.
     Cf. Stöckhardt. Jul. Ad. : Schule der Chemie.

Descriptive List of Experiments on the Fundamental Principles of Chemistry.   Cambridge, 1886.
Elements (The) of Chemical Physics.   London, 1886.
Laboratory Practice.   A Series of Experiments on the Fundamental Principles of Chemistry.   A companion volume to "The New Chemistry."   New York, 1891.   12mo.
New Chemistry (The).   New York, 1874.   pp. 326, 12mo.   Also, London, 1874.   12mo.   The International Scientific Series.
          Eighth edition, remodelled and enlarged.   London, 1884.
          Die Chemie der Gegenwart.   Leipzig, 1875.   8vo.   Ill.
          La nuova chimica.   Milano, 1877.   8vo.
          Новая химія.   Перев. Бутлерова А. М.   С-Петербургъ, 1876.

COOKE, JOSIAH PARSONS, JR. [Cont'd.]

Numerical Relations (The) between the Atomic Weights, with some thoughts on the classification of the Chemical Elements. (With a Table and Supplement.) Cambridge and Boston, 1855. 4to.

Principles of Chemical Philosophy. Boston, 1871.

Revised edition, Boston, 1881. pp. xi–623. 8vo. Ill.

Religion and Chemistry, or Proofs of God's Plan in the Atmosphere and its Elements. Ten lectures delivered at the Brooklyn Institute, Brooklyn, N. Y. Second edition. New York, 1867. pp. viii–348, 8vo.

COOLEY, LEROY C.

Guide to Elementary Chemistry for Beginners. New York, 1887.

COOPER, THOMAS.

A Practical Treatise on Dyeing and Callicoe Printing. Exhibiting the processes in the French, German, English and American practice of fixing colours on woollen, cotton, silk and linen. Philadelphia, 1815, pp. xv–506, 8vo.

COP, M. J.

Verscheidenheid van ligchamen in betrekking tot atomen-leer. Deventer, 1845. 8vo.

COPPA, RAFFAELE.

Elementi di chimica. Napoli, 1864. 2 vols., 8vo.

CORDUS, VALERIUS.

Dispensatorium pharmacorum omnium. Norimbergæ, 1535.

Compiled at the request of the Council of Nürnberg; was the first legal pharmacopœia in Germany; taught, among other things, the preparation of ether.

CORLEO, SIM.

Ricerche sulla vera natura de' creduti fluidi imponderabili. Palermo, 1852. 8vo.

CORMACK, JOHN ROSE.

A Treatise on the chemical, medicinal and physiological properties of creosote, illustrated by experiments on the lower animals, with some considerations on the embalmment of the Egyptians. Being the Harveian Prize Dissertation for 1836. Edinburgh, 1836. 8vo.

CORMIER, TH.

Précis sur le gaz hydrogène. Havre. 8vo.

CORNELIANI.

Opusculo sulla cotennazione del sangue, e sull' applicazione di alcuni principî alla teoria delle flogosi. Padova, 1854. 8vo.

CORNELIUS, CARL SEBASTIAN.

Ueber die Bildung der Materie aus ihren einfachen Elementen. Oder, das Problem der Materie nach ihren chemischen und physikalischen Beziehungen mit Rücksicht auf die sogenannten Imponderabilien. Leipzig, 1856. 8vo.

CORNETTE, CLAUDE MELCHIOR.

Mémoire sur la formation du salpêtre . . . Paris, 1779.

CORNULAT, E.

Le gaz à Paris et à Londre. Nancy, 1885.

CORNWALL, HENRY B.

Chemistry (The) of Butter and its Imitations. The physiological and hygienic bearings of the question, the prevention of the fraudulent use of imitation butter. (In report of the Dairy Commissioner of the State of New Jersey for 1886.) Trenton, 1887.

Laboratory Notes for the exclusive use of the Senior Academic Elective Class in Laboratory Chemistry in the College of New Jersey. *n. p.* [Princeton, N. J.], *n. d.* [1879]. pp. 45, 8vo.

Manuel d'analyse qualitative et quantitative au chalumeau. Traduit sur la seconde édition américaine par J. Thoulet. Paris, 1874. 8vo.

       Cf. Plattner, C. F.

CORNWALLIS, CAROLINE F.

An Introduction to Practical Organic Chemistry. With reference to the works of Davy, Brande, Liebig [etc.]. London, 1843, pp. 90. 16mo.

Another edition, Philadelphia, 1846. 16mo.

       Second edition, revised, London, 1854. 8vo. (Small books on great subjects.)

    Published anonymously.

CORPUT, VAN DEN.

Des fécules et des substances propres à les remplacer au point de vue de l'alimentation et des applications techniques. Bruxelles, 1857. 8vo.

CORTEN, F. R.

    See Ville, Georges.

CORTENOVIS, ANGELO MARIA.
Che la platina Americana era un metallo conosciuto dagli antichi [etc.].
Bassano, 1790.

COSSIGNY DE PALMA, JOSEPH FRANÇOIS CHARPENTIER.
Essai sur la fabrication de l'indigo.   Isle de France, 1779.
Recherches physiques et chimiques sur la fabrication de la poudre à
canon.   Paris, 1806.

COSTEL, JEAN BAPTISTE LOUIS.
Analyse des eaux de Pougues.   Paris, 1769.

COTTEREAU.
Notions de chimie avec applications aux usages de la vie à l'usage des
séminaires, colléges, etc.   Troisième édition, revue et complétée.
Angers, 1876. 12mo. Ill.
Seconde édition. Angers, 1873.

COTTING, JOHN RUGGLES.
An Introduction to Chemistry, with practical questions, designed for
beginners in the science. From the latest and most approved
authors, to which is added : A Dictionary of Terms. Boston,
1822. pp. vii–420, 12mo.

COUNET, J., ET T. KINET.
Traité élémentaire de chimie. Deuxième édition. Bruxelles, 1881.
12mo.

COURS DE LA FACULTÉ DES SCIENCES DE PARIS publiés par l'association
amical des élèves et anciens élèves de la faculté des sciences. Conférences faites au laboratoire de M. Friedel. 1888–89. Conférences de L. Bouveault ; Maquenne ; Arnaud ; A. Behal ; O. Saint
Pierre ; Ad. Fauconnier ; A. Étard. Paris, 1889. pp. viii–142,
8vo.

COURTOIS, BERNARD.
This pharmacist and manufacturer of saltpetre, who discovered iodine in
1812, published only a note in *Annales de chimie*, vol. LXXXVIII (1814).
In 1804, with Séguin, he investigated opium, and the latter published the
results eleven years later in the same journal, vol. XCII (1815).

COUTANCE, A.
Empoisonneurs-empoisonnés. Venins et poisons, leur production et
leur fonctions pendant la vie—dangers et utilité pour l'homme.
Paris, 1888. pp. xvi–420, 8vo.

Cox, ——.

Standard Course of Elementary Chemistry. Five parts. London, 1892. 4to.

Cox, E. J.

Practical Inorganic Chemistry. Analysis and Sketches. London, 1890. 8vo.

Problems in Chemical Arithmetic. London, 1891. 8vo.

Cox, John.

Poisons, their effects, tests and antidotes. London, 1852. 12mo.

Coxe, John Redman.

Observations on Combustion and Acidification ; with a theory of those processes founded on the conjunction of the phlogistic and anti-phlogistic doctrines. Philadelphia, 1811. 12mo.

Cozzi, Andrea.

Nuovo processo economico per ottenere il creosoto. Firenze, 1838.

Sulle applicazioni della forza elettro-chimica della pila all' analisi dei sali metallici disciolti in liquidi organici vegeto-animali. Firenze, 1838, 8vo.

Trattato elementare di chimica medico-farmaceutica. Firenze. 1840. 8vo.

> *See also, in Section II*, Fasi della nomenclatura chimica, *and in Section III*, Storia della scienza elettrica.

Craanen, Daniel.

Scheikundige werking van sommige tegengiften. Amsterdam, 1808. 8vo.

Crace-Calvert, Frederick.

Dyeing and Calico-Printing, including an account of the recent improvements in the manufacture and use of aniline colours, edited by John Stenhouse and Charles E. Groves. Second edition. Manchester, 1876.

Crafts, J. M.

A Short Course in Qualitative Analysis. Sixth edition, revised by Charles A. Schaeffer. New York, 1888. 12mo.

Cramer, Carl.

Die nähern Bestandtheile und die Nahrungsmittel der Pflanzen. Vorgetragen am 3. November, 1855, zum Behuf der Habilitation an der Züricher Universität. Zürich, 1856. 8vo.

CRAMER, CASPAR.
  Collegium chymicum.   Francofurti et Lipsiæ, 1688.

CRAMER, JOHANN ANDREAS.
  Anfangsgründe der Metallurgie.   Blankenburg und Quedlinburg,
    1744–47. III Parts.
  Elementa artis docimasticæ duobus tomis comprehensa, quorum prior
    theoriam, posterior praxim ex vera fossilium indole continet.
    Lugduni Batavorum, 1730. 2 vols, 8vo.
            Zweite Auflage, 1744.
            Anfangsgründe der Probierkunst.   Aus dem lateinischen
              übersetzt von C. E. Gellert.   Stockholm, 1746. 8vo.
            Another edition, von J. F. A. Göttling, Leipzig, 1794.
            Elements of the Art of Assaying Metals, containing the
              theory and practice of the said art, the whole deduced
              from the true properties and nature of fossils, confirmed
              by the most accurate and unquestionable experiments,
              explained in a natural order, and with the utmost clear-
              ness, by J. A. C., with several notes and observations
              not in the original, particularly useful to the English
              reader; with an appendix, containing a list of the chief
              authors that have been published in English upon
              minerals and metals.   London, 1741. 8vo. Ill. 6
              folding plates.

CRAMER DE CLAUSBRUCH, JACOB JOACHIM GEORG.
  Dissertatio inauguralis medico-chemica de præcipitatione chemica
    generatim considerata.   Halæ Magdeburgicæ, n. d. [1754]. pp.
    59, 4to.

CRAMPTON, C. A.
  Baking Powders.   Part Fifth of Foods and Food Adulterants. Investi-
    gations made under the direction of H. W. Wiley.   Bulletin No.
    13, Division of Chemistry, U. S. Department of Agriculture.
    Washington, 1889. 8vo.
  Record of Experiments at Des Lignes Sugar Experiment Station,
    Baldwin, La., during the Season of 1888.   Bulletin No. 22, Divi-
    sion of Chemistry, U. S. Department of Agriculture.   Washington,
    1889. 8vo.
          Cf. in Section VII, Bulletins of the Division of Chemistry, U. S. Depart-
          ment of Agriculture.

CRANZ, HEINRICH JOHANN NEPOMUK.
  Examinis chemici doctrinæ Meyerianæ de acido pingui et Blackianæ

CRANZ, HEINRICH JOHANN NEPOMUK. [Cont'd.]
de aëre fixo respectu calcis rectificatione. Lipsiæ, 1770. pp. 212, 8vo.

> Cf. Jacquin, Nic. Jos., and see Meyer, Johann Friedrich.

CRAWFORD, ADAIR.
Experiments and Observations on Animal Heat and the inflammation of combustible bodies, etc. London, 1779.
Second edition, 1788.

> Contains his much-discussed theory of heat, and his experiments on the specific heat of gases.

CRELL, LORENZ FLORENZ FRIEDRICH VON.
De acidorum nitrosi imprimis et muriatici dulcificatione. Helmstadtii, 1782.

CRELL, LORENZ FLORENZ FRIEDRICH VON [Editor.]
See in Section VII.

CRESTI, L.
Analisi chimica qualitativa ; procedimento alla ricerca dei più importanti acidi inorganici. Pavia, 1877. 8vo.

CRETIER, H.
Handleiding voor analytische chemie ten gebruike bij het middelbaar onderwijs. Deventer, 1873-'74. 8vo. Ill.

CREUZBERG.
Katechismus der Stöchiometrie. Wien, 1834. 8vo.

CREYDT, R.
Die quantitative Bestimmung der Raffinage. Erlangen, 1888. pp. 37, 8vo.

CREW, B. J.
A Practical Treatise on Petroleum, comprising its origin, geology, geographical distribution, history, chemistry, mining, technology, uses and transportation. Together with a description of Gas-wells, the application of gas as fuel, etc. With an appendix on the product and exhaustion of the oil regions and the geology of natural gas in Pennsylvania and New York by Ch. Ashburner. Philadelphia, London, 1887. pp. xxiv-508, 8vo. Ill.

CREUZNACH, LEONARD STEFFAN DE.
Verhandelingen over de kalk. 1761. 8vo.

CRISTIANI, R. S.

A Technical Treatise on Soap and Candles, with a glance at the industry of fats and oils.   Philadelphia, 1881.   pp. xvi–17–581, 8vo. Ill.

CRIVELLI, ANTONIO.

Nuovo meccanismo per ottenere la più vantaggiosa combustione dell' idrogeno mediante l'ossigeno.   Discorso accademico.   Milano, 1818.   8vo.

CROFT, HENRY.

Course of Practical Chemistry as adopted at University College, Toronto.   Toronto, 1860.   8vo.

КРОКЕРЪ.

Руководство къ сельско-хозяйств анализу съ спец. указаніемъ важнѣйшихъ, сельско-хозяйственныхъ продуктовъ.   Перев, Энгельгардта.

С. Петербургъ, 1867.

> CROKER.   Manual of agricultural analysis.   Translated by Engelhardt. St. Petersburg, 1867.

CROLL, OSWALD.   [CROLLIUS.]
See in Section VI.

CROME, GEORG ERNST WILHELM.

Disputatio chemico-physiologica alimentorum hominis et animalium domesticorum.   Göttingæ, 1811.

CRONSTEDT, AXEL FREDRIC.

Försök till Mineralogiens eller Mineral-Rikets upställning.   Stockholm, 1758.

CROOKES, WILLIAM.

Dyeing and Tissue Printing.   Edited by H. Trueman Wood.   London, 1882.

Éléments et méta-éléments.   Mémoire lu à la Société chimique de Londres.   Traduit par Lewy.   Paris; 1888.

Genèse (La) des éléments.   Traduit par Richard.   Paris, 1887.

Genesis (Die) der Elemente.   Ein Vortrag gehalten in der " Royal Institution " zu London am 18. Februar, 1887.   In das deutsche übertragen von Alfred Delisle.   Braunschweig, 1888.   8vo.   Ill.

On Thallium.   London, 1863.   4to.

On the Manufacture of Beet-Root Sugar in England and Ireland. London, 1870.   pp. xvi–290–xii, 8vo.   Ill.

Practical (A) Handbook of Dyeing and Calico-Printing.   London, 1874. 8vo.   Ill.

CROOKES, WILLIAM.   [Cont'd.]
Another edition, London, 1883.
Contains specimens of dyed and printed fabrics, and a bibliography of the subject.

Select Methods in Chemical Analysis.   (Chiefly Inorganic.)   London, 1871.   pp. xvi–468, 8vo.
Second edition, London, 1886.   pp. xxii–725, 8vo.
Third edition, London, 1888.   8vo.

CROSNIER, L.
Elementos de química mineral.   Santiago de Chile, 1846.   8vo.

CROSS, C. F., AND E. J. BEVAN.
Cellulose, being a short account of the chemical properties typical members of the cellulose group with reference to their natural history.   Edited by W. R. Hodgkinson.   London, 1885.   pp. 28, 8vo.
Contains a bibliography.

CROSS, GEORGE N.
Elementary Chemical Technics.   A handbook of manipulation and experimentation for teachers of limited experience, and in schools where chemistry must be taught with limited appliances.   Boston, 1887.   pp. vi–123, 12mo.   Ill.

CRÜGNER, MICHAEL.
Chymischer Garten-Baw, das ist spagyrische Beschreibung Vier und Dreissigerley Gewächs und Kräuter, nach rechter Fundamental und Hermetischer Anleitung : Welche auss der Putrefaction und Transplantation sich generirn, vom stetssuchenden Autore fleissig observiret allen Liebhabern zum Anlass zur fernerer Speculation und Observation dem Menschen zum besten herausgegeben. Nebenst angehängter kleinen Haliographia.   Nürnberg, 1653. pp. 286, sm. 4to.

CRUSIUS, WILHELM [Editor].
See Agriculturchemische Untersuchungen.

CSETNEKY.
Physikalischer Beitrag zur Chemie.   Linz, 1849.   8vo.

CUMENGE ET FUCHS.
Or.   See, in Section II, Fremy : Encyclopédie chimique, vol. III.

CURAUDAU, FRANÇOIS RÉNÉ.
Traité sur le blanchissage à la vapeur.   Paris, 1806.

CURIE ET JOLY.
> Bor, Silicium et silicates.   *See, in Section II*, Fremy : Encyclopédie chimique, vol. II, 3e Section.

CURTMAN, CHARLES O.
Uses, Tests for Purity and Preparation of Chemical Reagents employed in qualitative, quantitative, volumetric, docimastic, microscopic and petrographic analysis, with a supplement on the use of the spectroscope. St. Louis, Mo., 1890. 8vo.

CURTMAN, CHARLES O.
> *See* Beilstein (F.) : Lessons in Qualitative Chemical Analysis. Second edition. St. Louis, Mo., 1886. 8vo.

CUTBUSH, JAMES.
A System of Pyrotechny, comprehending the theory and practice, with the application of chemistry.   Designed for exhibition and for war. In four parts : containing an account of the substances used in fire-works, the instruments, utensils and manipulations ; fire for exhibition ; and military pyrotechny.   Adapted to the military and naval officer, the man of science and artificer. Philadelphia, 1825. pp. xliv–610, 8vo. Ill.
Philosophy (The) of Experimental Chemistry. Philadelphia, 1813. 12mo.

CUTLER, CONDICT W.
The Medical-Students' Essentials of Physics. New York, 1884. pp. 179, 24mo.
> The Medical-Students' Essentials of Physics and Chemistry. New York, 1887. pp. 468, 32mo.
> Essentials of Physics and Chemistry written especially for the use of students in medicine. Third edition, enlarged and revised. New York and London, 1889. pp. ix–296.
Introductory Lessons in Organic Chemistry. New York, 1886. pp. 121, 32mo.

CZUMPELIK, E.
Die Chemie als Mechanik der Atome. Wien, 1883. 8vo.

CZYRNIAŃSKI, EMIL.
Ein Beitrag zur chemisch-physikalischen Theorie. Krakau, 1887.
Słownictwo Polskie chemiczne. w Krakowie, 1852. 8vo.
Theorie der chemischen Verbindungen auf der rotirenden Bewegung der Atome basirt. Krakau, 1863. 8vo.

CZYRNIANSKI, EMIL. [Cont'd.]

    Dritte Auflage   Krakau, 1872

    Chemische Theorie auf der rotirenden Bewegung der Atome basirt, kritisch entwickelt. Vierte Auflage. Krakau, 1873. pp. [viii]–62, 8vo.

Theorya chemiczno-fizyczna. w Krakowie, 1884.

    Chemisch-physische Theorie, aus der Anziehung und Rotation der Uratome abgeleitet. Krakau, 1885, 8vo.

DABNEY, CHARLES W. [Editor].

    See, in Section VII, Association of Official Agricultural Chemists.

DACHAUER, G.

Haupt-Grundlehren der Chemie zur Einführung in diese Wissenschaft. München, 1863. 8vo.

    Hoofd-grondstellingen der scheikunde. Naar het Hoogduitsch door R. J. Opwijrda. Utrecht, 1863. 8vo.

DAHLEN, HEINRICH WILHELM.

    Die Weinbereitung. *See* Otto-Birnbaum's Lehrbuch.

DALENCÉ.

* Traittez des baromètres, thermomètres et notiomètres ou hygromètres. Amsterdam, 1688. pp. [x]–139–[iv], 16mo. Illustrated title-page. Ill.

    Published anonymously.

DALLAPORTA, N.

Trattato elementare dei gas. Padova, 1793.

DALLAS, A.

Outlines of Chemico-Hygiene and Medicine, or the application of chemical results to the preservation of health and cure of disease. Toronto, 1860. 8vo.

DALTON, JOHN.

Essay on the Phosphates and Arseniates. Manchester, 1840.

Essay on the Quantity of Acids, Bases and Water in the different varieties of salts, with a new method of measuring the water of crystallisation, etc. Manchester, 1840.

* New (A) System of Chemical Philosophy. Manchester, printed by S. Russell, 125 Deansgate, for R. Bickerstaff, Strand, London. 2 vols., 8vo. 1808–10. Vol. I, pp vi–[i]–220 ; II, viii, *from* 221–560. 8 plates.

DALTON, JOHN.   [Cont'd.]

> An epoch-making work, in which the immortal author established the atomic theory of chemistry.   For an early abstract of Dalton's theory, see Thomson's System of Chemistry (1807).

Ein neues System des chemischen Theiles der Naturwissenschaft.   Aus dem englischen von Friedrich Wolff. Berlin, 1812–13.   2 vols., 8vo.   8 plates.

On a New and Easy Method of Analysing Sugar.   Manchester, 1840.

DALTON, JOHN, UND W. WOLLASTON.

Die Grundlagen der Atomenlehre.   Herausgegeben von W. Ostwald. Leipzig, 1889.   Ill.

> Ostwald's Klassiker der exakten Wissenschaften.

Δαμβέργης, Αναστάσιος Κ.

Χημικαὶ ἀνάλύσεις τῶν ἐν Αἰγίνῃ ἰαματικῶν ὑδάτων.   Ἐν Ἀθήναις, 1884.

Χημικὴ ἀνάλυσις τοῦ ἐν Ἄνδρῳ ἰαματικοῦ ὕδατος.   Ἐν Ἀθήναις, 1885.

Νοθεύσεις ἐδωδίμων καὶ ἐξέλεγξις αὐτῶν.   Ἐν Ἀθήναις, 1877.

Περὶ τῶν ἐκρηκτικῶν οὐσιῶν.   Ἐν Ἀθήναις, 1888.

Πραγματεία ἐπὶ ὑφηγεσία περὶ χημικῆς συνθέσεως καὶ τῆς τεκυητῆς τῶν ἀλκαλοειδῶν παρασκευῆς.   n. p.   1880.   4to.

Στοιχεία χημείας.   Ἐν Ἀθήναις.   1890.

DAMMER, O.

Handbuch der anorganischen Chemie unter Mitwirkung von Gadebusch, Haitinger, Lorenz, Nernst, Philipp, Schellbach, von Sommaruga, Stavenhagen, Zeisel, herausgegeben von O. D.   Stuttgart, 1892. 3 vols., 8vo.   *In progress.*

DANA, JAMES FREEMAN.

*An Epitome of Chymical Philosophy ; being an extended syllabus of the lectures on that subject, delivered at Dartmouth College, and intended as a text-book for students.   Concord, N. H., 1825.   pp. 231, 8vo.

DANCY, F. B., AND H. B. BATTLE.

Chemical Conversion Tables for use in the analysis of commercial fertilizers.   Raleigh, 1885.   pp. 42, 8vo.

DANDOLO, VICENZO.

Fundamenti della scienza chimico-fisica applicati alla formazione dei corpi e dei fenomeni della natura . . . Venezia, 1795.

DANGER ET FLANDIN.

De l'arsenic, suivi d'une introduction propre à servir de guide aux experts dans les cas d'empoisonnement, etc.   Paris, 1841.   8vo.

DANIELL, JOHN FREDERIC.
    Chemistry. [London, 1838.]
        Published anonymously ; no title-page ; Society for the Diffusion of Useful
            Knowledge. Library of Useful Knowledge.
    Introduction (An) to the Study of Chemical Philosophy ; being a
        preparatory view of the forces which concur to the production of
        chemical phenomena. London, 1839. pp. xvi–565, 8vo. Ill.

DANNEHL, GUSTAV.
    Die Verfälschung des Bieres. Berlin, 1878. 8vo.
        Deutsche Zeit und Streit-Fragen, Jahrgang vii, Heft 100–101.

DAQUIN, JOSEPH.
    Analyses des eaux thermales d'Aix en Savoie. Chambéry, 1771.
    Analyses des eaux de la Boisse. Chambéry, 1784.

DARAPSKY, L.
    Las aguas minerales de Chile. Valparaiso, 1890.

DARCET, JEAN PIERRE JOSEPH. [D'ARCET.]
    Description des appareils à fumigation. Paris, 1818.
            Deuxième édition, 1830.
    Mémoire sur l'action d'un feu égal, violent et continué plusieurs jours
        sur un grand nombre de terres, de pierres et de chaux métalliques
        essayées pour la plupart telles qu'elles sortent du sein de la terre.
        Lu à l'Académie Royale des Sciences les 16 et 28 Mai, 1766.
        Paris, 1766. pp. 170. 8vo.
    Mémoire sur l'art de dorer le bronze au moyen de l'amalgame d'or et
        mercure. Paris, 1818.
    Résultats de l'emploi alimentaire de la gélatine des os. Paris, 1833.
    Sur les substances nutritives qui renferment les os . . . Paris, 1829.

DA-RIO, NICOLÓ.
    Introduzione alla chimica. Padova, 1798.

DAUBENY, CHARLES.
        An introduction to the Atomic Theory. *See in Section III.*

DAUBRAWA, FERDINAND.
    Der populäre technische Chemiker. Ein treuer leichtfasslicher Rath-
        geber für Oekonomen, Handels- und Gewerbsleute, angehend die
        Prüfung der Reinheit, der Güte und des Werthes der im Verkehr
        häufig vorkommenden Stoffe, als der verschiedenen Gewebe,
        des Mehles, der Milch etc. Prag, 1854. 8vo.

DAUBRÉ, A.

Recherches expérimentales sur le striage des roches au phénomène erratique, sur la formation des galets, des sables et du limon, et sur les décompositions chimiques produites par les agents mécaniques. Paris, 1858. 8vo.

DAVIDSON, WILLIAM [DAVISSON].

Philosophia pyrotechnica seu curriculus chymiatricus [etc.]. Paris, 1635.

Élémens de la philosophie de l'art du feu, ou chimie, traduits du latin par Jean Hellot. Paris, 1651.

Élémens de la philosophie de l'art du feu, ou cours de chimie, traduit en français par lui même. Paris, 1675. 8vo.

> The author, though born in Scotland, was Professor of Chemistry at the Jardin du Roi, Paris, in 1606, said to be the first in France. In the fourth part of this work he treats of crystallography in a scientific way, being the earliest to do so. French biographers, in accordance with their habit of misspelling proper names, write his name Dawisson, Davisson, and d'Avissone, Guillaume.

DAVIDOWSKY, F.

Die Leim- und Gelatine-Fabrikation.   Wien, 1883.

DAVIS, FLOYD.

Elementary Handbook of Potable Water.   New York, 1891.   8vo.

DAVIS, C. F.

The Manufacture of Leather ; being a description of all the processes for the tanning, currying and finishing of leather, including the various raw materials and the methods of determining their values . . . London, 1885.

DAVREUX, CHARLES JOSEPH.

Leçons sur la minéralogie et la chimie . . . Liège, 1828–29.   7 parts.

DAVY, SIR HUMPHRY.
       See Beddoes, Thomas.

DAVY, SIR HUMPHRY.

Account (An) of some new Analytical Researches on the nature of certain bodies ; particularly the alkalies, phosphorus and sulphur, carbonaceous matter, and the acids hitherto undecomposed ; with some general observations on chemical theory. London, 1809. 4to.

Electro-chemical Researches on the Decomposition of the Earths, with observations on the metals obtained from the alkaline earths, and on the amalgam procured from ammonia. London, 1808.   4to.

Davy, Sir Humphry. [Cont'd.]

Elements of Agricultural Chemistry in a course of lectures before the
Board of Agriculture.   London, 1813.   4to.

> Second edition, London, 1814.
>
> American edition : New York, 1815.   8vo.
>
> [Another edition,] to which is added a treatise on soils and
> manures by a practical Agriculturist.   Philadelphia,
> 1821.   2 parts.   8vo.
>
> Fifth edition.   London, 1836.   8vo.
>
> Elements of Agricultural Chemistry.   A new edition, with
> instructions for the analysis of soils and copious notes
> embracing the recent discoveries in agricultural chem-
> istry, by Liebig, Boussingault and others, by John
> Shier.   London, 1846.   pp. ix–293.
>
> Éléments de chimie agricole, en un corps de leçons pour le
> comité d'agriculture ; traduit de l'anglais, avec un
> traité sur l'art de faire le vin et de distiller les eaux-
> de-vie par A. Bulos.   Paris, 1819.   2 vols., 8vo.   Ill.
>
> Nouveau manuel de chimie agricole, traduit sur la cinquième
> édition anglaise des Éléments de chimie agricole de Sir
> H. D. avec les notes de J. Davy sur des faits connus
> seulement depuis 1826 ; par A. D. Vergnaud. Paris, 1838.
>
> Elementi di chimica agraria, tradotti da A. Targioni-Tozzetti.
> Firenze, 1815.   2 vols.
>
> Основанія земледѣльческой химіи, изложенныя серомъ Гумфри Деви.
> С.-Петербургъ, 1832.   8°

* Elements of Chemical Philosophy.   London, 1812.   pp. xiv–[i]–507,
8vo.   10 plates.

> American edition : Philadelphia and New York, 1812.
>
> Contains an Introduction reviewing the history of chemistry.
>
> Élémens de philosophie chimique traduit de l'anglais avec
> des additions par J. B. van Mons.   Paris et Amsterdam,
> 1813.   2 vols., 8vo.   Another edition, 1826.
>
> The author's name is given on the title-page as " Homfrede " Davy.
>
> Elemente des chemischen Theils der Naturwissenschaft, aus
> dem englischen übersetzt von Friedrich Wolff.   Berlin,
> 1814.   8vo.
>
> Elementi di filosofia chimica, tradotti con note di Brugna-
> telli e Configliachi.   Milano, 1814.   2 vols.
>
> Elementi di filosofia chimica, tradotti e commentati da G.
> Moretti e G. Primo.   Milano, 1814.   2 vols.   Plates.

Experimental Researches in Electro-chemistry : containing his Bakerian

DAVY, SIR HUMPHRY. [Cont'd.]
lectures on the chemical agencies of electricity, and on the metals and earths. Glasgow and London, 1842. 8vo.

Lecture (A) on the Objects of the Royal Institution. London, 1810.

On some New Phenomena of Chemical Changes produced by Electricity, particularly the decomposition of the fixed alkalies, and the exhibition of the new substances which constitute their bases . . . London, 1808. 4to.

On the Fire-damp of Coal Mines, and on methods of lighting the mines so as to prevent its explosion. London, 1815. 8vo.

On the Safety Lamp for Coal Miners, with some researches on flame. London, 1818. 8vo.

> [Another edition] with additions to the Appendix. London, 1825. 8vo.

* Outlines of a Course of Lectures on Chemical Philosophy. London, 1804. pp. [ii]-54, 12mo.

> Contains : Explanation of Terms used in Chemistry by John Sadler, with independent pagination. London, 1804. pp. 22, 12mo.

Researches Chemical and Philosophical, chiefly concerning Nitrous Oxide or dephlogisticated nitrous air, and its respiration. London, 1800. pp. xvi-580, 8vo. Plates.

> The first treatise on the effects of so-called laughing gas. These researches were made between April, 1799, when Davy first breathed nitrous oxide, and the summer of 1800, when this work was issued. Dr. Thomson writes of this work : "It gave him at once a high reputation as a chemist, and was really a wonderful performance when the circumstances under which it was produced are taken into consideration." Davy was but twenty-one years of age at the time.

> Chemische und physiologische Untersuchungen über das oxydirte Stickgas und das Athmen desselben. Aus dem englischen übersetzt. Lemgo, 1812–14. 2 parts. 8vo.

* Six Discourses Delivered before the Royal Society at their anniversary meetings on the award of the Royal and Copley medals ; preceded by an address to the Society on the progress and prospects of science. London, 1827. pp. xi–148. 4to.

Syllabus of a Course of Lectures on Chemistry. London, 1802.

The Collected Works of Sir Humphry Davy, edited by his brother John Davy. London, 1839. 9 vols., 8vo.

> Vol. I.   Memoirs of the Life of H. D., by John Davy. pp. viii–475; portrait and facsimile.

> II.   Early Miscellaneous Papers. pp. xii–466.

> III.   Researches chemical and philosophical, chiefly concerning nitrous oxide. pp. xvi–343 ; plate.

DAVY, SIR HUMPHRY. [Cont'd.]

IV. Elements of Chemical Philosophy. pp. xvi–376 ; plates.

V. Bakerian Lectures, and miscellaneous papers, 1806–15. pp. xv–527.

VI. Miscellaneous papers and researches, especially on the safety lamp and flame, and on the protection of the copper sheathing of ships, 1815–28. pp. xi–364; plates.

VII. Discourses delivered before the Royal Society. Agricultural Lectures, part I. pp. xiii–391.

VIII. Agricultural Lectures, part II, and other lectures. pp. viii–365.

IX. Salmonia, or Days of Fly-fishing. Consolation in Travel. pp. xi–388.

DAVY, JOHN.

Lectures on the Study of Chemistry in Connexion with the Atmosphere, the Earth and the Ocean and Discourses on Agriculture ; with introductions on the present state of the West Indies, and on the Agricultural Societies of Barbados. London, 1849. pp. xxv–291, 12mo.

DAXHELET, AUGUSTE.

Cours de chimie organique et inorganique d'après la théorie typique de Gerhardt, avec tableaux. Paris, 1869. 2 vols., 8vo. Ill.

DAY, GEORGE E.

Chemistry, in its Relations to Physiology and Medicine. London, 1860. pp. xvi–526, 8vo. Plates.

DE.

For French proper names beginning with DE not joined to following word, see next succeeding word.

DEBAT, L.

Essai sur la constitution de la matière et l'essence des forces dans l'ordre physique. Paris, 1873. 8vo.

DEBAY, A.

Nouveau manuel du parfumeur-chimiste. Les parfums de la toilette et les cosmétiques les plus favorables à la beauté, sans nuire à la santé. Suivis d'un grand nombre de produits hygièniques nouveaux complètements ignorés de la parfumérie. Paris, 1856. 18mo.

DEBRAY.

Leçon (etc.). *See* Société chimique de Paris.

DEBRAY ET JOLY.

Platine et métaux qui l'accompagnent. *See, in Section II*, Fremy: Encyclo-pédie chimique, vol. III.

DEBRAY, HENRI.

Cours élémentaire de chimie. Paris, 1863. 8vo. Ill.

Troisième édition, Paris, 187–. 2 vols., 8vo.

Quatrième édition, par A. Joly. Paris, 1884. 2 vols., 8vo. Ill.

Du glucium et de ses composés. Thèse de chimie. Paris, 1855. 4to.

DECANDOLLE, AUGUSTIN PYRAMUS.

Physiologie végétale, ou exposition des forces vitales des végétaux . . . Paris, 1832. 3 vols., 8vo.

ДЕШАНЪ И ГРО.

Химическая часть товаровѣдѣнія. Изслѣдованіе москательныхъ товаровъ. Москва, 1864.

DECHAMPS ET GROS. Chemical investigation of drugs. Moscow, 1864.

DECREMPS.

Diagrammes chimiques, ou Recueil de 360 figures . . . qui rendent sensible la théorie des phénomènes. Paris, 1823. 4to.

DEGUIN.

Cours élémentaire de chimie à l'usage des collèges et des autres établissements d'instruction publique. Paris, 1845. 8vo.

Deuxième édition, Paris, 1847.

Troisième édition, Paris, 1849.

Quatrième édition, Paris, 1853.

Curso de química por D. Traducido . . . por Mariano de Rementería. Madrid, 1848. 8vo. Ill.

DEHÉRAIN, PIERRE PAUL.

Nutrition de la plante. *See, in Section II*, Fremy: Encyclopédie chimique. vol. x.

Sur l'assimilation des substances minérales par les plantes. *See* Société chimique de Paris. Leçons . . . 1870.

DEHÉRAIN, PIERRE PAUL.

Cours de chimie agricole, professé à l'École d'agriculture de Grignon. Paris, 1873. pp. xii–616. 8vo. Ill.

Traité de chimie agricole. Développement des végétaux ; terre arable ; amendements et engrais. Paris, 1892. 8vo. Ill.

DEHÉRAIN ET TISSANDIER.

Éléments de chimie. Ouvrage rédigé conformément aux programmes officiels de 1866 pour l'enseignement secondaire spécial. Paris, 1869. 18mo.

DEHERRYPON, MARTIAL.

Les merveilles de la chimie. Paris, 1872. 8vo. Ill.
    Deuxième édition. Paris, 1873. 18mo. Ill.
    Quatrième édition. Paris, 1889. Ill.
    Bibliothèque des merveilles.

DEICKE, H.

Sammlung von Aufgaben aus der Chemie. Zum Gebrauche für Real-
und Gewerbeschulen, polytechnischen Lehranstalten und chemischen Laboratorien. Iserlohn, 1861. 8vo.

DEIDIER, ANTOINE.

Chymie raisonnée où l'on tache de découvrir la nature et la manière
d'agir des remèdes chymiques les plus en usages en médecine et
en chirurgie. Lyon, 1715.

DEIMAN, JOAN RUDOLPH.

Recherches physico-chymiques. Amsterdam, 1792–94. 3 parts. 4to.

DEITE, C.

    Die Darstellung der Seifen, Parfümerien und Cosmetica. *See* Bolley's
    Handbuch.

DEITE, C.

Die Industrie der Fette, enthaltend die Gewinnung und Reinigung der
Fette, sowie die Darstellung der Seifen, der Talg- und Wachs-
lichte, der Wagenfette und anderer Schmiermaterialen und der
Kunstbutter. Braunschweig, 1878. 8vo. Ill.
    Cf. Otto-Birnbaum's Lehrbuch der landwirthschaftlichen Gewerbe. Vol. x.

Handbuch der Parfümerie- und Toiletteseifen-Fabrikation. Unter
Mitwirkung von L. Borchert, F. Eichbaum, E. Kugler, H. Töffner
und anderen Fachmännern. Berlin, 1891. Ill.

Handbuch der Seifenfabrikation. Unter Mitwirkung von Fachmännern herausgegeben. Berlin, 1887. Ill.

DEJEAN, FERDINAND.

Dissertatio chemico-œconomico-practica, qua proponitur historia,
analysis chymica, origo et usus œconomicus sodæ hispanicæ.
Lugduni Batavorum, 1773. 4to.

Traité raisonné de la fabrication des liqueurs françaises et étrangeres
sans distillation. Troisième édition. 12mo.

DE LA GARAY.

    See La Garay, De.

DELAFOSSE.

Mémoire sur le plésiomorphisme des espèces minérales, c'est-à-dire sur
les espèces dant les formes offrent entre elles, le degré de ressem-
blance qu'on observe dans le cas d' isomorphisme sans que leur
compositions atomiques puissent se ramener à une même formule.
Lu à l'Académie, le 14 Avril, 1851.   Paris, 1851.   8vo.

DE LA MÉTHÉRIE.
   See La Méthérie, De.

DELAURIER, E.

Essai d'une théorie générale supérieure de philosophie naturelle et de
thermo-chimie, avec une nouvelle nomenclature binaire notative
pour la chimie minérale et organique.   Paris, 1883.   pp. 272, 8vo.   Ill.

  The author himself characterizes his work as an audacious attempt to sound
  the most hidden secrets of nature.   His system of nomenclature is more
  novel than practical.

DELAVAUD, H. C.

Revue analytique de la chimie contemporaine.   Paris, 1866.   8vo.

DEL BUL, G. C.

Trattato filosofico-chimico sopra la teoria atomistica.   Napoli, 1829.
pp. 369, 8vo.

DELEFOSSE, E.

Procédés pratiques pour l'analyse des urines, des dêpots et des calculs
urinaires.   Troisième édition.   Paris, 1885.

DELFFS, FRIEDRICH WILHELM HERRMANN.

Organische (Die) Chemie in ihren Grundzügen dargestellt.   Kiel, 1840.
8vo.

Reine (Die) Chemie in ihren Grundzügen dargestellt.   Zweite vermehrte
und verbesserte Auflage.   Kiel, 1845.   2 parts, 8vo.   Part 1:
Anorganische Chemie.   Part 11 : Organische Chemie.

   Dritte Auflage.   Erlangen, 1854.   2 parts, 8vo.   Part 1:
   Anorganische Chemie.   Dritte vermehrte und verbes-
   serte Auflage.   Part 11 : Lehrbuch der organischen
   Chemie.   Dritte umgearbeitete Auflage.

DELFOS, F. C.

Beginselen der scheikunde, ten dienste van onderwijzers, kweek- en
normaalscholen   Groningen, 1881.   8vo.

   Andere druk.   Groningen, 1884.

DELLINGSHAUSEN, N.
Die rationellen Formeln der Chemie auf Grundlage der mechanischen
Wärmetheorie entwickelt.   Heidelberg, 1876.   pp. 164, 8vo.

DELMART, A.
Die Echtfärberei der losen Wolle in ihrem ganzen Umfange.   Reichenbach, 1891.   8vo.   With samples.

DELOYERS, ÉMILE.
Les premiers éléments de la chimie minérale avec ses applications à
l'agronomie, à l'usage des écoles moyennes, des sections agricoles
et des cultivateurs.   Fleurus, 1888.   8vo.

DELUC, A. D.
La chimie pour tous.   Vol. 1: Métalloïdes et métaux, complètement
pratique.   Bruxelles, 1879.   12mo.

DEMACHY, JACQUES FRANÇOIS.
Art (L') du destillateur d'eau forte et du liquorist.   Paris, 1776.
Art (L') du vinaigrier.   Paris, 1785.
Examen chimique des eaux de Passy.   Paris, 1756.
Examen chimique des eaux de Verberie.   Paris, 1757.
Instituts de chymie, ou Principes élémentaires de cette science présentés
sous un nouveau jour.   Paris, 1766.   2 vols., 8vo.
Procédés chimiques rangés méthodiquement et définies.   Paris, 1769.
Recueil des dissertations physico-chimiques.   Paris, 1774.
Laborant (Der) im Grossen.   Aus dem französischen von Hahnemann.
Leipzig, 1801.   2 vols.

DEMEL, JOSEPH EUSTACHIUS.
Analysis plantarum.   Viennæ, 1782.   8vo.

DENIGÉS, G.
Formules générales pour servir au calcul rapide des analyses élémentaires.   Bordeaux, 1885.   8vo.

DENIS, PROSPER SYLVAIN.
Essai sur l'application de la chimie à l'étude physiologique du sang de
l'homme et à l'étude physiologico-pathologique, hygiénique et
thérapeutique des maladies de cette humeur.   Ouvrage présenté
à l'Académie des Sciences le 2 janvier, 1838.   Paris, 1838.   pp.
366, 8vo.
Études chimiques, physiologiques et médicales, faites de 1835 à 1840,
sur les matières albumineuses, etc.   Commercy, 1843.   8vo.

DENIS, PROSPER SYLVAIN. [Cont'd.]

Nouvelles études chimiques, physiologiques et médicales sur les substances albuminoïdes qui entrent comme principes immédiats dans
la composition des solides et des fluides organiques, tant animaux
que végétaux. Études faites en suivant la méthode d'expérimentation par les sels, la seule qui, dans l'état actuel de la science, semble
pouvoir être appliqué avec fruit à des recherches sur ces substances.
Mémoire présenté à l'Académie des Sciences en juin 1856. Paris,
1856. pp. 236, 8vo.

Mémoire sur le sang considéré quand il est fluide, pendant qu'il se
coagule et lorsqu'il est coagulé. Suivi d'une notice sur l'application de la méthode d'expérimentation par les sels à l'étude des
substances albuminoïdes. Mémoire présenté à l'Académie des
Sciences le 20 décembre, 1858. Paris, 1859. pp. viii–208, 8vo.

Recherches expérimentales sur le sang humain, considéré à l'état sain,
faites pour déterminer les modifications aux quelles est sujette dans
l'économie, la composition de cette humeur, et apprécier les phénomènes physiologiques qui s'y rapportent. Mémoire présenté à
l'Institut, Académie des Sciences en 1828. Commercy, 1830. pp.
xvi–358, 8vo.

DENNSTEDT.

Anweisung wie der Landwirth die Runkelrübe und andere Rübenarten
auf die vortheilhafteste Art anbauen kann. Leipzig, 1836.

DENRÉES (LES) ALIMENTAIRES. Leurs altérations et leurs falsifications.
Conférences données au grand concours international de Bruxelles
en 1888 à l'occasion de l'exposition d'un laboratoire pour l'analyse
des denrées alimentaires, précédées d'une notice sur la participation
de l'administration du service de santé au grand concours et à l'Exposition. Bruxelles, 1889. pp. xxviii–310, 8vo.

Денисовъ Ѳедоръ.

Рѣчь о вліяніи химіи на успѣхи мануфактурной промышленности.
Москва, 1822.

    DENISOF, THEODORE. Discourse on the Influence of Chemistry upon In-
      dustry. Moscow, 1822.

DÉON, PAUL HORSIN.

    See Horsin-Déon, Paul.

DÉPIERRE, J.

Traité de la teinture et de l'impression des matières colorantes artificielles. Partie I. Les couleurs d'aniline. Paris, 1890. pp. iv–
558, 8vo. 12 plates and 221 specimens.

DESAIGNES, V.

Travaux de chimie organique.   Vendôme, 1886.

DESCHAMPS.

Manuel pratique d'analyse chimique.   Avec 39 [41] figures intercalées
dans le texte.   Paris, 1859.   2 vols., 8vo.   Vol. I, pp. xvi–416 ;
vol. II, pp. 602.

DESCHARMES, C.

Mémoire sur l'opium indigène.   Amiens, 1855.   8vo.

DES CLOISEAUX.

Leçon (etc.).   *See* Société chimique de Paris.

DESCRIPTION DE DIVERS PROCÉDÉS POUR EXTRAIRE LA SOUDE DU SEL MARIN.

Imprimerie du Comité de Salut public.   An III [1795].   4to.

DESCROIZILLES, FRANÇOIS ANTOINE HENRI.

Description et usage du Berthollimètre.   . . .   Paris, 1802.

Notice sur l'alcalimètre et autres tubes chimico-métriques, ou sur le
polymètre chimique.   . . .   Paris, 1810.

Deuxième édition, 1818.   Troisième édition, 1818.

Quatrième édition, 1824.   Cinquième édition, 1839.

Der Alkalimeter, Acetimeter und Polimeter oder genaue Beschreibung der Anfertigung und vielseitige Anwendung
genannter Instrumente nebst Bemerkungen über Alca-
lien, Säuren, Seifen, Brandewein und Beschreibung
eines kleinen leicht transportabel Destillirapparats.   Ein
Handbuch für Apotheker, etc.   Aus dem französischen
übersetzt von C. E. Schaumburg.   Nach der vierten
verbesserten Ausgabe.   Eisenach, 1833.   8vo.

Notice sur la fermentation vineuse.   . . .   Paris, 1822.

DESMAREST, E.

Chimie récréative.   . . .   Paris, 1829.

Précis de chimie.   . . .   Paris, 1824.

Traité élémentaire de chimie avec les applications de cette science aux
arts et aux manufactures.   Quatrième édition.   Paris, 1843.   12mo.

Химія для всѣхъ сословій, примѣненная къ ремесламъ и искусствамъ,
сообразно понятіямъ всякаго незнакомаго съ ея основаніями и
содержащая въ себѣ руководство къ устроенію небольшой и не-
дорогой лабораторіи у себя, которой однакожъ, желающій зани·
маться этою отраслью, можетъ производить всѣ извѣстные химическіе опыты.   Москва, 1839.   2 части.   8°

Traité des falsifications.   . . .   Paris, 1827.

Desmazures, C.
>   Analyse chimique minérale, d'après Fresenius.   Paris, 1888.   Fol., 2 pp. and 11 plates.

Despretz, César Mansuète.
>   Élémens de chimie théorique et pratique, avec l'indication des principales applications aux sciences et aux arts ; ouvrage dans lequel les corps sont classés par familles naturelles.   Paris, 1830.   2 vols., 8vo.   I, pp. ix-717 ; II, 826.   Plates.

Detharding, Georg.
>   Chymischer Probierofen des Joh. Agricolæ.   Stettin, 1648.
>   Chymischer Processofen.   Discurs vom Auro potabili, was es sey und was es für Eigenschaften an sich haben müsse.   Stettin, 1642.

Deusing, Antonius.
>   Dissertatio de manna et saccharo.   Groningen, 1659.   12mo.

Deville, H. Sainte-Claire.
>   See Sainte-Claire Deville, H.

Dewar, J.
>   Chemical Notes : 1. On the atomic volume of solid substances.   2. On inverted sugar.   Edinburgh, 1870.   8vo.

Dewilde, P.
>   Traité élémentaire de chimie générale et descriptive.   Deuxième édition, revue corrigée et augmentée.   Vol. II, Chimie organique.   Bruxelles, 1878.   12mo.

Dexter, W.
>   The Chemical Tables for the Calculation of Quantitative Analyses of H. Rose.   Recalculated for the more recent determinations of atomic weights, and with other alterations and additions.   Boston, 1850.   8vo.

Diacon, É.
>   Recherches sur la solubilité des mélanges salins.   Thèse de physique soutenue le 12 mai, 1864, devant la Faculté des Sciences de Montpellier.   Montpellier, 1864.   pp. 65, 4to.   Plate.

Dibbits, H. C.
>   Eerste beginselen der qualitatieve analytische scheikunde ten dienste van het middelbaar onderwijs.   Amsterdam, 1869.   8vo.
>   Eenige scheikundige werkingen als bewegings-verschijnselen opgevat.   Inwijdingsrede bij de aanvaarding van het hoogleersambt in de

DIBBITS, H. C.   [Cont'd.]

scheikunde aan de Hoogeschool te Utrecht, uitgesproken den
23sten Juni, 1876.   Amsterdam, 1876.   8vo.

De Spectraal-Analyse.   Academisch Proefschrift.   Rotterdam. 1863.

> "A complete treatise, giving an historical sketch of the discoveries, with
> chromolithographs of the carbon and other spectra."

DIBBITS, J. E.

De Wet van Berthollet, getoetst aan de draaiing van het polarisatievlak
bij cinchonine-zouten.   Na zijn overlijden uitgegeven door H. C.
Dibbits.   Haarlem, 1873.   pp. vi–74, 8vo.

DICKMANN, F.

Ueber die Bestimmung des Glycerins in Form von Nitroglycerin.
Erlangen, 1887.   8vo.

DIETRICH, TH., UND J. KÖNIG.

Zusammensetzung und Verdaulichkeit der Futtermittel nach vorhandenen Analysen und Untersuchungen zusammengestellt.   Zweite
völlig umgearbeitete und sehr vermehrte Auflage.   Berlin, 1891.
2 vols.

DIETZSCH, OSCAR.

Kuhmilch (Die), ihre Behandlung und Prüfung im Stall und in der
Käserei.   Mit einem Anhang über Markt-Milch und Rahm.   Zürich,
1889.   pp. vi–68, 8vo.   Folding table.

Wichtigsten (Die) Nahrungsmittel und Getränke, deren Verunreinigung
und Verfälschung.   Dritte Auflage.   Zürich, 1879.
Vierte Auflage.   Zürich, 1884,   8vo.

DIMOND, E. W.

The Chemistry of Combustion applied to the Economy of Fuel, with
special reference to the construction of fire-chambers for steam
boilers.   Worcester, 1867.   8vo.

DIONISIO, MICHELE.

Compendio di chimica farmaceutica secondo il programma d'insegnamento adottato nella R. università di Torino.   Torino, 1874.
2 vols., 8vo.

DIOS DE LA RADA, D. J. DE.

Principios elementales de química.   Granada, 1839–40.   Three parts.
8vo.

DIRCKS, V.

Vejledning i kvalitativ analyse til brug ved landbrugsskoler.   Anden
omarbejdede udgave.   Christiania, 1886.   8vo.

DIRUF, OSCAR.

Historische Untersuchungen über das Chinoidin in chemischer, pharmaceutischer und therapeutischer Beziehung, nebst Beobachtungen
über seine Wirksamkeit in Krankheiten und Versuchen über dessen
Verhalten zum thierischen Organismus in toxicologischer Hinsicht.
Erlangen, 1851.   8vo.

DITTE, A.

Exposé de quelques propriétés générales des corps.   *See, in Section II*, Fremy:
Encyclopédie chimique.

DITTE, A.

Traité d'analyse chimique qualitative.   Avec tableaux en couleurs
d'analyse spectral.   Paris, 1880.
Traité élémentaire de chimie fondée sur les principes de la thermochimie, avec emploi des données calorimétriques.   Paris, 1884.
12mo.

Kurzes Lehrbuch der anorganischen Chemie, gegründet auf
die Thermo-Chemie.   Uebersetzt von H. Böttger.   Berlin, 1886.

DITTE ET GUNTZ.

Uranium, Étain et Antimoine.   *See, in Section II*, Fremy : Encyclopédie
chimique.

DITTEN, H. S.

Veiledning ved den qualitativ-chemiske Analyse med et Anhang, hvori
en kortfattet Beskrivelse over Elementernes og de vigtigste For-
bindelsers Egenskaber.   Christiania, 1853.   8vo.

DITTMAR, WILLIAM.

Analytical Chemistry, a series of laboratory exercises, constituting a
preliminary course of qualitative chemical analysis.   Compiled for
the use of beginners in his laboratory.   New edition.   London,
1886.   12mo.

Chambers's Educational Course.

Chemical Arithmetic.   Part I. Collection of Tables, Mathematical,
Chemical and Physical, for the use of Chemists and others.
Glasgow, 1890.   8vo.
Exercises in Quantitative Chemical Analysis.   With a short treatise on
gas analysis.   London, 1887.

Second edition.   London, 1888.   8vo.

Manual (A) of Qualitative Chemical Analysis.   London, 1877.   8vo.

Handleiding bij de qualitatieve scheikundige analyse.   Naar

DITTMAR, WILLIAM. [Cont'd.]
het Engelsch bewerkt door L. J. van der Harst. Amsterdam, 1878. 8vo.
Report on Researches into the Composition of Ocean Water, made during the voyage of the *Challenger*. London, 1884. Imp. 4to.

DITTMAR, WILLIAM, WITH THE ASSISTANCE OF J. McARTHUR, A. KLING AND T. BARBOUR.
Tables to facilitate chemical calculations. Glasgow, 1884.

DIXON, H. B.
Conditions of Chemical Change in Gases. London, 1885. 4to. 2 plates.

DOBBIN, L.
Arithmetical Exercises in Chemistry. A series of elementary lessons on chemical calculation, with preface by C. Brown. Edinburgh, 1891. 8vo.

ДОБРЕВЪ. Н.
Краткое руководство по качественному химическому анализу. Софія, 1891.
DOBREF, N. A Brief Textbook of Qualitative Chemical Analysis. Sophia, 1891.

DOCUMENTS SUR LES FALSIFICATIONS DES MATIÈRES ALIMENTAIRES et sur les travaux du laboratoire municipal. Paris, 1885. pp. 812, 4to.
Deuxième rapport. Paris, 1887. 4to. Ill.
Nouveau tirage augmenté des lois étrangères sur les falsifications. Paris, 1889. 4to. Ill.

DODD, GEORGE.
Curiosities of Industry. The Chemistry of Manufactures. London, 1853. 8vo.

DÖBEREINER, FRANZ.
Cameralchemie für Land- und Forstwirthe, Techniker, Sanitäts-, Cameral- und Justiz-Beamte. Dessau, 1851. 3 vols., 8vo. Ill.
Chemie in Beziehung auf Leben, Kunst und Gewerbe. In Form von Vorlesungen bearbeitet. Zweite völlig umgearbeitete und vermehrte Auflage. Stuttgart, 1850. 8vo.
Also under the title : Der angehende Chemiker, oder Einleitung in die angewandte Chemie mit Angabe der interessantesten Experimente. Zum Gebrauch für alle welche sich mit der Chemie nach ihren Gesetzen und deren Anwendung im Leben beschäftigen oder bekannt machen wollen.
Dünger- und Bodenbestandtheile (Die), oder chemische Lehre über die Nahrungsmittel der Pflanzen. Dessau, 1854. 8vo.

DÖBEREINER, FRANZ. [Cont'd.]
Lehre (Die) von den giftigen und explosiven Stoffen der unorganischen
Natur welche im gewerblichen und häuslichen Leben vorkommen.
Populär bearbeitet.   Dessau, 1858.   8vo.

DÖBEREINER, JOHANN WOLFGANG.
Anleitung und Darstellung zum Gebrauch aller Arten der kräftigsten
Bäder und zur künstlichen Bereitung der wirksamsten Heilwasser,
welche von Gesunden und Kranken getrunken und als Bäder ge-
braucht werden.   Jena, 1816.   4to.
Anleitung zur Bereitung verschiedenen Essige.   Jena, 1816.
Anfangsgründe der Chemie und Stöchiometrie.   Jena, 1819.   pp. xiv–
424, 8vo.   4 plates.
    Zweite Auflage, 1826.   pp. x–358, 8vo.   4 folding plates,
    and tables.

Also under the title :   Grundriss der allgemeinen Chemie zum Gebrauche
bei seinen Vorlesungen.   Zweite Auflage, Jena, 1819.
    Dritte Auflage, Jena, 1826.
    Supplement zu J. W. D.'s Grundriss der Chemie, von Franz
    Döbereiner.   Stuttgart, 1837.   4to.
        See, in Section II, Döbereiner, Franz.

Beiträge zur chemischen Proportions-Lehre als Anhang zur Darstellung
der Verhältnisszahlen der irdischen Elemente zu chemischen Verbindungen und zum Grundrisse der allgemeinen Chemie.   Jena,
1816.   8vo.

Also under the title :   Döbereiner's neueste Stöchiometrische Untersu-
chungen und chemische Entdeckungen.
Beiträge zur physikalischen Chemie.   Jena, 1824–35.   3 parts.   8vo.
    Part I *also under the title :* Zur pneumatischen Chemie, Theil 4, 1824.
    Part II *also under the title :* Zur pneumatischen Chemie, Theil 5, 1825.
    Part III *also under the title :* Döbereiner's neuesten Erfahrungen und
    Beobachtungen im Gebiete der physikalischen und technischen
    Chemie.   1835.
[Beiträge] zur pneumatischen Chemie.   Jena, 1821–25.   5 parts.   8vo.   Ill.
    Cf. Beiträge zur physikalischen Chemie, *above.*

Elemente der pharmaceutischen Chemie.   Jena, 1815.   Zweite Auflage,
1819.
    Grundriss der allgemeinen Chemie zum Gebrauch bey seinen Vorlesungen.
    Jena, 1816.
    Cf. Anfangsgründe der Chemie.

Lehrbuch der allgemeinen Chemie.   Jena, 1811–'12.   3 vols., 8vo.

DÖBEREINER, JOHANN WOLFGANG. [Cont'd.]
Neueste stöchiometrische Untersuchungen und chemische Entdek-
kungen. Jena, 1816.
Ueber neu entdeckte höchst merkwürdige Eigenschaften des Platins
und die pneumatisch-capillare Thätigkeit gesprungener Gläser.
Ein Beitrag zur corpuscular Philosophie. Jena, 1823. 4to.
Also under the title : Die neuesten und wichtigsten physikalisch-che-
mischen Entdeckungen.
Zur Chemie des Platins in wissenschaftlicher und technischer Bezie-
hung. Für Chemiker, Metallurgen, Platinarbeiter, Pharmaceuten,
Fabrikanten und die Besitzer der Döbereinerschen Platinfeuer-
zeuge. Stuttgart, 1836. pp. 102, 12mo. Folding plate.
Zur Gährungschemie. Jena, 1822. Zweite Auflage, 1844.

DOELTER, C.
Allgemeine chemische Mineralogie. Leipzig, 1890. pp. 277, 8vo.

DOERING, ADOLFO.
La química del carbono, ó tratado de química orgánica segun las teorias
modernas con aplicacion á las artes, industria, medicina y farmacia.
Texto de las lecciones dadas en . . . la universidad nacional
de San Cárlos en Córdoba [República Argentina]. Buenos Aires,
1877. 8vo.

DOIJER VAN CLEEF, G.
Leerboek der scheikunde. Haarlem, 1886. 8vo.

DOLLFUS-AUSSET.
Matériaux pour la coloration des étoffes. Paris, 1865. 2 vols., roy.
8vo. Vol. I, Auteurs ; vol. II. Coloration.
> Volume I is essentially a biographical and bibliographical dictionary of
> writers on dyeing.

DOMBASLE, CHRISTOPHE JOSEPH ALEXANDRE MATTHIEU DE.
Agriculture (L') pratique et raisonnée. Paris, 1824. 2 vols.
Essai sur l'analyse des eaux par les réactifs. Paris, 1810.
Faits et observations sur la fabrication du sucre des betteraves. Paris,
1818.
> Deuxième édition. Paris, 1823.
> Troisième édition. Paris, 1831. pp. 216, 16mo.

DOMESTIC (THE) CHEMIST : comprising instructions for the detection of
Adulteration, in numerous articles employed in domestic economy, medicine, and the arts. To which are subjoined the art of

DOMESTIC (THE) CHEMIST. [Cont'd.]
> detecting poisons in food and organic mixtures and a popular introduction to the principles of chemical analysis. London, 1831. pp. x–340, 16mo.

DOMEYKO, IGNACIO.
> Tratado de ensayes, tanto por la via seca como por la via húmeda, de toda clase de minerales y pastas de cobre, plomo, plata, etc., con descripcion de los caracteres de los principales minerales y productos de las artes en America y en particular en Chile. Segunda edicion. Valparaiso, 1858. 4to.

DONATH, ED.
> Monographie der Alkohol-Gährung als Einleitung in das Studium der Gährungstechnik. Brünn, 1874. 8vo.

DONOVAN, MICHAEL.
> Treatise on Chemistry. London, 1832. 8vo.
>> Lardner's Cabinet Clyclopædia.
>>> Beginselen der scheikunde. Vrij naar het Engelsch bewerkt, door S. J. van Roijen. Amsterdam, 1854. 12mo.

DORÉ FILS.
> Leçons de chimie élémentaire appliquées aux arts industriels. Paris, 1857. 3 vols., 8vo. Ill.

DORN, GERHARD [DORNÆUS].
> Clavis totius philosophiæ chymisticæ. Lugduni Batavorum, 1567. Francoforti, 1583. Herbornæ, 1594.
>> *See also in Section VI.*

Δόσιος, Λέανδρ.
> Στοιχειώδη μαθήματα τεχνολογικῆς χημείας. Ἐν Ἀθήναις, 1871. 8vo.

DOSNE, P.
> Rapport sur les industries chimiques concernant le blanchîment, l'impression, la teinture des fibres textiles, les apprêts, la fabrication des matières colorantes, etc. Rouen, 1877. 8vo.

DOSSIE, ROBERT.
> Elaboratory (The) laid open, or the Secrets of Modern Chemistry and Pharmacy revealed. London, 1758.
> Institutes of Experimental Chemistry. London, 1759. 2 vols.

DOUGLAS, SILAS HAMILTON, AND ALBERT BENJAMIN PRESCOTT.
Qualitative Chemical Analysis, a guide in the practical study of chemistry and in the work of analysis. Ann Arbor, 1874. 8vo.
> Second edition. Ann Arbor, 1877.
> Third edition wholly revised. With a study of oxidation and reduction by Otis Coe Johnson. New York, 1881. pp. 305, 8vo.

DOVERI, LEONARDO.
Trattato elementare di chimica organica. Livorno, 1849. pp. 379, 8vo.

DOZY, FRANÇOIS.
Expositio concinna et perspicua elementorum stochiometricæ. Lugduni Batavorum, 1827. 4to.

DRAGENDORFF, GEORG.
Analyse chimique de quelques drogues actives et de leurs préparations pharmaceutiques par J. Morel. Gand, 1876. 8vo.
Beiträge zur gerichtlichen Chemie einzelner organischer Gifte. Untersuchungen aus dem pharmaceutischen Institute in Dorpat. St. Petersburg, 1872. pp. 312, 8vo.
Gerichtlich-chemische (Die) Ermittelung von Giften in Nahrungsmitteln, Luftgemischen, Speiseresten, Körpertheilen, etc. St. Petersburg, 1868. pp. xxiv–426, 8vo.
> Zweite völlig umgearbeitete Auflage. St. Petersburg, 1876.
> Dritte Auflage. Göttingen, 1888. pp. 567, 8vo.
> Manuel de toxicologie. Traduit avec des nombreuses additions et augmenté d'un précis des autres questions de chimie légale par E. Ritter. Avec gravures dans le texte et une planche chromo-lithographiée représentant l'analyse spectrale du sang. Paris, 1873. 8vo.
> Manuel de toxicologie. Deuxième édition française revue et très augmentée par L. Gautier. Paris, 1886. pp. xx–736, 8vo. Ill.
> Судебно-химическое открытіе ядовъ въ пищевыхъ веществахъ воздухѣ, остаткацъ пищи, частяхъ тѣла и т. д.
> Перев. Капустинъ М. и Ментинъ Н. С.-Петербургъ 1875.
Qualitative (Die) und quantitative Analyse von Pflanzen und Pflanzentheilen. Göttingen, 1882. pp. xvi–288, 8vo. Ill.
> Plant Analysis, Quantitative and Qualitative, translated by H. Greenish. London, 1887.

DRAGGENDORFF [sic] ET SCHLAGDENHAUFFEN.
> Analyse des végétaux. See, in Section II, Fremy: Encyclopédie chimique, vol. x.

DRAPER, JOHN CHRISTOPHER.
  A Practical Laboratory Course in Medical Chemistry. New York, 1882. pp. 71, oblong 12mo.

DRAPER, JOHN WILLIAM.
  Chemical (The) Organization of Plants. With an appendix, containing several memoirs on capillary attraction, electricity, and the chemical action of light. With engravings. New York, 1844. 4to.
  Scientific Memoirs, being experimental contributions to a knowledge of radiant energy [xxx Memoirs]. New York, 1878. pp. xxii–473, 8vo. Illustrations and portrait.
  Text-book (A) of Chemistry for the use of Schools and Colleges. Sixth edition. New York, 1848. pp. xi–408, 8vo. Ill.

DREBBEL, JACOBSZ. CORNELIS.
  Kort (Een) tractaet van de natuere der elementen, en hoe zy veroorsaecken den wint, regen, blixem, donder en waaromme dienstig zijn. Rotterdam, 1621.
        * Gründliche Auflösung von der Natur und Eigenschafft der Elementen, und was die Ursache dass Donner und Blitz, Hitz und Kälte, Winde, Regen, Hagel und Schnee, sich in der obern und untern Region erzeugen, und worzu selbige Anlass geben? Mit einem Anhang und klaren Beweiss, die von so vielen gesuchte Quint-Essenz aus allen dreyen Reichen zu haben, auch herrlichen Dedication vom Primo Mobili, sambt andern raren physikalischen Fragen, von einem Liebhaber der Hermetischen Kunst, herausgegeben. Franckfurt am Mayn; verlegts Margaretha Gertraud Isingin im Jahr Christi 1715. pp. 118, 16mo.
      The author uses the word Elements with the Aristotelian meaning.

DRECHSEL, EDMUND.
        Chemie der Absonderungen und Gewebe. See Handbuch der Physiologie.

DRECHSEL, EDMUND.
  Leitfaden in das Studium der chemischen Reactionen. Leipzig, 1874. pp. 94, 18mo.
        Zweite Auflage, —— und zur qualitative Analyse. Leipzig, 1888. 134 pp.
        Introduction to the Study of Chemical Reactions. Translated with permission of the author and of the publisher, and specially adapted to the use of American Students, by notes, etc., by N. Fred. Merrill. Second edition. New York, 188–. 12mo.

DREHER, E.

Beiträge zur modernen Atom- und Moleculartheorie auf kritischer Grundlage. Halle, 1882. 8vo.

DREYFUS, GASPARD.

Recueil de procédés chimiques appliquées aux arts et métiers divisé en deux parties. Toutes les recettes sont éprouvées et garanties par l'auteur, Joseph Ferdinand Charles. Paris, 1835. 12mo.

DRIESSEN, JAN CONSTANTYN.

Dissertatio de auro fulminante. Groningen, 1814. 4to.

DRIESSEN, PETRUS.

Natuur- en scheikundige waarnemingen over eenige gewigtige onderwerpen der geneeskunde en oeconomie. Leyden, 1791. 8vo.

Scheikundige verhandeling over de magnesia alba . . . Amsterdam, 1786.

DRINCOURT, E.

Cours de chimie. Paris, 1887. Ill.

DRINKWATER, T. W.

Synopsis of Chemistry, Inorganic and Organic ; to assist students preparing for examinations. Edinburgh and London, 1882.

DROEZE, J. HAVER.

Leerboek der scheikunde. 1ste deel, Metalloiden. Nieuwediep, 1878. 8vo. Tweede deel, Metalen. Nieuwediep, 1880. Ill.

DROHOJOWSKA.

Berthollet, Chaptal, oeuvres diverses. Paris, 1886. 8vo. Ill.

DROUX, L.

Chimie industrielle. Les produits chimiques et la fabrication des savons. Première partie : Produits chimiques ; acide sulphurique, soude et potasse, acides gras, bougies stéariques. Seconde partie : La savonnerie, savons de Marseille, savons mous à base de potasse, la fabrication en Hollande, fabrication parisienne, etc., etc. Troisième partie : Alcalimétrie, essais des potasses, essais des soudes, les différentes méthodes, etc. Paris, 1874. 8vo. Ill.

Deuxième édition par V. Larue. Paris, 1887. 2 vols. and atlas of 14 plates.

DUBRUNFAUT, AUGUSTIN PIERRE.

Art (L') de fabriquer le sucre de betteraves, contenant, 1 : La description des meilleures méthodes usitées pour la culture et la conserva-

DUBRUNFAUT, AUGUSTIN PIERRE.   [Cont'd.]

    tion de cette racine. 2 : L'exposition détaillée des procédes et appareils utiles pour en extraire le sucre avec de grands avantages ; suivi d'un essai d'analyse chimique de la betterave propre à éclairer la théorie des opérations qui ont pour objet d'en séparer la matière sucrée. Paris, 1825. pp. xvi–559, 8vo.

Mémoire sur la saccharification des fécules, présenté en 1822 à la Société centrale d'agriculture de Paris pour le concours qu'elle a ouvert sur la culture de la pomme de terre et l'emploi de ses produits, suivi de diverses notes et mémoires depuis l'année 1822 et de plusieurs travaux inédits sur la saccharification des matières amylacées par le malte. Deuxième édition. Paris, 1822. 8vo.

Notice historique sur la distillation des betteraves, rédigée à l'occasion de deux procès en contrefaçon intentés : 1° contre la Société Bocquet et compagnie de Sermaise (Marne) ; 2° contre MM. Lenfrey, Lefèvre et compagnie, d'Aubencheul (Nord) ; suivie de pièces justificatives. Paris, 1856. 8vo.

Osmose (L') et ses applications industrielles. Paris, 1873. 8vo.

Sucrage des vendages avec les sucres raffinés de canne, de betterave, etc., ou Vues sur cette méthode industrielle de vinification considérée comme moyen de régulariser la qualité des vins au niveau des grandes années, et d'en augmenter, au besoin, la quantité dans les années de recoltes mauvaises ou insuffisantes. Paris, 1854. 8vo.

    Sucrage des vendages à l'aide de sucres bruts blancs en grains, ou la betterave-canne de Nord, pour la production du sucre et auxiliaire de la vigne pour la production du vin. Paris, 1874. 8vo.

    Sucrage des vendages avec les sucres purs de cannes ou de betteraves, ou Méthode rationelle de régulariser la qualité des vins et d'en accroître, au besoin, la quantité. Troisième édition, revue et corrigée. Paris, 1880. 8vo.

Sucre (Le) dans ses rapports avec la science, l'agriculture, l'industrie, le commerce, l'économie publique et administrative . . . ou Études faites depuis 1866 sur la question des sucres. Paris, 1873–'78. 2 vols., 8vo.

DUCATEL, JULIUS TIMOLÉON.

Manual of Practical Toxicology, condensed from Dr. Christison's treatise on poisons. With notes and additions. Baltimore, 1833. 12mo.

DUCLAUX.

    Conférence (etc.). *See* Société chimique de Paris.

    Microbiologie. *See, in Section II*, Fremy : Encyclopédie chimique. Vol. IX.

DUCLAUX.

Le lait ; études chimiques et microbiologiques. Paris, 1887. Ill.

DUCLOS, SAMUEL COTTEREAU.

Dissertation sur les principes des mixtes naturels. Amsterdam, 1680.

Observations sur les eaux minérales de plusieurs provinces de France.
Paris, 1675.

DUCOIN-GIRARDIN.

Entretiens sur la chimie et ses applications les plus curieuses, suivis de
notions de manipulation et d'analyses chimiques. Tours, 1841. 8vo.

Deuxième édition. Tours, 1843. 8vo.

Nouvelle édition. Avec portrait. Tours, 1855. 8vo.

Sixième édition. Tours, 1865. 8vo.

Septième édition, 1869. Huitième édition, 1874.

Trattenimenti sulla chimica e sue più curiose applicazioni.
Versione di Carlo A. Valle. Torino, 1842. 16mo.

Дудинъ, Николай.

Химическія забавы. Москва, 1797.

DUDINE, NICOLAS. Chemical Amusements. Moscow, 1797.

DUDLEY, WILLIAM L.

Iridium. (From Mineral Resources of the United States.) Washington, 1885. 8vo.

DUEHRING, EUGEN CARL AND ULRICH DUEHRING.

Neue Grundgesetze zur rationellen Physik und Chemie. Leipzig,
1878–86. 2 vols., 8vo.

DUFF, A. P.

Introduction to Chemistry. London, 1885. 8vo.

DUFLOS, ADOLF.

Anweisung zur Prüfung chemischer Arzneimittel, als Leitfaden bei
Visitation der Apotheken, wie bei Prüfung chemisch-pharmaceutischer Präparate überhaupt. Ein Anhang zu den verschiedenen Ausgaben des chemischen Apothekerbuches. Breslau,
1849. 8vo.

Zweite Ausgabe. Breslau, 1862.

Die Prüfung chemischer Arzneimittel und chemisch-pharmaceutischer Präparate. Ein Leitfaden bei analytischen
Untersuchungen wie bei Visitation der Apotheken.
Dritte wesentlich verbesserte und vermehrte Bearbeitung. Breslau, 1866. 8vo.

DUFLOS, ADOLF. [Cont'd.]

    Esame chimico dei medicinali e dei prodotti chimici farma-
ceutici, ovvero guida per le ricerche analitiche e per
la visita delle farmacie; versione di Luigi Gabba.
Milano, 1867. 8vo.

Chemie (Die) in ihrer Anwendung auf das Leben und die Gewerbe.
Breslau, 1852. 2 vols., 8vo.

    De scheikunde toegepast op het dagelijksch leven en de
nijverheid. Uit het Hoogduitsch vertaald door J. H.
Giltay. Dordrecht, 1854–1855. 2 vols., 8vo.

[Chemisches Apothekerbuch.] Theorie und Praxis der pharmaceutischen Experimental-Chemie oder erfahrungsmässige Anweisung
zur richtigen Ausführung und Würdigung der in den pharmaceutischen Laboratorien vorkommenden pharmaceutisch- und analytisch-chemischen Arbeiten. Mit specieller Berücksichtigung der
Pharmacopoea Austriaca, Borussica, etc. Nebst einem Anhange,
die wichtigsten chemischen Hülfstabellen enthaltend. Mit in den
Text gedruckten Holzschnitten. Breslau, 1841. 8vo.

    The cover bears the title : Chemisches Apothekerbuch mit specieller Berück-
sichtigung aller gültigen Landespharmacopöen.

    Chemisches Apothekerbuch. Theorie und Praxis der phar-
maceutischen Experimental-Chemie. Zweite durchaus
umgearbeitete Ausgabe. Breslau, 1843. 2 vols., 8vo.

    Dritte Ausgabe. Breslau, 1847. 2 Bände und Ergänzungs-
band. 8vo.

    Ergänzungsband, also under the title : Grundriss der pharmaceutischen
Chemie.

    Chemisches Apothekerbuch. Theorie und Praxis der in
pharmaceutischen Laboratorium vorkommenen phar-
maceutisch-technischen und analytisch-chemischen Arbeiten. Fünfte Bearbeitung nebst Hülfstabelle für die
Praxis in pharmaceutischen Laboratorien und vergleichende Uebersicht der Nomenclatur der arzneilich
angewandten chemischen Präparate der Pharmacopoea
Germaniæ, der Pharmacopöen von der Schweiz, England, etc. Breslau, 1867. 8vo. Ill.

    Sechste Auflage, Leipzig, 1880.

    Chemisches Apothekerbuch. Theorie und Praxis der in
pharmaceutischen Laboratorien vorkommenden chemischen Arbeiten. Kleinere Ausgabe in einem Bande in
völlig neuer Bearbeitung. Breslau, 1857. 8vo. Ill.

Handbuch der angewandten, pharmaceutisch- und technisch-chemischen Analyse, als Anleitung zur Prüfung chemischer Arzneimittel

DUFLOS, ADOLF. [Cont'd.]

und zur Visitation der Apotheken, wie als Wegweiser zur Untersuchung und Beurtheilung von der Pharmacie, den Künsten, den
Gewerben und der Landwirthschaft angehörenden chemischen
Präparaten und Fabrikaten. Unter Berücksichtigung der älteren
und neuen Pharmakopöen Deutschlands, Oesterreichs, der Schweiz,
Englands, Frankreichs und Russlands, wie der Ergebnisse der
neuesten Forschungen im Gebiete der technischen Chemie, in
vierten Auflage neu bearbeitet. Ein Ergänzungs-Band zu den
verschiedenen Ausgaben von des Verfassers Werk : Chemisches
Apothekerbuch. Breslau, 1871. pp. xxiv–432, 8vo. Ill.

Handbuch der angewandten gerichtlich-chemischen Analyse der
chemischen Gifte. Leipzig, 1873.

> A second supplementary volume to the Author's Chemisches Apothekerbuch.

Pharmacologische Chemie. Die Lehre von den chemischen Arzneimitteln und Giften ; ihre Eigenschaften, Erkennung, Prüfung und
therapeutische Anwendung. Ein Handbuch für academische
Vorlesungen und zum Gebrauche für praktische und gerichtliche
Aerzte und Wundärzte. Breslau, 1842. 8vo.

> Zweite Ausgabe. Mit besonderer Berücksichtigung der
> neuesten Pharmacopöen. Breslau, 1848. 8vo. Er-
> gänzungsband (zur 1. Auflage).

> *Also under the title :* Grundriss der pharmacologischen Chemie.

> Handleiding tot de kennis der scheikundige geneesmiddelen
> en vergiften, gevolgd naar het Hoogduitsch, omgewerkt
> en met aanmerkingen, door P. J. Haaxman. Met eene
> plaat en vergelijkende Tabellen over geneeskundig-
> scheikundige nomenclatuur, temperatuur-herleidingen,
> gewigten, oplosbaarheid van zouten, enz. Schoonhoven,
> 1842. 8vo.

Prüfung (Die) chemischer Gifte, ihre Erkennung im reinen Zustand
und Ermittelung in Gemengen. Ein Leitfaden bei gerichtlich-
chemischen Untersuchungen für Aerzte, Apotheker, gerichtliche
Chemiker und Criminalrichter. Breslau, 1867. 8vo. Ill.

DUFLOS, ADOLF.

> Prüfung (Die) chemischer Arzneimittel (etc.) *See* Anweisung zur Prüfung,
> dritte Bearbeitung, *above.*

DUFLOS, ADOLF, UND ADOLF G. HIRSCH.

Das Arsenik, seine Erkennung und sein vermeintliches Vorkommen in
organisirten Körpern. Leitfaden zur Selbstbelehrung und zum
praktischen Gebrauche bei gerichtlich-chemischen Untersuchungen

DUFLOS, ADOLF, AND ADOLF G. HIRSCH.   [Cont'd.]
> für Aerzte, Physiker, Apotheker und Rechtsgelehrte. Breslau,
> 1842. 8vo.
>> Het arsenik, deszelfs herkenning en vermoedelijk voorkomen
>> in bewerktuigde ligchamen. Leiddraad tot zelfonder-
>> richt en tot practisch gebruik bij geregtelijk scheikundige
>> onderzoekingen voor artsen, natuurkundigen, enz. Naar
>> het Hoogduitsch door P. van Genderen Stort. Gro-
>> ningen; 1843. 8vo.
>
> Oekonomische Chemie. Breslau, 1846. 8vo. 2 parts. Part I : Die
> wichtigsten Lebensbedürfnisse ; ihre Aechtheit und Güte, ihre
> zufälligen Verunreinigungen und ihre absichtlichen Verfälschungen
> auf chemischen Wege erläutert. Zur Selbstbelehrung für Jeder-
> mann, wie auch zum Handgebrauche bei polizeilich-chemischen
> Untersuchungen. Erste Auflage, 1842.
>> Zweite neu bearbeitete und bereicherte Auflage. Breslau,
>> 1846.
>
> Part II : Die chemischen Bedürfnisse des Ackerbaues, ihre Eigen-
> schaften, Erkennung, Prüfung und ihr Einfluss auf die Productivi-
> tät des Bodens. Breslau, 1843.
>> Huishoudkundige scheikunde, de echtheid en goede hoeda-
>> nigheid, benevens de toevallige verontreinigingen en
>> opzettelijke vervalschingen der gewigtigste levensbe-
>> hoeften, scheikundig opgehelderd ; tot eigen onder-
>> rigt voor een ieder, alsmede tot handleiding bij Staats-
>> scheikundige onderzoekingen. Uit het Hoogduitsch.
>> Schoonhoven, 1843. 8vo.
>>
>> Landbouwkundige scheikunde, de eigenschappen, het onder-
>> zoek en de herkenning van de scheikundige benoodigd-
>> heden voor den landbouw en derzelver invloed op het
>> voortbrengingsvermogen van den grond ; tot eigen
>> onderrigt voor landbouwers, alsmede ten gebruike bij
>> landbouwkundig-scheikundige onderzoekingen. Uit
>> het Hoogduitsch vertaald met eenige wijzingen en
>> aanteekeningen door J. P. C. van Tricht. Arnhem,
>> 1844. 8vo.
>>
>> Potrzeby chemiczne rolnictwa, ich własności, ocenienie,
>> badanie i wpływ na żyzność gruntów . . . przełożył J.
>> S. Zdzitowiecki. Warszawa, 1844. 8vo.

DUFOUR, LOUIS.
> Essai sur quelques points de l'état actuel de la physique et de la chimie.
> Paris, 1853. 4to.

DUHAMEL DU MONCEAU, HENRI LOUIS.

Art (L') de faire de l'amidon.   Paris, 1775

Art (L') de faire différentes sortes de colles.   Paris, 1771.

Art (L') de raffiner le sucre.   Paris, 1764.

> Deuxième édition, Paris, 1790.

> Nouvelle édition augmentée de tout ce qui à été écrit de mieux sur ces matières . . . par J. E. Bertrand. Paris, 1812.   4to.   Ill.

> Die Kunst des Zuckersiedens.   Aûs dem französischen der Descriptions des Arts et Métiers übersetzt von Johann Heinrich Gottlob von Justi.   Königsberg und Mitau, 1765.   4to.   Ill.

Art (L') du savonnier.   Paris, 1774.

> Nouvelle édition par J. E. Bertrand.   Paris, 1812.

DUJARDIN, J.

L'essai commercial des vins et vinaigres.   Paris, 1892.   pp. viii–368, 8vo.   Ill.

> Bibliothèque des connaissances utiles.

DUJARDIN–BEAUMETZ.

Lezioni di chimica terapeutica, raccolte da E. Carpentier, Mericourt. Versione italiana, di V. Cozzolino.   Parte prima : Malattie del cuore e dell' aorta.   Napoli, 1879.   8vo.

DULK, FRIEDRICH PHILIPP.

De lucis effectibus chemicis.   Regiomontani, 1831.

Handbuch der Chemie.   Berlin, 1833–34.   2 vols., 8vo.

> Lehrbuch der Chemie.   Zum Gebrauch bei seinen Vorlesungen und zum Selbstunterricht.   Zweite verbesserte Auflage.   Berlin, 1842.   2 vols., 8vo.

DULK, FRIEDRICH PHILIPP.

> Synoptische Tabellen über die Atomgewichte [etc.].   *See in Section II.*

DUMAS, JEAN BAPTISTE.

> Leçon (etc.).   *See* Société chimique de Paris.

DUMAS, JEAN BAPTISTE.

De l'action du calorique sur les corps organiques, applications aux opérations pharmaceutiques.   Paris, 1838.   4to.

Dissertation sur la densité de la vapeur de quelques corps simples. Thèse . . . Paris, 1832.   8vo.

Leçon sur la statique chimique des êtres organisés, professée . . . pour la clôture de son cours à l'École de médecine.   Paris, 1841.   8vo.

DUMAS, JEAN BAPTISTE. [Cont'd.]
Second edition under the title :

> Essai de statique chimique des êtres organisés. Leçon.
> Deuxième édition augmentée de documens numériques.
> Paris, 1842. 8vo.
>
> Troisième édition par J. Dumas et J. B. Boussingault.
> Paris, 1844. 8vo.
>
> Les over de scheikundige evenwigtsleer der bewerktuigde
> wezens. Uitgesproken tot sluiting mijner lessen aan
> de hoogeschool te Parijs. Naar het Fransch. Gouda,
> 1843. 8vo.
>
> The Chemical and Physiological Balance of Organic Nature.
> Third edition, with new documents. London, 1844.
> pp. xii–166. 12mo.
>
> Versuch einer chemischen Statik der organischen Wesen.
> Zweite mit den nöthigen Zahlenbelegen vermehrte
> Auflage. Aus dem französischen von Carl Vieweg.
> Leipzig, 1844. 8vo.
>
> Storia chimica degli esseri organizzati, ossia la chimica
> applicata alle leggi che governano la vita degli animali
> e de' vegetabili. Prima versione italiana. Milano,
> 1847. 8vo.
>
> Ensayo de estática química de los séres organizados. Ter-
> cera edicion aumentada con nuevos documentos, por J.
> Dumas y J. B. Boussingault. Traducido por Ramon
> Corres Muñoz. Madrid, 1846. 8vo.

Leçons sur la philosophie chimique professées au Collège de France
[en 1836]. Recueillies par M. Bineau. Paris, 1837. pp. 430, 8vo.
  * Seconde édition, Paris, 1878. pp. [ii]–470, 8vo.

The second edition is a reprint of the first, and retains its inaccuracies.

> Die Philosophie der Chemie. Vorlesungen gehalten im
> Collège de France, gesammelt von Bineau und ins
> deutsche übertragen von C. Rammelsberg. Berlin,
> 1839. 8vo.
>
> Lezioni di filosofia chimica, pronunciate nel Collegio di
> Francia e raccolte da M. Bineau. Prima versione
> italiana di Giuseppi Orosi. Livorno, 1842. 12mo.
>
> Lecciones sobre la filosofía química esplicadas en el colegio
> de Francia. Dadas á luz por Bineau. Madrid, 1844. 8vo.

This work well illustrates the singular lack of accuracy on the part of French
historians of chemistry. The author, writing of Priestley, names Shee-
field (Sheffield) ; Mitt-Hill (Mill-Hill) ; Waringthon (Warrington) ;
Susqueannah (Susquehanna) ; he says Priestley sought in America the

DUMAS, JEAN BAPTISTE. [Cont'd.]

hospitality of *Peaux-rouges!* And he adds : " Priestley was poisoned at a meal with his entire family by an accident never explained. No one, however, succumbed except himself." This wholly fictitious statement has been copied by French authors again and again, so that it now forms an ineradicable part of French biographical sketches of the English chemist. *See* Chem. News, Feb. 14, 1890, page 84.

* Mémoires de chimie. Paris, 1843. pp. [x]–412, 8vo. 5 folding plates.

Contains : Mémoires sur les types chimiques ; Recherches sur la composition de l'eau, and other classical essays.

Précis de l'art de la teinture. Paris, 1846. 8vo.

Traité de chimie appliquée aux arts. Paris, 1828–46. 8 vols., 8vo. Ill.

Grundzüge der Chemie, angewendet auf Künste und Gewerbe. Weimar, 1829–30. 2 vols., 4to. Ill.

Handbuch der angewandten Chemie. Ein nöthiges Hülfsbuch für technische Chemiker, Künstler, Fabrikanten und Gewerbtreibende überhaupt. Aus dem französischen übersetzt und mit Anmerkungen versehen von G. Alex, F. Engelhart und [Ludwig Andr.] Buchner, Jr. Nürnberg, 1830–50. 8 vols., 8vo. Ill.

Vols. 6–8 *also under the title :* Handbuch der Chemie in ihrer Anwendung auf Künste und Handwerke. Deutsch von Buchner.

Trattato di chimica applicata alle arti. Milano, 1840–44. Vols. I–VI. 8vo. Ill.

Tratado de química aplicado á las artes, traducido por Luciano Martinez y Enrique Mieg. Madrid, 1845–48. 11 vols. of text and 1 atlas in folio. Ill.

Resúmen de las lecciones de química pronunciadas en la Escuela central de artes manufactureras de Francia. Traducidas de la coleccion litografiada de la indicada escuela. Madrid, 1848. 8vo. Ill.

Sur la loi de substitution de la théorie des types. Paris, 1841. 4to.

DUMAS-GUILIN, MAX.

Manuel du dynamiteur. La dynamite de guerre et le coton-poudre ; leur fabrication, leur conservation, leur transport et leur emploi. Paris, 1887. 12mo.

DU MÊNIL, AUGUST PETER JULIUS.

Analyse (Die) der thierischen Concretionen, oder Anleitung diese abnormen Erzeugnisse nach ihren physikalischen Merkmalen kennen zu lernen. Altona, 1837. 8vo.

Chemische Analysen anorganischer Körper, als Beitrag zur Kenntniss ihrer innern Natur. Schmalkalden, 1823. 8vo.

Chemische Forschungen im Gebiete der anorganischen Natur, enthaltend

Du Mênil, August Peter Julius.   [Cont'd.]
über 50 Analysen der bisher am wenigsten bekannten Fossilien, Mineralquellen, etc.   Hannover, 1825.   8vo.

Handbuch der Reagentien- und Zerlegungslehre, oder chemisch-analytische Studien nach einem neuen erprobten Plan, vornämlich zum Selbstunterricht, bündig und mit sorgfältiger Benutzung älterer und jüngster analytischer Schriften, wie auch eigener Erfahrung, für Freunde der praktischen Chemie, als Pharmaceuten, Aerzte, Mineralogen, Fabrikanten, Landwirthe.   Lemgo, 1836.   2 vols., 8vo.

Leitfaden zur chemischen Untersuchung der Naturkörper für alle welche die practische Chemie auf Wissenschaft, Künste und Gewerbe anwenden, als Pharmaceuten u. s. w., den neuesten Erfahrungen und besten Methoden eines Arfwedson, Bergman, Berthier gemäss.   Gotha, 1829.   2 vols., 8vo.

Neue physikalische Untersuchungen des Schwefelwassers zu Eilsen.   Hannover, 1826.

Reagentien-Lehre (Die) für die Pflanzen-Analyse, nebst einer Anzeige der Folge, in welcher die gegenwirkenden Mittel bei der chemischen Prüfung und weiteren Untersuchung verschiedener Pflanzen-Körper angewandt werden.   Zweite sehr vermehrte Auflage.   Celle, 1841.   8vo.
     First edition, 1834.

Treuer Wegweiser für arbeitende Chemiker und Freunde der analytischen Chemie.   Nürnberg, 1842.   8vo.

Du Mênil, August Peter Julius.
     See also in Section III.

Du Monceau, Henri Louis Duhamel.
     See Duhamel du Monceau, Henri Louis.

Dumont, J. S.
Tableau synoptique, ou toxicologie populaire, contenant les poisons les plus répandus, sous leur dénomination la plus vulgaire avec une classification spéciale, adoptée pour son usage, qui consiste à faciliter à tout le monde des moyens d'administrer le contre-poison.   Cambrai et Paris, 1855.   Fol.

Du Moulin, N.
La toxicologie du cuivre.   Recueil des discours prononcés devant l'Académie royale de médecine de Belgique.   Bruxelles, 1886.   pp. viii–285, 8vo.

Duncan, Daniel [of Montpellier].
Chymiæ naturalis specimen.   Hag, 1707.

DUNCAN, DANIEL.   [Cont'd.]
Chymiæ rationalis specimen.   Hag, 1707.
Chymie (La) naturelle ou l'explication chymique et méchanique de la
  nourriture de l'animal.   Paris, 1682.   3 vols., 12mo.
    Printed at Montauban.

DUNDONALD, EARL OF.
    See Cochrane, Archibald.

DUPASQUIER, ALPHONSE.
Des eaux de source et des eaux de rivière comparées sous le double rapport hygiènique et industriel, et spécialement des eaux de sources
  de la rive gauche de la Saône, près Lyon . . . Lyon, 1840.   8vo.
Histoire chimique, médicale et topographique de l'eau minérale sul-
  fureuse de l'établissement d'Allevard.   Lyon, 1841.
Traité élémentaire de chimie industrielle.   Lyon et Paris, 1844.   8vo.
    Only one part published.

DUPLAIS, P., AÎNÉ.
Traité de la fabrication des liqueurs et de la distillation des alcools.
  Cinquième édition par Duplais jeune.   Paris, 1891.   2 vols., 8vo.

DUPONT, F.
État actuel de la fabrication du sucre en France au point de vue
  technique.   Paris, 1889.   Ill.

DUPORTAL, A. S.
Anleitung zur Kenntniss des gegenwärtigen Zustandes der Branntwein-
  brennerei in Frankreich.   Aus dem französischen übersetzt von
  Hermbstädt.   Berlin, 1812.   8vo.   Ill.

DUPRÉ, A., AND H. W. HAKE.
A Short Manual of Chemistry.   Vol. 1 : Inorganic Chemistry.   London,
  1886.

DUPUY, B.
Alcaloïdes, histoire, propriétés chimiques et physiques, extraction, action
  physiologique, effets thérapeutiques, toxicologie, observations,
  usages en médecine, formules etc.   Avec préface de Dujardin-
  Beaumetz.   Paris, 1889.   2 vols., 8vo.

DUPUY, EDMOND.
Recherches sur la solubilité.   Paris, 1884.   8vo.

DUQUESNAY.
    Mortiers et ciments.   *See, in Section II*, Fremy : Encyclopédie chimique,
    vol. v.

DURAND, A. [de Cherbourg].
Considérations sur la nitrification et les effets de la propagation lente du calorique considéré comme identique avec l'électricité naturelle. À Toulouse, 1824.   8vo.

DURAND, F. AUGUSTE.
Nouvelle théorie physique, ou études analytiques et synthétiques sur la physique et sur les actions chimiques fondamentales.   Paris, 1854. 8vo.

DURAND-CLAYE, CHARLES LÉON.
Chimie appliquée à l'art de l'ingénieur.   Paris, 1885.   pp. viii–299, roy. 8vo.
        Encyclopédie des travaux publics.

DURANDE, JEAN FRANÇOIS.
Élémens de chymie, rédigés dans un nouvel ordre.   Dijon, 1778.

DUREAU, B.
De la fabrication du sucre de betterave dans ses rapports avec l'agriculture et l'alimentation publique, avec des considérations sur la partie économique de la législation de cette industrie.   Paris, 1858.   8vo.

DUREAU, GEORGES.
Le procédé à la strontiane pour l'extraction du sucre des mélasses; théorie, pratique et avantages de ce procédé.   Paris, 18—.

DURO Y GARCES, JOSÉ.
Discurso sobre los diferentes métodos de ensayar y afinar los metales preciosos y sus aleaciones mas usuales.   Madrid, 1854.   4to.

DUROCHER, J.
Recherches sur les roches et les minéraux des îles Féroë.   Thèse de chimie.   Paris, 1841.   8vo.   Ill.

DUROY, J. L. P.
Expériences et considérations nouvelles pour servir à l'histoire de l'iode.   Paris, 1854.   8vo.

DÜRR, MICHAEL.
        See Müller, Johannes.

DURRANT, REGINALD G.
Laws and Definitions connected with Chemistry and Heat.   With notes on physical and theoretical chemistry; also special tests and examples for practical analysis.   London, 1887.   8vo.

DUSSAUCE, H.

General (A) Treatise on the Manufacture of Soap, theoretical and practical ; comprising the chemistry of the art, a description of all the raw materials and their uses, directions for the establishment of a soap factory, with the necessary apparatus, instructions in the manufacture of every variety of soap, the assay and determination of the value of alkalies, fatty substances, soaps, etc., etc. With an Appendix. Philadelphia and London, 1869. pp. xxviii–17–807, 8vo.

Practical (A) Treatise on the Fabrication of Matches, Gun-Cotton, Colored Fires and Fulminating Powders. Philadelphia, 1864. pp. 336, 8vo. Ill.

DUTTENHOFER, F. M.

Die Lehre von der Hauswirthschaft mit besonderer Rücksicht auf technische und chemische Grundsätze. Stuttgart, 1846. 8vo.

DYBONSKI, A.

Memento de chimie. Paris, 1887.
Deuxième édition. Paris, 1889.

DYER (THE) AND COLOR-MAKER'S COMPANION. New edition. Philadelphia, 1891. 12mo.

DYRENFURTH, M.

Vergiften en tegengiften. Uit het Duitsch vertaald door A. W. J. Zubli en A. Arn. J. Quanjer. 's Gravenhage, 1880. 8vo.

DWORŽAK, HUGO.

Ueber Malz und dessen Verwerthung. Programm der Landes-Ober-Realschule. Kremsier, 1886.

EARL, A. G.

The Elements of Laboratory Work. A course of natural science. London, 1890. pp. xii–179.

EATON, AMOS.

Chemical Instructor. Fourth edition. Troy, 1833. 12mo.

EBELMEN, JACQUES JOSEPH.

Chimie, céramique, géologie, métallurgie. Revue et corrigé par M. Salvétat, suivi d'une notice sur la vie et les travaux de l'auteur . . . par M. E. Chevreul. Paris, 1861. 3 vols., 8vo. Ill.
First edition, 1855-61.

EBELMEN, JOSEPH JACQUES, ET SALVÉTAT.

Recherches sur la composition des matières employées dans la fabrication et la décoration de la porcelaine de Chine, exécutées à la
manufacture nationale de porcelaine de Sèvres et présentées à
l'Académie des sciences.   Paris, 1852.   8vo.

EBERLE, J. N.

Physiologie der Verdauung nach Versuchen auf natürlichem und
künstlichem Wege.   Würzburg, 1838.   pp. xviii–408, 8vo.

EBERMAIER, JOHANN ERDWIN CHRISTOPH.

Physikalisch-chemische Geschichte des Lichts und dessen Einfluss auf
den menschlichen Körper.   Osnabrück, 1799.

Zweite Auflage, Leipzig, 1810.

Tabellarische Uebersicht der Kennzeichen der Aechtheit und Güte
sämmtlicher Arzeneimittel.   Leipzig, 1804.

Vierte Auflage, 1819.

EBERMAYER, ERNST.

Beschaffenheit (Die) der Waldluft und die Bedeutung der atmosphärischen Kohlensäure für die Waldvegetation.   Zugleich eine übersichtliche Darstellung des gegenwärtigen Standes der Kohlen-
säurefrage.   Aus dem chemisch-bodenkundlichen Laboratorium
der königlichen bayerischen forstlichen Versuchs-Anstalt.   Stuttgart, 1885.   8vo.

Naturgesetzliche Grundlagen des Wald- und Ackerbaues.   Erster
Theil: Physiologische Chemie der Pflanzen. Zugleich Lehrbuch
der organischen Chemie und Agriculturchemie für Forst- und
Landwirthe, Agriculturchemiker, Botaniker, etc.   Berlin, 1882.
2 vols., 8vo.

EBERT, H.

Die Methode der hohen Interferenzen in ihrer Verwendbarkeit für
Zwecke der quantitiven Spectralanalyse.   Erlangen, 1888.   pp.
71.   4to.   Ill.

EBOLI, C. W.

Curso de química organica analitica.   Trad. por José A. de los Rios.
Lima, 1865.   Fol.

ECCLES, R. G.

The Evolution of Chemistry.   New York, 1891.   12mo.

ECKHARD, C.

Physiologie des Rückenmarks.   *See* Handbuch der Physiologie.

ECKHARTSHAUSEN, KARL VON.

*Entwurf zu einer ganz neuen Chimie durch die Entdeckung eines
allgemeinen Naturprinzips wodurch sich das phlogistische System
der alten, und das antiphlogistische der neuen Chimisten als zwey
Extrema in ein Mittelsystem vereinigen lassen, worinn allein die
Wahrheit liegt, und die höhere Chimie der ältesten Vorzeit mit der
gemeinen Schulchimie der jetzigen Zeit vereiniget wird. Regensburg, 1800, bey Montag und Weiss.

> Vagaries of a mystical mind, unworthy of the period at which written. The
> author discourses of fire-matter, heat-matter and light-matter and
> shows how phosphorus produces " *Naturschwefel* " and " *Sonnenstoff.*"

EDE, ROBERT BEST.

Practical Facts in Chemistry exemplifying the rudiments and showing
with what facility the principles of the science may be experi-
mentally demonstrated at a trifling expense by means of simple
apparatus and portable laboratories, more particularly in reference
to those.   A new and enlarged edition, to which is added a distinct
chapter on agricultural analysis. London, 1845. 18mo.

> First edition, London, 1837.

EDER, JOSEF MARIA.

Ausführliches Handbuch der Photographie. Halle, 1884. 3 vols., 8vo.
Ill.

> Zweite Auflage, 1891.

Die chemischen Wirkungen des Lichtes (Photochemie), Spectralphoto-
graphie, die Photographie im Zusammenhang mit klimatischen
Verhältnissen und die Actinometrie. Zweite gänzlich umgearbeitete und vermehrte Auflage. Halle, 1891. 8vo. Ill.

> The Chemical Effect of the Spectrum.   Translated and
> edited by W. de W. Abney. London, 1883. 12mo.
> *Also* New York, 1885.

Ueber die Reactionen der Chromsäure und der Chromate auf Gelatine
Gummi, Zucker und andere Substanzen organischen Ursprungs in
ihren Beziehungen zur Chromatphotographie. Preisgekrönt von
der Photographischen Gesellschaft in Wien. Wien, 1878. pp. viii-
96, 8vo.

EDSON, HUBERT.

Record of Experiments at the Sugar Experiment Station on Calumet
Plantation, Pattersonville, La. Bulletin No. 23, Division of Chemistry, U. S. Department of Agriculture. Washington, 1889. 8vo.

EGGER, E.

Zweiter Rechenschaftsbericht des chemischen Untersuchungsamts für die Provinz Rheinhessen. Mainz, 1885. 8vo.

EGGERTZ, V.

Om kemisk pröfning af jern, jernmalmer och brännmaterialier. Falun 1870. 8vo.

Руководство къ химическимъ пробамъ желѣза, желѣзныхъ рудъ и горючихъ матеріаловъ Перев. Хирьякова М. С.-Петербургъ 1872.

EGLESTON, THOMAS.

The Metallurgy of Silver, Gold and Mercury in the United States. New York, 1887–90. 2 vols., 8vo.

Егоровъ, П.

Пачальныя основанія химіи. С.-Петербургъ 1862.

Another edition, 1854.

EGOROF, P. The first principles of chemistry according to Regnault. St. Petersburg, 1862.

EHRENREICH, JOHANN EBERHARD LUDWIG.

Abhandlung vom concentrirten Essiggeist. Königsberg, 1778.

EHRLICH, EDMUND

Wasserglass (Das). Seine Fabrikmässige Darstellung und rationelle Anwendung für Industrie, Gewerbe und Landwirthschaft. Nebst einem Anhang über Stereochromie. Nach den neuesten Erfahrungen zusammengestellt. Quedlinburg, 1858. 8vo.

Zünd-Waaren-Fabrikation (Die) in ihrem neuesten Stadium der Vollkommenheit. Nach praktischen Erfahrungen gesammelt. Würzburg, 1858. 8vo.

EHRMANN, MARTIN S.

Handbuch der populären Chemie in deren vielseitiger Beziehung zum gemeinen Leben und in der mannigfachen Benützung chemischer Grundsätze und Thatsachen gemeinfasslich zusammengestellt. Wien, 1842. 2 parts, 8vo.

Lehrbuch der Chemie. Dritte Auflage, Leipzig, 1840. 8vo.

Populäre Darstellung der neueren Chemie mit Berücksichtigung ihrer technischen Anwendung. Leipzig, 1828.

Stöchiometrie (Die), auf eine leichtfassliche Weise ohne Beihülfe algebraischer Berechnung erläutert. Wien, 1828–29. 8vo. Table in 4to.

EICHWALD, E. (JUN.).

Beiträge zur Chemie der gewebbildenden Substanzen und ihren Abkömmlinge. Berlin, 1873. 8vo.

EIDHERR, ED., UND ALOIS SCHÖNBERG.

Der chemisch-technische Brennereileiter. Populäres Handbuch der Spiritus- und Presshefe-Fabrikation. Vollständige Anleitung zu Erzeugung von Spiritus und Presshefe aus Kartoffeln, Kukuruz, Korn, Gerste, Hafer, Hirse und Melasse ; auf Grundlage vieljähriger Erfahrungen ausführlich und leichtfasslich geschildert. Dritte vollständig umgearbeitete Auflage. Leipzig, 1886. Ill.

    Cf. Schönberg, Alois.

EIMBKE, GEORG.

Analysis chemica fontium muriaticorum Oldesloënsium. Kiliæ, 1794.

    Versuch einer systematischer Nomenclatur. *See in Section II.*

Эйнбродтъ, Павелъ.

Объ атомическомъ вѣсѣ азота. Харьковъ, 1846. 8⁰

    EINBRODT, PAUL. On the Atomic Weight of Nitrogen. Charkow, 1846.

EINHOF, HEINRICH, UND ALBRECHT THAER.

Grundriss der Chemie für Landwirthe aus Heinrich Einhof's hinterlassenen Dictaten. Berlin, 1808. 8vo.

EISSLER, MANUEL.

Handbook (A) on Modern Explosives, being a practical treatise on the manufacture and application of dynamite, gun-cotton, nitroglycerine and other explosive compounds, including the manufacture of collodion-cotton. With about one hundred illustrations. London, 1890. pp. xviii–318, 8vo.

Nitroglycerine and dynamite ; their manufacture, their use, and their application to mining and military engineering ; pyroxyline, or gun-cotton ; the fulminates, picrates and chlorates ; also the chemistry and analysis of the elementary bodies, which enter into the manufacture of the principal nitro-compounds. New York, 1884. 8vo.

EKEBERG, ANDERS GUSTAF.

De calce phosphorata. Upsaliæ, 1793.

De topazio. Upsaliæ, 1796.

EKEBERG, ANDERS GUSTAF, ET J. J. BERZELIUS.

De nova analysi aquarum medeviensium. Upsaliæ, 1800.

ELBS, KARL.
 Die synthetischen Darstellungsmethoden der Kohlenstoff-Verbindungen.
      Leipzig, 1889–91.   2 vols., 8vo.
        A valuable key to the subject treated, written in a scientific spirit.

ELDERHORST, WILLIAM.
  A Manual of Blowpipe Analysis.   New York, 1856.
                A Manual of Blowpipe Analysis and Determinative Miner-
                   alogy.   Second edition revised and greatly enlarged.
                   New York, 1861.
          For later editions *see* Nason, Henry B.   *Cf. also* Landauer, F.

ELHUYAR, JUAN JOSÉ Y DON FAUSTO DE.
  Análisis químico del volfram y exámen de un nuevo metal que entra
     en su composicion.   Extractos de las Juntas de la Real Soc.
     Bascongada, 1783.   4to.
                A Chemical Analysis of Wolfram and examination of a new
                   metal entering into its composition.   Translated from
                   the Spanish by Ch. Cullen, to which is prefixed a trans-
                   lation of Mr. Scheele's analysis of the Tungsten or
                   heavy stone, with Mr. Bergmann's supplemental re-
                   marks.   London, 1785.
                Chemische Zergliederung des Wolframs von Don Fausto de
                   Luyart [*sic*].   [Translated by Gren.]   Halle, 1786.   8vo.

ELIOT, CHARLES WILLIAM, AND FRANCIS HUMPHREYS STORER.
  Compendious (A) Manual of Qualitative Chemical Analysis.   New York,
     1873.   12mo.
                New edition revised with the co-operation of the Authors by
                   William Ripley Nichols.   New York, 1880.   12mo.
                Sixteenth edition.   Newly revised by W. B. Lindsay.   New
                   York, 1892.
  Elementary (An) Manual of Chemistry.   Abridged from Eliot and
     Storer's Manual by Wm. Ripley Nichols.   New York, 1872.   12mo
     Ill.
                Other editions, 1873, 1880.
  Manual of Inorganic Chemistry, arranged to facilitate the experimental
     demonstration of the facts and principles of the science.   Second
     edition, revised.   New York and London, 1868.   8vo.
  Manual of Organic Chemistry, arranged to facilitate the experimental
     demonstration of the facts and principles of the science..   London,
     1868.   8vo.
        Compare Nichols, W. R., and L. M. Norton.

ELLIOT, J.

Elements of the Natural Philosophy connected with Medicine—viz : chemistry, optics, acoustics, hydrostatics, electricity and physiology. Including the doctrine of the atmosphere, fire, phlogiston, water, etc. Together with Bergman's Tables of Elective Attractions with explanations and improvements. Second edition, corrected, with additions. London, 1786. pp. xvi–338. 8vo.

ELLIS, G. E. R.

An Introduction to Practical Organic Analysis. Adapted to the requirements of the first M. B. examination. London, 1885. 8vo.

Papers on Inorganic Chemistry. With numerical answers progressively arranged. London, 1886.

Papers in Organic Chemistry. London, 1886.

ELLIS, ROBERT.

The Chemistry of Creation : being an outline of the chemistries of the earth, the air, the ocean. Published under the direction of the Committee of General Literature and Education, appointed by the Society for Promoting Christian Knowledge. London, 1850. pp. x–512, 8vo.

ELSHOLTZ, JOHANN SIGISMUND.

Distillatoria curiosa seu ratio ducendi liquores coloratos per alembicum. Berolini, 1674.

De phosphoris observationes. Berolini, 1671.

ELSNER, FRANZ CARL LEO.

Galvanische (Die) Vergoldung, Versilberung und Verkupferung. Berlin, 1843.

Neue Erfahrungen bei der galvanischen Vergoldung. Berlin, 1845.

Leitfaden der qualitativ-chemischen Analyse, oder Lehre von den Reagentien und dem Verhalten der am häufigsten vorkommenden Körper gegen Reagentien ; nebst specieller Anleitung zu qualitativ-chemischen Untersuchungen. Mit einem Anhange, welcher die quantitativen Bestimmungs-Methoden der gewöhnlicheren, bei Analysen vorkommenden Körper andeutet. Für diejenigen, welche mit chemischen Untersuchungen sich zu beschäftigen anfangen, bearbeitet Berlin, 1844–45. 2 parts, 8vo.

Zweite Auflage, 1851. 8vo.

ELSNER, FRITZ.

Grundriss der pharmaceutischen Chemie, gemäss den modernen An-

ELSNER, FRITZ. [Cont'd.]

    sichten. Ein Leitfaden für den Unterricht, zugleich als Handbuch zum repetiren für Pharmaceuten und Mediciner. Zweite gänzlich umgearbeitete Auflage. Berlin, 1875. 8vo. Ill.

        Dritte Auflage. Berlin, 1883. 8vo.

        Schets der pharmaceutische scheikunde, volgens de nieuwere begrippen. Leiddraad bij studie en practijk. Naar het Hoogduitsch, vrij en in overeenstemming met de pharmacopœia Neerlandica. Ed. II bewerkt door R. J. Opwyrda. Sneek, 1878. 8vo.

        Tweede uitgave. Sneek, 1882. 8vo.

    Neuere Nahrungs- und Genussmittel aus dem Pflanzenreiche sowie deren Surrogate und Verfälschungsmittel. Halle, 1885.

    Praxis (Die) des Nahrungsmittel-Chemiker. Anleitung zur Untersuchung von Nahrungsmitteln und Gebrauchsgegenständen, sowie für hygienische Zwecke. Für Apotheker, Chemiker und Gesundheits-Beamte. Dritte umgearbeitete und vermehrte Auflage. Hamburg, 1885. 8vo. Ill.

        Die Praxis des Chemikers bei Untersuchung von Nahrungsmitteln und Gebrauchsgegenständen, Handelsproducten, Luft, Boden, Wasser, bei bakteriologischen Untersuchungen, sowie in der gerichtlichen und Harn-Analyse. Ein Hülfsbuch für Chemiker, Apotheker und Gesundheitsbeamte. Vierte umgearbeitete und vermehrte Auflage. Hamburg, 1889. 8vo. Ill.

        Erste Auflage. Leipzig, 1880.

    Untersuchungen von Lebensmitteln und Verbrauchsgegenständen. Berlin, 1878.

ELTOFT, THOMAS.

    A Systematic Course of Practical Qualitative Analysis, specially arranged for Students preparing for the Science and Art Department, Medical Schools, Preliminary, Scientific and First B. Sc. (London), Oxford and Cambridge Local Practical Chemistry Examinations. London, 1882(?). 8vo.

EMBECH.

    Appunti di lezioni di chimica forestale ed agraria, date alle guardie della Selva di Schattenthal. Prima versione italiana. Firenze, 1884.

EMMET, JOHN PATTEN.

    An Essay on the Chemistry of Animated Matter. New York, 1822. pp. 125, 8vo.

EMMET, JOHN PATTEN. [Cont'd.]

> An Inaugural Dissertation at the College of Physicians and Surgeons of the
> University of New York for the degree of M.D. on the 2d April, 1822.

EMSMANN, A. H., AND OTTO DAMMER.

Des deutschen Knaben Experimentirbuch. Praktische Anleitung zum
unterhaltenden und belehrenden Experimentiren auf den Gebieten
der Physik und Chemie. Zwei Theile. Bielefeld und Leipzig,
1874. 8vo. Ill.

EMY, C. J.

Cours des sciences physiques et chimiques appliquées aux arts militaires.—Application de la métallurgie du fer, au service de l'artillerie, comprenant la fabrication des projectiles, des flaques d'affûts,
des mortiers, des essieux et des ancres. Metz et Paris, 1849.
8vo. Ill.

ENDLICH, F. M.

Manual of Qualitative Blowpipe Analysis and Determinative Mineralogy.
New York, 1891. Ill.

ENGEL, R.

Nouveaux éléments de chimie médicale et de chimie biologique avec
les applications à l'hygiène, à la médecine légale et à la pharmacie.
Paris, 1878. pp. vii–768, 12mo. Ill.

> Deuxième édition. Paris, 1883. 12mo.
>
> Troisième édition. Paris, 1888. 12mo.
>
> Nuevos elementos de química médica y biológica, con las
> aplicaciones á la higiene, medicina legal y farmacia.
> Traducción española y considerablemente aumentada
> por G. Saenz Diez y M. de Tolosa Latour. Madrid,
> 1882. 4to.
>
> Nuevos elementos de química médica y biológica, con las
> aplicaciones á la higiene. Tercera edición, traducida
> por V. M. de Argenta. Madrid, 1891. 8vo.

ENGELHARDT, ALBIN.

Handbuch der praktischen Seifen-Fabrikation. Die in der Seifen-
Fabrikation angewendeten Rohmaterialien, Maschinen und Geräthschaften. Wien, Pest, Leipzig, 1886. 8vo. Ill.

Handbuch der praktischen Seifen-Fabrikation. Die gesammte Seifen-
Fabrikation nach dem neuesten Standpunkte der Praxis und
Wissenschaft. Wien, Pest, Leipzig, 1886. 8vo. Ill.

Handbuch der practischen Kerzen-Fabrikation. Wien, Pest, Leipzig,
1887. 8vo. Ill.

ENGELHARDT, ALBIN. [Cont'd.]

Handbuch der praktischen Toilette-Seifen-Fabrikation. Praktische Anleitung zur Darstellung aller Sorte von deutschen, englischen und französischen Toiletteseifen, sowie der medicinischen Seifen, Glycerinseifen und der Seifenspecialitäten. Unter Berücksichtigung der hierzu in Verwendung kommenden Rohmaterialen, Maschinen und Apparate. Wien, Pest, Leipzig, 1888. 8vo. Ill.

ENGELMANN, Th. W.

Flimmer- und Protoplasmabewegung. *See* Handbuch der Physiologie.

ENGESTRÖM, GUSTAV VON.

Beschreibung eines mineralogischen Taschen-Laboratoriums, und insbesondere des Nutzens des Blaserohrs in der Mineralogie. Aus dem schwedischen übersetzt und mit Anmerkungen versehen von Christian Ehrenfried Weigel. Greifswald, 1774. 8vo. Ill.

Zweite Auflage. Greifswald, 1782.

Laboratorium chymicum. 1 Stück, Gold und Silver. 2. Stück, om Krasser. Stockholm, 1781. 8vo.

ENGLER, C.

Die deutschen Erdöle. Berlin, 1887. 4to.

Das Erdöl von Baku, Geschichte, Gewinnung, Verarbeitung. Stuttgart, 1886.

ENGLER, C.

Handbuch der chemischen Technologie. *See* Bolley's Handbuch, etc.

ENGRAIS (DES), DE LEUR COMPOSITION, de leur emploi, de leur action fertilisante dans l'agriculture, dans la culture maraîchère et dans l'horticulture. Recueil des meilleures et des plus récentes publications qui ont paru en France et en Angleterre sur l'usage pratique des engrais. Ouvrage publié sous les auspices de la Société d'Horticulture de Liège. Liège, 1848. 8vo. Ill.

ENKLAAR, J. E.

Handleiding bij het verrichten der proeven vermeld in de " eerste beginselen der scheikunde." Groningen, 1881. 8vo. Ill.

Leerboek der anorganische scheikunde op proefondervindelijken grondslag bewerkt. Deventer, 1873. 2 parts, 8vo.

EPSTEIN, J.

Einführung in das electro-technische Maassystem. Frankfurt am Main, 1891. 8vo.

ERCKER, LAZARUS.

Beschreibung aller fürnemsten mineralischen Ertzt und Berckwercks-
arten . . . mit schönen Figuren und Abriss der Instrumente
trewlich und fleissig an Tag geben. Prag, 1574. Fol.

> Other editions : Frankfurt am Mayn, 1580. Fol. Also 1598.
> Fol. Also 1629. Fol. Also as follows : Aula subter-
> ranea domina dominantium subdita subditorum : das
> ist unterirdische Hofhaltung . . . oder gründliche
> Beschreibung derjenigen Sachen so in der Tieffe der
> Erden wachsen. [Edited by J. H. Cardilucius.] In-
> terpres phraseologiæ metallurgicæ . . . zusammen-
> getragen durch E. Berwardum. Franckfurt, 1684. 3
> parts, 4to.

Another edition, 1703. Fol.

Fünffte Auflage. Franckfurt am Mayn, 1736. Fol.

For an English version see Pettus, John.

ERDMANN, CARL GOTTLIEB HEINRICH.

Lehrbuch der Chemie und Pharmakologie für Aerzte und Thierärzte.
Zum Gebrauch bei Vorlesungen und zum Selbstunterricht entworfen Mit Abbildungen. Berlin, 1836–54. 3 vols., 8vo.

ERDMANN, H.

Anleitung zur Darstellung chemischer Präparate. Ein Leitfaden für
den praktischen Unterricht in der anorganischen Chemie. Frankfurt a. M., 1890. 8vo. Ill.

ERDMANN, OTTO LINNÉ.

Lehrbuch der Chemie. Leipzig, 1828.

> Vierte völlig umgearbeitete und vermehrte Auflage. Leipzig,
> 1851.

> Algemeen overzigt der nieuwere scheikunde. Uit het
> Hoogduitsch door W. S. Swart. Amsterdam, 1836.
> 2 vols., 8vo.

De natura affinitatis chemicæ. Dissertatio academica. Lipsiæ, 1825.

ERLENMEYER, E.

Lehrbuch der organischen Chemie. Leipzig, 1883–92. 3 vols., 8vo.
Vol. I. Bearbeitet von E. E. und O. Hecht. Vol. II. Unter Mit-
wirkung von F. Carl und A. Lippe, von Otto Hecht. Vol. III.
Herausgegeben von Otto Hecht. In progress.

ERXLEBEN, JOHANN CHRISTIAN POLYCARP.

* Anfangsgründe der Chemie. Göttingen, 1775. pp. [xxxii]–472–[lii],
8vo.

ERXLEBEN, JOHANN CHRISTIAN POLYCARP. [Cont'd.]
>   Zweite Auflage vermehrt von J. C. Wiegleb. Göttingen, 1784.
>   Dritte Auflage. Göttingen, 1793. 8vo.
>   Начальныя основанія химіи. С.-Петербургъ 1788.
> Physikalisch-chemische Abhandlungen. Göttingen, 1777.
> Versuche über den Anbau der Zuckerrübe. Prag, 1818.

ERRERA, L.
> Pourquoi les éléments de la matière vivante ont-ils des poids atomiques
>   peu élevés ? Messina, 1886. 8vo.

ESBACH.
> Analyse complète du lait. Rapidité, précision. Paris, 1881. 8vo.

ESCHENBACH, CHRISTIAN GOTTHOLD.
> De quibusdam auri calcibus et salibus mercurialibus observationes.
>   Lipsiæ, 1785.

ESCHENBACHER, AUG.
> Die Feuerwerkerei oder die Fabrikation der Feuerwerkskörper. Eine
>   Darstellung der gesammten Pyrotechnik, enthaltend die vorzüg-
>   lichsten Vorschriften zur Anfertigung sämmtlicher Feuerwerks-
>   objecte, als aller Arten von Leuchtfeuern, Sternen, Leuchtkugeln,
>   Raketen, der Luft- und Wasser-Feuerwerke sowie einen Abriss der
>   für den Feuerwerker wichtigen Grundlehren der Chemie. Zweite,
>   sehr vermehrte und verbesserte Auflage. Wien, 1885. Ill.

ESCOSURA, LUIS DE LA.
> Química analítica. De la volumetria. Bol. ofic. de minas, 1845. pp.
>   213–249.

ESSAY ON CULINARY POISONS. With observations on the adulteration of
>   bread and flour, and the nature and properties of water. London,
>   1781. 8vo.

ETTMÜLLER, MICHAEL.
> Chemia rationalis ac experimentalis curiosa. Leiden, 1684.
>   Edited by J. Ch. Ausfeld after the author's death.
> Collegium chymicum habitum anno 1671. [*In* Opera omnia of M.
>   Ettmüller edited by his son Michael Ernst Ettmüller.] Francofurti,
>   1708.

ETTMÜLLER, MICHAEL, ET H. WARNATIUS.
> Medicina Hippocratis chymica. Lipsiæ, 1679. 4to.

EUGUBIUS, HIERONYMUS ACOROMBONUS.

Tractatus de lacte, nunc primum impressus.   Venetiis, 1536.   pp. [92],
16mo.

EULENBERG, HERMANN.

Die Lehre von den schädlichen und giftigen Gasen.   Toxikologisch,
physiologisch, pathologisch, therapeutisch mit besonderer Berücksichtigung der öffentlichen Gesundheitspflege und gerichtlichen
Medicin, systematisch und nach eigenen Versuchen bearbeitet.
Braunschweig, 1865.   8vo.   Ill.

EVANS, JOHN CASTELL.
        See Castell-Evans, John.

EVRARD, A.

Chimie.   Résumé du cours.   Paris, 1877.   16mo.

EWELL, THOMAS [of Virginia].

Plain Discourses on the Laws of Properties of Matter.   Containing the
elements or principles of modern chemistry.   New York, 1806.   8vo.

EXNER, SIGM.
        Physiologie der Grosshirnrinde.   See Handbuch der Physiologie.

EYTELWEIN, I. A.

Beschreibung der Erbauung und Einrichtung einer vereinigten Brauerei
und Branntweinbrennerei auf dem Lande.   Berlin, 1802.   pp. 68,
4to.   Plates.

F., F. C.

Die neue deutsche Zuckerbereitung aus Runkelrüben.   Tübingen, 1800.

FABBRO, T. E MARCO F.

Nozioni di storia naturale, d'igiene e di fisico-chimica per le scuole normali, secondo i programmi del 17 settembre, 1890.   Elementi di fisica
sperimentale per la terza preparatoria.   Torino, 1891.   8vo.   Ill.

Nozioni di storia naturale, d'igiene e di fisico-chimica per le scuole normali.   Organi del corpo umano, acustica e calore, per la prima
classe normale.   Torino, 1891.   8vo.

Nozioni di storia naturale, d'igiene e di fisica sperimentale e nozioni di
chimica per la terza preparatoria.   Torino, 1891.   8vo.

FABBRONI, GIOVANNI VALENTINO MATTIA.

Dell' arte di fare il vino.   Firenze, 1787.   Seconda edizione, 1790.
        Translated into French and German.

Dell' azione chimica de' metalli nuovamente avvertita.   Firenze, 1793.

FABBRONI, GIOVANNI VALENTINO MATTIA. [Cont'd.]

Sulla natura dell' arsenico e preparazione dell' acido arsenicale. Milano, 1780.

Sur l'action chimique des différens métaux entr'eux à la température de l'atmosphère et sur l'explication de quelques phénomènes galvaniques. Paris, 1796.

FABER, E.

Hwa hsioh chih leo. [Canton, 187–.]

> Synopsis of Chemistry in Chinese, by Rev. E. Faber, D.D. *Compare* Kerr, J. G., *and* Fryer, John.

FABER, PETRUS JOHANNES.

Alle in zwei Theile verfasste chymische Schriften. Hamburg, 1713.

FABER, PETRUS JOHANNES.

> One hundred and twelve chymical Arcanums. *See* Salmon, William : Polygraphice.

FABER, RUDOLF.

Die chemischen Elemente des gesunden Blutes. München, 1852. 8vo.

FABRE, J. HENRI.

Chimie (La) de l'oncle Paul. Paris, 1881. 12mo.

Notions élémentaires de chimie à l'usage des classes de lettres. Classe de sixième. Paris, 1880. 12mo.

Simples notions sur la chimie. Paris, 1880. 12mo.

FABRE, PIERRE JEAN.

L'abrégé des secrets chymiques où l'on voit la nature des animaux, des végétaux et minéraux entièrement découverte. Paris, 1636. 8vo.

FABULET, ADOLPHE.

Éléments de chimie théorique et pratique . . . Deuxième edition, Paris, 1813. 2 vols., 8vo.

FACEN, A.

Chimica bromatologica, ossia guida per riconoscere la bontà, le alterazioni e le falsificazioni delle sostanze alimentari. Firenze, 1872. 16mo.

FAIDEAU, F.

La chimie amusante. Expériences mises à la porté de tous. Paris, 1892. 8vo. Ill.

FAIRBAIRN, WILLIAM.

Iron, its History, Properties and Processes of Manufacture. Third edition revised and enlarged. London, 1869. 8vo. Plates.

FALCK, FERDINAND AUGUST.

Lehrbuch der praktischen Toxikologie für praktische Aerzte und Studirende mit Berücksichtigung der gerichts-ärztlichen Seite des Faches. Stuttgart, 1880. pp. vii–340, 8vo.

FALCKNER, J. L.

Beyträge zur Stöchiometrie und chemischen Statik. Basel, 1824. pp. 206, 12mo.

> An interesting contribution to the history of chemical equivalents.

Ueber die Verhältnisse und Gesetze wonach die Elemente der Körper gemischt sind. Basel, 1819. 8vo.

FALCONER, WILLIAM.

Observations and Experiments on the Poison of Copper. London, 1774. 8vo.

FALKNER, FREDERICK.

The Farmer's Manual; treatise on manures. With an account of the most recent discoveries in agricultural chemistry. New York, 1843. 12mo.

FARADAY, MICHAEL.

Chemical Manipulation, being Instructions to students in chemistry on the methods of performing experiments of demonstration or of research with accuracy and success. London, 1827. 8vo.

> New edition. London, 1830. pp. 656, 8vo.
>
> Third edition, revised. London, 1842. 8vo.
>
> First American edition, edited by J. K. Mitchell. Philadelphia, 1831. 8vo.
>
> Manipulations chimiques ; traduit de l'Anglais par Maiseau . . . et revu pour la partie technique par Bussy. Paris, 1827. 2 vols., 8vo.
>
> Chemische Manipulation, oder das eigentlich Praktische der sichern Ausführung chemischer Arbeiten und Experimente. Aus dem englischen. Weimar, 1828. pp. vi–810, 8vo. Folding plates.

Course (A) of six lectures on the Various Forces of Matter and their relations to each other. . . . Edited by W. Crookes. With numerous illustrations. London and Glasgow, 1860. 8vo.

> Second edition, 1873.

Course (A) of six lectures on the Chemical History of a Candle ; to which is added a lecture on Platinum . . . delivered during the Christmas Holidays of 1860–61. Edited by W. Crookes. London, 1861. 8vo.

FARADAY, MICHAEL. [Cont'd.]

[Another edition.] The Chemical History of a Candle. London, [1873]. 8vo. Ill.

Histoire d'une chandelle . . . avec une notice biographique et des notes complémentaires sur l'acide stéarique, les lampes, l'éclairage au gaz, et les lumières éblouissantes par Henri Sainte-Claire Deville. (Traduction par W. Hughes.) Paris, *n. d.* [1865]. pp. 310, 12mo.

Naturgeschichte einer Kerze. Sechs Vorlesungen für die Jugend aus dem englischen übertragen von Lüdicke. Berlin, 1871. 8vo. Ill.

[Another edition], herausgegeben von R. Meyer. Berlin, 1884. 8vo.

Storia chimica di una candela; traduzione dall' inglese. Milano, 1866. 16mo.

Seconda edizione. Milano, 1871

Исторія свѣчки. Перев. Бекетова Н. Харьковъ 1866.

Исторія свѣчки съ библіографіею. Перев. Зайцева. С.-Петербургъ, 1866.

Experimental Researches in Chemistry and Physics. Reprinted from the Philosophical Transactions of 1821–1857, the Journal of the Royal Institution . . and other publications. London, 1859. 8vo. Ill.

On the Liquefaction and Solidification of Bodies generally existing as gases. London, 1845. 4to.

Subject Matter (The) of a Course of Six Lectures on the Non-metallic Elements. Arranged by John Scoffern. London, 1853. 8vo.

FASBENDER, F.

Mechanische Technologie der Bierbrauerei und Malzfabrikation. Supplementband I. Leipzig, 1891. 4to. Ill.

FASOLI, G. B.

Sull' opinione d'identitá del bromo e dell' jodio. Considerazione. Venezia, 1856. 8vo.

FATO, ANTONIO.

Manuale del chimico clinico. Verona, 1878. 8vo.

FAULHABER, ALBERT FRIEDRICH.

Dissertatio inauguralis medica sistens theoriam solutionis chemicæ. Tubingæ, *n. d.* [1765]. pp. 36, 4to.

FAULKNER, FRANK.

The Theory and Practice of Modern Brewing, A re-written and much

FAULKNER, FRANK. [Cont'd.]

enlarged edition of "The Art of Brewing." With a complete and fully illustrated Appendix specially written for the present period. Second edition. London, 1888. pp. xx-396, 8vo. Ill.

FAUVELLE.

La physico-chimie, son rôle dans les phénomènes naturels astronomiques, géologiques et biologiques. Précédée d'une lettre de l'auteur à M. Berthelot sur l'unité de la science. Paris, 1889. pp. xxiv-512, 12mo.

> Bibliothèque des sciences contemporaines.

FEBURE, NICOLAS LE.

> *See* Le Febure, Nicolas.

FECHNER, GUSTAV THEODOR.

Repertorium der neuen Entdeckungen in der unorganischen Chemie. Leipzig, 1830-33. 3 vols., 8vo.

> *Also under the title :* Lehrbuch der theoretischen und praktischen Chemie von L. J. Thénard. Uebersetzt und vervollständigt von G. Th. F. Ersten Supplements erste Abtheilung.
>
> *See* Thénard, L. J.

Repertorium der organischen Chemie. Leipzig, 1826-28. 2 vols., 8vo.

Repertorium der neuesten Entdeckungen in der organischen Chemie. Leipzig, 1830-33. 2 vols., 8vo.

Resultate der bis jetzt unternommenen Pflanzenanalysen, nebst ausführlich chemisch-physikalischer Beschreibung des Holzes, der Kohle, der Pflanzensäfte und einiger andern wichtigen Pflanzenkörper. Leipzig, 1829.

Ueber die physikalische und philosophische Atomenlehre. Leipzig, 1855. 8vo.

FEHLING, HERMANN VON.

Chemische Untersuchung der Soolen, des Stein- und Kochsalzes, sowie der Mutterlaugen der königlich württembergischen Salinen. Mit besonderer Rücksicht der concentrirtén Mutterlaugen. Stuttgart, 1847. 8vo.

FEHLING, HERMANN VON [Editor].

> *See, in Section II,* Handwörterbuch der reinen und angewandten Chemie.

FEHRMANN, A.

Das Ammoniakwasser und seine Verarbeitung. Auf Grund selbstständiger Erfahrungen und mit Berücksichtigung der neuesten Verbesserungen. Braunschweig, 1887. 8vo. Ill.

FEICHTINGER, G.
>   Die chemische Technologie der Mörtelmaterialien. *See* Bolley's Handbuch der chemischen Technologie.

FEILBERG, C.
Analytisk Kemi til Brug ved Undervisning i Landbrugsskoler. Kjøbenhavn, 1888. 8vo.
Kortfattet uorganisk Kemi for Folkehöjskoler og Landbrugsskoler. Kjøbenhavn, 1876. 8vo.
Organisk Kemi. Odense, 1875. 8vo.

FELIU Y PEREZ, B.
Lecciones de química general inorgánica con applicaciones á la ciencia, á la industria y á las artes. Segunda edicion. Madrid, 1880. 4to.

FELKER, P. H.
What the Grocers sell us. New York, 1860.

FELLÖCKER, P. SIGMUND.
Die chemischen Formeln der Mineralien in geometrischen Figuren dargestellt. Lenz, 1879. pp. xxxii–158, 8vo.

FENNEL, CHARLES T. P.
Principles of General Chemistry, pursuant to a course by Adolphus Fennel. Cincinnati, 1886.

FENOGLIO, G.
Metodo di analisi chimico-pratiche pel riconoscimento delle fibbre tessili, materie coloranti sui tessuti, generi commestibili e combustibili. Ricettario per tinture in ogni colorazione su lana, seta e cotone. Torino, 1880. 16mo.

FENTON, HENRY JOHN HORSTMANN.
Notes on Qualitative Analysis, concise and explanatory. Cambridge, 1883. 4to.

FEOKTISTOW, A.
Experimentale Untersuchung über Schlangengift. Dorpat, 1888.

FERBER, JOHANN JAKOB.
Nachrichten und Beschreibungen einiger chemischen Fabriken, nebst Joh. Chr. Fabricius' mineralogischen und technologischen Bemerkungen auf einer Reise durch verschiedene Provinzen in England und Schottland, mit Anmerkungen und Zusatzen von J. J. Ferber. Mit Kupfern. Halberstadt, 1793. pp. [iv]–166, 8vo. 2 plates.

FERBER, JOHANN JAKOB. [Cont'd.]

* Untersuchung der Hypothese von der Verwandlung der mineralischen Körper in einander. Aus den Akten der kaiserlichen Akademie der Wissenschaften zu St. Petersburg übersetzt, mit einigen Anmerkungen vermehrt und herausgegeben von der Gesellschaft naturforschender Freunde zu Berlin. Berlin, 1788. pp. [viii]–72, 12mo.

FERET.

Étude comparée sur le lait de la femme, de l'ânesse, de la vache et de la chèvre, suivie de tableaux d'analyse. Paris, 1884. 8vo.

FERMOND, CH.

Monographie du tabac, comprenant l'historique, les propriétés thérapeutiques, physiologiques et toxicologiques du tabac, la description des principales espèces employées, sa culture, sa préparation et l'origine de son usage ; son analyse chimique, ses falsifications, sa distribution géographique, son commerce et la législation qui le concerne. Paris, 1857. 8vo. Portrait.

FERNBACH, A.

Recherches sur la sucrase, diastase inversive du sucre de canne. Sceaux, 1890. 8vo.

FERRARA, MICHELE.

Istituzioni di farmacia chimica. Napoli, 1805–11. 3 vols.

FERRARIO, OTTAVIO.

Corso di chimica generale. Milano, 1837–46. 10 vols., 8vo, with atlas in 4to.

FERREIRA LAPA, J. J.

Chimica agricola. Lisboa, 1875. 8vo.

FERREIRA DA SILVA, ANTONIO JOAQUIM.

Estudo sobre as classificaçãoes dos compostos organicos. Coimbra, 1877. pp. xviii–132, 8vo.

FERVILLE, E.

L'industrie laitière, le lait, le beurre et les fromages. Paris, 1888. 16mo. Ill

    Bibliothèque des connaissances utiles.

FESER, J.

Lehrbuch der theoretischen und praktischen Chemie für Aerzte, Thierärzte und Apotheker. Berlin, 1872–73. 8vo.

FESER. J. [Cont'd.]

Polizeiliche (Die) Controle der Markt-Milch. Zwei Vorträge. Leipzig,
1878. pp. 99, 8vo. Ill Folding table.

FICINUS, HEINRICH DAVID AUGUST

Anfangsgründe der medicinischen Chemie. Zum Gebrauch bei seinen
Vorlesungen, Leipzig, 1815.

Chemie allgemeinfasslich dargestellt. Dresden, 1829–30. 8vo, 2 vols.
in 4 parts.

Neue Ausgabe. Quedlinburg, 1843.

Allgemeine Taschenbibliothek der Naturwissenschaft.

FICK, A.

Specielle Bewegungslehre ; Gesichtssinn, etc. See Handbuch der Physiologie.

FICK, JOHANN JACOB.

Chymicorum in pharmacopœia Bateana et Londinensi explicatio.
Francofurti, 1711.

De calce viva. Jenæ, 1726.

De saccharo lactis. Jenæ, 1713.

De salium natura, genere et usu. Jenæ, 1713.

FIEDLER, CARL WILHELM.

Gründliche Anweisung zur Salpeter-Erzeugung. Cassel, 1786.

Lehrbegriff der grundsätzlichen Färber- und Zeugdruckerkunst. Cassel,
1826. 2 vols.

FIEDLER, JOHANN ALEXANDER.

De lucis effectibus chemicis in corpora inorganica. Vratislav, 1845.

FIGUIER, LOUIS GUILLAUME.

Il gas e le sue applicazioni. Milano, 1888.

De l'application méthodique de la chaleur aux composés organiques
définis. Paris, 1853.

See also in Sections III and IV.

FILACHOU, J. E.

Prodrome de chimie rationnelle (avec planches). Paris, 1879. 12mo.

FILETI, M.

Tavole di analisi chimica qualitativa ; terza edizione accresciuta.
Torino, 1883. 8vo.

Quarta edizione. Torino, 1891.

FINCKE, JACOB.

De elementis. Copenhagen, 1632.

FINOT, E. [Translator].
>   *See* Sutton, Francis.

ФИРСТОВА, Г.
>   Сѣра и ея неорган. соединеніе съ металлоидями, имѣющ. примѣненіе въ про-
>   мышленности. С,-Петербургъ 1857.
>>   FIRSTOF, H. Sulphur and its inorganic combinations with metalloids used
>>   in industry. St. Petersburg, 1857.

FISCHER, ANTON.
>   Die Likör-Fabrikation in ihrem ganzen Umfange. Vollständiges Hand-
>   und Hülfsbuch für Branntweinbrennerei, Conditoren, Destillateure,
>   Gast- und Schenkwirthe, sowie für Kaufleute. Bestehend in 1170
>   Rezepten zur Bereitung aller Sorten einfacher und Doppel-Brannt-
>   weine, der Usquebaugh, der Magentropfen, Essenzen und Tincturen,
>   Extracte, Rum, Punch, Arrac, Cognac, Franzbranntwein, der Alko-
>   holate und Wässer, sowie der Huiles. Nebst Anleitung zur Dar-
>   stellung derselben auf warmem und kaltem Wege. Leipzig, 1863.
>   pp. xx–314, 8vo.
>>   Otto Spamer's Bibliothek des Wissenswürdigsten aus der technischen
>>   Chemie und Gewerbskunde. Zweite Serie : Praktisches technisch-
>>   chemisches Haus- und Hülfsbuch.

FISCHER, BERNHARD.
>   Lehrbuch der Chemie für Pharmaceuten. Mit besonderer Berück-
>   sichtigung der Vorbereitung zum Gehülfen-Examen. Stuttgart,
>   1885. 8vo. Ill.

FISCHER, E.
>   Anleitung zur Darstellung organischer Präparate. Zweite Auflage.
>   Würzburg, 1887.
>>   Dritte Auflage. Würzburg, 1889.
>>   Exercises in the preparation of Organic Compounds. Trans-
>>   lated by A. Kling. Glasgow and London, 1889.

FISCHER, FERDINAND.
>   Industrie (Die) der Mineralöle. *See* Bolley's Handbuch.

FISCHER, FERDINAND.
>   Chemische (Die) Technologie des Wassers. Braunschweig, 1880. 8vo.
>>   *Cf.* Bolley's Handbuch [etc.]. Neue Folge.
>   Leitfaden der Chemie und Mineralogie. Hannover, 1880. 8vo. Ill.
>   Stöchiometrie. Mit 150 Aufgaben. Angabe der Resultate und An-
>   deutungen zur Auflösung. Für Studirende, Pharmaceuten und
>   Realschüler. Hannover, 1875. 8vo.

FISCHER, FERDINAND. [Cont'd.]

Wasser (Das), seine Verwendung, Reinigung und Beurtheilung mit besonderer Berücksichtigung der gewerblichen Abwässer. Zweite
umgearbeitete Auflage. Berlin, 1891.

FISCHER, FERDINAND, UND H. KRAUSE.

Leitfaden der Chemie und Mineralogie. Mit 224 in den Text eingedruckten Abbildungen. Dritte Auflage. Hannover, 1891. pp.
viii–280, 8vo.

FISCHER, H.

Der praktische Seifensieder, oder gründliche Anleitung zur Fabrikation
aller im Handel vorkommenden Riegel-, Schmier-, Textil- und
Toilette-Seifen. Unter Berücksichtigung der neuesten Erfindungen
und Fortschritte, nach dem jetzigen Standpunkte der Seifenfabrikation. Sechste Auflage des Werkes : "Die Kunst des Seifensiedens, etc.," in vollständiger Neubearbeitung herausgegeben.
Weimar, 1889. pp. xii–248, 8vo.

Also under the title · Neuer Schauplatz der Künste und Handwerke mit
Berücksichtigung der neuesten Erfindungen. Vierter Band.

FISCHER, JUSTUS WILHELM CHRISTIAN.

Chemische Grundsätze der Gewerbskunde, oder Handbuch der Chemie
für Fabrikanten . . . mit einer Vorrede begleitet von Sigismund Friedrich Hermbstädt. Berlin, 1802. Part I. Ill. 8vo.

Neue chemische Erfindungen für Fabriken und Manufacturen. . . .
Wien, 1802.

FISCHER, NICOLAUS WOLFGANG.

Chemische Untersuchungen der Heilquellen zu Salzbrunn. Breslau,
1821.

Systematischer Lehrbegriff der Chemie, in Tabellen dargestellt. Berlin,
1836. 4to.

Ueber die Natur der Metallreduction auf nassem Wege. . . . Breslau,
1828.

Ueber die Wirkung des Lichts auf das Hornsilber. Nürnberg, 1813. 8vo.

Ueber die chemischen Reagentien. Breslau, 1816. 8vo. .

Also under the title : Versuche zur Berichtigung und Erweiterung der Chemie.

Verhältniss (Das) der chemischen Verwandtschaft zur galvanischen
Electricität, in Versuchen dargestellt. Berlin, 1830. 8vo.

FISCHER, R.

Repetitorium der Chemie für Studirende der Naturwissenschaften,
Medicin und Pharmacie. Berlin, 1891. 8vo.

FISCHER, X.

Ueber die chemische Zusammensetzung altägyptischer Augenschmin-
ken. Erlangen, 1892. 8vo.

FISHER, WALTER WILLIAM.

A Class Book of Elementary Chemistry. London, 1888. Ill.

FITTIG, RUDOLF.

   See Wöhler, Fr.: Outlines of Organic Chemistry.

FIUMI, G.

Guida alla analisi chimica qualitativa. Rovereto, 1891. 8vo.
Trattato di chimica inorganica ed organica, ad uso delle scuole reali
superiori ed inferiori. Rovereto, 1885. 8vo.

FIZES, ANTOINE.

Leçons de chymie. Montpellier, 1750.

FLACHAT, E., BARRAULT, A., ET PITIET, J.

Traité de la fabrication de la fonte de fer, envisagée sous les trois
rapports, chimique mécanique, et commercial. 2 parties. Paris,
1846. 4to. Ill. Atlas of 92 plates.

FLANDIN, CHARLES (called des Aubues).

De l'arsenic, suivi d'une instruction propre à servir de guide aux experts dans les cas d'empoissonnement, etc. Paris, 1841.
Principes et philosophie de la chimie moderne, fondés sur la doctrine
des équivalents. Paris, 1864. pp. iv–701, 8vo.
Traité des poisons, ou toxicologie appliquée à la médecine légale, à la
physiologie et à la thérapeutique. Paris, 1846–'53. 3 vols., 8vo.

FLASCHNER, JOHANN.

De elemento aëris ; tractatus physico-experimentalis, in quo natura,
proprietates et effectus ejusdem elementi rationum et experimentorum serie demonstrantur. Pragæ, 1748.

FLECK, HUGO.

Fabrikation (Die) chemischer Producte aus thierischen Abfällen. Auf
Grund selbstständiger Erfahrungen und mit Berücksichtigung der
neuesten Verbesserungen bearbeitet. Braunschweig, Zweite Auflage, 1878.
   Cf. Bolley's Handbuch.
Ueber die Chemie in ihrer Bedeutung für die Gesundheitspflege. Berlin, 1883. 8vo.

FLEISCHER, EMIL.

Kurzgefasstes Lehrbuch der Maasanalyse, nebst Anleitung zu den geeignetsten Trennungsmethoden für analytische Bestimmungen und zur quantitativen Untersuchung technisch wichtiger Stoffe. Leipzig, 1867. 8vo.

Titrir-Methode (Die) als selbständige quantitative Analyse. Leipzig, 1871. 8vo.

> Zweite Auflage, Leipzig, 1876. 8vo.
> Dritte Auflage, Leipzig, 1884. 8vo.
> A System of Volumetric Analysis. Translated, with notes and additions, from the second German edition, by M. M. Pattison Muir. London, 1877. 8vo.

FLEISCHMANN, WILHELM.

> Das Molkereiwesen. *See* Otto-Birnbaum's Lehrbuch.

FLEISCHMANN, WILHELM.

Untersuchung der Milch von 16 Kühen des in Ost-Preussen rein gezüchteten holländischen Schlages während der Dauer einer Lactation. Mitgetheilt aus der Versuchsmolkerei zu Kleinhof-Tapian. Berlin, 1891. 8vo.

FLEURY, E., ET E. LAMAIRE.

Manuel pratique de diffusion. Paris, 1888. Ill.

FLEURY.

Méthode générale pour l'analyse organique immédiate. Paris, 1872. 8vo.

FLINT, AUSTIN.

Manual of Chemical Examination of the Urine in disease, with brief directions for the examination of the most common varieties of urinary calculi. New York, 1870. 12mo.

FLINT, TIMOTHY.

Lectures upon Natural History, Geology, Chemistry, the applications of steam and interesting discoveries in the arts. Boston, 1832.

FLÓREZ, J. F.

Nociones de quimica general y aplicada á los servicios militares y navales. Segunda edicion. Ferrol, 1891.

FLOURENS, G.

Étude sur la cristallisation du sucre et sur la fabrication du sucre candi. Paris, 18—.

FLÜCKIGER, FRIEDRICH AUGUST.
  Pharmaceutische Chemie. Berlin, 1878–79.
            Zweite Auflage, Berlin, 1887. 2 vols. in 1, 16mo.
            Chimica farmaceutica. Tradotta e corredata di numerose
                aggiunte e note da Torquato Gigli. Torino, 1882. 8vo.
  Pharmakognosie des Pflanzenreiches. Zweite Auflage. Berlin, 1888.
    8vo.
  Reactionen. Eine Auswahl pharmaceutisch wichtiger Präparate der
    organischen Chemie in ihrem Verhalten zu den gebräuchlichsten
    Reagentien. Berlin, 1892. 8vo.

FLÜCKIGER, F. A., UND A. TSCHIRCH.
  Grundlagen der Pharmakognosie. Einleitung in das Studium der
    Rohstoffe des Pflanzenreiches. Zweite Auflage. Berlin, 1885. 8vo.
            Elementi di farmacognosia. Versione di P. Giacosa. Tori-
                no, 1886. 8vo.

FLUHRER, WILHELM.
  Die Diastase. Eine ausführliche Zusammenstellung der Untersuchung
    über die Vorgänge beim Maischen. München, 1870. pp. vi–290.
    8vo.

        Perhaps this is a valuable work, but as it has no table of contents and no
        indexes, the time to ascertain its value is lacking.

FOCILLON, AD.
  Chimie. Manuel d'études pour la section des sciences dans les lycées ;
    à l'usage des aspirants aux écoles spéciales du gouvernement et
    des candidats au baccalauréat ès sciences. Paris, 1854. 8vo.
  Premiers enseignements de chimie. Deuxième édition. Tours, 1880.
    8vo. Ill.

FOCILLON, AD., ET PRIVAT-DESCHANEL.
  Cours élémentaire de chimie, rédigé conformément aux programmes
    des lycées et aux programmes pour les examens du baccalauréat
    ès sciences et du baccalauréat ès lettres. Première partie. Chimie
    minérale ; par Ad. Focillon. Paris, 1868. 18mo. Seconde partie :
    Métallurgie et chimie organique par Privat-Deschanel. Paris, 1866.
    18mo.

FOCK, A.
  Krystallographisch-chemische Tabellen. Leipzig, 1890. 8vo.
  Ueber die physikalischen Eigenschaften der Elemente und ihre
    anschauliche Erklärung. Ein Vortrag gehalten in der deutschen
    chemischen Gesellschaft am 26. Januar, 1891. Berlin, 1891. pp.
    16, 8vo.

FOL, FRÉDÉRIC.
Guide du teinturier. Paris, *n. d.* 18mo. Ill.

FOLWARCZNY, C.
Handbuch der physiologischen Chemie, mit Rücksicht auf pathologische Chemie und analytische Methoden. Wien, 1863. pp. xii–340, 8vo.

FONTAINE, HIPPOLYTE.
Électrolyse. Renseignements pratiques sur le nickelage, le cuivrage, la dorure, l'argenture, l'affinage des métaux et le traitement des minérais au moyen de l'électricité. Paris, 1885. pp. xv–296, 8vo, Ill.

> Deuxième édition. Paris, 1892. 8vo. Ill.
>
> Electrolysis, a Practical Treatise on Nickeling, Coppering, Gilding, Silvering, the refining of metals and treatment of ores by electricity. Translated from the French by J. A. Berly. New York, 1885. 8vo.

FONTANA, FELICE.
Descrizioni ed usi di alcuni stromenti per misurar la salubrità dell' aria Firenze, 1774.

> Contains his invention of the nitrous acid eudiometer.

Opuscules physiques et chymiques. Traduit de l'Italien par Gibelin. Paris, 1784. 8vo.
Recherches physiques sur la nature de l'air nitreux et de l'air déphlogistiqué. Paris, 1776. pp. 184, 8vo.

> Physische Untersuchungen über die Natur der Salpeterluft, der vom Brennbaren beraubten Luft, und der fixen Luft. Aus dem französischen und italienischen übersetzt von F. X. von Wasserberg. Wien, 1777. 8vo.

Ricerche fisiche sopra il veneno della vipera. Lucca, 1767.

> Treatise on the Venom of the Viper, the American Poisons, the Cherry, Laurel, and other Vegetable Poisons. Translated from the French by Joseph Skinner. London, 1787. 2 vols., 8vo. Plates.
>
> Traité sur le venin de la vipère, sur les poisons américains, etc. Florence, 1781. 2 vols., 4to.

FONTENELLE, JULIA.
> *See* Julia-Fontenelle, Jean Simon Étienne.

FONVIELLE, W. DE.
Le pétrole. Paris, 1887. Ill.

FORCHAMMER, C.

Lærebog ; Stoffernes almindelige Kemi. Förste Del : De enkelte Radi-
kalers almindelige Kemi. Kjøbenhavn, 1842. 8vo.

FORCHHAMMER, GEORG.

De mangano. Hafniæ, 1820. 4to.

Erindringsord til Forelæsninger over anvendt Kemi, fornemlig for
saa vidt den angaaer Stoffer af Mineralriget. Kjøbenhavn, 1826.

FORCRAND, DE.

Argent. *See, in Section II*, Fremy : Encyclopédie chimique, vol. III.

FORCRAND, DE, ET VILLIERS.

Lithium et Ammonium. *See, in Section II*, Fremy : Encyclopédie chimique, vol. III.

FORMOND, CH.

Monographie du tabac. Comprenant l'historique, les propriétés thérapeutiques, physiologiques et toxicologiques du tabac, la description
des principales espèces employées, sa culture, sa préparation et
l'origine de son usage ; son analyse chimique, ses falsifications, sa
distribution géographique, son commerce et la législation qui la
concerne. Paris, 1857. 8vo. Portrait.

FORNARI, F.

La chimica nelle arti, nelle industria nell' igiene e nell' economia
domestica spiegata alla buona al popolo ed ai giovanetti. Milano,
1872. 16mo.

FORNARI, P.

La piccola chimica nelle arti, nelle industrie, nell' igiene e nella
economia domestica spiegata alla buona al popolo ed ai giovinetti.
Milano, 1882. 16mo.

FORNARI, U.

La fabbricazione delle vernici e prodotti affini, lacche, inchiostri da
stampa, mastici, ceralacche. Milano, 1891. 12mo.

FORSYTH, J. S.

First (The) Lines of Philosophical and Practical Chemistry as applied
to Medicine and the Arts, including the recent discoveries and
doctrines of the science. London, 1828. pp. xxxvi-326, 12mo.
Plates.

Contains pp. 301-326, A Glossary of Chemical Terms.

FOSSEK, W.

Einführung in das Studium der Pharmacie. Vol. I : Allgemeine und anorganische Chemie. Vol. II : Organische Chemie. Vol. III : Pharmaceutische Botanik und Pharmacognosie. Vol. IV : Pharmaceutische Operation und Utensilien in Verbindung mit Physik. Wien, 1891. 4 vols., 8vo. Ill.

FOSTER, W.

First Principles of Chemistry illustrated by a series of the most recently discovered and brilliant experiments known to the science. Adapted specially for classes. Second edition. New York, 1868. pp. vii–136, 8vo. Ill.

> Of this work I have a Japanese version, translated by Sougida and printed in Japanese characters. Tokio, 1871. 4 vols.

FOUCHÉ, J. M. L.
> *See* Ajasson de Grandsagne, J. B. F.

FOURCROY, ANTOINE FRANÇOIS DE.

Leçons élémentaires d'histoire naturelle et de chimie, dans lesquelles on s'est proposé : 1° de donner un ensemble méthodique des connoissances chimiques . . . 2° d'offrir un tableau comparé de la doctrine de Stahl et de celle de quelques modernes. Paris, 1782. 2 vols., 8vo.

Élémens d'histoire naturelle et de chimie. Troisième édition. Paris, 1789, 5 vols., 12mo.

> Cinquième édition, Paris, An II [1793]. 5 vols., 12mo,
>
> Elementary Lectures on Chemistry and Natural History. Translated from the French by Thomas Elliot. With many additions, notes . . . by the translator. Edinburgh, 1785. 2 vols., 8vo.
>
> Elements of Natural History and of Chemistry : being the second edition of the elementary lectures on those sciences . . . enlarged and improved by the author. Translated into English with . . . notes and an historical preface by the translator [W. Nicholson]. London, 1788. 4 vols., 8vo.
>
> *The same :* Translated from the Paris edition 1789. To which is prefixed by the translator [W. Nicholson] a preface containing strictures on the history and present state of chemistry ; with observations on the positions, facts and arguments urged for and against the antiphlogistic theory . . . London, 1790. 3 vols., 8vo.
>
> *The same :* * Translated from the fourth edition of the origi-

Fourcroy, Antoine François de. [Cont'd.]

nal French work by R. Heron. London, 1796. 4 vols., 8vo.

Fifth edition, with notes by John Thomson [W. Nicholson and W. Allen]. Edited by John Thomson. Edinburgh, 1800. 3 vols., 8vo.

Handbuch der Naturgeschichte und der Chemie mit erläuternden Anmerkungen und einer Vorrede vorsehen von Johann Christian Wiegleb. Ins deutsche übersetzt von Ph. Loos. Erfurt, 1788–1791. 4 vols., 8vo.

* Mémoires et observations de chimie pour servir de suites aux Élémens de chimie publiés en 1782 par l'auteur. Paris, 1784. pp. xvi–448, 8vo. 3 plates.

Chemische Beobachtungen und Versuche. Nach dem französischen herausgegeben von Hebenstreit. Leipzig, 1785. 8vo.

Philosophie chimique, ou vérités fondamentales de la chimie moderne disposées sans un nouvel ordre. Paris, 1792–An IV. pp. 128, 12mo.

Nouvelle édition, augmentée de notes et d'axiomes tirés des dernières découvertes, par J. B. Van Mons. Bruxelles, An III [1795]. 8vo.

Troisième édition, Paris, 1806.

The new order is as follows:
    I. Action of light.
    II. Action of heat.
    III. Action of air in combustion.
    IV. The nature and action of water.
    V. The nature of earths, alcalies, etc.
    VI. The nature and properties of combustible bodies.
    VII. The formation and decomposition of acids.
    VIII. The union of acids with earths and alkalies.
    IX. The oxidation and solution of metals.
    X. The nature and formation of vegetable substances.
    XI. The passage of vegetable into animal matter, and the nature of the latter.
    XII. Spontaneous decomposition of vegetable and animal matter.

See, in Section III, Lamarck, J. B., for an attack on this work.

Chemisk Philosophie eller den nyere Chemies Grundsandheder, fremsatte i en nye Orden. Fordansket ved Carsten Ludvig Schiødt. Kjøbenhavn, 1797.

The Philosophy of Chemistry, or Fundamental Truths of modern Chemical Science. Arranged in a new order. Translated from the French of the second edition signed by the author. London, 1795. pp. viii–192, 8vo.

FOURCROY, ANTOINE FRANÇOIS DE.   [Cont'd.]

Philosophy of Chemistry, translated from the French second
edition [by W. Nicholson].   London, 1800.

Chemical Philosophy, translated from the French [third edi-
tion] by W. Desmond.   London, 1807.

Chemische Philosophie, oder Grundwahrheiten der neuern
Chemie auf eine neue Art geordnet.   Aus dem fran-
zösischen übersetzt von Johann Samuel Traugott
Gehler.   Leipzig, 1796.   8vo.

Cf. Link, H. F.

Χημικὴ φιλοσοφία ἤ στοιχειώδεις ἀλήθειαι τῆς νεωτέρας χιμικῆς,
νεωτέρᾳ τινὶ μεθόδῳ τετραγμέναι . . . Ἐκτρατκισθεῖσα
[*sic*] μετὰ προσθήκης καὶ . . . σημειωμάτων ὑπὸ Θ. Μ.
Ἡλιάδου. Ἐπιδιορθωθεῖσα καὶ τύποις ἐκδοθεῖσα ὑπὸ Ἀ.
Τάξη. Βιέννη τῆς Ἀουστρίας αω β´ [1802].   8vo.

Filosofia chimica.   Venezia, 1794.

Filozofia chemiczna, czyli fundamentalne prawdy teraz-
niejszej chemii, przez . . . przekładania ks. I Bys-
trzyckiego S. P.   Warszawa, 1808.   8vo.

Химическая философія или основательныя истины новѣйшей химіи.
Владиміръ 1799. pp. 140, 12mo.

Filosofia chímica, ó verdades fundamentales de la química
moderna.   Traducidas por Francisco Piguillon.   Madrid, 1801 (?).

Philosophia chemica, eller Grundsanningar of den nya
chemien.   Til nyttjande vid enskilte föreläsningar från
Fransyskan öfversatt . . . af Andr. Sparrman.
Stockholm, 1795.   8vo.

Système des connaissances chimiques et de leurs applications aux
phénomènes de la nature et de l'art.   Paris, An IX [1801].   11
vols., 8vo.   Also in 6 vols., 4to.

Section I, Article 3, is entitled : Esquisse historique de la chimie. Vol. I,
pp. 10–49. Vol. XI, also under the title

Table alphabétique et analytique des matières contenues
dans les dix tomes du Système des connaissances
chimiques, rédigée par Madame Dupiery et revue par la
Contesse Fourcroy.   Paris, An X.

A General System of Chemical Knowledge and its application to the phenomena of nature and art, translated
from the French by Wm. Nicholson.   London, 1804.
11 vols., 8vo, with folio atlas of Synoptic Tables.

System des chemischen Kenntnisse und Darstellung ihrer
Anwendung auf die Erscheinungen der Natur und zu

FOURCROY, ANTOINE FRANÇOIS DE. [Cont'd.]
> den Zwecken der Kunst. Aus dem französischen von einer Gesellschaft teutscher Gelehrten. Braunschweig, 1801. Vols. I and v, 8vo.

> Sistema de los conocimientos químicos y de sus aplicaciones á los fénomenos de la naturaleza y del arte. . . . Transladado al castellano por Pedro María Oliva. Madrid, 1803–09. 10 vols., 4to.

> > Incomplete. Vol. x was translated by Gregorio Gonzalez Azaola.

Elements of Chemistry, also Philosophy of Chemistry, with notes by John Thomson. London, 1798. 3 vols., roy. 8vo.

FOURCROY, A. F. DE.
> Tableaux synoptiques de chimie [etc.]. *See in Section II.*

FOWNES, GEORGE.

Chemical Tables, chiefly for the use of junior students. London, 1846. Fol.

Chemistry as Exemplifying the Wisdom and Beneficence of God. London, 1844. 12mo. *Also* Philadelphia, 1844.
> Second edition, London, 1849. 8vo.

> > The Actonian Prize Essay.

Introduction (An) to Qualitative Analysis. London, 1846. 8vo.

Manual (A) of Elementary Chemistry, theoretical and practical. With numerous wood-engravings. London, 1844. pp. xiv–566, 12mo.
> Second edition, London, 1848. 12mo.
> Third edition, London, 1850. 12mo.
> Fourth edition, London, 1852. 12mo.
> Fifth edition, by H. B. Jones and A. W. Hofmann, London, 1854. 12mo.
> Ninth edition, London, 1863. 12mo.
> Eleventh edition, by Henry Watts, London, 1873. pp. xxii–1026. 12mo.
> Twelfth edition, by H. Watts, London, 1877. 2 vols. 12mo.

> > Thirteenth and fourteenth editions. *See* Watts, Henry.

> A New American from the twelfth English edition. Edited by Robert Bridges. Philadelphia, 1878. pp. xxviii–*from* 25–1027, 8vo. Ill.

> Elementi di chimica esposti popolarmente, coll' aggiunta di un saggio sulle applicazioni della chimica alla agricoltura. Seconda edizione. Milano, 1871. 16mo.

Fox, Cornelius B.
  Ozone and Antozone, their history and nature.

> When \
> Where \
> Why \
> How   is ozone observed in the   atmosphere

  Illustrated with wood engravings, lithographs, and chromo-lithographs. London, 1873. pp. xvi–329, 8vo.

  Sanitary Examinations of Water, Air and Food. A Handbook for the Medical Officer of Health. London, 1878. pp. xix–508. Ill.

Foye, J. C.
  Chemical Problems. New and enlarged edition. New York, 1883. 16mo.

Fränkel, N.
  Przyczynek do znajomości tiodwufenilaminn. Kraków, 1886.

Franchimont, A. P. N.
        Conférence (etc.). *See* Société chimique de Paris.

Franchimont, A. P. N.
  Beginselen der Chemie. Anorganische Chemie. Tiel, 1884–86. 3 parts, 8vo.

  Handleiding bij praktische oefeningen in organische chemie voor eerstbeginnenden. Leiden, 1880. 8vo.

  Kort leerboek der organische chemie als leiddraad bij middelbaar onderwijs. Leiden, 1880. Roy. 8vo.

  Leiddraad bij de studie van de koolstof en hare verbindingen. Leiden, 1878. Roy. 8vo.

Francis, Ernest.
  Practical Examples in Quantitative Analysis, forming a concise guide to the analysis of water. London, 1873. 12mo.

Francis, G.
  Chemical Experiments illustrating the theory, practice and application of the science of chemistry, and containing the properties, uses, manufacture, purification and analysis of all inorganic substances, with numerous engravings of apparatus, etc. London, 1849. Fifth edition. pp. [iv]–250–[ii], 8vo.

        Chemistry for Students : being an abridgement of chemical experiments. Abridged and revised by W. White. London, 1851. 12mo. Ill.

Franck, Charles François.
  Tableaux et cahiers de chimie. Paris, 1854. 4to.

FRANCK, V.
Introduction à la chimie inorganique.   Huy, 1864.   8vo.

FRANCKEN, V.
Manuel de chimie générale théorique.   Bruxelles, 1880.   2 vols., 8vo.

FRANCOLIN, GUSTAVE.
Chimie.   Paris, 1866.

FRANK, JOSEPH.
Handbuch der Toxicologie, oder der Lehre von Giften und Gegengiften.
   Nach den Grundsätzen der Brown'schen Arzneylehre und der
   neuern Chemie bearbeitet.   Wien, 1800.   8vo.

FRANKE, E.
Chemie der Küche für Töchterschulen, sowie zum Selbstunterrichte.
   Zweite vermehrte und verbesserte Auflage.   Eisleben, 1860.   8vo.

FRANKE, O. BRUNO.
Chemische Abhandlungen.   Leipzig, 1889.   2 parts.   Ill.
Exacte Principien der Chemie.   Leipzig, 1891.   8vo.

FRANKEL, J., AND R. HUTTER.
A Practical Treatise on the Manufacture of Starch, Glucose, Starch
   Sugar and Dextrine.   Philadelphia, 1881.

FRANKLAND, EDWARD.
Experimental Researches in pure, applied, and physical chemistry.
   London, 1877.   pp. xliv–1047, 8vo.
How to Teach Chemistry ; Hints to Science Teachers and Students.
   Six Lectures delivered at the Royal College of Chemistry by E. F.
   Summarized and edited by George Chaloner.   London and Phila-
   delphia, 1874.   8vo.
Lecture Notes for Chemical Students ; embracing mineral and organic
   chemistry.   London, 1866.   pp. xx–422, 8vo.
          Second edition.   London, 1870–72.   2 vols., 8vo.
          Third edition, revised by F. R. Japp.   London, 1881.
          This work is noted for the high development of so-called graphic formulæ
          applied to inorganic and organic substances.

Researches on Organometallic Bodies.   London, 1852–59.   4 parts.   Ill.
Water Analysis for Sanitary Purposes ; with hints for the interpretation
   of results.   London, 1880.   8vo.
          Second edition, revised.   London, 1890.   8vo.

FRANKLAND, EDWARD, AND F. R. JAPP.
Inorganic Chemistry . . . With illustrations. London, 1884. 8vo.
Ill.

FRANKLAND, PERCY FARADAY.
Agricultural Chemical Analysis, founded upon Leitfaden für die agricultur-chemische Analyse von F. Krocker. London, 1883.
Cf. Krocker, F.

FRAXNO Y PALACIO, CLAUDIO DE.
Tratado de química aplicada á las artes y á las funciones peculiares del artillero. Madrid, 1844. 3 vols., 8vo.

FREBAULT, A.
Manipulations de chimie. Analyse volumétrique. Cours de travaux pratiques professé à l'École de médecine et de pharmacie de Toulouse. Paris, 1879. pp. ix–329, 8vo.

FREESE, C.
Beziehungen zwischen den physikalischen Eigenschaften und der Zusammensetzung chemischer Verbindungen. Brieg, 1884.

FREIND, JOHN.
Prælectiones chymicæ . . . anno 1704 in Museo Ashmoleano habitæ . . . Amstelodami, 1709.

FREITAG, JOSEPH.
Die Zündwaaren-Fabrikation. Anleitung zur Fabrikation von Zündhölzchen, Zündkerzchen, Cigarren-Zünder und Zündlunten, der Fabrikation der Zündwaaren mit Hilfe von amorphem Phosphor und gänzlich phosphorfreier Zündmassen, sowie der Fabrikation des Phosphors. Zweite Auflage. Wien, 1887. Ill.

FREMY, EDMONDE,
Discours préliminaires sur le développement . . . de la chimie. *See,* *in Section II,* Fremy : Encyclopédie chimique.

FREMY, EDMONDE.
Chimie végétale. La ramie. Paris, 1886. 8vo.
Sur l'oxygène et l'ozone. Deuxième conférence du 10 avril, 1866, stenographiée et publiée par Boillot. Conférences scientifiques sous le patronage de S. M. l'Impératrice, au bénéfice de la Société de secours des amis des sciences fondée par le baron Thénard. Paris, 1866. pp. 20, 8vo.

FREMY, EDMONDE, ET A. TERREIL.

Le guide du chimiste. Répertoire de documents théoriques et pratiques à l'usage des laboratoires de chimie pure et de chimie industrielle. Paris, 1885. pp. viii–988, roy. 8vo. Ill.

FRENCH, JOHN.

The Art of Distillation, or a Treatise of the choicest spagyrical preparations, experiments and curiosities performed by way of distillation . . . London, 1651. 4to.

Yorkshire (The) Spaw, or a Treatise of four famous medicinal wells . . . London, 1652. 12mo.

        Another edition, 1654.

FRENCH, N. S.

Experiment Blanks for a Short Course in Elementary Chemistry. Boston, 1884. 8vo.

FRESENIUS, C. REMIGIUS.

    See Alwens, Friedrich : Stöchiometrische Schemata.

FRESENIUS, C. REMIGIUS.

Analyse der Elisabethen–Quelle zu Homburg vor der Höhe. Wiesbaden, 1864. 8vo.

Analyse der Felsenquelle No. II. in Bad Ems. Wiesbaden, 1866. 8vo.

Analyse der fünf Eisenquellen in Bad Neudorf in Böhmen. Wiesbaden, 1876. 8vo.

Analyse der im Jahre 1856 erbohrten Louisenquelle zu Bad Homburg. Wiesbaden, 1859. 8vo.

        Analysis of the Water of the Louisenquelle at Homburg . . . London, 1859. 8vo.

Analyse der Mineralquelle bei Birresborn in der Eifel. Wiesbaden, 1876. 8vo.

Analyse der Römer-Quelle in Bad Ems . . . Wiesbaden, 1870. 8vo.

Analyse der Trinkquelle, der Badequelle und der Helenenquelle zu Pyrmont. Arolsen, 1865. 8vo.

Analyse der Trinkquelle zu Driburg, der Herster Mineralquelle, sowie des zu Bädern benutzten Satzer Schwefelschlammes. Wiesbaden, 1866. 8vo.

Analyse des Julianenbrunnens und des Georgenbrunnens im fürstlichen Bade Eilsen. Nebst Anhang : Analyse des Eilser Badeschlammes von R. Fittig. Wiesbaden, 1891. 8vo. Ill.

Analyse des Kaiser-Brunnens und des Ludwigs-Brunnens zu Homburg vor der Höhe. Wiesbaden, 1863. 8vo.

Analyse des Stahl-Brunnens zu Homburg vor der Höhe. Wiesbaden, 1873. 8vo.

FRESENIUS, C. REMIGIUS.   [Cont'd.]

Analyse des Tönnissteiner Heilbrunnens und des Tönnissteiner Stahl-
brunnens im Brohl-Thale. Wiesbaden, 1869. 8vo.

Chemische Analyse der Caspar-Heinrich-Quelle zu Bad Driburg.
Wiesbaden, 1889. 8vo.

Chemische Analyse der Natronlithionquelle zu Offenbach am Main.
Wiesbaden, 1888. pp. 20, 8vo.

Chemische Analyse der Marienquelle zu Oelheim. Wiesbaden, 1886.

Chemische Analyse der Mineralquelle bei Biskirchen im Lahnthale.
Wiesbaden, 1876. 8vo.

Chemische Analyse der Soolquelle im Admiralsgarten Bad zu Berlin.
Wiesbaden, 1888. pp. 20.

Chemische Analyse der warmen Soolquelle zu Werne in Westphalen.
Wiesbaden, 1877. 8vo.

Chemische Analyse der Wilhelmsquelle im Neuen Soolbaden zu Kolberg. Wiesbaden, 1882. 8vo.

Chemische Analyse des Oberbrunnens zu Salzbrunn in Schlesien.
Dritte Auflage. Wiesbaden, 1883. 8vo.

Chemische Untersuchung der Hunyadi János Bittersalz-Quellen des
Hrn. Andr. Saxlehner in Budapest. Wiesbaden, 1878. 8vo.

Chemische Untersuchung der Stettiner Stahlquelle. Wiesbaden, 1884.
8vo.

Chemische Untersuchung der Trink- oder Bergquelle des königlichen
Bades Bertrich, unter Mitwirkung von E. Hintz ausgeführt.
Wiesbaden, 1891. 8vo.

Chemische Untersuchung des Lamscheider Mineral-Brunnens. Wiesbaden, 1869. 8vo.

Chemische Untersuchungen der wichtigsten Mineralwasser des Herzog-
thums Nassau. Wiesbaden, 1850–68. 8vo.

    I. Der Kochbrunnen zu Wiesbaden. 1850.

    II. Die Mineralquellen zu Ems. 1851.

    III. Die Quellen zu Schlangenbad. 1852.

    IV. Die Quellen zu Langenschwalbach. 1855.

    V. Die Schwefelquelle zu Weilbach. 1856.

    VI. Die Mineralquelle zu Geilnau. 1857.

    VII. Die neue Natronquelle zu Weilbach. 1861.

    VIII. Die Mineralquelle zu Niederselters.

    IX. Die Mineralquelle zu Fachingen, 1868.

Neue chemische Untersuchung des Kochbrunnens zu Wiesbaden und
Vergleichung der Resultate mit den in 1849 erhaltenen. Wiesbaden,
1886.

Anleitung zur qualitativen chemischen Analyse, oder Systematisches

FRESENIUS, C. REMIGIUS. [Cont'd.]

Verfahren zur Auffindung der in der Pharmacie, den Künsten und Gewerben häufiger vorkommenden Körper. Für Anfänger bearbeitet. Bonn, 1841. 8vo. Ill.

Anleitung zur qualitativen chemischen Analyse, oder die Lehre von den Operationen, von den Reagentien und von dem Verhalten der bekannteren Körper zu Reagentien, so wie systematisches Verfahren zur Auffindung der in der Pharmacie, den Künsten und Gewerben und der Landwirthschaft häufiger vorkommenden Körper in einfachen und zusammengesetzten Verbindungen. Für Anfänger bearbeitet. Mit in den Text eingedruckten Holzschnitten und einem Vorworte von Justus Liebig. Zweite vermehrte und verbesserte Auflage. Braunschweig, 1843. 8vo.

Dritte vermehrte und verbesserte Auflage. Mit einem Vorwort von Justus Liebig. Braunschweig, 1843. 8vo.

Vierte Auflage, 1844. Fünfte Auflage, 1847.

Sechste Auflage, 1850. Siebente Auflage, 1852.

Achte Auflage, 1853. Neunte Auflage, 1856.

Zehnte Auflage, 1860. Elfte Auflage, 1864.

Zwölfte Auflage, 1866. Dreizehnte Auflage, 1869–70.

Vierzehnte Auflage, 1874. Fünfzehnte Auflage, 1885–86.

Hwa hio kao chih. [Translated into Chinese by John Fryer.] Shanghai, 187–. 6 vols.

Handleiding bij qualitatieve scheikundige ontledingen, of de leer der scheikundige bewerkingen, der herkenningsmiddelen, en van derzelver verhouding ten opzigte van de meest bekende ligchamen alsmede eene stelselmatige handelwijze ter opsporing der in de artsenijmengkunde en in de overige kunsten en bedrijven veelvuldig voorkomende ligchamen, wanneer zij in min of meer zamengestelde verbindingen zijn bevat. Uit het Hoogduitsch, volgens de tweede vermeerderde en verbeterde uitgave. Amsterdam, 1843. 8vo.

Handleiding tot de qualitatieve chemische analyse. Naar de negende uitgave uit het Hoogduitsch vertaald door C. F. Donnadieu. Met vele tusschen den tekst gedrukte houtsnee-figuren. Delft, 1856. 8vo.

Handleiding tot de qualitatieve scheikundige analyse. Naar de twaalfde uitgave uit het Hoogduitsch vertaald door W. F. Koppeschaar. Arnhem, 1870. 8vo. Ill.

Elementary Instructions in Chemical Analysis, qualitative,

FRESENIUS, C. REMIGIUS. [Cont'd.]

as practised in the Laboratory of Giessen. With a Preface by Liebig. Edited by J. Lloyd Bullock. London, 1843. 8vo.

Second edition, London, 1846. 8vo.

Third edition, London, 1850. 8vo.

Fourth edition under the title :

A System of Instruction in Qualitative Chemical Analysis. Edited by J. Lloyd Bullock. [Fourth edition.] London, 1855. 8vo.

Fifth edition, 1859. Sixth edition, 1864.

Seventh edition. Edited by A. Vacher. London, 1869.

Eighth edition. Translated from the thirteenth German edition by A. Vacher. London, 1872. 8vo.

Ninth edition. Translated from the fourteenth German edition by A. Vacher. London, 1876.

Tenth edition. Translated from the fifteenth German edition, and edited by C. E. Groves. London, 1887.

Précis d'analyse chimique qualitative, ou Traité des opérations chimiques, des réactifs et de leur action sur les corps les plus répandus, suivi d'un procédé systématique d'analyse appliquée aux corps le plus fréquemment employés en pharmacie et dans les arts. Édition française publiée sur la troisième édition allemande, par Sacc fils. Paris, 1845. 12mo. Ill.

Deuxième édition française publiée sur la cinquième édition allemande par F. Sacc. Corbeil, 1850. 12mo.

Traité d'analyse chimique qualitative. Des opérations chimiques, des réactifs et de leur action sur les corps les plus répandus . . . Traduit de l' Allemand sur la onzième édition par C. Forthomme. Édition revue et augmentée par l'auteur. Paris, 1866. 12mo.

Analyse chimique minérale, d'après Frésénius, traduit par C. Desmazures. Paris, 1879. 4to.

Septième édition traduit de l'allemand par L. Gautier. Paris, 1885.

Huitième édition, par L. Gautier. Paris, 1891.

Guida all' analisi chimica qualitativa. Opera scritta ad uso dei principianti. Con una prefazione di Giusto Liebig. Prima versione Italiana sulla terza edizione Tedesca del 1844 per cura di A. Sobrero. Con incisione in legno, note ed addizioni del traduttore. Torino, 1845. 8vo.

FRESENIUS, C. REMIGIUS.   [Cont'd.]

Veiledning i den qualitative chemiske Analyse, med specielt Hensyn til de i Pharmacien, Tekniken og Agronomien hyppigst forekommende Legemer i enkelte og sammen-satte Forbindelser.   Med en Fortale af Justus Liebig. Oversat efter fjerde forogede Oplag af C. G. Dietrichs. Christiania, 1846.   8vo.

Аналитическая качественная химія или ученіе о производствахъ, реагенціяхъ отношеніи извѣстныхъ тѣлъ къ реактивамъ и систематическое руководство къ открытію въ простыхъ и сложныхъ соединеніяхъ тѣлъ весьма часто встрѣчающихся въ аптекѣ, художествахъ, промыслахъ и хозяйствѣ, для начинающихся. Перев. съ нѣмецкаго 5-го изданія. Въ 2-хъ частяхъ. С.-Петербургъ. 1848.   8°

Руководство къ качественному химическому анализу К. Р. Фре-зеніуса. Въ 2-хъ частяхъ. Часть первая. Переводъ и до-полненъ М. Е. Тереховъ. Съ 9-го (послѣдняго) вновь исправленнаго и увеличеннаго нѣмецкаго изданія. Мос-ква, 1859.

Руководство къ качеств. анализу. Москва, 1865.

Compendio de análisis química cualitativa ; ó tratado de las operaciones químicas, de los reactivos y de su modo de obrar con los cuerpos mas esparcidos ; acompañada de un procedimiento sistemático de análisis aplicado á los cuerpos mas generalmente empleados en medicina, far-macia, y en las artes.   Traducido por Magin Bonet y Bonfil.   Madrid, 1846.   8vo.

Análisis química cualitativa, ó tratado de las operaciones químicas, de los reactivos y su accion sobre los cuerpos mas generalmente esparcidos ; seguido de un procedi-miento systemático de análisis aplicado á los cuerpos que mas se usan en farmacia y en las artes.   Traducido de la segunda edicion francesa que publicó con arreglo à la quinta alemana por Ramon Ruiz.   Madrid, 1859. 4to.

Tratado de análisis química cualitativa.   Vert. y adicion por V. Peset.   Valencia, 1884.   4to.

Anleitung zur quantitativen chemischen Analyse oder die Lehre von der Gewichts-Bestimmung und Scheidung der in der Pharmacie, den Künsten, Gewerben und der Landwirthschaft häufiger vor-kommenden Körper in einfachen und zusammengesetzten Verbin-dungen.   Für Anfänger und Geübte bearbeitet.   Braunschweig, 1846.   8vo.   Ill.

Zweite Auflage. Braunschweig, 1847.   8vo.   Ill.

FRESENIUS, C. REMIGIUS.   [Cont'd.]

Dritte Auflage.   Braunschweig, 1854.   8vo.   Ill.

Vierte Auflage.   Braunschweig, 1858.   8vo.   Ill.

Fünfte Auflage.   Braunschweig, 1862.   8vo.   Ill.

Sechste Auflage.   Braunschweig, 1873–77.   2 vols., 8vo.   Ill.

Hwa hio chiu sho.   [Translated into Chinese by John Fryer.]   Shanghai, 187–.   14 vols.

Handleiding bij quantitatieve ontledingen, voor aanvangers en meergeoefenden.   Uit het Hoogduitsch vertaald en met aanteekeningen vermeerderd door C. F. Donnadieu. 's Gravenhave, 1846, '47.   8vo.   Ill.

Instruction in Chemical Analysis (Quantitative).   Edited by J. Lloyd Bullock.   London, 1846.   pp. xix–626, 8vo.   Ill.

A System of Instruction in Quantitative Chemical Analysis. Second edition, edited by J. Lloyd Bullock.   London, 1854.   pp. xvi–624, 8vo.   Ill.

Fourth edition by J. L. B. and Arthur Vacher.   London, 1865.   pp. vi–791, 8vo.   Ill.

Another edition, translated by Ch. E. Groves.   London, 1886.

A System of Instruction in Quantitative Chemical Analysis. From the last English and German editions, edited by O. D. Allen, with the co-operation of Samuel W. Johnson.   New York, 1881.   8vo.

Précis d'analyse chimique quantitative, ou Traité du dosage et de la séparation des corps simples et composés les plus usités en pharmacie dans les arts et en agriculture. Édition française publiée par F. Sacc.   Paris, 1846. 12mo.

Cinquième édition, traduit de l'allemand par C. Forthomme. Paris, 1885.

Traité d'analyse chimique quantitative.   Sixième édition française par L. Gautier.   Paris, 1891.   8vo.

Минеральный количественный анализъ.   Перев. Тавилдарова, С.-Петербургъ. 1875.

Tratado de análisis química cuantitativa, traducido por Peset y Cervera.   Valencia, 1886–89.   2 vols., 4to.

Good wine needs no bush, and comments on Fresenius' Qualitative and Quantitative Chemical Analysis are superfluous.   We may, however, quote the testimony of one who appreciates their character :

" Fresenius' works are of the highest order.   The arrangement is simple, methodical and consecutive ; the theoretical explanations are appropriate, clear and intelligible ; the language plain, and the directness and

FRESENIUS, C. REMIGIUS.   [Cont'd.]

> honesty of purpose, together with the just appreciation of the labors of others displayed throughout, commend them to every one engaged in studying, teaching or practising Chemistry."—J. L. B. (1865).

Hwa hio chih nan. [Translated into Chinese by M. Billequin.] Peking. 8 vols.

> An abridged edition.

Lehrbuch der Chemie für Landwirthe, Forstmänner und Cameralisten. Braunschweig, 1847.

> Leerboek der scheikunde, voor land- en boschbouwers en staathuishoudkundigen. In het Nederduitsch bewerkt door F. A. Enklar. Gorinchem, 1852. 12mo. Ill.

FRESENIUS, C. R., UND H. WILL.

Neue Verfahrungsweisen zur prüfung Potassche und Soda, der Aschen, der Säuren, insbesondere des Essigs, sowie des Braunsteins, auf ihren wahren Gehalt und Handelswerth. Für Chemiker, Pharmaceuten, Techniker und Kaufleute, lediglich nach eigenen Versuchen bearbeitet. Heidelberg, 1843. 8vo.

> New Methods of Alkalimetry, and of determining the commercial value of acids and manganese. Edited by J. Lloyd Bullock. London, 1843. 12mo.

> Nouvelle méthode pour reconnaître et pour déterminer le titre véritable et le valeur commerciale des potasses, des soudes, des cendres, des acides et particulièrement de l'acide acétique et des manganésies. Traduit de l'allemand par C. W. Bichon. Paris, 1845. 12mo.

FREY, HEINRICH.

Histologie und Histochemie des Menschen. Lehre von den Form- und Mischungs-Bestandtheilen des Körpers für Aerzte und Studirende. Leipzig, 1859. pp. xviii–626, 8vo. 388 woodcuts.

> Dritte Auflage, Leipzig, 1870.

> Handbuch der Histologie und Histochemie des Menschen. Fünfte umgearbeitete und vermehrte Auflage. Leipzig, 1876. pp. viii–747, 8vo. 634 woodcuts.

> The Histology and Histo-Chemistry of Man. A Treatise on the Elements of Composition and Structure of the Human Body. Translated from the Fourth German Edition by Arthur E. J. Barker. London, 1875. pp. ix–683, 8vo. Ill.

FREYTAG, FERDINAND.

Atmosphäre (Die), oder das Wissenswertheste für das gewöhnliche

FREYTAG, FERDINAND. [Cont'd.]
Leben Unentbehrlichste aus der Physik und Chemie. Wernigerode, 1853.
Lehrbuch der Chemie. Quedlinburg, 1839.

FRICKHINGER, ALBERT.
Katechismus der Stöchiometrie, für Pharmaceuten, studirende Mediciner, Chemiker und Techniker. Nördlingen, 1844. pp. 101-[iii]. 8vo.

> Zweite Auflage. Nördlingen, 1853. 8vo.
> Dritte Auflage. Nördlingen, 1858. 8vo.
> Vierte Auflage. Nördlingen, 1865. 8vo.
> A text-book in questions and answers. 68 questions.
>
> Katechismus der Stöchiometric voor beoefenaars der scheien artsenijbereidkunde, enz. Naar het Hoogduitsch door T. J. van der Veer. Amsterdam, 1849. 8vo.

FRIEDBERG, HERMANN.
Die Vergiftung durch Kohlendunst klinisch und gerichtsärztlich dargestellt. Berlin, 1866. pp. xii–187.

FRIEDBERG, WILHELM.
Die Fabrikation der Knochenkohle und des Thieröles. Eine Anleitung zur rationellen Darstellung der Knochenkohle oder des Spodiums und der plastischen Kohle, der Verwerthung aller sich hierbei ergebenden Nebenproducte und zur Wiederbelebung der gebrauchten Knochenkohle. Wien, Pest, Leipzig, 1877. 8vo. Ill.
Die Verwerthung der Knochen auf chemischen Wege. Eine Darstellung der Verarbeitung von Knochen auf alle aus denselben gewinnbaren Producte, insbesondere von Fett, Leim, Düngemitteln und Phosphor. Wien, 1884. 8vo. Ill.

FRIEDEBERG, O.
Ueber Glycerinbestimmung in vergohrenen Getränken. Berlin, 1890. 8vo.

FRIEDEL, CH.
Cours de chimie organique professé à la faculté des sciences de Paris pendant le deuxième semestre, 1886–87. Rédigé par M. Mansard. Paris, 1887. 2 vols. Vol. I. Série grasse. Vol. II. Série aromatique.
Leçons (etc.), see Société chimique de Paris.

FRIEDRICH, HERMANN AUGUST.
Handbuch der animalischen Stoechiologie, oder der thierische Körper,

FRIEDRICH, HERMANN AUGUST. [Cont'd.]
> seine Organe und die in ihnen enthaltenen Substanzen in Hinsicht ihrer chemischen Bestandtheile, ihrer physischen und chemischen Eigenschaften. Helmstädt, 1828. 8vo.

FRIESIUS, GEORGIUS.
> *See* Müller, Johannes.

FRITSCH, J.
> Fabrication de la fécule et de l'amidon. Paris, 1892. 12mo. Ill.
> Nouveau traité de la fabrication des liqueurs d'après les procédés les plus récents. Paris, 1891. 8vo. Ill.

FRITSCH, J., ET E. GUILLEMIN.
> Traité de la distillation des produits agricoles et industriels. Paris, 1890. 2 vols., 8vo. Ill.

FRITZE, L.
> Anfangsgründe der unorganischen Chemie. Brandenburg, 1864. 8vo.

FROMHERZ, CARL.
> Lehrbuch der medizinischen Chemie zum Gebrauche bei Vorlesungen, für praktische Aerzte und Apotheker. Freiburg, 1832–36. 2 vols., 8vo. Vol. I. Pharmaceutische Chemie und Chemische Arzneimittel-Lehre. Vol. II. Physiologische, pathologische und gerichtliche Chemie.
> Anleitung zur chemischen Analyse der Arzneimittel des Pflanzenreichs. Freiburg, 1829.
> Ueber die elektro-chemische Theorie der Verwandtschaft. Freiburg, 1822.

FRUTIGER, G.
> Cours élémentaire de chimie inorganique d'après les théories modernes. Genève, 1885. 8vo.

FRÜHLING, R., UND J. SCHULZ.
> Anleitung zur Untersuchung der für die Zuckerindustrie in Betracht kommenden Rohmaterialien, Producte, Nebenproducte und Hülfs-substanzen. Zum Gebrauche zunächst für die Laboratorien der Zuckerfabriken, ferner für Chemiker, Fabrikanten, Landwirthe und Steuerbeamte, sowie für landwirthschaftliche und Gewerbeschulen. Dritte vermehrte und verbesserte Auflage. Braunschweig, 1885. 8vo. Ill.
>> Vierte vermehrte und verbesserte Auflage. Braunschweig, 1891. 8vo. Ill.

FRÜHLING, R., UND J. SCHULZ.   [Cont'd.]
  Appendix :
  Anleitung zur Ausführung der wichtigsten Bestimmungen bei der
    Bodenuntersuchung, zum Gebrauch im Laboratorium zusammen-
    gestellt. Braunschweig, 1892. 8vo.

FRYER, JOHN.
  Hwa hio i chih. [Manual of chemistry for schools.] Shanghai, 187–.
  Hwa hio pu pien. [A treatise on practical inorganic chemistry, adapted
    from Bloxam's text-book and translated into Chinese by John
    Fryer.] Shanghai. 6 vols.
  Hwa hio hsu chih. [Outlines of chemistry for elementary schools.]
    Shanghai.
  Hwa hio su pien. [A treatise on practical organic chemistry, adapted
    and translated into Chinese from Bloxam's text-book by John
    Fryer.] Shanghai, 1875. 6 vols.
        Cf. Bloxam, C. L.
  Seh hsiang liu chen. [A treatise on photography, compiled and trans-
    lated into Chinese by J. F.] Shanghai. 2 vols.
  Tsao hwang chiang shui fa. [The manufacture of sulphuric acid,
    compiled and translated into Chinese by J. F.]. Shanghai.
        Dr. John Fryer has also enriched Chinese scientific literature by translations
        of works on materia medica, mineralogy and mineralogical terms ; works
        not within the scope of this bibliography.

FUCHS, GEORG FRIEDRICH CHRISTIAN.
  Beiträge zu den neuesten Prüfungen der Bleiglasur. Jena, 1794–95.
    2 parts.
  Chemischer Lehrbegrif nach Spielmanns Grundsätzen ausgearbeitet
    und mit den neuesten Erfahrungen bereichert. Leipzig, 1787.
    8vo. Ill.
  Versuch einer natürlichen Geschichte des Boraxes und seiner Bestand-
    theile wie auch von dessen medicinischen und chymischen Ge-
    brauch Jena, 1784. pp. [xii]–96. 12mo.
  Versuch einer natürlichen Geschichte des Spiesglases, dessen che-
    mischer Zerlegung, arzneiischen und ökonomischen Gebrauch.
    Halle, 1786. pp. x–388, 8vo.
        The first chapter treats of the " natural history."
        Repertorium der chemischen Literatur (etc.). *See, in Section I.*

FUCHS, JOHANN NEPOMUK VON.
  Gesammelte Schriften des Johann Nepomuk v. Fuchs. Zum ehrenden
    Andenken herausgegeben von dem Central-Verwaltungs-Aus-
    schusse des polytechnischen Vereins für das Königreich Bayern

FUCHS, JOHANN NEPOMUK VON.   [Cont'd.]

Redigirt und mit einem Nekrologe versehen von seinem Schüler Cajetan Georg Kaiser. Mit dem Bildnisse sammt Facsimile und einer Abbildung des Geburtshauses des Verewigten. München, 1856. 4to.

> Contains necrology and portrait of Fuchs.

FUCHS, M. J.

Die wichtigsten Thatsachen der Chemie der Carbonide. Eine Grundlage für die Unterrichtszwecke der deutschen Mittelschulen. München, 1878. pp. iii–118. 8vo.

> Zweite Auflage, München, 1880.

FUCHS ET CUMENGE.

> Or. *See, in Section II*, Fremy : Encyclopédie chimique. Vol. III.

FUGGER, EB.

Anfangsgründe der Chemie, zunächst für den Gebrauch an Realgymnasien bearbeitet. Wien, 1868. 8vo.

FÜLLEBORN, F. L.

Das Uebereinstimmende und Abweichende der Grundregeln der Chemie und Logik aus der Einheitslehre als Grundwissenschaft entwickelt. Berlin, 1850. 8vo.

FULHAME, MRS.

Essay on Combustion, with a view to a new art of dyeing and painting. Wherein the Phlogistic and Antiphlogistic hypotheses are proved erroneous   First American edition, Philadelphia, 1810. 12mo.

FUNARO, A.

Chimica agraria. Milano, 1884. 2 vols., 8vo.

FUNARO, A., E PITONI.

Corso di fisica e chimica pei licei. Livorno, 1891. 3 vols., 8vo.

FUNKE, OTTO.

Atlas der physiologischen Chemie. Zugleich als Supplement zu C. G. Lehmann's Lehrbuch der physiologischen Chemie. 15 Tafeln enthaltend 90 Abbildungen, sämmtlich nach dem Mikroscop gezeichnet und erläutert. Leipzig, 1853. 4to.

> Zweite gänzlich neu gezeichnete Auflage. 18 Tafeln enthaltend 180 Abbildungen, sämmtlich nach dem Mikroscop gezeichnet und erläutert. Lithographie und Farbendruck von J. G. Bach. Leipzig, 1858. 8vo.

FUNKE, OTTO. [Cont'd.]

> Atlas of Physiological Chemistry, supplement to Lehmann's Physiological Chemistry. London, 1853. 15 plates.
>
> *Cf.* Cavendish Society ; *also*, Lehmann, C. G.
> Tastsinn und Gemeinfühle. *See* Handbuch der Physiologie.

FURNEAUX, WILLIAM S.

> Elementary Chemistry, inorganic and organic. Adapted to the requirements of the "Alternative" elementary syllabus of the Science and Art Department. London, 1888.
>
> Longmans' Elementary Science Manuals.

FÜRNROHR, AUG. EM.

> Lehrbuch der technischen Chemie für den ersten Unterricht an Gewerbschulen. Regensburg, 1842. 8vo.
>
> > Zweite Auflage, 1846.

FÜRSTENAU, E.

> Das Ultramarin und seine Bereitung nach dem jetzigen Stande dieser Industrie. Wien, 1880. Ill.

FÜRSTENAU, JOHANN FRIEDRICH.

> De alumine. Rinteln, 1748.
>
> De antimonio crudo. Rinteln, 1748.
>
> Desiderata physico-chemica. Rinteln, [1750?].

FUSS, K.

> Grundzüge der Chemie aus Versuchen entwickelt. Der neueren Anschauung gemäss und nach methodischen Grundsätzen als Leitfaden für den Unterricht in der Chemie zunächst an Lehrerbildungsanstalten bearbeitet. Nürnberg, 1878. 8vo.

FYFE, ANDREW.

> Manual (A) of Chemistry. Edinburgh, 1826. 12mo.
>
> Elements of Chemistry. Edinburgh, 1827. 2 vols., 8vo.
>
> > Another edition, with additions by J. W. Webster. Boston, 1827. 12mo.

GABBA, LUIGI.

> Manuale del chimico e dell' industriale. Milano, 1889.
>
> Sopra alcuni studj di chimica organica e sull' applicazione dei loro risultati all'arte tintoria. *n. p., n. d.*
>
> Sull' indirizzo dell' insegnamento nelle scuole di chimica applicata all' industria : relazione alle Societá d'incoraggiamento d'arti e mestieri in Milano. Milano, 1886. 8vo.
>
> Trattato di analisi chimica generale ed applicata, ad uso delle scuole d'applicazione degli ingegneri, delle universitá, etc. Parte prima :

GABBA, LUIGI. [Cont'd.]

ricerche chimiche generali qualitative e quantitative. Parte seconda: ricerche speciali qualitative e quantitative ad uso delle arti e delle industrie. Milano, 1880. 8vo. Ill.

Trattato di chimica inorganica ed organica ad uso degli istituti tecnici, delle università, delle scuole d'applicazione e professionali. Milano, 1883. 16mo.

Seconda edizione riveduta. Milano, 1888. 16mo. Ill.

GABER, AUGUST.

Die Fabrikation von Rum, Arrak und Cognac und allen Arten von Obst- und Früchtenbranntweinen, sowie die Darstellung der besten Nachahmungen von Rum, Arrak, Cognac, Pflaumenbranntwein (Slibowitz), Kirschwasser u. s. w. Nach eigenen Erfahrungen geschildert. Wien, Pest, Leipzig, 1886. Ill.

GADD, PETER ADRIAN.

Disquisitio chemica hypotheseos de transmutatione aquæ in terram. Aboæ, 1763. 4to.

Observationes chemico-physicæ de originaria corporum mineralium electricitate. Aboæ, 1769. 4to.

Tentamen speciminis chemicæ opticæ. Aboæ, 1772. 4to.

GADOLIN, JOHANN.

Inledning till Chemien. Åbo, 1798. 8vo.

GÄNGE, C

Lehrbuch der angewandten Optik in der Chemie. Spectralanalyse, Mikroskopie, Polarisation. Praktische Anleitung zu wissenschaft-lichen und technischen Untersuchungen mit Hülfe optischer In-strumente nebst theoretischer Erklärung der beobachteten Erschei-nungen. Mit Tabellen der Emissions- und Absorptionsspectra in Wellenlängen, zahlreichen Abbildungen im Text und 24 Spectral-tafeln. Braunschweig, 1886. 8vo. Ill.

GAHN, JOHANN GOTTLIEB

Underrättelse om Uppställning och Nyttjandet af en förbättrad Appareil för Vattens aërerande. Upsala, 1804.

GALE, L. D.

Elements of Chemistry. Especially for the use of Schools and Academies. New York, 1835. pp. viii–323, 16mo. Ill.

Second edition. New York, 1837.

GALIPPE, L. M. V.

Étude toxicologique sur le cuivre et ses composés. Paris, 1875. 8vo.

GALL, L

Die Schnellgerberei in Nord-Amerika.   Trier, 1824.

GALLINI, S.

Summa observationum anatomicarum ac physico-chymicarum quæ usque ab anno MDCCXCII expositæ præcurrerunt nova elementa physicæ corporis humani.   Patavii, 1824.   8vo.

GALLISCH, FRIEDRICH ANTON.

De acido salis ejusque dephlogisticatione.   Lipsiæ, 1782.

GALLO, G.

Memoria di filosofia chimica.   Torino, 1869.   8vo.

GALLOWAY, ROBERT.

First (The) Step in Chemistry ; A new method of teaching the elements of the science.   Third edition.   London, 1860.   12mo.

> Fourth edition.   London, 1868.   pp. xxiv–478, 12mo.   Ill.

Second (The) Step in Chemistry ; or, The Student's Guide to the Higher Branches of the Science.   London, 1864.   pp. xvii–771, 12mo.   Ill.

Manual of Qualitative Analysis.   London, 1850.

> Fourth edition.   London, 1864.   Fifth edition.   London, 1870.
>
> Handleiding tot de qualitatieve analyse voor eerstbeginnenden, bijzonder tot zelfœfening geschikt.   Naar de Hoogduitsche vertaling van Th. Gerding, door T. J. van der Veer.   Utrecht, Amsterdam, 1854.   12mo.

Fundamental (The) Principles of Chemistry practically taught by a new method.   London, 1888.

GALTIER, C. P.

Traité de toxicologie générale, ou des poisons et des empoisonnements en général.   Paris, 1855.   8vo.

Traité de toxicologie médicale, chimique et légale et de la falsification des aliments, boissons, condiments.   Paris, 1845–55.   3 vols., 8vo.

> Vol. I. under the title :

Traité de toxicologie médico-légale, et de la falsification des aliments, des boissons et des médicaments, poisons inorganiques ou minéraux.   Paris, 1845.   8vo.

GALVANI, LUIGI.

Opere edite ed inedite raccolte e pubblicate per cura dell' Accademia dell scienze dell' Istituto di Bologna.   (Aggiunta alla collezione delle opere, etc.) ... Con appendice ... Bologna, 1841–42.   4to.

> Portrait and facsimile.

GAMGEE, ARTHUR.

A Text-book of the Physiological Chemistry of the Animal Body including an account of the chemical changes occurring in disease.
London, 1880. [?] 2 vols., 8vo.

GANSWINDT, A.

Lehrbuch der Baumwollengarn-Färberei. Abtheilung IV; Die Gerbstoff als Beizen. München, 1892. 8vo. Ill.

    In progress; parts I-III are to follow part IV.

GARAY, DE LA.
        See La Garay, De.

GARCIA LÓPEZ, M.

Manual del tintorero ó arte de teñir toda clase de tejidos y fieltros
nuevos ó usados, seguido del arte del quita-manchas. Madrid,
1881. 4to.

GARCÍA SOLÁ, EDUARDO.

Manual de microquímica clínica ó diagnóstico médico fundado en las
exploraciones microquímicas. Madrid, 1876. 8vo.

GARDNER, D. P.

Medical Chemistry, for the use of students and the profession; being a
manual of the science, with its applications to toxicology, physiology,
therapeutics, hygiene. Philadelphia, 1848. pp. xii–13–396, 12mo. Ill.

GARDNER, JOHN.

Acetic Acid, Vinegar, Alum, Ammonia, etc. Their Manufacture, etc.
Philadelphia, 1885.

Bleaching, Dyeing and Calico Printing. With formulæ, a chapter on
dye stuffs. Philadelphia, 1884. 12mo. Ill.
        Also London, 1884.

The Brewer, Distiller and Wine Manufacturer. Giving full directions
for the manufacture of beers, spirits, wines, liquors, etc., etc. A
handbook for all interested in the manufacture and sale of alcohol
and its compounds. Philadelphia, 1883. Ill.

GARNETT, THOMAS.

Experiments and Observations on the Crescent Water at Harrowgate.
Leeds, 1791.

Outlines of a Course of Lectures on Chemistry. Liverpool, 1797.
        Another edition : London, 1801. 8vo.

GARNIER, J. (JEUNE).

Guide pratique de chimie élémentaire. Ouvrage mis à la portée des gens du monde, des lycées et des institutions, contenant les principes de cette science et leur application aux arts et aux questions usuelles de la vie. Suivi d'une série de problèmes avec leurs solutions, de la synonymie chimique et d'un vocabulaire de chimie de la description des appareils et de la nomenclature des réactifs nécessaires. Paris, 1862. 8vo.

GARNIER, JEAN JOSEPH JULES.

Précis élémentaire de chimie ; ouvrage mis à la portée des gens du monde, des candidats au baccalauréat ès sciences, des écoles normales primaires supérieures, des collèges et des institution ; contenant les principes de cette science et leur application aux arts et aux questions usuelles de la vie ; suivi d'une série de problèmes, de la synonymie, d'un vocabulaire, de la description des appareils et réactifs. Paris, 1841. 12mo.

GARNIER, JEAN JOSEPH JULES, ET CH. HAREL.

Des falsifications des substances alimentaires et des moyens chimiques de les reconnaître. Paris, 1844. 12mo.

Falsificaciones de las sustancias alimentiaras y medios de reconocerlas. Traducidas compendiadas y adicionadas por Magin Bonet y Bonfill. Barcelona y Madrid, 1846. 8vo.

GARNIER ET SCHLAGDENHAUFFEN.

Analyse chimique des liquides et des tissus de l'organisme. *See, in Section II*, Fremy : Encyclopédie chimique, vol. IX.

GARNIER, LÉON.

Ferments et fermentations. Paris, 1887. Ill.

GARREAU, L.

Des attractions moléculaires que les gaz chimiquement inertes exercent entre eux, et de leurs effets comme agents de dissociations. Lille, 1886. 8vo.

GARRETT, J. H.

The Action of Water on Lead, being an inquiry into the cause and mode of the action and its prevention. London, 1891.

GASC, A.

Manuel pratique des analyses chimiques des matières phosphatées. Liège, 1892. 8vo.

GASPARIN, PAUL JOSEPH DE.
Traité de la détermination des terres arables dans le laboratoire. Paris, 1872. 16mo.
> Troisième édition, revue et augmentée. Paris, 1877. pp. xii–270, 8vo.

GASSICOURT, L. C. CADET DE.
See Cadet de Gassicourt, L. C.

GAUBERT, P.
Étude sur les vins et les conserves, suivie du compte rendu de la séance de dégustation tenue par les membres de la onzième classe de l'Exposition universelle. Paris, 1857. pp. ix–467, 8vo.

GAUDIN.
> *See, in Section II,* Fremy : Encyclopédie chimique, vol. II.

GAUDIN, A.
Recherches sur le groupement des atomes dans les molécules et sur les causes les plus intimes des formes cristallines. Avec une planche. Paris, 1847. 8vo.
L'architecture du monde des atomes, dévoilant la structure des composés chimiques et de leur cristallogénie. Paris, 1873. 12mo. Ill.

GAUDIN, G.
Notions de chimie générale. Déterminations des nombres proportionnels ; théorie atomique ; dissociation et transformation allotropiques. Paris, 1891.

GAUDRY, JULES.
Guide pratique pour l'essai des matières industrielles d'un emploi courant dans les usines, les chemins de fer, les bâtiments la marine, etc. Paris, *n. d.* [1876]. 16mo. Ill.

GAULTIER DE CLAUBRY, HENRI FRANÇOIS.
Rapport sur la fabrication des poudres fulminantes. Paris, 1838.
> *See also, in Section VII,* Répertoire de chimie.

GAUTSCH, K.
Das chemische Feuerlöschwesen in allen seinen Theilen. München, 1891. 8vo.

GAUTIER (ÉMIL JUSTIN), ARMAND.
> Leçon (etc.). *See :* Société chimique de Paris.

GAUTIER, ÉMIL JUSTIN ARMAND.

Chimie appliquée à la physiologie, à la pathologie et à l'hygiène, avec les analyses et les méthodes de recherches les plus nouvelles. Paris, 1874. 2 vols., 8vo. Ill.

Cours de chimie. Paris, 1887–91. 3 vols., 8vo. Vol. I, Chimie minérale, pp. xx–644. Vol. II, Chimie organique, pp. 683. Vol. III, Chimie biologique.

Sophistication (La) des vins. Coloration artificielle et mouillage, moyens pratiques de reconnaître la fraude. Paris, 1877. 18mo.

> Sophistication (La) des vins. Méthodes analytiques et procédés pour reconnaître les fraudes. Paris, 1884. Troisième édition. pp. viii–268, 12mo. Ill.

> Sophistication et analyse des vins. Quatrième édition, entièrement refondue accompagnée de 4 planches noires et coloriées et de figures dans le texte. Paris, 1891. pp. vi–356.

Sur les alcaloïdes dérivés de la destruction bactérienne ou physiologique des tissus animaux, ptomaïnes et leucomaïnes. Paris, 1886.

GAUTIER, H., ET G. CHARPY.

Leçons de chimie à l'usage des élèves de mathématiques spéciales. Paris, 1892. 8vo. Ill.

GAUTIER, L.

Manuel pratique de la fabrication et du raffinage du sucre de betterave. Paris, 1880. pp. iv–207.

GAUTIER, L. [Translator].

> See Wagner et L. Gautier, Nouveau traité de chimie industrielle. See Dragendorff, Manuel; also Fresenius, C. R.; also Lunge, G.; also Post, Julius.

GAVARE, EMIL.

Wegweiser in die Chemie. Eine Vorschule dieser Wissenschaft für Studiengenossen und Freunde der Naturkunde. Leipzig, 1869. 16mo.

GAY LUSSAC, JOSEPH LOUIS.

Cours de chimie comprenant l'histoire des sels, la chimie végétale et animale. Paris, 1828. 2 vols., 8vo.

> Vol. I has eighteen chapters each with separate pagination. Vol. II has fifteen chapters each with independent pagination. The lectures were written out from shorthand notes and revised by Gaultier de Clanbry.

Instructions pour l'usage de l'alcoolomètre centesimal. Paris, 1824.

Instructions sur l'essai des matières d'argent par la voie humide. Paris, 1833.

GAY LUSSAC, JOSEPH LOUIS. [Cont'd.]
> Vollständiger Unterricht über das Verfahren Silber auf
> nassem Wege zu probiren. Braunschweig, 1833. 8vo.
> Ill.

Instructions sur l'essai de chlorure de chaux. Paris, 1824.

Untersuchungen über das Jod. (1814). Herausgegeben von W. Ostwald.
Leipzig, 1889. 8vo.

> Ostwald's Klassiker der exakten Wissenschaften.

GAY-LUSSAC, J. L., ET L. J. THÉNARD.
> Recherches physico-chimiques faites sur la pile ; sur la préparation
> chimique et les propriétés du potassium et du sodium ; sur la
> décomposition de l'acide boracique ; sur les acides fluorique,
> muriatique et muriatique oxigéné ; sur l'action chimique de la
> lumière ; sur l'analyse végétale et animale, etc.   Avec six planches
> en taille-douce. Paris, 1811. 2 vols., 8vo. Vol. I, pp. xv–405.
> Vol. II, pp. 443.
>
> > Udtog af en Afhandling over den chemiske Undersøgelse
> > af Dele henhørende til Plante- og Dyr-Rigerne, forelæst
> > i det franske Instituts første Klasse d. 15 Januarii, 1810.
> > oversat af Ole Hieronimus Mynster. Kjøbenhavn, 1813.

GAZZERI, G.
Compendio d'un trattato elementare di chimica. Firenze, 1728. 2 vols.,

GEBER [Arabian alchemist].
> *See, in Section VI.*, Geber.

GEDANKEN ÜBER DIE ANZIEHENDEN KRÄFTE welche bey den chemischen
Auflösungen, und der Erzeugung der sogenannten fixen Luft
können in Betrachtung gezogen werden. Verfasset in einem Send-
schreiben an einen Freund. Prag, 1778. pp. 33, 8vo. Folding
plate.

GEERTS, A. J. C.
> Les produits de la nature japonaise et chinoise et leurs applications
> aux arts, à l'industrie, à la médicine, etc. Yokohama, 1878–83. 2
> vols., 8vo.

GEERTS, P.
> Beginselen der quantitatieve analytische scheikunde ten gebruike bij
> het hooger en middelbaar onderwijs. Tweede gedeelte toegepast op
> industrie, pharmacie, physiologie en landbouw. Utrecht, 1867. 8vo.

GEISSLER, E.
Grundriss der pharmaceutischer Maas-Analyse. Berlin, 1884.

GEIST, R.

Methoden der qualitativen chemischen Analyse von Substanzen welche die häufiger vorkommenden Elemente enthalten. Für den Schulgebrauch zusammengestellt. Halle, 1863. 8vo.

GEITNER, ERNST AUGUST.

Briefe über Chemie. Leipzig, 1808. 2 vols.

Fabrikmässige Bereitung des Syrups und Zuckers aus Kartoffelmehl. Leipzig, 1811.

Familie (Die) West, oder Unterhaltungen über Chemie und Technologie. Leipzig, 1805–'06. 2 vols.

Versuche über das Blaufärben wollener Zeuge ohne Indigo. Leipzig, 1809.

GELLÉE, A.

Précis d'analyses pour la recherche des altérations et falsifications des produits chimiques et pharmaceutiques. Paris, 1860. 8vo.

GELLERT, CHRISTLIEB EHREGOTT.

Anfangsgründe der metallurgischen Chimie in einem theoretischen und practischen Theile nach einer in der Natur gegründeten Ordnung. Leipzig, 1751–55. 2 vols., 8vo.

> Vol. II *also under the title :* Anfangsgründe zur Probierkunst.

> Metallurgic Chymistry, being a system of mineralogy in general and of all the arts arising from this science. Translated from the original German by I. S. with plates. London, 1776. 8vo.

GENTELE, J. G.

Lehrbuch der Farbenfabrikation. Anweisung zur Darstellung, Untersuchung und Verwendung der im Handel vorkommenden Malerfarben, zum Gebrauche für Farben-, Tusch- und Tapetenfabrikanten, Chemiker, Techniker, Kaufleute, Maler, Coloristen, Anstreicher und andere Farbenconsumenten. Zweite umgearbeitete und stark vermehrte Auflage. Braunschweig, 1880. 8vo. Ill.

> First edition 1860.

GENTH, FREDERICK A.

Tabellarische Uebersicht der wichtigsten Reactionen welche die Basen in ihren Salzen zeigen. Mit einer Zugabe. Marburg, 1846. Roy. fol.

Tabellarische Uebersicht der wichtigsten Reactionen, welche Säuren in ihren Salzen zeigen. Marburg, 1846. Roy. fol.

GEORGI, M.

Mittheilungen über die theoretischen Bewerthung und praktische Untersuchung der Sprengstoffe. Freiberg, 1887. Ill.

GEORGIEVICS, G. VON.

Der Indigo vom praktischen und theoretischen Standpunkte dargestellt. Wien, 1892. 8vo. Ill.

GEPPERT, J.

Die Gasanalyse und ihre physiologische Anwendung nach verbesserten Methoden. Berlin, 1885. 8vo.

GERBER, NICLAUS.

Chemisch-physicalische Analyse der verschiedenen Milch-Arten und Kindermehle, unter besonderer Berücksichtigung der Hygiene und Marktpolizei. Ein Buch aus der Praxis für Chemiker, Apotheker, Aerzte, Sanitätsbeamte und Unterrichts-Anstalten. Bremen, 1880. pp. vi–vi–90, 8vo. Ill. Folding table.

> Chemical and Physical Analysis of Milk, condensed milk, and infants' milk-foods, with special regard to hygiene and sanitary milk inspection. A laboratory guide developed from practical experience, intended for chemists, physicians, sanitarians, students, etc. Translated from the revised German edition, and edited by Hermann Endemann. New York and London, 1882.

Praktische (Die) Milchprüfung mit Einschluss der Centrifugalmilchprüfung. Vierte Auflage. Bern, 1887.

> Fünfte sehr vermehrte und verbesserte Auflage mit 8 Abbildungen. Bern, 1890. pp. [iv]–80, 12mo.

GERBER, P.

Qualitative chemische Analyse in tabellarischer Uebersicht. Bern, 1885. 8vo.

GERDING, THEODOR.

Allgemeinen (Die) Grundlehren des wissenschaftlich- chemischen Lehrgebäudes, mit besonderer Rücksicht auf Physik und Stöchiometrie, oder die Theoreme der physikalischen, reinen und mathematischen Chemie. Wiesbaden, 1873. pp. xv–627, 8vo.

> Zweite Auflage, 1874.

Einführung in das Studium der Chemie. . . . Leipzig, 1852.

Gewerbe-Chemie, oder die Chemie in ihrer Beziehung zur allgemeinen Kunst- und Gewerbethätigkeit. Ein Handbuch der technischen Chemie und chemischen Technologie für Fabrikanten, Techniker, Künstler, Gewerbetreibende, Berg- und Hüttenleute, Cameralisten, Chemiker, Eleven der technischen Lehranstalten. Göttingen, 1860–64. 3 vols., 8vo. Ill.

Illustrirte Chemie der Hauswirthschaft und der Gewerbe. Für Haus-

GERDING, THEODOR. [Cont'd.]
>   frauen und Gewerbsleute allgemein verständlich dargestellt. Frankfurt am Main, 1869. 8vo. Ill.
> Vorschule der qualitativ-chemischen Analyse . . . Leipzig, 1853.

GERDING, THEODOR.
>   Handbuch der organischen Chemie. *See* Gregory, William.

GERHARD, CARL ABRAHAM.
> Beiträge zur Chemie und Geschichte des Mineralreichs. Berlin, 1773–76. 2 vols., 8vo.
>> *See also in Section III.*

GERHARDT, CHARLES FRÉDÉRIC.
> Aide-mémoire pour l'analyse chimique, contenant les caractères des acides, etc., à l'usage des élèves des laboratoires de chimie. Paris, 1852. 12mo.
>> Poradnik do rozbiorów chemicznych. Tłomaczony przez Wincentego Karpiúskiego. Warszawa, 1854. 8vo.
>> Памятная книжка химическаго анализа для руководства занимающ въ химич. лабораторіяхъ. Перев. Варавина П. С.-Петербургъ 1865.
> Précis de chimie organique. Paris, 1844–45. 2 vols., 8vo.
>> Another edition : Traité de chimie organique. Paris, 1853–56. 4 vols., 8vo.
>> A meritorious work once regarded as indispensable. An edition in German, with the co-operation of Rudolph Wagner, was published at Leipzig, 1854–55. 3 vols.
> Введеніе къ изученію химіи по унитарной системѣ. Съ біографіею автора. Перев. Алексѣева. С.-Петербургъ 1865. Съ 6 чертеж. и портр. Жерара и Лорана.
>> With biography of the author.
> Précis d'analyse chimique qualitative. Paris, 1855. 8vo.

GERHARDT, CHARLES FRÉDÉRIC.
>   Traité de chimie organique. *See* Précis de chimie, of which the Traité is a second edition.

GERHARDT, CHARLES FRÉDÉRIC [Translator].
>   *See* Liebig, Justus von.

GERHARDT, CHARLES FRÉDÉRIC, ET G. CHANCEL.
> Précis d'analyse chimique qualitative. Paris, 1859. 12mo.
>> Deuxième édition par G. Chancel. Ouvrage contenant la description des opérations et des manipulations générales

GERHARDT, CHARLES FRÉDÉRIC, ET G. CHANCEL.   [Cont'd.]

de l'analyse qualitative, la préparation et l'usage des
réactifs, les caractères des acides et des bases, les essais
au chalumeau, la marche de l'analyse qualitative, la
détermination des sels, l'essai des eaux potables et
l'analyse des eaux minérales, l'analyse des mélanges
gazeux, l'analyse immédiate des matières végétales et
animales, la recherche des poisons, l'exposition de
l'analyse spectrométrique. Paris, 1862. pp. iii–699,
12mo. Ill.

Trattati d'analisi chimica qualitativa e quantitativa, tradotto
con aggiunte da Giorgini e Ant. Gibertini.   Parma,
1867.   2 vols., 12mo.

Аналитическая химія. Качественный анализъ. Пер. Менделѣева.
С -Петербургъ 1864.

Précis d'analyse chimique quantitative.   Ouvrage contenant la description des appareils et des opérations générales de l'analyse quantitative, les méthodes de dosage et de séparation des bases et des
acides, l'analyse par les liqueurs titrées, l'analyse organique,
l'analyse des gaz, l'analyse des eaux minérales, des cendres, des
terres arables, l'exposition du calcul des analyses, à l'usage des
médecins, des pharmaciens, des aspirants aux grades universitaires
et des élèves de laboratoire de chimie. Paris, 1859. pp. [iii]–710,
12mo. Ill.

Аналитическая химія количественная. Перев. съ добавл. Рехтера,
Менделѣева и Бородина. С.-Петербургъ 1866—1867.

GERICHTEN, ED. VON.

Die Theorie der Säuren- und Salzbildung und die electro-chemische
Theorie. Erlangen, 1875. 8vo.

GERICKE, PETRUS.

Fundamenta chymiæ rationalis. Lipsiæ et Guelferbytum, 1740. 8vo.
Praelectiones chymicæ extraordinariæ. Helmstädt, 1745.

GERLACH, G. TH.

Specifische Gewichte der gebräuchlichsten Salzlösungen bei verschiedenen Concentrationsgraden. Nebst Beiträgen zur Kenntniss
der Volumenveränderungen, welche beim Verdünnen wässriger
Salzlösungen sowie beim Lösen der Salze in Wasser stattfinden,
und Beobachtungen über die Ausdehnung mehr und minder con-
centrirter gleichnamiger Lösungen durch die Wärme. Für Chemiker und Physiker. Freiberg, 1859. 8vo. Ill.

GERNER, RAIMUND.

Die Glas-Fabrikation. Eine übersichtliche Darstellung der gesammten Glasindustrie mit vollständiger Anleitung zur Herstellung aller Sorten von Glas und Glaswaaren. Zum Gebrauche für Glasfabrikanten und Gewerbetreibende aller verwandten Branchen auf Grund praktischer Erfahrungen und der neuesten Fortschritte bearbeitet. Wien, Pest, Leipzig, 188–. Ill.

GERSON, G. H.

Die Verunreinigung der Wasserläufe durch die Abflusswässer aus Städten und Fabriken und ihre Reinigung. Berlin, 1889. Ill.

GERSTENHÖFER, MORITZ.

Hülfsbuch für den gewerblichen Chemiker. *See in Section II.*

GESNER, ABRAHAM.

A Practical Treatise on Coal Petroleum and other distilled oils. Second edition, revised and enlarged, by George Weltden Gesner. New York, 1865. pp. vi–181, 8vo. Ill.

GESNER, CONRAD.

De thermis et fontibus medicatis Helvetiæ et Germaniæ. Venetiis, 1553.

GESNER, JOHANN ALBRECHT.

Historia cadmiæ fossilis metallicae sive cobalti. . . .   Berolini, 1743. 4to.

GEUBEL, HEINRICH CARL.

Die physiologische Chemie der Pflanzen, mit Rücksicht auf Agricultur. Zugleich eine wissenschaftliche Widerlegung der Ansichten Liebig's und Schleiden's. Frankfurt a. M., 1845. pp. xii–312, 8vo.

Grundriss der zoophysiologischen Chemie. Zugleich eine kritische Beleuchtung aller neueren physiologisch-chemischen Theorien. Frankfurt am Main, 1845. 8vo.

Grundzüge der wissenschaftlichen Chemie der unorganischen Verbindungen. Frankfurt am Main. 1847. 8vo.

GEUTHER, A.

Beispiele zur Erlernung der quantitativen chemischen Analyse. Jena, 1887.

Erste Uebung in der chemischen Analyse. Dritte verbesserte Auflage. Jena, 1881. 8vo.

Kurzer Gang in der chemischen Analyse. Zweite verbesserte Auflage. Jena, 1867. 8vo.

GEUTHER, A. [Cont'd.]

Dritte Auflage. Jena, 1872. Vierte Auflage. Jena, 1881.

Lehrbuch der Chemie gegründet auf die Werthigkeit der Elemente. Jena, 1870. pp. xii–742, 8vo.

GIANELLI, C.

Lezioni di chimica date nella scuola militare in Modena. Modena, 1876. 8vo.

GIANETTO, SALO.

Dei fenomeni della combustione in rapporto con l'azione chimica commutativa. Messina, 1878. 8vo.

GIANNETTI, C.

Lavori di chimica eseguiti nel triennio 1876–78 nel laboratorio di chimica generale nella regia università di Sassari. Sassari, 1879. 8vo

GIBBS, WOLCOTT, AND GENTH, F. A.

Researches on the Ammonia-cobalt Bases. Smithsonian Contributions to Knowledge. Washington, 1856. 4to.

GIBSON, ALFRED.

Agricultural Chemistry. New edition, London, 1892.

GIBSON, RICHARD H.

American Dyer (The) . . . embracing over 400 recipes. Boston, 1878. 8vo.

> Contains 70 samples of dyed fabrics.

Art (The) of Dyeing all Colors on raw cotton, or cotton waste for the purpose of working with raw wool. The system and science of colors. The principles and practice of woolen dyeing. The properties and composition of the dyestuffs and chymical compounds which enter into the constitution of colors. Willimantic, Connecticut, 1861. 8vo. Ill.

GIESE, JOHANN EMMANUEL FERDINAND.

Chemie der Pflanzen- und Thier-Körper in pharmazeutischer Rücksicht mehrentheils nach eigenen Erfahrungen bearbeitet. Leipzig, 1811. 5 parts, 8vo.

> *Also under the title:* Lehrbuch der Pharmacie zum Gebrauche bei öffentlichen Vorlesungen. . . .

Classification des substances végétales et animales selon leur propriétés chimiques. Moscou, 1810.

Giese, Johann Emmanuel Ferdinand. [Cont'd.]

Darstellung der allgemeinen Chemie, zum Behufe für Vorlesungen. Dorpat, 1820. 8vo.

Von den chemischen Processen den dabei sich darbietenden Erscheinungen . . Riga, 1806–11. Two parts in one vol.

Всеобщая химія для учащихъ и учащихся. Харьковъ 1813—1817, 5 частей.
General Chemistry for teachers and pupils. Charkof, 1813–17. 5 vols.

Gigli, Torquato.

Latte, cacio, burro, olii grassi alimentari. Milano, 1885. 8vo.

Prodotti chimici inorganici usati come medicamenti. Manuali dell' alterazione e falsificazione delle sostanze alimentari e di altre importanti materie di uso comune. Milano, 1891. 4to.

Giglioli, J.

Chimica agraria, campestre e silvana, ossia chimica della piante coltivate, dell' aria, del terreno, dei lavori rurali e dei concimi. Napoli, 1884.

Gil, Emmanuele Gervasio.

Disquisitio in causam physicam recentiorum chemicorum pro elasticitate aëris atmosphaerici et aliorum fluidorum elasticorum, quæ gas nuncupatur, cum appendice de causa fluiditatis. Placentiæ, 1799.

Gilkinet, Alfred.

Traité de chimie pharmaceutique. Liège, 1885. 8vo. Ill.

Gill, C. H.

Chemistry for Schools : an introduction to the practical study of chemistry. London, 1875. 8vo.

Gille, Norbert.

Falsifications des médicaments qui doivent se trouver dans l'officine du vétérinaire Belge. Bruxelles, 1852. 8vo.

Gillet, Achille.

Traité pratique du dégraissage et du blanchîment des tissus, des toiles, des écheveaux, de la flotte, enfin de toutes les matières textiles, ainsi que du nettoyage et du détachage des vêtements et tentures. Paris, 1883. pp. iv–102, 8vo. Ill.

Girard, Ch.

Les laboratoires de chimie [and other articles]. *See, in Section II*, Fremy : Encyclopédie chimique, vol. I and vol. x.

GIRARD, C.

Recherches sur le développement de la betterave à sucre.　Paris, 1887.
10 plates.

GIRARD, CH., ET J. DE BREVANS.

Margarine (La) et le beurre artificiel.　Procédés de fabrication, dangers
au point de vue de la santé, procédés chimiques et physiques employés pour la reconnaître, législation française et étrangère.　Par
Ch. Girard et J. de Brevans.　Avec figures intercalées dans le texte.
Paris, 1889.　pp. 172, 8vo.

GIRARD, CHARLES, ET G. DE LAIRE.

Traité des dérivés de la houille applicables à la production des matières
colorantes.　Paris, 1873.　8vo.　12 plates.

GIRARD, CH., ET PABST.

Série aromatique des matières colorantes et ses applications industrielles.
Paris, 1883.　8vo.

GIRARDIN, A.

Chimie organique.　Troisième année.　Ouvrage rédigé conformément
au programme officiel du 3 août, 1881, et aux instructions ministéri-
elles du 18 october, 1881.　Paris, 1886.　18mo.

GIRARDIN, JEAN PIERRE LOUIS.

Leçons de chimie élémentaire faites le dimanche à l'école municipal de
Rouen.　Paris, 1837.　8vo.　Ill.
　　　Deuxième édition.　Paris, 1839.
　　Third edition under the title :
　　　Leçons de chimie élémentaire appliquées aux arts industriels,
　　　　et faites le dimanche à l'école municipale de Rouen.
　　　　Troisième édition, revue, corrigée et augmentée avec
　　　　200 figures et échantillons d'Indiennes intercalées dans
　　　　le texte.　Paris, 1845.　2 vols., 8vo.
　　　Cinquième édition.　Paris, 1874.　5 vols., 8vo.
　　　Leçons de chimie élémentaire appliquée aux arts industriels.
　　　　Sixième édition avec 1403 figures et 50 échantillons
　　　　dans le texte, augmentée d'un supplement.　Paris,
　　　　1880.　5 vols., 8vo.
　　Illustrated with engravings and samples.
　　　Scheikunde voor den beschaafden stand en het fabrieks-
　　　　wezen.　Gouda, 1845.　2 vols.　Tweede druk, 1850.
　　　　Derde druk door J. B. Peters, 1856.
　　　Vierde nieuwe volksuitgave uit het Fransch door J. B.
　　　　Peters.　Gouda, 1867.　2 vols., 8vo.　Ill.

GIRARDIN, JEAN PIERRE LOUIS. [Cont'd.]

    Zesde uitgave. Gouda, 1868.

    Lecciones de química elemental, con figuras repartidas por el contesto, esplicadas los domingos en la escuela municipal de Ruan. Traducidas de la segunda edicion Francesa dada á luz en el año de 1839, y adicionadas por Francisco Carbonell y Font. Barcelona, Madrid, 1841. 2 vols., 4to.

    Lecciones de química elemental hechas los domingos en la Escuela municipal de Ruan y traducidas al castellano por J. Bermudez de Castro. Paris, 1843. 2 vols., 8vo.

    Populära Föreläsningar i Kemien. Øfversättning. Stockholm, 1850. 8vo.

Mémoires de chimie appliquée à l'industrie, à l'agriculture, à la médecine et à l'économie domestique. Rouen, 1839. 8vo.

Manuel de chimie appliquée. Bruxelles, 1851. 18mo. Ill.

    Handboek van toegepaste scheikunst. Brussel, 1854. 12mo. Ill.

    Algemeene scheikunde. Voor Nederland bewerkt door F. H. van Moorsel. Amsterdam, 1854. 8vo.

Chimie agricole. Du sol arable, de ses variétés et des moyens d'en apprécier les qualités. Fragmens de leçons faites à l'École d'agriculture et d'économie rurale du département de la Seine-Inférieure. Caen, 1842. 8vo.

    O gnojach używanych jako nawozy. Wydanie V. dzieła uwieúczonego przez Towarzystwo agronomicane w Cher i. t. d. Tłumaczenie z Franzuzkiego z przedmową obejmującą tréśćiowy rys chemii agronomicznéj. Poznań, żupański, 1853. 8vo.

Notices sur diverses questions de chimie industrielle, médicale et agricole. Rouen, 1847. 8vo.

GIRARDIN, JEAN PIERRE LOUIS, ET HENRI LECOQ.

Éléments de minéralogie appliquée aux sciences chimiques. . . . Paris, 1826. 2 vols., 8vo. *Cf.* Lecoq, Henri.

GIRTANNER, CHRISTOPH.

Anfangsgründe der antiphlogistischen Chemie. Berlin, 1792. 8vo.

    With portrait of the author.

    Zweite, verbesserte und stark vermehrte Auflage. Berlin, 1795. pp. 16–466, 8vo.

    Dritte Auflage. Berlin, 1801. 8vo.

    Начальныя основанія химіи, горючее существо опровергающей. С.-Петербургъ 1801.

GIRTANNER, CHRISTOPH. [Cont'd.]

De terra calcarea cruda et calcinata. Dissertatio inauguralis. Göttingæ, 1783.

   See also in Section II.

GLADSTONE, J. H., AND ALFRED TRIBE.

The Chemistry of the secondary batteries of Planté and Faure. London, 1883. 12mo.

GLAESER, CHRISTIAN GOTTLOB.

Experimenta chemica cum tribus mineris stanniferis. . . . Vitebonæ, 1798.

GLASER, CHRISTOPH.

Traité de chimie, contenant une méthode claire et facile d'obtenir les préparations de cet art les plus nécessaires à la médecine. Paris, 1663.

    Another edition. Paris, 1667.

   The author was demonstrator of chemistry at the Jardin du Roi, Paris, and apothecary to Louis XIV. His little treatise, being very clearly written, passed through many editions.

   The edition of 1673 *bears the title :*

    * Traité de la chymie enseignant par une briève et facile méthode toutes les plus nécessaires préparations. Nouvelle édition, revuë et augmentée en toutes ses parties, principalement dans la troisième, que la mort de l'Autheur avoit empêché de mettre en sa perfection. Paris, 1673. pp. [x]–439–[xi], 18mo. Illuminated title-page and three folding plates.

   On the title-page the author's name is given as Christophle Glaser.

    Other editions : Brussels, 1676, 12mo.; Lyon, 1676 and 1679.

    * Chimischer Wegweiser, das ist sichere Anweisung zur chimischen Kunst, darinnen durch einen kurtzen Weg und leichte Handgriffe gewiesen wird, wie man allerley Artzneyen durch die Chimie bereiten kann. Erstlich in französischer Sprach beschrieben von C. G., anjetzo aber auf Begehren in unsere teutsche Sprache übersetzt von einem Philochimico, [Fr. Menudier ?] Jena und Helmstadt, 1696. pp. [x]–528–[xii], 18mo. Illustrated with folding plates and frontispiece.

    Other German editions : Jena, 1684 ; Uebersetzt von Marschalk, Nürnberg, 1677. 8vo.

GLASSER, H.

Anleitung zu stöchiometrischen Rechnungen, besonders für angehende Chemiker und Pharmaceuten. Mit einem Anhange enthaltend Tabellen und Beispiele aus der praktischen Chemie. Stuttgart, 1837.

GLAUBER, JOHANN RUDOLPH.
  Opera. Amstelodami, 1650-'70.   4 vols., 8vo.   Ill.

Contents :
> Pharmacopoeia spagyrica—De auri tinctura—Descriptio qua ratione ex
> vini fecibus tartarum sit extrahendum—Operis mineralis partes tres—
> Descriptio artis destillatoriae cum appendice et annotationibus—Mi-
> raculum mundi cum continuatione, explicatione et annotationibus—
> Prosperitas Germaniae—Consolatio navigantum—De natura salium—
> De signatura salium, metallorum et planetarum—Libellus dialogorum—
> Novum lumen chymicum—De tribus principiis metallorum—De medi-
> cina universa—Explicatio verbi Salomonis : In herbis, verbis et lapidi-
> bus magna est virtus—Appendices—Apologia contra Farnerum.

Opera chymica, Bücher und Schrifften so viel deren von ihme bisshero
  an Tag gegeben worden . . .   Franckfurt am Main, 1658.   2 vols.
  Vol. I, pp. [xxii]-574 ; vol. II, ⌊xii]-444-[xviii], sm. 4to.

  *Works (The) of the highly experienced and famous John
    Rudolph Glauber.   Containing great variety of choice
    secrets in medicine and alchymy, in the working of
    metallick mines, and the separation of metals.   Also,
    various cheap and easie ways of making saltpetre, and
    improving of barren land and the fruits of the earth.
    Together with many other things very profitable for all
    the lovers of art and industry.   Translated into English,
    and published for publick good by the labour, care and
    charge of Christopher Packe, Philochymico-Medicus.
    London, printed by Thomas Milbourn, for the author,
    and are to be sold at his house next door to the Gun in
    Little-Moorfields ; by D. Newman at the King's Arms
    in the Poultry, and W. Cooper at the Pellican in Little-
    Britain, 1689.   3 Parts and an Index.   pp. [xii]-440-
    [iv]-220-92-[xii], folio.   Plates.

Contains the following treatises :
  Part I. (a) Philosophical furnaces : in 5 parts and an Appendix.
           (b) Of the tincture of gold, or the true Aurum Potabile.
           (c) The mineral work.   3 parts.
           (d) The apology of J. R. G. against the lying calumnies of
               Christopher Farner.
           (e) Miraculum mundi, etc., in two parts.
           (f) A treatise of the nature of salts, in one part.
           (g) A treatise of the signature of salts, metals and planets.
           (h) The consolation of navigators.
           (i) A true and perfect description of extracting good tartar from
               the lees of wine.
           (j) The prosperity of Germany, in 6 parts.
  Part II. (a) The first century of Glauber's weathy store-house of treasures.
           (b) The second century.

GLAUBER, JOHANN RUDOLPH. [Cont'd.]

  (c) The third century.
  (d) The fourth century.
  (e) The fifth century.
  (f) Novum lumen chymicum.
  (g) A spagyrical pharmacopoeia in 7 parts and 3 appendices.
  (h) Libellus ignium, or book of fires.
Part III. (a) Of the three principles of metals.
  (b) A short book of dialogues.
  (c) The goddess of riches.
  (d) Of Elias the artist.
  (e) Of the three fire-stones.
  (f) De purgatoria philosophorum.
  (g) Of the sacred fire of philosophers.
  (h) Of the animal-stone.

Consolation (La) des navigateurs dans laquelle est enseigné à ceux qui voyagent sur mer un moyen de se garantir de la faim et de la soif, voire mesme les maladies qui leur pourraient survenir durant un longue voyage ; traduit par Du Teil. Paris, 1659. 8vo.

Première (La) (2e, 3e, 4e et 5e) partie de la description des nouveaux fourneaux philosophiques, ou art distillatoire, par le moyen duquel sont tirez les esprits, huiles, fleurs et autres médicaments ; traduit par Du Teil. Paris, 1659. 8vo. Ill.

Première (La) (2e et 3e) partie de l'oeuvre minéral, où est enseignée la séparation de l'or des pierres à feu, sable, argile, et autres fossiles par l'esprit de sel ce qui ne se peut faire par autre voye ; traduit par Du Teil. Paris, 1659. 8vo.

Teinture (La) de l'or, ou véritable or potable, sa nature et sa différence d'avec l'or potable, faux et sophistique, sa préparation spagirique et son usage dans la médecine ; traduit par Du Teil. Paris, 1659. 8vo.

Tractatus de natura salium, oder ausführliche Beschreibung deren bekannten Salien. Amsterdam, 1658.

  Contains the first account of Glauber's Salt, Sal mirabile Glauberi, sulphate of soda.

Глинскій, Г.
  О радикалахъ въ органической химіи. Казань 1872.
   GLINSKY, H. On Radicals in Organic Chemistry. Kazan, 1872.

GLOGERUS, JOHANNES.
  See Müller, Johannes.

GLOVER, ROBERT MORTIMER.
  A Manual of Elementary Chemistry. London, 1855.
  On Mineral Waters, their physical and medicinal properties, with descriptions of the different mineral waters of Great Britain and

GLOVER, ROBERT MORTIMER. [Cont'd.]
the Continent and directions for their administration. London, 1857. 8vo.

GLÜCKSMANN, CARL.
Kritische Studien im Bereiche der Fundamentalanschauungen der theoretischen Chemie. Theil I, Ueber die Quantivalenz. Wien, 1891. pp. 63, 8vo.

GMELIN, CHRISTIAN GOTTLIEB.
Einleitung in die Chemie. Tübingen, 1835–37. 2 vols. Vol. I, pp. xxviii–950; vol. II, pp. 983–2131, 8vo.

GMELIN, JOHANN FRIEDRICH.
Chemische Grundsätze der Gewerbkunde. Hannover, 1795. 8vo.
Chemische Grundsätze der Probier- und Schmelzkunst. Halle, 1786. 8vo.
Einleitung in die Chemie. Nürnberg, 1780. 8vo.
Another edition, 1786.
Einleitung in die Mineralogie. Nürnberg, 1780.
Einleitung in die Pharmacie. Nürnberg, 1781.
Grundriss der allgemeinen Chemie zum Gebrauch bei Vorlesungen. Göttingen, 1789. 2 vols., 8vo.
Zweite Auflage, Göttingen, 1804. 2 vols., 8vo.
Grundriss der Mineralogie. Nürnberg, 1790.
Grundriss der Pharmacie. Göttingen, 1792.
Grundsätze der technischen Chemie. Halle, 1786. pp. xvi–750. Anhang, pp. viii–402, 12mo.
The Appendix treats of assaying and analysis.
Handbuch der technischen Chemie. Göttingen, 1795–96. 2 vols.
Ueber die neueren Entdeckungen in der Lehre von der Luft; und deren Anwendung auf Arzneikunst. Berlin, 1784. 8vo.
See also in Section III.

GMELIN, LEOPOLD.
Handbuch der theoretischen Chemie zum Behuf seiner Vorlesungen entworfen. Frankfurt-am-Main, 1817–19. 3 vols., 8vo.
Zweite Auflage. F. a. M., 1821–'22.
Dritte Auflage. F. a. M., 1827–'29. 3 vols.
Continued under the title : Handbuch der Chemie. Vierte umgearbeitete und vermehrte Auflage. Heidelberg, 1843–70. 10 vols., 8vo. Register zum ersten bis fünften Band bearbeitet von K. List.
Vol. VII is in two Abtheilungen. Vol. IX and X are Supplements.

GMELIN, LEOPOLD. [Cont'd.]

Fünfte Auflage, Heidelberg, 1852.

Handbuch der anorganischen Chemie. Sechste umgear-
beitete Auflage. Herausgegeben von Karl Kraut.
Heidelberg, 1871–'86. Three volumes in five parts, 8vo.
Ill.

Vol. I, Part 1 : Allgemeine und physikalische Chemie.
Bearbeitet von Alexander Naumann. 1877.

Vol. I, Part 2 : Nichtmetallische einfache Stoffe und ihre
Verbindungen untereinander. Bearbeitet von H. Ritter
und Karl Kraut. 1872.

Vol. II, Part 1 : Metalle. Bearbeitet von Karl Kraut. 1886.

Vol. II, Part 2 : Metalle. Bearbeitet von S. M. Jörgensen.
1876. [Incomplete.]

Vol. III : Metalle. Bearbeitet von S. M. Jörgensen. 1875.

Also under the title : Gmelin-Kraut : Handbuch der Chemie.

Handbook of Chemistry. Translated by Henry Watts.
Printed for the Cavendish Society. 19 vols. London,
1848–72. 8vo. Ill.

Second edition of Vol. I, 1861.

Cf. Cavendish Society.

Lehrbuch der Chemie zum Gebrauche bei Vorlesungen auf Universitäten, in Militärschulen, polytechnischen Anstalten, Realschulen,
etc., sowie zum Selbstunterricht. Erste Abtheilung, Unorganische
Chemie. Heidelberg, 1844. pp. x–380. 8vo.

Chimie organique appliquée à la physiologie et à la médecine ; contenant l'analyse des substances animales et végétales. Traduit de
l'allemand d'après la seconde édition par J. Ineichen. Avec des
notes et des additions sur diverses parties de la chimie et de la
physiologie par Virey. Paris, 1823. pp. 488, 12mo.

GMELIN, LEOPOLD UND FRIEDRICH TIEDEMANN.
See Tiedemann, Friedrich.

GMELIN-KRAUT.
Handbuch der Chemie. *See* Gmelin, Leopold : Handbuch der anorganischen Chemie.

GODEFFROY, RICHARD.
Tabellen und Formulare für qualitative und quantitative Analysen,
Titrirmethoden, Harnanalysen, etc. Wien, *n.d.* [1884]. 8vo.

GODEFROY, L.
Bismuth. *See, in Section II,* Fremy : Encyclopédie chimique, Vol. III.

GODEFROY, L.
Fabrication de l'alcool d'industrie chimiquement pur.   Paris, 1888.

GODET, A.
Contribution à l'étude des alcaloïdes de l'urine.   Paris, 1889.

GÖBEL, CARL CHRISTIAN TRAUGOTT FRIEDEMANN.
Grundlinien der pharmaceutischen Chemie   .   .   .   Jena, 1821.
Grundzüge der analytischen Chemie.   Erlangen, 1845.   8vo.
Vol. III of the Grundlehren der Pharmacie.

GÖPPERT, HEINRICH ROBERT.
Ueber die chemischen Gegengifte.   Zweite Auflage.   Breslau, 1843.

GÖRZ, J.
Handel und Statistik des Zuckers.   Berlin, 1884.

GÖTTLING, JOHANN FRIEDRICH AUGUST.
Anweisung zum Gebrauch seines chemischen Probierkabinets für
Scheidekünstler.   Jena, 1790.   8vo.
Description of a Portable Chest of Chemistry ; or complete
collection of chemical tests.   Translated from the Ger-
man.   London, 1791.   8vo.
Beschreibung verschiedener Blasemaschinen zum Löthen.   Erfurt,
1784.
Beytrag zur Berichtigung der antiphlogistischen Chemie auf Versuche
gegründet.   Weimar, 1794–98.   8vo.   Ill.
Chemische Bemerkungen über das phosphorsaure Quecksilber und
Hahnemann's schwarzen Quecksilberkalk.   Jena, 1795.
Chemische Versuche über eine verbesserte Methode den Salmiak zu
bereiten.   Weimar, 1782.
Einleitung in die pharmazeutische Chemie.   Mit einer Vorrede von
W. H. S. Buchholz.   Altenburg, 1778.   8vo.
Elementarbuch der chemischen Experimentirkunst.   Jena, 1808.   2
parts.   8vo.
Handbuch der theoretischen und praktischen Chemie.   Jena, 1798–
1800.   3 vols., 8vo.   I. Systematischer Theil.   II. Praktischer
Theil.   III. Pharmaceutischer Theil.
Praktische Anleitung zur prüfenden und zerlegenden Chemie.   Jena,
1802.
Praktische Vortheile und Verbesserungen verschiedener pharmaceutisch-
chemischer Operationen.   Weimar, 1783.
Syrup- (Die) und Zuckerbereitung aus Runkelrüben.   Jena, 1808.
Tabella über die Lehre von den Salzen   .   .   .   Weimar, 1784.

GÖTTLING, JOHANN FRIEDRICH AUGUST. [Cont'd.]
Versuch einer physischen Chemie. Jena, 1792.
Vollständiges chemisches Probir-Cabinet zum Handgebrauche für
Scheidekünstler, Aerzte, Mineralogen, Metallurgen, Technologen,
Fabrikanten, Oekonomen und Naturliebhaber. Jena, 1790. 8vo.
Part I, Untersuchungen auf dem nassen Wege.
Zuckerbereitung aus Mangoldarten. Jena, 1799.
See also in Sections II and VII.

GÖTTLING, JOHANN FRIEDRICH AUGUST, UND JOACHIM DIETERICH
BRANDIS.
Technologisches Taschenbuch für Künstler, Fabrikanten, und Metal-
lurgen. Göttingen, 1786.

GOHREN, CARL THEODOR VON.
Agricultur-Chemie (Die) nach dem heutigen Standpunkte . . .
Leipzig, 1872–77. 2 vols., 8vo Ill. Vol. I : Die naturgesetzlichen Grundlagen des Pflanzenbaues. Vol. II : Die Naturgesetze
der Fütterung der landwirthschaftlichen Nutzthiere.
Anleitung zu chemischen Untersuchungen mit besonderer Beziehung
auf Landwirthschaft und landwirthschaftliche Industrie. Prag,
1867. 8vo. Ill.
Methodischer Leitfaden für den chemischen Unterricht an landwirthschaftlichen Fachschulen. Wien, 1882. pp. xv–271. 8vo.

GOLDBERG, ALWIN, AND OSKAR GOLDBERG.
Die natürlichen und künstlichen Mineralwässer. Ein Handbuch enthaltend eine kurze Zusammenfassung der wichtigsten Kapitel der
Mineralquellenlehre und Darlegung der Prinzipien der Herstellung
künstlicher Mineralwässer, insbesondere der Nachbildung natürlicher Mineralwässer. Weimar, 1892. pp. v–219. 8vo.

GOMEZ, BERNARDIN.
Diascepseon de sale libri quatuor . . . denuo revisi per Petrum
Uffenbackium. Ursell, 1605. 8vo.

GOMEZ, PAMO, J. R.
Manual de análisis química aplicada á las ciencias médicas. Obra
ilustrada con grabados intercalados en el texto. Madrid, 1870.
4to.
Another edition. Madrid, 1882.

GONZALES VALLEDOR, VENANCIO, Y JUAN CHAVARRI.
Programa de un curso elemental de física y nociones de química para
el uso de los alumnos de quinto año de filosofía. Madrid, 1848.
8vo.

GOPPELSROEDER, FRIEDRICH.
Ueber die Darstellung der Farbstoffe sowie über deren geichzeitige Bildung und Fixation auf den Fasern mit Hilfe der Electrolyse. Reichenberg, 1885.

GORDON, D.
Methode das Gaslicht tragbar und dadurch brauchbar zu machen. Ein Supplement zu Accum's Werk über die Gasbeleuchtung. Weimar, 1820. 8vo.
>*See* Accum, Fr. Praktische Abhandlung [etc.].

GORDON, J. H.
Aids to practical Chemistry, especially arranged for the analysis of substances containing a single base and acid radicle. London, 1887.

GORE, GEORGE.
The Art of Electro-Metallurgy. Including all known processes of electro-deposition. London, second edition, 1884. pp. xx–397. 12mo. Ill.
>Text-books of Science.

The Art of Electrolytic Separation and Refining of Metals. London, 1890. 8vo.

GORE, GEORGE, MARCUS SPARLING, AND JOHN SCOFFERN.
Practical Chemistry, including the theory and practice of electro-deposition, photographic art, the chemistry of food, with a chapter on adulteration ; and the chemistry of artificial illumination. London, 1856. 8vo.

GORHAM, JOHN.
The Elements of Chemical Science. In two volumes, with plates. Boston, 1819–20. 8vo.
>" Not surpassed by any one with which we are acquainted, as a perspicuous, chaste, and philosophical treatise."—BENJ. SILLIMAN, 1822.

GORINI, G.
Olii vegetali, animali e minerali. Milano, 1878. 8vo.

GORUP BESANEZ, EUGEN C. FRANZ VON.
Anleitung zur qualitativen und quantitativen zoochemischen Analyse, enthaltend die Lehre von den Eigenschaften und dem Verhalten der im Thierreich vorkommenden oder aus diesem entstehenden chemischen Verbindungen gegen Reagentien, sowie systematisches Verfahren zur qualitativen und quantitativen chemischen Unter-

GORUP-BESANEZ, EUGEN C. FRANZ VON. [Cont'd.]
    suchung thierischer Secrete, Excrete und Gewebe, zum Gebrauch
    im Laboratorium und zum Selbstunterricht. Nürnberg, 1850. pp.
    xxiv–363–iv, 8vo.   2 folding plates.
        A pioneer work in a difficult field, which later discoveries made more feasible,
            as shown in the subsequent editions.
        Zweite Auflage. Nürnberg, 1854. 8vo.
        Anleitung zur qualitativen und quantitativen zoochemischen
            Analyse, für Mediciner, Pharmaceuten, Landwirthe
            und Chemiker, zum Gebrauche im Laboratorium und
            zum Selbstunterrichte bearbeitet.   Dritte vollständig
            umgearbeitete und vermehrte Auflage. Braunschweig,
            1871. 8vo. Ill.
        Traité d'analyse zoochimique qualitative et quantitative
            traduit par L. Gautier. Paris, 1875. 8vo.
    Lehrbuch der Chemie für den Unterricht auf Universitäten und mit
        besonderer Berücksichtigung des Standpunktes studirender Medi-
        ciner. Braunschweig, 1859–60. 2 vols., 8vo.
        Lehrbuch der Chemie für den Unterricht auf Universitäten,
            technischen Lehranstalten und für das Selbststudium.
            Braunschweig, 1873–'75. 3 vols., 8vo. Ill.
        Vol. I. Anorganische Chemie. Fünfte mit besonderer Be-
            rücksichtigung der neueren Theorien vollständig umge-
            arbeitete und verbesserte Auflage.   Mit zahlreichen in
            den Text eingedruckten Holzstichen und einer farbigen
            Spectraltafel. 1873.
        Vol. II. Organische Chemie. Fünfte auf Grundlage der
            neueren Theorien vollständig umgearbeitete und ver-
            besserte Auflage. 1875.
        Vol. III. Physiologische Chemie.   Dritte vollständig umgear-
            beitete und verbesserte Auflage.   Mit einer Spectraltafel
            und drei Tafeln in Holzstich, den Respirations-Apparat
            darstellend. 1875.
        Lehrbuch der Chemie für den Unterricht auf Universitäten
            technischen Lehranstalten und für das Selbststudium.
            Braunschweig, 1878–85. 3 vols., 8vo. Ill.
        Vol. I. Anorganische Chemie. Mit Einschluss der experi-
            mentellen Technik.   Siebente vollständig neu bear-
            beitete Auflage. Von Albrecht Rau. 1885. Ill.
        Vol. II. Organische Chemie. Sechste Auflage, neu bearbeitet
            von Hermann Ost. 1881. Ill.
        Vol. III. Physiologische Chemie. Vierte, vollständig umge-
            arbeitete und vermehrte Auflage. 1878. Ill.

GORUP-BESANEZ, EUGEN C. FRANZ VON. [Cont'd.]

Leerboek der organische scheikunde. In het Nederduitsch overgebragt door P. A. Huet. Utrecht en Amsterdam, 1860. 1. gedeelte, 8vo. Ill.

Руководство къ органической химіи. Перев. Бордюгова. Москва, 1861.

Traité de chimie physiologique, traduit de l'allemand par Schlagdenhaufen. 2 vols., 8vo.

Руководство къ физіологической химіи. Перев. Волкова и Манасеина. Дерптъ 1863.

Tafeln zur Erläuterung der Typentheorie, und der Ableitung Formeln organischer Verbindungen von den Typen. Braunschweig, 1860. pp. vi–37, 8vo.

The tables are made clear by the use of red ink for certain elements.

Vergleichende Untersuchungen im Gebiete der zoochemischen Analyse. Erlangen, 1850. 4to.

GOSSART, J.

Le contrôle chimique de la fabrication du sucre. Lille, 1886. 8vo.

GOTTLIEB, JOHANN.

Kurze Anleitung zur qualitativen chemischen Analyse. Für Anfänger bearbeitet. Wien, 1866. 8vo.

Lehrbuch der pharmaceutischen Chemie mit besonderer Berücksichtigung der oesterreichischen, preussischen und sächsischen Pharmakopöen. Berlin, 1857–9. 2 vols., 8vo. I, pp. 525 ; II, 621.

Lehrbuch der reinen und technischen Chemie. Braunschweig, 1853. 8vo.

Zweite Auflage. Braunschweig, 1861. 8vo.

Lehrbuch der reinen und angewandten Chemie. Zum Gebrauche an Real- und Gewerbeschulen, Lyceen, Gymnasien, technischen Lehranstalten etc., und zum Selbstunterrichte. Dritte verbesserte Auflage. Braunschweig 1868. 8vo. Ill.

Polizeilich-chemische Skizzen. 1. Ueber die Zusammensetzung, Werthbestimmung und Verfälschungen von Milch, Butter, Wachs, Walrath, Honig und Seife. Leipzig, 1853. 8vo.

Vollständiges Taschenbuch der chemischen Technologie zur schnellen Uebersicht bearbeitet. Leipzig, 1852. 16mo. Ill.

Volledig handboek der scheikundige technologie. Ten gebruike voor zelfonderrigt en op industrie-scholen, alsmede voor een algemeen overzigt bewerkt. Naar het Hoogduitsch bewerkt door J. R. F. Nievergeld. 's Gravenhage, 1855. Eerste stuk. Ill.

Химія и химическая технологія. Перев. Алексѣева.

С.-Петербургъ 1861.

GOUZY, P.
Promenade d'une fillette autour d'un laboratoire.   Paris, 1887.   8vo.

GOWER, ALFRED ROLAND.
An Elementary Textbook of Practical Metallurgy.   London, 1888.   8vo.

GRABFIELD, JOSEPH P., AND P. S. BURNS.
Chemical Problems.   Boston, 1888.   pp. 96, 12mo.

GRAEGER, NICOLAUS.
Fabrikmässige (Die) Darstellung chemischer Produkte, umfassend die
   Fabrikation aller, für Industrie und Gewerbe wichtigen, einfachen
   und zusammengesetzten Körper der anorganischen Chemie ; mit
   Angabe ihrer Eigenschaften und titrimetrischen Prüfung auf ihre
   Reinheit und Güte.   2 vols. and atlas.   Vol. I : Sauerstoff, Wasser-
   stoff, Stickstoff, Chlor, Jod und Kohlenstoff und deren technisch
   wichtigsten Verbindungen.   Vol. II : Bor, Schwefel und Phosphor,
   Natrium, Aluminium, Blei, Zink, Zinn, Kupfer, Uran, Quecksilber
   und deren technisch wichtigsten Verbindungen ; als Anhang : die
   Ultramarinfabrikation.   Weimar, 188–.   2 vols., 8vo.
Mass-Analyse (Die) oder die Bestimmung der chemisch wichtigen
   Körper auf volumetrischen Wege.   Ein Handbuch für Apotheker,
   Fabrikanten, Chemiker und solche die sich mit der Chemie be-
   schäftigen.   Nebst einer Anleitung zur Untersuchung auf Trink-
   wässer, hauptsächlich mit Rücksicht auf ihre hygienische
   Beschaffenheit.   Zweite vermehrte und verbesserte Ausgabe.
   Weimar, 1872.   8vo.   Ill.
Ueber die fabrikmässige Darstellung der Weinsäure und deren Ent-
   färbung mittelst eines neuen Verfahrens.   Halle, 1844.
         Zweite Auflage.   Halle, 1848.
            Based on Boussingault's treatise in his Économie rurale.

GRAFTIAU, FIRMIN.
Guide pratique pour l'analyse de la betterave à sucre, par la méthode
   alcoolique.   Paris, 1887.

GRAHAM, CHARLES.
The Chemistry of Bread Making.   Cantor Lectures, London Society of
   Arts.   London, 1880.
            La chimie de la panification.   Traduit de l'anglais.   Paris,
         1882.   12mo.
            Bibliothèque biologique internationale, No. 8.

GRAHAM, D. A.
Coal Analysis : a Treatise on the comparative commercial Values of
   Gas Coals and Cannels.   London, 18—.   8vo.

GRAHAM, THOMAS

Chemical and Physical Researches. Collected and printed for presentation only. Preface and analytical contents by R. Angus Smith. Edinburgh, 1876. pp. lvi–660. Plates and portrait.

Elements of Chemistry, including the applications of the science in the arts. London, 1842. 8vo.

> Second edition by T. G. and Henry Watts. London, 1847. 2 vols., 8vo.

> > Traité de chimie organique. Traduit de l'Anglais par E. Mathieu-Plessy. Paris, 1843. 8vo.

GRAHAM, THOMAS [Editor].

Chemical Reports and Memoirs, on atomic volume ; isomorphism ; endosmosis ; the simultaneous contrast of colors ; the latent heat of steam at different pressures ; the artificial formation of alkaloids ; and volcanic phenomena. London, 1848. pp. vi–[ii]–370, 8vo. Plates.

> Printed for the Cavendish Society. The essay on atomic volumes is by Prof. Otto ; on isomorphism by M. Filhol, and by Prof. Otto ; on endosmosis by Julius Vogel ; on colors by M. E. Chevreul ; on latent heat of steam by M. Regnault ; on alkaloids by M. E. Kop ; on volcanic phenomena by Prof. Bunsen. The translations were by G. E. Day.

GRAHAM, THOMAS, AND FRIEDRICH JULIUS OTTO.

Ausführliches Lehrbuch der Chemie. Vierte umgearbeitete Auflage. Braunschweig, 1863–78. 5 vols. in 9 parts, 8vo. Ill.

> For earlier editions, see Graham, Thomas, above, and compare Otto, Fr. Jul.

Vol. I : Lehrbuch der physikalischen und theoretischen Chemie, von H. Buff, H. Kopp und F. Zamminer. Zweite Auflage. Part I : Physikalische Lehren von H. Buff, H. Kopp und F. Zamminer. 1863.

Part II : Theoretische Chemie und Beziehungen zwischen chemischen und physikalischen Eigenschaften von Hermann Kopp. Vierte Auflage. 1863.

Vol. II : Ausführliches Lehrbuch der anorganischen Chemie. Part I. 1863. Part II von Robert Otto, 1863–72. Part III. 1863.

Vol. III (Dritte Auflage). Ausführliches Lehrbuch der organischen Chemie. 3 Parts. Part I, 1854. Part II, 186–. Part III, section 1, bearbeitet von E. Meyer und A. Weddige, 1878 ; section 2, bearbeitet von H. von Fehling, 1868.

> [Another edition.] Ausführliches Lehrbuch der Chemie. Braunschweig. 5 vols., 8vo. Ill.

> > Vol. I, in 3 parts : Physikalische Lehren. Von A. Winkelmann. Dritte gänzlich umgearbeitete Auflage des in

GRAHAM, THOMAS, AND FRIEDRICH JULIUS OTTO. [Cont'd.]
den früheren Auflagen von Buff, Kopp und Zamminer
bearbeiteten Werkes. Part 1 : Physikalische Lehren.
Von A. Winkelmann. Mit zahlreichen Holzstichen
und einer farbigen Tafel, 1885. Part 11 : Theoretische
Chemie einschliesslich der Thermochemie. Von A.
Horstmann, 1885 Part 111 (unpublished).
Vol. 11 : Anorganische Chemie. Neu bearbeitet von A.
Michaelis. Fünfte umgearbeitete Auflage In vier
Abtheilungen. 1879–89.
Vol. 111 : Organische Chemie. Zweite umgearbeitete und
vermehrte Auflage von Ernst v. Meyer.
Vol. 1v : Organische Chemie. Zweite umgearbeitete und
vermehrte Auflage von Ernst v. Meyer. In 3 parts.
1881–'84.
Vol. v : Organische Chemie. Bearbeitet von E. v.
Meyer und A. Weddige in Leipzig und H. v. Fehling
in Stuttgart. In 2 parts. 1868–'78.
Leerboek der onbewerktuigde scheikunde ; gedeeltelijk naar
Th. Graham's Elements of Chemistry, bewerkt door F.
J. Otto. Uit het Hoogduitsch vertaald en met aan-
teekeningen voorzien door J. P. C. von Tricht. Met
talrijke houtsneefiguren. Tiel, 1861.
Another edition, Tiel, 1867–70. 3 vols., 8vo.

> *Cf.* Otto, Friedrich Julius.

GRANDEAU, LOUIS.
> Leçon (etc.). *See :* Société chimique de Paris.

GRANDEAU, LOUIS.
Chimie et physiologie appliquées à l'agriculture et à la sylviculture. I.
La nutrition de la plante. Paris, 1879. pp. xvi–624, 8vo. Plates.
> With additional cap-title : Cours d'agriculture de l'école forestière.

Traité d'analyse des matières agricoles. Paris, 1877. 16mo.
Deuxième édition. Nancy, 1883. 16mo.
Handbuch für agriculturchemische Analysen. Berlin, 1879.
[Another edition :] Mit einem Vorwort von F. Henneberg.
Berlin, 1884.
Trattato di chimica analitica applicata all' agricoltura.
Prima traduzione italiana rifatta ed aumentata dall'
autore, eseguita da E. Mingioli e L. Paparelli. Firenze,
1888. pp. 547. Ill.
Instruction pratique sur l'analyse spectrale. Paris, 1863. Ill.

GRANDVAL, A., ET H. LAJOUX.
Nouveaux procédés pour la recherche et le dosage rapide de faibles
quantités d'acide nitrique dans l'air, l'eau, le sol, etc. Reims,
1886. 8vo.

GRANT, E. B.
Beet-Root Sugar and cultivation of the beet. Boston, 1867. pp. 158,
12mo.

GRANVILLE, AUGUSTUS BOZZI.
An Account of the Physical and Chemical Properties of the Malambo-
bark. . . . London, 1816. 8vo.
On a new Compound Gas resulting from animal decomposition taking
place in the human body. London, 1818.
On the Chemical Composition of two Liquids lately proposed as disin-
fectants of great power. London, 1827.
Spas (The) of England. London, 1838. 3 vols., 8vo.
Spas (The) of Germany. London, 1837. 2 vols., 8vo.
> Granville's true name was Bozzi, being of Italian birth.

GRÄUPNER.
Electrolyse und Katalyse, ihre Theorie und Praxis. Breslau, 1891.
8vo. Ill.

GRAUVOGEL.
Ueber die Zuckerbereitung aus Runkelrüben. Augsburg, 1811.

GRAVENHORST, J. L. C.
Die anorganischen Naturkörper nach ihren Verwandtschaften und
Uebergängen. Breslau, 1816. 8vo.

GRAY, ALONZO.
Elements of Chemistry. Intended as a Text-book for Academies. . . .
Second edition. New York, 1841. 12mo.
> Fourth edition. New York, 1842. 12mo. Ill.

GRAY, SAMUEL FREDERICK.
Operative (The) Chemist ; being a practical display of the arts and
manufactures which depend upon chemical principles. London,
1828. 8vo. Ill.
>> Traité pratique de chimie appliquée aux arts et manufac-
>> tures. Traduit de l'anglais. Paris, 1828–29. 3 vols.,
>> 8vo. 100 plates.
>> Der praktische Chemiker und Manufacturist. Aus dem
>> englischen übersetzt. Weimar, 1829–30. 8vo. 115 plates.

GREBNER, TH.

Runkelrüben-Zuckerfabrikation nach eigenen Erfahrungen und den besten französischen Schriften. Wien, 1830.

GREEN, JACOB.

Chemical Diagrams, or concise views of many interesting changes produced by chemical affinity. Philadelphia, 1837. pp. iv-90, 24mo.

A Text-book of Cnemical Philosophy, on the basis of Dr. Turner's Elements of Chemistry. Philadelphia, 1829. 8vo.

GREENE, WILLIAM H.

A Manual of Medical Chemistry. For the Use of Students. Based upon Bowman's Medical Chemistry. Philadelphia, 1883. 12mo. Ill.

A Practical Handbook of Medical Chemistry applied to clinical research and detection of poisons. Philadelphia, 1880. 8vo.

Lessons in Chemistry. Philadelphia, 1884. 12mo.

GREENWOOD, WILLIAM HENRY.

A Manual of Metallurgy. New York, *n. d.* [1875]. 2 vols., 12mo.

GREGORY, WILLIAM.

Elementary Treatise on Chemistry. Edinburgh, 1855. 8vo.

Outlines of Chemistry. London, 1845. 12mo.

> Handbook of Organic Chemistry [new edition of the Outlines]. Third edition. London, 1852.
>
> Handbook of Inorganic Chemistry [new edition of the Outlines]. Third edition. London, 1853.
>
> Handbook (A) of Chemistry, Inorganic and Organic ; for the use of students. Fourth edition. London, 1857. 8vo.
>
> Gregory-Gerding's organische Chemie, oder kurzes Handbuch der organischen Chemie, nach der dritten Auflage der " Outlines of organic chemistry " von William Gerding. Mit zahlreichen Zusätzen und Rücksicht auf technische Anwendung selbständig bearbeitet von Theodor Gerding. Braunschweig, 1854. 8vo. Ill.

> *Also under the title :*
>
> Handbuch der organischen Chemie für Universitäten, Real- und Gewerb-Schulen, sowie zum Selbstunterricht.

GREGORY, WILLIAM [Editor].

> *See* Liebig, Justus von.

GRÉHANT, N.

Les poisons de l'air, l'acide carbonique et l'oxyde de carbone. Asphyxie et empoisonnement par les puits, le gaz de l'éclairage, le tabac à

GRÉHANT, N.   [Cont'd.]

fumer, les poêles, les voitures chauffées, etc. Avec 21 figures intercalées dans le texte. Paris, 1890. pp. 320, 8vo.

Bibliothèque scientifique contemporaine.

GREINER, T.

Practical Farm Chemistry. A handbook of profitable crop feeding. New York, *n. d.* [1891]. 12mo.

GREN, FRIEDRICH ALBERT CARL.

See also, *in Section VII*, Journal der Physik.

GREN, FRIEDRICH ALBERT CARL.

Betrachtungen über die Gährung . . . Halle, 1784.

Published over the pseudonym G. F. J. v. P. (Jaspen von Pirch).

Grundriss der Chemie. Nach den neuesten Entdeckungen entworfen. Halle, 1796–97. 2 vols., 8vo.

Zweite verbesserte Auflage. Halle, 1800. 2 vols., 8vo.

Dritte Ausgabe, von Christian Friedrich Bucholz. Halle und Berlin, 1809. 2 vols., 8vo.

Vierte Ausgabe. Halle und Berlin, 1818. 2 vols., 8vo.

Observationes et experimenta circa genesin aëris fixi et phlogisticati ; dissertatio inauguralis. Halæ, 1786. 8vo.

Systematisches Handbuch der gesammten Chemie. Halle, 1787–90. 2 vols., 8vo.

Zweite Auflage. Halle, 1794–96.

Dritte Auflage, umgearbeitet von Martin Heinrich Klaproth. Halle, 1806–07. 3 vols., 8vo.

Haandbog i Chemien efter de nyeste Erfaringer udarbeidet of F. A. C. G. I Udtog oversat af Andr. Svendsen. Kjøbenhavn, 1801.

*Principles of Modern Chemistry systematically arranged. Translated from the German with notes and additions concerning later discoveries by the translator, and some necessary tables. Illustrated by plates. London, 1800. 2 vols., 8vo. Vol. I, pp. xviii–448 and 6 plates ; vol. II, pp. vi–498 and one plate.

This work begins with a brief historical introduction, and closes with a bibliography or list of the best books in chemistry systematically arranged.

GREVILLE, H. L.

Students' Handbook of Chemistry. With tables for chemical calculations. Second edition. Edinburgh, 1887.

GRIESSMAYER, VICTOR.
Verfälschung (Die) der wichtigsten Nahrungs- und Genussmittel vom
chemischem Standpunkte in populärer Darstellung. Augsburg,
1880. pp. vi–7–120, 16mo. Four folding tables.
Zweite Auflage. Augsburg, 1881.

GRIFFIN, JOHN JOSEPH.
Chemical Experiments. London, 1864. 8vo.
Chemical Handicraft. London, 1877.
Of this a Chinese version exists bearing the title :
Hwa hio ch'i chu tsai liao. [Translated by John Fryer.]
Shanghai, 187–. 2 vols.
Reprinted from the Chinese Scientific and Industrial Magazine.
Chemical Recreations and Romance of Chemistry. Seventh edition.
Glasgow, 1834. 12mo.
Thirteenth edition. London, 1853. 8vo.
Chemical (The) Testing of Wines and Spirits. London, 1866.
Another edition. London, 1872.
Radical (The) Theory in Chemistry. London, 1858. 8vo.
Treatise (A) on Chemical Manipulation and on the Use of the Blow-
pipe in Chemical Analysis. Glasgow, 1837.

GRIFFITHS, A. B.
A Treatise on Manures, or the Philosophy of Manuring. A practical
handbook for the agriculturist, manufacturer and student. London, 1889.

GRIFFITHS, W.
The Principal Starches used in Food, illustrated by photo-micrography
with a short description of their origin and character. Cirencester,
1892. 25 photo-micrographs.

GRIL, L'ABBÉ.
Petites leçons de chimie élémentaire, rédigées à l'usage de toutes les
maisons d'éducation. Paris, 1862. 18mo.

GRIMAUX, EDOUARD.
Conférence (etc.). See Société chimique de Paris.

GRIMAUX, EDOUARD.
Équivalents, atomes, molécules. Paris, 1866. 8vo.
Chimie inorganique élémentaire. Leçons professées à la Faculté de
médecine. Deuxième édition. Paris, 1879. pp. 513, 12mo. Ill.
Elementi di chimica inorganica. Napoli, 1878. 16mo.

GRIMAUX, EDOUARD. [Cont'd.]
  Chimie organique élémentaire. Quatrième édition. Paris, 1886. 8vo.
      Cinquième édition. Paris, 1889. 8vo.
      Chimie organique et inorganique élémentaire. Sixième édi-
        tion revue. Paris, 1892. 2 vols., 12mo. Ill.
      Elementi di chimica organica. Napoli, 1878. 16mo.
  Introduction à l'étude de la chimie. Théories et notations chimiques.
    Premières leçons du cours professé à l'École polytechnique. Paris,
    1883. pp. iii–244, 12mo.

GRIMELLI, G.
  Storia scientifica ed artistica dell' elettrometallurgia originale italiana,
    con un saggio teoretico-pratico di elettrometallurgia piana e solida,
    e un appendice lessicologica relativa alle diverse materie elettro-
    fisiche, elettrochimiche, elettrometallurgiche. Modena, 1844. 8vo.

GRINDEL, DAVID HIERONYMUS.
  Allgemeine Uebersicht der neuern Chemie zur Einleitung für Anfänger
    dargestellt. Riga, 1799. pp. (viii)–144, 12mo.
        A clear exposition of the (then) new theories of chemistry, admirably written.
  Handbuch der theoretischen Chemie zu akademischen Vorlesungen.
    Dorpat, 1808. 8vo.
  Organischen (Die) Körper chemisch betrachtet. Riga, 1811–12. 2
    vols., 8vo.
        Zweite Auflage. Riga, 1818.
  Ueber die verschiedenen Mittel die atmosphärische Luft zu reinigen.
    Riga, 1802. 8vo.
  Versuche über die Natur der Blausäure. Riga, 1804. 8vo.
  Wenzel's Lehre von der Verwandtschaft der Körper, mit Anmerkungen.
    Dresden, 1800.

GRINDEL, D. H., UND F. GIESE.
  Briefe über die Chemie, zur belehrenden Unterhaltung für Dilettanten.
    Riga, 1814. 8vo. Ill.

GRISCHOW, CARL CHRISTOPH.
  Physikalisch-chemische Untersuchungen über die Athmungen der
    Gewächse und deren Einfluss auf die gemeine Luft. Leipzig,
    1819. 8vo.

GRISCOM, J.
  Syllabus of a Select Course of Lectures on Chemistry. New York,
    1818. 12mo.

GRISON, THÉOPHILE.

Teinturier (Le) au XIXe siècle, en ce qui concerne les tissus où la laine
est la substance textile prédominante.   Déville-les-Rouen, 1860.
8vo.

> Die Färberei der feinen wollenen, wie der gemischten Mo-
> dezeuge mit baumwollener, oder seidener Kette und
> wollenen Einschuss, in den glänzendsten Farben, nach
> den neuesten und vortheilhaftesten zu Rouen gebräuch-
> lichen und in Deutschland bis jetzt noch wenig be-
> kannten Verfahrungsarten.   Ins deutsche übertragen
> von Chr. Schmidt.   Weimar, 188–.   8vo.

GRIVEAUX, F.

Traité de chimie à l'usage des élèves des lycées et collèges des jeunes
filles, des écoles normales primaires et des candidats au brevet
supérieur, rédigé conformément aux programmes.   Paris,   1884.
18mo.

GROLO, A.

Nozioni di chimica secondo le teorie moderne : libro di testo per i licei.
Seconda edizione.   Messina, 1885.   8vo.

GROOME, W.

Brief Notes on Chemistry.   (Bedford's Middle Class School Series.)
London, 1870.   8vo.
Concise Tables for Chemical Analysis.   (Bedford's Middle Class School
Series.)   1870.   8vo.

GROSHANS, J. A.

De la nature des éléments de la chimie.   Paris, 1875.
Des combinaisons chimiques $Cp\ Hq\ Or$ et des nombres de densité des
éléments.   Berlin, 1888.   pp. 83, 8vo.
Des dissolutions aqueuses par rapport aux nombres de densité des
éléments.   Berlin, 1888.   8vo.

> Ueber wässrige Lösungen.   Nach den Untersuchungen von
> Gerlach, Kremers und J. Thomsen. . . . Deutsch
> von Fr. Roth.   Leipzig, 1884.   pp. 50, 8vo.

Études et considérations sur la nature des élémens (corps non décom-
posés) de la chimie.   Leide, 1866–67.   3 parts.
Neues (Ein) Gesetz analog dem Gesetz von Avogadro.   Deutsch von
F. Roth.   Leipzig, 1882.   pp. 80, 8vo.

GROSMAN, J. (of the University of Prague).

Treatise (A) for the Service of Chemistry in General ; exhibiting the

GROSMAN, J. [Cont'd.]

universal and specific principles of body ; the simple and uniform procedure of nature, in petrification, in producing minerals, and the generation of gold. To which is added the most accurate process for ducifying corrosives. The medicine of Wedelius and Paracelsus for the gout. Medicines for the scurvy, the stone and the palsy. Considerations on the lues venerea, with its cure without mercury, together with several curious philosophic experiments, the reason why the fulminant gold strikes downward, and the true Bohemian paste for precious stones. London, 1766. pp. 106, 4to.

Partly in Latin, partly in English.

GROSOURDY, R. DE.

Chimie médicale. Paris, 1838–39. 8vo.

GROTHUSS, THEODOR FREIHERR VON [*really* CHRISTIAN JOHANN DIETRICH].

Mémoire sur la décomposition de l'eau et des corps qu'elle tient en dissolution à l'aide de l'électricité galvanique. Rome, 1805. *Also,* Mitau, 1806.

Contains his theory of the galvanic decomposition of water.

Physisch-chemische Forschungen. Nürnberg, 1840. 4to. Ill.

Verbindungsverhältniss-, oder chemische Aequivalenten-Tafeln. . . . Nürnberg, 1821. Fol.

GROUVEN, HUBERT.

Vorträge über Agricultur-Chemie mit besonderer Rücksicht auf Thier-Physiologie. Dritte ganz umgearbeitete Auflage. Köln, 1872. 8vo.

GROVE, WILLIAM ROBERT.

On the Correlation of Physical Forces. London, 1847. 8vo.

Third edition. London, 1855.

In French by Moigno with additions by Séguin, sen., Paris, 1856. *Cf.* Youmans, Edward L.

GROVES, CHARLES EDWARD AND WILLIAM THORP.

Chemical Technology, or Chemistry in its applications to arts and manufactures. Vol. I. Fuel and its applications, by E. J. Mills and F. J. Rowan. London, 1889. pp. xx–802, 8vo. Ill. *Also* Philadelphia. 1889.

A revised edition of Richardson and Watts' Chemical Technology, *q. v.*

GRUBE, JOHANNES GERHARDUS.

Disputatio chemico-physica prior et posterior de natura et usu salis. Hafniæ, 1703–4. 4to.

Cf. Licht, J. G.

GRÜNWALD, H.

Ueber einige Methoden zur quantitativen Bestimmung des Glycerins. Jena, 1889. 8vo.

GRÜTZNER, P.

Physiologie der Stimme und Sprache. *See* Handbuch der Physiologie.

GRUND, FRANCIS JOSEPH.

Elements of Chemistry. Boston, 1838. 12mo. Ill.

GRUNDY, CUTHBERT C.

An Introduction to the Study of Chemistry, written for the people. Manchester, 1870. 8vo.

GRUNER, L.

Traité de métallurgie. Première partie. Métallurgie générale. Vol. 1: Agents et appareils métallurgiques. Principes de la combustion. Paris, 1875. 18mo.

GRUNER ET ROSWAG.

Métallurgie et cuivre. *See, in Section II,* Fremy : Encyclopédie chimique, vol. v.

GUARESCHI, J., ET A. MOSSO.

Les ptomaïnes, recherches chimiques, physiologiques et médico-légales. Turin, 1883.

GUCKEISEN, AUGUST.

Lehrbuch der Chemie mit Berücksichtigung der Mineralogie für höhere Lehranstalten. Köln, 1879. 8vo.

GUÉGEN, A.

Mémoire sur la théorie chimique de la production du gas d'éclairage. Mémoire II. Paris, 1885. 8vo.

GÜLICH, JEREMIAS FRIEDRICH.

Anweisung zur Färberei auf Schaafwolle. . . . Ulm, 1786. 8vo.
Genaue Beschreibung und Vorschriften zur Manchester-Piqué-, Mousselin- und Casimir-Druckerey. Ulm, 1800. 8vo.
Vollständiges Färbe- und Bleichbuch. . . . Ulm, 1779–99. 7 vols., 8vo.

GÜNTHER, JOHANN JACOB.

Atmosphäre (Die) und ihre vorzüglichsten Erscheinungen. . . Mannheim, 1835.
Darstellung einiger Resultate die aus der Anwendung der pneumatischen Chemie, auf die praktische Arzneikunde hervorgehen. Mit einer Vorrede begleitet von Ferd. Wurzer. Marburg, 1801. 8vo.

GUERIN-VARRY, R. T.
Nouveaux élémens de chimie théorique et pratique, à l'usage des établissemens de l'Université ; précédés des notions de physique nécessaires à l'intelligence des phénomènes chimiques. Paris et Strasbourg, 1833. 8vo.
>Deuxième édition. Paris, 1840. 8vo.

GÜRTLER, G.
Der Chlorkalk in chemischer technischer und anderweitiger Beziehung. Wien, 1829. 8vo. Ill.

GUEYMARD, E.
Recueil d'analyses chimiques, à l'usage de l'agriculture moderne, comprenant toutes les analyses des substances végétales des fumiers naturels ou artificiels, des amendements de toute espèce d'eaux domestiques et d'eaux d'irrigation. Grenoble, 1869. 8vo.

GUIBOURT, NICOLAS JEAN BAPTISTE GASTON, ET BÉRAL.
Observations de pharmacie, de chimie et d'histoire naturelle pharmaceutique. Paris, 1838. 8vo.

GUIDOTTI, GALGANO.
Analisi chimica delle orine umane : studi compendiati. Milano, 1874. 16mo.

GUIGNET,
>Fabrication des couleurs. *See, in Section II*, Fremy : Encyclopédie chimique, vol. x.

GUILLAUME, E.
Fabrication de l'amidon, description des diverses opérations. Paris,1886.

GUILLEMIN, E.
Guide pratique du chimiste du distillerie et du sucraterie. Paris, 1890. pp. 392, 12mo.

GUILLOUD.
Traité de chimie appliquée aux arts et métiers. Paris, 1830. 2 vols., 8vo.
>Deuxiéme edition. Paris, 1835. 8vo.

GULDBERG, C. M., OG P. WAAGE.
Studier over Affiniteten. Christiania, 1864.
>>Études sur les affinités chimiques. Christiania, 1867. 4to.
>>Studien über die chemische Affinität. Leipzig, 1879.

GUMPRECHT, O.
Wie studirt man Chemie und die beschreibenden Naturwissenschaften ? Mit Berücksichtigung der sächsischen, preussischen und bayrischen Prüfungsordnungen. Leipzig, 1884. 8vo.

GUNARO, A.

Chimica dei concimi. Milano, 1884. 8vo.

GUNNING, J. W.

Beginselen (De) der algemeene scheikunde. 1e afdeeling : Experimen-
teele chemie. Schoonhoven, 1870. 8vo. Ill.

Leerboek der scheikunde. Ten gebruike aan inrigtingen van lager en
middelbaar onderwijs, en tot zelfonderrigt. I. deel : De scheikunde der nietmetalen. Schoonhoven, 1858. 8vo.

Leerboek der scheikunde. III. deel : Organische scheikunde. Schoonhoven, 1864. 2 parts, 8vo. 2 druk. Schoonhoven, 1866.

Onderzoek naar den oorsprong en de scheikundige natuur van eenige
Nederlandsche wateren. Utrecht, 1853. 8vo.

GUNTZ ET DITTE.

Uranium, Étain, Antimoine. *See, in Section II*, Fremy : Encyclopédie
chimique, vol. III.

GUSENBURGER, H.

Die Untersuchungen der Schmieröle und Fette mit specieller Berücksichtigung der Mineralöle. Luxemburg, 1886.

GUSSEROW, CARL AUGUST.

Die Chemie des Organismus abgeleitet aus Betrachtungen über die
elektro-chemischen Wirkungen der organischen und der diesen
ähnlich wirkenden Grundstoffe. Berlin, 1832. 8vo.

Die gerichtlich-chemischen Untersuchungen. Berlin, 1836.

ГУСТАВСОНЪ, Г.

Опытъ изслѣдованія реакцій взаимнаго обмѣна въ отсутствіи воды.
С.-Петербургъ 1873.

GUSTAVSON, H. An Essay upon the reactions of mutual exchange in
absence of water. St. Petersburg, 1873.

GUTHRIE, F.

Elements of Heat and of Non-metallic Chemistry. London, 1868. 12mo.

GUTIERREZ BUENO, PEDRO.

Curso de química dividido en lecciones para la enseñanza del real colegio
de San Carlos. Madrid, 1802. 8vo.

GUYE, PHILIPPE A.

Conférence [etc.], *see* Société chimique de Paris.

GUYTON DE MORVEAU, LOUIS BERNARD.

Allgemeine theoretische und practische Grundsätze über die sauren

GUYTON DE MORVEAU, LOUIS BERNARD. [Cont'd.]

Salze oder Säuren, zum Gebrauche für Chemisten und Aerzte. Aus dem französischen übersetzt und mit Anmerkungen versehen von David Ludewig Bourguet. Mit einer Vorrede begleitet von Sigismund Friedrich Hermbstädt. Berlin, 1796–1804. 3 vols., 8vo.

Allgemeine theoretische und praktische Grundsätze der chemischen Affinität oder Wahlanziehung, zum gemeinnützigen Gebrauch für Naturforscher, Chemisten, Aerzte und Apotheker. Aus dem französischen übersetzt von David Joseph Veit. Mit Anmerkungen begleitet und herausgegeben von Sigismund Friedrich Hermbstädt. Berlin, 1794. 8vo.

>These are translations of Guyton de Morveau's articles, Adhésion et Affinité, in the Encyclopédie méthodique.

Digressions académiques, ou Essais sur quelques sujets de physique, de chymie et d'histoire naturelle. Dijon et Paris, 1762. pp. xvi–417, 12mo.

>Annexed is : Défense de la volatilité du phlogistique. pp. [iv]–39. The first essay in this collection is : " Sur le phlogistique considéré comme corps grave," etc. Of this the second chapter is entitled : " Précis historique et critique des differens systèmes proposés pour l'explication de ce phénomène." The second essay is : " Sur la dissolution et la crystallisation."

Traité de désinfecter l'air [etc.]. Paris, 1801, 1802, 1803, 1805. 8vo.

>Contains his discovery of fumigation with chlorine described in detail.

GUYTON DE MORVEAU, MARET, ET DURANDE.

Élémens de chymie théorique et pratique, rédigés dans un nouvel ordre, d'après les découvertes modernes pour servir aux Cours publics de l'Académie de Dijon. Dijon, 1777. 3 vols., 12mo. Folding tables.

>Anfangsgründe der theoretischen und practischen Chemie, zum Gebrauch der öffentlichen Vorlesungen auf der Academie zu Dijon, nach den neuern Entdeckungen in eine neue Ordnung gebracht. Aus dem französischen übersetzt von Christian Ehrenfried Weigel. Leipzig, 1779. 3 vols., 8vo.

GUYTON DE MORVEAU, L. B., EN ABRAHAM VAN STIPRIAAN LUISCIUS.

Over de middelen om de lucht te zuiveren, de besmetting voortekomen, en derzelver voortgang te stuiten van L. B. Guyton Morveau. 1802. 8mo.

GUYTON DE MORVEAU ET PIOYER.

Clef (La) de la chimie, ou la nomenclature chimique mis à la portée de toutes les intelligences, inventée 1782. Publiée par L. M. Pioyer. Mans, 1855.

GUYTON DE MORVEAU, LOUIS BERNARD, LAVOISIER, BERTHOLLET, ETC.
Méthode de nomenclature chimique. *See in Section II.*

GYLLENBORG, GUSTAVUS ADOLPHUS.
The Natural and Chemical Elements of Agriculture.   Translated from
the Latin by John Mills. London, 1770. pp. xvi–198, 12mo.
One of the earliest scientific treatises on agricultural chemistry.

HAAF, C.
Ueber ein Verfahren zum Nachweis und zur Bestimmung flüchtiger
Fettsäuren. Bern, 1890. 8vo.

HAAN, ANDREAS LEOPOLD.
Libellus, in quo demonstratur quod non solum vegetabilia, animalia et
mineralia menstruo simplici possint solvi, verum etiam extracta
educi. Vindobonæ, 1766. 8vo.

HAAS, P. R. VON.
Tabellen zur qualitativen chemischen Analyse. Wien, 1892. 8vo.

HAAXMAN, P. J.
Handwoordenboekje van vervalschingen van scheikundige genees-
middelen en droogerijen, en verwisselingen van geneeskrachtige
plantendeelen. Voorburg, 1851. 12mo.

HABERMANN, J.
Mittheilungen aus dem Laboratorium für allgemeine und analytische
Chemie an der k. k. technischen Hochschule in Brünn.   6 Theile.
(Brünn, Verh. Naturf. Ver.), Brünn, 1889. 8vo.

HABICH, G. C.
Schule der Bierbrauerei.   Illustrirtes Hand- und Hülfsbuch für Brauer,
insbesondere ein Leitfaden für die Besucher von Brau-Lehranstal-
ten.   Nebst einer Vorschule : Die nöthigsten Vorkenntnisse in der
Braukunde.   Auf Grund eigener Erfahrungen sowie mit Benutzung
der neuesten in- und ausländischen Literatur neu herausgegeben
von Conrad Schneider.   Dritte, vielfach vermehrte und verbesserte Auflage. Leipzig, 1875. Two parts in one volume. pp.
xvi–330–456, 8vo. Ill.
Handbuch der Bierbrauerei auf Grundlage von Habich, Schule der
Bierbrauerei.   Herausgegeben von Conrad Schneider und Gottlieb
Behrend.   Fünfte, vollständig umgearbeitete und vermehrte Auflage
herausgegeben von E. Ehrich. Halle a. S., 1891. pp. xv–552.
8vo. Ill.
Химическая технологія по Боллею. Т. 5-й. Практич. руководство къ пиво-
вареніню. Перев. Усова и Яцуковича. С.-Петербургъ 1869.
Cf. Bolley's Handbuch.

HADLEY, JOHN.
Course of Chemical Lectures.   Cambridge, 1758.   8vo.

HÄNLE, CHRISTIAN FRIEDRICH.
Stöchiometrische Schemata, oder Darstellung des chemischen Prozesses pharmaceutisch-chemischer Präparate in atomistischen Formeln. Stuttgart, 1836.   pp. viii–204, 12mo.   Folding tables.

> Stechiometria chimico-farmaceutica, ossia etiologia e rappresentazione dei processi e dei fenomeni relativi ai preparati chimico-farmaceutici in formole anatomiche. Traduzione dal tedesco con aggiunte di G. B. Sembenini.   Verona, 1840.   8vo.

Chemisch-technische Abhandlungen.   Frankfurt am Main, 1808–21. 4 vols., 8vo.

HÄRING, A.
Repetitorium zu Stöckhardt's Schule der Chemie.   Braunschweig, 1864.   8vo.

> *Cf.* Stöckhardt, J. A.

HAGEMANN, G. A.
Aggregatzustände (Die) des Wassers.   Uebersetzt von P. Knudsen. Berlin, 1888.   8vo.
Chemische (Die) Energie.   Ausgearbeitet nach einem am 27. April, 1889, im chemischen Verein zu Kopenhagen gehaltenen Vortrag. Berlin, 1890.   pp. 40, 8vo.
Chemischen (Die) Kräfte.   Aus dem dänischen übersetzt von P. Knudsen. Berlin, 1888.   8vo.
Chemische (Die) Schwingungs-Hypothese und einige thermochemische Daten.   Berlin, 1888.   8vo.
Einige kritische Bemerkungen zur Aviditätsformel.   Aus dem dänischen übersetzt von P. Knudsen.   Berlin, 1887.   pp. 12, 8vo.
Studien über das Molekularvolumen einiger Körper.   Aus dem dänischen übersetzt von P. Knudsen.   Berlin, 1886.   pp. 58, 8vo.
Ueber die Energie und ihre Umwandlungen.   Einleitungsvortrag gehalten im dänischen Ingenieurverein zu Kopenhagen.   Berlin, 1892.   8vo.
Ueber Wärme- und Volumänderung bei chemischen Vorgängen.   Aus dem dänischen übersetzt von P. Knudsen.   Berlin, 1887.   8vo.

HAGEN, CARL GOTTFRIED.
Chemische Zergliederung der Thurenschen Wassers in Preussen. Königsberg, 1789.
De stanno.   Dissertationes III.   Regiomontanus, 1776.

HAGEN, CARL GOTTFRIED.   [Cont'd.]
Dissertatio de natura partis inflammabilis spiritus vini. Regiomontanus,
    1785.
Grundriss der Experimental-Chemie. Königsberg, 1786. 8vo.
        Zweite Auflage. Königsberg und Leipzig, 1790. 8vo.
        Vierte Auflage. 1815.
Isagoge in chemiam forensem. Regiomontanus, 1789.
Programma de similitudine salium alcalinorum cum terris absorbentibus.
    Regiomontanus, 1784.

HAGEN, JOHANN HEINRICH.
Chemisch-mineralogische Unterhaltungen einer merkwürdigen blauen
    Farberde aus den preussischen Torfbrüchen. Königsberg, 1772.
    4to.
Physikalisch-chemische Betrachtungen über die Herkunft und Ab-
    stammung des feuerbeständigen vegetabilischen Laugensalzes.
    Königsberg, 1768. 4to.
Physikalisch-chemische Betrachtungen über den Torf in Preussen.
    Königsberg, 1761. 4to.

HAGER, HERMANN.
Chemische Reactionen zum Nachweise des Terpentinöls in den äthe-
    rischen Oelen, in Balsamen, etc.   Für Chemiker, Apotheker,
    Drogisten und Fabrikanten ätherischer Oele. Berlin, 1885. pp.
    iv-166, 8vo.
Chemisch-pharmaceutischer Unterricht in 103 Lectionen.   Dritte Auf-
    lage. Berlin, 1877. 8vo.
        Vierte Auflage, 125 Lectionen. Berlin, 1885. 8vo.
Hager's Untersuchungen.   Ein Handbuch der Untersuchung, Prüfung
    und Wertbestimmung aller Handelswaren, Natur- und Kunster-
    zeugnisse, Gifte, Lebensmittel, Geheimmittel, etc. Zweite um-
    gearbeitete Auflage herausgegeben von H. Hager und E. Holder-
    mann. (Mit zahlreichen Holzschnitten.) Leipzig, 1885-'88. 2
    vols., 8vo.
        Het chemisch onderzoek.   Een handboek bij de beproeving
            der waarde van allerlei handelswaren, natuur- en
            kunstvoortbrengselen, levens- en geneesmiddelen, enz.
            Bew. naar het Hoogduitsch door G. C. W. Bohnensieg.
            Haarlem, 1875. 2 vols., 8vo. Ill.

HAHNEMANN, SAMUEL.
Ueber die Arsenikvergiftung, ihre Hülfe und gerichtliche Ausmittelung.
    Leipzig, 1786.   pp. xx-276, 12mo.
        The author was founder of the fallacious Homœopathic medical theory.

HAIDINGER, W.

Ueber die von Herrn Herapath und Herrn Stokes in optischer Beziehung untersuchte Jod-Chinin-Verbindung.   Wien, 1853.   8vo.

HAINES, ISAAC S.

Catechism on Chemistry.   Second edition.   Philadelphia.   1839.   12mo.

HAKE, H. WILSON.

Coloured Analytical Tables showing the behaviour of the more common metals, with special reference to the colour of the various oxides, salts, precipitates, flames, borax beads and blowpipe reactions. London, 1889.   8vo.

Practical (A) Examination Manual for Students of Chemistry.   Coloured analytical Tables, showing the behaviour of the more common metals and acids to the ordinary reagents, with special reference to the colour of the various oxides, salts, etc.   London, 1891. Roy. 8vo.   Coloured plates.

HALDAT DU LYS, CHARLES NICOLAS ALEXANDRE DE.

Recherches chimiques sur l'encre   .   .   .   Troisième édition.   Strasbourg, 1805.

HALES, STEPHEN.

* Philosophical Experiments containing useful and necessary instructions for such as undertake long voyages at sea. Shewing how sea-water may be made fresh and wholesome : And how freshwater may be preserv'd sweet. How biscuit, corn, etc., may be secured from the weevil, maggots, and other insects, and flesh preserv'd in hot climates by salting animals whole. To which is added an account of several experiments and observations on Chalybeate or Steel-Waters : With some attempts to convey them to distant places, preserving their virtue to a greater degree than has hitherto been done. Likewise a proposal for cleansing away mud out of rivers, harbours, and reservoirs. London, printed for W. Innys and R. Manby, at the West End of St. Paul ; and T. Woodword, at the Half-Moon between the Temple-Gates, in Fleet-Street, 1739.   pp. xxx–163–[viii], 12mo.

> Papers read before the Royal Society at several meetings.   Contain analyses of several chalybeate springs.

Statical Essays, containing Vegetable Statics, or an account of some statical experiments on the sap in vegetables. Being an essay towards a natural history of vegetation, of use to those who are curious in the culture and improvement of gardening ; also a specimen of an attempt to analyse the air by a great variety of

HALES, STEPHEN. [Cont'd.]

chymio-statical experiments, which were read at several meetings
before the Royal Society. Vol. I. London, 1769. Fourth Edition.
pp. x–[iv]–376, 8vo.

Statical Essays containing Hæmastatics, or an account of some
hydraulic and hydrostatical experiments made on the blood and
blood vessels of animals. Also an account of some experiments
on stones in the kidneys and bladder, with an enquiry into the
nature of those anomalous concretions. To which is added an
appendix containing observations and experiments relating to
several subjects in the first volume. The greatest part of which
were read at several meetings of the Royal Society. With an
Index to both volumes. Third edition corrected. Vol. II. London, 1769. pp. xxii–[xxvi]–354–[xxii], 8vo.

* Vegetable Staticks ; or an account of some statical experiments on
the sap in vegetables ; being an essay towards a natural history of
vegetation. Also a specimen of an attempt to analyse the air by a
great variety of chymio-statical experiments, which were read at
several Meetings of the Royal Society. London, 1727. pp. [vii]–
ix–376, 8vo. Twenty plates.

> Hales, coming after Mayow and before Priestley, contributed much to the
> method of handling gases, inventing improved apparatus and attempting
> quantitative analysis of the air. But he failed to differentiate the many
> gases he prepared, believing them all merely common air.

HALL, THOMAS WRIGHT.

A Correlation Theory of Chemical Action and Affinity. London, 1888.
pp. ii–360, 8vo.

HALLECK, H. W.

Bitumen ; its varieties, properties, and uses. Washington, 1841. 8vo.

HALLER, A.

Conférence (etc.). See : Société chimique de Paris.

HALLER, ALBIN.

Théorie générale des alcools. Paris, 1879. 8vo.

HALLIBURTON, WILLIAM DOBINSON.

A Text-Book of Chemical Physiology and Pathology. London, 1891.
8vo. Ill.

HALLIER, ERNST.

Gährungserscheinungen. Untersuchungen über Gährung, Fäulniss
und Verwesung mit Berücksichtigung der Miasmen und Contagien

HALLIER, ERNST.   [Cont'd.]
   sowie der Desinfection, für Aerzte, Naturforscher, Landwirthe und
   Techniker.   Leipzig, 1867.   pp. vi–116, 8vo.   Plate.

HALMALE, I. F.
   Mercurius verheerlykt, of Verhandeling over het Kwikzilver.   Amsterdam, 1707.   8vo.

HAMM, WILHELM.
   Katechismus der Ackerbauchemie, Bodenkunde und Düngerlehre.
      Ein Buch für alle Landwirthe, Lehrer und Schulen.   Mit zu
      Grundlegung der 17. Auflage von Johnston's Catechism of Agri-
      cultural Chemistry and Geology.   Leipzig, 1847.   8vo.   Ill.
         Zweite verbesserte und vermehrte Auflage, Leipzig, 1851.
         Dritte vielfach vermehrte und verbesserte Auflage, Leipzig,
            1854.   8vo.

      Cf. Johnston, James F. W.

         Le catechisme de la chimie de l'agriculture, de la science des
            terrains et de l'enseignement des engrais, contenant:
            1, tout ce qui a rapport à la composition à la nutrition
            aux éléments et principes organiques des plantes, etc.
            Traduit de l'Allemand sur la troisième édition, con-
            sidérablement augmentée avec 33 figures sur bois, suivi
            du tableau des principales maladies internes et externes
            qui affectent les animaux domestiques par Aimé
            Jacquot.   Besançon, 1855.   12mo.   Ill.
         Catechismus i Agerdyrknings-Chemie, Jordbrugs-Kundskab
            og Gjödningslære, en Bog for Landmænd, Lærere og
            Skoler.   Efter det Tyske.   Christiania, 1848.   8vo.
         Ручная книжка земледѣльч. химіи почвознанія и ученія о
            наземахъ.   С.-Путербургъ 1855.

HAMMERDAHL, J. G.
   De första grundämner uti Kemien jemte några korta Uppgifter på
      de allmännast förekommande tekniska Beredningar.   1 Delen:
      Oorganiska Kemi.   Åbo, 1847.   8vo.

HAMMARSTEN, O.
   Lärobok i farmaceutisk Kemi.   Upsala, 1886.
   Kortfattad Lärobok i farmaceutisk Kemi med Hänsyn till Svenska Far-
      makopéns Preparat jemte Handledning i Titreringsanalysen.   Up-
      sala, 1886.   8vo.

HAMMERSCHMIED, JOHANN.

Das Ozon und seine Wichtigkeit im Haushalte der Natur und des menschlichen Körpers. Mit einem Anhange. Wien, 1873. 8vo.

HAMPE, W.

Tafeln zur qualitativen chemischen Analyse. Clausthal, 1868. 8vo.
Ueber die Analyse der Sprengstoffe. Berlin, 1883. 4to.

HANAFORD, WILLIAM G.

Lectures on Chemistry, with familiar directions for performing experiments with a small apparatus. Boston, 1831, 12mo.

HANDBOEK DER TECHNOLOGIE.
        See Jacobsen, G. J.

HANDBUCH DER CHEMISCHEN TECHNOLOGIE.
        See Bolley's Handbuch.

HANDBUCH DER PHYSIOLOGIE.

Bearbeitet von H. Ambert, E. Drechsel, C. Eckhard, Th. W. Engelmann, S. Exner, A. Fick, O. Funke, P. Grützner, R. Heidenhain, V. Hensen [etc.]. Herausgegeben von L. Hermann. Leipzig, 1879–82. 6 vols. (in several parts), 8vo. Ill.

Vol. I. Handbuch der Physiologie der Bewegungsapparate. Part I : L. Hermann, Allgemeine Muskelphysik ; O. Nasse, Chemie und Stoffwechsel der Muskeln ; Th. W. Engelmann, Flimmer- und Protoplasmabewegung. Part II : P. Grützner, Physiologie der Stimme und Sprache ; A. Fick, Specielle Bewegungslehre.

Vol. II. Handbuch der Physiologie des Nervensystems. Part I : L. Hermann, Allgemeine Nervenphysiologie ; Sigm. Mayer, Specielle Nervenphysiologie. Part II : C. Eckhard, Physiologie des Rückenmarks ; Sigm. Exner, Physiologie der Grosshirnrinde.

Vol. III. Handbuch der Physiologie der Sinnesorgane. Part I: A. Fick, Gesichtssinn, Dioptrik, Nebenapparate des Auges, Lehre von der Lichtempfindung ; W. Kuhne, Chemische Vorgänge in der Netzhaut ; E. Hering, Raumsinn des Auges. Part II : V. Hensen, Gehör ; M. von Vintschgau, Geschmacksinn, Geruchsinn ; O. Funke, Tastsinn und Gemeingefühle ; E. Hering, Temperatursinn.

Vol. IV. Handbuch der Physiologie des Kreislaufs der Athmung und der thierischen Wärme. Part I : A. Rollett, Blut und Blutbewegung ; A. Aubert, Innervation der Kreislaufsorgane. Part II : N. Zuntz, Blutgase und Respirator. Gaswechsel ; J. Rosenthal, Athembewegungen und Innervation derselben. Thierische Wärme.

Vol. V. Part I : R. Heidenhain und B. Luchsinger, Physiologie der Absonderungsvorgänge ; Richard Maly, Chemie der Verdauungs-

HANDBUCH DER PHYSIOLOGIE.   [Cont'd.]
 säfte und Verdauung.   Part II, division 1 : W. von Wittich, Aufsaugung, Lymphbildung, Assimilation ; Sigm. Mayer, Bewegung der
 Verdauungs- Absonderungs- und Fortpflanzungsapparate.   Anhang
 über glatte Muskeln.   Part II, division 2 : E. Drechsel, Chemie der
 Absonderungen und Gewebe.   Sachregister zum 5 Bände und
 General-Register zu sämmtlichen Bänden des Handbuches.
 Vol. VI. Part I : C. von Voit, Physiologie des allgemeinen Stoffwechsels
 und der Ernährung.   Part II. V. Hensen, Physiologie der Zeugung.

HANDL, ALOIS.
 Beiträge zur Moleculartheorie.   Wien, 1875.   5 parts.
 Kleines Lehrbuch der Chemie.   Für Schulen und zum Selbstunterrichte.
 Teschen, 1875.   8vo.

HANDS, T.
 Numerical Exercises in Chemistry.   (Inorganic.)   London, 1884.   8vo.

HANKEL, E.
 Laboratoriumsversuche über die Klärung der Abfallwässer der Färbereien.   Hygienische Studie.   Glauchau, 1884.   8vo.

HANKEL, WILHELM GOTTLIEB.
 Enleitung zur Experimental-Chemie.   Halle, 1842.   8vo.

HANRIOT, MAURICE.
 Hypothèses actuelles sur la constitution de la matière.   Paris, 1880.   8vo.

HANSEN, ADOLPH.
 Die Farbstoffe des Chlorophylls, Kritik der Litteratur und experimentelle
 Untersuchungen.   Darmstadt, 1889.   pp. 88, 8vo.   Colored plates.

HANSEN, E. CH.
 Organismer i øl og ølurt.   Kjøbenhavn, 1879.   2 plates.
 Untersuchungen aus der Praxis der Gährungsindustrie.   München,
 1888, 8vo.
          Zweite vermehrte und verbesserte Auflage.   München, 1890.
    8vo.   Ill.

HANSSEN, AUGUST.
 Studien über den chemischen Nachweis fremder Fette im Butterfette.
 Erlangen, 1884.   pp. 34, 8vo.

HARAUCOURT, CÉLESTIN.
 Cours élémentaire de chimie à l'usage des établissements d'enseignement
 secondaire.   Paris, 1882.   8vo.

HARAUCOURT, CÉLESTIN. [Cont'd.]
>   Deuxième édition. Paris, 1883.
>   Troisième édition, Paris, 1887.
Leçons élémentaires de chimie à l'usage des écoles primaires supérieures. Paris, 1885.
>   Deuxième édition, Paris, 1886.
Leçons de chimie à l'usage des candidats au brevet supérieur. Paris, 1888. Ill.
Notions de chimie. Paris, 1878–80. 3 vols.
Premières leçons de chimie à l'usage des élèves de la classe de sixième et des classes primaires supérieures. Paris, 1881. 12mo.

HARCOURT, A., G. VERNON, AND H. G. MADAN.
Exercises in Practical Chemistry. London, 1870.
>   Fourth edition, Oxford, 1887. pp. xvi–590, 12mo.

HARDER, ALBERT.
Die wichtigsten Lehren der Ackerbauchemie zur Belehrung für die ländliche Jugend in Schule und Haus. In Fragen und Antwort zusammengestellt. Braunschweig, 1869. 8vo. Ill.

HARDER, P. E.
Das Molekulargesetz, mit besonderer Anwendung auf das Wasser, den Wasserdampf und die Luft. Hamburg, 1866. pp. 168, 8vo.

HARDWICH, T. FREDERICK.
A Manual of Photographic Chemistry, including the practice of the collodion process. Fifth edition. London, 1859 pp. xvi–516, 12mo.
>   Sixth edition, London, 1861. pp. xvi–571, 12mo.
>   Ninth edition, edited by J. Traill Taylor, London, 1883. 12mo.

HARDY, A.
Des recherches chimiques appliquées à l'étude des maladies. Paris, 1847. 4to.

HARDY, ERNEST.
Principes de chimie biologique. Avec figures dans le texte et une planche chromolithographiée représentant l'analyse spectrale du sang. Paris, 1871. pp. iv–563, 12mo.

HARE, ROBERT.
Compendium (A) of the Course of Chemical Instruction in the Medical Department of the University of Pennsylvania. In two parts . . . Fourth edition. With amendments and additions. Philadelphia, 1840–'43. Pt. 1, pp. xii–370 ; pt. 2, pp. vi–371–605–xx, 8vo.

HARE, ROBERT. [Cont'd.]

Description of a Part of the Apparatus used in the Chemical Course of the University of Pennsylvania. Philadelphia, 1826. 8vo.

Effort (An) to Refute the Arguments advanced in favour of the existence in the amphide salts or radicals, consisting, like cyanogen, of more than one element. Philadelphia, 1842. pp. 23, 8vo.

Essays, Chemical, Electrical, and Galvanic. Philadelphia, 1825. 8vo.

Lecture ; Introductory to a Course on Chemistry. Philadelphia, 1843. 8vo.

On the Explosiveness of Nitre, with a view to elucidate its agency in the tremendous explosion of July, 1845, in New York. [Washington, 1851.] 4to.

Radicals of more than one Element in the Amphide Salts. Philadelphia, 1842. 8vo.

Some Encomiums upon the excellent Treatise of Chemistry by Berzelius, also objections to his nomenclature, and suggestions respecting a substitute deemed preferable, in a letter to Professor Silliman, June, 1834. [Philadelphia, 1834.] 8vo.

Another edition : Objections to the Nomenclature of the celebrated Berzelius, with suggestions respecting a substitute, in a letter to Professor Silliman's Journal for 1835, vol. XXVII. Also a letter from the distinguished chemist above mentioned in reply, with a concluding examination of the suggestions in that letter by the author of the objections ; republished from the Journal of Pharmacy for April, 1837. Philadelphia, 1840. pp. 23, 8vo.

HARKORT, VON.

Die Probirkunst mit dem Löthrohre. Freyberg, 1827. 2 parts, 8vo.

HARMAIN, G.

Course of Qualitative Analysis. London, 1889. 8vo.

HARMENS, GUSTAV.

De elementis aquarum mineralium. Lundæ, 1734.

De genuino docimasiæ metallicæ fundamento. Lundæ, 1761.

De nitro. Lundæ, 1745.

De lapide calcareo. Lundæ, 1751.

De ortu et generatione salis alkali fixi et volatilis. Lundæ, 1760.

De sale communi. Lundæ, 1748.

HARMSEN, W.

Die Fabrikation der Theerfarbstoffe und ihrer Rohmaterialien. Berlin, 1889. pp. vii–317, 12mo. Ill.

HARNACK, E.

Die Hauptthatsachen der Chemie. Für das Bedürfniss des Mediciners, sowie als Leitfaden für den Unterricht zusammengestellt. Hamburg, 1887.

> Fatti principali di chimica. Tradotta di L. Trimani. Milano, 1888. 16mo.

HARRINGTON, ROBERT.

Chemical Essays. London [1793]. 8vo.

Death-Warrant (The) of the French Theory of Chemistry, signed by truth, reason, common-sense, honour and science, with a theory fully and rationally accounting for all the phenomena, also a full and accurate investigation of all the phenomena of galvanism, and strictures upon the chemical opinions of Messrs. Wiegleb, Cruickshanks, Davy, Leslie, Count Rumford and Thompson, likewise remarks upon Mr. Dalton's late theory and other observations. London, Oct., 1804. pp. 312, 8vo.

> The prejudice, shortsightedness and over-confidence exhibited by the author in his book is reflected on the title-page.

HARRIS, ELIJAH P.

A Manual of Qualitative Chemical Analysis. Fourth edition, thoroughly revised and corrected. Amherst, Mass., 1892. pp. 220, 12mo. Published by the author.

HARRISON, J. B.

Certain Points in Agricultural Chemistry considered in reference to the selection and application of manures for the sugar-cane in the Island of Barbadoes. Barbadoes, 1886. 8vo.

HARRISON, W. J.

Elementary Chemistry. London, 1890. 8vo.

HART, EDWARD.

A Handbook of Volumetric Analysis, designed for the use of classes in colleges and technical schools. New York, 1878. 12mo.

HART, H. MARTYN.

Elementary Chemistry. London, 1870. 8vo.

HARTLEY, F. W.

The Gas-Analyst's Manual. London, 1879. pp. viii–146, 12mo. Ill.

HARTLEY, W. N.

Absorption (The) Spectra of the Alcaloids. London, 1886. 4to.

Air and its Relations to Life. Being with some additions the substance

HARTLEY, W. N. [Cont'd.]

of a course of lectures delivered in the summer of 1874 at the Royal Institution of Great Britain. London, 1875. 16mo.

Course (A) of Quantitative Analysis for Students. London, 1887. pp. 230.

HARTMANN, C.

Die Runkelrüben-Zuckerfabrikation auf ihrem neuesten Standpunkte. Quedlinburg, 1850.

HARTMANN, CARL FRIEDRIC ALEXANDER.

Populäres Handbuch der allgemeinen und speciellen Technologie, oder der rationellen Praxis des chemischen und mechanischen Gewerbwesens . . . Berlin, 1841. 2 vols., 8vo.

Probirkunst, (Die) oder Anleitung die wichtigsten Metallgemische auf dem trockenen und nassen Wege zu untersuchen . . . Nach Chaudet, "l'Art de l'essayeur." Weimar, 1838. 8vo.

Zweite Ausgabe. Weimar, 1847. 8vo.

HARTMANN, F.

Das Verzinnen, Verzinken, Vernickeln, Verstählen und das Ueberziehen von Metallen mit anderen Metallen überhaupt. Zweite verbesserte und sehr vermehrte Auflage. Wien, 1886.

HARTMANN, HUGO.

Die Fette. Lehre von den natürlichen Fettkörpern, welche technische Anwendung finden. Vorkommen, Gewinnung, Handel, Eigenschaften, Veränderungen und Verfälschungen, sowie die Mittel zur Erkennung und Nachweisung der letzteren. Nach Theod. Chateau bearbeitet. Leipzig, 1864. 8vo.

Untersuchungen mit dem Löthrohr. Uebersicht der pyrognostischen Eigenschaften der unorganischen Substanzen. Tafeln über das Verhalten der Mineralkörper vor dem Löthrohre, nebst Angaben zu qualitativen Untersuchungen technisch-wichtiger Mineralien, Erze und Hüttenprodukte mittelst des Löthrohres. Leipzig, 1862. 4to.

HARTMANN, JOHANN.

Opera omnia medico-chymica. Francofurti, 1664, *also* 1690. Fol.

Hartmann held a chair of Chemistry at the University of Marburg from 1609, and was therefore the first Professor of Chemistry in Germany.

Praxis chymiatrica. Lipsiæ, 1633.

For an English translation, *see, in Section VI*, Croll, Oswald : Bazilica chymica.

HARTRODT, A.
>   Die Alkaloide. *See in Section II.*

HARTSEN, F. A. VON.
Chemie (Die) der Zukunft. Heidelberg, 1877. 8vo.
Philosophischen (Die) Grundlagen der Chemie. Als Einleitung zu den
    Lehrbüchern der Chemie. Heidelberg, 1876. pp. viii–56, 8vo.
Was heisst ein chemisches Aequivalent? Kritik der heutigen Chemie
    und Vorschlag zur Berichtigung. Heidelberg, 1876. 8vo.
>   Qu'appelle-t-on un équivalent chimique? Critique de la
>   chimie et moyen d'en rectifier la nomenclature. Paris,
>   1877. 8vo.
>   Que se llama un equivalente químico? Version de Manuelo
>   de Telosa y Gust. Saenz Diez. Madrid, 1877. 4to.

HARTUNG-SCHWARZKOPF, HEINRICH CARL.
Chemie der organischen Alkalien. München, 1855. pp. xii–452, 8vo.

HASERICK, E. C.
The Secrets of the Art of Dyeing Wool, Cotton and Linen, including
    bleaching and coloring wool and cotton hosiery and random-yarns.
    Cambridge, Mass., 1869. 8vo.

HASSALL, ARTHUR HILL.
Adulterations detected, or plain instructions for the discovery of frauds
    in food and medicine. London, 1857. 8vo.
>   Second edition. London, 1861. pp. xvi–[6]–712, 8vo.
Food; its Adulterations and the Methods for their Detection. London,
    1876. pp. vi–896, 8vo.
>   *See* Klencke, Hermann : Die Verfälschung der Nahrungsmittel.

HASSELT, A. W. VAN.
>   *See* Husemann, Th., und A. Husemann : Handbuch der Toxicologie.

HASSELT, A. W. VAN.
Handleiding der bijzondere vergiftleer, ten gebruike bij het onderwijs
    aan's Rijk's kweekschool voor militaire geneeskundigen. Utrecht,
    1850. 8vo.
Handleiding der bijzondere vergiftleer, ten gebruike bij het onderwijs
    aan's Rijks kweekschool voor militaire geneeskundigen. 3 stukken.
    Utrecht, 1853–'54. 8vo. 1 en 2 stuk : Vergiften uit het planten-
    rijk. 3 stuk : Vergiften uit het dierenrijk.
>   Handbuch der Giftlehre für Chemiker, Aerzte, Apotheker
>   und Gerichtspersonen. Aus dem holländischen nach
>   der zweiten Auflage frei bearbeitet und mit Zusätzen

HASSALT, A. W. VAN. [Cont'd.]

versehen von J. B. Henkel. Braunschweig, 1862. 8vo. 2 parts. Erster Theil : Allgemeine Giftlehre und die Gifte des Pflanzenreichs. Zweiter Theil : Die Thiergifte und die Mineralgifte.

HATCHETT, CH.

An Analysis of the Carinthian Molybdate of Lead, with experiments on the Molybdic Acid. London, 1796. 4to.

HAUBOLDUS, G. G.

De usu instrumentorum physico-mathematicorum recte æstimando. Lipsiæ, 1771. 4to.

HÄUSSERMANN, C.

Die Industrie der Theerfarbstoffe. Stuttgart, 1881. 8vo.

HAUSHOFER, K.

Beiträge zur mikroskopischen Analyse. I. Ueber die Anwendung der concentrirten Schwefelsäure ·in der mikroskopischen Analyse. München, 1886. 8vo.

Constitution (Die) der natürlichen Silicate. Auf Grundlage ihrer geologischen Beziehungen nach den neueren Ansichten der Chemie. Braunschweig, 1874. 8vo.

Leitfaden für die Mineralbestimmung. Braunschweig, 1892. 8vo.

Ueber einige mikroskopisch-chemische Reactionen. München, 1886. 8vo.

HAUSSKNECHT, O.

Lehrbuch der Chemie und chemischen Technologie für Realgymnasien. Hamburg, 1883.

HAVINGA, JOHANNES RUDOLPHUS.

Dissertatio de arsenico. Groningen, 1793. 8vo.

HEATON, CHARLES WILLIAM.

Experimental Chemistry, founded on the work of J. A. Stöckhardt. London, 1872.

New edition, revised. London, 1886. 8vo.

See Stöckhardt, J. A.

The Threshold of Chemistry ; an Experimental Introduction to Science. London, 1861. 8vo.

HECKENAST, WILHELM.

Desinfectionsmittel, oder Anleitung zur Anwendung der praktischsten

HECKENAST, WILHELM. [Cont'd.]
und besten Desinfectionsmittel, um Wohnräume, Krankensäle,
Stallungen, Transportmittel, Leichenkammern, Schlachtfelder, u. s.
w. zu desinficiren. Wien, Pest, Leipzig, 18—. Ill.

HECTOR, D. S.
Undersökningar öfver Svafvelurinämmens Förhållande till Oxidations-
medel. Upsala, 1892. 8vo.

HEHNER, OTTO, AND ARTHUR ANGELL.
Butter, its Analysis and Adulteration. Second edition entirely re-
written. London, 1877. 8vo.

HEIDEKAMPF.
Anleitung zur Zuckerfabrikation und Rübenbau. Weimar, 1842.

HEIDENHEIN R., UND B. LUCHSINGER.
Physiologie der Absonderungsvorgänge. *See* Handbuch der Physiologie.

HEINTZ, WILHELM HEINRICH.
De acido saccharico ejusque salibus ; dissertatio inauguralis . . .
Berolini, *n. d.* [1844]. 8vo.
Lehrbuch der Zoochemie. Berlin, 1853. pp. xviii–1107, 8vo. Ill.
Leitfaden für die qualitative chemische Analyse. Halle, 1875.

HEINZE, ROBERT.
Anleitung zur chemischen Untersuchung und rationellen Beurtheilung
der landwirthschaftlich wichtigsten Stoffe. Ein den praktischen
Bedürfnissen angepasstes analytisches Handbuch für Landwirthe,
Fabrikanten künstlicher Düngemittel, Chemiker, Lehrer der
Agriculturchemie und Studirende höherer landwirthschaftlicher
Lehranstalten. Nach dem neuesten Stande der Praxis verfasst.
Wien, Pest, Leipzig, 1883. Ill.

HEINZERLING, CHR.
Die Fabrikation der Kautschuk- und Guttaperchawaaren, sowie des Cellu-
loids [etc.]. *See* Bolley's Handbuch.
Grundzüge der Lederbereitung. *See* Bolley's Handbuch.

HEINZERLING, CHR.
Abriss der chemischen Technologie mit besonderer Rücksicht auf
Statistik und Preisverhältnisse. Cassel und Berlin, 1887. 8vo.
Gefahren (Die) und Krankheiten in der chemischen Industrie und die
Mittel zu ihrer Verhütung und Beseitigung. Halle a. S., 1884. 8vo.

HEISS, PH.

Die Bierbrauerei mit besonderer Berücksichtigung der Dickmaisch-
brauerei. Siebente vermehrte und verbesserte Auflage. Nach den
neuesten Fortschritten bearbeitet von E. Leyser. Nebst einem
Anhange enthaltend : Die einfache und doppelte Buchführung für
Braugeschäfte. Augsburg, 1880. 8vo.

HEIZLER, F., UND N. HOFMANN.

Chemie für die vierte Classe der Gymnasien und Real-Gymnasien, nach
methodischen Grundsätzen bearbeitet. Prag, 1881. 8vo. Ill.

HELDT, WILHELM.

Die Fundamental-Eigenschaften des Sauerstoffs und Wasserstoffs.
Experimental-Untersuchungen von W. H. Berlin, 1861. 4to.

HÉLÈNE, M.

La poudre à canon et les nouveaux corps explosifs. Deuxième édition.
Paris, 1886. 18mo.

HELLER, JOHANN FLORIAN.

Ueber die Rhodizonsäure, eine neue oxydationsstufe des Kohlenstoffs
und die Krokonsäure, dann die Salze beider. Prag, 1837. 8vo. Ill.

HELLMANN, A.

Vorlesungen der allgemein technischen Chemie. Gotha, 1851. Zwei
Theile, 8vo.

HELLOT, JEAN.

L'Art de la teinture des laines et étoffes de laine [etc.]. Paris, 1750.
    Nouvelle édition. Paris, 1786.
    Färbekunst, oder Unterricht Wolle und wollene Zeuge zu
        färben, nebst Vorschriften zu Prüfungen derselben
        durchs Absieden. Aus dem französischen übersetzt
        von Abraham Gotthelf Kästner. Dritte Auflage von
        Carl August Hoffmann. Altenburg, 1790. 8vo.

HELMHOLTZ, H.

Wissenschaftliche Abhandlungen. Leipzig, 1882. 2 vols., 8vo. Ill.
Die Thermodynamik chemischer Vorgänge. Berlin, 1882. 8vo.

HELMONT, JAN BAPTISTA VAN.

Opera omnia . . . Francofurti, 1682. 2 parts in 1 vol., 4to.
Opera omnia, novissima hac editione . . . repurgata . . . una cum
    introductione atque clavi M. B. Valentini . . . Francofurti, 1707.
    4to.
        Contains a life of the author by his son, Franciscus Mercurius van Helmont.

HELWIG, A.

Das Mikroskop in der Toxikologie. Beiträge zur mikroskopischen und mikrochemischen Diagnostik der wichtigsten Metall- und Pflanzen-gifte . . . mit einem Atlas photographirter mikroskopischer Prä-parate. Mainz, 1865. 8vo.

HÉMENT, EDGARD.

Histoire d'un morceau de charbon. Paris, 1868. 18mo.

>   Bibliothèque de la science pittoresque.

HEMMER, MORITZ.

Experimentale Studien über die Wirkung faulender Stoffe auf den thierischen Organismus. (Gekrönte Preisschrift.) München, 1866. pp. 170, 8vo.

HEMPEL, I. G.

Pharmaceutisch-chemische Abhandlung über die Natur der Pflanzen-säuren und die Mortificationen denen sie unterworfen sind, nebst einer chemischen Abhandlung der Winter- und Sommerreiche. Berlin, 1794. pp. 176, 8vo.

HEMPEL, WALTHER.

Neue Methoden zur Analyse der Gase. Braunschweig, 1880. 8vo.

>   Gasanalytische Methoden. Zweite Auflage. Braunschweig, 1890. 8vo. 100 ill.
>
>   Methods of Gas-Analysis. Translated from the second German edition by L. M. Dennis. London and New York, 1892. pp. xv-384, 12mo. Ill.

Ueber den Einfluss der chemischen Technik auf Leben und Sitte. Dresden, 1891.

HENDESS, HERMANN.

Allgemeine Giftlehre. Uebersichtliche Darstellung der gewöhnlichsten Giftstoffe in ihrer chemischen Zusammensetzung, ihrem Verhalten gegen Reagentien, ihren Gegengiften, sowie der besten Methoden zur Ausmittelung derselben. Mit einem Anhange enthaltend die neuesten gesetzlichen Bestimmungen über den Verkehr mit Giften. Ein praktisches Handbuch für Aerzte, Apotheker und Juristen, wie für Gebildete aller Stände. Berlin, 1880. pp. 107, 8vo.

HENKEL.

Allgemeine Waarenkunde. Eine systematische Darstellung der wich-tigsten im Handel erscheinenden Natur- und Kunstproducte. Erlangen, 1870. 8vo.

HENKEL, JOHANN FRIEDRICH.

De mediorum chymicorum appropriatione, in argenti cum acido salis communis communicatione. Dresden, 1727.

Kleine mineralogische und chemische Schriften, herausgegeben von Ch. F. Zimmermann. Dresden und Leipzig, 1744, 1747 [and] Wien, 1769. 8vo.

HENNEGUY, L. FÉLIX.

Étude physiologique sur l'action des poisons. Montpellier, 1875. pp. 168–3–3, 8vo.

HENRICH, F.

Tabellen zur qualitativen chemischen Analyse. Wiesbaden, 1886.

HENRICI, J.

Kleiner Grundriss der Elementar-Chemie. Leipzig, 1886.

HENRIVAUX.

Les laboratoires de chimie [and other articles]. *See, in Section II,* Fremy: Encyclopédie chimique.

HENRY, NOËL ÉTIENNE, ET ÉTIENNE OSSIAN HENRY.

Manuel d'analyse chimique des eaux minérales. Paris, 1825. 8vo.

Traité pratique d'analyse chimique des eaux minérales potables et économiques avec leurs principales applications à l'hygiène et à l'industrie. Considérations générales sur leur formation, leur thermalité, leur aménagement, etc., etc. Fabrication des eaux minérales artificielles. Paris, 1858. pp. xv–662, 8vo. Ill.

Manual de la análisis química de las aguas minerales, medicinales y de las destinadas á la economía doméstica . . . Traducido al castellano por Manuel Diez Moreno. Madrid, 1829. 8vo.

HENRY, ÉTIENNE OSSIAN, ET E. HUMBERT.

Recherches chimiques et médico-légales sur l'acide cyanhydrique et ses composés employés dans les arts, présentées et lues à l'Académie Impériale de médecine, dans la séance du 6. mai, 1856. Suivies du rapport fait à l'Académie par Wurtz et Boutron. Paris, 1857. 8vo.

HENRY, WILLIAM.

Epitome (An) of Chemistry. London, 1800. 12mo.

Another edition. Edinburgh, 1806. 8vo.

Second American from the fifth English edition. With

HENRY, WILLIAM.   [Cont'd.]

    additions and notes by Benjamin Silliman.   Boston, 1810.   8vo.

    Another edition : New York, 1808.

Continued under the title :

    Elements (The) of Experimental Chemistry.   London, 1810. 2 vols., 8vo.   Plates.

Dedicated to Mr. John Dalton, of Manchester.

    * Seventh edition greatly enlarged.   London, 1815.   2 vols., 8vo.   9 plates.

    Ninth edition.   London, 1823.   2 vols., 8vo.   Ill.

    Tenth edition.   London, 1826.   2 vols., 8vo.   Ill.

    The first American from the eighth London edition, comprehending all the recent discoveries.   Together with an account of Dr. Wollaston's Scale of chemical equivalents.   Also a substitute for Woulfe's or Nooth's apparatus ; and a new theory of galvanism by Robert Hare.   Philadelphia, 1819.   2 vols., 8vo.

    Second American edition.   Philadelphia, 1822.   2 vols., 8vo.

    Another edition.   Philadelphia, 1831.

    Élémens de chimie expérimentale.   Traduit de l'anglois sur la sixième édition, par H. F. Gaulthier-Claubry.   Paris, 1812.   2 vols., 8vo.   Ill.

    Chemie für Dilettanten, oder Anweisung die wichtigsten chemischen Versetzungen ohne grosse Kosten und ohne weitläufige Apparate anzustellen.   Aus dem englischen nach der zweiten Originalausgabe übersetzt und mit Anmerkungen versehen, von J. B. Trommsdorff.   Gotha, 1807.   8vo.

    Grundriss der theoretischen und praktischen Chemie, sowohl zum Selbstunterrichte als zu Vorlesungen eingerichtet.   Aus dem englischen nach der fünften Ausgabe übersetzt, von Friedrich Wolff.   Berlin, 1812. 2 vols., 8vo.

A Danish translation by Sever Christian Salling was published at Odense in 1805.

HENSEN, V.

    Das Gehör.   Physiologie der Zeugung.   *See* Handbuch der Physiologie.

HENSING, JOHANN THOMAS.

Meditationes et experimenta circa acidulas Svalbacenses . . .   Francofurti, 1711.   8vo.

HENSMANS, PIERRE JOSEPH.
Mémoire sur le proportionnement chimique pésé et mésuré des corps. Louvain, 1824.
Mémoire sur les esprits alcooliques.   Bruxelles, 1824.

HEPPE, GUSTAV.
>   Die chemische Reactionen . . . Tabellen [etc.].   *See in Section II.*

HEPPE, GUSTAV.
Hauswirthschaftliche Chemie.   Dritte Auflage.   Hamburg, 1889.
Katechismus der Chemikalienkunde.   Eine kurze Beschreibung der wichtigsten Chemikalien des Handels.   Leipzig, 1880.   8vo.

HERBERGER, JOHANN EDUARD.
Die menschliche und thierische Milch . . .   Nürnberg, 1839.
>   *See also in Section II.*

HERBIG, W.
Beiträge zur Glycerinbestimmung.   Jena, 1890.   8vo.

HERING, E.
>   Raumsinn des Auges. Temperatursinn.   *See* Handbuch der Physiologie.

HERRBURGER, A.
Allgemeines chemisch-technisches Recept-Handbuch.   Zweite Ausgabe. Leipzig, 1891.   8vo.

HERITIER, L'., S. D.
>   *See* L'Heritier, S. D.

HERLANT, ACHILLE.
Précis du cours de chimie usuelle professé aux sections d'infanterie et de cavalerie à l'École militaire de Belgique.   Deuxième édition, avec figures, corrigée, augmentée et imprimée en deux sortes de caractères pour servir aux élèves des athénées et des universités. Bruxelles, 1863.   8vo.

HERMANN.
Die Cultur der Runkelrübe in Russland.   1830.

HERMANN, BENEDICT FRIEDRICH JOHANN.
Ueber die allgemeinen Eigenschaften des Kupfers und über die Kenntniss seiner Erze.   Leipzig, 1812.   8vo.

HERMANN, FELIX.
Die Glas-, Porzellan- und Email-Malerei in ihrem ganzen Umfange. Ausführliche Anleitung zur Anfertigung sämmtlicher bis jetzt zur

HERMANN, FELIX. [Cont'd.]

Glas-, Porzellan-, Email-, Fayence- und Steingut-Malerei gebräuchlichen Farben und Flüsse, nebst vollständiger Darstellung des Brennens dieser verschiedenen Stoffe. Unter Zugrundelegung der neuesten Erfindungen und auf Grund eigner in Sèvres und anderen grossen Malereien und Fabriken erworbenen Kenntnisse bearbeitet und herausgegeben. Wien, Pest, Leipzig, 1882. Ill.

HERMANN, L.

See Handbuch der Physiologie.

HERMANN, L.

Lehrbuch der experimentellen Toxicologie. Berlin, 1874. pp. x–306, 8vo.

HERMANN, R.

Untersuchungen über die Zusammensetzung der Tantalerze. Moskau, 1850. 8vo.

Untersuchungen über Ilmenium, Niobium und Tantal. Moskau, 1855. 8vo.

HERMBSTÄDT, SIGISMUND FRIEDRICH.

Bibliothek der neuesten . . . Literatur. *See in Section I.*

HERMBSTÄDT, SIGISMUND FRIEDRICH.

Allgemeine Grundsätze der Bleichkunst [etc.]. Berlin, 1804. 8vo.

Anleitung zur Kunst wollene, seidene, baumwollene und leinene Zeuge zu bleichen [etc.]. Berlin, 1815.

Anleitung zur Fabrikation des Syrups und Zuckers aus Stärke [etc.]. Berlin, 1814.

Anleitung zur praktisch-ökonomische Fabrikation des Zuckers aus den Runkelrüben. Berlin, 1811.

Anleitung zur Zergliederung der Vegetabilien nach physisch-chemischen Grundsätzen. Berlin, 1807. 8vo.

Chemische Grundsätze der Destillirkunst oder Liqueurfabrikation [etc.]. Berlin, 1819.

Chemische Grundsätze der Kunst Bier zu brauen, oder Anleitung zur theoretisch-praktischen Kenntniss und Beurtheilung der neuesten und wichtigsten Entdeckungen und Verbesserungen in der Bierbrauerei; nebst einer Anweisung zur praktischen Darstellung der wichtigsten engländischen und deutschen Biere, so wie einiger ganz neuer Arten derselben. Berlin, 1814. 8vo. Ill.

Chemische Grundsätze der Kunst Brantwein zu brennen. Berlin, 1817.
Zweite Auflage. Berlin, 1823. 2 vols.

HERMBSTÄDT, SIGISMUND FRIEDRICH.   [Cont'd.]

Chemische Zergliederung des Wassers aus dem todten Meere, des aus dem Jordan, des bituminösen Kalks und eines andern Fossils aus der Nachbarschaft des todten Meeres.   Nürnberg, 1822.   8vo.

Chemisch-technologische Grundsätze der gesammten Ledergerberei [etc.].   Berlin, 1805–'07.   2 vols., 8vo.

Elemente der theoretischen und practischen Chemie für Militärpersonen ; besonders für Ingenieur- und Artillerie-Officiere.   Zum Gebrauche bei Vorlesungen und zur Selbstbelehrung.   Berlin, 1823.   3 vols., 8vo.   Vol.I, pp. xii–440–[iv], 2 plates.   Vol. II, pp. 441–836–[iv].   Vol. III, pp. 837–1143–[v].

> Each volume has a separate title-page in addition to the general one, only differing with respect to the Abtheilungen.

Grundriss der Färbekunst.   Berlin, 1802.   8vo.

> Dritte Auflage.   Berlin, 1824.   2 vols., 8vo.

Grundlinien der theoretischen und experimentellen Chemie, zum Gebrauche beim Vortrage derselben.   Berlin, 1814.   8vo.

Grundriss der Technologie oder Anleitung zur Kenntniss und Beurtheilung der Künste, Fabriken, Manufakturen und Handwerke. Berlin, 1814.   8vo.

> Zweite Auflage, Abtheilung I und II, 1830.

Grundsätze der experimentellen Kammeral-Chemie für Kammeralisten, Agronomen, Forstebediente und Technologen.   Berlin, 1808.   8vo.

> Dritte durchaus umgearbeitete Auflage.   Berlin, 1833.   8vo.
>
> Grundrids af den experimentale Kameral-Chemie for Kameralister, Agronomer, Forstbetiente og Technologer Oversat fra det Tydske ved Jo. Fr. Bergsøe.   Kjøbenhavn, 1812.

Grundsätze der Kunst alle Arten Leder zu gerben.   Berlin, 1805–06.

Physikalisch-chemische Versuche und Beobachtungen.   Berlin, 1786–89.   2 vols., 8vo.

Systematischer Grundriss der allgemeinen Experimental-Chemie zum Gebrauch seiner Vorlesungen.   Berlin, 1791.   3 vols., 8vo.

> Zweite Auflage, 1800–01.   2 vols.
>
> Systematischer Grundriss der allgemeinen Experimental-Chemie, zum Gebrauch der Vorlesungen und zur Selbstbelehrung beym Mangel des mündlichen Unterrichts ; nach der neuesten Entdeckungen entworfen. Dritte verbesserte und vermehrte Auflage.   Basel und Leipzig, 1816–27.   5 vols., 8vo.   Vol. I, pp. xxviii–446, 1812 ; vol. II, pp. xvi–592, 1813 ; vol. III, pp. xx–475, 1819 ; vol. IV, pp. xxii–475, 1823 ; vol. v [with an index], pp. xxxviii–610, 1827.

HERMBSTÄDT, SIGISMUND FRIEDRICH. [Cont'd.]
Wissenschaft (Die) des Seifensiedens [etc.]. Berlin, 1808. 8vo.
Zweite Auflage. Berlin, 1824.

HERMBSTÄDT, SIGISMUND FRIEDRICH [Translator].
See Duportal ; also Scheele ; also Chaptal ; also Lavoisier.

HERPIN, JEAN CHARLES.
De l'acide carbonique, de ses propriétés physiques, chimiques et physi-
ologiques ; de ses applications thérapeutiques comme anesthésique,
désinfectant, cicatrisant, résolutif, etc. . . . Paris, 1864. pp.
xii–564. 12mo.
De la graisse des vins . . . Paris, 1819. 8vo.
Notice sur l'art de cultiver la vigne et de faires les vins . . . Metz,
1821. 8vo.
Tableaux analytiques comparatifs des principales eaux minérales. Paris,
1855. 12mo.
Récréations chimiques . . . Paris, 1824. 2 vols., 8vo.
Recherches sur l'emploi des divers procédés nouveaux pour la conservation des substances animales. Metz, 1822. 8vo.

HERRING, WALTER RALPH.
The Construction of Gas-works practically described. With specially
prepared plates, illustrations and numerous useful tables. London, 1892.

HERZ, FRANZ JOSEPH.
Die gerichtliche Untersuchung der Kuhmilch sowie deren Beurtheilung.
Berlin und Neuwied, 1889. pp. vi–178. 8vo. Ill.

HERZBERG, CARL.
Vollständiges Handbuch der chemischen Fabrikenkunde oder Darstellung des Schwefels, der Schwefelsäure, des Kochsalzes, Natrons
(Soda), der Pottasche, des Boraxes, Salpeters, der Salpeter-, Salzsäure, des Alauns, Vitriols, Salmiaks, Phosphors und der gashaltigen
Wasser. Zweite sehr vermehrte und verbesserte Auflage. Weimar,
1858. pp. xx–879. Plates.

HERZFELD, J.
Farben (Die) und Bleichen von Baumwolle, Wolle, Seide, Jute, Leinen,
etc., im unversponnenen Zustande, als Garn und als Stückwaare.
Part 1. Berlin, 1889.
Praxis (Die) der Färberei von Baumwolle im losen Zustand im Strang,
als Kette, in Copsform und als Stückwaare ; Leinen-, Jute- und
Nesselfärberei ; Färberei der losen Wolle und des Wollgarns ;
Seidenfärberei im Strang und im Stück u. s. w. Berlin, 1892. 8vo.
Ill. In progress.

HESEKIEL, ADOLPH.
Die Pyridinbasen in der chemischen Litteratur. Hamburg, 1886. 8vo.

ГЕССЪ.
Основанія чистой химіи. С.-Петербургъ 1849.
> HESSE. Principles of pure chemistry. St. Petersburg, 1849.

HESS, GERMAIN HENRI.
Thermochemische Untersuchungen, 1839–1842. Herausgegeben von
W. Ostwald. Leipzig, 1890. 8vo.
> Ostwald's Klassiker der exakten Wissenschaften.

HESSEL.
Löthrohrtabellen für mineralogische und chemische Zwecke. Marburg,
1847. 4to.

HÉTET, FRÉDÉRIC.
Manuel de chimie organique élémentaire avec ses applications à la
médecine, a l'hygiène et à la toxicologie. Paris, 1880. pp. vii–767.
8vo. Ill.

HEUMANN, KARL.
Anleitung zum Experimentiren bei Vorlesungen über anorganische
Chemie. Zum Gebrauch an Universitäten und technischen Hoch-
schulen, sowie beim Unterricht an höheren Lehranstalten.
Braunschweig, 1876–79. pp. xxxiv–668, 8vo. Ill.
Anilinfarben (Die) und ihre Fabrikation. Braunschweig, 1888. 2 parts,
8vo. Ill. Part I, Triphenylmethan-Farbstoffe.

HEUSEL, E.
Eine neue Theorie der Lebens-Chemie in typischen Figuren veranschaulicht. Für Aerzte, Apotheker und Chemiker. Christiania,
1886. 8vo.

HEURICH, F.
Tabellen zur qualitativen chemischen Analyse. Wiesbaden, 1886. 8vo.

HEYER, C.
Ursache und Beseitigung des Bleiangriffs durch Leitungswasser.
Chemische Untersuchungen aus Anlass der Dessauer Bleivergiftun-
gen im Jahr 1886. Dessau, 1888. Plates.

ГЕЙМАНЪ.
Под. чтенія общей химіи, приложенной къ фабричному и заводскому дѣлу.
Составлено стенографически по публичному курсу, въ 1840 году пре-
поданному для Московскихъ фабрикантовъ. Съ пояснительными рисун-
ками, вырѣзанными на деревѣ. Москва 1845—49. 8°

Гейманъ. [Cont'd.]
>   Heymann, Rod. Chemistry for Manufacturers. Moscow, 1845–49.

О добываніи сѣры и сѣрной кислоты изъ колчедановъ. Москва 1854. 8°
>   On Sulphuric Acid and Sulphurous Acid. Moscow, 1854.

Hicks, J. W.
Inorganic Chemistry. London, 1877. 12mo.

Hielbig, Carl.
Kritische Beurtheilung der Methoden welche zur Trennung und
>   quantitativen Bestimmung der verschiedenen Chinaalkaloide be-
>   nutzt werden. Dorpat, 1880. 8vo.

Hierne, Urban. *See* Hjärne, Urban.

Higgins, Bryan.
*Experiments and Observations relating to Acetous Acid, Fixable Air,
>   Dense Inflammable Air, Oils and Fuel ; the matter of fire and
>   light, metallic reduction, combustion, fermentation, putrefaction,
>   respiration and other subjects of chemical philosophy. London,
>   1786. pp. xvi–353-[1].
>>   *See also, in Section VII*, Minutes of the Society for Philosophical Experi-
>>   ments.

Higgins, William.
*A Comparative View of the Phlogistic and Antiphlogistic Theories
>   with Inductions. To which is annexed an Analysis of the human
>   calculus with observations on its origin. London, 1789. pp. xiv-
>   [i]-316, 8vo.
An Essay on the Theory and Practice of Bleaching, wherein the sul-
>   phuret of lime is recommended as a substitute for potash. London,
>   1799.
*Experiments and Observations on the Atomic Theory and Electrical
>   Phenomena. Dublin, 1814. pp. 180, 8vo.

Hildebrandt, Georg Friedrich.
Anfangsgründe der allgemeinen dynamischen Naturlehre. Erlangen,
>   1807. 2 vols., 8vo.
>>   Zweite Auflage, 1821.
Anfangsgründe der Chemie, zum Grundriss akademischer Vorlesungen
>   nach dem neuen Systeme abgefasst. Erlangen, 1794–'95. 3 vols.,
>   8vo.
De metallorum nobilium puritate, arte paranda. Erlangen, 1796. 8vo.
Lehrbuch der Chemie als Wissenschaft und Kunst. Mit einem voll-
>   ständigen Register. Erlangen, 1816. 8vo.

HILDEBRANDT, GEORG FRIEDRICH.   [Cont'd.]
> *Supplement, under the title :*
>> Bischof, Carl Gustav : Lehrbuch der Stöchiometrie oder Anleitung die Verhältnisse zu berechnen, nach welchen sich die irdischen Körper mit einander verbinden.   Erlangen, 1819.   8vo.

HILGER, A.
>> *See* Husemann, A., und A. Hilger.

HILL, HENRY BARKER.
> Lecture Notes on Qualitative Analysis.   New York, 1876.   16mo.
> Another edition, New York, 1879.   16mo.

HILLAIRET, J. E.
> Notice historique sur l'empoisonnement par l'arsenic, sur l'emploi de l'appareil de Marsh et des autres moyens de doser ce toxique. Paris, 1846.   8vo.

HILLER, ARNOLD.
> Die Lehre von der Fäulniss.   Auf physiologischer Grundlage einheitlich bearbeitet.   Berlin, 1879.   pp. xii–547, 8vo.

HILLER, FERDINAND.
> Lehrbuch der Chemie.   Leipzig, 1861–'63.   8vo.   Ill.

HIMLY, FRIEDRICH CARL.
> De Caoutchouk ejusque destillationis siccæ productis et ex his de caoutchino novo corpore ex hydrogenio et carbono composito. Gottingæ, 1835.   8vo.

HINRICHS, GUSTAVUS.
> Beiträge zur Dynamik des chemischen Moleküls.   Leipzig, 1892.
>> *Contains :*
>>> I. Das Molekül als System materieller Punkte.
>>> II. Die Energie des Moleküls.
>>> III. Graphische Structurformeln.
>>> IV. Die Trägheitsmomente der Moleküle.
>>> V. Die Bewegungen der Moleküle.
>>> VI. Die Siedpunkte isomerer Körper bestimmt durch das Trägheitsmoment der Moleküle.

> Contributions to Molecular Science, or Atomechanics.   Iowa City, and Salem, Mass., 1868–70.   4 nos., 8vo.
> Principles (The) of Chemistry and Molecular Mechanics.   Davenport, Iowa, 1874.   8vo.
>> *Also under the title :* The Principles of Physical Science, Vol. II.
> Programme der Atommechanik oder die Chemie eine Mechanik der Panatome.   Iowa City, 1867.   4to.
>> With the text also in French.

HINTERBERGER, FRIEDRICH.

Lehrbuch der technischen Chemie für Ober-Realschulen.   Wien, 1855.
3 vols., 8vo.   Ill.

Lehrbuch der Chemie für Realschulen, sowie zum Selbstunterrichte.
Mit zahlreichen in den Text eingedruckten Holzschnitten.
Zweiter Theil : Organische Chemie.   Wien, 1852.   8vo.
Zweite verbesserte Auflage.   Wien, 1854.   8vo.
Vierte verbesserte Auflage.   Wien, 1857.   8vo.

Repetitorium der anorganischen Chemie.   Bearbeitet von R. Jama.
Wien, 1871.   8vo.

HINTERBERGER, FRIEDRICH, UND EDMUND SCHREINZER.

Kurze Anleitung zur qualitativen und quantitativen chemischen Analyse.
Erste Abtheilung : Qualitative Analyse.   Zweite verbesserte und
vermehrte Auflage.   Wien, 1856.   8vo.

First edition, Wien, 1852.

HINTZE, C.

Ueber die Bedeutung krystallographischer Forschung für die Chemie.
Bonn, 1885.

HIORNS, ARTHUR H.

Mixed Metals or Metallic Alloys.   London, 1891.

" The best manual of alloys to be met with in the English language."—
Chem. News.

Practical Metallurgy and Assaying.   A text-book.   London, 1888.   8vo.
Second edition, completely revised.   London, 1892.

Text-book (A) of Elementary Metallurgy for the Use of Students.
Added an Appendix of Examination Questions.   London, 1888.
8vo.

HIRSCH, B.

Die Fabrikation der künstlichen Mineralwässer und anderer moussiren-
der Getränke.   Zweite vermehrte Auflage.   Braunschweig, 1876.
4to.   Ill.

Separat-Abdruck aus Muspratt's technischer Chemie, q. v.

HIRSCHFELD, J., UND WILHELM PICHLER.

Die Bäder, Quellen und Curorte Europa's.   Stuttgart, 1875–76.   2 vols.,
roy. 8vo.   I, pp. 546 ; II, pp. iv–651.

Contains a large number of analyses of mineral waters.

HIRSCHWALD, J.

Löthrohr-Tabellen.   Ein Leitfaden zur chemischen Untersuchung auf
trockenem Wege, für Chemiker, Hüttenleute und Mineralogen.

HIRSCHWALD, J. [Cont'd.]

Nebst einer Uebersicht über die Zusammensetzung technisch-wichtiger Minerale und Hüttenproducte, sowie einem Schema der wichtigsten quantitativen Löthrohrproben und deren Beschickung. Leipzig und Heidelberg, 1875. 4to. Plates.

Anleitung zur systematischen Löthrohranalyse für Chemiker, Mineralogen und Hüttenleute. Zweite Auflage der "Löthrohr-Tabellen." Leipzig, 1891.

HIRZEL, CHRISTOPH HEINRICH.

Führer (Der) in die organischen Chemie. Leipzig, 1855.

Führer (Der) in die unorganischen Chemie. Leipzig, 1852.

Grundzüge der Chemie. Leipzig, 1857.

This forms vol. 1 of Rossmässler's Bücher der Natur.

Katechismus der Chemie. Fünfte Auflage. Leipzig, 1884.

Sechste Auflage. Leipzig, 1889.

Nux Vomica (Die) und ihre Bestandtheile. Eine Zusammenstellung der bis zum heutigen Tage hierüber gesammelten Erfahrungen. Leipzig, 1851. 8vo.

Opium (Das) und seine Bestandtheile. Eine Zusammenstellung der bisher gesammelten Erfahrungen. Leipzig, 1851. 8vo.

Steinöl (Das) und seine Producte. Nach A. Norman Tate's "The Petroleum and its Products." Leipzig, 1864. pp. xii–172, 12mo.

Cf. Tate, A. Norman, Petroleum.

Toiletten-Chemie. Leipzig, 1857.

Zweite sehr vermehrte und verbesserte Auflage. Leipzig, 1866. pp. viii–538. 8vo.

Vierte Auflage. Leipzig, 1892. 8vo.

Ueber das Aluminium und einige seiner Legirungen. Leipzig, 1858. 4to.

Ueber die Einwirkung des Quecksilbers auf das Ammoniak und die Ammoniaksalze. Leipzig, 1852.

HJÄRNE, URBAN.

Actorum laboratorii Stockholmensis parasceue elle Förberedelse. Stockholm, 1706. 4to.

Acta et tentamina chymica in regio laboratorio Stockholmiensi elaborata et demonstrata, in decades redacta atque divisa, una cum præmissa parasceue seu prævia manuductione ad experimenta rite perficienda. Holmiæ, 1712. pp. [xviii]–204–[ii], 4to.

With portrait of the author, folding plate, and frontispiece.

Actorum chemicorum Holmiensium tomus primus, hoc est parasceue sive præparatio ad tentamina in regio laboratorio Holmiensi peracta, ut et compendiosa manuductio ad elementa et principia chemica

HJÄRNE, URBAN. [Cont'd.]

rite investiganda, cum annotationibus Joh. Gotschalk Wallerii. Stockholmiæ, 1753. pp. xviii–194–[x]. 2 folding plates.

> The "Tomus secundus," mentioned by Poggendorff, I have not seen.
>
> This chronicles the work done in a chemical laboratory established in 1679 by the order of Charles XI., King of Sweden, who cultivated chemistry even in the midst of burdensome affairs of state and foreign wars.

Den lilla Vattenprofvaren. Stockholm, 1683.

Tractatus de acidulis medeviensibus. Linkjøping, 1679.

Utförlig Berättelse om de nyss uppfundne Surbrunnar i Medevij. Stockholm, 1680.

HJELM, PETTER JACOB.

Anvisning till bästa sättet, at tillverka Salpeter. Stockholm, 1799.

Om Indigo 's Tillverkning af Veideörten. Stockholm, 1801.

HJELT, EDUARD.

Grunddragen af allmänna organiska Kemien. Helsingfors, 1883. 8vo.

> Grunderna af allmänna organiska Kemi. Andra omarbetete upplagen. Helsingfors, 1887.
>
> Grundzüge der allgemeinen organischen Chemie. Berlin, 1887. 8vo.
>
> Principles of General Organic Chemistry, translated from the German by J. Bishop Tingle. London, 1890. 8vo.

Intramoleculare (Die) Wasserabspaltung bei organischen Verbindungen, Helsingfors, 1886.

HJORTDAHL, TH.

Begyndelsesgrundene af den kvalitative Analyse. Kortfattet Veiledning for de studerende ved Universitetets Laboratorium. Christiania, 1871. 8vo.

> Anden Udgave. Christiania, 1886. 8vo.

Kortfattet Lærebog i Kemi. Christiania, 1870. 8vo.

Lærebog i anorganisk Kemi. Fjerde Oplag. Christiania, 1888. 8vo. Ill.

Oversigt over de vigtigste Dele af den chemiske Fabrik-Industri. Forelæsninger ved Universitetet, 1870 og 1871. Første Hefte. Christiania, 1871. 8vo.

HLASIWETZ, HEINRICH HERMANN.

Anleitung zur qualitativen chemischen Analyse zum Gebrauche bei den praktischen Uebungen im Laboratorium. Vierte Auflage. Wien, 1883.

> Achte Auflage, durchgesehen von P. Weselsky. Wien, 1883.
>
> Neunte Auflage. Wien, 1888.

HLASIWETZ, HEINRICH HERMANN. [Cont'd.]

Zehnte Auflage, von R. Benedikt. Wien, 1892. 8vo.

Introduzione alla analisi chimica qualitativa ad uso degli esercizi pratici nei laboratori. Versione autorizzata sulla quarta edizione tedesca con note ed aggiunte per Paolo Matcovich. Wien, 1879. 8vo.

HLUBEK, FRANZ XAVER WILHELM.

Beleuchtung der organischen Chemie durch Dr. Liebig. Gratz, 1842. 8vo.

Runkelrübe (Die), ihr Anbau und die Gewinnung des Zuckers aus derselben. Laibach, 1839.

HOBLYN, RICHARD D.

A Manual of Chemistry, with glossary and index. London, 1846. 12mo. Also New York, 1860.

HOBSON, BENJAMIN.

Po wu hsin pien. Shanghai, 1858.

> Natural philosophy, including chemistry, in Chinese. The author has published also a medical vocabulary in English and Chinese which includes chemical terms. *Cf.* Fryer, John.

HOCHHEIMER, CARL FRIEDRICH AUGUST.

Allgemeine ökonomisch- und chemisch-technologisches Haus- und Kunstbuch. Leipzig, 1794–1810. 6 vols. Vols. III to VI by J. C. Hoffmann.

Chemische Mineralogie, oder vollständige Geschichte der analytischen Untersuchung der Fossilien in systematischer Ordnung aufgestellt. Leipzig, 1792–93. 2 vols., 8vo.

Handbuch zur chemischen Praxis für Apotheker, Mineralogen und Scheidekünstler. Leipzig, 1792. 8vo.

HODGES, JOHN F.

The First Book of Lessons in Chemistry in its Application to Agriculture. For the use of farmers and teachers. London, 1848. 12mo.

Second edition. London, 1849. 12mo.

Third edition. Belfast, 1856. 12mo.

Fourth edition. London, 1858. 12mo.

HODGKINSON, W. R.

Exercises in Practical Chemistry. An Introduction to Qualitative and Quantitative Analysis. Third Issue. London, 1890. 8vo.

HÖDL, E. J.

Die praktische Anwendung der Theerfarben in der Industrie. Praktische Anleitung zur rationellen Darstellung der Anilin-, Phenyl-, Naphtalin- und Anthracen-Farben in der Färberei, Druckerei, Buntpapier-, Tinten- und Zündwaaren-Fabrikation. Wien, Pest, Leipzig, 1885. Ill.

HOEFER, FERDINAND.

Éléments de chimie minérale, précédés d'un abrégé de l'histoire de la science et suivis d'un exposé des éléments de chimie organique. Paris, 1841. 8vo.

> Elementi di chimica minerale. Versione con note di Giovanni Giorgini. Modena, 1845. 2 vols., 8vo.

> *See also, in Sections II, III and IV,* Hoefer, Ferdinand.

HÖFER, HANS, UND FERDINAND FISCHER.

> Die Industrie der Mineralöle. Das Erdöl (Petroleum) und seine Verwandten. *See* Bolley's Handbuch.

HÖFER, HUBERT FRANZ.

Nachricht von dem in Toskana entdeckten natürlichen Sedativsalze und von dem Boraxe welche daraus bereitet wird. Aus dem italienischen übersetzt von B. S. Hermann. Wien, 1781. 8vo.

HÖVINGHOFF, HENR.

Dissertatio de incremento ponderis corporum calcinatorum. Hafniæ, 1772. 4to.

> *Cf.* Rey, Jean.

HÖVER, K.

Kemi til Skolebrug. Tredie Oplag. Kjøbenhavn, 1870. 8vo. Ill.
Fjerde aldeles omarbejdede Oplag. Kjøbenhavn, 1876. 8vo. Ill.
Nogle Bemærkninger om Kemien. Kjøbenhavn, 1860. 8vo.

HOFF, J. H. VAN'T.

Ansichten über die organische Chemie. Braunschweig, 1878–81. 3 parts. 8vo.
Études de dynamique chimique. Amsterdam, 1884. pp. iv.–214, 8vo.
Voorstel tot uitbreiding der tegenwoordig in de scheikunde gebruikte structuurformules in de ruimte ; benevens een daarmee samenhangende opmerking omtrent het verband tusschen optisch actief vermogen en chemische constitutie van organische verbindingen. Utrecht, 1874. 8vo.

> Chimie (La) dans l'espace. Rotterdam, 1875. pp. 44, 12mo. 2 plates.

HOFF, J. H. VAN'T. [Cont'd.]

Dix années dans l'histoire d'une théorie. (Deuxième édition de " La chimie dans l'espace.") Rotterdam, 1887. pp. 102, 12mo.

Chemistry in Space. Translated and edited by J. E. Marsh. Oxford, 1891. pp. vi–128, 12mo. Clarendon Press Series.

Die Lagerung der Atome im Raume. Nach des Verfassers Broschüre, " La chimie dans l'espace," deutsch bearbeitet von F. Hermann. Nebst einem Vorwort von Johannes Wislicenus. Braunschweig, 1877. 8vo. Ill.

Stereochemie. Nach van't Hoff's " Dix années dans l'histoire d'une théorie " neu bearbeitet von W. Meyerhoffer. Wien, 1892.

HOFFER, RAIMUND.

Kautschuk und Guttapercha. Eine Darstellung der Eigenschaften und der Verarbeitung des Kautschuks und der Guttapercha auf fabrikmässigem Wege, der Fabrikation des vulcanisirten und gehärteten Kautschuks, der Kautschuk- und Guttapercha-Compositionen, der wasserdichten Stoffe, elastischen Gewebe u. s. w. Für die Praxis bearbeitet. Wien, Pest, Leipzig, 188–.

HOFFMANN, FRIEDRICH.

Chemia rationalis et experimentalis, sive collegium physico-chemicum curiosum. Lugduni-Batavorum, 1748. 8vo.

Demonstrationes physicæ curiosæ experimentis et observationibus curiosis mechanicis ac chymicis illustratæ. Halæ, 1700. 4to.

De panis grossioris Westphalorum vulgo bompournickel natura, elementis chymicis et virtute. Halle, 1695. 4to.

Observationum physico-chymicarum selectiorum libri III, in quibus multa curiosa experimenta et lectissimæ virtutis medicamenta exhibentur. . . . Halæ, 1722. 4to.

Observations physiques et chymiques, dans lesquelles on trouve beaucoup d'expériences curieuses . . . Traduites du Latin de F. H. A Paris, 1754. 8vo. 2 parts.

Opera omnia physico-medica. Genevæ, 1740. 6 vols., fol.

Zweite Auflage, nebst 5 Supplementbänden, 1748, 1749, 1753 und 1760.

Gmelin, in his Geschichte der Chemie, enumerates 122 chemical treatises by Friedrich Hoffmann.

HOFFMANN, FREDERICK, AND FREDERICK B. POWER.

A Manual of Chemical Analysis as applied to the Examination of Medicinal Chemicals. A Guide for the determination of their

HOFFMANN, FREDERICK, AND FREDERICK B. POWER.   [Cont'd.]
  identity and quality and for the detection of impurities and adulterations.   For the use of pharmacists, physicians, druggists,
  manufacturing chemists and pharmaceutical and medical students.
  Third edition thoroughly revised and greatly enlarged.   Philadelphia, 1883.   pp. 624, 8vo.   Ill.

    First edition.   New York and London, 1873.

HOFFMANN, JOHANN MORITZ.
  * Acta laboratorii chemici Altdorfini, chemiæ fundamenta, operationes
    præcipuas et tentamina curiosa, ratione et experientia suffulta, com-
    plectentia.   Norimbergæ et Altdorfii, apud hæredes Joh. Dan.
    Tauberi, 1719.   pp. [iv]–288–54, sm. 4to.   Portrait.
  [Appendix I] : Auctuarium, notas, observationes et experimenta ad
    actorum sectionem primam.   Declarationem ulteriorum necessaria
    una cum programmato invitatorio ad inaugurationem laboratorii
    chemici Altdorfini præmisso, et monumento ad memoriam posteri-
    tatis publice erecto, ac indice rerum ac verborum, exhibens. pp. 42.
  [Appendix II] : Laboratorium novum chemicum apertum medicina
    cultoribus, cum amicâ ad orationem inauguralem invitatione.   A. C.
    1683.   pp. *from* 43–54–[xiv].   Norimbergæ et Altdorfii, 1719.

      One of the first works issued dealing with researches conducted in a
      chemical laboratory.   Compare, however, Hjärne, Urban.

HOFFMANN, KARL BERTHOLD, UND R. ULTZMANN.
    See Ultzmann, R.

HOFFMANN, MIKULÁŠ.
  Chemie organická pro vyšši školy realné.   V Praze, 1880.

HOFFMANN, MIKULÁŠ A FRANT. HEJZLAR.
  Chemie zkušebná pro čtvrtou školu gymnasíí a realných gymnasíí.   V
  Praze, 1880.

HOFFMANN, ROBERT.
  Theoretisch-praktische Ackerbau-Chemie nach dem heutigen Standpunkte der Wissenschaft und Erfahrung.   Dritte Auflage bearbeitet
  von Th. von Gohren.   Leipzig, 1876–77.

      Земледѣльческая химія.   Съ дополн. проф. Энгельгардта.
      С.-Петербургъ 1868.

ГОФМАНЪ, В.
    Руководство къ изученію качествен. и химич. анализа, составленное для на-
    чинающихъ.   Москва 1858.

        HOFFMANN, V.   Manual for the study of qualitative chemical analysis for
        beginners.   Moscow, 1858.

HOFMANN, ——.

Der Reactionair in der Westentasche oder rhythmischer Gang der qualitativen chemischen Analyse. Didactisches Poëm mit elegisch-spectral-analytischem Epilog und einer Gedanken-pyramide der Analyse. In chemische Knüttelreime gebracht. Sechste Auflage. Breslau, 1862.

HOFMANN, AUGUST WILHELM VON [Editor].
> *See* Liebig, Justus von.

HOFMANN, AUGUST WILHELM VON.

Einleitung in die moderne Chemie. Nach einer Reihe von Vorträgen gehalten in dem Royal College of Chemistry zu London. Braunschweig, 1866. pp. xvi–257, 8vo.

> Zweite Auflage. Braunschweig, 1866. 8vo.

> Vierte Auflage. Braunschweig, 1869. 8vo.

> Fünfte gekürzte und verbesserte Auflage. Braunschweig, 1871. 8vo.

> Sechste mit der fünften übereinstimmende Auflage. Braunschweig, 1877. 8vo.

> Inleiding tot de nieuwere scheikunde, bewerkt naar eene reeks van voorlezingen, in het Royal College of Chemistry te London gehouden. Naar het Engelsch door H. F. R. Hubrecht. Utrecht, 1866. 8vo. Ill.

> Introduction to modern Chemistry, experimental and theoretic. Embodying twelve lectures delivered in the Royal College of Chemistry. London, 1865. pp. xvi–233, 8vo.

> Introduzione alla chimica moderna. Prima versione italiana eseguita sotto gli occhi dell'autore sulla quarta edizione tedesca da Luigi Gabba. Torino, Firenze e Milano, 1870.

> Wstęp do nowoczesnéj chemji. Przetłomaczył . . . L. Masłowski. Kraków, 1875. 8vo.

> Введеніе къ изученію соврем. химіи. С.-Петербургъ 1866.

Organische (Die) Chemie und die Heilmittellehre. Berlin, 1871. 8vo.

Молекулярная химія. О силѣ присоединенія атомовъ. Перев. Редингъ С. Москва 1873.

> Molecular Chemistry. Moscow, 1873.

Органическая химія и фармакологія. Рѣчь сказан. 2 авг. 1871. Перев. Базаровъ А. Кіевъ 1873.

> Organic Chemistry and Pharmacology. A Discourse delivered on August 2, 1871. Kiev, 1873.

HOFMANN, CARL AUGUST.

Systematische Uebersicht der Resultate von 242 chemischen Untersuchungen von Mineralwässer . . .   Berlin, 1815.

Tabelle über einige 40 Mineralwässer.   Weimar, 1789.

Tabellarischer Entwurf der pharmaceutischen Scheidekunst.   Weimar,
1791.

HOFMANN, CARL B.

Lehrbuch der Zoochemie.   Wien, 1876.   pp. xvii–729, 8vo.   Ill.
Neue Ausgabe.   Wien, 1883.

HOFMANN, GOTTFRIED AUGUST.

Chymie zum Gebrauch des Haus-, Land- und Stadtwirthes, des Künstlers, Manufacturiers, Fabrikanten und Handwerkers.   Leipzig,
1757.   8vo.

Another edition : Chemie für Künstler und Fabrikanten . . .
mit Anmerkungen von J. C. Wiegleb.   Leipzig, 1779.

HOFMEISTER, A.

Die Fabrication des Alauns, sowie der Handel mit diesem Producte und
die über ihn erschienene Literatur.   Nach den besten Quellen,
neuesten Forschungen und Entdeckungen dargestellt zum Gebrauch für Techniker, Künstler und Professionalisten, ganz besonders aber für Besitzer von Alaunwerken, Färber, Kaufleute, etc.
Leipzig, 1840.   8vo.   Ill.

HOLBROOK, [DR.].

Tung chow.   [Handbook of chemistry in Chinese.]
Cf. Fryer, John.

HOLDERMANN, E.

See Hager's Untersuchungen . . .

HOLGER, PHILIPP ALOIS VON.

Chemie für Damen.   Wien, 1843.   16mo.

HOLLAND, JOHN.

A Treatise on the Progressive Improvement and Present State of the
Manufactures in Metals.   London, 1831–34.   3 vols., 16mo.

New edition.   London, 1838–42.

Lardner's Cabinet Cyclopædia.

HOLLAND, J. W.

The Urine.   Memoranda, chemical and microscopical, for laboratory
use.   Philadelphia, 1888.

HOLLAND, J. W. [Cont'd.]
>   The Urine, the common poisons and the milk.   Memoranda,
>   chemical and microscopical, for laboratory use.   Third
>   edition.   Philadelphia, 1889.   pp. 84, oblong 12mo.   Ill.

HOLLANDUS, ISAAK.
>   Rariores chemiæ operationes.   Lipsiæ, 1714.   8vo.

HOLLUNDER, CHRISTIAN FÜRCHTEGOTT.
>   Anleitung zur mineralischen Probirkunst.   Nürnberg, 1826.   2 parts.
>   Beiträge zur Begründung einer analytischen Chemie auf trocknem
>   Wege.   Nürnberg, 1827.
>   Versuch einer Anleitung zur mineralurgischen Probirkunst auf trock-
>   nem Wege.   Nürnberg, 1826.   3 vois., 8vo.
>   Beiträge zur Farben-Chemie und chemischen Farben-Kunde.   Leipzig,
>   1827.   8vo.
>>   *Also with the title :* Handbuch des technisch-koloristischen
>>   Theiles der chemischen Fabrikenkunde [etc.].

HOLM, F.
>   Svovlsyrefabrikationens Udvikling i dette Aarhundrede og dens Ind-
>   flydelse paa den övrige chemiske Industri.   Afhandling ved Kon-
>   kurrencen til en Lærerplads i technisk Chemi ved den polytechniske
>   Læreanstalt.   Kjøbenhavn, 1867.   8vo.

HOLTZAPFEL, S. E. G.
>   Populaire voorlezingen over scheikunde.   Tweede goedkoope uitgave.
>   Amsterdam, 1866.   8vo.

HOLVERSCHEIT, R.
>   Ueber die quantitative Bestimmung des Vanadins und die Trennung
>   der Vanadinsäure von Phosphorsäure.   Berlin, 1890.   8vo.

HOLZNER.
>   *See* Sechs Vorträge aus dem Gebiete der Nahrungsmittel-Chemie.

HOMANN, H.
>   Das Gewichtsalkolometer und seine Anwendung.   Mit Einleitung von
>   L. Loewenherz.   Berlin, 1889.   8vo.

HOME, FRANCIS.
>   Versuche im Bleichen.   Leipzig, 1777.   pp. vi–352, 8vo.
>   Translated from English.

HOMOLLE, E., ET QUEVENNE.
>   · Mémoire sur la digitaline.   Rapport fait à l'Académie de médecine par
>   Bouillaud.   Paris, 1851.   8vo.

HOOKER, WORTHINGTON.

First Book in Chemistry for the Use of Schools and Families. New
York, 1862. pp. 231, 12mo. Ill.

Revised Edition. New York, 1877. pp. 176, 12mo. Ill.

Science for the School and Family. Part II: Chemistry. New York,
1863. pp. 435, 8vo. Ill.

Second edition, revised and corrected. New York, 1876.
pp. 430, 8vo.

HOPFF, GUSTAV WILHELM LUDWIG.

Das Bier in geschichtlicher, chemischer, medizinischer, chirurgischer,
und diëtetischer Beziehung, mit Rücksicht auf seine Verschiedenheiten, seine Verfälschungen und deren Entdeckungen. Zwei-
brücken, 1846. 8vo.

HOUT, F. D'.

Contribution à l'étude du lait. Courtrai, 1891. 8vo.

HOPPE, FELIX [afterwards HOPPE-SEYLER].

Anleitung zur pathologisch-chemischen Analyse für Aerzte und Studirende. Berlin, 1858. pp. xiv–281, 12mo. Ill.

HOPPE-SEYLER, FELIX.

Handbuch der physiologisch- und pathologisch-chemischen Analyse
für Aerzte und Studirende. Berlin, 1858. 8vo.

Zweite Auflage. Berlin, 1865. 8vo.

Dritte Auflage. Berlin, 1870. pp. xii–420, 8vo.

Vierte Auflage. Berlin, 1875. pp. viii–486, 8vo.

Fünfte Auflage. Berlin, 1883.

Traité d'analyse chimique appliquée à la physiologie et à la
pathologie. Guide pratique pour les recherches cliniques, traduit de l'allemand sur la quatrième édition
par Schlagdenhauffen. Paris, 1877. 8vo. Ill.

Medicinisch-chemische Untersuchungen. Berlin, 1866–71. 4 parts,
8vo. Ill.

Physiologische Chemie. In vier Theilen. Berlin, 1877–1881. pp. vi–
1036, 8vo. 1. Allgemeine Biologie. 2, 3, 4, Specielle physiologische Chemie.

Ueber die Einwirkung des Sauerstoffs auf Gährungen. Festschrift zur
Feier des fünfundzwanzig-jährigen Bestehens des pathologischen
Instituts zu Berlin. Strassburg, 1881. 8vo.

HOPSON, CHARLES RIVINGTON.

A General System of Chemistry theoretical and practical, digested and
arranged with a particular view of its application to the arts;
chiefly taken from the German by Wiegleb. London, 1789.

HORATIUS, TH.

Die Fabrikation der Aether und Grundessenzen. Die Aether, Fruchtäther, Fruchtessenzen, Fruchtextracte, Fruchtsyrupe, Tincturen
zum Färben und Klärungsmittel. Nach den neuesten Erfahrungen
bearbeitet. Wien, Pest, Leipzig, 18—. Ill.

HORN VAN DEN BOS, VAN DER.

De nederlandsche scheikundigen van het laatst der vorige eeuw.
Utrecht, 1881. 4to.

HORN, F. M.

Anleitung zur chemisch-technischen Analyse organischer Stoffe. Wien,
1890. pp. xv–244, 8vo. Ill.

HORN, JOHANN BERNHARDUS.

Synopsis metallurgica, oder kurtze, jedoch deutliche Anleitung zu der
höchst nütz- und ergötzlichen edlen Probierkunst, verfasset in 15
Tabellen . . . zum Druck befördert durch Kellnern. 1690. 8vo.

HORNE, HENRY.

Essays concerning Iron and Steel; the first, containing observations
on American sand-iron. The second, observations founded on
experiments on common iron-ore, with the method of reducing it
first into pig or sow-metal, and then into bar-iron; on the sort of
iron proper to be converted into good steel, and the method of
refining that bar-steel by fusion, so as to render it fit for the more
curious purposes. With an account of Mr. Réaumur's method of
softening cast-iron, and an Appendix discovering a more perfect
method of charring pit-coal so as to render it a proper succedaneum
for charred wood-coal. London, 1773. pp. [vi]–223, 16mo.

HORNIG, E.

Lehrbuch der technischen Chemie für Ober Realschulen und technische
Lehranstalten. I Theil, Unorganische Chemie. Wien, 1860. pp.
vi–428, 8vo. Plates.

HORSFORD, EBEN NORTON.

Chemical Essays relating to Agriculture. [Analyses of grains and
vegetables, distinguishing the nitrogenous from the non-nitrogenous
ingredients for the purpose of estimating their separate values for
nutrition. Also on ammonia found in glaciers, and on the action
and ingredients of manures.] Boston, 1846. 12mo.

HORSIN-DÉON, PAUL.

Traité théorique et pratique de la fabrication du sucre. Guide du
chimiste-fabricant. Paris, 1882. pp. xvi–640, roy. 8vo. 5 folding plates.

HORSLEY, JOHN.

A Catechism of Chemical Philosophy : being a familiar exposition of the principles of chemistry and physics, in their application to the arts and comforts of life. London, 1835. 8vo. Ill.

> Catechismo di chimica elementare, ossia espozione dialogica dei principii di chimica e di fisica colle loro applicazioni alle arti ed al bisogni della vita illustrato con 134 incisioni sul legno intercalate nel testo. Traduzione dall' inglese di G. Gorini. Milano, 1858. 8vo.

The Toxicologist's Guide : a new manual on poisons, giving the best methods of manipulation to be pursued for their detection, postmortem or otherwise. London, 1866. pp. viii–73, 12mo. Ill.

HORSTMANN, A.

> Physikalische und theoretische Chemie. *See* Graham-Otto.

HORSTMANN, A.

Theoretische Chemie einschliesslich der Thermochemie. Braunschweig, 1885.

Ueber den Zusammenhang zwischen dem Wärmewerth und dem Verlauf chemischer Reactionen. Heidelberg, 1884.

HOSAEUS, A.

Grundriss der Chemie. Nach methodischen Grundsätzen unter Berücksichtigung gewerblicher und . landwirthschaftlicher Verhältnisse sowie der neueren Ansichten der Wissenschaft zum Schulgebrauche zusammengestellt. Zweite vermehrte Auflage. Hannover, 1878. 8vo. Ill.

> Dritte Auflage. Hannover, 1884. 8vo. Ill.
>
> Методическій курсъ неорганической химіи. Перев. Крыловъ А. Москва 1873—75.

Grundzüge der Agriculturchemie. Heidelberg, 1878.

Leitfaden für praktisch-chemische Uebungen. Helmstadt, 1875.

Vorschule der Chemie. Leipzig, 1876.

HOULSTON, THOMAS.

Observations on Poisons . . . New edition, with additions, amendments and an appendix. Edinburgh and London, 1787. 8vo.

HOUSSAYE, J. G.

Monographie du thé, Description botanique, torrefaction, composition chimique, propriétés hygiéniques de cette feuille. Paris, 1843. 8vo.

HOUSTON, E. J.
The Elements of Chemistry; for the use of schools, academies and colleges. Philadelphia, 1883. 8vo.

HOWARD, J.
Practical Chemistry for use in science classes and higher and middle-class schools. New York, 1873. pp. 140, 12mo. Ill.
Putnam's Elementary Science Series.

HOYER, EGBERT.
Die Fabrikation des Papiers. *See* Bolley's Handbuch.

HRUSCHAUER, FRANZ.
Elemente der medicinischen Chemie und Botanik. Gratz, 1839. 8vo.

HUARD, AUGUSTE.
Traité comparé de chimie organique . . . Paris, 1852. 12mo.
Epitome (An) of Chemistry, theoretical and practical, intended for the use of candidates for the degree of "Bachelor in Science" and serving also as a summary of chemistry for the professor and student. Translated with additions by John B. Fessenden. Boston and Cambridge, 1857. pp. xii–127, 12mo.

HUBERT, ALOYS VON.
Anleitung durch Colorimetrie den Kupferhalt von Erzen- und Hütten-Producten schnell und genau zu ermitteln. Wien, 1852. pp. xvi-58, 8vo.

HUBERT, P.
Phosphates (Les) de chaux naturels. Recherche des gisements, essais chimiques, extraction, emplois dans l'industrie, phosphates indus-triels, superphosphates. Paris, 1892. 8vo. Ill.

HUE, F.
Le pétrole, son histoire, ses origines, son exploitation dans tous les pays du monde. Paris, 1885. 12mo.

HUEBBENET, C. DE.
De acido arsenicoso maximeque ejus cum toxicologia et medicina publica ratione. Dorpati, 1847. 8vo.

HÜFNER, G.
Ueber die Entwicklung des Begriffs Lebenskraft und seine Stellung zur heutigen Chemie. Tübingen, 1873.

HÜLSMANN, J.
De suiker uit een natuurkundig, technisch en œconomisch oogpunt beschouwd. 's Gravenhage, 1885.

HÜNEFELD, FRIEDRICH LUDWIG.

Chemie (Die) der Rechtspflege oder Lehrbuch der polizeigerichtlichen
Chemie.   Berlin, 1832.   pp. xxxii–603, 8vo.

Chemie und Medicin in ihrem engeren Zusammenwirken, oder Bedeutung der neueren Fortschritte der organischen Chemie für erfahrungsmässige und speculative ärztliche Forschung, als vollständige
Lehrschrift für die Studien der organischen Chemie überhaupt,
insbesondere aber für die im Gebiete der Medicin und Pharmacie
so wie für die Fortschritte der Heilmittellehre.   Berlin, 1841.   2
vols., 8vo.

Chemismus (Der) in der thierischen Organisation.   Physiologisch-
chemische Untersuchungen der materiellen Veränderungen oder
des Bildungslebens im thierischen Organismus, insbesondere des
Blutbildungsprocesses, der Natur der Blutkörperchen und ihrer
Kernchen.   Ein Beitrag zur Physiologie und Heilmittellehre.
Gekrönte Preisschrift.   Leipzig, 1840.   pp. xvi–269, 8vo.   Plate.

Physiologische Chemie des menschlichen Organismus, zur Beförderung
der Physiologie und Medicin.   Leipzig, 1826–27.   2 parts, 8vo.

HUGHES, SAMUEL.

Gas-Works, their construction and arrangement.   London, 1853.
Seventh edition, with important additions.   London, 1885.
12mo.
Eighth edition, revised, with notices of recent improvements.
London, 1892.   12mo.   Ill.

HUGOUNENQ, L.

Les alcaloïdes d'origine animale.   Paris, 1886.

Traité des poisons ; hygiène industrielle, chimie légale.   Paris, 1890.
8vo.

HUGUENY, F.

Recherches expérimentales sur la dureté des corps et spécialement sur
celle des métaux.   Strasbourg, 1865.   8vo.   6 plates.

Recherches sur la composition chimique et les propriétés qu'on doit
exiger des eaux potables.   Strasbourg, 1865.   pp. xiii–166, 8vo.

HUIZINGA, D.

Handleiding bij het eerste onderwijs in scheikunde aan burgerscholen.
Groningen, 1869.   8vo.
Tweede druk.   Groningen, 1878.

Handleiding bij de chemische oefeningen in het physiologisch laboratorium der Rijks-Universiteit te Groningen.   Groningen, 1891.
8vo.

HUMBOLDT, ALEXANDER VON.
   *Versuche über die chemische Zerlegung des Luftkreises und über
      einige andere Gegenstände der Naturlehre. Braunschweig, 1799.
      pp. [iv]–258, 12mo. 2 plates.

HUME, GILBERT LANGDON.
   Chemical Attraction, an Essay in five chapters, with an historical intro-
      duction. Cambridge, 1835. pp. viii–175–[ii], 8vo.
         The history occupies pp. 1–84.

HUMMEL, J. J.
   The Dyeing of Textile Fabrics. London, 1885. pp. xii–534, 12mo.
         Die Farberei und Bleicherei der Gespinnstfasern. Uebersetzt
            von E. Knecht. Berlin, 1888.

HUNT, ROBERT.
   Photography ; a treatise on the chemical changes produced by solar
      radiation, and the production of pictures from nature, by the
      Daguerreotype, calotype, and other photographic processes, with
      additions by the American editor. New York, 1851. 12mo.
   Researches on Light ; an examination of all the phenomena connected
      with the chemical and molecular changes produced by the influence
      of the solar rays, embracing all the known photographic processes
      and new discoveries in the art. With a plate and woodcuts.
      London, 1844. 8vo.
         Researches on light and its chemical relations, embracing a
            consideration of all the photographic processes. Second
            edition. London, 1854. 8vo.

HUNT, THOMAS STERRY.
   A New Basis for Chemistry : A Chemical Philosophy. Boston, 1887.
      pp. viii–165, 12mo.
         Un système de chimie nouveaux. Traduit de l'anglais par
            Spring. Paris, 1889.
   On the Chemistry of the Earth. Annual Report of Smithsonian Insti-
      tution for 1869. Washington, D. C., 1871. 8vo.

HURCOURT, E. ROBERT D'.
   De l'éclairage au gaz. Développements sur la composition des gaz
      destinés a l'éclairage sur la construction des fournaux et des
      cheminées sur la pose des tuyaux, sur les phénomènes de la
      lumière, etc. Paris, 1863. pp. xi–521, 8vo. Plates.

HUREAUX, JEAN PIERRE.

Histoire des falsifications des substances alimentaires et médicamenteuses précéde d'une instruction élémentaire sur l'analyse.   Paris,
1855.   8vo.

HUSBAND, H. A.

Aids to the Analysis of Food and Drugs.   London, 1885.   12mo.

HUSEMANN, AUGUST.

Grundriss der reinen Chemie.   Als Lehrbuch für Realschulen, Lyceen
und technische Lehranstalten sowie als Repetitorium für Studirende
der Medicin und Pharmacie bearbeitet.   Berlin, 1868.   8vo.
Grundriss der unorganischen Chemie.   Zweite Auflage.   Berlin, 1877.

HUSEMANN, AUGUST, A. HILGER UND TH. HUSEMANN.

Die Pflanzenstoffe in chemischer, physiologischer, pharmacologischer
und toxikologischer Hinsicht.   Zweite Auflage.   Berlin, 1882–84.
2 vols., 8vo.

HUSEMANN, TH., UND A. HUSEMANN.

Handbuch der Toxicologie.   Im Anschlusse an die zweite Auflage von
A. W. M. van Hasselts Handleiding tot de vergiftleer für Aerzte
und Apotheker.   Berlin, 1862.   pp. x–978, roy. 8vo.

> *See* Hasselt, A. W. M. van.

Supplementband zu Th. und A. Husemann's Handbuch der Toxikologie.   Bearbeitet von Th. Husemann.   Berlin, 1867.   pp. 187, 8vo.

HUSSON, C.

Le lait, la crème et le beurre, au point de vue de l'alimentation de
l'allaitement naturel, de l'allaitement artificiel et de l'analyse
chimique.   Paris, 1878.   pp. vii–252, 12mo.
Du vin, ses propriétés, sa composition, sa préparation, ses maladies et
les moyens de les guérir, ses falsifications et les procédés usités
pour les reconnaître.   Paris, 1877.   pp. 204, 12mo.

HUSSON, H. M., ADELON, PELLETIER, CHEVALLIER, ET CAVENTOU.

Rapports sur les moyens de constater la présence de l'arsenic dans les
empoissonnements par ce toxique, au nom de l'Académie royale
des sciences sur le même sujet, par MM. Thénard, Dumas, Boussingault et Regnault, et d'une réfutation des opinions de MM.
Magendie et Gerdy sur cette question par Orfila.   Paris, 1842.   8vo.

HUTH, E.

Das periodische Gesetz der Atomgewichte und das natürliche System
der Elemente.   Frankfurt a. O., 1884.   8vo.

HUTH, E. [Cont'd.]
> Zweite Auflage. Breslau, 1887.
>> Sammlung naturwissenschaftlicher Vorträge.

HYATT, JAMES.
> The Elements of Chemistry ; embracing the general principles, applications of the science, recent discoveries and present theories. New York, 1855. 8vo.

HYLTÉN-CAVALLIUS, CARL E.
> Elementar-Kurs i oorganisk Kemi. Stockholm, 1854. pp. xxii–519.

ИЛЬЕНКОВЪ, В.
> Курсъ химич. технологіи. С.-Петербургъ 1862. 2 тома.
>> IL'ENKOF V. Course of chemical technology. St. Petersburg, 1862. 2 vols.

ИЛЬИНСКИ, А.
> Таблицы справочныя Фармакологическія и токсикологическія, составленныя по Эстерлену, Trousseau а Pictoux и Giacomini съ присовокупленіемъ нѣкоторыхъ собственныхъ наблюденій А. Ильинскимъ. Въ двухъ частяхъ. С.-Петербургъ 1858. 16°
>> IL'INSKY, A. Toxicological Tables. St. Petersburg, 1858. 16mo.

IMHOF, MAXIMUS VON.
> Anfangsgründe der Chemie zum Gebrauche für Vorlesungen. München, 1803. 8vo.

IMISON, JOHN.
> Elements of Science and Art, being a familiar introduction to natural philosophy and chemistry, together with their application to a variety of elegant and useful arts. A new edition considerably enlarged and adapted to the improved state of science, by Thomas Webster. London, 1822. 2 vols., 8vo.

INDAGINE, INNOCENTIUS LIBORIUS AB.
> A pseudonym of Johann Ludolph Jaeger, *q.v.*

INGENHOUSZ, JAN.
> Account of a new kind of Inflammable Air or Gas. London, 1779. 4to.
> Easy Methods of measuring the diminution of bulk taking place on the mixture of common and nitrous air, with experiments on platine. London, 1776. 4to.
> Experiments upon Vegetables discovering their great power of purifying the Common Air in the sunshine and of injuring it in the shade and at night ; to which is joined a new method of examin-

INGENHOUSZ, JAN. [Cont'd.]

    ing the accurate degree of salubrity of the atmosphere.   London,
    1779.   pp. lxviii–302–[xvii].   One plate.

        Uitslag der proefnemingen op de planten, strekkende ter
        ontdekking van derselver zonderlingen eigenschap om
        de gemeene lugt te zuiveren op plaatsen, waar de zon
        schynt, en dezelve te bederven in de schaduwe, en
        gedurende den nagt. Amsterdam, 1780. 8vo.

INTRODUCTION TO PRACTICAL ORGANIC CHEMISTRY.   With references
    to the works of Davy, Brande, Liebig.   London, 1844.   12mo.

Введеніе къ изученію химіи. С.-Петербургъ 1876.

    Introduction to the study of Chemistry.   St. Petersburg, 1876.

IRVINE, WILLIAM.

    Essays, chiefly on Chemical Subjects.   London, 1805.   pp. xxxi–490.
    8vo.

IRVING.

    Catechism of Chemistry.   Exhibiting a concise view of the present
    state of science.   New edition revised by R. J. Mann.   London,
    1876.   18mo.

ISAMBERT, FERDINAND.

    Précis de chimie à l'usage des élèves de l'enseignement secondaire spé-
    cial, des candidats aux baccalauréats et des élèves de l'enseigne-
    ment primaire supérieur.   Paris, 1885.

Изложеніе, Полное, гальванопластики, гальванической позолоты и сбереженія
    составленное по новѣйшимъ источникамъ съ присовокупленіемъ способа
    золоченія и серебренія электро-химики, безъ посредства гальваноп ла-
    стической батареи. Съ чертежами и таблицею показывающею отношеніе
    французскихъ метрическихъ мѣръ къ русскимъ. С. Петербургъ 1844.
    2 т.

    Complete Treatise on Electro-metallurgy.   St. Petersburg, 1844.

ИВАНОВЪ.

    Начальныя основанія аналитической химіи. Учебное руководство. 3 части.
    С.-Петербургъ 1854. 8°

    IVANOF. Elements of Analytical Chemistry.   St. Petersburg, 1854.

ИВАНОВЪ, С.

    Къ вопросу о сѣроводородѣ въ крови. Харьковъ 1873.

    IVANOF, S. Contribution to the question of sulphuretted hydrogen in the
    blood. Kharkov, 1873.

ISRAEL, JOHANNES FRIDERICUS CHRISTIANUS.
De chemicorum instrumentis mechanicis, errorum et dissensus fontibus.
Vitebergæ, *n.d.* [1783]. pp. 24, 4to.

ITALY. MINISTERO DELL' INTERNO. DIREZIONE DELLA SANITÀ PUBBLICA.
Regolamento speciale per la vigilanza igienica sugli alimenti, sulle be-
vande e sugli oggetti d'uso domestico. Roma, 1890. 8vo.

ITTNER, I. von.
Beiträge zur Geschichte der Blausäure mit Versuchen über ihre Ver-
bindungen und Wirkungen auf den thierischen Organismus.
Freyburg und Konstanz, *n. d.* [1809]. 8vo. pp. [viii]–150.

IZARN, JOSEPH.
Explication du nouveau langage des chimistes pour tous ceux qui, sans
s'occuper de la science, voudroient profiter de ses découvertes.
Paris, An XII, [1803.] pp. xi–148. 8vo.

JACKSON, CHARLES LORING.
Laboratory Experiments in Chemistry. (A course given at Harvard
College.) 1885–86. Cambridge, 1885. 16mo.

JACOB, J.
Traité élémentaire de chimie expérimentale et appliquée, suivi d'une
méthode d'analyse pour reconnaître les métaux, les bases, les
acides, les sels et les principaux corps que l'on rencontre dans
l'industrie et dans la nature. Illustré de 200 belles gravures sur
bois, intercalées dans le texte, accompagné d'une notice biographique et bibliographique sur les principaux chimistes auxquels
la science doit ses progrès . . . Paris, 1867. 8vo.

JACOBI, V.
Bewährte und umfassende Anleitung die Runkelrübe zum Behufe der
Zuckerfabrikation zu cultiviren. Leipzig, 1837.

JACOBS, WILLIAM STEPHEN.
The Student's Chemical Pocket Companion. Philadelphia, 1807.
12mo.

JACOBSEN, G. J.
Handboek der technologie, of wetenschappelijke beschrijving en ver-
klaring van alle fabriekmatige verrigtingen. Ten gebruike bij
technologische lessen en tot eigen onderrigt. Naar het Hoogduitsch van Barentin, Wagner en anderen, voor Nederland bewerkt door G. J. Jacobsen met een voorberigt van M. J. Cop.
Deventer, 1859. 8vo.

JACOBSEN, O.

Die Glycoside. Breslau, 1887. 8vo.

JACOBSOHN, H.

Untersuchungen über Muawin. Dorpat, 1892. 8vo.

JACQUEMIN, E. T.

De la putréfaction au point de vue de la chimie, de la physique, de la toxicologie. Thèse. Strasbourg, 1854. 4to.

JACQUEMYNS, E.

Handboekske der chemie. Gent, 18—. 12mo.

JACQUIN, JOSEPH FRANZ VON.

Lehrbuch der allgemeinen und medicinischen Chymie. Zum Gebrauche seiner Vorlesungen. Wien, 1793. 2 vols., 8vo.
>   Vierte umgearbeitete Auflage. Wien, 1810–12.
>   Elementa chemiæ universalis et medicæ ex lingua germanica versa. Viennæ, 1793. 2 vols.
>   Elements of Chemistry. Translated from the German [by Henry Stutzer]. London, 1799. pp. xi–405, 8vo.

JACQUIN, NIKOLAUS JOSEPH VON.

Anfangsgründe der medicinisch-practischen Chymie, zum Gebrauche seiner Vorlesungen. Zweyte Auflage. Wien, 1785. pp. [xvi]–526–[xviii]. 8vo.

Examen chemicum doctrinæ Meyerianæ de acido pingui et Blackianæ de aëre fixo, respectu calcis. Vindobonæ, 1769. 8vo.
>   Cf. Cranz, H. J. N., and see Meyer, Johann Friedrich.

JAEGER, JOHANN LUDOLPH [Indagine, Innocentius Liborius ab].

Chemisch-physikalische Nebenstudien oder Betrachtungen über einige nicht gemeine Materien . . . Hof, 1780.

Memorabilia bismuthi, das ist chemisch-physikalische Abhandlung zu näherer Kenntniss des annoch ziemlich unbekanten Minerals welches Wissmuth und Magnesia wie auch antimonium foemininum gennenet wird . . . Nürnberg, 1782. pp. 358–[ii], 12mo.

Trifolium chemico-physico-salinum, oder Dreyfache chemisch-physikalische Abhandlung, worinnen Drey berühmte Salze, namentlich : Salmiac, Salpeter und Borax nach ihrer Natur und Wesenheit . . . betrachtet werden. Amsterdam und Leipzig, 1771. 8vo.

JAGNAUX, RAOUL.

Aide-Mémoire du chimiste. Chimie organique et chimie inorganique. Documents chimiques, documents physiques, documents minéralogiques, documents mathématiques. Paris, 1890. 8vo. Ill.

JAGNAUX, RAOUL. [Cont'd.]

Analyse chimique des substances commerciales, minérales et organiques. Liège, *n. d.* [1888]. pp. 946, 8vo. Ill.

Traité de chimie générale analytique et appliquée. Paris, 1887. 4 vols., pp. 2200, 8vo. Ill.

Traité pratique d'analyses chimiques et d'essais industriels. Paris, 1884. 12mo.

JAGO, WILLIAM.

Inorganic Chemistry, theoretical and practical. With an introduction to the principles of chemical analysis, inorganic and organic. Eighth edition, London, 1886. 12mo.

Ninth edition, rewritten and greatly enlarged. London and New York, 1889. 12mo. Ill.

The Chemistry of Wheat, Flour, and Bread ; and technology of bread-making. Brighton, 1886. pp. 474, 8vo.

JAHN, HANS.

Elektrolyse (Die) und ihre Bedeutung für die theoretische und angewandte Chemie. Wien, 1883. pp. viii–206, 8vo.

Grundsätze (Die) der Thermochemie und ihre Bedeutung für die theoretische Chemie. Wien, 1882. pp. viii–238, 8vo.

Zweite Auflage. Wien, 1892.

JAMES, J. WILLIAM.

Notes on the Detection of the Acids (Inorganic and Organic) usually met with in analysis for the use of laboratory students. London, 1883. 12mo.

JAMIESON, T.

Inorganic Chemistry, theoretical and practical. First Course. Aberdeen, 1874.

JAMIN.

Leçon (etc.). *See* Société chimique de Paris.

JANEČEK, GUSTAV.

Leitfaden für die praktischen Uebungen in der qualitativen chemischen Analyse unorganischer Körper. Wien, 1879. 8vo.

JANNASCH, PAUL.

Gesammelte chemische Forschungen. Göttingen, 1888. Band I. pp. x–270, 8vo.

"Nicht für den Buchhandel bestimmt."

JANOVSKY, J. V.

Anleitung zur qualitativen Analyse unorganischer und organischer Körper.   Hilfsbuch bei den praktischen Arbeiten in gewerblichen und technischen Laboratorien.   Prag, 1882.
> Zweite Auflage.   Prag, 1891.   8vo.

JAPING, EDUARD.

Darstellung (Die) des Eisens und der Eisenfabrikate.   Handbuch für Hüttenleute und sonstige Eisenarbeiter, für Techniker, Händler mit Eisen und Metallwaaren, für Gewerbe- und Fachschulen.   Wien, Pest, Leipzig, 188-.   Ill.

Elektrolyse, (Die) Galvanoplastik und Reinmetallgewinnung.   Mit besonderer Rücksicht auf ihre Anwendung in der Praxis.   Wien, Pest, Leipzig, 1883.   pp. viii-260, 8vo.   Ill.
> > L'Électrolyse, la galvanoplastie et l'électrométallurgie.   Édition française par Ch. Baye et G. Fournier.   Paris, 1885.   16mo.   Ill.

JARMAIN, GEORGE.

Systematic Course of Qualitative Analysis arranged in tables.   Fifth edition.   London, n. d. [1884].   8vo.

JAYS, L.

Problèmes de physique et de chimie choisis parmi les sujets de compositions proposés dans les concours et par les diverses facultés dans ces dernières années.   Paris, 1886.   8vo.

JEAN, FERDINAND.

Chimie analytique des matières grasses.   Paris, 1892.   8vo.   Ill.

Méthodes chimiques pour la recherche des falsifications, l'essais, l'analyse des matières fertilisantes.   Paris, 1874.   18mo.

JEFE.

Tratado elemental de química analítica, precedido de algunas ideas sobre la filosofía química.   Lecciones explicadas en la Escuela Especial de Minas, por el Ingeniero Jefe.   Madrid, 1867.   8vo.   Ill.

JENNINGS, CHARLES GODWIN.

Practical Urine Testing.   A guide to office and bedside urine analysis for physicians and students.   Detroit, 1887.   pp. 124, 12mo, Ill.

JERVIS, G.

Guida alle acque minerali d'Italia, cenni storici e geologici coll' indicazione delle proprietà fisiche, chimiche e mediche delle singole sorgenti, corredata di analisi chimiche, raccolte ed ordinate in 12 specchi.   Provincie centrali.   Torino e Firenze, 1868.   8vo.

JESPERSEN, H.
    Kortfattet Lærebog i uorganisk Kemi.   Et Grundlag for den mundt·
    lige Undervisning.   Kjøbenhavn, 1874.   8vo.

JESSEN, P.
    Kortfattet uorganisk Kemi.   Andet Oplag.   Trykt som Manuskript.
    Kjøbenhavn, 1878.   8vo.

JETTEL, W.
        Die Zundwaaren-Fabrikation.   *See* Bolley's Handbuch [etc.].

JIMINEZ, E.
    Escuela de labradores.   Nociones de química agricola.   Madrid, 1878.
    8vo.

JOANNIS.
        See, in Section II, Fremy: Encyclopédie chimique, vol. II, vol. III, etc.

JOCLÉT, VICTOR.
    Chemische (Die) Bearbeitung der Schafwolle oder das ganze der Fär-
        berei von Wolle und wollenen Gespinnsten.   Ein Hilfs- und Lehr-
        buch für Färber, Färberei-Techniker, Tuch und Garn-Fabrikanten
        und solche die es werden wollen.   Dem heutigen Standpunkte
        der Wissenschaft entsprechend und auf Grund eigener langjährigen
        Erfahrungen im In- und Auslande vorzugsweise praktisch darge-
        stellt.   Wien, Pest, Leipzig, 188–.   Ill.
    Kunst- (Die) und Feinwäscherei in ihrem ganzen Umfange.   Enthaltend:
        Die chemische Wäsche, Fleckenreinigungskunst, Kunstwäscherei,
        Hauswäscherei, die Strohhut-Bleicherei und Färberei, Handschuh-
        Wäscherei und Färberei.   Zweite Auflage.   Wien, Pest, Leipzig,
        1888.   Ill.
    Woll- (Die) und Seidendruckerei in ihrem ganzen Umfange.   Ein prak-
        tisches Hand- und Lehrbuch für Druckfabrikanten, Färber und
        technische Chemiker.   Enthaltend: das Drucken der Wollen-,
        Halbwollen- und Halbseidenstoffe, der Wollengarne und seidenen
        Zeuge.   Unter Berücksichtigung der neuesten Erfindungen und
        unter Zugrundelegung langjähriger praktischer Erfahrungen.
        Wien, Pest, Leipzig, 188–.   Ill.
    Vollständiges Handbuch der Bleichkunst oder theoretische und prak-
        tische Anleitung zum Bleichen der Baumwolle, des Flachses, des
        Hanfes, der Wolle und Seide, sowie der daraus gesponnenen Garne
        und gewebten oder gewirkten Zeuge.   Nebst einem Anhange
        über zweckmässiges Bleichen der Hadern des Papieres, der Wasch-
        und Badeschwämme, des Strohes und Wachses.   Nach den neues-
        ten Erfahrungen durchgängig praktisch bearbeitet.   Wien, Pest
        und Leipzig, 188–.   Ill.

JÖRGENSEN, S. M.

Kortfattet Kemi til Brug for Skoler.   Kjøbenhavn, 1874.   8vo.

      Kemiens Begyndelsesgrunde til Brug for Skoler.   For-
kortet Udgave af samme Forfatters "Kortfattet Kemi
til Brug for Skoler."   Kjøbenhavn, 1876.   8vo.

Vejledning i de uorganiske Stoffers quantitative Analyse.   Kjøbenhavn,
1870.   8vo.

Das Thallium.   Eine Zusammenstellung der vorhandenen Beobachtungen.   Heidelberg, 1871.   8vo.

Lærebog i organisk Kemi.   Kjøbenhavn, 1880.

Mindre Lærebog i uorganisk Kemi.   Kjøbenhavn, 1888.   8vo.

    See Gmelin, Leopold.

JOFFROY, J.

Leçons élémentaires de chimie.   Enseignement secondaire spécial.
Paris, 1868.   8vo.

JOHN, JOHANN FRIEDRICH.

Chemische Tabellen der Pflanzenanalysen, oder Versuch eines systematischen Verzeichnisses der bis jetzt zerlegten Vegetabilien nach
den vorwaltenden näheren Bestandtheilen geordnet und mit Anmerkungen versehen.   Nürnberg, 1813.   Fol.

Chemische Tabellen des Thierreichs.   Oder systematische Uebersicht
der Resultate aller bis jetzt zerlegten Animalien mit Rücksicht auf
die wichtigsten medicinischen Thatsachen, welche aus der Chemie
entlehnt sind ; einige wichtige chemische Erscheinungen der
Zoochemie und Eigenschaften der animalischen Körper, und die
Literatur.   Berlin, 1814.   Fol.

Chemische Untersuchungen mineralischer, vegetabilischer und animali-
scher Substanzen.   Fortsetzung des chemischen Laboratoriums.
Berlin, 1810–1821.   5 vols., 12mo.

    Vols. III–V *also with the title :* Chemische Schriften.   Vol. I, 1810 ; II,
1811 ; III, 1813 ; IV, 1816 ; V, 1821.

Chemisches Laboratorium, oder Anweisung zur chemischen Analyse
der Naturalien.   Nebst Darstellung der nöthigsten Reagenzien.
Mit einer Vorrede von Martin Heinrich Klaproth.   Berlin, 1808.
pp. xii–522, 12mo.   2 folding plates.

    For continuation, *see* Chemische Untersuchungen, *above.*

Naturgeschichte des Succins . . .   Cöln, 1816.   2 vols., 8vo.

Ueber die Ernährung der Pflanzen und den Ursprung der Potassche.
Berlin, 1819.   8vo.

Versuch einer Methode zur Untersuchungen der Mineralwasser.
Moskau und Leipzig, 1805.

    See also in Section II.

JOHN, JOSEPH.
Die Schule der Gährungschemie in Anwendung auf Bierbrauerei, Bieruntersuchungen und Spirituserzeugung. Zum wissenschaftlichen Selbstunterrichte für Bierbrauer, Branntweinbrenner und Steuerbeamte leichtfasslich dargestellt. Zweite vermehrte Auflage. Prag, 1856. 8vo.

JOHNSON, A. E.
The Analysts' Laboratory Companion. London, 1888. 8vo.

JOHNSON, CUTHBERT WILLIAM.
Agricultural Chemistry for young farmers. London, 1843. 12mo.
Use of Salt for agricultural purposes. London, 183-. 8vo.
      Observations sur l'emploi du sel en agriculture et en horticulture, avec les conseils fondés sur l'expérience. Treizième édition. Londres, 1838. 12mo.
      Deuxième édition française. Paris, 1846. 12mo.

JOHNSON, G.
On the Various Modes of Testing for Albumen and Sugar in the Urine Two lectures. London, 1884. 12mo.

JOHNSON, GEORGE WILLIAM.
Chemistry (The) of the World, being a popular explanation of the phenomena daily occurring in and around our persons, houses, gardens and fields. London, 1858. 8vo. Ill.

JOHNSON, SAMUEL W.
Chemical Notation and Nomenclature, old and new, with rules for converting old system formulæ into new. New York, 1871. 8vo.
How Crops Grow. New York, 1868.
      How Crops Grow, the Chemical Composition, Structure and Life of the Plant, with Tables of Analysis, revised with numerous additions, and adapted for English use by Church and Thisleton Dyer. London, 1869. 8vo. Ill.
      Préceptes de chimie agricole, à l'usage des jeunes fermiers. Traduits de l'anglais, d'après la cinquième édition, par Louis Léonzon. Paris, 1870. 8vo.
      Also translated into Russian by N. R. Temashev. St. Petersburg, 1873.
Lectures on Agricultural Chemistry delivered at the Smithsonian Institute [sic] December, 1859. n. p. n. d. pp. 72, 8vo.

JOHNSTON, JOHN.
A Manual of Chemistry on the basis of Dr. Turner's Elements of Chemistry, containing, in a condensed form, all the most important

JOHNSON, JOHN. [Cont'd.]

facts and principles of the science. Designed for a text-book in colleges and other seminaries of learning. Middletown, 1840. pp. 453, 8vo. Ill.

A new edition, Philadelphia, 1842. 8vo.

Cf. Turner, Edward, Elements of Chemistry.

JOHNSTON.

Illustrations of Chemistry, edited by T. J. Menzies. Edinburgh, 1884. Fol.

JOHNSTON, JAMES FINLAY WEIR.

Catechism of Agricultural Chemistry and Geology. Edinburgh and London, 1844. 8vo.

Seventh edition, Edinburgh, 1845.

[The same.] New York, 1846.

New edition, enlarged, London, 1849.

Twenty-third edition, Edinburgh, 1854. 8vo.

A new edition, with an introduction by John Pitkin Norton. New York, 1856.

Katechismus i Agerdyrkningschemie og Jordbundslære. Oversat efter 24nde Udgave af E. Thomsen. Udgivet af det kongelige Landhusholdningsselskab, 1850. Andet Oplag. Kjøbenhavn, 1854. 8vo.

Femte Oplag, ved C. Barfoed, 1867.

Catechismus voor landbouwkundige scheikunde en aard- kunde, naar de 16 uitgave vertaald door J. R. E. von Laer. Rotterdam, 1848. 8vo.

Naar de 22 druk bewerkt ; tweede vermeerderde uitgave, Rotterdam, 1849. 8vo.

Catechismus der Agricultur-Chemie und Geologie für Landwirthe und Lehrer an Volksschulen, etc. Aus dem englischen übersetzt nach der 14. Original-Auflage. Dresden, 1847. 8vo.

Catechismo di geologia e di chimica agraria, tradotto sulla 14 edizione inglese da Giovenale Vegezzi-Ruscalla. Torino, 1847. 16mo.

Catechismus i Agerdyrkningens Chemie og Geologie oversat ved C. T. Christensen. Drammen, 1846. 8vo.

Katechizm rolniczy, oparty na zasadach chemii i geologii. Tłómaczony z angl. przez M. Oborskiego. Lwów, 1847. 8vo.

Katekes i Jordbrukets Kemi och Geologi. Öfversättning

JOHNSTON, JAMES FINLAY WEIR. [Cont'd.]

　　från femte engelska original upplagan af H. B. P.
　　Jönköping, 1846. 8vo.

　　Elfenau Amaethyddiaeth a Daeardraith . . . Rhan 1.
　　Dosbarth wyddorol. Dinbych, 1851. 12mo.

Chemistry (The) of Common Life. Edinburgh and London [1853–55].
　　2 vols., 8vo. Also, New York, 1855. 2 vols., 12mo.

　　New edition, revised and brought down to the present time
　　by G. H. Lewis. Edinburgh, 1859. 2 vols., 8vo.

　　Another edition, revised and brought down to the present
　　time by A. H. Church. Edinburgh and London, 1879.
　　8vo. Also, New York, 1880. pp. xxvi–592. 8vo. Ill.

　　Hwa hio wei sheng lun. [Translated into Chinese by John
　　Fryer.] Shanghai. 4 vols.

　　Reprinted from the Chinese Scientific and Industrial Magazine. *Cf.*
　　Fryer, John.

　　De scheikunde in het dagelijksch leven. Voor Nederlan-
　　ders bewerkt door J. W. Gunning. Sneek, 1855–56.
　　3 parts, 8vo. Tweede uitgave, Sneek, 1865.

　　Die Chemie des täglichen Lebens. Deutsch bearbeitet von
　　Th. O. G. Wolff. Berlin, 1854. 2 vols.

　　Neu bearbeitet von Dornblüth. Stuttgart, 1887.

　　Chemische Bilder aus dem täglichen Leben. Für Frauen
　　bearbeitet von S. Augustin. Leipzig, 1856. 2 vols.,
　　8vo. Zweite Ausgabe. Leipzig, 1858. 8vo.

　　Die Chemie des gewöhnlichen Lebens. Aus dem englischen
　　übersetzt. Cassel, 1855. 16mo.

　　Chemische Bilder aus dem täglichen Leben. Aus dem
　　englischen von Wilhelm Hamm. Leipzig, 1855. 8vo.

　　Chemische Bilder aus dem Alltagsleben. Nach der eng-
　　lischen dritten Ausgabe. Leipzig, 1870. 8vo.

　　CSENGERY ANTAL. Vegytani képek a közéletböl Johnston
　　nyomán. Elsö és második kötet. Pest, 1857. 8vo.

　　Hvardagslifvets Kemi. Öfversättning af Gustav Thomée.
　　Stockholm, 1855. 8vo.

　　Kemiska Bilder ur dagliga iifvet. Efter J. F. W. Johnston's
　　"Chemistry of common life" dels öfversättning, dels
　　bearbetning af A. Första delen. Med flera kartor och
　　en mängd fina träsnitt. Ørebro, 1854. 8vo.

　　Химическія свѣдѣнія предметовъ изъ вседневной жизни. Перев.
　　Ходнева. С.-Петербургъ 1858.

Elements of Agricultural Chemistry and Geology. Edinburgh and
　　London, 1842. 8vo.

JOHNSTON, JAMES FINLAY WEIR. [Cont'd.]

Seventh edition, London, 1856. 12mo. Also, Boston, 1853.

Twelfth edition, by J. F. W. J. and C. A. Cameron. Edinburgh, 1881. 8vo.

Thirteenth edition, by J. F. W. J. and C. A. C. London, 1883.

Grondbeginselen der landbouwkundige scheikunde en geologie. Naar den derden druk vertaald door P. F. H. Fromberg. Met eene voorrede van G. J. Mulder. Rotterdam, 1844. 8vo.

Grondbeginselen der landbouwkundige scheikunde en geologie. Naar den vierden druk, bewerkt door J. R. E. van Laer. Met eene voorrede van G. J. Mulder, tweede geheel omgewerkte en vermeerderde uitgave. Rotterdam, 1849. 8vo.

Éléments de chimie agricole et de géologie. Traduits de l'anglais par F. Exschaw, et revus par J. Rieffel. Paris, 1845. 12mo. Deuxième édition. Paris, 1849. 12mo. Troisième édition. Liège, 1849. 16mo.

Anfangsgründe der praktischen Agrikultur-Chemie und Geologie. Aus dem englischen . . . mit einem Vorworte von . . . F. Schulze. Neubrandenburg, 1845. 8vo. .

De første Grunde i Agerbrugs-Kemi og Geologi, letfattelingen fremstillet. Oversat efter det Svenske. Fredrikshald, 1848. 8vo.

Experimental Agriculture, being the results of past, and suggestions for future experiments in scientific and practical agriculture. Edinburgh and London, 1849. 8vo.

Instructions for the Analysis of Soils. Edinburgh and London, 1847. 8vo.

Third edition, London, 1855. Also, Cleveland, Ohio, 1855.

Handboek vor schei- en aerdkunde, op den landbouw toegepast. Uit het engelsch vertaeld. Antwerpen, 1847. 12mo.

Handboek over de landbouw-stofscheiding en over de kennis der aerde. Uit het engelsch vertaeld en vermeerderd met eenen oagslag op de aerdkundige gesteldheit van Belgie door M. Dumont. Brüssel, 1851. 18mo.

Lectures on Agricultural Chemistry and Geology. With an appendix containing suggestions for experiments in practical agriculture. Edinburgh and London, 1844. pp. xvi–116–xx, 8vo.

JOHNSTON, JAMES FINLAY WEIR. [Cont'd.]

> Second edition, Edinburgh and London, 1847. 8vo.
>
> [The same :] New York, 1842. pp. 10–255–40, 8vo.
>
> Forelæsninger over Agerdyrkningskemi og Jordbundslære. Oversat efter anden Udgave, 1847, af E. Thomsen. Kjøbenhavn, 1854. 8vo.
>
> Leerwyze van landbouwkundige chemie en geologie. Vrye navolging der fransche vertaling van F. André. Luik, 18—.
>
> Manuel de chimie agricole et de géologie. Traduit de l'Anglais. Augmenté d'un coup d'œil sur la constitution géologique de la Belgique, par M. Dumont. Bruxelles, 1850. 18mo.
>
> Landbrugets naturvidenskabelige Grunde, med specielt Hensyn til Chemi og Geologi. Oversat af P. Dørum, Første Del. Christiania, 1850. 8vo.
>
> Практическое руководство къ употребленію всѣхъ до нынѣ извѣстныхъ землеудобрительныхъ веществъ или туковъ. Съ 2-мя литографированными таблицами. С.-Петербургъ, 1846. 8°
>
> Landbrukets naturvetenskapliga Grunder, med hufvudsakeligt afseende paå Kemi och Geologi. Öfversättning af J. Th. Nathhorst. Stockholm, 1848–'52. 8vo.

JOHNSTON, JOHN.

> Manual (A) of Chemistry, on the basis of Turner's Elements of Chemistry. Seventh edition, revised. Philadelphia, 1870. 12mo.
>
> > *Cf.* Turner, Edward.

JOLLY, L.

> Les phosphates, leurs fonctions chez êtres vivants. Paris, 1887. pp. xii–584, 4to.

JOLY, A.

> > *See, in Section II,* Fremy : Encyclopédie chimique, Vol. II and Vol. III.

JOLY, A.

> Éléments de chimie. Troisième année. Paris, 1888.
>
> Cours élémentaire de chimie et de manipulations chimiques. Paris, 1888. 16mo.
>
> > Quatrième année. Paris, 1886. 8vo.
> >
> > Cinquième année. Paris, 1887.
> >
> > Sixième année. Paris, 1888.
> >
> > Huitième année. Paris, 1890. 8vo. Ill.
>
> Éléments de chimie pour les classes de rhétorique et de philosophie. Paris, 1885.

JOLY, ET FILHOL, E.

Recherches sur le lait.   Bruxelles, 1856.   4to.

JONAS, L. E.

Handbuch der Chemie, in welchem die anorganischen, organischen und organisirten Verbindungen dem neuesten Standpunkt der Wissenschaft entsprechend und des leichtern Ueberblicks wegen nach den Grundsätzen der dualistischen Ansicht, in zwei neben einander verlaufenden Rubriken deren eine die basischen die andere die sauren Verbindungen enthält, abgehandelt sind. Zum Gebrauch bei Lehrvorträgen sowie auch zum Selbststudium für Aerzte, Pharmaceuten, Techniker, Oekonomen, u. s. w.   Leipzig, 1846.   8vo.

Katechismus der Chemie.   Dritte gänzlich umgearbeitete dem gegenwärtigen Standpunkt der Wissenschaft entsprechende stark vermehrte Auflage.   Leipzig, 1840.   8vo.

> *Also under the title:* Lehrbuch der Chemie, in katechetischer Form methodisch-systematisch abgefasst.

JONES, CHAPMAN H.

Text-Book of Experimental Organic Chemistry for Students.   Reprinted from the last English edition.   New York, 1881.   pp. 145, 16mo.

JONES, EDWIN GODDEN.

Chemical Analysis of the Mineral Waters of Spa.   Liège, 1816.

JONES, FRANCIS.

The Owen's College junior Course of Practical Chemistry.   With a preface by Prof. Roscoe.   Third edition, revised and enlarged. London, 1875.   pp. viii–179, 16mo.   Ill.

JONES, HENRY BENCE.

Lectures on Some of the Applications of Chemistry and Mechanics to Pathology and Therapeutics.   London, 1867.   pp. vi–314.

On Animal Chemistry in its application to stomach and renal diseases. London, 1850.   8vo.

JONES, THOMAS P.

New Conversations on Chemistry.   With engravings, questions, experiments and a glossary.   On the foundation of Mrs. Marcet's "Conversations on Chemistry."   Philadelphia, 1834.   12mo.

> *Cf.* Marcet, Jane.

JORBAN, N.
Vergleichende Untersuchungen der wichtigeren zum Nachweis von Arsen in Tapeten und Gespinnsten empfohlenen Methoden. Dorpat, 1889.

JORDAN, SAMSON.
Album du cours de métallurgie professé à l'École centrale des arts et manufactures. 140 planches in-folio et un volume de texte. Paris, 1874–75. Fol. and 8vo.

JORGENSEN, ALFRED.
Die Mikroorganismen der Gährungsindustrie. Berlin, 1886. 8vo.
Zweite vermehrte und verbesserte Auflage. Berlin, 1890. 8vo.

JOULIE, HENRI.
Conférences sur les engrais chimiques appliqués à la culture de céréales et des betteraves. Paris, 1885.

JOULIN.
Les potasses et les soudes de Stassfurt (Prusse et Anhalt). Paris, 1866. 8vo.

JOYCE, FR.
Praktische Anleitung zur chemischen Analytik und Probierkunst der Erze, Metallgemische, Erden, Kalien, Mineralwasser, oder Grundzüge der mineralogischen Chemie. Aus dem englischen des F. J. mit Anmerkungen von J. Waldauf von Waldenstein. Wien, 1827. Ill.

JOYCE, JEREMIAH.
Dialogues in Chemistry intended for the Instruction and Entertainment of Young People ; in which the first principles of that science are fully explained. To which are added questions and other exercises for the examination of pupils. London, 1807. 18mo.
Fourth edition, corrected and newly arranged. London, 1822. 2 vols., 18mo. Vol. I, pp. viii–352 ; 4 plates. Vol. II, pp. 332 ; 4 plates.

JUCH, CARL WILHELM.
Europäische (Die) Zuckerfabrikation aus Runkelrüben. Augsburg, 1811.
Handbuch der Chemie für Fabrikanten . . . Nürnberg, 1807. 8vo.
Ideen zu einer Zoochemie systematisch dargestellt. Mit Zusätzen und einer Vorrede versehen von Joh. Bartholomä Trommsdorff. Erfurt, 1800. 8vo.

JUCH, CARL WILHELM. [Cont'd.]

System der antiphlogistischen Chemie nach den neuesten Entdeckungen
entworfen. Nürnberg, 1803. 8vo.

> Part I deals with theoretical chemistry. The work was not continued.

JUCH, JULIUS CARL.

Angewandte (Die) Chemie für Leser aus allen Ständen, insbesondere
für Pharmaceuten, Fabrikanten aller Art [etc.]. Augsburg, 1836.
8vo.

Anleitung zur Ausführung qualitativer chemisch-analytischer Untersuchungen anorganischer Körper. Für Studirende der Medicin,
Pharmacie und Chemie. Nach H. Rose's Handbuch in Tabellenform ausgeführt und mit der nothwendigen Erläuterung versehen.
Zweite unveränderte Ausgabe. Augsburg, 1836. 8vo.

JÜNEMANN, F.

Die Fabrikation des Alauns, der schewefelsauren und essigsauren
Thonerde, des Bleiweisses und Bleizuckers. Wien, 1882.

JÜPTNER VON JONSTORFF, H.

Praktisches Handbuch für Eisenhütten-Chemiker. Wien, 1885. 8vo.

> Traité pratique de chimie métallurgique. Traduit de
> l'allemand par E. Vlasto. Édition française revue et
> augmentée par l'auteur. Paris, 1891. 8vo.

JÜRGENSEN, C.

Procentische chemische Zusammensetzung der Nahrungsmittel des
Menschen graphisch dargestellt. Berlin, 1888. pp. 16, 4to.

JUGEL, JOHANN GOTTFRIED.

* Freyentdeckte Experimental-Chymie oder Versuch den Grund natürlicher Geheimnisse durch die Anatomie und Zerlegungskunst,
in dem astralischen, animalischen, vegetabilischen und mineralischen Reiche durch systematische Grundsätze, Lehrsätze, Beweise,
Gegensätze, Gegenbeweise, Anmerkungen, Versuche, Erfahrungen und darauf folgende Schlüsse nebst dem deutlichen Natur-
begriffe der metallischen Generation wie solche täglich in der Erde
getrieben wird, durch eine lange Untersuchung, also vorzustellen,
dasz es ein jeder Naturforschender einsehen und erkennen kann ;
In zwey Theile abgefasset, und zu jedermanns Nutzen und Vergnügen dem Drucke überlassen. Leipzig, verlegts Johann Paul
Krausze, Buchhändler, 1766. pp. [xiv]-368, sm. 8vo. Illustrated
title-page.

Sehr rare und wahrhafte chymische experimentirte Kunststücke. Berlin, 1758-63. 3 vols., 8vo.

JULIA-FONTENELLE, ÉTIENNE.
  Manuel de chimie médicale.   Paris, 1824.   8vo.
              Manuale di chimica medica.   Tradotto per cura e con
                aggiunte di G. Caproni.   Milano, 1825.   16mo.
              Compendio elemental de química aplicada á la medicina
                . . . traducido al castellano con muchas y nuevas adi-
                ciones por Jose Benito y Lentijo.   Madrid, (?)   2 vols.,
                4to.
  Manuel portatif des eaux minérales [etc.].   Paris, 1825.   8vo.
  Theoretisch-praktisches Handbuch der Oelbereitung und Oelreinigung
    nebst einer Darstellung der Gasbeleuchtung.   Aus dem franzö-
    sischen übersetzt von Gustav Heinrich Haumann.   Ilmenau, 1828.
    pp. xvi–344, 8vo.

JULIUS, PAUL.   *See* Noelting, E.

JULLIEN, C. E.
  La chimie nouvelle, ou le crassier [?] de la nomenclature chimique de
    Lavoisier.   Paris, 1870.   8vo.

JUNCKEN, JOHANN HELFRICUS.
  *Chimia experimentalis curiosa, ex principiis mathematicis demon-
    strata ; in qua ex triplici regno remedia generosiora a neotericis
    et aliis hactenus inventa fideliter exhibentur, adjunctis singulari-
    orum remediorum formulis . . .   Francofurti, 1681.   pp. [xx]–
    898, 12mo.   Plate and engraved title-page.
              Another edition, 1702.
        See also in Section II.

JUNCKER, JOHANNES.
  Conspectus chemiæ theoretico-practicæ in forma tabularum repræsenta-
    tus, in quibus physica, præsertim subterranea, et corporum natu-
    ralium principia, habitus inter se, proprietates, vires et usus itemque
    præcipua chemiæ pharmaceuticæ et mechanicæ fundamenta e
    dogmatibus Becheri et Stahlii potissimum explicantur, eorundemque
    et aliorum celebrium chemicorum experimentis stabiliuntur.   Halæ
    Magdeburgicæ, 1730.   2 vols., sm. 4to.   Vol. I, pp. [x]–1086–[42 of
    Index].   Vol. II, pp. [xii]–598–[30 of Index].   Portrait of Juncker.
              Second edition, 1744.
          "Tabula I, De chemia in genere," contains a list of authors, with notes
            biographical and bibliographical, but imperfectly prepared.   In Vol.
            II the Index definitionem is a useful addition.
        Conspectus chemiæ theoretico-practicæ.   Vollständige Ab-
          handlung der Chemie nach ihrem Lehr-Begriff und der

JUNCKER, JOHANNES.   [Cont'd.]

Ausübung, darin die Naturlehre, besonders von den
Mineralien, der natürlichen Cörper ersten Bestandtheile,
Verhalten gegen einander, Eigenschaften, Kräfte und
Gebrauch, zur wohlgegründeten und nützlichen Anwendung in der Apotheckerkunst, andern Künsten und
Handwercken, der Hauswirthschaft und gemeinen Leben, vornehmlich nach Bechers und Stahls Grundlehren
ausgeführt, und mit eben dieser, wie auch anderer
berühmten Chemicorum Erfahrungen bestätigt werden.
Halle, 1749-53. 3 vols., 4to. Vol. I, pp. [xvi]-724-
[54], 1749. Vol. II, pp. [viii]-568-[44], 1750. Vol. III,
pp. [xx]-789-[51], 1753.

Élémens de chymie suivant les principes de Becker et de
Stahl traduits du Latin sur la deuxième édition de
Juncker avec des notes par M. Demachy. Paris, 1757.
6 vols., 12mo.

Section II of Part I, Chapter I, treats of the history of chemistry (pp. 23-87).

JUNG, H.

Leitfaden für den Unterricht in der Chemie und Technologie. I. Anorganische Chemie. Weimar, 1887.

JUNGFLEISCH, ÉMILE.

Les laboratoires de chimie [and other articles]. *See, in Section II*, Fremy:
Encyclopédie chimique.

JUNGFLEISCH, ÉMILE.

Manipulations de chimie. Guide pour les travaux pratiques de chimie
de l'École supérieure de Pharmacie de Paris. Avec 372 figures
intercalées dans le texte. Paris, 1886. 2 vols., 8vo. Ill.

JURISCH, KONRAD WILHELM.

Die Vereinigung der Gewässer. Eine Denkschrift im Auftrage der
Flusskommission des Vereins zur Wahrung der Interessen der
chemischen Industrie Deutschlands. Berlin, 1888 [?].

Fabrikation von chlorsaurem Kali und anderen Chloraten. Berlin,
1888. 5 plates.

JURITZ, CHARLES F.

Chemical (The) Constitution of some Colonial Fodder Plants and Woods.
Wynberg [South Africa], 1892.

Systematic (A) Course of Qualitative Chemical Analysis of Inorganic
Substances. With tests for the more important metals and acids.
For the use of students of analytical chemistry. Cape Town and
Johannesburg, 1892.

JUSTI, JOHANN HEINRICH GOTTLOB VON.

Gesammelte chymische Schriften worinnen das Wesen der Metalle und die wichtigsten chymischen Arbeiten von dem Nahrungstand und das Bergwesen ausführlich abgehandelt werden. Berlin und Leipzig, 1760–71. 3 vols., 8vo. Ill.

Vols. I and II, Zwote Auflage. Berlin und Leipzig, 1773–74.

KABLOUKOF, J.

The present Theories of Solution according to Van't Hoff and Arrhenius, together with the doctrine of chemical equilibrium. Moscow, 1891. 8vo.

In the Russian language, of which the above is a translation. The original work I have not seen.

KAEPPELIN, R.

Cours élémentaire des sciences physiques. Cours de chimie théorique et pratique. Paris, 1838.

Seconde édition. Paris, 1841. 12mo.

Curso elemental de química teórico y práctico . . . traducido de la segunda edicion y adicionado por Rafael Saez Palacios y Carlos Ferrari y Scardini. Madrid, 1843. 8vo.

KAHLBAUM, GEORG W. A.

Siedetemperatur und Druck in ihren Wechselbeziehungen. Studien und Vorarbeiten. Mit 15 lithographirte Tafeln. Leipzig, 1885. 8vo.

KAISER, C. G.

Chlorkalk und Chlornatron in ihrer Anwendung im öffentlichen Leben. Landshut, 1831.

KAISER, R.

Chemisches Hilfsbuch für die Metall-Gewerbe. Würzburg, 1885. 8vo.

KALLE, F.

Nahrungsmittel-Tafel. Nebst erläuterndem Text für den Lehrer. Wiesbaden, 1892. 8vo.

KALMANN, W., UND TH. MORAWSKI.

Technisch-chemische Rechenaufgaben. Wien, 1889. pp. viii–44, 8vo.

KANE, SIR ROBERT.

Elements of Chemistry, including the most recent discoveries and applications of the science to medicine and pharmacy and to the arts. London and Dublin, 1841. 8vo. Ill.

An American edition by John William Draper. New York, 1846.

KAPP, GEORG CHRISTIAN FRIEDRICH.

Ueber die Wirkung der Lebensluft auf die thierischen Körper. Erlangen, 1799. 8vo.

See also in Section III.

КАРАЧАРОВЪ, П.

О способахъ очищенія воды. С.-Петербургъ 1853. 8°

KARACHAROF, P.   Methods of purifying water.   St. Petersburg. 1853.

KARMARSCH, CARL.

Grundriss der Chemie nach ihrem neuesten Zustande.   Wien, 1823. 8vo.   Ill.

KARSTEN, CARL JOHANN BERNHARD.

Grundriss der Metallurgie und der metallurgischen Hüttenkunde. Berlin, 1818.

Lehrbuch der Salinenkunde.   Berlin, 1846–47.   2 vols., 8vo.

Philosophie der Chemie.   Berlin, 1843.   pp. viii–327, 8vo.

* Revision der chemischen Affinitätslehre mit beständiger Rücksicht auf Berthollets neuer Theorie.   Leipzig, 1803.   pp. 278, 8vo.

System der Metallurgie, geschichtlich, statistisch, theoretisch, und technisch.   Berlin, 1831.   5 vols., 8vo.   Plates and Atlas.   Vol. I, pp. x–549 ; II, pp. iv–523 ; III, vi–490 ; IV, vi–629 ; V, viii–704.

Ueber die chemischen Verbindungen der Körper.   Four parts.   Berlin, 1824–41.

See also in Sections III and VII.

KARSTEN GUSTAV.

Untersuchungen über das Verhalten der Auflösungen des reinen Kochsalzes in Wasser.   Berlin, 1846.   8vo.   Ill.

KARSTEN, WENCELAUS JOHANN GUSTAV.

Grondbeginselen der scheikunde.   Zutphen.   1803.   8vo.

Wencelaus Johann Gustav Karsten died in 1787 ; if the above title be correct it must be a translation and posthumous.

KASSNER, G.

Repetitorium der Chemie.   Breslau, 1887.

KASTELEYN, PETRUS JOHANNES [also CASTELEIN].

Beschouwende en werkende, pharmaceutische, oeconomische en natuurkundige chemie.   Amsterdam, 1786–94.   3 parts, 8vo.

Chemische en physische oefeningen voor de beminnaars der scheikunst in het algemeen, de Apothekers in't byzonder.   Amsterdam, 1783–88.   5 parts, 8vo.

KASTELEYN, PETRUS JOHANNES, AND J. P. SCHONCK.
Over het aanzien der scheikunde. Utrecht, 1786. 8vo.

KASTNER, KARL WILHELM GOTTLOB.
Beiträge zur Begründung einer wissenschaftlichen Chemie. Frankfurt
und Heidelberg, 1806. 12mo.

> *Also under the title :* Physikalisch-chemisch-mineralogische und pharma-
> ceutische Abhandlungen.

Chemie (Die) zur Erläuterung der Experimental-Physik. Erlangen,
1850.

Einleitung in die neuere Chemie behufs seiner Vorlesungen und zum
Selbstunterrichte für Anfänger. Halle und Berlin, 1814. pp. xxiv–
696, 8vo.

> Pages 13–26 contain a classified bibliography ; and pages 523–696 a Chrono-
> logische Uebersicht der Geschichte der Chemie, in form of tables.
> Compare note on this work *in Section III.*

Grundriss der Chemie. Heidelberg, 1807. 8vo.

> [Another edition.] Grundriss der Chemie nach ihrem
> neuesten Zustande, besonders in technischer Beziehung.
> Wien, 1823. 8vo.

Theorie der Polytechnochemie. Ein Versuch. I. Einleitung oder
Darstellung der Grundwahrheiten der gesammten reinen Chemie,
nach ihrem neuesten Zustande. II. Ausführung oder Nachweisung
der Gesetze der Chemie und deren Anwendung zur Erklärung der
Polytechnochemie. Eisenach, 1828. 2 vols., 8vo. I, xvi–544 ;
II, xxiv–824.

Vergleichende Uebersicht des Systems der Chemie. Halle, 1821. 4to.

> *See also in Sections II and VII.*

KAUER, A.
Elemente der Chemie gemäss den neueren Ansichten für Real-Gymnasien und Unter-Realschulen. Wien, 1869. 8vo.

> Siebente Auflage. Wien, 1884. 8vo.
>
> Achte Auflage. Wien, 1888.
>
> Elementi di chimica per le classi inferiori delle scuole medie.
> Prima versione italiana sulla settima edizione tedesca
> di E. Girardi. Wien, 1884. 8vo.

KAUSLER, CHRISTIAN FRIEDRICH.
Die Kunst rohe und calcinirte Potasche zu verfertigen. Tübingen,
1780. 8vo.

KAUZMANN.
Einige Worte über Runkelrübenbau und Zuckerfabrikation. Nürnberg,
1838.

KAWAMOTO, KOMIN.

Kagaku Tsu. Tokio, 1876.   7 vols., 8vo.

> General chemistry in Japanese.

KAY-SHUTTLEWORTH, U. J.

First Principles of Modern Chemistry. A Manual of Inorganic Chemistry for Students and for Use in Schools and Science Classes. London, 1868. pp. viii–214, 12mo.

> Second edition, 1875[?]. 8vo.

KAYSER.

> *See :* Sechs Vorträge aus dem Gebiete der Nahrungsmittel-Chemie.

KEIR, JAMES.

Treatise on the various Kinds of Permanently Elastic Fluids or Gases. London, 1777. 8vo.

> *See also in Section II.*

KEIR, WILLIAM.

Disputatio physica inauguralis de attractione chemica.   Edinburg, 1778. 8vo.

KEKULÉ, AUGUST.

Chemie der Benzolderivate oder der aromatischen Substanzen.   Mit in den Text eingedruckten Holzschnitten.   Erlangen, 1867–87. 8vo. 2 vols. and vol. III, pt. I.   Vol. I, pp. viii–502, 1867 ; vol. II, pp. iv–ii–ix–587 [pt. I, 1880 ; pt. II, 1881 ; pt. III, 1882] ; vol. III, pt. I, pp. ii–240, 1887.

Lehrbuch der organischen Chemie, oder der Chemie der Kohlenstoff-Verbindungen. [From vol. III] fortgesetzt unter Mitwirkung von R. Anschütz und G. Schultz. [From vol. IV], und W. La Coste. Erlangen, [from vol. III] Stuttgart, 1859–87.   3 vols. and 1 part, roy. 8vo.

Органическая химія или химія углерод. соединеній. Москва 1863—1864.

Руководство къ органич. химіи или химіи углеродныхъ соединеній. Перев. Струговщикова М. дополненный. С.-Петербургъ 1864.

KEKULÉ, AUGUST, UND O. WALLACH.

Tableaux servant à l'analyse chimique.   Bonn, 1885.   2 vols., 8vo.

KELBE, W.

Grundzüge der Maassanalyse.   Karlsruhe, 1888.   pp. viii–136, 8vo. Ill.

KELLER, WILHELM.

Branntweinbrennerei (Die) aus Kartoffeln und Getreide. . . . Berlin, 1840.

KELLER, WILHELM. [Cont'd.]

Neue Auflage. Berlin, 1849. 2 vols., 8vo.

Fermentations-Prozess (Der) enthaltend die bisher unermittelten Veränderungen der künstlichen Hefen, welche dieselben bei Aus-
gährung der Getreide- und Kartoffelmaischen eingehen ; oder die
Haupt- und Nebenbedingungen welche allein zur richtigen Führung
des Brennereibetriebes durchaus erforderlich sind. Berlin, 1842.
8vo.

Forst- und Ackerbau-Chemie. Erklärungen der wichtigsten chemischen Vorgänge so weit sie auf das Wachsthum der Pflanzen Bezug
haben, und der Wirkung der verschiedenen Düngerarten, nach
dem neuesten Stande der Wissenschaft und geläuterter Erfahrung
dargestellt. Nördlingen, 1847. 8vo.

Handbibliothek für angehende Chemiker und Pharmazeuten, Schüler
an technischen Anstalten und Dilettanten, sowohl zum Lehrvor-
trage, als zum Selbstunterrichte. Kempten, 1838–'45. 5 vols., 12mo.

Vol. I. Die Stöchiometrie oder die rechnende Chemie. . . . 1838.

Vol. II. Kurze Anleitung zur chemischen Analyse. . . . 1838.

Vol. III. Die chemischen Grundstoffe und ihre Verbindungen. I.
Abtheilung, Anorganische Verbindungen. 1842.

Vol. IV. Organische Stoffe und Verbindungen. 1844.

Vol. V. Die Beschreibung und Erklärung der Naturerscheinungen.
. . . 1845.

Neueste und vollständigste Farben- und Lack-Kunde für Künstler und
Handwerker ; oder theoretisch-praktische Anleitung zur rationellen
Kenntniss und Fabrikation aller Arten Farben und Lackfirnisse, so-
wie zur Prüfung der Güte oder Verfälschung der hierzu verwendeten
Materialen und deren Wirkung auf den menschlichen Organismus.
Berlin, 1841. 8vo. Ill.

Theoretisch-praktische Anleitung zur Destillirkunst und Liqueurfabrikation. Oder vollständige Anweisung zum Darstellen aller einfachen und doppelten Branntweine und Liqueure durch Extraktion
und durch Destillation auf gewöhnlichem Wege und auf kaltem
mittelst ätherischer Oele, so wie der Cremes, Oele, Ratafia's und
der verschiedenen Elixire. Nebst Angabe der allein richtigen auf
eigene Erfahrung begründeten Methoden, einen fuselfreien Spirit
darzustellen, vom die auf künstlichem Wege gewonnen Rum's,
Cognak's, Franzbranntweine, etc., den ächten am ähnlichsten zu
machen. Berlin, 1842. 8vo.

KELLNER, DAVID.

Anleitung zur Probirkunst. Gotha, 1690. 8vo.

KELLNER, DAVID. [Cont'd.]

Ars metallica curiosa, oder curiose angestellte und experimentirte Schmelzproben. Nordhausen, 1701. 8vo.

Ars separatoria oder Scheidekunst. Leipzig, 1693. 12mo.

Von den edlen Bierbrauerkunst. Gotha, 1690. 8vo.

Vorstellung der zur edlen Chymie gehörigen Wissenschaften. Nordhausen, 1702. 8vo.

Weg der Natur zur Verbesserung der Metalle. Nordhausen, 1704. 8vo.

KELS, HEINRICH WILHELM.

Dissertatio de carbone vegetabili. Helmstadii, 1791. 4to.

KELS, HEINRICH WILHELM.

Onomatologia chymica practica. *See in Section II.*

KEMP, THOMAS LINDLEY.

The Phasis of Matter; being an outline of the discoveries and applications of modern chemistry. London, 1855. 2 vols., 12mo.

KEMP, W. J.

Chemical Tables for Elementary Students. London, 1884. 8vo.

KEMSHEAD, W. B.

Inorganic Chemistry, adapted for students in the elementary classes of the science and art department. (Collins' elementary series.) Enlarged edition. London, 1877. 12mo.

KENSINGTON, E. T.

Chemical Composition of Foods, Waters, Soils, minerals, manures and miscellaneous substances. London, 1877. pp. x–306, 12mo.

KENT, J. E.

The Fireside Companion, or guide to knowledge. (First lessons in chemistry and geology applied to agriculture.) Boston, 1858. 12mo.

KENTISH, T.

The Pyrotechnist's Treasury. Complete Art of making Fireworks. Second edition. London, 1887.

KERBERT, COENRAAD.

Acidi carbonici nativi historia naturalis et chemica. Lugduni-Batavorum, 1836. 4to.

KERCKHOFF, PETRUS JOHANNES VAN.

Blik (Een) op den tegenwoordigen toestand der scheikunde. Groningen, 1858.

KERCKHOFF, PETRUS JOHANNES VAN. [Cont'd.]
Scheikunde (De) in verhouding tot maatschappelyke belangen. Groningen, 1851.
Over chemische verbinding. Utrecht, 1868. 4to.

KEREKES, FRANZ.
Betrachtungen über die chemischen Elemente. Pest, 1819. 8vo.

KERL, BRUNO.
Grundriss der allgemeinen Hüttenkunde. Zweite vermehrte und verbesserte Auflage. Leipzig, 1879. 8vo.
Grundriss der Salinenkunde (Separat-Abdruck aus Muspratt's technische Chemie). Braunschweig, 1868. 8vo. Ill.
Handbuch der metallurgischen Hüttenkunde zum Gebrauche bei Vorlesungen und zum Selbststudium. Freiberg, 1855. 4 vols.
> Zweite umgearbeitete und vervollständigte Auflage. Freiberg und Leipzig, 1861–65. 4 vols. and atlas in 3 vols., 8vo.
> Practical (A) Treatise on Metallurgy adapted from the last German edition by William Crookes and Ernst Röhrig. London, 1868–70. 3 vols., 8vo.
Leitfaden beim Löthrohrprobierunterrichte an der Bergschule zu Clausthal. Clausthal, 1851. 8vo.
> Leitfaden bei qualitativen und quantitativen Löthrohruntersuchungen. Zweite Auflage. Clausthal, 1862. 8vo.
> Leitfaden bei qualitativen und quantitativen Löthrohruntersuchungen. Zum Gebrauche beim Unterrichte und zum Selbststudium für Chemiker, Pharmaceuten, Mineralogen, Berg- und Hüttenleute und sonstige Techniker. Zweite Auflage. Zweite durch Nachträge vervollständigte Ausgabe. Klausthal, 1877. 8vo.
Metallurgische Probierkunst zum Gebrauche bei Vorlesungen und zum Selbststudium. Leipzig, 1866. 8vo.
> The Assayer's Manual. An abridged treatise on the docimastic examination of ores, and furnace and other artificial products. Translated from the German by William T. Brannt. Edited by William H. Wahl. Philadelphia, 1883. 8vo. Ill.
> Second edition. Philadelphia, 1889. 8vo. Ill. Also London, 1889.

KERNER, JUSTINUS.
Das Fettgift, oder die Fettsäure und ihre Wirkungen auf den thierischen Organismus, ein Beytrag zur Untersuchung des in verdorbenen

KERNER, JUSTINUS. [Cont'd.]
Würsten giftig wirkenden Stoffes. Stuttgart und Tübingen, 1822.
pp. xxiii–368, 12mo.

KERR, J. G.
Hwa hsioh chu chei. Translated by Ho Lieo Yan. Canton, 1875.
New and enlarged edition. 4 vols. Ill.

> Principles of Chemistry in Chinese. Vol. II, the only one I have seen, has
> 120 pages. Each elementary body is represented by a Chinese charac-
> ter. Dr. Kerr has published also : Hsi yao leo tseh. Canton, 1875.
> 4 vols.; a work on Materia Medica. *See, in Section II*, Kerr, J. G.,
> and in Section V, Fryer, John.

KERSTEN, CARL MORITZ.
> Handbuch der metallurgisch-analytischen Chemie. *See* Berthier, P.

KERSTEN, CARL MORITZ.
Kreuz- und Ferdinandsbrunnen (Der) von neuem chemisch untersucht.
Leipzig, 1845.

KERTÉSZ, A.
Die Anilinfarbstoffe, Eigenschaften, Anwendung und Reactionen. Auf
praktischer Grundlage bearbeitet für Chemiker, Coloristen, Färber
des Woll-, Baumwoll-, und Seidenfaches, sowie für sonstige Interes-
senten der Anilinfarbstoffe. Braunschweig, 1888. 8vo.

KESSLER, FR.
Is das Atomgewicht des Antimons 120 oder 122 ? Bochum, 1879. 4to.

KEYSER, C. JOH.
Kemien. Bibliothek i populär Naturkunnighet VII. Andra afdelning.
Metaller och deras Föreningar. II. Tunga Metaller. Landskrona,
1875. 8vo.
Kort Lärobok i oorganisk Kemi till tekniske elementarskolarnes tjenst
utgifven. Stockholm, 1878. 8vo.
Kurs i Laborationsöfningar, afsedd at förbereda den oorganiska kemiska
Analysen. Ørebro, 1865. 8vo.
Praktiska Förstudier till den kemiska Analysen. Stockholm, 1879. 8vo.
Tillägg til C. W. Blomstrands kort Lärobok i oorganisk Kemi. (Tryckt
som manuskript.) Norrköping, 1878. 8vo.
Wårt Tidhwarfs Kemi. Svensk Bearbetning efter Fr. Schödler's
Chemie der Gegenwart. Stockholm, 1856. 8vo.

ХОДНЕВЪ, А.
Курсъ технич. химіи. С.-Петербургъ 1856. 2 части.
> KHODNEF, A. Course of technological Chemistry. St. Petersburg, 1856.

Ходневъ, Л. [Cont'd.]
Курсъ физіологической химіи. Харьковъ 1847.

> KHODNEF, A.   A Course of physiological Chemistry.   Kharkov, 1847.

Химическая часть товаровѣдѣнія. Изслѣдованіе съѣстныхъ припасовъ и напитковъ. С-Петербургъ 1861.

> Chemical investigations of food and drinks.   St. Petersburg, 1861.

Хомяковъ, Н.
Къ вопросу о гніеніи. Казань 1876.

> KHOMYAKOF, N.   On the question of putrefaction.   Kazan, 1876.

Хозеусъ.   *See* Hosaeus, A.

KIELMAIER, or KIELMEYER.
> *See* Kielmayer, Carl Friedrich.

KIELMAYER, CARL FRIEDRICH.
De reductione metallorum via humida ope combustilium stricte sic dictorum perficienda.   Tubingæ, 1804.   4to.

Examen experimentorum quorundam effectus magnetis chemicos spectantium.   Tubingæ, 1813.

Examen mineralogico-chemicum strontianitarum in monte Jura juxta Aroviam obviarum.   Tubingæ, 1813.   8vo.

Physisch-chemische Untersuchungen des Schwefelwassers von Stachelberg im Canton Glarus.   Tubingæ, 1816.

KIMBALL, ARTHUR L.
The Physical Properties of Gases.   London, 1891.

> "A general view of existing knowledge concerning gases adapted to the requirements of the non-specialist."   (*Chem. News.*)

KING.
Treatise on the Science and Practice of the Manufacture and Distribution of Coal-Gas, edited by Thos. Newbigging and W. T. Fewtrell, with copious indexes to all the volumes.   London.   3 vols., roy. 8vo.

KINGZETT, CHARLES THOMAS.
Animal Chemistry, or the relations of chemistry to physiology and pathology.   A manual for medical men and scientific chemists.   London, 1878.   pp. xx-494, 8vo.

History, (The) Products and Processes of the Alkali Trade, including the most recent Improvements.   With 23 illustrations.   London, 1877.   pp. xvi-247, 8vo.

> The historical sketch is brief.

KINKELIN, F., UND G. KREBS.

Leitfaden der Chemie für Mittelschulen insbesondere für höhere Töchterschulen. Leipzig, 1881. 8vo. Ill.

KINNERUS, SAMUEL.
   See Müller, Johannes.

KIRCHHOF.

Zucker- und Syrupfabrikation aus Runkelrüben und Kartoffeln. Leipzig, 1836.

KIRCHHOFF, GUSTAV ROBERT.

Untersuchungen über das Sonnenspectrum und die Spectren der chemischen Elemente. Berlin, 1861. 4to.
   Zweite Auflage. Berlin, 1862.
   On the Solar Spectrum and the Spectra of the Chemical Elements. 2 parts. London, 1861–62.

KIRCHHOFF, GUSTAV ROBERT, UND R. BUNSEN.

Chemische Analyse durch Spectralbeobachtungen. Leipzig, 1861. 8vo.

KIRSTEN, MICHAEL.

Non-entia chymica, seu catalogus eorum operum operationumque, quæ, cum non sint in rerum natura, nec esse possint, magno cum strepitu a vulgo chemicorum passim circumferuntur et orbi obtruduntur (sub larva Utis Udenii edit.). Frankfurt, 1645 et 1650. 12mo.

KIRWAN, RICHARD.

Essay (An) on Phlogiston and the Constitution of Acids. London, 1787. 8vo.
   * A New Edition, to which are added Notes exhibiting and defending the Antiphlogistic Theory and annexed to the French edition of this work by de Morveau, Lavoisier, de la Place, Monge, Berthollet and de Fourcroy, translated into English with additional remarks and replies by the author. London, 1789. pp. xxiii–317.

   This work is of much interest to the student of chemical history, dealing with the transition period from the phlogistic to antiphlogistic theories.

   Essai sur le phlogistique et sur la constitution des acides, traduit de l'anglais de M. Kirwan ; avec des notes de MM. de Morveau, Lavoisier. . . . A Paris, 1788. 8vo.

   Translated by Madame Lavoisier.

KIRWAN, RICHARD. [Cont'd.]

> Antiphlogistische Anmerkungen der Herren de Morveau, Lavoisier, de la Place, Monge, Berthollet und Fourcroy zu Kirwan's Abhandlung über das Phlogiston. Nebst Kirwans Gegenerrinerungen und Adets Beantwortung derselben. Aus dem französischen und englischen übersetzt von Friedrich Wolff. Berlin, 1791. 8vo.

Essay (An) on the Analysis of Mineral Waters. London, 1799. pp. vii–279, 8vo. Folding tables.

> Versuch einer Zerlegung der Mineralwasser. Nebst einigen anderen Abhandlungen. Aus dem englischen übersetzt von Lorenz von Crell. Berlin und Stettin, 1801. 8vo.

On the Composition and proportion of Carbon in Bitumens and Mineral Coal. Dublin, 1796. 4to.

Of the Strength of Acids and the Proportion of ingredients in neutral salts. Dublin, 1791. 4to.

Physisch-chemische Schriften, aus dem englischen übersetzt und mit einer Vorrede versehen von Lorenz Crell. Berlin und Stettin, 1783. 5 vols., 8vo.

Versuche und Beobachtungen über die specifische Schwere und der Anziehungskraft verschiedener Salzarten, und über die wahre neuentdeckte Natur des Phlogistons. Uebersetzt und mit Vorrede versehen von Lorenz Crell. Berlin, 1783–85. 2 vols. in 1.

What are the Manures most advantageously applicable to the various sorts of soils and what are the causes of their beneficial effect. . . . Dublin, 1794. 4to.

> Second edition. The Manures most advantageously applicable to the various soils, and the causes of their beneficial effect. . . . London, 1796. 8vo.
>
> Third edition. London, 1801.
>
> Sixth edition. London, 1806.

KITAIBEL, PAUL.

De aqua soteria thermarum Budensium. Budæ, 1804.

Examen thermarum Budensium. Neosolii, 1804.

Ueber die Bardfelder Mineralwasser. Kaschau, 1801.

KITAIBEL, PAUL, ET J. SCHUSTER.

Hydrographica Hungariæ. Budæ, 1829.

KLAPROTH, MARTIN HEINRICH.

Beiträge zur chemischen Kenntniss der Mineralkörper. Berlin und Stettin, 1795–1815. 6 vols., 8vo.

> Of the first two volumes an English translation was published :

KLAPROTH, MARTIN HEINRICH.   [Cont'd.]

> *Analytical Essays towards promoting the Chemical Knowledge of Mineral Substances.   London, 1801–04. 2 vols., pp. 591 and 267, 8vo.
>
> Mémoires de chimie contenant des analyses de minéraux. Traduit de l'allemand par B. M. Tassaert.   Paris, 1807. 2 vols., 8vo.

Chemische Untersuchung der Mineralquellen zu Karlsbad.   Berlin, 1790.   8vo.

KLEIN, A.

Studien über die gerichtlich-chemische Nachweisung von Blut.   Dorpat, 1889.

KLEIN, JOSEPH.

Elemente der forensisch-chemischen Analyse.   Ein Hilfsbuch für Studirende und kurzes Nachschlagebuch.   Hamburg, 1891.

KLEIST, F. W.

Betrachtungen über die schädlichen Wirkungen arsenikhaltiger Farben auf den menschlichen Organismus.   Berlin und Cassel, 1854.   8vo.

KLEMENT, C., AND A. RENARD.

Réactions microchimiques à cristaux et leurs applications en analyse qualitative.   Bruxelles, 1886.   pp. 126, 8vo.

KLENCKE, PHILIPP FRIEDRICH HERMANN.

Reagentien-Tabelle.   Alphabetisch-tabellarische Zusammenstellung der auf einander wirkenden chemischen Körper und ihrer Reactionserscheinungen.   Leipzig, 1858.   8vo.

Verfälschung (Die) der Nahrungsmittel und Getränke der Kolonialwaaren, Droguen und Manufacte, der gewerblichen und landwirthschaftlichen Producte.   Nach Arthur Hill Hassall und A. Chevalier und nach eigenen Untersuchungen des Verfassers. Leipzig, 1858.   pp. xx–1099, 8vo.   Ill.

> See Hassall, Arthur Hill : Food, etc.

KLETZINSKY, VINCENZ.

Chemischen (Die) Grundstoffe oder Elemente.   Eine atomistische Skizze.   Wien, 1875.   8vo.

Compendium der Biochemie.   Wien, 1858.   2 parts, 8vo.   Part I, pp. xi–223 ; eleven folding tables.   Part II, pp. xvi–84.

Graphischen (Die) Formeln der chemischen Verbindungen.   Wien, 1865.

Kurzer Abriss der chemischen Analyse, als Leitfaden bei praktischen Arbeiten im Laboratorium.   Wien, 1867.   8vo.

KLEVER, A.

Die Chemie in ihrer Gesammtheit bis zur Gegenwart und der chemischen Technologie der Neuzeit.   Stuttgart, 1888.

KNAPE, CHRISTOPH.

Theoria metamorphosis chemico-philosophicis rationibus superstructa.
Halæ, 1773.   4to.

KNAPP, FRIEDRICH LUDWIG.

Lehrbuch der chemischen Technologie zum Unterricht und Selbststudium.   Mit zahlreichen in den Text eingedruckten Holzschnitten.   Braunschweig, 1844.   2 vols., 8vo.

    Zweite Auflage.   Braunschweig, 1848–1853.   2 vols., 8vo.
    Vol. I, pp. xiv–655.   Vol. II, pp. xix–895.

    Dritte   umgearbeitete   und   vermehrte   Auflage.   Braun-
schweig, 1865–1875.   3 vols., 8vo.

Of this edition only vol. I and parts 1 and 2 of vol. II have been published.

    Leerboek der scheikundige technologie, ten dienste van het
onderwijs en tot zelfonderrigt.   Uit het Hoogduitsch.
Gouda, 1846.   8vo.   Part I, Licht en lichtstoffen.

    Chemical Technology, or chemistry applied to the arts and
to manufactures.   Edited, with numerous notes and
additions, by Edmund Ronalds and Thomas Richardson.   London, 1848–51.   3 vols., 8vo.   Vol. I, . . .
Vol. II [Glass and Earthenware], pp. vii–491.   Vol. III,
by Edmund Ronalds and Thomas Richardson.   [Foods],
pp. vi–458–iv.   Plates.

For a second revised edition, *see* Richardson, Th., and Henry Watts.

    First American Edition with Notes and Additions by
Walter R. Johnson.   Philadelphia, 1848–49.   2 vols.,
8vo.   Ill.

    Traité de chimie technologique et industrielle ; traduction
revue et augmentée par Merijot et A. Debize.   Paris,
n. d.   2 vols., 8vo.

    Lärobok uti kemisk Teknologi.   Öfversättning af J. F.
Bahr.   Stockholm, 1849–50.   8vo.

Nahrungsmittel (Die) in ihren chemischen und technischen Beziehungen.
Braunschweig, 1848.   8vo.   Ill.

KNAUER, A.

Elemente der Chemie gemäss den neueren Ansichten für Real-Gymnasien und Unter-Realschulen.   Vierte verbesserte Auflage.   Wien,
1876.   8vo.   Ill.

KNAUER, K.

Chemischer Rathgeber für Gewerbtreibende. Ein höchst nützliches
Handbuch, enthaltend eine auserlesene Sammlung praktischer und
vielfach geprüfter Anweisungen zur Auflösung aller in das Gebiet
der technischen Chemie einschlagenden Operationen für Maurer,
Zimmerleute, Tüncher, etc. Nach den besten französischen
Quellen bearbeitet. Altenburg, 1848. 8vo.

KNAUST, HEINRICH.

Fünff Bücher von der Göttlichen und Edlenn Gabe der philosophischen
. . . Kunst Bier zu brawen. Jetzo auffs newe übersehen. Erffurdt,
1614. 4to.

KNÖRR, JOHANN GOTTFRIED.

Observationes chemicæ miscellaneæ. Gottingæ, 1768. 4to.

KNOP, WILHELM.

Handbuch der chemischen Methoden. Leipzig, 1859. 8vo.
Kreislauf (Der) des Stoffs. Lehrbuch der Agricultur-Chemie. Leipzig,
1868. 2 vols.
Kreislauf (Der) des Wassers. Lehrbuch der Agricultur-Chemie. Leipzig, 1868. 2 vols., 8vo.
Körpermolecule. Nachweisung der Thatsache dass die Molecule der
modernen Chemie durch zusammenlegen von Tetraedern und
Oktaedern atomistisch nachgebildet werden können. Leipzig,
1876. 8vo.

KNUTH, P.

Lehrbuch der Chemie für Maschinisten und Torpeder. Kiel, 1883. 8vo.

KOBELL, FRANZ VON.

Galvanographie (Die), eine Methode gemalte Tuschbilder durch galvanische Kupferplatten im Drucke zu vervielfältigen. München,
1842. 4to. Ill.
        Zweite Auflage. 1846. Roy. 8vo.
        Гальванографія или способъ производить гальванически мѣдныя
            доски для печатанія кистью работанныхъ рисунковъ. Соч.
            Ф. Кобеля, съ рисунками. С.-Петербургъ 1843. 8º
Tafeln zur Bestimmung der Mineralien mittelst einfacher chemischer
Versuche auf trockenem und nassem Wege. München, 1833. 8vo.
        Zweite Auflage, 1835. Dritte Auflage, 1838.
        Vierte Auflage, 1846. Fünfte Auflage, 1853.
        Sechste Auflage, 1858. Siebente Auflage, 1861.
        Achte Auflage, 1864. Neunte Auflage, 1870.

KOBELL, FRANZ VON. [Cont'd.]

Zehnte vermehrte Auflage, 1873. Elfte Auflage, 1878.

Zwölfte neu bearbeitete und vermehrte Auflage von K. Oebbeke. München, 1884.

Instructions for the Discrimination of Minerals by simple Chemical Experiments. Glasgow and London, 1841. 8vo.

Tableau pour reconnaître les minéraux au moyen d'essais chimiques simples par la voie sèche et la voie humide. Traduit par E. Melly sur la deuxième édition allemande. Genève, 1836. 8vo.

Tableaux pour la détermination des minéraux. Traduit de l'allemand par J. Gilon. Bruxelles, 1851. 8vo.

Les minéraux ; guide pratique pour leur détermination sûre et rapide au moyen de simples recherches chimiques par voie sèche et par voie humide, à l'usage des chimistes, ingénieurs, industriels, etc. Traduit sur la dixième édition allemande par le comte Ludovic de la Tour-du-Pin. Refondu et augmenté d'un avant-propos et d'additions par F. Pisani. Paris, 1872, 12mo.

Tavole per riconoscere i minerali per mezzo di saggi chimici semplici per la via secca e per la via umida. Tradotta dalla seconda edizione Tedesca in Francese da E. Melly e dal Francese in Italiano conforme alla terza edizione originale da Adriano de Bonis. Firenze, 1842. 8vo.

Nuova edizione. Firenze, 1850. 8vo.

KOBERT, R.

Ueber Cyanmethaemoglobin und den Nachweis der Blausäure. Stuttgart, 1891. pp. viii–64, 8vo, and one colored plate.

KOCH, A.

Küchen- und Haushaltungschemie. Leitfaden der Chemie mit besonderer Berücksichtigung der Gesundheitslehre für Lehrerinnen-Seminare, höhere' Töchter- und Mädchenschulen, sowie für das Haus nach methodischen Grundsätzen bearbeitet. Hannover, 1884. 8vo. Ill.

KÖHLER, FRIEDRICH WILHELM.

Die Chemie in technischer Beziehung. Leitfaden für Vorträge in Gewerbsschulen. Berlin, 1834.

Zweite umgearbeitete und erweiterte Ausgabe. Berlin, 1837.

Dritte Ausgabe. Berlin, 1840. 8vo.

KÖHLER, FRIEDRICH WILHELM. [Cont'd.]
Fünfte Ausgabe. Berlin, 1846.
Siebente völlig umgearbeite Ausgabe. Berlin, 1854. 8vo.
De scheikunde met betrekking tot het fabriekwezen. Eene
handleiding bij industriëel onderwijs. Naar de vierde
uitgave uit het Hoogduitsch vertaald door C. F. Donnadieu. Delft, 1844-46. 2 parts, 8vo.

Cf. Loppens, P.

Rys chemii organicznej i jej zastosowanie podług dzieła
Koehlera przez J. Seweryn. Zdzitowieckiego. Warszawa,
1840. 8vo.

KÖHLER, H.
Carbolsäure und Carbolsäure-Präparate, ihre Geschichte, Fabrikation,
Anwendung und Untersuchung. Berlin, 1891. 8vo.

KÖHLER, HIPPOLYT.
Die Fabrikation des Russes und der Schwärze aus Abfällen [etc.]. *See*
Bolley's Handbuch.

KÖHLER, J. G.
Die Salze aus dem elektro-chemischen Gesichtspunkte betrachtet.
Prag, 1839. 8vo.

KÖHLER, JOSEPH.
Anleitung zur Photographie auf Glas. *n. p.*, July 4, 1855.
A lithographic copy of a MS. by a teacher and photographer of Reichen-
berg. Two parts, 61 and 35 pages.

KOENE, CORNEILLE JEAN.
Mémoires de chimie. Paris, 1857. 18mo.
The author was Professor of chemistry and toxicology at the University of
Bruxelles, from 1840.

KÖNIG, ARNOLD.
Zur Theorie und Geschichte der fünfgliedrigen Kohlenstoffringe, enthaltend eine Berechnung des relativen Abstandes von Kohlen-
stoffatomen bei doppelter und einfacher Bindung sowie einen
Beitrag zur Kenntniss der Hydrindenderivate. Leipzig, 1889. pp.
80, 8vo.

KÖNIG, EMANUEL.
Chymia physica circa corporum naturalem et artificialem statum.
Basileæ, 1693. 4to.

KÖNIG, F.

Composizione chimica e valore nutritivo degli alimenti umani rispetto al loro prezzo con la razione giornaliera e la digestibilita di alcuni cibi. Con una grande tavola grafica. Milano, 1884. 8vo.

KÖNIG, J.

Bestand und Einrichtungen der Untersuchungsämter für Nahrungs- und Genussmittel in Deutschland und ausserdeutschen Staaten. Nebst Vorschlägen zur einheitlichen Organisation. Berlin, 1882. pp. viii–163, 8vo.

Chemische Zusammensetzung der menschlichen Nahrungs- und Genuss- mittel. Nach vorhandenen Analysen mit Angabe der Quellen zusammengestellt und berechnet. Zweite sehr vermehrte und ver- besserte Auflage. Berlin, 1882. pp. xxi–351, 4to.

> Erste Auflage, 1878–80.
> Dritte Auflage, Berlin, 1889, pp. xxviii–1161.
>> *The first part of :* Chemie der menschlichen Nahrungs- und Genussmittel. Berlin, 1882.

Menschlichen (Die) Nahrungs- und Genussmittel, ihre Herstellung, Zu- sammensetzung und Beschaffenheit, ihre Verfälschungen und deren Nachweisung, mit einer Einleitung über die Ernährungs- lehre. Zweite Auflage, Berlin, 1883. pp. xviii–820. 4to.

> Erste Auflage, 1879.
>> *Second part of :* Chemie der menschlichen Nahrungs und Genussmittel. Berlin, 1883.

Procentische Zusammensetzung und Nährgeldwerth der menschlichen Nahrungsmittel. Chromolith. Tabelle. Fünfte Auflage. Berlin, 1888. 4to.

Verunreinigung (Die) der Gewässer, deren schädliche Folgen nebst Mitteln zur Reinigung der Schmutzwässer. Berlin, 1887. 10 plates.

KÖSTLIN, CARL HEINRICH.

Von der Methode die mineralischen Wasser vermittelst der fixen Luft . . . nachzumachen. Stuttgart, 1780. 4to.

> *Cf.* Priestley, Joseph.

KOFAHL, H.

Ueber einige Methoden zur Bestimmung und Trennung von Eisen, Mangan, Nickel und Kobalt. Berlin, 1890. 8vo.

KOHLMANN, B., UND F. FRERICHS.

Rechentafeln zur quantitativen chemischen Analyse. Leipzig, 1882. 8vo.

KOHLRAUSCH, FRIEDRICH.

Gegenwärtigen (Die) Anschauungen über die Electrolyse von Lösungen. Vortrag gehalten in der Sitzung des Electrotechnischen Vereins am 29 März, 1887. Berlin, 1887. 8vo.

KOHLRAUSCH, O.

Physiologie und Chemie in ihrer gegenseitigen Stellung beleuchtet durch eine Kritik von Liebig's Thierchemie. Göttingen, 1844. 8vo.

KOLB, J.

Produits chimiques. *See in Section II*, Fremy : Encyclopédie chimique, vol. v.

KOLB, J.

Étude sur la fabrication de l'acide sulphurique, considérée au point de vue théorique et technologique. Lille, 1865. 4to.

Sur l'évolution actuelle de la grande industrie chimique. Lille, 1883. 8vo.

KOLBANI, PAUL.

Gifthistorie des Thier-, Pflanzen- und Mineralreichs nebst den Gegengiften und der medicinischen Anwendung der Gifte. Zweite vermehrte und verbesserte Auflage. Wien, 1807. pp. 489, 8vo.

KOLBE, HERMANN.

Ausführliches Lehrbuch der organischen Chemie. Braunschweig, 1854–78. 3 vols., 8vo. Ill.

Also under the title : Graham-Otto's ausführliches Lehrbuch der Chemie, dritter bis fünfter Band.

Ausführliches Lehr- und Handbuch der organischen Chemie. (Zugleich als dritter, vierter und fünfter Band zu Graham-Otto's ausführlichem Lehrbuch der Chemie.) Braunschweig, 1868–84. 3 vols., 8vo.

Vol. I. Zweite umgearbeitete und vermehrte Auflage. Von Ernst von Meyer. 1880.

Vol. II. Zweite umgearbeitete und vermehrte Auflage. Von Ernst von Meyer. Three parts. 1881–1884.

Vol. III, bearbeitet von E. von Meyer und A. Weddige in Leipzig, und H. von Fehling in Stuttgart. Two parts. 1868–1878.

Uitvoerig leerboek der organische scheikunde. Naar het Hoogduitsch door J. F. L. Rendler. Met tusschen den tekst geplaatste houtsneefiguren. Haarlem, 1854–'55. 8vo.

KOLBE, HERMANN.   [Cont'd.]
Kurzes Lehrbuch der anorganischen Chemie.   Braunschweig, 1877.
pp. xxiv–679.   8vo.   Ill.

    Kurzes Lehrbuch der Chemie.   Braunschweig, 1883–1884.
    2 vols., 8vo.   Ill.

    Part I.   Anorganische Chemie.   Zweite verbesserte Auflage.
    1884.

    Part II.   Organische Chemie.   1883.

    A Short Text-book of Inorganic Chemistry.   Translated
    from the German, and edited by T. S. Humpidge.
    London, 1888.   8vo.   Ill.   *Also* New York, 1888.

    Third edition revised by H. Lloyd Snape.   London, 1892.
    8vo.   Ill.

Ueber die chemische Constitution der organischen Kohlenwasserstoffe.
Vortrag nebst Eröffnungsrede, gehalten bei Einweihung des neuen
chemischen Laboratoriums der Universität Leipzig am 16 Nov.,
1868.   Braunschweig, 1869.   8vo.

KOLLER, THEODOR.
Chemische Präparatenkunde.   Handbuch der Darstellung und Gewinnung der am häufigsten vorkommenden chemischen Körper.
Wien, Pest und Leipzig.   1890.   12mo.
Praktische (Die) Herstellung von Lösungen.   Ein Handbuch zum
raschen und sicheren Auffinden der Lösungsmittel aller technischen
und industriellen wichtigen festen Körper, sowie zur Herstellung
von Lösungen solcher Stoffe.   Wien, Pest und Leipzig, 1888.
12mo.

Колли, А.
Процессы броженія.   Москва 1876.

    KOLLI, A.   Processes of fermentation.   Moscow, 1876.

KOLLMYER, A. H.
Chemia Coartata, or the Key to Modern Chemistry.   London and
Philadelphia *n. d.* [1875].   Oblong 12mo.

Κομνῆνος, Τελέμαχος.
    Ἡ ἀτομικότης τῶν στοιχείων εἰς τὰς ἐνώσεις τῶν πραγμάτων ἐπὶ
    ὑφηγεσίᾳ.   Ἐν Ἀθήναις, 1884.

KONINCK, LAURENT GUILLAUME DE.
Éléments de chimie inorganique.   Liège, 1839.   8vo.
Résumé de la théorie chimique des types.   Liège et Paris, 1865.   12mo.

KONINCK, LAURENT GUILLAUME DE, AND E. DIETZ.
Manuel pratique d'analyse chimique appliquée à l'industrie du fer. Liège, 1871. 12mo.

> A Practical Manual of Chemical Analysis and Assaying as applied to the manufacture of iron from its ores, and to cast iron, wrought iron and steel, as found in commerce. Edited with notes by Robert Mallet. London, 1872. pp. xviii–210, 12mo. Ill.

KONINCK, LAURENT GUILLAUME DE, ET E. PROST.
Exercises d'analyse chimique qualitative. Liège, 1886. 8vo.

KONING, B.
Verhandeling over de zelfontvlamming van den phosphorus in het luchtledige. Te Middelburg, 1825. 8vo.

KONKOLY, N. von.
Handbuch für Spectroscopiker im Cabinet und am Fernrohr. Praktische Winke für Anfänger auf dem Gebiete der Spectralanalyse. Halle, 1889. 8vo.

Константиновичъ, М.
Химическія бесѣды. Общепонятное изложеніе химич. свѣдѣній. Кіевъ 1874.

> KONSTANTINOVICH, M. Chemical Conversations. Popular exposition of chemical science. Kiev, 1874.

Κοπέλλα, Ανδ9.
> Χρωματολογία, ἤτοι περὶ φύσεως, ονομασίας καὶ τῆς χημικῆς συστάσεως τῶν χρωμάτων παρά τε τοῖς ἀρχαίοις καὶ τοῖς νεωτέροις. Ἐν Ἀθήναις, 1886.

KOPFER, FERD.
Die quantitative Bestimmung des Kohlenstoff- und Wasserstoffgehaltes der organischen Substanzen. Wiesbaden, 1877. 8vo. Ill.

KOPP, E.
Sur la préparation et les propriétés du verre soluble ou des silicates de potasse et de soude. Analyse de tous les travaux publiés jusqu' à ce jour sur ce sujet. Paris, 1857. 4to.

KOPP, HERMANN.
> Theoretische Chemie. *See* Graham–Otto.

KOPP, HERMANN.
Bemerkungen zur Volumtheorie. Braunschweig, 1844. 8vo.
Physikalisch-chemische Beiträge. Frankfurt a. M. 1841. 8vo.

KOPP, HERMANN. [Cont'd.]

Also under the title : Ueber die Modification der mittleren Eigenschaft, oder über die Eigenschaft von Mischungen in Rücksicht auf ihrer Bestandtheile. Part I.

Ueber das specifische Gewicht der chemischen Verbindungen. Frankfurt a. M., 1841. 8vo.

See also in Section III.

KOPP, JOHANN HEINRICH.

Ausführliche Darstellung der Selbstverbrennung des menschlichen Körpers. Frankfurt am Main, 1811. 8vo.

Grundriss der chemischen Analyse mineralischer Körper. Frankfurt am Main, 1805. 8vo.

KOPPE, SIEGFRIED WALTER.

Das Glycerin, seine Darstellung, seine Verbindung und Anwendung in den Gewerben, in der Seifen-Fabrik, Parfumerie und Sprengtechnik. Für Chemiker, Parfumeure, Seifen-Fabrikanten, Apotheker, Sprengtechniker und Industrielle geschildert. Wien, Pest und Leipzig, 1883. 12mo. Ill.

KOPPER, FERDINAND.

Die quantitative Bestimmung des Kohlenstoff und Wasserstoffgehaltes der organischen Substanzen. Wiesbaden, 1877. pp. 75, 8vo. Plates.

KOPPESCHAAR, W. F.

Leerboek der chemie en van eenige harer toepassingen. Deel I., Anorganische chemie. Vijfde druk. Leiden, 1887. Deel II. Anorganische chemie. Metalen. Vierde druk. Leiden, 1886.

KOPPI, VON.

Die Runkelrüben-Zucker-Fabrikation. Breslau und Liegnitz, 1810. 8vo.

KOSMANN, B.

Die Darstellung von Chlor- und Chlorwasserstoffsäure aus Chlormagnesium. Berlin, 1891. Ill.

KOSSAKOWSKI, L.

Die Beziehung der Verdampfungszeiten der Fettsäurereester und anderer Substanzen zum Moleculargewicht. Zürich, 1890. 8vo.

KOSSEL, A.

Leitfaden für medicinisch-chemische Curse. Zweite vermehrte Auflage. Berlin, 1888.

KOTIKOVSKY.

Ueber die Nicht-Einfachheit der Metalle, des Schwefels, der Kohle, des
Chlors und überhaupt über die Nicht-Einfachheit der gegen-
wärtig sogenannten " Einfachen Stoffe " mit Angabe ihrer nächsten
Bestandtheile wie diese aus den Grundsätzen echter Naturfor-
schung von selbst und zwar zunächst sich ergeben. Sammt nächsten
Folgerungen über das Wesen der nachweislich-zusammengesetzten
Stoffe. Auszug aus dem 1 Buche einer umfassenderen noch nicht
veröffentlichten Abhandlung. Wien, 1854. pp. x-118, 8vo.

> A curious example of misdirected energy. A single example of his reason-
> ing will suffice. " Water (= 18) cannot contain Oxygen (= 32) as a
> constituent because no part can weigh more than the whole." (Page 29.)

Котовщиковъ, Н.

Способы опредѣленія азотной и азотистой кислотъ (въ разведенныхъ раство-
рахъ) индиго-сѣрной кислотою. Казань 1875.

> KOTOVSHCHIKOFF, N. Means of detecting the presence of azotic and
> azotous acids in diluted solutions by indigo-sulphuric acid. Kasan,
> 1875.

Ковалевскій, С.

Элементарныя свѣдѣнія изъ химіи. С.-Петербургъ 1873

> KOVALEVSKY, S. Elementary notions of Chemistry. St. Petersburg, 1873.

Учебникъ химіи, Съ 34 политипажами. С.-Петербургъ 1874

> Handbuch of Chemistry. St. Petersburg, 1874.

KRÄTZER, HERMANN.

Chemische Unterrichts-Briefe. Für das Selbst-Studium Erwachsener.
Mit besonderer Berücksichtigung der neuesten Fortschritte der
Chemie und unter Mitwirkung hervorragender Fachmänner und
Gelehrten herausgegeben. 1 Curs : Die anorganische Chemie.
Leipzig, 1883–84. 8vo.

II Curs. Die organische Chemie, oder die Chemie der Kohlenstoff-
verbindungen. Mit besonderer Berücksichtigung der chemischen
Technologie. Unter Mitwirkung der Herren Landerer, Herrburger, W. A. Herrmann, Alwin Engelhardt, F. Eichbaum, [etc.,]
bearbeitet. Leipzig, 1884–87. 8vo.

Wasserglas und Infusorienerde, deren Natur und Bedeutung für Industrie, Technik und die Gewerbe. Wien, Pest, Leipzig. 1887. Ill.

KRAFFT, F.

Kurzes Lehrbuch der Chemie. Anorganische Chemie. Wien, 1891.
8vo. Ill.

KRAMER, ANTON JOHANN VON, ET LAUGIER, E.
See, in Section II, Laugier, E.

KRAUCH, C.

Die Prüfung der chemischen Reagentien auf Reinheit, mit einem Vorwort von J. König.   Darmstadt, 1888.   8vo.

Zweite Auflage.   Berlin, 1891.

Essais de pureté des réactifs chimiques.   Guide pratique à l'usage des laboratoires de chimie et de microbiologie. Édition française annotée par J. Delaite, avec préface par L. de Koninck.   Liège, 1892.   8vo.

Examen de la pureza de los reactivos químicos.   Traducido de la segunda edición alemana por E. Mascarenas y Hernández.   Barcelona, 1892.   4to.

KRAUS, GREGOR.

Grundlinien zu einer Physiologie der Gerbstoffe.   Leipzig, 1888.

Zur Kenntniss der Chlorophyllfarbstoffe und ihrer Verwandten.   Spectralanalytische Untersuchungen.   Stuttgart, 1872.   pp. 132.   8vo. Plates.

KRAUSE, G.

Darstellung des Zuckers aus Runkelrübe.   Wien, 1834.

KRAUSE, G.

Die Industrie von Stassfurt und Leopoldshall und die dortigen Bergwerke.   In chemisch-technischer und mineralogischer Hinsicht betrachtet.   Cöthen, 1876.   8vo.

KRAUSS, C.

Kritische Studien über die Trennung von Nickel und Kobalt.   Erlangen, 1889.

KRAUT, CARL.

See Gmelin, Leopold.

KRECHEL, GEORGES.

Choix de méthodes analytiques des substances qui se rencontrent le plus fréquemment dans l'industrie.   Paris, 1887.   pp. 477–v, 8vo. Ill.

KRECKE, F. W.

Handleiding der chemische technologie.   Gorinchem, 1880.   8vo.   Ill.

КРЕЙНИГЪ, А.

Задачи по химіи. Перев. Левъ Раерманъ.   Одесса 1875.

KREINING, A.   Problems of Chemistry.   Odessa, 1875.

KREMERS, P.

Physikalisch-chemische Untersuchungen. Wiesbaden, 1869–78. 8vo. Ill.

KREUSLER, U.

Atomgewichtstafeln, enthaltend die neueren Atomgewichte der Elemente, nebst multiplen Werthen. Bonn, 1884. 8vo.

Lehrbuch der Chemie nebst einem Abriss der Mineralogie. Berlin, 1880. 17 plates.

KREUSSER, H.

Das Eisen, sein Vorkommen und seine Gewinnung. Kurze gemeinfassliche Darstellung der Eisen-Erzeugung bearbeitet für das Verständniss eines grösseren Leserkreises, zum Gebrauche für Techniker, Metallarbeiter, Kaufleute, sowie an Gewerbe- und Industrie-Schulen. Weimar, 1886. 8vo.

KRIEG, LUDWIG.

Theorie und praktische Anwendung von Anilin in der Färberei und Druckerei nebst Bemerkungen über die Anilin-Surrogate. Berlin-1860. Roy. 8vo.

> Zweite Auflage. Berlin, 1862.

> Dritte umgearbeitete und verbesserte Auflage. Nach dem Tode des Verfassers bis auf die jüngste Zeit vermehrt, bearbeitet, und herausgegeben von Theodor Oppler. Berlin, 1866. pp. xl–385, 8vo. Plates. (Die Chemie und Industrie der Mineralöle.)

Κρίνος, Γεόργιος.

Σχετικὰ πρὸς τὴν γνωμοδότησιν τῆς φιλοσοφικῆς σχολῆς τοῦ ἐθνικοῦ πανεπιστημίου περὶ ἀναπληρώσεως τῆς ἕδρας τῆς φαρμακευτικῆς χημείας ἐν ἔτει 1882. Ἐν Ἀθήναις, 1883.

KROBATIN, A. VON.

Lehrbuch der Chemie für die kaiserlichen königlichen Cadettenschulen. Wien, 1886.

KROCKER, F.

Leitfaden für die agricultur-chemische Analyse mit specieller Anleitung zur Untersuchung landwirthschaftlich-wichtiger Stoffe. Breslau, 1861. 8vo.

> Zweite Auflage. Breslau, 1862. 8vo.

> Vierte vermehrte und verbesserte Auflage. Breslau, 1878. 8vo.

For an English version, see Frankland, P. F.

KRÖNIG, A.
Die Werthlosigkeit einer grossen Anzahl von chemischen Formeln, dargethan durch die Grösse der Fehler in Liebig's Analysen . . . Berlin, 1866. 8vo.

KRONBERG, H.
Wasserglas und Infusorienerde, deren Natur und Bedeutung für Industrie, Technik und Gewerbe. Wien, 1887.

KRÜGER, JUL.
Ausführliches Lehrbuch der Destillation. Die Bereitung von Branntweinen, Liqueuren, Essig, u.s.w. nebst einem Anhang werthvoller Recepte. Durchgesehen und ergänzt von G. Wettendorfer. Zweite Auflage. Leipzig, 1891. 8vo.
Leitfaden zu qualitativen Untersuchungen mittelst des Löthrohrs. Berlin, 1851. 8vo.

KRÜSS, GERHARD.
Spezielle Methoden der Analyse. Anleitung zur Anwendung physikalischer Methoden in der Chemie. Hamburg und Leipzig, 1892. 12mo. Ill.
Untersuchungen über das Atomgewicht des Goldes. München, 1886. 8vo.

KRÜSS, G. UND H.
Colorimetrie und quantitative Spectralanalyse in ihrer Anwendung in der Chemie. Hamburg, 1891. 8vo. Ill.

KRUKENBERG, C. Fr. W.
Chemische Untersuchungen zur wissenschaftlichen Medicin. Jena. Two parts. 1888.
Eigenartigen (Die) Methoden der chemischen Physiologie. Vortrag. Heidelberg, 1884. 8vo.
Grundriss der medicinisch-chemischen Analyse. Heidelberg, 1884. 8vo.
Compendio di analisi chimica medica, che ha per base il corso dato nel laboratorio chimico-fisiologico di Würzburg. Tradotto corredata di note ed aggiunte da Gibertini. Parma, 1885. 8vo.

KRUPP, A.
Die Legirungen. Handbuch für Praktiker. Enthaltend : Die Darstellung sämmtlicher Legirungen, Amalgame und Lothe für die Zwecke aller Metallarbeiter, insbesondere für Erz-Giesser, Glockengiesser, Bronzearbeiter, Gürtler, Sporer, Klempner, Gold und Silberarbeiter, Mechaniker, Techniker u.s.w. Wien, 1879. Ill.

KRUTWIG, JEAN.
Exercises d'analyse chimique quantitative. Liège et Paris, 1885. 12mo.
> *Cf.* Tableaux servant à l'analyse chimique.

KRUTZSCH, K. L.
Das A B C der Chemie, enthaltend das gemeinnützigste aus der chemischen Wissenschaft für Nichtchemiker und die zum Verständniss eines besondern chemischen Unterrichts erforderlichen allgemeinen chemischen Vorkenntnisse. Zweite vermehrte und verbesserte Auflage. Aus des Verfassers " Bodenkunde " besonders abgedruckt. Dresden und Leipzig, 1845. 8vo.
> Reprinted from : Populärer Abriss der wissenschaftlichen Bodenkunde. Of this the first edition was published in 1842.

> A B C der scheikunde. Eene handleiding voor den landbouwer, en voor ieder die zich op eene gemakkelijke wijze de grondbeginselen dezer wetenschap wenscht eigen te maken. Vrij vertaald naar het Hoogduitsch door F. A. Enklaar. Zwolle, 1851. 8vo.
> Jordbundslære med Chemiens A B C. En Haandbog for Landmænd, Forstmænd og Gartnere. Frit oversat efter anden Udgave af C. Dalgas. Kjøbenhavn, 1850. 8vo.

KRYSZKA, A.
Chemiczne sprawy żywotne. Warszawa, 1855. pp. v–322, 8vo.

KUBEL, WILHELM.
Anleitung zur Untersuchung von Wasser, welches zu gewerblichen und häuslichen Zwecken oder als Trinkwasser benutzt werden soll. Zum Gebrauche für Techniker, Fabrikanten Pharmaceuten, etc. Braunschweig, 1866. 8vo.

> Anleitung zur Untersuchung von Wasser, welches zu gewerblichen und häuslichen Zwecken oder als Trinkwasser benutzt werden soll. Zweite vollstandig umgearbeitete und vermehrte Auflage von Ferdinand Tiemann unter Mitwirkung des Verfasser der ersten Auflage. Braunschweig, 1874. 8vo. Ill.
> Die chemische und mikroskopisch-bakteriologische Untersuchung des Wassers zum Gebrauche für Chemiker, Aerzte, Medicinalbeamte, Pharmaceuten, Fabrikanten und Techniker bearbeitet. Zugleich als Dritte vollständig umgearbeitete und vermehrte Auflage von Kubel-Tiemann's Anleitung zur Untersuchung von Wasser welches zu gewerblichen und häuslichen

KUBEL, WILHELM. [Cont'd]

    Zwecken sowie als Trinkwasser benutzt werden soll. Braunschweig, 1889. pp. xxxii–705. 10 plates.

       Prepared with the co-operation of F. Tiemann and A. Gärtner.

KUBLIN, S.

    Die Bewegung der Elemente. Eine kosmisch-tellurische Studie. Fünf-kirchen, 1892. 8vo.

KÜHLMANN, CARL FRIEDRICH.

    Applications des silicates alcalins et solubles au durcissement des pierres calcaires poreuses. Paris, 1855. 8vo.

    Expériences chimiques et agronomiques. Lille, 1847.

KÜHN, OTTO BERNHARD.

    Cyan (Das) und seine anorganischen Verbindungen nebst dem Mellon. Eine Zusammenstellung aller darüber bekannt gewordenen Erfahrungen. Für Aerzte, Apotheker, Techniker und Chemiker. Leipzig, 1863. pp. viii–320, 8vo.

    Lehrbuch der Stöchiometrie. Leipzig, 1837. 2 tables.

    Praktische Chemie für Staatsärzte. Erster Theil : Praktische Anweisung die in gerichtlichen Fällen verkommenden chemischen Untersuchungen anzustellen. Leipzig, 1829. 8vo.

    System der organischen Chemie als Leitfaden zum Studium der theoretischen Chemie bearbeitet. Göttingen, 1848. 8vo.

    Versuch einer Anthropochemie. Leipzig, 1824. 8vo.

KÜHNE, W.

    Chemische Vorgänge in der Netzhaut. *See* Handbuch der Physiologie.

KÜHNE, W.

    Lehrbuch der physiologischen Chemie. Leipzig, 1868. pp. viii–605, roy. 8vo. Ill.

       Учебникъ физіолог. химіи. Перев. Сѣченова П.   С.-Петербургъ 1866—67, 3 выпуска.

KÜNSTLE, GUIDO.

    Kohlenstoffskizzen. München, 1877. pp. 59, 8vo.

KÜTZING, FRIEDRICH TRAUGOTT.

    Die Chemie in ihrer Anwendung auf das Leben. Nordhausen, 1838.

    Phycologia germanica. Nordhausen, 1845. 8vo.

KUHLMANN, CARL FRIEDRICH.

    See Kühlmann, C. F.

KUNCKEL, JOHANN VON LÖWENSTERN.

* Ars vitraria experimentalis, oder vollkommene Glasmacher-Kunst,
lehrende als in einem, aus unbetrüglicher Erfahrung, herfliessendem
Commentario, über die von dergleichen Arbeit beschriebene sieben
Bücher P. Antonii Neri, von Florenz, und denen darüber gethanen
gelehrten Anmerckungen Christophori Meretti (so aus den Italien
und Lateinischen beyde mit Fleysz ins Hochteutsche übersetzt).
Die allerkurtz-bündigsten Manieren, das reineste Chrystall-Glas ;
alle gefärbte oder tingirte Gläser ; künstliche Edelstein oder
Flüsse ; Amausen, oder Schmeltze, Doubleten, Spiegeln, das
Tropff-Glas ; die schönste Ultramarin, Lacc- und andere nützliche
Mahler-Farben ; in gleichen wie die Saltze zu den allerreinesten
Chrystallinen Gut, nach der besten Weise an allen Orten Teutsch-
lands mit geringer Müh und Unkosten copieus und compendieus
zu machen, auch wie das Glas zu mehrer Perfection und Härte zu
bringen.   Nebst ausführlicher Erklärung aller zur Glaskunst
gehörigen Materialien und Ingredientien ; sonderlich der Zaffera
und Magnesia, etc.   Anzeigung der nöthigsten Kunst- und Handgriffe ; dienlichsten Instrumenta ; bequemsten Gefässe, auch nebst
anderen des Autoris sonderbaren Ofen, und der gleichen mehr,
nützlichen in Kupffer gestochenen Figuren.   Samt einem II.
Haupt-Theil.   So in drey unterschiedenen Büchern, und mehr
als 200 Experimenten bestehet, darinnen vom Glasmahlen, ver-
gulden und Brennen ; vom holländischen Kunst- und Porcellan-
Töpfferwerck ; vom kleinen Glasblasen mit der Lampen ; von
einer Glas-Flaschen-Forme, die sich viel 1000 mal verändern lässet ;
wie Kräuter und Blumen in Silber abzugiessen ; Gypsz zu tractiren ;
Rare Spick- und Lacc-Fürnisse ; Türckisch Papier : Item der vortreffliche Nürnberg Gold-Sträu-Glantz ; und viel andere umgemeine
Sachen zu machen, gelehret werden.   Mit einem Anhange von
denen Perlen und fast allen natürlichen Edelsteinen ; Wobey auch
in gewissen Tabellen eigentlich zu sehen, wie sich die künst-
lichsten derselben nach dem Gewicht an ihren Preisz verhöhen, und
einem vollständigen Register.   Alles hin und wieder in dieser
dritten Edition um ein merckliches vermehret.   Nürnberg, in
Verlegung Christoph Riegels, Buchhändlers unter der Vesten,
1743.   pp. [xii]-472-[xx], 4to.   Portrait, illustrated title-page and
plates.

>   First published in 1689 ; a French translation was issued at Paris in 1752,
>   4to.   The work contains Antonio Neri's Arte vitraria, with additions by
>   Kunckel and others.   *Cf.* Neri, Antonio.

Oeffentliche Zuschrifft von dem Phosphoro mirabili und dessen leuch-
tenden Wunder-Pilulen.   Sammt angehängten Discurs von dem

KUNCKEL, JOHANN, VON LÖWENSTERN.   [Cont'd.]
weyland recht benahmten Nitro, jetzt aber unschuldig genandten
Blut der Natur.   Wittemberg, 1678.   pp. [xii]–51–[iv], 16mo.

Philosophia chemica experimentis confirmata, in qua agitur de principiis
chymicis, salibus acidi et alcalibus, fixis et volatilibus, in tribus illis
regnis, minerali, vegetabili et animali, itemque de odore et colore,
etc.; accedit perspicilium chymicum contra nonentia chymica.
Amstelædami, 1694.   pp. [xiv]–333, 16mo.

> An Experimental Confirmation of Chymical Philosophy,
> treating of the principles of the chymists in the animal
> and mineral kingdoms ; as also of smell and colour
> [etc.].   London, 1730.   8vo.

*Vollständiges Laboratorium chymicum worinnen von den wahren
principiis in der Natur, der Erzeugung, den Eigenschaften und der
Scheidung der Vegetabilien, Mineralien und Metalle, wie auch von
Verbesserung der Metalle gehandelt wird.   4 Theile.   Vierte verbesserte Auflage.   Berlin, in der Rüdigerschen Buchhandlung,
1767.   pp. [xx]–671–[xvii], 8vo.   1 Plate.

> First edition, 1716.

Contains Kunckel's account of the discovery of elementary phosphorus.

KUNEMAN, J.
Étude sur la fermentation spontanée, procédé de fermentation sans
levure de bière ; travail de la mélasse de betteraves.   Paris, 1892.
8vo.

KUNTZE, O.
Die wichtigsten Lehren der Chemie.   Für den Gebrauch an höheren
Schulen, besonders an Gymnasien zur Ergänzung physikalischer
Lehrbücher und speciell als Anhang an die 20 Auflage des Brett-
nerschen Leitfadens für den Unterricht in der Physik bearbeitet.
Stuttgart, 1884.   8vo.   Ill.

KUPFFERSCHLÆGER, IS.
Appréciation des nouvelles théories chimiques.   Bruxelles, 1880.   8vo.
Éléments de chimie toxicologique à l'usage des pharmaciens et des
médecins experts.   Liège et Bruxelles, 1879.   pp. 180, 8vo.
Tableaux d'analyse chimique résumant l'exécution de la méthode
d'élimination générique.   Liège, 1879.   8vo.
Traité d'analyse chimique quantitative par la voie humide.   Bruxelles,
1878.   8vo.

KURELLA, ERNST GOTTFRIED.
Chymische Versuche und Erfahrungen.   Berlin, 1756–59.   Five parts.
8vo.

KURRER, JACOB WILHELM HEINRICH VON.

Druck- und Farbekunst (Die) in ihrem ganzen Umfang.   Wien, 1848–
      1850.   3 vols., 8vo.

Kunst (Die) vegetabilische und andere Stoffe zu bleichen.   Nürnberg,
      1831.

Neuesten (Die) Erfahrungen in der Bleichkunst.   Nürnberg, 1838.

Ueber das Bleichen der Leinwand und der leinenen Stoffe.   Braun-
      schweig, 1850.

         See also in Section III.

KURZES REPETITORIUM DER CHEMIE.   Zur Preparation zu den Prü-
      fungen der Mediciner, Pharmaceuten und Lehramtskandidaten,
      bearbeitet nach den Werken und Vorlesungen von Fresenius,
      Gorup-Besanez, Graham-Otto, Hager u. A.   Theil II : Organische
      Chemie.   Wien, 1891.   8vo.

KURZES REPETITORIUM DER CHEMISCHEN ANALYSE für Mediciner, Phar-
      maceuten, Chemiker und Agronomen, gearbeitet nach den Werken
      und Vorlesungen von Weidel, Fresenius, Lieben, Will, Medicus
      und Andere.   Abtheilung I ; Qualitative Analyse.   Wien, 1892.   8vo.

KURZGEFASSTE ANLEITUNG ZUR QUALITATIVEN CHEMISCHEN ANALYSE.
      Giessen, 1886.   8vo.

Куторга, С.

   Химическіе законы образованія, разрушенія и превращенія минеральныхъ
      формъ.   С.-Петербургъ 1851.

         KUTORGA, S.   The chemical laws of the formation, solution and trans-
            formation of mineral substances.   St. Petersburg, 1851.

KWEEKSCHOOL DER SCHEYKUNDE.   Zynde verzamelingen en waarnee-
      mingen uyt de beste en nieuwste schryvers getrokken.   Ten
      dienste der geene, die zich in deeze weetenschappen willen œfenen.
      Amsterdam, 1773.   8vo.

L———.

Stellvertreter des indischen Zuckers, oder der Zucker aus Runkelrüben.
      Berlin, 1799.

L., F. C.

Sammlung acht hundert und sieben und fünfzig chymischer Experi-
      mente einer Gesellschaft in dem Ertzgebürge, darinnen alle die
      Erscheinungen, welche man bey chymischer Bearbeitung verschie-
      dener Körper wahrgenommen treu und aufrichtig angezeiget
      werden, nebst einer Vorrede begleitet von E. G. Kurella.   Berlin,
      1759.   6 parts.   pp. [viii]–404, 8vo.

         The preface is signed " von dem Verfasser, F. C. L."   Each part has a
            distinct title-page.

LAACHE, S.
Harn-Analyse für praktische Aerzte.   Leipzig, 1885.   8vo.   Ill.

LABARRAQUE, ANTOINE GERMAIN.
Emploi (L') de chlorures de sodium et de chaux.   Paris, 1825.
Manière de se servir du chlorure d'oxide de sodium.   Paris, 1825.

LABAT, A.
Du degré de certitude de l'analyse des eaux.   Paris, 1885.   8vo.

ЛАБЕНСКІЙ, П.
Химическіе анализы винъ, встрѣчающихся въ продажѣ въ Петербургѣ и
ихъ полмѣси. С.-Петербургъ 1876.
> LABENSKY, P.   Chemical analyses of wines sold in St. Petersburg, and their
> adulterations.   St. Petersburg, 1876.

LABORATORIES (in general).
> See Eggert, Körner, Fremy, Carnot and Kolbe.

LABORATORIES (special, description of).
> See, in Section III, under the following towns : Berlin, Chemnitz, Freiburg,
> Giessen, Göttingen, Graz, Greifswald, Heidelberg, Klausenburg, Leip-
> zig, Manchester, Marburg, München, Paris, Pest, Prag, Strassburg,
> Wien, Wiesbaden, Zürich.

LABORDE, J. V., ET H. DUQUESNIL.
Des aconits et de l'aconitine, histoire naturelle, chimie et pharmacolo-
gie, physiologie et toxicologie, thérapeutique.   Paris, 1883.   8vo.
Ill.

LACOMBE.
Cours de chimie analytique à l'Institut industriel du Nord de la France,
1888–'89 (Deuxième année d'études).   Lille, 1889.   4to.   Ill.

LADAME, HENRI.
Essai sur la composition et la constitution de l'atmosphère.   Neufchatel,
1846.   8vo.

LADD, E. F.
Report of the Assistant Chemist to the New York Agricultural Experiment Station.   Elmira, N. Y., 1887.   8vo.

LADENBURG, A.
Theorie der aromatischen Verbindungen.   Braunschweig, 1876.   8vo.

LADREY, C.
Chimie appliquée à la viticulture et à l'œnologie, leçons professées en
1856 [à Dijon].   Paris, 1857.   12mo.

LADUREAU, A. [Editor].
>   *See* Congrès betteravier   .   .   .   Paris.

LAGARAYE, CLAUDE TOUSSAINT MAROT.
Chymie hydraulique, pour extraire les sels essentiels des végétaux, animaux et minéraux avec l'eau pure.   Paris, 1746.   12mo.
>   Nouvelle édition, 1775.   pp. [xxiv]-512-[viii], 12mo.
>   *Chymia hydraulica, oder Neuentdeckte Handgriffe, vermittelst welcher man das wesentliche Saltz aus Vegetabilien, Animalien und Mineralien mit schlechtem Wasser ausziehen kan. Erfunden und anfänglich in französischer Sprache bekannt gemacht, nunmehro aber wegen Vortrefflichkeit der Sache ins Teutsche übersetzt von einem Liebhaber der Naturlehre. Zweyte verbesserte Auflage.   Frankfurt und Leipzig, 1755. pp. [xxxii]-364-[iv], 12mo.   2 folding plates.

LAGRANGE, EDMONDE JEAN BAPTISTE BOUILLON.
Essai sur les eaux minérales naturelles et artificielles.   Paris, 1811.
Manuel d'un cours de chimie, ou principes élémentaires théoriques et pratiques de cette science.   Paris, 1799.   3 vols., 8vo.
>   Troisième édition, Paris, An. XI, [1802.]   3 vols.
>   Cinquième édition, avec 25 planches et des tableaux.   Paris, 1812.   3 vols., 8vo.   Vol. I, pp. 4-xx-504 ; vol. II, pp. 746 ; vol. III, pp. 658.
>   A Manual of a Course of Chemistry, or a series of experiments and illustrations necessary to form a complete course of that science.   With seventeen plates.   London, 1800.   2 vols., 8vo.

>   *See also in Section II.*

LAHOUSSE, E.
La vie et l'affinité chimique.   Gand, 1891.   8vo.

LAINER, ALEXANDER.
Lehrbuch der photographischen Chemie und Photochemie.   Erster Theil.   Anorganische Chemie.   Wein, 1889.

LALIEU, A.
Manuel d'oxalimétrie, ou méthode de titrages fondée sur l'emploi combiné de l'acide oxalique et du permanganate de potasse applicable à l'essai de substances médicamenteuses, alimentaires, etc.   Bruxelles, 1881.   12mo.   Ill.

LAMARCK, JEAN BAPTISTE PIERRE ANTOINE DE MONNET, CHEVALIER DE.
Réfutation de la théorie pneumatique . . . Paris, 1796. *See in Section III ; compare in Section V*, Fourcroy, A. F. de : Philosophie chimique.

LAMBE, WILLIAM.
Researches into the Properties of Spring Water, with medical cautions against the use of lead in the construction of pumps, waterpipes, cisterns, etc. London, *n. d.* [1803]. 8vo.

LAMBERTI, ANDREAS VON.
Allerneuesten (Die) Fortschritte der Destillirkunst.   Dorpat, 1821. Part I, Alcoholometrie ; Part II, Pyrometrie.
Dampf- (Der) Destillir-Apparat . . . Dorpat, 1811.

LAMBOTTE, HENRI.
Établissements de produits chimiques. Considérations sur les émanations qui s'en échappent, sur la manière dont elles se disséminent dans l'atmosphère suivant les conditions météorologiques et sur la part d'influence qu'elles peuvent exercer sur les êtres exposés à leur action. Bruxelles, 1856. 12mo.
Nouvelle théoire de chimie organique basée sur les lois de la composition binaire. Bruxelles, 1840. 8vo.

LA MÉTHÉRIE, DE.
Essai analytique sur l'air pur et les différentes espèces d'air. Paris, 1785. 8vo.
Deuxième édition, Paris, 1788.   2 parts, 8vo.

LAMPADIUS, WILHELM AUGUST.
Beiträge zur Atmosphärologie . . . Freyberg, 1816. 8vo.
Beyträge zur Erweiterung der Chemie und deren Anwendung auf Hüttenwesen, Fabriken und Ackerbau. Freyberg, Graz und Gerlach, 1804. 8vo.
Chemische Briefe für Frauenzimmer. Freiberg, 1818. 8vo.
Erfahrungen über den Runkelrübenzucker. Freiberg, 1800.
Erläuternde Experimente über die Grundlehren der allgemeinen und Mineral-Chemie, welche in dem Freyberger akademischen Lehrkurse in 1808–1809 angestellt wurden, nach eigenen Beobachtungen gesammelt und herausgegeben von Johannes Breisig. Freyberg, Graz und Gerlach, 1809–1810. 2 vols., 8vo.
Experimente über die technische Chemie . . . Freyberg, 1815. 8vo.
Grundriss der Elektrochemie. Freyberg, 1817. 8vo.
Grundriss der Hüttenkunde. Göttingen, 1827. 8vo.
Manuel de métallurgie générale, suivi d'additions extraites du Supplément de Lampadius, traduit, revue, augmen-

LAMPADIUS, WILHELM AUGUST. [Cont'd.]
        tée et mis au niveau des connaissances actuelles, par G.
        A. Arrault. Paris, 1840. 2 vols., 8vo. Plates.
Grundriss der technischen Chemie zum Gebrauch bey Vorlesungen
    und Selbstunterricht. Freyberg, 1815. 8vo.
Handbuch der chemischen Analyse der Mineralkörper. Freyberg, 1801.
    8vo. Plate.
        Nachträge . . . Freyberg, 1818. 8vo.
Kurze Darstellung der vorzüglichsten Theorien des Feuers, dessen
    Wirkungen und verschiedenen Verbindungen. Göttingen, 1793.
    8vo.
Neue Erfahrungen im Gebiete der Chemie und Hüttenkunde gesam-
    melt im chemischen Laboratorio zu Freyberg und in den Hütten-
    werken und Fabriken Sachsens in den Jahren 1808–1815, von W.
    A. L. Weimar, 1816. 2 vols., 8vo.
Sammlung practisch-chemischer Abhandlungen und vermischter Be-
    merkungen. Dresden, 1795–1800. 3 vols., 8vo. Plates.
Systematische Darstellung der einfachen Naturkörper . . . Freiberg,
    1806. Fol.

LAMPERT, A.
Leitfaden der unorganischen Chemie mit Berücksichtigung der neues-
    ten Ansichten der Wissenschaft. Brieg, 1875. 8vo.

LAMY, A.
        Leçon (etc.) : See Société chimique de Paris.

LAMY, H.
Unité de la matière, nouvelle théorie chimique, basée sur le calcul et
    confirmée par l'expérience. Paris, 1872. 8vo.

LANCELLOTTI, FRANCESCO.
Elementi di chimica. Napoli, 1835. 8vo.
Institutione di farmacia galenica. Institutione di chimica. Napoli,
    1817 [?]
Novella chimica con tutte le applicazioni. Napoli, 1819 [?] 4 vols.
Saggi analitici sulle acque minerali del territorio di Pozzuoli . . .
    Napoli, 1819. 8vo.

LANCILLOTTI, CARLO.
Guida alla chimica, che per suo mezo conduce gl' affezionati alle opera-
    zioni sopra ogni corpo misto animale, minerale o vegetabile. Dimo-
    strando come s' estraggono i loro sali, ogli, essenze, magisterii,
    mercurii, etc. Con il modo di fare varii colori, belletti et altri rari

LANCILLOTTI, CARLO. [Cont'd.]
   secreti. Opera utilissima a medici, speciali, alchimisti, pittori,
   orefici et altre persone curiose. Data in luce da C. L. Divisa in sei
   libri. Venetia, 1674. pp. [viii]–326–[xviii].

LANDAUER, F.
   Die Löthrohranalyse. Anleitung zur qualitativen chemischen Untersuchung auf trockenem Wege. Mit freier Benutzung von William Elderhorst's Manual of Qualitative Blowpipe Analysis.
   Braunschweig, 1876.
         Zweite Auflage, Berlin, 1881. 8vo.
         Analisi al cannello. Introduzione alla richerche chimiche
            qualitative per via secca. Tradotta da V. Fino. Tori-
            no, 1878. 8vo. Ill.
      Cf. Elderhorst, William.

LANDAUER, JOHANN.
   Systematischer Gang der Löthrohr-Analyse. Wiesbaden, 1877.
         Blowpipe Analysis. Authorized English edition. By
            James Taylor and William E. Kay. London, 1879.
            12mo.
         Second edition, revised and enlarged. London, 1892. 8vo.

LANDGREBE, GEORG.
   Ueber die chemischen und physiologischen Wirkungen des Lichtes.
      Ein Versuch. Marburg, 1834. pp. x–602, 8vo.
      An extensive monograph for the period.

LANDOLT, H.
   Das optische Drehungsvermögen organischer Substanzen und die prak-
      tische Anwendung desselben. Für Chemiker, Physiker und
      Zuckertechniker. Braunschweig, 1879. 8vo. Ill.
         Handbook of the Polariscope and its practical applications.
            Adapted from the German edition of H. Landolt . . .
            By D. C. Robb and V. H. Veley. With an appendix
            by J. Steiner. London, 1882. pp. xiv–262, 8vo. Ill.

LANDRIANI, MARSIGLIO.
   Opuscoli fisico-chimici. Milano, 1781. 8vo. Ill.
   Ricerche fisiche intorno alle salubrità dell' aria. Milano, 1775. 4to.
      A German translation was published at Basel in 1778.

LANDRIN, E. ET H.
   Nouveau manuel complet de la fabrication et de l'application des en-
      grais animaux, végétaux et minéraux et des engrais chimiques.
      Nouvelle édition. Paris, 1888. Ill.

LANG, VICTOR.

Die Fabrikation von Kunstbutter, Sparbutter und Butterine. Eine Darstellung der Bereitung der Ersatzmittel der echten Butter nach den besten Methoden. Allgemein verständlich geschildert. Wien, Pest, Leipzig, 18—. Ill.

ЛАНГЪ, В. фонъ.

Динамическая теорія газовъ. Перев. Левицкій М. С.-Петербургъ 1874.

LANG, V. VON. The dynamic theory of gases. St. Petersburg, 1874.

LANGBEIN, G.

Die Genussmittel. Leipzig, 1869. 8vo.

LANGE, CHRISTIAN.

Collegium chymicum [edited by Rivinus]. Lipsiæ, 1704.

LANGE, JOHANN JOACHIM.

Grundlegung zu einer chemischen Erkenntniss der Körper. Halle, 1770. 8vo.

LANGER, CARL, UND VICTOR MEYER.

Pyrochemische Untersuchungen von C. L. und V. M. mitgetheilt von Letzterem. Mit 17 in den Text eingedruckten Holzstichen. Braunschweig, 1885. 8vo.

LANGER, THEODOR.

Grundriss der Chemie für Brauer und Mälzer. Zweite Auflage. Leipzig, 1890.

Lehrbuch der Chemie, mit besonderer Berücksichtigung der Gährungsgewerbe. Verfasst nach paedagogischen Grundsätzen. Leipzig, 1878. pp. xii–451, 8vo. Ill.

LANGGUTH, CHRISTIAN AUGUST.

De chemiæ recentioris præstantia. Vitebergæ, 1779.

LANGHOFF, F.

Lehrbuch der Chemie. Zum Gebrauche an Schullehrer-Seminarien, höheren Bürgerschulen, Mittelschulen, höheren Knaben- und Töchterschulen, Gymnasien, Ackerbauschulen, Gärtner-Lehranstalten, Handwerker-Fortbildungsschulen. Vierte nach der neueren Theorie und neuer Rechtschreibung umgearbeitete und vermehrte Auflage. Leipzig, 1883. 8vo.

LANGLEBERT, J.

Chimie. Avec 125 gravures dans le texte. Paris, 1857. 8vo.

LANGLEBERT, J. [Cont'd.]

Trente-quatrième édition tenue au courant des progrès de la science les plus rècents (1882) et augmentée d'un appendice de la mécanique chimique. Paris, 1882. 12mo.

Trente-neuvième édition. Paris, 1888.

Química. Traduccion de la última edicion francesa por L. Elizaga. Paris, 1886. 12mo.

LAPA, J. I. F.

Chimica agricola, ou estudo analytico dos terrenos, das plantas e dos estrumes. Lisboa, 1875. 8vo.

LAPOSTOLLE [of Amiens].

Plan d'un cours de chimie expérimentale. Amiens, 1777.

LARBALÉTRIER, A.

Alcool (L') au point de vue chimique, agricole, industriel, hygiénique et fiscal. Paris, 188–. 12mo.

Bibliothèque scientifique contemporaine,

Analyse chimique des matières agricoles. Notions élémentaires simplifiées. Paris, 1887. 2 plates.

Engrais (Les) chimiques et les matières fertilisantes d'origine minérale ; emploi pratique des engrais chimiques. Paris, 1885.

LARDNER, DIONYSIUS.

Anorganische und organische Chemie für Schulen. Deutsche Uebersetzung von G. Tröbst. Neue wohlfeile Titel-Ausgabe. Weimar, 1866. 8vo. Ill.

LARKIN, F. C., AND LEIGH, R.

Outlines of practical Physiological Chemistry. Second edition, enlarged and revised. London, 1891. 8vo.

LARRIEU, F.

De l'analyse chimique et microscopique de l'urine. Toulouse, 1875. 8vo.

LASSAIGNE, JEAN LOUIS.

Abrégé élémentaire de chimie considérée comme science accessoire à l'étude de la médecine, de la pharmacie et de l'histoire naturelle. Paris, 1829.

Deuxième édition. Paris, 1839.

Troisième édition. Paris, 1842. 2 vols., 8vo. I, pp. x– 704 ; II, pp. 754. Ill.

LASSAIGNE, JEAN LOUIS. [Cont'd.]

Quatrième édition. Paris, 1846.

Atlas [de ,la chimie]. Paris, 1846. Seven plates and 16 colored tables.

Compendio elementare di chimica, considerata come scienza accessoria della medicina, farmacia, e della storia naturale. Tradotto da A. Buffini. Milano, 1839. 8vo.

Tratado completo de química, considerada como ciencia accesoria al estudio de la medicina, de la farmacia y de la historia natural. . . . Traducido de la tercera y última edicion francesa por Francisco Alvarez Alcalá. Madrid, 1844. 3 vols., 8vo. 15 plates and 4 tables.

LASSAR-COHN.

Arbeitsmethoden für organisch-chemische Laboratorien. Ein Handbuch für Chemiker, Mediciner und Pharmaceuten. Hamburg, 1891. pp. ix–339, 8vo. Ill.

Méthodes de travail pour les laboratoires de chimie organique. Traduit par F. Ackermann. Paris, 1892. 12mo. Ill.

Moderne Chemie. Zwölf Vorträge vor Aerzten gehalten. Hamburg, 1891. 8vo.

LÁSZLÓ, E. D.

Chemische und mechanische Analyse ungarländischer Thone mit Rücksicht auf ihre industrielle Verwendbarkeit. Ungarisch und Deutsch. Budapest, 1886. 8vo.

LATSCHENBERGER, J.

Kurze Anleitung zur qualitativen chemischen Mineralanalyse für Mediciner. Freiburg, 1883.

LAU, L., UND A. HAMPE.

Praktischer Unterricht in der heutigen Wollenfärberei. Wien, 1892. 8vo.

LAUBENHEIMER, AUGUST.

Grundzüge der organischen Chemie. Heidelberg, 1884. 8vo.

LAUBER.

Handbuch des Zeugdrucks. Mit Abbildungen und Zeugproben. Unter Mitwirkung von A. Steintheil und M. Kohn. Leipzig, 1884. 8vo.

Praktisches Handbuch des Zeugdrucks. Herausgegeben unter Mitwirkung von J. Langer und M. Zelakowski. Leipzig, 1888. 8vo. With samples.

LAUGA, A.

Précis élémentaire de chimie, formant une collection de douze tableaux synoptiques représentant un cours en douze leçons. Bordeaux et Paris, 1838. 4to.

LAUGIER, A.

Cours de chimie générale. Paris, 1829. 4 vols., 8vo. Vol. I, Première à dix-huitième leçon [each separately paginated]. Vol. II, pp. xvi–511. Vol. III, pp. xxiv–541. Vol. IV, plates and indices, pp. 64 and 8 plates.

LAUGIER, E., ET A. DE KRAMER.

Tableaux synoptiques, ou abrégé des caractères chimiques des bases salifiables. Paris, 1828. 8vo.

ЛАУПМАНЪ, А.

Карболовая кислота съ точки зрѣнія химической, гигіенической, медицинской, сельско-хозяйственной и физіологической. С.-Петербургъ 1872.

> LAUPMAN, A. Carbolic acid from the standpoint of chemistry, hygiene, medicine, agriculture and physiology. St. Petersburg, 1872.

LAURENT, AUGUSTE.

Méthode de chimie. Paris, 1854. pp. xxii–464, 8vo.

> Chemical Method, notation, classification and nomenclature translated by William Odling. Printed for the Cavendish Society, London, 1855. pp. xxiii–382, 8vo.

LAVÉN, A. W.

Oorganisk Kemi. Andra upplagan. Stockholm, 1855. 8vo.

LAVOISIER, ANTOINE LAURENT.

Mémoires de chimie. Paris, 1805. 2 vols., 8vo.

> Published by Madame Lavoisier.

Œuvres, publiées par les soins de son Excellence le Ministre de l'Instruction publique et des cultes. Paris, 1864–92. 5 vols., 4to. Vol. I, pp. xi–728, 16 plates, 1864. Vol. II, pp. 828, 8 plates, 1862. Vol. III, pp. 795, 12 plates, 1865. Vol. IV, pp. 775, 4 plates, 1868. Vol. V, 1892.

* Opuscules physiques et chymiques. Tome premier. Paris, 1774. pp. xxx–436, 8vo. Three folding plates.

> This classical work deserves a full analysis. It consists of two distinct parts, the first bearing the title : " Précis historique sur les émanations élastiques qui se dégagent des corps pendant la combustion, pendant la fermentation et pendant les effervescences." In this historical essay the author, beginning with Van Helmont and his "gas sylvestre," traces the successive discoveries of Boyle, Hales, Black, Cavendish and Priestley and joins with them the theoretical views of the chemists

LAVOISIER, ANTOINE LAURENT.   [Cont'd.]

named and of Boerhaave, Stahl, Meyer, Rouelle and others of less fame.   To the experiments of Priestley he devotes about fifty pages, and everywhere gives him credit and praise.   In the concluding paragraph Lavoisier says : "These nineteen chapters contain all that I could obtain of interest on fixed air," and refers to papers by Rutherford and Weigel.

The second part contains Lavoisier's original researches on fixed air.   The book concludes with a Report commending Lavoisier's essays to the Academy of Sciences, signed by Trudaine, Macquer, Le Roy and Cadet and dated December 7, 1773, eight months before the discovery of oxygen by Priestley.

Essays, Physical and Chemical.   Translated from the French with notes, and an Appendix by Thomas Henry.   London, 1776.   pp. xxxii–475, 8vo.   Plates.

Physikalisch-chemische Schriften.   Aus dem französischen übersetzt von Christian Ehrenfried Weigel.   Greifswald, 1783–94.   5 vols., 8vo.

Vols. IV and V : fortgesetzt von H. F. Link.

* Traité élémentaire de chimie présenté dans un ordre nouveau et d'après les découvertes modernes.   Avec figures.   Paris, 1789.   2 vols., 8vo.   Vol. I, pp. xliv–322.   Vol. II, pp. viii, from 323 to 653.   13 folding plates [engraved by Madame Lavoisier].

*Deuxième édition.   Paris, 1793.   Vol. I, xliv–322 ; II, viii–331.

Troisième édition, corrigée et augmentée de plusieurs mémoires nouveaux.   Paris, An IX, 1801.   2 vols., 8vo.   I, pp. xliv–386 ; II, pp. vii–377.

Elements of Chemistry in a new systematic order, containing all the modern discoveries, illustrated by thirteen copperplates.   Translated from the French by Robert Kerr.   Edinburgh, 1790.   8vo.

Second edition, Edinburgh, 1793.

Fourth edition, 1799, pp. 592, 8vo.   Plates.

Fifth edition, with considerable additions.   Edinburgh, 1802.   2 vols., 8vo.   Also, New York, 1806.   2 vols., 8vo.

System der antiphlogistichen Chemie.   Aus dem französischen übersetzt und mit Anmerkungen und Zusätzen begleitet von Sigismund Friedrich Hermbstädt.   Zweite Ausgabe.   Berlin und Stettin, 1803.   2 vols., 8vo.

Trattato elementare di chimica presentato in un ordine nuovo.   Con figure.   Recato dalla francese nell' italiana favella e corredato di annotazioni da Vincenzo Dandolo.   Edizione seconda.   Veneto, 1792.   4 vols., 8vo.

LAVOISIER, ANTOINE LAURENT. [Cont'd.]
> *In Dutch*, translated by Hemery and Van Werkhoven.
> Utrecht, 1789, 1792, 1795, and 1801.
>
> It is hardly possible to estimate too highly the influence which Lavoisier's work exerted on the history of chemistry. The first treatise on chemistry in which the phlogistic theory was dropped and the new theory of combustion was clearly promulgated, in conjunction with the "Méthode de nomenclature chimique," published two years before, it changed the existing language of chemistry and shaped the course of progress still pursued.

LAVOISIER, ANTOINE LAURENT.
> *See* Nieuwland, Pieter.

LAWES, J. B., AND J. H. GILBERT.
> Agricultural Chemistry in relation to the mineral theory of Baron Liebig. London, 1851. pp. 41, 8vo. Folding plate.
>> Entgegnung auf Baron Liebig's Grundsätze der Agricultur-Chemie, mit Rücksicht auf die in England angestellten Untersuchungen. Leipzig, 1856. 8vo.

ЛАВРОВЪ, Н. И.
> Неорганическая химія. С.-Петербургъ, 1865.
>> LAVROFF, N. I. Inorganic chemistry. St. Petersburg, 1865.

ЛАВРОВЪ, Н.
> Конспектъ свѣдѣній изъ химіи и товаровѣдѣнія. С.-Петербургъ, 1872.
>> Summary of industrial chemistry. St. Petersburg, 1872.

LEAPER, CLEMENT J.
> Outlines of Organic Chemistry. London, 1892. 8vo.

LEBEL, J. A.
> Conférence (etc.). *See* Société chimique de Paris.

LEBLANC, ——.
> Cours de chimie analytique. École centrale des arts et manufactures. Deuxième année. Année scolaire 1870–71. Paris, 1871. 8vo.

LEBLANC, NICOLAS.
> Mémoires sur la fabrication du sel ammoniac et de la soude. Paris, 1798.

LEBLANC, NICOLAS, ET LA SOUDE ARTIFICIELLE.
> *See, in Section III*, Scheurer-Kestner ; *also in Section IV*, Anastasi, Aug.

LEBLANC, R.
> Manipulations de chimie. Troisième édition. Paris, 1883.

LE CANU, LOUIS RENÉ.
Études chimiques sur le sang humain.   Paris, 1837.   8vo.
Nouvelles études chimiques sur le sang.   Paris, 1852.
                Nuevos estudios químicos sobre la sangre por L. R. L.
                acompañados del informe de Thénard, Dumas y Andral.
                Traducidos por Luis Fernandez Molina.   Madrid, 1854.
                4to.
Recherches sur les corps gras.   Paris, 1843.   8vo.

LE CAT, CLAUDE NICOLAS.
Dissertation sur la nature du fluide des nerfs et son action pour le
    mouvement musculaire.   Berlin, 1753.   8vo.
        Crowned by the Berlin Academy.

LECHARTIER, G.
Cours de chimie agricole, professé en 1873.   Rennes, 1874.   12mo.

LECHAT, F.
Notions élémentaires de chimie.   Paris, 1875.

LE CHATELIER, H.
        Conférence (etc.).   *See* Société chimique de Paris.

LE CHATELIER, H.
Recherches expérimentales et théoriques sur les équilibres chimiques.
    (Extraits des Annales des Mines, livraison de Mars-Avril, 1888.)
    Paris, 1888.   pp. 230, 8vo.

LECLERC, J.
Cours complet de chimie.   Paris, 1887.

LEÇONS DE CHIMIE.
        See Société chimique de Paris.

LECOQ DE BOISBAUDRAN.
        Gallium [and other articles].   *See, in Section II*, Fremy : Encyclopédie
        chimique, vol. III.

LECOQ DE BOISBAUDRAN.
Spectres lumineux.   Spectres prismatiques et en longueurs d'ondes
    destinés aux recherches de chimie minérale.   Paris, 1874.   2 vols.,
    roy. 8vo.   The second volume is an " Atlas " of 29 plates.

LECOQ, HENRI.
Tableau synoptique de chimie minérale.   Toulouse, 1840.   Fol.

LECOQ, HENRI, ET J. P. L. GIRARDIN.
Éléments de chimie appliquée aux sciences.   Paris, 1826.   2 vols., 8vo.
> *Cf.* Girardin, J. P. L.

LEDEBUR, A.
> Die Metallverarbeitung.   *See* Bolley's Handbuch [etc.].

LEDEBUR, A.
Leitfaden für Eisenhütten-Laboratorien.   Braunschweig, 1880.   8vo.
> Zweite durch einem Nachtrag vermehrte Ausgabe.   Braun-
> schweig, 1885.   8vo.
> Dritte neu bearbeitete Auflage.   (Separat-Abdruck aus der
> " Chemisch-technischen   Analyse,"   zweite   Auflage,
> herausgegeben von Julius Post.)   Braunschweig, 1889.
Legierungen (Die) in ihrer Anwendung für gewerblichen Zwecke.   Ein
Hand- und Hülfsbüchlein für sämmtliche Metallgewerbe.   Berlin,
1890.
Metalle (Die), ihre Gewinnung und ihre Verarbeitung allgemein fasslich
dargestellt.   Mit 64 in den Text eingedruckten Holzschnitten.
Stuttgart, 1886.   8vo.
Handbuch der Eisenhüttenkunde.   Leipzig, 1884.

LE DOCTE, A.
Contrôle chimique de la fabrication du sucre.   Tableaux numériques,
etc.   Bruxelles, 1883.   4to.

LEDUC, A.
Cours élémentaire de chimie, rédigé conformément au programme de
1885 pour la classe de rhétorique et le baccalauréat ès lettres.
Paris, 1885.   12mo.

LEE, ALEXANDER.
Chemical Diagrams accompanied with a concise description of each
decomposition : the vegetable alkalies ; the urine and urinary
calculi ; tables of chemical equivalents, etc., intended to facilitate
the progress of the medical student.   London, 1833.   pp. xix–182,
16mo.

LEEDS, ALBERT R.
> *See* Morton, Henry, and Albert R. Leeds.

LEEDS, ALBERT R.
Method of Separation of Coloring Matters in Butter, Imitation Butter,
and so-called Butter Colors.   (In Report of the Dairy Commissioner of the State of New Jersey for 1886.)   Trenton, N. J., 1887.
8vo.

LEERSTELLINGEN (DE) DER MODERNE CHEMIE. Bevattelijk voorgesteld voor studenten in de chemie. Uit het Hoogduitsch door J. D. H. V. Utrecht, Amsterdam, 1870.

LEFEBURE, NICOLAS.
Traicté de la chymie. Paris, 1660. 2 vols., 12mo.
> *Deuxième édition, corrigé de plusieurs fautes. Leyde,1669. 2 vols., 24mo. Vol. I, pp. [lxvi]–556 ; II, [i]–from 557 to 1216–[xxi]. Engraved title-page. Plates.
> Nouvelle édition fort augmentée. Paris, 1674. 12mo.
> Cinquième édition, par Du Moustier. Paris, 1751. 5 vols., 12mo.
> A Compleat Body of Chymistry : wherein is contained whatsoever is necessary for the attaining to the curious knowledge of this art ; comprehending in general the whole practise thereof : and teaching the most exact preparation of animals, vegetables and minerals, so as to preserve their essential vertues. Laid open in two books, and dedicated to the use of all apothecaries, &c. Renderd into English, by P. D. C., Esq. ; one of the Gentlemen of His Majesties Privy-Chamber, with additions. London, 1670. 2 vols., 4to. Vol. I, pp. [xii]–286–[vi] ; vol. II, pp. 320–[viii]. Folding plate.

On the title-page the author is called Nicasius Le Febure.

> Chymischer Handleiter und Guldenes Kleinod, das ist, richtige Anführung . . . wie man die chymische Schrifften . . . recht verstehen . . . möge. Aus dem französischen von Arnold Doude. Nürnberg, 1676. 12mo.
> [Another edition], von Cardilucio. Nürnberg, 1688. pp. [lii]–1149–[xix], 12mo.

Nicolas Lefebure was demonstrator of chemistry at the Jardin des Plantes under Louis XIV. In 1664 he was called to London and became an honored member of the newly founded Royal Society. He died in 1674. His Traicté de la chymie is logical, systematic and free from affectation of mystery.

LEFÈVRE, NICOLAS.
See Lefebure.

LEFÈVRE.
> Teinture et apprêts des tissus de coton. *See, in Section II*, Fremy : Encyclopédie chimique, vol. x.

LEFFMANN, HENRY.
Compendium of Chemistry, Inorganic and Organic. With full explanations of difficult points. Philadelphia, 1881. 16mo.

LEFFMANN, HENRY. [Cont'd.]
> Second edition. Philadelphia, 1888.
> Compend of Chemistry, inorganic and organic. Including urine analysis. Third edition. Revised and adapted for Students of Medicine and Dentistry. Philadelphia, 1890. 12mo.
>> ? Quiz-Compends ? No. 10.

Elements of Chemistry. Philadelphia, 1882. pp. 227. 8vo. Ill.
> Butler's Science Series.

First steps in Chemical Principles ; Introduction to modern chemistry, intended especially for beginners. Philadelphia, 1880. 16mo.

LEFFMANN, HENRY, AND W. BEAM.
Examination of Water for sanitary and technical purposes. Philadelphia, 1889. 8vo. Ill.
> Second edition. Philadelphia, 1891.

LEFORT, JULES.
Chimie des couleurs pour la peinture à l'eau et à l'huile comprenant l'historique, la synonymie, les propriétés physiques et chimiques, la préparation, les variétés, les falsifications, l'action toxique et l'emploi des couleurs anciennes et nouvelles. Paris, 1855. 8vo.
Traité de chimie hydrologique comprenant des notions générales d'hydrologie et l'analyse chimique qualitative et quantitative des eaux douces et des eaux minérales. Paris, 1859. 8vo.
> Deuxième édition. Paris, 1873. 8vo. Ill.

LE GIVRE, PIERRE.
L'anatomie des eaux minérales de Provins. Paris, 1654.
Le secret des eaux minérales acides nouvellement découvert par le moyen des principes chimiques. Paris, 1667.

LEHMANN, AUGUST.
Gründliche Anweisung zur Schnell-Essig Fabrikation, oder die Kunst in Zeit von zwei Stunden einen guten, scharfen, chemisch-reinen Essig ohne bedeutende Kosten zu bereiten, so wie die Fabrikation des Doppel-Essigs. Theoretisch und praktisch dargestellt. Zweite verbesserte Auflage. Quedlinburg, 1854.
> Dritte Auflage. Quedlinburg, 1857. 8vo.

LEHMANN, CARL GOTTHELF.
Handbuch der physiologischen Chemie mit besonderer Berücksichtigung der zoochemischen Dokimastik. Leipzig, 1854. 8vo.
> Zweite Auflage. Leipzig, 1859. 8vo. Ill.

LEHMANN, CARL GOTTHELF. [Cont'd.]

Handboek der physiologische scheikunde. Onder medewerking van den schrijver vertaald door L. Egeling.
Utrecht en Amsterdam, 1856. 8vo.

Manual of Chemical Physiology. Translated with notes
and additions by J. Cheston Morris, with an introductory essay on vital force by Samuel Jackson.
Philadelphia, 1856. 8vo.

Précis de chimie physiologique animale. Traduite de
l'Allemand par Ch. Drion. Paris, 1855. 18mo. Ill.

Краткая животно-физiологич. химiя перев. Ходнева А.
С.-Петербургъ 1860.

Lehrbuch der physiologischen Chemie. Leipzig, 1842–52. 3 vols., 8vo.

Zweite Auflage. Leipzig, 1853. 3 vols., roy. 8vo.

Physiological Chemistry. Translated from the second edition
by George E. Day. London, 1851. 3 vols., 8vo. I, pp.
xii–455 ; II, pp. xi–465 ; III, pp. xii–579.

> Printed for the Cavendish Society. *For* Supplementary Atlas *see* Funke,
> Otto.

[*The same :*] Edited by R. E. Rogers. With illustrations
selected from Funke's Atlas of physiological chemistry
and an appendix of plates. Philadelphia, 1855. 8vo.

Vollständiges Taschenbuch der theoretischen Chemie, zur schnellen
Uebersicht und leichten Repetition bearbeitet. Leipzig, 1840. 16mo.

Zweite wesentlich verbesserte und vermehrte Auflage. Leipzig, 1842. 8vo.

Dritte verbesserte und vermehrte Auflage. Leipzig, 1846.

Vierte vollkommen umgearbeitete Auflage. Leipzig, 1850.

Fünfte Auflage. Leipzig, 1851. Sechste Auflage. Leipzig,
1854.

Volledig zakboek der theoretische scheikunde. Naar het
Hoogduitsch door N. W. de Voogt. II deelen. Utrecht
en Amsterdam, 1840–41. 12mo. 1. deel, Onbewerktuigde scheikunde. 2. deel, Bewerktuigde scheikunde.

Полное карманное руководство къ теоретической химiи К. Г.
Лемана. Съ шестаго нѣмецкаго изданiя. Перев. Петръ Бо-
борыкинъ. III тома. С.-Петербургъ, Н. О. Рольфа, 1855—56.
16°.

> Third edition, 1867.

Zoochemie, in Verbindung mit Huppert bearbeitet und herausgegeben.
Heidelberg, 1858. pp. iv–734, 8vo.

> Reproduced from C. G. Gmelin's Handbuch der organischen Chemie, vol.
> VIII, part 2.

LEHMANN, JOHANN GOTTLOB.

Abhandlung von den Metallmüttern und der Erzeugung der Metalle aus der Naturlehre und Bergwerkswissenschaft hergeleitet und mit chymischen Versuchen erwiesen. Berlin, 1753. 8vo.

Abhandlung von Phosphoris deren verschiedener Bereitung, Nutzen, u. s. w. Dresden und Leipzig, 1749. 4to.

Разсужденіе о фосфорѣ. С.-Петербургъ, 1780.

Physikalisch-chemische Schriften. Berlin, 1761. 8vo.

LEHMANN, O.

Die Krystallanalyse, oder die chemische Analyse durch Beobachtungen der Krystallbildung mit Hülfe des Mikroskops. Leipzig, 1891.

LEHNER, SIGISMUND.

Die Tintenfabrikation, und die Herstellung der Tusche, der Stempel-Druckfarben, sowie der Waschblaues. Ausführliche Darstellung der Anfertigungen aller Schreib-, Comptoir- und Copirtinten, aller farbige und sympathetische Tinten, der chinesische Tusche, lithographische Stifte und Tinten unauslöschliche Tinten zum Zeichnen der Wäsche, sowie zur Ausführung von Schriften von jedem beliebenen Materiale, der Bereitung des besten Waschblaues und der Stempeldruckfarben. Nebst eine Anleitung zum Lesbarmachen alter Schriften. Ein Handbuch für Fabrikanten chemischer Producte, Apotheker, Kaufleute u. A. Nach eigenen Erfahrungen dargestellt. Wien, Pest und Leipzig, 1876. 8vo.

Zweite sehr vermehrte und verbesserte Auflage. Wien, Pest und Leipzig, 1880.

Dritte Auflage. Wien, 1885.

LEIDENFROST, IVAN GOTTLOB.

De aquæ communis nonnullis qualitatibus tractatus. Dinsburgi ad Rhenum, 1756. 8vo.

Contains the so-called " Leidenfrost's Experiment."

Opuscula physico-chemica et medica . . . Lemgoviæ, 1797-98. 4 vols., 8vo.

Published after his death in 1794.

LELLMANN, EUGEN.

Principien der organischen Synthese. Berlin, 1887. pp. xi-511, 8vo.

LE MARCHAND, E. V.

Chimie de l'unité ; étude comparative des mathématiques cosmiques par la science de l'arithmétique naturelle. Caën, 1886.

LEMERY, NICOLAS.

Cours de chimie contenant la manière de faire les opérations qui sont en usage dans la médecine, par une méthode facile ; avec les raisonnements sur chaque opération, pour l'instruction de ceux qui veulent s'appliquer à cette science. Paris, 1675. 8vo.

> Deuxième édition. Paris, 1677.
>
> [Other Paris editions.] Paris, 1679, 1682, 1683, 1687, 1690, 1696, 1697, 1698, 1701.
>
> Dixième édition. Paris, 1713.
>
> *Onzième édition, revûë, corrigée et augmentée par l'auteur. Paris, chez Jean Baptiste Delespine, 1730. pp. [xxiv]–938–[lviii], 8vo. Portrait and plates.
>
> [Another edition] Paris, 1738.
>
> [Other editions.] Leyde, 1697, 1716, 1730. 8vo. Bruxelles, 1744, 1747. 8vo. Avignon, 1751. 4to. Amsterdam, 1682 and 1698. 8vo.
>
> * Nouvelle édition revue, corrigée et augmentée d'un grand nombre de notes et de plusieurs préparations chymiques qui sont aujourd'hui d'usage et dont il n'est fait aucune mention dans les éditions de l'auteur, par [Theodore] Baron. Paris, 1756. pp. [iv]–xxiv–945, 4to. 9 folding plates.

Het philosoophische laboratorium, of der chymisten stook-huis, leerende . . . alle de gebruikelykste medica-menten op de chymische wyse bereiden. Na het laatste France exemplaar en met noodige aanteikeningen ver-rijkt. Amsterdam, 1683. 8vo.

> * A Course of Chymistry, containing the easiest manner of performing those operations that are in use in physik. Illustrated with many curious remarks and useful dis-courses upon each operation. Together with additional remarks to the former operations, the process of the volatile salt of tartar and some other useful prepara-tions by way of appendix. Translated by Walter Harris. London, printed for Walter Kettilby at the Bishop's Head in St. Paul's Church-Yard, 1680. pp. [xxxiv]–323–[xv]–[xv]–140–[xii], 12mo.

> The title-page to "A Course of Chemistry" is dated 1677, and to the "Appendix" 1680.

> A Course of Chymistry . . . Second edition inlarged. Translated from the fifth edition in the French by W. Harris. London, 1686. 8vo. Also, London, 1698.

LEMERY, NICOLAS. [Cont'd.]

A Course of Chemistry, containing an easie method of preparing those chymical medicines which are used in
physick, with curious remarks upon each preparation,
for the benefit of such as desire to be instructed in the
knowledge of this art. The fourth edition translated
from the eleventh edition in the French which has been
revised, corrected and much enlarged beyond any of
the former. London, 1720. 8vo.

Cours de Chymie, oder der vollkommene Chymist, welcher
die in der Medizin gebräuchlichen Processe machen
lernt. Dresden, 1734. 8vo. Ill.

Also Dresden, 1754. 8vo. Another edition, 1698.

Corso di chimica. Venezia, 1732. 8vo.

Du Fresnoy names an Italian edition dated 1700, also at Venezia, but gives
no particulars. Hoefer assigns the date 1763.

Cursus chymicus, continens modum parandi medicamenta
chymica usitatoria brevi et facili methodo una cum
notis et dissertationibus super unamquamque præparationem. Ex ultima editione Gallica Latine versus.
Geneva, 1681. pp. [x]–664. 16mo.

The numerous editions and translations attest the popularity of this work.
" Lemery avait le talent de décrire les choses les plus obscures et les
plus arides avec une simplicité et une précision remarquables. Ce
talent est la pierre de touche d'un esprit qui sait apprécier l'importance
des détails."—(Hoefer.)

Traité de l'antimoine, contenant l'analyse chymique de ce minéral, et
un recueil d'un grand nombre d'opérations rapportées à l'Académie
royale des sciences, avec les raisonnemens qu'on a crus nécessaires.
Paris, 1707. 12mo.

Neue curieuse chymische Geheimnisse des Antimonii. Aus
dem französischen durch Johann Andreas Mahler.
Dresden, 1709. 8vo.

Trattato dell' antimonio, che contiene l'analisi di questo
minerale, e una raccolta di gran numero di operazioni.
. . . Traduzione dal linguaggio francese di Selvaggio
Canturani. In Venezia, 1732. 8vo.

LEMERY, LOUIS.

Sur les précipitations chimiques. Paris, 1711.

Traité des aliments . . . Paris, 1702.

Deuxième édition. 1705.

LEMME, G.

Ueber die Trennung von Wismuth und Blei. Berlin, 1889. 8vo.

LEMOINE, G.

Essais sur les équilibres chimiques [and other articles]. *See, in Section II,*
Fremy : Encyclopédie chimique.

LE MORT, JACOB.

See Collectanea chymica Leidensia.

LE MORT, JACOB.

Chymia, rationibus et experimentis auctioribus, iisque demonstrationibus superstructa, in qua malevolorum calumniæ modeste simul
diluuntur. Lugduni-Batavorum, 1688. 8vo.
Chymiæ veræ nobilitas ut utilitas. Lugduni-Batavorum, 1696. 4to.
Compendium chymiæ. Lugduni-Batavorum, 1682. 12mo.

LENG, HEINRICH.

Vollständiges Handbuch der Zuckerfabrikation. Ilmenau, 1834. 8vo.

LENGERKE, A. H. VON.

Die gesammte Chemie der Gutswirthschaft in fünffarbigen Bildern
nebst kurzem Text. Fasslich für Jedermann zur Aufklärung und
Anregung. Görlitz, 1875. 8vo. Ill.

LENGYEL, BELA.

Chemia. Tankönyv a felsö tanintézetek számára Számos ábrával és
egy színkep-táblával. Budapest, 1889. Elsö kötet Szervetlen
chemia. pp. xiii–536. 8vo.

LENOBLE.

Manual de química elemental, para los alumnos de la Universidad de
farmacia, industriales, mineros, etc. Paris, 1857. 18mo. Ill.
Segunda edicion. Paris, 1861. 18mo.
[Another edition.] Paris, 1877.

LE NOIR.

Chimie élémentaire. Paris, 1881.
Deuxième édition. Paris, 1888. Ill.

LE NORMAND, LOUIS SÉBASTIEN.

Art (L') du distillateur des eaux-de-vie. . . . Paris, 1817. 2 vols., 8vo.
Essai sur l'art de la distillation . . . Paris, 1811. 2 vols., 8vo.
Manuel de l'art du dégraisseur . . . Paris, 1819. 12mo.
[Another edition.] Paris, 1826.

LENZ, LEOPOLD.

Kurze Anleitung der qualitativen chemischen Analyse. Iglau, 1873.
8vo.

LEONHARDI, JOHANN GOTTFRIED.
   De pyrophoro aluminari theoriæ vindiciæ. Vitebergæ, *n. d.* [1789].
   Two parts. pp. 14–14, 4to.

LEONHARDT, J. H.
   Handbuch der pharmazeutischen Chemie, oder Darstellung und Prü-
      fung der sämmtlichen chemisch-pharmazeutischen Präparate. Zum
      praktischen Gebrauche für Physiker, Aerzte und Apotheker. Mit
      einer Vorrede von A. du Mênil. Hannover, 1825. 8vo.

LEPLAY, H.
   Chimie théorique et pratique des industriels du sucre. La mélasse
      dans la fabrication et le raffinage des sucres de betteraves et de
      cannes. Paris, 1883. 8vo.
   Études chimiques sur la formation du sucre et de tous les principes
      organiques définis contenus dans les différentes parties des plantes
      sucrées et de leurs fonctions dans la végétation. Paris, 1889.
   Osmomètre, endosmose de Dubrunfaut, dialyse de Graham. Paris,
      1887. 8vo. Ill.

LE ROUX, F. P.
      Leçon (etc.). *See* Société chimique de Paris.

LERSCH, B. M.
   Einleitung in die Mineralquellenlehre. Ein Handbuch für Chemiker
      und Aerzte. Erlangen, 1860. 8vo.
         Hydrochemie, oder Handbuch der Chemie der natürlichen
            Wässer. Zweite Auflage der beträchtliche Theiles der
            "Einleitung in die Mineralquellenlehre." Berlin, 1864.
            8vo. Ill.
         Zweite Auflage durch Zusätze vermehrt. Bonn, 1860. 8vo. Ill.

LE SAGE, GEORGE LOUIS.
   Essai de chimie mécanique. 1758. 4to.
      Awarded a prize by the Academy of Sciences at Rouen.

LETHEBY, H.
   On Food : its varieties, chemical composition, nutritive value, com-
      parative digestibility, physiological functions and uses, preparation,
      culinary treatment, preservation, adulteration, etc. Being the sub-
      stance of four Cantor lectures, delivered before the Society for the
      Encouragement of Arts, Manufactures and Commerce, in the
      months of January and February, 1868. London, 1870. pp. xiii–
      277–vi, 8vo.

LEUCHS, EHRHARD FRIEDRICH.

Pottaschenfabrikant (Der) . . . Nürnberg, 1834. 8vo.

Vollständige Anleitung zur Fabrikation des Natrons . . . Nürnberg, 1834. 8vo.

LEUCHS, JOHANN CARL.

Beschreibung der färbenden und farbigen Körper, mit genauer Angabe ihrer Eigenschaften und ihres Gebrauchs. Ein unentbehrliches Handbuch für Färber, Katundrucker, Maler, Lakirer, Farbenbereiter, Gerber und Künstler die mit diesen Waren handeln. Nürnberg, 1825. 2 vols. Ill.

> *Also under the title :* Vollständige Farben- und Färbekunde. Erster Theil.
>
> The second part contains the preparation of colors. Contains a bibliography of dyeing, etc., pp. 644–677.

> Traité complet des propriétés, de la préparation et de l'emploi des matières tinctoriales et des couleurs. Par J. C. Leuchs. Traduit de l'allemand. Revue pour la partie chimique par E. Peclet. Paris, 1829. 2 vols.

Holzessig (Die) Fabrikation. Gründliche Anweisung zur Bereitung, Reinigung und Benutzung der Holzsäure, oder des Holzessigs nach den neuesten und besten Verfahrungsarten. Nürnberg, 1834. 8vo.

Neueste Farben- und Färbekunde, oder Beschreibung und Anleitung zur Bereitung und zum Gebrauch aller färbenden und farbigen Körper. Nürnberg, 1825. 2 vols. Ill.

> Vol. I also under the title :
>
> Beschreibung der färbenden und farbigen Körper, *q. v.*
>
> Vol. II also under the title :
>
> Anleitung zur Bereitung aller Farben und Farbflüssigkeiten, so wie zur Verfertigung der künstlichen Edelsteine, der Zeichenstifte, Pastellfarben, Tusche und zur Malerei auf Glas, Porzelan und Email.
>
> A Supplement to this work bears the title :
>
> 250 Entdeckungen und Verbesserungen in der Färberei und Druckerei. Gemacht in den Jahren 1838–1839.
>
> A second Supplement bears the title :
>
> 100 neue Vorschriften zur Farbenbereitung. Auch als zweiter Nachtrag zu J. C. Leuchs Anleitung zur Bereitung aller Farben. Nürnberg, 1839.

Vollständige Braukunde, oder wissenschaftlich-praktische Darstellung der Bierbrauerei in ihrem ganzen Umfange. Nürnberg, 1831. 8vo. Ill.

LEUCHS, JOHANN CARL. [Cont'd.]

Vollständige (Die) Essigsiederei, wissenschaftlich und praktisch dargestellt. Nürnberg, 1829. 8vo.

> Also under the title :

Vollständige Essigfabrikation vermehrt mit einem Geheimniss Essig
aus jeder Essig gebenden Flüssigkeit binnen zwei Stunden . . .
zu machen.

> Dritte Ausgabe. Nürnberg, 1833. 8vo.

> Vollständige Essigfabrikation. Vierte ganz umgearbeitete
> Ausgabe, enthaltend alle Arten der Essigbereitung,
> nebst dem verbesserten Geheimniss haltbaren, klaren
> Essig aus Branntwein, Wein, Most, Zucker, Simp, Malz,
> Obst, Molken, etc., binnen 12 Stunden ohne Zusatz mit
> grossem Vortheil im Grossen und Kleinen in einem mit
> Hobelspänen angefülltem Fass zu machen, worin oben
> die Flüssigkeit eingegossen wird und unten als fertiger
> Essig abläuft. Nürnberg, 1840.

> Contains a bibliography of the subject.

Vollständige Runkelrüben-Zucker-Fabrikation, nebst Anleitung zur
Abscheidung und Raffination des Zuckers aus Aepfeln, Ahorn,
Honig, Kastanien, Mais, Milch, Möhren, Pflaumen, Süssholz, Wein-
trauben, Zuckerrohr und dreissig andern Körpern, und Beschreibung der besten Geräthe und Einrichtungen. Nürnberg, 1836. 8vo.

Zucker-Fabrikation (Die), oder Anleitung zur Erzeugung des Zuckers
aus Stärkmehl, Kartoffeln, Gummi, Papier, Stroh und Holz, mit
und ohne Schwefelsäure, mit und ohne Sieden. Zweite ganz umgearbeitete und sehr vermehrte Auflage. Nürnberg, 1835. 8vo.

Zusammenstellung der in letzten 30 Jahren in der Gerberei und Lederfabrikation gemachten Beobachtungen und Verbesserungen.
Zweite Auflage. Nürnberg, 1832.

LEURET ET LASSAIGNE.

Recherches physiologiques et chimiques pour servir à l'histoire de la
digestion. Ouvrage mentionné honorablement par l'Académie
royale des sciences, dans sa séance publique du 20 Juin, 1825.
Paris, 1825. pp. 220, 12mo. Plates.

LEVAUX, P. F.

Traité de teinture sur laine et sur étoffes de laine. Liège, 1890. pp.
244, 12mo.

> The preface is dated at Constantinople, the author being director of the
> Imperial Ottoman Manufactories.

LEVERRIER, V.

La métallurgie. Paris, 1892. 16mo.

> Bibliothèque des connaissances utiles.

LEVIN, L.

Lehrbuch der Toxicologie für Aerzte, Studirende und Apotheker.
Berlin, 1885. pp. vi–456, 8vo. Ill.

LEVY, S.

Anleitung zur Darstellung organischer Präparate. Stuttgart, 1887. Ill.

LEWENAU, VON.

Chemische Abhandlungen über das Selen. Wien, 1823. 8vo.

LEWES, VIVIAN B.

Inorganic Chemistry, with a short account of its more important applications. London, 1890. 8vo. Ill.
Second edition. London, 1892. 8vo. Ill.
Service Chemistry : being a short manual of chemistry and its applications in the naval and military services. London, *n. d.* [1889].
pp. xvi–521, 8vo.

LEWIN, GEORG.

Toxicologische Tabellen. Uebersichtliche Darstellung der gewöhn-
lichen Giftstoffe in ihrer chemischen Zusammensetzung, ihrem
Verhalten gegen die Reagentien, ihren Wirkungen und ihren
Gegengiften, sowie der besten Methoden, sie aufzufinden. Für
Gerichts-, Militär- und Civil-Aerzte, Pharmaceuten und Studirende
bearbeitet. Berlin, 1856. 4to.

LEWIS, E. R.

Chemical Analysis. Published at the American Mission Press.
Beirût, Syria, 1876.
    In Arabic language.
Chemistry of the Air and Water. Published at the American Mission
Press. Beirût, Syria, 1879.
    In Arabic language.

LEWIS, POLYDORE.

Philosophical Inquiry into the Nature and Properties of Common Water,
with Observations on its medicinal qualities . . . London, 1790.
8vo.

LEWIS, WILLIAM.

A Course of Practical Chemistry. In which are contained all the
operations described in Wilson's "Complete Course of Chemistry,"
with many new and several uncommon processes. To each
article is given the chemical history, and to most, an account of

LEWIS, WILLIAM. [Cont'd.]
the quantities of oils, salts, spirits yielded in distillation, etc.,
from Lémery, Hoffman, the French Memoirs, Philosophical Transactions and from the Author's own experience. London, 1746.
pp. [xlii]–432, 8vo. Plates.

Historie des Goldes und der Gewerbe so davon abhangen. Zürich, 1764.

A second edition bears the title :

Geschichte des Goldes und verschiedene damit sich be-
schäftigende Künste und Arbeiten. Grätz, n. d. [1780].
pp. [vi]–325. 4 folding plates.

The title is misleading as the work is not a history but a treatise on the
metallurgy and technology of gold. Lewis was F.R.S.

LEYDE, EDUARD.
Anleitung für den ersten Unterricht in der qualitativen chemischen
Analyse. Berlin, Posen und Bromberg, 1836. 8vo.

Zweite Ausgabe vermehrt durch eine Anleitung zur
Uebung in der gerichtlich-chemischen Analyse. Berlin, 1842. 8vo.

Dritte Ausgabe. Berlin, 1851. 8vo.

Ueber die Constitution organischer Verbindungen. Berlin, 1845.

LEYDIÉ.
Éthers. *See, in Section II*, Fremy : Encyclopédie chimique, vol. VII.

LEYSER, E.
Die Bierbrauerei mit besonderer Berücksichtigung der Dickmaisch-
brauerei. Achte Auflage. Stuttgart, 1887. 8vo. Ill.

Neunte vollständig umgearbeitete Auflage. Stuttgart, 1892.

L'HERITIER, S. D.
Traité de chimie pathologique ou recherches chimiques sur les solides
et les liquides du corps humain dans leurs rapports avec la physiologie et la pathologie. Paris, 1842. 8vo. Plate.

LI SHIH CHEN.
Pun tsao kang muh tseuen shoo. Edited by Woo Yuh Chang. 52 vols.
1603.

This Chinese work on materia medica, compiled in the Ming dynasty, contains
eighteen hundred and ninety-two different substances classified and arranged under sixteen divisions according to their qualities. All early
notions of chemistry are found in this work. Abridgments of the 52
vols. also exist. (Communicated by John Fryer, LL.D., of Shanghai.)

LIBAVIUS, ANDREAS [LIBAU, ANDREAS].
Alchemia e dispersis passim optimorum auctorum . . . collecta, adhibi-
tisque ratione et experientia quanta potuit esse methodo accurata

LIBAVIUS, ANDREAS.   [Cont'd.]

  explicata et in integrum corpus redacta.   Francofurti, 1595.   Fol.
  Plates.

> Sometimes styled the First Textbook of Chemistry.   Second edition under
>   the title :

> *Alchymia recognita, emendata et aucta, tum dogmatibus
>   et experimentis nonnullis, tum commentario medico
>   physico chymico ; qui exornatus est variis instrumen-
>   torum chymicorum picturis, partim aliunde translatis,
>   partim plane novis ; in gratiam eorum qui arcanorum
>   naturalium cupidi, ea absque involucris elementarium
>   et ænigmaticarum sordium intuere gaudent.   Franco-
>   furti, 1606.   pp. 196–402–192, fol.   Ill.

  Commentationum metallicarum libri quatuor . . . Francofurti ad
      Mœnum, 1597.   4to.

  Opera chymica.   Francofurti.   7 vols., folio.

  *Praxis alchymiæ, hoc est doctrina de artificiosa præparatione præ-
      cipuorum medicamentorum chymicorum duobus libris explicata ;
      quorum primus de destillatione aquarum et oleorum de salium et
      extractorum quintar. essentiarum, aquarum vitæ, floram et balsam-
      orum etc. confectione, ab autore anonymo, propria experientia con-
      scriptus est ; alter de lapide philosophorum agit in quo ut recte is
      comparandus sit, remotis omnibus figuris et parabolis dilucide
      docetur uterque correctus et declaratus opera A. Libavii . . .
      nunc ex Germanico idiomate in Latinum traductus [by L. Doldius].
      Annexus est libellus J. Bessoni de absoluta ratione extrahendi olea
      et aquas a medicamentis simplicibus, etc.   Francofurti, 1604.   pp.
      680–[xx], 8vo., sm. 12mo.   Ill.

  Rerum chymicarum epistolica forma ad philosophos et medicos quosdam
      in Germania excellentes descriptarum.   Liber ı, in quo tum rerum
      quarundam naturalium continentur explicationes ingeniosæ, tum
      chymiæ disciplina pyronomica scenastica et vocabularia . . .
      declarantur fideliter.   Liber ıı, continens operationes chymicas
      artificum præceptis, naturæ documentis et experientia declaratas
      . . . Francofurti, 1595–99.   3 parts, 8vo.

> " Il y a peu d'Auteur moderne qui ait autant écrit sur la science hermétique
>   que Libavius, il n'y en a peut-être pas qui ait moins réussi.   C'est une
>   consolation pour ceux qui ne sçauroient réussir de s'entretenir du moins
>   dans une agréable folie et dans les espérances chimériques."—*Lenglet
>   Du Fresnoy.*

LIBRO ELEMENTARE DI CHIMICA, contenente i principii generali di questa scienza, la proporzione chimica, la teoria atomistica, l'iso-
  morfismo etc di Liebig ; e capitoli tre di Carlo Matteucci cioé :

LIBRO ELMENTARE DI CHIMICA. [Cont'd.]

    1. Dell' elettro-chimica ; 2. Della relazione fra i pesi atomistici e alcune proprietà fisiche dei corpi ; 3. Delle generalità della chimica organica ; e più una tavola in cui sono al vero rappresentate le reazioni che servono a scoprire i metalli nella loro combinazione e l'apparecchio recentemente immaginato da Marsh per iscoprire l'arsenico. Pisa, 1842. 8vo.

LICHT, JOHANNES GOTTFRIDUS.

Meditationes de sale communi ejusque existentia in omnibus rebus mundanis, unde problema sal primigenium hoc contineri probabile reddi videtur. Hafniæ, 1749. 4to.

LICHTENBERG, CARL.

    Fettwaaren (Die) und fetten Oele. Ausführliche Erörterungen über Herkunft, Eigenschaften, Verhalten und Zusammensetzung, Gewinnung und Herstellung, Verfälschung, Reinigung und Bleichung, Aufbewahrung und Verwendung aller für den Handel, ingleichen für den Haushalt, für Gewerbe und Künste, für Kosmetik und Pharmacie wichtigen animalischen, wie auch vegetabilischen Fettsubstanzen. Nach den neuesten wissenschaftlichen Erfahrungen bearbeitet. Weimar, 1880. 8vo.

    Seifenfabrikation (Die). Handbuch bei Darstellung aller Arten von Kern-, Leim-, Schmier-, Harz-, und Luxus-Seifen nach den neuesten Vervollkommnungen in diesen Industriezweige, nebst einer kurzen Beschreibung der Rohmaterialen und deren Prüfung auf ihre Reinheit und Güte. Dritte Auflage, herausgegeben von A. Steinheil. Weimar, 1879. 8vo.

LICHTENBERGER, G. E.

    Die Ultramarin-Fabrikation, beschrieben nach ihrer Entstehung, Geschichte und gegenwärtigen Ausbildung, mit Angabe der für die Hauptfabrikorte und Systeme eigenthümlichen Einrichtungen und Bereitungsarten. 1 vol., 8vo, with Atlas in folio.

LIEBER, OSCAR MONTGOMERY.

    The Assayer's Guide ; or practical directions to assayers and smelters for the tests by heat and by wet processes of the ores of all the principal metals, and of gold and silver coins and alloys. Philadelphia, 1852. 12mo.

        Another edition : Philadelphia, 1884. 12mo.

LIEBERMANN, LEO.

Anleitung zu chemischen Untersuchungen auf dem Gebiete der Medi-

LIEBERMANN, LEO. [Cont'd.]

cinalpolizei, Hygiene und Forensischen Praxis, für Aerzte, Medicinalbeamte und Physikats-Candidaten. Stuttgart, 1877. pp. xii–274,
8vo. Ill.

Chemische (Die) Praxis auf dem Gebiete der Gesundheitspflege und
gerichtlichen Medicin, für Aerzte, Medicinalbeamte und Physikats-
Candidaten, sowie zum Gebrauch in Laboratorien. Stuttgart.
Zweite Auflage, 1883. 8vo.

Grundzüge der Chemie des Menschen für Aerzte und Studirende.
Stuttgart, 1880. 8vo.

LIEBIG, JUSTUS VON.

Chemie (Die) in ihrer Anwendung auf Agricultur und Physiologie. *See*
Organische (Die) Chemie [etc.].

Criticism on J. von L. *See* Mulder, G. J.

LIEBIG, JUSTUS VON.

Anleitung zur Analyse organischer Körper. Braunschweig, 1837. 8vo.

Zweite umgearbeitete und vermehrte Auflage. Braunschweig,
1853. 8vo.

Instructions for the Chemical Analysis of Organic Bodies.
Translated from the German by W. Gregory. Glasgow,
1839. 8vo.

Handbook of Organic Analysis, containing a detailed account
of the various methods used in determining the elementary composition of organic substances. Edited by
A. W. Hofmann. Second edition. London, 1853. pp.
viii–135. 8vo.

Instruction sur l'analyse des corps organiques. Traduit de
l'allemand par A. Schmersahl. Avec deux planches
gravées et une table dressée avec beaucoup d'exactitude
pour exécuter tous les calculs de la manière la plus
simple et la plus expéditive. Paris, 1838. 8vo.

Manuel pour l'analyse des substances organiques . . . Traduit de l'allemand par A. J. L. Jourdan, suivie de
l'examen critique des procédés et des résultats de l'analyse des corps organisés par F. V. Raspail. Paris, 1838.
8vo. Ill.

Руководство къ анализу органич. тѣлъ. Перев. Терехова М. Е.
Москва 1858.

Bemerkungen über das Verhältniss der Thier-Chemie zur Thier-Physiologie. Heidelberg, 1844. 8vo.

Briefwechsel zwischen J. von Liebig und T. Renning über landwirthschaftliche Fragen aus den Jahren 1854 bis 1883. Dresden, 1884.
8vo.

LIEBIG, JUSTUS VON. [Cont'd.]

Chemische Briefe. Heidelberg, 1844. 12mo.

   Zweite Auflage. Heidelberg, 1845.

   Dritte Auflage. Heidelberg, 1851.

   Vierte umgearbeitete und vermehrte Auflage. Leipzig, 1859.
      2 vols., 8vo.

   Fünfte wohlfeile Volksausgabe. Leipzig, 1865-8.

   Sechste Auflage, 1878.

   Chemiske Breve. I Dansk Oversættelse af Jacob Davidsen.
      Andet med Forfatterens nyeste Skrift om "den nyop-
      fundne Patentgjödning" forögede Oplag. Kjøbenhavn,
      1846. 8vo.

   Chemiske Breve. Ny Samling. Ved. F. Möller-Holst.
      Kjøbenhavn, 1854. 8vo.

   Rationelle Breve for Landmænd og Veterinairer ; eller :
      Grundtræk af Dyr- og Plantelivets chemiske Functioner
      og Fordringer til praktisk Benyttelse. En Afdeling af
      Liebigs "chemische Briefe" frit bearbeidet af G.
      Michelsen. Kjøbenhavn, 1844. 8vo.

   Brieven over scheikunde en de betrekking waarin deze
      wetenschap staat tot koophandel, physiologie en land-
      bouw. Uit het Engelsch vertaald door J. C. Kruse-
      mann. Haarlem, 1844. 8vo.

   Scheikundige brieven. Naar het Hoogduitsch door J. M.
      van Bemmelen. Tweede druk. Leeuwarden, 1864.
      3 parts, 8vo.

   Familiar Letters on Chemistry and its relation to commerce,
      physiology and agriculture. Edited by John Gardner.
      London, 1843. pp. vii–179, 12mo.

         Contains 16 letters.

   [*The same.*] Second Series. The Philosophical Principles
      and General Laws of the Science. Edited by John
      Gardner. London, 1844. pp. vii–218, 12mo.

         Contains 11 letters.

 * Familiar Letters on Chemistry in its relations to physiology,
      dietetics, agriculture, commerce and political economy.
      London, 1851. pp. xx–536, 12mo.

         This edition contains 35 letters and an Appendix of six numbers.

   Lettres sur la chimie considérée dans ses rapports avec
      l'industrie, l'agriculture et la physiologie. Traduites de
      l'allemand sur la deuxième édition par F. Bertel Dupiney
      et E. Dubreuil Helion. Paris, 1845. 12mo.

LIEBIG, JUSTUS VON.   [Cont'd.]

Lettres sur la chimie et sur ses applications à l'industrie, à la physiologie et à l'agriculture. Traduites de l'allemand par G. W. Bichon. Paris, 1845. 12mo. Portrait of Liebig.

Lettres sur la chimie considérée dans ses applications à l'industrie, à la physiologie et à l'agriculture. Nouvelle édition française publiée par Charles Gerhardt. Paris, 1847. pp. 330, 12mo.

> With lithographic portrait of Liebig by Artus from a drawing by Engel.

Nouvelles lettres [etc.] Paris, 1852. 12mo.

Lettere prime e seconde sulla chimica e sue applicazioni all' agricoltura, alla fisiologia, alla patologia, all' igiene ed alle industrie. Nuova edizione condotta sull' originale tedesco da E. Leone, ed annotata da F. Selmi. Torino, 1857. 16mo.

Lettere chimiche, serie terza. Versione dal tedesco di A. Passerini. Parma, 1858. 16mo.

Sei nuove lettere chimiche sull' agricoltura. Compendiate e annotate da Gustavo Dalgas. Firenze, 1858. 8vo.

Listy o chemii o jéj zastósowaniach w przemyśle, fizyologii i w rolnictwie przełożył J. S. Zdzitowiecki. Warszawa, 1845. 12mo.

Nowe listy o chemii zastósowanéj do przemysłu fizyologii i rolnictwa. Przełożył Ludwik Natanson. Warszawa, 1854. 8vo.

Письма о химіи и ея приложеніяхъ къ промышленности, физіологіи и земледѣлія. Перев. Дымчевичъ. Съ примѣч. переводчика. С.-Петербургъ 1847. 8º

Письма о химіи. Переводъ съ послѣдняго исправленнаго ʏ допол неннаго изданія. Москва 1855. 8º

Письма о химіи. Перев. Алексѣева. С.Петербургъ 1861.

Новыя письма о химіи, въ ея приложеніяхъ къ промышленности. физіологіи и земледѣлію. Переводъ Адама Іохеро. Въ 16 д. л. до 500 страницъ убористаго, но весьма четкаго шрифта, па хорошей бумагѣ. Москва. 8º

Cartas sobre la química y sobre sus aplicaciones á la industria, á la fisiologia y á la agricultura, . . . Traducidas del frances al castellano por José Villar y Macias. Salamanca, 1845. 8vo.

Nuevas cartas sobre la química considerada en sus aplicaciones á la industria, á la fisiología y á la agricultura.

Liebig, Justus von. [Cont'd.]

Edicion española por D. Ramon Torres Muñoz y Luna. Madrid, 1853.

Cartas químicas. Traducidas al castellano. Barcelona, 1850. 4to.

Kemisk bref. Öfversättning från tredje upplagan af Georg Schentz. Stockholm, 1854. 8vo.

See Moleschott, Jacob : Der Kreislauf des Lebens.

Chemische Untersuchung der Schwefel-Quellen Aachen's. Aachen und Leipzig, 1851. 8vo.

Chemische Untersuchung über das Fleisch und seine Zubereitung zum Nahrungsmittel. Heidelberg, 1847. 8vo.

Chimica (La) alla portata di tutti, ossia la chimica organica. Compilata da Lorenzo Agostino Ghisi. Col ritratto litografico di Liebig. Mouza, 1846. 16mo.

Einleitung in die Naturgesetze des Feldbaues, besonderer Abdruck aus Liebig's : Die Chemie in ihrer Anwendungen auf Agricultur und Physiologie. Achte Auflage. Braunschweig, 1862.

Eenige hoofdstukken der nieuwere scheikunde. Uit het Hoogduitsch door J. M. van Remmelen. Groningen, 1863. 8vo.

Handbuch der organischen Chemie mit Rücksicht auf Pharmacie. Aus der neuen Bearbeitung des ersten Bandes von Geiger's Handbuch der Pharmacie. Heidelberg, 1843. 8vo.

Complete Works on Chemistry. Comprising his agricultural chemistry, or organic chemistry in its application to agriculture and physiology ; animal chemistry, or organic chemistry in its application to physiology and pathology ; familiar letters on chemistry and its relations to commerce, physiology and agriculture ; the origin of the potato disease and researches into the motion of the juices in the animal body, and evaporation of plants, etc. Philadelphia, 1856. 8vo.

Grundsätze (Die) der Agricultur-Chemie mit Rücksicht auf die in England angestellten Untersuchungen. Braunschweig, 1855. 8vo.

Zweite Auflage. Braunschweig, 1855. pp. 152. 8vo.

Principles of Agricultural Chemistry, with special reference to the late researches made in England. Translated by W. Gregory. London, 1855. 8vo. Also New York, 1855.

The Relations of Chemistry to Agriculture, and the agricultural experiments of Mr. J. B. Lawes. Translated by Samuel W. Johnson at the author's request. Albany, N. Y., 1855. 12mo.

Question des engrais. Principes de chimie agricole et

Liebig, Justus von. [Cont'd.]

critique des essais d'application de quelques praticiens Anglais et Allemands. Traduit, sous les yeux de l'auteur, sur la deuxième édition allemande par Paul Picard. Paris, 1856. 8vo.

I principii fondamentali della chimica agraria in relazione alle richerche istituite in Inghilterra. Prima traduzione italiana, eseguita sulla seconda tedesca, per cura di Alfonso Cossa. Milano, 1856. 16mo.

Åkerbruks-Kemiens Grundsatser med Hänseende til de i England utförda Undersökningar. Öfversättning af C. Juhlin-Dannfelt. Stockholm, 1856. 8vo.

Introduction à l'étude de la chimie, contenant les principes généraux de cette science, les proportions chimiques, la théorie atomique, le rapport des poids atomiques et les formules chimiques, les combinaisons isomériques, les corps catalytiques et accompagnée de considérations détaillés sur les acides, les bases et les sels. Traduite de l'allemand par Ch. Gerhardt et augmentée d'une table analytique des matières. Paris, 1837. 12mo.

Introduzione allo studio della chimica filosofica, contenente i principii generali di questa scienza, le proporzioni chimiche, la teoria atomica, il rapporto dei pesi atomici e delle formole chimiche, le combinazioni isomeriche, i corpi catalitici, ecc., e particolari considerazioni sugli acidi, sulle basi e sui sali. Traduzione dal tedesco di Giovanni Battista Sembenini. Verona, 1839. 8vo.

Moderne (Die) Landwirthschaft als Beispiel der Gemeinnützigkeit der Wissenschaften. Rede gehalten in der öffentlichen Sitzung der königlichen Akademie der Wissenschaften zu München am 28 November, 1861. Braunschweig, 1862. 8vo.

Det nyere Agerbrug, et Exempel paa Videnskabens Almeennytte. Oversættelsen gjennemseet og udgivet af Th. Segelcke. Kjøbenhavn, 1862. 8vo.

Agerbrugets Naturlove. Kjøbenhavn, 1864. 8vo.

Onbewerktuigde (De) natuur en het bewerktuigde leven. Eene redevoering tegen het materialisme dezer dagen. Naar het Hoogduitsch door H. Kloete Nortier. Rotterdam, 1856. 8vo.

Organische (Die) Chemie in ihrer Anwendung auf Agricultur und Physiologie. Braunschweig, 1840. 8vo.

Zweite Auflage [in which the word "Organische" is dropped]. Braunschweig, 1841.

Fünfte Auflage, 1843. Sechste Auflage, 1846.

LIEBIG, JUSTUS VON. [Cont'd.]

Neunte Auflage im Auftrage des Verfassers herausgegeben von Ph. Zöller. Braunschweig, 1876. 8vo.

Agricultur-Chemi, eller Chemien anvendt paa Agerdyrkning og Physiologie. Oversat af Jacob Davidsen. Aalborg, 1846. 8vo.

De bewerktuigde scheikunde, toegepast op landbouwkunde en physiologie. Naar de tweede uitgave uit het Hoogduitsch vertaald, door J. P. C. van Tricht. Assen, 1843. 8vo.

De scheikunde in hare toepassing op landbouw en physiologie, naar den zevenden Hoogduitschen druk, door E. C. Enklar. Zwolle, 1863-'67. 8vo.

Organic Chemistry in its applications to Agriculture and Physiology. Edited from the manuscript of the author by Lyon Playfair. London, 1840. 8vo.

Second edition [in which the first word, "Organic," is dropped], with very numerous additions. London, 1842. 8vo.

Third edition, revised. London, 1843.

Fourth edition, revised and enlarged. Edited . . . by Lyon Playfair and W. Gregory. 2 parts. London, 1847. pp. xii-418, 8vo. Also, New York, 1846.

Organic Chemistry in its applications to Agriculture and Physiology. Edited from the manuscript of the author by Lyon Playfair. First American edition, with an introduction, notes and appendix by John Webster. Cambridge [Mass.], 1841. pp. xx-435. 8vo.

Chimie organique appliquée à la physiologie végétale et à l'agriculture. Suivie d'un essai de toxicologie. Traduction faite sur les manuscrits de l'auteur par Charles Gerhardt. Paris, 1842. 8vo.

Deuxième édition, revue et considérablement augmentée, traduction faite sur la quatrième édition allemande par Ch. Gerhardt, et revue par J. Liebig. Paris, 1844. 8vo.

Chimie appliquée à la physiologie végétale et à l'agriculture. Abrégé de cet ouvrage par J. B. Bivort. (Extrait du Moniteur des compagnes.) Bruxelles, 1850. 8vo.

La chimica applicata all' agricoltura ed alla fisiologia. Versione sulla quinta edizione originale Tedesca, dall' autore rifusa e di molto ampliata da Giuseppe Netwald. Vienna, 1844. 8vo.

LIEBIG, JUSTUS VON. [Cont'd.]

Chemia z zastósowaniern do rolnictwa i fizyologii, z piątego
    przerobionego i pomnożonego wydania przełożył J. S.
    Zdzitowiecki. Warszawa, 1846. 8vo.

Органическая химія, приложенная къ растительной физіологіи
    и земледѣлію. С.-Петербургъ 1842.

Organische (Die) Chemie in ihren Beziehungen zu Dr. Gruber und Dr.
    Sprengel. Heidelberg, 1841.

Principes de chimie agricole. Paris, 1856. 12mo.

Reden und Abhandlungen. Leipzig und Heidelberg, 1874. pp. viii–
    334, 8vo.

Researches on the Chemistry of Food . . . edited from the manuscript
    of the author by W. Gregory. London, 1847. 8vo.

Suppe für Säuglinge, mit Nachträgen in Beziehung auf ihre Bereitung
    und Anwendung. Zweite Auflage. Braunschweig, 1866. 8vo.

    Dritte vermehrte Auflage. Braunschweig, 1877. 8vo.

Thier-Chemie (Die) oder die organische Chemie in ihrer Anwendung
    auf Physiologie und Pathologie. Braunschweig, 1842. 8vo.

    Zweite unveränderte Auflage. Braunschweig, 1843. 8vo.

    Dritte umgearbeitete und sehr vermehrte Auflage. Erste
        Abtheilung. Braunschweig, 1846. 8vo.

    No more published. *See* Kohlrausch, O : Physiologie und Chemie.

    Dierlijke scheikunde of bewerktuigde scheikunde, toegepast
        op physiologie en pathologie. Vertaald uit het Engelsch
        naar de van Hoogduitsche manuscript bezorgde uitgave
        van W. Gregory, door E. C. Donders. s' Hage, 1842.
        8vo.

    Chemistry and Physics in relation to Physiology and Path-
        ology. London, 1846. 8vo.

        A translation of Theil 2 of Abtheilung 1 of " Die Thier-Chemie."

    Animal Chemistry, or organic chemistry in its applications to
        physiology and pathology. Edited from the author's
        manuscript by William Gregory. London, 1842. pp.
        xix-354, 8vo.

        Second edition, 1843. Third edition, revised and enlarged.
            Part I. London, 1846. 8vo.

            No more published.

    Animal Chemistry, or organic chemistry in its application to
        physiology and pathology. Edited by William Gregory,
        with additions, notes and corrections by John Webster.
        Cambridge [Mass.], 1842. 8vo.

    La chimie organique appliquée à la physiologie animale et à
        la pathologie. Traduction faite sur les manuscrits de

LIEBIG, JUSTUS VON. [Cont'd.]

l'auteur par Ch. Gerhardt, et revue par J. Liebig. Paris, 1842. 8vo.

Chemia organiczna w zastósowaniu do zoofizyologii i patologii. Przełoźne z niem. i dodatkiem o budowie i Znaczeniu organów zywienia powiększył Jan Pankiewicz. Warszawa, 1844. 8vo.

Química orgánica aplicada á la fisiologia animal y á la patologia. Traducida al francés de sus manuscritos por Carlos Gerhardt, y vertida al español por Manuel José de Porto. Cadiz y Madrid, 1845. 4to.

Traité de chimie organique. Édition française, revue et considérablement augmentée par l'auteur et publiée par Ch. Gerhardt. Paris, 1841–44. 3 vols., 8vo.

Trattato di chimica organica. Prima versione sull' ultima edizione francese di Brusselles, di Giovanni Vanzani. Milano, 1844. 8vo.

Tratado de química orgánica, revisado y considerablemente aumentado por el autor, publicado en francés por Ch. Gerhardt y vertido de este idioma al español por Rafael Saez y Palacios y Carlos Ferrari y Scardini. Madrid, 1847–48. 4 vols., 4to.

Ueber die Constitution der organischen Säuren (1838). Herausgegeben von H. Kopp. Leipzig, 1891. 8vo.

Ostwald's Klassiker der exakten Wissenschaften.

Ueber Theorie und Praxis in der Landwirthschaft. Braunschweig, 1856. 8vo.

De la théorie et de la pratique en agriculture. Lille, 1857. 8vo.

La teoria e la pratica della agricoltura. Prima edizione italiana con note eseguita sull' originale tedesco per cura di Alfonso Cossa. Milano, 1857. 16mo.

Untersuchungen über einige Ursachen der Säftebewegung im thierischen Organismus. Braunschweig, 1848. 8vo. Ill.

Onderzoekingen over eenige oorzaken van de beweging der vochten in het dierlijk organismus. Uit het Hoogduitsch. Groningen, 1849. 8vo.

Researches on the motion of juices in the animal body. Edited from the manuscripts of the author by William Gregory. London, 1848. 8vo.

Researches on the Chemistry of food and the motion of the juice in the animal body. Edited by William Gregory.

LIEBIG, JUSTUS VON. [Cont'd.]
>Edited from the English edition by Eben W. Horsford.
>Lowell [Mass.], 1848. 8vo.

Vera (La) maniera facile ed economica di fare senz' uva diverse qualità
di vino di ottimo gusto e non di pregiudizio alla salute. Palermo,
1853. 32mo.

Zur Beurtheilung der Selbstverbrennung des menschlichen Körpers.
Heidelberg, 1850. 8vo.
>Is de zelfverbranding van het menschelijk ligchaam al of niet
>mogelijk. Haarlem, 1850. 8vo.

Herr Emil Wolff in Hohenheim und die Agricultur-Chemie. Nachtrag
zu den Grundsätzen der Agricultur-Chemie. Braunschweig, 1855.
8vo.

LIEBIG, JUSTUS VON, AND WILLIAM GREGORY.
Oily Acids (The); being the first supplement to the seventh edition of
Turner's Chemistry. London, 1841. 8vo.
>*Cf.* Turner, Edward. Elements of Chemistry; seventh edition.

LIEBIG'S THIERCHEMIE UND IHRE GEGNER, ein vorzüglich für praktische
Aerzte berechneter ausführlicher Commentar, zu dessen physiologischen, pathologischen und pharmakologischen Ansichten. Nach
dem englischen des Henry Ancell bearbeitet und mit Anmerkungen vermehrt von A. W. King. Pest, 1844. 8vo. Ill.

LIELEGG, ANDREAS.
Erster Unterricht aus der Chemie an Mittelschulen. Ausgabe für Realschulen. Wien, 1871. 8vo.

LIÈS-BODART.
Conférences et manipulations chimiques. Paris, 1885. 12mo.

LILLEY, H. J.
Bench-book for Test-tube Work in Chemistry. London, 1890. 8vo.
Lecture Course (A) in Elementary Chemistry. London, 1892.

LIMBOURG, JEAN PHILIPPE DE.
Dissertation sur les affinités chymiques qui a remporté le prix de physique de l'an 1758 quant à la partie chymique, au jugement de
l'Académie royale des Sciences, Belles-Lettres et Arts de Rouen.
Liège, 1761, pp. 87, 12mo. One folding plate.
>Written for the prize offered in 1758 by the Academy in Rouen on the
>theme: " Déterminer les affinités, qui se trouvent entre les principaux
>mixtes, ainsi que l'a commencé M. Geoffroy, et trouver un système
>physico-mécanique de ces affinités."

Traité des eaux minérales de Spa. Liège, 1756.

LIMPRICHT, H.
Lehrbuch der organischen Chemie. Braunschweig, 1862. pp. [ii]–
1292, 8vo. Ill.

LIMOUSIN–LAMOTHE.
Éléments pratiques de chimie agricole, suivis d'une notice sur roque-
fort. Arras, 1878. 12mo.

LINCOLN, MRS. ALMIRA H.
See Phelps, Mrs. A. H. Lincoln.

LINDAUER, G.
Compendium der Hüttenchemie. Prag, 1861. 8vo.

LINDES, AUGUST WILHELM.
Beiträge zur gerichtlichen Chemie ; enthaltend : I. Die Ermittelung der
Mahlsteuer-Defraudationen. II. Die Unterscheidung von Leinen
und Baumwolle in gemischten Geweben. III. Ueber die Auffin-
dungen des Arseniks in Leichen. Berlin, 1853. 8vo.
Chemische Farbenlehre für Maler, Tapetenfabrikanten und Farbwaaren-
händler. Eine auf Grundsätze der Chemie gestützte Anleitung zur
Prüfung der im Handel vorkommenden Malerfarben auf Echtheit
und Güte, sowie ein Rathgeber bei Farbenmischungen, namentlich
bei solchen welche stehen oder sich vergänglich zeigen. Weimar,
1861. 8vo.
> Neuer Schauplatz der Künste und Handwerke.

Gründliche Anleitung zur praktischen Ausführung der wichtigsten
agronomisch-chemischen Untersuchungen. Ein zeitgemässer, un-
entbehrlicher Rathgeber für Landwirthe. Berlin, 1849. 8vo. Ill.
Gründliche Anleitung zur richtigen Beurtheilung und chemischen Prüfung aller, in den verschiedenen Zweigen der Färberei und Zeugdruckerei erforderlichen Materialen auf Aechtheit und Güte.
Nebst einer Anleitung zur chemischen Untersuchung gefärbter und
bedruckter Zeuge. Berlin, 1849. 8vo.
Handbuch der Chemie. Zur Erleichterung von Repetitionen bearbeitet. Berlin, 1854. 8vo.
Praktische Anleitung zu den wichtigsten gerichtlich-chemischen und
sanitäts-polizeilichen Untersuchungen. Für Physiker, Aerzte und
Apotheker. Berlin, 1849. pp. x–238, 12mo. Plates.
Praktische Anleitung zur Prüfung und Werthbestimmung der wichtigsten im Handel vorkommenden Düngemittel, zunächst für Landwirthe. Berlin, 1855. 8vo.

LINDET, L.
La bière. Paris, 1892. 8vo.

LINK, CHRISTIAN.
Chemistry. Buffalo, N. Y., 185–. 8vo.

LINK, HEINRICH FRIEDRICH.
Beiträge zur Physik und Chemie. Rostock, 1795–97. 3 parts.
Einige Bemerkungen über das Phlogiston. Göttingen, 1790. 8vo.
Grundwahrheiten (Die) der neueren Chemie, nach Fourcroy's Philoso-
phie chimique, herausgegeben mit vielen Zusätzen von H. F. L.
Zweite sehr verbesserte und vermehrte Auflage. Leipzig, Rostock
und Schwerin, 1815. 8vo.
    Cf. Fourcroy, A. F.

LINNELL, H. LEONARD.
Sewage Treatment, or the purification of water-carried sewage. Bolton
[England], 1891.

LINTNER, CARL.
Lehrbuch der Bierbrauerei.   Nach dem heutigen Standpunkte der
Theorie und Praxis, unter Mitwirkung der angesehensten Theoreti-
ker und Praktiker. (Zugleich als erster Theil zu Otto-Birnbaum's
Lehrbuch der landwirthschaftlichen Gewerbe, siebente Auflage.)
Braunschweig, 1878. 8vo.
    Cf. Otto-Birnbaum's Lehrbuch.

LIPPHARD, J. CH. L.
Handbuch der Chemie, nebst moralischer Bildung des Apothekers, in
Briefen für Lernende. Leipzig, 1800.

LIPPMANN, EDMUND O. VON.
Zucker (Der), seine Derivate und sein Nachweis. Eine Monographie.
Wien, 1878. 8vo.
Zuckerarten (Die) und ihre Derivate. Vom Vereine für die Rübenzucker-
Industrie des deutschen Reichs mit dem ersten Preise gekrönte
Monographie. Braunschweig, 1882. pp. xii–228–[ii]. 8vo.

LISSAJOUS.
    Leçon (etc.)   See Société chimique de Paris.

LIST, E. [Editor].
Sechs Vorträge aus dem Gebiete der Nahrungsmittel-Chemie, gehalten
bei Gelegenheit der ersten Versammlung bayerischer Chemiker zu
München von Holzner, Kayser, List, Prior, Sendtner, Vogel. Im
Auftrage der Referenten zusammengestellt von E. List. Würz-
burg, 1884. 8vo.

LIST, KARL GEORG ERNST.

Leitfaden für den ersten Unterricht in der Chemie, besonders auf Gewerbe- und Realschulen. Erster Theil ; Unorganische Chemie. Heidelberg, 1853. Zweiter Theil ; Organische Chemie. Heidelberg, 1855. 12mo.

Zweite verbesserte Auflage. Heidelberg, 1859. 2 parts, 12mo.

Theil 1 : Unorganische Chemie, nebst 300 Repetitionsfragen und einem Anhange : Zur Einführung in die neuere Chemie. Dritte vermehrte und verbesserte Auflage. Heidelberg, 1868. 8vo.

Fünfte Auflage. Heidelberg, 1880.

LISTE GÉNÉRALE DES FABRIQUES DE SUCRE, raffineries, distilleries de France, de Hollande, d'Allemagne, d'Autriche-Hongrie, etc., suivie d'un Traité d'analyse à l'usage des fabricants de sucre, revu pour 1891–92, de la législation des sucres et des usages commerciaux en France et dans les principaux pays [vingt-troisième année de publication], campagne de 1891–92. Paris, 1891.

LIVEING, D.

Chemical Equilibrium the Result of the Dissipation of Energy. Cambridge, 1885.

ЛИСЕНКО.

Руководство къ неорганич. химіи, теорет. описаніе и прикладной. С. Петербургъ 1867.

LISENKO. Manual of Inorganic Chemistry, theoretical, descriptive and practical. St. Petersburg, 1867.

LIVERANI, V.

La nuova chimica, opera di Giosia Cooke. Considerazioni critiche. Bologna, 1877. 4to.

Cf. Cooke, Josiah P.: The New Chemistry.

LLACH Y SOLIVA, JOSÉ.

Nociones de química para el quinto año de filosofía elemental . . . Gerona, 1847. 8vo.

LLOYD, FREDERICK JAMES.

The Science of Agriculture. London, 1884. pp. vi–365, 8vo.

LLOYD, J. U.

Chemistry (The) of Medicines ; Practical text- and reference-book for the use of students, physicians and pharmacists ; embodying the principles of chemical philosophy and their application to those chemicals that are used in medicine and pharmacy, including all those that are officinal in the pharmacopœia of the U. S. Cincinnati, 1881. 12mo.

LOBATO, JOSE G.

Estudio quimico-industrial de los varios productos del maguey Mexicano y analisis quimico del aguamiel y el pulque. México, 1884. pp. viii–9–191, 12mo.

LOCATELLI, LODOVICO.

Theatrum arcanorum chymicorum, sive de arte chimico-medica tractatus exquisitissimus. Francofurti, 1636. 8vo.

LOCHNER, ZACHARIAS.

Probier-Büchlein auff alle Metalle, so die Ertzt und Bergwerck des hochloblichen teutschen Landts geben . . . Augspurg, 1565. 8vo.

LOCK, CHARLES G. WARNFORD, BENJAMIN E. R. NEWLANDS, AND JOHN A. R. NEWLANDS.

Sugar ; a Handbook for planters and refiners, being a comprehensive treatise on the culture of sugar-yielding plants, and the manufacture, refining and analysis of cane, beet, palm, maple, melon, sorghum, milk and starch-sugars ; with copious statistics of their production and commerce and a chapter on the distillation of rum. Illustrated by 13 plates and 249 engravings. London and New York, 1888. pp. xxiv–920, 8vo.

　　　　Contains a short bibliography.

LOCKE, JOHN.

Lecture on Toxicology, delivered Jan. 15, 1841, before the Class of the Medical College of Ohio. Cincinnati, 1841. 8vo.

LOCKYER, J. N.

The Chemistry of the Sun. London, 1887. 8vo.

LOCQUES, NICOLAS DE.

* Les rudimens de la philosophie naturelle, touchant le système du corps mixte. Paris, chez Geoffroy Marcher, ruë Saint Jacques à la ville de Rome. 1664–68. 8vo. Frontispiece.

　　　　Five treatises, the first in two books, each having independent pagination.

　　　　[I.] *Traité premier, Livre premier :* Cours théorique, où sont clairement expliquez les preceptes et les principes de la chymie, qui ont esté jusques icy cachez des anciens philosophes. 1665. pp. [xxii]–184–[viii].

　　　　[II.] *Livre second :* Cours pratique, où il est traité des operations suivant la doctrine de Paracelse ; qui n'ont pas jusqu'icy esté connuës, que de fort peu de personnes. pp. [xxiv]–196. Des termes de l'alchymie vulgaire et commune. pp. *from* 199–214–[ii].

LOCQUES, NICOLAS DE.   [Cont'd.]

[III.] *Traité second :* De la fermentation, où on void ce qui se passe intérieurement dans le mouvemens divers des substances.   Avec le Traitté du sang et les propositions de la chymie résolutive.   1665.   pp. [xvi]–146.

[IV.] Les vertus magnétiques du sang, de son usage interne et externe.   Pour la guarison des maladies.   Paris, de l'imprimerie de Jacques le Gentil, ruë de Noyers, et se vend chez l'Autheur, ruë des Mauvais Garçons, à l'Image Saint Martin.   1664.   pp. [xvi]–54–[ii].

[V.] Propositions touchant la physique résolutive.   Paris, chez Geoffroy Marcher, ruë S. Jaques, à la ville de Rome.   1665.   pp. 39.

[VI.] Élémens philosophiques des arcanes et du dissolvant général ; de leurs vertus, propriétez, et effets.   Où sont ponctuellement expliquées en général leurs sécrettes compositions, et les expériences qui en ont esté faites ; l'ordre et la manière de s'en servir, pour les usages de la médicine.   Paris, 1668.   pp. [xx]–87.

LODIN.

Zinc et Plomb ; Étain.   *See, in Section II*, Fremy : Encyclopédie chimique, vol. v.

LOEBISCH, W. F.

Anleitung zur Harn-Analyse für praktische Aerzte, Apotheker und Studirende.   Wien, 1878.   pp. 238, 8vo.   Ill.

LOEFFLER, KARL.

Tabellen der pyrognostischen Merkmale welche die allein oder mit Reagentien behandelten mineralischen Substanzen darbieten.   Berlin, 1863.   8vo.

Популярное изложеніе началъ химіи и физики въ примѣненіи къ сельскому хозяйству.   С·Петербургъ 1863

Popular exposition of the principles of chemistry and physics in their application to agriculture.   St. Petersburg, 1863.

LÖSCHER, MARTIN GOTTHELF.

De sale ammoniaco ejusque usu chimico et curioso.   Vitebergæ, 1726.

LOEW, OSCAR.

Leitfaden durch die anorganische, organische und physiologische Chemie, für Brauer, Landwirthe und sonstige Techniker.   München 1889.

LOEW, OSCAR, UND THOMAS BOKORNY.

Die chemische Kraft-Quelle des lebenden Protoplasma.   München, 1883. 8vo.

Die chemische Ursache des Lebens.   München, 1881.   8vo.

LÖWIG, C.

Arsenikvergiftung und Mumification.   Gerichtlich-chemische Untersuchung.   Breslau, 1889.

LÖWIG, CARL JACOB.

Brom (Das) und seine chemischen Verhältnisse. Heidelberg, 1829. 8vo.

Chemie der organischen Verbindungen.   Zürich, 1839–40.   2 vols., 8vo.

    Zweite Auflage.   Braunschweig, 1845–46.   2 vols., 8vo.

Chemische Untersuchung der Mineralwasser zu Seewen, Canton Schwyz. Zürich, 1834. 8vo.

Grundriss der organischen Chemie.   Braunschweig, 1852. 8vo.

    Principles of Organic and Physiological Chemistry.   Translated by Daniel Breed.   London, 1853.   pp. xl–33–481, 8vo.   Also Philadelphia, 1853.

Lehrbuch der Chemie, mit besonderer Berücksichtigung des technischen und medicinischen Theils.   Heidelberg, 1832.   8vo.

Mineralquellen (Die) von Baden im Aargau.   Zürich, 1837.   8vo.

Repertorium für organische Chemie.   (Supplement zu des Verfassers Chemie der organischen Verbindungen.)   Erster und zweiter Jahrgang.   1840 und 1841.   Zürich, 1841.   8vo.

    Dritter Jahrgang, 1842.   Zürich, 1843.   8vo.

Theoretische Betrachtung über die sauren und basischen Eigenschaften der nichtmetallischen Körper.   Zürich, 1835.   4to.

Ueber Bildung und Zusammensetzung der organischen Verbindungen. Zürich, 1843.   4to.

Untersuchung der Schwefelquelle zu Schinznach.   Aarau, 1844.   8vo.

LOHMANN, JUSTUS HEINRICH FRIEDRICH.

Ueber den gegenwärtigen Zustand der Zuckerfabrikation in Deutschland.   Magdeburg, 1818.   8vo.

LOMAS, J.

A Manual of the Alkali Trade, including the manufacture of sulphuric acid, sulphate of soda and bleaching powder.   London, 1880.   8vo.   Second edition, London, 1886.

LONDON, ROYAL COLLEGE OF CHEMISTRY.

Reports. and Researches concluded in the Laboratories in 1845–47.   Vol. I, London, 1849.   8vo.

    No more published.

LONGCHAMP.
Exposition d'une loi à laquelle sont soumises toutes les combinaisons de la chimie inorganique, ou Nouvelle doctrine chimique.   Paris, 1833. 8vo.
Sur les produits de la combustion du soufre. Paris, 1833. 8vo.

LOOS, DIRK DE.
Historisch-kritische beschouwing der glucosiden. Rotterdam, 1858. 8vo.
Handleiding bij de practische oefeningen in de scheikunde. Ten gebruike op hoogere burgerscholen. Leiden, 1882. 8vo.

LOPPENS, P.
Handboek van scheikunde tot het fabriekwezen toegepast. Dienstig voor industrie- en lagere-schoolen. Vry gevolgd naer het Hoogduitsch van F. Koehler, en byzonderlyk tot het binnenlandsch fabriekwezen toegepast aenmerkelyk vermeerdert en met platen voorzien. Gent, 1848-'54. 2 vols., 8vo.

LORENZ, F. A.
Chemisch-physikalische Untersuchung des Feuers. Kopenhagen und Leipzig, 1789. 8vo.

LORIDAN, J.
Cours élémentaire de chimie contenant les matières du programme du baccalauréat ès lettres. Paris, 1884. 12mo.

LORSCHEID, J.
Lehrbuch der anorganischen Chemie. Freiburg i. Br., 1870.
    Zweite Auflage, nach den neuesten Ansichten der Wissenschaft. Freiburg i. B. 1872.
    Dritte verbesserte und vermehrte Auflage. Freiburg in B., 1874. Vierte Auflage, 1876.
    Fünfte Auflage, 1877. Sechste Auflage, 1877.
    Siebente Auflage, 1878. Achte Auflage, 1880.
    Neunte Auflage, 1882. Zehnte Auflage, 1884.
    Lehrbuch der anorganischen Chemie mit einem kurzen Grundriss der Mineralogie. Elfte Auflage, bearbeitet von H. Hovestadt. Freiburg, 1887. 8vo. Ill.
    Zwölfte Auflage bearbeitet von H. Hovestadt. Freiburg, 1892. 8vo. Ill.
Lehrbuch der organischen Chemie. Freiburg i. B., 1874. 8vo.
    Zweite Auflage, 1877.

LOSANITSCH, S. M.

Chemische Technologie. I, Ueber Wasser und Brennstoffe. Belgrad, 1887. 8vo.

LOSCHMIDT, J.

Chemische Studien. I, Constitutions-Formeln der organischen Chemie in graphischer Darstellung.—Das Mariotte'sche Gesetz. Wien, 1861. Roy. 8vo.

LOTH, JULIUS.

Anorganische (Die) Chemie auf Grundlage methodisch-geordneter Versuche für den Unterricht an höheren Lehranstalten und zur Selbstbelehrung. Braunschweig, 1876. 8vo.

Grundriss der systematischen Chemie. Entwurf einer Statik und Dynamik der chemischen Kräfte. Chemie der anorganischen Körper. Leipzig, 1847. 8vo.

      Zweite Auflage. Leipzig, 1854. 8vo.

Lehrbuch der Chemie und Mineralogie auf Grundlage der neueren Theorien, für den Unterricht an höheren Lehranstalten. Leipzig, 1872. pp. xvi-388.

LOTH, JULIUS J.

Der Wegweiser durch das Gebiet der Chemie. Erfurt, 1852. 8vo.

LOTMAN. G.

Handboek voor het onderzoek van grondstoffen en producten der suiker-industrie. Andere serie. Amsterdam, 1886.

Praktische handleiding tot het onderzoek van alle suikerhoudende stoffen, bewerkt naar de nieuwste en beste bronnen en toegelicht met een aantal resultaten der fabrikage en raffinage. Amsterdam, 1867. pp. vi.-185, 8vo.

LOUISE, E.

Synthèse d'hydrocarbures, d'acétones, d'acides, d'alcools, d'éthers et de quinones dans la série aromatique. Paris, 1885.

LOUYET, P.

Cours élémentaire de chimie générale, inorganique, théorique et pratique à l'usage des universités et des écoles industrielles. Bruxelles, 1842-'44. 8vo.

LOW, DAVID.

An Inquiry into the nature of the Simple Bodies of Chemistry. Third edition. Edinburgh, 1856. pp. xiv-386, 8vo.

LOWN, CLARENCE, AND HENRY BOOTH.
  Fossil Resins, a compilation. New York, 1891. pp. xv–119, 12mo. Plates.

LOWRIE, W. F.
  Toxicologia, or a Treatise on internal poisons in their relation to medi-
    cal jurisprudence, physiology and the practice of physic. New
    York, 1832. 8vo.

LUANCO, J. R.
  Compendio de las lecciones de química general, explicadas en la uni-
    versadad de Barcelona. Barcelona, 1882. 4to.
          Segunda edicion. Barcelona, 1884.

LUBARSCH, O.
  Elemente der Experimental-Chemie. Ein methodischer Leitfaden für
    den chemischen Unterricht an höheren Lehranstalten. Theil 1:
    Die Metalloide, Berlin, 1888. pp. xvi–128. Theil II : Die Metalle.
    Berlin, 1888. pp. viii–184, 8vo.
  Technik des chemischen Unterrichts auf höheren Schulen und gewerb-
    lichen Lehranstalten. Eine kurze Anleitung zur Ausführung der
    grundlegenden chemischen Demonstrationsversuche. Für den
    praktischen Schulgebrauch bearbeitet. Berlin, 1889. 8vo.

LUCA, G. DE.
  Elementi di chimica industriale, compilati sull' edizione francese del
    dizionario di chimica industriale pubblicato da Barreswil e A. Girard,
    con la cooperazione di G. de Luca. Milano, 1884. 2 vols., 16mo.
          Cf. Barreswil et Girard : Dictionnaire de chimie industrielle.

LUDOLFF, HIERONYMUS VON.
  Die in der Medicin siegende Chemie . . . Erfurt, 1746–49. Seven
    parts and a supplement dated 1750.

LUDWIG, ERNST.
  Medicinische Chemie in Anwendung auf gerichtliche, sanitätspolizeiliche
    und hygienische Untersuchungen sowie auf Prüfung der Arznei-
    präparate. Ein Handbuch für Aerzte, Apotheker, Sanitätsbeamte
    und Studirende. Wien, 1885. 8vo. Ill.

LUDWIG, HANS.
  Algebra (Die) der Chemie. Eine ausführliche Bearbeitung der anor-
    ganischen Zersetzungsgleichungen in übersichtlicher Form. Frei-
    burg i. B., 1876. 8vo.
  Natürlichen (Die) Wässer in ihren chemischen Beziehungen zu Luft und
    Gesteinen. Stuttgart, 1862. 8vo.

LUDWIG, HERMANN.

Grundzüge der analytischen Chemie unorganischer Substanzen, zum Gebrauche in landwirthschaftlich-chemischen Laboratorien entworfen. Jena, 1851. 8vo.

LÜDERSDORF, FRIEDRICH WILHELM.

Fabrikation (Die) des Runkelrübenzuckers. Berlin, 1836.

LUFF, ARTHUR P.

Introduction (An) to the Study of Chemistry, specially designed for Medical and Pharmaceutical Students. London, 1885.

Manual (A) of Chemistry, Inorganic and Organic, with an Introduction to the Study of Chemistry. London, Paris and Melbourne, 1892. pp. xvi–525, 12mo. Ill.

> Prepared as a guide for students of medicine.

LUHMANN, E.

Die Kohlensäure. Eine ausführliche Darstellung der Eigenschaften, des Vorkommens, der Herstellung und der technischen Verwendung dieser Substanz. Handbuch für Chemiker, Apotheker, Fabrikanten künstlicher Mineralwässer, etc., Bierbrauer und Gastwirthe. Wien, 1885. 8vo.

LUIGI, MASCHI.

Storia naturale del chimismo animale fisiologico e patologico. Parma, 1863. 8vo.

LUISCIUS-STIPRIAAN, ABRAHAM VAN.

> See Stipriaan, Abraham van Luiscius.

LULL, RAMÓN.

> See, in Section VI, Lullius, Raymundus.

LUND, CANUT.

Tentamen physico-chymicum de causis principalibus solutionis. Hafniæ 1772. 4to.

LUNGE, GEORG.

Destillation (Die) des Steinkohlentheers und die Verarbeitung der damit zusammenhängenden Nebenproducte. Braunschweig, 1867. 8vo. Ill.

Industrie (Die) des Steinkohlentheers und Ammoniaks. Dritte vermehrte und verbesserte Auflage. Braunschweig, 1888. pp. 195, 8vo. Ill.

> Cf. Bolley's Handbuch [etc.], Neue Folge.

> A Treatise on the Distillation of Coal-tar, and ammoniacal liquor, and the separation from them of valuable products. London, 1882.

LUNGE, GEORG. [Cont'd.]

> Second edition. London, 1887.
>
> Traité de la distillation du goudron de houille et du traite-ment de l'eau ammoniacale, traduit de l'allemand par L. Gautier. Paris, 1885. 8vo. Ill.

Handbuch der Soda-Industrie und ihrer Nebenzweig für Theorie und Praxis. Braunschweig, 1879–80. 2 vols., 8vo. Ill.

> *Cf.* Bolley's Handbuch.
>
> A Theoretical and Practical Treatise on the Manufacture of Sulphuric Acid and Alkali, with the collateral branches. London, 1879–80. 8vo. Vol. I, pp. xv–658 ; vol. II, pp. xiv–708 ; vol. III, pp. xvi–422. Ill.
>
> New and enlarged edition. London, 1891. 8vo. Ill.

Taschenbuch für die Soda- Pottasche- und Ammoniak-Fabrikation. Herausgegeben im Auftrage des Vereins Deutscher Sodafabri-kanten und unter Mitwirkung der Commissions-Mitglieder J. Stroof (Griesheim), Vorsitzender, J. Dannien (Buckau), H. Kunheim (Berlin), Dr. Mayer (Heufeld), Dr. Nobiling (Schöningen), Dr. Richters (Saarn), E. Schott (Heinrichshall). Berlin, 1883.

> The Alkali-maker's Pocket-book. Tables and analytical methods for manufacturers of sulphuric acid, nitric acid, soda, potash and ammonia. London, 1884.
>
> Edited by F. Hurter.
>
> Second edition. London, 1891, [with the title "Handbook" in place of "Pocket-book."]
>
> Vade-mecum du fabricant de produits chimiques. Traduit sur la deuxième édition par V. Hassreidter et E. Prost. Paris, 1892. 12mo. Ill.

LUNGE ET NAVILLE.

La grande industrie chimique. Traité de la fabrication de la soude et de ses branches collatérales. Vol. II : Sulfate de soude ; acide chlorhydrique, soude brute. Paris, 1880. 8vo.

LUPTON, SIDNEY.

Chemical Arithmetic. With 1200 examples. Second edition. London, 1886. pp. xii–169, 12mo.

LUTOSŁAWSKI, W.

Das Gesetz der Beschleunigung der Esterbildung. Beitrag zur che-mischen Dynamik. Halle a. S., 1885. pp. 12, 4to.

LUYART, DON FAUSTO DE.

> *See* D'Elhujar, Fausto.

LUYNES, DE.

Leçons (etc.). *See* Société chimique de Paris.

Львовъ, Э. Н.

Изъ исторіи великихъ открытій и изобрѣтеній. Открытіе кислорода. С.-Петербургъ 1874.

LVOF, E. N. From the History of great discoveries and inventions. Discovery of Oxygen. St. Petersburg, 1874.

Лясковскій, Н.

Проростаніе тыквенныхъ сѣмянъ въ химич. отношеніи. Москва 1874

LYASKOVSKY, N. On the development of pumpkin seeds and their chemical relations. Moscow, 1874.

О химическомсъ оставѣ пшеничнаго зерна. Москва 1865

On the chemical composition of the seed of wheat. Moscow, 1865.

LYNEN, J. M.

Aanhangsel tot de Catechismus der Apothekerskunst van S. F. Hermbstädt. Behelzende eene beknopte verklaaring van de grondslaagen, beginselen en de daarop gegronde bewerkingen, welke in dat werkje vervat zyn, volgens het Lavoisieriaansche system. Amsterdam, 1800. 8vo. Ill.

LYON, FACULTÉ DE MÉDECINE DE. Travaux du laboratoire de chimie organique. Année 1889. Lyon, 1890.

M, R.

Chemische Trennungs-Methoden. Uebungen zum Gebrauche beim Laboratoriums-Unterrichte. Graz, 1880. 8vo.

MAANEN, FLORENTIUS JACOBUS VAN.

Over de eenvoudigheid der tegenwoordige bespiegelende en beoefenende scheikunde. s'Hage, 1806. 8vo.

MACADAM, STEVENSON.

Practical Chemistry. London and Edinburgh, 1866. pp. x–147, 12mo. Ill.

Chambers's Educational Course.

Практическая хи iя. Перев. Булыганскій А. Д. Москва 1874.

MACBRIDE, DAVID.

Experimental Essays on the following subjects. I. On the Fermentation of Alimentary Mixtures. II. On the Nature and Properties of Fixed Air . . . Illustrated with Copper Plates. London, 1764. 4to.

MACDONALD, J. D.

A Guide to the Microscopical Examination of Drinking Water. With an Appendix on the Microscopical Examination of Air. With twenty-five lithographic plates. Second edition. London, 1882. pp. xi–83, 8vo.

McGILL, J. T.
  An Introduction to the Study of Qualitative Chemical Analysis.   Nash-
    ville, Tenn., 1889.   8vo.

M'GOWAN, GEORGE [Translator].
        See, in Section III, Meyer, Ernst von.

MACKENZIE.
  Mémorial pratique du chimiste-manufacturier ou recueil de procédés
    d'arts et de manufactures.   Traduit de l'anglais.   Paris, 1824–25.
    3 vols., 8vo.

MACLAREN, J. J.
  Some Chemical Difficulties of Evolution.   London, 1877.

MACLEAN, JOHN.
  Two Lectures on Combustion . . . containing an examination of
    Priestley's considerations on the doctrine of phlogiston and the
    decomposition of water.   Philadelphia, 1797.   8vo.

McMILLAN, WALTER G.
  Treatise on Electro-Metallurgy, embracing the application of electrolysis
    to the plating, depositing, smelting and refining of various metals,
    and to the reproduction of printing surfaces, art-work, etc.   Lon-
    don, 1890.   pp. xvi–387, 8vo.   Ill.

McMULLEN, THOMAS.
  Handbook of Wines, practical, theoretical and historical ; with a descrip-
    tion of foreign spirits and liqueurs.   New York, 1852.   pp. xii–
    327, 8vo.

McMUNN, C. A.
  Outlines of the Clinical Chemistry of Urine.   London, 1889.

MACNEVEN, WILLIAM JAMES.
  Chymical Exercises in the Laboratory of the College of Physicians and
    Surgeons, by the pupils of the laboratory.   New York, 1819.   8vo.
  Exposition of the Atomic Theory of Chymistry, and the doctrine of
    definite proportions.   With an appendix of chymical exercises.
    New York, 1819.   8vo.

MACH, EDMUND.
  Die Gährung und die Technologie des Weines.   (Aus Schwackhöfer's
    Lehrbuch der landwirthschaftlich-chemischen Technologie).   Wien,
    1884.   8vo.
        Cf. Schwackhöfer, F.

MACHADO, ANTONIO.

Resúmen de las lecciones de química orgánica explicádas en la Facultad de ciencias médicas de Cádiz. Cádiz, 1845. 4to.

MACK, EDUARD.

Lehrbuch der Chemie für Realschulen, mit besonderer Berücksichtigung der Anwendung derselben auf die Gewerbe. Erster Theil: Unorganische Chemie. Pressburg, 1853. 8vo.

MACQUER, PIERRE JOSEPH.

Art (L') de la teinture en soie. Paris, 1763. 8vo.

Élémens de chymie pratique, contenant la description des opérations fondamentales de la chymie aves les explications et des remarques sur chaque opération. Paris, 1751. 2 vols., 16mo. Vol. I, pp. xvi-[viii]-517.

>Nouvelle édition. Paris, 1756. 3 vols., 12mo.

Élémens de chymie théorique. Paris, 1749. pp. xxiv-336-[xxiv], 16mo. 3 folding plates.

>Nouvelle édition. Paris, 1755. 8vo.

>* Elements of the Theory and Practice of Chymistry. Translated from the French [by Andrew Reid]. London, printed for A. Millar, and J. Nourse, in the Strand, 1758. 2 vols., 8vo. Vol. I, 6 plates.

>Third édition, 1775, 2 vols.

>Anfangsgründe der theoretischen Chymie aus dem französischen ins Deutsche übersetzt. Leipzig, 1752. 8vo.

>Zweyte Auflage. 1768.

>Elementi di chimica teorica e pratica. Venezia, 1781. 4 vols., 8vo. Ill.

>Начальныя основанія умозрительной химіи. С.-Петербургъ 1791.

See also in Section II.

MACQUER, PIERRE JOSEPH, ET ANTOINE BAUMÉ.

Plan d'un cours de chimie expérimentale et raisonnée. Paris, 1757.

MADAC, PETER.

Theoria affinitatum chemicorum. Tyrnau, 1774. 8vo.

MADAN, H. G.

Tables of Qualitative Analysis. London, 1881.

MAERCKER, MAX.

Handbuch der Spiritusfabrikation. Berlin, 1877. 8vo. Ill.

>Fünfte Auflage. Berlin, 1889.

MAERCKER, MAX. [Cont'd.]
> Traité de la fabrication de l'alcool.   Traduit de la quatrième
> édition allemande par Bosker et Warnery.   Lille, 1889.
> 2 vols.

MAERCKER, M., UND MORGEN, A.
Wesen und Verwerthung der getrockneten Diffusionsrückstände der
Zuckerfabriken.   Nach Untersuchungen im Laboratorium der
Versuchsstation Halle a. S. . . .   Fütterungsversuchen und
Beobachtungen der Praxis bearbeitet.   Berlin, 1891.   8vo.

MAETS, CARL LUDWIG VAN.
Prodromus chymiae rationalis.   Lugduni-Batavorum, 1684.   8vo.
> [Second edition,] Chymia rationalis et praxis chymiatricæ
> rationalis.   Lugduni-Batavorum, 1687.   4to.

MAETS, CARL LUDWIG VAN.
> *See* Collectanea chymica Leidensia.

MAGELHAENS, JOAO HYAZINTHE DE.
> *See* Magellan, J. H.

MAGELLAN, J. H. [MAGELHAENS, J. H.].
Description of a Glass Apparatus for making the best mineral waters of
Pyrmont, Spa, Seltzer, Seydschütz etc., together with the description
of two new eudiometers, etc.   London, 1779.   8vo.
> Third edition.   London, 1783.
> Beschryving van een glazen werktuig om in weinige minuten
> en met geringe kosten minerale wateren te maken.
> Utrecht, 1779.   8vo.
> Beschreibung eines Glasgeräthes vermittelst dessen man
> mineralische Wasser in kurzer Zeit und mit geringem
> Aufwande machen kann ; wie auch einiger neuen
> Eudiometer . . . in einem Sendschreiben an Priestley.
> Uebersetzt aus dem englischen von G. J. Wenzel und
> mit Zusätzen, besonders in Rücksicht der Eudiometer
> erläutert von C. F. Wenzel.   Dresden, 1780.   8vo.
Nouvelle construction d'alembic.   Utrecht, 1781.   4to.   4 plates.
Essai sur la nouvelle theorie du feu élémentaire et de la chaleur des
corps.   London, 1780.
> Proeve eener nieuwe beschouwing over het elementaire
> vuur en de warmte der ligchamen . . . Nieuwe ther-
> mometer.   Utrecht, 1780.   8vo.   Ill.

MAGENDIE, FRANÇOIS.

Recherches physiologiques et chimiques sur l'emploi de l'acide
prussique . . . Paris, 1819. 8vo.

Précis élémentaire de physiologie. Paris, 1816. 2 vols., 8vo.
Deuxième édition. Paris, 1836.

MAGISTRETTI, C.

Elementi di chimica e mineralogia ad uso delle scuole normali, tecniche,
ecc. Torino, 1891. 8vo. Ill.

MAGNIER DE LA SOURCE, L.

Analyse des vins. Paris, 1892. 8vo. Ill.

Chimie appliquée à l'hygiène et aux falsifications. Composition et
analyse du vin ; recherche des altérations frauduleuses de ce
liquide. Paris, 1881. 8vo.

MAHNERUS, ADAMUS.
See Müller, Johannes.

MAHRENHOLTZ, A.

Die praktisch-chemischen Uebungen an Landwirthschaftsschulen. Zum
Gebrauch bei den analytischen Arbeiten im Laboratorium zusammengestellt. Liegnitz, 1886. 8vo.

MAIER, JULIUS.

Die aetherischen Oele   Ihre Gewinnung, chemischen und physikalischen
Eigenschaften, Zusammenzetzung und Anwendung. Zum Ge-
brauche für Aerzte, Chemiker, Droguisten und Pharmaceuten.
Stuttgart, 1862. pp. ix–201, 8vo. Ill.

MAI, J.

Vademecum der Chemie. Repetitorium der unorganischen, organischen
und analytischen Chemie. Bearbeitet für Studirende denen die
Chemie als Hülfswissenschaft dient. Mannheim, 1890. 8vo.

MAISSEN.

Appunti di chimica. Modena, 1892. 8vo.
Lithographed.

MAJOR, H.

Inorganic Chemistry (advanced). Manchester and London, 1875. 12mo.

MAKINS, G. H.

A Manual of Metallurgy, more particularly of the precious metals ;
including the methods of assaying them. Philadelphia, 1865.
12mo. Ill.

MALACARNE, INNOCENZE.

Elementi di chimica applicata alle arti. Milano, 1848. 8vo. Ill.

MALAGUTI, FAUSTINO JOVITA.

Chimie appliquée à l'agriculture, précis des leçons professées depuis 1852 jusqu'à 1862 sur différents sujets d'agriculture. Nouvelle édition. Corbeil, 1862. 2 vols., 18mo.

Cours de chimie agricole professé en 1858 à la Faculté des sciences de Rennes, sous les auspices de M. le Ministre de l'agriculture et du commerce, et publié par décision du conseil général d'Ille-et-Vilaine. Paris, 1858. 8vo. Ill.

The same for 1859. Paris, 1859.

Leçons de chimie agricole. Paris, 1848. 12mo.

Nouvelle édition. Paris, 1855. 12mo.

Lezioni di chimica agraria. Edizione Italiana per cura di F. Selmi. Torino, 1850. 12mo.

Nuove lezioni di chimica agraria, dette nel 1853. Versione Italiana di P. Carlevaris con note del traduttore. Torino, 1854. 12mo.

Nuove lezioni di chimica agraria, tradotte da A. Selmi. Piacenza, 1857.

Lezioni di chimica applicata all' agricoltura. Tradotte da A. Selmi. Torino, 1865. 3 vols., 12mo.

Chimica agraria elementare. Traduzione di S. Cavallero. Schio, 1884. 16mo.

Leçons élémentaires de chimie. Paris, 1853. 2 vols., 12mo.

Deuxième édition entièrement refondue et augmentée. Paris, *n. d.* [1858–59]. 2 vols., 12mo. Ill. I, pp. 1,052 ; II, pp. 828.

Troisième édition, corrigée et augmentée. Paris, 1863–64. 2 vols., 12mo.

Hwa hio shen yuan. [Translated into Chinese by A. Bille-quin]. Peking. 4 vols.

Petit cours de chimie agricole à l'usage des écoles primaires. Paris, 1857 (?). 18mo. Ill.

Первоначальный учебникъ химіи. Перев. Савченкова. С.-Петербургъ 1872.

MALATRET, P.

See Bayen, Pierre.

MALEPEYRE, FRÉDÉRIC.

Nouveau manuel complet de la fabrication des acides gras concrets employés dans les arts, et de celle des bougies stéariques, marga-

MALEPEYRE, FRÉDÉRIC. [Cont'd.]

    riques, élaidiques, palmétiques et concéniques. Paris, 1849. 18mo. Plates.

        Handbuch der Fabrikation der Stearinkerzen, sowie der Margarin-, Elaidin-, Palmitin- und Cocinin-Kerzen. Eine vollständige und auf die neuesten Erfahrungen in der Chemie begründete theoretisch-praktische Anweisung zur zweckmässigsten Gewinnung des Stearins, Margarins, Palmitins, Cocinins und Elaidins, sowie zur Anfertigung der beliebtesten und in jeder Hinsicht vorzüglichsten Arten von Kerzen aus den genannten Stoffen. In Verbindung mit Mehreren herausgegeben. Nach der neuesten Auflage aus dem französischen. Quedlinburg, 1853. 8vo. Ill.

MALEPEYRE, F., ET A. PETIT.

    Manuel d'alcoolométrie, contenant la description des appareils et des méthodes alcoolométriques, les tables de force et mouillage des alcools, d'après Gay-Lussac et Küpffer. Paris, 1887. 8vo.

MALLARD, E.

        Conférence (etc.). *See* Société chimique de Paris.

MALLE, P.

    Considérations médico-légales sur les empoisonnements simples et complexes, suivies d'une nouvelle méthode d'analyse générale et d'un nouveau mode d'isolement de l'arsenic. Strasbourg, 1838. 8vo.

    Essai d'analyse toxique générale. Strasbourg, 1839. 8vo.

MALLET, J. W.

    Chemistry Applied to the Arts. Lynchburg, 1868. 8vo.

MALOUIN, PAUL JACQUES.

    Chymie médicinale . . . Paris, 1750.

        Troisième édition. Paris, 1756. 2 vols., 8vo.

        Medizinische Chemie, die Arzneien zu bereiten. Altenburg, 1768. 2 parts, 8vo.

    Traité de chymie. Paris, 1734. 12mo.

        Deuxième édition. Paris, 1755. 2 vols., 12mo.

MALY, RICHARD.

        Chemie der Verdauungssäfte und der Verdauung. *See* Handbuch der Physiologie herausgegeben von L. Hermann. Leipzig, 1880.

MALY, R., UND R. ANDREASCH.

    Studien über Caffeïn und Theobromin. Wien, 1883. 8vo.

MAMONE-CAPRIA, DOMENICO.

Trattato de' reagenti ed uso di essi con quattro esemplari pratici di analisi, dei quali uno riguarda l'analisi dell' arsenico ; altro l'analisi di terreni aratorii. Napoli, 1864. 24mo.

[MANCHERLEI.] Allgemein nützliches chemisch-physikalisches Mancherlei. Mit Kupfern. Berlin, 1781–82. 2 vols. Vol. I, pp. viii–244, 2 plates, 1781 ; vol. II, pp. [iv]–372, 2 plates.

> Containing 22 treatises, among which are : "Geschichte der künstlichen Luft, von J. R. Spielmann und J. F. Corvinus," and "Ueber die Zerlegung von Wasser, von T. Bergmann."

MANDELBLÜH, CLEMENT.

Leitfaden zur Untersuchung der verschiedenen Zuckerarten sowie der in der Zuckerfabrikation vorkommenden Producte. Ein Hilfsbuch für Fabriksdirectoren, Techniker und Siedmeister, zusammengestellt von Mandelblüh. Brünn, 1867. pp. [vi]–106, 8vo. Ill.

MANG, A.

Leitfaden der Chemie, Mineralogie und Gesundheitslehre für Bürger- und Realschulen und verwandte Anstalten, unter Berücksichtigung des practischen Lebens methodisch bearbeitet. Weinheim, 1883. 8vo.

MANGIN, ANTIDE.

Notions mathématiques de chimie et de médecine, ou Théorie du feu, où l'on démontre par les causes, la lumière, les couleurs, le son, la fièvre, nos maux, la clinique, etc. Paris, 1800. 8vo.

MANGOLD, CHRISTOPH ANDREAS.

Chymische Erfahrungen und Vortheile in Bereitung einiger sehr bewährter Arzeneimittel . . . Erfurt, 1748.
　　　　Neue chymische Erfahrungen [etc.]. Erfurt, 1749.

MANN, FRIEDRICH.

Elementarcurs der Chemie in inductorischer Methode. Für untere Industrieschulen, Secundarschulen, höhere Bürgerschulen, Lehrerseminare etc. Frauenfeld, 1857. 8vo.

MANN, L.

Atomaufbau (Der) in den chemischen Verbindungen, und sein Einfluss auf die Erscheinungen. Berlin, 1884. 8vo.

Atomgestalt (Die) der chemischen Grundstoffe. Berlin, 1883. 8vo.

MANSFIELD, CHARLES BLACKFORD.

Benzole, its nature and utility. London, 1849. 8vo.

MANSFIELD, CHARLES BLACKFORD. [Cont'd.]

Theory (A) of Salts : a treatise on the constitution of bipolar (two-membered) chemical compounds. London, 1865. pp. lii–608, 8vo.

Edited by N. S. M., *i.e.*, N. S. Mansfield?

MANUEL DE CHIMIE INORGANIQUE à l'usage des élèves du pensionnat des frères des écoles chrétiennes de Reims. Reims, 1886. 8vo.

MAQUENNE.

Conférence (etc.). *See* Société chimique de Paris.

MARABELLI, F.

Lezioni di chimica-farmaceutica. Pavia, 1805. 3 vols.

MARAVIGNA, CARMELO.

Prime linee di chimica inorganica applicata alla medicina ed alla farmacia. Messina, 1826, 1827, 1828. 3 vols., 8vo.

MARCANO, V.

Elementos de filosofía química segun la teoria atómica. Precedido de una carta de A. Naquet. Carácas, 1881. 8vo.

MARCET, JANE.

Conversations on Chemistry, in which the elements of that science are familiarly explained and illustrated by experiments. London, 1806. 2 vols., 12mo.

This work passed through more than twenty editions, having a success now difficult to comprehend. *Compare* Jones, Thomas P., and Comstock, John Lee.

MARCET, W.

On the Composition of Food and how it is adulterated, with practical directions for its analysis. London, 1856. 8vo.

MARCHAND, ANTON.

* Neue Theorie der Gährung. Nebst zwei Abhandlungen über die nützlichste Art Brandtwein zu brennen und Essig zu sieden. Mannheim, 1787. pp. 120, 8vo.

* Ueber Phlogiston, elektrische Materie, Licht, Luft und die unmittelbare Ursache der Bewegung. Mannheim, 1787. pp. 132, 8vo.

MARCHAND, EUGÈNE.

Eaux (Les) potables en général considérées dans leur constitution physique et chimique et dans leurs rapports avec la physique du globe, la géologie, la physiologie générale, l'hygiène publique, l'industrie et l'agriculture ; et en particulier des eaux utilisées dans

MARCHAND, EUGÈNE. [Cont'd.]
les arrondissements du Havre et d'Yvetot, avec la carte géologique de ces arrondissements. Paris, 1855. 4to.

Étude sur la force chimique contenue dans la lumière du soleil, la mesure de sa puissance et la détermination des climats qu'elle caractérise. Paris, *n. d.* [1873]. pp. xx-196, 8vo.

MARCHAND, RICHARD FELIX.
Chemische Tafeln zur Berechnung der Analysen. Leipzig, 1847. 8vo.

Grundriss der organischen Chemie. Leipzig, 1839. 8vo.

Schets der bewerktuigde scheikunde. Uit het Hoogduitsch. Amsterdam, 1840. 8vo.

Lehrbuch der organischen Chemie. Leipzig, 1838.

Lehrbuch der physiologischen Chemie. Berlin, 1844. 8vo.

MARCKWALD, W.
Ueber die Beziehungen zwischen dem Siedepunkte und der Zusammensetzung chemische Verbindungen welcher bisher erkannt worden sind. Berlin, 1887.

MARCO, FELICE.
Nozioni di chimica moderna ad uso specialmente dei Licei. Torino, 1870. 16mo.

Quarta edizione. Torino, 1879.

Settima edizione. Torino, 1887. 12mo. Ill.

[Another edition.] Torino, 1891.

MARECK, FRIEDRICH.
Leitfaden für den chemischen Unterricht an Unterrealschulen, Bürger-, Handels- und Gewerbeschulen. Wien, 1867. 8vo. Ill.

MARET, HUGUES.
Analyse de l'eau de Pont-de-Vesle. Dijon, 1779. 8vo.

Élémens de chimie théorique et pratique. Dijon, 1777. 3 vols.

MARGGRAF, ANDREAS SIGISMUND.
* Chymische Schriften. Berlin, 1761–1767. 2 vols., 12mo. Second edition of vol. I in 1768, pp. [xxii]–330–[vi], 2 plates. Vol. II, pp. xiv-206, 1 plate.

The preface to vol. I is signed Johann Gottlob Lehmann, and that to vol. II, L. von Beausobre.

Opuscules chymiques. Paris, 1762. 2 vols., 12mo.

Chymische Untersuchung eine s sehr merkwürdigen Salzes, welches das Saure des Phosphors in sich enthält. Leipzig, 1757. 4to.

MARGGRAVIUS, CHRISTIAAN.
See Collectanea chymica Leidensia, Maetsiana et Marggraviana.

MARGOTTET, J.
Cours élémentaire de chimie.   Troisième année.   Paris, 1884.   18mo.

MARGOTTET, J.
See, in Section II, Fremy : Encyclopédie chimique, vol. II, vol. III, etc.

MARGUERITTE ET CAMUS.
Industrie du gaz.   *See, in Section II.,* Fremy : Encyclopédie chimique, vol. V.

MARHERR, PHILIPP AMBROSIUS.
Dissertatio chemica de affinitate corporum . . . publicæ disquisitioni
committit . . . die 2 mensis Aprilis, 1762.   Vindobonæ, *n. d.*
[1762].   pp. [xvi]–99, 12mo.   4 Tables.
> Chemische Abhandlung von der Verwandtschaft der Körper
> aus dem lateinischen übersetzt von E. G. Baldinger.
> Leipzig, 1764.   pp. 144, 12mo.   4 Tables.
>> The tables indicate relative affinities of acids to various substances, aided by
>> alchemical symbols.   Bibliography is not neglected.

MARICHAUX.
Ueber den gegenwärtigen Stand der Zuckerfabrikation.   Nürnberg, 1812.

MARINO, SALVATORE SALOMONE.
See Salomone-Marino, S.

MARIOTTE, EDMONDE.
*Traité du mouvement des eaux et des autres corps fluides.   Divisé en
cinq parties.   Mis en lumière par les soins de M. de la Hire.   Paris,
chez Estienne Michallet, rue saint Jaques, à l'Image S. Paul, 1686.
pp. [xii]–408–[xx], 16mo.   Ill.

MARIUS.
Korte beschriiving over de aanwending en het nut van het waterglas.
Bijeenverzameld uit wetenschappelijke tijdschriften en opgaven
van scheikundigen.   Arnhem, 1856.   8vo.

MARLIN, PAUL.
Examen comparatif de la fabrication des produits chimiques en Belgique
et en Angleterre.   Fabrication de l'acide sulfurique.   Fabrication
du sulfate de soude.   Fabrication du carbonate de soude (sel de
soude).   Liège, 1865.   8vo.   Ill.
Exposé des principaux procédés analytiques à l'usage des fabricants et
de consommateurs de produits chimiques.   Liège, 1869.   8vo.

MARQUISAN, H.
Goudrons (Des) de houille et leurs dérivés des goudrons de pétrole. Paris, 1885. 8vo.
Saccharine (La).   Marseille, 1887.

MARSHALL, JOHN.
Notes on the Chemical Lectures in the Medical Department of the University of Pennsylvania for second-year students.   Published by authority of Theodore G. Wormley.   Philadelphia, 1889.   pp. 107, 8vo.

MARTEAU, PIERRE ANTOINE.
Analyse des eaux minérales d'Aumale.   Paris, 1756.   12mo.
Analyse des eaux de Forges.   Paris, 1756.   12mo.
Traité de l'analyse des eaux minérales.   Paris, 1778.   4to.

MARTIN, E. N.
Études de chimie philosophique, exposé de principes de chimie d'une nouvelle école, conduisant à la solution de toutes les grandes questions de chimie et de physique, à l'explication de la vie végétale et animale de la constitution des corps organisés, des fermentations et des divers phénomènes naturels les plus intéressans ; avec des applications à l'hygiène la médecine et l'agriculture.   Première partie, Principes de chimie d'une nouvelle école.   Paris, 1842.   8vo.

MARTIN, JEAN ÉMILE.
L'Art de dégraisser . . . les tissus.   Paris, 1828.   18mo.
L'Art de la teinture de la soie, du coton . . .   Paris, 1828.   18mo.
L'Art de la conservation des substances alimentaires.   Paris, 1829.   18mo.
L'Art de préparer la chaux . . .   Paris, 1829.
L'Art de l'amidonnier . . .   Paris, 1836.   8vo.
Chimie du teinturier.   Paris, 1828.   18mo.
Éléments de chimie appliquée aux manufactures . . .   Paris, 1829.   2 vols., 12mo.
Étude de chimie philosophique . . .   Paris, 1842.   8vo.
Nouvelle école électro-chimique, ou Chimie des corps pondérables et impondérables, découverte des véritables corps simples et d'une théorie chimique générale à l'aide de laquelle les actions chimiques et électrochimiques sont devoilées, la science aggrandie et simplifié.   Paris, 1858.   8vo.

MARTIN, W. A. P.
Ke chih ju men.   1865.   New edition, enlarged and revised.   Peking, 1891.   7 vols.
Natural Philosophy and Chemistry in Chinese.

MARTIN Y CASTRO, FLORENCIO.

Nociones elementales de química para el uso de los aspirantes al grado de bachiller en filosofía. Cáceres, 1847. 8vo.

MARTINY, BENNO.

Die Milch, ihr Wesen und ihre Verwerthung. Danzig, 1871. 2 vols., 8vo.

MARTIUS, C. A.

Verzeichniss der chemischen Fabriken Deutschlands. Im Auftrage des Vereins zur Wahrung der Interessen der chemischen Industrie Deutschlands herausgegeben. Berlin, 1880. 4to.

MARTIUS-MATZDORFF, J.

Katechismus der Galvanoplastik. Leipzig, 1868. pp. viii–104, 12mo.

MARUM, MARTINUS VAN.

Beschryving van chemische Werktuigen behoorende aan Teyler's Stichting. Haarlem, 1798. 4to.

> Description de quelques appareils chimiques nouveaux ou perfectionner de la fondation Teylerienne et des expériences faites sur ces appareils. Haarlem, 1798. 4to.

Natuurkundige verhandeling over de gephlogisteerde en dephlogisteerde luchten. Haarlem, 1781. 4to.

MARVIN, CH.

The Petroleum Industry of Southern Russia and the Caspian Region. London, 1884. Roy. 4to.

MARX, CARL MICHAEL.

Chemische Tabellen. Braunschweig, 1831.

MARX, K. F. H.

Die Lehre von den Giften in medizinischer, gerichtlicher und polizeylicher Hinsicht. Göttingen, 1827. 1 vol., 2 parts, 8vo.

MARX, LOUIS.

Le laboratoire du brasseur. Traité analytique des eaux, des orges, des malts, des houblons, des poix et des bières, leurs qualités et leurs falsifications. Les levures et les levures pures. Trouble des bières, comment on les reconnaît, leurs causes, leurs remèdes et le moyen de les éviter. Troisième édition considérablement augmentée, entièrement relue et corrigée. Couronné par la Société d'encouragement pour l'industrie nationale. Paris, 1890. pp. vi–421, roy. 8vo. Ill.

> Printed at Valence, 1889.

MARZUCCHI, JOSEPH (ABBAS).

Nova et vera chymiæ elementa. Patavii, 1751. pp. [xvi]-237. Folding table.

> A work on theoretical chemistry; chapter VI is entitled : De legibus attractionis ubi de propositionibus geometrice expositis.

MASCAREÑAS, EUGENIO, Y JAIME SANTOMA.

Estudio analítico de vinos catalanes. Barcelona, 1885.

MASCHI, L.

Storia naturale del chimismo animale fisiologico e patologico. Parma, 1863. 8vo.

MASING.

Elemente der pharmaceutischen Chemie. St. Petersburg, 1885.

MASON, WILLIAM P.

Notes on Qualitative Analysis arranged for the use of students of the Rensselaer Polytechnic Institute. Troy, N. Y. Second edition, 1889. pp. 44, 8vo.

MASSEROTTI, V.

Rudimenti di chimica organica. Milano, 1860. 8vo.

MASSON, HERBERT.

Original Research. The Governing Principles of the Elements, explaining and supplementing Dalton's doctrine of definite reciprocal and multiple proportions, and the substitution hydrogen-replacement and compound radical theories of modern chemistry. Part I. London, 1878.

> " The key to the Universe, the grand *arcanum* which shall forever disperse the clouds of ignorance from the fields of science, is here offered us in imposing form and in language which rises in magniloquence until it even bursts into exceedingly blank verse. . . . Turning to the body of the work we read it again and again, each time with a more profound sense of bewilderment until it seemed as a kind of scientific nightmare."—*Chem. News.*

MASURE, F.

Cours public de chimie organique. Deuxième année. La Rochelle, 1861. 8vo.

Éléments de chimie appliquée à l'agriculture, à l'économie domestique et à l'industrie, à l'usage des lycées et des établissements d'instruction secondaire, des écoles normales et des écoles primaires. Paris, 1870. 18mo. Ill.

MATA, PEDRO.
Sinopsis filosófica de la química. Obra escrita para facilitar y abreviar
el estudio de esta ciencia. Madrid, 1849. 8vo.

MATCOVICH, P.
Avviamento alla analisi chimica qualitativa pei corpi organici. Milano,
1880. 8vo.

MATECKI, T. T.
Słownictwó chemiczne Polskie. Poznań, 1855. 8vo.

MATHIEU, L.
Vade-mecum des travaux pratiques au laboratoires de chimie, à l'usage
des lycées, collèges, écoles professionnelles, écoles normales, écoles
primaires, etc. Paris, 1888. 4to.

MATHIEU, L., ET J. MORAUX.
Caractérisation des fuchsines et autres couleurs de la houille dans les
vins à l'aide d'un procédé simple et rapide. Paris, 1885.

MATIGNON, C.
Recherches sur les uréides. Paris, 1892. 8vo.

MATTEUCCI, CARLO.
Fenomeni fisico-chimici dei corpi viventi. Pisa, 1844. 8vo.
        Seconda edizione con molte correzione ed aggiunte. Pisa,
            1846. 8vo.
        Leçons sur les phénomènes physiques et chimiques des corps
            vivants, professées à Pise en 1844. Édition française
            publiée sous les yeux de l'auteur avec des additions
            considérables par Leblanc. Paris, 1845. 8vo.
        Seconde édition, publiée sur la seconde édition italienne.
            Paris, 1846. 12mo.

MAUMENÉ, E. F.
Théorie générale de l'action chimique. Paris, 1880. 8vo.
Traité théorique et pratique de la fabrication du sucre, comprenant la
culture des plantes saccharines, l'extraction du sucre brut, le raffinage,
le traitement des mélasses, la distillation et ses opérations relatives
au travail des salins et potasses, l'analyse des matières utiles à la
culture et à la fabrication, etc., etc. Paris, 1876. 2 vols., 8vo. Ill.

MAURO, F., R. NASINI, E E. A. PICCINI.
Analisi chimica dell' acque potabili della città di Roma, eseguite per in-
carico del municipio. Roma, 1884. 8vo.

MAYBURY, A. C.

The Student's Chemistry. Part I. Non-metallic Elements. London, 1885. 12mo.

MAYER. *See also* Meyer.

MAYER, ADOLPH.

Chemische Technologie des Holzes als Baumaterial. *See* Bolley's Handbuch.

MAYER, ADOLF.

Lehrbuch der Agriculturchemie in vierzig Vorlesungen zum Gebrauch an Universitäten und höheren landwirthschaftlichen Lehranstalten, sowie zum Selbststudium. Heidelberg, 1870. 8vo. Ill.

Zweite Auflage. Heidelberg, 1876.

Dritte Auflage in drei Abtheilungen. Heidelberg, 1885–86.

Учебникъ земледѣльческой химіи. Пыраніе зеленыхъ растеній. С.-Петербургъ 1871.

Учебникъ земледѣльческой химіи въ сорока лекціяхъ. Перев. Со-овѣтова. С.-Петербургъ 1876

Lehrbuch der Gährungschemie in elf Vorlesungen, als Einleitung in die Technologie der Gährungsgewerbe. Heidelberg, 1874. 8vo.

Dritte umgearbeitete Ausgabe. Heidelberg, 1879. 8vo.

Lehre (Die) von den chemischen Fermenten oder Enzymologie. Auf Grund von vorhandenen und eigenen Versuchen. Heidelberg, 1882. pp. viii–124, 8vo.

La chimica delle fermentazioni, tradotta dal prof. A. Pavesi. Milano, 1874. Ill.

Новый способъ опредѣленія удѣльнаго вѣса. С.-Петербургъ 1855.

A new method for the determination of specific gravities. St. Petersburg, 1855.

MAYER, J. R.

Die organische Bewegung in ihrem Zusammenhange mit dem Stoffwechsel. Heilbronn, 1845. 8vo.

MAYER, MORITZ.

Grundzüge der Militair-Chemie. Berlin, 1834. 8vo.

MAYER, SIGMUND.

Specielle Nervenphysiologie. Bewegung der Verdauungs-, Absonderungs- . . . Apparate. *See* Handbuch der Physiologie.

MAYOW, JOHANNES.

Tractatus quinque medico-physici quorum primus agit de salnitro et spiritu nitro-aereo, secundus de respiratione, tertius de respiratione fœtus in utero et ovo, quartus de motu musculari et spiritibus

MAYOW, JOHANNES. [Cont'd.]

> animalibus, ultimus de rhachitide. Oxonii, 1674. pp. [xl]–335 and 1–152, 12mo. Portrait and plates.
>
>> Traduction des œuvres chimiques et physiologiques de Jean Mayow ; par Leopold Ledru et H. C. Gaubert. Paris, 1840. 8vo.
>>
>> Chemisch-physiologische Schriften. Aus dem lateinischen übersetzt von Joh. Koellner. Nebst einer Vorrede von Alex. Nicol Scherer. Jena, 1799. pp. xxii–456–[xiv], 8vo. Ill.
>>
>> *Cf. in Section III*, Scherer, Johann Andreas von.

MAZZONI, A.

> I corpi considerati come chimiche individualità. Faenza, 1868. 8vo.

MEAD, RICHARD.

> A Mechanical Account of Poisons in several essays. The second edition revised, with additions. London : printed by J. M. for Ralph Smith, at the Bible, under the Piazzas of the Royal Exchange, Cornhill. 1708. pp. [xiv]–189, 12mo.

MEADS, S. P.

> Chemical Primer ; an elementary work for use in high schools, academies and medical colleges. Fourth edition. Oakland, California, 1885.

MECKLENBURG, J. M., UND J. FR. SIMON.

> Grundzüge der Chemie in Tabellen-Form. Zunächst als Repertorium für angehende Aerzte und Pharmaceuten bearbeitet. Berlin, 1835. 4to.

MEDICUS, LUDWIG.

> Einleitung in die chemische Analyse. Kurze Anleitung zur technisch-chemische Analyse. Tübingen, 1891.
>
> Gerichtlich-chemische Prüfung von Nahrungs- und Genussmittel. Methoden und Daten zur Beurtheilung zusammengestellt. Würzburg, 1881. pp. vii–152, 8vo. Folding tables.
>
> Kurze Anleitung zur Massanalyse. Mit specieller Berücksichtigung der Pharmakopöe bearbeitet. Vierte Auflage. Tübingen, 1888. pp. 154.
>
> Kurze Anleitung zur Gewichtsanalyse. Tübingen, 1886.
>
> Kurze Anleitung zur qualitativen Analyse. Dritte Auflage. Tübingen, 1886. 8vo.
>
>> Fünfte Auflage. Tübingen, 1888.
>
> Mitteilungen aus dem Laboratorium der Untersuchungsanstalt zu Würzburg. Würzburg, 1886.

MEERTEN, L. A. VAN.

Iets over het door distilleren en filtreren drinkbaar maken van zeewater. Amsterdam, 1829. 8vo.

Over het verbranden van metalen en aether in het overzuurd zoutzuurgas. Amsterdam, 1805. 8vo.

MÈGE-MOURIÈS.

Fabrication des acides gras propres à la confection des bougies et fabrication de savons. Paris, 1864. 4to.

MEHNER, H. D.

Die Fabrikation chemischer Düngemittel in Leipzig. Leipzig, 1885. 8vo.

MÉHU, C.

Traité pratique et élémentaire de chimie médicale appliquée aux recherches cliniques. Paris, 1870. 12mo.

　　　　Deuxième édition. Paris, 1878. 12mo.

MEIDINGER, K. F. VON.

Nähere Beleuchtung der Erfindung Zucker aus Runkelrüben zu erzeugen. [Wien], 1799.

MEIGS, ARTHUR V.

Milk Analysis and Infant Feeding. A practical treatise on the examination of human and cows' milk, cream, condensed milk, etc. ; and directions as to the diet of young infants. Philadelphia, 1885. pp. ix–10–102, 12mo.

MEIJERRINGH, W.

Beknopt leerboek der scheikunde. Zwolle, 1884. 8vo. Ill.

MEINECKE, ANDREAS HEINRICH.

Chemischer Catechismus mit besonderer Rücksicht auf die Bedürfnisse der Landwirthe, der Gewerbtreibenden und überhaupt aller jener, welche die Chemie nicht blos als Studium sondern auch zur Anwendung im Leben sich eigen zu machen wünschen. Unter stäter Beobachtung der neuesten Entdeckungen der Engländer, Franzosen und Deutschen. Prag, 1820. 8vo.

MEINECKE, JOHANN LUDEWIG GEORG.

Die chemische Messkunst, oder Anleitung die chemischen Verbindungen nach Mass und Gewicht auf eine einfache Weise zu bestimmen und zu berechnen. Auf Versuche gegründet und durch Beyspiele erläutert. Halle und Leipzig, 1815–17. 2 vols., pp. 278 and 204, 8vo.

MEINEKE, F. L. G.
Ueber das Schiesspulver, eine chemisch-technische Abhandlung. Halle,
1814.

MEISSAS, ALEXANDRE ANDRÉ DE.
Notions élémentaires de chimie. Paris, 1850. 18mo.

MEISSAS, NICOLAS DE.
Nouveaux élémens de chimie destinés à l'enseignement. Chimie organique et analyse. Chimie inorganique. Paris et Lyon. 1839–40.
2 vols., 8vo.

MEISSNER, CARL FRIEDRICH WILHELM.
    See, in Section VII, Almanach für Scheidekünstler.

MEISSNER, G.
Neue Untersuchungen über den electrischen Sauerstoff. Göttingen,
1869. 4to. Plates.
Untersuchungen über den Sauerstoff. Hannover, 1863. pp. ix–370,
8vo. Folding plate.
    A contribution to the theory of antozone.

MEISSNER, G., UND CHARLES U. SHEPARD.
Untersuchungen über das Entstehen der Hippursäure im thierischen
Organismus. Hannover, 1866. 8vo. Ill.

MEISSNER, PAUL TRAUGOTT.
Aräometrie (Die) in ihrer Anwendung auf Chemie und Technik. Nürnberg, 1816. 2 parts. Ill.
Chemische Aequivalenten- oder Atomenlehre. Zum Gebrauche für
Chemiker, Pharmaceuten und Techniker, gemeinfasslich dargestellt.
Neue unveränderte Ausgabe. Wien, 1838. 2 vols., 8vo. Vol. I,
pp. x–355 ; vol. II, pp. vi–329.
    First edition, 1834. 2 vols., 8vo.
Handbuch der allgemeinen und technischen Chemie. Zum Selbstunterricht, und zur Grundlage seiner ordentlichen und ausserordent-
lichen Vorlesungen entworfen von P. T. Meissner. Wien, 1819–33,
8 vols., 8vo.
    Vol. 5 is in 3 "Abtheilungen," the third of these again is in two parts,
    with second title-page reading "Anfangsgründe des chemischen
    Theiles der Naturwissenschaft."
Neues System der Chemie. Zum Leitfaden eines geregelten Studiums
dieser Wissenschaft. Nebst einem Anhange, enthaltend ein alphabetisches Repertorium der neuesten Entdeckungen und Fortschritte
der Chemie. Neue unveränderte Ausgabe. Wien (1835–38)
1841. 8vo.

MEISSNER, PAUL TRAUGOTT.   [Cont'd.]
       Also under the title :
         Neues System der Chemie organischer Körper.   Mit steter
         Berücksichtigung der Functionen in der organischen
         Natur und Medicin.   1838.   3 vols., 8vo.

MEITZ, OSKAR.
  Die Fabrikation der moussirenden Getränke.   Praktische Anleitung
    zur Fabrikation aller moussirenden Wässer, Limonaden, Weine,
    etc., und gründliche Beschreibung der hierzu nöthigen Apparate.
    Wien, 1881.   Ill.

MEITZEN, E.
  Plan einer chemischen Lehrmethode für Industrielle.   Oder wie erlernt
    der Industrielle in möglichst kürzester Frist die Chemie derart, dass
    er sie selbstandig zum Nutzen seines Faches verwenden kann?
    Leipzig, 1867.   8vo.

MELANDRI-CONTESSI, GIROLAMO.
  Elementi di chimica generale.   Padova, 1809–10.   2 vols., 8vo.
  Nuove ricerche fisico-chimiche ed analisi delle acque di Staro e di
    Civillina.   Padova, 1830.
  Trattato elementare di chimica generale e particolare.   Padova, 1829.
    8vo.

MELDOLA, RAPHAEL.
  The Romance of Science.   Coal and what we get from it.   A romance
    of applied science.   Expounded from the notes of a lecture de-
    livered in the Theatre of the London Institution January 20, 1890.
    Published under the direction of the Committee of General Litera-
    ture and Education appointed by the Society for Promoting Chris-
    tian Knowledge.   London, 1891.
        The author undertakes to give an intelligible account of the development of
        the coal-tar industry without assuming any knowledge of chemical
        science on the part of the reader.   This attempt is " meritorious and
        signally successful."
  Three Lectures on Photographic Chemistry.   London, 1891.   Roy. 8vo.

MÉMOIRE SUR LA SACCHARIFICATION DES FÉCULES.   Deuxième édition.
  Paris, 1882

МЕНДЕЛѢЕВЪ, Д.
    Основы химіи или общедоступное и подробное изложеніе свѣдѣній не-
    органической химіи, ея теорій и приложеній. С.-Петербургъ 1868.
    Other editions, 1872 and 1889.

Менделѣевъ Д   [Cont'd.]

MENDELÉEFF, DMITRI.   The Principles of Chemistry, or a popular and complete exposition of the theories and applications of inorganic chemistry.

Добавленіе къ „Основамъ химіи“. С.-Петербургъ 1889. 8⁰

Supplement to the Principles of Chemistry, embracing two lectures delivered in London, 1889.

The Principles of Chemistry. Translated by George Kamensky and edited by A. Greenaway. London, 1892. 2 vols., 8vo.

Grundlagen der Chemie, aus dem russischen übersetzt von L. Jawen und A. Thillot. St. Petersburg, 1890-91. Roy. 8vo.

Изоморфизмъ въ связи съ другими отношеніями крестал. формы, С.-Петербургъ 1856

Isomorphism in relation to the crystalline form. St. Petersburg, 1856.

Обзоръ парижской всемірной выставки 1867 г. О современномъ развитіи нѣкот. химическихъ производствъ. С.-Петербургъ 1867.

Report on the universal exhibition of 1867 in Paris. Modern development of some chemical productions. St. Petersburg, 1867.

О соединеніи спирта съ водою. С.-Петербургъ 1865.

On the chemical combination of alcohol and water. St. Petersburg, 1865.

Химическія изслѣдованія почвъ и продуктовъ съ опытныхъ полей Симбирской, Смоленской, Московской и Петербургской губерній, произеденныя въ химической лабораторіи С.-Петербургскаго Университета. Москва 1871.

Chemical investigations of soils and of the production from the experimental field of Sembersk, Smolensk, Moscow and Petersburg governments, executed at the chemical laboratory of St. Petersburg University. Moscow, 1871,

Органическая химія. С.-Петербургъ 1863.

Organic Chemistry. St. Petersburg, 1863.

Объ упругости газовъ. С -Петербургъ 1875.

On elasticity of gases. St. Petersburg, 1875.

Менделѣевъ, Д. Густавсонъ, Г. Яцуковичъ, И.

Аналитическая химія. С.-Петербургъ 18—.

MENDELÉEFF, GUSTAVSON AND YATZUKODICH. Analytical Chemistry. St. Petersburg, 18—.

Mène, Ch.

Guide pratique pour l'analyse chimique des engrais. Paris, 1874. 8vo.

MÉNIL, AUGUST PETER JULIUS DU.

See Du Ménil, A. P. J.

Меншуткинъ, Н.

Аналитическая химія. С.-Петербургъ 1871.

Another edition, 1874.

MENSCHUTKIN, N.
> Analytische Chemie. Für den Gebrauch im Laboratorium
> und für Selbststudium. Uebersetzt von O. Bach. Leip-
> zig, 1877.
> Zweite Auflage. Leipzig, 1888.

MERCER, NATHAN.
The Chemistry of Gold ; with a sketch of its natural history and geological distribution, more especially with reference to information
valuable to Australian emigrants. Liverpool, 1853. 12mo.

MERING, J. VON.
Das chlorsaure Kali, seine physiologischen, toxischen und therapeu-
tischen Wirkungen. Berlin, 1885. pp. 142, 8vo.

MERMET, A.
Manipulations de chimie. Métalloides. Paris, 1885.

MERTENS, WILHELM.
Die Fabrikation und Raffinirung des Glases. Genaue, übersichtliche
Beschreibung der gesammten Glasindustrie, wichtig für den Fabrikanten, Raffineur, als auch für das Betriebsaufsichtspersonal, mit
Berücksichtigung der neuesten Errungenschaften auf diesem Gebiete
und auf Grund eigener vielseitiger, praktischer Erfahrungen bearbeitet. Wien, Pest und Leipzig, 1888.

MÉTHÉRIE, DE LA.
> See La Méthérie, De.

METHODS OF ANALYSIS OF COMMERCIAL FERTILIZERS, Feeding Stuffs,
and Dairy Products adopted at the fourth annual convention of the
Association of Official Agricultural Chemists, August 16, 17 and 18,
1887. Bulletin No. 16, U. S. Department of Agriculture, division
of chemistry. Washington, 1887. 8vo.
> Cf. in Sections VII, Bulletins of the Division of Chemistry, U. S. Depart-
> ment of Agriculture.

METZGER, SIGMUND.
Pyridin, Chinolin und deren Derivate. Gekrönte Preisschrift der . . .
Universität Würzburg. Braunschweig, 1885. pp. iv–91, 8vo.

METZNER.
Résumé d'analyse minérale. Paris, 1892. 8vo.
> Cf. in Section II, Fremy : Encyclopédie chimique, vol. IV.

MEUNIER, STANISLAS.
Traité pratique de chimie et de géologie agricoles. Paris, 1880.

MEUNIER, STANISLAS.
> *See* Thorpe, T. E.
>
> Combustibles minéraux ; *see*, *in Section II*, Fremy : Encyclopédie chimique. vol. II. Météorites ; *see the same*, vol. II. Cobalt et Nickel ; *see the same*, vol. III.

MEUSEL, E.

Die Quellkraft der Rhodonate und die Quellung als Ursache ferment-artiger Reactionen. Gera, 1886.

MEUSING, W.

Leichtfassliche Anleitung zu stöchiometrischen Rechnungen für ange-hende Chemiker und Pharmaceuten. Mit Vorrede von J. B. Trommsdorff. Erfurt, 1824. 8vo.

MEYER. *See also* Mayer.

MEYER, AD.

Das Chlorophyllkorn in chemischer, morphologischer und biologischer Beziehung. Leipzig, 1883. 4to. Ill.

Handbuch der qualitativen chemischen Analyse anorganischer Substan-zen, nebst Anleitung zur volumetrischen Analyse. Bearbeitet für Apotheker und Gerichts-Chemiker, sowie zum Gebrauch beim Un-terricht in chemischen Laboratorien. Berlin, 1884. 8vo. Ill.

MEYER, E.

Einige chemische Beobachtungen und Versuche über die Zubereitung des Zuckers aus einheimischen Pflanzen, welche als Zuckersurro-gate statt der ostindischen Zuckerpflanze dienen können. Düssel-dorf, 1799.

MEYER, E., UND WEDDIGE.
> *See* Graham-Otto.

MEYER, E. VON.
> Explosivkörper und die Feuerwerkerei. *See* Bolley's Handbuch [etc.].

MEYER, FRANZ EDUARD MORITZ.

Feuerwerkerei (Die) in ihrer Anwendung auf Kunst, Wissenschaft und Gewerbe. Leipzig, 1833. 8vo.

Grundzüge der Militairchemie. Leipzig, 1834.

MEYER, JOHANN FRIEDRICH.

Chymische Versuche zur näheren Erkenntniss des ungelöschten Kalchs, der elastischen und elektrischen Materie des allerreinsten Feuer-wesens und der ursprünglichen allgemeinen Säure [acidum pingue] nebst einem Anhange von den Elementen. Hannover und Leipzig, 1764. pp. 418–[xxviii], 8vo.

MEYER, JOHANN FRIEDRICH. [Cont'd.]

Zwote nach dem eigenhändig verbesserten Exemplar des Verfassers und mit dessen alchimistischen Briefen vermehrte Ausgabe. Hannover, 1770. 8vo.

MEYER, LOTHAR.

Modernen (Die) Theorien der Chemie und ihre Bedeutung für die chemische Mechanik. Breslau, 1864. 8vo.

Zweite Auflage. Breslau, 1872. Dritte Auflage. Breslau, 1876.

Fünfte Auflage. Breslau, 1884.

Modern Theories of Chemistry. Translated from the German (5th edition) by P. Phillips Bedson and W. Carleton Williams. London, 1888. pp. xliii–587, 8vo. Folding table.

Les théories modernes de la chimie et leur application à la mécanique chimique. Ouvrage traduit de l'allemand sur la cinquième édition par Albert Bloch. Paris, 1887. Vol. I, pp. viii–452, roy. 8vo. Tome II, précédé d'une préface de M. Friedel. Paris, 1889.

Повѣйшія теоріи химіи и ихъ значеніе для химич. статистики. Перев. Афанасьина П. С.-Петербургъ 1863.

Ueber die neuere Entwickelung der chemischen Atomlehre. Tübingen, 1886. 8vo.

Outlines of Theoretical Chemistry, translated by P. Phillips Bedson and W. Carleton Williams, with a preface by the author. London, 1892. 8vo.

MEYER, LOTHAR, UND KARL SEUBERT.

Die Atomgewichte der Elemente. *See, in Section II*, Meyer, Lothar, *and compare in Section II*, Clarke, F. W.

MEYER, L., UND K. Seubert.

Das natürliche System der Elemente. Nach den zuverlässigsten Atomgewichten zusammengestellt. Leipzig, 1889. Fol. Plates.

MEYER, MRS. L. J. (R.).

Real Fairy Folks. Explorations in the World of Atoms. Boston, 1887. 12mo. Ill.

MEYER, RICHARD.

Einleitung in das Studium der aromatischen Verbindungen. Leipzig, 1882.

MEYER, VICTOR.

Ergebnisse und Ziele der stereochemischen Forschung. Voutrag gehalten in der Sitzung der deutschen chemischen Gesellschaft zu Berlin am 28 Januar, 1890. Heidelberg, 1890. pp. 92, 8vo.

Thiophengruppe (Die). Braunschweig, 1888. pp. viii–302, 8vo.

MEYER, VICTOR, UND PAUL JACOBSON.

Lehrbuch der organischen Chemie. Leipzig, 1891–92. 2 vols., 8vo. Ill.

MEYER, VICTOR, AND F. P. TREADWELL.
          See Treadwell, F. P.

MEYER, W. CHR.

Praktische chemische Tabellen für Aerzte, Apotheker und Liebhaber der Chemie, zur leichtern Uebersicht der natürlichen, einfachen und zusammengesetzten Körper. Gotha, 1806. Fol.

MEYERHARDT, S.

Studien über die hygienische Bedeutung des Kupfers. Würzburg, 1890. 8vo.

MEYLINK, BERNARD.

Allereerste beginselen der scheikunde. Eene handleiding voor allen die eene oppervlakkige kennis dezer wetenschap wenschen te ver-krijgen. Deventer, 1837. 2 vols., 12mo.

Analytische (De) scheikunde. Deventer, 1844. 8vo.

Landbouw-scheikunde. De scheikunde toegepast op de landhuishoud-kunde onzer dagen. Met in den tekst gedrukte afbeeldingen. Amsterdam, 1855. 8vo.

Schei- artsenymeng- en natuurkundige bibliotheek. Deventer, 1824–34, 18 vols., 8vo.

Nieuwe schei-artsenymeng- [etc.] bibliotheek. Deventer, 1835–40. 3 vols., 8vo.

MIAHLE, LOUIS.

Chimie appliquée à la physiologie et à la thérapeutique. Paris, 1856. pp. xxxii–703, 8vo.

  Chimica applicata alla fisiologia ed alla terapeutica. Versione Italiana di Antonio Bianchi. Venezia, 1864.

  Tratado de química aplicado á la fisiología y á la terapéutica. Vertido por Félix Borrell y Font. Quinta edicion. Madrid, 1876. 4to.

Du rôle chimique de l'acide carbonique dans l'économie animale. Paris, 1856. 8vo.

Essai de propositions et observations pharmaceutiques. Paris, 1856. 8vo.

MICÉ, L.

De la notation atomique et de sa comparaison avec la notation en équivalents. Bordeaux, 1871. 8vo.

MICHAELIS, ——.

Neueste (Der) deutsche Stellvertreter des indischen Zuckers oder Zucker aus Runkelrüben. 1799.

MICHAELIS, A.

Anorganische Chemie, *see* Graham-Otto.

MICHAELIS, A.

Einführung in die allgemeine Chemie und die physikalisch-chemischen Operationen. Besonderer Abdruck der Einleitung zur fünften Auflage von Graham-Otto's ausführlichem Lehrbuche der anorganischen Chemie. Braunschweig, 1879. 8vo. Ill.

MICHAELIS, JOHANN.

Praxis chymiatrica generalis. [1650 ?]

MICHAUD, L.

La chimie des dames, ou cours de chimie élémentaire appliquée aux usages domestiques. Genève, 1851. 8vo.

MICHEL, G.

Des levures pures et de leur emploi dans les fermentations alcooliques industrielles. Marseilles, 1889.

MICHEL DU TENNETAR.

Élémens de chimie . . . Metz, 1779. 12mo.

MICHELET, L.

Handboek van aanvankelijke scheikunde ten gebruike der normaalscholen en middelbare scholen. Namur, 1888. 12mo. Ill.

Traité de chimie élémentaire à l'usage des écoles normales et des écoles primaires. Namur, 1882. 12mo.

MICHOTTE, FÉLICIEN, ET E. GUILLAUME.

Traité de la fabrication industrielle des eaux gazeuses et des boissons qui s'y rattachent. Paris, *n. d.* pp. v–311, 12mo. 14 folding plates.

Bibliothèque des professions industrielles commerciales et agricoles.

MIELE, S.

Compendio di analisi chimica per lo studio dei minerali. Napoli, 1879.

MIERZIŃSKI, STANISLAUS.
Desinfectionsmittel (Die). Berlin, 1878. 8vo.

Erd-, Mineral- und Lack-Farben (Die), ihre Darstellung, Prüfung und Anwendung. Zum Gebrauche für Fabrikanten, Techniker, Farb-waarenhändler, Maler, Anstreicher, etc. Vierte Auflage von Schmidt's "Farbenlaboratorium" in vollständiger Neubearbeitung. Weimar, 1881. pp. xii–500, 8vo. Ill.

Fabrikation (Die) der aetherischen Oele und Riechstoffe. Zum Ge-brauche für Fabrikanten, Chemiker, Apotheker, Droguisten und Parfümisten. Berlin, 1872. pp. v–185, 8vo. Ill.

Fabrikation (Die) des Aluminiums und der Alkalimetalle. Wien, 1885. Ill.

Gerb- und Farbstoff-Extracte (Die). Wien, Pest, Leipzig. 1887. 8vo. Ill.

Theerfarbstoffe (Die) ihre Darstellung und Anwendung. Leipzig, 1878. 8vo.

MIJERS, J.
Leerboek der scheikunde. 1e deel. Amsterdam, 1882. pp. 200, 8vo.

MIKAN, JOHANN CHRISTIAN.
Ueber Zucker-Erzeugung aus Ahornsaft. Prag, 1811. 4to.

MILANI, G.
La chimica in famiglia con cinquanta disegni originali di E. Marzanti. Firenze, 1886. 16mo.

MILLAR, JAMES.
A new Course of Chemistry, in which the theory and practice of that art are delivered in a familiar and intelligible manner. The fur-naces, vessels and instruments are described, and the preparations of the several medicines are laid down according to the most easy and certain processes. Together with a succinct account of the several drugs used in the preparation of chemical medicines as to their nature, production and country. London, 1754. 8vo.

Elements of chimistry. Edinburgh, 1820. 8vo.

Éléments de chimie pratique appliquée aux arts et aux manufactures. Traduits par Coulier. Paris, 1822. 8vo.

МИЛЛЕРЪ, О. ЗАРУБИНЪ, И.
Пособіе къ изученію химіи. Москва 1871

MILLER, O., AND ZARUBINE. Introduction to the Study of Chemistry. Moscow, 1871.

MILLER, WILLIAM ALLEN.

Elements of Chemistry, theoretical and practical. London, 1855–57.
3 parts, 8vo.

> Second edition, with additions. London, 1860–62. 3 vols.,
> 8vo. Ill. Vol. I, Chemical Physics, pp. xvi–482 ; vol.
> II, Inorganic Chemistry, pp. xxiv–882 ; vol. III, Organic
> Chemistry, pp. xxxii–950.

> Vol. I, sixth edition, edited by McLeod, 1888. Vol. II, sixth
> edition, edited by Groves. London, 1888. Vol. III,
> fifth edition, edited by Armstrong and Groves. London.

Introduction to the Study of Inorganic Chemistry. With Questions
for Examination. London [also New York], 1872. pp. xi–304,
12mo. Ill.

> Completed after the author's death by C. Tomlinson in 1870.

MILLER, W. VON, AND KILIANI, H.

Kurzes Lehrbuch der analytischen Chemie. Zweite vermehrte und
verbesserte Auflage. München, 1891. pp. xii–609, 8vo. Ill.

MILLON, NICOLAS AUGUSTE EUGÈNE.

Éléments de chimie organique comprenant les applications de cette
science à la physiologie animale . . . Paris, 1845. 2 vols., 8vo.
Ill.

Recherches sur l'acide nitrique. Paris, 1842. 8vo.

> See also, in Section VII, Annuaire de chimie.

MILLOT, A.

> Conférence (etc.). See Société chimique de Paris.

MILLS, E. J.

Destructive Distillation. A manuallette of the paraffin, coal-tar, resin,
oil and kindred industries. London, 1886. 8vo.

MILLS, J.

Alternative Elementary Chemistry, being a course of lessons adapted to
the requirements of the new Syllabus of Chemistry recently in-
stituted by the Science and Art Department. With a preface by
Thorpe. London, 1891. 8vo.

MILLS, JOHN, AND BARKER NORTH.

Handbook (A) of Quantitative Analysis. London, 1889. pp. viii–212,
8vo. Ill.

> An expansion of the Introductory Lessons by the same authors.

Introductory Lessons on Quantitative Analysis, suitable for the examinations of the Science and Art Department and the University of
London. London, 1889. 8vo.

MINET, ADOLPHE.

L'aluminium, fabrication, emploi. Paris, *n. d.* [1892]. pp. 312, 12mo.

> Bibliothèque des actualités industrielles, No. 48. A specimen of aluminium foil obtained by electrolysis is inserted in the paper cover of each copy.

MINGUEZ, C. F.

El vino antí la química analítica. Métodos de análisis y reconocimiento de las adulteraciones, con un nuevo procedimiento para averiguar las materias colorantes artificiales. Madrid, 1889. 4to.

MINKELERS, JOHANN PETER.

Mémoire sur l'air inflammable tiré de différentes substances. Louvain, 1784. pp. 49, 8vo. 11 folding tables.

> Contains a : Table des gravités spécifiques de différentes espèces d'air par J. F. Thysbaert, de Louvain. The whole is a contribution to aëronautical chemistry.

MITCHELL, CLIFFORD.

The Physicians' Chemistry. Chicago, 1886. pp. 301.

MITCHELL, JOHN.

Manual (A) of Agricultural Analysis. London, 1845.
> New edition. London, 1851.

Manual (A) of Practical Assaying. London, 1849. 8vo. Ill.
> Second edition, 1854.
>> Third edition, edited by William Crookes. London, 1868. 8vo.
>> Practical Assaying, with copious Tables, edited by William Crookes. New edition revised with the recent discoveries. London, 1888. 8vo.

> A portion of this work, especially the analysis of coal and iron, has been translated and published in Chinese by John Fryer. *Cf.* Fryer, John.

Treatise on the Falsifications of Food, and the chemical means employed to detect them. Containing water, flour, bread, milk, cream, beer, cider, wines, spirituous liquors, coffee, tea, chocolate, sugar, honey, lozenges, cheese, vinegar, pickles, anchovy sauce and paste, catsup, olive (salad) oil, pepper, mustard. London, 1848 pp. xviii–334. 12mo.

MITCHELL, T. D.

Elements of Chemical Philosophy on the basis of Reid, comprising the rudiments of that science and the requisite experimental illustrations. Cincinnati, 1832. 8vo.

Medical Chemistry, or a Compendious View of the various substances employed in the practice of medicine that depend on chemical

MITCHELL, T. D. [Cont'd.]
  principles for their formation. Designed for the use of medical
  students. To which is appended a discourse on the medical
  character. *n. p.* 1819. 18mo.

MITCHILL, SAMUEL LATHAM.
  Outlines of the Doctrines in Natural History, Chemistry and Economy.
    New York, 1792.
  Remarks on the Gaseous Oxyd of Azote, and of its effects . . . New
    York, 1795. 12mo.

MITSCHERLICH, ALEXANDER.
  Chemische Abhandlungen veröffentlicht in den Jahren 1859–1865.
    Berlin, 1865–75. 3 parts, 8vo.

MITSCHERLICH, EILHARDT.
  Lehrbuch der Chemie ; mit Holzschnitten von Unzelmann. Berlin,
    1829–30. 2 vols., 8vo.
        Zweite Auflage. Berlin, 1833–35. 2 vols.
        Vierte Auflage. Berlin, 1844–47.
        Éléments de chimie, traduits par B. Valérius. Bruxelles,
          1835. 3 vols. Ill.
        Elementi di chimica, tradotti da F. du Pré. Venezia,
          1835–38.

MITTEREGGER, J.
  Lehrbuch der Chemie für Oberrealschulen. Erster Theil : Anorga-
    nische Chemie. Zweite Auflage. Dem Normallehrplane für Real-
    schulen entsprechend umgearbeitet. Wien, 1883.
        Dritte Auflage. Wien, 1886.
        Trattato di chimica inorganica pelle scuole reali superiori
          Tradotto da E. Gerardi. Wien, 1885.

MIXTER, W. G.
  An Elementary Text-book of Chemistry. New York and London,
    1889.
        Third edition. New York, 1890.

MODEL, JOHANN GEORG.
  Chymische Nebenstunden. St. Petersburg, 1762. 8vo.
        Fortsetzung seiner chymischen Nebenstunden. St. Peters-
          burg, 1768. 8vo.

MÖCKERN, VERSUCHSSTATION IN. *See* Agriculturchemische Untersuchungen.

MOEBIUS, GOTTFRIED.

Anatomia camphoræ, ejusque originem, qualitates, præparationes chymicas ac vires . . . demonstrans. Jenæ, 1660. 4to.

MÖHLAU, R.

Organische Farbstoffe welche in der Textilindustrie Verwendung finden. Uebersicht ihrer Zusammensetzung, Gewinnung, Eigenschaften, Reactionen und ihrer Anwendung zum Färben und Bedrucken von Seide, Wolle und Baumwolle. Dresden, 1890. Roy. 8vo. With 175 specimens of dyed fabrics.

MÖLLER, C. F. C.

Lærebog i uorganisk Kemi. Udarbeidet til Skolebrug. Prisbelönnet af Selskabet for Naturlærens Udbredelse. Kjøbenhavn, 1874. 8vo.

MÖLLER, J.

Ueber den Alcohol. Berlin, 1883. 8vo.

MÖNCH, CONRAD.

Beschreibung und chymische Untersuchung der Dorf Geismarschen Mineralbrunnens. Cassel, 1778. 8vo.

MOEWES, A. L.

Destillir-Kunst (Die) der geistingen Getränke, nach den neuesten und praktisch-bewährten Erfahrungen, mit wissenschaftlichen Erläuterungen und besonderer Rücksicht auf den gegenwärtigen Standpunkt des Gewerbes, bearbeitet. Zur Benutzung für den praktischen Destillateur, etc. Nebst einen Abhandlung der Essig- und Schnellessigfabrikation. Berlin, 1834.

> Zweite vermehrte und umgearbeitete Auflage. Berlin, 1841. 8vo. Ill.
> Dritte Auflage. Berlin, 1850.
> Die Destillirkunst der geistigen Getränke auf warmem wie auf kaltem Wege. Ein vollständiges Handbuch der Liqueurfabrikation, nebst einer praktischen Anleitung zur Essig- und Schnell-Essigfabrikation.
> Vierte Auflage. Berlin, 1854. 8vo. Ill.
> Fünfte Auflage. Berlin, 1857. 8vo. Ill.
> Sechste verbesserte Auflage. Berlin, 1864. pp. xx–311, 8vo. Ill.

MOHR, CARL FRIEDRICH.

Allgemeine Theorie der Bewegung und Kraft als Grundlage der Physik und Chemie. Ein Nachtrag zur mechanischen Theorie der chemischen Affinität. Braunschweig, 1869. 8vo.

MOHR, CARL FRIEDRICH. [Cont'd.]

Chemische Toxicologie. Anleitung zur chemischen Ermittelung der Gifte. Braunschweig, 1874. 8vo.

>Toxicologie chimique, guide pratique pour la détermination chimique des poisons. Traduit par L. Gautier. Paris, 1875. 8vo.

Lehrbuch der chemisch-analytischen Titrirmethode. Nach eigenen Versuchen und systematisch dargestellt. Für Chemiker, Aerzte [etc]. Braunschweig, 1855–56. 2 parts, 8vo.

>Zweite Auflage. Braunschweig, 1862. 8vo. Ill.

>Vierte durchaus umgearbeitete Auflage. Braunschweig, 1874, pp. xxii–772, 8vo. Ill.

>Sechste Auflage neu bearbeitet von Alexander Classen. Braunschweig, 1886. 8vo. Ill.

>Traité d'analyse chimique à l'aide des liqueurs titrées. Traduit de l'allemand par C. Forthomme. Paris, 1857. 8vo. Ill.

>Deuxième édition. Paris, 1875.

>Troisième édition, revue et augmentée par A. Classen. Paris, 1888. 8vo. Ill.

>Руководство къ химическому анализу мѣрою (метода титрированiя). Перев. Ходнева. С.-Петербургъ 1859. 2 тома

Lehrbuch der pharmaceutischen Technik nach eigenen Erfahrungen bearbeitet. Fur Apotheker, Chemiker, chemische Fabrikanten, Aerzte und Medicinalbeamte. Braunschweig, 1847. 8vo.

>Zweite vermehrte und verbesserte Auflage. Braunschweig, 1853. 8vo.

>Dritte Auflage. Braunschweig, 1866. 8vo.

>Also translated into English and Dutch.

Mechanische Theorie der chemischen Affinität und die neuere Chemie. Braunschweig, 1868. pp. viii–364, 8vo.

MOHR, CARL FRIEDRICH.

>Weinbau (Der) und die Weinbereitungskunde [etc.]. *See* Bolley's Handbuch [etc.].

MOIGNO, L'ABBÉ.

L'Ozone, ce qu'il est, ses propriétés physiques et chimiques. Analyse des recherches dont il a été l'objet. Paris, 1878. 12mo.

Saccharimétrie optique chimique et mélassimétrique. Paris, 1869. pp. xxviii–xxviii–256, 12mo.

MOISSAN.

>Conférence (etc.). *See* Société chimique de Paris.

>Manganèse [and other articles]. *See in Section II*, Fremy: Encyclopédie chimique, vol. III, etc.

MOISSAN, H.

Recherches sur l'isolement du fluor. Paris, 1887. 8vo.

MOISSON, A.

Pyrodynamique. Théorie des explosions dans les canons et les torpilles.
Paris, 1887. 8vo.

MOJON, GIUSEPPE.

Analisi delle acque termali di Voltri. Genova, 1805.
Analyses des eaux sulfureuses et thermales d'Acqui. Gênes, 1808. 8vo.
Corso analitico di chimica. Genova, 1806. 2 vols.
          Seconda edizione, 1811.
          Terza edizione. Milano, 1815.
          [Other editions.] Livorno, 1815, 1816.
          Curso analitico de química escrito en italiano . . . traducido
               en castellano por Francisco Carbonell y Bravo. Bar-
               celona, 1818. 4to.
Memorie di chimica. Genova, 1804. 8vo.
Osservazioni sopra la tavola delle espressioni numeriche di affinità del
     cittadino Guyton Morveau. [n. p. 1802.] 8vo.

MOLDENHAUER, FRIEDRICH.

Chemische Reagentien, oder wie prüft man einen Körper auf Verfäl-
     schung und benutzt ihn, chemisch rein, selbst wieder als Reagens.
     Ilmenau, 1830. 8vo.
Lehrbuch der Chemie. Carlsruhe, 1835. 8vo.
Reihenfolge und Eintheilung der chemisch-einfachen Körper oder
     Grundstoffe nebst ihren chemischen Zeichen, Aequivalenten . . .
     Carlsruhe, 1843.

MOLESCHOTT, JACOBUS ALBERTUS WILLEBRORDUS.

Kreislauf (Der) des Lebens. Physiologische Antworten auf Liebig's
     Chemische Briefe. Mainz, 1852, 12mo.
          Dritte vermehrte und verbesserte Auflage. Mainz, 1857.
               pp. xii–534, 12mo.
          Vierte Auflage. Mainz, 1862–63.
Kritische Betrachtung von Liebig's Theorie der Pflanzenernährung, mit
     besonderer Angabe der empirisch constatirten Thatsachen. Eine
     von der Teyler'schen Gesellschaft im Jahre 1844 gekrönte Preis-
     schrift. Harlem, 1845. pp. viii–122, 4to.
          Also under the title: Verhandeling bevattende een antwoord op de vraag
               betraffende Liebig's Theorie der Plantenvoeding.
Lehre der Nahrungsmittel. Für das Volk. Zweite Auflage. Erlangen,
     1853. pp. xx–256, 8vo.

MOLESCHOTT, JACOBUS ALBERTUS WILLEBRORDUS.   [Cont'd.]
> Dritte Auflage, 1858.
> Leer der voedingsmiddelen, voor het volk.   Utrecht, 1850.
> 8vo.

Physiologie (Die) der Nahrungsmittel.   Ein Handbuch der Diätetik.
Friedrich Tiedemann's "Lehre von dem Nahrungsbedürfniss, dem
Nahrungsbetrieb und den Nahrungsmittel des Menschen" nach
dem heutigen Standpunkte der physiologischen Chemie völlig
umgearbeitet.   Darmstadt, 1850.   8vo.

> De physiologie der voedingsmiddelen, een handboek der
> diaetetica, volgens Friedrich Tiedemann's leer der
> voedingsmiddelen, naar het tegenwoordig standpunt der
> physiologische scheikunde geheel omgewerkt.   Amster-
> dam, 1849.   8vo.
> Chemistry of Food and Diet.   Translated by E. Bronner.
> London, 1867–68.   3 vols., 8vo.
> De l'alimentation et du régime.   Traduit de l'allemand sur
> la troisième édition par Ferd. Flocon.   Paris, 1858.   8vo.

Physiologie des Stoffwechsels in Pflanzen und Thieren.   Erlangen,
1851.   8vo.

MOLINARI, F.
Laterizi, gesso, pozzolane, calci e cementi ad uso degli industriali, ingegneri, architetti, capimastri e costruttori, con appendice sugli
stessi materiali in rapporto all'igiene.   Milano, 1887.   8vo.   Ill.

MOLITOR, NICOLAUS CARL.
Darstellung des chymischen Laboratoriums.   Mainz, 1791.

MOMBERGER, KONRAD PHIL. KONST.
Die Gasarten, zur Erleichterung ihrer Kenntniss für angehende Che-
miker und Pharmaceuten systematisch zusammengestellt.   Mit einer
Vorrede und einigen Anmerkungen begleitet von Ferd. Wurzer.
Cassel und Marburg, 1810.   8vo.

MONAVON, MARIUS.
La Coloration artificielle des vins.   Avec figures intercalées dans le
texte.   Paris, 1890.   pp. 164, 12mo.
> Petite bibliothèque scientifique.

MONCKMAN, JAMES.
Combined Hand-Book and Note-Book for Students of Inorganic Chemistry.   London, [188-].   8vo.

MONDIÈRE, ——.
Traité comparé de chimie théorique et pratique, à l'usage des candidats
aux baccalauréats ès-lettres et ès-sciences à l'école polytechnique.
Paris, 1851. 8vo.

MONHEIM, JOHANN PETER JOSEPH.
Analyses des eaux thermales de Borcette, suivie de l'examen du gaz
azote sulfuré dégagé des sources sulfureuses, tant d'Aix-la-Chapelle
que de Borcette. Aix-la-Chapelle, 1812. 8vo.
Heilquellen (Die) von Aachen, Burtscheit, Spaa [etc.].   Aachen, 1829.
8vo.

MONHEIM, JOHANN PETER JOSEPH, ET REUMONT.
Analyses des eaux sulfureuses d'Aix-la-Chapelle.   Aix-la-Chapelle,
1810. 8vo.

MONHEIM, JOHANN PETER JOSEPH, ET SARTORIUS.
Untersuchung einer Arsenikvergiftung.   Cöln und Aachen, 1826. 8vo.
Untersuchungen zweier Zinkvergiftungen. Cöln und Aachen, 1826. 8vo.

MONIER, ÉMILE.
Guide pour l'essai et l'analyse des sucres indigènes et exotiques à l'usage
des fabricants de sucre. Résultats de deux cents analyses de
sucres classés d'après leur nature. Paris, 1865. pp. 94, 18mo.
        The eminent chemist Payen praised this work at a meeting of the Academy
        of Sciences.
Mémoires sur l'analyse du lait et des farines par les méthodes volumé-
triques, présenté à l'Académie des sciences. Paris, 1858. 8vo.

MONNET, ANTOINE GRIMOALD.
Démonstration de la fausseté des principes des nouveaux chimistes, pour
servir de supplément au Traité de la dissolution des métaux. Paris,
An VI, [1798]. 8vo.
Dissertation et expériences relatives aux principes de la chimie pneu-
matique, ou à la théorie des chimistes pneumatistes, pour servir de
supplément au Traité des métaux. Turin, 1789. 4to.
        Extrait du neuvième volume des Mémoires de l'Académie de Turin.
Dissertation sur l'arsenic, qui a remporté le prix proposé par l'Académie
de Berlin, 1774. 4to.
Traité de la dissolution des métaux. Paris, 1775. 12mo.
Traité de la vitriolisation et de l'alunation, ou l'Art de fabriquer l'alun
et le vitriol. Paris, 1769. 12mo.
Traité des eaux minérales avec plusieurs mémoires de chimie relatifs à
cet objet. Paris, 1768. 12mo.

MONNIÈRES, A. H. DE.
Histoire, analyse et effets du guano du Pérou. Paris, 1845. 8vo.

MONRO, DONALD.
Medical (A) and Pharmaceutical Chemistry and the Materia Medica.
Edinburgh, 1788–90. 3 vols., 8vo.
On Mineral Waters. Edinburgh, 1770. 2 vols., 8vo.

MONS, JEAN BAPTISTE VAN.
Abrégé de chimie, à l'usage des leçons. Louvain, 1831–35. 5 vols., 12mo.
Chimie (La) des éthers. Louvain, 1837.
Conspectus mixtionum chemicarum. Louvain, 1827. 12mo.
Essai sur les principes de la chimie antiphlogistique. Bruxelles, 1785.
8vo.
Essai sur une théorie chimique modifiée. Bruxelles, 1806–07. 4 vols., 8vo.
Principes élémentaires de chimie philosophique avec des applications
générales de la doctrine des proportions déterminées. Bruxelles,
1818. pp. iv–380, 12mo.
Théorie de la combustion. Bruxelles, An XII [1804]. 8vo.

MONSELISE, A.
L'ambra primaticcia o sorgo zuccherino del Minnesota. Seconda edi-
zione. Mantova, 1884.
La chimica moderna. Verona, 1881. 2 vols.

MONTANUS, AUGUST SCHULZE.
            See Schulze-Montanus, August.

MONTEFERRANTE, R.
·Manuale pratico di ricerche tossicologiche. Napoli, 1872. 16mo.

MONTELLS Y NADAL, FRANCISCO DE PAULA.
Curso elemental de química aplicada á las artes. Granada y Madrid,
1841. 2 vols., 8vo.
Compendio de física esperimental y algunas nociones de química . . .
Granada, 1848. 8vo.

MONTFERRIER, ALEXANDRE ANDRÉ VICTOR SARRAZIN DE.
Précis de physique et de chimie. Paris, 1839.

MOORE, S. W.
Notes of Demonstrations on Physiological Chemistry. London, 1874.
12mo.

MOORMAN, J. J.
The Mineral Waters of the United States and Canada, with a map and
plates and general directions for reaching mineral springs. Baltimore, 1867. pp. 507, 8vo.

MORALES, JOSÉ MARIA PEREZ, Y BENITO TAMAYO.

Curso de química general, arreglado á las esplicaciones del Vicente Santiago de Masarnau, y comprendiendo toto lo mandado en el plan vigente de estudios. Madrid, 1848. 8vo. Ill.

MOREL, J.

Handboek der anorganische scheikunde. Tweede uitgave. Gand, 1880. 12mo.

MORELL, CARL FRIEDRICH.

Chemische Untersuchungen einiger Gesundbrunnen und Bäder der Schweiz . . . Berne, 1788.

MORELOT, SIMON [or MORELLOT].

Cours élémentaire théorique et pratique de pharmacie-chimique, ou Manuel du pharmacien-chimiste. Paris, 1803.

Histoire naturelle appliquée à la chimie . . . Paris, 1809. 2 vols., 8vo.

MORFIT, CAMPBELL.

Chemical and Pharmaceutical Manipulations : a manual of the mechanical and chemico-mechanical operations of the laboratory ; containing a complete description of the most approved apparatus, with instructions as to their application and management both in manufacturing processes and in the more exact details of analysis and research. Assisted by A. Muckle. For the use of chemists, druggists, teachers and students. Philadelphia, 1851. 8vo. Ill.

Second and enlarged edition by Campbell and Clarence Morfit. Philadelphia, 1856. 8vo. Ill.

Perfumery, its manufacture and use ; with instructions in every branch of the art, and recipes for all the fashionable preparations : the whole forming a valuable aid to the perfumer, druggist and soap-manufacturer. From the French of Celnart and other late authorities, with additions and improvements. Second edition revised and improved. Philadelphia, 1853. 8vo.

Practical Treatise on Pure Fertilizers ; and the chemical conversion of rock guanos, marlstones, coprolites, and the crude phosphates of lime and alumina generally into various valuable products. London, 1873. Royal 8vo. Ill.

Treatise (A) on Chemistry applied to the manufacture of soap and candles ; being a thorough exposition, in all their minutiæ, of the principles and practice of the trade, based upon the most recent discoveries in science and art. A new and improved edition. Philadelphia, 1856. 8vo. Ill.

Cf. Booth, James C.

MORIDE, ÉDOUARD.
   Traité pratique de savonnerie. Matières premières—Matériel—Procédés de fabrication des savons de toute nature. Paris, 1888. pp. 402, roy. 8vo. Ill.
       See, in Section III, the same author.

MORLEY, CHRISTOPH LOVE [Editor].
       See Collectanea chymica Leidensia.

MORLEY, H. F.
   Outlines of Organic Chemistry. London, 1886. 8vo.

MORRILL, ARTHUR B.
   Outlines of a Short Elementary Course in General Chemistry. London, 1883. 12mo.

MORRIS, D.
   A Class-Book of Inorganic Chemistry with tables of chemical analysis, and directions for their use : compiled specially for pupils preparing for the Oxford and Cambridge middle-class examinations, and the matriculation examination of the University of London. London, 1870. pp. vii–157, 12mo.

MORT, JACOB LE.
   Chymia medico-physica, rationibus et experimentis instructa ; cui annexa est metallurgia contracta. Lugduni-Batavorum, 1684. 4to.
       Second edition, 1699.
       Translated into Dutch by Jacob Roman. Amsterdam, 1696. 8vo.

MORTON, HENRY, AND ALBERT R. LEEDS.
   The Student's Practical Chemistry. A text-book on chemical physics and inorganic and organic chemistry. Philadelphia, 1877. pp. viii–312, 8vo. Ill.

MORUS, JOHANN GOTTFRIED.
   Versuch einer physisch-chemische Untersuchung der Mineralquelle bei Homburg vor der Höhe. Homburg, 1811.

MORUZZI, G. B.
   Saggio di chimica popolare. Piacenza, 1868. 16mo.

MORVEAU. *See* Guyton de Morveau, Louis Bernard.

MOSELEY, BENJAMIN.
   Abhandlung über den Zucker, aus dem englischen mit Anmerkungen und einem Anhange von Karl August Nöldechen. Berlin und Stettin, 1800. 8vo.

MOSER, A., UND J. K. STRAHL.

Handbuch der physiologischen und pathologischen Chemie nach den neuesten Quellen bearbeitet. Leipzig, 1851. 12mo.

MOSER, E.

Ueber die organischen Substanzen des Mainwassers bei Würzburg. Beitrag zur Frage der Flussverunreinigung. Würzburg, 1887.

MOSER, H.

Chemische Abhandlung über das Chrom. Wien, 1824. 8vo.

MOSER, J.

Grundzüge der Agricultur-Chemie. Wien, 1857. 8vo.

Lehrbuch der Chemie für Land und Forstwirthe. Wien, 1870. 8vo. Ill.

Leitfaden zur qualitativen und quantitativen agricultur-chemischen Analyse. Wien, 1855. 8vo.

> Handleiding bij qualitatieve en quantitatieve landbouw-scheikundige ontledingen. Uit het Hoogduitsch door G. Duffer Blom. Utrecht, 1857. 8vo.

MOSMANN, GEORG.

Aequivalente (Die) der Grundstoffe und ihre specifischen Gewichte nach den neuesten Bestimmungen zusammengestellt. Chur, 1849.

> A lithographed sheet in imperial folio.

Regeln (Die) der Bildung chemischer Namen, Zeichen und Formeln. Zur leichten Uebersicht und Repetition besonders für Anfänger bearbeitet. Zweite vermehrte Auflage. Schaffhausen, 1858. 8vo.

MOTT, HENRY A.

The Chemist's Manual : a practical treatise on chemistry, qualitative and quantitative analysis, stoichiometry, blowpipe analysis, mineralogy, assaying, toxicology, etc. New York, 1877. 8vo.

MOUTIER, F.

Five Hundred Chemical Experiments for One Shilling. Arranged for Young Beginners in Chemistry. London, 1874.

MOUTIER, J.

La thermodynamique. Paris, 1885. 8vo.

MOWBRAY, GEORGE M.

Tri-nitro-glycerine, as applied in the Hoosac Tunnel, and to submarine blasting, torpedoes, quarrying, etc. Being the result of six years' observation and practice during the manufacture of five hundred thousand pounds of this explosive, mica blasting powder, dyna-

MOWBRAY, GEORGE M. [Cont'd.]

    mite ; with an account of the various systems of blasting by electricity, priming compounds, explosives, etc., etc. Third edition. New York, 1874. 8vo. Ill.

        First edition, North Adams, Mass., 1872.

MOYRET, MARIUS.

    Traité de teinture et impression, blanchiment et apprêt de fils et tissus. Lyon, 1887. 8vo.

    Traité de la teinture des soies, précédé de l'histoire chimique de la teinture de la soie. Lyon, 1877. 8vo.

MUCK, F.

    Einfachere gewichts-analytische Uebungsaufgaben in besonderer Anordnung nebst Einleitung als Vorwort : Einiges über Unterricht in chemischen Laboratorien. Breslau, 1887. 8vo.

    Elementarbuch der Steinkohlenchemie für Praktiker. Zweite Auflage. Essen, 1887.

    Grundzüge und Ziele der Steinkohlenchemie. Bonn, 1881.

        Chemie (Die) der Steinkohle, für Lehrende und Lernende an höheren und mittleren technischen Schulen, insbesondere Montanlehranstalten, sowie zum Selbstunterricht für Chemiker, Berg- und Hüttenleute und Ingenieure. Zweite grösstentheils umgearbeitete und vermehrte Auflage der "Grundzüge und Ziele der Steinkohlenchemie." Leipzig, 1891. pp. viii–284, 8vo.

MUDIE, ROBERT.

    The Air, a popular account of the atmospheric fluid. Second edition. London, 1847. 12mo.

MÜHLBERG, F.

    Der Kreislauf der Stoffe auf der Erde. Basel, 1886. 8vo.

MÜHLHÄUSER, O.

    Die Technik der Rosanilinfarbstoffe. Stuttgart, 1889.

MÜLLER, ALEXANDER.

    Chemische (Die), Zusammensetzung der gebräuchlichsten Nahrungsmittel und Futterstoffe bildlich dargestellt. Dresden, 1861. Fol.

        Zweite Auflage. Dresden, 1864.

        Dritte Auflage. Dresden, 1868.

        Vierte verbesserte und vermehrte Auflage. Dresden, 1875. Folio. Ill.

        La composition chimique des aliments représentés en tableaux coloriés. Bruxelles, 1862.

MÜLLER, ALEXANDER. [Cont'd.]

Gasbeleuchtung (Die) im Haus und die Selbsthilfe des Gas-Consumenten. Praktische Anleitung zur Herstellung zweckmässiger Gasbeleuchtungen, mit Angabe der Mittel eine möglichst grosse Gasersparniss zu erzielen. Wien, Pest, Leipzig, 18—. Ill.

MÜLLER, CARL GEORG.

Die trockene Destillation und die hauptsächlichsten auf ihn beruhenden Industriezweige. Leipzig, 1858. 8vo.

MÜLLER, CHRISTIAN.

Anleitung zur Prüfung der Kuhmilch. Zweite Auflage. Bern, 1868. pp. [iv]–35. 8vo. Folding table.
> Vierte Auflage. Bern, 1877. 8vo.

MÜLLER, E.

Generalitá di fisica e nozioni sperimentali di chimica per la prima liceale. Torino, 1891. 8vo.

MÜLLER, FERDINAND ALBRECHT.

Berzelius' Ansichten. Ein Beitrag zur theoretischen Chemie. Breslau, 1846. 8vo.

Lehrbuch der theoretischen Chemie. Berlin, 1850.

MÜLLER, JOHANNES.

ΣΤΟΙΧΕΙΟΛΟΓΙΑ, hoc est, elementorum ratione sui et mixta physica contemplatio, in academia Wittebergensi proposita. Wittebergæ, 1622. 4to.
> This contains eight Disputationes as follows :
> (1) Georgius Friesius, De elementis in genere et in specie de igne.
> (2) Martinus Ringius, De aere.
> (3) Johannes Glogerus, De aqua.
> (4) Lucas Scultetus, De terra.
> (5) Andreas Otto, De primis qualitatibus.
> (6) Adamus Mahnerus, De secundis qualitatibus.
> (7) Samuel Kinnerus, De elementorum actione, passione ac mistione.
> (8) Michael Dürr, De generatione et corruptione.

MÜLLER, JOHANNES.

Gerichtlich-chemische Untersuchungen für Juristen und Mediciner. Berlin, 1848. 8vo.

Gerichtlich-chemisch Untersuchung einer Arsenikvergiftung. Cleve, 1845. 8vo.

Gifte (Die), ihre Wirkung auf den Organismus, sowie Anleitung wie man sich zu verhalten hat, um bei Vergiftungszufällen, Erfrorenen, Ertrunkenen, etc., schnelle Hülfe leisten zu können, für Geistliche, Lehrer, Familienväter. Nürnberg, 1840. 8vo.

MÜLLER, JOHANNES. [Cont'd.]
> Over de vergiften. Derzel verwerking op het organismus, en handleiding hoedanig men zich te gedragen heeft, om by vergiftigingen, dadelijke hulp te kunnen verleenen. Voor den beschaafden stand. Duisburg, 1841[?]. 8vo.

Tabak (Der) in geschichtlicher, botanischer, chemischer, medicinischer und diätetischer Hinsicht. Emmerich, 1842. 8vo.

MÜLLER, J. P.
> Leitfaden für den Unterricht in der Chemie. Für höhere Lehranstalten bearbeitet. Zweiter Theil : Organische Chemie. Liegnitz, 1869. 8vo.

MÜLLER, LUDWIG.
> Commentationes de isomerismo, metamerismo et polymerismo particula prior ; dissertatio quam . . . die 1 M. Septembris anno 1837 publice defendet L. M. Vratislaviæ, n. d. [1837]. 8vo.

MÜLLER, L.
> Lehrbuch der theoretischen Chemie. Berlin, 1850. 8vo.

MÜLLER, P.
> Handbuch für Bierbrauer. Eine wissenschaftlich-praktische Anleitung zum Bierbrauen im ganzen Umfange des Gewerbes. Mit Rücksicht auf die neuesten Erfahrungen und Verbesserungen im Braufache, und unter Beifügung der verschiedenen Braumethoden in Baiern und anderen Ländern. Nach den besten Quellen und vieljährigen eigenen Erfahrungen bearbeitet. Mit einem Vorworte von Fr. Jul. Otto. Braunschweig, 1854. 8vo. Ill.

MÜLLER, PHILIPP.
> Tyrocinium chymicum : miracula chymica et mysteria medica. Lipsiæ, 1611. 12mo.

MÜLLER, W., OG J. G. OTTO.
> Medicinsk-kemisk Praktikum. Kristiania, 1884. 8vo.

MÜLLER, W. VON, UND H. KILIANI.
> Kurzes Lehrbuch der analytischen Chemie. München, 1884. 8vo.

MÜNTZ, ACHILLE.
> Méthodes chimiques analytiques appliquées aux substances agricoles. Paris, 1888. 8vo.

MÜNTZ, ACHILLE.
> Résumé de chimie analytique appliquée spécialement à l'industrie et à l'agriculture. *See, in Section II*, Fremy : Encyclopédie chimique, vol. IV.

MUIR, MATTHEW MONCRIEFF PATTISON.
Practical Chemistry for Medical Students, specially arranged for the
first M. B. Course. London, 187–. 18mo.
Treatise (A) on the Principles of Chemistry. Cambridge, 1884. pp.
xxii–488, 8vo.

MUIR, M. M. PATTISON, AND DOUGLAS CARNEGIE.
Practical Chemistry. A Course of Laboratory Work. A companion
Volume to Muir and Slater's Elementary Chemistry. Cambridge,
1887. pp. viii–224, 12mo. Ill.

MUIR, M. M. PATTISON, AND C. SLATER.
Elementary Chemistry. A companion Volume to Muir and Carnegie's
Practical Chemistry. London, 1887.

MUIR, M. M. PATTISON, AND DAVID MUIR WILSON.
Elements (The) of Thermal Chemistry. London, 1885. pp. xvi–312. 8vo.

MULDER, CLAAS.
Chemiæ (De) usu in illustrandis vitæ phænomenis. Groningen, 1841.
4to.
Handleiding tot de scheikunde. Amsterdam, 1824–25. 2 vols., 8vo.
Populaire scheikundige lessen gehouden bij het genootschap ter bevor-
dering van de natuurkundige wetenschappen te Groningen. Amsterdam, 1845. 8vo.

MULDER, E.
Leerboek der zuivere en toegepaste scheikunde. 1 afdeeling : Zuivere
scheikunde. 2 deelen, 3 stukken. Delft, 1866. 8vo.
Scheikundige aanteekeningen, uitgegeven door E. Mulder. Utrecht,
1865–74. 3 vols., 8vo. Vol. I, 4 pts., 1865–67. Vol. II, pp. v–228,
1871. Vol. III, pp. 258, 1874.

MULDER, GERRIT JAN.
Bijdragen tot de geschiedenis van het scheikundig gebonden water.
Rotterdam, 1864.
Matériaux pour servir à l'histoire de l'eau en combinaison
chimique. Haarlem. 8vo. Ill.
Chemie (Die) der austrocknenden Oele, ihre Bereitung und ihre tech-
nische Anwendung in Künsten und Gewerben. Nach der holländ-
ischen Original-Ausgabe bearbeitet von J. Müller. Berlin, 1867.
8vo.
Chemie (Die) des Bieres. Aus dem holländischen übersetzt von Chr.
Grümm. Leipzig, 1858. 8vo.

Mulder, Gerrit Jan. [Cont'd.]

De la bière ; sa composition chimique, sa fabrication, son emploi comme boisson. Traduit du hollandais par Aug. Delondre avec le concours de l'auteur. Corbeil, 1861. 8vo.

Elementen (De). Rotterdam, 1844   8vo.

Geregtelijke scheikundige onderzoekingen, gedaan in het laboratorium der hoogeschool te Utrecht, 1843–1844. Uitgegeven voor jonge lieden. 1. Stuk : irriterende vergiften. Rotterdam, 1845. 8vo.

Gerichtlich-chemische Untersuchungen ausgeführt unter G. J. Mulder's Leitung im Laboratorium zu Utrecht. Aus dem holländischen für deutsche Juristen, Aerzte und diejenigen bearbeitet, welche sich mit diesem Zweige der Chemie beschäftigen müssen, von Johannes Müller. Berlin, 1848. 16mo.

Handboek voor scheikundige werktuigkunde. Rotterdam, 1834–35. 2 vols., 8vo. 24 Plates.

Proeve eener algemeene physiologische scheikunde. Rotterdam, 1844– 46. 2 vols., 8vo. Ill.

Versuch einer allgemeinen physiologischen Chemie. Aus dem holländischen übersetzt von Jac. Moleschott. Heidelberg, 1844–47. 8vo.

Versuch einer allgemeinen physiologischen Chemie. Mit eigenen Zusätzen des Verfassers für diese deutsche Ausgabe seines Werkes. Mit 8 colorirten und 12 schwarzen Kupfertafeln. Braunschweig, 1844–51. 2 vols., 8vo. Vol. I, pp. xii–532. Vol. II, pp. 533–1289-15.

The Chemistry of Vegetable and Animal Physiology. Translated from the Dutch by P. F. H. Fromberg, with an introduction and notes by James F. W. Johnston. Edinburgh, 1845–49.

First authorized American edition with notes and corrections by B. Silliman, jr. New York, 1845. 12mo.

Scheikunde (De) der bouwbare aarde. Rotterdam, 1860. 4 vols., 8vo. Vol. I, pp. xiv–482 ; vol. II, pp. vi–484 ; vol. III, pp. viii–502 ; vol. IV, pp. viii–435.

Chemie (Die) der Ackerkrume. Aus dem holländischen unter mitwirkung des Verfassers von Chr. Grimm. Leipzig, 1861–62. 2 vols., roy. 8vo.

Chemie (Die) der Ackerkrume. Nach der holländischen Original-Ausgabe deutsch bearbeitet und mit Erläuterungen versehen von Johannes Müller in Berlin. Berlin, 1861–63. 3 vols., 8vo.

MULDER, GERRIT JAN. [Cont'd.]

Scheikundig onderzoek van koper tot dubbeling der schepen gebruikt. Amsterdam, 1836. 4to.

Scheikundig onderzoek van Chinesche en Java-Thee. Rotterdam, 1836. 8vo.

Scheikundige onderzoekingen gedaan in het laboratorium der Utrechtsche hoogeschool. Uitgegeven door G. J. M. Rotterdam, 1842–47. 5 parts, 8vo.

> Chemische Untersuchungen. Unter des Verfasser's Mitwirkung übersetzt von A. Völker. Frankfurt a. M., 1847–8. 3 parts, 8vo.

Scheikundige verhandelingen en onderzoekingen. Rotterdam, 1862. 3 parts, 8vo.

Silber-Probirmethode (Die). Leipzig, 1859. 8vo.

Streven (Het) der stof naar harmonie. Eene voorlezing. Rotterdam, 1844. 8vo.

> Das Streben der Materie nach Harmonie. Eine Vorlesung. Aus dem holländischen übersetzt. Braunschweig, 1844.

Vergelijkend onderzoek van suiker, met en zonder stoom bereid. Rotterdam, 1850. 8vo.

Vraag (De) van Liebig, aan de zedelijkheid en de wetenschap getoetst. Rotterdam, 1846. 8vo.

> Liebig's questions to Mulder, tested by morality and science. Translated by P. F. H. Fromberg. London, 1846. 8vo.

> Reply to Liebig on the chemistry of animal and vegetable physiology. Translated by P. F. H. Fromberg. London, 1846. 8vo.

Wet (De) op het zout aan de scheikunde getoetst. Rotterdam, 1851. 8vo.

Wijn (De) scheikundig beschouwd. Rotterdam, 1855. 8vo.

> The Chemistry of Wine. Edited by H. Bence Jones. London, 1857. pp. xii–390, 12mo.

> Die Chemie des Weines. Aus dem holländischen von Karl Arentz. Leipzig, 1856. 8vo.

Zeewater (Het) en het zout in verband tot nijverheid en wetgeving beschouwd. Rotterdam, 1851. 8vo. Ill.

MUNIN.

Chimie expérimentale et théorique appliquée aux arts industriels et agricoles. Paris, 1845. 2 vols., 8vo.

MUNK, PH. ET E. LEYDEN.

Die acute Phosphor-Vergiftung. Mit besonderer Rücksicht auf Pathologie und Physiologie experimentell bearbeitet. Berlin, 1865. pp. 188, 8vo.

MUÑOZ Y AMADOR, BERNARDO.

Arte de ensayar oro y plata con breves reglas para la teoria y práctica. Segunda edicion . . . Madrid, 1845. 8vo.

MUNOS Y LUNA, RAMON TORRES.

Guia del químico práctico ó compendio de análysis química. Madrid, 1852. pp. xv–332, 12mo. Ill.

Prontuario de química general para complemento de la instruccion preparatoria de los institutos de segunda enseñanza seminarios y colegios. Madrid, 1865. 16mo. Ill.

Quimica (La) en sus principales aplicaciones à la agricultura. Obra publicada bajo la proteccion del Ministerio de Fomento. Madrid, 1856. 8vo.

Lecciones elementales de química general, para uso de los alumnos de medicina, ciencias, farmacia, ingenieros industriales, agrónomos, de minas, etc., etc. Madrid, 1861. 2 vols., 8vo. Vol. 1, pp. 507–[v.]. Vol. 11, pp. 458-[iii.].

Tratado de química general y descriptiva : obra de texto escrita con destino á la enseñanza de los alumnos de ciencias, medicina, farmacia y escuelas especiales. Quinta edicion, notablemente corredada y aumentada con arreglo á los últimos adelantos de la ciencia. Madrid, 1885. 4to.

Urinometria. Nuevo método normal para analizar la orina, descubierto per el célebre químico Justo Liebig. Leccion pronunciada en la cátedra de química de San Isidro el dia 1 Julio de 1853. Madrid, 1853. 4to.

MUNNER Y VALLE, VICENTE.

Análisis químico de aplicacion á las ciencias médicas. Segunda edicion. Barcelona, 1874. 4to.

MURPHY, J. G.

Review of Chemistry, for students adapted to courses as taught in the principal medical schools in the United States. Philadelphia, 1852. 12mo.

MURRAY, JOHN.

A Sketch of Chemistry, practical and applied. London, 1839. pp. vii–337, 12mo. Ill.

> Inorganic Chemistry only.

Elements of Chemistry. Edinburgh, 1801. 2 vols., 8vo. 4th edition, 1816. Cf. Wolff, Fr. 6th edition, Edinburgh, 1828. 2 vols.

System (A) of Chemistry. Edinburgh, 1806–7. 4 vols., 8vo. Supplement, 1812. 4th edition, 1819.

MURRAY, R. M.

Chemical Notes and Equations for use of students. Third edition.
Edinburgh, 1888.

MUSAIO, G.

Nozioni elementari di analisi chimica qualitativa delle sostanze minerali.
Caserta, 1884. 8vo.

MUSPRATT, JAMES SHERIDAN.

Chemistry, theoretical, practical and analytical as applied and relating
to the arts and manufactures. Edinburgh, London and New York.
Glasgow, *n. d.* [1853–61]. 2 vols., 4to.

> Vol. I contains portraits of: Muspratt, Brande, Berzelius, Black, Bunsen,
> Chaptal, Chevreul, Christison, Davy, Dumas, Dalton, Faraday, Fownes,
> Gregory, Gay Lussac, Graham, Gmelin, Horsford, Hofmann, Kane,
> Lavoisier, Morfit, Mitscherlich, Priestley, Playfair, Rose, Thomson,
> Ure and Wöhler.
>
> There have been many later editions.

Muspratt's theoretische, praktische und analytische Chemie in Anwendung auf Künste und Gewerbe. Frei bearbeitet von F. Stohmann.
Braunschweig, 1854–58. 2 vols., 8vo.

> Zweite verbesserte und vermehrte Auflage. Braunschweig,
> 1865–1870. 6 vols., 4to.

> *Also under the title:* Encyklopädisches Handbuch der technischen Chemie.
> From Vol. III, fortgesetzt von Bruno Kerl.

> Dritte verbesserte und vermehrte Auflage. Frei bearbeitet
> von Bruno Kerl und F. Stohmann. 7 vols., 4to. Ill.

> *Also under the title :* Encyklopädisches Handbuch der technischen Chemie.
> Von Bruno Kerl und F. Stohmann.

> Vierte verbesserte und vermehrte Auflage. Frei bearbeitet
> von F. Stohmann und B. Kerl. Braunschweig, 1886–92.
> 7 vols., 4to. [In progress.]

> *Also under the title :* Encyclopädisches Handbuch der technischen Chemie.
> Von F. Stohmann und B. Kerl.

> Теоретическая, практич. и аналитич. химія въ приложеніи къ
> искусствамъ и промышленности; перев. Калиновска Москва
> 1869. 4 части.

Outlines of Qualitative Analysis. London, 1849. 8vo.

MUTER, JOHN.

Introduction to Theoretical, Pharmaceutical and Medical Chemistry.
London, 1879. Roy. 8vo.

Manual (A) of Analytical Chemistry, qualitative and quantitative, inorganic and organic. Philadelphia, 1888. 8vo.

MUTER, JOHN. [Cont'd.]

Practical and Analytical Chemistry, quantitative and qualitative. Fourth edition. Revised and Illustrated. Philadelphia, 1891. 8vo.

Manual of Analytical Chemistry for Laboratory Use, including testing
for metals, acids, poisons, etc., volumetric, air, drugs and gas
analysis. London, 1889. 8vo. Ill.

Short (A) Manual of Analytical Chemistry ; qualitative, quantitative,
inorganic and organic. Third edition. London, 1887.

Fourth edition. London, 1890. 8vo. Ill.

Fifth edition. London, 1892. 8vo.

First American edition from the Fourth English edition.
Edited by Claude C. Hamilton. Philadelphia, 1891. 8vo.

MUYKENS, THEODORUS.

See Collectanea chymica Leidensia.

MYLIUS, J. D.

Opus medico-chymicum, continens tres tractatus sive basilicas, quorum
prior inscribitur Basilica medica, secundus Basilica chymica, tertius Basilica philosophica. Francofurti, 1618. 2v., 4to.

NÄGELI, C. VON.

Theorie der Gährung. Ein Beitrag zur Molekularphysiologie. München,
1879. 8vo.

NAGY DE NYIR, FRANZ.

Darstellung der neuern Ansichten über die Natur der Salzsäure. Wien,
1819. 8vo.

NAHUYS, ALEXANDER PETRUS.

De aquæ origine ex basibus aeris puri et inflammabilis. Harderov, 1791.
8vo.

[Another edition.] Trajectæ ad Rhenum, 1789.

Chymische Abhandlung von der Entstehung des Wassers aus
der Verbindung der Grundstoffe der reinen und brenn-
baren Luft. Aus dem lateinischen mit Erläuterungen
und Zusätzen herausgegeben von Johann Andreas
Scherer. Wien, 1790. pp. xii–276, 8vo.

Is het Phlogiston een waar beginsel der ligchamen. Utrecht, 1789. 8vo.

Tractatus chemicus, continens nova quædam experimenta cum basi
salis marini, nitri et aluminis. Pars 1 Amstelodami, 1761. 8vo.

NAPIER, JAMES.

Manual (A) of Electro-Metallurgy : including the applications of the
art to manufacturing processes. London, 1851. 8vo.

NAPIER, JAMES. [Cont'd.]

> Second edition. London, 1852. 8vo.
>
> [*The Same*] From the second London edition, revised and enlarged. Philadelphia, 1853. pp. xv–356, 8vo. Ill.

Manual (A) of Dyeing and Dyeing Receipts, composing a system of elementary chemistry, as applied to dyeing, with receipts for the general reader for dyeing any colour on cotton, silk and wool, with coloured pattern of cloth of each fabric. London, 1853. 8vo.

> New edition. London, 1858. 8vo.
>
> Third edition. London, 1875. pp. xxviii–420, 8vo. Ill.

NAPOLÉON, LOUIS.

Considérations sur la question des sucres. Paris, 1842.

> The author afterwards became Emperor of France.

NAPOLI, RAFFAELE.

Prontuario di chimica elementare moderna. Napoli, 1867–68. 2 vols., 8vo. Vol. I, pp. 283 ; vol. II, xi–328.

> A third volume dealing with organic chemistry is announced as in preparation.

NAQUET, ALFRED.

Application de l'analyse chimique à la toxicologie. Paris, 1859. 4to.

De l'allotropie et de l'isomérie. Paris, 1860. 8vo.

Della sintesi in chimica organica. Palermo, 1866.

> *Cf.* Naquet, A. : Principes de chimie.

Des sucres. Paris, 1863. pp. 82, 8vo.

Manuel de toxicologie. Paris, 1867. 18mo. Ill.

Précis de chimie légale, guide pour la recherche des poisons, l'examen des armes à feu, l'analyse des cendres, l'altération des écritures, des monnaies, des alliages, des denrées et la détermination des taches dans les expertises chimico-légales, à l'usage des médecins, pharmaciens, chimistes, experts, avocats. Paris, 1872. 18mo.

> Legal Chemistry. A guide to the detection of poisons, examination of tea, stains, etc., as applied to chemical jurisprudence. Translated with additions from the French by J. P. Battershall. Second edition, revised, with additions. New York, 1884. pp. 190, 12mo.
>
> Compendio de química legal. Guía para la investigacion de los venenos, exámen de las armas de fuego, análisis de las cenizas, alteracion de escritos, de monedas, de aleaciones, de las sustancias alimenticias y determinacion de las manchas, etc., en las operaciones químico-legales

NAQUET, ALFRED. [Cont'd.]

para uso de los médicos, farmacéuticos, abogados, químicos, peritos, etc. Traducido y adicionado con trabajos especiales de los célebres químicos Fresenius, Wurtz, Odling, Dragendorff y Bolley, por Vicente Martin Argenta. Madrid, 1876. 4to.

Судебная химія. Москва, 1874

Principes de chimie fondées sur les théories modernes. Paris, 1864. 12mo.

Deuxième édition revue et considérablement augmentée. Paris, 1867. 2 vols., 12mo. Vol. I, pp. iv–444 ; vol. II, pp. 630. Ill.

Vol. II contains a valuable " Histoire de la synthèse en chimie organique." pp. 555–594.

Principles of Chemistry founded on modern theories. Translated by W. Cortis. London, 1867.

New edition revised by Stevenson. London, 1868. 8vo.

Grundzüge der modernen Chemie. Nach der zweite Auflage von A. Naquet's " Principes de chimie." Berlin, 1868–70. 2 vols., 8vo.

Translated and edited by Eugen Sell.

Principii di chimica fondati sulle teorie moderne, tradotti da C. Parenti. Firenze, 1879–81. 2 vols., 8vo.

Translated and edited by Hauriot.

Курсъ химіи, основан. на современ. теоріяхъ пер. Лесгафтъ Фр. С.-Петербургъ, 1867. 3 тома.

NASH, J. A.

The Progressive Farmer ; a scientific treatise on agricultural chemistry, the geology of agriculture, on plants, animals, manures, and soils, applied to practical agriculture. New York, 1853. 8vo.

NASINI, R., E VILLAVECCHIA.

Relazione sulle analisi e sulle ricerche eseguite durante il triennio 1886–'89 nel laboratorio chimico centrale delle gabelle diretto da Stan. Cannizzaro. Roma, 1890. 4to.

NASON, HENRY BRADFORD.

Manual of Qualitative Blowpipe Analysis and Determinative Mineralogy. Elderhorst's Manual rewritten and revised. Philadelphia, n. d. [1881]. pp. 371, 12mo.

Cf. Elderhorst, William ; also Landauer, F. See Wöhler, F.: Handbook of Mineral Analysis.

NASSE, JOHANN FRIEDRICH WILHELM.
Ueber die Aetherbildung im Allgemeinen. Leipzig, 1809.
Ueber Naturphilosophie in Bezug auf Physik und Chemie. Freiberg,
    1809. 8vo.

NASSE, O.
        Chemie und Stoffwechsel der Muskeln. *See* Handbuch der Physiologie.

NATANSON, JACÓB.
Krótki rys chemii organicznéj ze szczególnym względem na rolnictwo,
    technologie i medycyne. Warszawa, 1860.

NAUDIN, L.
Désinfection des alcools du mauvais goût par électrolyse des flegmes ;
    historique de l'industrie des alcools ; progrès accomplis depuis le
    XVIII siècle. Paris, 1881. 8vo.

NAUMANN, ALEXANDER,
        See Gmelin, Leopold.

NAUMANN, ALEXANDER.
Grundlehren der Chemie. Heidelberg, 1879.
Grundriss der Thermochemie, oder der Lehre von den Beziehungen
    zwischen Wärme und chemischen Erscheinungen vom Standpunkt
    der mechanischen Wärmetheorie dargestellt. Braunschweig, 1869.
    8vo.
Lehr- und Handbuch der Thermochemie. Braunschweig, 1882. pp.
    xi–606, 8vo.
        Основанія термахиміи или ученіе о зависимости между тепловыми
            и химическими явленіями, изложенное на основаніи
            механической теоріи тепла; перев. Лисенко К. С.-Петербургъ
            1871.
Ueber Molekülverbindungen nach festen Verhältnissen. Heidelberg,
    1872. 8vo.

NAVIER, PIERRE TOUSSAINT.
Contrepoisons de l'arsenic, du sublimé corrosif, du verd-de-gris, et du
    plomb. Paris, 1777. 2 vols., 12mo.
        Gegengifte des Arseniks, ätzenden Sublimats, Spangrüns
            und Bleies aus dem französischen von Weigel. Greifs-
            wald, 1782. 2 vols., 4to.

NAYUDU, P. LAKSHMI NARASU.
Notes on Qualitative Chemical Analysis. Madras, 1892. 8vo.

NEALE, [DR.].

Chi nan foo. [Practical Chemistry in Chinese.]

Cf. Fryer, John.

NEAT, J. W.

A Catechism of Chemistry, including heat, magnetism and electricity. Arranged chiefly with reference to the new army and other examinations. London, 1858. 12mo.

NEBBIEN, C. H.

Wie ist der grösste und reinste Zuckergehalt in der Runkelrübenwirthschaft zu erzeugen? Leipzig, 1836.

НЕЧАЕВЪ, Н. П.

Очеркъ новѣйшей химіи. Москва 1868.

NECHAEFF, N. P. Sketch of modern chemistry.

Формулы химич. реакцій по унитарной системѣ для тѣлъ чаще другихъ встрѣч. приходѣ качеств анализа. Москва 1864.

Formula of chemical reactions on the unitary system.

НЕДАТСЪ, Константинъ.

Сравнительныя таблицы приблизительнаго химическаго состава болѣе употребительныхъ веществъ и напитковъ. С.-Петербургъ 1876.

NEDATS, KONSTANTINE. Comparative tables of approximate composition of the food and drink most in use.

NEES VON ESENBECK, C. G. (and others).

Die Entwickelung der Pflanzensubstanz, physiologisch, chemisch und mathematisch dargestellt mit combinatorischen Tafeln der möglichen Pflanzenstoffe, und den Gesetzen ihrer stöchiometrischen Zusammensetzung. Herausgegeben von C. G. Nees von Esenbeck, C. G. Bischof, H. A. Rothe. Erlangen, 1819. 4to.

NEILL, J. H., AND F. G. SMITH.

A Handbook of Chemistry with nineteen illustrations. Being a portion of an analytical compend of the various branches of medicine. Philadelphia, 1848. 8vo.

НЕЛЮБИНЪ, А.

Общая судебно-медицинская и полицейская химія, съ присовокупленіемъ общей токсикологіи или науки о ядахъ и противуядныхъ средствахъ. 2 части. С.-Петербургъ 1851. 8°

NELUBINN, A. General medico-legal chemistry. St. Petersburg, 1851.

NENDTVICH, KÁROLY MAXIMILIAN.

Az életmütlen müipari vegytannak alapismeretei. Az iparegyesületi

NENDTVICH, KÁROLY MAXIMILIAN. [Cont'd.]
> felolvasásokhoz alkalmazva. Fába metszett rajzokkol. Pesten,
> 1844. 8vo.

Grundriss der allgemeinen technischen Chemie in drei Abtheilungen,
> (Metalloide, Metalle, Organische Chemie). Pest, 1854. 3 vols., 8vo.
> Zweite Auflage. Pest, 1859. 8vo.

Необходимѣйшія свѣдѣнія изъ описательной химіи. С.-Петербургъ 1876.

> The essentials of descriptive chemistry. St. Petersburg, 1876.

NERI, ANTONIO.
> L'arte vetraria distinta in libri sette ; ne' quali si scoprono maravigliosi
> effetti e s'insegnano segreti bellissimi del vetro nel fuoco ed altre
> cose curiose. Firenze, 1612. 4to.

> > L'art de la verrerie de Neri, Merret et Kunckel, auquel a
> > ajouté le Sol sine veste d'Orschall, traduit de l'allemand
> > [par le baron d'Holbach]. Paris, 1754. 4to.

> > > One of the earliest treatises on glass manufacture. A Latin translation was
> > > issued, with notes by Ch. Merrett, at Amsterdam in 1681. 12mo. A
> > > German version by Kunckel is found in his Ars vitraria. *Cf.* Kunckel,
> > > Johanan.

NESSLER, J.
> Kemisk-teknisk Handbok for Vinhandlande och Vinfabrikanter, inne-
> hållende ett urval af nyaste metoder för viners behandling samt
> botemedel mot deras sjukdomer, jemte beskrifning paa nödiga
> apparater. Stockholm, 1872. 8vo. Ill.

NETTL, ANTON S.
> Grundriss der unorganischen Chemie nach den neuesten Ansichten der
> Wissenschaft. Frankenberg, 1876. 8vo.

NEUBAUER, CARL THEODOR LUDWIG.
> Anleitung zur qualitativen und quantitativen Analyse des Harns. Ent-
> haltend die Lehre von den Eigenschaften und dem Verhalten der
> im Harn vorkommenden Bestandtheile zu Reagentien und unter
> dem Microscop, zum Gebrauch für Mediciner und Pharmaceuten
> bearbeitet. Mit 3 lithographirten Tafeln und 20 Holzschnitten.
> Bevorwortet von R. Fresenius. Wiesbaden, 1854. 8vo.

> > Anleitung zur qualitativen und quantitativen Analyse des
> > Harns, sowie zur Beurtheilung der Veränderungen dieses
> > Secrets mit besonderer Rücksicht auf die Zwecke des
> > praktischen Arztes. Zweite Auflage. Wiesbaden, 1856.
> > 8vo.

> > > Published with the co-operation of Julius Vogel.

NEUBAUER, CARL THEODOR LUDWIG.   [Cont'd.]

        Dritte verbesserte Auflage. Wiesbaden, 1858. 8vo.

        Fünfte vermehrte und verbesserte Auflage. Wiesbaden,
            1867. 8vo. Ill.

        Achte umgearbeitete und vermehrte Auflage von H. Huppert
            und L. Thomas, bevorwortet von R. Fresenius. Wies-
            baden, 1881–85. 2 vols., 8vo. Ill.

        Neunte umgearbeitete und vermehrte Auflage von H. Hup-
            pert und L. Thomas. Wiesbaden, 1890. Two parts.
            I, pp. x–581 ; II, pp. vi–288. 8vo. Ill.

        Guide to the qualitative and quantitative analysis of the
            urine. Translated by William O. Markham. London,
            1863.

        De l'urine et des sédiments urinaires. Propriétés et
            caractères chimiques et microscopiques des éléments
            normaux et anormaux de l'urine. Analyse qualitative
            et quantitative de cette sécrétion. Description et valeur
            séméiologique de ses altérations pathologiques, etc.
            Précédé d'une introduction par R. Frésénius. Traduit
            sur la cinquième édition allemande par L. Gautier.
            Paris, 1869. 8vo. Ill.

Systematischer Gang der qualitativen und quantitativen Analyse des
    Harns. Achte Auflage, bearbeitet von E. Borgmann. Wiesbaden,
    1882–3.

        Neunte Auflage. Wiesbaden, 1890. 8vo.

Ueber die Chemie des Weines. Drei Vorträge gehalten im Winter
    1869–70 in Mainz, Oppenheim und Oestrich am Rhein. Wiesbaden,
    1870.

NEUBAUER, CARL THEODOR LUDWIG, UND JULIUS VOGEL.

    Farbentabelle für den Urin. Wiesbaden, 1872. 8vo. Ill.

NEUBER, AUGUST WILHELM.

    De natura acidorum ac basium placitarum Winterli disquisitio. Kiliæ.
    1809.

    Ueber die Materie und den Urstoff in seinem vierfachen chemischen
    Grundverhältniss und seiner fünffachen Erscheinungsform. Ham-
    burg, 1830. 8vo.

NEUMANN, CARL AUGUST.

    Betrachtungen der chemischen Elemente, ihrer Qualitäten, Aequivalente
    und Verbreitung. Prag, 1858. 8vo.

    Betrachtungen über die Wichtigkeit, Möglichkeit und Nützlichkeit der

NEUMANN, CARL AUGUST. [Cont'd.]

Zuckererzeugung aus europäischen Gewächsen, mit besonderer
Rücksicht auf Böhmen. Prag, 1810.

Chemie (Die) als natürliche Grundlage wissenschaftlicher Natur- und
Gewerbskunde . . . Prag und Frankfurt, 1842. Fol.

Lehrbuch der Chemie, mit besonderer Hinsicht auf Technologie. Prag,
1810.

Vergleichung der Zuckerfabrikation aus in Europa einheimischen Gewächsen mit der aus Zuckerrohr in Tropenländern, mit Bezug auf
Staats- und Privatwirthschaft. Prag, 1837.

NEUMANN, CARL C. O.

Repetitorium der Chemie für Chemiker, Pharmaceuten, Mediciner, etc.,
sowie zum Gebrauch an Realschulen und Gymnasien. Düsseldorf,
1884.

NEUMANN, CASPAR.

Chymia medica, dogmatico-experimentalis, das ist gründliche und mit
experimenten erwiesene medicinischen Chymie. Herausgegeben
von J. Chr. Zimmermann. Züllichau, 1749–55. 4 vols., 4to.

        [A second edition abridged and condensed.] Züllichau, 1756.
        2 vols., 4to.

        A French translation by Roux. Paris, 1781. 4to.

        For other editions of Neumann's Lectures *see* Prælectiones chemicæ.

        * The Chemical Works of C. Neumann, abridged and
        methodized. With large additions, containing the later
        discoveries and improvements made in chemistry and
        the arts depending thereon, by William Lewis. London,
        1759. pp. [xvi]–586–[xxxviii], 4to.

        With a critical list of Neumann's works.

        Second edition. London, 1773. 2 vols., 8vo.

        Grondlyke en met proeven bewezene medicinale en natuur-
        kundige chymie. Leeuwarden, 1766. Vol. 1, 8vo.

        No more published.

Chymische Untersuchung der meisten zum Pflanzenreiche gehörigen
Materien. Züllichau, 1751–52. 4 parts.

Lectiones publicæ von vier subjectis pharmaceutico-chemicis, nehmlich
vom gemeinem Saltze, Weinstein, Salmiac und der Ameise . . .
Leipzig, 1737. pp. [viii]–379, sm. 4to.

Lectiones chymicae von salibus alkalino-fixis, und von camphora. Als
zwey Proben, umb daraus zu sehen wie alle übrige lectiones bey
dem in Berlin gestifteten Kgl. Collegio Medico Chirurgico publice

NEUMANN, CASPAR. [Cont'd.]

  abgehandelt werden . . . Berlin, 1727. pp. [viii]–164, sm.
  4to. Folding title-page.

Prælectiones chemicæ seu chemia medico-pharmaceutica experimentalis
  et rationalis. Oder grundlicher Unterricht der Chemie worinnen
  nicht nur eine gründliche Anleitung zur Chemie, sondern auch die
  aus allen dreyen Natur-Reichen vorkommende Præparationes et
  Producta chemica medico-pharmaceutica und nützliche auch ver-
  werffliche Medicamenta, mit lauter Experimentis und vernünfftigen
  Demonstrationibus gewiesen und erläutert, dersgleichen einige ad
  Fontem diæteticum gehörige Esculenta und Potulenta abgehandelt
  seyn, nebst beygefügten Annotationibus und dem Usu Medico-
  Chirurgico. Denen Medicis und Apotheckern, theils auch denen
  Chirurgis zum höchstnützlichen Gebrauch herausgegeben von
  Johann Christian Zimmermann. Berlin, 1740. pp. [xxx]–1872–
  [lxiii], sm. 4to. Portrait of Neumann.

        Allgemeine Grundsätze der theoretisch practischen Chemie,
        das ist, gründlicher und vollständiger Unterricht in der
        Chemie ; in welchem nicht nur überhaupt eine gründ-
        liche Anleitung zu allen Theilen der Chemie, sondern
        auch die Operationes und Producta mit Demonstra-
        tionibus und Experimentis gelehret werden ; nebst bey-
        gefügten medicinischen, chirurgischen, œconomischen,
        metallurgischen, etc., Gebrauch und Anwendung.
        Herausgegeben von Johann Christian Zimmermann.
        Dresden, 1755. 2 vols., 4to.

      The " Prælectiones" and the German translation "consist of notes taken
        down by one of Neumann's pupils, intermixed with a number of inco-
        herent compilations from different authors." The edition entitled
        Chymia medica, etc., Züllichau, 1749–55, is said to be genuine.

NEVOLE, MILAN A BOH. RAÝMAN.

  Chemie organická pro vysoké učení české. V Praze, 1881.

NEWELL, LYMAN C.

  Experiments in General Chemistry, systematically and progressively
    arranged. Pawtucket [Rhode Island], n d. [1891].

NICHOLS, JAMES ROBINSON.

  Chemistry of the Farm and the Sea ; with other familiar chemical essays.
    Boston, 1867. 8vo.

NICHOLS, WM. RIPLEY.

  Water Supply considered mainly from a Chemical and Sanitary Stand-
    point. New York, 1883. 8vo. Plates.

NICHOLS, WILLIAM RIPLEY, AND LEWIS M. NORTON.
Laboratory Experiments in General Chemistry, compiled from Eliot
and Storer's Manual, and other sources. For the use of Students
of the Massachusetts Institute of Technology. Printed, not published. Boston, Mass., 1884. pp. 58–viii, 16mo. Ill.
[Another edition.] Revised by Fred. L. Bardwell. Boston,
Mass., 1891. pp. 56–viii, 12mo.

NICHOLSON, WILLIAM.
Controversy (The) between Kirwan and the French Academicians on
Phlogiston. London, 1787.
First (The) Principles of Chemistry. London, 1790. 8vo.
Second edition. London, 1792.
Anfangsgründe in der Scheidekunst. Aus dem englischen
übersetzt von C. H. Spohr. Leipzig, 1791. 8vo.
* Introduction (An) to Natural Philosophy. Illustrated with copper-
plates. The third edition, with improvements. Philadelphia,
printed for Thomas Dobson at the Stone house, in Second Street,
between Market and Chestnut Street, 1788. pp. xvi–554 [570]–
[viii], 8vo. 25 plates.

NICKEL, EMIL.
Die Farbenreactionen der Kohlenstoffverbindungen. Für chemische,
physiologische, mikrochemische, botanische, medicinische und phar-
makologische Untersuchungen bearbeitet. Zweite umgearbeitete,
vermehrte und erweiterte Auflage. Berlin, 1890. 8vo.

NICOLAS, PIERRE FRANÇOIS.
Cours de chimie théorique-pratique. Paris, 1777.
Dissertation chimique sur les eaux minérales de St. Diez. Nancy, 1780.
12mo.
Dissertation sur les eaux minérales de la Lorraine. Nancy, 1784.
Manuel du destillateur .   .   .   Nancy, 1787.

NIELSEN, P.
Kemi for Jordbrugere. Veiledning for Landmænd, Gartnere og andre
Havebrugere, i uorganisk Kemi. Randers, 1881. 8vo.

NIEPCE DE SAINT-VICTOR, ABEL.
Traité pratique de gravure héliographique sur acier et sur verre. Paris,
1856. 8vo.

NIETZKI, R.
Organische Farbstoffe. Breslau, 1886.
Chemie der organischen Farbstoffe. Berlin, 1889. 8vo.

NIETZKI, R. [Cont'd.]

From the : Encyclopaedie der Naturwissenschaften, Abtheilung, Handwörterbuch der Chemie, herausgegeben von A. Ladenburg.

Chemistry of the Organic Dye-Stuffs. Translated, with
additions, by A. Collin and W. Richardson. London,
1892. 8vo.

NIEUWLAND, PIETER.

Schets van het scheikundige leerstelsel van Lavoisier. Amsterdam,
1792. 8vo.

NIKOLAI.

Was ist für und wieder den Zuckerbau in den preussischen Staaten zu
sagen ? Berlin, 1799.

NILSON, L. F.

Handledning i bruket af Blåsrör och Lamplåga vid qualitativ kemisk
analys. Upsala, 1873. 8vo.

Schematisk Öfversigt af qualitativa kemiska Analyser. Upsala, 1873.
8vo.

NIVET.

Dictionnaire des eaux minérales du Puy-de-Dôme. Clermont-Ferrand,
1846. 8vo.

NIVOIT.

[Produits chimiques.] *See, in Section II,* Fremy : Encyclopédie chimique,
vol. v.

NIVOIT ET MARGOTTET.

Calcium, Baryum, Strontium, Magnesium et Aluminium. *See, in Section II,*
Fremy : Encyclopédie chimique, vol. III.

NOAD, HENRY M.

Chemical Manipulation and Analysis Qualitative and Quantitative.
(Intended to form part of a series of treatises on chemistry in the
library of useful knowledge.) London, 1848. 8vo. Philadelphia,
1849.

Chemical Manipulation and Analysis, qualitative and quantitative, explanatory of the general principles of chemical
nomenclature. New edition. London, 1852.

Lectures on Chemistry, including its applications in the arts and the
analysis of organic and inorganic compounds. London, 1843. 8vo.

A Manual of Chemical Analysis, qualitative and quantitative. For the
use of students. London, 1863–64. 8vo. Pt. I, 1863, pp. 211 ;
pt. II, 1864, pp. x–213–663.

Fourth edition. London, 1871.

NOBLE, CAPTAIN, AND F. A. ABEL.

Researches on Explosives. Fired Gunpowder. 2 parts. London, 1875 and 1880. 4to.

Recherches sur les substances explosives. Paris, 1877.

See Abel, F. A.

NÖLDEKE.

Vorkommen (Das) des Petroleums in nordwestlichen Deutschland, insbesondere in der Lüneburger Heide. Vortrag . . . Celle, 1881. 8vo. Ill.

Vorkommen und Ursprung des Petroleum, neu bearbeitet. Celle, 1883. 8vo. Ill.

NÖLTING, E.

Conférence, (etc.), *see* Société chimique de Paris.

NÖLTING, E.

Künstlichen (Die) organischen Farbstoffe. Unter Zugrundelegung von sechs Vorlesungen gehalten von E. N. bearbeitet von Paul Julius. Berlin, 1887. pp. vii–235, 12mo.

NÖLTING, E., UND A. LEHNE.

Anilinschwarz und seine Anwendung in Färberei und Zeugdruck. Berlin, 1892. 8vo. Ill.

NOLDECHEN, K. A.

Ueber den Anbau der Runkelrüben und die damit angestellten Zuckerversuche. Berlin, 1799. 3 parts, 8vo.

NORDENSKJÖLD, NILS GUSTAV.

Försök till Framställning af kemiska Mineralsystemet. Stockholm, 1827. [Another edition.] Helsingfors, 1833.

NORMANDY, ALPHONSE RÉNÉ LE MIRE.

Chemical (The) Atlas; or, tables showing at a glance the operations of qualitative analysis, with practical observations and copious indices of tests and reagents. London, 1857. Fol.

Tableaux d'analyse chimique. Ouvrage présentant toutes les opérations de l'analyse qualitative accompagné de nombreuses observations pratiques. Paris, 1858. 4to.

Таблицы химич. анализа; перев. Федорова и Виннера. С.-Петерб. 1865.

Commercial (The) Hand-Book of Chemical Analysis, or practical instructions for the determination of the intrinsic or commercial value of substances used in manufactures, in trades and in the arts. London, 1850. 8vo. Ill.

NORMANDY, ALPHONSE RÉNÉ LE MIRE. [Cont'd.]
> Second edition. London, 1865. pp. xii–640, 12mo. Ill.
> New edition, enlarged and to a great extent rewritten by H.
>   M. Noad. London, 1875. 8vo.
> Manuel commercial d'analyse chimique, ou instructions
>   pratiques pour déterminer la valeur intrinsique ou com-
>   merciale des substances employées dans les manufac-
>   tures, le commerce et les arts. Nouvelle édition, revue
>   et augmentée. Traduit et remis au courant des connais-
>   sances scientifiques actuelles, par L. Quéry et L. Debacq.
>   Paris, 1884. 18mo. Ill.

Farmers' (The) Manual of Agricultural Chemistry . . . London, 1853.
Practical Introduction to H. Rose's Treatise on Chemical Analysis.
> London, 1849. 8vo. Ill.

NORTHCOTE, A. B., AND A. H. CHURCH.
> A Manual of Qualitative Chemical Analysis. London, 1858. 8vo.

NORTON, JOHN P.
> Elements of Scientific Agriculture, or the connection between science
>   and the art of practical farming. Albany, 1850. pp. v–208, 12mo.

NORTON, SIDNEY AUGUSTUS.
> The Elements of Chemistry, inorganic and organic. Cincinnati and
>   New York, 1884. pp. vi–504, 8vo. Ill.

NORWEGIAN NORTH ATLANTIC EXPEDITION, Chemistry of. *See*, Schmelck, L.

NOSOTTI, J.
> Carni fresche, carni salate, o in altro modo preparate e conservate,
>   grassi animali. Milano, 1886. 8vo. Ill.
> Chimica clinica. Pavia, 1883. 8vo.

NOTT, ELIPHALET.
> Syllabus of a Course of Lectures on Chemistry. In four parts. Part I.
>   Schenectady, 1825. 16mo.
>> Published anonymously.

NOUVEAU COURS DE CHYMIE. *See* Senac, J. B.

NOVARIO.
> Nouveaux élémens de chimie. Paris, 1823. 8vo.

NOYES, WM. A.
> The Elements of Qualitative Analysis. Terre Haute, 1887. 8vo.
> Qualitative Analysis. New York, 1891. 12mo.

Nüys, T. C. van.
  Chemical Analysis of healthy and diseased Urine.   Philadelphia, 1888.
    8vo.

Nyquist, J.
  Populär Kemi.   Nyköping, 1870.

Nysten, P. H.
  Recherches de physiologie et de chimie pathologiques.   Paris, 1811.   8vo.

O химическихъ паяхъ. . . . . С.-Петербургъ 1852.   8º
          On the use of the blowpipe in chemistry.   St. Petersburg, 1852.

O'Brine, David.
  Laboratory Guide to Chemical Analysis.   Second edition, revised.   New
    York, 1889.   8vo.
  Practical (The) Laboratory Guide in Chemistry.   Columbus, Ohio, 1883.

Ochse, W.
  Experimentelle Untersuchungen über den Einfluss der Concentration
    und der Temperatur einiger wässeriger Salzlösungen auf ihre Ober-
    flächenspannung.   Rostock, 1890.

Ochsenius, C.
  Die Bildung des Natronsalpeters aus Mutterlaugensalzen.   Mit einer
    Karte und vier Profilen der mittleren südamericanischen West-
    küste.   Stuttgart, 1887.   8vo.

Odling, William.
  Course (A) of Practical Chemistry arranged for the use of medical
    students.   Second edition.   London, *n. d.* [1863].   8vo.
          Fifth edition.   London, 1876.
          Курсъ практической химіи; перев. Сивченкова Р.   С.-Петерб.
            1867.
  Course (A) of Six Lectures on the Chemical Changes of Carbon.   Re-
    printed from the Chemical News, with notes by William Crookes.
    London, 1869.   pp. xii–162, 12mo.
          Métamorphoses (Les) chimiques du carbone.   Paris, 1870.
  Lectures on Animal Chemistry delivered at the Royal College of Phy-
    sicians.   London, 1866.
          Животпая химія.   Перев. Бакета. П.   С.-Пъ., 1867.
  Manual (A) of Chemistry descriptive and theoretical.   London, 1861.
    pp. xiv–380, 8vo.
          Manuel de chimie théoretique et pratique.   Édition française
            publiée avec l'autorisation de l'auteur par Edm. Willm.
            Paris, 1868.

ODLING, WILLIAM. [Cont'd.]

Beschreibendes und theoretisches Handbuch der Chemie, übersetzt von A. Oppenheim. Erlangen, 1865.

Руководство къ химіи описат. и теоретической; перев. Савченкова Ф С -Петербургъ 1863,

Outlines of Chemistry, or brief notes of chemical facts. London, 1870. 8vo.

Tables of Chemical Formulæ. London, 1864. 8vo.

Что такое озонъ? перев. Шарапова С. С.-Петербургъ 1874.

What is ozone? St. Petersburg, 1874.

OECHSNER DE CONINCK.

Cours de chimie organique. Vol. I, Chimie organique générale. Série grasse. Vol. II, Série aromatique. Paris, 1892. 8vo. *In progress.*

OEFFINGER, H.

Die Ptomaine oder Cadaver-Alcaloide, nach einem für den badensischen staatsärtzlichen Verein ausgearbeiteten Vortrag dargestellt. Wiesbaden, 1885. pp. 42, 8vo.

OEHME, JULIUS.

Die Fabrikation der wichtigsten Antimon-Präparate. Mit besonderer Berücksichtigung des Brechweinsteines und Goldschwefels. Wien, Pest, Leipzig, 1884. 8vo. Ill.

OERSTED, HANS CHRISTIAN.

Ansicht der chemischen Naturgesetze, durch den neueren Entdeckungen gewonnen. Mit einer Kupfertafel. Berlin, 1812. 8vo.

In French by Marcel de Serres. Paris, 1813.

Erindringsord af Forelæsninger over Chemiens almindeligste Grundsætninger. Kjøbenhavn, 1825.

Second edition, 1826.

Materialien zu einer Chemie des neunzehnten Jahrhunderts. Regensburg, 1803. 8vo.

OERSTED, HANS CHRISTIAN [Translator].

See Chaptal, J. A.

OERTEL, CHRISTIAN THEODOR.

Programma IV de præstantia systematis chemiæ antiphlogisticæ. Baruthi, 1795–98. 4to.

OETTEL, F.

Hülfstafel zur Berechnung der Analysen. Dresden, 1888. 12mo.

OETTGEN, GODEFRID WOLFFGANG.

Tentamen inaugurale chemicum de theoria solutionum. Duisburgi ad Rhenum, *n. d.* [1782]. pp. 28, 4to.

OGIER.

> Analyse des gaz [and other articles]. *See, in Section II*, Fremy : Encyclo-pédie chimique, vol. II, vol. IV, etc.

OGLIALORO TODARO, A.

Trattato di chimica generale inorganica ed organica esposta sotto il punto di vista delle dottrine moderne per cura del Prof. Marco Sbriziolo. Napoli, 1882. pp. 204, 8vo.

OGONOWSKI, E.

La photochromie. Tirages d'épreuves photographiques en couleurs. Paris, 1891. pp. 30, 16mo.

Околовъ, Э.

О вліяніи салициловой и бензойной кислотъ на гніеніе и броженіе. С·Петербургъ 1876

> OKOLOF, E. On the influence of salicylic and benzoic acids upon putrefac-tion and fermentation. St. Petersburg, 1876.

OLDBERG, OSCAR, AND JOHN H. LONG.

A Laboratory Manual of Chemistry, Medical and Pharmaceutical. Chicago, 1887. 8vo. Ill.

OLEWINSKY, L.

Sur l'atomicité des éléments et sur les limites des combinaisons chimiques. Paris, 1861. 4to

OLMEDILLA, J.

Estudio químico de las generalidades de alcaloides. Edicion del Semanario Farmacéutico. Madrid, 1880. 4to.

OLMSTED, DENISON.

Elements of Chemistry. New Haven, 1851. 12mo.

O'NEILL, CHARLES.

Chemistry of Calico-Printing, Dyeing and Bleaching, including silken, woollen and mixed goods, with copious references to original sources of information and abridged specifications of the patents connected with these subjects. London, 1860.

Practice (The) and Principles of Calico Printing, Bleaching, Dyeing, etc. Manchester, 1878. 2 vols., 8vo.

OPOIX, CHRISTOPHE.
Analyse des eaux minérales de Provins. Paris, 1770.
Moyen de suppléer la potasse pour la fabrication de la poudre. Paris,
1793.
Traité des eaux minérales de Provins. Paris, 1824.

OPPERMANN, CARL FRIEDRICH.
Considérations sur les poisons végétaux. Strasbourg, 1845. 8vo.

OPPIO, L. DALL'.
L'ozóno, appunti critici ed alcune esperienze. Bologna, 1875.

OPPLER, THEODOR.
Handbuch der Fabrikation mineralischer Oele aus Steinkohlen, Braun-
kohlen, Holz, Torf, Petroleum und anderen bitumösen Substanzen,
so wie der Gewinnung von künstlichen Farbstoffen des Anilins, und
verwandter Producte des Steinkohlentheers. Berlin, 1862. pp.
xiii–284, 8vo. Ill.

ORFILA, MATHÉO JOSÉ BONAVENTURA.
Élémens de chimie médicale. Paris, 1817. 2 vols., 8vo. 2 plates.
Éléments de chimie appliquée à la médecine et aux arts.
Troisième édition revue, corrigée et augmentée. Paris,
1824. 2 vols., 8vo. Vol. I, 12 plates ; vol. II, 2 plates.
Cinquième édition. Paris, 1831.
Sixième édition. Paris, 1835.
Septième édition. Paris, 1843.
Huitième édition. Paris, 1851.
Practical Chemistry or a description of the processes by
which the various articles of chemical research in the
animal, vegetable and mineral kingdoms are produced.
Together with the best mode of Analysis. Translated
from the French by John Redman Coxe. To which is
added a variety of subjects of practical utility, and a
copious glossary of chemical terms and synonimes.
Philadelphia, 1818. pp. xv–355–lxxxvi, 8vo.
Handbuch der medizinischen Chemie, in Verbindung mit den
allgemeinen und technischen Theilen der chemischen
Wissenschaft nach ihrem neuesten Standpunkt. Aus
dem französischen übersetzt von Friedrich Tromms-
dorff . . . und mit Anmerkungen begleitet von Johann
Bartholmä Trommsdorff. Erfurt, 1819. 2 vols., 8vo.
Leçons de médecine légale. Paris, 1823. 3 vols., 8vo.
Deuxième édition. Paris, 1828. 3 vols., 8vo.

ORFILA, MATHÉO JOSÉ BONAVENTURA.   [Cont'd.]
    Fourth edition under the title :

> Traité de médecine légale. Quatrième édition, revue, corrigée
> et considérablement augmentée, contenant en entier le
> Traité des exhumations juridiques par Orfila et Lesueur.
> Paris, 1848.   3 vols. in 4, and one vol., plates, 8vo.

> Vorlesungen über gerichtliche Medicin. Uebersetzt mit
> Anmerkungen von J. Hergenröther. Leipzig, 1829.
> 3 vols., 8vo.

> Lehrbuch der gerichtlichen Medicin. Uebersetzt von G.
> Krupp. Leipzig, 1848–50. Three parts in 4 vols.

> Tratado de medicina legal, traducido de la cuarta edicion y
> arreglado á la legislacion española por Enrique Ataide.
> Madrid, 1847.   4 vols., 8vo.

Leçons de toxicologie. Paris, 1858. 8vo.

Leçons de chimie appliquée à la médecine pratique et à la médecine
légale. Nouvelle édition. Bruxelles, 1836. 24mo.

> Lecciones de química de Orfila aplicadas á la medicina
> práctica y á la medicina legal . . . Traducidas del
> francés por F. D. J. y aumentadas con notas por T. D.
> Barcelona, 1840. 8vo.

Leçons faisant partie d'un cours de médecine légale. Paris, 1821.

Rapport sur les moyens de constater la présence de l'arsenic dans les
empoisonnements par ce toxique. Paris, 1841. 8vo.

> Informe acerca de los medios para probar la presencia del
> arsénico en los envenenamientos por este tóxico . . .
> traducido del francés por Juan de Mata Castro. Madrid,
> 1842. 4to.

Recherches médico-légales et thrérapeutiques sur l'empoisonnement par
l'acide arsénieux, précédées d'une histoire de l'arsenic métallique,
etc. Recueillies et rédigées par Beaufort. Paris, 1841. 8vo.

> Zwölf Vorlesungen über die Lehre von den Vergiftungen im
> Allgemeinen und über die mit Arsenik insbesondere.
> Uebersetzt von F. Händel. Weimar, 1858.

Résumé des leçons de chimie appliquée à la médecine pratique et à la
médecine légale par Alexandre Pichon. Paris, 1828. 18mo.

Secours à donner aux personnes empoisonnées et asphyxiées. Suivis
des moyens propres à reconnaître les poisons et les vins frelatés et
à distinguer la mort réalle de la mort apparente. Paris, 1818. 12mo.

> Deuxième édition. Paris, 1821.

> Troisième édition. Paris, 1825.

> Quatrième édition, revue, corrigée et augmentée. Paris,
> 1830. 12mo.

ORFILA, MATHÉO JOSÉ BONAVENTURA. [Cont'd.]

Directions for the Treatment of Persons who have taken Poison, and those in a state of apparent death . . . Translated from the French by R. H. Black. With an Appendix on suspended animation. London, 1818. 8vo.

Second edition with additions and corrections. London, 1820.

A Popular Treatise on the Remedies to be employed in cases of poisoning and apparent death, including the means of detecting poisons, of distinguishing real from apparent death and of ascertaining the adulteration of wines. Translated from the French, under the inspection of the author, by William Price. London, 1818. 8vo.

A Practical Treatise on Poisons and Asphyxies adapted to general use, by M. P. Orfila . . . Translated from the French, with notes and additions, by J. G. Stevenson. With an Appendix containing the principles of medical jurisprudence . . . Boston, 1826. 8vo.

Rettungsverfahren bei Vergiftungen und im Scheintode, nebst den Mitteln zur Erkennung der Gifte und der verfälschten Weine und zur Unterscheidung des wahren Todes vom Scheintode. Aus dem französischen übersetzt von P. G. Brosse. Berlin, 1819.

Traité des poisons tirés des règnes minéral, végétal et animal, ou Toxicologie générale considérée sous les rapports de la physiologie de la pathologie et de la médecine légale. Paris, 1814–15. 4 vols. in 2, 8vo.

Deuxième édition. Paris, 1818.

Troisième édition, revue, corrigée et augmentée. Paris. 1827. 2 vols., 8vo.

Cinquième édition. Paris, 1852. 2 vols., 8vo.

Traité de toxicologie. Quatrième édition. Paris, 1843. 2 vols., 8vo. Plates.

A General System of Toxicology, or a treatise on Poisons . . . considered as to their relations with physiology, pathology and medical jurisprudence. Translated from the French. London, 1815. 8vo.

A General System of Toxicology, or a Treatise on Poisons found in the mineral, vegetable and animal kingdoms, considered in their relations with physiology, pathology and medical jurisprudence, abridged and partly translated from the French of P. Orfila þy Jos. G. Nancrede. Philadelphia, 1817. 8vo.

ORFILA, MATHÉO JOSÉ BONAVENTURA. [Cont'd.]

A Treatise on Mineral, Vegetable and Animal Poisons, considered as to their relations with physiology, pathology,
and medical jurisprudence. Translated from the French
by John Waller. London, 1817. *2 vols., 8vo.*

A secondary title on p. 1 reads : A General System of Toxicology.

Second edition corrected, by John Augustine Waller.
London, 1818. *2 vols., 8vo.*

Allgemeine Toxicologie oder Giftkunde, worin die Gifte des
Mineral-Pflanzen- und Thierreichs aus dem physiolo-
gischen, pathologischen und medizinisch-gerichtlichen
Gesichtspunkte untersucht werden. Nach dem fran-
zösischen . . . von Sigismund Friedrich Hermbstädt.
Berlin, 1818–19. *4 vols., 8vo. Ill.*

Seemann und Karls' Toxikologie oder die Lehre von den
Giften und Gegengiften. Nach der dritten Auflage des
" Traité des poisons " von Orfila durchaus frei bearbeitet
von J. A. Seemann und A. O. S. F. Karls. Berlin, Posen
und Bromberg, 1829–31. *2 vols., 8vo.*

Allgemeine Toxicologie. Uebersetzt von O. B. Kühn.
Leipzig, 1839. *2 vols.*

Lehrbuch der Toxicologie. Nach der fünften umgearbeiteten,
verbesserten und vielfach vermehrten Auflage aus dem
französischen mit selbstständigen Zusätzen bearbeitet
von G. Krupp. Braunschweig, 1852–53. *2 vols., 8vo.*

Tratado completo de toxicologia. Cuarta edicion revisada
corregida y aumentada. Traducido al castellano por
Pedro Calvo Asensio. Madrid, 1845–48. *4 vols., 4to.*

ORFILA, M. J. B.

Traité de médecine légale.   Under this title was published the fourth edition
of Orfila's Leçons de médecine légale, *q. v.*

Traité de toxicologie.   Under this title was published the fourth edition
of Traité des poisons. Paris, 1843. *2 vols., 8vo.*

ORME, F.

Rudiments of Chemistry. London, 1886.

ORME, TEMPLE.

Matriculation Chemistry ; chemistry of the non-metallic elements and
their compounds. London, 1892. *12mo.*

OROSI, GIUSEPPE.

Dottrine (Le) chimiche intorno agli equivalenti e agli atomi. Introduzione allo studio della chimica odierna. Firenze, 18—. *16mo.*

OROSI, GIUSEPPE. [Cont'd.]
  Manuale di chimica analitica inorganica qualitativa e quantitativa.
    Firenze, 1871. 8vo.

ORSCHIEDT, H. R.
  Lehrbuch der anorganischen Chemie und Mineralogie an der Hand des
    Experiments. Nichtmetalle. Strassburg, 1879.
        Neue Ausgabe. Strassburg, 1884. 8vo.

ORSONI, FRANCESCO.
  Note scientifiche. Sommario: 1. Sopra delle polveri esplosive. 2.
    Nomenclatura chimico-quantitativa di alcune sostanze minerarie.
    3. Nomenclatura chimico-quantitativa di alcune sostanze vegetali.
    4. Alcuni esempî di nomenclatura quantitativa vegetale. 5. Con-
    clusione. Noto, 1873. pp. 23, 16mo.

OSANN, EMIL.
  Physikalisch-medicinische Darstellung der bekannten Heilquellen der
    vorzüglichsten Länder Europa's. Berlin, 1829–32. 2 vols., 8vo.
        Zweite Auflage, 1839–43. 3 vols.

OSANN, GOTTFRIED WILHELM.
  Beiträge zur Chemie und Physik. Jena, 1822–24. 2 parts, 8vo.
        Neue Beiträge zur Chemie und Physik. Würzburg, 1843.
          8vo.
  De natura affinitatis chemicæ. Jenæ, 1821. 4to.
  Handbuch der theoretischen Chemie. Jena, 1827.
  Messkunst der chemischen Elemente. Zweite Auflage. Jena, 1830.
    8vo.

OSBORN, HENRY STAFFORD.
  The Metallurgy of Iron and Steel, theoretical and practical, with special
    reference to American materials and processes. Philadelphia, 1869.
    8vo. Ill.

OSBURG, JOHANN JACOB.
  Chemische Untersuchungen der Alacher Mineralwassers. Erfurt, 1786.
    4to.
  Chemische Versuche über die Frage: Ob mineralisches Alkali und
    Laugensalz als Arten oder Varietäten verschieden seyen? Erfurt,
    1786. 4to.

O'SHAUGHNESSY, W. B.
  A Manual of Chemistry, arranged for native general and medical
    students and the subordinate medical department of the service.
    Second edition. Calcutta, 1842. 8vo.

OSMOSE (L') ET SES APPLICATIONS INDUSTRIELLES. Paris, 1873.

OST, H.

Lehrbuch der technischen Chemie. Berlin, 1889. Ill.

OSTERMAYER, EUGEN.

Die organischen Farbstoffe der Steinkohlen-Theer-Industrie. Wissen-schaftlich-technisch kurz zusammengestellt und bearbeitet. Lörrach, 1879. 8vo.

OSTHUES.

Ueber die Fabrikation und Verwendung des Wassergases zu Heizungs-und Beleuchtungszwecken. Vortrag gehalten im Casino-Saale zu Dortmund am 28. März, 1885. Dortmund, 1885.

Остряковъ, П.

Вспомогательная книжка для занимающихся въ химич. лабораторіяхъ. С.-Петербургъ 1867.

OSTRYAKOF, P. Aids for laboratory students. St. Petersburg, 1867.

OSTWALD, WILHELM.

Energie (Die) und ihre Wandlungen. Antrittsvorlesung gehalten am 23 November, 1887, in der Aula der Universität Leipzig. Leipzig, 1888. pp. 25, 8vo.

Grundriss der allgemeinen Chemie. Leipzig, 1889. pp. ix–402, 8vo. Ill.

Lehrbuch der allgemeinen Chemie. Leipzig, 1885–86. 2 vols., roy. 8vo. Vol. I, Stöchiometrie, pp. xv–855 ; vol. II, Verwandtschafts-lehre, pp. 909.

Zweite gänzlich umgearbeitete Auflage. Leipzig, 1891–92. 2 vols.

Outlines of General Chemistry. Translated, with the author's sanction, by James Walker. London, 1890. pp. xii–396, 8vo.

Of this a Russian translation by Jegorof and Masing, and edited by J. Kabloukof, was published at Moscow in 1891.

Solutions. (Being the fourth book, with some additions, of the second edition of Ostwald's " Lehrbuch der allgemeinen Chemie.") Trans-lated by M. M. Pattison Muir. London, 1891. pp. xiii–316, 8vo.

Verwandtschaftslehre. Leipzig, 1887. 8vo.

This forms also vol. II of Ostwald's Lehrbuch der Chemie, q. v.

Volumchemische Studien über Affinität. Dorpat, 1877.

OSTWALD'S KLASSIKER DER EXAKTEN WISSENSCHAFTEN. Leipzig, 1889.

No. 3. Die Grundlagen der Atomtheorie. Abhandlungen von J. Dalton und W. H. Wollaston. (1803–1808.)

No. 4. Untersuchungen über das Jod von Gay Lussac. (1814).

OTA YUNEI.

Shinshiki Kagaku. Tokio, 1879. 10 vols.

A new system of chemistry, compiled from Barker, Fownes, Cooke, Roscoe, Chandler and others. In Japanese text.

OTT, ADOLF.

Das Petroleum, seine Entdeckung, Ausbeutung und Verwerthung in den Vereinigten Staaten nebst Mittheilungen über die Prüfung auf seine Feuergefährlichkeit. Zürich, 1874. 8vo. Ill.

OTTO, ANDREAS.

See Müller, Johannes.

OTTO, FRIEDRICH JULIUS.

Bierbrauerei (Die) [etc.], *See* Bolley's Handbuch der chemischen Technologie.

Essig- (Die), Zucker und Stärke-Fabrikation [etc.]. *See* Bolley's Handbuch der chemischen Technologie.

OTTO, FRIEDRICH JULIUS.

Anleitung zur Ausmittelung der Gifte. Ein Leitfaden bei gerichtlich-chemischen Untersuchungen, enthaltend die Ausmittelung des Arsens, Kupfers, Bleis, Quecksilbers, Antimons, Zinns, Zinks, der Blausäure, des Phosphors, des Alkohols und Chloroforms, der Alkaloide sowie die Erkennung der Blutflecken. Für Chemiker, Apotheker, Medicinalbeamte und Juristen. Braunschweig, 1856. 8vo.

Zweite durch einen Nachtrag vermehrte Auflage. Braunschweig, 1857.

Anleitung zur Ausmittelung der Gifte und zur Erkennung der Blutflecken bei gerichtlich-chemischen Untersuchungen. Vierte Auflage. Braunschweig, 1870.

Fünfte Auflage bearbeitet von Robert Otto. Braunschweig, 1875. 8vo. Ill.

Sechste Auflage. Braunschweig, 1884. 8vo. Ill.

Handleiding ter opsporing der vergiften. Een leiddraad bij geregtelijk scheikundige onderzoekingen. Uit het Hoogduitsch vertaald door J. P. C. van Tricht. Rotterdam, 1856. 8vo.

A Manual of the Detection of Poisons by Medico-Chemical Analysis. Translated from the German, with notes and additions, by William Elderhorst. London and New York, 1857. 12mo.

Instruction sur la recherche des poisons et détermination des taches de sang dans les expertises chimico-légales, à l'usage des pharmaciens, des médecins et des avocats.

Otto, Friedrich Julius. [Cont'd.]

> Traduit sur la troisième édition allemande, par G. E. Strohl. Paris, 1869. 8vo.

Lehrbuch der Chemie, zum Theil auf Grundlage von Thomas Graham's Elements of Chemistry. Braunschweig, 1840. 8vo. Ill.

> Zweite Auflage. Braunschweig, 1844–46. 3 vols., 8vo. Ill.

> Scheikunde der bewerktuigde ligchamen. Naar de Hoogduitsche omwerking van F. J. Otto in het Nederduitsch overgebragt door C. de Bordes. Amsterdam, 1845. 8vo. Ill.

> For later editions, *see* Graham, Thomas, and F. J. Otto.

Lehrbuch der Essigfabrikation, enthaltend die Anleitung zur rationellen Bereitung aller Arten von Essig, sowohl nach der älteren langsamen Methode, als auch nach der neueren schnellen Methode ; zur Darstellung der Kräuteressige ; zur Prüfung des Essigs auf seinen Säuregehalt und zur Anlage von Essigfabriken. Für Essigfabrikanten, Kaufleute, Landwirthe, Cameralisten und Techniker. Zweite umgearbeitete Auflage. Braunschweig, 1857. pp. xii–279, 12mo. Ill.

Химическая технологія по Боллею. Производство уксуса, свекловичнаго сахара, крахмала, камеди, крахм. сахара и крахмальной патоки. Производство сыра и масла ; перев. Алексѣева. С.-Петербургъ 1869. 2 тома.

> Chemical Technology according to Bolley. Preparation of vinegar, beet-root sugar, starch, resin, starch-sugar and starch-treacle. Preparation of cheese and butter. St. Petersburg, 1869. *Cf.* Bolley, P. A.

Lehrbuch der rationellen Praxis der landwirthschaftlichen Gewerbe. Die Bierbrauerei und Branntweinbrennerei, die Spirit-, Hefe-, Liqueur-, Essig-, Stärke-, Stärkezucker- und Runkelrübenzuckerfabrikation, die Cider- oder Obstmostbereitung, die Kalk-, Gyps- und Ziegelbrennnerei, Potaschesiederei, Oelraffinerie, Butter- und Käsebereitung, das Brotbacken und Seifesieden umfassend. Zum Gebrauch bei Vorträgen über die landwirthschaftlichen Gewerbe und zum Selbstunterrichte für Chemiker, Landwirthe, Fabrikanten, Architekten, Ingenieure und Steuerbeamte. Sechste revidirte Aufflage. Braunschweig, 1865–68. 2 vols., 8vo. Ill.

> Lehrbuch der rationellen Praxis der landwirthschaftlichen Gewerbe [Encyklopädie]. (Zugleich als 7. Auflage von Fr. Jul. Otto's Lehrbuch der landwirthschaftl. Gewerbe.) Herausgegeben in Gemeinschaft mit E. Birnbaum, Bronner, Dahlen, Deite, Fleischmann, Lintner, Richard, Rühne, Stammer, v. Wagner, P. Wagner u. A. und redigirt von K. Birnbaum. Braunschweig, 1875–1887. 14 vols., 8vo., and Register. Ill.

OTTO, FRIEDRICH JULIUS. [Cont'd.]

> Also called Otto-Birnbaum's Lehrbuch . . .
>
> Vol. I. Die Bierbrauerei. Von C. Lintner.
>
> Vol. II. Die Branntweinbrennerei und deren Neben-zweige. Von K. Stammer.
>
> Vol. III. Die Zuckerfabrikation. Von K. Stammer. (Zweite Auflage, 1887.)
>
> Vol. IV. Das Molkereiwesen. Von Wilhelm Fleisch-mann.
>
> Vol. V. Die Stärke-, Dextrin- und Traubenzuckerfabri-kation. Von Ladislaus von Wágner. Zweite vermehrte Ausgabe.
>
> Vol. VI. Die Weinbereitung. Von Heinr. Wilh. Dahlen.
>
> Vol. VII. Die Essigfabrikation. Von Paul Bronner.
>
> Vol. VIII. Das Brotbacken. Von K. Birnbaum.
>
> Vol. IX. Die Gewinnung der Gespinnstfasern. Von H. Richard.
>
> Vol. X. Die Industrie der Fette. Von C. Deite.
>
> Vol. XI. Die Torf-Industrie und die Moor-Cultur. Von E. Birnbaum und K. Birnbaum.
>
> Vol. XII. Die Düngerfabrikation. Von P. Wagner.
>
> Vol. XIII. Die Kalk-, Cement-, Gyps und Ziegelfabrika-tion. Von J. F. Rühne.
>
> Vol. XIV. Kurzes erläuterndes Wörterbuch. Von K. Birnbaum.
>
> General-Register. Zusammengestellt von K. Birnbaum.

OTTO, P. C.

Einleitung in die wissenschaftliche Chemie im Geiste von Kant's und Berthollet's Lehren und mit kritisch-philosophischer Berücksichti-gung der damit im Widersprüche stehenden Hypothesen. Mit einer Vorrede begleitet von C. W. Snell. Wiesbaden, 1816. 8vo.

> *Also under the title :*
>
> Beiträge zur chemischen Statik oder Versuch eines kritisch-philosophischen Commentars über Berthollet's und An-derer neue chemische Theorien.

OTTO, ROBERT. *See* Graham-Otto.

OUDEMANS, A. C., JR.

> Histor-kritisch overzigt . . . *See, in Section III,* Mulder, L., *also* Mulder, E.

OVERDUIN, J.

Leer der vergiften en tegengiften (toxicologia), nuttig voor alle huisge-zinnen. Breda, 1848. 12mo.

OVERMAN, FREDERICK.

Practical Mineralogy, Assaying and Mining, with a description of the useful minerals, and instructions for assaying and mining according to the simplest methods. Philadelphia, 1851. 12mo.

Metallurgy, or Chemistry of Metals. New York, 1852. 8vo. Ill.

Sixth edition. New York, 1865. 8vo.

OZONE (L'), ce qu'il est, ses propriétés physiques et chimiques, son existence et son rôle dans la nature ; analyse des recherches et des travaux dont il à été l'objet. Saint Denis, 1882. 18mo. Ill.

PABST.

Les laboratoires de chimie [and other articles]. *See, in Section II*, Fremy : Encyclopédie chimique.

PAETS VAN TROOSTWIJK, A.

Nader onderzoek over de verandering van water tot stikstof. Amsterdam, 1798. 4to.

Schets der nieuwe ontdekkingen omtrent het water. *n. p., n. d.* [Amsterdam, 1790]. 8vo.

PAGNOUL, A.

Premiers éléments de chimie, comprenant l'étude des corps non métallique et quelques notions sur les métaux, les sels et les matières organiques. Arras, 1862. 8vo.

PALAGI, F.

Elementi di chimica inorganica secondo le moderne teorie, ad uso delle scuole liceali e superiori, corredati di note etimologiche e di problemi. Milano, 1877. 16mo. Ill.

Dodici lezioni di chimica inorganica ed organica. Torino, 1887. 8vo.

Nozioni elementari di chimica inorganica ed organica. Seconda edizione. Torino, 1890. 8vo. Ill.

PALISSY, BERNARD.

Discours admirables de la nature des eaux et fontaines, tants naturelles qu' artificielles ; des métaux, des sels et salines, des pierres, des terres, du feu et des émaux ; avec plusieurs autres excellents secrets des choses naturelles. Plus, un traité de la marne fort utile et nécessaire pour ceux qui se meslent de l'agriculture. Le tout dressé par dialogues, esquels sont introduits la théorique et la pratique. Par Bernard Palissy, inventeur des rustiques figulines du Roy et de la Royne, sa mère. Paris, chez Martin le jeune, à l'enseigne du serpent, 1580. 8vo.

PALISSY, BERNARD. [Cont'd.]

* Moyen (Le) de devenir riche et la manière véritable par laquelle tous les hommes de la France pourront apprendre à multiplier et augmenter leurs thresors et possessions. Avec plusieurs autres exellens secrets des choses naturelles, desquels iusques à présent l'on n'a ouy parler. Paris, chez Robert Foüet, rue St. Jaques, à l'Occasion devant les Mathurins, 1636. pp. [xvi]–255–526. 16mo.

Oeuvres, revues . . . par Faujas de St. Fond et Gobet. Paris, 1777.

Recepte véritable par laquelle tous les hommes de la France pourront apprendre à multiplier leurs trésors . . . La Rochelle, 1564. Sm. 4to.

PALM, R.

Grundriss der qualitativen und quantitativen chemischen Analyse nebst einer General-Tabelle der wichtigsten Pflanzen-Alkaloïde und einer Spectraltafel. Leipzig, 1885. 8vo.

Wichtigsten (Die) und gebräuchlichsten menschlichen Nahrungs- und Genussmittel, ihre chemische Zusammensetzung, Verfälschungen, etc. St. Petersburg, 1882. Ill.

PALMERI, P., e E. CASORIA.

Vini adulterati. Analisi e determinazioni quantitative, delle materie coloranti estranee e specialmente dell' oricello. Due memorie. *n.p.* [Portici], 1886. 8vo.

PANCIROLI, GUIDO.

Rerum memorabilium libri duo. *See in Section III.*

ПАНТЮХОВЪ, И.

Химическія свѣдѣнія : неорганическія соединенія, органическія соединенія, почва. Кіевъ 1873.

PANTYOUCHOV, I. A text-book of inorganic and organic chemistry. Kiev, 1873.

PAPACINO D'ANTONI, ALESSANDRO VITTORIO.

See Antoni, Alessandro Vittorio Papacino d'.

PAPIN, DENIS.

A new digestor or engine for softening bones, containing the description of its make and use in cookery, voyages at sea, confectionary, making of drincks, chymistry and dyeing. London, 1681. 4to.

New edition. London, 1687. 4to.

La manière d'amolir les os, et de faire cuire toutes sortes de viandes en fort peu de temps, et à peu de frais. Avec une description de la machine dont il se faut servir pour cet effet, ses propriétez et ses usages confirmez par plusieurs expériences nouvellement inventé. Paris, 1682. 8vo. *Also*, Amsterdam, 1688.

PARACELSUS [AUREOLUS PHILIPPUS THEOPHRASTUS PARACELSUS BOMBAST VON HOHENHEIM].

* Opera omnia medico-chemico-chirurgica, tribus voluminibus comprehensa. Editio novissima et emendatissima, ad germanica et latina exemplaria accuratissimè collata : Variis tractatibus et opusculis summa hinc inde diligentia conquisitis, ut in voluminis primi præfatione indicatur, locupletata, indicibusque, exactissimis instructa. Geneva, sumptibus Joan. Antonij et Samuelis De Tournes, 1658. 3 vols., pp. 34-828-39, 22-718-32, 12-212-37, folio. Portrait.

> The editions of Paracelsus are numbered by scores, and would of themselves form a library. They have been catalogued by Dr. Friedrich Mook (Würzburg, 1870), who is, however, justly criticised by Prof. John Ferguson, of Glasgow (1877 and 1885). The collection of Paracelsus's works formed by Dr. Constantine Hering of Philadelphia is preserved in the library of the Homœopathic Medical College in that city. For other bibliographical data, *see*, *in Section IV*, Paracelsus.

PARA, ABBÉ.

Théorie des nouvelles découvertes en genre de physique et de chymie, pour servir de supplément à la théorie des êtres sensibles ; ou au cours complet et concours élémentaire de physique de l'Abbé Para par l'auteur de ces deux ouvrages. Paris, 1786. 8vo. Ill.

PARIS, JOHN AYRTON.

The Elements of Medical Chemistry, embracing only those branches of chemical science which are calculated to illustrate or explain the different objects of medicine, and to furnish a chemical grammar to the author's Pharmacologia. New York, 1825. 8vo.

> *Also*, London, 1825. 8vo.

PARKES, SAMUEL.

Chemical Catechism, with Experiments ; Vocabulary of Chemical Terms. London, 1806. 8vo. Ill.

> Ninth edition. London, 1819. Also New York, 1821.
> Another edition. London, 1830.
> New edition by W. Barker. London, 1854.
> *Cf.* Marcet, Jane.

>> Chemischer Katechismus. Mit Noten, Erläuterungen und Anleitungen zu Versuchen. Aus den englischen dritten, nach der zehnten und elften englischen Ausgabe berichtete Auflage revidirt und zum Theil umgearbeitet von J. B. Trommsdorff. Weimar, 1826. 8vo. Ill.

Chemical Essays, principally relating to the arts and manufactures of the British dominions. London, 1815. 5 vols., 16mo. Ill. Portrait of Francis Bacon in vol. 1. Second edition. London, 1823.

PARKES, SAMUEL. [Cont'd.]

> These vols. contain many historical notes, especially vol. v, which also has a full index to the five vols.

Essais chimiques sur les arts et les manufactures de la Grande Bretagne, traduit de l'anglais par Delaunay. Paris, 1820. 3 vols., 8vo. Ill.

Elementary (An) Treatise on Chemistry upon the basis of the Chemical Catechism, with illustrations, notes and experiments. A new edition. London, 1852. pp. xii–411, 12mo.

Rudiments (The) of Chemistry illustrated by experiments, and copperplate engravings of chemical apparatus. Third edition. London, 1822. pp. xvi–376, 18mo. Plates.

> The Rudiments of Chemistry . . . with Notes by James Renwick. The first American edition. New York, 1824. 8vo.

> Grundsätze der Chemie, durch Versuche und fünf Kupfern erläutert. Nebst einem Anhang die neuesten chemischen Entdeckungen enthaltend. Leipzig, 1822. 8vo.

PARKINSON, JAMES.

The Chemical Pocket-book ; or, Memoranda Chymica. To which is now added an Appendix by J. Woodhouse. From the second London edition. Philadelphia, 1802. 8vo. Plates.

> Fourth edition, with the latest discoveries. London, 1807, pp. xii–368–[ii].

PARMENTIER.

> Molybdène, Vanadium et Titane. *See, in Section II*, Fremy : Encyclopédie chimique, vol. III.

PARMENTIER, ANTOINE AUGUSTIN.

Dissertation sur la nature des eaux de la Seine . . . Paris, 1787.

Examen chimique de la pomme de terre. Paris, 1773. 12mo.

Expériences et reflexions relatives à l'analyse du blé et des farines. Paris, 1778. 8vo.

Parfait (Le) boulanger, ou traité complet sur la fabrication et le commerce du pain. Paris, 1778. 8vo.

Recherches sur les végétaux nourrissants . . . Paris, 1781. 8vo.

PARMENTIER, ANTOINE AUGUSTIN, ET N. DEYEUX.

Précis d'expériences et observations sur les différentes espèces de lait. Paris, 1790. 8vo.

> Deuxième édition. Paris et Strasbourg, 1799.

> Neueste Untersuchungen und Bemerkungen über die ver-

PARMENTIER, ANTOINE AUGUSTIN, ET N. DEYEUX. [Cont'd.]

schiedenen Arten der Milch in Beziehung auf die Chemie, die Arzneikunde, und die Landwirthschaft. Aus dem französischen übersetzt. Herausgegeben von A. N. Scherer. Jena, 1800. 8vo.

PARMENTIER [Editor].

See La Garaye, De.

PARMENTIER, F.

Guide élémentaire pour les premières recherches d'analyse qualitative des matières minérales. Paris, 1887. 8vo.

PARNELL, E. A.

Elements of Chemical Analysis, Inorganic and Organic. London, 1842.

Practical (A) Treatise on Dyeing and Calico-Printing. Also a description of the origin, manufacture . . . of the substances employed in those arts. With Appendix containing definitions of chemical terms, tables of weights, measures, thermometers, hydrometers, etc. New York, 1846. 8vo. 10 plates.

Published anonymously.

PARONE, S.

Corso di nozioni fisico-chimiche e di materie prime. 2 vols. Torino, 1886. 8vo.

PARROT, FRIEDRICH.

Ueber Gasometrie nebst einigen Versuchen über die Verschiebbarkeit der Gase. Preisschrift. Dorpat, *n. d.* [1812]. 8vo.

PARRY, JOSEPH.

Water : its composition, collection and distribution. A practical Hand-book for domestic and general use. London, 1881. 16mo. Ill.

PASCAL, BLAISE.

* Traitez de l'équilibre des liqueurs et de la pesanteur de la masse de l'air. Contenant l'explication des causes de divers effets de la nature qui n'avaient point esté bien connus jusques-icy, et particulièrement de ceux que l'on avoit attribuez à l'horreur du vuide. Paris, chez Guillaume Desprez, ruë S. Jaques, à S. Prosper et aux trois vertus vis-à-vis la porte du cloistre des Mathurins, 1698. pp. [xxvi]–238, 16mo. Plates.

PASQUALINI, A.

Generalità (Le) della chimica. Rimini, 1884. 16mo.

Guida alla analisi chimica qualitativa dei corpi inorganici. Forli, 1880. 8vo.

PASTEUR, LOUIS.
Leçon (etc.). *See* Société chimique de Paris. *Also* Laugel, A.

PASTEUR, LOUIS.
Études sur la bière, ses maladies, causes qui les provoquent, procédé pour la rendre inaltérable, avec une théorie nouvelle de la fermentation. Paris, 1876. pp. viii–387, roy. 8vo. 12 plates.

Études sur le vinaigre, sa fabrication, ses maladies, moyen de les prévenir. Nouvelles observations sur la conservation des vins par la chaleur. Paris, 1868. 8vo.

Études sur le vin, ses maladies, causes qui les provoquent et procédés nouveaux pour le conserver et pour le veillir. Deuxième édition, revue et augmentée. Paris, 1873. 8vo. Ill.

Première édition. Paris, 1866.

Examen critique d'un écrit posthume de Claude Bernard sur la fermentation. Paris, 1879. pp. xxiv–153–[ii], 8vo.

Mémoire sur la fermentation alcoolique. Paris, 1857–60. 4to.

Die Alkohol-Gährung. Uebersetzt von Victor Griessmayer. Augsburg, 1871.

Zweite Titel-Ausgabe. Augsburg, 1878.

Studies on Fermentation, the diseases of beer, their causes and the means of preventing them, translated by Frank Faulkner and D. Constable Robb. London, 1879. 8vo. Ill.

Nouveau procédé industriel de fabrication du vinaigre. Paris, 1862. 4to.

Der Essig, seine Fabrikation und Krankheiten, sowie Mittel den letzteren vorzubeugen. Neue Beobachtungen über Conservirung der Weine durch die Wärme. Autorisirte deutsche Ausgabe, übersetzt und mit Anmerkungen versehen von Eugen Borgmann. Braunschweig, 1878. 8vo. Ill.

Ueber die Asymetrie bei natürlich vorkommenden organischen Verbindungen (1860). Herausgegeben von M. und A. Ladenburg. Leipzig, 1891. 8vo.

Ostwald's Klassiker der exakten Wissenschaften.

PATCHETT, J.
Elementary Chemical Calculations. Leeds, 1888. 8vo.
Qualitative Chemical Analyses (Inorganic and Organic). Part I. Elementary stage for schools and science classes. Leeds, 1886. 12mo.

PATISSIER, PHILIBERT, ET BOUTRON-CHALARD.
Manuel des eaux minérales naturelles . . . Paris, 1819. 8vo.
Deuxième édition. Paris, 1837.

PATTISON MUIR, M. M.
    See Muir, M. M. Pattison.

PATTONE, CARLO DEODATO.
Trattato elementare di chimica generale. Alessandria, 1865–68. 2 vols.,
    8vo.

PATZIER, M. J.
Anleitung zur metallurgischen Chemie. Ofen, 1805. 4 vols., 8vo.

PAUGUY, C.
Nouvelle méthode naturelle chimique, ou disposition des corps simples
    et composés propre à rendre l'étude de cette science plus facile et
    plus courte. Paris, 1828. 8vo.

PAULA, FRANCISCO DE, MONTELLS Y NADAL.
    See Montells y Nadal, Francisco de Paula.

PAULET, MAXIME.
Chimie agricole. Théorie et pratique des engrais, précédées d'anatomie
    et de physiologie végétales. Paris, 1846. 8vo.

PAULI, JOHANN WILHELM.
Dissertationum chymico-physicarum prima de corporum dissolutione
    . . . Lipsiae, *n. d.* [1679]. pp. [18], 4to.

PAULY, C.
Einführung in die quantitative chemische Analyse. Zum praktischen
    Gebrauche in chemischen, pharmaceutischen und technischen La-
    boratorien. Braunschweig, 1880. 8vo.

PAUPAILLE, J. J.
Résumé complet de la chimie inorganique, contenant l'exposé des
    principes généraux de la science, et l'étude des corps inorganisés
    simples et composés, précédée d'une introduction historique. Paris,
    1825. pp. viii–v–276, 24mo. Ill. Folding plate.
        Encyclopédie portative, sous la direction de C. Bailly de Merlieux. The
        historical introduction occupies pages 1 to 32.

Résumé complet de la chimie organique, contenant la chimie végétale et
    animale, un précis d'analyse chimique et un aperçu sur les princi-
    paux poisons ; terminé par la biographie des plus illustres Chimistes
    et par une bibliographie et un vocabulaire chimiques. Orné de
    planches. Paris, 1825. pp. viii–314, 24mo.
        Encyclopédie portative, sous la direction de C. Bailly de Merlieux. The
        biographies are limited to 52 names. Under *Bayen* the author endeav-
        ors to wrest from PRIESTLEY the honor of discovering oxygen. The
        bibliography embraces 60 titles, with a brief analysis of each work ; the
        books are almost exclusively in the French language.

PAUPAILLE, J. J. [Cont'd.]

 Chemie. Erster Theil : Die unorganische Chemie. Eine Darstellung der allgemeinen Grundsätze der Chemie und Beschreibung der einfachen und zusammengesetzten unorganischen Körper nebst einer historischen Einleitung. Zweiter Theil : Organische Chemie. Eine Darstellung der chemischen Untersuchungen der Pflanzen und Thiere, und der vorzüglichsten Gifte, Biographien ausgezeichneter Chemiker, chemische Bibliographie und Wörterbuch. Nach den französischen des J. J. Paupaille von C. G. Ch. Hartlaub. Leipzig, 1828. 2 vols., 16mo. Ill.

PAVESI, A.

Guida allo studio dell'analisi chimica qualitativa : manuale pratico con tavole ad uso degli studenti degli istituti tecnici e delle scuole universitarie. Terza edizione. Milano, 1884. 16mo.

PAVY, F. W.

Food and Dietetics. Second edition. London, 1875.

PAWLEWSKI, BRONISLAW.

Sposoby oceniania wartości nafty. Warszawa, 1885.

PAYEN, ANSELME.

 Chimie (La) enseignée en vingt-deux leçons. Paris, 1825. 8vo.

  Chimie (La) enseignée en vingt-six leçons. Paris, 1827. 12mo.

  Die Chemie in 26 Vorlesungen, enthaltend die Entwickelung der Theorie sowie Versuche und praktische Anwendung auf Künste und Handwerke. Stuttgart, 1829. 12mo.

  La chimica insegnata in ventisei lezioni. Opera tradotta sulla nona edizione parigina da A. L. P. Milano, 1846. 2 vols.

  Química (La) enseñada en 26 lecciones. Contiene la manifestacion de las teorías de esta ciencia puestas al conocimiento de todos, y en cada leccion experiencias físicas con su aplicacion á las artes. Obra traducida al castellano de la nona edicion inglesa y de la cuarta francesa por D. Barcelona, 1830. 4to.

Cours de chimie inorganique appliquée. Analyse des leçons données et description des planches par Knab et Schmersahl. (Premières leçons.) Paris, 1843. 8vo.

PAYEN, ANSELME. [Cont'd.]

Cours de chimie organique appliquée. Description des appareils de chimie appliquée, légende des lithographies du cours ; par Knab et Leblanc. Paris, 1842. 8vo.

Cours de chimie élémentaire et industrielle, destiné aux gens du monde. Paris, 1830. 8vo.

> [Another edition.] Bruxelles, 1840. 2 vols., 18mo.

> Curso de química elemental é industrial, dedicado á toda clase de personas e splicado, traducido y augmentado con las últimas lecciones dadas por el mismo por Antonio Gonzalez Bustamante y Mina. Madrid, 1842. 8vo.

Des engrais. Théorie actuelle de leur action sur les plantes principaux, moyens d'en obtenir le plus d'effet utile. Paris, 1839. 12mo.

Des substances alimentaires, et des moyens de les améliorer, de les conserver, et d'en reconnaître les altérations. Paris, 1853. 16mo.

> Troisième édition. Précis historique et pratique. Paris, 1856.

> Quatrième édition. Paris, 1865. 8vo.

Manuel du cours de chimie organique appliquée aux arts industriels et agricoles. Par Jules Rossignon et Jules Garnier. Paris, 1841–'42. 2 vols., 8vo.

Manuel du cours de chimie organique appliquée aux arts industriels et agricoles. Paris, 1842–43. 2 vols., 8vo, and atlas folio.

Mémoire sur l'amidon, la dextrine et la diastase considérés sous les points de vue anatomique, chimique, et physiologique. Paris, 18—. 4to. 8 plates.

> Vollständiges Handbuch der Stärke-Fabrikation aus Kartoffeln und Waizen. Nach dem französischen bearbeitet und mit Zusätzen vermehrt. Quedlinburg und Leipzig, 1852. 8vo. Ill.

Mémoire sur le houblon . . . Paris, 1823. 8vo.

> Troisième édition. Paris, 1825. 12mo.

Mémoire sur les bitumens . . . Paris, 1824. 8vo.

Précis de chimie industrielle à l'usage des écoles préparatoires, aux professions industrielles, des fabricants et des agriculteurs. Paris, 1849. 8vo.

> Deuxième édition, augmentée. Paris, 1850.

> Troisième édition, où l'on a introduit les derniers perfectionnements apportés aux applications chimiques, et plusieurs chapitres sur les industries nouvelles. Paris, 1855. 8vo.

> Quatrième édition. Paris, 1859.

> Sixième édition. Paris, 1877–78.

PAYEN, ANSELME. [Cont'd.]

Industrial Chemistry, a Manual for Technical Colleges and manufacturers, based upon a translation, partly by T. D. Barry, of Stohmann and Engler's German edition of Payen's Précis de chimie industrielle, edited throughout and supplemented with chapters on the Chemistry of the Metals, by B. H. Paul. London, 1878. 8vo.

Populäres Handbuch der industriellen Chemie. Für Künstler, Fabrikanten und Gewerbtreibende aller Art. Uebersetzt und bearbeitet von J. F. Hartmann und C. G. Meerfeld. Quedlinburg, 1838–'40. 2 vols., 8vo.

Gewerbs-Chemie. Ein Handbuch für Gewerbschulen, sowie zum Selbstunterricht für Gewerbtreibende, Kameralisten, Landwirthe, etc. Nach dem französischen Originale bearbeitet von Hermann Fehling. Zweite vielfach vermehrte Auflage. Stuttgart, 1852. 8vo.

Handbuch der technischen Chemie. Frei bearbeitet von F. Stohmann und C. Engler. Stuttgart, 1872–74. 2 vols., 8vo. Ill.

Traité complet de la distillation des principales substances qui peuvent fournir de l'alcool : vins, grains, betteraves, fécule, tiges, fruits, racines, tubercules, bulbes, etc. Paris, 1857. 8vo. 14 plates.

Deuxième édition, revue et augmentée. Paris, 1861. 16 plates. Ill.

Vollständiges Handbuch der Branntweinbrennerei aus den Hauptsubstanzen welche Weingeist liefern können, und zwar Weine, Getreidearten, Runkelrüben, . . . Nebst Anweisung zur Rectification des Branntweines. Ins Deutsche übertragen von E. O. Fromberg. Quedlinburg, 1858. 8vo. 13 plates.

Traité de la fabrication des diverses sortes de bière . . . Paris, 1829. 12mo.

Traité de la fabrication et du raffinage des sucres de canne de betterave . . . Paris, 1832. 8vo.

Traité de la pomme de terre. . . . Paris, 1826. 8vo.

PAYEN, ANSELME.

Précis théorique et pratique des substances alimentaires. *See* Des substances alimentaires.

PAYEN, ANSELME, ET A. CHEVALLIER.

Traité élémentaire des réactifs, leurs préparations, leurs emplois spéciaux et leur application à l'analyse. Paris, 1822. 8vo.

Deuxième édition. Paris, 1829–30. 2 vols. 4 plates.

PAYEN, ANSELME, ET A. CHEVALLIER.   [Cont'd.]

Troisième édition.   Paris, 1841.   3 vols., 8vo.

Unterricht über die Reagentien, ihre Bereitung, ihre Anwendung, etc   Aus dem französischen von L. Cerutti. Leipzig, 1823.   8vo.

PAYEN, ANSELME, UND CARTIER.

Fabrikation (Die) der Schwefelsäure nach den neuesten französischen und englischen Methoden und Verbesserungen ; nebst Beschreibung und Abbildung der dazu erforderlichen Apparate.   Nach dem französischen bearbeitet und mit Zusätzen vermehrt.   Quedlinburg, 1832.   8vo.   Ill.

PAYEN, ANSELME, ET GAUTIER.

Résumé du cours pratique de fabrication du sucre indigène.   Paris, 1838.   8vo.

PAYEN, ANSELME, ET RICHARD.

Précis d'agriculture théorique et pratique à l'usage des écoles d'agriculture, des propriétaires et des fermiers.   Paris, 1851.   2 vols., 8vo.

PEACOCK, JAMES.

A Short Account of a New Method of Filtration by Ascent.   With explanatory sketches upon six plates.   London, 1793.   pp. 21, 4to.

The author was an architect, and describes a gravel-filter for purifying water.

PEALE, ALBERT C.

Lists and Analyses of the Mineral Springs of the United States.   (A preliminary study.)   Bulletin No. 32 of the U. S. Geological Survey, Department of the Interior.   Washington, 1886.   pp. 235, 8vo.

PEART, EDWARD.

Anti-phlogistic (The) Doctrine of Lavoisier critically examined and demonstratively confuted.   London, 1795.

On the Properties of Matter, the Principles of Chemistry, and the nature and construction of aëriform fluids.   London, 1792.

PECKHAM, S. F.

Elementary Chemistry, a Text-Book for beginners designed as an introduction to Barker's Chemistry.   Louisville, 1877.   pp. 254, 8vo. Ill.

PÉCHINEY.

[Produits chimiques.]   *See, in Section II.* Fremy : Encyclopédie chimique, vol. v.

PECKSTON, THOMAS S.

Theory (The) and Practice of Gas-lighting ; in which is exhibited an historical sketch of the rise and progress of the science, and the theories of light, combustion and formation of coal, with descriptions of the most improved apparatus for generating, collecting and distributing coal-gas for illuminating purposes.   London, 1819. pp. xx–438, 8vo.   14 plates.

> Practical (A) Treasure on Gas-Lighting, in which the Gas-Apparatus generally in use is explained and illustrated. Third edition. London, 1841. pp. vii-472, 8vo. Plates.

PÉCLET, JEAN CLAUDE EUGÈNE.

Cours de chimie.   Marseille, 1823.   pp. xviii–776, 4to.   9 folding plates.

PEDRONI, P. M.

Nouveau manuel complet des falsifications des drogues simples et composées. Paris, 1848. 24mo. Ill.

PÉLIGOT, EUGÈNE MELCHIOR.

Études chimiques et physiologiques sur les vers à soie.   Paris, 1853. 8vo.

Recherches sur la composition chimique de la canne à sucre de la Martinique. Paris, 1840. 8vo.

Sur la composition des sucres bruts.   Paris, 1851.   8vo.

Traité de chimie analytique appliquée à l'agriculture.   (Méthodes générales d'analyse, la terre arable, les eaux, les engrais, les cendres végétales, les céréales, les fourrages, etc.) Paris, 1883. 8vo. Ill.

> Première édition. Paris, 1843.

PÉLIGOT, EUGÈNE MELCHIOR, ET DECAISNE.

Recherches sur l'analyse et la composition chimique de la betterave à sucre, et sur l'organisation anatomique de cette racine.   Paris, 1839. 8vo.

PELLEREAU, F.

Chimie minérale, ou Traité complet des métaux, des oxydes et des acides, d'après une nouvelle méthode, avec l'indication de tous les réactifs qui servent à faire reconnaître ces substances et des secours ou contre-poisons à administrer en cas d'empoisonnement par ces corps. Boulogne, 1839. 8vo.

PELLET, H.

Chimie sucrière.   Études nouvelles sur les jus et pulpes de diffusion. Paris, 1880. 8vo.

PELLET, H., ET L. BIARD.

Agenda et calendrier de poche du fabricant de sucre.   *See*, Stammer, Carl:
Wegweiser in der Zuckerfabrikation.

PELLET, H., ET SENCIER.

Fabrication (La) du sucre.   Vol. 1 : Historique ; les principes sucrés ;
saccharimétrie chimique et physique ; analyse des sols ; les terres à
betteraves.   Paris, 1883.   8vo.   Ill.

PELLETIER, BERTRAND.

Description de divers procédés pour extraire la soude du sel marin.
Paris, 1794.   4to.

Mémoires et observations de chimie.   Recueillis et mis en ordre par
Charles Pelletier et Sedillot jeune.   Paris, An. VI [1798].   2 vols.,
12mo.   Folding plates.

PELLETIER, JOSEPHE.

Mémoire sur un nouveau alcali, la strychnine.   Paris, 1818.   8vo.

Notice sur les recherches chimiques.   Paris, 1829.

PELLETIER, JOSEPHE, ET CAVENTOU.

Analyse chimique des quinquina.   Suivie d'observations médicales sur
l'emploi de la quinine et de la cinchonine.   Paris, 1821.   pp.
viii–88, 8vo.

PELLETIER, JOSEPHE, ET ADELON, CHEVALLIER [AND OTHERS].

Rapport sur les moyens de constater la présence de l'arsenic dans l'empoisonnement .   .   .   Paris, 1841.   8vo.

PELLEW, CHARLES E.

Lessons in Medical Chemistry.for Laboratory Use.   New York, 1889.
8vo.

Manual of Practical Medical and Physiological Chemistry.   New York,
1892.   pp. xiv–314, 8vo.   Ill.

PELLIZZARI, V.

L'Ellettrolisi.   Milano, 1891.   12mo.

PELOUZE, EDMOND.

Art (L') de fabriquer les couleurs et les vernis· .   .   .   Paris, 1828.
2 parts, 18mo.·

Traité méthodique de la fabrication du coke et du charbon de tourbe.
Paris, 1841.   8vo.

Traité de l'éclairage au gaz tiré de la houille, des bitumes, des lignites,
de la tourbe, des huiles, des résines, des graisses, etc., précédé d'un
examen approfondi de la teneur de ces combustibles en hydrogène

PELOUZE, EDMOND. [Cont'd.]

et en carbone, de leur comparaison sous le point de vue des facultés illuminantes, de considérations sur la préférence à donner à chacun d'eux respectivement, selon les localités, pour la fabrication du gaz, et du tableau statistique de la production en France, avec une statistique particulière des houilles de la Belgique et de l'Angleterre et le tableau des importations en France de ces deux provenances, les circonstances et les frais de transport, etc. Revu, quant aux principes théoriques et à l'analyse de matières par Pelouze fils. Paris, 1841. 2 vols., 8vo, and atlas of 28 plates, 8vo.

Deuxième édition. Paris, 1859. 2 vols., 8vo.

PELOUZE, THÉOPHILES JULES, ET E. FREMY.

Abrégé de chimie. Paris, 1848, 8vo.

Deuxième édition. Paris, 1855. 8vo.

Septième édition, entièrement refondue. Paris, 1876. 8vo.

Cours de chimie générale. Paris, 1847–48. 3 vols. and an atlas of 49 plates, 8vo.

Continued under the title:

Traité de chimie générale, comprenant les applications de cette science à l'analyse chimique, à l'industrie, à l'agriculture et à l'histoire naturelle. Deuxième édition. Paris, 1855–57. 6 vols., 8vo, and an atlas of 53 plates, 4to.

Traité de chimie générale, analytique, industrielle et agricole. Troisième édition. Paris, 1867. 7 vols., 8vo. Ill.

Handboek der scheikunde. Voor Nederland bewerkt door Louis Mulder. Utrecht, 1851. 2 vols., 8vo. Ill.

Handboek der scheikunde. Nederlandsch door J. N. de Rijk. Utrecht, 1856-57. 2 vols., 8vo.

Handbuch der Chemie für Gewerbtreibende, Künstler und Ackerbautreibende. Leipzig, 1854. 8vo. Ill.

Notions générales de chimie. Paris, 1853. 8vo.

Deuxième édition. Paris, 1855.

General Notions of Chemistry. Translated from the French by Edmund C. Evans. Philadelphia, 1854. 12mo. Ill.

PELTZ UND HABICH.

Pratisches Handbuch und Hülfsbuch für Bierbrauer und Mälzer. Zweite Auflage. Braunschweig, 1888.

PEMBERTON, HENRY.

Course (A) of Chemistry. Published by J. Wilson. London, 1771. 8vo.

PENNETIER, GEORGES.

Leçons sur les matières premières organiques.   Origines—Provenance—
Caractères—Composition—Sortes commerciales—Altérations naturelles—Falsifications et moyens de les reconnaîtres—Usages. Paris,
1881. pp. xi-1018, 8vo. Ill.

PENOTUS, BERNARDUS GEORGIUS.

Abditorum chymicorum tractatus varii. Francofurti, 1595. 8vo.
Apologia chemiæ transmutatoriæ. Bernæ, 1608. 8vo.
* Tractatus varii de vera præparatione et usu medicamentorum chymicorum nunc primum editi. Francofurti, apud Joannem Feyrabend,
impensis Petri Fischeri, 1594. pp. 256, 16mo.

PENROSE, F.

Essays Physiological and Practical, founded on the modern Chemistry
of Lavoisier, Fourcroy, etc. London, 1794. 8vo.

PENSAURO, SANTES DE ARDOYNIS DE.   *See* Santes de Ardoynis de Pensauro.

PEÑUELAS Y FORNESA, LINO.

Tratado elemental de química analítica, precedido de algunas ideas
sobre la filosofía química. Lecciones esplicadas en la escuela especial
de minas. Madrid, 1867. pp. xci-1,003, 12mo. Ill.

PENZOLDT, F.

Aeltere und neuere Harnproben und ihr praktischer Werth. Dritte
Auflage. Jena, 1890. 8vo.

PEPE, VINCENZO.

Elementi di chimica sperimentale ragionata, e corredata delle più recenti
scoperte. Per uso del suo studio privato. Seconda edizione. Napoli,
1811. 4 vols., 8vo. Ill.

> *Accompanied by :*
>
>> Tavole di affinità chimiche da servire di Supplemento alla
>> sua opera titolata Elementi di chimica, etc. Napoli,
>> 1811. 8vo. 21 tables.

PEPPER, JOHN HENRY.

Boy's Play-Book of Science, including the various manipulations and
arrangements of chemical and philosophical apparatus required for
the successful performance of scientific experiments in illustration
of the elementary branches of chemistry and natural philosophy.
Second edition. London, 1860. 8vo. Ill.
Chemistry, Electricity, Light. With illustrations.   London [1876].
8vo. Ill.

PEPPER, JOHN HENRY. [Cont'd.]

Playbook (The) of Metals : including personal narratives of visits to coal, lead, copper, and tin mines. With a large number of interesting experiments relating to alchemy and the chemistry of the fifty metallic elements. A new edition. London and New York, 1866. pp. viii–502. 8vo. Ill.

PERCIVAL, THOMAS.

Observations and Experiments on the Poison of Lead. London, 1774. 8vo.

PERCY, JOHN.

Metallurgy : the art of extracting metals from their ores. New edition, greatly enlarged. Vol. I, containing Introduction, refractory materials, and fuel (including slags, crucibles, fire-bricks). London, 1875. 8vo. Ill.

> The original edition including Copper, Zinc, and Brass. London, 1861. 8vo. Ill.

Metallurgy of Gold and Silver. Part I. London, 1880. 8vo. Ill.

Metallurgy of Iron and Steel. London, 1864. 8vo. Ill.

Metallurgy of Lead, including Desilverisation and Cupellation. London, 1870. 8vo. Ill.

> Traité complet de métallurgie. Traduit sous les auspices de l'auteur, avec introduction, notes et appendice par E. Petitgand et A. Ronna. Paris et Liège. 1864–67. 5 vols., roy. 8vo. Ill.
>
> Die Metallurgie. Gewinnung und Verarbeitung der Metalle und ihrer Legirungen, in praktischer und theoretischer, besonders chemischer Beziehung. Uebertragen und bearbeitet von F. Knapp, H. Wedding und C. Rammelsberg. Braunschweig, 1862–88. 4 vols., 8vo.
>
> Vol. I. Die Lehre von den metallurgischen Processen im Allgemeinen und den Schlacken, die Lehre von den Brennstoffen und den feuerfesten Materialien als Einleitung, und die Metallurgie des Kupfers, des Zinks und der Legirung aus beiden. Bearbeitet von F. Knapp. 1862–1863. Ill.
>
> Vol. II. Ausfürliches Handbuch der Eisenhüttenkunde. Gewinnung des Roheisens und Darstellung des Schmiedeeisens und Stahls, in praktischer und theoretischer Beziehung, unter besonderer Berücksichtigung der englischen Verhältnisse. Bearbeitet von H. Wedding. In drei Abtheilungen. 1864–1878. Ill.

PERCY, JOHN. [Cont'd.]

Vol. III. Die Metallurgie des Bleies. Bearbeitet von C. Rammelsberg. 1872. Ill.

Vol. IV. Die Metallurgie des Silbers und Goldes. Uebertragen und bearbeitet von C. Rammelsberg. Erste Abtheilung. 1881. Ill.

I. Ergänzungsband : Der basische Bessemer- oder Thomas-Process von Hermann Wedding. 1884. Ill.

II. Ergänzungsband. Von Hermann Wedding. Die Berechnungen für Entwurf und Betrieb von Eisenhochöfen. Berechnungen behufs Beaufsichtigung und Veränderungen des Betriebes eines Hochofens. 1887–1888. 8vo. Ill.

PEREIRA, JOAO FELIX.

Principios de chymica. Lisboa, 1864. 8vo.

PEREZ MORALES, JOSÉ MARIA, Y BENITO TAMAYO.

Curso de química general arreglado á las esplicaciones de Vincente Santiago de Masarnau . . . Madrid, 1848. 8vo.

PERKIN, W. H.

Course of Chemistry. Edinburgh, 1890. 8vo.

PERL, EDUARD.

Die Beleuchtungsstoffe und deren Fabrikation. Eine Darstellung aller zur Beleuchtung verwendeten Materialien thierischen und pflanzlichen Ursprungs, des Petroleums, des Stearins, der Theeröle und des Paraffins. Enthaltend die Schilderung ihrer Eigenschaften, ihrer Reinigung und praktischen Prüfung in Bezug auf ihre Reinheit und Leuchtkraft, nebst einem Anhange über die Verwerthung der flüssigen Kohlenwasserstoffe zur Lampenbeleuchtung und Gasbeleuchtung im Hause, in Fabriken und öffentlichen Localen. Wien, Pest und Leipzig, 18—. Ill.

PERSOZ, JEAN FRANÇOIS.

Introduction à l'étude de la chimie moléculaire. Paris, Strasbourg, 1839. pp. xv–894, 8vo.

Traité théorique et pratique de l'impression des tissus. Paris, 1846. 4 vols., 8vo, and an atlas, 4to, of 20 plates.

The 8vo volumes have 165 engravings and 429 specimens of dyed fabrics intercalated in the text.

PERUTZ, H.

Die Industrie der Fette und Oele. Enthaltend die Seifen- und Glycerin-

PERUTZ, H. [Cont'd.]

Fabrikation, die Darstellung der Paraffin-, Stearin-, Wachs- und Wallrathskerzen, ferner die Rüböl-Raffinerie, Palmöl-Bleicherei, Toilett-, Seifen- und Parfümerie-Fabrikation. Nach eigenen Erfahrungen und mit Benutzung der neuesten deutschen, französischen und englischen Literatur. Berlin, 1866. pp. xxii–397, 8vo. Ill.

Die Industrie der Mineralöle, des Petroleums, Paraffins und der Harze, nebst sämmtlichen damit zusammenhängenden Industriezweigen. Wien, 1868. Part I, pp. xii–348. Part II, pp. x–206. 8vo. Ill.

PÉTERMANN, A.

Recherches de chimie et de physique appliquées à l'agriculture. Analyse des matières fertilisantes et alimentaires. Deuxième édition, revue et augmentée. Bruxelles, 1886. 8vo.

PÉTERMANN, A., ET GRAFTIAN, J.

Recherches sur la composition de l'atmosphère. 1. Acide carbonique contenue dans l'air atmosphérique. Bruxelles, 1892, 8vo.

PETERSEN, E.

Kortfattet Fremstilling af de vigtigste Fremgangsmaader og Prøver i den kvalitative uorganiske Analyse. Til Brug for Studerende ved den polytekniske Læreanstalt. Kjøbenhavn, 1882. 8vo.

PETERSEN, E. P. F.

Pharmaceutisk Teknik. En Lærebog. Kjøbenhavn, 1884. 8vo.

PETERSEN, THEODOR.

Die chemische Analyse. Ein Leitfaden für die qualitative und quantitative Analyse. Berlin, 1863. Three parts in two vols., 8vo.

Die Typentheorie und die Molekularformeln. Eine Uebersicht für Studirende der Chemie. Berlin, 1862. 8vo.

PETIT, OTHON.

Des emplois chimiques du bois dans les arts et l'industrie. Paris, 1888. 8vo.

PETRI, F.

Leitfaden für den chemischen Unterricht der anorganischen Chemie. Berlin, 1870.

Zweite Auflage. Berlin, 1876.

ПЕТРОВЪ, ВАСИЛІЙ.

Собраніе физико-химическихъ новыхъ опытовъ и наблюденій. С. Петербургъ 1801.

PETROF, VASILY. Collection of Physico-chemical Experiments and Observations. St. Petersburg, 1801.

PETTENKOFER, MAX.
   Die Chemie in ihrem Verhältnisse zur Physiologie und Pathologie.
      München, 1848. 4to.

PETTERSON, O.
   On the Properties of Water and Ice. Stockholm, 1883. Roy. 8vo.

PETTUS, JOHN.
   *Fleta Minor. The Laws of Art and Nature in knowing, judging,
      assaying, fining, refining and inlarging the bodies of confin'd Metals.
      In two parts. The first contains Assays of Lazarus Erckern, Chief
      Prover (or Assay-Master General of the empire of Germany) in v.
      Books : originally written by him in the Teutonick language, and
      now translated into English. The second contains Essays on
      Metallick Words, as a Dictionary to many pleasing Discourses.
      London, printed for and sold by Stephen Bateman at the Sign of
      the Bible over against Furnivals-Inn Gate in Holbourn, 1683–'86.
      pp. [xlii]–345–132 [no pagination before p. 81], folio. Ill.
         The dictionary includes many chemical terms. *Cf.* Ercker, Lazarus.

PETZHOLDT, ALPH.
         Die Erzeugung der Eisen- und Stahlschienen. *See* Bolley's Handbuch [etc.],

PETZHOLDT, GEORG PAUL ALEXANDER.
   Populäre Vorlesungen über Agriculturchemie, in der ökonomischen
      Gesellschaft für das Königreich Sachsen während des Winterhalb-
      jahres 1843–44 gehalten. Mit Sachregister und eingedruckten
      Holzschnitten. Leipzig, 1844. 8vo.
         Agriculturchemie (Die) in populären Vorlesungen. Leipzig,
            1846. 8vo.
         Almenfattelige Forelæsninger over Agerdyrkningschemien.
            Oversat af P. D. Broager. Roeskilde, 1847. 8vo.
         Lectures to Farmers on Agricultural Chemistry. (Translated
            by William Gregory.) London, 1844. 8vo. New edi-
            tion, 1846. 8vo.

PETZOLD, WILHELM.
   Die Grundzüge der chemisch-physikalischen Atomtheorie. Programm.
      Neubrandenburg, 1877. 4to.

PEW, RICHARD.
   Observations on the Art of making Gold and Silver ; or, the probable
      means of replenishing the nearly exhausted mines of Mexico, Peru
      and Potosi, in a letter to a friend. To which are added some obser-
      vations on the structure and formation of metals, and an attempt to

PEW, RICHARD.   [Cont'd.]

prove the existence of the ὀξὺ σελασφόρον, the phlogiston of Stahl, the volatilizing principle, or the principle of inflammability. Salisbury, London, 1796.   pp. 18, 4to.

PEXA, J. L.

Darstellung der erfolgreichsten Art und Weise des Anbaues, der Pflege und Aufbewahrung der Runkelrübe und der Zuckerbereitung aus derselben.   Wien, 1842.

PEYRONE.

Lezioni sulla chimica agraria.   Torino, 1869.

PFAFF, CHR. HEINRICH.

Handbuch der analytischen Chemie für Chemiker, Staatsaerzte, Apotheker, Oekonomen und Bergwerkskundige.   Altona, 1821–22.   2 vols., 8vo.   Tables.

Zweite vermehrte und verbesserte Ausgabe.   Altona, 1824–25. 8vo.

Ueber das chemische Gebläse mit explosiven Gasgemengen, oder den sogenannten Neumann'schen Apparat.   Eine Zusammenstellung der bis jetzt darüber bekannt gewordenen Arbeiten, nebst eigenen Experimental-Untersuchungen.   Nürnberg, 1819.   8vo.   Ill.

PFAUCIUS, CHRISTOPH.

De elementis.   Lipsiæ, 1666.   4to.

PFEIFFER, A.

Lærebog i Kemi til Skolebrug.   Kjøbenhavn, 1876.   8vo.

PFEIFFER, EMIL.

Die Stassfurter Kali-Industrie.   *See* Bolley's Handbuch (etc.).

PFEIFFER, EMIL.

Analyse (Die) der Milch.   Anleitung zur qualitativen und quantitativen Untersuchung dieses Secretes für Chemiker, Pharmazeuten und Aerzte.   Wiesbaden, 1887.   pp. v–84, 8vo.   Ill.

Handbuch der Kali-Industrie.   Die Bildung der Salzlager von Stassfurt und Umgegend, sowie von Kaluss und Beschreibung dieser Salzlager.   Die technische Gewinnung der Kalisalze aus den natürlich vorkommenden Salzen mit ihren Nebenzweigen und Anwendung der Kalisalze in der Landwirthschaft.   Braunschweig, 1887.   8vo. Ill.

Cf. Bolley's Handbuch der chemischen Technologie.

PFINGSTEN, JOHANN HERMANN.

Bibliothek ausländischer Chemisten, Mineralogen und mit Mineralien beschäftigter Fabrikanten nebst derley biographischen Nachrichten. Nürnberg, 1781–84. 4 vols., 12mo. Vol. I, pp. xlvi–618–[iv]. Portrait of Joh. Ernst. Imman Walch. Vol. II, pp. xxx–568–[vi]. Portrait of Joh. Albr. Gesner, 1782. Vol. III, pp. xxii–523. Portrait of Franz Ernst Brückmann, 1783. Vol. IV, pp. xiv–623. Portrait of Carl Friedr. Hundertmark.

> This is a collection of treatises by Walch, von Borch, Nicolas of Nancy, Gesner, Brongniart, Oesterreicher, Bergman, Buchoz and Salchow. Biographies are given of those whose portraits have been named.

PHÆDRO, GEORGE.

* The Art of Chemistry, written in Lattin. Done into English by Nicholas Culpepper. The third edition. London, printed for Simon Neale at the Sign of the three Pidgeons in Bedfordstreet in Covent Garden, 1674. pp. 133, 24mo. Portrait.

PHARMACY (THE) AND POISON LAWS of the United Kingdom. Their history and interpretation, with a brief account of the pharmacy laws in force in Australasia, Canada and Cape Colony. London and Melbourne, 1892.

PHELPS, MRS. ALMIRA H. LINCOLN.

Chemistry for beginners. New York, 1852. 12mo.

Chemistry for schools, families and private students. New York, 1850. 8vo.

Family lectures on chemistry. New York, 1850. 12mo.

PHILADELPHIA EXHIBITION. *See* Goldschmidt, G.

PHILIP, M.

Das Pyridin und seine nächsten Derivate. Stuttgart, 1889.

PHILIPPS, JOS.

Der Sauerstoff. Vorkommen, Darstellung und Benutzung desselben zu Beleuchtungszwecken nebst einem neuen Verfahren der Sauerstoffbeleuchtung. Berlin, 1871.

PHILLIPS, H. JOSHUA.

Engineering Chemistry, a practical treatise for the use of Analytical Chemists, Engineers, Ironmasters, Iron Founders, students and others, comprising methods of analysis and valuation of the principal materials used in engineering work, with numerous analyses, examples and suggestions. London, 1891. pp. xii–312, 8vo. Ill.

PHILLIPS, H. JOSHUA. [Cont'd.]

Fuels : solid, liquid and gaseous ; their analysis and valuation for the use of Chemists and Engineers. London, 1890. 8vo.

    Second edition. London, 1892.

PHILLIPS, JOHN ARTHUR.

Elements of Metallurgy. A practical treatise on the art of extracting metals from their ores. London, 1874. 8vo. Ill.

    New edition, revised and enlarged by the author and H. Bauerman. London, 1887. 8vo.

PHILLIPS, SIR RICHARD. [*Pseudonym :* D. BLAIR.]

A Grammar of Chemistry, wherein the principles of the science are explained and familiarized by a variety of examples and illustrations and by numerous useful and entertaining experiments, to which are added interrogatory exercises, and a glossary of terms used in chemistry. Corrected and revised by Benjamin Tucker. Third edition. Philadelphia, 1819. 24mo.

    Fourth edition. Philadelphia, 1823. pp. 196, 24mo. Plate.

PHILLIPS, S. E.

Old or New Chemistry, which is fittest for survival? And other essays in chemical philosophy. London, 1887.

PHIPSON, THOMAS LAMB.

Essay on the Uses of Salt in Agriculture. Prize essay, Salt Chamber of Commerce. London, 1863.

Force (La) catalytique. Harlem, 1858. 4to. Société hollandaise des Sciences.

Outlines of a New Atomic Theory. Fourth edition. London, 1886. pp. 4, roy. 8vo.

PIALAT, R.

Caractères des sels métalliques. Deuxième édition. Paris, 1888.

PIANCIANI, GIAMBATTISTA.

Elementi di fisico-chimica. Napoli, 1840.

    Seconda edizione, riveduta. Napoli, 1842. 2 vols., 8vo. Ill.

    Terza edizione. Roma, 1844.

Istituzioni fisico-chimiche. Roma, 1833–1835. 4 vols., 8vo. Ill.

PIAZZA, PIETRO.

Formole atomistiche e tipi chimici. Alcune nozioni elementari premesse al corso di chimica organica. Bologna, 1864. 8vo.

    Prima edizione, 1862.

PIAZZA, PIETRO.   [Cont'd.]

Lezioni di chimica organica generale e chimica animale date nella R.
Università di Bologna.   Bologna, 1873.   8vo.

   Another edition, 1865.

PICALAUSA, O.

Chimie des écoles moyennes.   Namur, 1884.   8vo.

PICK, S.

Alkalien (Die).   Darstellung der Fabrikation der gebräuchlichsten Kali-
und Natron-Verbindungen, der Soda, Potasche, des Salzes, Salpeters,
Glaubersalzes, Wasserglases, Chromkalis, Blutlaugensalzes, Wein-
steins, Laugensteins, u. s. w., und deren Anwendung und Prüfung.
Ein Handbuch für Färber, Bleicher, Seifensieder, Fabrikanten von
Glas, Zündwaaren, Lauge, Papier-Farben, überhaupt von chemischen
Producten, für Apotheker und Droguisten.   Wien, Pest, Leipzig,
1877.   Ill.

Künstlichen (Die) Düngemittel.   Darstellung der Fabrikation des
Knochen-, Horn-, Blut-, Fleisch-Mehls, der Kalidünger, des schwe-
felsauren Ammoniaks, der verschiedenen Arten Superphosphate,
der Poudrette, u. s. w., sowie Beschreibung des natürlichen Vorkommens der concentrirten Düngemittel.   Ein Handbuch für Fabrikanten künstlicher Düngemittel, Landwirthe, Zucker-Fabrikanten,
Gewerbetreibende und Kaufleute.   Wien, Pest, Leipzig, 1886.   Ill.

Mineralsäuren (Die).   Nebst einem Anhang : Der Chlorkalk und die
Ammoniak-Verbindungen.   Darstellung der Fabrikation von Schwe-
flichersäure, Schwefel-, Salz-, Salpeter-, Kohlen-, Arsen-, Bor-,
Phosphor-, Blausäure, Chlorkalk und Ammoniaksalzen, deren Un-
tersuchung und Anwendung.   Ein Handbuch für Apotheker,
Droguisten, Färber, Bleicher, Fabrikanten von Farben, Zucker,
Papier, Düngemittel, chemischen Producten und für Gastechniker.
Wien, 1879.   Ill.

Untersuchung (Die) der im Handel und Gewerbe gebräuchlichsten
Stoffe einschliesslich der Nahrungsmittel.   Wien, 1881.

PICKNY, J.

Aufforderung zum Anbau der Runkelrübe behufs der Futtergewinnung
oder Ablieferung derselben an in der Nähe befindliche Zucker-
fabriken, nebst einer Anleitung über die zweckmässigste Cultur der
Rübe.   Saaz, 1850.

PICTET, AMÉ.

La constitution chimique des alcaloïdes végétaux.   Paris, 1888.   pp.
vii–310, 8vo.   Ill.

PICTET, AMÉ. [Cont'd.]
>   Die Pflanzenalkaloide und ihre chemische Konstitution. In
>   deutscher Bearbeitung von Richard Wolffenstein. Ber-
>   lin, 1891, pp. v–382, 8vo.

PICTET, RAOUL.

Mémoire sur la liquéfaction de l'oxygène, la liquéfaction et la solidification de l'hydrogène et sur les théories des changements des corps.
Tiré des Archives des sciences de la Bibliothèque universelle.
Genève, Paris et Neuchatel, 1878. pp. 108, 8vo. Ill. 3 folding
plates.

Nouvelles machines frigorifiques basées sur l'emploi de phénomènes
physico-chimiques. Deuxième édition. Basel, 1885. Ill.

>   Neue Kälte-Erzeugungsmaschinen auf Grundlage der An-
>   wendung physikalisch-chemischer Erscheinungen. Nebst
>   Aufgaben über neue Einrichtung von Kälte-Erzeugungs-
>   maschinen. Uebersetzt von R. Schollmayer. Leipzig,
>   1885. Ill.

PIEPENBRING, GEORG HEINRICH.

Physikalisch-chemische Nachrichten von dem sogenannten neuen mine-
ralische Salzwasser auf der Saline bei Pyrmont. Leipzig, 1793.
8vo.

PIÉQUET, O.

La chimie des teinturiers.   Nouveau traité théorique et pratique de l'art
de la teinture et de l'impression des tissus. Paris, 1891. 8vo.

PIERRE, JOACHIM ISIDORE.

Chimie appliquée à l'agriculture.   Manuel élémentaire pour l'analyse
des terres, des amendements, des engrais liquides et solides, des
eaux d'irrigation ; des fourrages, des tourteaux et d'autres substances destinées à l'alimentation des hommes et des animaux.
Leçons professées à la faculté des sciences de Caen. Deuxième
édition des notions élémentaires d'analyse chimique appliquée à
l'agriculture. Paris, 1875. 8vo.

Chimie agricole, ou l'Agriculture considérée dans ses rapports principaux
avec la chimie. Deuxième édition. Paris, 1858. 18mo.

Considérations chimiques sur l'alimentation du bétail, au point de vue
de la production du travail, de la viande, de la graisse, de la laine
et du lait. Résumé des leçons faites à la Faculté des sciences de
Caen pendant l'année scolaire 1855–56. Paris, 1857. 8vo.

Études de chimie agricole. Caen, 1870. 8vo.

PIERRE, JOACHIM ISIDORE. [Cont'd.]

Introduction à l'étude de la chimie. Exposé sommaire des principaux difficultés qui peuvent embarrasser les personnes qui commence l'étude de cette science. Caen, 1853. 12mo.

Notions élémentaires d'analyse chimique. Paris, 1863. 18mo.

Recherches analytiques sur le thé de foin. Caen, 1858. 8vo.

Résumé de quelques leçons . . . sur les substances alimentaires. Caen, 1854. 12mo.

PIESSE, CHARLES H.

Chemistry in the Brewing-Room. A course of lectures to practical brewers, with tables of alcohol-extract and original gravity. London, 1877. 8vo.

PIESSE, G. W. SEPTIMUS.

Laboratory (The) of Chemical Wonders, a scientific mélange intended for the instruction and entertainment of young people. London, 1860. pp. xiii–256, 8vo.

Chimie des parfums et fabrication des savons, odeurs, essences, sachets, eaux aromatiques, pommades, etc. Paris, 1890. 16mo. Ill.

> Bibliothèque des connaissances utiles.

Histoire des parfums et hygiène de la toilette, poudres, vinaigres, œntifrices, fards, teintures, cosmétiques, etc. Édition française par F. Chardin-Hadancourt et H. Massignon et G. Halphen. Paris, 1889. 16mo. Ill.

> Bibliothèque des connaissances utiles.

> > Piesse's Art of Perfumery and the methods of obtaining the odours of plants, the growth and general farm system of raising fragrant herbs; with instructions for the manufacture of perfumes for the handkerchief, scented powders, odorous vinegars and salts, snuff, dentifrices, cosmetics, perfumed soap, etc. Edited by C. H. Piesse. London, 1891. pp. xvi–498, 8vo. Ill. Folding plate.

> Contains a sketch of the history of the art of perfumery.

PIFFARD, HENRY G.

A Guide to Urinary Analysis for the use of physicians and students. New York, 1873. pp. 88, 8vo. Ill.

PILLEY, J. J.

Chemistry of Common Objects. London, 1892. 8vo.

Elementary Inorganic Chemistry. London, 1892. 12mo.

Inorganic and Organic Qualitative Analysis; with equations and notes. London, 1888. 12mo.

PINK, WM. W., AND GEORGE E. WEBSTER.

A Course of Analytical Chemistry, qualitative and quantitative; to which is prefixed a brief treatise upon modern chemical nomenclature and notation. London, 1874. 12mo.

PINNER, ADOLF.

Einführung in das Studium der Chemie. Berlin, 1887.

Imidoäther (Die) und ihre Derivate. Berlin, 1892. 8vo.

Repetitorium der anorganischen Chemie. Mit besonderer Rücksicht auf die Studirenden der Medicin und Pharmacie bearbeitet. Berlin, 1874. 8vo.

    Dritte Auflage, 1878. Fünfte Auflage, 1883.

    Achte Auflage, 1889. 8vo.

Repetitorium der organischen Chemie. Mit besonderer Rücksicht auf die Studirenden der Medizin und Pharmacie. Berlin, 1872. 8vo.

    Zweite Auflage, 1874. Fünfte Auflage, 1881.

    Neunte Auflage. Berlin, 1890. 8vo.

    An Introduction to the Study of Organic Chemistry. Translated and Revised from the fifth German Edition by Peter T. Austen. New York, 1883. pp. xix–403, 12mo. Ill.

PINOLINI, G. D.

Adulterazioni (Le) del vino. Metodi per riconoscerle e leggi che le riguardano. Torino, 1889. 16mo.

Nozioni di fisica e di chimica per la terza normale preparatoria. Torino, 1891. 8vo.

PIQUE, EDWARD.

A Practical Treatise on the Chemistry of Gold, Silver, Quicksilver and Lead, tracing the crude ores from the mines through the various mechanical and metallurgical elaborations. San Francisco, 1860. 12mo.

PIRCH, JASPEN VON.

    A pseudonym of F. A. C. Gren. *See* Gren, Friedrich Albert Carl.

PIRIA, RAFFAELE.

Elementi di chimica inorganica [ecc.]. Vol. 1 in due parti. Napoli, 1841. 8vo.

Lezioni elementari di chimica organica. Seconda edizione. Torino, 1870. 12mo.

Trattato elementare di chimica inorganica. Seconda edizione. Pisa, 1845. pp. xvi–355, 8vo.

PISANI, F.

Traité pratique d'analyse chimique qualitative et quantitative à l'usage des laboratoires de chimie. Troisième édition. Paris, 1889. 12mo.

PISANI, F., ET DIRWELL.

La chimie du laboratoire. Paris, 1888.

PIZZI, A.

I pesi atomici e le proprietà fisiche principali dei corpi indecomposti, od elementi chimici che nello state presente della scienza si conoscono. Reggio nell' Emilia, 1886. 3 vols., 4to.

PIZZIGHELLI, G.

Die Actinometrie, oder die Photometrie der chemisch-wirksamen Strahlen für Chemiker, Optiker, Photographen, etc., in ihrer Entwickelung bis zu Gegenwart zusammengestellt. Wien, 1884. 8vo.

Handbuch der Photographie für Amateure und Touristen. Vol. I : Die photographischen Apparate und die photographischen Processe. Dargestellt für Amateure und Touristen. Vol. II : Die Anwendung der Photographie für Amateure und Touristen. Halle, 1887. 2 vols., 8vo. Ill.

PLACOTOMUS, JOHANN. [BRETSCHNEIDER].

De destillationibus chymicis. Francofúrti ad Viadrum, 1553.

De natura cerevisiarum et de mulso. Vitebergæ, 1551.

PLATTNER, CARL FRIEDRICH.

Beitrag zur Erweiterung der Probirkunst durch ein systematisches Verfahren bei Ausmittelung eines in Erzen, Hütten- und Kúnstprodukten befindlichen Gehalts an Kobalt, Nickel, etc., auf trockenem Wege. Freiberg, 1849. 8vo.

Probirkunst (Die) mit dem Löthrohre, oder Anleitung, Mineralien, Erze, Hüttenproducte und verschiedene Metallverbindungen mit Hülfe des Löthrohrs qualitativ auf ihre sämmtlichen Bestandtheile und quantitativ auf Silber, Gold, Kupfer, Blei, Zinn, Nickel, Kobalt und Eisen zu untersuchen. Dritte grösstentheils umgearbeitete und verbesserte Auflage. Leipzig, 1853. 8vo.

> First edition, Leipzig, 1835 ; second edition, 1847.

> > Probirkunst mit dem Löthrohre oder Vollständige Anleitung zu qualitativen und quantitativen Löthrohr-Untersuchungen. Vierte Auflage neu bearbeitet und vermehrt von Theodor Richter. Mit 86 in den Text eingedruckten Holzschnitten und einer Steindrucktafel. Leipzig, 1865. pp. xvi–680, 8vo.

PLATTNER, CARL FRIEDRICH. [Cont'd.]

Fünfte Auflage. Leipzig, 1878. 8vo.

Het onderzoek met het blaaspijp. Of handleiding om metalen, ertsen, huttenproducten, verschillende metaalmeng-
sels enz . . . met behulp van de blaaspijp, qualitatief,
naar hunne gezamentlijke bestanddeelen, en quantitatief,
met betrekking tot hun zilver, goud, koper, lood, tin,
nikkel, kobalt en ijzergehalte te onderzoeken. Naar de
tweede geheel omgewerkte en vermeerderde uitgaaf uit
het Hoogduitsch vertaald door P. J. Kipp. 'sGravenhage,
1849. 8vo.

Use of the Blowpipe in the Examination of Minerals, Ores,
Furnace Products and other Metallic Combinations,
translated with notes by J. S. Muspratt, with Preface by
Liebig. London, 1845. 8vo. Ill.

Third edition. London, 1854.

Plattner's Manual of Qualitative and Quantitative Analysis
with the Blowpipe. From the last German edition, revised and enlarged by Th. Richter. Translated by
Henry B. Cornwall, assisted by John H. Caswell. New
York, 1871. 8vo.

Third edition. New York, 1875. pp. xvii–548–[i].

For a French translation of Plattner, *see* Cornwall, Henry B. The French
translator ignores the German author and credits the work to the American translator.

Arte de ensayar con el soplete, cualitativa y cuantitativamente,
los minerales, aleaciones y productos metalúrgicos . . .
traducido al castellano de la segunda edicion inglesa
. . . por Ignacio Fernandez de Henestrosa. Madrid,
1853. 4to.

Tableaux des caractères que présentent au chalumeau les alcalis, les
terres et les oxydes métalliques, soit seuls, soit avec les réactifs.
Extraits du traité des essais au chalumeau. Traduits de l'Allemand
par A. Sobrero. Paris. 1843. 4to.

Tabellarisch overzigt van de kenteekenen der alkaliën-
aarden- en metaal-oxyden in het vuur voor de blaaspijp, uit het werk : Onderzoek met de blaaspijp getrok-
ken door P. J. Kipp. 'sGravenhage, 1849. 8vo.

Vorlesungen über allgemeine Hüttenkunde. Nach dem hinterlassenen
Manuscript herausgegeben und vervollständigt von Theodor Richter.
Freiberg, 1860. 2 vols., 8vo.

PLAYFAIR, SIR LYON.

On the Chemical Principles involved in the manufactures of the Exhibi-

PLAYFAIR, SIR LYON. [Cont'd.]
    tion as indicating the necessity of industrial instruction.   London,
    1852.   8vo.
        Society for the Encouragement of Arts, Manufactures and Commerce.
            Lectures, Series 1.
    On the Chemical Properties of Gold . . . Lectures on Gold for the
        Instruction of Emigrants about to proceed to Australia [etc.].
        London, 1853.   8vo.

PLAYFAIR, LYON [Editor].
    See Liebig, Justus von.

PLENCK, JOSEPH JACOB VON.
    Elementa chymiæ.   Viennæ, 1800.   8vo.
            Anfangsgründe der Chemie.   Wien, 1801.   8vo.
    Elementa chymiæ pharmaceuticæ, sive doctrina de præparatione ac
        compositione medicamentorum.   Viennæ, 1802.   8vo.
            Anfangsgründe der pharmaceutischen Chemie, oder Lehre
                von der Bereitung und Zusammensetzung der Arznei-
                mittel.   Wien, 1803.   8vo.
    Hygrology (The) or chemico-physiological doctrine of the fluids of the
        human body, translated from the Latin by Robert Hooper.   Lon-
        don, 1797.   8vo.
    Toxicologia, seu doctrina de venenis et antidotis.   Viennæ, 1785.   8vo.
            Toxikologie, oder Lehre von den Giften und Gegengiften ;
                aus dem lateinischen.   Wien, 1785.   pp. 302–[vii], 8vo.

PLEVANI, SILVIO.
    Chimica del pensiero.   Roma, 1880.   16mo.

PLINIUS, CAIUS, SECUNDUS.
    Plinii Secundi, naturalis historiæ libri xxvii.   Venetiis, Joannes de
        Spira, 1469.   355 ff., fol.
            This, the first edition of Pliny, contains (ff. 1 to 18) a life of the author.
                See Brunet, 4467, and Hain, 13087.
    C. Plinii Secundi naturalis historiæ libri xxxvii cum epistola dedicatoria
        Joh. Andreæ ad Paulum Papam II.   Venetiis, Nic. Jenson, 1472.
        ff. 358
            See Panzer, iii, 88, 90 ; Hain, 13089* ; Dibdin, Bibl. Spenc., ii, 258 ;
                Brunet, iv, 714 ; Graesse, v, 337. Of the very large number of editions
                of this work in various languages we can only mention the following :
                Lemaire, Paris, 1827–32, 11 vols. Sillig, Gotha and Hamburg, 1851–
                54, 6 vols. Külb, Stuttgart, 1840–47, 7 vols., 8vo. *Philemon Hol-
                land, London, 1601, 2 vols., folio. Brotier, Londini, 1826, 13 vols.,
                8vo.
            The "Father of Natural History" can hardly be accounted a chemist, but
                no bibliography of any branch of science, conceived on a historical basis,
                can afford to omit the Natural History of Pliny.

PLYMPTON, GEORGE W.

The Blow-pipe : A guide to its use in the determination of salts and minerals. Compiled from various sources. New York, 1877. 12mo.

PÖHLMANN, R.

Repetitorium der Chemie für Studirende. Theil I, Anorganische Chemie. Leipzig, 1888. Theil II, Organische Chemie. Leipzig, 1889.

PÖRNER, CARL WILHELM.

Anleitung zur Färbekunst . . . Leipzig, 1785. 8vo.

Chymische Versuche zum Nutzen der Färbekunst. Leipzig, 1772–73. 3 vols., 8vo.

Delineatio pharmaciæ chemico-pharmaceuticæ. Lipsiæ, 1764. 8vo.

POGGIALE, A. B.

Action des alcalis sur le sucre dans l'économie animale. Paris, 1856. 8vo.

Analyse de l'eau minérale acidule ferrugineuse d'Orezza. Paris, 1854. 8vo.

Du pain de munition distribué aux troupes de puissances Européennes et de la composition chimique du son. Paris, 1854. 8vo.

Recherches sur la composition chimique et les équivalents nutritifs des aliments de l'homme. Paris, 1856. 8vo.

Recherches sur la composition de l'eau de la Seine . . . Paris, 1856. 8vo.

Recherches sur les eaux des casernes des forts . . . de Paris. Paris, 1853. 8vo.

Traité d'analyse chimique par la méthode des volumes, comprenant l'analyse des gaz et des métaux, la chlorométrie, la sulfhydrométrie, l'acidimétrie, l'alcalimétrie, la saccharimétrie, etc. Avec 171 figures intercalées dans le texte. Faris, 1858. pp. 606, 8vo.

POHL, JOSEPH JOHANN.

Lehrbuch der chemischen Technologie. Einleitung zur chemischen Technologie. Wien, 1865. 8vo. Ill.

POIRÉ, P.

Leçons de chimie appliquée à l'industrie, l'usage des industriels, des écoles normales primaires, des établissements d'instruction primaire supérieure, des écoles professionnelles, etc. Paris, 1868. 18mo.

Cinquième édition. Paris, 1887.

Notions de chimie appliqués aux arts, à l'hygiène et à l'économie domestique, à l'usage des demoiselles. Paris, 1869. 8vo. Ill.

Huitième édition. Paris, 1883.

POLECK, THEODOR.

Beiträge zur Kenntniss der chemischen Veränderungen fliessender Gewässer. Breslau, 1869. 8vo.

Chemische (Das) Atom und die Molekul. Rede bei dem Antritt des Rektorats der königlichen Universität zu Breslau am 15. Ocktober 1888 gehalten. Breslau, 1889. 8vo.

Chemische (Die) Natur der Minengase und ihre Beziehung zur Minenkrankheit. Berlin, 1867.

Chemische Analyse des Oberbrunnens zu Flinsberg in Schlesien. Breslau, 1883. 8vo.

POLIN, H., ET H. LABIT.

Étude sur les empoisonnements alimentaires (Microbes et Ptomaines). Paris, 1890. pp. 226, 8vo.

Examen des aliments suspects. Avec une préface par J. Arnould. Paris, 1892. 8vo. Ill.

POLIS, ALFRED.

Grundzüge der theoretischen Chemie. Aachen, 1887.

Précis de chimie théorique à l'usage de étudiants, traduit de l'allemand par Ad. Lecrenier. Aix-la-Chapelle, 1888. 12mo.

POLLACCI, E.

Chimica teoretica, ovvero principii fondamentali di chimica generale basati sulle odierne dottrine chimiche. Milano, 1891. 8vo. Ill.

Corso di chimica medico-farmaceutica, scritto per uso degli studenti e degli esercenti la medicina e la farmacia. Parte organica. Milano, 1891. 8vo. Ill.

Metalloidi. Milano, 1892. 8vo. Ill.

POLLEYN, F.

Die Appreturmittel und ihre Verwendung. Darstellung aller in der Appretur verwendeten Hilfsstoffe, ihrer speciellen Eigenschaften, der Zubereitung zu Appreturmassen und ihrer Verwendung zu Appretiren von leinenen, baumwollenen, seidenen und wollenen Geweben ; feuersichere und wasserdichte Appreturen und der hauptsächlichen maschinellen Vorrichtung. Ein Hand- und Hilfsbuch für Appreteure, Drucker, Färber, Bleicher, Wäschereien. Wien, Pest, Leipzig, 1886. 8vo. Ill.

POLONI, G.

Lezioni elementari di chimica teorica, compilata ad uso dei licei. Milano, 1878. 16mo.

POMMIER.

[Produits chimiques.] *See, in Section II*, Fremy : Encyclopédie chimique, vol. v.

PONCELET, ABBÉ.

Chimie du goût et de l'odorat, ou Principes pour composer à peu de frais les liqueurs à boire et les eaux à senteur. Paris, 1755. 8vo.

Nouvelle chimie du goût [etc.]. Paris, 1774.

Nouvelle édition, 1800. 2 vols., 8vo.

PONCI, LUIGI.

Chimica tecnologica. Tintura della seta. Studio fisico-chimico della seta e delle materie coloranti che si impiegano a tingerla, macchine e manipolazioni tintorie. Milano, 1876. 8vo.

Eight tables and specimens of dyed silk.

Introduzione alla teoria atomica. Como, 1872. 16mo.

PONZONI, A. B.

Nozioni di chimica e fisica preparatoria all'esame d'ammissione nei corsi uffiziali telegrafici, in base al programma ufficiale. Milano, 1877. 16mo.

POORE, G. V.

Coffee and Tea. London, 1883.

POPE, GEORGE.

A Class Book of Rudimentary Chemistry. London, 1864. 18mo.

Поповъ, А.

Сборникъ работъ химической лабораторіи иИмператорскаго Варшавскаѣо. Университета (1870—1876). Варшава, 1876. pp. x–219, 8vo.

POPOFF, A. Contributions from the Chemical Laboratory of the Imperial University of Warsaw. Warsaw, 1876.

POPPER, HEINRICH.

Die Fabrikation der nichttrübenden ätherischen Essenzen und Extracte. Vollständige Anleitung zur Darstellung der sogenannten extra-starken, in 50 %-igem Sprit, löslichen ätherischen Oele, sowie der Mischungs-Essenzen, Extract-Essenzen, Frucht-Essenzen und der Fruchtäther. Nebst einem Anhange : Die Erzeugung der in der Liqueur-Fabrikation zur Anwendung kommenden Farbtincturen. Ein Handbuch für Fabrikanten, Materialwaarenhändler und Kauf-leute. Auf Grundlage eigener Erfahrungen praktisch bearbeitet. Wien, Pest, Leipzig, 18—. Ill.

PORTEFEUILLE FÜR GEGENSTÄNDE DER CHEMIE UND PHARMACIE.   Erstes
Stück.   Hamburg, 1784.   8vo.

PORTER, ARTHUR LIVERMORE.

The Chemistry of the Arts, being a practical display of the arts and
manufactures which depend upon chemical principles.   On the
basis of Gray's "Operative Chemist" adapted to the United States,
with treatises on calico printing, bleaching and other large additions.
Philadelphia, 1830.   2 vols., 8vo.

See Gray, Samuel Frederick.

PORTER, JOHN A.

First book of Chemistry and allied Sciences ; including an outline of
agricultural chemistry.   New York, 1857.   12mo.

Principles of Chemistry, embracing the most recent discoveries in the
science, and the outlines of its application to agriculture and the
arts.   New York, 1856.   12mo.

POSADSKY, S.

Praktische Modification der Pettenkofer-Nagorsky'schen Methode zur
Bestimmung des Kohlensäuregehalts der Luft.   Mit Tabellen :
I. Der Bestimmung des Kohlensäure-Volumens der Luft nach
Abnahme der alkalischen Reaction der Barytlösung, und II : Zur
Reduction eines Gas-Volumens auf 0° Temperatur und auf 760^{mm}
Luftdruck.   St. Petersburg, 1886.   8vo.

POSSELT, LOUIS.

Die analytische Chemie tabellarisch dargestellt.   Heidelberg, 1846.   Fol.
        De analytische scheikunde in tafels.   Uit het Hoogduitsch
        vertaald en vermeerderd onder opzigt van S. Bleekrode.
        Utrecht en Delft, 1848.   4to.

Tabellarische Uebersicht der qualitativen chemischen Analyse.   Zum
Gebrauche bei den praktischen Arbeiten im Laboratorium.   Heidelberg, 1845.   8vo.

POST, H. VON.

Grundlinier til Åkerbrukskemien, efter föreläsninger hållna vid Ultuna
landtbruksinstitut.   Utgifvet af Rob.   Stockholm, 1877.   8vo.

POST, JULIUS.

Chemisch-technische Analyse.   Handbuch der analytischen Untersuchungen zur Beaufsichtigung des chemischen Grossbetriebes und zum
Unterrichte.   Unter Mitwirkung von L. Aubry [and others] herausgegeben von Jul. Post.   Zweite vermehrte und verbesserte Auflage.

POST, JULIUS. [Cont'd.]

Mit zahlreichen Holzstichen. Braunschweig, 1888–90. 2 vols. Vol. I, pp. xxxiii–870.

First edition Braunschweig, 1881–'82.

Traité complet d'analyse chimique appliqué aux essais industriels. Publié avec la collaboration de 22 chimistes, traduit de l'allemand par L. Gautier. Paris, 1884. pp. viii–1143, 8vo. Ill.

Grundriss der chemischen Technologie. Berlin, 1877–'79. 8vo. Ill. 2 Parts. Vol I : Fabrikation der Rohproducte. Vol. II : Fabrikation der Endproducte.

POSTEL, EMIL.

Kleine Chemie, insbesondere für Seminaristen, sowie für angehende Landwirthe und Gewerbetreibende. Langensalza, 1860. 8vo.

Zweite Auflage. Langensalza, 1865.

Dritte Auflage. Langensalza, 1868.

Grondbeginselen der anorganische en organische scheikunde. Naar het Hoogduitsch door Th. H. A. J. Abeleven. Tweede uitgaaf. Sneek, 1864. 8vo.

Laien-Chemie, oder leichtfassliche, an einfache Versuche geknüpfte Darstellung der Hauptlehren der Chemie für Gebildete aller Stände insbesondere für Lehrer, Oeconomen und Gewerbetreibende. Zweite Auflage. Langensalza, 1860. 8vo. Ill.

Dritte Auflage. Langensalza, 1866. 8vo. Ill.

POSTHUMUS, L.

Handleiding bij het onderwijs in de scheikunde aan middelbare scholen voor meisjes. Dortrecht, 1878. 8vo. Ill.

POTT, JOHANN HEINRICH.

Chymische Untersuchungen welche fürnehmlich von der Lithogeognosia, oder Erkäntniss und Bearbeitung der gemeinen einfacheren Steine und Erden, ingleichen von Feuer und Licht handelt. Zweite Auflage. Berlin, 1757. pp. [xiv]–88-[viii]–120-[xiv]–148-[ii]–44-[xiii], sm. 4to. Folding plate.

The three parts have separate title-pages bearing the dates 1746, 1751 and 1754. First edition, Potsdam, 1746.

* Lithogéognosie, ou examen chymique des pierres et des terres en général, et du talc de la topaze, et de la stéatite en particulier, avec une dissertation sur le feu et sur la lumière. Ouvrages traduits de l'allemand. Paris, 1753. 2 vols., 12mo. Vol. I, pp. viii–431.

POTT, JOHANN HEINRICH.   [Cont'd.]

> Vol. II *is entitled :* Continuation de la lithogéognosie pyro-
> technique. Où l'on traite plus particulièrement de la
> connoissance des terres et des pierres, et de la manière
> d'en faire l'examen. pp. 267-[v]-ciii.
>
> > Pages ciii. consist of a series of tables for the determination of minerals by
> > action of solvents, etc.

Exercitationes chymicæ de sulphuribus metallorum, de auripigmento, de
solutione corporum particulari, de terra foliata tartari, de acido
vitrioli vinoso et de acido nitri vinoso, sparsim hactenus editæ, jam
vero collectæ, restitutæ, a mendis repurgatæ variisque notis experimentis et discussionibus ab auctore adauctæ, illustratæ. Berolini,
1738. pp. [vi]-220, sm. 4to.

Observationum et animadversionum chymicarum, præcipue circa sal
commune acidum salis vinosum et wismuthum versantium, collectio
prima. Berolini, 1739. pp. [viii]-197, sm. 4to.

> *Followed by a second collection :*

Observationum et animadversionum chymicarum, præcipue zincum,
boracem et pseudo-galenam tractantium, collectio secunda. Berolini, 1741. pp. [iv]-120, sm. 4to.

Physicalische chymische Abhandlung von dem sonderbahr feuerbe-
ständigen und zartflüssigen Urin-Saltz und dessen weitläuftige Anwendung und Nutzen. Ingleichen eine Untersuchung der Verbindungen eines Acidi vitrioli mit dem sauren Weinstein. Zweite mit
einem Anhang vermehrte Auflage. Berlin, 1761. 4to.

> First edition, Berlin, 1757.

> > Dissertations chimiques de Pott recueillies et traduites, tant
> > du latin que de l'allemand par Demachy. Paris, 1759.
> > 4 vols., 8vo.

POTT, ROBERT.

Kurzes Lehrbuch der anorganischen Chemie für Landwirthe. Berlin,
1878. 8vo.

POULSEN, V. A.

Botanische Mikrochemie. Eine Anleitung zu phytohistologischen Untersuchungen. Aus dem dänischen unter Mitwirkung des Verfassers
von Carl Müller. Cassel, 1881. 16mo.

> Contains a bibliography pp. xiii–xvi.

> > Botanical Microchemistry. An introduction to the study of
> > vegetable histology. Prepared for the use of students.
> > Translated with the assistance of the author and con-
> > siderably enlarged by William Trelease. Boston, 1884.
> > pp. xviii-118, 12mo.

POULSEN, V. A. [Cont'd.]
Microchimie végétal, guide pour les recherches phytohisto-
logiques. Traduit par J. P. Lachmann. Édition française augmentée. Paris, 1883. 16mo.
Bibliothèque biologique internationale, No. 11.

POURIAU, A. F.
Chimie inorganique suivie de l'étude des marnes, des eaux et d'une
méthode générale pour reconnaître la nature d'un des composés
minéraux intéressant l'agriculture ou la médecine vétérinaire.
Paris, 1862. 18mo.

POUTEAUX, A.
La poudre sans fumée et les poudres anciennes. Dijon, 1892. 8vo.

POUTET, JEAN JOSEPH ÉTIENNE.
Considérations générales sur l'art de faire le vin. Marseille, 1829. 8vo.
Considérations sur les savons de Marseille. Marseille, 1820. 8vo.
Instruction pour reconnaître la falsification de l'huile d'olive. Mar-
seille, 1819. 8vo.
Nouveau manuel du raffineur du sucre. Marseille, 1826.
Sur les sulfites dans les sels de soude. Marseille, 1821. 8vo.

POWER, FREDERICK B.
See Hoffmann, Frederick.

POZZI, GIOVANNI.
Materia medica chimico-farmaceutica. Milano, 1816. 8vo.
See also in Section II.

POZZO, LORENZO DEL.
Trattato elementare di chimica applicata. all' agricoltura. Vercelli,
1847. 8vo.
Trattato elementare di chimica applicata all' agricoltura, compilato
giusta i migliori autori dell' epoca. Torino, 1868.

PRAKTISCHE ANLEITUNG ZUM RÜBENBAU. Carlsruhe, 1836.

PRECHT, ——.
Die Salz-Industrie von Stassfurt und Umgegend. Stassfurt, 1883. 8vo.

PRECHTL, JOHANN JOSEPH.
Anleitung zur zweckmässigsten Einrichtung der Apparate zur Beleuch-
tung mit Steinkohlengas. Wien, 1817. 8vo.
Grundlehren der Chemie in technischer Beziehung. Für Kammeralisten,
Oekonomen, Techniker und Fabrikanten. Wien, 1813–15. 2 vols.,
8vo.
Neue vermehrte Ausgabe. Wien, 1817–19. 2 vols., 8vo.

PREIS, K.
Navedení ku chemickému rozboru. I, Analysa kvalitativná. v Praze,1881.

PRESCOTT, ALBERT BENJAMIN.
　　　　See Douglas, Silas H., and Albert B. Prescott.

PRESCOTT, ALBERT BENJAMIN.
　　Chemical Examination of Alcoholic Liquors. A manual of the con-
　　　　stituents of the distilled spirits and fermented liquors of commerce
　　　　and their qualitative and quantitative determinations. New York,
　　　　1873. 12mo.
　　First Book of Qualitative Chemistry. New York, 1879.
　　Guide in the Practical Study of Chemistry and in the work of Analysis.
　　　　Ninth edition, revised. New York, 1890. 8vo.
　　Organic Analysis : A manual of the descriptive and analytical chemistry
　　　　of certain carbon compounds in common use. For the qualitative
　　　　and quantitative analysis of organic materials ; commercial and
　　　　pharmaceutical assays ; the estimation of impurities under authorized
　　　　standards ; forensic examinations for poisons ; and elementary
　　　　organic analysis. New York, 1887. pp. 533, 8vo.
　　Outlines of Proximate Organic Analysis for the identification, separation
　　　　and quantitative determination of the more commonly occurring
　　　　organic compounds. New York, 1875. pp. 192, 12mo.

PRESL, JOHANN SWATOPLUK.
　　Experimental-Chemie. Prag, 1828. 8vo.

PRESTINARI, JOHANN NEPOMUK.
　　Handbuch der Cameralchemie. Heidelberg, 1827–28. 2 vols., 8vo.
　　Lehre (Die) von den Reagentien nach ihrem ganzen Umfang systema-
　　　　tisch bearbeitet für Chemiker, Staatsärzte, Apotheker, Metallurgen,
　　　　Mineralogen, Fabrikanten und Oekonomen. Heidelberg, 1823.
　　　　pp. xiv–viii–478, 12mo.

PRESTWICH, JOHN.
　　Dissertation on Mineral, Animal and Vegetable Poisons ; containing a
　　　　description of poisons in general, their manner of action, effects on
　　　　the human body. And respective antidotes with experiments and
　　　　remarks on noxious exhalations from earth, air and water, together
　　　　with several extraordinary cases, and elegant engravings of the
　　　　principal poisons of the different countries. London, 1775. pp.
　　　　ix–320, 8vo.

PREUSS, E.
　　Leitfaden für Zuckerfabrikchemiker zur Untersuchung der in der

PREUSS, E. [Cont'd.]
 Zuckerfabrikation vorkommenden Producte und Hilfsstoffe.   Berlin, 1892. 8vo. Ill.

PREYER, W.
 Blausäure (Die) physiologisch untersucht.   Bonn, 1868.   2 parts, 8vo.
  Cf. in Section I, Preyer, W.
 Blutkrystalle (Die) Untersuchungen.   Jena, 1871.   pp. viii–263, 8vo. Ill.
 Organischen (Die) Elemente und ihre Stellung im System.   Ein Vortrag gehalten in der deutschen chemischen Gesellschaft zu Berlin am 23. März, 1891.   Wiesbaden, 1891.   pp. 47, 8vo.

PRIBRAM, R., UND NEWMANN-WENDER.
 Anleitung zur Prüfung und Gehaltsbestimmung der Arzneistoffe für Apotheker, Chemiker, Aerzte und Sanitätsbeamte.   Wien, 1892. 8vo. Ill.

PRICE, R.
 Charte über pharmaceutische Chemie, eine Angabe verschiedener Artikel der Pharmacopoe nebst Gegenüberstellung derer, mit welchen jene chemisch unverträglich sind, wodurch die Kunst, wissenschaftlich zu verschreiben, erleichtert und jene Decompositionen vermeiden werden sollen, welche in ihren medicinischen Wirkungen oft die Absichten des Arztes vereiteln.   Aus dem englischen übertragen, vermehrt und berichtigt.   Weimar, 1823.   Fol. Ill.

PRIESTLEY, JOSEPH.
 * Directions for impregnating Water with Fixed Air in order to communicate to it the peculiar spirit and virtues of Pyrmont Water, and other mineral waters of a similar nature.   London, printed for J. Johnson, No. 72, in St. Paul's Churchyard, 1772.   pp. [i]–iii–22, 12mo.   Plates.
  This work describes the preparation of soda-water as it is now called ; it antedates Priestley's discovery of oxygen by two years.
 * Experiments on the generation of Air from Water, to which are prefixed experiments relating to the decomposition of dephlogisticated and inflammable air.   From the Philosophical Transactions, vol. LXXXI, p. 213.   With a dedication to the Members of the Lunar Society at Birmingham, 1793.   London, 1793.   pp. x–39, 8vo.
 * Experiments and Observations on Different Kinds of Air.   London, 1774-79.   3 vols., 8vo.   Vol. I, pp. xxviii-324, 2 plates, 1774. (Second edition, 1775 ; Third, 1781.)   Vol. II, pp. xliv-399-[xvi], 1 plate, 1775.   (Second edition, 1784.)   Vol. III, pp. xl-411-[viii], 1 plate, 1777.

PRIESTLEY, JOSEPH. [Cont'd.]
>   Proeven en waarnemingen op verschillende soorten van
>   lucht. [Translated by] Jacob Ploos van Amstel.
>   Amsterdam, 1778. 8vo.

* Experiments and Observations relating to various branches of Natural
Philosophy with a continuation of the observations on Air. London
and Birmingham, 1779–1786. 3 vols., 8vo. Vol. I, pp. xxxii–490,
1 plate, 1779. Vol. II, pp. xx–408, 1 plate. Birmingham, 1781.
Vol. III, pp. xxxii–454, 1 plate. Birmingham, 1786.

>   Versuche und Beobachtungen über verschiedene Gattungen
>   der Luft. Aus dem englischen übersetzt von C. Ludwig.
>   Wien und Leipzig, 1779–80. 3 vols., 8vo. Folding plates.

>   Expériences et observations sur différentes branches de la
>   physique, avec une continuation des observations sur
>   l'air, traduit de l'anglais par Jacques Gibelin. Paris,
>   1777–80. 5 vols., 12mo.

>   Versuche und Beobachtungen über verschiedene Theile der
>   Naturlehre ; nebst fortgesetzten Beobachtungen über die
>   Luft. Aus dem englischen. Leipzig, 1780–82. 2 vols.,
>   8vo. Ill.

Heads of Lectures on a Course of Experimental Philosophy particularly
including Chemistry, delivered at the New College in Hackney.
London, 1794. pp. xxviii–180, 8vo.

* Philosophical Empiricism : containing remarks on a charge of plagiarism
respecting Dr. H——s, interspersed with various observations relating to different kinds of air. London, 1775. pp. [iv]–86, 8vo.

PRIESTLEY, HENRY, AND BLACK.
* Auserlesene kleine Werke. Die Schwängerung des gemeinen Wassers
mit fixer Luft, die Magnesia und Kalkerde, die fäulungswidrige
Kraft gewisser Arztneyen und andere erhebliche Gegenstände betreffend. Kopenhagen und Leipzig, bey Johann Friedrich Heineck
und Faber, 1774. pp. 152, 16mo. 1 Plate.

PRIETO, C. F., DE LANDERO Y R.
Dinamica química. Guadalajara, 1886. 8vo.

PRINGLE, JOHN.
Redevoering over de verschillende soorten der lugt. Franeker, 1776.
8vo.
>   Treats of " fixed air."

PRINGSHEIM, N.
Ueber das Chlorophyll. Berlin, 1874–'76. 2 parts.

PRIOR.   *See :* Sechs Vorträge aus dem Gebiete der Nahrungsmittel-Chemie.

PRITCHETT, J.

Qualitative Chemical Analyses (Inorganic and Organic).   Part 1, Elementary stage for schools and science classes.   Leeds, 1886.   12mo.

PRIVAT-DESCHANEL.

Premières notions de chimie pour l'enseignement de la physique et de la chimie dans la classe de sixième.   Paris, 1881.   16mo.   Ill.

PROCTER, H. R.

A Text-book of Tanning.   A treatise on the conversion of skins into leather, both practical and theoretical.   London, 1885.

PROCTER, WILLIAM.

Notes on the Practical Chemistry of the non-metallic elements and their compounds.   London, 1876.

PROJECT DO SŁOWNICTWA CHEMICZNEGO.   Warszawa, 1853.   16mo.
      Project for a chemical terminology.

PROOST, A.

Manuel de chimie agricole et de physiologie végétale et animale appliquée à l'agriculture.   Bruxelles, 1884.   12mo.

Traité pratique de chimie agricole et de physiologie.   Bruxelles, 1881.   8vo.

PROUST, LOUIS JOSÉ.

Anales del real laboratorio de química de Segovia, ó coleccion de memorias sobre las artes, la artillería, la historia natural de España y Américas, la docimástica de sus minas, etc.   Segovia, 1791–95.   2 vols., 4to.

Análisis de una piedra meteórica caida en las immediaciones de Sixena en Aragon, el 17 de Noviembre de 1773.   Madrid, 1804.   8vo.

Mémoire sur le sucre des raisins.   Paris, 1808.   8vo.

PROUT, WILLIAM.

Chemistry, Meteorology, and the function of Digestion, considered with reference to natural theology.   London, 1833.
            Third edition.   London, 1845.   8vo.
            Fourth edition edited by J. W. Griffith.   London, 1855.
      Bridgewater Treatise No. 8.

PROVENZALI, FRANC. SAV.

Trattato elementare di chimica moderna.   Vol. 1, Chimica generale : metalloidi, metalli.   Roma, 1877.   8vo.

PRUNIER, L.

Tableaux d'analyse qualitative.   Paris, 1885.   8vo.

PRUNIER, L.

Tableaux analytiques. . . . Alcools et phénols . . . [and other articles].   *See, in Section II,* Fremy : Encylopédie chimique.

PUERTA RODENAS Y MAGANA, GABRIEL DE LA.

Tratado de química organica general y aplicada á la farmacia, industria y agricultura, con un tratado de química biológica vegetal y animal. Segunda edicion aumentada.   Madrid, 1879.   2 vols., 4to.   Ill.

PUILLE, D.

Leçons normales de chimie élémentaire théorique, appliquée.   Nouvelle édition.   Paris, 1864.

PURGOTTI, ENRICO.

Saggio di un processo sistematico per la ricerca del radicale acido organico ed inorganico che nei casi più comuni puó trovarsi salificato in un liquido.   Torino, 1878.   8vo.

PURGOTTI, SEBASTIANO.

Riflessioni sulla teoria degli atomi.   Perugia, 1835.   8vo.   pp. 234.

Trattato di chimica elementare applicata specialmente alla medicina. Perugia, 1839–41.   8vo.   Vol. I, pp. 815, 1839; vol. II, pp. 426, 1841.

Edizione riveduta, ampliata e corretta.   Perugia, 1864.

Trattato elementare di chimica.   Seconda edizione.   Perugia, 1845. 2 vols.

PUSCH, GEORG GOTTLOB, UND CH. BERCHT.

Erläuternde Experimente über die Grundlehren der allgemeinen und Mineralchemie.   Freiberg, 1810.   8vo.

QUADRAT, BERNHARD.

Anleitung zur qualitativen und quantitativen chemischen Analyse für Ober-Realschulen und höhere Gewerbschulen.   Brünn, 1855.   8vo.

Lehrbuch der Chemie für Ober-Realschulen und technische Anstalten sowie zum Selbstunterrichte.   Brünn, 1853–'54.   2 parts, 8vo.

Zweite Auflage.   Brünn, 1855–'57.   8vo.   Erste Abtheilung : Unorganische Chemie.   Zweite Abtheilung : Organische Chemie.

Lehrbuch der technischen Chemie.   Wien, 1862.   8vo.   Ill.

QUADRAT, BERNHARD, UND K. J. BÁDAL.

Elemente der reinen und angewandten Chemie für Unter-Realschulen. Brünn, 1860.   8vo.   Ill.

Dritte Auflage.   Brünn, 1876.

Qualitative chemische Analyse in tabellarischer Uebersicht. Bern, 1889.

Quarizius, C. G.
Die chemisch-künstliche Bereitung der moussirenden Weine überhaupt und insbesondere des französischen Champagners ingleichen die durchaus nicht anstössige und sichere Nachbildung der heilsamsten und gebräuchlichsten Mineralwässer, namentlich des Rakoczy, Eger, Pyrmonter, etc., Wassers in solcher Vollkommenheit dass sie hinsichtlich ihrer Konstitution den natürlichen analog sind, ja in gewissen Fällen letztere an Heilkraft noch übertreffen. Zweite vermehrte Auflage. Weimar, 1861. 8vo.
Populäres Handbuch der organisch-technischen Chemie. Berlin, 1842. 8vo.

Quatremère D'Isjonval, Denis Bernard.
Collection de mémoires chimiques et physiques. Paris, 1784. pp. 311, 4to.
> No more published. The volume includes essays on indigo, marls, dyeing of wool, etc.
Mémoires sur la découverte des sels triples. Paris, 1784.

Quelmalz, Samuel Theodor.
Utrum arsenicum sit primum principium metallorum. Lipsiæ, 1755. 4to.

Quensel, Conrad.
Anledning till kemiska och physiska försök. Stockholm, 1802. 8vo.

Quincke, H.
Balneologische Tafeln. Graphische Darstellung der Zusammensetzung und Temperatur der wichtigsten Heilquellen. Berlin, 1872. pp. 27, and 11 tables printed in colors, 8vo.

Quinquand, E.
Traité technique de chimie biologique avec applications à la physiologie, à la pathologie, à la clinique et à la thérapeutique. Paris, 1883. 8vo.

Quesneville, George.
Nouvelle méthode pour la détermination des éléments du lait et de ses falsifications. Paris, 1884. 4to.
> Neue Methoden zur Bestimmung der Bestandtheile der Milch und ihrer Verfälschungen. Deutsch von Vict. Griessmayer. Neuburg a D., 1885. 8vo.

[Rabinowitz, H. M.]

צבי בחרב מאיר הכהן ראבינאוויטץ.
אוצר החכמה והמדע כולל יסודי חכמת הטבע הכללית וגם ידיעת חרושה
המעשה וידיעות מדעים שונים.　יצא לאור בעזרת חברת מרבי=השכלה
בישראל אשר ברוסיא.　וווילנא בדפום והוצאת האלמנה והאחים ראם.
שנת תרל״ז לפ״ק.

With two additional title pages, in Russian and in German as below.
The text, however, is wholly in Hebrew.

Библіотека наукъ и знаній.　Общія основанія естествовѣдѣнія о химіидля
самоученія.

Bibliothek der gesammten Naturwissenschaften.　Buch der Chemie auf
einfache Experimente gegründet, in populärer Darstellung, zur
Selbstbelehrung für Jedermann.　Methodisch bearbeitet.　Mit 185
in den Text eingedruckten Abbildungen.　[Vilna, 1876.]　pp. 306,
8vo.　Illustrated.

Rabuteau, Antoine.
Éléments de toxicologie et de médecine légale appliquée à l'empoisonne-
ment.　Paris, 1874.　12mo.
Éléments d'urologie, ou analyse des urines, des dépôts et calculs urinaires.
Paris, 1875.　pp. vii–256, 12mo.　Ill.
Traité élémentaire de chimie médicale.　I. Chimie minérale.　Paris,
1878.　8vo.　Ill.

Rad.
Die Rübenzucker.　Wien, 1848.

Рачинскій, С.
О нѣкоторыхъ химическихъ превращеніяхъ раст. тканей.　Москва, 1866.
RACHINSKY, S.　On some Transformations in Plant Tissues.　Moscow, 1866.

Raffaele, Antonio.
La putrefazione sotto il rapporto della medicina legale.　Napoli, 1879.
pp. xix–220, 8vo.

Rahn, Johann Conrad.
Dissertatio de aquis mineralibus Fabariensis seu Piperinis.　Lugduni-
Batavorum, 1757.　4to.

Ralfe, Charles Henry.
Clinical Chemistry.　An account of blood, urine, morbid products, etc.,
with an explanation of some of the chemical changes that occur in
the body, in disease.　Philadelphia, 1883.　pp. viii–308, 16mo.　Ill.
Outlines of Physiological Chemistry, including the qualitative and
quantitative analysis of the tissues, fluids, and excretory products.
London, 1873.　pp. xxxii–236, 16mo.

RAMAER, JOHAN NICOLAAS.

Commentatio de gas oxygenii præparandi et colligendi rationibus deque ejusdem usu chymico et œconomico. Groningen, 1837. 4to. 5 plates.

RAMMELSBERG, CARL FRIEDRICH.

Anfangsgründe der quantitativen mineralogisch- und metallurgisch-analytischen Chemie durch Beispiele erläutert. Als Einleitung zu H. Rose's Handbuch der analytischen Chemie für Anfänger bearbeitet. Berlin, 1845.

> Leitfaden für die quantitative chemische Analyse besonders der Mineralien und Hüttenprodukte. Zweite umgearbeitete Auflage. Berlin, 1863. 8vo.
> Vierte Auflage. Berlin, 1886.
> Guide to a Course of Quantitative Chemical Analysis, especially of Minerals and Furnace Products. Illustrated by examples. Translated by J. Towler. New York, 1872.

Chemische Abhandlungen 1838–1888. Berlin, 1888. 8vo.

Chemische (Die) Natur der Mineralien. Berlin, 1886.

Chemische (Die) Natur der Meteoriten. Berlin, 1871–79 Two parts, 4to.

Grundriss der unorganischen Chemie gemäss den neueren Ansichten. Berlin, 1867. 8vo.

> Dritte Auflage. Berlin, 1873.
> Vierte Auflage : Grundriss der Chemie gemäss den neueren Ansichten. Berlin, 1874.
> Fünfte Auflage. Berlin, 1881. 8vo.
> Schets der anorganische scheikunde, naar de nieuwere beschouwingen. Uit het Hoogduitsch vertaald door H. Yssel de Schepper. Deventer, 1867. 8vo.
> Учебникъ неорганич. химіи по утилитарной системѣ. Пер. Савченкова Ф. С.-Петербургъ, 1867.

Handbuch der krystallographischen Chemie. Berlin, 1855. 8vo.

> Supplement. Leipzig, 1857.

> *Also under the title :* Die neuesten Forschungen in der krystallographischen Chemie.

Handbuch der krystallographisch-physikalischen Chemie. Mit Holzschnitten. Leipzig, 1881, 1882. 2 vols., 8vo. Vol. I, pp. xvi-615; vol. II, pp. xvi-532.

Lehrbuch der chemischen Metallurgie. . . . Berlin, 1850.

Lehrbuch der Stöchiometrie und der allgemeinen theoretischen Chemie. Berlin, 1842. pp. iv-386, 8vo.

> With a bibliography, pp. 383-386.

RAMMELSBERG, CARL FRIEDRICH. [Cont'd.]

Leitfaden für die qualitative chemische Analyse, mit besonderer Rücksicht auf Heinrich Rose's Handbuch der analytischen Chemie. Berlin, 1843. 8vo.

        Zweite Auflage. Berlin, 1847. Dritte Auflage, 1854.
        Vierte Auflage, 1860. Fünfte Auflage, 1867.
        Sechste Auflage, 1874. Siebente Auflage, 1885.
        Ledetraad i den qualitative chemiske Analyse. Oversat efter
            Originalens femte Oplag af C. Nebelong. Kjøbenhavn,
        1867. 8vo.

Neuesten (Die) Forschungen in der krystallographischen Chemie zugleich als Supplement zu dem Handbuch der krystallographischen Chemie. Leipzig, 1857. 8vo.

RAMMELSBERG, CARL FRIEDRICH.
        Handbuch der Mineralchemie. *See, in Section II*, Handwörterbuch des
        chemischen Theils der Mineralogie.

RAMOS Y LAFUENTE.

Elementos de química inorgánica. Obra de texto para la segunda enseñanza en los Institutos. Ilustrada con grabados intercalados y cinco láminas. Sexta edicion notablemente aumentada. Madrid, 1878. 8vo.

RAMSAUER, P.

Petroleum. Oldenburg, 1886. 8vo.

RAMSAY, WILLIAM.

Elementary Systematic Chemistry for the use of Schools and Colleges. London, 1891. 8vo.

Experimental Proofs of Chemical Theory for beginners. London, 1884. pp. xiv–134, 18mo.

System (A) of Inorganic Chemistry. London, 1891. 8vo.

        A systematic treatise based on the periodic classification of the elements. It
        includes all the elementary bodies and really teaches the science.

RAMSEY, WILLIAM (OR RAMESEY).

Life's Security, or the Names, Natures and Vertues of all Sorts of Venoms and Venemous Things. London, 1665. 8vo.

RANALLI, G.

Nozioni elementari di chimica scritte unicamente per uso dei propri scolari. Arezzo, 1875. 4to.

RAND, B. H.

Outline of Medical Chemistry for Students. Philadelphia, 1855. 8vo.

RANDAU, PAUL.

Fabrikation (Die) der Emaille und das Emailliren. Anleitung zur Darstellung aller Arten Emaille für technische und künstlerische
Zwecke und zur Vornahme des Emaillirens auf praktischem Wege.
Für Emaillefabrikanten, Gold- und Metallarbeiter und Kunstindus-
trielle. Wien, Pest, Leipzig, 18—. Ill.

RAOULT.

   Conférence (etc.). *See* Société chimique de Paris.

RASPAIL, FRANÇOIS VINCENT.

Essai de chimie microscopique appliquée à la physiologie ou l'art de
transporter le laboratoire sur le porte-objet dans l'étude des corps
organisés. Paris, 1830. 8vo.

Nouveau système de chimie organique fondé sur des nouvelles méthodes
d'observation et précédé d'un traité complet de l'art d'observer et
de manipuler en grand et en petit dans le laboratoire et sur le porte-
objet du microscope. Paris, 1833. pp. 96*–576, 8vo. 12 folding
plates.

   Deuxième édition, 1838. 8vo.

   A new System of Organic Chemistry. Translated from the
   French with notes and additions by William Henderson.
   London, 1843. pp. lxxvi–from 77–602, 8vo. 11 folding
   plates.

   Neues System der Chemie organischer Körper, auf neue
   Methoden der Beobachtungen gegründet. Aus dem
   französischen übersetzt und mit einigen Anmerkungen
   begleitet von Friedrich Wolff. Stuttgart, 1834. 8vo. Ill.

Recherches chimiques et physiques destinées à expliquer la structure et
le développement des tissus végétaux. Paris, 1827.

Tableau comparatif des caractères physiques de diverses fécules. Paris,
1826. 8vo.

RATH.

Die Rübenzucker-Industrie Oesterreichs. Prag, 1857.

RATHKE, BERNHARD.

Principien der Thermochemie und ihre Anwendung. Halle, 1881. 4to.

RATTON, J. J. L.

A Handbook of Common Salt. Madras, 1877. pp. xvii–281, 8vo.

RAU, ALBRECHT.

Bernsteinsäure (Die) als Product der alkoholischen Gährung zuckerhal-
tiger Flüssigkeiten, nebst Studien über die quantitative Bestimmung
derselben. Erlangen, 1892. 8vo.

Rau, Albrecht. [Cont'd.]
 Theorien (Die) der modernen Chemie. Braunschweig, 1877–84. 3 parts.
  8vo. Part I. Die Grundlage der modernen Chemie. Part II. Die
  Entwickelung der modernen Chemie. Part III. Die Entwickelung
  der modernen Chemie. Neue Folge.
   See also, *in Section III*, Rau, Albrecht.

Raulin, Joseph.
 Analyse des eaux minérales spatico-martiales de Provins. Paris, 1778.
  8vo.
 Traité analytique des eaux minérales en général . . . Paris, 1772–
  74. 2 vols., 12mo.

Ravaglia, G.
 Sunti di lezioni intorno agli elementi di chimica, destinati a facilitare lo
  studio ai giovani che frequentano i Licei, le Scuole Tecniche e
  Normali. Ravenna, 1886. 12mo.

Raydz-Kunheim.
 Die flüssige Kohlensäure und ihre Verwendung. Berlin, 1883. 4to.

Raynant, Ferdinando.
 La teoria atomica : lezione fatta nello studio di P. Palmieri. Napoli,
  1871. 8vo.

Reactions-Schema für die qualitative Analyse zum Gebrauche im
  chemischen Laboratorium zu Berlin. Berlin, 1870. Fol.

Reale, N.
 Lezioni di chimica farmaceutica e tossicologica. Napoli, 1877. 12mo.

Réaumur, Réné Antoine Ferchault de.
 L'Art de convertir le fer forgé en acier et l'art d'adoucir le fer fondu,
  ou de faire des ouvrages de fer fondu aussi fin que de fer forgé.
  Paris, 1722. 4to. 17 plates.
   This work brought the author a pension of 12,000 francs from the Duke of
    Orleans.

Recreaciones químicas, ó colleccion de experiencias curiosas é instructivas. . . . Barcelona, 1829. 2 vols., 4to. Ill.

Reddie, C.
 Chemical Atlas. Edinburgh, 1886. 8vo.

Redtel.
 Praktische Anleitung für den ersten Unterricht in der qualitativen
  chemischen Analyse der gewöhnlichen Verbindungen. Frankfurt
  am Main, 1843. 8vo.

REDTENBACHER, JOSEPH.
Der Sauerbrunnen zu Bilin in Böhmen chemisch untersucht . . .
Prag, 1845. 8vo.

REDWOOD, B.
Petroleum, its Production and Use.   New York, 1887.

REESE, JOHN R.
A Manual of Toxicology, including the consideration of the nature,
properties, effects and means of detection of poisons, more especially
in their medico-legal relations.   Philadelphia, 1874.   pp. xvi–507,
roy. 8vo.

REGNAULT, HENRI VICTOR.
Cours élémentaire de chimie à l'usage des facultés, des établissements
d'enseignement secondaire, des écoles normales et des écoles industrielles.   Paris, 1847–49.   2 vols. in 4 parts, 8vo.
> Troisième édition.   Paris et Bruxelles, 1850.
> Quatrième édition.   Paris, 1853.   4 vols., 8vo.
> Cinquième édition.   Paris, 1860.   4 vols., 8vo.
> Elements of Chemistry for the use of Colleges, Academies
> and Schools, translated from the French by J. Forrest
> Betton.   Edited by J. C. Booth and W. L. Faber.
> Second edition.   Philadelphia, 1853.   2 vols., 8vo.
> Lehrbuch der Chemie für Universitäten, Gymnasien, Real-
> und Gewerb-Schulen sowie für den Selbstunterricht.
> Uebersetzt von [Carl Heinrich Detlev] Boedeker. Berlin,
> 1849–51.   4 vols., 12mo.
> Corso elementare di chimica per uso delle scuole universitarie
> secondarie, normali ed industriali.   Prima traduzione
> Italiana (sulla seconda edizione Francese) di F. Selmi e
> G. Arpesani.   Con note dei traduttori e molti disegni
> intercalati nel testo.   Torino, 1851.   12mo.
> Curso elemental de química para el uso de las universidades,
> colegios y escuelas especiales escrito en Francés.   Tra-
> ducido de la segunda y ultima edicion Francesa, aumen-
> tado y publicado con annuencia y cooperacion del autor,
> por Gregorio Verdú.   Madrid, 1850–53.   4 vols., 8vo.
Premiers éléments de chimie.   Paris, 1850.   18mo.
> Quatrième édition.   Paris, 1861.
> Primi elementi di chimica.   Edizione Italiana per cura di
> Vincenzo Masserotti.   Milano, 1851.   16mo.   Ill.
> Quarto edizione.   Milano, 1862.

REGNAULT, HENRI VICTOR. [Cont'd.]

Primi elementi di chimica, settima edizione italiana con note e coll' aggiunta dei rudimenti di chimica organica per cura di Vincenzo Masserotti. Milano, 1868. 8vo.

REGNAULT, VICTOR, UND ADOLPH STRECKER.

Kurzes Lehrbuch der Chemie. In zwei Theilen. I. Anorganische Chemie. II. Organische Chemie. Theilweise nach Victor Regnault. Selbständig bearbeitet von Adolph Strecker. Braunschweig, 1851. 12mo. Ill.

Dritte Auflage. Braunschweig, 1855–57.• 2 vols., 12mo.

Vierte Auflage. Braunschweig, 1858. 2 vols., 12mo.

Siebente verbesserte Auflage. Braunschweig, 1866. 2 vols., 12mo. Ill.

Vol. I. Neunte verbesserte Auflage, 1877–81.

Vol. II. Sechste verbesserte Auflage, 1876. Bearbeitet von Johannes Wislicenus.

Beknopt leerboek der scheikunde. Naar de Hoogduitsche bewerking van Adolf Strekker. In het Nederduitsch overgebragt door J. J. Blekkingh. Utrecht, 1852. 12mo.

Beknopt leerboek der onbewerktuigde scheikunde. Naar het hoogduitsch door E. H. van Baumhauer en R. J. Opwijrda. Tweede en dere uitgave. Utrecht, 1862–64. 2 vols., 8vo.

A. Strecker's Short Text-Book of Organic Chemistry by J. Wislicenus. . . . Translated and edited with extensive additions by W. R Hodgkinson and A. J. Greenaway. London, 1881. 8vo.

Organic Chemistry. Translated and edited with extensive additions by W. R. Hodgkinson and A. J. Greenaway. Second edition. London, 1885. pp. 803, 8vo.

A szervéngi vegytan rövid kézikönyve. Után magyaritotta Oroszhegyi Jósa. A szövegbe foglalt 42 fametszvénynyel. Pest, 1857. 8vo.

Краткій Учебникъ органич. химіи. Пер Андреева. С.-Петербургъ, 1856.

REGNAULT, V., ET J. REISET.

Recherches chimiques sur la respiration des animaux. Paris, 1849.

REGNER, A. VON.

Bereitung (Die) der Schaumweine. Mit besonderer Berücksichtigung der französischen Champagner-Fabrikation. Genaue Anweisung und Erläuterung der vollständigen rationellen Fabrikationsweise

REGNER, A. VON. [Cont'd.]

aller moussirenden Weine und Champagner. Mit Benützung des Robinet'schen Werkes, auf Grund eigener praktischer Erfahrungen und wissenschaftlicher Kenntnisse dargestellt und erläutert. Wien, Pest, Leipzig, 18—. Ill.

REGNER, RICHARD VON.

Fabrikation (Die) des Rübenzuckers, enthaltend : Die Erzeugung des Brotzuckers, des Rohzuckers, die Herstellung von Raffinad- und Candiszucker nebst einem Anhange über die Verwerthung der Nachproducte und Abfälle. Zum Gebrauche als Lehr und Handbuch leichtfasslich dargestellt. Wien, 1879. Ill.

REGNIER, L. R.

L'intoxication chronique par la morphine et ses diverses formes. Paris, 1890. pp. 171, 8vo.

REGODT, HONORÉ.

Notions de chimie applicable aux usages de la vie. Deuxième édition. Paris, 1858. 12mo. Ill.

Vingt-neuvième édition. Paris, 1888. 12mo. Ill.

REHWALD, FELIX.

Die Stärke-Fabrikation und die Fabrikation des Traubenzuckers. Eine populäre Darstellung der Fabrikation aller im Handel vorkommenden Stärke sorten, als der Kartoffel-, Weizen-, Mais-, Reis-, Arrowroot-Stärke, der Tapioca, u. s. w. ; der Wasch- und Toilettestärke und des künstlichen Sago, sowie der Verwerthung aller bei der Stärke-Fabrikation sich ergebenden Abfälle, namentlich des Klebers und der Fabrikation des Dextrins, Stärkegummis, Traubenzuckers, Kartoffelmehles und der Zucker-Couleur. Ein Handbuch für Stärke- und Traubenzucker-Fabrikanten, sowie für Oekonomie-Besitzer und Branntweibrenner. Zweite sehr vermehrte und verbesserte Auflage. Wien, Pest, Leipzig, 1876. 8vo. Ill.

REIBENSCHUH, ANTON FRANZ.

Die neueren chemischen Theorien. Einleitung in das Studium der modernen Chemie. Graz, 1871. 8vo.

REICH, EDUARD.

Medicinische Chemie. Mit Berücksichtigung der österreichischen und preussischen Pharmakopoe. In zwei Bänden. Erlangen, 1858. 8vo. Vol. I, pp. xvi–314 ; vol. II, pp. xvi–465.

With an additional title-page running " Lehrbuch der Chemie," etc.

REICHARDT, EDUARD.

Ackerbauchemie, oder die Chemie in ihrer Anwendung auf Agricultur. Erlangen, 1861. pp. vi–636, 8vo.

Chemische Untersuchung der Mineralquelle zu Liebenstein. Hannover, 1859.

Chemischen (Die) Verbindungen der anorganischen Chemie, geordnet nach dem electro-chemischen Verhalten, mit Inbegriff der durch Formeln ausdrückbaren Mineralien. Erlangen, 1858. 8vo.

Grundlagen zur Beurtheilung des Trinkwassers, zugleich mit Berück- sichtigung seiner Brauchbarkeit für gewerbliche Zwecke nebst Anleitung zur Prüfung des Wassers. Für Behörden, Aerzte, Apo- theker und Techniker. Dritte sehr vermehrte Auflage. Jena, 1875. 8vo. Ill.

> Zweite Auflage. Jena, 1872.
> Vierte Auflage. Halle, 1880.

REICHEL, KARL GLI. WILHELM.

Handbuch der medizinischen Chemie nach den neuesten und besten Quellen, mit Berücksichtigung ihrer technischen Anwendung, be- arbeitet für Aerzte, Wundaerzte und Studirende, sowie zum Selbst- studium und zur Vorbereitung zum Examen. Bevorwortet von Heinrich Ficinus. Leipzig und Baltimore, 1837. 8vo.

REICHENBACH, KARL.

Das Kreosot in chemischer, physischer und medicinischer Beziehung. Zweite mit Nachträgen und Zusätzen von Schweigger-Seidel ver- mehrte Ausgabe. Leipzig, 1835. 8vo.

REID, DAVID BOSWELL.

Academical Examinations on the Principles of Chemistry, being an Introduction to the study of that science. Edinburgh, 1825. 2 vols., 12mo.

Elements of Practical Chemistry, comprising a systematic series of experiments, with directions, important tests and reagents. London, 1831. 8vo. Ill.

> Third edition. Edinburgh, 1839.

Rudiments of Chemistry. With illustrations of the chemistry of daily life. Fourth edition. London, 1851. 8vo. Ill.

> Vorschule der Chemie in einer Reihe von Versuchen. Mit Erläuterungen über chemische Erscheinungen im täg- lichen Leben. Für Schule und Haus gemeinfasslich behandelt. Aus dem englischen übersetzt von Kotten- kamp. Stuttgart, 1849. 8vo.

REID, HUGO.

Chemistry of Science and Art ; or elements of chemistry, adapted for reading along with a course of lectures for self-instruction, for use in schools and as a guide to teaching. Edinburgh, 1841. 12mo.

REIMANN, A.

Grundriss der Chemie. Ein Leitfaden für den ersten Unterricht in der Chemie. Saalfeld, 1854. 8vo.

> Grundriss der Chemie. Ein Leitfaden für den Unterricht in Realschulen und ihnen verwandten Lehranstalten. Nebst einem Anhang enthaltend Stöchiometrische Aufgaben zu den Nichtmetallen und ihren wichtigsten unorganischen Verbindungen. Zweite umgearbeitete Ausgabe. Saalfeld, 1870. 8vo.

REIMANN, M.

Färberei (Die) der Gespinnste und Gewebe. Practisches Handbuch der Färbekunst mit besonderer Berücksichtigung der modernen Farbstoffe. Berlin, 1867. 8vo. Ill.

Färberei (Die) der Wolle und der anderen wollenartigen Faserstoffe, für den Gebrauch des praktischen Färbers bearbeitet. Berlin, 1891. 8vo.

Leichtfassliche Chemie. Kurze allgemein verständliche Erklärung der chemischen Vorgänge in Färberei, Druckerei, Bleicherei, Appretur und den verwandten Industriebranchen. Dritte Auflage. Berlin, 1891.

Technologie des Anilins. Handbuch der Fabrikation des Anilins und der von ihm derivirten Farben. Mit Benutzung der neuesten Fortschritte der Wissenschaft. Berlin, 1866, 8vo. Ill.

> Aniline and its Derivatives, revised and edited by William Crookes. London and New York, 1868. 8vo.

REINDEL, FRANZ.

Lehrbuch der technischen Chemie. Vierte vollständig umgearbeitete Auflage des Fürnrohr'schen Lehrbuchs. Regensburg, 1863. 8vo.

> *Cf.* Fürnrohr, Aug. Em.

REINESIUS, THOMAS.

Chimiatria, hoc est, medicini nobili et necessaria sui parte chimia, instructa et exornata . . . Gerae Ruth, 1624. 4to.

> [Another edition.] Jenæ, 1678.

REINLEIN, JACOBUS.

Dissertatio physico-chemico-medica de phosphoris. Viennæ, 1768. pp. 62, 8vo.

REINSCH, HUGO.

Arsenik (Das). Sein Vorkommen, hauptsächlichsten Verbindungen, Anwendung und Wirkung, seine Gefahren für das Leben und deren Verhütung, seine Erkennung durch Reagentien, die verschiedenen Methoden zu dessen Ausmittelung, nebst einer neuen von Jedermann leicht ausführbaren zu dessen Auffindung. Zur allgemeinen Belehrung, so wie zum Gebrauche für Aerzte, Apotheker und Gerichtspersonen bearbeitet. Nürnberg, 1843. pp. 58, 8vo. Ill.

Grundriss der Chemie für den Unterricht an technischen Lehranstalten so wie zur Selbstbelehrung für angehende Pharmaceuten, Techniker, und Gewerbtreibende. Nebst einem Anhang über die chemische Technologie. Mannheim, 1854. 8vo.

Ueber die wahrscheinliche Zusammensetzung der chemischen Grundstoffe. Hof und Wunsiedel, 1839. 8vo.

Versuch einer von der atomistischen Ansicht abweichenden Erklärungsweise der chemischen Verbindungen. Erlangen, 1854. 8vo.

REISCHAUER, C.

Die Chemie des Bieres. Herausgegeben von Griessmayer. Augsburg und Zürich, 1878.

REISIG, F. W.

The Guide for Piece-Dyeing. Containing one hundred receipts with samples. New York, 1889. 8vo.

РЕЙСЪ Ф.

Способъ химическихъ препаратовъ и снарядовъ, въ пользу публичныхъ лекцій, сохраняемыхъ при Императорскомъ Московскомъ Университетѣ, Москва, 1826.

   REISS, F. Catalogue of chemical preparations and instruments used in the public lectures of Moscow University. Moscow, 1826.

REISSERT, ARNOLD.

Das Chinolin und seine Derivate. Braunschweig, 1889. pp. viii–228, 8vo.

REMER, WILHELM HERRMANN.

Lehrbuch der polizeilich-gerichtlichen Chemie. Helmstedt, 1803. pp. xxxii–454.

   Dritte Auflage. Helmstädt, 1827. 2 vols., 8vo.

  Translated into French by Bouillon-Lagrange, Paris, 1816 ; into Italian by Chiappari, Milano, 1818 ; into Russian, St. Petersburg, 1818.

Ueber die Definition der Salze und die Eintheilung der Säuren. Helmstädt, 1798.

REMLER, JOHANN CHRISTIAN WILHELM.

Chemische Untersuchung der Tamarindensäure, . . . Erfurt, 1787. 4to.

Salzchemie in Tabellen. Erfurt, 1789.

Tabellarische Uebersicht der festen und flüchtigen Bestandtheile, die in einem Pfunde der Mineralwässer enthalten sind . . . Erfurt, 1793. Folio.

Tabelle der Bestandtheile der genauer untersuchten Stein- und Erdarten. Erfurt, 1790.

Tabelle über den Gehalt der in neueren Zeiten untersuchten Mineralwässer. Erfurt, 1790.

Tabelle über die in Wasser und Weingeist löslichen Bestandtheile der Pflanzen. Erfurt, 1789.

Tabelle welche die Menge des wesentlichen Oels anzeigt das aus verschiedenen Gewächsen erhalten wird . . . Erfurt, 1789. 4to.

Taschenbuch für Tintenliebhaber . . . Leipzig, 1795.

See also in Section II.

REMSEN, IRA.

Elements (The) of Chemistry. A Text-book for Beginners. New York and London, 1887.

Inorganic Chemistry. New York and London, 1889. pp. xxi–827, 8vo.

Introduction (An) to the Study of Chemistry. New York and London, 1886. pp. xii–387, 12mo.

American Science Series, Briefer Course.

Introduction (An) to the Study of the Compounds of Carbon, or Organic Chemistry. Boston and London, 1885. pp. x–364, 12mo.

Einleitung in das Studium der Kohlenstoffverbindungen, oder organische Chemie. Zweite Auflage. Tübingen, 1891.

Laboratory (A) Manual, containing directions for a course of experiments in general chemistry. Systematically arranged to accompany the author's " Elements of Chemistry." New York, 1889. 12mo.

Second edition. New York, 1890.

Organic Chemistry. Boston, 1885.

Principles of Theoretical Chemistry, with special reference to the Constitution of Chemical Compounds. Philadelphia and London, 1877.

Second edition. Philadelphia, 1883. 12mo.

Grundzüge der theoretischen Chemie. Mit besonderer Rücksicht auf die Constitution chemischer Verbindungen. Tübingen, 1888.

Principi di chimica teorica, con speciale considerazione alla

REMSEN, IRA. [Cont'd.]

costituzione dei composti chimici. Traduzione eseguita
sulla terza edizione e corredata di aggiunte e note da A.
Alessi. Pisa, 1892. 8vo.

RENARD, ADOLPHE.

See Klement and Renard : Réactions microchimiques à cristaux [etc.].

RENARD, ADOLPHE.

Corps gras, huiles, graisses, beurres, cires. Ouvrage contenant l'indication
des lieux de provenance des corps gras, leurs fabrication, épuration,
propriétés, usages, ainsi que les meilleures méthodes employées
pour reconnaître leurs falsifications. Rouen, 1880. pp. 142, 8vo.

Traité de chimie appliquée à l'industrie. Avec 225 figures dans le
texte. Paris, 1890. pp. xxv–846, 8vo.

Traité des matières colorantes du blanchîment et de la teinture du
coton. Paris, 1883. 8vo. Avec un album de 83 échantillons.

RENNIE, JAMES.

Alphabet of Scientific Chemistry for the use of beginners. New edition,
much improved. London, 1834. pp. x–160, 16mo. Ill.

RENOTTE, F.

Petite chimie agricole à la portée des cultivateurs. Entretien familier
sur la vie des plantes et spécialement sur la culture rationnelle de
la betterave et sur les champs d'expériences. Louvain, 1886. 8vo.

Séparation (La), traitement des jus de betteraves par le saccharate de
chaux, derniers perfectionnements comprenant entre autres la suppression du lait de chaux et du noir animal. Paris, 1885.

RENWICK, JAMES.

First Principles of Chemistry, being a familiar introduction to the study
of that science. For use of Schools, Academies, and the lower
classes of Colleges. New York, 1852. pp. vi–444, 12mo. Ill.

Another edition. New York, 1840.

REPETITORIUM DER ORGANISCHEN CHEMIE nach den besten Quellen bearbeitet für Studierende der Medicin und Pharmacie. Augsburg,
1891. 12mo.

REPORT ON THE SCIENTIFIC RESULTS OF THE VOYAGE OF H. M. S.
CHALLENGER during the years 1873–76. . . . Prepared under the
superintendence of the late Sir C. Wyville Thomson . . . and now
of John Murray. . . . Physics and Chemistry, vol. II. London,
1889. 4to.

REUDLER, J. F. L.

Handboek der Chemie en practische pharmacie, in verband met de Pharmacopœa Neerlandica. (Physica, practische pharmacie en anorganische chemie). Amsterdam, 1854. 8vo.

REULEAUX, F.

Die Chemie des täglichen Lebens . . . Achte umgearbeitete und bedeutend erweiterte Auflage unter Mitwirkung von Gustav Heppe, Th. Schwartze [und] Jul. Zöllner. Leipzig, 1886. 4to. Ill.

> *Cf.* Johnston, James F. W. : Chemistry of Common Life.

REUSS, CHRISTIAN FRIEDRICH.

Sammlung einiger Abhandlungen aus der Oekonomie, Kameralwissenschaft, Arzeneykunde und Scheidekunst. Leipzig, 1777. 8vo.

Ueber des Salpeters vortheilhafteste Verfertigungsarten. Tübingen, 1783. Erster Supplement, 1785. Zweiter Supplement, 1786.

Untersuchung des Cyders oder Apfelweins. Tübingen, 1781. 8vo.

Untersuchungen und Nachrichten von des berühmten Selterwassers Bestandtheilen. Leipzig, 1775. 8vo.

> Zweite Auflage. Leipzig, 1780.

REUSS, FRANZ AMBROSIUS.

Chemisch-medicinische Beschreibung des Kaiser Franzensbad. Dresden, 1794. 8vo.

Saidschützer (Das) Bitterwasser, physikalisch, chemisch und medicinisch beschrieben. Prag, 1791. 8vo.

REUTER, CARL FRIEDRICH.

Gerichtliche Gutachten über Ausgrabung, Section und chemische Analyse von vier mit Arsenik vergifteten Leichen. Wiesbaden, 1846. pp. [vi]–140, 8vo.

RÉVEIL.

Du Lait. Paris, 1856. 4to.

REVERDIN, F., UND E. NÖLTING.

Ueber die Constitution des Naphtalins und seiner Abkömmlinge. Genf und Basel, 1880. 4to.

> Sur la constitution de la naphthaline et de ses dérivés. Mühlhausen i. E., 1888.

REY, A.

L'huile de pétrole ; exploitation, procédés pour l'extraire et la raffiner . . . Genève, 1865. 8vo.

Rey, Jean.
Essai sur la recherche de la cause pour laquelle l'estain et le plomb augmentent de poids quand on les calcine. Bazas, 1630. pp. 143.
> Nouvelle édition, revue sur l'exemplaire original et augmentée sur les manuscrits de la Bibliothèque du Roi et des Minimes de Paris. Avec des notes par Gobet. Paris, 1777. pp. xxxii–216, 8vo.
>> A classical work that might have overthrown the Phlogistic theory of combustion had its teachings been duly appreciated. *Compare* Høvinghoff, H.

Rey, J. A.
Ferments et fermentations ; travailleurs et malfaiteurs microscopiques. Paris, 1887.

Reyher, Samuel.
De auro et argento chymico. Gothæ, 1692.
> *See also, in Section III*, Reyher, Samuel.

Reynaud, Antoine André Louis.
Traité élémentaire de mathématiques de physique et de chimie . . . Paris, 1824. 8vo.
> Troisième édition. Paris, 1845. 2 vols., 8vo.

Reynolds, J. Emerson.
Experimental Chemistry for junior students. Second edition. London, 1885. Three parts, 16mo.
> Four parts. London, 1888. Part I, Introduction ; II, Non-metals ; III, Metals ; IV, Carbon Compounds.
> Leitfaden zur Einführung in die Experimental-Chemie. Uebersetzt von G. Siebert. 4 Theile. Leipzig, 1883–88.
Six Short Lectures on Experimental Chemistry. Second edition. Dublin and London, 1876. 8vo.
Qualitative Analyse. Mit specieller Berücksichtigung aller Erfordernisse, die Photographen nöthig haben. Aus dem englischen von Fr. Bollmann. Leipzig, 1863. 8vo.

Reynoso, Alvaro.
Recherches naturelles, chimiques et physiologiques sur le curare, poison des flèches des sauvages américains. Paris, 1855. 8vo.

Rhyn, Did. van.
Oratio de dogmatico chemiæ cultu a practico haud separando, deque causis quibus viros principes ad chemiam colendam impelli decet. Amstelodami, 1786. 4to.

RIBAN [ET BOURGOIN].
> Acides gras. *See, in Section II*, Fremy : Encyclopédie chimique, vol. VII.

RIBAUCOURT.
Élémens de chimie docimastique, à l'usage des orfèvres, essayeurs et affineurs. Paris, 1786. 8vo.
> Elementos de química docimástica . . . traducidos por Miguel Gerònimo Suarez y Nuñez. Madrid, 1791. 4to.

RIBBENTROP, HEINRICH GOTTLIEB FRIEDRICH.
Resultate chemischer und metallurgischer Erfahrungen in Absicht der Blei-Ersparung bei Schmelzprocessen, von Da Camera, übersetzt mit Anmerkungen von Lampadius. Dresden, 1797.

RICCA, FRANCESCO.
Trattato di chimica applicata alla mineralogia, alla botanica, alla fisiologia, alla igiene e alla patologia umana, alla farmacia, alla materia medica, alla giurisprudenza penale e civile, ed alla industria agricola ed artiera. Napoli, 1838. 4 vols., 8vo. Vol. I, pp. xx-613 ; vol. II, pp. 379 ; vol. III, pp. 260 ; vol. IV, pp. 251.

RICCI, A.
Chimico farmacista. Prontuario di tossicologia chimica. Seconda edizione, corretta ed aumentata. Milano, 1882. 8vo.

RICERCHE ESEGUITE NELL' ISTITUTO CHIMICO della R. Università di Roma nell' anno scolastico 1890–91. Roma, 1891. 8vo. Ill.

RICHARD, H.
> Die Gewinnung der Gespinnstfasern. *See* Otto-Birnbaum's Lehrbuch, vol. IX.

RICHARDS, EDGAR.
Principles and Methods of Soil-Analysis. Bulletin 10, Department of Agriculture, Division of Chemistry. Washington, 1886. 8vo.

RICHARDS, MRS. ELLEN H.
Chemistry (The) of Cooking and Cleaning. Boston, 1882.
Food Materials and their Adulterations. Boston, 1886. pp. iv–183, 12mo.
> Household Manuals, vol. II.

RICHARDS, JOSEPH W.
Aluminum, its History, Occurrence, Properties, Metallurgy and Applications, including its Alloys. Philadelphia, 1887. 8vo. Ill.

RICHARDSON, A. T.
Tables for Chemical Analysis. London, 1890. 8vo.

RICHARDSON, CLIFFORD [Editor].
> *See, in Section VII,* Association of Official Agricultural Chemists.

RICHARDSON, CLIFFORD.

Foods and Food Adulterants. Part II, Spices and Condiments. Bulletin No. 13, U. S. Department of Agriculture, Division of Chemistry. Washington, 1887. 8vo.

Investigation (An) of the Composition of American Wheat and Corn. Bulletin No. 1, Chemical Division, Department of Agriculture. Washington, 1883. pp. 69, 8vo.

Investigation (An) of the Composition of American Wheat and Corn. Second Report, Department of Agriculture, Bureau of Chemistry. Washington, D. C., 1884. 8vo.

Third Report on the Chemical Composition and Physical Properties of American Cereals. Bulletin No. 9, Division of Chemistry, Department of Agriculture. Washington, 1886. pp. 82, 8vo.

RICHARDSON, THOMAS, AND HENRY WATTS.

Chemical Technology, or chemistry in its applications to the arts and manufactures ; with which is incorporated a revision of Dr. Knapp's "Technology." Second Edition. London 1855–67. One vol. in 5 parts. 8vo Ill.

Parts I and II, By Edmund Ronalds and Thomas Richardson. Fuel and its applications. 1855. pp. xvii–799–32, 833 to 836, 8vo.

Part III, by Richardson and Watts. Acids, alkalies and salts. 1863. pp. xxvii–771, 8vo.

Part IV. Acids, alkalies and salts. 1865. pp. xxii–611.

Part V. Acids, alkalies and salts. 1867. pp. xxiii–894.

> *See* Knapp, F. *See also* Groves, Charles E., and William Thorpe.
> A portion of this work, relating to the manufacture of Gunpowder, has been translated into Chinese by John Fryer of Shanghai, and published in eight volumes. *Cf.* Fryer, John.

RICHE, ALFRED.

Leçons de chimie professées aux élèves de l'École Sainte Barbe qui se préparent à l'École polytechnique. Mesnil, 1863. 18mo.
> Troisième édition. Paris, 1887. 2 vols. Ill.

Manuel de chimie médicale et pharmaceutique. Paris, 1881. pp. viii–846, 12mo. Ill.

RICHE, A., ET A. GÉLIS.

L'art de l'essayer. Paris, 1888. Ill.
> Bibliothèque des connaissances utiles.

RICHTER, CARL.

Zink, Zinn und Blei.   Eine ausführliche Darstellung der Eigenschaften dieser Metalle, ihrer Legirungen unter einander und mit anderen Metallen, sowie ihrer Verarbeitung auf physikalischem Wege.   Für Metallarbeiter und Kunst-Industrielle geschildert.   Wien, Pest, Leipzig, 18—.   Ill.

RICHTER, CARL FRIEDRICH.

Kleine (Der) Chemiker, oder Anleitung zur Chemie zum Selbststudium. Leipzig, 1823.   8vo.

RICHTER, FRIEDRICH ADOLPH.

Lehrbuch der Chemie.   Halle, 1791.

RICHTER, JEREMIAS BENJAMIN.

Anfangsgründe der Stöchyometrie oder Messkunst chymischer Elemente. Breslau und Hirschberg, 1792–94.   3 vols., 8vo.

De usu matheseos in chymia.   Regiomontanus, 1789.

Ueber die neueren Gegenstände der Chymie.   Breslau, Hirschberg und Lissa, 1791–1802.   11 parts, 8vo.

RICHTER, M.

Tabellen der Kohlenstoff-Verbindungen.   *See in Section II.*

RICHTER, O.

General Outline of an Original System of Chemical Philosophy, comprising the determination of the volume-equivalents.   Edinburgh, 1869.   8vo.

Chemical Constitution of the Inorganic Acids, Bases and Salts, as viewed and interpreted from the standpoint of the "typo-nucleus theory."   Edinburgh, 1882.   8vo.

RICHTER, ROBERT.

Leitfaden zum Unterricht in der quantitativen analytischen Chemie. Freiberg, 1853.   8vo.   Ill.

Leiddraad tot onderrigt in de quantitatieve analytische scheikunde.   Uit het Hoogduitsch door T. J. van der Veer. Utrecht en Amsterdam, 1854.   12mo.

RICHTER, VICTOR VON.

Kurzes Lehrbuch der anorganischen Chemie, wesentlich für Studirende auf Universitäten und Polytechnischen Schulen so wie zum Selbstunterricht.   Bonn, 1875.   8vo.   Ill.

Zweite Auflage, 1878.   Vierte Auflage, 1884.

Sechste Auflage.   Bonn, 1889.

Inorganic Chemistry.   A Text-book for students.   Trans-

RICHTER, VICTOR VON. [Cont'd.]

lated by Edgar F. Smith.   Philadelphia, 1883.   pp. viii-
424, 12mo.   Ill.

Second edition.   Philadelphia, 1885.   pp. 432, 12mo.   Ill.

Third American edition from the fifth German edition.
Philadelphia, 1887.   12mo.   Ill.

Trattato di chimica inorganica, tradotto col consenso dell'autore sulla quarta edizione originale e corredato di note e
di un' appendice da A. Piccini.   Torino, 1884.   8vo.   Ill.

Учебникъ неор_аническои химіи по новѣйшимъ воззрѣніямъ.
Варшава, 1874.

Kurzes Lehrbuch der organischen Chemie, oder der Chemie der Kohlenstoffverbindungen.   Bonn, 1876.   8vo.

Zweite Auflage.   Bonn, 1880.   Dritte. Auflage.   1882.

Fünfte neu bearbeitete Auflage.   Bonn, 1888.

Sechste Auflage.   Bonn, 1891.

Chemistry of the Carbon Compounds, or Organic Chemistry.
Translated from the fourth German edition by Edgar
F. Smith.   Philadelphia, 1885.   12mo.   Ill.

Second American from the sixth German edition.   Philadelphia, 1891.   12mo.   Ill.

La chimica delle combinazioni del carbonio ovvero chimica
organica, tradotta sulla terza edizione originale da
G. Carnelutti.   Torino, 1883.   8vo.

RICKETTS, PIERRE DE PEYSTER.

Notes on Assaying and Assay Schemes.   New York, 1876.   pp. 169,
8vo.   Ill.   Four parts and an Appendix.

Part I :   Introduction—Apparatus—Reagents and Operations.

Part II :   Dry or Fire Assays.

Part III :   Wet Assays or Analyses.

Part IV :   Tables and References.

Appendix : Manipulation—Blowpipe Analysis—Apparatus and Reagents.

Fifth edition.   New York, 1880.

Fourteenth edition.   New York, 1891.

RICKETTS, PIERRE DE P., AND S. H. RUSSELL.

Skeleton Notes upon Inorganic Chemistry.   For the use of students at
lectures and as a guide to systematic study.   New York, 1888.   2
parts, 8vo.

RIDEAL, S.

Practical Organic Chemistry; the detection and properties of some of
the more important compounds.   London, 1889.   8vo.

RIDSDALE, C. H.
   Chemical Percentage Tables and Laboratory Calculations.   Manchester,
      1883.

RIEDEL, F. W.
   Einleitung zur Chemie nach der electro-chemischen- und der Typen-
      theorie.   Frankfurt a. O., 1873.   4to.

RIEMSDIJK, A. D. VAN.
   De scheikundige werking der warmte op anorganische verbindingen.
      Utrecht, 1864.

RIETH, R.
   Volumetrie (Die), oder chemische Maassanalyse zum Gebrauch im
      Laboratorium.   Bonn, 1871.   pp. ix–347, 8vo.   Ill.
   Volumetrische Analyse, ein Handbuch für Apotheker und Chemiker.
      Hamburg, 1883.

RIFFAULT DES HÊTRES, JEAN RENÉ DENIS.
   Art (L') du salpétrier.   Paris, 1813.
   Manuel complet du teinturier et du dégraisseur.   Paris, 1825.
   Manuel de chimie.   Paris, 1825.
   Traité de l'art de fabriquer la poudre à canon.   Paris, 1812.

RIGBY, EDWARD.
   Chemische Bemerkungen über den Zucker.   Aus dem englischen mit
      Anmerkungen von Samuel Hahnemann.   Dresden, 1791.   8vo.

RIGHINI, G.
   Prontuario per l'analisi chimica qualitativa dei corpi inorganici e dei
      principali composti organici.   Bologna, 1890.   12mo.

RIGG, ARTHUR, AND WALTER THOMAS GOOLDEN.
   An Easy Introduction to Chemistry.   London, 1875.   Roy. 8vo.
         New edition revised.   London, 1883.

RIGG, ROBERT.
   Experimental Researches, chemical and agricultural, shewing carbon to
      be a compound body. made by plants and decomposed by putre-
      faction.   London, 1844.   12mo.

RILLIET, AB.
   Notions de chimie.   Résumés du cours fait à l'École supérieure des
      jeunes filles.   Genève, 1884.   8vo.

RINGIUS, MARTINUS.   *See* Müller, Johannes.

RION, AD.
Éléments de chimie. Définition. Nomenclature. Analyse. Combinaisons. Métaux. Sels. Corps organiques, etc. Avec vignettes
dans le texte. Deuxième édition. Paris, 1855. 8vo.
Cinquième édition, Paris, 1875. 8vo.

RIPOLL Y PALOU.
Sustancias explosivas. Cartagena, 1886. 4to.

RIPPER, W.
Practical Chemistry, with notes and questions on theoretical chemistry,
adapted to the revised syllabus of the science and art department for
the elementary stage of inorganic chemistry. London, 1883. 8vo.

RISLER, JEAN. *See* Will, Heinrich ; Guide.

RITTER, E.
Cours de chimie ; leçons faites à la faculté de médecine à Nancy
recueillies par L. Garnier. Deuxième partie : chimie organique.
Troisième partie : chimie toxicologique. Nancy et Paris, 1880. 2
vols., 8vo.
Manuel de chimie pratique, analytique, toxicologique, zoochimique, à
l'usage des étudiants en médecine et en pharmacie. Paris, 1874. 8vo.

RITTER, FRIEDRICH.
Wasser und Eis. Eine Darstellung der Eigenschaften, Anwendung und
Reinigung des Wassers für industrielle und häusliche Zwecke und
der Aufbewahrung, Benützung und künstlichen Darstellung des
Eises. Für Praktiker bearbeitet. Wien, Pest, Leipzig, 1879. Ill.

RITTER, H. *See* Gmelin, Leopold.

RITTER, JOHANN WILHELM.
Physisch-chemische Abhandlungen in chronologischer Folge. Leipzig,
1806. 3 vols., 8vo. Ill.

RITTER, S.
Beiträge zur quantitativen Eiweissbestimmung. Breslau, 1887.

RITTHAUSEN, H.
Die Eiweisskörper der Getreidearten, Hülsenfrüchte und Oelsamen.
Beiträge zur Physiologie der Samen der Culturgewächse, der
Nahrungs- und Futtermittel. Bonn, 1872. pp. xi–252, 8vo.

РИТТИХЪ РУДОЛЬФЪ.
Сущность и цѣль химическихъ изслѣдованій и занятій. С.-Петербургъ 1871.
RITTIG, RUDOLPH. The scope and the aim of chemical investigations.
St. Petersburg, 1871.

RIVIÈRE, A. ET C.
Traité de manipulations de chimie à l'usage des établissements d'instruction secondaire, des écoles professionelles et des facultés.   Paris, 1882.   12mo.

RIVINUS, AUGUST QUIRINUS [BACHMANN].
Manuductio ad chemiam pharmaceuticam.   Lipsiæ, 1690.

RIVOT, LOUIS EDMOND.
Analyse des gaz et des eaux minérales.   Paris, 1862.   8vo.
Docimasie.   Traité d'analyse des substances minérales à l'usage des ingénieurs des mines et des directeurs de mines et d'usines.   Paris, 1861–66.   4 vols., roy. 8vo.   Vol. v, Paris, 1886.
> Handbuch der analytischen Mineralchemie.   Zum praktischen Gebrauche, insbesondere bei technischen und mineralogisch-chemischen Untersuchungen.   Unter specieller Autorisation und Mitwirkung des Verfassers ins Deutsche übersetzt und mit Anmerkungen versehen von Adolf Remelé.   Leipzig, 1863–'66.   2 vols., 8vo.
Principes généraux du traitement des minérais métalliques.   Traité de métallurgie, théorique et pratique.   Nouvelle édition.   Paris, 1871–73.   3 vols., 8vo.   Atlas oblong folio.

ROBERTS–AUSTEN, WILLIAM CHANDLER.
See Austen, William Chandler Roberts.

ROBIERRE, A.
Traité des manipulations chimiques.   Description raisonné de toutes les opérations chimiques et des appareils dont elles nécessitent l'emploi.   Paris, 1844.   8vo.

ROBIN, CHARLES, AND FR. VERDEIL.
Traité de chimie anatomique et physiologique normale et pathologique, ou des principes immédiats normaux et morbides qui constituent le corps de l'homme et des mammifères.   Accompagné d'un atlas de 45 planches gravées, en partie coloriées.   Paris, 1853.   8vo.   Vol. I, pp. xxxii–728 ; vol. II, pp. 584 ; vol. III, pp. 595 ; vol. IV, pp. 36 and 45 plates.

ROBIN, ÉDOUARD.
Chimie médicale raisonnée.   Paris, 1835–'37.   2 vols., 8vo.
Loi nouvelle régissant les différentes propriétés chimiques et permettant de prévoir, sans l'intervention des affinités, l'action des corps simples sur les composés binaires, spécialement par voie sèche. Nouvelle théorie de la fusion aqueuse et du mode d'action de la

ROBIN, ÉDOUARD.   [Cont'd.]
chaleur dans la fusion, la volatilisation et la décomposition. Propriétés chimiques fondamentales. Stabilité et solubilité. Paris, 1853. 18mo.
Précis élémentaire de chimie minérale et organique expérimentale et raisonnée. Première méthode par laquelle les faits se déduisent de lois générales au lieu d'être exposés comme des faits sans liaison qu'il fait apprendre de mémoire ou ignorer. Quatrième édition. Première partie: Lois qui régissent les propriétés physiques. Paris, 1853. 18mo.

ROBINET, ÉDOUARD (FILS).
Étude historique et scientifique sur la fermentation. Épernay, 1877. 8vo.
Manuel général des vins.   Fabrication des vins mousseux.   Paris, 1877. 12mo. Ill.
Manuel pratique d'analyse des vins, fermentations, alcoolisation, falsifications, procédés pour les reconnaitre. Troisième édition. 1879. 12mo.
    First edition, 1866.

ROBINET, STEPHANE.
Examen chimique des fruits du Lilas.   Paris, 1824.

ROBINSON, FRANKLIN CLEMENT.
Bowdoin College Course in Chemistry. Vol. i. *n. p., n. d.* [Brunswick, Maine, 1891]. 8vo.

ROCHLEDER, FRIEDRICH.
Anleitung zur Analyse von Pflanzen und Pflanzentheilen. Würzburg, 1858. pp. vii–112, 8vo.
Beiträge zur Phytochemie. Wien, 1847. 8vo.
Chemie und Physiologie der Pflanzen. (Abdruck aus C. G. Gmelin's Handbuch der organische Chemie.) Heidelberg, 1858. 8vo.
Genussmittel (Die) und Gewürze in chemischer Beziehung. Wien, 1852. 8vo.
Phytochemie. Leipzig, 1854. 8vo.

RODRIGUEZ BUSTILLO, JUAN ANTONIO.
Tratado de química analítica. Vigo, 1858. 4to.

RODRIGUEZ, T.
Elementos de física y química modernas. Madrid, 1892. 4to. Ill.

RÖHMANN, F.
Anleitung zum chemischen Arbeiten für Studirende der Medicin. Berlin, 1890. 8vo. Ill.

RÖSSIG, K. G.
  Abhandlungen über die vorzüglichsten einheimischen Zuckersurrogate.
    Leipzig, 1799.

RÖSSING, ADELBERT.
  Einführung in das Studium der theoretischen Chemie.   München und
    Leipzig, 1890.   pp. x–332, 8vo.
        Part I bears the title : Geschichtliche Entwicklung der chemischen Theorien,
          pp. 1 to 51.

ROGERS, SAMUEL BALDWYN.
  An Elementary Treatise on Iron Metallurgy up to the manufacture of
    puddled bars, built upon the atomic system of philosophy ; the
    elements operated upon being estimated according to Wollaston's
    hydrogen scale of equivalents.   London, 1857.   8vo.

ROGNETTA.
  Nouvelle méthode de traitement de l'empoisonnement par l'arsenic et
    documens médico-légaux sur cet empoisonnement.   Suivis de la
    déposition de Raspail devant la cour d'assises de Dijon.   Paris,
    1840.   pp. xxxii–108, 8vo.

ROHART, FRANÇOIS FERDINAND.
  Traité théorique et pratique de la fabrication de la bière.   Paris, 1848.
    2 vols., 8vo.

ROITI, A., E ALESSANDRI.
  Nozioni di fisica e chimica.   Libro di testo per i licei conforme ai pro-
    grammi scolastici.   Parte I. Chimica, compilata da G. Alessandri.
    Firenze, 1891.   8vo.   Parte II. Meccanica, Acustica, Cosmografia,
    compilata da A. Roiti.   Firenze, 1891.   8vo.

ROLFE, WILLIAM JAMES, AND JOSEPH ANTHONY GILLET.
  Handbook of Chemistry for school and home use.   Boston, 1869.   12mo.

ROLFINCK, WERNER.
  Chimia in artis formam redacta, sex libris comprehensa.   Genevæ, 1671.
    pp. [viii]–443–[x], sm. 4to.
          Editio secunda, Francofurti ad Mœnum, 1676.   Sm. 4to.
        Chapter II treats of the origin, antiquity and progress of chemistry, in which
          Adam and Tubal-cain receive full credit.
  De metallis imperfectis et mollibus.   Jenæ, 1638.
  De minera martis   Jenæ, 1668.
  De objecto chimiæ et de metallis perfectis sole et luna.   Jenæ, 1637.
  Dissertationes chymici sex :   De tartaro, sulphure, margaritis, auro,
    argento, antimonio, ferro et cupro.   Jenæ, 1691.

ROLFINCK, WERNER. [Cont'd.]
Non-entia chemica, mercurius metallorum et mineralium. Jenæ, 1670. 4to.
Scrutinium chimicum vitrioli. Jenæ, 1666.
> The author was the first to open a chemical laboratory at Jena.

ROLLETT, A.
> Blut und Blutbewegung. *See* Handbuch der Physiologie.

ROLOFF, JOHANN CHRISTOPH HEINRICH.
Anleitung zur Prüfung der Arzeneykörper . . . Magdeburg, 1812. 8vo.
> Dritte Auflage, 1820.

ROLOFF, J. F.
Die technische Chemie. Grundriss für Lehranstalten und Freunde der
> Naturwissenschaft. Erster Theil : Die anorganische Chemie.
> Neustrelitz, 1844. 8vo. Ill.

ROMEGIALLI, A.
Contribuzione alla teoria della fermentazione acetica e alla tecnologia
> dell' acetificazione. Roma, 1883. 8vo.

ROMEN, B.
Bleicherei, Färberei und Appretur der Baumwollen- und Leinenwaaren.
> Ein Lehr- und Handbuch, den Anforderungen der Gegenwart
> gemäss entworfen und unter Zugrundelegung der im praktischen
> Fabriksbetriebe gemachten Erfahrungen bearbeitet. Berlin, 1882.
> 8vo. Ill.

RONALDS, EDMUND, AND THOMAS RICHARDSON.
> *See* Richardson and Watts, Chemical Technology.

RONNA, A.
Chimie appliquée à l'agriculture. Travaux et expérience de A. Voelker.
> Paris, 1888. 2 vols., 8vo.

ROQUÉ Y PAGANI, PEDRO.
Curso de química industrial. Revisado por José Roura. Dedicado al
> ilustrisimo Sr. D. A. Gil de Zárate. Barcelona, 1851. 2 vols. Vol.
> I, pp. xxii–544, 7 plates ; vol. II, pp. vi–629, 5 plates.

ROSA, A. ED E. PERRONCITO.
Relazione sull' analisi chimica e biologica di alcune acque proposte per
> provvedere di acqua potabile la cittá di Aosta. Aosta, 1886. 16mo.

ROSCOE, SIR HENRY ENFIELD.
Chemistry. London, 1872. 16mo.
> Science Primers edited by Huxley, Roscoe and B. Stewart. No. 2.

ROSCOE, SIR HENRY ENFIELD. [Cont'd.]

Beginselen der scheikunde. Naar de vierde engelsche uitgave voor Nederland bewerkt door H. Wefers Bettink. Groningen, 1876. 8vo.

Derde druk. Groningen, 1884.

Kemia, suomentanut K. Suomalainen. Helsingissä, 1876. pp. vii–104, 16mo. Ill.

One of the series : Luonnontieteen Alkeiskirjoja, being Roscoe's Primer of Chemistry in Finnish.

Chemie. Uebersetzt von Rose. Dritte Auflage. Strassburg, 1882.

Vierte Auflage. Strassburg, 1886.

Elementarbuch der Chemie. Deutsche Ausgabe von F. Rose. Fünfte durchgesehene Auflage. Strassburg, 1892. 8vo. Ill.

Chimica, tradotta da A. Pavesi. Milano, 1891. 4to. Ill.

Kemi, öfversatt af P. T. Cleve. Stockholm, 1876. 8vo. Ill.

Lessons in Elementary Chemistry [inorganic and organic]. London, 1866. 16mo. Ill.

New editions in 1867, 1869, 1871, 1875, 1878, 1881, 1886, 1892.

Of the series known as " Macmillan's School Class Books."

Hwa hio ch'i meng. [Translated into Chinese by J. Edkins ; Canton ?]

Beknopt leerboek der scheikunde, volgens de nieuweste beschouwingen der wetenschap. Bewerkt naar de tweede Duitsche uitgave van Carl Schorlemmer, door H. J. Menalda van Schouwenburg. Dortrecht, 1870. 2 vols., 8vo. Ill. Vol. II : Scheikunde der koolstofverbinding (Organische scheikunde).

Vijfde geheel omgewerkte druk. Utrecht, 1879.

Kurzes Lehrbuch der Chemie. Uebersetzt von C. Schorlemmer. Braunschweig, 1871. 8vo.

Kurzes Lehrbuch der Chemie nach den neuesten Ansichten der Wissenschaften. Achte vermehrte Auflage. Braunschweig, 1886–89. 8vo. Two parts. Part I, Anorganische Chemie, 1886. Part II, section 1. Dritte Auflage, 1885. Section 2, 1889.

Cf. Schorlemmer, Carl, Lehrbuch der Kohlenstoffverbindungen.

Lezioni di chimica elementare inorganica ed organica. Seconda edizione sulla quinta edizione inglese di Orazio Silvestri. Milano, 1884. 16mo. Ill.

Prima edizione. Milano, 1874.

ROSCOE, SIR HENRY ENFIELD.   [Cont'd.]

> Краткій Учебникъ минеральной органич. химіи. Пер. Менде-
> лѣева. Д. С.-Петербургъ, 1868.

> [Other editions, 1873 and 1876.]

Spectrum Analysis.   Six Lectures delivered in 1868 before the Society
of Apothecaries of London.   London, 1869.   pp. xvi–348, 8vo.
Plates and Ill.

> Die Spectralanalyse in einer Reihe von 6 Vorlesungen mit
> wissenschaftlichen Nachträgen.   Autorisirte deutsche
> Ausgabe bearbeitet von C. Schorlemmer.   Zweite ver-
> mehrte Auflage.   Braunschweig, 1873.   8vo.   Ill.

Technical Chemistry.   Science Lectures at South Kensington.   London
and New York, 1877.   pp. 46, 12mo.   Ill.

ROSCOE, SIR HENRY ENFIELD, AND C. SCHORLEMMER.

Ausführliches Lehrbuch der Chemie.   Braunschweig, 1877–92.   6 vols.,
roy. 8vo.   Ill.   Vol. I, pp. ix–633, 1877 ; zweite Auflage, 1885.   Vol.
II, pp. xiii–866, 1879 ; zweite Auflage, 1888–89.   Vol. III, pp. xi–1179,
1882–1884.   Vol. IV, pp. xiv–1220, 1885–89.

> A Treatise on Chemistry.   London and New York, 1878–92.
> 9 vols., roy. 8vo.   Ill.

> > Vol. I. The Non-metallic Elements.   1878.   pp. vii–769.
> > Portrait of Dalton.

> > Vol. II, part I. Metals.   1879.   pp. viii–504.   Part II.
> > Metals.   1880.   pp. xiii–552.

> > Vol. III, part I. Organic Chemistry.   1882.   pp. x–724.
> > Part II. 1884.   pp. ix–655.   Part III. 1887.   pp. ix–388.
> > Part IV. 1888.   pp. xi–544.   Part V. 1890.   pp. x–523.
> > Part VI. 1892.   pp. xi–582.

> " As a systematic treatise on chemistry the work in its entirety has no equal
> in the English language."   *Chem. News.*

ROSE, HEINRICH.

Handbuch der analytischen Chemie.   Berlin, 1829.   8vo.

> Dritte Auflage, 1833.   2 vols., 8vo.

> Vierte Auflage.   Berlin, 1838.   2 vols., 8vo.

> Fünfte Auflage : Ausführliches Handbuch der analytischen
> Chemie.   Braunschweig, 1851.   2 vols., 8vo.

> Sechste Auflage.   Nach dem Tode des Verfassers vollendet
> von R. Finkener.   Leipzig, 1867–71.   2 vols., 8vo.
> Vol. I. Qualitative Analyse, pp. [viii]–863.   Vol. II.
> Quantitative Analyse, pp. [vi]–957.   Ill.

Handboek der analytische scheikunde.   Naar de derde

Rose, Heinrich.   [Cont'd.]

    uitgave vertaald door F. E. de Vry.   Het eene voorrede van G. J. Mulder.   Te Rotterdam, 1835–36.   2 vols., 8vo.

A Manual of Analytical Chemistry.   Translated from the German by John Griffin.   London, 1831.   pp. xvi–454, 8vo.

A Practical Treatise of Chemical Analysis, including Tables for calculations in analysis.   Translated from the French and from the fourth German edition with notes and additions by A. Normandy.   London, 1849.   2 vols., 8vo.

Traité d'analyse chimique, suivi de tables servant dans les analyses à calculer la quantité d'une substance d'après celle qui a été trouvée d'une autre substance.   Traduit de l'allemand sur la seconde édition par A. F. F. Jourdan.   Paris, 1832–38.   2 vols.

Revue et corrigé d'après la troisième édition allemande, et augmenté de tableaux synoptiques et d'une mémoire sur l'analyse chimique par B. Valérius.   Bruxelles, 1836–38.

Traduit sur la quatrième édition par A. F. F. Jourdan, accompagné de notes et additions par E. Péligot.   Paris, 1843.   2 vols., 8vo.

Traité complet de chimie analytique.   Édition française originale.   Paris, 1859–62.   2 vols., 8vo.   Vol. i. Analyse qualitative, pp. xvi–1063.   Vol. ii. Analyse quantitative. pp. vi–1251.   Ill.

Аналитическая химія С-Петербургъ, 1837.

Методическій ходъ качествен. анализа. Москва, 1862.

Tratado práctico de análisis química cualitativa traducido al francés de la cuarta edicion alemana por J. L. Jourdan . . . y del francés al castellano bajo la direccion de Pedro Mata.   Madrid, 1851.   8vo.

Tabulæ atomicæ.   The chemical tables for the calculation of quantitative analyses.   Recalculated . . . by William P. Dexter.   Boston, 1850.   8vo.

Roseleur, Alfred.

Manipulations hydroplastiques, guide pratique du doreur, de l'argenteur et du galvanoplaste.   Paris, 1855.   8vo.   Ill.

    Cinquième édition entièrement refondue et mise au niveau actuel des connaissances électro-chimiques.   Paris, 1884. Ill.

    Galvano-plastic Manipulations : a practical guide for the gold and silver electro-plater and the galvano-plastic

ROSELEUR, ALFRED. [Cont'd.]

> operator : comprising the electro-deposition of all metals by means of the battery and the dynamo-electric machine, as well as the most approved processes of deposition by simple immersion, with descriptions of apparatus, chemical products employed in the arts. Based largely on the "Manipulations hydroplastiques" of Alfred Roseleur. [Edited by William H. Wahl.] Philadelphia and London, 1883. pp. xxxviii–656, 8vo. Ill.

ROSENBERG, J. O.

Kemiska Kraften framstäld i dess förnämsta Verkningar. Stockholm, 1887. 8vo.

Lärobok i oorganisk kemi. Stockholm, *n. d.* [1887–88]. pp. iv–544, 8vo.

ROSENFELD, MAX.

Erster Unterricht in der Chemie. Für die unteren Classen der Mittel-schulen. Prag, 1880. 8vo. Ill.

> Eerste onderwijs in de scheikunde door B. van der Meulen. Groningen, 1881. 8vo. Ill.

Leitfaden der anorganischen Chemie. Freiburg, 1886. 8vo.

ROSENSTIEHL.

Recherches sur la formation du rouge d'aniline et sur quelques dérivés isomères du toluène. Strassbourg, 1871.

ROSENTHAL, J.

> Athembewegungen, etc. *See* Handbuch der Physiologie.

ROSS, WILLIAM ALEXANDER.

Blowpipe (The) in Chemistry, Mineralogy and Geology. Containing all known methods of anhydrous analysis, many working examples, and instructions for making apparatus. With one hundred and twenty illustrations by the author. Second edition, revised and enlarged. London, 1889. pp. xv–214, 8vo.

> *See, in Section II,* Ross, W. A.

> Das Löthrohr in der Chemie und Mineralogie. Uebersetzt von Kosmann. Leipzig, 1889. Ill.

Pyrology, or Fire Chemistry ; a science interesting to the general philosopher, and an art of infinite importance to the chemist, mineralogist, metallurgist, geologist, agriculturist, engineer (mining, civil, and military), etc., etc. London, New York, 1875. pp. xxviii–346, 4to. Ill., with folding plates.

> These works by a Major in the Royal Artillery reflect the interesting views of the author, and are too often neglected. He affects a peculiar and original terminology, which however, is explained by a glossary.

Rossi, S., e Malladra, A.
Nozioni di fisica, chimica, fisiologia umana ed igiene, ad uso delle scuole
tecniche. Milano, 1891. 16mo.

Rossignon, Jules.
Traité de chimie organique appliquée aux arts, à l'agriculture et à la
médecine. Paris, 1842. 3 parts, 8vo. Plates.

Roswag.
Cuivre: Argent; [Metallurgy of, and other articles]. *See, in Section II,*
Fremy: Encyclopédie chimique, vol. v.

Roth, E.
Die Chemie der Rothweine. Zweite Auflage. Heidelberg, 1884.

Roth, Gottfried.
Gründliche Anleitung zur Chymiæ. Vierte Auflage. Frankfurt und
Leipzig, 1750. 8vo.

Roth, J.
Allgemeine und chemische Geologie. Vol. i. Bildung und Umbildung
der Mineralien. Quell-, Fluss- und Meerwasser. Die Absätze.
Berlin, 1879. Vol. ii. Petrographie, Bildung, Zusammensetzung
und Veränderung der Gesteine. Drei Abtheilungen. Berlin,
1883–87.
Gesteins-Analysen in tabellarischer Uebersicht. Berlin, 1861. 4to.

Rothe, Carl.
Grundriss der Chemie. Wien, 1864. 8vo.
    Zweite Auflage. Wien, 1868.
        Handleiding bij beœfening van scheikunde. Bewerkt naar
        het Hoogduitsch door Corstiaan de Jong. Utrecht,
        1866. 8vo.
Ueber die Entdeckung von Elementen. Wien, 1885.

Rothe, Gottfried.
De salibus metallicis. Halæ, 1708.
Gründliche Anleitung zur Chymie . . . Leipzig, 1717. 8vo.
        This treatise, published after the author's death, passed through seven
        editions, the last in 1750, and was translated into French.

Rothe, Viggo.
Agerdyrknings-Chemie eller om Agerjordens Bestanddele, deres Egenskaber og Indflydelse paa Vegetationen. En Lærebog for Landmænd, Gartnere, Forstmænd og andre Jordbrugere. Kjøbenhavn,
1839. 8vo. Ill.

ROTHER, R.

The Chemistry of Pharmacy, an exposition of chemical science in its relations to medicinal substances according to a practical and original plan. Detroit, Michigan, 1888. pp. 71, 8vo.

ROTH-SCHOLTZ, FRIEDRICH.

Chymica curiosa variis experimentis adornata. Norimbergæ, 1720.

Cf. in Section I, Roth-Scholtz.

ROTTNER, F.

Chemie für Gewerbetreibende. Eine Darstellung der Grundlehren der chemischen Wissenschaft und deren Anwendung in den Gewerben. Wien, 1889. Ill.

ROUCHER, C.

Recherches toxicologiques. De la présence des poisons minéraux dans le système nerveux à la suite des empoisonnements aigus. Mémoire sur le traitement des matières organiques, en vue de la recherche des poisons. Paris, 1852. 8vo.

Toxicologie. Sur les empoisonnements par le phosphore, l'arsenic, l'antimoine et le plomb. Note et rapports présentés à la société de médecine légale. Paris, 1874. 8vo.

ROUELLE, GUILLAUME FRANÇOIS [called the Senior].

The "Cours de chimie" of this distinguished demonstrator of chemistry at the Jardin des Plantes, Paris, (1742–1768), exists only in manuscript. *Cf.* Hoefer's Histoire de la chimie, II, 389.

ROUELLE, HILAIRE MARTIN [called Junior].

Recherches chimiques sur l'étain publiées par ordre du gouvernement. Paris, 1781. 8vo.

Tableau de l'analyse chimique des procédés du cours de chimie. Paris, 1774. 12mo.

ROUGET DE LISLE.

Notice historique théorique et pratique, sur le blanchissage du linge de toile, de la flanelle et des divers vêtements. Paris, 1852. 1 vol., 4to.

ROUPPE, H. W.

Redevoering over den invloed der hedendaagsche scheikunde, op de œconomische wetenschappen en artsenijmengkunde. Te Rotterdam, 1796. 8vo.

ROURA, JOSÉ.

Tratado sobre los vinos, su destilacion y aceites. Nueva edicion. Barcelona, 1857. 8vo.

Rousseau.

    Généralités sur les métaux . . . Potassium, Sodium [and other articles].
    See, in Section II, Fremy: Encyclopedie chimique, vol. III.

Rousseau, Émile.

Introduction à l'étude de la chimie. Paris, 1844. 18mo.

Rousseau, Georg Ludwig Claudius.

Chemisch-mineralogische Abhandlungen. Nürnberg, 1790. 8vo.

Roux, Augustin.

Mémoires de chimie. Paris, 1764. 12mo.

Roux, J. P.

La fabrication de l'alcool. Distillation des betteraves. Paris, 1889.
La fabrication de l'alcool. Distillation des grains. Paris, 1888. Ill.
La fabrication de l'alcool. Distillation des vins. Paris, 1889.
La fabrication de l'alcool. Distillation du cidre. Paris, 1889.
La fabrication de l'alcool. La rectification. Paris, 1888.
La fabrication de l'alcool. Production du rhum. Paris, 1885.
La fabrication de l'alcool ; distillation des cannes et mélasses de cannes, production du rhum. Deuxième édition. Paris, 1892. 8vo. Ill.

Royal College of Chemistry, London Reports, etc. ; *see* London.

Rubien, Emil.

Kurzes Lehrbuch der Chemie nach den neueren Ansichten der Wissenschaft für Realschulen, höhere Bürgerschulen, Gewerbeschulen und Ackerbauschulen. Wriezen, 1875. 8vo.

Ruchte, S.

Ersten (Die) Anfangsgründe der Chemie mit besonderer Rücksichtnahme auf ihre Anwendung in den Gewerben. Für den Unterricht in der Chemie an höhern Bürger- und gewerblichen Fortbildung-Schulen, sowie zum Selbstunterrichte bearbeitet. München, 1869. 8vo.
Grundriss der Chemie. Ein Leitfaden für den Unterricht an Gewerbeschulen und verwandten Lehranstalten. Unter Berücksichtigung der Bestimmungen der Schulordnung für die technischen Lehranstalten bearbeitet. Rosenheim, 1866. 8vo.
Repetitorium der Chemie. 71 Fragen aus der Chemie für Chemiker, Mediciner und Pharmaceuten. München, 1863. 8vo.
Zusammenstellung der wichtigsten chemischen Processe. Neuburg, 1876.

Rüde, Georg Wilhelm.

Chemisches Probierkabinet. . . . Cassel, 1821.

RÜDE, GEORG WILHELM.   [Cont'd.]

Fassliche Anleitung, die Reinheit und Unverfälschtheit der vorzüglichsten chemischen Fabrikate einfach und doch sicher zu prüfen.
Cassel, 1806. 8vo.

Populäre Anweisung zur analytischen Prüfung der vorzüglichsten chemischen Heilmittel, oder Chemisches Probirkabinet für angehende
Aerzte und Apotheker.   Dritte wohlfeile Auflage.   Cassel, 1828.

RÜDIGER, ANTON.

Systematische Anleitung zur reinen und überhaupt applicirten oder allgemeinen Chymie.   .   .   .   Leipzig, 1756. 8vo.

RÜDINGER, HERMANN.

Die Bierbrauerei und die Malzextract-Fabrikation.   Eine Darstellung
aller in den verschiedenen Ländern üblichen Braumethoden zur
Bereitung aller Biersorten, sowie der Fabrikation des Malzextractes
und der daraus herzustellenden Producte.   Zweite vermehrte und
verbesserte Auflage.   Wien, 1887.   Ill.

RÜDORFF, FR.

Anleitung zur chemischen Analyse für Anfänger. Berlin, 1866.   4to.
      Zweite verbesserte Auflage.   Berlin, 1870.
      Vierte verbesserte Auflage.   Berlin, 1875.
      Siebente Auflage . . . für Anfänger.   Berlin, 1886.   8vo.
      Beginselen der scheikundige analyse ten dienste van hooger
         burgerscholen.   Naar den tweeden verbeterden druk
         bewerkt door R. Sinia.   Enkhuizen, 1871.   8vo.

Grundriss der Chemie für den Unterricht an höheren Lehranstalten.
Neunte Auflage.   Berlin, 1888.
      Schets der scheikunde ten dienste van het middelbaar
         onderwijs.   Met in den tekst gedrukte houtsneden.
         Uit het Hoogduitsch.   1e deel : Anorganische schei-
         kunde.   Amsterdam, 1868.   8vo.

RÜHNE, J. F.

      Die Kalk-, Cement-, Gyps-, und Ziegelfabrikation.   See Otto-Birnbaum's
      Lehrbuch.

RUETZ, OTTO.

Anleitung zur Prüfung von Trinkwasser und Wasser zu technischen
Zwecken nebst Methoden zur Beurtheilung des Trinkwassers.
Neuwied und Leipzig, 1882.   8vo.
      Zweite Auflage.   Neuwied, 1885.

Wie lassen sich Verfälschungen der Nahrungs-, Genussmittel und Con-
sumartikel leicht und sicher nachweisen ?   Eine Anleitung zur

RUETZ, OTTO. [Cont'd.]

Untersuchung derselben nach leichten Methoden und wissenschaftlichen Grundsätzen mit dem Gutachten der Reichsgesundheitsämter
und vielen Abbildungen für Apotheker, Aerzte, Gesundheitsämter
und Droguisten. Neuwied und Leipzig, 1883. pp. vi–125, 12mo.
Zweite Auflage. Neuwied und Leipzig, 1885.

RUDOLF, E.

Die gesammte Indigo-Küpenblau Färberei, Reservage- und Aetz-
Druckerei (Blaudruck) auf Baumwolle und Leinen. Aus den
Jahrgängen 1875–85 der Färberei-Muster-Zeitung gesammelt.
Leipzig, 1885. 8vo.

RUGGIERI.

La dynamite, histoire, fabrication, propriétés physiques, conservation,
emmagasinage et emploi commode facile et sûr. Paris, 1873. 8vo.

RUHLAND, REINHOLD LUDWIG.

System der allgemeinen Chemie, oder über den chemischen Process.
Berlin und Stettin, 1818. 8vo.

RUMFORD, SIR BENJAMIN THOMPSON, COUNT OF.

Essays political, economical and philosophical. Fourth edition. London, 1798. 3 vols., 8vo.
    A French translation was published at Geneva, 1799–1806, and a German
    translation at Weimar, 1800–'05.

Philosophical Papers ; being a Collection of Memoirs, dissertations and
experimental investigations relating to various branches of natural
philosophy and mechanics, London, 1803. 8vo.

RUNGE, FRIEDLIEB FERDINAND.

Bildungstrieb (Der) der Stoffe veranschaulicht in selbstständig ge-
wachsenen Bildern. Oranienburg, 1855. Fol.
Einleitung in die technische Chemie für Jedermann. Berlin, 1836.
pp. xiv–570, 8vo. Ill.
Farbenchemie. Berlin, Bromberg und Posen, 3 vols., 8vo, 1834, 1842,
1850.
Grundlehren der Chemie für Jedermann, besonders für Aerzte, Apo-
theker, Landwirthe, Fabrikanten und Gewerbtreibende und alle
Diejenigen welche in dieser nützlichen Wissenschaft gründliche
Kenntnisse sich erwerben wollen. Dritte vermehrte Ausgabe. Mit
82 Tafeln, worauf die chemischen Verbindungen befindlich sind.
Berlin, 1843. 8vo.
Grundriss der Chemie. Herausgegeben von dem unter Leitung Seiner

Runge, Friedlieb Ferdinand. [Cont'd.]
königlichen Hoheit des Kronprinzen Maximilian von Bayern
stehenden Vereine zur Verbreitung nützlicher Kenntnisse durch
gemeinfassliche Schriften. München, 1846. 2 vols., 8vo. Ill.
Neuer Abdruck. München, 1847–48. 2 vols., 8vo. Ill.

> Throughout these volumes the colors of chemical bodies, precipitates, etc.,
> are indicated by pigments inserted on squares in the text.

Neueste phytochemische Entdeckungen zur Begründung einer wissenschaftlichen Phytochemie. Berlin, 1820–21. 2 vols., 8vo. Ill.
Plates.

> *Also under the title :* Materialen zur Phytologie.

Technische (Die) Chemie der nützlichsten Metalle für Jedermann.
Berlin, 1838–39. 2 vols. Ill.
Zur Farben-Chemie. Musterbilder für Freunde des Schönen und zum
Gebrauch für Zeichner, Maler, Verzierer und Zeugdrucker. Dar-
gestellt durch chemische Wechselwirkung. Berlin, 1850. 4to. 21
tables.

Ruperti, I. O.
Das Probiren, in so weit diese Wissenschaft zum Münzwesen gehört.
Braunschweig, 1765. 8vo.

Ruprecht, Karl.
Die Fabrikation von Albumin und Eierconserven. Eine Darstellung
der Eigenschaften der Eiweisskörper und der Fabrikation von
Eier- und Blutalbumin, des Patent- und Naturalbumins, der Eier-
und Dotter-Conserven und der zur Conservirung frischer Eier
dienenden Verfahren. Wien, 1882. Ill.

Rybicki, T.
Zasady technologii chemicznéj, obejmujące wiadomósci treściwie zebrane
o fabrykacie i użytkach ważniejszych produktów mineralnych. Z
atlasem z 12 tablicami. Warszawa, 1846. 12mo.

Rydberg, J. R.
Om de kemiska Grundämnernas periodiska System. Stockholm, 1885.
8vo.

Rymsza, A.
Ein Beiträg zur Toxikologie der Pikrinsäure. Dorpat, 1889. 8vo.

Rzehak, A.
Ergebnisse der mikroskopischen Untersuchung des Wassers der Stadt
Brünn. Brünn, 1886.

S.

A System of Instruction in the Practical Use of the Blowpipe.   Being a graduated course of analysis for the use of students and all those engaged in the examination of metallic combinations.   New York, 1858.   pp. viii–269, 8vo.   Ill.

S., S.

Tesi per gli assistenti farmacisti, precedute dalla nomenclatura chimica ed alcune nozioni chimiche e botaniche.   Napoli, 1889.   12mo.

SAAVEDRA, FEDERICO.

Tratado de química y nociones de historia natural con aplicacion á los reconocimientos de aduanas, aprobado por el Consejo de la Direccion general del ramo y declarado de testo por la misma para la enseñanza de sus cátedras.   Barcelona, 1852.   pp. [iv.]–482, 8vo.

SABATIER.

Zinc, Cadmium et Thallium [and other articles].   *See, in Section II*, Fremy: Encyclopédie chimique, vol. III.

SACC, FRÉDÉRIC.

Éléments de chimie.   Paris, 1882.   2 vols.

Précis élémentaire de chimie agricole.   Paris, 1848.   18mo.

Deuxième édition.   Paris, 1855.

This treatise was honored by the award of a golden medal from the King of Prussia.

Tratado elemental de química agrícola escrito en francés, traducido al castellano por Balbino Cortés.   Madrid, 1853.   8vo.

Trabajos del laboratorio nacional de química en Cochabamba.   La Paz 1886.   12mo.

SACHSSE, ROBERT.

Chemie (Die) und Physiologie der Farbstoffe, Kohlenhydrate und Proteïnsubstanzen.   Ein Lehrbuch für Chemiker und Botaniker. Leipzig, 1877.   pp. viii–339, 8vo.   Ill.

Lehrbuch der Agrikulturchemie.   Leipzig, 1888.   pp. vi–628.

Phytochemische Untersuchungen.   Leipzig, 1880.   8vo.

SADEBECK, MORITZ.

Anfangsgründe der Chemie.   Leitfaden für den Unterricht an Gymnasien und Realschulen.   Breslau, 1841.   12mo.

SADTLER, SAMUEL P.

Chemical Experimentation, being a Hand-book of lecture experiments in inorganic chemistry.   Louisville, 1878.   8vo.

SADTLER, SAMUEL P. [Cont'd.]

Gewinnung (Die) des Theers und Ammoniakwassers. Vortrag. Aus dem englischen mit eine Nachtrage von G. Bornemann. Leipzig, 1886.

Handbook (A) of Industrial Organic Chemistry adapted for the use of Manufacturers, Chemists and all interested in the utilization of organic materials in the Industrial Arts. Philadelphia, 1891. pp. xiv–519, roy. 8vo. Ill.

> Contains short bibliographies at the close of each chapter, on the following topics, the one-line titles being arranged chronologically under each heading.
> 1. Petroleum and Mineral Oil Industry.
> 2. Industries of the Fats and Fatty Oils.
> 3. Industry of the Essential Oils and Resins.
> 4. The Cane-Sugar Industry.
> 5. The Industries of Starch and its Alteration Products.
> 6. Fermentation Industries (Malting, Brewing, Wines, Spirits, Vinegar, Flour and Bread).
> 7. Milk Industries.
> 8. Vegetable Textile Fibres and their Industries.
> 9. Textile Fibres of Animal Origin.
> 10. Leather, Glue and Gelatin.
> 11. Destructive Distillation Industries.
> 12. Artificial Coloring Matters.
> 13. Natural Dye Colors.
> 14. Bleaching, Dyeing and Textile Printing.

SAENZ DIEZ, MANUEL.

Historia y juicio crítico de la diálisis, considerada como procedimiento analítico. Madrid, 1871. 8vo.

SAEZ Y PALACIOS, RAFAEL.

Tratado de química inorgánica teórico y prático aplicada á la medicina y especialmente á la farmacia. Obra con figuras intercaladas en el texto. Madrid, Paris, Londres, Nueva York, 1868–69. 8vo. Tom. I, pp. 695 ; tom. II, pp. 751.
Segunda edicion. Madrid, 1875. 2 vols., 8vo.

SAFARIK, A.

Ueber die chemische Konstitution der natürlichen chlor- und fluorhaltigen Silikate. Prag, 1874. 4to. Ill.

ŠAFAŘÍK, V.

Rukovět chemie pro vysoké učení české. V Praze, 1881.

SAFFRAY.

La chimie des champs. Troisième édition. Paris, 1884. 16mo.

SAGE, BALTHAZAR GEORGES.
   Analyse chimique et concordance des trois règnes.   Paris, 1786.   3 vols.,
      8vo.   Ill.
   Analyse de l'eau de mer.   Paris, 1817.   8vo.
   Analyse des blés.   .   .   .   Paris, 1776.   8vo.
   Art (L') d'essayer l'or et l'argent.   Paris, 1780.   8vo.
         Die Kunst Gold und Silber zu probiren, oder Erfolg der
            Kapellirung.   Reval und Leipzig, 1782.   8vo.
   Art (L') d'imiter les pierres précieuses.   Paris, 1778.   8vo.
   De la formation de la terre végétale nommé humus . . . Paris, 1816.   8vo.
   Des mortiers et des ciments.   Paris, 1808.   Nouvelle édition, 1809.
   Éléments de minéralogie docimastique.   Paris, 1772.   8vo.
         Deuxième édition.   Paris, 1777.   2 vols., 8vo.
   Examen analytique des œufs de poule . . .   Paris, 1823.   8vo.
   Examen chymique de différentes substances minérales.   Essais sur le
      vin, les pierres, les bézoards, etc.   Traduction d'une lettre de M.
      Lehmann sur la mine de plomb rouge.   À Paris, 1769.   8vo.
         Des Herrn Sage chemische Untersuchung verschiedener
            Mineralien.   Aus dem französischen übersetzt.   Mit
            einigen Anmerkungen vermehrt von Johann Beckmann.
            Göttingen, 1775.   pp. 208-[iv], 12mo.
   Institutions de physique.   Paris, 1811.   3 vols., 8vo.
         Supplément.   Paris, 1812.
   Mémoires de chimie.   Paris, 1773.   pp. vii–262–xxxviii, 12mo.   Fold-
      ing table.
         This contains many analyses of minerals.   The memoirs are reprinted from
            the Mémoires de l'Académie des Sciences.
   Moyen de rémédier aux poisons. . . .   Paris, 1811.   8vo.
   Opuscules physico-chimiques.   Paris, 1818.   8vo.
   Précis historique des mémoires sur l'eau de mer.   Paris, 1817.   8vo.
         See also in Section IV.

SAGE, BALTHAZAR GEORGES, ET PERTHUIS DE LAILLEVAULT.
   Art (L') de fabriquer le salin et la potasse . . .   Paris, 1777.   8vo.
         [Another edition.]   Paris, 1794.

SAINTE-CLAIRE-DÉVILLE, HENRI.
      Leçon (etc.).   *See* Société chimique de Paris.   *Cf.* Faraday, M. : Histoire
         d'une chandelle.

SAINTE-CLAIRE-DÉVILLE, HENRI.
   De l'aluminium, ses propriétés sa fabrication et ses applications.   Paris,
      1859.   pp. ix–176, 8vo.   One folding plate.
   Recherches sur la décomposition des corps par la chaleur et la dissocia-
      tion.   Paris, 1860.   8vo.

SAINTE-CLAIRE-DÉVILLE, HENRI, ET LEBLANC.

Mémoire sur la composition chimique des gaz rejetés par les évents
volcaniques de l'Italie méridionale.   Paris, 1889.   8vo.

SALA, ANGELO.

Hydrelæologia, darinnen wie man allerley Wasser, Oliteten und bren-
nende Spiritus der vegetabilischen Dingen durch gewisse chymische
Regeln und Manualia in ihren besten Kräfften distilliren und rectifi-
ciren soll. . . .   Rostock, 1639.   12mo.

Opera medico-chymica.   Francofurti, 1647.   4to.

> A second edition was published by J. D. Thom. A., under the following
> title :

> > *Collectanea chimica curiosa, quæ veram continent rerum
> > naturalium anatomiam sive analisin et triplici regno tam
> > vegetabili animali quam et minerali, unde generosa hac-
> > tenus a neotericis hinc inde tradita resultant et traduntur
> > medicamina adversus omnes corporis morbos, cum usu
> > simul et applicandi modo accurate adjecto opera et
> > studio.   Francofurti, apud viduam Hermanni à Sande,
> > 1693.   pp. [viii]-927-[xxv], sm. 4to.   Ill.

> Contains 19 treatises on chemistry, medicine and alchemy.   The Saccharo-
> logia is noteworthy.

> In "Exegesis chymiatrica Andreæ Tentzelii," we note as a curiosity in
> practical chemistry an illustration of a method for obtaining a heat suf-
> ficient to melt the neck of a glass flask or to apply "Hermes' seal."   An
> eopile with a curved neck drawn to a small bore directs a current of
> steam against the flame of a candle.   This "Globus Æolius ad sigilla-
> tionem Hermeticam aptus" forms a substitute for the mouth blowpipe,
> and seems to anticipate the "hot blast."   (Page 769.)

Saccharologia, darinnen erstlich von der Natur qualiteten, nützlichem
Gebrauch und schädlichem Missbrauch des Zuckers.   Darnach wie
von demselben ein Wein mässiger starcker Getranck Brandwein
und Essig als auch unterschiedliche Art hochnützlicher Medicamenta damit können bereitet werden, beschrieben und angezeiget
wird.   Rostock, 1637.   pp. [xxiv]-190-[xliv], 12mo.

SALAZAR, A. E., Y C. NEWMAN.

Examen químico y bacteriológico de las aguas potables.   Con un capí-
tulo de Rafael Blanchard sobre los animales parasítos introducidos
por el agua en el organismo.   Obra ilustrada con 127 grabados, 16
fotomicrografías, y 5 fotogramas de cultivos.   Londres, 1890.   pp.
xix-513.   Plates.

> The preface is dated : Laboratorio de la Escuela Naval, Valparaiso, 1890.
> Blanchard's essay occupies pages 375 to 502.

SALER, HIERONIMUS.
    See Brunschwick, Iheronimus.

SALET, GEORGE.
Traité élémentaire de spectroscopie. Paris, 1887. 8vo.

SALMON, WILLIAM.
  * Polygraphice : or the arts of drawing, engraving, etching, limning, painting, washing, varnishing, gilding, colouring, dying, beautifying and perfuming. In seven books. Exemplified in the drawing of men, women, lanskips, countreys and figures of various forms ; The way of engraving, etching and limning, with all their requisites and ornaments ; The depicting of the most eminent pieces of antiquities ; The paintings of the antients ; Washing of maps, globes or pictures ; The dying of cloths, silk, horns, bones, wood, glass, stones, and metals ; The vernishing, colouring and gilding thereof, according to any purpose or intent ; The painting, colouring and beautifying of the face, skin and hair ; The whole doctrine of perfumes (never published till now), together with the original advancement and perfection of the art of painting ; And a discourse of perspective, chiromancy and alchymie. To which also is added :

    I. The one hundred and twelve chymical Arcanums of Petrus Johannes Faber, a most learned and eminent physician, translated out of Latin into English.

    II. An abstract of choice chymical preparations, fitted for vulgar use, for curing most diseases incident to humane bodies.

  The fifth edition, enlarged with above a thousand considerable additions. Adorned with xxv copper sculptures ; the like never yet extant. London, printed for Thomas Passinger at the three Bibles on London Bridge ; and Thomas Sawbridge at the Three Flower de Luces in Little-Britain, 1685. pp. [lxvi]–767, 8vo. Portrait of Salmon, plates and an illust. title-page.

SALOMON, F.
Systematischer Gang der qualitativen Analyse. Braunschweig, 1878. 8vo.

SALOMONE-MARINO, SALVATORE.
L'acqua tofana, notizie raccolte. Palermo, 1882. 8vo.

SALTONSTALL, W.
An Inaugural Dissertation on the chemical and medical history of Septon, Azote, or Nitrogene, and its combinations with the matter of heat and the principle of acidity. New York, 1796. 8vo.

SALVÉTAT, LOUIS ALPHONSE.

Leçons de céramique professées à l'École centrale des arts et manufactures, ou Technologie céramique comprenant les notions de chimie, de technologie, et de pyrotechnie, applicables à la fabrication, à la synthèse, á l'analyse, à la décoration des poteries. Paris, 1857. 2 vols., 18mo.

    Cf. Ebelmen, J. J.

SALZER, CARL FRIEDRICH.

Bleyzucker-Fabrikation (Die) in ihrem ganzen Umfange. Karlsruhe, 1820. 8vo.

Fabrication (Die) des Leims. Heilbronn, 1842. 8vo.

Neue Entdeckungen das Meerwasser . . . mittelst der Luftpumpe trinkbar zu machen. Heilbronn, 1834.

Neues Verfahren das blausaure Eisenkali fabrikmässig zu bereiten. Heilbronn, 1842. 8vo.

Neueste (Das) . . . Verfahren der Schnell-Essigfabrikation . . . Heilbronn, 1833.

Versuche über die Schiesspulver. . . . Karlsruhe, 1824.

Versuche zu einer neuen Verdunstung . . . nebst einer Abhandlung das Meerwasser auf eine ganz einfache Weise trinkbar zu machen. Heilbronn, 1833. 8vo.

SAMMLUNG ACHT HUNDERT UND SIEBEN UND FÜNFZIG CHYMISCHER EXPERIMENTE einer Gesellschaft in dem Ertzgebürge. *See* L., F. C.

SAMMLUNG CHEMISCHER EXPERIMENTE zum Nutzen der Künstler, Fabrikanten und überhaupt für alle Stände. Leipzig, 1793. 8vo.

SAMMLUNG NEUER ENTDECKUNGEN UND VERBESSERUNGEN in der Färberei, örtlichen Druckerei und Farbenbereitung. Zweite Auflage. Nürnberg [?], 1834.

SÁN CRISTÓBAL, JOSÉ MARÍA, Y JOSÉ GARRIGA Y BUACH.

Curso de química general aplicada á las artes. Paris and Madrid, 1804. 2 vols., 4to.

SANDERS, J. MILTON.

A System of Instruction in the Practical Use of the Blowpipe. New York, 1858. 12mo. Ill.

SANDRI, L.

Appunti di chimica organica, per la seconda classe. Parma, 1891. 4to.

SANSON, A.

Les principaux faits de la chimie. Deuxième édition. Paris, 1860. 12mo.

SANSONE, ANTONIO.
Dyeing, comprising the dyeing and bleaching of wool, silk, cotton, etc. With numerous plates and specimens. Manchester and London, 1888. 8vo. Ill.
Printing (The) of cotton fabrics, comprising calico-bleaching, printing and dyeing. Manchester and London, 1887. 8vo.

SANTES DE ARDOYNIS DE PĒSAURO.
Incipit liber de venenis quem Magister Santes de Ardoynis de Pēsauro phisicus Salvatoris nostri cõfisus auxilio edere cepit Venetiis die octavo Novẽbris, 1424. . . . Impressum Venetiis opera Bernardini Ricii de Novaria . . . impensa vero Dñi Magistri Joãnis Dominici de Nigro. 1492. Fol.
> *See* Hain, No. 1554 ; *and compare in* Graesse, *and in* Brunet, Ardoynis, Santes de.

SANTINI, S.
Lezioni di chimica inorganica, organica ed analitica ad uso degli Istituti Tecnici. Seconda edizione, completamente rifatta ed accresciuta. Torino, 1892. 8vo.

SANTOS DE CASTRO, FERNANDO.
Nociones elementales de química acomodadas á los alumnos del segundo año de filosofía . . . Sevilla, 1842. 8vo.

SANTOS E SILVA, JOAQUIM DOS.
Factoren-Tabellen. Zur Ausführung chemischer Rechnungen mittels der von L. Meyer und K. Seubert gegebenen Atomgewichte. Braunschweig, 1887. 8vo.

SAPORTA, A. DE.
Chimie (La) des vins. Les vins naturels, les vins manipulés et falsifiés. Paris, 1889.
Théories (Les) et les notations de la chimie moderne. Avec une introduction par C. Friedel. Paris, 1888. 12mo.

SARRAU, E.
> Conférence (etc.), *see* Société chimique de Paris.

SARREAU ET VIEILLE.
> Substances explosives. *See, in Section II*, Fremy : Encyclopédie chimique, vol. v.

SARTORI, G.
Analisi del latte : guida pratica (chimica agraria). Milano, 1886. 8vo.
Chimica agraria. Milano, 1887. 8vo.

SAUCEROTTE.

Petite chimie des écoles. Simples notions sur les applications les plus utiles de cette science à l'agriculture, à l'industrie et à l'économie domestique. Paris, 1868. 18mo.

Deuxième édition. Paris, 1874.

Sixième édition, revue et modifié. Paris, 1885. 18mo.

SAUNDERS, WILLIAM.

A Treatise on the chemical history and medical powers of some of the most celebrated Mineral Waters, with practical remarks on the aqueous regimen, to which are added observations on the use of cold and warm bathing. Second edition. London, 1805. pp. xx–570, 8vo.

SAUNIER, E.

Tableaux synthétiques de chimie atomique, théorique et pratique, avec les propriétés physiologiques et toxicologiques, les réactifs, contrepoisons, usages, détails de manipulations, etc. Métalloïdes. Métaux et chimie organique. Troisième édition. Bruxelles, 1883. 18mo.

SAUSSURE, NICOLAS THÉODORE DE.

Recherches chimiques sur la végétation. Paris, 1804. 8vo.

Chemische Untersuchung über die Vegetation. Uebersetzt von Wieler. Leipzig, 1890. 8vo.

SAUVAGES DE LA CROIX, FRANÇOIS BOISSIER DE.

Mémoires sur les eaux minérales d'Alais, . . . Montpellier, 1736. 4to.

САВЧЕНКОВЪ Ф.

Основнія понятія химіи. С.-Петербургъ, 1863.

SAVCHENKOF, F. The First Principles of Chemistry. St. Petersburg, 1863.

SBRIZIOLO, MARCO.

Conferenze di chimica moderna inorganica. Caltanisetta, 1878. 4to.

Trattato di chimica analitica qualitativa e quantitativa. Napoli, 1883. pp. xii–574.

Trattato di chimica generale inorganica ed organica esposto sotto il punto di vista della dottrina moderna. Napoli, 1883. 8vo.

Seconda edizione. Napoli. 1884.

Trattato teoretico-pratico di tossicologia generale e speciale medico-clinico-legale. Napoli, 1886. 8vo.

Cf. Oglialoro Todaro, A.

SCHABUS, J.
  Bestimmung der Krystallgestalten in chemischen Laboratorien erzeugter
    Producte. Eine von der kaiserlichen Akademie der Wissenschaften
    gekrönte Preisschrift. Wien, 1855. Roy. 8vo.

SCHAEDLER, CARL.
  Kurzer Abriss der Chemie der Kohlenwasserstoffe. Zugleich ein Re-
    petitorium für Studirende und praktische Chemiker, Techniker,
    Apotheker, etc. Leipzig, 1885 8vo.
  Technologie (Die) der Fette und Oele des Pflanzen- und Thierreichs.
    Berlin, 1882–87. 2 vols., 8vo. I, pp. xii–1108 ; II, Leipzig, pp.
    xii–1052. Ill. Plates.
          In accordance with the unhappy custom of Germany vol. II also bears two
            title-pages :
            (1) Die Technologie der Fette und Oele der Fossilien
              (Mineralöle) sowie der Harzöle und Schmiermittel.
              (2) Die Technologie der Fette und Oele. Erster Theil.
              Verseifbare Fette und Oele. Die Technologie der
              Fette und Oele des Pflanzen- und Thierreichs. Zweiter
              Theil. Nichtverseifbare Fette und Oele. Die Tech-
              nologie der Fette und Oele der Fossilien (Mineralöle)
              Harzöle und Schmiermittel. Leipzig, 1887.
            Zweite völlig umgearbeitete Auflage bearbeitet von P. Loh-
              man. Leipzig, 1892. 8vo. Ill.
  Untersuchung der Fette, Oele, Wachsarten und der technischen Fett-
    producte unter Berücksichtigung der Handelsgebräuche. Leipzig,
    1890. 8vo.

SCHAER, EDUARD.
  Das Zuckerrohr, seine Heimat, Kultur und Geschichte. Zürich, 1889 4to.

SCHAFFER, F.
  Einfache Methoden zur Prüfung der wichtigsten Nahrungsmittel.
    Bern, 1890. 8vo.

SCHAFFT, A.
  Uebersichtstafeln zum Unterricht in der anorganischen Chemie und
    Mineralogie. Bielefeld, 1886. 8vo.

SCHAFS, A.
  Notes des aides-chimistes pour sucreries. St. Trond, 1885. 8vo.

SCHARFFIUS, BENJAMIN.
  Τοξιϰολογία, seu tractatus de natura venenorum in genere. Jenæ,
    1682. 8vo.

SCHASCHL, JOSEF.

Die Galvanostegie mit besonderer Berücksichtigung der fabriksmässigen Herstellung dicker Metallüberzüge auf Metallen mittelst des galvanischen Stromes.   Wien. Pest, Leipzig, 1886.   pp. xv–224.

SCHATZ, E.

Reactions-Schema für die qualitative Analyse zum Gebrauche für chemische Laboratorien zusammengestellt.   Berlin, 1870.   Fol.

SCHAUB, JOHANN.

Systematisches Lehrbuch der allgemeinen Chemie, mit Hinsicht auf die neuesten Entdeckungen, zum Gebrauche für Vorlesungen und zum Selbstunterricht.   Nürnberg, 1804.   8vo.

SCHAUMANN, F.

Bestimmung von Glycerin im Wein, nebst Notizen über sächsisch-thüringische Weine.   Erlangen, 1892.   8vo.

SCHEELE, CARL WILHELM.

Chemical Essays, comprising fluor-mineral and its acid, arsenic, experiments upon molybdaena, plumbago, prussian blue, etc., translated from the Transactions of the Academy of Science in Stockholm. London, 1786.   8vo.

Chemische Abhandlung von der Luft und dem Feuer.   Nebst einem Vorbericht von Torbern Bergman.   Upsala und Leipzig, 1777. 12mo, pp. vi–155.

> Contains Scheele's discovery of oxygen, made independently of Priestley.

> Zweite verbesserte Ausgabe mit einer eigenen Abhandlung über die Luftgattungen, wie auch mit der Herren Kirwan und Priestley Bemerkungen und Herrn Scheele's Erfahrungen vermehrt und mit einem Register versehen von Joh. Gottfried Leonhardi.   Leipzig, 1782. 8vo.   Ill.

> *Chemical observations and experiments on Air and Fire, with a prefatory introduction by Torbern Bergman, translated from the German by J. R. Forster.   To which are added notes by Richard Kirwan with a letter to him from Joseph Priestley.   London, 1780.   pp. xl–260, 8vo.   Plate.

> Traité chimique de l'air et du feu.   Avec une introduction de Torbern Bergmann.   Ouvrage traduit de l'allemand par le Baron de Dietrich.   Paris, 1781.   pp. xliv–268, 12mo.

Opuscula chemica et physica ; Latine vertit Godofr. Henr. Schaefer ;

SCHEELE, CARL WILHELM. [Cont'd.]

    edidit et praef. est Em. Benjam. Gottl. Hebenstreit. Lipsiæ, 1788–89. 2 vols., 8vo. Ill.

        * Mémoires de chymie. Tirés des mémoires de l'Académie royale des Sciences de Stockholm. Traduits du Suédois et de l'Allemand. Première partie. Dijon, chez l'éditeur, Place Saint-Fiacre, No. 989, et se trouve à Paris chez Théophile Barrois jeune, Libraire, Quai des Augustins. Cuchet, Libraire, rue et hôtel Serpente, 1785. 2 vols.: pp. [iii]–vi–269 ; [iii]–vi–246, 16mo. Plates.

    Sämmtliche physische und chemische Werke, nach dem Tode des Verfassers gesammelt und in deutscher Sprache herausgegeben von Sigismund Friedrich Hermbstädt. Berlin, 1793. 2 vols., 8vo. Plates.

    Supplément au Traité chimique de l'air et du feu de M. Scheele, contenant un Tableau abrégé des nouvelles découvertes sur les diverses espèces de l'air par Jean Godofr. Leonhardy ; des notes de Rich. Kirwan ; traduit et augmenté de notes et du complement du Tableau abrégé par M. le Baron de Dieterich. Avec la traduction par M.M. de l'Académie de Dijon des expériences de M. Scheele sur la quantité d'air pur qui se trouve dans l'atmosphère. À Paris, 1785. 8vo.

SCHEERER, THEODOR.

    Isomorphismus und polymerer Isomorphismus. Braunschweig, 1850. 8vo.

        Scheerer's full name, Karl Johann August Theodor, is rarely used by him.

    Löthrohrbuch. Eine Anleitung zum Gebrauche des Löthrohrs nebst Beschreibung der vorzüglichsten Löthrohrgebläse, für Chemiker, Mineralogen, Metallurgen, Metallarbeiter und andere Techniker, sowie zum Unterrichte auf Berg-, Forst- und landwirthschaftlichen Academien, polytechnischen Lehranstalten, Gewerbeschulen u. s. w. [Mit Zusätzen versehener Abdrück zweier Aufsätze aus dem "Handwörterbuch der reinen und angewandten Chemie von Liebig, Poggendorff, Wöhler und Kolbe.] Braunschweig, 1851. 8vo. Ill.

        Löthrohrbuch. Eine Anleitung zum Gebrauche des Löthrohrs, sowie zum Studium des Verhaltens der Metalloxyde, der Metalle und der Mineralien vor dem Löthrohre, nebst Beschreibung der vorzüglichsten Löthrohrgebläse. Für Chemiker, Mineralogen, Metallurgen, Metallarbeiter und andere Techniker, sowie zum Unterrichte auf Berg-, Forst- und landwirthschaftlichen Akademien, polytechnischen Lehranstalten, Gewerbe-

SCHEERER, THEODOR. [Cont'd.]

schulen u. s. w. Zweite vermehrte Auflage. Braunschweig, 1857. 8vo. Ill.

Introduction (An) to the Use of the Mouth-Blowpipe.
Together with a description of the blow-pipe characters
of the important minerals. The whole translated and
compiled by Henry J. Blanford. London, 1856. 12mo.
Third edition revised. London, 1875. 12mo.

Руководство къ употреблению паяльной трубки по предметамъ
химіи, минералогіи, металлургіи и механики. Пер. Бекъ.
С Петербургъ, 1853. 8°

Paramorphismus (Der) und seine Bedeutung in der Chemie, Mineralogie
und Geologie. Braunschweig, 1854.

SCHEFFER, HENRIK THEOPHILUS.

Kemiska Föreläsningar rörande salter, jordarter, vatten, fetmor, metaller
och förgning . . . Upsala, 1775. 8vo.

Andra Upplagan. Stockholm, Upsala und Åbo, 1779.
Tredie Upplagan, 1796.
Chemische Vorlesungen über die Salze, Erdarten, Wässer,
entzündliche Körper, Metalle und das Färben ;
gesammelt in Ordnung gestellt und mit Anmerkungen
herausgegeben von Torbern Bergmann. Aus dem
schwedischen übersetzt von Christian Ehrenfried Weigel.
Greifswald, 1779. 8vo.

SCHEFFER, JOHAN DANIEL REINIER.

Over levulose en dextrose. Leiden, 1877. 8vo.

SCHEIBLER, C.

Festschrift zur Feier des 25. jährigen Bestehens des Vereins für die
Rübenzuckerindustrie des Deutschen Reiches. Berlin, 1875.

Gehaltsermittelung (Die) der Zuckerlösungen durch Bestimmung des
specifischen Gewichts derselben bei der Temperatur von + 15°
Celsius. Berlin, 1891.

Le titrage des solutions sucrées par la détermination de leurs
densités à la température de + 15° Centigrade. Traduit par D. Sidersky. Berlin, 1891.

SCHEIDING, F.

Neue Tabelle zur Berechnung der Phosphorsäure bei Anwendung von
5 g. Substanz. Cöthen, 1892. Fol.

SCHELHAMMER, GÜNTHER CHRISTOPH.

Disputationes tres de corporum per ignem resolutione chemica. Kiloniæ,
1701-03.

SCHELHAMMER, GÜNTHER CHRISTOPH. [Cont'd.]
  Tractatus de nitro, vitriolo, alumine et atramentis. Amstelodami, 1709.
    8vo.

SCHELLEN, H.
  Die Spectralanalyse in ihrer Anwendung auf die Stoffe der Erde und
    die Natur der Himmelskörper. Gemeinfasslich dargestellt Braun-
    schweig, 1870. pp. xi-452, 8vo. Ill.
        Zweite durchaus umgearbeitete und sehr vermehrte Auflage.
        Braunschweig, 1871. 8vo.
        Spectrum Analysis in its application to terrestrial substances
          and the physical constitution of the heavenly bodies
          familiarly explained Translated from the second en-
          larged and revised German edition by Jane and Caro-
          line Lassell, edited with notes by William Huggins.
          With numerous woodcuts and coloured plates and Ång-
          ström's and Kirchhoff's maps. London, 1872. pp.
          xxvi-662, 8vo. Ill.

SCHEMA DER QUALITATIVEN CHEMISCHEN ANALYSE. Zum Gebrauche
  bei den praktischen Uebungen im Laboratorium. Zweite vermehrte
  Auflage. Innsbruck, 1862. 8vo.

SCHERER, ALEXANDER NICOLAUS VON.
  Beiträge zur Berichtigung der antiphlogistischen Chemie. Weimar,
    1795. 8vo.
  Grundriss der Chemie. Für academische Vorlesungen. Stuttgart,
    1800. 8vo.
  Grundzüge der neuern chemischen Theorie. Mit dem Bildnisse La-
    voisier's. Jena, 1795.
        The Appendix contains : Nebst einigen Nachrichten von Lavoisier's Leben
          und einer tabellarischen Uebersicht der neuern chemischen Zeichen.
  In welchem Verhältniss stehen Theorie und Praxis der Chemie gegen
    einander? . . . Dorpat, 1803. 8vo.
  Kurze Darstellung der chemischen Untersuchungen der Gasarten. Für
    seine öffentlichen Vorlesungen entworfen. Weimar, 1799. 8vo.
    pp. x-62.
        Dritte Auflage. Berlin, 1809. 8vo.
        A Short Introduction to the Knowledge of Gaseous Bodies.
          Translated from the German. London, 1800. 8vo.
  Uebersicht der Untersuchungen über die Verwandlung des Wassers in
    Stickstoff-Gas. Halle, 1800. 8vo.
  Uebersicht der Zeichen für die neuere Chemie. Jena, 1811. Fol.

SCHERER, ALEXANDER NICOLAUS VON. [Cont'd.]
Versuch einer populairen Chemie. Mühlhausen, 1795. 8vo.
Versuch einer systematischen Uebersicht der Heilquellen des russischen
Reichs. St. Petersburg, 1820. 8vo.

ШЕРЕРЪ АЛЕКСАНДРЪ.
Руководство къ преподаванію химіи. С.-Петербургъ 1808.
> SCHERER, ALEXANDER NICOLAUS. Handbook of Chemistry. St. Peters-
> burg, 1808. Part I only was published.

SCHERER, ALEXANDER NICOLAUS VON, UND CARL CHRISTOPH FRIEDRICH
JÄGER.
Ueber das Leuchten des Phosphors im atmosphärischen Stickgas . . .
Nebst Chph. Heinr. Pfaff's Bemerkungen zu Göttling's Schrift :
Beytrag zur Berichtigung der antiphlogistischen Chemie. Weimar,
1795. 8vo. Ill.

SCHERER, JOHANN BAPTIST ANDREAS VON.
Eudiometria, sive methodus aëris atmosphærici puritatem salubritatemve
examinandi. Viennae, 1782. 8vo.
Genaue Prüfung der Hypothese vom Brennstoff. Prag, 1793. 8vo.
> Translated by R. Bretfeld from the Latin, in N. J. Jacquin's Collectanea
> ad . . . chemiam . . . spectantia, Vindobonæ, IV, 1790.
> *See also in Section III.*

SCHERER, JOHANN JOSEPH.
Chemische und mikroskopische Untersuchungen zur Pathologie, angestellt an der Kliniken des Julius-Hospitales zu Würzburg. Heidelberg, 1843. 8vo. Plates.
Lehrbuch der Chemie mit besonderer Berücksichtigung des ärztlichen
und pharmaceutischen Bedürfnisses. Wien, 1861. 8vo. Ill.

SCHERK, CARL.
Anleitung zur Bestimmung des wirksamen Gerbstoffgehaltes in den
Naturgerbstoffen. Nach den neuesten Erfahrungen für Inhaber und
Leiter von Lederfabriken und Gerbereien, sowie für Gerbstoff-
händler und Forstbeamte. Wien, Pest, Leipzig, 1891. pp. viii-
70, 12mo.

SCHERZER, JOSEPH.
Lehrbuch der Militär-Chemie als Leitfaden für die Vorlesungen im
K. K. Bombadier-Corps. Wien, 1846. pp. xiv-735, 8vo. Ill.

SCHEURER-KESTNER, A.
> Conférence (etc.). *See* Société chimique de Paris.

SCHEURER-KESTNER, A.
Principes élémentaires de la théorie chimique des types appliquée aux combinaisons organiques. Paris, 1862. 8vo.

SCHIBLER, J. J.
Lehrbuch der Agricultur-Chemie. Aarau, 1864. 8vo.

SCHIEL, J.
Anleitung zur organischen Analyse und Gasanalyse. Erlangen, 1860. 8vo. Ill.
Einleitung in das Studium der organischen Chemie. Erlangen, 1860. 8vo. Ill.

SCHIFF, HUGO.
Einführung in das Studium der Chemie, nach Vorlesungen gehalten am naturwissenschaftlichen Institut in Florenz. Berlin, 1876. pp. xi–328, 8vo.
Bibliothek für Wissenschaft und Literatur, vol. x.
Introduzione allo studio della chimica. Torino, 1876.

SCHIFF, R.
Ueber Aldehydadditionsproducte. Zürich, 1876.

SCHIK, J. C.
Repetitorium der Chemie mit besonderer Rücksicht auf Physiologie, Pathologie und Pharmacologie für Mediciner. Prag, 1864. 16mo.

SCHILD, H.
Tabellen zu Rauchanalysen. Berlin, 1888. 8vo.

SCHILLING, N. H.
Handbuch für Steinkohlengas-Beleuchtung. Mit einer Geschichte der Gasbeleuchtung von F. Knapp. München, 1860. 4to. Ill.
Zweite Auflage, 1865.
Dritte umgearbeitete und vermehrte Auflage. München, 1879. pp. xv–692, 4to. Atlas and tables.

SCHINDLER, CH. C.
Der geheimbde Münz-Guardein und Bergprobirer, welche zeiget und an den Tag giebet alle geheime Handgriffe, so bisshero sind verschwiegen und zurück gehalten worden. Franckfurth, 1705. 8vo.

SCHIØDT, C. L. [Translator].
See Fourcroy, A. F.

SCHLAGDENHAUFFEN, CHARLES FRÉDÉRIC.
Analyse des végétaux [and other articles]. See, in Section II, Fremy: Encyclopédie chimique, vols. IX and X.

SCHLAGDENHAUFFEN, CHARLES FRÉDÉRIC.

Rapports (Des) de la physique, de la chimie et de la toxicologie. Thèse. Strasbourg, 1854. 8vo.

Recherches sur le sulfure de carbone. Thèse pour le doctorat . . . Paris, 1857.

SCHLAGDENHAUFFEN, CHARLES FRÉDÉRIC [Translator].
See Hoppe-Seyler.

SCHLEGEL, L.

Théorie des équivalents chimiques. Paris, 1863. 8vo.

SCHLEICHER, E.

Zur Kenntniss der Tiophengruppe. Göttingen, 1887.

SCHLICHTER, CHRISTIAN.

Laien-Chemie. Eine populäre Darstellung der unorganischen Chemie für gehobene Volks-, Real- und Gymnasialschulen, sowie zum Selbstunterricht mit Rücksicht auf Künste, Gewerbe, Haushalt und Landwirthschaft bearbeitet. Mit einem kurzen Anhange über Electricität und Magnetismus, etc. Stuttgart, 1853. 8vo.

SCHLICHTING, M.

Chemische Versuche einfachster Art, ein erster Cursus in der Chemie, in der Schule und beim Selbstunterricht ausführbar ohne besondere Vorkenntnisse und mit möglichst wenigen Hülfsmitteln. Mit einem Vorwort von C. Himly. Kiel, 1862. Ill.

> Zweite Auflage. Kiel, 1862.
> Siebente Auflage. Bearbeitet von A. Wilke. Kiel, 1880.
> Achte Auflage. Kiel, 1885.

SCHLICKUM, O.

Beknopte methoden tot een pathologisch- en geregtelijkscheikundig onderzoek, enz. Gevolgd naar het Hoogduitsch en vermeerderd door R. J. Opwijrda. Utrecht, 1864. 8vo.

Chemische (Der) Analytiker. Gründliche Einführung in die qualitative chemische Analyse nebst abgekürzten Untersuchungsmethoden. Neuwied, 1864. 8vo.

> Zweite Auflage. Leipzig, 1875. 8vo.

Junge (Der) Chemiker. Gründliche Einführung in das Studium der Chemie für angehende Chemiker und Pharmaceuten, Gewerbe und Realschüler, Berg- und Hüttenleute. Neuwied, 1863.

> Dritte Auflage. Neuwied, 1867.
> [Another edition.] Leipzig, 1875.

Scheikunde (De) in proeven voorgedragen. Naar het Hoog-

SCHLICKUM, O.   [Cont'd.]
>   duitsch bewerkt door R. J. Opwijrda.   Utrecht, 1864.
>   12mo.

>   Beginselen (De) der scheikunde, in proeven voorgedragen.
>   Naar het Hoogduitsch bewerkt en vermeerderd door R.
>   J. Opwijrda. Tweede uitgave.   Utrecht, 1865.   8vo.
>   Ill.

SCHLOESING, Th.

Chimie agricole.   Paris, 1892.   8vo.

Sur la déperdition d'azote pendant la décomposition des matières organiques.   Nancy, 1889.

SCHLOESING, PÈRE ET FILS.
>   Contribution à l'étude de la chimie agricole.   *See, in Section II,* Fremy:
>   Encyclopédie chimique, vol. x.

SCHLOSSBERGER, JULIUS EUGEN.

Die Chemie der Gewebe des gesammten Thierreichs.   Leipzig und
Heidelberg, 1856.   2 vols., 8vo.

>   Also with a second title-page : Erster Versuch einer allgemeinen und ver-
>   gleichenden Thier-Chemie.

Lehrbuch der organischen Chemie mit besonderer Rücksicht auf Physiologie und Pathologie, auf Pharmacie, Technik, und Landwirthschaft.
Stuttgart, 1850.   pp. [vi]–662, 12mo.

>   Zweite Auflage.   Stuttgart, 1852.   8vo.
>   Fünfte Auflage.   Leipzig und Heidelberg, 1860.   pp. viii–
>   1047, 8vo.
>   Leerboek der organische scheikunde met toepassing op
>   physiologie en pathologie, pharmacie, techniek en land-
>   huishoudkunde.   Vrij naar het Hoogduitsch door J. W.
>   Gunning.   Utrecht, 1852–53.   8vo.

SCHLOSSER, JOHANNES ALBERTUS.

Tractatus de sale urinæ humanæ nativo.   Harlingen, 1760.   8vo.

SCHLUTTIG, Osw., UND G. S. NEUMANN.

Die Eisengallustinten.   Grundlagen zu ihrer Beurtheilung.   Dresden,
1890.

SCHMELCK, L.

Chemistry of the Norwegian North Atlantic expedition.   Christiania,
1882.   Roy. 4to.

SCHMELZER, AUGUST.

Vollständiges Handbuch der Schwefelsäure-Fabrikation, aus Vitriol, Schwefel, Schwefelkies, Gyps und Schwerspath. Nach allen vorhandenen Angaben, nebst Vorschlägen zu abweichenden Methoden zur Entbehrung der Bleikammern und Platinkessel, etc. Quedlinburg, 1849. 8vo.

SCHMID, E.

Ueber die Einwirkung von reiner nitroser und rauchender Schwefelsäure und Salpetersäure auf reines Blei und Legirungen von Blei mit Antimon und Kupfer. Ueber die Bestimmung des Sauerstoffs im Weichblei und dessen Einfluss auf die Angreifbarkeit durch Schwefelsäure. Basel, 1892. 8vo. Ill.

SCHMID, ERNST ERHARD.

Organische Chemie, Meteorologie, Geognosie, Bodenkunde und Düngerlehre. Für Landwirthe bearbeitet. Braunschweig, 1850. 8vo. Ill.
Encyclopädie der gesammten theoretischen Naturwissenschaften.

Physik, anorganische Chemie und Mineralogie. Für Landwirthe bearbeitet. Braunschweig, 1850. 8vo.
Encyclopädie der gesammten theoretischen Naturwissenschaften.

SCHMID, JOHANN ANDREAS.

De principiis chemicorum non chemiæ diversis. Helmstadii, 1720. 4to.

SCHMID, W.

Anleitung zu sanitäts- und polizeilich-chemische Untersuchungen. Zürich, 1878.

SCHMIDT, ALEXANDER.

Beitrag (Ein) zur Kenntniss der Milch. Dorpat, 1874. 4to.

Lehre (Die) von den fermentativen Gerinnungserscheinungen in den eiweissartigen thierischen Körperflüssigkeiten. Zusammenfassender Bericht über die früheren, die Faserstoffgerinnung betreffenden, Arbeiten des Verfassers. Dorpat, 1877. pp. 62, 8vo.

SCHMIDT, CARL.

Diagnostik (Die) verdächtiger Flecke in Criminalfällen. Mitau und Leipzig, 1848. 8vo.

Entwurf einer allgemeinen Untersuchungsmethode der Säfte und Excrete des thierischen Organismus. Basirt auf krystallonomische, histologische und mikrochemische Bestimmungen. Mitau und Leipzig, 1846. 8vo.

De microcrystallometria ejusque in chemia physiologica et pathologica momento. Dorpati, 1846. 8vo.

SCHMIDT, CARL WILHELM.
   Die Bierbrauerei in ihrem ganzen Umfange.   Züllichau, 1820.   8vo.
      Also under the title : Lehrbuch der speciellen Bierbrauerei.

ШМИДТЪ К.
   Химико-физіологическія основанія земледѣлія и скотоводство.
      SCHMIDT, CONSTANTIN.   Chemico-physiological basis of agriculture and
         cattle breeding.   [Moscow?]   2 parts.
   Руководсдво къ химическому анализу важнѣйшихъ сельско-хозяйственныхъ
   матеріаловъ и продуктовъ. Москва 1852.
      Manual of chemical analysis applied to agriculture.   Moscow, 1852.

SCHMIDT, CHRISTIAN HEINRICH.
   Branntweinbrennereibetrieb (Der) in seiner allerneuesten Vervollkomm-
      nung, besonders in Beziehung auf Spirituserzeugung aus Getreide,
      Kartoffeln, Runkelrübenmelasse mittelst Apparaten für ununter-
      brochene Arbeit und Dampfmaschine.   Weimar, 1854.   8vo.   Ill.
         Neuer Schauplatz der Künste und Handwerke.
   Fabrikation (Die) der für die Glasmalerei, Emailmalerei und Porcellan-
      malerei geeigneten Farben.   Weimar, 1843.   8vo.   Ill.
         Zweite vermehrte Auflage.   Weimar, 1848.
         Dritte sehr vermehrte Auflage.   Weimar, 1861.
         Neuer Schauplatz der Künste und Handwerke.
   Fabrikation (Die) der künstlichen Mineralwässer nebst Beschreibung
      der vorzüglichsten Brauverfahren in Oesterreich, am Rhein, in den
      Niederlanden.   Zweite völlig umgearbeitete und sehr vermehrte
      Auflage.   Weimar, 1853.   8vo.
         Neuer Schauplatz der Künste und Handwerke.
   Farbwaarenkunde (Die) und Farbenchemie für Färberei und Zeugdruck,
      oder instructive Anweisung, alle in der Färberei und Druckerei in
      Anwendung kommenden Farbwaaren, Säuren, Alkalien, etc., auf
      ihre Aechtheit oder Verfälschung zu prüfen, sowie ihren nutzbaren
      Gehalt, etc., auszumitteln, darzustellen und sie aus den besten
      Quellen zu beziehen.   Ein unentbehrliches Handbuch für Färber,
      etc., nach dem neuesten Standpuncte der Wissenschaft bearbeitet.
      Weimar, 1852. 8vo.
         Zweite vermehrte Auflage.   Weimar, 1856.
         Neuer Schauplatz der Künste und Handwerke.
   Grundsätze der Bierbrauerei nach den neuesten technisch-chemischen
      Entdeckungen.   Weimar, 1837.   8vo.
         Zweite völlig umgearbeitete und vermehrte Auflage.   Wei-
         mar, 1853.   Dritte Auflage.   Weimar, 1860.
         Neuer Schauplatz der Künste und Handwerke.

SCHMIDT, CHRISTIAN HEINRICH. [Cont'd.]
  Handbuch der gesammten Lohgerberei. Weimar, 1841, 8vo.
        Zweite vermehrte Auflage. Weimar, 1847.
        Dritte vermehrte Auflage. Weimar, 1855.
      Neuer Schauplatz der Künste und Handwerke.

  Handbuch der Zuckerfabrikation namentlich des Runkelrübenzuckers,
      des Rohrzuckers und des Stärkemehlzuckers nach ihrem Stand-
      punkte im Jahre 1840 in Deutschland. Weimar, 1841. 8vo. Ill.
        Zweite Auflage. Weimar, 1847.
        Dritte vermehrte Auflage. Weimar, 1850. Ill.
        Handbuch der Zuckerfabrikation namentlich des Runkel-
            rübenzuckers, des Rohrzuckers und des Stärkemehl-
            zuckers, nach ihrem Standpunkte im Jahre 1858 in
            Frankreich, Belgien und Deutschland. Nebst einem
            Anhange, die im Hause der preussischen Kammern
            Stattgefundenen Discussionen über die Besteuerung des
            Runkelrübenzuckers betreffend, besonders in ihren
            technischen Beziehungen. Vierte ganz umgearbeitete
            und vermehrte Auflage. Weimar, 1858. pp. xxiv–531,
            8vo. Plates.
      Neuer Schauplatz der Künste und Handwerke.

  Kerzen- (Die) und Seifenfabrikation nach den neuesten Vervollkomm-
      nungen dieses Industriezweiges, oder gründliche Anweisung alle
      Arten von Kerzen, etc., und Wachsstöcke zu fabriciren, ingleichen
      alle Arten der Kali- und Natronseifen, etc., darzustellen. Weimar,
      1852. 8vo.
        Zweite vermehrte Auflage. Weimar, 1856.
      Neuer Schauplatz der Künste und Handwerke.

  Kompendiöses Handbuch der Färberei wollener, seidener, baumwol-
      lener, leinener und hanfener Gewebe, sowie auch der Garne aus
      obigen Stoffen nach den bewährtesten ältern Verfahrungsarten, wie
      auch nach den neuesten Verbesserungen und Erfindungen im
      Gebiete der Färberei. Weimar, 1861. 8vo.
      Neuer Schauplatz der Künste und Handwerke.

  Lederfärbekunst (Die), oder chemische Grundsätze und Vorschriften,
      alle Ledergattungen in allen Farben ächt zu färben, etc. Zweite
      vermehrte Auflage. Weimar, 1848.
        Dritte Auflage. Weimar, 1861.
      Neuer Schauplatz der Künste und Handwerke.

  Lehrbuch der Chemie für Färber und Zeugdrucker. Leipzig, 1840. 8vo.
      Neuer Schauplatz der Künste und Handwerke.

SCHMIDT, CHRISTIAN HEINRICH. [Cont'd.]

Neuesten (Die) wichtigsten Fortschritte, Erfindungen und Verbesserungen in der Farbenfabrikation, sowohl in Bezug auf Erd- und
Oxydfarben, als auch Lackfarben und Farbstoffe aus Harnsäure
(Murexyd), wie aus Steinkohlentheer (Anilinfarben) zum Färben
und Drucken der Gewebe. Weimar, 1864. 8vo. Ill.

>   Neuer Schauplatz der Künste und Handwerke.

Verschiedenen (Die) Substanzen welche gegenwärtig zur Beleuchtung
angewendet werden, als Thran, Reps- oder Rüböl, Harzöl, Stein-
kohlenöl, Wallrathöl, Schieferöl, Talg, Braconnot's Ceromimem,
Elaidin, Elaidinsäure, Palmitin, Palmitinsäure, Stearin, Stearinsäure,
Corin, Corinsäure, Wachs, Wallrath Photogene, Paraffin, Naphtalin,
Alkohol, Holzgeist, Camphin, Mineralöl, Solaröl, Leuchtgas aus
Steinkohlen, Leuchtgas aus Oel, Leuchtgas aus Harz, Leuchtgas
aus Holz, Leuchtgas aus Torf, Wasserstoffgas, Wassergas, elec-
trisches Leuchtgas, Electricität,—ihre Gewinnung, Zubereitung und
ihr Nutzeffect. Weimar, 1856. 8vo. Ill.

>   Neuer Schauplatz der Künste und Handwerke.

Vollständiges Farben-Laboratorium, oder ausführliche Anweisung zur
Bereitung der in der Malerei, Staffirmalerei, Illumination, etc.
gebräuchliche Farben. Weimar, 1841. 8vo. Ill.

>   Zweite vermehrte Auflage. Weimar, 1847.
>
>   Dritte Auflage, 1857.
>
>   Neuer Schauplatz der Künste und Handwerke.

Vollständige (Der) Feuerzeugpraktikant. Weimar, 1840. 8vo. Ill.

>   Zweite vermehrte Auflage. Weimar, 1847.
>
>   Dritte vermehrte Auflage. Weimar, 1861.
>
>   Neuer Schauplatz der Künste und Handwerke.

SCHMIDT, ERNST.

Anleitung zur qualitativen Analyse. Zum Gebrauche im pharmaceutisch-chemischen Laboratorium zu Marburg. Zweite Auflage.
Halle, 1885.

>   Dritte vermehrte und verbesserte Auflage. Halle, 1890.
>   8vo.

Ausführliches Lehrbuch der pharmaceutischen Chemie. Braunschweig,
1879–82.

>   Zweite vermehrte Auflage. Braunschweig, 1887–1890. 2
>   vols., 8vo. Vol. I, Anorganische Chemie; vol. II, Or-
>   ganische Chemie.

SCHMIDT, F. X.

Lehrbuch der gewerblichen Chemie für den Unterricht an gewerblichen

SCHMIDT, F. X. [Cont'd.]

Fortbildungsschulen, Real- und Gewerbeschulen, technischen Lehranstalten, Realgymnasien, sowie für den Selbstunterricht. Stuttgart, 1864–70. 2 vols., 8vo.

> Leerboek der nijverheid-scheikunde, ten dienste van het
> onderrigt aan nijverheidscholen en inrigtingen van mid-
> delbaar onderwijs, alsmede tot huiselijk zelfonderrigt.
> Uit het Hoogduitsch in het Nederlandsch overgebragt
> door R. J. Opwijrda. Utrecht, 1864–65. 8vo.
>
> Only the inorganic part was published.

Шмидтъ Г. Р.

Основанія химіи въ примѣненіи ея къ сельск. хозяйству, технич. промышленности и домашн. быту. Москва 1854. 2 т.

> SCHMIDT, H. R. Principles of Chemistry in their application to Agriculture,
> technology, industry and household economy. Moscow, 1854. 2 vols.

SCHMIDT, M. VON.

Anleitung zur Ausführung agriculturchemischer Analysen. Wien, 1891.
8vo.

SCHMIEDER, CARL CHRISTOPH.

Dissertatio de affinitatibus chemicis. Halæ Magdeburgicæ, 1799. 8vo.

> For a notice of the author, *see, in Section III*, Schmieder, C. C. : Geschichte
> der Chemie.

Gemeinnützige (Das) der Chemie. Freiberg, 1804–05. 2 vols., 8vo.

SCHMITT, ——.

Mittheilungen aus der amtlichen Lebensmittel- Untersuchungs-Anstalt
und chemischen Versuchsstation zu Wiesbaden, über die geschäft-
liche und wissenschaftliche Thätigkeit in dem Betriebsjahre 1883–
84. Berlin, 1885. pp. [iv]–268, 8vo. Plates.

SCHMÖGER, VON.

Grundlinien der allgemeinen Chemie zum Gebrauche bei Vorlesungen.
Regensburg, 1842. 8vo.

SCHNABEL, CARL.

Die wichtigsten chemischen Prozesse. In anschaulicher Erklärungsweise graphisch dargestellt. Ein Hülfsbuch beim Studium und
Unterricht. Erster Theil : Die unorganische Chemie (Metalloide
und Metalle). Zweite umgearbeitete Auflage. Siegen, 1859. 8vo.

> Erste Auflage. Siegen, 1840.

SCHNAUBERT, LUDWIG.
Untersuchung der Verwandschaft der Metalloxyde zu den Säuren.
Nach einer Prüfung der neuen Berthollet'schen Theorie. Erfurt,
1803. 8vo.

SCHNEIDER, FRANZ COELESTIN.
Anfangsgründe der Chemie. Ein Leitfaden für Vorlesungen und zum
Selbststudium. Wien, 1853. 8vo.
Gerichtliche (Die) Chemie für Gerichtsärzte und Juristen. Bearbeitet
von F. C. Schneider. Wien, 1852. pp. 385, 8vo.
Proeve van eene geregtelijke scheikunde. Handboek voor
regtsgeleerden, geneesheeren en scheikundigen. Naar
het Hoogduitsch. Voor Nederland bewerkt door W.
M. Perk. Campagne, 1855. 8vo.
Grundzüge der allgemeinen Chemie mit besonderer Rücksicht auf die
Bedürfnisse des ärztlichen Studiums bearbeitet. Wien, 1851. 8vo.

SCHNEIDER JOSEPH.
Populäre Toxikologie, oder Lehre von den Giften und Gegengiften.
Ein Handbuch für höhere niedere Schulen, Lehrer und Jedermann.
Frankfurt am Main, 1838. pp. xiv–200, 12mo.

SCHOEDLER, FRIEDRICH.
Atlas der chemischen Technik. Leipzig, 1873.
Atlas voor de scheikundige technologie, met verklarenden
text, vrij bewerkt naar het Hoogduitsch door D. de Loos.
Leiden, 1874. 4to. Ill.

SCHÖDLER, FRIEDRICH.
Chemie (Die) als geistig bildendes Moment für den Unterricht in
Gymnasien. Eine Rede. Braunschweig, 1846. 8vo.
Chemie (Die) der Gegenwart in ihren Grundzügen und Beziehungen
zu Wissenschaft und Kunst, Gewerbe- und Ackerbau-Schule und
Leben. Für Gebildete aller Stände dargestellt. Leipzig, 1854.
8vo. Ill.
Dritte Auflage. Leipzig, 1859.
Elements of Chemistry. Edited by H. Medlock. London
and Glasgow, 1851. 8vo.

SCHÖNBEIN, CHRISTIAN FRIEDRICH.
Beiträge zur physikalischen Chemie. Basel, 1844. pp. xii–114, 8vo.
Chemische Beobachtungen über die langsame und rasche Verbren-
nung der Körper in atmosphärischer Luft. Basel, 1845. 8vo.
Denkschrift über das Ozon. Basel, 1849. 4to.
Ueber den Einfluss des Sonnenlichtes auf die chemische Thätigkeit

SCHÖNBEIN, CHRISTIAN FRIEDRICH.   [Cont'd.]
des Sauerstoffs und den Ursprung der Wolkenelektricität und des
Gewitters. Basel, 1850. 4to.

Ueber die Erzeugung des Ozons auf chemischen Wege. Basel, 1844.

Ueber die Häufigkeit der Berührungswirkungen auf dem Gebiete der
Chemie. Basel, 1843. 4to.

Verhalten (Das) des Eisens zum Sauerstoff. Ein Beitrag zur Erweiterung electro-chemischer Kenntnisse. Basel, 1837. 8vo. Ill.

SCHØNEMANN, CHRISTIAN WILHELM.
Praktisk Anvisning til at tilberede de meest brugelige chemiske Reagentier, tilligemed en kort Beskrivelse over disses vigtigste chemiske
Egenskaber samt Anvendelse, Nytte og Virkning som Reagentier
betragtede. Kjøbenhavn, 1824.

SCHÖPFER, CARL.
Der chemische Tausendkünstler. Eine reiche Sammlung der unter-
haltendsten und zugleich belehrendsten chemischen Experimente
von H. Herold. Nordhausen, 1845. 12mo.
> Neue Ausgabe: Der chemische Tausendkünstler oder
> famose Zauberer. Leipzig, 1849. 12mo

SCHOLL, HERMANN.
Die Milch, ihre häufigen Zersetzungen und Verfälschungen, mit speziel-
ler Berücksichtigung ihrer Beziehungen zur Hygiene. Mit einem
Vorwort von F. Hüppe. Wiesbaden, 1891. 8vo. Ill.

SCHOLZ, BENJAMIN.
Lehrbuch der Chemie. Wien, 1825.
> Zweite Auflage. Wien, 1829–'31. 2 vols., 8vo. Ill.

SCHORLEMMER, CARL.
Lehrbuch der Kohlenstoffverbindungen, oder der organischen Chemie.
Zugleich als zweiter Band von Roscoe's kurzem Lehrbuch der
Chemie. Braunschweig, 1872. 8vo.
> Zweite verbesserte Auflage, 1874. 8vo.
> Dritte verbesserte Auflage. Braunschweig, 1885–89.   2
> parts, 8vo.
> *Cf.* Roscoe, H. E., and Schorlemmer : Kurzes Lehrbuch der Chemie.
> Manual (A) of the Chemistry of the Carbon Compounds, or
> organic chemistry. London, 1874. 8vo.
> Trattato delle combinazioni del carbonio, o di chimica or-
> ganica. Prima traduzione di Maurizio Sella riveduta
> da Luigi Gabba. Milano, 188–. 16mo. Ill.
> Краткій учебникъ химіи углеродистыхъ соединеній. С.-Петер-
> бургъ, 1873.

SCHORN, P.
>Leitfaden der unorganischen Chemie. Nebst einer Darstellung der wichtigsten Lehren und Formeln der modernen Typentheorie, die den Gebrauch des Leitfadens bei dem Unterrichte der Chemie nach den neuesten Anschauungen ermöglichen soll. Erster Theil: Die Metalloide. Zweiter Theil: Die Metalle. Fünfte Auflage. Münster, 1867–68. 8vo.

>>Leiddraad der anorganische scheikunde. Naar den vijfden Hoogduitschen druk. Vrij vertaald door A. L. Lamers en J. Ringeling. 1e deel: Metalloiden. 'sHertogenbosch, 1869. 8vo.

>Leitfaden für die ersten analytisch-chemischen Arbeiten. Zweite Auflage. Münster, 1867.

SCHOTTEN, C.
>Kurzes Lehrbuch der Analyse des Harns. Leipzig, 1888. 8vo. Ill.

SCHRAMM, THEODOR.
>Examinatorium der Chemie. Tübingen, 1848–'49. 3 vols., 16mo.
>>I: Examinatorium der unorganischen Chemie. 1848. Zweite verbesserte und vermehrte Auflage, 1852.
>>II: Examinatorium der organischen Chemie. 1848. Zweite verbesserte und vermehrte Auflage, 1853.
>>III: Anleitung zur chemischen Analyse. 1849.
>>>Herinneringsregelen der scheikunde. Naar het Hoogduitsch bewerkt door P. J. Hollmann. Amsterdam, 1857. 8vo.

SCHREGER, CHRISTIAN HEINRICH THEODOR.
>Tabellarische Uebersicht der rohen und künstlichen Farben und Farbenmaterialen . . . Nürnberg, 1805. 2 vols., 4to.

SCHREGER, CHRISTIAN HEINRICH THEODOR.
>Kurze Beschreibung [etc.]. *See in Section III.*

SCHREIBER, A.
>Grundriss der Chemie und Mineralogie. Zweite verbesserte Auflage. Berlin, 1873.
>>Vierte vollständig umgearbeitete Auflage. Berlin, 1886. 8vo.

SCHREIBEREY ALLERHAND FARBEN und mancherley weyse Dinten zu bereyten . . . Zu Mentz, bey Peter Jordan, 1531. 4to.

SCHREITTMAN, C.

Probirbüchlin. Frembde und subtile Künst, vormals im Truck nie gesehen, von Woge und Gewicht, auch von allerhand Proben auff Ertz, Golt, Silber [etc.]. Franckfurt, 1580. 8vo.

SCHRODT, M.

Anleitung zur Prüfung der Milch im Molkereibetriebe. Bremen, 1891. 8vo. Ill.

SCHRÖDER, HEINRICH GEORG FRIEDRICH.

Molecularvolume (Die) der chemischen Verbindungen im festen und flüssigen Zustand. Mannheim, 1843. 8vo.

Siedhitze (Die) der chemischen Verbindungen als wesentliche Kennzeichen zur Ermittelung ihrer Componenten, nebst vollständigen Beweisen für die Theorie der Molekular-Volume der Flüssigkeiten. 1. Theil, enthaltend die Kohlenwasserstoffe und Kohlenwasserstoffoxyde. Mannheim, 1844. 8vo.

SCHRÖDER, JOHANN.

Pharmacopœia medico-chymica. Ulm, 1641. 4to.

> Editio quarta. Ulm, 1655.
>
> Editio septima. Coloniæ, 1687. Fol.
>
> [Other editions], 1746, 1748.
>
> The Compleat Chymical Dispensatory, treating of all sorts of Metals, Precious Stones, and Minerals. The like work never extant before. Englished by Wm. Rowland. London, 1669. Fol.

SCHRÖTTER, ANTON.

Chemie (Die) nach ihrem gegenwärtigen Zustande mit besonderer Berücksichtigung ihres technischen und analytischen Theiles. Wien, 1847–49. 2 vols., 8vo. Ill.

Ueber einen neuen allotropischen Zustand des Phosphors. Wien, 1858. Fol.

> Contains the discovery of amorphous phosphorus.

SCHUBARTH, ERNST LUDWIG.

Beiträge zur näheren Kenntniss der Runkelrübenzuckerfabrikation in Frankreich. Berlin, 1836. 4to.

> Nachtrag, von G. Reich. Berlin, 1837.

Elemente der technischen Chemie zum Gebrauch beim Unterricht im Königl. Gewerb-Institut und den Provinzial-Gewerbschulen. Berlin, 1831–33. 2 vols. in 8vo. Ill. And an Atlas, 1 vol., sm. fol.

> Zweite Auflage. Berlin, 1835.

SCHUBARTH, ERNST LUDWIG. [Cont'd.]
Continued under the title :
Handbuch der technischen Chemie. Dritte vermehrte Ausgabe. Berlin, 1839–40. 3 vols., 8vo, and Atlas. Ill.
Vierte Auflage. Berlin, 1851. 3 vols., 8vo, and Atlas.
Lehrbuch der theoretischen Chemie. Behufs seiner Vorträge und zum Selbstunterricht. Dritte Ausgabe. Berlin, 1827.
Vierte Auflage. Berlin, 1829. pp. xiv–818, 8vo.

SCHUBERT, EDUARD.
Praktisches Recept-Taschenbuch für Destillation ; 859 Recepte zur Bereitung aller Sorten Liqueure, der Doppel- und einfachen Branntweine auf warmem wie auf kaltem Wege ; Bereitung des Schweizer-Absynth, der Magen-Tropfen, -Essenzen und -Elixire, der Punsch- und Grog-Extrakte, der Fruchtweine, der Rums, Aracs, Cognacs und Franzbranntweine, der Fruchtsäfte, der aromatischen Essenzen, Sprite und Wasser, der wohlriechenden Essenzen, der Eau de Cologne, Toilettenwasser, Räuchermittel, Riechmittel, Zahnpulver, Zahntincturen, Pomaden, Haaröle und Toilettseifen. Mit Anleitung zur Destillation, etc., nebst Darstellung der gebräuchlichen Destillirapparate, sowie des in neuester Zeit construirten Fein-Sprit-Apparates. Zum Gebrauche für Branntweinbrenner, Destillateure, Kaufleute, Conditoren, Gast- und Schenkwirthe. Mit einem Vorwort von Fr. Jul. Otto. Dritte vermehrte und verbesserte Auflage von H. Beckurts. Mit Holzstichen und einer Reductions-Tabelle für österreichisches Mass und Gewicht. Braunschweig, 1877. 8vo.
Rationelle (Der) Brennereibetrieb. Enthaltend gründliche Anweisung zur Ausführung der besten Einmaischmethoden, wodurch der grösstmöglichste Vergährungsgrad der Maische, mithin der grösste Spiritusertrag und zwar von einigermassen gutem Materiale allermindestens 10 Procent Alkohol vom Quartmaischraum erzielt wird, sowie zur Bereitung bewährter Kunsthefen, des Pilz- und Schaufelmalzes, der Presshefe, etc. ; nebst Darstellung eines in neuester Zeit zweckmässig construirten Destillir-Apparates. Nach eigenen langjährigen Erfahrungen bearbeitet. Mit einem Vorwort von Fr. Jul. Otto. Dritte verbesserte Auflage. Braunschweig, 1865. 8vo. Ill.

SCHUBERT, F.
Lehrbuch der technischen Chemie. Erlangen, 1854. pp. viii–528, 8vo. Ill.
Zweite vermehrte und verbesserte Auflage. Stuttgart, 1866. 8vo. Ill.

SCHUBERT, M.
Praktisches Handbuch der Cellulosefabrikation. Berlin, 1892. 8vo. Ill.

SCHUCHARDT, B., HUSEMANN, TH., SEIDEL und SCHAUENSTEIN.
Die Vergiftungen in gerichtsärztlicher Beziehung. Tübingen, 1882.
pp. x–794.
Maschka's Handbuch der gerichtlichen Medicin, vol. II.

SCHÜBLER, GUSTAV.
Grundsätze der Agricultur-Chemie in näherer Beziehung auf Land- und
Forstwirthschaftliche-Gewerbe. Zweite Auflage durchgesehen und
vermehrt von K. L. Krutzsch. Leipzig, 1838. 8vo. Plates and
tables.
Cf. Schulze, Franz Ferdinand.

SCHUEREN, VAN DER.
Traité élémentaire de chimie industrielle d'après les travaux les plus
récents. Namur, 1880. 2 vols., 8vo. Ill.

SCHÜRMANN, E.
Ueber die Verwandschaft der Schwermetalle zum Schwefel. Tübingen,
1889.

SCHÜTZENBERGER, PAUL.
Leçon (etc.). *See* Société chimique de Paris.

SCHÜTZENBERGER, PAUL.
Chimie appliquée à la physiologie animale, à la pathologie et au diag-
nostic médical. Paris, 1864. pp. vii–515, 8vo.
Éléments dè chimie. Paris, 1881. Ill.
Farbstoffe (Die) mit besonderer Berücksichtigung ihrer Anwendung in
der Färberei und Druckerei. Bearbeitet von H. Schröder. Berlin,
1868–70. 2 vols., 8vo. Ill.
Fermentations (Les). Paris, 1875. 8vo.
On Fermentation. London, 1889. 8vo. Ill.
Die Gährungserscheinungen. Autorisirte Ausgabe. Leip-
zig, 1876. 8vo.
Le fermentazioni. Milano, 1876. Ill.
Traité de chimie générale comprenant les principales applications de la
chimie aux sciences biologiques et aux arts industriels. Paris,
1880. 2 vols., 8vo. Ill.
Deuxième édition. Paris, 1884–92. 6 vols., 8vo.

SCHULTZ, ADRIAN GOTTLOB.
Chymischer Wegweiser. Flensburg, 1757. 8vo.

SCHULTZ, GUSTAV.
Die Chemie des Steinkohlentheers mit besonderer Berücksichtigung
der künstlichen organischen Farbstoffe. Zweite Auflage. Braunschweig, 1886–90. 2 vols., 8vo. Vol. I, pp. 823 ; vol. II, pp. 1368.
First edition, Braunschweig, 1882. 8vo.

SCHULTZ-FLEETH, E.
Der rationelle Ackerbau in seiner Begründung durch die Ergebnisse der
neueren Naturforschung.   Berlin, 1856.   pp. xiv–455, 8vo.

SCHULTZER, F. F.
Anwendung der Grundsätze der antiphlogistischen Chemie auf die
Lehre von den Gasarten.   Neu-Ruppin, 1797.   8vo.

SCHULZ, C. G.
Die Fabrikation des Zuckers aus Rüben.   Theorie und Praxis.   Berlin,
1862–65.   2 vols., in 6 parts, 8vo.

SCHULZE, A.
Technologische Chemie und Materialenkunde.   Quedlinburg und Leipzig, 1826.   8vo.

SCHULZE, FRANZ FERDINAND.
Lehrbuch der Chemie für Landwirthe, zum Gebrauche bei Vorlesungen
an höheren landwirthschaftlichen Lehranstalten, und zum Selbstunterrichte.   (Als dritte Auflage von Schübler's Grundsätzen der
Agriculturchemie.)   Leipzig, 1846–'53.   2 vols., 8vo.   Vol. 1:
Unorganische Chemie, 1846.   Vol. II, Part 1: Organische Chemie,
1853.
      Dritte Auflage, von Th. Hübner.   Leipzig, 1877–78.   2 vols.
      For earlier edition, *see* Schübler, Gustav.

SCHULZE, JOHANN ERNST FERDINAND.
Toxicologia veterum, plantas venenatas exhibens Theophrasti, Galeni,
Dioscoridis, Plinii aliorumque auctoritate ad deleteria venena relatas.
Halæ, 1788.   4to.

SCHULZE, JOHANN HEINRICH.
Chemische Versuche, nach seinem Manuscript herausgegeben von K.
Strumpf.   Halæ, 1745.   8vo.
      Zweite Auflage.   Halæ, 1757.
De metallorum analysi per calcinationem.   Halæ, 1738.   4to.
De sale corporum mixtorum principio constitutivo.   Halæ, 1736.   4to.
De solutionibus corporum chemicæ fundamento.   Halæ, 1736.   4to.

SCHULZE-MONTANUS, KARL AUGUST.
Die Reagentien und deren Anwendung zu chemischen Untersuchungen
nebst zwei ausführlichen Abhandlungen über die Untersuchung der
mineralischen Wasser und die Prüfungen auf Metallgifte.   Zweite
Auflage.   Berlin, 1818.   12mo.
      Vierte Ausgabe, bearbeitet von Aug. Wilh. Lindes.   Berlin,
1830.   8vo.   Plates.
      Also under the title: A. W. Lindes, Versuch einer ausführlichen Darstellung
[u. s. w.].

SCHUMACHER, C. A.

Populaire Forelæsninger over Agerdyrknings-Chemiens Hovedlærdomme.
Trykt som Manuscript for den " Mellemslesvigske Landboforening."
Aabenraa, 1853. 8vo.

SCHUMANN, C.

Anleitung zur Untersuchung der künstlichen Düngemittel und ihrer
Rohstoffe, mit Berücksichtigung der unter den Agriculturchemikern
vereinbarten Untersuchungsmethoden, sowie der Prüfung auf Verunreinigungen und Verfälschungen. Für Chemiker, Techniker und
Fabrikanten. Braunschweig, 1876. 8vo. Ill.

SCHUMANN, GOTTHELF DANIEL.

Chemisches Laboratorium für Realschulen und zur Selbstbelehrung.
Anleitung zum chemischen Experimentiren in einer Auswahl der
wichtigeren und instructiveren chemischen Versuche. Mit einem
Vorworte von Fr. J. P. v. Rieche. Zweite Auflage. Esslingen, 1857.
pp. xii–355, 8vo. Plates.
        Erste Auflage, Esslingen, 1849.

SCHURER, FRIEDRICH LUDWIG.

Historia præcipuorum experimentorum circa analysin chemicam aëris
atmosphærici usumque principiorum ejus in componendis diversis
naturæ corporibus. Argentorati, 1784.

Synthesis oxygenii experimentis confirmata. Argentorati, 1789. pp.
126–[iv], 4to.
        Contains a bibliography of ''factitious airs.''
        Abhandlung vom Säurestoff und seiner Verbindung mit
        anderen Körpern. Aus dem lateinischen übersetzt und
        mit einigen Anmerkungen und Zusätzen vermehrt. Ber-
        lin, 1790. pp. xii–208, 16mo.

SCHUSTER, JOHANN.

De ferro. Pesth, 1829.
De iodo. Pesth, 1827.
Kleiner chemischer Apparat. Pesth, 1830.
System der dualistischen Chemie des Prof. Jeremias Joseph Winterl. . . .
Berlin, 1806–1807. 2 vols., 8vo.

SCHUTTE, C. H.

Uitvinding van een nieuwe manier om de vitriool-oly te maken. Haarlem, 1758. 8vo.

SCHWACKHÖFER, F.

Lehrbuch der landwirthschaftlich-chemischen Technologie. Wien, 1883.
8vo.
        Cf. Mach, Edmund.

SCHWANERT, HUGO.

Hülfsbuch zur Ausführung chemischer Arbeiten.   Braunschweig, 1866.
    4to.
        Zweite Auflage.   Braunschweig, 1874.
        Dritte umgearbeitete Auflage.   Braunschweig, 1891.
    Lehrbuch der pharmaceutischen Chemie.   Braunschweig, 1880–82.   3 vols.
        Trattato di chimica farmaceutica.   Brunswick, 1879.   8vo.

SCHWANNECKE, E.

Die Theorie der chemischen Structur.   Berlin, 1874.

SCHWARZ, A.

Brautechnische Reiseskizzen.   Neue Folge.   Eine ausführliche Darstel-
    lung der wichtigsten Neuerungen auf dem Gebiete der Brauereien.
    Stuttgart, 1889.

SCHWARZ, CARL LEONHARD HEINRICH.

Chemie (Die) und Industrie unserer Zeit, oder die wichtigsten chemi-
    schen Fabrikationzweige nach dem Standpunkte der heutigen
    Wissenschaft. In populären Vorträgen. Breslau, 1856–58. 3 vols.,
    8vo.
Ueber die Maassanalyse, besonders in ihrer Anwendung auf die Bestim-
    mung des technischen Werthes der chemischen Handelsproducte,
    wie Potasche, Soda, Chlorkalk, Braunstein, Säuren, Arsen, Chrom,
    Eisen, Kupfer, Zinn, Blei, Silber, u. s. w.   Braunschweig, 1850.
    8vo. Ill.
        Praktische Anleitung zu Maassanalysen (Titrir-Methode),
            besonders in ihrer Anwendung auf die Bestimmung des
            technischen Werthes der chemischen Handelsproducte,
            wie Potasche, Soda, Ammoniak, Chlorkalk, Jod, Brom,
            Braunstein, Säuren, Arsen, Chrom, Eisen, Kupfer, Zink,
            Zinn, Blei, Silber, Indigo, etc.   Zweite durch Nachträge
            vermehrte Auflage. Braunschweig, 1853. pp. xi–157.

SCHWARZ, R.

Memoranda der physiologischen Chemie, mit Rücksicht auf Pathologie.
    Leipzig, 1856.   12mo.

SCHWARZENBERG, PHILLIP, UND G. LUNGE.

    Die Technologie der chemischen Produkte (etc.).   *See* Bolley's Handbuch.

SCHWEIGGER, JOHANN SALOMO CHRISTOPH.

Ueber stöchiometrische Reihen im Sinne Richter's auf dem wissen-
    schaftlichen Standpunkte der neuesten Zeit.   Nachtrag zum Hand-
    wörterbuche der Chemie und Physik und zu den Lehrbüchern der
    Chemie überhaupt.   Halle, 1853.   8vo.

SCHWEINSBERG, G.
Uebersicht der wichtigsten chemischen Reagentien, nebst Angabe ihrer häufigsten Verwendung, für Anfänger und Dilettanten. Heidelberg, 1836.

SCHWEITZER, JOHANN CONRAD FRIEDRICH.
Chemische Versuche und Beschreibung eines vortrefflichen Stahlbrunnens zu Langenschwalbach. Wetzlar, 1770. 8vo.
De oleis essentialibus sive æthereis vegetabilium absque destillatione parandis. Giessæ, 1756. 4to.

SCHWEIZER, MATTHIAS EDUARD.
Practische Anleitung zur Ausführung quantitativer chemischer Analysen. Für Anfänger bearbeitet. Bevorwortet von Carl Löwig. Chur, 1848. 8vo.

> Practische Anleitung zur Ausführung quantitativer chemischer Analysen. In einer stufenmässig geordneten Reihe von Beispielen enthaltend Analysen von Salzen, Legierungen, Mineralien, Mineralwassern, Pflanzenaschen und technischen Producten. Zum Gebrauche im Laboratorium sowie zum Selbstunterrichte. Für Anfänger bearbeitet. Zweite vermehrte und verbesserte Auflage. Zürich, 1853. 8vo.

SCHWERDTFEGER, JULIUS.
Die chemische Maassanalyse, oder chemisch-anaiytische Titrirmethode und ihre Anwendung auf die Chlorometrie. Eine gründliche Anleitung zu volumetrischer Prüfung des Chlorkalks und anderer chlorhaltiger Bleichmittel; für Chlorkalk-, Kattun-, Papierfabrikanten, Kunstbleichereien, etc. Regensburg, 1857. 12mo.

SCIVOLETTO, PIETRO.
Principi fondamentali di chimica analitica con applicazioni alla tossicologia. Con tavole. Napoli, 1869. pp. 217, 8vo.

SCOFFERN, JOHN.
Chemistry of Inorganic Bodies ; their Compounds and Equivalents. London, 1856. 8vo.
Chemistry no Mystery ; or a Lecturer's Bequest. Being the Subject-Matter of a Course of Lectures, delivered by an old Philosopher and taken in short-hand by one cf the audience whose name is not known. Arranged from the original manuscript and revised by J. S. London, 1839. pp. xiii-310, 12mo. Ill.

> An amusing frontispiece, " Laughing Gas," by George Cruikshank.

SCOFFERN, JOHN. [Cont'd.]

Manual (A) of Chemical Analysis for the Young. Being a series of progressive lessons, illustrative of the properties of common chemical substances and of the means of separating them from each other. London, 1854. 12mo. Ill.

SCOPOLI, GIOVANNI ANTONIO.

Fundamenta chemiæ prælectionibus publicis accomodata. Pragæ, 1777. pp. [vi]–165, 12mo.

   Anfangsgründe der Chemie. Aus dem lateinischen über-
    setzt von Karl von Meidinger. Wien und Leipzig
    1786. 8vo.

SCOTT, ALEXANDER.

An Introduction to Chemical Theory. London and Edinburgh, 1891.

SCOTT, R. H.

A Handbook of Volumetrical Analysis. London, 1858. 8vo.

SCOTT-WHITE, A. H.

Chemical Analysis for schools and science classes. Qualitative-Inorganic. Adapted to meet the requirements of the London preliminary scientific and intermediate B. Sc. examinations. London, 1884.

SCOUTETTEN, H.

L'ozone, ou recherches chimiques, météorologiques, physiologiques et médicales sur l'oxygène électrisé. Paris, Metz, 1856. pp. 287, 8vo. 5 tables.

   Deuxième édition. Strasbourg, 1857. 8vo.

SCULTETUS, LUCAS. *See* Müller, Johannes.

SEBELIEN, JOHN.

Beiträge zur Geschichte der Atomgewichte. Eine von der Universität zu Kopenhagen gekrönte Preisschrift, mit einigen Veränderungen ins Deutsche übersetzt. Braunschweig, 1884. pp. vi–208, 8vo.

SEDNA, LUDWIG.

Das Wachs und seine technische Verwendung. Darstellung der natürlichen, animalischen und vegetabilischen Wachsarten, des Mineralwachses (Ceresin), ihrer Gewinnung, Reinigung, Verfälschung und Anwendung in der Kerzenfabrikation, zu Wachsblumen und Wachsfiguren, Wachspapier, Salben und Pasten, Pomaden, Farben, Lederschmieren, Fussbodenwichsen und vielen anderen technischen Zwecken. Wien, Pest, Leipzig, 1886. 8vo. Ill.

SEELHORST, G., UND H. MEIDINGER.
  Metallindustrie.   Braunschweig, 1875.

SEELIG, E.
  Organische Reactionen und Reagentien.   Stuttgart, 1892.   8vo.

SEFSTRÖM, NILS GABRIEL, UND J. J. BERZELIUS.
  Ueber das Vanadium, ein neues Metall.   Halle, 1831.

SÉGUIN, J. M.
  Cours de chimie conformé au programme du cours de philosophie.
    Paris, 1881.   12mo.

SÉGUR, OCTAVE DE.
  Lettres élémentaires sur la chimie.   Paris, 1803.   2 vols., 12mo.   Ill.

SELL, EUGEN.
  Ueber Branntwein, seine Darstellung und Beschaffenheit im Hinblick
    auf seinen Gehalt an Verunreinigungen, sowie über Methoden zu
    deren Erkennung, Bestimmung und Entfernung.   Berlin, 1888.
  Ueber Cognak, Rum und Arak ; das Material zu ihrer Darstellung, ihre
    Bereitung und nachher Behandlung unter Berücksichtigung der im
    Handel üblichen Gebräuche sowie ihrer Ersatzmittel und Nach-
    ahmungen sowie die Ergebnisse ihrer chemischer Untersuchung.
    Berlin, 1891.   4to.
  Ueber Kunstbutter, ihre Herstellung, sanitäre Beurtheilung und das
    Mittel zu ihrer Unterscheidung von Milchbutter.   Beiträge zur
    Kenntniss der Milchbutter und den zu ihrem Ersatz in Anwendung
    gebrachten anderen Fette.   Berlin, 1886.

SELL, EUGEN [Translator].
     See Naquet : Principes de chimie.

SELLA, VENANZIO GIUSEPPE.
  Polimetria chimica, ossia metodo comparativo per determinare gli acidi,
    gli alcali, i sali ed i corpi semplici nelle loro soluzioni.   Torino,
    1850.   12mo.   Ill.

SELLENATI, ANDREA CARLO.
  La chimica applicata all' agricoltura, divisa in tre parti.   Venezia, 1857.
    8vo.

SELMI, ANTONIO.
  Chimica applicata all' agricoltura.   Lezioni date nell' istituto tecnico
    provinciale di Mantova.   Il terreno, l'acqua e l'aria, ossia i mezzi
    nei quali vive la natura organica.   Parte I. Il terreno.   Parte II.

SELMI, ANTONIO. [Cont'd.]

L'acqua e l'aria. Parte III. I concimi e gli ingrassi. Parte IV. Le industrie rurali. Milano. 1871. 16mo.

Chimica applicata all' igiene ed alia economia domestica : lezioni date nell' istituto tecnico provinciale di Mantova. Milano, 1873. 2 vols., 16mo.

Chimica elementarissima, ossia nozioni facili e compendiose di chimica applicata all' igiene, all' economia domestica ed alle arti. Torino, 1856. 12mo.

Manuale di chimica inorganica, disposto secondo i programmi ministeriali ad uso degli istituti tecnici. Torino, 1866. 16mo.

Nozioni preliminari elementarissime di chimica generale per servire di introduzione allo studio di questa scienza. Firenze, 1879. 16mo.

SELMI, FRANCESCO.

Memorie sopra argomenti tossicologici. Con appendici di argomento agronomico. Bologna, 1878. pp. 178–52, 8vo.

Nuovo processo generale per la ricerca delle sostanze venefiche con appendici di argomenti tossicologici od affini. Bologna, 1875. pp. 120, 8vo.

Principii elementari di chimica agraria. Torino e Palermo, 1851. 24mo.

Principii elementari di chimica minerale. Torino, 1850. 16mo.

[Another edition.] Torino, 1856. 12mo. Ill.

Principii di chimica organica. Torino, 1852. 16mo.

Studi sperimentali e teoretici di chimica molecolare (dall' anno 1843 al 1846). Modena, n. d. 8vo.

Sulle ptomaine od alcaloidi cadaverici e loro importanza in tossicologia, osservazioni di F. S. aggiuntavi una perizia per la ricerca della morfina. Bologna, 1878. pp. v–110, 8vo.

Ptomaine od alcaloidi cadaverici e prodotti analoghi di certe malattie in correlazione colla medicina legale. Bologna, 1881. pp. iv–307, 8vo. Ill.

SEMBENINI, GIOVANNI BATTISTA.

Supplemento al trattato di chimica elementare teoretica e pratica del Barone di Thénard, contenente i progressi teorici delle scienze fisico-chimiche, ossiano fasti filosofico-chimici estratti dalle relazioni annuali di Berzelius. Mantova, 1844. 3 vols., 8vo.

See, in Section VII, Berzelius, J. J.: Jahresbericht.

SEMPLE, C. E. A.

Aids to Chemistry. London, Edinburgh, Glasgow and Dublin, 1878. pp. 114, 12mo.

New edition. London, 1888.

SEMPLE, C. E. A. [Cont'd.
Tablets of Chemical Analysis for the detection of one metal and one
    acid. London, 1884. 12mo.

SENAC, J. B.
Nouveau cours de chymie suivant les principes de Newton et de Sthall
    [*sic*]. Avec un discours historique sur l'origine et les progrez de la
    chymie. Paris, 1723. 2 vols., 12mo. Vol. I, pp. lxvii–246 ; vol. II,
    pp. 247–796.
Nouvelle édition revue et corrigée. Paris, 1737. 2 vols., 12mo.
        Published anonymously. The historical sketch occupies pp. i–c of vol. I,
        and has no especial value.

SENDTNER.
        See List, E.: Sechs Vorträge aus dem Gebiete der Nahrungsmittel-Chemie.

SENEBIER, JEAN.
Essai sur l'art d'observer et de faire des expériences. Genève, 1775.
    2 vols., 8vo.
            Deuxième édition, 1802. 3 vols., 8vo.
Expériences sur l'action de la lumière solaire dans la végétation.
    Genève et Paris, 1788. 8vo.
Mémoires physico-chimiques sur l'influence de la lumière solaire pour
    modifier les êtres des trois règnes de la nature et surtout du règne
    végétal. Genève, 1782. 3 vols., 8vo.
Physiologie végétale, contenant une description anatomique des organes
    des plantes et une exposition des phénomènes produits par leur
    organisation. Genève et Paris, 1800. 5 vols., 8vo.
        Based on Senebier's articles in the Dictionnaire des forêts et bois, which
        forms part of the Encyclopédie méthodique.
Recherches sur l'influence de la lumière solaire pour métamorphoser
    l'air fixe en air pure par la végétation. Genève, 1783. 8vo.
Recherches analytiques sur la nature de l'air inflammable. Genève,
    1784. 8vo.
            Analytische Untersuchungen über die Natur der brennbaren
                Luft. Aus dem französischen übersetzt mit einigen
                von R. Kirwan erhaltenen und eigenen Anmerkungen
                herausgegeben von Lorenz Crell. Leipzig, 1785. pp.
                [xxiv]–232, 12mo.

SENFTLEBEN, FELIX ADELBERT.
Erstes Lehrbuch der Chemie, nach ihrem gegenwärtigen Standpunkte.
    Frankfurt, 1834. 8vo.

SENGER, E.
Ueber den Werth der zur Erkennung fremder Fette (Oleomargarin, Kunstbutter, Kunstbutterschmalz, Schweinefett, u. s. w.) in dem Milchbutterfett in Anwendung befindlichen Methoden. Erlangen, 1892. 8vo. Ill.

SENNERT, DANIEL.
De chymicorum cum Aristotelicis et Galenicis consensu ac dissensu liber ; cui accessit appendix de constitutione chymiæ. Editio novissima. Venetiis, 1641. Fol.
Disputatio de natura chymiæ et chymicorum principiis. Wittenbergæ, 1629. pp. [xxvi], sm. 4to.
Hypomnemata physica de rerum naturalium principiis, de occultis qualitatibus, de atomis, de mistione . . . Francofurti, 1635. 4to.

SERTÜRNER, FRIEDRICH WILHELM.
Kurze Darstellung einiger über Elementar-Attraction, minder mächtige Säuren und Alkalien, Weinsteinsauren, Opium, Imponderabilien und einige andere chemische und physikalische Gegenstände . . . Göttingen, 1820. 8vo.
System der chemischen Physik, oder Entdeckungen und Berichtigungen im Gebiete der Chemie und Physik. Grundlinien eines umfassenden Lehrgebäudes der Chemie. Göttingen, 1820–22. 2 vols., 8vo.

SESTINI, FAUSTO.
Analisi chimica, con illustrazione medica di Decio Valentini, delle acque minerali di Loreto [Romagna] proprietà dei fratelli Brasini di Forlì. Forlì, 1868. 8vo.
Raccolta di problemi ad uso degli studenti di chimica e dei chimici pratici, compilata da Fausto Sestini. Forli, 1868. 8vo.
Saggio di analisi volumetrica. Torino, Firenze e Roma, 1871. 8vo.

SESTINI, FAUSTO, E A. FUNARO.
Elementi di chimica ad uso degli istituti tecnici secondo i nuovi programmi governativi del 21 giugno, 1885. Livorno, 1886. 16mo.
Seconda edizione. Livorno, 1888. 16mo.

SETSCHENOW, J.
Absorption der Kohlensäure durch Salzlösungen. St. Petersburg, 1876. 4to.
Ueber die Absorptionscoefficienten der Kohlensäure in den zu diesem Gase indifferenten Salzlösungen. St. Petersburg, 1886.
Weiteres über das Anwachsen der Absorptionscoefficienten von Kohlensäure in den Salzlösungen. St. Petersburg, 1887.

Севергинъ, Василій.

Пробирное искусство или руководство къ химическому испытанію металлическихъ рудъ и другихъ ископаемыхъ тѣлъ. С.-Петербургъ, 1801.

> Severghín, Vasíli. The Art of Assaying, or handbook of mineral analysis. St. Petersburg, 1801.

Способъ испытывать чистоту и неподложность химическихъ произведеній лѣкарственныхъ. С.-Петербургъ, 1800.

> Methods of testing the purity and genuineness of chemico-medical products. St. Petersburg, 1800.

Sexton, A. Humboldt.

Elementary Inorganic Chemistry, Theoretical and Practical, with a course of chemical analysis and a series of examples in chemical arithmetic. London, 1889.

Outlines of Qualitative Analysis. London, 1888. 8vo.

Outlines of Quantitative Analysis for the use of students. London, 1886. 8vo.

> Second edition. London, 1888.

Sharp, W. H.

Universal Attraction, its relation to the chemical elements. The key to a consistent philosophy. Edinburgh, n. d. [1884]. 8vo.

Shaw, Peter.

* Chemical Lectures publickly read at London in the years 1731 and 1732; And at Scarborough, in 1733; for the improvement of arts, trades and natural philosophy. The second edition, corrected. London, printed for T. and T. Longman, in Pater-Noster Row; J. Schuckburgh, in Fleet-Street; and A. Millar, in the Strand, 1755. pp. xxiv-467-[ix], 8vo.

> The preface is dated Scarborough, July 15, 1733.

Essays for the improvement of arts, manufactures and commerce by means of chemistry, containing . . . [three Essays]. London, 1741.

> Second edition. pp. xix-258, 8vo.

> Leçons de chymie propres à perfectionner la physique, le commerce et les arts. Traduit par Mme. d'Arconville. Paris, 1759. 4to.

Portable (A) Laboratory. London, 1731. 8vo.

* Secrets (Les) et les fraudes de la chymie et de la pharmacie modernes dévoilés, par l'exposition de plusieurs pratiques nouvelles et importantes pour tous ceux qui ont intérêt de s'assurer de la bonté des remèdes, et de pouvoir les fournir à un prix raisonnable. Ouvrage traduit de l'Anglois. À la Haye, chez Pierre Grosse, Junior. Libraire de Son Altesse Royale, 1759. pp. xii-370-[x], 8vo.

SHAW, SIMEON.

The Chemistry of the several natural and artificial heterogeneous compounds used in manufacturing porcelain, glass and pottery. London, 1837. pp. xliv–685, 8vo.

Pages 544–659 contain a tabular view of chemical bodies.

Tables of the Characteristics of Chemical Substances ; adapted to facilitate chemical analysis. London, 1843. 8vo.

SHAW-BREWSTER, M.

First Book of Chemistry. New York, 1887.

Щегловъ Н.

Химія, сост. Н. Щегловымъ. Учебное руководство для Военно-Учебныхъ. Заведеній съ чертежами. С.-Петербургъ, 1841 8vo.

SHCHEGLOF, N. Elements of Chemistry. St. Petersburg, 1841.

SHENSTONE, W. A.

Methods (The) of Glass Blowing. For the use of physical and chemical students. London, 1886.

Anleitung zum Glasblasen für Physiker und Chemiker. Nach dem englischen bearbeitet von H. Ebert. Leipzig, 1887.

Practical (A) Introduction to Chemistry, intended to give a practical acquaintance with the elementary facts and principles of chemistry. London, 1886. Ill.

New and revised edition. London and New York, 1892. Ill.

SHEPARD, CHARLES UPHAM.

Syllabus to Lectures on Chemistry. Charleston, S. C., 1841. 8vo.

[Another edition.] Charleston and New York, 1859. 8vo.

SHEPARD, JAMES H.

Elements of Inorganic Chemistry, Descriptive and Qualitative. Boston, 1885. pp. v–377, 12mo. Ill.

[Another edition.] Boston, 1887. 12mo.

SHERRERD, J. M.

Iron Analysis Record. With a complete table of atomic weights, their elements and symbols ; with the old and new systems. New York, 1883. 12mo.

SHILTON, A. J.

Household Chemistry for the Non-chemical. London, 1882.

Шлехтеръ Н.

Краткое руководство фармацевтической химіи для Московской военно-фельд-шерской школы. Москва, 1874.

> Shlechter, N. Short Manual of pharmaceutical chemistry for the Moscow
> Military Medical School. Moscow, 1874.

Шуреръ Фридерикъ.

Разсужденіе о соединеніи кислотворнаго вещества съ другими тѣлами. С.-Петербургъ 1800. 2 ч.

> Shurer, Frederic. Dissertation on combinations of oxygen with other
> bodies. St. Petersburg, 1800. 2 vols.

Siats, H.

Anleitung zur einfachen Untersuchung landwirthschaftlichwichtiger
Stoffe. Hildesheim, 1888.

> Zweite vermehrte und verbesserte Auflage. Hildesheim,
> 1892. 8vo. Ill.

Sibson, Alfred.

Agricultural Chemistry. With a preface by A. Voelcker, and an appendix. New edition. London, 1875. 12mo.

> Revised, extended and brought up to date by the author
> and A. E. Sibson. London and New York, 1892. 8vo.
> Beginselen der landbouwscheikunde, uit het engelsch voor
> Nederland bewerkt door S. J. van Roijen. Tweede ver-
> meerderde druk. Groningen, 1876. 8vo.

Sicard, Adrien.

Monographie de la canne à sucre de la Chine, dite sorgho à sucre.
Deuxième édition, revue, corrigée et considérablement augmentée.
Paris, 1858. 2 vols., 8vo.

Sidersky, D.

Le contrôle chimique du travail des mélasses. Paris, 18—.
Traité d'analyse des matières sucrées. Paris, 1890. pp. xxi–447, 8vo.
Ill.

Siebert, Georg.

Leitfaden für den Unterricht in der Chemie. Leipzig, 1877. 8vo.

Siegel, Oscar.

Anleitung zur Ausführung qualitativer chemischer Analysen. Für Gewerbeschulen, Bergschulen und landwirthschaftliche Anstalten.
Liegnitz, 1875. 8vo.

Siegmund, F.

Die Wunder der Physik und Chemie. Wien, 1880.

SIGAUD DE LA FOND.
Essai sur différentes especes d'air qu'on désigne sous le nom d'air fixe. Pour servir de suite et supplément aux Élémens de physique du même auteur. Paris, 1779. pp. xvi–400, 8vo. 5 folding plates.
　　　Nouvelle édition par Rouland. Paris, 1785. pp. xxviii–499, 8vo.
See also in Section III.

SILFVERBERG, A. E. W.
Lærebog i Chemien udgiven til Brug ved Skoleundervisningen. Kjøbenhavn, 1848. 8vo.

SILFVERBERG, G.
Chemisk Physik til Brug ved Skoleundervisningen. Tredie omarbejdede Oplag. Kjøbenhavn, 1865. 8vo. Ill.

SILLIMAN, BENJAMIN.
Elements of Chemistry, in the order of the lectures given in Yale College, New Haven, 1830. 2 vols., 8vo. Vol. I, pp. xii–518 ; II, viii–696–48–12. Ill.

SILLIMAN, BENJAMIN [JR.]
First Principles of Chemistry. Philadelphia, 1847. 12mo.
　　　Twenty-third edition. Philadelphia, 1853.
　　　Forty-eighth edition. Philadelphia, 1859.
Synopsis of Lectures on Chemistry delivered in Yale College. [New Haven, 186–.] 8vo.

SILLIMAN, BENJAMIN [JR.], AND HENRY WURTZ.
The Hydrocarbon Gas Process. Report of working results on a large scale, under the Gwynne-Harris Patents, Nov., 1868–May, 1869. New York, 1869. pp. 126, 8vo. Plates.

SILVA, R. D.
Traité d'analyse chimique. Rédigé par Engel. Paris, 1891. 8vo. Ill.

SILVERSMITH, JULIUS.
A Practical Handbook for Miners, Metallurgists and Assayers. Comprising the most recent improvements in the disintegration, amalgamation [etc.] of the precious ores, with a comprehensive digest of the mining laws. New York, 1866. 12mo.
　　　Fourth edition revised and corrected. 1867. 12mo.

SILVESTRINI, L.
Manuale pratico di preparazioni chimiche e guida allo studio dell' analisi chimica qualitativa. Novara, 1876. 16mo.

SIMMLER, R. THEODOR.

Chemie (Die) in ihren Beziehungen zur Landwirthschaft und zum Nationalhaushalt. Bern, 1863. 8vo.

Löthrohr-Chemie (Die) in Verbindung mit einigen Reactionen auf nassem Wege, angewendet auf land- und hauswirthschaftlich-chemische, sowie gewerbliche Untersuchungen. Eine Anleitung für angehende Chemiker, Studirende der Landwirthschaft, Polytechniker, und Gewerbtreibende, hauptsächlich zur Prüfung von Bodenarten, Düngern, Pflanzen- und Thieraschen, etc. Zürich, 1874. 8vo. Ill.

SIMMONDS, P. L.

Waste Products and undeveloped substances : or hints for enterprise in neglected fields. London, 1862. pp. vi–430.

SIMMONS, JAMES W.

A Practical Course in Qualitative Analysis for use in High Schools and Colleges. With additions for Students' work by La Roy F. Griffin. Chicago, 1888. pp. 88, 12mo. Ill.

SIMÕES DE CARVALHO, JOAQUIM AUGUSTO.

Lições de philosophia chimica. Segunda ediçao reformada. Coimbra, 1859. pp. xi–296.

SIMON, JOHANN FRANZ.

Beiträge zur physiologischen und pathologischen Chemie und Mikroscopie in ihrer Anwendung auf die praktische Medicin, unter Mitwirkung der Mitglieder des Vereins für physiologische und pathologische Chemie und anderer Gelehrten. Berlin, 1844. 2 vols., 8vo.

> Animal Chemistry with reference to the physiology and pathology of man. Translated by George E. Day. London, 1858. 2 vols., 8vo. Vol. I, pp. xx–359, 1 plate ; vol. II, xii–560.

De lactis muliebris ratione chemica et physiologica, dissertatio. Berolini, 1838. 8vo.

Handbuch der angewandten medizinischen Chemie nach dem neuesten Standpunkte der Wissenschaft und nach zahlreichen eigenen Untersuchungen. Zwei Theile. I. Medizinisch-analytische Chemie. II. Physiologische und pathologische Anthropochemie. Berlin, 1840–42. 2 vols., 8vo. Ill.

> > Vol I *also under the title :* Medizinisch-analytische Chemie, oder Chemie der näheren Bestandtheile des thierischen Körpers.
> > Vol. II *also under the title :* Physiologische und pathologische Anthropo-

SIMON, JOHANN FRANZ. [Cont'd.]

chemie mit Berücksichtigung der eigentlichen Zoochemie nach einer grossen Reihe eigener Untersuchungen und den Erfahrungen fremder Forscher.

SIMON, JOHANN FRANZ, UND J. F. SOBERNHEIM.

Handbuch der practischen Toxicologie.   *See* Sobernheim, J. F., und J. F. Simon.

SIMON, WILLIAM.

Manual of Chemistry. A Guide to lectures and laboratory work for beginners in chemistry. A textbook specially adapted for students of pharmacy and medicine. Philadelphia, 1885. pp. 479, 8vo. Ill. Second edition. Philadelphia, 1891.

SIMONS, GERRIT.

De aquæ vaporibus in atmosphæra contentis. Ultrajecti, 1823. 8vo.

SIMONS, GERRIT, EN P. L. RYKE.

Officieele stukken betrekkelijk het drinkbaar maken van zeewater van den heer L. Roulet. Amsterdam, 1850. 8vo.

SINGER, MAX.

Traité pratique pour reconnaître sans le secours de la chimie les fraudes, falsifications et sophistications des denrées alimentaires. Paris, 1876. 16mo.

SIRACUSA, F. P. C.

La clorofilla, sua natura, sua influenza nelle diversi funzioni vegetative. Palermo, 1878.

SKALWEIT.

Bericht über die fünfjährige Thätigkeit des hannoverschen Lebensmittel-Untersuchungsamts. Hannover, 1883. 8vo.

SKOMOROWSKI, T.

Chemia życia codziennego podług najnowszych żródeł. Warszawa, 1875. 8vo.

SKRIMSHIRE, FENWICK.

Series of Popular Chymical Essays. Second edition. London, 1804. 12mo.

SLATER, JOHN WILLIAM.

Handbook of Chemical Analysis for practical men. Containing directions for examining and valuing several hundreds of the most important

SLATER, JOHN WILLIAM. [Cont'd.]
articles of commerce, manufacturing products, residues, etc. Also
a brief systematic course of quantitative analysis, and a variety of
useful chemical tables. Second edition. London, 1861. pp. xvi–
384, 8vo. Ill.
    Third edition. London, 1871.
Manual (The) of Colours and Dye-wares ; their properties, applications,
valuation [etc.]. London, 1870. 8vo.

SLATTER, G. W.
Outlines of Qualitative Analysis. London, 1888. 8vo.

SLIASKY, S., AND F. WASSILTOFSKY.
Tablizi [etc.]. Kiew, 1891. 8vo.
    Tables for Chemists and Sugar-refiners, in the Russian language. I have
    not seen the original.

SMEE, ALFRED.
Elements of Electro-Metallurgy. Third edition, revised and illustrated.
London, 1851. 8vo.
        The first American from the third London edition. Revised
        and considerably enlarged by J. R. Chilton. New York,
        1853. 12mo. Ill.
        Nouvelle manuel complet de galvano-plastic, ou éléments
        d'électro-métallurgie, contenant l'art de réduire les
        métaux à l'aide du fluide galvanique, etc. Augmentée
        d'un grand nombre de notes, d'après Jacoby, Spencer,
        Elsner, etc. Ouvrage publié par E. de Valincourt.
        Nouvelle édition traduite sur la troisième édition de
        l'original Anglais entièrement refondue et mise au
        courant de toutes les découvertes nouvelles. Paris,
        1854. 2 vols., 18mo.
    Première édition. Paris, 1843.
        Elemente der Electro-Metallurgie. Deutsch bearbeitet nach
        der dritten vermehrten und verbesserten englischen
        Original-Ausgabe (von Otto Bernhard Kühn). Leipzig,
        1851. 8vo. Ill.

SMETH, DIDERICUS DE.
De aere fixo. Trajecta ad Rhœnum, 1772. 4to. Ill.

SMITH, DANIEL B.
The Principles of Chemistry ; prepared for the use of schools, academies,
and colleges. Second edition. Philadelphia, 1863. 12mo.

SMITH, DAVID.

Smith's Practical Dyer's Guide. Containing five hundred dyed patterns to each of which a genuine receipt is given. The work comprises practical instructions in the dyeing of silk, cotton and wool in a raw and manufactured state. Also instructions for dyeing plain and mixed fabrics in single and two colours, and a great variety of bronzes. Also receipts for making all the dye spirits with which to dye every colour in the work and brief explanation of the process of extracting "burrs from wool" and "cotton from rags" both by the liquid and the gas process. Second edition. Manchester, 1880. pp. 238, 8vo. Ill.

SMITH, EDGAR F.

Electro-Chemical Analysis. A practical Handbook. Philadelphia, 1890. 12mo. Ill.

SMITH, EDGAR F., AND HARRY KELLER.

Experiments arranged for Students in General Chemistry. Second edition, enlarged. Philadelphia, 1891. pp. 56, 8vo. Ill.

SMITH, EDWARD.

Foods. London, 1874. 12mo.
> Eighth edition. London and New York, 1883. 12mo.
> Die Nahrungsmittel. Leipzig, 1875. 2 vols.
The International Scientific Series.

SMITH, GODFREY.

* The Laboratory or School of Arts, in which are faithfully exhibited and fully explain'd : I. A Variety of curious and valuable experiments in refining, calcining, melting, assaying, casting, allaying, and toughening of gold, with several other curiosities relating to gold and silver. II. Choice secrets for Jewellers in the management of gold. . . . III. Several uncommon experiments for casting in silver, copper, brass, tin, steel and other metals. . . . IV. The Art of making glass. . . . V. A Collection of very valuable secrets for the use of Cutlers, Pewterers, Brasiers, Joiners, Turners, Japanners, Book-binders, Distillers, Lapidaries, Limnars, etc. . . . VI. A dissertation on the nature and growth of salt-petre. . . . VII. The art of preparing rockets, crackers, fire-globes, stars, sparks, etc., for recreative fireworks. VIII. The art and management of dyeing silks, worsteds, cottons, etc., in various colours. Compiled from German and other foreign authors. Third edition, with additions. . . . London, 1750. pp. [viii]–352–[viii], 12mo. 16 plates and illuminated frontispiece.

SMITH, HENRY ARTHUR.

The Chemistry of Sulphuric Acid Manufacture. London, New York, 1873. pp. xvi–81.

> Die Chemie der Schwefelsäure-Fabrikation. Aus dem englischen übersetzt und mit Anmerkungen versehen von F. Bode. Freiberg, 1874. 8vo. Ill.

SMITH, JOHN LAWRENCE.

* Mineralogy and Chemistry : Original researches. Louisville, Kentucky, 1873. pp. 401, 8vo. Ill.

* Original researches in Mineralogy and Chemistry. Printed for presentation only. Edited by J. B. Marvin. Louisville, Kentucky, 1884. pp. xl–630, 8vo. Ill.

> Portrait of J. L. Smith.

SMYTH, GEORGE A.

Entwicklung der Ansichten über die gepaarten Schwefelverbindungen. Berlin, 1876.

SNAITH, W. A.

Inorganic Chemistry for Elementary Classes. Designed chiefly for use in the elementary stage of classes in connection with the Science and Art Department. London, 1871. 18mo.

SNIADECKI, JEDRZ.

Początki chemii dla użycia słuchaczów akademickich ułożone. Edycya powtórna powiększona i poprawna. w Wilnie, 1807. 2 vols., 8vo. Ill. Vol. I, pp. xvi–520–[viii], 2 folding plates ; vol. II, pp. [x]–353–45.

> The appendix bears the title :
>
> Rosprawa o nowym metallu w surowey platynie odkrytym. w Wilnie, 1808. pp. 45, 8vo.

SNIJDERS, A. J. C.

Scheikundige brieven. Leercursus ter beoefening der chemie door zelfonderricht. Volgens de nieuweste uitkomsten der wetenschap, naar aanleiding van Krätzers chemische Unterrichts-Briefe. Zutphen, 1886. Roy. 8vo.

> Cf. Krätzer, Hermann.

SNIVELY, J. H.

Elements of Systematic Qualitative Chemical Analysis : a hand-book for beginners. Philadelphia, 1883. 12mo. Ill.

Tables for Systematic Qualitative Chemical Analysis. Nashville, 1879. 8vo.

SOAMES, PETER.

A Treatise on the Manufacture of Sugar from the Sugar Cane. London, 1872. pp. viii–136, 8vo. Folding plates.

SOBERNHEIM, J. F., UND JOHANN FRANZ SIMON.

Handbuch der praktischen Toxicologie. Nach dem neuesten Standpunkte dieser Wissenschaft und ihrer Hilfsdoctrinen; für angehende, praktische und Physikatsärzte, so wie für Kreiswundärzte und Apotheker. Berlin, 1838. pp. viii–734, 8vo. Plates.

SOBRERO, ASCANIO.

Lezioni di chimica docimastica fatte nella R. Scuola d'Applicazione per gli Ingegneri in Torino l'anno 1875–76. Raccolte stenograficamente dagli allievi Leosini V. e Pastore G., e rivedute dal professore. Torino, 1877. 8vo.

Manuale di chimica applicata all' arte del costruttore, della Regia Università di Torino, per l'anno scolastico 1850–51. Torino, 1850. 4to.

Manuale di chimica applicata alle arti. Torino, 1853–78. 12mo. Ill.

SOCIÉTÉ CHIMIQUE DE PARIS.

Leçons de chimie professées en 1860 par Pasteur, Wurtz, Cahours, Berthelot, Sainte-Claire-Déville, Barral et Dumas. Paris, 1861. 8vo.

[*Contents :*] Recherches sur la dissymmétrie moléculaire des produits organiques naturels.—Histoire des radicaux organiques.—Recherches sur les glycols.—De la synthèse en chimie organique.—Des lois de nombre en chimie et de la variation de leurs constantes.—De l'influence exercée par l'atmosphère sur la végétation.—Pièces historiques concernant Lavoisier et N. Le Blanc.

Leçons de chimie et de physique professées en 1861 par Jamin, Debray, Lissajous, Cloëtz, Edm. Becquerel et Pasteur. Paris, 1862. 8vo.

[*Contents :*] Loi de l'équilibre et du mouvement dans les corps poreux.—Sur la production des températures élévées et sur la fusion du platine.—Sur l'étude optique des sons.—Recherches sur la nitrification et considérations généraux sur le rôle des nitrates dans la végétation.—Effets lumineux qui résultent de l'action de la lumière sur les corps.—Sur les corpuscules organisés qui existent dans l'atmosphère, examen de la doctrine des générations spontanées.

Leçons de chimie professées en 1862 par Verdet et Berthelot. Paris, 1863.

[*Contents :*] Exposé de la théorie mécanique de la chaleur.—Sur les principes sucrés.

Leçons de chimie professées en 1863 par Wurtz, A. Lamy, Louis Grandeau. Paris, 1864.

[*Contents :*] Sur quelques points de la philosophie chimique.—Leçons sur le thallium.—Leçon sur le rubidium et caesium.

Société chimique de Paris.   [Cont'd.]

Leçons de chimie professées en 1864 et 1865 par Berthelot, De Luynes, Sainte-Claire-Déville et Des Cloiseaux.   Paris, 1866.   8vo.

> [*Contents :*]   Sur l'isomérie.—Sur la dissociation.—Sur l'étude des propriétés optiques biréfringuentes.—Sur les principes contenus dans les lichens à orseille.

Leçons de chimie professées en 1866 et 1867 par St.-Claire-Déville, Schützenberger et le Roux.   Paris, 1869.   8vo.

> [*Contents :*]   Sur l'affinité.—Sur les matières colorantes.—Sur les courants thermo-électriques.

Leçons de chimie professées en 1868 et 1869 par P. P. Dehérain, P. Schützenberger, Ch. Friedel, F. P. Le Roux, A. Gautier, B. Tollens. Paris, 1870.   pp. 240, 8vo.

> [*Contents :*]   Sur l'assimilation des substances minérales par les plantes.—Sur le rôle de l'acide hypochloreux en chimie organique et sur une nouvelle classe d'anhydrides mixtes.—Sur les composés organiques du silicium.—Sur les phénomènes électro-capillaires.—De l'acide cyanhydrique, de ses homologues et de leurs isomères.—Sur quelques hydrocarbures de la série aromatique.

Conférences faites à la Société chimique de Paris en 1883–1886 par Wurtz, Pasteur, Friedel, Scheurer-Kestner, Grimaux, Duclaux, Moissan, Raoult, Schützenberger.   Paris, 1886.   pp. 191.

> [*Contents :*]   Histoire chimique de l'aldol.—La dissymétrie moléculaire.—Synthèse des combinaisons aromatiques.—Nicolas Leblanc et la soude artificielle.—Les substances colloidales.—Le lait et sa composition chimique.—Les fluorures du phosphore.—Congélation des dissolutions, —Matières protéiques.

Conférences faites à la Société chimique de Paris en 1887–1888 par Henri Moissan, E. Mallard, A. Haller, Scheurer-Kestner, H. Millot, A. Combes, Maquenne.   Paris, 1889.   pp. 250, 8vo.

> [*Contents :*]   Le fluor.—Les groupements cristallins.—Le camphre et ses dérivés.—Recherches sur la combustion de la houille.—La fabrication du chlore.—Synthèses dans la série grasse au moyen du chlorure d'aluminium.—La constitution des sucres.

Conférences faites à la Société chimique de Paris en 1889–1892 par E. Duclaux, Franchimont, Cazeneuve, Sarrau, Lebel, Guye, Le Chatelier, Nœlting, Béchamp.   Paris, 1892.   pp. 313–[iii], 8vo.

> [*Contents :*]   Le lait au point de vue alimentaire, par E. Duclaux ; Action de l'acide azotique réel sur les composés de l'hydrogène, par Franchimont.—Le camphre et la série térébénique, par Cazeneuve.—La continuité des états liquide et gazeux, par E. Sarrau.—Le pouvoir rotatoire et la structure moléculaire, par J.-A. Lebel.—La dissymétrie moléculaire, par Philippe-A. Guye.—La mesure des hautes températures, par H. Le Chatelier.—Recherches sur les colorants dérivés du triphenylméthane, par E. Nœlting.—Sur la constitution histologique et la composition chimique comparées des laits de vache, de chèvre, d'ânesse et de femme, par A. Béchamp.

SOLER Y SANCHEZ, JOSE.
  Las teórias de la química.   Madrid, 1874.   pp. 142, 8vo.

SOLLY, EDWARD.
  Rural Chemistry : an elementary introduction to the study of the science
    in its relation to agriculture.   London, 1843.   8vo.
        Third edition.   London, 1851.   8vo.
        First American edition.   Philadelphia, 1852.   12mo.
        Agriculturchemie.   Aus dem englischen übersetzt und heraus-
            gegeben von der Redaction der Allgemeinen Gartenzei-
            tung in Berlin.   Berlin, 1844.   8vo.
  Syllabus of a Complete Course of Lectures on Chemistry ; including its
    applications to the arts, agriculture and mining prepared for the use
    of the Gentlemen Cadets at the Hon. East India Company's Military
    Seminary.   London, 1849.   8vo.

SON, VAN.
  Beknopte handleiding tot practische aanwending der blaaspijp, voor
    apothekers-leerlingen.   'sGravenhage, 1842.   8vo.

SONNENSCHEIN, FRANZ LEOPOLD.
  Anleitung zur chemischen Analyse für Anfänger.   Berlin, 1852.   8vo.
        Dritte Auflage.   Berlin, 1858.
  Anleitung zur quantitativen chemischen Analyse.   Berlin, 1864.   8vo.
  Handbuch der analytischen Chemie.   Mit Benutzung der neuesten
    Erfahrungen.   Qualitative Analyse.   Berlin, 1870.   8vo.
  Handbuch der analytischen Chemie.   Quantitative Analyse.   Berlin,
    1871.   8vo.   Ill.
  Handbuch der gerichtlichen Chemie.   Nach eigenen Erfahrungen
    bearbeitet.   Berlin, 1869.   pp. xii–564.   8vo.   6 plates.
        Zweite gänzlich umgearbeitete Auflage, neu bearbeitet von
            Alexander Classen.   Berlin, 1881.   pp. xii–560.   8vo.
            Ill.

SOREL.
  Fabrication de l'acide sulfurique et des engrais chimiques.   Paris, 1887.
    8vo.
        Cf. in Section II, Fremy : Encyclopédie chimique, vol. v.

SOTO Y MARRUGAN, F.
  El analisis volumétrico y electrolisis en la determinacíon del cobre.
    Madrid, 1890.   4to.

SOUBEIRAN, EUGÈNE.
  Manuel de pharmacie théorique et pratique.   Paris, 1826.   18mo.   Ill.

SOUBEIRAN, EUGÈNE. [Cont'd.]
Notice sur la fabrication des eaux minérales artificiales. Troisième
édition. Paris, 1843. 12mo.
  Première édition, Paris, 1839.

> Anleitung zu Verfertigung künstlicher Mineralwässer und
> ähnlicher Compositionen. Aus dem französischen über-
> setzt und durch Zusätze, so wie die Formeln der vor-
> züglichsten deutschen Mineralwässer vermehrt. Leipzig,
> 1840. 8vo.

Traité de pharmacie théorique et pratique. Paris, 1835–36. 2 vols.,
8vo.

> Cinquième édition. Paris, 1857. 2 vols., 8vo.
> Septième édition, par J. Regnauld. Paris, 1869. 8vo.
> Translated into German and Spanish.

SOULIÉ, ÉMILE, ET HIPP. HANDOÜIN.
Le pétrole, ses gisements, son exploitation, son traitement industriel, ses
produits dérivés, ses applications à l'éclairage et au chauffage. Paris,
1865. pp. viii–231.

SOUSA PINTO, ANTONIO JOSE DE.
Observações sobre a incerteza das analyses e reagentes, ou equivocação
em que cahem os que attribuem à cada reagente, hum caractér
particular para distinguir exclusivamente huma determinada sub-
stancia. Lisboa, 1819. pp. 32, 8vo.

SOUTHBY, E. R.
Systematic Book of Practical Brewing. Second edition. London, 1885.

SOUTHWICK, A. P.
Question-book of Chemistry ; with notes, queries, etc. Syracuse, N. Y.,
1883. 16mo.

SOXHLET, V. H.
Färberei (Die) der Baumwolle mit direct färbenden Farbstoffen. Stutt-
gart, 1891. 8vo. Ill.
Praxis (Die) der Anilin-Färberei und Druckerei auf Baumwoll-Waaren.
Enthaltend die in neuerer und neuester Zeit in der Praxis in
Aufnahme gekommenen Herstellungsmethoden. Echtfärberei mit
Anilinfarben, das Anilinschwarz und andere auf der Faser selbst
zu entwickelnde Farben, Anwendung der Anilinfarben zum Zeug-
druck. Wien, Pest, Leipzig, 18—. Ill.

SPALLANZANI, LAZZARO.
Chimico esame degli esperimenti del Sig. Göttling, sopra la luce del fosforo di Kunkel osservata nell' aria comune. Modena, 1796. 8vo.

SPARKES, GEORGE.
 An Easy Introduction to Chemistry. Second edition. London, 1846.
 12mo.

SPATARO, D.
 Igiene delle abitazioni. Vol. II. Igiene delle acque. Proprietá fisico-
 chimiche. Inquinamento. Analisi delle acque potabili. Idrologia.
 Generalitá sulla circolazione delle acque. Idrografia sotterranea
 d'Italia. Valore sanitario delle varie acque in natura. Milano,
 1891.

SPÄTH, JOHANN LEONHARD.
 Ueber die Natur der Gase oder die Gasometrie. Nach neuen und
 eigenen Ansichten vorgetragen. München, 1835. 8vo.

SPÉCZ, R. VON.
 Grundriss der technischen Chemie. Wien, 1837. 8vo.

SPENCER, GUILFORD L.
 Hand-Book (A) for Sugar Manufacturers and their Chemists. Contain-
 ing practical instruction in sugar-house control, the diffusion pro-
 cess, selected methods of analyses, reference tables, etc. New
 York, 1889. 12mo.
 Report of Experiments in the Manufacture of Sugar at Magnolia Station,
 Lawrence, La. Season of 1885–'86. Second Report, Bulletin No.
 11, Division of Chemistry, Department of Agriculture. Washing-
 ton, 1886. pp. 26. Folding plates.
   The First Report is contained in Bulletin No. 5.
 Third Report, Season of 1886–87, Bulletin No 15, Division of Chemistry,
 U. S. Department of Agriculture. Washington, 1887. pp. 35.
 Report of Experiments in the Manufacture of Sugar by Diffusion at
 Magnolia Station, Lawrence, La. Season of 1888–'89. Bulletin
 No. 21, Division of Chemistry, U. S. Department of Agriculture.
 Washington, 1889. pp. 67, 8vo.
 Tea, Coffee and Cocoa Preparations. Foods and Food Adulterants,
 investigations made under the direction of H. W. Wiley. Bulletin
 No. 13, Part VII, United States Department of Agriculture, Division
 of Chemistry. Washington, 1892. 8vo.
   See Wiley, H. W., Clifford Richardson [and others], *and see, in Section VII*,
   Bulletins of the Division of Chemistry, U. S. Department of Agriculture.

SPENCER, J.
 Elementary Practical Chemistry and Laboratory Practice ; with more
 than 200 experiments, 300 exercises in chemical calculations ; with

SPENCER, J. [Cont'd.]

answers and 250 examination questions set at the science examinations, the London university matriculation, the Cambridge local, and the College of Preceptors' examinations. Part 1. London, 1883. 12mo.

SPENCER, THOMAS.

Vital Chemistry. Lectures on Animal Heat. Published by request of the Class. Geneva, N. Y., 1845. pp. ix–114, 12mo.

SPENCER, W. H.

Elements of Qualitative Chemical Analysis. London and Cambridge, 1866. 4to.

SPENNRATH, J.

Die Chemie im Handwerk und Gewerbe. Ein Lehrbuch zum Gebrauche an technischen und gewerblichen Schulen sowie zum Selbstunterricht. Aachen, 1889.

SPICA, P.

Tavole di chimica analitica qualitativa. Padova, 1892. 8vo.

SPICE, R. P.

A Treatise on the Purification of Coal-gas and the advantage of Cooper's coal-liming process. London, 1884.

SPIELMANN, JACOB REINBOLD.

Cf. Wittwer, the editor of a collection of treatises by J. R. S.

SPIELMANN, JACOB REINBOLD.

Examen acidi pinguis. Argentorati, 1769. 4to.
Examen de compositione et usu argillæ. Argentorati, 1773. 4to.
De causticitate. Argentorati, 1779. 4to.
De principio salino. Argentorati, 1748. 4to.
Dissertatio de argilla. Argentorati, 1765. 4to.
Dissertatio sistens analecta de tartaro. Argentorati, 1780. 4to.
Dissertatio sistens commentarium de analysi urinæ et acido phosphoreo. Argentorati, 1781. 4to.
* Institutiones chemiæ prælectionibus academicis adcommodatæ. Argentorati, 1763. pp. [xiv]–309–[lix].

> Editio secunda, 1766.

>> Chemische Begriffe und Erfahrungen nach der lateinischen Urschrift und der französischen Uebersetzung mit Anmerkkungen von J. H. Pfingsten. Dresden, 1783. 8vo. Ill.

Translated into French by Cadet le jeune. 2 vols., 12mo. Paris, 1770. Contains a "Syllabus auctorum" embracing about 330 titles alphabetically arranged. *See also in Section III.*

SPIRGATIS, H.
Anleitung für die qualitative chemische Analyse. Dritte Auflage neu bearbeitet von E. Pieszczek. Königsberg, 1892. 8vo.

SPRENGEL, CARL.
Chemie für Landwirthe, Forstmänner und Cameralisten. Göttingen, 1831–32. pp. xxviii–793, 12mo. 2 vols.
Leuchs remarks of this work, "Ausgezeichnet."

SPRINGMÜHL, FERDINAND.
Die chemische Prüfung der künstlichen organischen Farbstoffe. Leipzig, 1874. 8vo.

STABEL, EDUARD.
Das Ozon und seine mögliche therapeutische Bedeutung. Kreuznach, 1883. pp. 36, 16mo. Ill.

STADIOT, FR.
Ueber das Wachsthum und den Nutzen der Runkelrübe. Prag, 1770.
Ueber den Nutzen der Anpflanzung der Zuckerrübe. Prag, 1775.

STÄDELER, GEORG ANDREAS.
Leitfaden für die qualitative chemische Analyse anorganischer Körper. Vierte Auflage. Zürich, 1870.
Fünfte Auflage. Zürich, 1871.
Sechste Auflage, durchgesehen und ergänzt von H. Kolbe. Zürich, 1873.
Siebente Auflage. Zürich, 1879.
Achte Auflage. Zürich, 1882.
Neunte Auflage, neu bearbeitet von H. Abeljanz. Zürich, 1890. 8vo.
Instruction sur l'analyse chimique qualitative des substances minérales. Revue par Hermann Kolbe. Traduite sur la sixième édition allemande par L. Gautier. Paris, 1873. pp. 68, 32mo.
Guida all' analisi chimica qualitativa dei corpi inorganici. Traduzione autorizzata sulla settima edizione originale, di Vincenzo Fino, seconda edizione corretta ed aumentata. Torino, 1878. 8vo.
Terza edizione. Torino, 1889. Ill.
Руководство къ качествен. химическому анализу неорганич. тѣлъ. Москва, 1862.
Another edition, 1872.

STAES, STEPHANUS I.
Στοιχειώδη μαθήματα χημείας. Σμυρνῆ, 1887. pp. 164, 8vo. Ill.
Contains portrait of Lavoisier.

STAHL, GEORG ERNST.
    Cf. Becher, J. J. : Physica Subterranea.

STAHL, GEORG ERNST.
    Ausführliche Betrachtung und zulänglicher Beweiss von den Saltzen das dieselben aus einer zarten Erde mit Wasser innig verbunden bestehen. Halle, 1723. pp. [xvi]–432, 12mo.

        Zweyte Auflage mit einem Vorbericht, Anmerkungen und einem Register versehen von Johann Joachim Langen. Halle, 1765. 8vo.

    * Billig Bedencken, Erinnerung und Erläuterung über J. Bechers Natur-Kündigung der Metallen. Franckfurth und Leipzig, 1723. pp. 443, 12mo. 2 parts, pagination continuous.

    Experimenta, observationes, animadversiones, CCC numero, chymicæ et physicæ. Editio secunda. Berlin, 1731. 8vo.

    Fundamenta chymiæ dogmaticæ et experimentalis, et quidem tum communioris physicæ, mechanicæ, pharmaceuticæ ac medicæ, tum sublimioris sic dictæ hermeticæ atque alchymicæ ; olim in privatos auditorum usus posita, jam vero, indultu autoris, publicæ luci exposita. Norimbergæ, 1723.

        * Philosophical Principles of Universal Chemistry, or the foundation of a scientifical manner of inquiring into and preparing the natural and artificial bodies for the uses of life ; both in the smaller way of experiment and the larger way of business. Design'd as a general introduction to the knowledge and practice of artificial philosophy or genuine chemistry in all its branches. Drawn from the Collegium Jenense of George Ernest Stahl by Peter Shaw. Printed for John Osborn and Thomas Longman at the Ship in Paternoster Row, London, 1730. pp. xxviii–424–[36], 8vo.

        * Chymia rationalis et experimentalis ; oder gründliche der Natur und Vernunfft gemässe und mit Experimenten erwiesene Einleitung zur Chymie ; darinnen hauptsächlich die Mixtion derer sublunarischen Cörper nebst deren Zerlegung und Relation gegen einander untersuchet, und mit vielen Experimenten gezeiget wird. Nebst einer Zugabe vor denen Mercuriis Metallorum, Mercurio animato, und Lapide Philosophorum. Leipzig, 1720. pp. [xvi]–520–[xxxi], 12mo.

        * Zweite Auflage, 1729. pp. [xvi]–560–[xxxii], 12mo.

        Contains an historical and experimental inquiry into the philosopher's stone.

    Fundamenta chymico-pharmaceutica generalia. Accessit manuductis

STAHL, GEORG ERNST.   [Cont'd.]

ad enchirises artis pharmaceuticæ specialis ; cura Benjamin Roth-
Scholtzii.   Herrnstadii, 1721.   pp. 48, 8vo.

> Contains a list of the works of Stahl.

* Gründliche und nützliche Schriften, von der Natur, Erzeugung,
Bereitung, und Nutzbarkeit des Salpeters, mit denen hieher gehörigen Kupfern, und vielen diensamen Anmerckungen vermehret,
und wegen ihres unbeschreiblichen Nutzens aus dem lateinischen
ins teutsche übersetzet.   Franckfurt und Leipzig, Verlegts Johann
Leopold Montag, Buchhandlern in Regenspurg, 1734.   pp. [x]–206,
12mo.   Plates.

> [Another edition.]   Stettin und Leipzig, 1748.

* Observationes physico-chymico-medicæ curiosæ, antehac observationibus Hallensibus selectis ad rem litterariam spectantibus sparsim
insertæ, nunc vero in unum fasciculum collecta et in gratiam
quorundam philiatrorum edita.   *n. p.*, 1719.   pp. 136, 16mo.

Observationum chymico-physico-medicarum curiosarum, mensibus singulis . . .   Francofurti et Lipsiæ, 1697.   pp. 352, 16mo.

> The parts bear the names of the months from July to December, both
> inclusive.

Opusculum chymico-physico medicum, seu schediasmatum a pluribus
annis variis occasionibus in publicum emissorum, nunc quadantenus
etiam auctorum et deficientibus passim exemplaribus in unum volumen jam collectorum, fasciculus publicæ luci redditus præmissa
præfationis loco authoris epistola ad Michaelem Alberti.   Halæ
Magdeburgicæ, 1740.   pp. [viii]–856-[xxxviii], 4to.   Portrait of
Stahl.

> [Another edition.]   Halæ Magdeburgicæ, 1715.   4to.

* Zufällige Gedanken und nützliche Bedencken über den Streit von
dem sogenannten Sulphure, und zwar sowol dem gemeinen, ver-
brennlichen, oder flüchtigen, als unverbrennlichen oder fixen.
Halle, 1718.   pp. [viii]–373, 8vo.

> Zweite Auflage.   Halle, 1747.   8vo.

Zymotechnia fundamentalis seu fermentationis theoria generalis [etc.].
Halæ, 1697.   8vo.

> Contains his theory of Phlogiston.

> * Zymotechnia fundamentalis, oder allgemeine Grund-Er-
> känntniss der Gährungs-Kunst, vermittelst welcher die
> Ursachen und Würckungen dieser alleredelsten Kunst,
> welche den nutzbahrsten und subtilesten Theil der gan-
> zen Chymie ausmacht.   Aus den wesentlichen mechan-
> isch-physischen Haupt-Gründen überhaupt mit höchstem

STAHL, GEORG ERNST. [Cont'd.]

Fleisz ans Licht gestellet, und mit einem neuen chymischen Experiment, wie ein wahrer Schwefel durch Kunst zum Vorschein zu bringen, wie auch mit andern nützlichen Erfahrungs-Proben und Anmerckungen dem Publico mitgetheilet werden. Wegen ihres unbeschreiblichen Nutzens aus dem lateinischen ins teutsche übersetzt. Franckfurth und Leipzig. Verlegts Johann Leopold Montag in Regenspurg, 1734. pp. [xxii]-304.

STAHL, LEOPOLD.

Allgemeiner Gang der qualitativen chemischen Analyse fester und tropfbarflüssiger anorganischer Körper mit Berücksichtigung der häufiger vorkommenden organischen Säuren. Berlin, 1862. 8vo. Ill.

STAHL, W.

Ueber Raffination, Analyse und Eigenschaften des Kupfers. Clausthal, 1886. 8vo.

STAHLSCHMIDT, C.

Die Gährungs-Chemie, umfassend die Weinbereitung, Bierbrauerei und Spiritusfabrikation. Nebst einem Anhang, die Essigfabrikation enthaltend. Nach dem heutigen Standpunkt der Wissenschaft und Praxis bearbeitet. Mit 93 in den Text eingedruckten Holzschnitten. Berlin, 1868. pp. vii-412, 8vo.

STAMM, ANTON.

Oxygen and Ozone, their Applications. F. C. Garbutt, Agent [etc.]. Leadville, 1884. pp. [vi]-81, 8vo.

STAMMER, CARL.

Die Branntweinbrennerei und deren Nebenzweige. *See* Otto-Birnbaum's Lehrbuch, vol. II.

STAMMER, CARL.

Chemisches Laboratorium. Anleitung zum Selbstunterrichte in der Chemie. Giessen, 1856–'57. 3 parts, 8vo. Ill.

Химическая лабораторія. Руководство къ практическому изученію химіи безъ помощи учителя. Перев. Вериго. С.-Петербургъ, 1865 3 ч.

Kurz gefasstes Lehrbuch der Chemie und chemischen Technologie. Essen, 1857. 8vo.

Zweite Auflage. Essen, 1869. 8vo.

Tabellen chemischer Schemata. Zum Gebrauche beim Unterricht in der unorganischen Chemie. Braunschweig, 1856.

STAMMER, CARL. [Cont'd.]

Leitfaden bei den praktischen Arbeiten im chemischen Laboratorium. Zum Gebrauche beim Unterrichte in der unorganischen Chemie an Gewerbe- und Realschulen. Braunschweig, 1854. 8vo.

Sammlung von chemischen Rechenaufgaben. Zum Gebrauche an Real- und Gewerbe-Schulen, an technischen Lehranstalten und beim Selbststudium für Studirende, Pharmaceuten, chemische Fabrikanten u. A. Zweite, vollständig neu bearbeitete Auflage. Braunschweig, 1878. 8vo.

Antworten und Auflösungen zu den Sammlungen von chemischen Rechenaufgaben. Zum Gebrauche beim Selbststudium für Studirende, Pharmaceuten, chemische Fabrikanten u. A., sowie für Lehrer an technischen Lehranstalten, Real- und Gewerbeschulen. Zweite, vollständig neu bearbeitete Auflage. Braunschweig, 1878. 8vo.

Chemical Problems, with explanations and answers, translated by W. S. Hoskinson. Philadelphia, 1885. 12mo.

Lehrbuch der Zuckerfabrikation. Zweite Auflage. Mit 2 Bildnissen, A. S. Marggraf und F. C. Achard, 4 Chromographien, 3 autographirten Plänen, 562 Holzstichen und 9 lithographirten Tafeln. Zwei Hälften und ein Atlas. Braunschweig, 1887. 8vo.

Ergänzungsband zur ersten Auflage des Lehrbuchs der Zuckerfabrikation. Mit 130 Holzstichen und zwei Bildnissen. Braunschweig, 1881. 8vo.

Wegweiser in der Zuckerfabrikation. Vorzugsweise zum Gebrauch für Fabrikbeamte, Techniker, Siedemeister, etc. Mit zahlreichen Holzstichen. Braunschweig, 1876. 8vo.

Agenda et calendrier de poche du fabricant de sucre ; à l'usage des fabricants de sucre, raffineurs, distillateurs, chimistes, chefs de fabrication, contre-maîtres et employés des régies. Traduit de l'allemand avec l'autorisation de l'auteur et suivi d'un Traité d'analyse chimique à l'usage des fabricants de sucre et des distillateurs. Nouvelle édition entièrement revue par les auteurs et augmentée. Paris, 1890. pp. 546, 16mo. Ill.

STANĚK, JOHANN.

Chemie všeobecná. Dil 1. O nekovech. Sepsal Jan Staněk. S 51 vyobrazeními. v Praze, 1858. 8vo.

STAS, JEAN SERVAIS.

Nouvelles recherches sur les lois des proportions chimiques, sur les poids atomiques et leurs rapports mutuels. [Bruxelles, 1865] 4to. Plate.

STAS, JEAN SERVAIS. [Cont'd.]

Untersuchungen über die Gesetze der chemischen Proportionen, über die Atomgewichte und ihre gegenseitigen Verhältnisse. Uebersetzt von L. Aronstein. Leipzig, 1867. pp. xii–347.

STEELE, J. DORMAN.

A Fourteen Weeks' Course in Chemistry. New York and Boston, 1869. pp. 268, 8vo. Ill.

[Another edition.] New York, 1887.

STEELE, JOHN H.

An Analysis of the Mineral Waters of Saratoga and Ballston. Albany, 1819. 12mo.

STEENBUCH, C.

Kortfattet Veiledning til Brug ved Fremstillinger af de vigtigste chemiske Præparater. Kjøbenhavn, 1876. 8vo.

Lærebog ; Chemie til Brug for Apotheksdisciple ved Forberedelsen til Medhjælperexamen. Kjøbenhavn, 1881. 8vo.

Titreranalysen, med særligt Hensyn til dens Anvendelse ved Undersögelsen af de i de Skandinaviske Pharmacopoeer officinelle Præparater. Kjøbenhavn, 1887. 8vo.

STEFANELLI E F. SESTINI.

Sommario degli studj di chimica si pura come applicata, ecc. Firenze, 1863. 8vo.

STEFFEN, W.

Lehrbuch der reinen und technischen Chemie nach System Kleyer bearbeitet. Stuttgart, 1892. 2 vols., 8vo. Ill.

Vol. I. Die Metalloide. 1889. Vol. II. Anorganische Experimental-Chemie. Die Metalle. 1892.

STEIN, GOTTLIEB.

Die Bleicherei, Druckerei, Färberei und Appretur der baumwollenen Gewebe. Ein praktisches Handbuch für Chemiker, Coloristen, Techniker, Leiter von Fabriken, Studirende der Chemie auf Universitäten, polytechnischen Hochschulen und anderen Anstalten, zum praktischen Gebrauche und zum Selbstunterricht. Nach den neuesten eigenen Erfahrungen. Braunschweig, 1884. 8vo. Illustrated with 16 woodcuts and 100 samples.

STEIN, WILHELM.

Die Fabrikation des Glases. *See* Bolley's Handbuch (etc.).

STEIN, WILHELM.
Prüfung (Die) der Zeugfarben und Farbmaterialen. Eutin, 1874. 8vo.
Folding tables.

STEINACH, HUBERT, UND GEORG BUCHNER.
Die galvanischen Metallniederschläge (Galvanoplastik und Galvanostegie) und deren Ausführung. Berlin, 1890. pp. vi–258, 8vo.

STEINMANN, FERDINAND.
Bericht über die neuesten Fortschritte auf dem Gebiete der Gasfeuer-
ungen. Berlin, 1879. pp. iv–49, 8vo. Plates.

STENBERG, S.
Chimie appliquée à la physiologie et à la thérapeutique par Mialhe.
Paris, 1856. 8vo. Stockholm, 1856. 8vo.

Стендеръ Ф.
Паяльная трубка. Руководство для химиковъ, минералоговъ, металлурговъ
и техниковъ. Казань 1872.

    STENDER, F. Blow-pipe Analysis. Handbook for chemists, mineralogists,
    metallurgists and technologists. Kasan, 1872.

STENZEL, G.
Anleitung zur Darstellung einfacher chemischer Präparate. Breslau,
1878. pp. xvi–271, 12mo.

STÉVART, A.
Des meilleures méthodes d'analyse des minérais, qui en Belgique servent
à l'extraction du fer, du cuivre du zinc et du plomb. Liège, 1863.
8vo.

STEVENSON, THOMAS.
A Treatise on Alcohol with Tables of Spirit-Gravities. Second edition.
London, 1888. pp. xxiv–73, 12mo.

STEVENSON, W. F.
The Non-Decomposition of Water Distinctly Proved : in answer to the
award of a medal by the Royal Society, whereby the contrary
doctrine is absolutely affirmed. London, 1848. 8v.

    The subject announced on the title contrasts curiously with the date of the
    same.

STEWART, F. L.
Sugar made from Maize and Sorghum. A new discovery. Washington,
1880. 8vo.

STIEREN, EDUARD.

Chemische Fabrik. Ein auf 33jährige durchaus eigene Erfahrungen gestütztes praktisches Handbuch zur fabrikmässigen Darstellung chemischer Präparate. Mit einer Vorrede von G. G. Wittstein. München, 1865. pp. vi–621, 8vo. Ill.

STIERLIN, R.

Das Bier, seine Verfälschungen und die Mittel solche nachzuweisen. Bern, 1878. pp. 130, 8vo. Plates.

STIPRIAAN LUISCIUS, ABRAHAM VAN.

Over de middelen om rottend, bedorven en stinkend water tot goed en drinkbaar water te herstellen. Haarlem, 1798. 8vo.

Over de oorzaaken der verrotting in plantaardige en dierlyke zelf-standigheden. Rotterdam. 1798. 4to. Ill.

Redevoering over het nut der scheikunde en derzelver invloed op de geneeskonst. Delft, 1789. 8vo.

STISSER, JOHANN ANDREAS.

* Actorum laboratorii chemici, autoritate atque auspiciis . . . ducum Brunsvic et Lyneburg. In academia Julia editorum specimen primum [secundum, tertium] medico-chemica nec non physico-mechanica observata quædam rariora exhibens. Helmstadii, typis et sumtibus Georg-Wolfgangi Hammii, Acad. typogr., 1693–1701. pp. [178], 4to. Frontispiece. Three parts. Part II, 1693 ; Part III, 1698 ; General title, 1701.

STITZINGER, E.

Scheikundige verhandeling over het met loodstof bezwangerde regen-water, of handleiding bij de beproeving van hetzelve met het liquor probatorius Wirtembergicus het liquor probatorius Hahnemanni en het aqua hydro-sulphurata. Amsterdam, 1807. 8vo.

STODDARD, JOHN T.

Outline of Lecture Notes on General Chemistry. The non-metals. Northampton, Mass., 1884. 12mo.

Outline (An) of Qualitative Analysis for beginners. Northampton, Mass., 1883. pp. iv–55, 12mo.

STÖCKHARDT, JULIUS ADOLPH.
    See Heaton, C. W.

STÖCKHARDT, JULIUS ADOLPH.

Chemische Feldpredigten für deutsche Landwirthe. Zwei Abtheilungen. Leipzig, 1851–'53.

STÖCKHARDT, JULIUS ADOLPH. [Cont'd.]

Theil I. Zweite unveränderte Auflage, 1853. Dritte verbesserte Auflage, 1854.

Theil II. Zweite unveränderte Auflage, 1854.

Nejhlavnèjṡi ; základ rolnictvi : Lučba rolnická. v Praze, 1851. 8vo.

Chemiske Markprædikener. I Dansk Oversættelse af E. Möller Holst. Anden Afdeling. Kjøbenhavn, 1852–'54. 8vo.

Algemeene landbouwscheikunde, bevattelijk voorgesteld. Naar het Hoogduitsch, door L. Mulder. Utrecht, 1854. 8vo.

Nieuwe proefnemingen en onderzoekingen op het gebied van landbouw. Vrij naar het Hoogduitsch bewerkt door N. Mouthaan. Rotterdam, 1856. 8vo.

Chemical Field Lectures. A familiar exposition of the chemistry of agriculture, addressed to farmers. . . . Translated from the German. Edited with notes by A. Henfrey. To which is added a paper on irrigating with liquid manure by J. J. Mechi. London, 1847. 8vo. Bohn's Scientific Library.

Chemical Field-Lectures for Agriculturists. Translated from the German edition with notes by J. E. Teschemacher. London, 1853. 8vo.

[Other editions ;] Cambridge, 1853. 12mo. New York, 1855. 12mo.

Agricultural Chemistry ; or chemical field lectures. A familiar exposition of the chemistry of agriculture, addressed to farmers. Translated from the German, with notes by Arthur Henfrey. To which is added a paper on liquid manure by J. J. Mechi. London, 1855. 8vo.

Landbrukskemien lättfattligt framställd. Öfversättning af C. J. J. Keyser. Sednare afdelingen : Gödsellära, jordartslära, landbruksstatistik och meteorologi. Örebro, 1854. 8vo.

Vigtigaste (Det) i Landbruksläran eller om åkerjordens gödning med afseende på betänklighetarne emot artificiella gödningsämnen. Öfversättning. Carlskrona, 1856. 8vo.

Guanobüchlein. Eine Belehrung für den deutschen Landwirth über die Bestandtheile, Wirkung, Prüfung und Anwendung dieses wichtigen Düngemittels. Leipzig, 1851. 8vo.

Dritte vollständige Auflage. Leipzig, 1854. 8vo.

Stöckhardt, Julius Adolph. [Cont'd.]

Guano. Nauka o częściach składowych, działaniu i skutkach tego nawozu tudzież wskazanie sposobów przekonania się o jego dobroci i użycia go w rólnictwie. Trzecie poprawne wydanie niemieckiego originalu. Poznań, 1856. 8vo.

Guano-boken. Underrättelse för landtmannen om guanons beståndsdelar des verkan samt sättet at pröfva och använda den samma. Øfversättning från 3 upplagan. Med tillägg inbefattande rön om detta gödningsämnes verkan vid des begagnande i Swerige. Stockholm, 1855. 8vo.

Schule (Die) der Chemie, oder erster Unterricht in der Chemie, versinnlicht durch einfache Experimente. Zum Schulgebrauch und zur Selbstbelehrung, insbesondere für angehende Apotheker, Landwirthe, Gewerbtreibende, etc. Braunschweig, 1846. 8vo. Ill.

Dritte verbesserte Auflage. Braunschweig, 1847. 8vo.

Vierte verbesserte Auflage Braunschweig, 1849. 8vo.

Fünfte verbesserte Auflage. Braunschweig, 1850. 8vo.

Sechste verbesserte Auflage. Braunschweig, 1851. 8vo.

Siebente verbesserte Auflage. Braunschweig, 1853. 8vo.

Achte verbesserte Auflage. Braunschweig, 1855. 8vo.

Neunte verbesserte Auflage. Braunschweig, 1857. 8vo. Ill.

Zehnte Auflage. Braunschweig, 1858.

Elfte Auflage. Braunschweig, 1860.

Zwölfte Auflage. Braunschweig, 1861.

Dreizehnte Auflage. Braunschweig, 1863.

Vierzehnte Auflage. Braunschweig, 1864.

Fünfzehnte Auflage. Braunschweig, 1868-'69.

Sechszehnte Auflage. Braunschweig, 1870.

Siebzehnte Auflage. Braunschweig, 1873.

Achtzehnte Auflage. Braunschweig, 1876.

Neunzehnte Auflage. Braunschweig, 1881. pp. 219. Ill.

Navedeni k lučebnictvi pro hospodáře řemeslníky. v Praze, 1853. 8vo.

De scheikunde van het onbewerktuigde en bewerktuigde rijk, bevattelijk voorgesteld en met eenvoudige proeven opgehelderd. Naar het Hoogduitsch door J. W. Gunning, met eene voorrede van G. J. Mulder. Met eene menigte af beeldingen. Schoonhoven, 1847. 12mo.

Tweede herziene en vermeerderde druk. Schoonhoven, 1850.

Derde herziene en veel vermeerderde uitgave bewerkt door J. W. Gunning. Schoonhoven, 1855.

STÖCKHARDT, JULIUS ADOLPH.   [Cont'd.]

The Principles of Chemistry exemplified in a series of simple experiments ; with upward of 200 diagrams and engravings. Translated from the German. London. 1851. 8vo.

The Principles of Chemistry illustrated by simple experiments. From the German by C. H. Peirce. Cambridge [Massachusetts], 1850.

La chimie usuelle appliquée à l'agriculture et aux arts. Traduit de l'allemand sur la onzième édition par F. Brustlein. Paris, 1860. 18mo. Ill.

A Chemia iskolája. Magány és iskolai használatra. Forditotta Berde Aron tanár. 160 a szövegbe nyomott ábrával. Kolozsváftt, 1849. 8vo.

Wykład chemii, czyli pierwsze zasady téj nauki, wsparte najprostszemi doświadczeniami. Dla szkolnego użycia i kształcenia się w niéj bez nauczyciela, szczególniéj zaś dla poczynających Farmaceutów, wiejskich gospodarzy, rękodzielników i. t. d Tłomaczenie J. Filipowicza i W. Tomaszewicza z ostatniego niemieckiego wydania. Z. 390 w tekscie odbitemi drzeworytami. Wilno, 1856. 12mo.

Kemi-Skola. Undervisning i första Grunderna af Kemien, åskådliggjordt genom enkla Experimenter ; för Skolor och Sjelfstudium. Øfversatt och bearbetad af Clemens Ullgren. Jönköping, 1851. 8vo.

Учебникъ химіи или первонач. изученіе химіи при помощи самыхъ простыхъ опытовъ безъ пособія наставника. Перев Ходнева. С.-Петербургъ, 1862.

Ueber die Zusammensetzung, Erkennung und Benutzung der Farben im Allgemeinen und der Giftfarben insbesondere, wie über die Vorsichtsmassregeln beim Gebrauche der letzteren. Zweiter vervollständigter Abdruck. Leipzig, 1844. 8vo.

STÖLZEL, CARL.

Die Metallurgie.   *See* Bolley's Handbuch [etc.].

STÖLZEL, CARL.

Enstehung (Die) und Fortentwicklung der Zuckerfabrikation. Braunschweig, 1851.

Metallurgie (Die). Gewinnung der Metalle. Braunschweig, 1863–86. 2 vols, 8vo. Ill.

STOHMANN, FRIEDRICH.

Handbuch der Zuckerfabrikation. Berlin, 1878. 8vo.

Zweite Auflage. Berlin, 1885. 8vo. Ill. Plates.

STOHMANN, F., UND CARL ENGLER.
    Handbuch der technischen Chemie. *See* Payen, Anselme : Précis de chimie
    industrielle.

STOHMANN, F., UND B. KERL.
    Encyclopädisches Handbuch der technischen Chemie. *See* Muspratt, J. S.

STOKER, G. N.
    Easy Lessons in Chemistry, organic and inorganic ; a complete course
    for young students. London, 1882. 8vo.

STOKER, G. N., AND E. G. HOOPER.
    Chemistry : organic and inorganic, elementary and advanced, worked
    out in full as models. (Stewart's educational series.) London,
    1881. 8vo.

STOKES, G. R.
    Inorganic Chemistry. Eleven years' papers worked out as models
    (Science and art department, South Kensington). London, 1881.
    12mo.

STOKLAS, EDVARD.
    Základové chemie pro ústavy učitelské. v Praze, 1881.

STORCH, V.
    Mikroskopiske og kemiske Undersögelser over Smördannelsen ved
    Kjærningen samt Smörrets fysiske og kemiske Sammensætning.
    Kjøbenhavn, 1883. 8vo.

STORER, FRANK H.
    Agriculture in some of its relations with Chemistry. New York, 1887.
    2 vols., 8vo. Vol. I, x–529 ; vol. II, vii–509.
        A most valuable treatise of modern scientific agriculture, an indispensable
        work for students in this department and for the library.

STRAHL, MORITZ.
    Handbuch der Naturwissenschaften, nach den neuesten und besten
    Quellen bearbeitet für Aerzte und Studirende. Leipzig, 1835. 8vo.
        Vol I *also under the title :* Grundriss der medicinischen Chemie nach Ber-
        zelius, Dumas, Mitscherlich, Rose, Schubarth, Dulk, etc.

STRASSER, K.
    Ueber die Anwendung von Ferricyankalium zur quantitativen Be-
    stimmung von Kobalt, Nickel, Mangan und Zink. Erlangen, 1888.
    8vo.

STRATINGH, SIBRANDUS.

Beknopt overzigt over de leer der stochiometrie door stochiometrische beweegbare cirkels. Groningen, 1827. 8vo. 4 plates

Uebersicht der Stöchiometrie, 1828. 8vo.

Chlorine-verbindingen (De) beschouwd in hare scheikundige fabryk-matige genees-en huishoudkundige betrekkingen. Groningen, 1827. 8vo.

Ueber die Bereitung, die Verbindungen und die Anwendung des Chlors in chemischer, medicinischer, oekonomischer und technischer Hinsicht. Frei aus dem holländischen mit Nutzung des neuesten Werkes von Chevallier und mit Anmerkkungen von C. G. Kaiser. Ilmenau, 1829. 8vo. Ill.

Scheikundig handboek voor essaijeurs, goud- en zilversmeden. Groningen, 1821.

Translated into German by J. H. Schultes. Augsburg und Leipzig, 1823.

Scheikundig verhandeling over de cinchonine en quinine . . . Groningen, 1823.

Scheikundig verhandeling over de morphine en andere hoofdbestand-deelen des opiums . . . Groningen, 1823.

Scheikundige verhandeling over eenige verbinding van den phosphorus. Groningen, 1809. 8vo.

STREATFEILD, FREDERICK WILLIAM.

Practical Work in Organic Chemistry. With a Prefatory Note of R. Meldola. London, 1891.

Finsbury Technical Manuals.

STRECKER, ADOLPH.

Kurzes Lehrbuch der Chemie. See Regnault, Victor, and Regnault-Strecker.

STRECKER, ADOLPH.

Theorien und Experimente zur Bestimmung der Atomgewichte der Elemente. Braunschweig, 1859. 8vo.

STRIPPELMANN.

Die Petroleumindustrie Oesterreich-Deutschlands. Leipzig, 1878.

STROMEYER, FRIEDRICH.

Grundriss der theoretischen Chemie. Göttingen, 1808. 2 vols., 8vo.

Untersuchungen über die Mischung der Mineralkörper und anderer damit verwandter Substanzen. Göttingen, 1821. 8vo.

STROTT.

Technische Chemie für das Bau- und Maschinenwesen. Holzminden, 1876.

STRUMPF, F. L.

Die Fortschritte der angewandten Chemie. Berlin, 1853-'54. 2 vols., 8vo.

> Vol. I, Part I, *also under the title :* Die Fortschritte der Chemie unter Anwendung auf Agricultur und Physiologie. [Neue Folge der neuesten Entdeckungen der angewandten Chemie.] 1853. Ill.
>
> Vol. II, Part 2 : Technische Chemie, *also under the title :* Die Fortschritte der Chemie in ihrer Anwendung auf Gewerbe, Künste und Pharmacie. [Neue Folge der neuesten Entdeckungen der angewandten Chemie.]

Die neuesten Entdeckungen der angewandten Chemie. Berlin, 1845. 2 vols., 8vo. Ill.

STRUVE, ERNST GOTTHOLD.

Paradoxum chymicum sine igne, i. e. operationes et experimenta physico-chymico-pharmaceutica, ipsaque medicamenta chymica ignis ope parari solita sine igne exhibet. Jenæ, 1715. 8vo.

STRUVE, FRIEDRICH ADOLPH AUGUST.

Remarks on an Institution for the preparation and use of artificial mineral waters in Great Britain. London, 1823.

Ueber die Nachbildung der natürlichen Heilquellen. Dresden, 1824 and 1826. 2 parts, 8vo.

Struve'schen (Die) Mineralwasseranstalten. Leipzig, 1853. 8vo.

> Zweite Auflage, 1858.

СТРУВЕ Г.

Химическія таблицы служащія для вычисленія количеств. разясненій. С.-Петербургъ, 1853.

> STRUVE, H. Chemical Tables for calculation of quantitative analyses. St. Petersburg, 1853.

STUTZER, A.

Der Chilisalpeter, seine Bedeutung und Anwendung als Düngemittel. Unter Berücksichtigung der Schrift von A. Damseaux bearbeitet und herausgegeben von P. Wagner. Berlin, 1886. 8vo.

> Nitrate of soda, its importance as manure. A prize essay. Rewritten and edited by Paul Wagner. London, 1887.
>
> Le nitrate de soude, son importance et son emploi comme engrais. Ouvrage couronné. Édité par P. Wagner. Paris, 1887.

STUTZER, ROBERT.
   Das Fahlberg'sche Saccharin. Anhydroorthosulfaminbenzoësäure.
      Braunschweig, 1890. pp. 67.
         Saccharin was discovered by Ira Remsen and his assistant, Constantine
            Fahlberg ; there are no reasons why it should bear Fahlberg's name, and
            good reasons for its bearing Remsen's name.

SUCKOW, GEORG ADOLPH.
   Analyses chemicæ aquarum Jenensium. Jenæ, 1772. 4to.
   Anfangsgründe der ökonomischen und technischen Chymie. Leipzig,
      1784. 8vo.
            Zweite Auflage. Leipzig, 1789.
            Zusätze zu der zweiten Auflage. Leipzig, 1798. 8vo.
   Anfangsgründe der Physik und Chemie. Augsburg, 1813–14. 2 vols.,
      8vo.
   Bemerkungen über einige chymische Gewerbe. Mannheim, 1791. 8vo.
   Von den Nützen der Chymie zum Behufe der bürgerlichen Lebens . . .
      Mannheim, 1775. 4to.

SUCKOW, GUSTAV.
   De lucis effectibus chemicis in corpora organica et organis destituta.
      Jenæ, 1828.
            Chemischen (Die) Wirkungen des Lichtes. Darmstadt,
               1832. 8vo.
   Drei Tabellen uber das Verhalten der Löthrohrproben gegen Reagentien.
      Jena, 1832. Fol.
   Lehrbuch der theoretischen und praktischen Chemie unorganischer
      Körper. Jena, 1851. 8vo.

SUCKOW, LAURENZ JOHANN DANIEL.
   Entwurf einer physischen Scheidekunst. Frankfurt und Leipzig, 1769.
      8vo. Ill.

SUCRE (LE) DANS SES RAPPORTS AVEC LA SCIENCE, l'agriculture, l'industrie,
   le commerce, etc. Paris, 1873–78. 2 vols.

SUNDELIN, CARL HEINRICH WILHELM.
   Handbuch der medizinischen Chemie für studirende und ausübende
      Aerzte. Berlin, 1823. pp. 324. 8vo.

SUNDSTRÖM, K. J.
   Traité général des matières explosives à base de nitroglycérine. Brux-
      elles, 1884. 8vo.

SUTTON, FRANCIS.

A Systematic Handbook of Volumetric Analysis, or the quantitative estimation of chemical substances by measure, applied to liquids, solids and gases. Adapted to the requirements of pure chemical research, pathological chemistry, pharmacy, metallurgy, manufacturing chemistry, photography, etc., and for the valuation of substances used in commerce, agriculture and the arts. London, 1871. pp. xx–378–6. 8vo. Ill.

> First edition. London, 1863. 8vo. Ill.
> Fifth edition. London, 1886. 8vo. Ill.
> Sixth edition. London, 1890. 8vo. Ill.
>> Manuel systématique d'analyse chimique volumétrique, ou le dosage quantitatif des substances chimiques par les mesures appliquée aux liquides, aux solides et aux gaz. Traduit sur la quatrième édition anglaise par C. Méhu. Paris, 1883. 8vo.
>> Traité pratique d'analyse chimique à l'aide des méthodes volumétriques, dosages des matières minérales, organiques et des substances employées dans la métallurgie, l'agriculture, la pharmacie, la médecine, la chimie, les arts, le commerce, l'industrie, par E. Finot. Traduction libre du "Handbook of Volumetric Analysis" de Francis Sutton. Strasbourg et Paris, 1884. 16mo. Ill.
> [Another edition,] par E. Finot et A. Bertrand. Paris, 1887.

SUVOONG, V. P. [Translator].

Pao yao chi yao. [Fulminates and explosives, translated into Chinese from various sources by V. P. Suvoong.]

> *Cf.* Fryer, John.

SVANBERG, LARS FREDRIK.

Har Kemien ei en Plats uti almänna Bildningen. Stockholm, 1847. 8vo.

SVENDSEN, ANDRAS [Translator].

> *See* Gren, F. A. C.

SVERDRUP, JACOB.

Chemie for Landmænd. Christiania, 1836. 8vo. Ill.

SWALVE, BERNHARDUS.

Alcali et acidum, sive naturæ et artis instrumenta pugilica. Amsterdam, 1670. 12mo.

> Editio secunda. Amsterdam, 1676.

SWART, WILLEM SIMON.
Oratio de chemia physicæ auxiliis ad scientiæ dignitatem evecta et
physicæ progressibus alterius perficienda. Amstelodami, 1835. 4to.

SWARTS BEVEL, MICHIEL ISAAK.
Verhandeling over de algemeene gasverlichting. Amsterdam, 1817.
8vo. Pl.
    Tweede druk, 1837.

SWARTZ, TH.
Introduction à l'étude de la chimie théorique. Gand, 1878. 12mo.
Notions élémentaires d'analyse chimique qualitative. Deuxième édition.
    Gand, 1878. 2 vols., 12mo.
Précis de chimie générale et descriptive, exposée au point de vue des
    doctrines modernes. Deuxième édition, refondue et augmentée.
    Gand, 1878. 2 vols., 12mo. Ill.
        Manuale di chimica generale e descrittiva, esposta sotto il
            punto di vista delle dottrine moderne, tradotta da Min-
            gioli. Napoli, 1880–81. 2 vols., 12mo.
Principes fondamentaux de chimie à l'usage des écoles. (Ouvrage cou-
    ronné par l'Académie royale de Belgique. Concours de Keyn.)
    Paris, 1884. pp. viii–280, 8vo.
    [Another edition.] Gand, 1884.

SWEDENBORG, EMMANUEL [SVEDBERG].
Nova observata et inventa circa ferrum et ignem, et præcipue circa
    naturam ignis elementarem, una cum nova camini inventione.
    Amsterdam, Johannes Osterwyk, 1721. 16mo. Ill.
        [Second edition.] Amsterdam, 1727.
        An English translation of this is contained in the work by C. E. Strutt
        named below.
Prodromus principiorum rerum naturalium sive novorum tentaminum
    chymiam et physicam experimentalem geometrice explicandi. Am-
    sterdam, Johannes Osterwyk, 1721. 16mo.
        [Second edition,] Johannes et Abraham Strander. Amster-
            dam, 1727.
        [Third edition,] J. G. Harrisch. Hildburghausen, 1754.
        For notice of this work see Acta eruditorum, February, 1722, pp. 83–87.
* Some Specimens of a Work on the Principles of Chemistry, with other
    Treatises. Translated from the Latin by Charles Edward Strutt.
    London, 1847. pp. xlii–253. Ill.
        Swedenborg attempts to explain the phenomena of chemistry and physics
        on geometrical principles, anticipating modern stereochemistry.

SYKORA, W., UND F. S. SCHILLER.
Kurzgefasste Chemie der Rübensaft-Reinigung. Zum Gebrauche für praktische Zuckerfabrikanten. Wien, 1881.

SYLVESTER, CHARLES.
Elementary Treatise on Chemistry, comprising the most important facts of the science, with tables of decomposition on a new plan. Liverpool, 1809. 8vo.

SYLVIUS, FRANÇOIS DE LA BOE [DUBOIS].
Opera omnia . . . in unum volumen redacta. Paris, 1671. 8vo.

SYMONDS, BRANDRETH.
Manual of Chemistry, for the special use of Medical Students. Philadelphia, 1889. 12mo.

SYMONS, H. H.
Chemical Laboratory Labels. Kentish Town, 1888. 8vo.

SZILÁGYI, GYULA.
Az erjedés chemiájának kézikönyve. Szeszés sörgyarosok, gazdákbortermelök és chemikusok részére. A szöveg közé nyomolt ábrákkal. Budapest, 1890. pp. 191, 8vo.

TABLEAUX SERVANT À L'ANALYSE CHIMIQUE. Deux parties. Traduits de l'allemand par J. Krutwig. Bonn, 1886. Roy. 8vo.

TADDEI, GIOACCHINO.
Elementi di farmacologia, sulle basi della chimica. Edizione seconda rivista, corretta ed ampliata dall' autore. Firenze, 1837–40. 4 vols., 8vo.
Lezioni orali di chimica generale pronunziate in un corso privato nell' anno 1849–50. Raccolte e publicate per cura dei dottori Balocchi, Landi, Minati, T. Taddei, e D. Casanti. Firenze, 1850-57. 7 vols. Vol. I, pp. 655, 1850 ; vol. II, pp. 640, 1851 ; vol. III, pp. 719, 1852 ; vol. IV, pp. 716, 1852 ; vol. V, pp. 763, 1855 ; vol. VI, pp. 816, 1855 ; vol. VII (addizioni e indici), pp. 200, 1857.
Prelezione al corso di chimica organica e fisica medica. Milano, 1848. 4to.
Saggio di metalloscopia . . . Firenze, 1844. 8vo.
Sistema di stechiometria chimica o teoria di proporzioni determinate. Firenze, 1824.
Sull' uffizio delle materie inorganiche nei corpi organici viventi. Milano, 1847. 8vo.

TAETS AB AMERONGEN, GERARDUS ARNOLDUS.
Dissertatio philosophica inauguralis de elementis . . . ad diem xxii Octobris, 1773. Trajecti ad Rhenum [1773]. pp. [vi]-92-[iv], 4to. One folding plate.

    A philosophical essay of no little interest considering the date.

TAFFE, A.
Cours de chimie appliquée aux matériaux. Deuxième édition. Paris, 1845. Fol.

TALBOT, WILLIAM HENRY FOX.
Some account of the art of photogenic drawing, or the process by which natural objects may be made to delineate themselves without the aid of the artist's pencil. Proc. Royal Soc., IV, pp. 120–121, 1839. London, 1839. 4to.

    Contains his discovery of photography on paper and of the art of making copies of photographs.

TALOTTI, G. B.
Manuale pratico di voltimetria. Terza edizione. Bologna, 1883. 12mo.

TAMARIT, EMILIO DE.
Nociones de química inorgánica y orgánica, necesarias para el mejor conocimiento de las primeras materias que se emplean en la construccion de los efectos del material de guerra, y simples que entran en la confeccion de los fuegos artificiales. Seguido de un tratado análitico sobre el modo de conservar las provisiones del ejército, manera de conocer su decomposicion y causas que la motivan. Madrid, 1858. 8vo.

TAMM, A.
Analyser af Jernmalmer utförda i åren 1871-'90. Stockholm, 1890. 4to.

ТАМОЧКИНЪ Д.
Основаніе химіи для хлѣбопашцевъ. С.-Петербургъ, 1864.

    TAMOCHKIN, D. Principles of Chemistry for Agriculturists. St. Petersburg, 1864.

TAPPARONE CANEFRI, C.
Nozioni semplici ed elementari di fisica chimica e storia naturale. Parte I. Fisica, chimica e mineralogia. Torino, 1891. 8vo.

TAPPEINER, H.
Anleitung zu chemisch-diagnostischen Untersuchungen am Krankenbette. München, 1885. 16mo.

TARDIEU, AMBROISE ET Z. ROUSSIN.
Étude médico-légale et clinique sur l'empoisonnement.   Paris, 1867.
   pp. xxii-1072.   8vo.

TARUGI, N.
La teoría atomica.. Modena, 1891.   12mo.

TASSART, LOUIS CHARLES.
L'industrie de la teinture avec 55 figures intercalées dans le texte.
   Paris, 1890.   8vo.   Ill.
        Bibliothèque des connaissances utiles.

TASSINARI, G.
Guida ad esercizii pratici di chimica per gli istituti tecnici.   Torino,
   1885.   8vo.

TASSINARI, P.
Avviamento allo studio della chimica, xxx lezioni.   Pisa, Torino,
   Milano e Firenze, 1868.   16mo.
Lezioni di chimica.   Introduzione.   Parte speciale.   Pisa, 1877.   8vo.
Manuale di chimica.   Chimica inorganica.   Seconda edizione.   Pisa,
   1868.   8vo.
Precetti di analisi chimica qualitativa.   Torino, 1892.
Sunto delle lezioni di chimica.   Pisa, 1867.   16mo.

TATE, A. NORMAN.
Petroleum and its Products : an Account of the history, origin, composition, properties, uses, and commercial value of petroleum.   The
   methods employed in refining it, and the properties, uses, etc., of
   its products.   London, 1863.   pp. iv-116.   12mo.
            Du pétrole et de ses dérivés ; histoire, origine, composition,
               propriétés, emplois, etc.   Traduit par Brandon.   Paris,
               1864.   8vo.

TATE, THOMAS.
Outlines of Experimental Chemistry : being a familiar introduction to
   the science of agriculture, designed for the use of schools and
   schoolmasters.   London, 1850.   12mo.
        [Another edition].   London, 1854.

TAUBA, KEIZO, TINICHIRO SHINOYAMA, AND KEIZO SHIBATA.
Yuki Kagaku.   Tokio, 1882.   2 vols., 8vo.
        Organic chemistry in Japanese.

TAYLOR, ALFRED SWAINE.
On Poisons in relation to medical jurisprudence and medicine.   Second
edition.   London, 1859.   pp. vi–863.   12mo.
>Third American from the third and thoroughly revised Eng-
>lish edition.   Philadelphia, 1875.   pp. xvi–788, roy. 8vo.

TAYLOR, R. L.
Analysis Tables for Chemical Students.   London, 1886.   8vo.
Chemistry for Beginners.   Adapted for the elementary stage of the
Science and Art Department's examinations in inorganic chemistry.
London, 1887.   8vo.
Students' Chemistry.   London, 1892.   8vo.   Ill.

TEGETMEIR, W. B.
A Catechism of Agricultural Chemistry.   London, 1853.   18mo.

TEICHMEYER, HERMANN FRIEDERICH.
Institutiones chimiæ dogmaticæ et experimentalis, in quibus chemicorum
principia, instrumenta, operationes et producta, simulque analyses
trium regnorum, succinta methodo traduntur.   Jenæ, 1729.   4to. Ill.
>Medicinisch und chemische Abhandlung vom seignettischen
>Saltze . . . ins deutsche übersetzt . . . von Gottfried
>Heinrich Burghart.   Breslau und Leipzig, 1749.   8vo.

TELLKAMPF, A.
Anfangsgründe der chemischen Naturlehre.   Hamm, 1831.   8vo.

ТИМИРЯЗЕВЪ К.
Спектральный анализъ хлорофила.   С.-Петербургъ, 1871.
>TEMERYAZEF, K.   Spectrum analysis of chlorophyll.   St. Petersburg, 1871.

TEPLOW, M. N.
Die   Schwingungsknoten-Theorie   der   chemischen   Verbindungen.
Uebersetzt von L. Jawein.   St. Petersburg, 1885.   pp. 136.   4to.

TERFOE, O.
Pétit guide pour les travaux pratiques au laboratoire de chimie.
Namur, 1891.

TERREIL, A.
>Les laboratoires de chimie [and other articles].   *See, in Section II*, Fremy:
>Encyclopédie chimique.

TERREIL, A.
Atlas de chimie analytique minérale.   Paris, 1861.   8vo.
Traité pratique des essais au chalumeau contenant tout ce qui est relatif

TERREIL, A. [Cont'd.]

à l'emploi de cet instrument et aux réactions que les corps de la chimie présentent dans les essais pyrognostiques, la description des propriétés des minéraux et les caractères chimiques que peuvent les faire reconnaître dans les essais au chalumeau. Paris, 1875. 8vo. Ill.

TESSARI, N.

Compendio di chimica generale compilato secondo lo spirito attuale della scienza. Firenze e Torino, 1869. 8vo.

TESSIER, P.

Chimie pyrotechnique, ou Traité pratique des feux colorés, contenant : 1° l'examen chimique, la description et la fabrication des matières pyrotechniques ; 2° des procédés nouveaux et faciles pour la préparation des divers composés tels que le chlorate de baryte, etc.; 3° des formules nombreuses et économiques pour la confection des lances, des étoiles et des feux de Bengale de toutes les couleurs, et suivi d'un petit traité special pour la fabrication des pastilles simples et des pastilles diamant de différents calibres. Paris, 1858. 8vo.

TEXEIRA MENDEZ, R.

La philosophie chimique d'après Aug. Comte. Indications générales sur la théorie positive des phénomènes de composition et de décomposition, suivie d'une appréciation sommaire de l'état actuel de la chimie. Paris, 1887. 18mo.

THAER, ALBRECHT. *See* Einhof, Heinrich.

THALÉN, R.

Spektralanalys exposé och Historie, med en Spectralkarta. Upsala, 1866.

THALMANN, F.

Die Fette und Oele. Darstellung der Eigenschaften aller Fette und Oele, der Fett- und Oelraffinerie und der Kerzenfabrikation. Nach dem neuesten Stande der Technik leichtfasslich geschildert. Wien, 1881. 8vo. Ill.

Zweite vermehrte und verbesserte Auflage. Wien, 1892. 8vo. Ill.

THAN, K. VON.

Az ásváyvizeknek [etc.]. Budapest, 1890. 8vo.

Chemical Constitution and comparison of Mineral Waters, in Hungarian.

THAULOW, HAROLD.
　　Veiledning ved qualitativ-chemiske Analyser.　For Lœger og Pharma-
　　ceuter.　Christiania, 1840.　8vo.

THAULOW, MORITZ CHR. JULIUS.
　　Chemiens Anvendelse i Agerdyrkningen.　Christiania, 1841.　8vo.
　　Den qvalitative chemiske Analyse.　Et Compendium til et praktisk
　　Cursus i Universitetets chemiske Laboratorium.　Christiania, 1847.
　　4to.

THAUSSING, J. E.
　　Die Theorie und Praxis der Malzbereitung und Bierfabrikation.　Dritte
　　Auflage.　Leipzig, 1888.　Atlas of 14 plates.
　　　　　　Preparation of Malt and the fabrication of beer, translated
　　　　　　　　by W. T. Brannt, A. Schwarz and A. H. Bauer.　Lon-
　　　　　　　　don, 1887.　Roy. 8vo.

THEIN, J.
　　Chemisch-technische Instruction.　Auf vieljähriger Erfahrungen ge-
　　　　stützte praktische Anweisungen für Fabrik, Gewerbe und Haus-
　　　　Industrie, nebst einem Anhang enthaltend Geheimmittel-Analysen.
　　　　Dritte sehr vermehrte und verbesserte Auflage.　Prag, 1872.　8vo.

THÉNARD, LOUIS JACQUES.
　　　　See Warvinsky, editor of a Russian version.

THÉNARD, LOUIS JACQUES.
　　Brom (Das) ein neuentdeckter einfacher Stoff.　Leipzig, 1828.
　　Traité de chimie élémentaire, théorique et pratique, suivi d'un essai sur
　　　　la philosophie chimique, et d'un précis sur l'analyse.　Paris, 1813-
　　　　16.　4 vols., 8vo.
　　　　　　Deuxième édition.　Paris, 1817-18.　4 vols.
　　　　　　Troisième édition.　Paris, 1821.　4 vols.
　　　　　　Sixième édition.　Paris, 1833-36.　5 vols., 8vo.
　　　　　　Also editions at Bruxelles, 1829 and 1836.　2 vols., 8vo.
　　　　　　Lehrbuch der theoretischen und praktischen Chemie.　Vierte
　　　　　　　　neu durchgesehene, vermehrte und verbesserte Ausgabe.
　　　　　　　　Uebersetzt und vervollständigt von Gustav Theodor
　　　　　　　　Fechner.　Leipzig, 1825-28.　6 vols., 8vo.　Vol. IV in
　　　　　　　　3 parts and vol. V in two parts.
　　　　　　Portrait of Thénard in Vol. I.　For a supplement, *see* Fechner, G. T.:
　　　　　　　Repertorium [etc.].
　　　　　　Trattato di chimica elementare teorica e pratica, seguito da
　　　　　　　　un saggio sulla filosofia chimica e da un compendio sull'
　　　　　　　　analisi.　Tradotto sulla sesta edizione Parigina da G. B.
　　　　　　　　Sembenini.　Mantova, 1838-41.　7 vols., 8vo.

THÉNARD, LOUIS JACQUES. [Cont'd.]

Trattato di chimica elementare teorica e pratica. Atlante e descrizione degli apparati e di tutti gli agenti meccanici dei quali deve essere fornito un laboratorio di chimica, ed in generale dei diversi apparecchi che rappresentano le tavole dell' opera. Verona, 1840–'41. Fol.

Tratado completo de química teórica y práctica traducido por la quinta y última edicion francesa. . . . Nantes, 1830. 6 vols., 8vo., and atlas.

Tratado elemental teórico-práctico de química, traducido de la sétima edicion francesa por una sociedad de profesores de química, farmacia, etc. Cádiz, 1840. 4to.

Essay (An) on Chemical Analysis, chiefly translated from the fourth volume of the last edition of the Traité de chimie élémentaire, with additions comprehending the latest discoveries and improvements in this branch of the science. By John George Children. London, 1819. pp. xvii–494, 8vo.

Anleitung zur chemischen Analyse dem gegenwärtigen Zustand der Wissenschaft gemäss. Nach L. J. Thénard's Handbuch der theoretischen und praktischen Chemie. Aus dem französischen übersetzt und mit Anmerkungen begleitet von Johann Bartholomä Trommsdorff. Erfurt, 1817. 8vo. Ill.

Tratado de análisis química escrito en francés . . . traducido al castellano de la quínta edicion. Madrid, 1828. 4to. Ill.

THÉNARD, LOUIS JACQUES, CHEVET ET A. CHEVALLIER.

Élémens de chimie, d'après M. le Baron L. J. Thénard. Troisième édition, revue, etc., par Blondeau. Paris, 1846. 2 vols., 18mo. Ill.

THÉNARD, LOUIS JACQUES, ET GAY-LUSSAC.

Recherches physico-chimiques faites sur la pile, sur la préparation chimique et les propriétés du potassium et du sodium, sur la décomposition de l'acide boracique, etc. Paris, 1811. 2 vols., 8vo.

THENIUS, GEORG.

Harze (Die) und ihre Producte. Deren Abstammung, Gewinnung und technische Verwerthung. Nebst einem Anhange : Ueber die Producte der trockenen Destillation des Harzes oder Colophoniums ; das Camphin, das schwere Harzöl, das Codöl und die Bereitung von Wagenfetten, Maschinenölen aus den schweren Harzölen, sowie

THENIUS, GEORG.   [Cont'd.]

   die Verwendung derselben zur Leuchtgas-Erzeugung.   Ein Handbuch für Fabrikanten, Techniker, Chemiker, Droguisten, Apotheker, Wagenfett-Fabrikanten und Brauer.   Nach den neuesten
   Forschungen zusammengesetzt.   Wien, 1879.   Ill.

Holz (Das) und seine Destillations-Producte.   Ueber die Abstammung
   und das Vorkommen der verschiedenen Hölzer.   Ueber Holz,
   Holzschleifstoff, Holzcellulose, Holzimprägnirung und Holzcon-
   servirung, Meiler- und Retorten-Verkohlung, Holzessig und seine
   technische Verarbeitung, Holztheer und seine Destillationsproducte, Holztheerpech und Holzkohlen nebst einem Anhange : Ueber
   Gaserzeugung aus Holz.   Ein Handbuch für Waldbesitzer, Forstbeamte, Lehrer, Chemiker, Techniker und Ingenieure, nach den
   neuesten Erfahrungen praktisch und wissenschaftlich bearbeitet.
   Wien, 1880.   Ill.

Fabrikation (Die) der Leuchtgase nach den neuesten Forschungen.
   Wien, 1891.   8vo.

Technische (Die) Verwerthung des Steinkohlentheers, nebst einem
   Anhange : Ueber die Darstellung des natürlichen Asphaltheers und
   Asphaltmastix aus den Asphaltsteinen und bituminösen Schiefern
   und Verwerthung der Nebenproducte.   Wien, 1878.   Ill.

THIBAUT, P.

   * The Art of Chymistry as it is now practised.   Now translated into
   English by a Fellow of the Royal Society.   London, printed for
   John Starkey, at the Miter near Temple-Bar in Fleet street, 1675.
   pp. [xxx]-279, 16mo.

THIEBOUT, C. H.

Korte leiddraad bij het geven van onderwijs in de allereerste beginselen
   der scheikunde.   Voor burger- en burgereravondscholen.   Eerste
   stukje.   De niet-metalen.   Arnheim, 1869.   8vo.

Scheikundige technologie ten dienste van het middelbaar onderwijs.
   Vrij bewerkt naar Knapp, Wagner, Wagner's Jahresbericht en
   anderen.   Eerste stuk.   Arnhem, 1868.   8vo.   Ill.

THIELAU, FRIEDRICH VON.

Der Kalk in seiner vielfachen Beziehung zum praktischen Leben ; mit
   Benutzung vorhandener grösserer Werke, namentlich Graham
   Otto's ausführlichen Lehrbuch der Chemie.   Dritte Auflage.   Bres-
   lau, 1873.   8vo.

THIEME, Fr. W.

Anfangsgründe der theoretischen und praktischen Chemie, nebst Anwendungen auf die Gewerbe.   Leipzig, 1839.   8vo.   Ill.

THOMPSON, FRANCIS BENJAMIN.

Fire ; its causes considered and explained on the basis of chemical and
electrical science, including the theory of " Spontaneous Combustion," together with scientific facts. London, 1857. 8vo.

THOMPSON, G.

An Essay on Manures, containing correct analytical tables of plants,
soils and fertilizers ; being a compendium of agricultural chemistry
adapted to the purposes of practical agriculturists ; written with the
express object of expounding a system whereby the farmer may
ascertain not only what his soil requires to render it fertile, but
what manure will afford him the necessary matter in the cheapest
form. Kidderminster, 1853. 8vo.

THOMPSON, SIR HENRY.

Food and Feeding. London, 1880. 8vo.
    Third edition. London, 1884.

THOMPSON, JOHN MILLAR [Editor].
        See Bloxam, Charles Loudon.

THOMS, G.

Die landwirthschaftlich-chemische Versuchsstation zu Riga. Riga, 1875.

THOMSEN, JULIUS.

Forelæsninger over teknisk Chemie ved den Polytechniske Læreanstalt,
    Tredie Halvaar. Kjøbenhavn, 1883. 8vo.
Forsög (Et) paa en almeenfattelig ‚Fremstilling af Chemiens vigtigste
    Resultater. Kjøbenhavn, 1853. 8vo. Ill.
        Andet Oplag. Kjøbenhavn, 1854. 8vo.
Kortfattet Lærebog i den uorganiske Chemie. Kjøbenhavn, 1850. 8vo.
Lærebog i uorganisk Chemie for Begyndere. Kjøbenhavn, 1875. 8vo.
        Tredie Udgave. Kjøbenhavn, 1885.
Thermochemiske Undersøgelser. 7 parts. Kjøbenhavn, 1869–73. 4to.
    Tables.
        Thermochemische Untersuchungen. Leipzig, 1882–86. 4
        vols., 8vo.
Vejledning i den præparative Chemie. Udarbejdet til Brug ved che-
    miske Øvelser. Kjøbenhavn, 1853. 8vo.

THOMSON, ANTHONY.

Anleitung zur Erkenntniss und Behandlung der Vergiftungen. Nebst
der chemischen Analyse und der Sectionsbefunde. In alphabetischer Ordnung. Nach dem englischen bearbeitet von Alexander
Reumont. Aachen, 1846, 8vo.

THOMSON, ANTHONY.   [Cont'd]
> Handleiding tot het leeren kennen en behandelen der ver-
> giftigingen.   Benevens de chemische analysen en de
> resultaten der lijkopeningen.   Naar het engelsch.
> Groningen, 1847.   8vo.

THOMSON, MURRAY.
Analytical Tables for the Use of Students of Practical Chemistry.
Edinburgh, 1861.   8vo.

THOMSON, ROBERT DUNDAS.
School Chemistry ; or practical rudiments of the science.   London,
1848.   12mo.

THOMSON, THOMAS.
Attempt (An) to Establish the First Principles of Chemistry by Experiment.   London, 1825.   2 vols., 8vo.   I, pp. xxiii–478 ; II, pp. vii–
532.
> Principes de la chimie établis par les expériences ; ou Essai
> sur les proportions définies dans la composition des
> corps.   Traduction de l'anglais publiée avec l'assenti-
> ment de l'auteur.   Paris, 1825.   2 vols., 8vo.

Brewing and Distillation.   With practical instructions for brewing porter
and ales according to the English and Scottish methods.   By W.
Stewart.   Edinburgh, 1849.   8vo.

Chemistry of Animal Bodies.   Edinburgh, 1843.   pp. ix–702, 8vo.

Chemistry of Inorganic Bodies.   London, 1831.   2 vols., 8vo.

Chemistry of Organic Bodies.   Vegetables.   London and Glasgow,
1838.   8vo.

Elements (The) of Chemistry.   Edinburgh, 1810.   8vo.

System (A) of Chemistry.   Edinburgh, 1802.   4 vols., 8vo.
> Second edition, 1804.   Third edition, 1807.   5 vols.
> Fourth edition, 1810.   5 vols.
> Seventh edition.   Inorganic bodies.   Edinburgh, 1831.   2
> vols., 8vo.
> *A new System of Chemistry, including Mineralogy and
> vegetable, animal and dyeing substances, comprehend-
> ing the latest discoveries and improvements of the
> science.   Philadelphia, 1803.   pp. 364, 4to.   Ill.
> [Another edition,] with Notes by Thomas Cooper, from the
> fifth London edition.   Philadelphia, 1818.   4 vols., 8vo.
> Système de chimie, traduit de l'anglois sur la dernière édition
> de 1807, par Jean Riffault, précédé d'une introduction
> de C. L. Berthollet.   Paris, 1809.   9 vols., 8vo.

THOMSON, THOMAS. [Cont'd.]
>   The full name of the translator is Jean René Denis Riffault des Hêtres.

>   System der Chemie in vier Bänden.   Nach der zweiten Aus-
>   gabe aus dem englischen übersetzt von Friedrich Wolff.
>   Bérlin, 1805.   4 vols., 8vo.

>   *Also* vol. v, *with the title :* Zusätze und Erweiterungen der Wissenschaft seit
>   1805. Berlin, 1811. 2 parts, 8vo.

THORPE, T. E.
A Series of Chemical Problems, with Key, for use in Colleges and
    Schools.   Adapted for the preparation of Students for the Govern-
    ment Science and Society of Arts examinations.   With a preface
    by Roscoe.   New edition.   London, 1881.   pp. 88, 18mo.

>   New edition, revised and enlarged by W. Tate, with preface
>   by H. E. Roscoe.   London, 1891.   8vo.

Quantitative Chemical Analysis.   London, 1873.   12mo.

>   Eighth edition.   London and New York, 1889.   12mo.

>   Traité pratique d'analyse chimique à l'aide des méthodes
>   gravimétriques, d'après Thorpe.   [Par S. Meunier.]
>   Paris, 1887.   Ill.

>   Die qualitative Analyse nebst Anleitung zu Uebungen im
>   Laboratorium.   Autorisirte deutsche Ausgabe von E.
>   Fleischer.   Zweite Ausgabe.   Berlin, 1879.   pp. viii–224,
>   8vo.   Ill.

THOUVENEL, PIERRE.
Mémoire chimique et médical sur le mécanisme et les produits de la
    sanguification, qui a remporté le prix de l'Académie impériale de
    Saint Pétersbourg pour l'an 1776.   Saint Pétersbourg, 1777.   4to.
Mémoire chimique et médical sur la nature, les usages et les effets de
    l'air et des airs, des aliments et des médicaments relatifs à l'écono-
    mie animale.   Ouvrage qui a remporté le prix double proposé par
    l'Académie des sciences . . . de Toulouse pour l'année 1778.
    Paris, 1780.   4to.
Précis chimique sur les principes de la formation de l'acide nitreux,
    ouvrage qui a remporté le prix proposé par la Société roïale des
    sciences de Copenhague en 1776.   Copenhague, 1784.   4to.

THUDICHUM, J. L. W.
Aids to Physiological Chemistry.   London, 1884.   12mo.
Manual (A) of Chemical Physiology, including its points of contact with
    pathology.   London, 1872.   pp. viii–195, 8vo.   Ill.

>   Grundzüge der anatomischen und klinischen Chemie.   Ana-

THUDICHUM, J. L. W. [Cont'd.]

    lecten für Forscher, Aerzte und Studirende. Berlin, 1886. 8vo.

Treatise (A) on the Chemical Constitution of the Brain. Based throughout upon original researches. London, 1884. 4to.

THÜRINGER, AMBRÓ.

Természettani ismeretik algymnasiumok, reál-s elemi iskolák számára és magán használatra az ui tanszervezet terve szerint. Szöveg közé nyomott 118 fametszvénynyel. Második javitott és tétemesen bövitett kiadás. Pest, 1853. 8vo.

THURNEISSER ZUM THURN, LEONHARD.

  * ΜΕΓΑΛΗ ΧΥΜΙΑ, vel magna alchymia. Das ist ein Lehr und Unterweisung von den offenbaren und verborgerlichen Naturen, Arten und Eigenschaften, allerhandt wunderlicher Erdtgewechssen, als Ertzen, Metallen, Mineren, Erdsäfften, Schwefelen, Mercurien, Saltzen und Gesteinen. Und was der Dingen zum Theil hoch in den Lüfften, zum Theil in der Tieffe der Erden, und zum Theil in den Wassern, welche aus dem Chaos oder der Confusion und Vermischung elementischer Substanzen, als geistlicher, und doch subtiler, noch unbestendiger weis verursacht, empfangen und radicirt. Aber von himelischer Zuneigung der influentischen Impression oder Eintruckung, seelischer und fixer oder bestendiger Weise, zu einer wesentlichen Materia digerirt, coagulirt, oder præparirt, und durch die natürliche Vermöglichkeit, Krafft und Forthtreibung, jedes in seiner Gestalt. Als ein greiffelichs, eintzigs, wesentlichs Ding, corporalischer, volkommener Weise, von seiner Radice abgelöset, an Tag auszgestossen, und in Gestalt einer sichtigen Massæ geboren : Und wie, oder welcher Gestalt, oder auff was Weisz und Wege deren ein jedes, mit Zusatz des andern, durch menschlichen Handgriff, oder den Usum (dieser sehr alten Kunst) eintweders in ein Liquorem, Oehl, Saltz, Stein, Wasser, Schwefel, Mercurium, oder andere Mineren und Metall verwandelt, oder sonst zum Nutz, Gebrauch und Wolstandt, menschlichs zeitlichs Lebens zugericht und bereitet wird. Gedruckt in Berlin durch Nicolaum Voltzen, 1583. pp. [xii]–144–[xii]. Portrait on reverse of title-page.

THWAITE, B. H.

Gaseous Fuel, including Water-Gas, its Production and Application. London, 1889.

TIDY, C. MEYMOTT.

Handbook of Modern Chemistry, Inorganic and Organic for the use of Students. London, 1878. 8vo.

    Second edition. London, 1887. 8vo.

TIDY, C. MEYMOTT, AND RAWDON MACNAMARA.
The Maybrick Trial : A Toxicological Study.　London, 1890.　8vo.

TIEBOEL, BOUDEWYN.
Verhandeling over de vaste lugt.　Rotterdam, 1781.　4to.

TIEDEMANN, FRIEDRICH, UND LEOPOLD GMELIN.
Verdauung (Die) nach Versuchen.　Heidelberg und Leipzig, 1826.
2 vols., 4to.
Zweite Auflage.　Heidelberg, 1831.
Versuche über die Wege auf welchen Substanzen aus dem Darmkanal in
das Blut gelangen.　Heidelberg, 1820.

TIEFTRUNK, FERDINAND.
Die Gasbeleuchtung.　Stuttgart, 1874.　8vo.　Ill.

TIELLE, C. H.
Theoretische (Die) Chemie, nebst Hinweisungen ihrer Anwendung auf
analytisch-chemische Untersuchungen.　Zunächst für Studirende
der Medicin bearbeitet.　Anorganische Chemie.　Kiel, 1842.　8vo.
Ill.

TIEMANN, F., UND A. GÄRTNER.
Die chemische　.　.　.　Untersuchung des Wassers.　*See* Kubel, Wilhelm.

TILDEN, WILLIAM A.
See Watts, Henry : Manual of Chemistry.

TILDEN, WILLIAM A.
Answers to Problems in the Introduction to Chemical Philosophy.
London, 1884.　8vo.
Introduction to the Study of Chemical Philosophy.　The Principles of
Theoretical and Systematic Chemistry.　New York, 1876.　pp. xvi–
279, 12mo.
Text-books of Science.

TILINGIUS, M.
Prodromus praxeos chimiatricæ.　.　.　.　Rintelii, 1674.　16mo.

TILLMANN, SAMUEL E.
Principles of Chemical Philosophv.　*n. p.* [West Point, N. Y.], *n. d.*
[1885].　12mo.

TISSIER, CHARLES, ET ALEXANDRE TISSIER.
Aluminium (L') et les métaux alcalins.　Recherches historiques et
techniques sur leurs propriétés, leur procédé d'extraction et leurs
usages.　Paris, 1858.　18mo.　Ill.

TISSIER, CHARLES, ET ALEXANDRE TISSIER. [Cont'd.]
Guide pratique de la recherche de l'extraction et de la fabrication de l'aluminium et des métaux alcalins. Recherches techniques sur leurs propriétés, leurs procédés d'extraction et leurs usages. Paris, 1864. 8vo. Ill.

The authors were manufacturers at an establishment near Rouen.

TISSIER.
Essai sur la théorie des trois élémens comparée aux élémens de la chimie pneumatique. Lyon. An XII [1804]. 8vo.

TJADEN MODDERMAN, R. S.
De leer der osmose. Akademisch proefschrift. Leeuw, 1857. 8vo.

TOIFEL, WILHELM F.
Glas-Industrie (Die) als Kunstgewerbe. Musterblätter zum praktischen Gebrauche nebst erläuterndem Text für Fabrikanten, Raffineure, Glasarbeiter und Händler. Herausgegeben unter Mitwirkung namhafter Fachmänner. Leipzig, 1877. 4to and folio.
Handbuch der Chemigraphie, Hochätzung in Zink für Buchdruck mittelst Umdruck von Autographien und Photogrammen und directer Copirung oder Radirung des Bildes an der Platte. (Photo-Chemigraphie und Chalco-Chemigraphie.) Wien, Pest, Leipzig, 1883. Ill.

TOLLENS, B. Leçon (etc.). *See* Société chimique de Paris.

TOLLENS, B.
Einfache Versuche für den Unterricht in der Chemie, für agricultur-chemische Laboratorien. Berlin, 1878. 8vo. Ill.
Kurzes Handbuch der Kohlenhydrate. Breslau, 1888.

Pages 331 to 360 contain a bibliography of carbohydrates.

TOMMASI, DONATO.
Traité théorique et pratique d'électrochemie. Electrolyse, galvanoplastie, dorure, argenture, nickelage, cuivrage, etc. Électrométallurgie, affinage électrolytique des métaux, application de l'électrolyse au blanchiment des matières textiles à la rectification des alcools, etc. Analyse électrolytique. Paris, 1889. pp. xi–1102, roy. 8vo.

Contains a translation of William Walter Webb's Index to the Literature of Electrolysis. *See in Section I.*

TONDI, MATTEO.
Istituzioni di chimica, per servire ad un corso d'operazioni, appartenenti alla medesima. Napoli, 1786. 8vo.

TOPHAM, JOHN.

Chemistry made easy for Agriculturists. Third edition. Droitwich, 1846. 16mo.

TORNÖE, H.

Norwegian North-Atlantic Expedition. Chemistry. (I. On the air in seawater. II. On the carbonic acid in seawater. III. On the amount of salt in the water of the Norwegian Sea.) Christiania, 1880. Fol. Plates.

Uebungen des Salzgehaltes des Nordmeer-Wassers. Wien, 1881. 8vo. Ill.

Veiledning i den qualitative uorganiske Analyse. Tredie Udgave. Kjøbenhavn, 1887. 8vo.

Anden Udgave. Kjøbenhavn, 1878.

TORRES, DIEGO ANTONIO.

Tratado elemental de quimica. Segunda edicion. Santiago de Chile, 1869. 8vo. Plates.

[Another edition.] Santiago de Chile, 1882. 8vo.

TORRES MUÑOZ DE LUNA, RAMON.

Guia del químico prático ò compendio de análisis química. Madrid, 1852. 8vo.

TOUR, JAN CAREL, BARON DU.

Beschryving van een werktuig om steden en groote gebouwen te verlichten door middel van vlamvatbaar gas uit steenkolen. 's Gravenhage, 1817. 8vo.

TOURNIER, ÉMILE.

Nouveau manuel de chimie simplifiée, pratique et expérimentale ; sans laboratoire, manipulations, préparations, analyses. Paris, 1868. 18mo. Ill.

Deuxième édition. Paris, 1880.

Товскій, Александръ.

Начальныя основанія химіи. Москва, 1822

TOVSKY, ALEXANDER. First Principles of Chemistry. Moscow, 1822.

О важности химическихъ изслѣдованій въ кругу наукъ и искуствъ. Москва, 1827.

Importance of Chemical Investigations in Science and Arts. Moscow, 1827.

TRAITÉ ÉLÉMENTAIRE ET PRATIQUE DE CHIMIE GÉNÉRALE appliquée aux arts, à la médecine et à l'agriculture. Chimie inorganique. Première partie : Métalloides. Paris, 1841. 18mo.

TRATADÒ COMPLETO DE LOS GASES, extraido de la obra de J. A. Saco y adicionado con las experiencias de Thenard, Gay Lussac, Berzelius, Dumas, Saussure, Biot, Arago, Davy, etc. Cádiz, 1841. 8vo.

TRAUTZSCH, CAROLUS FRIDERICUS.
Dissertatio inauguralis medico-chemico de elementorum quæ nunc exstant ponderabilium genuinitate. Lipsiæ, 1823. 4to,

TRAUZL, ISIDOR.
Dynamite. Ihre ökonomische Bedeutung und ihre Gefährlichkeit. Wien, 1876. 8vo.
Explosive Nitrilverbindungen, insbesondere Dynamit und Schiesswolle, deren Eigenschaften und Verwendung in der Sprengtechnik. Zweite Auflage. Wien, 1870. 8vo.

TREADWELL, F. P.
Tabellen zur qualitativen Analyse, bearbeitet unter Mitwirkung von V. Meyer. Zweite Auflage. Berlin, 1884. 8vo.
Dritte Auflage. Berlin, 1890, 8vo.
Guide pratique pour l'analyse qualitative. Traduit sur la troisième édition allemande par G. Jaubert. Berlin, 1892. 8vo. Ill.

TRICHT, J. P. C. VAN [Translator].
Leerboek der onbewerktuigde scheikunde. *See* Graham, Thomas, and F. J. Otto.

ТРИФАНОВСКІЙ, Дм.
Къ вопросу о химическомъ составѣ человѣческой жилчи. Москва, 1875.
TRIFANOVSKY, D. The Chemical Composition of human bile. Moscow, 1875.

TRIMBLE, HENRY.
Practical and Analytical Chemistry. Being a complete course in Chemical Analysis. Philadelphia, 1885.
Second edition. Philadelphia, 1886.
Third edition, enlarged. Philadelphia, 1889.

TRIMMER, JOSHUA.
Practical Chemistry for Farmers and Landowners. London, 1842. 8vo.

TRINKWASSER (Das) der Stadt Kiel auf Grundlage von Analysen aller Brunnenwasser Kiels, ausgeführt im Jahre 1883, im Auftrage der städtischen Gesundheits-Commission durch das agriculturchemische Laboratorium der landwirthschaftlichen Versuchsstation zu Kiel. Kiel, 1886. 4to.

TROMMSDORFF, JOHANN BARTHOLOMÄUS.

Allgemeine chemische Bibliothek des neunzehnten Jahrhunderts. Erfurt, 1802–05. 5 vols., 8vo.

Allgemeines theoretisches und praktisches Handbuch der Färbekunst [u. s. w.]. Erfurt und Gotha, 1805–20. 5 vols., 8vo.

Allgemeine Uebersicht der einfachen und zusammengesetzten Salze. Gotha und Weimar, 1789. 4to. 4 tables.

Anfangsgründe der Agriculturchemie. Gotha, 1816. 8vo.

Apothekerschule (Die), oder tabellarische Darstellung der gesammten Pharmacie. Gotha, 1803. Fol.

Zweite Auflage, 1810.

Chemisches Probierkabinet, oder Nachricht von dem Gebrauche und den Eigenschaften der Reagentien. Erfurt, 1801. 8vo.

Dritte Ausgabe. Erfurt und Gotha, 1818. 8vo.

Chemische Receptirkunst, oder Taschenbuch für praktische Aerzte, welche bei dem Verordnen der Arzneien Fehler in chemischer und pharmaceutischer Hinsicht vermeiden wollen. Fünfte neu bearbeitete Auflage. Hamburg, 1844. 8vo.

First edition, Erfurt, 1797.

Grundsätze (Die) der Chemie, mit Berücksichtigung ihrer technischen Anwendung in einer Reihe allgemein fasslicher Vorlesungen entwickelt und durch Versuche erläutert. Für Fabrikanten, Künstler und Gewerbtreibende. Erfurt, 1829. 8vo.

Lehrbuch der pharmaceutischen Experimental-Chemie. Altona, 1796. 8vo.

Dritte Auflage. Hamburg, 1822.

Systematisches Handbuch der gesammten Chemie zur Erleichterung des Selbststudiums dieser Wissenschaft. Gotha und Erfurt, 1800–1807. 8 vols., 8vo.

Zweite Auflage. Erfurt, 1805–1820. 8vo.

Also under the title : Die Chemie im Felde der Erfahrung.

Systematisches Handbuch der Pharmacie. Erfurt, 1792. 8vo.

Vierte Auflage, 1831.

Uebersicht der wichtigsten Entdeckungen in der Chemie. Weimar, 1792. Fol.

TROMMSDORFF, JOHANN BARTHOLOMÄUS.

Darstellung der Säuren [etc.]. *See in Section II.*

Tabelle über alle bis jetzt bekannten Gasarten u. s. w. *See in Section II.*

See also in Section III.

TROMMSDORFF, HUGO.

Statistik (Die) des Wassers und der Gewässer, ihre Wichtigkeit und bisherige Vernachlässigung . . . und Anleitung zur maass-analytischen Bestimmung der organische Stoffe. Erfurt, 1869. 8vo.

TROOST, LOUIS.

Précis de chimie. Seconde édition, rédigée conformément aux nouveaux programmes et suivie de quelques notions de chimie organique. Paris, 1867. 18mo.

> Troisième édition. Paris, 1869.
>
> Dixneuvième édition. Paris, 1886.
>
> Vingt et une édition. Paris, 1888.
>
> Leerboek der scheikunde, bevattende de toepassingen op het gebied van nijverheid en gezondheidsleer. Vrij bewerkt naar het Fransch : gewijzigden op vele plaatsen uitgebreid door de Loos. Ten gebruike aan inrichtingen van middelbaar onderwijs. Eerste deel : Nietmetalen. Tweede deel : Metalen. Rotterdam, 1866–'68. 8vo. Ill.
>
> Tratado elemental de química arreglado al programa official de la segunda enseñanza con las principales aplicaciones à las artes, industria, medicina e higiene. Version castellana con autorizacion por A. Sanchez de Bustamente. Tercera edicion. Paris, 1880. 8vo.
>
> Cuarta edicion. Paris, 1878.
>
> Sesta edicion. Paris, 1885.

Traité élémentaire de chimie, comprenant les principales applications à l'hygiène, aux arts et à l'industrie. Paris, 1864. 18mo.

> Deuxième édition. Paris, 1869.
>
> Neuvième édition. Paris, 1886.

TSCHELNITZ, S.

Farben-Chemie insbesondere der Oel und Wasserfarben, nach ihrem chemischen und physikalischen Verhalten, ihrer Darstellung und Verwendung, so wie ihren gewöhnlichen Verfälschungen ; für Fabrikanten, Maler, Techniker. Wien, 1857. pp. viii–302, 8vo.

TSCHEUSCHNER, E.

Handbuch der Glasfabrikation nach allen ihren Haupt- und Nebenzweigen. Fünfte Auflage, von Leng-Gräger's Handbuch. Weimar, 1885. Ill. Atlas of 34 plates, folio.

TSCHIRCH, A.

Untersuchungen über das Chlorophyll. Berlin, 1884. Roy. 8vo.

TUCKER, J. H.

Manual of Sugar Analysis, including the applications in general of analytical methods to the sugar industry, with introduction on the chemistry of cane sugar, dextrose, levulose, and milk sugar. New York, 1881. pp. 353, 8vo. London, 1890. 8vo.

Tunisi, Mohammed et, ibn Omar ibn Suleiman.
[General Chemistry in Arabic text. Published in Egypt (Cairo ?) in
    1842.]

Turner, Edward.
    * An Introduction to the study of the laws of chemical combination and
        the atomic theory . . . [With a table of atomic weights.] . . .
        London, 1825. pp. x–114, 18mo.
                Einleitung in die Gesetze der chemischen Verbindung und
                    die Atomen-Lehre. Zum Selbstunterricht. Uebersetzt
                    und erweitert von F. Steinbeiss. Tübingen, 1828. 8vo.
        Elements of Chemistry, including the recent discoveries and doctrines
            of the science. Edinburgh, 1827. 8vo. Plates.
                Third edition. London, 1831.
                Fourth edition, enlarged. London, 1833.
                Fifth edition. London, 1834.
                Sixth edition, enlarged and revised. Edited by J. Liebig,
                    Wilton George Turner, and W. Gregory. London, 1842.
                Seventh edition, including the actual state and prevalent
                    doctrines of the science. Edited by J. Liebig and W.
                    Gregory. London, 1842.
                Eighth edition. London, 1847.
            Cf. Johnston, John.
                Lehrbuch der Chemie, deutsch bearbeitet von Hartmann.
                    Leipzig, 1829. 8vo.
        Unterhaltungen über die Chemie, in welcher die Elemente dieser Wis-
            senschaft in Gesprächsform durchgegangen und durch Experimente
            erläutert werden. Aus dem englischen von K. G. Kühn. Leipzig,
            1822. 8vo. Ill.

Tuttle, D. K., and C. F. Chandler.
    A Manual of Qualitative Analysis. New York, 1873. pp. 46, 8vo.

Tuttle, H. A.
    Pocket Analysis Book. For ores, iron, coal, steel, coke and limestone.
        With tables for miners, dealers and iron smelters. New York,
        n. d. 18mo.

Tweede antwoord op de vraag : is het phlogiston een waar beginsel
    der ligchaamen. Utrecht, 1789. 8vo.

Twining, T.
    Familiar Lessons on Food and Nutrition. London, 1882.

TYCHSEN, NIC.
Chemisk Haandbog. Kjøbenhavn, 1794. 3 vols., 8vo.
     Kurzes chemisches Handbuch. Kopenhagen, 1787.

TYLER, HARRY W.
Entertainments in Chemistry. Easy lessons and directions for safe experiments. Chicago, 1886.

TYNDALL, JOHN.
The Forms of Water in Clouds and Rivers, Ice and Glaciers. London and New York, 1880. pp. xxiii–196, 12mo. Ill.
    The International Scientific Series.

TYSON, J.
A Guide to the Practical Examination of the Urine, for the use of physicians and students. Fifth edition. Philadelphia, 1886. 12mo.

UEBER DEN ANBAU DER RUNKELRÜBE und Einrichtungen der Zucker-fabriken. Prag, 1833.

UHLENHUTH, EDUARD.
Darstellung (Die) des Aluminiums, Kaliums, Natriums, Magnesiums, Bariums, Strontiums und Calciums und der Metalloide Bor und Silicium. Nach den neuesten Arbeiten von Sainte-Claire Deville, Wöhler, Heinrich Rose, Brunner, Bunsen und Andere. Quedlin-burg, 1858–59. 2 parts. I, iv–66 ; II, iv–64. Ill.
Handbuch der Photogen- und Paraffin-Fabrikation aus Torf, Braun-kohle und bituminösem Schiefer. Nach den neuesten Versuchen und Erfahrungen. Nebst einem Anhang : Ueber den Heizeffect des Torfes und seine künstliche Bearbeitung. Von Fischer. Quedlinburg, 1858. 8vo.
Junge (Der) Chemiker. Eine methodische Anleitung zur Anstellung chemischer Versuche und zur Begründung einer sichern wissen-schaftlich chemischen Erkenntniss. Ein Handbuch für Lehrer und Schüler der Real-, Gewerb- und höheren Bürgerschulen, Gewerbs-Institute. Berlin, 1859. pp. x–267, 8vo. Ill.
Ueber Leuchtgasbereitung aus Steinkohlen, Holz, Torf, u. s. w. Ueber Fabrication der Seifen. Frankfurt a. d. O., 1856. 8vo.

UILKENS, F.
Eenige aanmerkingen nopens het sal seignetti en het acidum essentiale tartari. Amsterdam (?), 1791. 4to.

ULBRICHT, R. M.
Leitfaden für die qualitative und quantitative Analyse in chemischen

ULBRICHT, R. M. [Cont'd.]

und technischen Laboratorien. Zunächst für die Studirenden höherer landwirthschaftlichen Lehranstalten bearbeitet. Erster Theil : Die qualitative Analyse. Wien, 1871. 8vo.

ULBRICHT, R., UND L. VON WÁGNER.

Handbuch der Spiritusfabrikation. Weimar, 1888. Ill.

ULE, OTTO.

Die Chemie der Küche oder die Lehre von der Ernährung und den Nahrungsmitteln der Menschen und ihren chemischen Veränderungen durch die Küche. Zweite vermehrte Auflage. Halle, 1871. 8vo.

Химія кухни. С.-Петербургъ, 1865.

ULLGREN, CLEMENS.

Organiska Kemien i Sammendrag, med Förord af Berzelius. Stockholm, 1839.

ULTZMANN, ROBERT.

Anleitung zur Untersuchung des Harnes mit besonderer Berücksichtigung der Erkrankung des Harnapparats. Unter Mitwirkung von Karl Berthold Hofmann. Wien, 1871.

Zweite Auflage. Wien, 1878.

Analysis of the Urine, with special reference to the diseases of the genito-urinary organs. Translated from the German by T. Barton Brune and H. Holbrook Curtis. New York, 1890. 8vo.

Guide to the Examination of Urine with special reference to the diseases of the Urinary Apparatus. Second edition, translated and edited by F. Forcheimer. Cincinnati, 1886. pp. [v]–251, 12mo.

UNCLE DAVY [Pseudonym].

First Lessons in Chemistry. Sixth edition. New York, 1846. 24mo. Ill.

UPMANN, J.

Das Schiesspulver, dessen Geschichte, Fabrikation, Eigenschaften und Proben. Braunschweig, 1874.

Traité sur la poudre, les corps explosifs et la pyrotechnie. Traduit par Desortiaux. Paris, 1878. 8vo. Ill.

URBAIN.

La conservation des substances alimentaires. Paris, 1892. 8vo. Ill.

URBAIN.   [Cont'd.]

> A part of Fremy's Encyclopédie chimique, vol. x, published since Section
> II of this work was printed.   *See, in Section II*, Fremy : Encyclopédie
> chimique, vol. II.

URBAIN ET ST. MEUNIER.

La chimie du charbon de bois, du noir de fumée et des combustibles
minéraux.   Paris, 1885.   8vo.

URE, ANDREW.   *See in Section II.*

URE, ANDREW.

The Revenue in Jeopardy from Spurious Chemistry, demonstrated in
researches upon wood-spirit and vinous-spirit.   London, 1843.   pp.
35, 8vo.

URECH, FRIEDERICH.

Itinerarium durch die theoretische Entwicklungsgeschichte der Lehre
von der chemischen Reactionsgeschwindigkeit.   Berlin, 1885.

URQUHART, J. W.

Electroplating.   A practical Handbook, including the practice of electro-
typing.   London, 1882.   pp. viii–216, 8vo.

UTNE, A.

Lærebog i Chemi for Borger- og Realskoler.   Christiania, 1860.   8vo.

VACHER, ARTHUR [Editor].
> *See* Fresenius, C. R.

VACHER, ARTHUR.

A Primer of Chemistry including Analysis.   London, 1877.   pp. viii–
108, 16mo.

VADEMECUM DER CHEMIE.   Ein nützliches Buch für Handwerker und
Oeconomen.   Stuttgart, 1854.   8vo.

VALENTIN, WILLIAM GEORGE.

Course (A) of Qualitative Chemical Analysis.   Second edition.   London,
1873.

> Third edition, 1874.   Fourth edition, 1876.
> Fifth edition, revised and corrected by W. R. Hodgkinson
> and H. M. Chapman.   London, 1880.
> Sixth edition.   London, 1884.
> Seventh edition edited by Hodgkinson, Chapman-Jones and
> Matthews.   London, 1888.   8vo.   Ill.

VALENTIN, WILLIAM GEORGE.   [Cont'd.]

Laboratory-Textbook (A) of Practical Chemistry, or introduction to
qualitative analysis, a guide to the course of practical instruction
given in the laboratories of the Royal College of Chemistry.   London, 1871.   8vo.

Twenty Lessons in Inorganic Chemistry : embracing the course of instruction in chemistry required for the first stage or elementary
classes of the science and art department.   London and New York,
1879.   pp. 186, 8vo.   Ill.

Valentin's Tables for the Qualitative Analysis of simple and compound
substances.   Third edition, revised by Hodgkinson, Chapman-Jones
and Matthews.   London, 1888.   8vo.

VALLHONESTA JOSÉ.

Colores derivados de la anilina.   Historia, fabricacion y aplicacion á la
tintorería y otros varios ramos de la industria.   Madrid, 1876.   4to.
Ill.

VALON, W. A.

De la fabrication de l'oxygène dans les usines de gaz et les résultats
pratiques de son emploi dans l'épuration du gaz de houille.   Paris,
1889.   Ill.

VAN, AND VAN DER.

For Dutch names beginning with VAN and VAN DER, see next succeeding
word.   A few Americanized names form exceptions, as here follows.

VANDENBERGH, F. P.

A Laboratory Guide in Pharmaceutical Chemistry, with two hundred
experiments.   Buffalo, 1888.   pp. 70, oblong 12mo.

VAN DYCK, C. V. A.

Elementary Chemistry [in Arabic text.]   Science Primer Series.   Published at the American Mission Press, Beirût, Syria. 1886.   Based
on Roscoe's Chemistry.

Larger Chemistry [in Arabic text.]   Published at the American Mission
Press, Beirût, Syria, 1869.

VANUCCINI, G.

Analisi chimica delle nuove polle d' acqua del risorgimento di Montecatini
in Val di Nievole.   Prato, 1885.   8vo.

Letame e concimi chimici ; considerazioni e raffronti, lettura, ecc.
Città di Castello, 1885.   8vo.

VAUBEL, WILHELM.

Das Stickstoffatom.   Giessen, 1891.   pp. 8, 8vo.   Folding plate.

VAUGHAN, VICTOR C.
Lecture Notes on Chemical Physiology and Pathology. Second edition.
Ann Arbor, 1879. 8vo.

VAUQUELIN, LOUIS NICOLAS.
Instruction sur la combustion des végétaux, la fabrication du salin, de
la cendre gravelée, etc. Tours, 1794, 1799 and 1803. 4to.
Manuel de l'essayeur, approuvé en l'an VII par l'administration des
monnaies. . . . Paris, 1812. 8vo.
        Manuel complet de l'essayeur suivi de l'instruction de Gay
           Lussac sur l'essai des matières d'argent par la voie
           humide, et des dispositions du laboratoire de la monnaie
           de Paris par Darcet. Nouvelle édition, entièrement
           refondue, augmentée de plusieurs tableaux d'essais et
           d'un grand nombre de figures par Ad. Vergnaud. Paris,
           1835. 18mo. Ill.
        Handbuch der Probirkunst aus dem französischen übersetzt
           von Fr. Wolff mit Anmerkungen von M. H. Klaproth.
           Königsberg, 1800. 8vo.
        Manual del ensayador de oro, plata y otros metales traducido
           de la última edicion francesa. Paris, 1826. 12mo.

VEJLEDNING TIL DEN CHEMISKE ANALYSE af de almindeligst forekommende
organiske Stoffer. Kjøbenhavn, 1854.

VELASCO Y PANO, BONIFACIO.
Análisis química fundada en las observaciones de los espectros lumino-
sos . . . Madrid, 1862. 4to.
Tratado de química orgánica aplicada á la farmacia y á la medicina,
escrito con arreglo á la teorías modernas. Granada, 1872. 8vo.

VELEZ DE PAREDES, E.
Manual de química divertida o sea recreaciones químicas. Cuarta
edicion. Paris, 1881. 16mo.

VENABLE, F. P.
A Course in Qualitative Chemical Analysis. New York, 1883. 12mo.
        Second edition. New York, 1892. pp. iv–53, 12mo.
        The author acknowledges the assistance of F. P. Dunnington.

VENABLES, ROBERT.
Elements of Urinary Analysis and Diagnosis, chemical and microscopical, being concise directions for a chemico-pathological examination
of the urine and urinary concretions, etc. Second edition. London, 1850. 8vo.

VERA Y LOPEZ, V.
  Breves nociones de química orgánica para uso de los institutos de
  segunda enseñanza.   Madrid, 1885.   8vo.

VERDEIL, F.   *See* Robin and Verdeil: Traité de chimie anatomique, etc.

VERDET.   Leçon (etc.).   *See* Société chimique de Paris.

VERGNIAUD, AMAND DENIS.
  Nouveau manuel complet de chimie inorganique et organique dans
  l'état actuel de la science, suivi d'un Dictionnaire de chimie.
  Nouvelle édition.   Paris, 1845.   18mo.
        La première édition, 1825.

  Nouveau manuel complet de chimie amusante, ou nouvelles récréations
  chimiques.   Nouvelle édition, revue, corrigée et considérablement
  augmentée.   Paris, 1845.   18mo.
        This work is based upon Accum's Chemical Amusements.   *Cf.* Accum,
        Frederick.

VERGUIN, E.
  Élémens de chimie générale.   Lyon et Paris, 1845.   12mo.
            Elementos de química general.   Traducidos del francés al
            castellano espresamente para servir de testo en el Real
            Colegio de San Cristobal y anotados por los alumnos de
            la clase de química de dicho Colegio.   Habana, 1848.
            2 vols., sm. 4to.

VERLOOREN, MARGARETHA CORNELIS.
  De corporum chemicorum organicorum constitutione.   Trajecta ad
  Rhenum, 1845.   8vo.

VERMEULEN, P. J. F.
  Over de chemische constitutie der aromatische verbindingen,   Delft,
  1871.   8vo.

VERMOREL, V.
  Le sulfure de carbone, ses propriétés, sa fabrication, moyens pratiques de
  vérifier sa pureté.   Tours, 1886.

VERNAY, F.
  Éléments de chimie.   Nouvelle édition.   Paris, 1880.   16mo.

VERNOIS, MAXIME, ET ALFRED BECQUEREL.
  Analyse du lait des principaux types de vache, chèvre, brebis, bufflesse,
  présentés au concours agricole universel de 1856.   Paris, 1857.
  pp. 35, 12mo.

VERNOIS, MAXIME, ET ALFRED BECQUEREL. [Cont'd.]
Du lait chez la femme dans l'état de santé et dans l'état de maladie.
Mémoire suivi de nouvelles recherches sur la composition du lait
chez la vache, l'annesse, la chèvre, la jument, la brebis, et la
chienne. Paris, 1853. 8vo.

VERSUS MEMORIALES ANALYTICO-CHEMICI, oder Anweisung wie die
trockene Substanz auf Platin und vor dem Löthrohre zu behandeln
sei. In artige Reimlein gebracht von Aureolus Ambica. München,
1871. 16mo.

VERVER, B.
Num publicæ sanitati nocere possint venena metallica, quibus conse-
rantur agri, ad occidenda animalia nociva. Groningæ, 1841. 8vo.

VIARD, ÉMILE.
Traité général des vins et de leurs falsifications d'après le mémoire
couronné, remanié, corrigé et considérablement augmenté. Deuxième édition. Étude complet de vue chimique, des procédés licites
de traitement des vins et des falsifications. Méthodes de recherches
et d'analyses précises et douteuses, description de tous les appareils
employés. Paris et Nantes, 1884. pp. 499, 8vo. Ill.

VICCAJEE, KAIKHOSRU RUSTAMJEE.
Rasayàna Castra   Bombay, 1875. 8vo. Ill.
    A Treatise of chemistry in the Gujerathi language.

VICKERS, B.
Genealogical Tree of Coal-tar Products. London and Derby, 1888. Fol.

VIELLE (ET SARREAU). Substances explosives. See, in Section II, Fremy : Encyclopédie
    chimique, vol. v.

VIERORDT, KARL.
Anwendung (Die) des Spectralapparates zur Photometrie der Absorp-
tionsspectren und zur quantitativen chemischen Analyse. Tübingen,
1873. pp. iv–169, roy. 4to. Plates.
Quantitative (Die) Spectralanalyse in ihrer Anwendung auf Physiologie,
Physik, Chemie und Technologie. Tübingen, 1876. pp. iv–125,
roy. 4to. Plates.

VIETH, P.
Die Milchprüfungs-Methoden und die Controle der Milch in Städten
und Sammelmolkereien. Bremen, 1879. pp. 116, 8vo.

VIGANUS, JOHANNES FRANCISCUS.
Chymia jam variis experimentis aucta multisque figuris illustrata. Londini, 1687. 8vo. Ill.

VILLE, GEORGES.
École (L') des engrais chimiques. Premières notions de l'emploi des agents de fertilité. Paris, 1871. 18mo.

> Engrais (Les) chimiques. Entretiens agricoles donnés au champ d'expériences de Vincennes. Paris, 1891. 3 vols., 12mo. I. Les principes et la théorie. II. Les cultures spéciales. III. Le fumier et le bétail.

> De chemische bemesting volgens G. Ville door F. R. Corten. Eenige beschouwingen over de doelmatigheid der chemische meststoffen, naar aanleiding van de beoordeeling van den Villemest door A. Mayer. Amsterdam, 1877. 8vo.

> School (The) of Chemical Manures : Or elementary principles in the use of fertilizing agents. Translated from the French by A. A. Fesquet. Philadelphia, 1872. 12mo. Ill.

> On Artificial Manures : Their chemical selection and scientific application to agriculture. A series of lectures given at the Experimental Farm at Vincennes during 1867 and 1874–75. Translated and edited by Wm. Crookes. London, 1879. 8vo. Ill.
>
> Second edition, revised. London, 1882.

> Letame, bestiame e concimi chimici. Torino, 1876.

VILLEJEAN, E.
Pigments et matières colorantes de l'économie animale. Paris, 1886.

VILLIERS, A.   *See, in Section II,* Fremy: Encyclopédie chimique, vol. III, vol. VI, etc.

VILLIERS, A.
Précis d'analyse quantitative. Paris, 1892. 8vo. Ill.
Recherche des poisons végétaux et animaux. Paris, 1882. 8vo.
Tableaux d'analyse qualitative des sels par voie humide. Paris, 1890.

VILLON, A. M.   Nickel et Cobalt [metallurgy of]. *See, in Section II,* Fremy : Encyclopédie chimique, vol. V.

VILLON, A. M.
Les corps gras, huiles, beurres, graisses, suifs, cires, pétrole, vaseline, minéraux, etc. Paris, 1890. pp. 334, 8vo.

VILLON, A. M. [Cont'd.]

Traité pratique des matières colorantes artificielles dérivées du goudron de houille. Paris, 1890. pp. x–560. Ill.

Traité pratique de photogravure au mercure, ou mercurographie. Paris, 1891. pp. 32, 16mo.

> With additional title : Bibliothèque photographique.

ВИЛУЕВЪ.

Разсужденіе о вѣсѣ пая высмута. С.-Петербургъ, 1849.

> VILUEF. On the Atomic Weight of Bismuth. St. Petersburg, 1849.

VINANT, MICHEL DE.

Traité pratique de la teinture l'impression sur étoffes et du blanchissage. Renfermant les procédés pour l'impression le blanchiment et l'apprêt des étoffes, suivi de la fabrication des principaux produits chimiques de la distillation des essences minérales, des alcools et des liqueurs. Paris, 1872. pp. xxiii–878, 8vo. Plates.

VINCENT, C.

Industrie des produits ammoniacaux. Paris, 1884. 8vo.

> *Cf., in Section II*, Fremy : Encyclopédie chimique, vol. x.

VINTSCHGAU, M. VON.

> Geschmacksinn, Geruchsinn. *See* Handbuch der Physiologie.

VINTSCHGAU, M. VON.

Ueber die Hoffmann'sche Tyrosin-Reaction und über die Verbindungen des Tyrosins mit Quecksilberoxyd. Wien, 1869.

VIOLLE. Éclairage électrique. *See, in Section II*, Fremy : Encyclopédie chimique, vol. v.

VIOLETTE, HENRI.

Notions élémentaires de chimie à l'usage des écoles. Nancy, 1838. 12mo.

Nouvelles manipulations chimiques simplifiées, contenant la description d'appareils, entièrement nouveaux, d'une construction simple et facile et suivie d'un cours de chimie pratique à l'aide de ces instruments. Paris, 1839. 8vo.

> Deuxième édition. Paris, 1847.
>
> Упрощенныя химич. манипуляціи или дешевая домашн. лаборатарія. Москва, 1864.

Полный курсъ практич. химіи. Москва, 1867.

> VIOLETTE, HENRI. Complete Course of Practical Chemistry. Moscow, 1867.

VIRCHOW, C.

Analytische Methoden zu Nahrungsmittel-Untersuchungen, nebst einem

VIRCHOW, C. [Cont'd.]

Anhang enthaltend die Untersuchungen einiger landwirthschaftlicher und technischer Producte und Fabrikate sowie die Harnanalyse. . . . Berlin, 1891. pp. xii–172, 8vo.

VITALI, D.

Manuale delle alterazioni e sofisticazioni dei principali preparati chimici e delle più importanti droghe medicinali. Piacenza, 1881. 8vo.

VITALIS, J. B.

Cours élémentaire de teinture sur laine, soie, lin, chanvre et coton, et sur l'art d'imprimer les toiles. Paris, 1823. pp. xx–459, 8vo.

Lehrbuch der gesammten Färberei auf Wolle, Seide, Leinen, Hanf und Baumwolle. . . . Nach dem französischen . . . frei bearbeitet. Dritte mit Beihaltung der S. Rennir'schen und H. Leng'schen Verbesserungen sehr vermehrte Auflage, bearbeitet von L. Bergmann.

Neuer Schauplatz der Künste und Handwerke.

[*The same.*] Nebst einem Anhang über Kattun-Druckerei. Nach dem französischen. Vierte sehr vermehrte . . . Auflage von Ch. H. Schmidt. Weimar, 1840. 8vo.

Sechste sorgfältig revidirte und mit Benutzung des Traité de l'impression des tissus par J. Persoz vermehrte, etc., Auflage. Bearbeitet von Chr. H. Schmidt. Weimar, 1854. 8vo.

VIVARELLI, A.

Lezioni di chimica applicata. Livorno, 1887. 16mo.

VLENTEN, A. VAN.

Bijdrage tot de kennis der geschiedenis van Avogadro's hypothese. Leiden, 1873. 8vo.

VOGEL, AUGUST.

See Sechs Vorträge aus dem Gebiete der Nahrungsmittel-Chemie.

VOGEL, AUGUST.

Lehrbuch der Chemie. Als Leitfaden bei seinen Vorlesungen für die Studirenden an der Universität und zum Selbstunterricht. Stuttgart, 1830–'32. 2 vols., 8vo. Ill.

Ueber den Chemismus der Vegetation. Festrede zur Vorfeier des Geburtstages Seiner Majestät Maximilian II. Königs von Bayern gehalten in der öffentlichen Sitzung der königlich bayerischen Akademie der Wissenschaften am 27 November 1852. München, 1852. 4to.

VOGEL, AUGUST [THE YOUNGER].

Bieruntersuchung (Die). Eine Anleitung zur Werthbestimmung und Prüfung des Bieres nach den üblichsten Methoden. Berlin, 1866. pp. vi–96, 8vo.

Fünfzig praktische Uebungsbeispiele in der quantitative-chemischen Analyse, mit besonderer Rücksicht auf die Werthbestimmung landwirthschaftlicher und technischer Produkte. Erfurt, 1861. 8vo. Ill.

Dritte Auflage. Erfurt, 1863. 8vo.

Metallische (Das) Zink. Eine Darstellung seines natürlichen Vorkommens, seiner Gewinnung, Eigenschaften. . . . München, 1861. 12mo.

VOGEL, AUGUST, UND E. WEIN.

Anleitung zur qualitativen Analyse landwirthschaftlich wichtiger Stoffe in praktischen Beispielen. Fünfte Auflage. Berlin, 1879. 8vo.

VOGEL, E.

The Atomic Weights and their variation. San Francisco, 1888. 8vo.

VOGEL, F., UND A. RÖSSING.

Handbuch der Electrochemie und Electrometallurgie. Stuttgart, 1891. 8vo.

VOGEL, HANS.

Ueber Milchuntersuchung und Milchkontrolle. Vortrag gehalten bei Gelegenheit der ersten Versammlung bayrischer Chemiker zu München. Würzburg, 1884. pp. 23, 8vo.

VOGEL, HEINRICH.

Die Verfälschung und Verschlechterung der Lebensmittel. Zweite Auflage. Schwelm, 1873. 16mo.

VOGEL, HERMANN W.

Chemischen (Die) Wirkungen des Lichts und die Photographie in ihrer Anwendung in Kunst, Wissenschaft und Industrie. Berlin, 1874.

Internationale wissenschaftliche Bibliothek.

The Chemistry of Light and Photography in its applications to art, science and industry. London, 1872. 8vo. Ill.

The International Scientific Series.

La photographie et la chimie de la lumière. Troisième édition. Paris, 1881. 8vo. Frontispiece and Ill.

Химическія дѣйствія свѣта и фотографія въ ихъ приложеніи къ искуству, наукѣ и промышленности. Пер. Чутковскаго Я.

Fortschritte (Die) der Photographie seit dem Jahre 1879 . . . mit . . .

VOGEL, HERMANN W. [Cont'd.]

einem Anhang ; Photographie für Amateure . . . Zugleich als Ergänzung zur dritten Auflage von des Verfassers Lehrbuch der
Photographie. Berlin, 1883. 8vo. Ill.

Lehrbuch der Photographie. . . . Berlin, 1870. 8vo.

Part I was published in 1867.

Zweite verbesserte Auflage. Berlin, 1874.

Dritte gänzlich umgearbeitete Auflage. Berlin, 1878. 8vo.

Ausführliches Handbuch der Photographie. Vierte vollständig umgearbeitete und vermehrte Auflage. Berlin,
1890. 4 vols., 8vo. Illustrated.

Practische Spectralanalyse irdischer Stoffe. Anleitung zur Benutzung
der Spectralapparate in der qualitativen und quantitativen chemischen Analyse organischer und unorganischer Körper, im Hüttenwesen, bei der Prüfung von Mineralien, Farbstoffen, Arzneimitteln,
Nahrungsmitteln bei physikalischen und physiologischen Untersuchungen, etc. Nördlingen, 1877. pp. viii–398, 8vo. Ill. Plates.

Zweite vollständig umgearbeitete Auflage. Theil I : Qualitative Spectralanalyse. Berlin, 1889. 8vo. Ill. Plates.

VOGEL, MAX.

Die Entwicklung der Anilin-Industrie. Die Anilinfarben, ihre Entstehung, Herstellung und technische Verwendung. Leipzig, 1866.
8vo.

Zweite stark erweiterte Auflage. Leipzig, 1870.

VOGEL, RUDOLF AUGUSTIN.

Experimenta chemicorum de incremento ponderis corporum quorundam
igne calcinatorum examinat. Göttingæ, *n. d.* [1753]. pp. 24, 4to.

The author was unacquainted with the work of Jean Rey, published in 1630.

Institutiones chemiæ ad lectiones academicas accommodatæ. Editio
nova polita et locupleta. Francofurti et Lipsiæ, 1762. 8vo.

[Earlier editions]. Göttingæ, 1755, and Leyden, 1757.

Lehrsätze der Chemie. Ins deutsche übersetzt und mit Anmerkungen versehen von Johann Christian Wiegleb.
Weimar, 1775. 8vo.

VOGL, A. E.

Die gegenwärtig am häufigsten vorkommenden Verfälschungen und
Verunreinigungen des Mehles und deren Nachweisung. Wien,
1880. 8vo. Ill.

VOGTHERR, M.

Maassanalytisches Uebungsbuch zu der Einführung in die Maassanalyse.
Neuwied, 1891. 4to.

VOIGT, JOHANN HEINRICH.

Versuch einer neuen Theorie des Feuers, der Verbrennung, der künstlichen Luftarten, des Athmens, der Gährung, der Electricität, der
Meteoren, des Lichts, und des Magnetismus.   Jena, 1793.   8vo.

VOIT, C. VON.

Physiologie des allgemeinen Stoffwechsels und der Ernährung.   *See* Handbuch der Physiologie.

VOLCKMAR, E.

Lehrbuch der anorganischen Chemie für den Unterricht an höheren
Lehranstalten.   Kassel, 1887.

VOLCKXSOM, E. W. V.

Catechism of Modern Elementary Chemistry, or solutions of the
questions set at the London matriculation examinations 1844-'82.
London, 1882.   8vo.

VOLLRATH, ANTON.

Die Fabrikation des Alauns . . .   Quedlinburg und Leipzig, 1834.
8vo.   Ill.

VOLTA, ALESSANDRO.

Nozioni di chimica compilate giusta il vigente programma d'insegnamento pei licei.   Milano, 1882.   8vo.

VOLTA, ALESSANDRO.

Lettres de M. A. V. sur l'air inflammable des marais, auxquelles on a
ajouté trois lettres du même auteur tirées du Journal de Milan.
Traduites de l'italien.   Strasbourg, 1778.   pp. 191, 12mo.   Folding
plate.

Briefe über die entzündbare Luft der Sümpfe.   Nebst drey
andern Briefen von dem nämlichen Verfasser die aus
dem mayländischen Journal genommen sind.   Aus dem
italienischen übersetzt von Carl Heinrich Köstlin.
Strassburg, 1778.   pp. [xii]–226, 12mo.   Folding plate.

VOLTENIUS, FLORIS JACOBUS.

De lacte humano ejusque cum asinino et ovillo comparatione observationes chemicæ; accesserunt Henrici Doorschodte de lacte adque
Joh. Georgii Greiselii de cura lactis in arthritide commentationes;
conjunctim edendas curavit Joh. Georgius Fridericus Franzius.
Lipsiæ, 1779.   8vo.

VORTMANN, G.

Anleitung zur chemische Analyse organischer Stoffe.   Wien, 1891.   pp.
xii–408, + 18 tables, 8vo.   Ill.

VOSMAER, A.

The Mechanical and other properties of Iron and Steel in connection
with their chemical composition. London, 1891. 8vo.

WAAGE, P.

Kemiens förste Grunde efter H. E. Roscoe. Christiania, 1878. 8vo.
Ill.
    Cf. Roscoe, H. E.

Kursus (Et) i den kvalitative kemiske Analyse. Christiania, 1866.
8vo. Three tables.
        Anden Udgave. Christiania, 1876.

Meddelelser fra Universitetets Kemiske Laboratorium. Særskilt Aftryk
af Forhandlinger i Christiania Videnskabs-Selskab, 1871. [Chris-
tiania, 1872.] 8vo. Ill.

Udvikling af de surstofholdige Syreradikalers Theorie. Christiania,
1859. 8vo.
        A prize thesis, from the "Nyt Magazin for Naturvidenskaberr.e." *See also*
        Guldberg, C. M., and P. Waage.

WAALS, VAN DER.

Die Continuität des gasförmigen und flüssigen Zustandes. Uebersetzt
und mit Zusätzen versehen von F. Roth. Leipzig, 1881. 8vo. Ill.

WABST, CHRISTIAN XAVIER.

De hydrargyro tentamen physico-chemico-medicum. Vindobonæ,
1754. pp. [viii]–218, 4to. Large margins.
        Contains numerous bibliographical references.

WACHTEL, A. VON.

Hilfsbuch für chemisch-technische Untersuchungen auf dem Gesammtgebiete der Zuckerfabrikation. Prag, 1884. 8vo.

WACKENRODER, B.

Anleitung zur chemischen Untersuchung technischer Producte. Leipzig, 1874.

WACKENRODER, HEINRICH WILHELM FERDINAND.

Anleitung zur qualitativen chemischen Analyse der unorganischen und
organischen Verbindungen. Jena, 1836. 2 vols., 8vo.

Ausführliche Characteristik der stickstofffreien organischen Säuren, nebst
Anleitung zur qualitativen chemischen Analyse, Jena, 1841.

Chemische Classification der einfachen und zusammengesetzten Körper,
nebst Tafeln über die Atomgewichte oder Aequivalente der einfachen Körper und über die wichtigsten Verbindungen derselben.
Zum Behuf seiner Vorlesungen entworfen. Jena, 1851. 8vo.

WACKENRODER, HEINRICH WILHELM FERDINAND. [Cont'd.]
Chemische Tabellen zur Analyse der unorganischen Verbindungen,
oder ausführliche Characteristik der unorganischen Salzbasen und
Säuren. Jena, 1829. Fol.

    Dritte Auflage. Jena, 1834.

        Fünfte vermehrte Auflage. Erster Theil: Ausführliche
Characteristik der unorganischen Salzbasen und Säuren.
Nebst einem Vorworte und einer kurzen Einleitung.
Jena, 1843. Zweiter Theil: Erste Abtheilung: Ausführliche Characteristik der wichtigern stickstofffreien
organischen Säuren. Nebst einer Anleitung zur qualitativen chemischen Analyse der chemischen Körper und
ihrer Verbindungen. Jena, 1841.

Commentatio de anthelminticis regni vegetabilis præmio regio ornata.
Göttingæ, 1826.

Commentatio de cerevisiæ vera mixtione et indole chemica . . . Jena,
1850. 8vo.

Kleine analytisch-chemische Tabellen zur Analyse der unorganischen
Verbindungen, bearbeitet nach der fünfte Auflage der grössern
chemischen Tabellen. Ein Handbuch in Tabellenform zum Gebrauch beim ersten Unterricht in der qualitativen chemischen
Analyse, insbesondere für pharmaceutische physiologische, polytechnische und landwirthschaftliche Institute, Realschulen und
Gymnasien. Jena, 1847. Fol.

WAEBER, R.
Leitfaden für den Unterricht in der Chemie für Präparandenanstalten,
Mittel- und Fortbildungsschulen. Leipzig, 1876. 8vo. Ill.

    Zweite Auflage. Leipzig, 1879. 8vo. Ill.

    Fünfte Auflage. Leipzig, 1885.

    Sechste Auflage. Leipzig, 1888.

Lehrbuch der Chemie mit besonderer Berücksichtigung der Mineralogie
und chemischen Technologie. Für Seminarien, landwirthschaftliche Mittelschulen, Gewerbschulen u. a. m. Leipzig, 1876. 8vo.
Ill.

    Zweite Auflage. Leipzig, 1879.

        Lehrbuch für den Unterricht in der Chemie mit besonderer
Berücksichtigung der Mineralogie und chemischen Technologie. Sechste Auflage. Leipzig, 1888.

    Siebente Auflage. Leipzig, 1890.

WAGNER, A.
Lehrbuch der organischen Chemie für Mittelschulen. München, 1886.

WAGNER, A. [Cont'd.]

Lehrbuch der unorganischen Chemie für Mittelschulen sowie zum
    Selbststudium. München, 1885.

WAGNER, DANIEL.

Ueber das Kalium, die Verbindungen der ersten Stufe, der Zusammen-
    setzung derselben und über das Aetzkali. Als Beytrag zum che-
    mischen Theil der Naturwissenschaft. Wien, 1825. 8vo.

WAGNER, F.

Lærebog i den almindelige Teknologi omfattende Metallernes og Træets
    mekaniske Teknologi samt Jærnets Metallurgie, udarbeidet til Brug
    ved Undervisningen i tekniske Skoler og til Selvstudium for Ingeni-
    örer og Fabrikanter. Udgivet med Understöttelse af det Reier-
    senske Fond. Kjøbenhavn, 1884. 8vo. Ill.

WAGNER, H.

Licht und Feuer oder die Feuerzeugfabrikation vom Standpunkte ihrer
    gegenwärtigen Entwickelung theoretisch und praktisch beschrieben.
    Nebst Anleitung zur Prüfung und Darstellung der hierbei ange-
    wandten Materialien, sowie auch zur Bereitung der verschiedenen
    Zündpräparate, etc., sowie einer Reihe der bewährtesten Vorschriften
    für Phosphor, Antiphosphor und phosphorfreie Zündhölzchen, mit
    besonderer Berücksichtigung der in den Phosphorzündholzfabriken
    vorkommenden Krankheiten und wie denselben zu begegnen ist.
    Weimar, 1869. 8vo. Ill.
        Neuer Schauplatz der Künste und Handwerke.

WAGNER, JOHANN RUDOLF.
        See Wagner, Rudolf.

WAGNER, JULIUS.

Tabellen der im Jahre 1882 bestimmten physikalischen Konstanten
    chemischer Körper. Leipzig, 1884.

WÁGNER, LADISLAUS VON.

Bierbrauerei (Die) nach dem gegenwärtigen Standpunkte der Theorie
    und Praxis des Gewerbes. Mit besonderer Berücksichtigung des
    Brauverfahrens in Ungarn-Oesterreich, Bayern, am Rhein, etc.
    Auf Grund eigener Erfahrungen sowie mit Benutzung der neuesten
    deutschen, englischen, und französischen Literatur bearbeitet.
    Vierte sehr vermehrte und gänzlich umgearbeitete Auflage von
    Chr. H. Schmidt's Grundsätze der Bierbrauerei. Weimar, 1869.
    8vo. Ill.

WÁGNER, LADISLAUS VON.   [Cont'd.]

Handbuch der Bierbrauerei nach dem heutigen Standpunkte der Theorie und Praxis. Mit besonderer Berücksichtigung des von Pasteur aufgestellten neuen Verfahrens der Bierfabrikation. Auf Grund eigener Studien und praktischen Erfahrungen, sowie mit Benutzung des vorhanden literarischen Materials und unter Mitwirkung hervorragender Theoretiker und Praktiker verfasst. Fünfte sehr vermehrte und gänzlich umgearbeitete Auflage. Weimar, *n. d.* 2 vols., 8vo. Ill.

Sechste Auflage. Weimar, 1884. 2 vols., 8vo, and atlas of 27 plates.

For earlier editions, *see* Schmidt, Christian Heinrich.

Handbuch der Stärkefabrikation, mit besonderer Berücksichtigung der mit der Stärkefabrikation verwandten Industriezweige, namentlich der Dextrin-, Stärkesyrup- und Stärkezuckerfabrikation. Zweite Auflage. Weimar, 1884. 8vo. Atlas of 11 plates.

A Practical Treatise on the Manufacture of Starch, Glucose, Starch-Sugar, and Dextrine. Translated from the German by Julius Frankel. Edited by Robert Hutton. London and Philadelphia, 1881. 8vo. Ill.

Cf. Otto-Birnbaum's Lehrbuch, vol. v.

WAGNER, P.

La question des engrais d'après des expériences récentes. Paris, 1885.

L'augmentation économique de la production agricole par l'emploi rationel des engrais azotés. Traduit de l'allemand par Gieseker. Paris, 1887. Ill.

Cf. Otto-Birnbaum's Lehrbuch. Die Düngerfabrikation.

WAGNER, RICHARD.

Harnuntersuchungen. Halle, 1888. Ill.

WAGNER, RUDOLF VON.

Chemie (Die) fasslich dargestellt nach dem neuesten Standpunkte der Wissenschaft, zum Schulgebrauch und Selbstunterricht, namentlich für Studirende der Naturwissenschaften. Leipzig, 1850. 2 parts, 8vo. Ill. Theil 1: Unorganische Chemie. Theil 11: Organische Chemie.

Zweite vermehrte und verbesserte Auflage. Leipzig, 1851. 8vo.

Chemie (Die) fasslich dargestellt nach dem neuesten Standpunkte der Wissenschaft, für Studirende und Freunde

WAGNER, RUDOLF VON. [Cont'd.]

der Naturwissenschaften, der Medicin und der Pharmacie, so wie zum Gebrauche für technische Lehranstalten.
Dritte umgearbeitete und vermehrte Auflage. Leipzig,
1854. 8vo. Ill.

Vierte umgearbeitete Auflage. Leipzig, 1858. 8vo. Ill.

Fünfte Auflage. Leipzig, 1864.

Sechste Auflage. Leipzig, 1873. pp. xiv–663, 12mo. Ill.

De scheikunde, volgens het nieuwste standpunt der wetenschap, bevattelijk voorgesteld aan beoefenaars en lief-
hebbers der natuurwetenschappen. Naar de derde
Hoogduitsche uitgave vertaald door D. van der Waal
Spruijt. Utrecht, 1856. 8vo. Ill.

Chemische (Die) Technologie fasslich dargestellt nach dem neuesten
Standpunkte des Gewerbewesens und der Wissenschaft, zum Schulgebrauche und Selbstunterrichte, namentlich für Kameralisten, Ge-
werbe- und Realschüler. Leipzig, 1850. 8vo. Ill.

Chemische (Die) Technologie fasslich dargestellt nach dem
jetzigen Standpunkte der Wissenschaft und des Gewerbewesens, als Leitfaden bei Vorlesungen an Universitäten,
Gewerbschulen und polytechnischen Anstalten, sowie
zum Selbstunterricht. Zweite umgearbeitete und vermehrte Auflage. Leipzig, 1853. 8vo. Ill.

Dritte Auflage. Leipzig, 1856. 8vo. Ill.

Vierte Auflage, 1859. Fünfte Auflage, 1863.

Sechste Auflage, 1866. Siebente Auflage, 1868.

Achte Auflage, 1870. Neunte Auflage, 1873.

Zehnte Auflage, 1875. Elfte Auflage, 1880.

Zwölfte Auflage, bearbeitet von F. Fischer, 1886.

Handbuch der chemischen Technologie. Dreizehnte stark
vermehrte Auflage. Neu bearbeitet von Ferdinand
Fischer. Leipzig, 1889. 8vo. Ill.

Handboek der fabriekscheikunde, of de leer der scheikunde
theoretisch verklaard en pract. toegepast op alle fabrie-
ken en takken van nijverheid, benevens aanwijzing der
middelen om de waarde en zuiverheid der handels- en
nijverheidsproducten te bepalen. Uit het Hoogduitsch.
Voor Nederland bewerkt door C. C. J. Teerling. Leiden, 1862–65. 2 parts in three volumes, 8vo. Ill.

Nouveau traité de chimie industrielle, à l'usage des chimistes,
des ingénieurs, des industriels, des fabricants de produits chimiques, des agriculteurs, des écoles d'arts et

WAGNER, RUDOLF VON.   [Cont'd.]

    manufactures et d'arts et métiers, etc. Deuxième édition française très augmentée, publiée sur la dixième édition allemande. Avec figures intercalées dans le texte. Paris, 1878–79. 8vo. Vol. I, pp. III–840 ; vol. II, pp. 956.

    With the co-operation of L. Gautier.

Nouveau traité de chimie industrielle. Troisième édition française très augmentée, publiée d'après la treizième édition allemande. Paris, 1891–92. 2 vols., 8vo. Ill.

Nuovo trattato di chimica industriale per uso dei chimici ingegneri, industriali, fabbricanti di prodotti chimici agricoltori, medici legali, manifatture d'arti e mestieri, scuole tecniche, ecc., con riguardo alla statistica industriale. Traduzione autorizzata, con aggiunte dell' autore, di Alf. Cossa e C. Morbelli. Torino, 1876. 2 vols., 8vo. Ill.

[Another edition.] Torino, 1883. 8vo.

Nuovo trattato di chimica industriale. Seconda edizione con 377 incisioni ed aggiunte ed un' Appendice contenente i progressi dal 1883 al 1889. Torino, 1889. 8vo. Ill.

    With the co-operation of A. Cossa and A. Romegialli. First edition. Torino, 1873.

A Handbook of Chemical Technology. Translated and edited from the eighth German edition, with extensive additions by William Crookes. London, 1872. 8vo. Ill. *Also* New York, 1881. pp. xvi–745–xvi, 8vo. Ill.

Manual of Chemical Technology. Translated and edited by William Crookes, from the thirteenth enlarged German edition as remodelled by Ferdinand Fischer. London, 1892. 8vo. Ill.

Grundriss der chemischen Technologie. Zweite vermehrte und verbesserte Auflage. Leipzig, 1874. pp. xiv–566, 8vo.

    Schets der chemische technologie voor het middelbaar onderwijs en tot zelfonderricht. Vrij bewerkt naar het Duitsch door S. P. Huizinga. Arnhem, 1872. 8vo.

Ueber die homologen Reihen in der organischen Chemie. Nürnberg, 1852. 4to.

    Program der technischen Lehranstalten in Nürnberg.

WAHL, WILLIAM H.

    Galvanoplastic Manipulations. *See* Roseleur, Alfred.

WAHNSCHAFFE, FELIX.

Anleitung zur wissenschaftlichen Bodenuntersuchung. Berlin, 1887.
8vo.

> Guide (A) to the Scientific Examination of Soils, comprising
> select methods of mechanical and chemical analysis and
> physical investigation. Translated from the German of
> F. W., with additions by Wm. T. Brannt. Philadelphia,
> 1892. pp. xii– from 17–177, 12mo.

ВАЛЬБЕРХЪ Н., ПАВЛЕНКОВЪ Ф

Краткій очеркъ популярной химіи. С.-Петербургъ. 1874

> WALBERG, N., AND PAVLENKOF, F. Short Sketch of Popular Chemistry.
> St. Petersburg, 1874.

WALCHNER, FRIEDRICH AUGUST.

Chemie, volksfasslich und in Beziehung auf die Gewerbe und das
bürgerliche Leben bearbeitet. Karlsruhe, 1843. 8vo.

Unorganische (Die) Chemie. Zweite Ausgabe. Stuttgart, 1853. Sm. 8vo.

WALCHNER, F. H.

Die Nahrungsmittel des Menschen, ihre Verfälschungen und Verunreinigungen. Berlin, 1875.

WALENN, W. H.

Little Experiments for Little Chemists. London, 1866. 12mo.

WALKER, J.

Zur Affinitätsbestimmung organischer Basen. Leipzig, 1890. 8vo.

WALKHOFF, LOUIS.

Der praktische Rübenzuckerfabrikant und Raffinadeur. Ein Lehr- und
Hülfsbuch für Rübenzuckerfabrikanten, Betriebsdirigenten, Siede-
meister, Maschinenbauer, Ingenieure, Landwirthe und Studirende
an landwirthschaftlichen Lehranstalten. Nach eigenen langjährigen
Erfahrungen bearbeitet. Mit einem Vorwort vom Fr. Jul. Otto.
Vierte neu bearbeitete und vermehrte Auflage. Mit zahlreichen
Holzstichen, nach Originalzeichnungen der neuesten und besten
Constructionen aller Apparate der Rübenzuckerfabrikation. Braun-
schweig, 1872. 2 parts, 8vo.

> Dritte Auflage. Braunschweig, 1871.

> Traité complet de la fabrication et du raffinage du sucre de
> betterave. Deuxième édition française, traduite de la
> quatrième édition allemande, et considérablement aug-
> mentée par E. Merijotet J. Gay-Lussac. Paris, 1870.
> 2 vols., 8vo. Ill.

WALLACH, OTTO.
>> *See* Kekulé, August, und O. Wallach.

WALLACH, OTTO.
Hülfstafeln für den chemisch-analytischen Unterricht. Bonn, 1879.
Tabellen zur chemischen Analyse, zum Gebrauch im Laboratorium und
>> bei der Repetition. Erster Theil: Verhalten der Elemente und
>> ihrer Verbindungen. Zweiter Theil: Methoden zur Auffindung
>> und Trennung der Elemente. Zweite veränderte Auflage. Bonn,
>> 1889. 8vo.

>> First edition, Bonn, 1880.

>> Tableaux servant à l'analyse chimique, traduits de l'alle-
>> mand par J. Krutwig. Première partie: Caractères des
>> éléments et de leurs combinaisons. Bonn, 1880. 8vo.

WALLENIUS, A. W.
Dissertatio doctrinas de affinitatibus chemicis exhibens. Aboæ, *n. d.*
>> [1819]. 4to.

WALLER, ELWYN [Editor].
>> *See* Cairnes, F. A.

WALLER, RICHARD [Translator].
>> *See* Academie del Cimento.

WALLERIUS, JOHAN GOTTSCHALK.
Åkerbrukets chemiska Grunder. Holmiæ, 1761.
>> In Swedish and Latin. Another edition in Swedish, Stockholm, 1778.

Bref om Chemiens rätta Beskaffenhet och Nytta. Stockholm, 1751.
Chemia physica, första delen, föreställande chemiens natur och beskaf-
>> fenhet i genom dess historia, characterer, instrumenter, operationer
>> och producter. Med Kopparstick. Stockholm, 1759. 3 vols.,
>> 12mo.

>> Andra Delen. 2 vols., 12mo, 1765, 1768.

>> Vol. 1 only also in Latin. Holmiæ, 1760.

>> Physische Chemie, welche von der Natur und Beschaffenheit
>> der Chemie überhaupt, von ihrer Geschichte, Zeichen,
>> sowol leidenden als wirkenden Werkzeugen, und end-
>> lich von den Arbeiten und hervorgebrachten chemischen
>> Cörpern, auf systematische Art handelt. Aus dem
>> schwedischen ins lateinische übersetzt und vermehrt
>> herausgeben von J. G. W., und nummehr ins deutsche
>> übersetzt und mit einigen nöthigen Anmerkungen ver-
>> sehen von Christ. Andr. Mangold. Mit Kupfern.

WALLERIUS, JOHAN GOTTSCHALK. [Cont'd.]
>Schleusingen, 1772. pp. [xii]–470 [and Index], 8vo.
>Four folding plates.

>>In both editions of this work Chapter II treats of the history of chemistry
>>and of its name, pp. 7–21 (and 15–44 of the German edition). Chapter
>>III discusses the symbols and language used by chemists. Chapter XXVI
>>also contains historical notes on the transmutation of metals.

>>A second German edition by Christian Ehrenfried Weigel was published at
>>Leipzig, 1780.

Elementa metallurgiæ speciatim chemicæ. Holmiæ, 1768.
>Zweite Auflage, 1779.

De causa chymificationis. Lund, 1733.

De origine et natura nitri. Holmiæ, 1749.

De origine salium alcalinorum. Holmiæ, 1753.

Dissertatio sistens quæstione, an et quævisque chemia resolvat corpora
naturalia in illas a quibus fuerunt composita partes ? Holmiæ, 1748.

Hydrologia eller Beskrifning af Vatten-Riket. Stockholm, 1748.
>Hydrologie oder Wasserreich, von ihm eingetheilet und
>beschrieben. Nebst einer Anleitung zur Anstellung von
>Wasserproben wie auch dessen Gedanken vom Danne-
>marks Gesundbrunner. Ins deutsche übersetzt von
>Johann Daniel Denso. Berlin, 1751. 8vo.

WALLQUIST, ELOF.

Dissertatio chemica de salibus nonnullis duplicibus ex acido tartarico,
oxydo stibico et oxydis magis electropositivis. Upsaliæ, 1822.

Dissertatio chemicæ præparandi methodi æquationibus explicatæ.
III partes. Upsaliæ, 1822. 4to.

WALQUE, F. DE.

Manuel de manipulations chimiques ou de chimie opératoire. Troisième
édition. Louvain, 1886.

WALTER, ARWED.

Untersuchungen über Molecularmechanik nach analytisch-geometrischer
Methode als matematische Grundlage der chemischen Statik. Berlin, 1873. 8vo.

WANDTAFELN DER ATOMGEWICHTE DER CHEMISCHEN ELEMENTE. $H = 1$.
Litograph. Wien. 2 Nos. Large folio.

WANKLYN, JAMES ALFRED.

Milk-Analysis ; a practical treatise on the examination of milk and its
derivatives, cream, butter and cheese. London, 1874. 8vo.
>Second edition. London, 1886.

WANKLYN, JAMES ALFRED. [Cont'd.]

Tea, Coffee and Cocoa; a practical treatise on the analysis of tea, coffee, cocoa, chocolate, maté (Paraguay tea), etc. London, 1874. 8vo.

The Gas Engineer's Chemical Manual. Second edition. London, 1888. 8vo.

WANKLYN, JAMES ALFRED, AND ERNEST THEOPHRON CHAPMAN.

Water-Analysis; a practical treatise on the examination of potable water. London, 1868. 8vo. Ill.

Second edition, edited by E. T. Chapman. London, 1870.
Third edition, entirely rewritten by J. A. W. London, 1874.
Fourth edition. London, 1876.
Fifth edition. London, 1879.
Sixth edition. London, 1884.
Seventh edition. London, 1889.
Eighth edition. London, 1891. 8vo. Ill.

WANKLYN, JAMES ALFRED, AND W. J. COOPER.

Air Analysis; a practical treatise on the examination of air. With an appendix on illuminating gas. London, 1890. 8vo.

Bread Analysis; a practical treatise on the examination of flour and bread. London, 1881. 8vo.

New edition. London, 1886.

WARD, GEORGE MASON.

A Compend of Inorganic Chemistry, with table of elements. Philadelphia, 1883. pp. 111, 12mo.

WARD, T. F.

First Lessons in Inorganic Chemistry. Manchester, 1870. 8vo.
Outline Facts of Chemistry, with exercises. Manchester, 1866. 2 parts, 8vo.

WARDEN, C. J. H., AND L. A. WADDELL.

The Non-bacillar Nature of Abrus-poison, with observations on its chemical and physiological properties. Calcutta, 1884.

WARE, LEWIS S.

The Sugar Beet; including a history of the beet sugar industry in Europe, varieties of the sugar beet, examination, soils, tillage, seeds and sowing, yield and cost of cultivation, harvesting, transportation, conservation, feeding qualities of the beet and of the pulp, etc. Philadelphia, 1880. 12mo. Ill.

WARINGTON, R.

The Chemistry of the Farm.   London and New York, 1882.   pp. 120,
12mo.

> Second edition.   London, 1882.
> Fourth edition.   London, 1886.
> Sixth edition.   London, 1891.
>
> Of the series known as Morton's Handbooks of the Farm.

WARLTIRE, J.

Synopsis of Natural and Artificial Chemistry.   *n. p., n. d.*   Fol.

WARNFORD, C. G., B. E. R. AND J. A. E. NEWLANDS.

Sugar, a Handbook for Planters and Refiners.   London, 1888.   Ill.

WARTHA, VINCENZ.

Die qualitative Analyse mit Anwendung der Bunsen'schen Flammen-
reactionen.   Zürich, 1867.   8vo.

> Précis d'analyse qualitative, voie humide et réactions de la
> flamme selon Bunsen . . .   1877.   12mo.
>
> *Cf.* Bunsen, Robert Wilhelm.

ВАРВИНСКІЙ Г.

Начальныя основанія всеобщей химіи.   Сост. по системѣ Тенаре.   С.-Петер-
бургъ, 1832, 3 части.

> WARVINSKY, I.   The First Principles of General Chemistry according to
> the System of Thénard.   St. Petersburg, 1832.   3 vols.

WASSERBERG, FRANZ AUGUST XAVER VON.

Beiträge zur Chemie.   Wien, 1791.   8vo.

Chemische Abhandlung vom Schwefel.   Wien, 1788.   8vo.

Institutiones chemicæ.   Vindobonæ, 1773–80.   3 vols., 8vo.   Vol. 1:
Regnum animale.   1773.   Vol. II: Regnum minerale.   1778.   Vol.
III : Inflammabilia.   1780.

Von dem Nutzen die Luft rein u. s. w. zu halten.   Wien, 1772.   8vo.

WATSON, RICHARD [BISHOP OF LLANDAFF].

Essay (An) on the Subjects of Chemistry and their general divisions.
London, 1771.   8vo.

Institutiones metallurgicæ.   London, 1768.   8vo.

Chemical Essays.   London, 1782.

> Second edition.   5 vols., 12mo.   Vol. I, [x]–349 ; II, 368 ;
> III, 376 ; IV, [xxiv]–354–[50] ; v, v–375.
> Fourth edition.   London, 1787.
> Fifth edition.   London, 1789.
> Seventh edition.   London, 1800.
>
> Vol. I contains an Essay on the Rise and Progress of Chemistry, pp. 1–48.

WATSON, WILLIAM H.

Science Teachings in Living Nature. Introduction to the study of Physiological Chemistry and Sanitary Science. London, 1879. 8vo.

WATT, ALEXANDER.

Art (The) of Leather Manufacture. Second edition. London, 1887.

Art (The) of Soap-making ; a practical handbook of the manufacture of hard and soft soaps. Second edition. London, 1885. 8vo.

> Practical (A) Handbook of the manufacture of hard and soft soaps, etc., including many new processes, and a chapter on the recovery of glycerine from waste leys, with illustrations. Third edition, revised. London, 1889. 8vo.

Electro-Deposition. Practical treatise on the electrolysis of gold, silver, copper, nickel, and other metals and alloys. With descriptions of voltaic batteries, magneto- and dynamo-electric machines, thermopiles, and the materials and processes used in every department of art, and several chapters on electro-metallurgy. London, 1886. 8vo.

> Third edition. London, 1889.

> Tien chi tu chin leo fa. [Translated into Chinese by John Fryer.] 2 vols.

> *Cf.* Fryer, John.

Scientific Industries explained, showing how some of the important articles of commerce are made. Edinburgh, 1881. 8vo.

WATT, JAMES.

* Supplement to the description of a Pneumatic Apparatus, for preparing factitious airs ; containing a description of a Simplified Apparatus, and of a Portable Apparatus. Birmingham, 1796. pp. 47, 8vo. Plate.

> *Cf.* Beddoes, Thomas, and James Watt.

WATTS, HENRY.

Chemistry of Carbon Compounds, or Organic Chemistry. Thirteenth edition, revised. Philadelphia, 1887. 12mo.

> This is an American edition of the second volume of Watts's Manual of Chemistry. *Cf.* Fownes, George.

Manual of Chemistry. Thirteenth edition of Fownes'. London, 1886. 2 vols., 12mo.

> Second edition. Fourteenth of Fownes'. By William A. Tilden. London, 1889. 2 vols., 12mo.

> *See* Fownes, George. *Also, in Section II*, Watts, Henry.

WATTS, W. MARSHALL.
  Index of Spectra. With a preface by H. E. Roscoe. London, 1872.
    pp. xvi–74, 8vo. 9 plates.
  Practical (A) Introduction to the Elements of Chemistry. London,
    1891.
      Nisbet's Elementary Science Manuals.

WEBER, C.
  Leitfaden für den Unterricht in der landwirthschaftlichen Chemie.
    Stuttgart, 1888.
          Dritte vermehrte Auflage. Stuttgart, 1892. 8vo. Ill.

WEBER, CHRISTOPH.
  Tractatio chemica de pyrophoro. Göttingæ, 1758. 4to.

WEBER, JACOB ANDREAS.
  Beschreibung einiger zum Gebrauch der dephlogistirten Luft beim
    Blaserohr eingerichteten Maschinen . . . Tübingen, 1785. 8vo.
  Chemische Erfahrungen bei meinen und anderen Fabriken . . . Neu-
    wied, 1793.
  Kurze Anweisung für einen Anfänger der Apothekerkunst und Chemie.
    Berlin, 1779. 8vo.
          Zweite Auflage. Berlin, 1785.
  Leichtfassliche Chemie für Handwerker . . . Tübingen, 1790.
          Zweite Auflage, 1793.
  Neuentdeckte Natur und Eigenschaften des Kalkes und der äzenden
    Körper, nebst einer œkonomisch-chemischen Untersuchung des
    Kochsalzes und dessen Mutterlauge. Berlin, 1778. 8vo.
  Physikalisch-chemische Untersuchungen der thierischen Feuchtigkeiten.
    Tübingen, 1781. 8vo.
  Vollständige theoretische und praktische Abhandlung von dem Salpeter
    und der Zeugung derselben, nebst einer Abhandlung von der
    Gährung durch physische und chemische Grundsätze und Er-
    fahrungen bestätigt. Tübingen, 1779. 8vo.

WEBER, K.
  Die Malz-Fabrikation. Wien, 1886.

ВЕБЕРЪ К. К.
  Для вседневной жизни. Знакомство съ химіею при помощи домашнихъ
    средствъ. С.-Петербургъ, 1876.
          WEBER, K. K. Household Chemistry. St. Petersburg, 1876.

WEBER, R.
  Atomgewichts-Tabellen zur Berechnung der bei analytisch-chemischen
    Untersuchungen erhaltenen Resultate. Braunschweig, 1852. 8vo.

WEBSTER, JOHN.

* Metallographia, or an history of Metals. Wherein is declared the signs of ores and minerals both before and after digging, the causes and manner of their generations, their kinds, sorts, and differences ; with the description of sundry new metals, or semi-metals, and many other things pertaining to mineral knowledge. As also, the handling and shewing of their vegetability, and the discussion of the most difficult questions belonging to mystical chemistry ; as of the Philosophers gold, their mercury, the liquor alkahest, aurum potabile and such like. Gathered forth of the most approved that authors have written in Greek, Latine, or High Dutch ; with some observations and discoveries of the author himself. London, 1671. pp. (xvi)–388, sm. 4to.

> Not so prolix a work as the title suggests. The author is, however, credulous, a follower of Paracelsus and a believer in hermetic mysteries.

WEBSTER, JOHN WHITE.

A Manual of Chemistry, containing the principal facts of the science in the order in which they are discussed and illustrated in the lectures at Harvard University, and several other colleges and medical schools in the United States. Compiled and arranged as a text-book for the use of students, and persons attending lectures on chemistry. Third edition, comprising a summary of the latest discoveries as contained in the works of Brande, Turner, Thomson and other distinguished chemists, illustrated with upwards of two hundred engravings on wood. Boston, 1839. pp. xxi–554, 8vo.

> First edition, Boston, 1826. Second edition, Boston, 1828. pp. viii–619.
> The author was hanged Nov. 23, 1849, for the murder of Dr. Parkman.

WEBSTER, N. B.

Outlines of Chemistry for Agricultural Colleges, public and private schools, and individual learners. New York, 1883. 32mo.

WEDDERBURN, ALEX. J.

Popular (A) Treatise on the Extent and Character of Food Adulterations. Bulletin No. 25, Division of Chemistry, U. S. Department of Agriculture. Washington, 1890. 8vo.

Special Report on the Extent and Character of Food Adulterations, including State and other laws relating to foods and beverages. Bulletin No. 32, Division of Chemistry, U. S. Department of Agriculture. Washington, 1892. 8vo.

WEDDING, HERMANN.

Ausführliches Handbuch der Eisenhüttenkunde. *See* Percy, John: Die Metallurgie, vol. II.

WEDDING, HERMANN.

Der Phosphor im Haushalte der Natur und der Menschen. Vortrag.
Berlin, 1884. 4to.

WEDEL, GEORG WOLFGANG.
See also in Section VI.

WEDEL, GEORG WOLFGANG.

Compendium chymiæ theoreticæ et practicæ, methodo analytica
propositæ. Jenæ, 1715. 4to.

Experimenta chymica de sale volatili plantarum. Jenæ, 1672. 12mo.

De anil, indico, glasto. Inaug.-dissert. Jenæ, 1689. 4to.

De clave principiorum chimicorum. Jenæ, 1685. 4to.

Non-entia chymica. Francofurti, 1670.

Tabulæ chymicæ XV in synopsi universam chymiam exhibentes. Jenæ,
1692. 4to.

WEDEL, JOHANN ADOLPH.

De fermentis chimicis, dissertatio. Jenæ, 1695. 4to.

Dissertatio inauguralis chimico-medica de principiis chimicorum. *n. p.*
[Jenæ], 1716. 4to.

WEEREN, J. M.

Atomgewichtsbestimmung des Mangans. Halle, 1890. 8vo.

WEGSCHEIDER, R.

Zur Regelung der Nomenclatur der Kohlenstoffverbindungen. Wien,
1892. 8vo.

WEHNER.

Leitfaden der Chemie mit besonderer Berücksichtigung der landwirthschaftlichen Gewerbe. Zum Gebrauche an Real- und landwirthschaftsschulen bearbeitet. Berlin, 1883. 8vo. Ill.

WEIGEL, CHRISTIAN EHRENFRIED.
Grundriss der reinen und angewandten Chemie. 1777. *See in Section I.*
Beiträge zur Geschichte der Luftarten. 1784. *See in Section III.*

WEIGEL, CHRISTIAN EHRENFRIED.

Einfluss (Der) chemischer Kenntniss in der Oekonomie. Greifswald,
1775. 4to.

*Einleitung zur allgemeinen Scheidekunst. Leipzig, 1788-94. Drei
Theile. 8vo.
Part III is in two volumes. Valuable for its bibliography.

Observationes chemicæ et mineralogicæ. Göttingæ, 1771-73. 2 parts.
4to.

WEIGEL, CHRISTIAN EHRENFRIED.   [Cont'd.]
>   Chemisch-mineralogische Beobachtungen, aus dem lateinischen übersetzt und mit vielen Zusätzen vermehrt von Johann Theodor Pyl. Breslau, 1779. pp. [vi]-182-[ii], 8vo. Ill. 2 folding plates.
Vom Nutzen der Chemie. . . . Greifswald, 1774. 4to.

WEIGMANN, H.
>   Die Methoden der Milchconservirung speciell das Pasteurisiren und Sterilisiren der Milch. Bremen, 1893. 8vo. Ill.

WEIHRICH, G.
>   Die Ansichten der neueren Chemie. Mainz, 1872. 8vo.

WEILL-GOETZ, L., ET F. DESOR.
>   Traitement des eaux ammoniacales et matières épurantes épuisées provenant des usines à gaz. Strasbourg, 1889. pp. 251, roy. 8vo.
>>   Contains a brief bibliography, consisting of one-line titles without places or dates. Such a bibliography [?] is a mere aggravation.

WEIN, ERNST.
>   Agriculturchemische Analyse. Stuttgart, 1889. 8vo.
>   Tabellen zur quantitativen Bestimmung der Zuckerarten. Nebst erläuterndem Text. Stuttgart, 1888. pp. xi-55, 12mo.

WEINHOLZ, WILHELM.
>   Anweisung zum Gebrauch der von ihm entworfenen verschiebbaren chemischen Aequivalenten-Scale, für Aerzte, Apotheker, Hüttenleute, für theoretische und practische Chemiker überhaupt. Braunschweig, 1830. 8vo.
>   Technisch-chemisches Handbuch der Erforschung, Ausscheidung und Darstellung des in den Künsten und Gewerben gebräuchlichen metallischen Gehalts der Mineral-Körper, unter steter Berücksichtigung sämmtlicher bis jetzt in der Chemie gemachten Erfahrungen ; zum Selbststudium, besonders für angehende Hüttenbeamte, Cameralisten, etc. Hannover, 1830. 8vo.
>   Verschiebbare chemische Aequivalenten-Skala für Aerzte, Apotheker, Hüttenleute, Fabrikanten. Eine Tafel nebst erläuterndem Text. Braunschweig, 1830.

WEINLIG, CHRISTIAN ALBERT.
>   Examen theoriæ electro-chemico-atomisticæ. Lipsiæ, 1840. 8vo.
>   Lehrbuch der theoretischen Chemie. Zum Gebrauche bei Vorlesungen zur Repetition für Studierende. Leipzig, 1840-41.

WEINLIG, CHRISTIAN ALBERT. [Cont'd.]

Pflanzenchemie (Die), ein Handbuch für Aerzte und Apotheker. Unter theilweiser Zugrundelegung von Thomson's Organic Chemistry . . . bearbeitet. Leipzig, 1839. 8vo.

WEINRICH.

Kurze Anleitung zum Anbau der Runkelrüben. Prag, 1835.

WEINSTEIN, B.

Capillaritäts-Untersuchungen und ihre Verwendung bei der Bestimmung der alkoholometrischen Normale. Berlin, 1889. 4to.

Ueber die Bestimmung von Aräometern mit besonderer Anwendung auf die Feststellung der deutschen Urnormale für Alkoholometer. Berlin, 1890. 4to.

WEINZIERL, THEODORE VON.

Die qualitative und quantitative mikroscopische Analyse. Eine neue Untersuchungsmethode der Mahlproducte auf deren Futterwerth und Verfälschungen. Wien, 1887. 8vo.

ВЕЙРИХЪ.

Воззрѣнія современной химіи, Перев. Лесгафта Р. С.-Петербургъ, 1874.

WEIRICH. Views of Modern Chemistry. St. Petersburg, 1874.

WEISS, TH.

Die Darstellung künstlicher Mineralwasser. Friedrichshafen, 1871. 8vo.

WELL, JOHANN JACOB.

Rechtfertigung der Blackischen Lehre von der figirten Luft gegen die vom Herrn Wiegleb . . . darwider gemachten Einwürfe. Wien, 1771. 8vo.

Cf. Wiegleb, Johann Christian.

WELLS, DAVID A.

Principles and Applications of Chemistry. New York, 1859. 8vo. Ill.

Passed through many editions.

Hwa hio chien yuan. [Translated into Chinese by John Fryer.] 1870. 4 vols.

WELTZIEN, CARL.

Systematische Zusammenstellung der organischen Verbindungen. *See in Section II.*

WELTZIEN, CARL.

Grundriss der theoretischen Chemie insbesondere für Artillerie- und Ingenieur-Officiere bearbeitet. Carlsruhe, 1854. 8vo. Ill.

WELTZIEN, CARL. [Cont'd.]
Vorträge über Chemie. Gehalten im Gewerbverein zu Carlsruhe. Nach einem nachgeschriebenen Heft herausgegeben vom Gewerbverein. Carlsruhe, 1846. 8vo.

WENDT, GUSTAV.
Entwickelung (Die) der Elemente. Entwurf zu einer biogenetischen Grundlage für Chemie und Physik. Berlin, 1891. pp. 49, 8vo. Folding table.

> The author collects the evidence in favor of the non-primordial character of the elements, and constructs a genealogical table of the same. "Herr Wendt's work is entitled to the most careful study of every chemist who is really concerned about a truly rational development of his science." —*Chem. News.*

WENGHÖFFER, L.
Kurzes Lehrbuch der Chemie der Kohlenstoffverbindungen. Stuttgart, 1882. 8vo.
Lehrbuch der anorganischen, reinen und technischen Chemie. Stuttgart, 1883–84. 2 parts.

WENZEL, CARL FRIEDRICH.
Chymische Untersuchung des Flussspathes. Dresden, 1783. 8vo.
Chymische Versuche die Metalle vermittelst die Reverberation in ihre Bestandtheile zu zerlegen. Kopenhagen, 1781. 4to.

> This Essay was awarded a prize by the Royal Danish Academy of Sciences.

Einleitung zur höheren Chemie. Leipzig, 1774.
* Lehre von der Verwandschaft der Körper. Dresden, 1777. pp. 491, 12mo.

Zweite Auflage. Dresden, 1782.
[Another edition] mit Anmerkungen herausgegeben von David Hieronimus Grindel. Dresden, 1800. 8vo.

> For a critical analysis of this important work in chemical theory, see R. A. Smith's Memoir of John Dalton, pp. 160–166.

WERBER, ANTON.
Lehrbuch der praktischen Toxicologie zum Selbstudium und zum Gebrauch für Vorlesungen. Erlangen, 1869. pp. viii–140, 8vo.

WERKHOVEN, PIETER VAN.
Beknopt handboek der scheikunde. Utrecht, 1809. 8vo.
Nieuwe chemische en physische oefeningen voor de beminnaars der schei- en natuurkunde. Amsterdam, 1797–1800. 2 vols, 8vo.

WERNER, A.
Ueber räumliche Anordnung der Atome in stickstoffhaltigen Molekülen. Zürich, 1890. 8vo.

WERNERUS, ABRAHAM.
Oratio de confectione ejus potus qui Germaniæ usitatus veteri vocabulo, secundum Plinium, cerevisia vocatus. Witebegæ, 1567. 8vo.

WERNICH, A.
Grundriss der Desinfectionslehre, zum praktischen Gebrauch. Auf kritischer und experimenteller Grundlage bearbeitet. Wien, 1880. 8vo. Ill.
Zweite Auflage. Wien, 1882. 8vo. Ill.

WERTHER, GUSTAV.
Die unorganische Chemie, ein Grundriss für seine Vorlesungen an der Artillerie- und Ingenieur-Schule in Berlin. Berlin, 1850–'52. 8vo, Ill. Section I : Die unorganische Chemie mit Ausnahme der zur Artillerie- und Ingenieur-Technik gehörigen Gegenstände. Section II : Zur Artillerie und Ingenieur-Technik gehörigen Gegenstände.
Zweite Auflage. Berlin, 1863. 8vo.

WESELSKY, P., UND P. BENEDIKT.
Dreissig Uebungsaufgaben als erste Anleitung zur quantitativen Analyse. Wien, 1883. 8vo.
Zweite Auflage. Wien, 1892. 8vo. Ill.

WESENER, FELIX.
Lehrbuch der chemischen Untersuchungs-Methoden zur Diagnostik innerer Krankheiten. Berlin, 1890. pp. viii–280, 8vo. Ill.
Wreden's Sammlung kurzer medicinischer Lehrbücher, vol. XV.

WEST, GRATIEN.
Statistique des volumes des équivalents chimiques et d'autres données relatives à leurs propriétés physiques ; suivie d'un mémoire sur quelques questions moléculaires. Paris, 1873. 8vo.

WESTERGAARD, H. B.
Kortfattet Lærebog i den uorganiske Chemie. Kjøbenhavn, 1853. 8vo. Ill.

WESTRUMB, JOHANN FRIEDRICH.
Geschichte der . . . Metallisirung der . . . Erdarten. *See in Section III.*

WESTRUMB, JOHANN FRIEDRICH.
Bemerkungen und Vorschläge für Bleicher. Hannover, 1800.
Also under the title : Kleine physikalisch-chemische Abhandlungen, vol. VI, part 2.
Bemerkungen und Vorschläge für Branteweinbrenner. Hannover, 1793. 8vo.
Dritte Auflage. 1803.

WESTRUMB, JOHANN FRIEDRICH. [Cont'd.]

Bemerkungen und Vorschläge für Fruchtbranteweinbrenner. Hannover, 1821. 8vo.

Beschreibung des Gesundbrunnens in Selters, dem Herrn Wurzer zur Prüfung vorgelegt. Marburg, 1813. 8vo.

Beschreibung einer sehr vortheilhafter Essigfabrik. Hannover, 1818. 8vo.

Kleine physikalisch-chemische Abhandlungen. Leipzig, 1785–1800. 6 vols., 8vo.

Kleine Schriften physikalischen, chemischen und technischen Inhalts. Hannover, 1805. 8vo. Ill.

> *Also under the title :* Beschreibung der Gesundbrunnen und der Schwefel-bäder zu Eilsen in der Grafschaft Lippe-Schaumburg.

Materialen für Branntweinbrenner, oder Bemerkungen und Vorschläge über die Verbesserung des Brenngeschäfts und über die Veredlung des gemeinen Fruchtbranntweins zu Weinbranntwein, Rum, Arak und Liqueure. Aus des Verfassers nachgelassenen Papieren herausgegeben von A. H. L. Westrumb. Hannover, 1826. 8vo.

Physikalisch-chemische Beschreibung der Mineralquellen zu Pyrmont. Leipzig, 1789. 8vo.

Ueber das Bleichen mit Säuren nach französischen und englischen Vorschriften. Nebst Beschreibung des besten Bleichverfahrens. Berlin, 1819. 8vo.

Ueber Glasbereitung, dessen Verbesserung, Verwohlfeilung, etc. Hannover, 1818. 8vo.

Versuch eines Beitrages zu den Sprachbereicherungen für die deutsche Chemie. Hannover, 1793. 8vo.

WETHERILL, CHARLES M.

A Brief Sketch of the Modern Theory of Chemical Types. Smithsonian Institution. Washington, 1863. pp. 153–168.

Lecture-Notes on Chemistry. Bethlehem, Pennsylvania, 1868. 8vo. In three parts. Part I : The Metalloids. pp. 73. Part II : The Metals. pp. 112. Part III : Organic Compounds. pp. 111.

WEYDE, FRANZ.

Anleitung zur Herstellung von physikalischen und chemischen Apparaten. Wien, 1882. 36 plates.

WEYL, THEODOR.

Analytisches Hilfsbuch für die physiologisch-chemischen Uebungen in Tabellenform. Berlin, 1882. 8vo.

Lehrbuch der organischen Chemie für Mediciner. Berlin, 1891. pp. xii–587, 8vo. Ill.

WEYL, THEODOR. [Cont'd.]

Theerfarben (Die) mit besonderer Rücksicht auf Schädlichkeit, hy-
gienisch- und forensisch-chemisch untersucht.   Berlin, 1889.

The Coal-tar Colors, with especial reference to their in-
jurious qualities and the restriction of their use.   A
sanitary and medico-legal investigation.   With a preface
by [Eugen] Sell.   Translated with permission of the
author by H. Leffmann.   Philadelphia and London,
1893.

WHARTON, FRANCIS, AND ALFRED STILLÉ.

Medical Jurisprudence.   Philadelphia, 1873.   *2 vols.*

Fourth edition, edited by Robert Amory and Edward S.
Wood.   Philadelphia, 1882–84.   3 vols., roy. 8vo.

Vol. I also under the title : A Treatise on Mental Unsoundness.   By F. W.

WHEELER, C. GILBERT.

Medical Chemistry.   Third edition, revised.   Chicago, 1891.   8vo.

Outlines of Modern Chemistry, organic.   Based in part upon Riche's
Manuel de chimie.   New York and Chicago, 1877.   pp. 231,
12mo.

WHITE, A. H. S.

Chemical Analysis for Schools and Science-Classes.   Qualitative-Inorganic, adapted to meet the requirements of the London preliminary
scientific and intermediate B. Sc., the locals and the South-Kensing-
ton practical chemistry.   London, 1883.   8vo.

WHITE, ROBERT.

A Summary of the Pneumato-Chemical Theory, with a table of its
nomenclature intended as a supplement to the analysis of the new
London Pharmacopœia.   London, 17—.   8vo.

WHITELEY, R. LLOYD.

Chemical Calculations, with Explanatory Notes.   Problems and answers,
specially adapted for use in Colleges and Science Schools.   With a
Preface by F. Clowes.   London, 1892.   8vo.

WICKE, WILHELM.

Anleitung zur chemischen Analyse.   Nebst Beispielen für Anfänger und
Geübtere.   Braunschweig, 1857.   pp. 467, 8vo.   Ill.

WICKERSHEIMER [*sic*].

Aluminium [Industry].   *See, in Section II,* Fremy: Encyclopédie chimique,
vol. v.

WIECHMANN, FERDINAND G.

Sugar Analysis. For refineries, sugar-houses, experimental stations, etc., and as a handbook of instruction in schools of chemical technology. New York, 1890. pp. viii–187, 8vo. Ill.

ВИКЕ В.

Руководство къ химическому анализу съ примѣрами. Пер. Ломоносова А. С.-Петербургъ, 1863.

WIECKE, W. Manual of Chemical Analysis, with illustrations. St. Petersburg, 1863.

WIEDEMANN, G.

Das Beleuchtungswesen. *See* Bolley's Handbuch, etc.

WIEGLEB, JOHANN CHRISTIAN.

Chemische Versuche über die alkalischen Salze. Berlin und Stettin, 1774. 8vo.

Neue verbesserte und vermehrte Auflage. Berlin und Stettin, 1781. pp. 264, 8vo.

Handbuch der allgemeinen Chemie. Berlin und Stettin, 1781. 2 vols., 8vo.

Zweite Auflage. Berlin und Stettin, 1786. 2 vols., 8vo.

Dritte Auflage. Berlin und Stettin, 1796. 2 vols., 8vo.

A General System of Chemistry, theoretical and practical. Digested and arranged with a particular view to its application to the arts. Taken chiefly from the German of M. Wiegleb. By C. R. Hopson. London, 1789. pp. viii–670–[xlvi]. Plates.

Kleine chymische Abhandlungen von dem grossen Nutzen der Erkenntniss des Acidi pinguis bey der Erklärung vieler chymischen Erscheinungen. Nebst einer Vorrede worinnen Meyers Leben erzählt und von dessen Verdiensten gehandelt wird von E. G. Baldinger. Langensalza, 1767. 8vo.

Fortgesetzte kleine Abhandlungen. Langensalza, 1770.

Neue Begriffe von der Gährung und den ihr unterwürfigen Körpern. Weimar, 1776. 8vo.

Revision der Grundlehren von der chemischen Verwandschaft der Körper. Erfurt, 1780. 4to.

Vertheidigung der Meyerischen Lehre vom Acidi pingui gegen verschiedene darwider gemachten Einwürfe. Altenburg, 1770. 8vo.

See also in Section III.

WIEGMANN, A. F., UND L. POLSTORFF.

Ueber die anorganischen Bestandtheile der Pflanzen, oder Beantwortung der Frage : sind die anorganischen Elemente, welche sich in der

WIEGMANN, A. F., UND L. POLSTORFF.   [Cont'd.]

Asche der Pflanzen finden, so wesentliche Bestandtheile des vegetabilischen Organismus, dass dieser sie zu seiner völligen Ausbildung
bedarf und werden sie den Gewächsen von Aussen dargeboten?
Eine in Göttingen im Jahre 1842 gekrönte Preisschrift, nebst
einem Anhange über die fragliche Assimilation des Humusextractes.
Braunschweig, 1842.   8vo.

WIENER, FERDINAND.

Lederfärberei (Die) und die Fabrikation des Lackleders.   Ein Handbuch für Lederfärber und Lackirer.   Anleitung zur Herstellung
aller Arten von färbigem Glacéleder nach dem Anstreich- und
Tauchverfahren, sowie mit Hilfe der Theerfarben, zum Färben von
schwedischem sämischgarem und lohgarem Leder, zur Saffian-,
Corduan-, Chagrinfärberei, etc., zur Fabrikation von schwarzem
und färbigem Lackleder.   Wien, Pest, Leipzig, 1881.   8vo.   Ill.

Lohgerberei (Die), oder die Fabrikation des lohgaren Leders.   Ein
Handbuch für Lederfabrikanten.   Enthaltend die ausführliche
Darstellung der Fabrikation des lohgaren Leders nach dem gewöhn-
lichen und dem Schnellgerberverfahren, nebst der Anleitung zur
Herstellung aller Gattungen Maschinen-riemenleder, des Juchten-,
Saffian-, Corduan-, Chagrin- und Lackleders.   Wien, Pest, Leipzig,
1877.   8vo.   Ill.

Weissgerberei (Die), Sämischgerberei und Pergamentfabrikation.   Ein
Handbuch für Lederfabrikanten, enthaltend die ausführliche Darstellung der Fabrikation des weissgaren Leders nach allen Verfahrungsarten, des Glacéleders, Seifenleders, etc. ; der Sämischgerberei,
der Fabrikation des Pergamentes und der Lederfärberei, mit besonderer Berücksichtigung der neuesten Fortschritte auf dem
Gebiete der Lederindustrie.   Wien, Pest, Leipzig, 1877.   8vo.   Ill.

WIENER, J.

Compendium der Chemie für Mediciner und Pharmaceuten vorzüglich
zur Repetition für die strengen Prüfungen.   Wien, 1863.   pp. xvi-
164.   8vo.

WIESNER, JULIUS.

Rohstoffe (Die) des Pflanzenreichs.   Leipzig, 1873.   8vo.   Ill.

Technisch (Die) verwendeten Gummiarten, Harze und Balsame.   Ein
Beitrag zur wissenschaftlichen Begründung der technischen Waarenkunde.   Erlangen, 1869.   8vo.   Ill.

WIGGERS, AUGUST.

Chemische Untersuchung der Pyrmonter Eisensäuerlinge.   Hannover,
1857.   8vo.

WIGGERS, AUGUST. [Cont'd.]

Grundriss der Pharmacognosie. Göttingen, 1840. 8vo.

Vierte Auflage. Göttingen, 1857.

Trennung (Die) und Prüfung metallischer Gifte aus verdächtigen organischen Substanzen ; mit Rücksicht auf Blausäure und Opium. Göttingen, 1833. 8vo.

WILBRAND, FERDINAND.

Grundzüge der Chemie nach inductiver Methode. Hildesheim, 1885.

Leitfaden für den methodischen Unterricht in der anorganischen Chemie. Zweite umgearbeitete Auflage. Hildesheim, 1875. 8vo.

Fünfte Auflage. Hildesheim, 1882. Ill.

Sechste Auflage. Hildesheim, 1892. 8vo. Ill.

Ziel und Methode des chemischen Unterrichts. Hildesheim, 1881.

WILBRAND, JULIUS, UND FERDINAND WILBRAND.

Leitfaden für die ersten Uebungen im chemischen Laboratorium. Zum Gebrauch an höheren Mittelschulen zusammengestellt. Neuwied, 1867. 16mo.

Leiddraad bij de eerste oefeningen in het scheikundig laboratorium. Ten gebruike aan scholen voor middelbaar onderwijs. Naar het Hoogduitsch door R. J. Opwijrda. Utrecht, 1867. 12mo.

WILDE, H.

On the Origin of Elementary Substances, and on some new relations of their atomic weights. [Also with the German title :] Ueber den Ursprung der elementaren Körper und über einige neue Beziehungen ihrer Atomgewichte. London, 1892. Imp. 4to.

The same in English and French. London, 1892.

WILDE, P. DE.

Traité élémentaire de chimie générale et descriptive. Troisième édition. Bruxelles, 1884. 2 vols., 12mo.

WILDT, E.

Katechismus der Agricultur-Chemie. Sechste Auflage, neu bearbeitet unter Benutzung der fünften Auflage von Hamm's Katechismus der Ackerbauchemie, der Bodenkunde und Düngerlehre. Leipzig, 1884. 8vo.

For earlier editions see Hamm, Wilhelm.

WILEY, HARVEY W.

Analytical Processes used in the general qualitative examination of liquids and solids in the Laboratories of Purdue University and Medical College of Indiana. Indianapolis, 1875. pp. 23, 12mo.

WILEY, HARVEY W.   [Cont'd.]

Diffusion, its application to sugar-cane and record of experiments with sorghum in 1883.   Bulletin of the Department of Agriculture, Chemical Division ; No. 2.   Washington, D. C., 1884.   8vo.

Economical (The) aspects of Agricultural Chemistry.   An Address before the American Association for the Advancement of Science, at the Buffalo meeting, August, 1886.   Cambridge, 1886.   8vo.

Experiments with diffusion and carbonatation at Ottawa, Kansas.   Campaign of 1885.   Bulletin No. 6, Division of Chemistry, Department of Agriculture.   Washington, 1885.   pp. 20, 8vo.

Experiments with Sugar Beets in 1890.   Bulletin No. 30, Division of Chemistry, U. S. Department of Agriculture.   Washington, 1891. pp. 93, 8vo.

Foods and Food Adulterants.   Part 1, Dairy Products.   Bulletin No. 13, U. S. Department of Agriculture, Division of Chemistry. Washington, 1887.   8vo.

Methods and Machinery for the application of diffusion to the extraction of sugar from sugar-cane and sorghum and for the use of lime, and carbonic and sulphurous acids in purifying the diffusion juices.   Bulletin No. 8, Division of Chemistry, Department of Agriculture.   Washington, 1886.   pp. 85, 8vo.   24 folding plates.

Northern (The) Sugar Industry : A Record of its progress during the season of 1883.   Bulletin No. 3, Chemical Division, Department of Agriculture.   Washington, 1884.   pp. 118, 8vo.   11 folding plates.

WILEY, HARVEY W. [Editor.]

Record of Experiments conducted by the Commissioner of Agriculture in the manufacture of sugar from sorghum and sugar-canes at Fort Scott, Kansas ; Rio Grande, New Jersey and Lawrence, Louisiana, 1887–88.   Bulletin No. 17, Division of Chemistry, U. S. Department of Agriculture.   Washington, 1888.   pp. 118, 8vo.   Plates.

Record of Experiments at Fort Scott, Kansas, in the manufacture of sugar from sorghum and sugar-canes in 1886.   U. S. Department of Agriculture, Division of Chemistry, Bulletin No. 14.   Washington, 1887.   8vo.

Record of Experiments conducted by the the Commissioner of Agriculture in the manufacture of sugar from sorghum at Rio Grande, New Jersey ; Kenner, Louisiana ; Conway Springs, Douglass and Sterling, Kansas, 1888.   Bulletin No. 20, Division of Chemistry, U. S. Department of Agriculture.   Washington, 1889.   pp. 162, 8vo.

Record of Experiments in the production of sugar from sorghum in 1889 at Cedar Falls, Iowa ; Rio Grande, New Jersey ; Morrisville, Virginia ; Kenner, Louisiana ; College Station, Maryland ; and

WILEY, HARVEY W. [Cont'd.]

Conway Springs, Attica, Medicine Lodge, Ness City, Liberal, Arkalon, Meade, Minneola, and Sterling, Kansas. Bulletin No. 26, Division of Chemistry, U. S. Department of Agriculture. Washington, 1890. pp. 112, 8vo.

Record of Experiments with sorghum in 1890. Bulletin No. 29, Division of Chemistry, U. S. Department of Agriculture. Washington, 1891. pp. 125, 8vo.

Sugar Beet (The) Industry. Culture of the sugar beet and manufacture of beet sugar. Bulletin No. 27, Division of Chemistry, U. S. Department of Agriculture. Washington, 1890. pp. 262, 8vo. Ill. Folding plates.

Sugar Industry (The) of the United States. Bulletin No. 5, Division of Chemistry, U. S. Department of Agriculture. Washington, 1886. 8vo.

Sugar producing Plants. Record of analyses made by authority of the Commissioner of Agriculture under direction of the Chemist, 1887–'88. Sorghum : Fort Scott, Kansas ; Rio Grande, New Jersey. Sugar-Cane : Lawrence, Louisiana. Together with a study of the data collected on sorghum and sugar-cane. Bulletin No. 18, Division of Chemistry, U. S. Department of Agriculture. Washington, 1888. pp. 132, 8vo.

See also, in Section VII, Association of Official Agricultural Chemists.

WILEY, HARVEY W., CLIFFORD RICHARDSON, C. A. CRAMPTON, AND GUILFORD L. SPENCER.

Foods and Food Adulterants. By direction of the Commissioner of Agriculture. Bulletin No. 13, Division of Chemistry, U. S. Department of Agriculture. Washington, 1887 to 1892. Seven parts, 8vo.

Part I. Dairy Products, by H. W. Wiley. 1887. pp. 132.

Part II. Spices and Condiments, by Clifford Richardson. 1887. pp. 130.

Part III. Fermented Beverages, by C. A. Crampton. 1887. pp. 140.

Part IV. Lard and Lard Adulterants, by H. W. Wiley. 1889. pp. 184.

Part V. Baking Powders, by C. A Crampton. 1889. pp. 73.

Part VI. Sugar, Molasses and Sirup, Confections, Honey and Beeswax. 1892. pp. *from* 633–874–ix, 8vo.

Part VII. Tea, Coffee and Cocoa Preparations, by Guilford L. Spencer, with the collaboration of Ervin E. Ewell. 1892. pp. viii, *from* 875–1013–vi, 8vo. Ill.

Contains a Bibliography of the Literature of Tea, Coffee and Cocoa Preparations. Appendix A, pp. 991–1009. To each method for estimation of Tannin, is attached an analytical summary. And the same is true of Caffeine.

WILFERT, ADOLF.

Kartoffel (Die) und Getreidebrennerei. Handbuch für Spiritusfabrikanten, Brennereileiter, Landwirthe und Techniker. Enthaltend
die praktische Anleitung zur Darstellung von Spiritus aus Kartoffeln, Getreide, Mais und Reis, nach den älteren Methoden und
nach dem Hochdruckverfahren. Dem neuesten Standpunkte der
Wissenschaft und Praxis gemäss populär geschildert. Wien, Pest,
Leipzig, 1885. 8vo. Ill.

WILHELMY, L.

Ueber das Gesetz nach welchem die Einwirkung der Säuren auf den
Rohrzucker stattfindet (1850). Herausgegeben von W. Ostwald.
Leipzig, 1891. 8vo. Ill.
      Ostwald's Klassiker der exakten Wissenschaften.

WILL, HEINRICH.

Anleitung zur [qualitativen] chemischen Analyse zum Gebrauche im
chemischen Laboratorium zu Giessen. Heidelberg, 1846. pp. 148,
8vo. 5 folding tables.
          Zweite Auflage, 1851. Fünfte Auflage, 1859.
          Achte Auflage, 1869. Zwölfte Auflage, Leipzig, 1883.
          Handleiding bij de scheikundige analyse. Bewerkt naar den
            achtsten hoogduitschen druk. Leiden, 1870. 8vo. Ill.
          Outlines of the Course of Qualitative Analysis followed in
            the Giessen Laboratorium. With a preface by Baron
            Justus Liebig. London, 1846. 8vo.
          Outlines of Chemical Analysis, prepared for the chemical
            laboratory at Giessen. From the German by Daniel
            Breed and Lewis H. Steiner. Boston, 1854. 8vo. Ill.
          Outlines of Chemical Analysis, edited by E. N. Hosford.
            Boston, 1854. 12mo.
          De l'analyse qualitative. Instruction pratique à l'usage des
            laboratoires de chimie. Traduit de l'allemand, par G.
            W. Bichon. Paris, 1847. 12mo.
          Guide pratique d'analyse qualitative, instruction élémentaire
            et pratique à l'usage des laboratoires de chimie. Traduit
            de l'allemand par G. W. Bichon. . . . Paris, 1864.
          Guide pour l'analyse chimique à l'usage des médecins, des
            pharmaciens et des étudiants en chimie et en minéralogie.
            Traduit d'après la troisième édition allemande par Jean
            Risler. Paris, 1857. 8vo.
          Guide de l'analyse chimique et tableaux d'analyse qualitative.
            Deuxième édition française. Paris, 1858. pp. 282, 8vo,
            and eleven tables.

WILL, HEINRICH. [Cont'd.]

Guida per l'analisi chimica ed atlante d'analisi quantitativa. Traduzione di N. Reale. Napoli, 1864 8vo.

Tratado de análisis química . . . trasladada por Ramon Botet y Fonsellá. Lérida, 1856. 4to.

Руководство къ качеств. химич. анализу. Перев. Шталь И. С.-Петербургъ, 1847.

Руководство къ химич. анализу для употреб ленія Гесенск. химич. лабораторіи. Перев. Гейхеля. С.-Петербургъ, 1856.

Tafeln zur qualitativen chemischen Analyse Heidelberg, 1846. 8vo.

Zweite Auflage, 1851. Dritte Auflage, 1854.

Vierte Auflage, 1857. Fünfte Auflage, 1859.

Sechste Auflage, 1862. Siebente Auflage, 1866.

Achte Auflage, 1869. Zwölfte Auflage. Leipzig, 1883. 8vo.

Tabellen tot het qualitatief scheikundig onderzoek. Uit het Hoogduitsch vertaald en vermeerderd door W. M. Perk. Delft, 1851. 4to.

Tables for Qualitative Chemical Analysis with an introductory chapter on the course of Analysis. Edited by Charles F. Himes. Second American from the ninth German edition. Philadelphia, 1874. pp. 12, 8vo, and 14 tables.

Tableaux pour l'analyse chimique qualitative. Traduits par Jean Risler. Mulhouse, 1856. 8vo.

Tavole per l'analisi chimica qualitativa ; traduzio ne di G. Carnelutti sull' ultima edizione tedesca. Milano, 1885. 8vo.

Clave de análisis química, ò sea cuadros para el estudio de la análisis química cualitativa . . . traducidos . . . por Magin Bonet y Bonfill. Madrid, 1855. 4to.

For an appendix on legal-chemistry *see* Botet y Jomellà, Ramon.

Таблицы качеств. анализа. С.-Петербургъ, 1867.

WILL'S HAND-BOOK TO PRACTICAL ANALYSIS. London, 1877.

WILL, HEINRICH, F. WÖHLER AND J. LIEBIG.

Nouveau manuel complet de chimie analytique, contenant des notions sur les manipulations chimiques, les éléments d'analyse inorganique qualitative et quantitative, et des principes de chimie organique. Traduit de l'allemand sur les dernières éditions, par F. Malepeyre. Paris, 1855. 2 vols., 18mo. Ill.

WILL, HEINRICH, UND C. R. FRESENIUS.

See Fresenius, C. Remigius.

Вилль, Ж.

Химическія удобренія. Сельско-хозяйственныя бесѣды на Весенскомъ опытномъ полѣ С. Петербургъ, 1871.

> WILL, G. Chemical fertilizers. Agricultural conversations. St. Petersburg, 1871.

WILLGERODT, C.

Ableitung, Entwickelung und Construktion der Kohlenstoffkerne. Freiberg, 1880.

Allgemeinsten (Die) chemischen Formeln, ihre Entwicklung und Anwendung zur Ableitung chemischer Verbindungen. Heidelberg, 1878. pp. viii–208, 8vo.

Ueber Ptomaïne (Cadaveralkaloïde) mit Bezugnahme auf die bei gerichtlich chemischen Untersuchungen zu berücksichtigenden Pflanzengifte. Freiberg, 1890. 8vo.

WILLIAMS, C. GREVILLE.

A Handbook of Chemical Manipulation. London, 1857. pp. xv–580, 8vo. Ill.

> Handbuch der chemischen Manipulationen. Aus dem englischen übersetzt von A. von Hammerl. Mit einem Vorworte von G. C. Wittstein. München, 1860. pp. xx–567, 8vo. Illustrated with 407 woodcuts.

WILLIAMS, C. W.

The Combustion of Coal and the Prevention of Smoke chemically and practically considered. Part the first. Second edition, illustrated with coloured diagrams. London, 1841. 8vo ; and atlas of 12 plates. Liverpool, 1841. 4to.

> Considérations chimiques et pratiques sur la combustion du charbon et sur les moyens de prévenir la fumée. Traduit de l'anglais par D. Bona Christave, publié sous les auspices de S. Exc. l'amiral Hamelin . . . avec l'autorisation de l'auteur. Paris, 1858. 8vo.

WILLIAMS, G. HUNTINGDON.

Elements of Crystallography for Students of Chemistry, Physics and Mineralogy. Second edition, revised. London, 1891.
> Contains a bibliography of the subject.

WILLIAMS, RUFUS PHILLIPS.

Introduction to Chemical Science. London, 1889. 4to.
> Also Boston. 12mo.

Laboratory Manual of General Chemistry. Including general directions for performing one hundred of the more important experiments in general chemistry and metal analysis [etc.]. Specially adapted to accompany " Introduction to Chemical Science." Boston, 1889. 12mo.

WILLIAMS, W. MATHIEU.
    Chemistry (The) of Cookery.   London and New York, 1885.
    Chemistry (The) of Iron and Steel Making.   London, 1890.   pp. x–420, 8vo.
    Framework (The) of Chemistry.   Part I. Typical facts and elementary theory.   London, 1892.   8vo.

WILLIAMSON, ALEXANDER W.
    Chemistry for Students.   Oxford, 1865.   12mo.
            New edition.   Oxford, 1868.   pp. xxii–479, 12mo.   Ill.

WILLIGK, ERWIN.
    Lehrbuch der Chemie für Real- und höhere Bürgerschulen.   Zweite Auflage.   Prag, 1864–'66.   8vo.   Ill.
    Lehrbuch der Chemie.   Gemäss den neueren Ansichten mit Rücksicht auf die Bedürfnisse des Technikers.   Prag, 1872.   2 vols., 8vo.
            Kemien, utarbetad af C. J. Keyser.   Tredie Afdeling: Organisk Kemi.   Landskrona, 1876.   8vo.

WILLIS, TIMOTHY.
    Propositiones tentationum, sive propædeumata de vitis et fæcunditate compositorum naturalium, quæ sunt elementa chymica.   Excusum per Johannem Legatt.   Londini, 1615.   8vo.

WILLM, E., ET HANRIOT.
    Traité de chimie minérale et organique comprenant la chimie pure et ses applications.   Paris, 1883–1889.   4 vols., 8vo.

WILLS, G. S. V.
    Chemistry.   Vol. 1. Inorganic.   London, 1888.
    Handbook to Practical Analysis.   London, 1878.   8vo.
            Manual (A) of Practical Analysis.   Eighth edition.   London, 1885.

WILSON, A. RIVERS.
    Chemical Notes for Pharmaceutical Students, including the Chemistry of the Additions to the Pharmacopœia.   London, 1886.   8vo.

WILSON, BENJAMIN.
    A Series of Experiments relating to Phosphori and the prismatic colors they are found to exhibit in the dark.   Together with a translation of two Memoirs from the Bologna Acts upon the same subject by J. Beccari.   London, 1775.   4to.

WILSON, GEORGE.
    * A Compleat Course of Chymistry, containing not only the best chymi-

WILSON, GEORGE. [Cont'd.]

cal medicines, but also great variety of useful observations. The
fourth edition carefully corrected, very much enlarged and illustrated with copper plates. To which are added the Author's experiments upon metals by way of Appendix. London, 1721. pp.
[xxxi]–[xxiv]–383–[xii], 12mo. Portrait of author and 8 plates.

> Contains recipes for Mathew's and for Starkey's pills. Also describes a
> series of alchemical operations which were interrupted by a mob who
> took the author 'for a "conjurer or something worse and broke my
> glasses and Athanor, saying I was preparing the Devil's Fireworks."

WILSON, GEORGE [of Edinburgh].

Inorganic Chemistry. Revised and enlarged by Stevenson Macadam.
With new notation added. London and Edinburgh, 1866. pp. x–
385, 8vo. Ill.

* Religio chemici. Essays. London and Cambridge, 1862. pp. viii–
386, 12mo.

> *Contents :* Chemistry and Natural Theology ; Chemistry of the stars ;
> Chemical final causes ; Robert Boyle ; Wollaston ; Life and discoveries
> of Dalton ; Thoughts on the resurrection.

WILTNER, FRIEDRICH A.

Fabrikation (Die) der Toilette-Seifen. Praktische Anleitung zur Darstellung aller Arten von Toilette-Seifen auf kaltem und warmen
Wege, der Glycerin-Seife, der Seifenkugeln, der Schaumseifen und
der Seifen-Specialitäten. Mit Rücksicht auf die hierbei in Verwendung kommenden Maschinen und Apparate geschildert. Wien,
1884. Ill.

Seifenfabrikation (Die). Handbuch für Praktiker. Enthaltend die
vollständige Anleitung zur Darstellung aller Arten von Seifen im
Kleinen wie im Fabriksbetriebe mit besonderer Rücksichtnahme
auf warme und kalte Verseifung und die Fabrikation von Luxus-
und medicinische Seifen. Wien, Pest und Leipzig, 1885. 12mo. Ill.

WIMMER, ANTON.

Grundriss der Chemie mit besonderer Berücksichtigung der Mineralogie.
Nach den neueren Theorien für Anfänger in der Chemie, zunächst
für Schüler an Realschulen und mit diesen auf gleicher Stufe
stehende Unterrichts-Anstalten bearbeitet. Landshut, 1877. 8vo.

WINCKLER, EMIL. *See* Bibliothek des Wissenswürdigsten aus der technischen Chemie und
Gewerbskunde.

WINDISCH, KARL.

Die Bestimmung des Molekulargewichts in theoretischer und praktischer

WINDISCH, KARL. [Cont'd.]

    Beziehung. Mit einem Vorwort von Eugen Sell. Berlin, 1892. 8vo. Ill.

        Contains, pp. 1–9: Geschichtliche Entwickelung der Molekulartheorie.

WINKELBLECH, C.

    Elemente der analytischen Chemie. Marburg und Leipzig, 1840. 8vo.

WINKELMANN, A. Physikalische Lehren. *See* Graham-Otto.

WINKLER, CLEMENS.

    Anleitung zur chemischen Untersuchung der Industrie-Gase. Erste Abtheilung : Qualitative Analyse. Mit 31 in den Text getruckten Holzschnitten und einer lithographirten Tafel. Freiberg, 1876. pp. vi–166, 8vo.

    Lehrbuch der technischen Gasanalyse. Kurzgefasste Anleitung zur Handhabung gasanalytischer Methoden bewährter Brauchbarkeit. Freiburg, 1884.

        Handbook of Technical Gas-analysis, containing concise instructions for carrying out gas-analytical methods of proved utility. Translated with a few additions by G. Lunge. London, 1885.

        Manuel pratique de l'analyse industrielle des gaz. Traduit de l'allemand par C. Blas. Paris, 1886.

        Технико-химическій газовый анализъ Перев Ахматовъ П. и Шеинъ П Москва, 1873.

    Maassanalyse (Die) nach neuern titrimetrischem System. Kurzgefasste Anleitung zur Erlernung der Titrirmethode. Freiberg, 1883. pp. viii–98, 8vo.

    Praktische Uebungen in der Maassanalyse. Anleitung zur Erlernung der Titrirmethode. Unter Zugrundelegung des ursprünglichen titrirmetrischen Systems der chemischen Anschauung der Neuzeit gemäss bearbeitet. Freiburg, 1888. 8vo. Ill.

WINTERL, JACOB JOSEPH.

    Kunst (Die) Blutlauge zu bereiten. Wien, 1790.

    Methodus analyseos aquarum mineralium. Viennæ et Budæ, 1781.

        Editio secunda, 1784.

    Prolusiones ad chemiam sæculi decimi noni. Budæ, 1800. 8vo.

        Accessiones novæ ad prolusionem suam primam et secundam. Budæ, 1803.

        Darstellung der vier Bestandtheile der anorganischen Natur. Eine Umarbeitung des ersten Theiles seiner Prolusionen und Accessionen. Aus dem lateinischen übersetzt von Johann Schuster. Jena, 1804. pp. xlvi–528, 8vo.

WINTERL, JACOB JOSEPH.   [Cont'd.]

System der dualistischen Chemie dargestellt von Johann Schuster. Berlin, 1807. 2 vols., 8vo.

Systematis chemici ex demonstrationibus Tirnav. pars rationalis et experimentalis. Viennæ, 1773.

WINTERL, JACOB JOSEPH, UND J. G. KAIM.

De metallis dubiis. Viennæ, 1770.

> Contains the first account of metallic manganese.

WIPACHER, DAVID.

De phlogisto, unionis rerum metallicarum menstruo dissertatio. Lipsiæ, 1752. 4to.

WIRTH, G.

Wiederholungs- und Hülfsbuch für den Unterricht in der Chemie. Für die Hand der Schüler bearbeitet. Berlin, 1875. 8vo.

WISLICENUS, JOHANN.

Theorie der gemischten Typen. Berlin, 1859. 8vo.

Ueber die räumliche Anordnung der Atome in organischen Molekulen und ihre Bestimmung in geometrisch-isomeren ungesättigten Verbindungen. Mit 186 Figuren. (Aus Abhandlungen der königlichen sächsischen Gesellschaft der Wissenschaften.) Leipzig, 1887.

Руководство къ органической химіи. Перев Гемиліана В. и Гудыва В. С.-Петербургъ, 1875–1876.

> Handbook of Organic Chemistry. St. Petersburg, 1875–76.

WISLICENUS, J., UND ADOLPH STRECKER.   *See* Regnault-Strecker.

WISSER, John P.

Chemical Manipulations. Course of Sciences applied to military art. Printed at the United States Artillery School, Fort Monroe, Va. 1883. Roy. 8vo.

Виттъ II.

Промышленная химія. Публычныя бесѣды о важнѣйшихъ химическихъ производствахъ, происходившія въ залѣ Императорскаго Вольнаго Экономическаго Общества. 2 части. С.-Петербургъ, 1847–48. 8o

> WITT, J. Technological Chemistry. Public Lectures. St. Petersburg, 1847–48.

Собраніе чертежей промышленной химіи, съ подробными описаніями. С.-Петербургъ, 1847–8.

> Collection of chemico-technological Formula. St. Petersburg, 1847–48.

Виттъ М.

Органическая химія. С.-Петербургъ, 1849.

> WITT, M. Organic Chemistry. St. Petersburg, 1849.

WITT, OTTO N.

Chemische Homologie und Isomerie in ihrem Einflusse auf Erfindungen aus dem Gebiete der organischen Chemie. Berlin, 1889.

Chemische Technologie der Gespinnstfasern, ihre Geschichte, Gewinnung, Verarbeitung und Veredlung. (Zugleich als fünften Bandes, zweite Gruppe, erste Lieferung des Handbuchs der chemischen Technologie.) Braunschweig, 1888. 8vo. Ill.

> Contains a Bibliographie der Faserstoffe and a Geschichtlicher Ueberblick über die Entwickelung der Textilgewerbe. See Bolley's Handbuch [etc.].

WITTHAUS, RUDOLPH AUGUST.

Essentials of Chemistry and Toxicology. [Wood's Pocket Manuals.] New York, 1878.

Second edition. New York, 1888. 16mo.

Eleventh edition. New York, 1890. pp. 4-299, 16mo.

General Medical Chemistry for the Use of Practitioners of Medicine. [Wood's Library.] New York, 1881. pp. vii-442, roy. 8vo.

Laboratory (A) Guide in Urinalysis and Toxicology. New York, 1886. Oblong 12mo.

Second edition. New York, 1889.

Medical Student's (The) Manual of Chemistry. New York, 1883. pp. xi-370.

Second edition, 1889. pp. xi-391.

Third edition. New York, 1890. pp. xii-528, 8vo.

WITTICH, W. VON.

Aufsaugung, Lymphbildung und Assimilation. See Handbuch der Physiologie.

WITTING, ERNST.

Beiträge für die pharmaceutische und analytische Chemie. Schmalkalden, 1821-22. 8vo.

WITTSTEIN, GEORG CHRISTOPH.

Anleitung zur Analyse der Asche von Pflanzen oder organischen Substanzen überhaupt. München, 1862. 8vo. Ill.

Anleitung zur chemischen Analyse von Pflanzentheilen auf ihre organischen Bestandtheile. Nördlingen, 1868. pp. 355, 8vo.

Organic (The) Constituents of Plants and Vegetable Substances and their chemical Analysis. Authorized translation from the German original, enlarged with numerous additions by Ferd. von Mueller. Melbourne, 1878. pp. xviii-332.

Anleitung zur Darstellung und Prüfung chemischer und pharmaceutischer Präparate. München, 1845. 8vo.

WITTSTEIN, GEORG CHRISTOPH.   [Cont'd.]

>Zweite Auflage, 1851.   Dritte Auflage, 1857.

>Anleitung zur Darstellung und Prüfung chemischer und pharmaceutischer Präparate.   Ein auf eigene Erfahrungen gegründetes insbesondere den Apothekern gewidmetes praktisches Hülfsbuch.   Vierte vermehrte und verbesserte Auflage.   München, 1867.   8vo.   Ill.

>Practical Pharmaceutical Chemistry ; An Explanation of chemical and pharmaceutical processes with the methods of testing the purity of the preparations deduced from original experiments.   Translated and edited from the second German edition by Stephen Darby.   London, 1853.   pp. xiv–624, 12mo.

Grundriss der Chemie.   Zunächst bearbeitet für technische Lehranstalten.   München, 1852.   2 parts.   8vo.   Part 1 : Allgemeiner Theil und unorganische Chemie.   Part 11 : Organische Chemie.

>Zweite Auflage.   München, 1868.

Taschenbuch der Chemikalienlehre.   München, 1879.

Widerlegung der chemischen Typenlehre.   München, 1862.   8vo.

>*See also in Section II.*

WITTWER, ———.

Delectus dissertationum medicarum Argentoratensium.   Norimbergæ, 1777–81.   4 vols., 8vo.

>Contains chemical treatises by Jacob Reinbold Spielmann, q. v.

>Kleine practische medicinische und chemische Schriften.   Leipzig, 1786.   8vo.

WITTWER, WILHELM CONSTANTIN.

Grundzüge der Molecular-Physik und der mathematischen Chemie.   Stuttgart, 1885.   8vo.   3 folding plates.

Versuch einer Statik der chemischen Verbindungen.   München, 1854.   8vo.

WÖHLER, FRIEDRICH.

Grundriss der Chemie.   Berlin, 1833.   2 vols., 8vo.

>Vol. 1 *also under the title :* Grundriss der unorganischen Chemie.

>Vol. 11 *also:* Grundriss der organischen Chemie.

>Each volume passed through many editions independently, of which the following are a few :

>Vol. 1. Zweite Auflage.   Berlin, 1833.

>Fünfte Auflage.   Berlin, 1838.

>Achte Auflage.   Berlin, 1845.

Wöhler, Friedrich. [Cont'd.]

Dreizehnte Auflage. Berlin, 1863.

Vierzehnte Auflage, mit einer Einleitung von Hermann Kopp. Leipzig, 1868.

Fünfzehnte Auflage. Leipzig, 1873.

Vol. ii. Dritte Auflage. Berlin, 1844.

Vierte Auflage. Berlin, 1848.

Sechste Auflage. Bearbeitet von Rudolph Fittig. Berlin, 1863.

Siebente Auflage. Berlin, 1868.

Zehnte Auflage. Leipzig, 1876.

Grundrids af Chemien. Paa Dansk bearbejdet af Simon Groth. Kjøbenhavn, 1854, '55. 2 Dele, 8vo.

Förste Del : Grundrids af den uorganiske Chemie.

Anden Del : Grundrids af den organiske Chemie.

Schets der onbewerktuigde scheikunde. Naar de vierde Hoogduitsche uitgave vertaald. Gouda, 1839. 8vo.

Schets der onbewerktuigde scheikunde. Naar de laatste Hoogduitsche vertaald en met bijvoegsels vermeerderd door N. W. de Voogt. Utrecht, 1845. 8vo.

Schets der bewerktuigde scheikunde. Naar de derdie onlangs verschenen geheel omgewerkte Hoogduitsche uitgave met bijvoegsels en aanmerkingen, door P. J. Knipp. Utrecht, 1844. 8vo.

Schets der organische scheikunde. Naar de achtste Hoogduitsche uitgave van Wöhler's schets, in het Nederlandsch bewerkt door F. W. Krecke. Utrecht, 1872. 8vo. Ill.

Wöhler's Outlines of Organic Chemistry by R. Fittig. . . . Translated from the eighth German edition with additions by I. Remsen. Philadelphia, 1873. 12mo.

Cours de chimie générale, traduit sur la neuvième édition allemande et augmenté de notes par Mareska et H. Valerius. Bruxelles, 1848. 2 vols., 8vo.

Éléments de chimie inorganique. Traduit de l'allemand sur les éditions deuxième et cinquième par L. Grandeau, avec le concours de F. Sacc, et des additions par H. Sainte Claire Deville. Paris, 1858. 8vo.

Éléments de chimie organique et inorganique. Traduit de l'allemand sur la onzième édition par L. Grandeau. Paris, 1872. 8vo.

Lærebog i Chemien efter Wöhler. [By] M. W. Sinding. Christiania, 1838. 8vo.

Wöhler, Friedrich. [Cont'd.]

    Utkast till den organiska Kemien. Öfversatt och bearbetad af N. J. Berlin. Lund, 1848. 8vo.

    Oorganiska Kemien i Sammandrag. Från sidsta Tyska upplagan öfversatt af Clemens Ullmann. Andra upplagan. Ny bearbetning. Stockholm, 1851. 8vo.

    Органичвсаая химія по Веллеру и Фиттиху. Справочная книга при занятіяхъ въ лабораторіи и для практиковъ. Перев. Кульбергъ А. и Тавильдаровъ И. С.-Петербургъ, 1871.

    Wöhler and Fittig. Organic chemistry according to Wöhler and Fittig. St. Petersburg, 1871.

Practische Uebungen in der chemischen Analyse. Göttingen, 1853. pp. xiv–218, 8vo. Ill.

    The second edition bears the title :

    Die Mineral-Analyse in Beispielen. Zweite umgearbeitete Auflage. Göttingen, 1861. pp. xiv–234, 8vo.

    Handbook of Inorganic Analysis ; one hundred and twenty-two examples illustrating the most important processes for determining the elementary composition of mineral substances. Edited by A. W. Hofmann. London, 1854. 8vo.

    The Analytical Chemist's Assistant ; a Manual of chemical analysis, both quantitative and qualitative, or natural and artificial inorganic compounds ; to which are appended the rules for detecting arsenic in a case of poisoning. Translated from the German, with an introduction, illustrative and copious additions, by Oscar M. Lieber. Philadelphia, 1853. 12mo.

    Handbook of Mineral Analysis. [Translated from the German.] Edited by H. B. Nason. Philadelphia, 1871. 8vo.

    Traité pratique d'analyse chimique. Paris, 1865. 12mo. Ill.

    Минеральный анализъ въ примѣрахъ. Перев. Отто. С.Петербуръг. 1863.

Wöhler, Friedrich, und E. Von Siebold.

Das forensisch-chemische Verfahren bei einer Arsenik-Vergiftung. Berlin, 1847. pp. 32, 12mo. Ill.

Wöhler, Friedrich, und Justus von Liebig.

Untersuchungen über das Radikal der Benzoësäure. Herausgegeben von H. Kopp. Leipzig, 1891. 8vo.

    Ostwald's Klassiker der exakten Wissenschaften, No. 22.

WOJCZYNSKI, F.

Wiadomości chemiczno-gospodarskie, czyli : opisanie naiważniejszych
potrzeb życia jako to : pokarmów, napojów i niektórych przypraw,
ich wyrabianie, dobroć i czystość ; tudzież dochodzenie przypad-
kowego zanieczyszczenia i umyślnego zafałszowania, oraz o naczy-
niach kuchennych i stołowych. Warszawa, 1845. 8vo.

WOLF, RUDOLPH.

Ueber den Ozongehalt der Luft und seinen Zusammenhang mit der
Mortalität. Vorträge gehalten in der bernischen naturforschenden
Gesellschaft. Bern, 1855. 8vo. Ill.

WOLFF, EMIL THEODOR.

Anleitung zur chemischen Untersuchung landwirthschaftlich wichtiger
Stoffe. Zum Gebrauch bei quantitativ-analytischen Arbeiten im
chemischen Laboratorium und bei Vorträgen über landwirthschaftlich-chemische Analyse. Zweite durchaus neu bearbeitete Auflage.
Mit steter Berücksichtigung der unter den Agricultur-Chemikern
gebräuchlichen und vereinbarten Untersuchungsmethoden. Stuttgart, 1867. 8vo.

Aschen-Analysen von landwirthschaftlichen Producten, Fabrik-Abfällen
und wildwachsenden Pflanzen. Einheitlich berechnet und mit
Nachweisung der Quellen systematisch geordnet nebst Notizen
über das untersuchte Material und verschiedene Uebersichts-
tabellen. Berlin, 1871. 4to.

Chemischen (Die) Forschungen auf dem Gebiete der Agricultur- und
Pflanzen-Physiologie. Leipzig, 1847. 8vo.

La chimica agraria congiunta alla pratica agricola. Dal
tedesco tradotta in italiano e corredata di riflessioni
preliminari, da T. H. Ohlsen. Napoli, 1868. 8vo.

Руководство къ химич. изслѣдованію важнѣйшихъ сельско-хозяйственныхъ продуктовъ. Перев. Похписева П. Москва, 1860

Naturgesetzlichen (Die) Grundlagen des Ackerbaues. Leipzig, 1851.
2 vols., 8vo.

Dritte Auflage, 1856.

Praktische Düngerlehre mit einer Einleitung über die allgemeinen
Nährstoffe der Pflanzen. Gemeinverständlicher Leitfaden der
Agricultur-Chemie. Vierte verbesserte und vermehrte Auflage.
Berlin, 1872. 8vo.

Praktisk Gjødningslære med en Indledning om Planternes
almindelige Næringsstoffer og den dyrkede Jords
Egenskaber. Letfattelig Ledetraad i Agerdyrknings-
kemien. Anden Udgave. Paa Dansk ved J. V. T.
Hertel. Kjøbenhavn, 1876. 8vo.

WOLFF, EMIL THEODOR. [Cont'd.]

Praktische bemestingleer, met eene inleiding over de algemeene voedingsstoffen der planten en de eigenschappen van den bouwgrond. Een algemeen verstaanbare leiddraad tot de kennis der landbouwscheikunde. Tweede druk. Naar den negenden verbeterden Hoogduitschen druk bewerkt door F. J. van Pesch. Zwolle, 1886. Roy. 8vo. Ill.

Manuale pratico per l'uso dei concimi e degli ingrassi secondo i risultati delle teorie moderne ; traduzione da Antonio Selmi, con note ed aggiunte di alcuni studii pratici di scienza applicata all' agricoltura del traduttore. Milano, 1872. 16mo.

See also, in Section I, Quellen-Literatur der . . . Chemie.

WOLFF, FRIEDRICH BENJAMIN.

Lehrbuch der Chemie nach den neuesten Werken von Murray, Thénard und Thomson frei bearbeitet. Berlin, 1820–21. 3 vols., 8vo.

Vorlesungen über Chemie. Berlin, 1829–30. 2 vols., 8vo.

See also, in Section II, Chemisches Wörterbuch, and, in Section VII, Annalen der chemischen Literatur.

WOLFF, H.

Die Beizen, ihre Darstellung, Prüfung und Anwendung. Für den praktischen Färber und Zeugdrucker bearbeitet. Wien, Pest, Leipzig, 18—.

WOLFF, H., UND J. BAUMANN.

Tabellen zur Berechnung der organischen Elementaranalyse. Berlin, 1886. 8vo.

WOLFF, JACOB.

Chemische Analyse der wichtigsten Flüsse und Seen Mecklenburgs, mit ausführlicher Angabe des eingehaltenen Untersuchungsganges. Wiesbaden, 1872. 8vo. Ill.

WOLFF, K.

Chemische Analyse der am meisten bekannten Elemente nach einem neuen System zusammengestellt, für Mediciner, Apotheker, Chemiker und Landwirthe. Göttingen, 1872. 4to.

WOLFF, LAWRENCE.

Applied Medical Chemistry, containing a description of the apparatus and methods employed in the practice of medical chemistry, the chemistry of poisons, physiological and pathological analysis, urinary and fecal analysis, sanitary chemistry, and the examination of medicinal agents, foods, etc. Philadelphia, 1886. 8vo.

WOLFF, LAWRENCE. [Cont'd.]

 Essentials of Medical Chemistry, organic and inorganic. Containing also questions of medical physics, chemical philosophy, analytical processes, toxicology, etc. Prepared especially for students of medicine. Third and revised edition. Philadelphia, 1891. pp. xi-from 17-218, 12mo.

 Saunders' Question-Compends, No. 4.

Questions and Answers on the Essentials of Medical Chemistry. Philadelphia, 1889.

WOLFF, LEOPOLD.

Das Wasserglas. Seine Darstellung . . . in den technischen Gewerben. Quedlinburg und Leipzig, 1846. 8vo.

WOLFRING, M. C.

Verhältniss des Organischen zum Anorganischen. Erlangen, 1848.

WOLLASTON, WILLIAM HYDE.

 This eminent chemist does not appear to have published any independent book, but, contrary to my plan, and to introduce his name, I give two references to papers announcing important discoveries in the Philosophical Transactions. *See* Catalogue Scientific Papers by the Royal Society of London, vol. VI.

On a new metal found in crude platina [Palladium]. Phil. Trans., 1804. pp. 419–430.

On the discovery of Palladium ; with observations on other substances found with Platina [Rhodium]. Phil. Trans., 1805. pp. 316–330.

WOLPERT, H.

Eine einfache Luftprüfungsmethode auf Kohlensäure mit wissenschaftlicher Grundlage. Jena, 1891. 8vo. Ill.

WOOD, THOMAS.

Chemical Notes for the Lecture Room, on heat, laws of chemical combination, and the chemistry of the non-metallic elements. Second edition. London, 1868. 8vo.

Elementary Questions in Chemistry for Students. First Series. Thirty Papers on Heat and Metalloids. London, 1870.

Notes on the Metals. A second series of chemical notes for the lecture room. London, 1868. 8vo.

WOODLAND, J.

Laboratory Work for Students. London, 1889. 12mo.

WOODWARD, C. J.

Arithmetical Chemistry, Part I. New edition, entirely re-written. London, 1890. pp. viii–83, 8vo. Part II. New edition. London and Birmingham, 1892. 8vo.

Questions in Chemistry and Natural Philosophy given at the matriculation examination of the University of London. New edition containing answers. London, 1886.

WOODY.

Essentials of Medical Chemistry. Third edition, enlarged. Philadelphia, 1890. 12mo. Ill.

WOOTEN, H.

Three Hundred Problems in Chemical Physics and Specific Gravities with Key. London, 1886. 8vo.

WORMLEY, THEODORE G.

    Chemical Lectures, 1889. *See* Marshall, John.

WORMLEY, THEODORE G.

Micro-Chemistry of Poisons, including their physiological, pathological and legal relations : Adapted to the use of the medical jurist, physician and general chemist. New York, 1867. pp. xxxi–668, 8vo. Ill. and Plates.

        New edition revised and enlarged. With an appendix on the detection and microscopic discrimination of the blood. Philadelphia, 1885. 8vo.

WORTHINGTON, LEWIS NICHOLAS.

Chimie inorganique et organique, botanique, zoologie. Notes servant à la préparation de l'examen du premier doctorat. Paris, 1889. 8vo.

WOUWERMANS, ALWIN VON.

Farbenlehre. Für die praktische Anwendung in den verschiedenen Gewerben und in der Kunstindustrie bearbeitet. Wien, Pest, Leipzig, 1879. 8vo. Ill.

WRIGHT, CHARLES ROMLEY ALDER.

Metals, and their chief industrial applications. London, 1878. 8vo. Ill.

Oils, Fats, Waxes and allied materials, and the manufacture therefrom of candles, soaps and other products. London, 1893. 8vo.

WRIGHT, HERBERT EDWARDS.

A Handy Book for Brewers, being a practical guide to the art of brewing and malting, embracing the conclusions of modern research which bear upon the practice of brewing. London, 1892.

WROBLEWSKI, S. VON, UND K. OLSZEWSKI.
Ueber die Verflüssigung des Sauerstoffs, Stickstoffs und Kohlenoxyds. Leipzig und Wien, 1883. Ill.
Ueber die Verflüssigung des Sauerstoffs und die Erstarrung des Schwefelkohlenstoffes und Alcohols. Wien, 1883.

WUCHERER, GUSTAV FRIEDRICH.
Leitfaden zum Gebrauch bei Vorlesungen über die Stöchiometrie der unorganischen Körper. Carlsruhe, 1820. 8vo.

WUNDER, GUSTAV.
Die Vorbereitung für den Eintritt in die chemische Technik. Eine Schrift zur Orientirung für künftige Techniker, nebst Beschreibung des neuen Laboratoriums der technischen Staatslehranstalten in Chemnitz. Chemnitz, 1879. 8vo. Ill.

WUNDERLICH, A.
Configuration organischer Molekule. Leipzig, 1886.

WURTZ, ADOLPHE. Leçons (etc.). *See* Société chimique de Paris.

WURTZ, ADOLPHE.
Atomic Theory (The). Translated by E. Cleminshaw. London, 1880. 8vo.
> International Scientific Series. For French and German editions *see in Section III.*

Introduction à l'étude de la chimie. Avec 60 figures dans le texte. Paris, 1885. pp. v–276, 8vo.
Leçons élémentaires de chimie moderne. Paris, 1866. 12mo. Ill.
> Deuxième édition, revue et augmentée. Paris, 1871.
> Troisième édition. Paris, 1875.
> Cinquième édition. Paris, 1883.
> Elements of Modern Chemistry. Translated and edited with the approbation of the author from the fourth French edition by W. H. Greene. With one hundred and thirty illustrations. Philadelphia, 1880. pp. ix–687, 8vo.
> [*The same*], from the fifth French edition by W. H. Greene. Philadelphia, 1887. pp. ix–770, 8vo.
> Lezioni elementari di chimica inorganica. Traduzione di R. Monteferrante. Napoli, 1885. 12mo.
> Lezioni elementari di chimica organica moderna. Traduzione di R. Monteferrante. Napoli, 1882. 12mo.
> Элементарный учебникъ химіи. Кіевъ, 1867.
> Уроки новѣйшей химіи. Перев. Алексѣева. Кіевъ, 1869. 2 части.

WURTZ, ADOLPHE. [Cont'd.]

Traité de chimie biologique. Paris, 1880–85. 8vo.

> Tratado de química biológica. Versión española con adiciones por V. Peset y Cervera. Madrid, 1892. 4to. Ill.

Лекціи по нѣкоторымъ вопросамъ теоретич. химіи. Перев. Алексѣева П. С.-Петербургъ, 1865

> Lectures upon some questions of theoretical chemistry. St. Petersburg, 1865.

WURTZ, ADOLPHE, A. LAMY, LOUIS GRANDEAU.

Leçons de chimie professées en 1863. Société chimique de Paris. Sujets des leçons : Sur quelques points de philosophie chimique. Leçon sur le thallium. Leçon sur le rubidium et le cæsium. Paris, 1864. pp. 314, 8vo.

WURZER, FERDINAND.

Handbuch der populären Chemie zum Gebrauche bei Vorlesungen und zur Selbstbelehrung. Leipzig, 1806.

> Dritte Auflage. Leipzig, 1820.
> Vierte umgearbeitete Auflage. Leipzig, 1826. 8vo.

Physikalisch-chemische Beschreibung der Mineralquelle zu Godesberg. Bonn, 1790.

Ueber das Gemeinnützige der chemischen Kenntnisse. Marburg, 1805. 8vo.

WUTTIG, JOHANN FRIEDRICH CHRISTIAN.

Anleitung Metallgemische durch ein neues Verfahren zu probiren. Berlin, 1819. 8vo.

Gründliche Anleitung zur Fabrikation der Schwefelsäure. Berlin, 1812. Ill.

Kurze Anleitung Messing, Kanonenmetall und viele andere Metallmischungen durch ein neues Verfahren auf das quantitative Verhältniss ihres Inhalts schnell und genau zu probiren. Berlin, 1820. 8vo. Ill.

Uebersicht meiner Systeme der Hylognosie und chemischen Fabrikenkunde. Berlin, 1821.

Versuche über die Gallussäure. Dorpat, 1806.

WYATT, FRANCIS.

The Phosphates of America. Where and how they occur ; How they are mined ; and what they cost. With practical treatises on the manufacture of sulphuric acid, acid phosphate, phosphoric acid and concentrated superphosphates, and select methods of chemical analysis. New York, 1892. 8vo.

YARKOVSKI, IVAN.

Hypothèse cinétique de la gravitation universelle en connexion avec la formation des éléments chimiques. Moscow, 1888. 8vo.

YOUMANS, EDWARD L.

Alcohol and the Constitution of Man ; being a popular scientific account of the chemical history and properties of alcohol, and its leading effects upon the healthy human constitution. New York, 1854. 16mo.

Chemical Atlas, or the Chemistry of Familiar Objects ; exhibiting the general principles of the science in a series of beautifully colored diagrams, and accompanied by explanatory essays embracing the latest views of the subjects illustrated. Designed for the use of students and pupils in all schools where chemistry is taught. New York and London, 1856. pp. 106, 4to.

Classbook of Chemistry. New York, 1851. 12mo. Ill.

> This passed through many editions and reached a sale of over 144,000 copies.

Elementos de química, para uso de los colegios y escuelas. Libro que contiene los ultimos descubrimientos de la ciencia, y en que se indican sus aplicaciones a las artes, y a la mejor inteligencia de los fenomenos de la naturaleza. Traducido de la última edicion inglesa por Marco A. Royas. Segunda edicion corregida. Nueva York, 1879. pp. 500, 8vo. Ill.

YOUMANS, EDWARD L. [Editor].

The Correlation and Conservation of Forces : A Series of Expositions by Grove, Helmholtz, Mayer, Faraday, Liebig and Carpenter. With an introduction and brief biographical notices of the chief promoters of the new views. New York, 1871. pp. xlii–438, 8vo.

YPEY, ADOLPHUS.

Disputatio philosophica de igne. Franeker, 1767. 4to.

Waarnemingen over eenige stoffen die de verrotting bevorderen of tegenstaan. 1776. 8vo.

YVON, P.

De l'analyse chimique de l'urine normale et pathologique au point de vue clinique. Paris, 1875. 8vo.

> Manuale pratico per le analisi delle urine. Traduzione italiana sulla seconda edizione originale con note di G. de Luca. Napoli, 1886. 8vo. Ill.

ZABOROWSKI.

Les boissons hygièniques. Paris, 1889. 16mo.

> Petite bibliothèque scientifique.

ZAENGERLE, MAX.

Grundriss der Chemie nach den neuesten Ansichten der Wissenschaft für den Unterricht an Mittelschulen, besonders Gewerbe-, Handels- und Realschulen. Dritte Auflage. Braunschweig, 1886. 2 parts, 8vo. Ill.

> Erste Auflage. München, 1872.

Grundriss der Chemie und Mineralogie. Nach den neuesten Ansichten der Wissenschaft für den Unterricht an Mittelschulen, besonders Gewerbe-, Handels-, und Realschulen bearbeitet. 1· part. Braunschweig, 1883. 8vo. Ill.

Grundzüge der Chemie und Naturgeschichte für den Unterricht an Mittelschulen. München, 1887.

Lehrbuch der Chemie nach den neuesten Ansichten der Wissenschaft für den Unterricht an technischen Lehranstalten bearbeitet. Dritte vermehrte Auflage. Braunschweig, 1885. 2 vols., 8vo. Ill.

Kemian alkeet. Porvoosa, 1885.

ЗАГУМЕННЫЙ, АЛЕКСАНДРЪ.

О нѣкоторыхъ производныхъ дезоксибензоина. С.-Петербургъ, 1875·

> ZAGUMENNUI. On some derivatives of desoxybenzoin. St. Petersburg, 1875.

ZALIESKI, S.

Значеніе химіи для культуры и человѣчества. Метода ея преподаванія. Томскъ, 1888. 16º.

> Significance of chemistry for culture and for mankind. Methods of teaching. Tomsk, 1888. 16mo.

ZANGENMEISTER, O., UND E. VON SCHWARTZ.

Die wichtigsten Futter- und Düngemittel in ihrer chemischen Zusammensetzung graphisch dargestellt, nebst leichfasslicher Darlegung ihrer Verwendung für den praktischen Landwirth. Gotha, 1889.

ZANON, G.

La spettroscopia e le sue conseguenze : esame degli studii di analisi spettrale del signor Norman Lockyer. Bologna, 1880. 16mo.

ZANTEDESCHI, FRANCESCO.

Della correlazione delle forze chimiche molecolari colla rifrangibilità delle irradiazioni luminose e calorifiche oscure. Padova, 1857. 8vo.

Elenco delle principali opere scientifiche dell' Abate Francesco Zantedeschi. Venezia, 1849. 8vo.

Ricerche fisico-chimiche-fisiologiche sulla luce. Venezia, 1846. 4to. Ill.

Zantedeschi, Francesco [Editor].
Raccolta fisico-chimica italiana, ossia collezione di memorie originali edite ed inedite di fisici, chimici e naturalisti italiani. Venezia, 1846–48. 3 vols., 8vo. Ill.

Zasady technologii chemicznéi gospodarskiéj (przez Józefa Bełze). Warszawa, 1840. 12mo.

Zauschner, Johann Baptista Joseph.
Vindiciæ phlogisti. Pragæ, 1794. 8vo.

Zdeborsky.
Anleitung zum Anbau der Runkelrüben. Prag, 1836.

Zdzitowiecki, Józef Seweryn.
Wykład początkowy chemii o metalloidach i ich związkach. Warszawa, 1850. 12mo. Ill.
> [*The same*] o metalach i ich związkach. z siedmiu tablicami figur. Warszawa, 1851.

Zehle, L.
Ein Beiträg zur Trinkwasserfrage. Olmütz, 1890. 8vo. 6 plates.

Zeise, Willhelm Christoph.
Erindringsord til Forelæsninger over anvendt Chemie. Kjøbenhavn, 1825.
Udførlig Fremstilling af Chemiens Hovedlærdomme saavel i theoretisk som practisk Henseende. Kjøbenhavn, 1829. Ill.

Zeisel, S.
Chemie ; eine gemeinfassliche Darstellung der chemischen Erscheinungen und ihrer Beziehungen zum praktischen Leben. Wien, 1892. 8vo. Ill.

Zenneck, Ludwig Heinrich.
Anleitung zur Untersuchung des Biers nach seinen sowohl erlaubten als unerlaubten Bestandtheilen, für Polizeibehörden, Chemiker und Bierbrauer. München, 1834. pp. vi–142, 12mo. Folding plate.
Grundlinien einer populären Chemie. Stuttgart, 1829.
Physikalisch-chemisches Hülfsbuch, die verschiedenen Eigenschaften und chemischen Prozesse aller ökonomisch, pharmaceutisch und technisch wichtigen Körper auf anschauliche Art zu bequemer Uebersicht darstellend. Berlin, 1842. 8vo.

Zenoni, E.
Nuova guida elementare allo studio della analisi chimica qualitativa dei principali composti inorganici, compendiata in xvi tavole ad uso degli Istituti tecnici, ecc. Pavia, 1884. 8vo.

ZETTNOW, EMIL.

Anleitung zur qualitativen chemischen Analyse ohne Anwendung von
Schwefelwasserstoff und Schwefelammonium. Berlin, 1867. pp.
x–158, 8vo. Ill.

ZIEGELER, G. A.

Die Analyse des Wassers, nach eigenen Erfahrungen bearbeitet. Stuttgart, 1887. pp. viii–117, 8vo. Ill.

ZIERL, LORENZ.

Bayerische (Die) Braunbier-Fabrication . . .   München, 1843.
Agriculturchemie. München, 1830. 8vo.
Grundriss der Chemie, als Propädeutik der chemischen Technologie,
Pflanzen- und Thierphysiologie, zum Behufe von Vorträgen an
technischen, land- und forstwirthschaftlichen Lehranstalten. Die
unorganische Chemie. München, 1842. 8vo.
Landwirthschaftlich-vegetabilische (Die) Productionslehre. München,
1830. 2 vols., 8vo.

ZIMMERMANN, A. F.

Ausführliches Lehrbuch der Bierbrauerei. Zweite Auflage. Berlin,
1852. 8vo. Ill.

ZIMMERMANN, W. F. A.

Chemie für Laien.   Eine populäre Belehrung über die Geheimnisse der
Chemie, deren Aufschlüsse über das innere Leben der Natur, sowie
ihre Bedeutung und praktische Nutzung für das Leben. Mit Abbildungen. Berlin, 1858–63. 9 vols., 8vo.
Vol. I, pp. viii–518, 1858 ; vol. II, pp. vi–504, 1858 ; vol. III, pp. xii–
660, 1859 ; vol. IV, pp. vi–569, 1859 ; vol. v, pp. viii–759, 1860 ;
vol. VI, pp. vi–794, 1860 ; vol. VII, pp. vi–568, 1861 ; vol. VIII, pp.
[xvi]–720, 1861 ; [Pt. 1], pp. vi–721–1381, 1862 ; [Pt. 2], [Index] ;
vol. IX, pp. 190, 1863.

ZINCKEN, C.

Das Naturgas Amerikas, nach A. Williams, C. Zincken, C. A. Ashburner,
etc. Leipzig, 1887. 4to.

ZINNO, SILVESTRO.

Elementi di chimica generale a sistema moderno. Seconda edizione.
Vol. 1 : Chimica dei corpi inorganici. Napoli, 1876. 2 vols., 16mo.
Elementi di chimica inorganica ed organica adattati all' insegnamento
delle università, istituti, licei, ecc. Libro secondo. Chimica or-
ganica. Napoli, 1871. 16mo.
Memoria sull' ozono (premiata). Napoli, 1874. 8vo.

ZINNO, SILVESTRO. [Cont'd.]
Riassunto di chimica generale a sistema moderno, compilato per cura di
S. Emilio Liebler. Napoli, 1880. 16mo.

ZIPPERER, P.
Untersuchungen über Kakao und deren Präparate. Gekrönte Preisschrift. Hamburg, 1887. Ill.

ZORN, W.
Ueber Cinchonin. Leipzig, 1873. 8vo.

ZOTOS, Z. N.
Ein Beitrag zur Kenntniss des Cerberins. Dorpat, 1892. 8vo.

ZUCCHI, CARLO, E ALLESSANDRO RANZOLI.
Prontuario di farmacia coll' aggiunta di nozioni di chimica legale e di
chimica medica e della raccolta delle leggi vigenti nel regno Lom-
bardo-Veneto sull' esercizio farmaceutico. Edizione a spesa degli
autori. Milano, 1859. 8vo.

ZUELZER, W.
Lehrbuch der Harnanalyse. Berlin, 1880.

ZUNE, AUGUSTE J.
Analyse des eaux potables et détermination rapide de leur valeur hygiènique. Paris et Bruxelles, 1889. pp. 144, 8vo. Ill.
Traité général d'analyse des beurres, préparation, caractères, composition,
altérations et falsifications, méthode générale d'analyse, discussion
et appréciation des résultats. Première partie, avec 83 figures et 63
tableaux intercalés dans le texte. Paris et Bruxelles, 1892. pp.
xii–5–490. 8vo. Ill.

> This extensive work is published by the author and is not in the trade. The
> preface is dated Arcachon, 25 November, 1891. The second part may
> be expected.

ZUNTZ, N. Blutgase und respirator. Gaswechsel. *See* Handbuch der Physiologie.

ZWICK, HERMANN.
Hydraulischer Kalk und Portland-Cement nach Rohmaterialien, physikalischen und chemischen Eigenschaften, Untersuchung, Fabrikation und Werthstellung unter besonderer Rücksicht auf den
gegenwärtigen Stand der Cement-Industrie. Wien, Pest, Leipzig,
1879. 8vo. Ill.
Kalk und Luftmörtel. Auftreten und Natur des Kalksteines, das Brennen desselben und seine Anwendung zu Luftmörtel. Nach gegen-

ZWICK, HERMANN. [Cont'd.]

wärtigem Stande der Theorie und Praxis dargestellt.   Wien, Pest,
Leipzig, 1879.   8vo.   Ill.

Lehrbuch der chemischen Technologie, zum Gebrauch beim Unterrichte
an technischen Lehranstalten, sowie zum Selbststudium für Chemiker, Techniker, Apotheker, etc.   München, 1870.   8vo.   Ill.

Wasserglas (Das) seine Natur und seine Bedeutung für die Industrie
und Technik.   Zürich, 1877.

For additional titles see ADDENDA at close of Section VII.

SECTION VI.

ALCHEMY.

For works on the History of Alchemy, see Section III; for Biographies of Alchemists, see Section IV. Section VI is purposely restricted to a small selection of representative works; those marked with asterisks are in my private library.

---

*A B C vom Stein der Weisen. Berlin, 1782. Bey Friedrich Maurer. 4 parts. pp. [xi]–7–318–[i]; 348, 100, 299, 325–64–[iii], 12mo. Illustrations and folding plates. Dedication signed by C. U. Ringmacher.

For an earlier edition, *see* Hermetisches A B C.

[*Contents :*]

PART I.

1. *Trismegistus, Hermes :* Die Smaragdtafel. p. 57.
2. *Baruch, Samuel :* Gabe Gottes. p. 61.
3. *Eleazar, Abraham :* Schrifft. p. 71.
4. Nodus sophicus enodatus. p. 73.
5. Kinderbett des Steins der Weisen d'un chevalier françois. p. 86.
6. *Barcius,* oder *Johann von Sternberg :* Gloria Mundi, kleine Paradeis-Tafel, oder Beschreibung der uralten Wissenschaft Lapidis Philosophorum. p. 95.
7. *Grashofer, Johann* [*Chortalasseus*] : Geheimnisse des grossen und kleinen Bauers und Cabala chemica. p. 121.
8. *Siebmacher, Johann :* Das güldene Vlies. p. 145.
9. *S., J. :* Wasserstein der Weisen. p. 156.
10. *Hautnorton, Josaphat Friedrich :* Vom philosophischen Salz. p. 157.
11. *Leona, Constantia* (Johanna Leade) : Sonnenblume der Weisen. p. 172.
12. *Brotoffer, Ratich :* Elucidarius major, oder Erleuchtung über die Reformation der ganzen Welt. p. 176.

943

A B C vom Stein der Weisen. [Cont'd.]

13. *Suchten, Alexander von :* De explicatione tincturæ physicorum Theophrasti Paracelsi. p. 195.

14. *S., A. v. :* De tribus facultatibus. p. 214.

15. *Nuysement, de :* Tractat vom wahren geheimen Salz der Weisen und dem allgemeinen Geiste der Welt. p. 230.

16. *Antonius de Abbatia :* Bericht von Verwandlung der Metallen. p. 261.

17. Der Aufrichtige deutsche Wegweiser zum Licht der Natur von dem Authore : Domino in limo, non malo malo. p. 263.

18. *Ventura, Laurentius :* Liber unus de lapide philosophorum. p. 268.

19. *Johannes von Padua :* Vollendete heilige Weisheit. p. 283.

20. *Ficinus, Marsilius :* Ein Büchlein vom Stein der Weisen ; aus dem Welschen Original ausgezogen und übersetzt. p. 294.

21. *Daustenius, Joannes :* Rosarium vom Stein der Weisen. p. 300.

22. *Trismegistus, Hermes :* Ein güldenes Tractätlein von der Zusammensetzung des Steins der Weisen. p. 311.

23. *Lullius, Raymundus :* Apertorium de compositione lapidis philosophorum. p. 312.

24. *L., R. :* Elucidarius über sein Testament und Codicill. p. 315.

25. *Aristoteles :* Alchymische Schrift an Alexandrum Magnum de lapide philosophorum. p. 316.

PART II.

1. *Spiess :* Concordantz über des Nuysement Sal coeleste : ex arcano hermeticæ philosophiæ. p. 15.

2. *Gutwasser, Benedict :* Aus dem aufrichtigen Glaubensbekenntniss. p. 41.

3. Das eröfnete philosophische Vaterherz. p. 56.

4. Das Buch " Amor Proximi," geflossen aus dem Oel der göttlichen Barmherzigkeit, geschärft mit dem Wein der Weisheit, bekräftiget mit dem Salz der göttlich- und natürlichen Wahrheit. p. 72.

5. Das aus der Finsternis von sich selbst hervor brechende Licht. p. 106.

6. Fama mystica hermetica. p. 163.

7. Der rechte Weg zu der hermetischen Kunst, nebst Anmerckungen über die Irrwege. p. 175.

8. Die geheime Naturlehre der hermetischen Wissenschaft, nach dem System des Sendivogii. p. 193.

9. *Welling, Georg von :* Opus mago-cabalisticum et theosophicum. p. 231.

A B C vom Stein der Weisen. [Cont'd.]

PART III.

PART IV.

A B C VOM STEIN DER WEISEN. [Cont'd.]

  2. Die Fürstlich-Monarchische Rose von Jericho, oder Moses Testament. p. 39.

  3. *Elias Artista :* Aus dem Geheimnis vom Salz. p. 47.

  4. Einzelne Zeugnisse aus *Hermann Fictuld's* beiden Classen des Probirsteins. p. 71.

  5. *Frydan, Johann Ferdinand von :* Zeugniss. p. 75.

  6. *Frydan, F. F. :* Das Licht des Lichts. p. 95.

  7. *Altenburg, Leonhard von :* Delarvatio tincturæ philosophorum, oder eine kurze einfältige Erklärung des Lapidis benedicti. p. 122.

  8. *Loen, von :* Geheimniss der Verbrennung und Verwesung aller Dinge in der grossen und kleinen Welt. p. 137.

  9. Aus dem microcosmischen Vorspiele eines neuen Himmels und der neuen Erde. p. 167.

  10. *Marsciano, Franciscus Onuphrius de :* Die Sendschreiben und die hermetische Untersuchung. p. 14.

ÆGYDIUS DE VADIS. *See* Theatrum chemicum.

ÆNIGMA EX VISIONE ARISLEI PHILOSOPHI ET ALLEGORIIS SAPIENTUM.
      See Artis auriferæ ; *also* Manget, J. J.

AGRIPPA VON NETTESHEIM, HEINRICH CORNELIUS.
  Of the Vanitie and Uncertaintie of Artes and Sciences. Englished by Ja. San[ford]. London, H. Wykes, 1569.
      * The Vanity of Arts and Sciences. "Vanity of Vanities, all is Vanity." London : Printed by J. C. for Samuel Speed, and sold by the Booksellers of London and Westminster, 1676. pp. [xviii]-368, 8vo. Portrait.
      In Chapter 90 the Author writes of Alchemy and the Vanity thereof. The first edition in Latin was published at Cologne in 1533. Editions and translations are numerous.

ALANUS AB INSULIS [ALAIN DE LISLE].
  Dicta Alani philosophi de lapide philosophico, e germanico idiomate latine reddita per Justum Balbian. Lugduni-Batavorum, 1599.
      * Dicta Alani das ist : Kurtze Lehr- und Unterricht-Sprüche von der Bereitung des grossen Stein der Weisen. Von einem alten Philosopho Alano de Insulis oder von Issle aus Flandern, beschrieben und hinterlassen. Nürnberg, Verlegts Johann Paul Krauss, Buchhändler in Wienn, nächst der Kayserlichen Burg. *n. d.* pp. from 308-324, 12mo.
      Alanus ab Insulis, or Alain de Lisle, born 1114, died 1202. The so-called Universal Doctor was the earliest Flemish alchemist.

ALANUS PHILOSOPHUS. *See* Theatrum chemicum.

ALBERTUS MAGNUS. *See* Theatrum chemicum ; *also* Petrus Bonus Ferrariensis, Pretiosa margarita.

ALBERTUS MAGNUS.

* De mineralibus et rebus metallicis libri quinque. Coloniæ, apud Joannem Birckmannum et Theodorum Baumium, 1569. pp. 39 [xi], 16mo.

De secretis mulierum, de virtutibus herbarum, lapidum et animalium, de mirabilibus mundi, ac de quibusdam effectibus causatis a quibusdam animalibus, etc. Argentorati, 1600. 16mo.

* Philosophie naturalis isagoge sive introductiones, emendate nuper et impresse summa diligentia, in libros phisicorum : De celo et mundo ; De generatione ; Metheororum ; De anima, Aristotelis ; cum annotaciunculis marginalibus. [Colophon :] Impressum Vienne Pannonie, per Joannem Singrenium ; expensis vero Leonhardi Alantse, liuis Viennensis, idibus decembribus anno 1514. ff. lxxvi, sm. 4to.

ALBINEUS, NATHAN. *See* Manget, J. J.

ALBINEUS, NATHAN.

* Bibliotheca chemica contracta ex delectu et emendatione N. A. in gratiam et commodum artis chemicæ studiosorum. Genevæ, 1654. 12mo.

Second edition, 1673.

Contains four alchemical treatises separately paged. Bibliographically of no value.

1. *Augurellus, Joannes Aurelius :* Chrysopœia et vellus aureum, seu chrysopœia major et minor, cum Nathanis Abinei carmine aureo. pp. 77.

2. *Sendivogius, Michael* (Divi leschi genus amo) : Novum lumen chemicum e naturæ fonte et manuali experimentis depromptum, cui accessit tractatus de sulphure. pp. 11–175.

3. Enchiridion physicæ restitutæ, in quo verus naturæ concentus exponitur, plurimique antiquæ philosophiæ errores per canones et certas demonstrationes dulcide aperiuntur. pp. 179.

4. Arcanum Hermeticæ philosophiæ opus, in quo occulta naturæ et artis circa lapidis philosophorum materiam et operandi modum canonice et ordinate sunt manifesta. pp. 83.

Nos. 3 and 4 are by the same anonymous author, who conceals his name in No. 3 under the anagram *Spes mea est in agno*, and in No. 4 under *Penes nos unda Tagi*. The anagrams are, however, not identical.

ALCHYMIÆ COMPLEMENTUM ET PERFECTIO.
    See Dreyfaches hermetisches Kleeblatt ; also Besondere Geheimnisse . . .

ALCHYMIST, DER VON MOSE UND DENEN PROPHETEN ÜBEL URTHEILENDE.
    See Der von Mose, etc.

ALETHOPHILUS. See Roth-Scholtz, Fr., Deutsches Theatrum chemicum ; also Trisme-
    gistus, Hermes.

ALEXANDER REX. See Artis auriferæ.

ALI PULI. See Quadratum alchymisticum.

ALLEGORIÆ SAPIENTUM SUPRA LIBRUM TURBÆ PHILOSOPHORUM.
    See Manget, J. J.; also Theatrum chemicum.

ALLEGORIÆ SUPER LIBRUM TURBÆ. See Artis auriferæ ; also Manget, J. J.

* ALLERHÖCHSTE (DAS), edelste, kunstreicheste Kleinod und der urälteste
    verborgene Schatz der Weisen in welchem die allgemeine Materia
    Prima, derselben nothwendige Präparation und überaus reiche
    Frucht des philosophischen Steins augenscheinlich gezeiget und
    klärlich dargethan wird. Frankfurt und Leipzig, 1755. pp.
    [xvi]-209, 12mo. Frontispiece and plates.

ALLERLEY PARTICULARIA. See Aureum Vellus.

ALPHONSUS REX CASTELLÆ. See Theatrum chemicum.

ALTENBURG, LEONHARD VON. See A B C.

ALVETANUS, CORNELIUS ARNSRODIUS. See Theatrum chemicum.

AMOR PROXIMI. See A B C.

ANCIENNE (L') GUERRE DES CHEVALIERS, OU LE TRIOMPHE HERMÉTIQUE.
    See R[ichebourg], J. M. D.

ANDER (EIN) TRACTÄTLEIN SEHR NUTZLICH ZU LESEN.
    See Roth-Scholtz, Fr., Deutsches Theatrum chemicum.

ANDREÄ, JOHANN VALENTIN.
    * Chymische Hochzeit Christiani Rosencreütz, Anno 1459. Arcana
    publicata vilescunt : et gratiam prophanata amittunt. Ergo : ne
    margaritas obijce porcis, seu asino substerne rosas. Strassburg,
    in Verlägung Lazari Zetzners S. Erben, 1616. pp. 143, 16mo.
        Published anonymously. According to George Soane this curious work is an
        attack on the pretended brethren of the Rosy Cross, written by Johann
        Valentin Andreä, the reputed author of the Fama Fraternitatis (pub-
        lished at Frankfurt in 1617). For an English translation of this work
        consult A. E. Waite's Real History of the Rosicrucians, London, 1887.

ANDRENAS, PHIL., SIEUR D'AUBIGNY.

Extrait d'un livre intitulé or potable Levain, ou Discours de l'or potable Levain. Et l'offre fait au public d'en faire de très parfait en presence de témoins, aux conditions de deux millions de livres de recompense. Paris, 1674. 12mo.

> Of this Lenglet du Fresnoy remarks : " Ho ! c'est mettre le secret à un trop haut prix ; et je doute que l'auteur ait été pris au mot ; he ! que le pourroit faire."

ANDREWES, ABRAHAM. *See* Ashmole, E. : Theatrum chemicum Britannicum.

* ANNULUS PLATONIS, oder physikalisch-chymische Erklärung der Natur nach ihrer Entstehung, Erhaltung und Zerstöhrung, von einer Gesellschaft ächter Naturforscher aufs neue verbessert und mit vielen wichtigen Anmerkungen herausgegeben. Berlin und Leipzig, bei George Jacob Decker, 1781. pp. xxviii–551, 8vo. Ill.

ANTHONIUS DE ABBATIA.

See Roth-Scholtz, Fr., Deutsches Theatrum chemicum ; *also* A B C.

ANTONIE, FR. *See* Collectanea chymica.

AQUINAS, THOMAS. *See* Theatrum chemicum.

AQUINAS, THOMAS.

Secreta alchimiæ magnalia, de corporibus supercœlestibus, et quod in rebus inferioribus inveniantur, quoque modo extrahantur de lapide minerali, animali et plantali ; item thesaurus alchimiæ secretissimus . . . Opera Danielis Brouchouisii, editio tertia. Lugduni-Batavorum, 1612. 12mo.

ARIA VON DER WEISEN STEIN. *See* A B C.

ARISLEUS. *See* Manget, J. J. ; *also* Theatrum chemicum ; *also* R[ichebourg], J. M. D. ; *also* Artis auriferæ.

ARISTOTELES. *See* A B C ; *also* Manget, J. J. ; *also* Theatrum chemicum ; *also* Arnaldus de Villa Nova : Schrifften ; *also* Artis auriferæ.

ARNALDO DE VILLA NOVA.

Also called Arnaldus Villanovanus, Arnald Bachuone, Arnauld de Villeneuve, Arnaldus Novicomensis. *See* Arnaldus de Villanova.

See Hermetischer Rosenkranz ; *also* Petrus Bonus Ferrariensis : Pretiosa Margarita ; *also* Manget, J. J. ; *also* Theatrum chemicum ; *also* Artis auriferæ ; *also* Roth-Scholtz, Fr. : Deutsches Theatrum chemicum.

ARNALDUS DE VILLA NOVA.

* Chymische Schriften. Allen Liebhabern der wahren Alchimie zu Gefallen aus dem latein mit höchstem Fleisz in teutscher Sprache

ARNALDUS DE VILLA NOVA. [Cont'd.]

übersetzet durch Johannem Hoppodamum. Verlegts Johann Paul Krauss, Kayserlicher und Königlicher priviligirter Niederlags-Verwandter, Buchhändler in Wienn, 1742. pp. [xvi]–411–[v], 16mo.

[*Contents :*]

1. Rosarius philosophorum, das ist ein gründlicher Bericht von der wahren Zusammensetzung der natürlichen Philosophie dadurch alle unvollkommene Metallen zu wahrem Gold und Silber verwandelt werden. p. 1.

2. Ein Buch welches Novum Lumen oder das neue Licht genennet wird. p. 119.

3. Flos florum ; Das aller vollkommenste Magisterium und die hertzlichste Freude Magistri Arnoldi de Villa Nova an den Grossmächtigen König Arragoniæ geschrieben : Welches dann eine Blume aller Blumen, ein Schatz aller Schätze und ein edle Perle ist. Darinnen des wahren Elixirs Composition oder Zusammensetzung, beydes zum Weissen und zum Rothen, nemlich auf Gold und Silber, kürtzlich verfasset, und in guter Erklärung zu befinden. p. 145.

4. Spiegel der Alchemie, in welchem auch die allerverborgensten Geheimnüsse der Kunst deutlich eröffnet und so hell und klar, als vergönnet, und immer möglich ist, erkläret werden. Nun zum ersten mahl in hoch-teutscher Sprache herausgegeben durch J. L. M. C. p. 173.

5. Gedichte des Arnoldi de Villa Nova. p. 249.

6. Eine Epistel wegen der Alchimie an den Neapolitanischen König geschrieben. p. 253.

7. Der Prophetin *Mariæ* Moysis Schwester, Practica in der Kunst der Alchimie. p. 263.

8. *Calid filius Jazichi :* Ein Buch von der Alchimie Geheimnüssen. Aus dem hebräischen in die arabische und aus derselbigen in die lateinische Sprach durch einen unbekandten Authorn ; nunmehr aber auch aus solcher in unsere teutsche versetzet. p. 273.

9. *Kallid, Rachaidibus :* Ein Buch von den dreyen Worten. p. 321.

10. *Aristoteles :* Ein Tractätlein von der Practica des philosophischen Steins. p. 337.

11. Der Tractat so der Weiber Arbeit und der Kinder-Spiel genennet wird. p. 359. [Running title : Ludus puerorum.]

ARTEFIUS. *See* Manget, J. J. ; *also* Theatrum chemicum ; *also* R[ichebourg], J. M. D. ; *also* Ripley, George : Schrifften ; *also* Hermetischer Rosenkranz.

ARTIS AURIFERAE QUAM CHEMIAM VOCANT. Basileæ, 1572. 2 vols., 16mo.
    Second edition under the title :

* Artis auriferæ quam chemiam vocant volumen primum, quod continet
    turbam philosophorum, aliosque antiquissimos autores ; cum indice
    rerum et verborum locupletissimo. Basileæ, excudebat Conrad
    Waldkirch, expensis Claudii de Marne et Joan Aubry, 1593. 2
    vols., 16mo. Vol. I, pp. [xvi]–631–[xxxii] ; vol. II, pp, 525–[xxii].

[*Contents :*]

VOL. I.

1. Propositiones seu maximæ artis Chymicæ. p. [x].
2. [*Arisleus.*] Turba philosophorum in secunda philosophia longe
    diversa et copiosior quam reliquæ quæ passim circumferun-
    tur. p. 1.
3. Turbæ philosophorum alterum exemplar. p. 66.
4. Allegoriæ super librum turbæ. p. 139.
5. Ænigmata ex visione Arislei et allegoriis sapientum. p. 146.
6. In turbam philosophorum exercitationes. p. 154.
7. Aurora consurgens, quæ dicitur aurea hora. p. 185.
8. Rosinus ad Euthiciam. p. 246.
9. Rosinus ad Saratantam episcopum. p. 277.
10. Liber divinarum interpretationum et definitionum. p. 316.
11. Practica Mariæ prophetissæ in artem alchimicam. p. 319.
12. *Calid :* Liber secretorum alchemiæ.
13. *Kalid, Rachaidibus :* Liber trium verborum.
14. *Aristoteles :* De lapide philosophorum. p. 361.
15. *Avicenna :* De congelatione et conglutinatione lapidis. p. 374.
16. Expositio epistolæ Alexandri regis. p. 382.
17. Autor ignotus : De secretis lapidis philosophorum. p. 389.
18. *Merlinus :* Allegoria de arcano lapidis. p. 392.
19. *Kalid, Rachaidibus Veradianus :* De materia philosophici lapidis.
    p. 397.
20. *Avicenna :* Tractatulus de alchimia. p. 405.
21. *Arnoldus de Villa Nova :* Semita semitæ. p. 437.
22. Clangor buccinæ. p. 448.
23. Correctio fatuorum. p. 545.
24. Liber de arte chymica. p. 575.

VOL. II.

25. *Morienus Romanus :* Sermo. p. 7.
26. *Trevirensis, Bernardus.* [*Trevisanus*] : Responsio ad Thomam de
    Bononia medicum regis Caroli octavi. p. 55.
27. Scala philosophorum. p. 107.

ARTIS AURIFERAE QUAM CHEMIAM VOCANT.   [Cont'd.]

28. Opus mulierum et ludus puerorum.   p. 171.

29. Rosarium philosophorum.   p. 204.

30. *Arnaldus de Villa Nova :* Rosarium cum figuris.   p. 385.

31. *A. d. V. N. :* Novum lumen.   p. 456.

32. *A. d. V. N. :* Flos florum.   p. 470.

33. *A. d. V. N. :* Nova epistola super alchymia ad regem Neapolitanum.   p. 488.

34. *Baco, Rogerius :* De mirabili potestate artis et naturæ.   p. 494.

> Of this work a third edition in three volumes, Basileæ, 1610, is named by
> Lenglet du Fresnoy.   According to Schmieder this collection was made
> and edited by Gul. Gratarolus (b. 1516, d. 1568), Professor of Medicine at
> Marburg, and afterwards at Basel.   *Cf.* Gratarolus, G. : Veræ alchemiæ,
> etc.   For a German translation, *see* Morgenstern, Philippus Islebiensis.

ASHMOLE, ELIAS.

Arcanum, or the great secret of the Hermetical Philosophy.   London,
1650.   8vo.

* Theatrum chemicum Britannicum ; containing severall poeticall Pieces
of our Famous English Philosophers, who have written the Her-
metique Mysteries in their owne Ancient Language.   Faithfully
collected into one volume with Annotations thereon by E. A., qui
est Mercuriophilus Anglicus.   The first Part.   London, printed by
J. Grismond for Nath. Brooke, at the Angel in Cornhill, 1652.
pp. [xvi]-486-[viii], 4to.   Plates.   Portrait inserted.

[Contents :]

1. *Norton, Thomas :* Ordinall of Alchemie.   p. 1.

2. *Ripley, George :* Compound of Alchemie.   p. 107.

3. Pater Sapientiæ.   p. 194.

4. Hermes's Bird.   p. 211.

5. *Chaucer, Geoffry.*   p. 227.

6. *Dastin, John :* Dastin's Dreame.   p. 257.

7. *Pearce the Black Monke,* upon the Elixir.   p. 269.

8. *Richard Carpenter's* Work.   p. 275.

9. *Andrewes, Abraham :* Hunting of the Greene Lyon.   p. 278.

10. *Charnock, Thomas :* Breviary of Naturall Philosophy.   p. 291.

11. *Charnock, Thomas :* Ænigmas.   p. 303.

12. *Bloomefield, William :* Bloomefield's Blossomes.   p. 305.

13. *Sir Edward Kelley's* Worke.   p. 324.

14. *Sir Edward Kelley* to G. S. Gent.   p. 332.

15. *John Dee's* Testament.   p. 334.

16. *Thomas Robinson* of the Philosophers' Stone.   p. 335.

17. Experience and Philosophy.   p. 336.

ASHMOLE, ELIAS. [Cont'd.]

18. The Magistery. W. B. p. 342.
19. *Gower, John*, upon the Philosophers' Stone. p. 368.
20. *George Ripley's* Scrowle. p. 375.
21. Mystery of Alchymists. p. 380.
21. *Ripley, George :* Preface to the Medulla. p. 389.
22. *Ripley, G. :* A Short Work. p. 393.
23. *Lydgate, John :* Secreta Secretorum. p. 397.
24. Hermit's Tale. p. 415.
25. Description of the Stone. p. 420.
26. The Standing of the Glass. p. 421.
27. *Redmann, W. :* Ænigma Philosophicum. p. 423.
28. Fragments. p. 424.

> A minute account of this work and of its compiler Ashmole will be found in Dr. Kipp's Biographica Britannica, I, 298 (1778). Part II was never issued. Ashmole was the well-known English antiquary who founded the Ashmolean Museum, Oxford. Lenglet du Fresnoy says of him : " Il avoit cette folie en tête sans peut-être la pratiquer, ou du moins sans y réussir."

* Way (The) to Bliss ; in three Books. London, printed by John Grismond for Nath. Brook, at the Angel in Corn-hill, 1658. pp. [vi]–220, sm. 4to.

AUBERTUS, JAC.

Explicatio de ortu et causis metallorum contra chemicos. Lugduni-Batavorum, 1575. 8vo.

> The author, a Lausanne physician, attacks the current alchemical views, and was himself answered by Joseph Quercetanus (Lyons, 1600).

AUBIGNÉ DE LA FOSSE, NATHAN. *See* Albineus.

AUFRICHTIGE (DER) DEUTSCHE WEGWEISER ZUM LICHT DER NATUR. *See* A B C.

AUGURELLUS, JOANNES AURELIUS.

> *See* Manget, J. J.; *also* Theatrum chemicum ; *also* Trois anciens traictez de la philosophie naturelle.

AUGURELLUS, JOHANNES AURELIUS.

* Vellus aureum et Chrysopœia, seu Chrysopœia major et minor, das ist Gülden-Vliess, und Golderzielungs-Kunst, oder grosse und kleine Golderzielungs-Kunst ; an ihre Päbstliche Heiligkeit Leonem den Zehenden. Aus dem lateinischen ins teutsche übersetzet von Valentino Weigelio. Hamburg, zu finden bey Samuel Heyl, 1716. pp. [xii]–112, 16mo. Illustrated title-page.

> It is related that when Augurellus presented this poem to Pope Leo X the latter rewarded the poet with an empty purse, saying that he who knew so well how to make gold could easily fill it.

* AUREUM VELLUS, oder güldin Schatz und Kunst-Kammer : darinnen der aller fürnembsten, fürtreffenlichsten, auserlesenesten, herrlichsten und bewehrtesten Auctorum, Schrifften, Bücher, aus dem gar uhralten Schatz der überbliebnen, verborgenen, hinterhaltenen Reliquien und Monumenten der Ægyptiorum, Arabum, Chaldæorum und Assyriorum Königen und Weisen. Von dem edlen, hocherleuchten, fürtreffenlichen bewehrten Philosopho Salomone Trismosino (so des grossen Philosophi und Medici Theophrasti Paracelsi Præceptor gewesen) in sonderbare unterschiedliche Tractetlein disponirt und in das deutsch gebracht. Sampt anderen philosophischen alter und newer Scribenten sonderbaren Tractetlein, alles zuvor niemalen weder erhört noch gesehen wie der Catalogus zu verstehen gibt. Durch einen der Kunst-Liebhabern mit grossen Kosten, Mühe, Arbeit und Gefahr die Originalia und Handschrifften zusammen gebracht, und auffs trewlichste und fleissigst an Tag geben. Erstlich gedruckt zu Rorschach am Bodensee, anno MDXCIX. Part I, pp. 215 ; II, pp. 165 ; III, pp. 701. 16mo. Ill.

> Contains 96 treatises—portrait (profile) of Paracelsus and rude cuts. Compare Morgenstern.

> *** Aureum Vellus**, oder guldin Schatz und Kunst-Kammer, darinnen der aller fürnemisten, fürtreffenlichsten, auserlesenesten, herrlichisten und bewehrtesten Auctorum, Schrifften und Bücher aus dem gar uralten Schatz der überbliben, verborgnen, hinterhaltenen Reliquien und Monumenten der Ægyptiorum, Arabum, Chaldæorum und Assyriorum Königen und Weysen. Von dem edlen, hocherleuchten fürtreffenlichen bewehrten Philosopho Salomone Trismosino (so dess grossen Philosophi und Medici Theophrasti Paracelsi Præceptor gewesen) in sonderbare underschiedliche Tractätlein disponiert und in das teutsch gebracht. Sampt anderen philosophischen alter und newer Scribenten sonderbaren Tractätlein, alles zuvor niemalen weder erhört noch gesehen wie der Catalogus gleich nach der Vorrede zuverstehen gibt. Durch einen der Kunst-Liebhabern mit grossen Kosten, Mühe, Arbeyt und Gefahr die Originalia und Handschrifften zusammen gebracht, und auffs trewlichest und fleissigst an Tag geben. Vormahls gedruckt zur Rorschach am Bodensee anno M.D.XCVIII und zu Basel, 1604, in fünff verschiedene Tractaten ; itzo aufs neue auffgelegt und in ein Volumen gebracht. Hamburg, bey Christian Liebezeit, 1708. pp. [xvi]-816. Ill.

Aureum Vellus. [Cont'd.]

[*Contents :*]

TRACTATUS I.

TRACTATUS II.

AUREUM VELLUS. [Cont'd.]

B. Korndorffer, Bartholomaus :

1. Die geistlich Sonn auss dem Buch Carneson.   p. 129.

2. Liber tincturarum particularum.   p. 132.

     a. Particular mit dem brennenden Himmel.   133.

     b. Zu einbringen vier Loth ☉ in die Marck ☽.   p. 134.

     c. Coagulatio Mercurii.   p. 135.

     d. Ein geheim suginolisch Stuck.   p. 136.

     e. Zwo philosophische und coloffonische Figierungen.   p. 138.

     f. Wie man die sieben Metall clarificieren soll.   142.

     g. Der viscosisch Bronn oder Gradierwasser.   p. 145.

     h. Gradierwasser auss dem ungarischen Coagulat.   p. 146.

     i. Process eines geheimen Gradierwassers.   p. 147.

     j. Zwey geheyme fürtreffenliche hohe Augment in der Geistlich-
         heit mit Sol und Luna.   p. 148.

     k. Tinckturisch Gradieröl mit dem schwarzen Adler.   p. 150.

     l. Process mit dem Silber.   p. 150.

     m. Tinckturisch Zimentpulfer.   p. 153.

     n. Particularstuck.   p. 153.

     o. Ignis perpetuus.   p. 155.

     p. Tinctura patris Gregorii.   p. 158.

     q. Oleum Mercurii.   p. 156.

     r. Zwey ewige . . . Liechter.   p. 159.

TRACTATUS III.

1. Splendor Solis mit seinen Figuren.   p. 166.   [22 plates.]

2. *Poyselius, Ulrichus :* Spiegel der Alchymey.   p. 214.

3. Clavis sambt seiner Declaration der chymischen Handgriffen. **p. 229.**

     a. Aqua regis philosophorum.   p. 231.

     b. Mercurius essensiuicatus.   p. 231.

     c. Das gefrorne Eysswasser.   p. 231.

     d. Aqua mercurialis philosophorum.   p. 232.

     e. Mercurius vitæ communis.   p. 232.

     f. Lac virginis.   p. 233.

     g. Quinta essentia tartarisata.   p. 234.

     h. Præparatio et sublimatio Solis in Mercurium.   p. 234.

     i. Compositio et solutio Mercurii Solis cum Mercurio vitæ com-
         munis.   p. 235.

     j. Ablutio cum quinta essentia tartarisata.   p. 235.

     k. Processus et fixatio lapidis philosophorum.   p. 236.

     l. Augmentatio lapidis philosophorum.   p. 236.

     m. Usus lapidis philosophorum.   p. 236.

4. Allerley Particularia.   p. 237.

AUREUM VELLUS. [Cont'd.]

14. Ein schönes Tractätlein. p. 554.
15. Schenken : Ein schön Stuck. p. 563.
16. Vom Antimonio Philosophorum. p. 570.
17. Oleum antimonii. p. 579.

TRACTATUS V.

1. Von Offenbarum der philosophischen Materien und Dingen. p. 593.
2. Thesaurus philosophiæ Euferarii. p. 606.
3. Tractatus, darinnen das ganze Secret der Alchymey von dem Stein der Weisen begriffen ist. p. 623.
4. Lux lucens in tenebris. p. 636.
5. Tractatus de vitriolo philosophorum. p. 642.
6. [*Baco, Rogerius*] : Oleum vitrioli. p. 651.
7. *Lullius, Raimundus :* Experimenta. p. 664.
8. [*Zacharias, Dionysius*] : Das Büchlein der Philosophi der Metallen. p. 727.
9. Ein Streit und Gespräch des Golds und Mercurii wider den Stein der Weisen. p. 765.
10. Korndorfferische Particularia. p. 773.
11. Spiegel der Philosophey. p. 787. [12 illustrations.]

AURORA CONSURGENS. *See* Artis auriferæ.

AUS DEM MICROCOSMISCHEN VORSPIELE EINES NEUEN HIMMELS. *See* A B C.

AUS (DAS) DER FINSTERNISS VON SICH SELBST HERVORBRECHENDE LICHT. *See* A B C.

AVICENNA. *See* Manget, J. J.; *also* Theatrum chemicum; *also* Aureum Vellus; *also* Artis auriferæ; *also* S[chröder, Fr. J. W.].

B., H. C. A. V. T.

Explicatio chimica characteris spiritus mundi in quo occultantur scientiarum omnium arcana; das ist : chimische Erklärung der Abbildung oder Bezeichnung des Welt-Geistes Mercurii in welchem alle Geheimnisse natürlicher Wissenschafften verborgen liegen, auff unterschiedliches Ansuchen aus dem lateinischen in das teutsche übersetzet an vielen Orten vermehret und deutlicher erkläret. *n. p.*, 1690. pp. 66, 18mo.

BACO, ROGERIUS.

See Manget, J. J.; *also* Collectanea chymica; also Theatrum chemicum; *also* S[chröder, Fr. J. W.]; *also* Artis auriferæ; *also* Roth-Scholtz, F. (Deutsches Theatrum chemicum).

BACON, ROGER.

Frier Bacon, his Discovery of the Miracles of Art, Nature and Magick, faithfully translated out of Dr. Dee's own copy by T. M. and never before in English. London, 1659. 12mo.

* Miroir (Le) d'alquimie. Traduit de latin en français par un gentilhomme du D'aulphiné. Le page suivant déclare le contenu en cest œuvre. À Lyon, par Macé Bonhomme, 1557. Avec privilege du Roy. pp. 134, 24mo.

[*Contents :*]

1. Bacon, Rogier : Miroir d'alquimie. p. 5.
2. Table d'esmeraude de Hermes Trimegiste. p. 35.
3. Hortulain : Petit commentaire dict des jardins maritimes, sur la table d'esmeraude d'Hermes Trimegiste. p. 39.
4. Calid filz de Iazic Juif : Le livre des secretz d'alquimie. p. 57.
5. Le miroir de maistre Jean de Mehun. p. 109.

Mirror (The) of Alchimy composed by R. B. Also a most excellent and learned Discourse of the admirable Force and Efficacie of Art and Nature written by the same author, with certaine other worthie Treatises of the like Argument. (The Smaragdine Table of Hermes Trismegistus, a commentarie of Hortulanus, the booke of the secrets of alchemie by Calid the son of Jazich, London, 1597. 4to.)

Opera quædam hactenus inedita. Vol. I, Opus tertium ; vol. II, Opus minus ; III, Compendium philosophiæ ; [and in Appendix] Epistola . . . de secretis operibus artis et naturæ . . . Edited by J. S. Brewer. London, 1859. 8vo.

Scripta de arte chymiæ. Francofurti, 1603. 12mo.

BALBIAN, JUSTUS À. *See* Theatrum chemicum.

BALDUINUS, CHRISTIANUS ADOLPHUS. *See* MANGET, J. J.

BALDUINUS, CHRISTIANUS ADOLPHUS.

* Aurum superius et inferius auræ superioris et inferioris hermeticum. Francofurti et Lipsiæ, 1675. pp. [xxxii]–174, 24mo. Ill.

BARCIUS, ODER JOHANN VON STERNBERG.

See A B C ; *also* Musæum hermeticum ; *also* Roth-Scholtz, Fr. (Deutsches Theatrum chemicum).

BARLET, ANNIBAL.

Le vray et méthodique cours de la physique resolutive vulgairement dite chymie, représenté par figures générales et particulaires pour connoistre la théotechnie ergocosmique, c'est à dire l'art de Dieu en l'ouvrage de l'univers. Paris, 1653. pp. (viii)–626, 4to. Frontispiece and plates.

BARNAUDUS NICOLAUS À CRISTA ARNAUDI DELPHINATIS. *See* Manget. J. J.

BARUCH, SAMUEL. *See* A B C.

BARUCH, SAMUEL.

* Donum Dei Samuelis Baruch des Juden Rabbi, Astrologi und Philosophi
  gebohren aus dem Stamm Abrahams, Isaacs, Jacobs und Juda.
  Welcher erlernet das grosse Geheimniss des grossen Meisters Tubal-
  kains aus dessen Tabell gefunden von Abrahamo Eliazare dem
  Juden. *n. p.* I. N. U. CXI. pp. 86–[xiv], 12mo.

    Cf. Eleazar, Abraham.

BASILIUS VALENTINUS. *See* Valentinus, Basilius.

BAUER, A.

* Chemie und Alchymie in Oesterreich bis zum beginnen des XIX. Jahrhundert. Eine Skizze. Wien, 1883. pp. 86, 16mo.

    A readable sketch. Contains authentic portraits of Paracelsus and Rudolph
    II., and a facsimile of Wenzel Seiler's medallion, 1677.

BAUER, GEORG. [Editor.]

* Nützliche Versuche und Bemerkungen aus dem Reiche der Natur, allen
  Erz- und Naturkündigern, Bergwerksverwandten, wie auch denen
  Liebhabern der Alchimie zum Gebrauch und Nutzen herausgegeben.
  Mit Kupfern. Nürnberg, *n. d.* [1760]. pp. 214, 12mo.

    Contains 23 treatises, and facsimiles of alchemistic coins.

BECHER, JOHANN JOACHIM.

    See Manget, J. J.; *also* Roth-Scholtz, Fr. (Deutsches Theatrum chemicum).

BEDENCKEN ÜBER DIE FRAGE : OB DIE TRANSMUTATIO METALLORUM MÖGLICH ?

    See Roth-Scholtz, Fr. (Deutsches Theatrum chemicum).

BENIAM, MUTAPHIA.

Epistola hebraica et latina de auro potabili. *In his* " Sententiis sacro
  medicis." Hamburg, 1640. 8vo.

    The author, a Hebrew physician, who died at Amsterdam in 1674, ascribes
    the origin of alchemy to the earliest period in the history of the Israelites
    as related by Moses and others.

BERICHT WAS DIE WAHRHAFTIGE KUNST ALCHEMEY INNHALT. *See* Aureum Vellus.

BERNARDUS TREVISANUS. *See* Trevisanus, Bernardus.

BERRY, F. HABERT DE. [Translator.]

    See Trois anciens traictez de la philosophie naturelle.

BERTHELOT, ANDRÉ.

Rapport sur les manuscrits alchimiques de Rome. Archives des Missions scientifiques et littéraires. III Série, Tome XIII. Paris, 1887. pp. 819–854.

BERTHELOT, M., AND CH.-EM. RUELLE. [Editors.]

* Collection des anciens alchimistes grecs publiée sous les auspices du ministère de l'instruction publique. Paris, 1887–88. 4to.

> Published in three livraisons, the introduction, Greek text, and the translation each having separate pagination : Introduction, xxviii–268. Greek text, 1–459. Translation, 1–428. Illustrated.
>
> This fountain-head of information on the earliest authentic manuscripts of chemistry displaces the earlier works of Reuvens and of Leemans, *q. v.*

BESEKE, JOHANN MELCHIOR GOTTLIEB.

* Ueber Elementarfeuer und Phlogiston als Uranfänge der Körperwelt insbesondere über electrische Materie. In einem Schreiben an Herrn Director Achard in Berlin. Leipzig, in der Johann Gottfried Müllerschen Buchhandlung, 1768. pp. 52, 8vo.

* BESONDERE GEHEIMNISSE EINES WAHREN ADEPTI VON DER ALCHYMIE zum Gebrauch und Nutzen deren Liebhabern herausgegeben und mit Figuren erläutet von C. G. H. Dresden, bey Johann Nicolaus Gerlach, 1757. pp. [xii]–276–[xx], 12mo. Frontispiece and plates.

> Contains the same treatises as Dreyfaches Hermetisches Kleeblatt, *q. v.*

[*Contents :*]

1. Mercurius redivivus, oder Unterricht wie man den philosophischen Stein sowohl den weissen als den rothen aus dem Mercurio machen solle. p. 1.

2. Catholicon physicorum, oder Unterricht wie man die Tincturam physicam und alchymicam, so von den alten Philosophis sehr ämsig gesucht worden machen solle, aber noch von keinem Auctore, weder von Alten noch neuen gänzlich und vollkommentlich ans Licht gegeben worden benebst dieser Tinctur Abkürzung. Bey welchem Tractätlein mit zu finden ist, der dreyfache Zweig von der Composition der Jungfrauen Milch, oder des philosophischen Essigs. Auctor inartus. Dieser Stein ist dreyeckigt in Esse oder Wesen, viereckigt in Qualitate oder Beschaffenheit. p. 25.

3. Venus vitriolata in elixir conversa, nec non Mars victoriosus seu elixirizatus, oder Unterricht wie man den philosophischen Stein sowohl aus dem Venere oder dem Kupfer, als aus dem Marte oder dem Eisen machen solle. p. 43.

BESONDERE GEHEIMNISSE, etc. [Cont'd.]

4. Elixir seu medicina vitæ, oder Unterricht wie man das wahre Aurum et Argentum potabile machen soll, samt derselben Kraft und Wirkung, nach der Meynung der alten und neuen Philosophen ; bey welchen auch zu finden, der Modus, wie man das Glas möge zurichten, das sichs ziehen, treiben und hämmern lasse. p. 63.

5. Saturnus saturatus, dissolutus et cœlo restitutus, oder Unterricht wie man den philosophischen Stein, so wohl den weissen als den rothen, aus dem Bley machen soll. Wie den auch eben auf solche Weise aus dem Jove oder Zinn. Bey welchem zu finden, die Abkürzung des Wercks Saturni, zugleich auch der Modus das Quecksilber aus dem Bley zu extrahiren. p. 85.

6. Metamorphosis lapidum ignobilium in gemmas quasdam pretiosas, oder Unterricht wie man aus kleinen Perlen, grosse und gute machen solle : Auch wie man die künstlichen Carfunkeln, und andere köstliche Steine zurichten solle, welche weit edler als die natürlichen seyen sollen. Bey welchen auch der Modus gefunden wird, wie man das Electrum artificiale, als das höchste aller Elixirium mit Andeutung des natürlichen, und metallischen Electri (so den Alten ganz unwissend gewesen) bereiten solle. p. 117.

7. Alchymiæ complementum et perfectio, oder Unterricht wie man alle Steine und Elixiria augmentiren oder multipliciren solle, sowohl in virtute als in qualitate, auch wie man zugleich die Projection verrichten solle. Bey welchen auch zu finden ist, die Erläuterung der Philosophen Meynung, wenn sie von der zehenden Zahl reden, in welcher das Werck solle gemacht werden. Ingleichen das Wundergeheimniss des animalischen Steines aus derer Menschen-Blute nach der Meynung Georgii Riplei. Endlich ist auch dazu gesetzt der Beschluss des Buchs welches alles denen 6 vorhergehenden alchymistenschen Büchern, sehr dienlich, wie auch diesen, so sonsten zuvor herausgegangen sind, seyen wird. p. 135.

8. Ein Tractätlein welches von denen philosophischen Schriften handelt, so die Alten über die Alchymie haben lassen ausgehen. Und begreift in sich die Auslegung der dunklen Wörter, Namen und seltsamen Reden, so in und bey dieser Kunst geführet werden. p. 163.

9. *Ficinus, Marsilius :* Vom Stein der Weisen Lapis Philosophorum genannt. p. 183.

Besondere Geheimnisse, etc. [Cont'd.]

    10. *Nuysement, de :* Von dem wahren geheimen Salz der Philosophen und allgemeinen Weltgeiste. p. 233.

      De Nuysement's treatise is more complete in Dreyfaches Hermetisches Kleeblatt.

Beuther, David.

    * Universal und Particularia, worin die Verwandelung geringer Metalle in Gold und Silber klahr und deutlich gelehret wird, nebst einem Anhange von unvergleichlich curieusen alchymischen Kupffern, darin die Kunst von Anfang bis zum Ende vorgemahlet ist, und einer Vorrede von Beuthers Person und Schrifften, Joh. Christoph Sprögels. Hamburg, bey Samuel Heyl in der St. Johannis Kirche, 1718. pp. [xxx]–140–[iv], 12mo. Plates.

      Beuther was alchemist to Elector Augustus of Saxony, 1575 to 1582. His adventures, imprisonment, and suicide are graphically told by Kunkel and by Wiegleb.

* Beyden (Die) Hauptschriften der Rosenkreuzer, die Fama und die Confession. Kritisch geprüfter Text mit Varianten und den seltenen lateinischen Original der zweyten Schrift ; nebst Einleitung und angehängten Verzeichniss einiger andern Rosenkreuzerschriften. Frankfurt a. M., 1827. pp. xii–95, 8vo.

Bibliothèque des philosophes chimiques.

    See R[ichebourg] ; *also* Salmon : Bibliothèque.

Biltdorffer, Georg. *See* Aureum Vellus.

Birckholtz, Adam Michael. [AdaMah Booz.]

    * Sieben (Die) heiligen Grundsäulen der Ewigkeit und Zeit, in deutlichen Sinnbildern zum Besten aller Weisheit-Suchenden. Nebst dem Brunnen der Weisheit und Erkenntniss der Natur. Den Grundsätzen der wahren Alchemie und vier merkwürdigen Briefen eines Adepten. Leipzig, bey Paul Gotthelf Kummer, 1783. pp. 132, 8vo.

Blawenstein, Salomon de.

    Interpellatio brevis ad philosophos veritatis tam amatores, quam scrutatores pro lapide philosophorum, contra Antichymisticum Mundum Subterraneum Athanasii Kircheri Jesuitæ ; qua non solum antichymistica ejus putatitia argumenta subnervantur, sed et ars ipsa quantum fieri potest intelligentibus manifestatur. Biennæ, apud Bernates, 1667. pp. [30], 4to.

      Polemical ; a reply to the attacks of Kircher. Also in Manget's Bibliotheca chemica curiosa, I, 113.

Bloomefield, William. *See* Ashmole, E.: Theatrum chemicum Britannicum.

Bodenstein, Adam von.

    Tractatus de veritate alchemiæ. Basel, 1560.

BOLNEST, EDWARD.

* Aurora chymica, or a rational way of preparing animals, vegetables and minerals for a physical use ; by which preparations they are made most efficacious, safe, pleasant medicines for the preservation and restoration of the life of Man. London, printed by Tho. Ratcliffe, and Nat. Thompson for John Starkey at the Miter within Temple-Bar, 1672. pp. [xvi]–148, 16mo.

BOLTON, HENRY CARRINGTON.

* Contributions of Alchemy to Numismatics. Read before the New York Numismatic and Archæological Society, Dec. 5, 1889. New York, 1890. pp. 44, 4to. Ill.

    Cf. Reyher, Samuel.

BONUS, PETRUS, FERRARIENSIS. *See* Petrus Bonus Ferrariensis.

BOOZ, ADAMAH. *See* Birckholtz, Adam Michael.

BORCH, OLE. *See* Borrichius, Olaus.

BORRI, FRANCESCO GIUSEPPI [or BURRHI, FRANCISCUS JOSEPH].

* Relatio fidei, actionum ac vitæ Burrhianæ, das ist : Eine Erzehlung des Glaubens, Thaten und Leben des berühmten Italiäners F. J. Burrhi. Welcher umb der Gleichheit der Materie halber beygefüget ist die Historia de tribus hujus seculi famosis impostoribus, nemlich Padre Ottomanno, Mahomed Bei, oder Johann Michael Cigala, und Sabatai Sevi. Gedruckt Anno 1670. *n. p.* pp. [xl], 24mo.

    Contains the judgment of the Inquisition on Borri.

BORRICHIUS, OLAUS.

    De ortu et progressu chemiæ dissertatio. *See* Manget, J. J., *and Section III.*
    Hermetis Ægyptiorum et chemicorum sapientia. *See in Section III.*

BOYLE, ROBERT.

Of a degradation of Gold made by an anti-elixir. A strange chymical narrative. London. pp. vi–17, sm. 4to.

    A clever satire on alchemical pretensions ; published anonymously.

BRACESCHUS, JOANNES BRIXIANUS. *See* Manget, J. J.

BRANDE, W. T.

A Sketch of the History of Alchymy. The Quarterly Journal, July, 1820. pp. 225–239 of vol. IX.

BRENTZIUS, ANDREAS. *See* Theatrum chemicum.

BREVE OPUS AD RUBEUM CUM SOLE PER AQUAS FORTES. *See* Theatrum chemicum.

BROTOFFER, RATICH. *See* A B C.

BRUNN DER WEISHEIT. *See* A B C.

BUCH (DAS) VON DEN ANFÄNGEN DER NATUR UND CHYMISCHEN KUNST.
　　See S[chröder, Fr. Jos. W.].

BUDDEUS, JOHANNES FRANCISCUS.
　　Quaestio politica : An Alchymistæ in republica sint tolerandi ?　Hallæ,
　　1702.　4to.
　　　　Historisch- und politische Untersuchung von der Alchemie
　　　　　　und was davon zu halten sey ?　Aus dem lateinischen
　　　　　　ins teutsche übersetzet, nun aber zum Druck befördert
　　　　　　durch Friedrich Roth-Scholtzen.　Nürnberg, 1727.　pp
　　　　　　146, 12mo.
　　　　　　A critical study and defence of alchemy, in which bibliography has not been
　　　　　　　　neglected.

BURRHI, FRANCISCUS JOSEPH.　*See* Borri, Francesco Giuseppi.

CALID FILIUS JAICHI.
　　　　See Manget, J. J. ; *also* Theatrum chemicum ; *also* Artis auriferæ ; *also*
　　　　　　Arnaldus de Villa Nova : Schrifften ; *also* Bacon, Roger : Le miroir
　　　　　　d'alquimie.

CAMBRIEL, L. P. FRANÇOIS.
　　*Cours de philosophie hermétique ou d'alchimie en dix-neuf leçons.
　　　　Traitant de la théorie et de la pratique de cette science, ainsi que
　　　　de plusieurs autres opérations indispensables, pour parvenir à trou-
　　　　ver et à faire la Pierre Philosophale, ou transmutations métalliques
　　　　lequelles ont été cachées jusqu'à ce jour dans tous les écrits de
　　　　philosophes-hermétiques ; suivies des explications de quelques arti-
　　　　cles des cinq premiers chapitres de la Genèse, par Moïse ; et de
　　　　trois additions prouvant trois vies en l'homme, animal parfait ;
　　　　Ouvrage nouveau, curieux et très nécessaire pour éclairer tous ceux
　　　　qui désirent pénétrer dans cette science occulte, et qui travaillent à
　　　　l'acquérir ; ou chemin ouvert à celui qui veut faire une grosse for-
　　　　tune.　Ouvrage fini en janvier, 1829, et du règne de Charles X, roi
　　　　de France, la cinquième.　Première édition.　Paris, 1843.　pp.
　　　　215, 12mo.

CAMPBELL, JOHN.
　　*Hermippus redivivus, or the Sage's Triumph over Old Age and the
　　　　Grave.　Wherein, a method is laid down for prolonging the life and
　　　　vigour of man.　Including a Commentary upon an antient inscrip-
　　　　tion in which this great secret is revealed, supported by numerous
　　　　authorities.　The whole interspersed with a great variety of remark-
　　　　able and well attested relations.　The second edition, carefully

CAMPBELL, JOHN. [Cont'd.]
corrected and much enlarged. London ; Printed for J. Nourse, at
the Lamb, against Catherine-Street in the Strand, 1749. pp. [viii]–
248.

> This work was published anonymously by Dr. John Campbell in imitation
> of another work under the same title by Dr. J. H. Cohausen. Cohau-
> sen's work was published at Frankfurt in 1742 (in Latin). The first
> edition of Campbell's book was published in 1743.

CARERIUS, ALEXANDER.
Dissertatione ; An possint arte simplicia veraque metalla gigni. Patau,
1579. 4to.

> *Also in* Karl Weisenstein's Quinta essentia chemicorum.

CARINI, ISIDORO.
Sulle scienze occulte nel medio evo e sopra un codice della famiglia
Speciale. Discorso letto all' accademia di scienze e lettere in
Palermo. Palermo, 1872. pp. 98–xxxii, 8vo.

> The 71 brief extracts from the Codex Speciale are only sufficient to stimulate
> the desire for more. It consists of a manual of alchemy on finest parch-
> ment and beautifully written. Probably of the XVIth century. It con-
> tains two chemical glossaries, one of 12 and one of 38 pages.

CARPENTER, RICHARD. *See* Ashmole, E. : Theatrum chemicum Britannicum.

CATHOLICON PHYSICORUM.
See Dreyfaches hermetisches Kleeblatt ; *also* Besondere Geheimnisse.

CATO CHEMICUS. *See* Manget, J. J.

CHARNOCK, THOMAS. *See* Ashmole, E. : Theatrum chemicum Britannicum.

CHARTIER, JOANNES. *See* Theatrum chemicum.

CHARTIER, JOANNES.
* La science du plomb sacré des sages, ou de l'antimoine, où sont
décrites ses rares et particulières vertus, puissances et .qualitez.
À Paris, chez I. de Senlecque en l'Hostel de Bauieres proche la
porte de S. Marcel et François le Cointe, ruë Saint Jaques à l'Image
Saint Remy prés le College du Plessis, 1651. pp. [iv]–56, sm. 4to.

CHAUCER, GEOFFRY. *See* Ashmole, E. : Theatrum chemicum Britannicum.

CHEVREUL, MICHEL-EUGÈNE.
Du traité alchimique d'Artéfius intitulé : " Clavis majoris sapientiæ." A
series of reviews in Journal des Savants. 1867, p. 767 ; 1868, pp.
45, 153, 209, 664.

CHEVREUL, MICHEL-EUGÈNE. [Cont'd.]

* Examen critique au point de vue de l'histoire de la chimie d'un écrit alchimique intitulé Artefii Clavis majoris sapientiæ, et preuve que cet écrit est identique avec l'écrit publié sous le nom d'Alphonse X, Roi de Castille et de Leon auquel l'astronomie doit les Tables Alphonsines. Présenté à l'Académie des sciences le 2 avril 1867. [Paris, 1867]. pp. 82, 4to.

CHRISTOPHORUS PARISIENSIS. *See* Theatrum chemicum.

CHRYSOGONUS DE PURIS. *See* Roth-Scholtz, Fr. (Deutsches Theatrum chemicum).

CHRYSOPOIEA : being a dissertation on the Hermetical Science. *See* Vazquez, Jablada.

* CHYMISCHE EXPERIMENTE EINER GESELLSCHAFFT IN DEM ERTZGEBÜRGE. Erstes Stück. Berlin, zu finden bey Seel Joh. Jac. Schützens Wittwe, 1753. pp. 80, 12mo.

* CHYMISCHE VERSUCHE aus alten raren Manuscripten gezogen in welchen die Kunst Gold und Silber, Gesundheit und ein langes Leben zu finden enthalten : mit vortreflichen Anmerkungen welche den nechsten Weg nach Colchos führen und die Ehre der Kunst befördern. Zwei Theile. Frankfurt und Leipzig, 1767. Vol. I, pp. [xxiv]–260–[iv] ; II, pp. [xxiv]–224–[viii], 12mo. Plate.

CLANGOR BUCCINÆ. *See* Artis auriferæ ; *also* Manget, J. J.

CLAUDER, GABRIEL. *See* Manget, J. J. ; *also* S[chröder, Fr. Jos. W.].

CLAUDER, GABRIEL.

Dissertatio de tinctura un.versali (vulgo lapis philosophorum dicta) in qua 1, Quid hæc sit ; 2, Quod detur in rerum natura; 3, an Christiano consultum sit immediate in hanc inquirere ; 4, e qua materia, et 5, quomodo præparetur [etc.]. Ad normam Academiæ naturæ curiosorum. Altenburgi, 1678. pp. [xii]–272[–xxiv], 4to.

Schediasma de tinctura universali, vulgo lapis philosophorum dicta, cum Petri Johannis Fabri manuscripto, res alchymicorum obscuras extraordinaria perspicuitate explanante, nec non D. Adami Gottlob Berlichii dissertatione de medicina universali, quin et D. Emanuelis Koenigii epistola de elixirio sophorum, ejusque observationibus de modo exaltationis et vitrificatione metallorum revisum et hinc inde auctum, promisso simul blati Clauderi vitæ curriculo sistit D. Gabriel Frid. Clauderus Nepos. Norimbergæ, 1736. pp. [vii]–304–[xlviii], 4to. 13 folding tables.

> These tables, which have a distinct title-page running " Tabulæ metallurgico-docimasticæ, etc.," consist of a list of minerals (" Fossilien"), with information classified under the headings " Nomen, Substantia, Color, Pondus, Natura, Præparatio, Tractatio, Contenta."

CLAUDER, GABRIEL. [Cont'd.]

Tractatus de tinctura universali, ubi in specie contra R. P. Athanasium Kircker pro existentia lapidis philosophici disputatur. *In* Manget's Bibliotheca chemica curiosa, vol. I, p. *119*.

> A reply to Athanasius Kircher, *q. v.*

CLAUDIUS GERMANUS. *See* Manget, J. J.; *also* S[chröder, Fr. Jos. W.].

* CLAVICULA HERMETICÆ SCIENTIÆ ab hyperborӕo quodam horis subsecivis calamo consignata anno millesimo septingentesimo triagesimo secundo. Amstelӕdami, apud Petrum Mortier, 1751. pp. 73, 16mo.

> Also with the title in French.

> > Clavicule (La) de la Science hermétique écrite par un habitant du Nord dans ses heures de loisir. L'an 1732. À Amsterdam, chez Pierre Mortier, 1751. pp. 73, 16mo.

> > Printed in Latin and French on opposite pages.

* CLAVIER DU PLESSIS. Mytho-Hermetisches Archiv. Ein periodisches Werk. Aus dem französischen. Erster Band. Gotha bey Carl Wilhelm Ettinger, 1780. pp. 160–32, 16mo.

> This periodical devoted to alchemy died with the first number, which is not surprising considering the date.

CLAVIS PHILOSOPHIÆ CHEMICÆ. *See* Aureum Vellus.

CLAVIS SAMBT SEINER DECLARATION DER CHYMISCHEN HANDGRIFFEN.
> *See* Aureum Vellus.

CNÆFFELLIUS, ANDREAS. *See* Manget, J. J.

Собраніе разныхъ достовѣрныхъ химическихъ книгъ, а именно: Іоанна Исаака Голанда, рука Философовъ, о Сатурнѣ, о растеніяхъ, минералахъ, кабала и о камнѣ философическомъ, съ пріобщеніемъ сочиненія о заблужденіяхъ Алхимистовъ. С.-Нетербургъ, 1787.

> Collection of various authentic books on chemistry, viz.: *John Isaak Holland's* Philosopher's Hand; on Saturnus; on plants, minerals and the Cabala; on the Philosopher's Stone; with addition of an essay on the errors of alchemists. St. Petersburg, 1787.
> "The errors of alchemists" ought to make a huge volume.

* COLLECTANEA CHYMICA: A Collection of Ten several Treatises in Chymistry, concerning the Liquor Alkahest, the Mercury of Philosophers, and other curiosities worthy the perusal. Written by Eir. Philaletha, Joh. Bapt. Van-Helmont, Fr. Antonie, Bernhard of Trevisan, Geo. Ripley, Rog. Bacon, Geo. Starkey, Hugh Platt and the Tomb of Semiramis, see more in the Contents. London, 1684. Printed for William Cooper at the Pelican in Little Britain. pp. 193-32-16, 16mo.

COLLECTANEA CHYMICA.   [Cont'd.]
[Contents :]

1. *Philalethes.* The Secret of the Immortal Liquor called Alkahest or Ignis-Aqua, communicated to his friend, a son of art and now a philosopher. By question and answer. p. 5.

2. The Practice of Lights, or an excellent and ancient Treatise of the Philosophers' Stone. p. 27.

3. *Helmont, Joh. Baptist van:* Præcipiolum, or the Immature Mineral-Electrum. The first Metall, which is the Minera of Mercury. p. 45.

4. *Antonie, Fr.:* Aurum potabile, or the receit shewing the way and method how he made and prepared that most excellent medicine for the body of man. p. 71.

5. *Trevisan, Bernard:* A Treatise of the Philosophers' Stone. p. 81.

6. *Riply, George:* The Bosome-Book. Containing his philosophical accurtations in the makeing the Philosophers' Mercury and Elixirs. p. 101.

7. *Bacon, Roger:* Speculum Alchymiæ; the true Glass of Alchemy. London, 1683. p. 125.
   [The first Latin edition of Speculum Alchymiæ dates 1614 in Nürnberg. For a French one *see* Bacon, Roger.]

8. *Starkey, George:* The admirable Efficacy and almost incredible Virtue of true Oyl which is made of Sulphur-vive, set on fire, and called commonly Oyl of Sulphur per Campanam, to distinguish it from that rascally sophisticate Oyl of Sulphur which instead of this true Oyl is unfaithfully prepared and sold by Druggists and Apothecaries, to the dishonour of art, and unspeakable damage of their deluded patients. Faithfully collected out of the writings of the most acute Philosopher, and unparalleled Doctor of this last Age, John Baptist Van-Helmont, of a noble extraction in Belgia. p. 139.

9. *Platt, Sir Hugh:* New and artificial Remedies against Famine. Written upon the occasion of the great dearth in the year 1596. p. 155.

10. H. V. D. [Pantaleon]. The Tomb of Semiramis Hermetically sealed, which if a Wise-man open (not the ambitious covetous Cyrus) he shall find the Treasures of Kings, inexhaustible Riches to his content. p. 7.
    Contains several lists of books on chemistry sold by William Cooper.

COLLESSON, JOANNES. *See* Theatrum chemicum.

COMBACH, LUDWIG.
   Scripta chymica Riplæi.   Cassel, 1649.
      Cf. Ripley, George.

* COMPASS (DER) DER WEISEN, von einem Mitverwandten der innern Verfassung der ächten und rechten Freymäurerey beschrieben ; herausgegeben mit Anmerkungen einer Zueignungschrift und Vorrede, in welcher die Geschichte dieses erlauchten Ordens, vom Anfang seiner Stiftung an, deutlich und treulich vorgetragen, und die Irrthümer einiger ausgearteter französischer Freymäurer-Logen entdeckt werden, von Ketmia Vere. Berlin und Leipzig, bey Christian Ulrich Ringmacher, 1779. pp. 386, 12mo. Plates.

COMITIBUS, LUDOVICUS DE. *See* Manget, J. J.

CONSILIUM CONJUGII SEU DE MASSA SOLIS ET LUNÆ.
  See Manget, J. J. ; *also* Theatrum chemicum.

COOPER, WILLIAM. [Editor.]
 * The Philosophical Epitaph of W. C. Esquire, for a Memento Mori on his Tomb-stone. With three Hieroglyphical Scutcheons, and their Philosophical Mottos and Explanation with the Philosophical Mercury, Nature of Seed, and Life, and Growth of Metalls ; and a Discovery of the Immortal Liquor Alchahest. The Salt of Tartar volatized, and other Elixirs with their Differences. Also : A Brief of the Golden Calf (the Worlds Idol), discovering the rarest Miracle in Nature, how by the smallest proportion of the Philosophers-Stone a great piece of common lead was totally transmuted into the purest transplendent Gold at the Hague, 1666, by Fr. Helvetius. And : The Golden Ass well managed, and Midas restored to Reason ; Or, A new Chymical Light, Demonstrating to the blind world that good Gold may be found as well in Cold as Hot Regions, and be profitably extracted out of Sand, Stones, Gravel, and Flints, to be wrought by all Sorts of People ; written by Jo. Rud. Glauber. —With Jehior [Aurora Sapientiæ,] or the Daydawning or Light of Wisdom containing the three Principles or Original of all things ; whereby are discovered the Great and many Mysteries in God, Nature, and the Elements, hitherto hid, now revealed. All published by W. C., Esquire. With a Catalogue of Chymical Books. London, Printed by T. R. and N. T. for William Cooper, at the Pellican in Little Britain, Anno Domini 1673. pp. [xvii]–16–[vi] –[x]–41–[xi]–36–56–[xviii]–78–[v]–[lxxxvii], 12mo. Plates. Engraved title.
  The date of the catalogue is 1675. Published under the initials W. C. (= William Cooper).

CORRECTIO FATUORUM. *See* Artis auriferæ ; *also* Manget, J. J.

CREILING, JOHANN CONRAD.
 * Dissertatio academica de aureo vellere aut possibilitate transmuta-

CREILING, JOHANN CONRAD. [Cont'd.]

tionis metallorum ; Præside Joanno Conrado Creilingio. Tubingæ, Litteris Roebelianis, 1737. pp. 88, 4to.

> *Abhandlung vom goldenen Vliess, oder Möglichkeit der Verwandlung der Metalle ; aus dem lateinischen übersetzt. Tübingen, bei Jacob Friedrich Heerbrandt, 1787. pp. [xvi]–176, 16mo.

CREMERUS. *See* Musæum hermeticum.

CRINOT, HIERONYMUS. *See* Aureum Vellus.

CROLL, OSWALD. [CROLLIUS.]

Basilica chymica, continens philosophicam propria laborum experientia confirmatam descriptionem et usum remediorum chymicorum selectissimorum et lumine gratiæ et naturæ desumptorum. Frankfort, 1609.

> [Other editions :] Francofurti, 1619, 1634, 1647 ; Genevæ, 1638, 1658 ; Venetiis, 1642.
>
> According to Poggendorff, 18 editions were published before 1658.

> *Bazilica chymica and praxis chymiatricæ, or Royal and Practical Chymistry. In three treatises wherein all those excellent medicines and chymical preparations are fully discovered, from whence all our modern chymists have drawn their choicest remedies. Being a translation of Oswald Crollius, his Royal Chemistry, augmented and inlarged by John Hartman. To which is added his treatise of Signatures of Internal Things, or a true and lively Anatomy of the greater and lesser world. As also : The practice of chymistry of John Hartman, augmented and inlarged by his son. All faithfully Englished by a lover of chymistry. London, printed for John Starkey at the mitre in Fleetstreet near Temple-Bar, and Thomas Passinger at the Three Bibles on London-Bridge, 1670. pp. [vi]–180–[xvi]–37–[xi]–186–[xix], sm. folio.

> *La royale chymie de Crollius. Traduite en Français par I. Marcel de Boulene. À Rouen, 1634. pp. 460–[lxi], 12mo.
>
> *To this edition is added :* Traicté des signatures, ou vraye et vive anatomie du grand et petit monde. pp. 126–[xxxii], 12mo.

> *Hermetischer Probier-Stein, darauf nicht allein alle und jede in dess Osvvaldi Crollii intitulirten alchymistischen Königlichen Kleynod befindliche Process und chymische

CROLL, OSWALD. [Cont'd.]

Artzneyen examiniret und auff die Prob gesetzet, sondern dieselbe auch mit unterschiedlichen andern schönen
und nützlichen durch selbst eygnen Handgriff und
tägliche Erfahrung approbirten Artzneyen vor diesem
in lateinischer Sprach vermehret und verbessert worden,
von Johann Hartmann. Neben angehengten Crollischen
Tractatlein von den innerlichen Signaturen oder Zeichen
aller Dinge und dem Hermetischen Wunderbaum, dem
gemeinen Nutzen zum besten ins teutsche versetzet.
Franckfurt am Mayn, in Verlegung Johann Gottfried
Schönwetters, 1647. pp. [viii]–392–[xvi]–61–[x]–83,
sm. 4to. Ill. Plates.

    The engraved title-page reads : Basilica chymica, oder Alchymistisch könig-
    lich Kleynod . . . The half-title : Chymisch Kleynod . . .

CRÜGNER, MICHAEL. Chymischer Garten-Baw. *See in Section IV*.

D., W. G. L.

  * Des Englischen Grafens von S——— experimentirte Kunst-Stücke,
oder Sammlung einiger rarer curieuser und geheimer chymischer
Processe und andere höchst-nützliche Arcana in welchen die Kunst
Gold zu machen mehr als auf einen Weg ohne dunckle Worte und
Allegorien gantz deutlich gezeiget und mit allen Umständen be-
schrieben und denen Liebhabern der edlen Chymie zu sonderbahren
Nutzen ans Licht gegeben worden. Braunschweig, 1731. Zu fin-
den in der Rengerischen Buchhandlung. pp. [xii]–84, 12mo.

DASTIN, JOHN. *See* Ashmole, E. : Theatrum chemicum Britannicum.

DAUSTENIUS, JOANNES. *See* A B C ; *also* Manget, J. J.

* DE ALCHEMIA DIALOGI DUO : quorum prior genuinam librorum Gebri
sententiam, de industria ab autore celatam et figurato sermone
involutam, retegit et certis argumentis probat ; alter Raimundi
Lullii Maioricani mysteria in lucem producit. Quibus pramittun-
tur propositiones centum viginti novem, idem argumentum compendiosa brevitate complectentes, ex Tuscanico idiomate traductæ.
Lugduni, excudebant Godefridus et Marcellus Beringi fratres, 1548.
pp. 147, 12mo.

    Of this Lenglet du Fresnoy says : " Recueil assez rare, mais cela ne veut pas
    dire qu'il soit meilleur que les autres ouvrages."

    And in another place this original bibliographer writes : " Geber et Raymond
    Lulle sont deux grands maitres en cette science ; mais pour les bien
    commenter, il faut les bien entendre ; et quand on les entend, on ne
    cherche pas à les faire entendre aux autres ; on réserve ce secret pour
    soi."

DE LAPIDE IN AQUA MERCURIALI. *See* Aureum Vellus.

DE LAPIDE PHILOSOPHORUM. *See* Aureum Vellus.

DE MAGNI LAPIDIS COMPOSITIONE ET OPERATIONE. *See* Theatrum chemicum.

DE PRIMA MATERIA LAPIDIS PHILOSOPHORUM. *See* Aureum Vellus.

DE QUINTA ESSENTIA PULCHERRIMUS TRACTATUS. *See* Aureum Vellus.

DEANUS, EDMUNDUS. *See* Dreyfaches hermetisches Kleeblatt.

DEE, ARTHUR.
 Fasciculus chymicus, abstrusæ scientiæ hermeticæ ingressum, progressum, coronidem explicans. Paris, 1631.
   * Fasciculus Chemicus : or Chymical Collections. Expressing the ingress, progress, and egress of the secret Hermetick Science, out of the choisest [*sic*] and most famous authors. Collected and digested in such an order, that it may prove to the advantage, not only of the beginners, but proficients of this high art, by none hitherto disposed in this method. Whereunto is added the Arcanum, or grand secret of hermetick philosophy. [By Arthur Dee.] Both made English by James Hasolle, qui est Mercuriophilus Anglicus. London, printed by J. Fletcher for Richard Mynne at the sign of St. Paul in Little Britain, 1650. pp. [xlviii]–268, 12mo. Additional engraved title-page.

    The preface is dated March, 1629. The "Arcanum" is by an author who conceals his name in the anagram *Penes nos unda Tagi ;* and is said to be "third edition amended and enlarged."

DEE, JOHN. *See* Theatrum chemicum ; *also* Ashmole, E. : Theatrum chemicum Britannicum.

DEE, JOHN.
 * The Private Diary of Dr. John Dee and the Catalogue of his Library of Manuscripts, from the original manuscripts in the Ashmolean Museum at Oxford, and Trinity College Library, Cambridge. Edited by James Orchard Halliwell. London, printed for the Camden Society, 1842. pp. viii–102, 4to.

DEMOCRITUS.
 * De rebus sacris naturalibus et mysticis cum notis Synesii et Pelagii. Norimbergæ, apud Hæredes Johannis Danielis Tauberi, 1717. pp. 42, 16mo.

* DER VON MOSE UND DENEN PROPHETEN ÜBEL URTHEILENDE ALCHYMIST
wird fürgestellet in einer Schrifft-gemässen Erweisung ; Dass Moses
und einige Propheten, wie auch David, Salomon, Hiob, Esra und
dergleichen, keine Adepti Lapidis Philosophorum gewesen sind.
Ingleichen dass die Lehre und alchymistisch Vorgeben, von Verwandlung der geringen Metalle in Gold, eine lautere Phantasie und
schädliche Einbildung sey. Von einem Liebhaber der Wahrheit,
der sich tröstet dass der Allmächtige sein Gold sey, Hiob xx, v. 25,
und nichts im Golde sucht. Chemnitz, bey Conrad Stöffelen, 1706.
pp. [xiv]–144, 12mo. Frontispiece and plates.

> Contains a portrait of the unfortunate rogue George Honauer, and a repre-
> sentation of his execution on gilded gallows, April 2, 1597, he being at
> the time only 24 years old.

DESCRIPTION OF THE STONE. *See* Ashmole, E. : Theatrum chemicum Britannicum.

DEUTSCHES THEATRUM CHEMICUM. *See* Roth-Scholtz, Fr.

DIALOGUS MERCURII ALCHYMISTÆ ET NATURÆ. *See* Theatrum chemicum.

DIALOGUS PHILOSOPHIÆ. *See* Aureum Vellus.

DICKINSON, EDMUND. *See* S[chröder, Fr. J. W.].

DICKINSON, EDMUND.
De chrysopœia, sive de quintessentia philosophorum. Oxoniæ, 1725.
pp. [iv]–224, 8vo.

> The essay bears the date August, 1683. Contains : Quæstiones propositæ
> Theodoro Mundano (pp. 93–224).

DIGBY, SIR KENELME.
* A late Discourse made in a Solemne Assembly of Nobles and Learned
Men at Montpellier in France, touching the Cure of Wounds by the
Powder of Sympathy ; With instructions how to make the said
Powder ; whereby many other Secrets of Nature are unfolded.
Rendered faithfully out of French into English. By R. White,
Gent. The third edition corrected and augmented with the addi-
tion of an Index. London, printed for R. Lowndes at the White
Lion, and T. Davies at the Bible in S. Pauls Churchyard, over
against the little North-Door, 1660. pp. [x]–152–[iv], 16mo.

* DIVERS TRAITEZ DE LA PHILOSOPHIE NATURELLE. Scavoir, la turbe des
philosophes ou le code de vérité en l'art ; la parole délaissée de
Bernard Trevisan ; les deux traitez de Corneille Drebel, Flaman ;
avec le très-ancien duel des chevaliers. Nouvellement traduits en
François, par un Docteur en Medecine. À Paris, chez Jean d'Houry

DIVERS TRAITEZ DE LA PHILOSOPHIE NATURELLE. [Cont'd.]
à l'Image S. Jean au bout du Pont-neuf sur le Quay des Augustines, 1672. pp. [viii]-298-[v].

DONATO D' EREMITA (FRA).

* Dell' elixir vitæ di Fra Donato d'Eremita di Rocca d'Evandro, dell' Ord[ine] de' Pred[icatori] libri quattro. Al Serenissimo Ferdinando Secondo Gran Duca di Toscana. Napoli, 1624. pp. [xii]-182, sm. fol., and 19 plates. Illuminated title-page.

> Contains : Tavola delle varie infirmita secondo il linguaggio Arabico, Greco e Latino, pp. 133-165 ; also : Tavola de gli scrittori di medicina e di chirurgia Arabi, Greci, Latini e Italiani . . . della presente opera.

* Antidotario di Fra D. d'E. Diviso in libri tre. Napoli, 1639. pp. [vi]-142, sm. fol. Ill.

> Very little seems to be known of this author. He is not mentioned by Schmieder, Borel, Kopp's Geschichte, Kopp's Alchemie, Hoefer, Gmelin, Roth-Scholtz or Fuchs. Lenglet du Fresnoy catalogues only the Elixir Vitæ and adds : " On voit que jusques à ces derniers temps, les Religieux ne sont pas difficulté de se mêler de la science Hermétique. Et il vaut mieux qu'ils s'y appliquent que de faire d'autres choses, qui ne leur conviennent pas " (vol. III, p. 147). Of the Antidotario only one Book was published. It is really a treatise on pharmacy.

DORNÆUS, GERARDUS. *See* Theatrum chemicum ; *also* Manget, J. J.

DRECHSLER, JOHANN GABRIEL.

Disputationes duo de metallorum transmutatione et in primis de chrysopœia. Lipsiæ, 1673. 4to.

DREBBEL, CORNELIUS.

> *See* S[chröder, Fr. Jos. W.] ; *also* Divers Traitez de la philosophie naturelle.

DREBBEL VAN ALKMAR, CORNELIS.

* Grondige oplossinge van de Natuur en Eygenschappen der Elementen, En hoe sy veroorsaaken Donder, Blixem, Hitte, Koude, Wind, Regen, Hagel, Sneeuw &c. En waar toe sy dienstig zyn. Als mede een klare beschryving van de Quinta Essentia, Noyt voor desen gedrukt. Noch een Dedicatie van 't Primum mobile. Alles gedaan door den grooten Hollandschen Philosooph Cornelis Drebbel van Alkmaar. Vermeerdert met het Leven van den zelve ; als mede een Brief van Nacha Ree. Te Amsteldam ; by Samuel Lamsveld Boekverkooper, 1732. pp. 24-108. Illustrated title-page.

> The engraved title-page is dated 1709. Contains a wood-cut and description of an experiment that has caused the author to be erroneously credited with the invention of the thermometer. The first edition in Latin was published in 1621. For a German edition, *see in Section V.*

DREBBEL, JACOBSZ CORNELIS.

Van de Elementen, quinta essentia en primum mobile. Rotterdam, 1761. 12mo. Ill.

* DREY CURIEUSE CHYMISCHE TRACTÀTLEIN ; Das erste betitult : Güldene Rose, das ist : Einfältige Beschreibung des allergrössesten von dem Allmächtigsten Schöpffer Himmels und der Erden Jehova in die Natur gelegten und dessen Freunden und Auserwählten zugetheilten Geheimnisses, als Spiegels der Göttlichen und Natürlichen Weissheit. Das andere : Brunn der Weissheit und Erkänntnis der Natur, von einem unvergleichlichen Philosopho gegraben. Das Dritte : Blut der Natur, oder Entdeckung des allergeheimsten Schatzes derer Weisen, seyende nichts anders als der rothe Lebens-Safft davon alle Geschöpfe nach dem Willen des Allmächtigen herstammen, erhalten und fortgepflanzet werden. Frankfurt und Leipzig, zu finden in dem Kraussischen Buchladen, 1774. pp. 206, 12mo. Plate.

* DREYFACHES HERMETISCHES KLEEBLATT in welchem begriffen dreyer vornehmen Philosophorum herrliche Tractätlein. Das erste von dem geheimen waaren Saltz der Philosophorum, und allgemeinen Geist der Welt, H. Nuysement aus Lothringen. Das andere Mercurius Redivivus, Unterricht von dem Philosophischen Stein sowol den weissen als rothen aus dem Mercurio zu machen, Samuelis Nortoni sonsten Rinville. Und das dritte von dem Stein der Weisen Marsilii Ficini Florentini welche ehedessen von denen Authoribus in frantzösischer und lateinischer Sprach beschrieben, nunmehro aber allen Liebhabern, so der lateinischen Sprach unkündig, zum besten, in unser teutsche Muttersprach übersetzet und mit einem zweyfachen Register zum Druck verfertiget. Durch Vigilantium de Monte Cubiti. Nürnberg, in Verlegung Michael und Johann Friedrich Endtern, 1667. pp. [xxii]–448–[xxx], 16mo. Illustrated title-page. Ill.

> Contains the same treatises as '' Besondere Geheimnisse eines wahren Adepti,'' but differently placed. De Nuysement's tract is more complete here.

DAS ERSTE TRACTÄTLEIN.

Nuysement, de : Tractat von dem waaren geheimen Saltz der Philosophorum, und von dem allgemeinen Geist der Welt. Zu Ergäntzung des lang-begehrten dritten Principii, Michaelis Sendivogii, welches er vom Salze verheissen hat. p. 1.

DAS ZWEITE TRACTÄTLEIN.

1. Mercurius Redivivus, oder Unterricht wie man den philosophischen Stein, sowol den weissen, als den rothen aus dem Mercurio

Dreyfaches Hermetisches Kleeblatt.   [Cont'd.]

machen solle.   Erstlich von weiland *Samuele Nortono*, sonsten *Rinvillo Briszollensi* angefangen.   Nachmals mit Fleisz verbessert und vermehret durch *Edmundum Deanum* von Oxonien, Doctorem der Artzney zu Eborach in Engelland.   Nunmehr aber wegen seiner grossen Nutzbarkeit dieser Kunst Liebhabern, die der lateinischen Sprach unkündig sind, zum besten aufs fleissigste in die teutsche Muttersprach versetzet worden.   Bey welchem Tractätlein zu finden, wie man beide Fermenta, so wol zum weissen aus der Luna oder dem Silber, als zum rothen aus Sole oder dem Golde machen solle.   p. 207.

2. Catholicon physicorum, oder Unterricht wie man die Tincturam physicam und alchymiam so von den alten Philosophis sehr emsig gesuchet worden, machen solle ; aber noch von keinem Authore weder den alten noch neuen, gäntzlich und vollkommenlich ans Licht gegeben worden, benebens dieser Tinctur Abkürtzung.   Bey welchem Tractätlein zu finden, der dreyfache Zweige von der Composition der Jungfrauen-Milch oder des philosophischen Essigs.   p. 231.

3. Venus vitriolata in elixir conversa ; nec non Mars victoriosus seu elixiratus, oder Unterricht wie man den philosophischen Stein sowol aus der Venere oder dem Kupffer, als aus dem Marte oder den Eisen machen solle.   p. 247.

4. Elixir seu medicina vitæ, oder Unterricht wie man das ware Aurum und Argentum potabile machen solle, samt derselben Krafft und Würckungen, nach Meinung der alten und neuen Philosophen.   Bey welchem auch zu finden der Modus wie man das Glas möge zurichten dass sichs ziehen und treiben lasse.   p. 263.

5. Saturnus saturatus dissolutus et coelo restitutus, oder Unterricht wie man den philosophischen Stein, sowoln den weissen, als den rothen, aus dem Blei machen soll, wie dann auch eben auf solche Weise aus dem Jove oder Zinn.   Bey welchen zu finden die Abkürzung des Wercks Saturni, zugleich auch der Modus das Quecksilber aus dem Bley zu extrahiren.   Ueber das ist auch dabey ein Tractätlein von der Philosophorum Methodo in dem opere Saturni nach dem Georg Riplæo vermehret und verbessert, zugleich mit der Abkürzung des Riplæi von dem Mercurio sublimato auch vermehret und verbessert.   p. 283.

6. Metamorphosis lapidum ignobilium in gemmas quasdam pretiosas, oder Unterricht wie man aus kleinen Perlein grosse und gute machen solle ; auch wie man die künstlichen Carfunkeln und andere köstliche Stein solle zurichten welche weit edler, als die

DREYFACHES HERMETISCHES KLEEBLATT. [Cont'd.]

natürlichen seyen sollen. Bey welchem der Modus gefunden
wird, wie man das Electrum artificiale, als das höchste aller
Elixirium, mit Andeutung des natürlichen und metallischen
Electri, so den alten gantz unwissend, bereiten solle. p. 303.

7. Alchymiæ complementum et perfectio, oder Unterricht und Processus, wie man alle Steine und Elixira augmentiren oder
multipliciren solle, so wol in virtute oder qualitate, zugleich
auch wie man die Projection verrichten solle. Bey welchem
zu finden die Erläuterung der Philosophorum Meinung wann
sie von der zehenden Zahl reden in welcher das Werck solle
gemachet werden. In welchem auch angezeiget wird, das
Wunder-Geheimniss des animalischen Steins aus dem Menschen-
Blut, nach der Meinung Georgii Riplæi. Endlich ist darzu
gesetzt der Beschluss des Buchs, welches alles den sechs
alchimistischen Büchern, wie auch diesem dienstlich ist, so
zuvor herausgangen sind. p. 329.

8. Ein Tractätlein welches von den philosophischen Schrifften
handlet, so die Alten über die Alchymei haben lassen ausgehen. Begreifft in sich die Auslegung der duncklen Wörter,
Namen und seltzamen Reden, so in dieser Kunst geführet
werden. p. 355.

DAS DRITTE TRACTÄTLEIN.

Ficinus, Marsilius : Büchlein vom Stein der Weisen erstlich von dem
Authore selbsten in lateinischer Sprach beschrieben, an jetzo
aber allen Liebhabern und des Lateins unkündigen zulieb in
die teutsche Muttersprach übersetzt. p. 373.

DU CHESNE, JOSEPH. [QUERCETANUS].

A Briefe Aunswere of Iosephus Quercetanus Armeniacus to the exposition of Jacobus Aubertus Vindonis, concerning the original and
causes of Mettalles ; set foorth against Chimists. Another ex-
quisite and plaine treatise of the same Josephus, concerning the
spagericall preparations, and use of minerall, animall, and vegitable
medicines. Whereunto is added divers rare secretes, not heretofore
knowne of many. By John Hester. London, 1591. pp. [vi] ff. 61-[iv].

DU PLESIS, SCIPIO.

* The Resolver or Curiosities of Nature. Usefull and pleasant for all.
London, printed by N. & I. Okes. Anno Domini 1635. pp. [vi]-
408, 16mo. Illustrated title-page.

A very quaint work on early science, in question and answer, by a Counsellor
and Historiographer to the King of France.

DUVAL, ROBERT. *See* Theatrum chemicum : Vallensis, Robertus ; *also, in Section III*, Duval, Robert.

E. I., I. N. VON.

* Alchymia denudata revisa et aucta, oder das bisanhero nie recht ge-
glaubte, durch die Erfahrung nunmehro aber würcklich beglaubte,
und aus allen Zweifel gesetzte, neu übersehene und vermehrte, oder
in vielen besser erklärte Wunder der Natur, nebst angehängter ausführlichen Beschreibung der unweit Zwickau, in Meissen, zu Nieder-
hohendorf und anderen umliegenden Orten gefundenen goldischen
Sande ; vorstellend Welchergestalt aus unterschiedenen allhier aufrichtig mit Namen genannten Materien, wie auch auf unterschiedene
Art und Weise, in der That und Wahrheit eine Universal-Medicin,
auf menschlichen Leib und zur Verbesserung der Metallen zu
bereiten, wie auch dass ausser dem Fonte universali aller philosophorum Schriften ungeachtet, dennoch ein höchst nutzbares und
grossen Profit tragendes Particulare zu erlangen sey alles nach
langwierigem dem Studio Chymico obgelegenen Fleisse, theils mit
Augen gesehen, theils mit Händen selbst gemacht, und des von
vielen sich darauf beruffenden philosophischen Fluchs ungeachtet,
um erheblicher Ursachen willen, und andern zu einem guten Exempel, in dergleichen mit Experimenten nachzufolgen, an Tag
gegeben.   Leipzig, bey Johann George Löwen, 1769. pp. 224, 8vo.
Anderer Theil, vorstellend worinnen der Vortheil der im ersten Theile
erwiesenen Wahrheit der Transmutation, oder geringerer Metallen
in bessere und vollkommene beruhe, und wie aus denen alldorten
unterschiedenen aufrichtig mit Namen genannten Materien, wie
auch auf unterschiedene Art und Weise nicht nur in der That und
Wahrheit eine rechte und weit höhere Universal-Medicin auf
menschlichen Leib, sondern auch zur Verbesserung der Metallen
als alldorten angewiesen worden, zu bereiten, indem er specialiter
anweiset, wie effective oder würklich zum Fonti Universali zu gelangen.   Ebenfalls alles nach langwierigem und in die etliche
dreyssig Jahre dem Studio chymico obgelegenem Fleisse, sowohl
mit Augen gesehen, als mit Händen, durch die Gnade Gottes dem
ewig Lob und Dank dafür gesaget sey, nun selbst gemacht und aus
Ursachen, wie in der Vorrede gedacht werden wird, an Tag gegeben.
1769. pp. 376.

ECK DE SULTZBACH. *See* Theatrum chemicum.

ECKHARTSHAUSEN, VON.

* Ueber die Zauberkräfte der Natur. Eine freie Uebersetzung eines
egyptischen Manuscripts in coptischer Sprache. Mit einem An-

ECKHARTSHAUSEN, VON. [Cont'd.]
hange eines aus magischen Characteren entzifferten Manuscripts.
Ein nachgelassenes Werck von Hofrath von E. . . . München,
1819. pp. 66, 8vo.

EDELGEBORNE JUNGFER ALCHYMIA. *See in Section III.*

EFFERARIUS MONACHUS. *See* Theatrum chemicum ; *also* Aureum Vellus.

EHREN-RETTUNG DER ALCHYMIE. *See, in Section III,* Edelgeborne Jungfer Alchymia.

* EHRENRETTUNG DER HERMETISCHEN KUNST durch solche chymisch-
physikalische Beweise dargethan die jeder, auch nur mittelmässige
Kenner und Künstler leicht einsehen, selbst nachmachen, und
dadurch zugleich überzeugt werden kann und soll : dass Alchymie
und Chrysopœia keine leere Einbildung müssiger Köpfe sey und
noch weniger in die Zauber-Höhle gelehrter Windmacher gehöre.
Drey Theile. Erfurt, 1785. I, pp. 61; II, pp. 72; III, pp. 116, 12mo.

EINZELNE ZEUGNISSE aus Hermann Fictulds beiden Classen des Probiersteins. *See* A B C.

ELEAZAR, ABRAHAM.
* Uraltes chymisches Werk welches ehedessen von dem Autore theils in
lateinischer und arabischer, theils auch in chaldäischer und
syrischer Sprache geschrieben, nachmals von einem Anonymo in
unsere deutsche Mutter-Sprache übersetzet, nun aber nebst zugehörigen Kupffern, Figuren, Gefässen, Oefen, einer kurtzen Vorrede,
nöthigen Registern, wie auch beygefügten Schlüssel derer in selbi-
gen vorkommenden fremden Wörter, mit gewöhnlicher Approbation
zu Nutz und Gebrauch aller Liebhaber der edlen hermetischen
Philosophie, in zwei Theilen zum öffentlichen Druck befördert
worden durch Julium Gervasium Schwartzburgicum. Erfurt, Verlegts Augustinus Crusius, 1735. pp. [xxx]–122–[xvi]–87–[xxvii],
12mo. Ill.

> Second edition, 1760. Contains : Kurtzer doch deutlicher Schlüssel der-
> jenigen fremden Wörter welche in Abraham dem Juden enthalten sind.
> The vocabulary embraces German, Latin, Greek and Hebrew.
>
> Nothing seems to be known of Abraham Eleazar personally, nor of Samuel
> Baruch, nor of Julius Gervasius the publisher. This work is of great
> rarity, not being mentioned in Schmieder's Geschichte, Bauer's Biblio-
> theca, Brunet's Manuel nor Graesse's Trésor. Heinsius, in his Bücher-
> Lexikon, mentions only a second edition dated 1760. J. F. Gmelin
> names it, however. For an account of this work and testimony to its
> exceeding rarity *see* Kopp : Die Alchemie, Heidelberg, 1886, II, 315.
> Kopp had not seen this, the editio princeps.
>
> *Cf.* Baruch, Samuel, *and see* A B C.

ELIAS ARTISTA. *See* A B C ; *also* S[chröder, Fr. J. W.].

ELIXIR SEU MEDICINA VITÆ.
See Dreyfaches hermetisches Kleeblatt ; *also* Besondere Geheimnisse.

ENTRETIEN DU ROI CALID ET DU PHILOSOPHE MORIEN. *See* R[ichebourg], J. M. D.

EPISTEL ODER SEND-BRIEF DES KAYSERS ALEXANDRI.
See Roth-Scholtz, Fr. (Deutsches Theatrum chemicum).

ERASTUS, THOMAS.
Explicatio quæstionis famosæ : Utrum ex metallis ignobilibus aurum verum et naturale arte conflari possit. Basiliæ, 1572. 4to.
Also in his Disputatione de medicina nova Paracelsi. Basiliæ, 1572. 4to.
Cf. Claveus, Gaston Dulco, who attacks the views of Erastus.

ERÖFNETE (DAS) PHILOSOPHISCHE VATERHERZ.
See A B C ; *also* Ripley, George : Schriften.

ESPAGNET, DOMINUS D'.
See Manget, J. J. ; *also* Roth-Scholtz, Fr. (Deutsches Theatrum chemicum).

ETTNER, HANS CHRISTOPH VON. [ETTNER VON EITERITZ].
* Rosetum chymicum, oder chymischer Rosen-Garten, aus welchem der vorsichtige Kunst-Beflissene, vollblühende Rosen, der unvorsichtige Laborant aber Dornen und vervaulte Knospen abbrechen wird, in sonderliche Garten-Better abgetheilet und vorgestellet. Frankfurt und Leipzig, bey Michael Rohrlachs Wittib und Erben, 1724. pp. [xii]–564, 16mo.
* Vade et occide Cain, oder gehe und schlage den Cain todt. Franckfurt und Leipzig, bey Michael Rohrlachs Wittib und Erben, 1724. pp. 70, 16mo.

EXEMPLUM ARTIS PHILOSOPHICÆ. *See* Aureum Vellus.

EXPERIENCE AND PHILOSOPHY. *See* Ashmole, E.: Theatrum chemicum Britannicum.

EXPLICATION TRÈS CURIEUSE DES ÉNIGMES et figures hyéroglifiques qui sont au portail de l'église Cathédrale de Nottre-Dame. *See* R[ichebourg], J. M. D.

FABER, PETRUS JOHANNES.
See Manget, J. J. ; *also* Roth-Scholtz, Fr. (Deutsches Theatrum chemicum).

FABRE, PIERRE JEAN.
* L'Abrégé des secrets chymiques, où l'on void la nature des animaux, végétaux, et minéraux, entièrement découverte ; avec les vertus et propriétez des principes qui composent et conservent leur estre, et un Traitté de la médecine générale. À Paris, chez Pierre Billaine, ruë S. Jacques à la Bonne-Foy devant S. Yues, 1636. Avec privilége du Roy. pp. [xv]–392, 12mo.

FABRE, PIERRE JEAN.
Res alchymicorum obscuras extraordinariâ perspicuitate explanans. *See* Manget, J. J. Bibliotheca chemica curiosa, vol. I, p. 291.
The explanations are scarcely more intelligible than the subjects treated.

* FAMA MYSTICA HERMETICA von dem grossen Universal-Stein, oder
  Lapide Philosophorum der uralten Weisen, ein abgenöthigter Beweiss von desselben wahrhaftigem Daseyn als eine Antwort auf
  dasjenige Avertissement, das eine unbekannte, aber sehr erlauchte
  Feder, in dem Monat Hornung des Jahrs 1765, durch das Frank-
  furter und Erlanger Wochenblatt an die erlauchten Hohen Socie-
  täten London, Paris, Berlin abgegeben, und öffentlich bekannt
  gemacht hat. Frankfurt und Leipzig, bey Johann Paul Krauss,
  1772. pp. 88, 12mo.

  Cf. A B C.

FANIANUS, JOHANNES CHRYSIPPUS.

De jure artis alchemiæ, hoc est, variorum authorum et præsertim jurisconsultorum judicia et responsa ad quæstionem ; an alchemia sit
  ars legitima.

  See Manget, J. J.   Bibliotheca chemica curiosa, vol. I, p. 210.

Liber de arte metallicæ metamorphoseos ; accedunt variorum ICtor
  judicia et responsa de jure artis alchemiæ, an sit ars legitima.
  Basiliæ, 1576.   8vo.

  Reprinted in Theatrum chemicum, *q. v.*

FAUST, JOHANN MICHAEL.

Compendium alchymisticum novum, sive Pandora explicata et figuris
  illustrata ; das ist die edelste Gabe Gottes, oder ein güldener
  Schatz, mit welchem die alten und neuen Philosophi, die unvollkommene Metall, durch Gewalt des Feuers vebessert und allerhand
  schädliche und unheylsame Kranckheiten innerlich und äusserlich
  durch deren Würckung vertrieben haben. Diese Edition wird
  annoch nebst vielen Kupffern und über 800 philosophischen Anmerckungen, ein vollkommenes Lexicon alchymisticum novum und
  ein vollständiges Register rerum et verborum beygefüget. Franckfurt und Leipzig, 1706. pp. [xxvi]-1071-[cxcii]-104-236, 12mo.
  Ill.

  The Lexicon alchymisticum fills 104 pages, and forms a pretty full glossary.

FICINUS, MARSILIUS.

  See Manget, J. J, ; *also* A B C ; *also* Dreyfaches hermetisches Kleeblatt ;
  also Besondere Geheimnisse.

FICTULD, HERMANN.   *See* A B C.

FICTULD, HERMANN.

  * Abhandlung von der Alchymie und derselben Gewissheit. Erlangen,
  1754. pp. 226, 12mo.

  Contains a curious presentation of reasons for a belief in the verity of transmutation based on Holy Scripture and early authors.

FICTULD, HERMANN. [Cont'd.]

* Azoth et Ignis, das ist das wahre elementarische Wasser und Feuer, oder Mercurius Philosophorum, als das einige nothwendig der Fundamental-Uranfänge und Principiorum des Steins der Weisen. Aureum Vellus, oder Goldenes Vliess, was dasselbe sey, sowohl in seinem Ursprunge, als erhabenen Zustande. Denen Filiis Artis und Liebhabern der Hermetischen Philosophie dargelegt, auch dass darunter die Prima Materia Lapidis Philosophorum samt dessen Praxi verborgen eröffnet von H. F. Leipzig, bey Michael Bloch-berger, 1749. pp. 379, 12mo. Frontispiece and plates.

* Chymische Schrifften darinnen in zwölf königlichen Palästen von dem Stein der Weisen gehandelt wird. Samt einer kurtzen Vorrede ans Licht gestellet durch Friederich Roth-Scholtzen, Herenstadio-Silesium. Franckfurt und Leipzig, bey Johann Christoph Göpner Buchhändler in Nürnberg, 1734. pp. [viii]–230, 16mo.

[1.] Das edle Perlein und theurer Schatz der Himmlischen Weisheit, in zwölff königlichen Palästen vorgestellet und beschrieben, nehmlich wie der Stein der Weisen vom Anfang zum Ende gemacht und bereitet werde. p. 1.

[2.] Anhang und Anweisung vom allgemeinen natürlichen Chaos der Naturgemässen Alchymie und Alchymisten, daraus alles herkommt so zu unserer Kunst gehöret. p. 153.

[3.] Ein Gespräche zwischen dem König Mascos, seiner Schwester Agos und einem Hermetischen Lehr-Jünger von dem Stein der Weisen zu fernerer Erklärung des Chaos und der Arbeit. p. 199.

FLAMELLUS, NICOLAUS. [NICOLAS FLAMELL.]
See Theatrum chemicum ; also Musæum hermeticum ; also R[ichebourg], J. M. D. ; also Wasserstein der Weisen ; also Métallique (La) transforma-tion ; also Trois Traictez de la philosophie.

FLAMELLUS, NICOLAUS. [NICOLAS FLAMELL.]

* Chymische Werke als 1. Das güldene Kleinod der hieroglyphischen Figuren. 2. Das Kleinod der Philosophiæ. 3. Summarium Phi-losophicum. 4. Die grosse Erklärung des Steins der Weisen zur Verwandelung aller Metallen. 5. Schatz der Philosophiæ. Den Liebhabern der Kunst aus dem französischen in das teutsche übersetzt von J. L. M. C., zu finden bey Johann Paul Kraus, Buch-händler in Wien, 1751. pp. 290, 16mo. Frontispiece and plates.

* Zwey ausserlesene chymische Büchlein. 1. Das Buch der Hierogly-phischen Figuren Nicolai Flamelli des Schreibers, wie dieselben stehen unter dem vierdten Schwiebbogen auf dem Kirchhofe der Unschuldigen Kinder zu Paris, wann man zur Pforten von S.

FLAMELLUS, NICOALAUS. [Cont'd.]

> Dionysii Strassen hinein gehet zur rechten Handwerts, sampt der-
> selben Bedeutung oder Erklährung durch gemeldten Flamell.
> Worinnen gehandelt wird von Transmutation oder Verwandelung
> der Metallen. II. Das wahrhafte Buch des gelährten griechischen
> Abts Synesii, vom Stein der Weisen, welches aus der kaiserlichen
> Bibliothek herkommen. Zuvor noch nie im teutschen gesehen
> nun aber den Liebhabern der Kunst zu gutem aus dem franzö-
> sischen ins hochteutsche übersetzt. Anno 1751.
>> Another edition, Anno 1673.

> * His Exposition of the Hieroglyphical Figures which he caused to bee
> painted upon an Arch in St. Innocens Church-yard, in Paris.
> Together with the Secret Booke of Artephius, and the Epistle of
> John Pontanus : Concerning both the Theoricke and the Practicke
> of the Philosophers Stone. Faithfully and (as the Maiesty of the
> thing requireth) religiously done into English out of the French
> and Latine Copies, by Eirenæus Orlandus, qui est, vera veris
> enodans. Imprinted at London by T. S. for Thomas Walkley, and
> are to bee solde at his Shop, at the Eagle and Childe in Britans
> Bursse, 1624. pp. [xii]–139, 24mo. 1 folding plate.

FLORETUS A BETHABOR. *See* A B C ; *also* Mehun, J. von : Der Spiegel der Alchymie.

FLUDD DE FLUCTIBUS, ROBERT.

> * Schutzschrift für die Aechtheit der Rosenkreutzergesellschaft. Wegen
> seiner überaus grossen Seltenheit und Wichtigkeit auf Begehren
> aus dem lateinischen ins deutsche, zugleich mit einigen Anmer-
> kungen übersetzt von AdaMah Booz [Adam Michael Birkholtz].
> Leipzig, verlegts Adam Friedrich Böhme, 1782. pp. [xviii]–320.

FRAGMENTS [ALCHEMICAL]. *See* Ashmole, E. : Theatrum chemicum Britannicum.

* FREYMÄURERISCHE VERSAMMLUNGSREDEN DER GOLD UND ROSENKREUT-
ZER DES ALTEN SYSTEMS. Mit zwölf eingedruckten Vignetten. Amsterdam, 1779. pp. [xvi]–304, 8vo.

FRITSCHIUS, JOHAN CHRISTIAN. *See* Pyrotechnical Discourses.

FRYDAN, JOHANN FERDINAND VON. *See* A B C.

FÜRSTLICH-MONARCHISCHE (DIE) ROSE. *See* A B C.

GALLUS, FRIEDRICH. *See* Mehun, J. von : Der Spiegel der Alchymie.

* GANZE (DIE) HÖHERE CHEMIE UND NATURWISSENSCHAFT in allgemeinen
Grundsätzen, nach den drei Uranfängen und Grundkräften der
ganzen Natur. Aus dem lateinischen übersetzt, mit beygefügten
Anmerkungen von AdaMah Booz [Adam Michael Birkholz]. Leipzig, bei Johann Friedrich Junius, 1787. pp. [xvi]–366, 8vo.

GARLANDUS, JOHANNES, SEU HORTULANUS.

See Roth-Scholtz, Fr. (Deutsches Theatrum chemicum); *also* Hermetischer
Rosenkranz; *also* Bacon, Roger: Le Miroir d'alquimie.

GASTON DE CLAVIUS, DULCO.

Apologia chrysopœiæ et argyropœiæ contra Thomam Erastum. 1598.
pp. xvi–216, 12mo.

Also in Theatrum chemicum, vòl. II. *Cf.* Erastus, Thomas.

* Traité philosophique de la triple préparation de l'or et de l'argent. À
Paris, chez Laurent d'Houry, ruë Saint Jaques, devant la Fontaine
Saint Severin, au Saint Esprit, 1595 [1695]. Avec privilege du
Roy. pp. 119, 16mo.

> Of this writer Lenglet du Fresnoy remarks : " Cet auteur est estimé et à
> l'entendre, la manière d'arriver au grand œuvre est facile ; il paroîtroit
> même qu'il l'auroit euë. Les préparations qu'il enseigne sont assez
> simples, c'est un des plus clairs des auteurs . . . ; supposé néanmoins
> qu'il dise vrai : car il ne faut pas se fier à tous ces messieurs qui parlent
> si clairement."

GEBER. [JÁBIR IBN HAYYÁN (ABÚ MÚSÁ).]

Geberi philosophi ac alchimistæ maximi, de alchimia libri tres. [Co-
lophon :] Argentoragi [*sic*] arte et impensa solentis viri Johannis
Grieninger anno a virgineo partu M.D.XXIX decimo die Martij. Fol.
66, 4to.

> The earliest Latin version.

Geberis philosophi perspicacissimi summa perfectionis magisterii in sua
natura ex bibliothecæ Vaticanæ exemplari undecunque ; emendatis-
simo nuper edita, cum quorundam capitulorum, vasariorum et fornacum, in volumine alias mendosissime impresso omissorum,
librique investigationis magisterii, et testamenti ejusdem Geberis,
ac aurei trium verborum libelli, et Avicennæ summi medici et
acutissimi philosophi mineralium additione castigatissima. Venetiis,
apud Petrum Schæffer Germanum, Maguntinum, Anno 1542. ff.
126, 16mo.

Geber Arabis . . . de alchemia traditio summæ perfectionis in duos
libros . . . Item : liber investigationis magisterii ejusdem. Ab
. . . mendis repurgata . . . [Strassburg], 1598. 8vo.

* Chimia sive traditio summæ perfectionis et investigatio magisterii
innumeris locis emendata à Caspare Hornio. Accessit ejusdem :
Medulla alchimiæ Gebricæ. Omnia edita à Georgio Hornio.
Lugduni-Batavorum, apud Arnoldo Doude, 1668. pp. [xviii]–279,
32mo. Illustrated title-page.

* Curieuse vollständige chymische Schriften worinnen in den vier
Büchern das Quecksilber, Schwefel, Arsenicum, Gold, Silber, Bley,

GEBER. [Cont'd.]

Zinn, Kupfer, Eisen, Oefen, Instrument, Sublimationen, Descension, Destillationen, Calcination, Solution, Coagulation, Fixation, Ceration, Test, Cement, Feurung, Schmelzung, etc., ferner deren Anfänge, Præparationen, Essenzen, Salze, Alaune, Atramente, Salpeter, Salarmoniæ, Vitriol, Antimonium, Bolus, Cinnober, Glass, Borax, Essig, etc., abgehandelt werden ; Wie auch das Testament, Güldene Buch der Dreyen Wörter Kallid Rachaidibi, und andere chymische Tractätgen. Summa die ganze Kunst die unvollkommenen Metalle, als Kupfer, Zinn, Bley, Eisen, etc., in Vollkommene, als Silber und Gold zu verwandeln ; das ist : Wie man Silber und Gold machen soll, enthalten. Alles aus einem uhralten Manuscripto genommen, nach dem vorhandenen Exemplar in der Vaticanischen Bibliothec eingerichtet, mit gehörigen Figuren und Register versehen, und an Tag gegeben von Philaletha. Wienn, verlegts Joh. Paul Krauss, 1753. pp. [xx]–332–[iv], 12mo.

> *The volume contains also :* Avicenna : Tractat von der Congelation und Conglutination der Steine. p. 319.

* Works (The) of Geber, the most famous Arabian Prince and Philosopher, faithfully Englished by R[ichard] R[ussel], a Lover of Chemistry. London, printed for N. E. by Thomas James, Mathematical Printer to the Kings most Excellent Majesty, at the Sign of the Printing-press in Mincing-lane ; and are to be sold by Booksellers, 1678. pp. [xvi]–302, 12mo.

> Geber, properly Jábir ibn Hayyán (Abú-Músá), was an Arabian physician of the eighth century. His Arabic manuscripts are preserved in Leyden, Paris and Rome. Versions in modern languages are numerous. *See* Manget, J. J. ; *also* R[ichebourg], J. M. D.

GEDICHT (EIN) VON DEN BRÜDERN DES ROSEN-KREUTZES. *See* A B C.

GEGEN ANTWORT DES ALCHYMISTEN. *See* Wasserstein der Weisen.

GEHEIME NATÜRLEHRE DER HERMETISCHEN WISSENSCHAFT. *See* A B C.

* GEHEIMNISSE EINIGER PHILOSOPHEN UND ADEPTEN AUS DER VERLASSENSCHAFT EINES ALTEN MANNES. Erster Theil. Leipzig, bey Christian Gottlob Hilscher, 1780. pp. 187, 12mo.

GENTY, ACHILLE. [Editor.]

* La Fontaine des amoureux de science composée par Jehan de la Fontaine de Valenciennes en la Comte de Herault. Poème hermétique du xve siècle. Paris, 1861. pp. 93, 12mo.

> Contains also " Balade du secret des philosophes." With an historical introduction and bibliographical notes. My copy was presented by the editor to Alphonse de Lamartine.

GERHARD, JOHANN CONRAD.

Extractum chymicarum quæstionum, sive responsionis ad theoriam lapidis philosophici editam in academia Regiomontana a quodam ibidem antichymista ; ubi veritas artis chymicæ etiam contra principia negantem asseritur, et multæ difficiles et jucundæ quæstiones discutiuntur. Argentorati 1616. pp. [xvi]–134, 12mo.

GERHARDUS, JOANNES. *See* Manget, J. J.

GESPRÄCH ZWISCHEN DEM SATURN DER WEISEN UND EINEM CHEMISTEN. *See* A B C.

ГИЛБОА.

Алхимистъ безъ маски. Москва, 1789.

GILBOA. Alchymist without the masque. Moscow, 1789.

GLAUBER, JOHANN RUDOLPH.
See C[ooper], W[illiam], *and cf. in Section V.*

GLAUBER, JOHANN RUDOLPH.

* De auri tinctura sive auro potabili vero ; was solche sey und wie dieselbe von einem falschen und sophistischen Auro potabili zu unterscheiden und zu erkennen. Auch wie solche auff spagirische weise zugerichtet und bereitet werde ; und wozu solche in Medicina könne gebraucht werden. Amsterdam, gedruckt bey Johann Fabeln, im Jahr 1646. pp. 39, 16mo.

* GOLDMACHER CATECHISMUS IN FRAG UND ANTWORT ; zum Nutzen und Vergnügen aller derjenigen welche in diesem Hospital krank darnieder liegen, lehrend, wie sie wieder zur wahren Erkäntniss gelangen können. Aufrichtig beschrieben von einem Liebhaber in Philadelphia. Berlin und Leipzig, 1776, bey Christian Ulrich Ringmacher. pp. vi–80, 12mo.

GOWER, JOHN. *See* Ashmole, E. : Theatrum chemicum Britannicum.

GRASHOFER, JOHANN. *See* Grasseus Chortalasseus.

GRASSEUS, ALIAS CHORTALASSEUS JOANNES.
See Manget, J. J. ; *also* Theatrum chemicum ; *also* A B C.

GRASSHOF, JOHANN.

* Dyas chymica Tripartita, das ist : Sechs herrliche teutsche philosophische Tractätlein. Deren II, Von an itzo noch im Leben ; II, Von mitlern Alters ; und II, Von ältern Philosophis beschrieben worden. Nunmehr aber allen Filiis Doctrinæ zu Nutz an Tag geben und mit schönen Figuren gezieret durch H. C. D. [Dr. Hermannus Condeesyanus]. Franckfurt am Mayn, bey Luca Jennis zu finden, 1625. pp. 87–150, sm. 4to. Ill.

GRASSHOF, JOHANN. [Cont'd.]

[*Contents :*]

1. Ein güldener Tractat vom philosophischen Steine. Von einem noch lebenden, doch ungenannten Philosopho den Filiis Doctrinæ zur Lehre den Fratribus auræ crucis aber zur Nachrichtung beschrieben. p. 13

2. *Madathanus, Hinricus :* Aureum seculum redivivum, das ist : Die uhralte entwichene güldene Zeit, so nunmehr wieder auffgangen, lieblich geblühet, und wohlriechenden güldenen Samen gesetzet. Welchen teuren und edlen Samen allen wahren Sapientiæ et Doctrinæ Filiis zeigt und offenbahret. p. 67.

3. *Valentinus, Basilius :* Vier Tractätlein. p. 1.
   a. Handgriffe über die Bereitung des grossen Steins.
   b. Handgriffe wie er seine Artzneyen gemacht hat.
   c. Schlussreden vom Sulphure, Vitriola, und Magnete.
   d. Supplementum oder Zugabe.

4. *Lambspring :* Ein herrlicher teutscher Tractat vom philosophischen Steine, welchen für jahren ein adelicher teutscher Philosophus so Lampert Spring geheissen, mit schönen Figuren beschrieben hat. p. 83.

5. Vom philosophischen Steine ein schöner Tractat, von einem teutschen Philosopho im Jahre 1423, beschrieben mit diesem Titul : Ein wahrhafftige Lehr der Philosophie von der Gebehrung der Metallen, und ihrem rechten Beginne an Tag gegeben. p. 119.

6. Vom philosophischen Steine, ein kurtzes Tractätlein, so von einem unbekanten teutschen Philosopho beynahe für zweihundert Jahren beschrieben, und "Liber Alze" genennet worden, jetzo aber an Tag geben. p. 137.

The publisher, L. Jennis, issues with this :

Hermetico-Spagyrisches Lustgärtlein : Darinnen hundert und sechtzig unterschiedliche, schöne kunstreiche, chymicosophische Emblemata, oder geheymnussreiche Sprüche der wahren Hermetischen Philosophen ; Sampt beygefügten noch vier grossen, schönen unnd tiefsinnigen theosophischen Figuren. Nicht allein sehr dienstlich, Augen und Gemüt dardurch zu erlüstigen, sondern zugleich ein scharffes nachdencken der Natur, bey allen Filiis Doctrinæ zuerwecken. Franckfurt am Mayn, bey Luca Jennis, zu finden, 1625. pp. 24, sm. 4to. Plates.

GRATAROLUS, GUL.

Veræ alchemiæ artisque metallicæ, citra ænigmata, doctrina certusque modus, scriptis tum novis tum veteribus nunc primum et fideliter majori ex parte editis, comprehensus : quorum elenchum a præfatione reperies. Basiliæ, 1561. pp. [xvi]–244–299, folio.

> Also under the title : Veræ alchymiæ scriptores aliquot collecti et una editi. (Schmieder.) Another edition, 1572. 8vo. *Cf.* Artis auriferæ, for another collection by this editor.
>
> Lenglet du Fresnoy writes of this collection : "D'ailleurs quand ces auteurs disent qu'ils vont parler sans énigmes, qu'on ne s'y fie pas ; ce n'est que pour mieux tromper."

GREVVERUS, JODOCUS. *See* Theatrum chemicum.

GÜGLER. *See* Lutz von Laufelfingen.

GÜLDENE (DIE) ROSE. *See* A B C.

GÜLDENER TRACTAT (EIN) VOM PHILOSOPHISCHEN STEINE. *See* Grasshof, Johann.

*GÜLDENE (DAS) VLIESS oder das allerhöchste, edelste, kunstreichste Kleinod, und der urälteste Schatz der Weisen. In welchem da ist die allgemeine Materia Prima derselben nothwendige Præparation und überaus reiche Frucht des philosophischen Steins augenscheinlich gezeiget und klarlich dargethan. Philosophischer und theologischer Weise beschrieben und zusammen verfasset durch einen ungenannten, doch wohl bekannten, etc. *Ich sags nicht.* Nürnberg, bey Johann Adam Schmidt, 1737. pp. [xiv]–208, 12mo. Frontispiece and Plates.

> Contains also :
>
> Duodecim articuli philosophici, oder zwölf gedoppelte Stücke so einem jeden rechten theosophischen, wie auch wahren philosophischen Artisten nothwendig zu wissen, zu kennnen und zu erkennen sind, und ohne welche sonst im Lapide Philosophorum alles Studiren und Laboriren nicht allein beschwerlich, sondern auch gefährlich, vergeblich und unnützlich fürgenommen wird. Aufs allerkürtzeste einfältig angedeutet. p. 197.

GÜLDENFALK, SIEGMUND HEINRICH.

*Die himmlische und hermetische Perle, oder die göttliche und natürliche Tinctur der Weisen. Frankfurt und Leipzig, in der Fleischerischen Buchhandlung, 1785. pp. [iv]–172, 12mo.

> *See also in Section III.*

GUIBERTUS, NICOLAUS.

Alchymia ratione et experientia impugnata et expugnata. Argentorati, 1603. 8vo.

> The author attacks current views concerning transmutation.

GUIBERTUS, NICOLAUS. [Cont'd.]

De interitu alchymiæ metallorum transmutatoriæ tractatus aliquot mul-
tiplici eruditione referti. Adjuncta est ejusdem apologia in sophis-
tam Libaviam, alchymiæ refutatæ furentem calumniatorem, quæ loco
præfationis in eosdem tractatus esse possit. Tulli, 1614. pp. [xvi]
–141, 12mo.

GUTWASSER, BENEDICT. *See* A B C.

HALLIWELL, JAMES ORCHARD [Editor]. *See* Dee, John (The private Diary of J. **Dee**).

HAPPELIUS, NICOLAUS NIGER. *See* Theatrum chemicum.

HAUTNORTHON, JOSAPHAT FRIEDRICH.
        See Roth-Scholtz, Fr. : Deutsches Theatrum chemicum ; *also* A B C.

HARTMANN, JOSEPH.

* Alchemie und Arkanologie im Gegensatze zur Schulmedizin. Die Ar-
kana die Remedia divina der alten Alchemisten. Zürich, 1888.
pp. 32, 8vo.
        Dedicated to Gottlieb Latz, "meinem hochverehrten Lehrer"; see notes to
        Latz, G. A fountain cannot rise higher than its source.

HELIAS ARTISTA. *See* Theatrum chemicum

HELLWIG, JOHANN OTTO DE.

* Curiosa physica, oder gründliche Lehre von unterschiedlichen Natur-
Geheimnissen, sonderlich das philosphische Meisterstück oder so
genandten Lapidem Philosophorum betreffend, gleichsam als sein
letztes Testament ; zum andermal herausgegeben und mit unterschiedlichen curiösen Stücken vermehret von L. Christoph Hellwig,
i. z. Pr. Erffurt. Franckfurt und Leipzig, verlegts Michael Käyser,
Buchhändler. Mühlhausen, druckts Tob. Dav. Brückner, im Jahr
1714. pp. [x]–154–[x], 16mo. Portrait.

HELMONT, JOHN BAPTIST VAN. *See* Collectanca chymica.

HELMONT, JOHANN BAPTIST VAN.

* Delineamenta Catarrhi, or the Incongruities, Impossibilities and Ab-
surdities couched under the vulgar opinion of Defluxions. The
Translator and Paraphrast, Dr. Charleton, Physician to the late
King. London, Printed by E. G., for William Lee at the signe of
the Turks-head in Fleet-street, 1650. pp. [x]–75, sm. 4to.
        Bound with the Ternary of Paradoxes.

* Ternary (A) of Paradoxes. The Magnetick Cure of Wounds ; the
Nativity of Tartar in Wine ; the Image of God in Man. Translated, illustrated and ampliated by Walter Charleton. London,
Printed by James Flesher for William Lee, dwelling in Fleetstreet,
at the sign of the Turks-head, 1650. pp. [xlv]–144, sm. 4to.
        See also in Section V.

HELVETIUS, JOHANNES FRIEDERICUS.

 See Manget, J. J. ; *also* Musæum hermeticum ; *also* Roth-Scholtz, Fr. (Deutsches Theatrum chemicum) ; *also* Cooper, William.

HELVETY, JOHAN FRED. [HELVETIUS].

 * Theatridium Herculis Triumphantis, ofte kleyn Schouw-Tooneel, van den Triumpherenden Hercules. Met volkomen kennisse der natuyr-lijke dingen, bestaende in Sympathia, ende Antipathia, Magicè ende Magnetice Midstsgaders ; grondighe Weder-legginghe der Schriften van Sijn Excell. Digby, ængæende Poudre Sympathie daer door Wonden sonderverhin deringhe der tusschen-komende distantien, van salven, plaesters, etc., te genesen, oock smerten, te veroorsaken ende wegh te nemen. Alles wat uwe handt vindt om te doen, doet dat met uwe macht, want daer en is geen werck noch versinninge, noch wetenschap noch wijsheydt in het graf dar ghy henen gaet. In s'Graven-Hage, By Johannes Tongerloo, Boek-verkoper in de Hoogh-staet, Anno 1663. pp. [xvi]–200, 12mo. Portrait.

  The author, a Dutch physician, became a zealous alchemist under circumstances narrated in his " Vitulus Aureus." *Cf.* Cooper, Wm.

HENTSCHEL, CARL.

 Chemie, Alchemie und Botanik. Lehr- und Lobgedicht in vier Abtheilungen. Hof und Wunsiedel, 1840. 8vo.

HERMAPHRODITISCHE (DAS) SONN UND MONDS-KIND. *See* A B C.

HERMES'S BIRD. *See* Ashmole, E. : Theatrum chemicum Britannicum.

HERMES TRISMEGISTUS. *See* Trismegistus, Hermes.

HERMETICO-SPAGYRISCHES LUSTGÄRTLEIN. *See* Grasshof, Johann.

 * HERMETISCHER ROSENKRANZ, das ist vier schöne anserlesene chymische Tractätlein ; allen Liebhabern dieser edlen Kunst zum besten aus dem lateinischen ins teutsche übersetzt und nun zum zweytenmal in Druck gegeben. Franckfurt am Mayn, bey Johann Friedrich Fleischer, 1747. pp. 112, 16mo.

[*Contents :*]

 1. *Artefius :* Geheimes Buch von der geheimen Kunst und Stein der Weisen. p. 7.

 2. *Garlanäius, Johannes,* seu Hortulanus Anglus : Compendium Alchimiæ, oder Erklärung der smaragdischen Tafel Hermetis Trismegisti von der Chimia. p. 51.

 3. *Arnaldus de Villa Nova :* Erklärung über den Commentarium Hortulani. p. 81.

HERMETISCHER ROSENKRANZ.   [Cont'd.]
  4. Trevirensis, Bernhardus : Ein schöner absonderlicher Tractat vom
     Stein der Weisen. Aus dem latein ins teutsche übersetzet.
     p. 99.

HERMETISCHES A B C derer ächten Weisen alter und neuer Zeiten vom
   Stein der Weisen. Ausgegeben von einem wahren Gott- und Men-
   schenfreunde. Berlin, 1778–9. 4 parts, 12mo. I, pp. 318 ; II,
   348 ; III, 299 ; IV, 326.

> For contents see A B C vom Stein der Weisen, which is a later edition of the
> above.

HERMETISCHES MUSEUM.   Allen Liebhabern der wahren Weisheit gewidmet von dem Herausgeber. Reval und Leipzig, 5782–5783. 2
   vols., 8vo.

* HERMETISCHE TRIUMPH (DER), oder der siegende philosophische Stein ;
   ein Tractat völliger und verständlicher eingerichtet als einer jemals
   bissher gewesen handelnde von der Hermetischen Meisterschaft.
   Hiebevor in französischer Sprache gedruckt zu Amsterdam bey
   Heinrich Wetstein, Anno 1689. Nunmehro gegenwärtig ins deutsche
   versetzt. Frankfurt und Leipzig, bey Johann Paul Krauss, 1765.
   pp. 224, 12mo.   Plate.

> *Contains :* I. Uralter Ritter-Krieg : L'ancienne guerre des chevaliers. [In
> German and in French on opposite pages.] II. Gespräch des Eudoxi
> und Pyrophile über den uralten Ritter-Krieg. III. Sendschreiben an die
> wahren Schüler Hermetis.

HERMIT'S TALE.   *See* Ashmole, E. : Theatrum chemicum Britannicum.

HERTOOT A TODTENFELD, JOANNES FERDINANDUS.   *See* Manget, J. J.

HERVERDI, JOSEPH FERDINAND.
   * Erklärung des mineralischen Reichs ; ein Beytrag zur Geschichte der
     Alchymie. Berlin, bey Arnold Wever, 1783. pp. 124, 16mo.

HIRSCHING, W. S. C.
   * Versuch physicalisch-chymischer Lehrbegriffe zu möglicher Prüfung
     des Wesens, des Beständnisses, und der Wirkungsart des so berüch-
     teten metallverwandelnden Meisterstückes und dessen vorgeblicher
     Nutzanwendung zu einem allgemeinen Genessmittel in Absicht
     einiger Vergnügung einer natur- und grundforschenden Wissbe-
     gierde entworfen. Leipzig, verlegts Carl Ludwig Jacobi, 1754. pp.
     [xxiv]–488–[xiv], 8vo.

HITCHCOCK, ETHAN ALLEN.

* Remarks on the Sonnets of Shakespeare ; with the Sonnets. Showing
that they belong to the Hermetic class of writings, and explaining
their general meaning and purpose. Second edition enlarged.
New York, 1867. pp. xxvi–366, 8vo.

> Published anonymously.

* Remarks upon Alchemy and the Alchemists indicating a method of
discovering the true nature of Hermetic Philosophy, and showing
that the search after the Philosophers' Stone had not for its object
the discovery of an agent for the transmutation of metals. Being
also an attempt to rescue from undeserved opprobrium the reputa-
tion of a class of extraordinary thinkers in past ages. Boston, 1857.
pp. xv–304, 8vo.

> Published anonymously. The author claims that the works of the alchemists
> are treatises upon religious education. This work of 304 pages is not
> divided into chapters, has no table of contents and no index !

* Swedenborg, a Hermetic Philosopher. Being a sequel to Remarks on
Alchemy and the Alchemists. Showing that Emanuel Swedenborg
was a hermetic philosopher and that his writings may be interpreted
from the point of view of Hermetic Philosophy. With a chapter
comparing Swedenborg and Spinoza. New York, 1858. pp. 352,
8vo.

> Mystical rather than alchemical ; Anonymous.

HOGHELANDE, THEOBALDUS DE.
> *See* Manget, J. J.; *also* Theatrum chemicum ; *and cf. in Section III.*

HOLLANDUS, ISAAK.

Libellus rarissimus, dictus secreta revelatio veræ operationis manualis
pro universali opere et lapide sapientium, sicut filio suo M. Johanni
Isaaco Hollandus e Flandria paterno animo fidelissimo manu
tradidit. 8vo. 16—.

Mineralia opera, seu de lapide philosophico duo libri. Middelburg,
1600. 8vo.

Opera mineralia et vegetabilia, sive de lapide philosophorum quæ
reperiri potuerunt omnia. Arnhem, 1617. 8vo.

Rariores chemiæ operationes. Lipsiæ, 1714. 8vo.

> *See also* Theatrum chemicum.

HOLLANDUS, JOHANNES ISAACUS.

* Hand (Die) der Philosophen, mit ihren verborgenen Zeichen. Wie
auch desselben Opus saturni mit annotationibus. Geheimer und
bis dato verborgen gehaltener trefflicher Tractat. Item Opera
vegetabilia, so viel davon biss dato hat können erforschet werden.

HOLLANDUS, JOHANNES ISAACUS. [Cont'd.]

Mit grosser Müh und Fleiss auss erforscheten niederländischen Manuscriptis verhochdeutschet, nebenst fleissigster Nachreissung aller darin enthaltenen Figuren, von einem geübten Liebhaber der hermetischen Philosophy. Dabey mit angehenckt worden, ein zwar kleiner, aber überauss herrlicher Tractat so von Michaelis Sendivogii Diener herkommt, und ohne Zweiffel vom Sendivogio selber gestellet worden, darin der Grund der metallischen meliora-tion mit Fingern gleichsam gezeiget wird. Franckfurt, in Verlegung Thomæ Matthiæ Goetzens, 1667. pp. 384, 12mo. Ill.

* Sammlung unterschiedlicher bewährter chymischer Schriften, nament-lich : Hand der Philosophen, opus saturni, opera vegetabilia, opus minerale, cabala, de lapide philosophico ; Nebst einem Tractat von den Irrgängen derer Alchymisten, Auctoris incerti, neue und ver-besserte Auflage, mit gehörigem Fleisse übersehen und mit einem Verzeichniss derer in jeglichem Tractat befindlichen wichtigsten Materien vermehret, wie auch mit nöthigen Kupfern gezieret. Wien, in Verlag bey Joh. Paul Krauss, Buchhändler, 1773. pp. [xvi]–762, 12mo.

HORNIUS, CHRISTOPHORUS. *See* Theatrum chemicum.

HORTULANUS, JUNIOR.

* The Golden Age : or the Reign of Saturn review'd. Tending to set forth a true and natural way, to prepare and fix common mercury into silver and gold. Intermix'd with a discourse vindicating and explaining, that famous universal medicine of the ancients, vul-garly called, the Philosophers' Stone, built upon four natural principles. An Essay. Preserved and published by R. G. Lon-don, printed by J. Mayos, for Rich Harrison, at New-Inn, without Temple-Bar, 1698. pp. [xxiv]–215, 12mo.

HORTUS DIVITIARUM. *See* Aureum Vellus.

HYDROLITHUS SOPHICUS SEU AQUARIUM SAPIENTUM. *See* Musæum hermeticum.

* INTRODUCTION À LA PHILOSOPHIE DES ANCIENS. Par un amateur de la vérité. À Paris, chez la veuve de Claude Thiboust et Pierre Esclassan, Libraire-Juré et ordinaire de l'Université, ruë Saint Jean de Latran, vis-à-vis le Collége Royal, 1689. Avec privilége du Roy. pp. [xii]–395, 12mo.

IN TURBAM PHILOSOPHORUM EXERCITATIONES. *See* Artis auriferæ ; *also* Manget, J. J.

ISAACUS HOLLANDUS. *See* Hollandus, Isaak.

JÁBIR IBN HAYYÁN (ABÚ MÚSÁ). *See* Geber.
> The incomparable Catalogue of the British Museum gives Geber's name as
> above. For convenience I adopt the form more commonly occurring.

JACOB, P. L.
> * Curiosités des sciences occultes ; alchimie, médecine chimique et
> astrologique, talismans, amulettes, baguette divinatoire, astrologie,
> chiromancie, magie, sorcellerie, secrets d'amour, etc. Paris, 1885.
> pp. 391, 8vo.

JAMSTHALER. *See* A B C.

JANITOR PANSOPHUS. *See* Musæum hermeticum.

JOANNES DE PADUA. *See* Aureum Vellus ; *also* A B C.

JOHNSONIUS, GULIELMUS. *See* Manget, J. J.

JOLY, GABRIEL [Translator]. *See* Trois anciens traictez de la philosophie naturelle.

KALID, RACHAIDIBUS VERADIANUS.
> *See* Artis auriferæ ; *also* Manget, J. J. ; *also* Arnaldus de Villa Nova:
> Schrifften.

KELLÆUS, EDOUARDUS.
> * Tractatus duo egregii de lapide philosophorum, una cum theatro
> astronomiæ terrestri, cum figuris ; in gratiam filiorum Hermetis
> nunc primum in lucem editi, curante J. L. M. C. Hamburgi, apud
> Gothofredum Schultzen. Anno 1676. pp. 125, 16mo. Ill.

KELLEY, SIR EDWARD.
> *See* Ashmole, Elias : Theatrum chemicum Britannicum ; *also* Roth-Scholtz,
> Fr. (Deutsches Theatrum chemicum).

KERENHAPUCH.
> Deutsches Fegefeuer der Scheidekunst, worinnen nebst den neugierigsten
> und grössten Geheimnissen die wahren Besitzer der Kunst, wie
> auch die Ketzer, Betrüger, Pfuscher, Stümpler, Bönhasen und
> Herren Gerngrose vor Augen gestellt werden ; mit gar vielen Orten
> aus der Schrift und anderen Urkunden erörten, von einem Feind
> der Vizlipuzli der ehrlicher Leute, Ehre und der aufgeblasenen
> Schande entdecken will. Hamburg, 1702. 8vo.

KHUNRATH, HEINRICH.
> * Alchymisch-philosophisches Bekenntniss vom universellen Chaos der
> naturgemässen Alchymie, mit beygefügter Warnung und Ver-
> mahnung an alle wahre Alchymisten. Neue von den deutschen
> Sprachfehlern ohne Verletzung des Sinnes gesäuberte, und mit des
> Verfassers Anmerkungen versehene Auflage. Leipzig, 1786. pp.
> 348, 12mo.

KHUNRATH, HEINRICH. [Cont'd.]

* De igne magorum philosophorumque secreto, externo et visibili, das
    ist, Philosophische Erklärung des geheimen, äusserlichen, sichtbaren
    Glut und Flammenfeuers der uralten Weisen und anderer wahren
    Philosophen. Nebst Johann Arndts philosophisch-kabalistischen
    Judicio über die vier ersten Figuren des grossen Khunrathischen
    Amphitheaters. Neue und mit Anmerkungen versehene Auflage.
    Leipzig, bey Adam Friedrich Böhme, 1783. pp. 109, 12mo.

* Magnesia Catholica Philosophorum, oder eine in der Alchymie höchst
    nothwendige und augenscheinliche Anweisung die verborgene
    catholische Magnesia des geheimen Universalsteins der ächten
    Philosophen zu erlangen. Im Jahr 1599 gründlich geschrieben
    und zu Magdeburg herausgegeben. Neue von den Sprach und
    Druckfehlern gesäuberte Auflage. Leipzig, bey Adam Friedrich
    Böhme, 1784. pp. vi–112, 12mo.

* Wahrhafter Bericht vom philosophischen Athanor und dessen Ge-
    brauch und Nutzen. Wegen seiner überaus grossen Seltenheit
    nach der dritten im Jahr 1615 zu Magdeburg im Verlag des Ver-
    fassers gedruckten Ausgabe aufs neue von den deutschen Sprach-
    fehlern ohne Verletzung des Sinnes gesäubert, und mit einem
    historischen Vorberichte von seinen sämmtlichen Schriften, nebst
    dem in Kupfer gestochenen Athanor auf Begehren herausgegeben.
    Leipzig, bey Adam Friedrich Böhme, 1783. pp. 58, 12mo.

KINDERBETT DES STEINS DER WEISEN. *See* A B C.

KIRCHERUS, ATHANASIUS. *See* Manget, J. J.

KIRCHMAYER, GEORG CASPAR.
    Metallico-metamorphosis, principiis ac experimentis curiosis metal-
    lurgicis asserta. Publicabitur ad d. Decembr. 1693, in auditorio
    majori à respondente Ludovico Caspare Mayero, Neukirchensi,
    Franco. Wittenbergæ, typis Christiani Schrödteri, Acad. Typ.
    1693.

KIRCHWEGER, ANT. JOSEPHUS DE FORCHENBRONN.
* Microscopium Basilii Valentini, sive commentariolum et cribrellum über
    den grossen Kreuzapfel der Welt ♂, ein Euphoriston der ganzen
    Medicin, ex Theoria et Praxi Gravinii, allen Philo-Medicis, Chymicis,
    Pharmacopaeis, Chirurgis et singulis Medicinæ Amatoribus Chymi-
    caeque artis praeprimis Fautoribus, zu ihrem freundlichen Faveur
    und Benevolenz. Ein Compendium der ganzen chymischen Scienz
    und Physica Hermetica concentrata; ein Werk so noch nie gesehen
    worden, höchst nützlich zur Praxi und der jetzigen Welt höchst
    nöthig. Berlin, 1790. pp. x–172, 12mo.

KOFFSKHIUS, VINCENTIUS.

* Hermetische Schriften denen wahren Schülern und Nachfolgern unserer geheimen spagierischen Kunst zum Nutz beschrieben und hinterlassen, den 4ten October Ao. D. 1478. Zwey Theile. Nürnberg, 1786. pp. 119, 12mo. Plates.

> Vincenz Koffsky, a Polish monk, died at Danzig in 1488.

KORNDORFFER, BARTHOLOMAUS.

> *See* Aureum Vellus; *also* Roth-Scholtz, Fr. (Deutsches Theatrum chemicum).

KORTUM, KARL ARNOLD.

* Noch ein Paar Worte über Alchimie und Wiegleb, oder erster Anhang der Vertheidigung der Alchimie wider die Einwurfe der neuesten Gegner. Duisburg, in Commission der Helwingschen Universitäts-Buchhandlung, 1791. pp. 80, 8vo.

* Verteidiget die Alchimie gegen die Einwürfe einiger neuen Schriftsteller besonders Herrn Wieglebs. Duisburg, in der Helwingschen Universitäts-Buchhandlung, 1789. pp. [iv]-360, 8vo.

KRIEGSMANN, WILHELM CHRISTOPH. *See* Manget, J. J.

KRIEGSMANN, WILHELM CHRISTOPH.

* Taaut, oder Ausslegung der chymischen Zeichen; damit die Metallen und andere Sachen von Alters her bemerckt werden. Auff Begehren beschrieben. Franckfurt, bey Thoma Matthia Götzen, 1665. pp. [iv]-75-[iv], 18mo. Plates.

> My copy is bound with Andrëa's "Chymische Hochzeit."

KUNCKEL VON LÖWENSTERN, JOHANN.

* Nützliche Observationes, oder Anmerckungen von den fixen und flüchtigen Salzen, Auro und Argento potabili, Spiritu Mundi und dergleichen, wie auch von den Farben und Geruch der Metallen, Mineralien und andern Erdgewächsen; durch vieljährige, eigene Erfahrung, Mühe und Arbeit mit Fleiss untersuchet, angemercket, und nun auff vieler der edlen Chimie beflissenen und unverdrossener Naturforscher inständiges Begehren zu dero Nutz und Gefallen an den Tag gegeben. Hamburg, auff Gottfried Schultzens Kosten, im Jahr 1676. pp. [100], 12mo.

KUNCKEL, JOHANN. *See* Pyrotechnical Discourses; *and cf. in Section V.*

* KURTZER DOCH DEUTLICHER SCHLÜSSEL derjenigen fremden Wörter, welche in Abraham dem Juden enthalten sind, wodurch denen Liebhabern der edlen hermetischen Wissenschaft ein sonderbahres Licht gegeben wird. *n. p., n. d.* [1735(?)]. pp. [xii].

> My copy is bound with Baruch, Samuel, *q. v.*

KURTZES TRACTÄTLEIN VOM PHILOSOPHISCHEN STEINE. *See* Grasshof, Johann.

LA BROSSE, DE. *See* Theatrum chemicum.

LACINI, GIOVANNI. *See* Petrus Bonus Ferrariensis : Pretiosa Margarita.

LACINI, GIOVANNI. [LACINIUS, JANUS.]
* Præciosa ac nobilissima artis chymiæ collectanea de occultissimo ac
præciosissimo philosophorum lapide ; nunc primum in lucem edita
cum totius libelli capitum indice. Norimbergæ, apud Gabrielem
Hayn, Joann. Petrei generum, 1554. pp. [xvi]–124, 4to. Ill.

LA FONTAINE, I. DE. *See* Métallique (La) transformation.

LA FONTAINE, DE. *See* Genty, Achille.

LAGNEUS, DAVID.
Harmonia sive consensus philosophorum chemicorum. Paris, 1611.
12mo.
> *Also in* Theatrum chemicum, vol. IV.

LAMBSPRINCK. *See* Theatrum chemicum ; *also* Musæum hermeticum ; *also* Grasshof,
Johann.

LANGELOT, JOEL. *See* Roth-Scholtz, Fr. (Deutsches Theatrum chemicum).

LANGELOT, JOEL, MORHOF ET TILEMANN.
Virorum clarissimorum epistolarum circa utilissima aliquot chymica
experimenta conscriptarum trias : 1. Joel Langelottus de quibusdam in chimia prætermissis ; 2. D. G. Morhoff de transmutatione
metallorum ; 3. Jo. Tilemann experimenta circa veras et irreduci-
biles auri solutiones. Hamburgi, 1673. 8vo.
> *Cf*. Morhof, Daniel Georg.

LASNIORO, JOANNES DE. *See* Theatrum chemicum.

LATZ, GOTTLIEB.
* Die Alchemie, das ist die Lehre von den grossen Geheim-Mitteln der
Alchemisten und den Speculationen welche man an sie knüpfte.
Ein Buch welches zunächst für Aerzte geschrieben ist, zugleich
aber auch jedem gebildeten Denker geboten wird. Bonn, 1869.
pp. vi–570, 4to, double columns. Privately printed.
> Mystical, cabalistic, occult, inscrutable, whimsical and valueless. An extra-
> ordinary production for the year 1869. The bibliography omits places
> and dates of publication. Compare Hartmann, Joseph.

LAVINUS, WENCESLAUS MORAVUS.
> *See* Theatrum chemicum ; *also* R[ichebourg], J. M. D.

LE CROM.

* Vade mecum philosophique en forme de dialogue, en faveur des enfans de la science, nouvellement mis au jour, où l'on fait voir ce que c'est que la vraye quintessence. Avec un petit traité des dissolutions et coagulations naturelles et artificielles. À Paris, chez Daniel Joliet, Imprimeur-Libraire au bout du Pont Saint Michel, du costé du Marché-Neuf, au Livre Royal. Et chez la veuve Papillon, près des Augustins à la descente du Pont-Neuf aux Armes d'Angleterre. 1719. pp. [viii]-107. Le petit traité. pp. 37.

* Dissertation philosophique sur le sel Arabe et la poudre solaire. *n. p.* [Paris], *n. d.* [1718]. [pp. xxviii], 16mo.

> Bound with Vade mecum philosophique.

LEHRSÄTZE VON DER WEISEN STEIN. *See* A B C.

LEONA CONSTANTIA (JOHANNA LEADE). *See* A B C.

LERMINA, JULES [Editor].

* Collection d'ouvrages relatifs aux sciences hermétiques sous la direction de Jules Lermina. L'or et la transmutation des métaux par G. Théodore Tiffereau. Mémoires et conférences précédes de Paracelse et l'alchimie au XVI Siècle par M. Franck. Paris, 1889. pp. x–184, sq. 12mo.

> *See* Tiffereau, G. T. This is virtually a new edition of Tiffereau's curious work ; with the addition of a lecture given by Tiffereau March 16, 1889 (in which he reiterates his claim to make artificial gold), and other essays and letters.

LETTRE PHILOSOPHIQUE DE PHILOVITE À HÉLIODORE. *See* R[ichebourg], J. M. D.

LIBAVIUS, ANDREAS. *See* Manget, J. J. *See also in Section V.*

* LIBER MUTUS. *n. p., n. d.* pp. [i]–43, sm. 4to. 17 plates. Manuscript.

> The fifteen engraved plates of the " Liber Mutus " in this French manuscript, colored by hand, were first published by Denis Tollé of Rochelle, 1677, and were the work of Jacob Saulat Démarest, who anagrammatized his name as Altus. The fifteen plates pretend to portray the secrets of transmutation, but are intelligible only to adepts. Pages 7–43 of this MS. contain : " Description du fourneau ou athanor philosophique," with two plates. This neatly written manuscript dates probably about the close of last century.

* LIBER SAPIENTIÆ. [Manuscript.] 1745. pp. 168, 8vo. Illustrated title-page.

LOEN VON. *See* A B C.

LULL ; LULLI ; LULLY ; LULLUS. *See* Lullius, Raymundus.

LULLIUS, RAYMUNDUS.

  See Aureum Vellus ; *also* A B C ; *also* Manget, J. J.; *also* Theatrum chemi-
  cum ; *also* R[ichebourg], J. M. D. ; *also* Neue Sammlung, [etc.] ; *also*
  Paracelsus : Of the chymical Transmutation [etc.] ; *also* Petrus Bonus
  Ferrariensis : Pretiosa margarita.

LULLIUS, RAYMUNDUS.

 * Arbor scientiæ venerabilis et cælitus ; liber ad omnes scientias utilissimus. [Lugdunum], 1515. pp. [iv]–681–[xix], sm. 4to. Plates.

   Contains annexed : Introductorium magnæ artis generalis . . . R. L. ad
   omnes scientias utilissimum.

 * Clavicule (La) ou la science de Raymond Lulle. Avec toutes les
  figures de rhétorique. Par le Sieur Jacob. Et la vie du mesme
  Raymond-Lulle, par Monsieur Colletet. À Paris, *n. d.* [1646]. pp.
  [xxviii]–252–[iv], 12mo.

   Of the " Clavicula " Lenglet du Fresnoy remarks : " Raymond Lulle assure
   lui-même que ce Traité est nécessaire pour bien entendre ce qu'il a écrit
   sur la philosophie hermétique, cependant quand on l'a lû, on n'en sçait
   pas beaucoup plus qu' auparavant."

 * Codicillus, seu Vade mecum in quo fontes alchimicæ artis ac philosophiæ reconditioris uberrime traduntur. Secunda editio, in qua in-
  numerabiles loci multorum exemplarium collatione restituuntur, et
  multa prius omissa supplentur. Coloniæ, apud hæredes Arnoldi
  Birckmanni, 1572. Cum privilegio Cæsaris Maiestatis ad decennium.
  pp. 248, 16mo.

 * Libelli aliquot chemici ; nunc primum, excepto Vade mecum, in lucem
  opera Doctoris Toxitæ editæ ; quorum omnium nomina versa pagina
  dabit. Cum privilegio Cæsaris Maiestatis ad decennium. Basileæ,
  typis Conradi Waldkirchii, 1600. pp. [xii]–393–[xxvi], 12mo.

  1. Testamentum nȯvissimum integrum. p. 1.

  2. Elucidatio vocabulorum eius. p. 144.

  3. Vade mecum. p. 258.

  4. Compendium de transmutatione animæ metallorum ; pro media
   parte ex antiquo exemplari auctum. p. 274.

  5. De compositione gemmarum et lapidum preciosorum. p. 298.

  6. Epistola accurtatoria ad regem Neapolitanum. p. 319.

  7. Medicina magna. p. 330.

  8. Dialogus Demogorgon, qui Lullianis scriptis multam præclare
   lucem adfert. p. 375.

   [Another edition :] Basiliæ, 1572.

 Obras de Ramón Lull, texto original publicado con notas, variantes,
  ilustraciones y estudios biográficos y bibliográficos por Jerónimo
  Roselló. Biografía. Palma, 1886. pp. 28–12, 8vo. Illuminated
  initials.

   I have seen only one small fragment of this work.

LULLIUS, RAYMUNDUS. [Cont'd.]
    Opera ea quæ ad adinventam ab ipso artem universalem, scientiarum
        artiumque omnium brevi compendio firmaque memoria apprehen-
        dendarum, locupletissimaque vel oratione ex tempore pertractan-
        darum, pertinent. Argentorati, Zetznerus, 1598. pp. 992, besides
        several not numbered. Ill.
  * Opera ea quæ ad adinventam ab ipso artem universalem, scientiarum
        artiumque omnium brevi compendio firmaque memoria apprehen-
        dendarum, locupletissimaque ; vel oratione ex tempore pertractan-
        darum pertinent ; ut et in eandem quorundam interpretum scripti
        commentarii ; quæ omnia sequens indicabit pagina ; et hoc demum
        tempore conjunctim emendatiora locupletioraque non nihil edita
        sunt. Accessit huic editioni Valerii de Valeriis Patricii Veneti
        aureum in artem Lullii generalem opus : adjuncto indice cum
        capitum, tum rerum ac verborum locupletissimo. Argentorati,
        sumptibus hæredum Lazari Zetzneri, 1617. pp. [xvi]-1109-[xl],
        8vo. Plates.

[*Contents :*]

1. Ars brevis.
2. De auditu Kabbalistico seu Kabbala.
3. Duodecim principia philosophiæ Lullianæ.
4. Dialectica seu Logica.
5. Rhetorica.
6. Ars magna.
7. *Brunus Jordanus :* De specierum scrutinio.
8. *Brunus Jordanus :* De lampade combinatoria Lulliana.
9. *Brunus Jordanus :* De progressu et lampade venatoria logico-
   rum.
10. Commentaria Agrippæ in artem brevem Lullian.
11. Articuli fidei.
12. *Valerii de Valeriis* tam in arborem scientiarum quam artem
    generalem opus aureum.

    Opera omnia, quinque seculorum vicissitudinibus illæsa et integra ser-
        vata. [Ivo Salzinger, Sacerdos, edita.] Moguntiæ, 1721-42. Vols.
        I-VI and IX-X ; fol.
        Vols. VII and VIII were never published.

LUMEN JUVENIS EXPERTI NOVUM. *See* Theatrum chemicum.

LUTZ ZU LAUFELFINGEN, MARCUS.
  * Chemische Analyse und Synthese, ein alchymistischer Versuch von
        einem Mystiker des 19ten Jahrhundert [Gügler]. Luzern, 1816.
        pp. 151, 12mo.

LUX LUCENS IN TENEBRIS. *See* Aureum Vellus ; *also* R[ichebourg], J. M. D.

LYDGATE, JOHN. *See* Ashmole, E.: Theatrum chemicum Britannicum.

MADATHANUS, HENRICUS. *See* Musæum hermeticum ; *also* Grasshof, Johann.

* MAGAZIN FÜR DIE HÖHERE NATURWISSENSCHAFT UND CHEMIE. Tübingen, bey Jacob Friedrich Heerbrandt, 1784.

MAGISTERY, THE. *See* Ashmole, E.: Theatrum chemicum Britannicum.

MAIER, MICHAEL. *See* Musæum hermeticum.

MAIER, MICHAEL.

> * Arcana arcanissima, hoc est Hieroglyphica Ægyptio-Græca vulgo necdum cognita, ad demonstrandam falsorum apud antiquos deorum, dearum, heroum, animantium et institutorum pro sacris receptorum, originem, ex uno Ægyptiorum artificio, quod aureum animi et corporis medicamentum peregit, deductam, unde tot poëtarum allegoriæ scriptorum narrationes fabulosæ et per totam encyclopædiam errores sparsi clarissima veritatis luce manifestantur, suæque tribui singula restituuntur, sex libris exposita. *n. p.*, *n. d.* [London, 1614]. pp. [xii]–285–[xiv], 4to. Illustrated title-page.
>
>> The author was physician to Rudolph II., Emperor of Germany. His mystical and well illustrated works are much sought by bibliophiles. He gave an alchemical interpretation to Greek and Roman mythology. Lenglet Du Fresnoy says of this : " Très rare, très curieux, et recherché des amateurs."
>
> * Atalanta fugiens, hoc est Emblemata nova de secretis naturæ chymica, accommodata partim oculis et intellectui, figuris cupro incisis, adjectisque sententiis, epigrammatis et notis, partim auribus et recreationi animi plus minus 50 fugis musicalibus trium vocum, quarum duæ ad unam simplicem melodiam distichis canendis peraptam correspondeant, non absque singulari jucunditate videnda, legenda, meditanda, intelligenda, dijudicanda, canenda et audienda. Oppenheimii, ex typographia Hieronimi Galleri, sumptibus Joh. Theodore de Bry, 1618. pp. 211–[iii], 4to. Ill.
>
>> The Atalanta fugiens was republished in 1687, but without the music, under the title " Scrutinium chymicum," *q. v.*
>
> * Cantilenæ intellectuales de phœnice redivivo ; ou chansons intellectuelles sur la resurrection du Phenix. Traduites en François sur l'original Latin par M. L. L. M. À Paris, chez Debure l'aîné, Quai des Augustins, à l'Image S. Paul, 1758. Avec approbation et privilége du Roi. pp. [viii]–129, 16mo.
>
>> The preface is dated 1622. Printed in Latin and French on opposite pages.

MAIER, MICHAEL. [Cont'd.]

* Chymisches Cabinet, derer grossen Geheimnussen der Natur durch
wohl ersonnene sinnreiche Kupfferstiche und Emblemata, auch zu
mehrerer Erleuchterung und Verstand derselben, mit angehefften
sehr dienlich und geschickten Sententien und poëtischen Ueber-
schrifften, dargestellet und ausgezieret. Welches nachdeme es
wegen vieler darinn entdeckten raren Geheimnussen und Erläuterung der philosophischen Subtilitäten, von verschiedentlichen
hocherleuchtenden und zu grossen Künsten sich applicirenden
Liebhabern zum öffteren begehret und verlanget worden ; der
chymischen Republic und dero Liebhabern, zur Speculation, Be-
tracht- und Untersuchung aus wohlmeinender Veneration und
Liebe zum zweyten mahl in der lateinischen Sprach ausgefertiget,
vor jetzo aber zum ersten mahl in das hochteutsche übersetzet ist ;
von G. A. K. Der philosophischen Künsten Liebhabern. Deme
beygefüget ist, eine Application des hohen Lied Salomonis auff die
Universal-Tinctur der Philosophorum. Franckfurt, verlegts Georg
Heinrich Oehrling, Anno 1708. pp. [iv]–153, 4to.

> A German translation of Scrutinium chymicum, *q. v.*

*Examen fucorum pseudo-chymicorum detectorum et in gratiam veritatis
amantium succincte refutatorum. Francofurti, typis Nicolai Hoff-
manni, sumptibus Theodori de Bry. Anno 1617. pp. 47, 4to.
Illustrated title-page.

* Scrutinium chymicum per oculis et intellectui accuratè accommodata,
figuris cupro appositissimè incisa, ingeniosissima emblemata, hisque
confines et ad rem egregie facientes sententias, doctissimaque item
epigrammata, illustratum ; opusculum ingeniis altioribus et ad
majora natis, ob momenta in eo subtilia, augusta, sancta, rara, et alio,
qui nimium quantum abstracta, quam maxime expetitum, desideratum ; iterata vice amplissimæ reipublicæ chymicæ bono et emolu-
mento, non sine singulari jucunditate legendum, meditandum,
intelligendum, dijudicandum, depromptum. Francofurti, impensis
Georgii Heinrici Oehrlingii, Bibliopolæ, typo Johannis Philippi
Andreæ. 1687. pp. [vi]–147, 4to. Ill.

> This is a second edition of Atalanta fugiens, omitting the music of the latter.

* Symbola Aureæ Mensæ duodecim nationum ; hoc est, Hermæa seu
mercurii festa ab heroibus duodenis selectis, artis chymicæ usu,
sapientia et authoritate paribus celebrata ad Pyrgopolynicen seu
adversarium, illum tot annis iactabundum, virgini chemiæ injuriam
argumentis tam vitiosis quam convitiis argutis inferentem, confun-
dendum et exarmandum, artifices verò optimè de ea meritos suo
honori et famæ restituendum, ubi et artis continuatio et veritas

MAIER, MICHAEL. [Cont'd.]

invicta 36 rationibus et experientia librisque authorum plus quam trecentis demonstratur; opus ut chemiæ sic omnibus aliis antiquitatis et rerum scitu dignissimarum percupidis, utilissimum 12 libris explicatum et traditum, figuris cupro incisis passim adjectis. Francofurti, typis Antonii Hummij, impensis Lucæ Jennis, 1617. pp. [xx]-621-[xlii], 4to. Portraits and illustrations.

* Tripus aureus, hoc est tres tractatus chymici selectissimi, nempe :

I. *Valentinus, Basilius :* Practica una cum duodecim clavibus et appendice, ex Germanico ;

II. *Norton, Thomas :* Crede mihi, seu ordinale, ante annos 140 ab authore scriptum, nunc ex Anglicano manuscripto in Latinum translatum, phrasi cuiusque authoris ut et sententia retenta.

III. *Cremerus [Johannes]* : Testamentum, hactenus nondum publicatum : nunc in diversarum nationum gratiam editi, et figuris cupro affabre incisis ornati, opera et studio.

Francofurti, ex chalcographia Pauli Jacobi, impensis Lygæiennis. Anno 1618. pp. 196, 4to. Portrait of Michael Maier. Ill.

* Viatorium, hoc est, de montibus planetarum septem seu metallorum ; tractatus tam utilis quam perspicuus, quo, ut indice mercuriali in triviis, vel Ariadneo filo in Labyrintho, seu Cynosura in Oceano chymicorum errorum immenso, quilibet rationalis veritatis amans ad illum qui in montibus sese abdidit de Rubeapetra Alexicacum, omnibus Medicis desideratum investigandum, uti poterit. Oppenheimii, ex typographia Hieronymi Galleri, sumptibus Joh. Theodori de Bry, 1618. pp. 136, 4to.

[Another edition.] Rothomagi, sumptibus Joannis Berthelin in area Palatii. Anno 1651. pp. 224, 12mo.

"Tous les traités de M. M. sont recherchés, quelques-uns mêmes sont extrêmement rares, et ils renferment beaucoup de curiosités ; mais pour dire ce que je pense, cet habile Médecin a trop écrit sur la Philosophie Hermétique pour croire qu'il ait jamais été grand Praticien, ou qu'il ait réussi." *Lenglet du Fresnoy.*

MALVASIUS, CAROLUS CÆSAR. *See* Manget, J. J.

MANGET, JEAN JACQUES. [MANGETUS.]

* Bibliotheca chemica curiosa, seu rerum ad alchemiam pertinentium thesaurus instructissimus, quo non tantum artis auriferæ ac scriptorum in ea nobiliorum historia traditur, lapidis veritas argumentis et experimentis innumeris, immo et jurisconsultorûm judiciis evincitur, termini obscuriores explicantur, cautiones contra impostores, et

MANGET, JEAN JACQUES. [Cont'd.]

difficultates in tinctura universali conficienda occurrentes, declarantur ; verum etiam tractatus omnes virorum celebriorum qui in magno sudarunt elixyre, quique ab ipso Hermete, ut dicitur, Trismegisto, ad nostra usque tempora de chrysopœia scripserunt, cum præcipuis suis commentariis, concinno ordine dispositi exhibentur ; ad quorum omnium illustrationem additæ sunt quamplurimæ figuræ aënæ. Genevæ, 1702. 2 vols., fol. Vol. I, pp. [xviii]–938 ; vol. II, 904. Portrait of Manget and plates.

VOLUMEN I. LIBER I.

SECTIO PRIMA.

§ I.

1. *Borrichius, Olaus :* De ortu et progressu chemiæ dissertatio. p. 1.

§ II.

2. *Borrichius, O. :* Conspectus scriptorum chemicorum celebriorum. p. 38.

SECTIO SECUNDA.

§ I.

3. *Kircherus, Athanasius :* De lapide philosophorum dissertatio. p. 54.

4. *Kircherus, A. :* De alchymia sophistica. p. 82.

5. *K., A. :* . . . an aurum chymicum sit licitum. p. 101.

6. *K., A. :* Consectarium antitheticum, sive contradictorium. p. 109.

§ II.

7. *Blawenstein, Salomon de :* Interpellatio brevis ad philosophos pro lapide philosophorum, contra antichimisticum Mundum Subterraneum Athanasii Kircheri. p. 113.

§ III.

8. *Clauder, Gabriel :* Tractatus de tinctura universali, ubi in specie contra R. P. Athanasium Kircherum, pro existentia lapidis philosophici disputatur. p. 119.

§ IV.

9. *Morhof, Daniel Georg :* Epistola de metallorum transmutatione ad virum nobilissimum et amplissimum Joelem Langelottum, serenissimi Cimbrici principis archiatrum celeberrimum. p. 168.

§ V.

10. *Sachs a Levvenheimb, Philip Jacob :* Aurum chymicum. p. 192.

§ VI.

11. *Helvetius, Johannes Fridericus :* Vitulus aureus quem mundus

MANGET, JEAN JACQUES.   [Cont'd.]

adorat et orat, in quo tractatur de rarissimo natura miraculo transmutandi metalla, nempe quomodo tota plumbi substantia, vel intra momentum, ex quavis minima lapidis veri philosophici particula in aurum obryzum commutata fuerit. Hagæ Comitis. p. 196.

§ VII.

12. *Fanianus, Johannes Chrysippus :* De jure artis alchymiæ, hoc est variorum autorum et præsertim jurisconsultorum judicia et responsa ad quæstionem, an alchemia sit ars legitima. p. 210.

SECTIO TERTIA.   UBI TERMINI ARTIS CHEMICA EXPLICANTUR.

§ I.

13. *Johnsonius, Gulielmus :* Lexicon chymicum. p. 217.

§ II.

14. *Faber, Petrus Joannes :* Manuscriptum ad serenissimum Holsatiæ ducem Fridericum olim transmissum, res alchymicorum obscuras extraordinaria perspicuitate explanans. p. 291.

§ III.

15. *Becher, Johann Joachim :* Œdipus chimicus, obscuriorum terminorum et principiorum chimicorum mysteria aperiens et resolvens. p. 306.

SECTIO QUARTA.   DE ALCHIMIÆ ARTIS DIFFICULTATIBUS.

§ I.

16. *Hoghelande, Theobaldus de :* De alchimiæ difficultatibus liber, in quo docetur quid scire quidque vitare debeat veræ chemiæ studiosus ad perfectionem aspirans. p. 336.

§ II.

17. *Cato Chemicus :* Tractatus quo veræ ac genuinæ philosophiæ hermeticæ, et fucatæ ac sophisticæ pseudo-chemiæ, et utriusque magistrorum characterismi accurate delineantur. p. 368.

LIBER II.

SECTIO PRIMA.   TRACTATUS CHYMICI AUTORUM ÆGYPTIORUM.

§ I.

18. *Trismegistus, Hermes :* Tabula smaragdina, cui titulus Verba secretorum Hermetis Trismegisti, W. Chr. Kriegsmanni et Gerardi Dornei commentariis illustrata. p. 380.

§ II.

19. *T., H. :* Tractatus aureus de lapidis physici secreto, in septem capitula divisus, cum scoliis anonymi. p. 400.

MANGET, JEAN JACQUES.　[Cont'd.]

SECTIO SECUNDA.

§ I.

20. [*Arisleus :*] Turba philosophorum ex antiquo manuscripto codice excerpta, qualis nulla hactenus visa est editio.　p. 445.

21. In turbam philosophorum sermo unus anonymi.　p. 465.

22. Allegoriæ sapientum super librum turba philosophorum xxix distinctiones.　p. 467.

23. Turbæ philosophorum aliud exemplar.　p. 480.

24. Allegoriæ super librum turba.　p. 494.

25. Ænigma ex visione Arislei philosophi et allegoriis sapientum. p. 495.

26. In turbam philosophorum exercitationes quindecem.　p. 497.

§ II.

27. *Artefius :* Liber qui clavis majoris sapientiæ dicitur.　p. 503.

§ III.

28. *Morienus Romanus :* Liber de compositione alchemiæ.　p. 509.

§ IV.

29. *Geber :* Summa perfectionis magisterii in sua natura, ex Bibliothecæ Vaticana exemplari undecunque emendatissime edita, cum vera genuinaque delineatione vasorum et fornacum, denique libri investigationis magisterii et testamenti ejusdem additione, castigatissima. p. 519.

30. *Geber :* Liber investigationis magisterii.　p. 558.

31. *Geber :* Testamentum.　p. 562.

32. *Braceschus, Joannes Brixianus :* Dialogus veram et genuinam librorum Gebri sententiam explicans.　p. 565.

33. *Gerhardus, Joannes :* Exercitationes perbreves in Gebri libros duos summæ perfectionis.　p. 598.

SECTIO TERTIA.

§ I.

34. *Baco, Rogerius :* Libellus de alchemia, cui titulum fecit, Speculum alchemiæ.　p. 613.

35. *Baco, R.* Epistola de secretis operibus artis et naturæ et de nullitate magiæ.　p. 616.

§ II.

36. *Avicenna :* Tractatus de alchemia.　p. 626.

37. *Avicenna :* De congelatione et conglutinatione lapidum.　p. 636.

§ III.

38. *Aristoteles :* De perfecto magisterio.　p. 638.

39. *Aristoteles :* Tractatulus de practica lapidis philosophici.　p. 658.

MANGET, JEAN JACQUES. [Cont'd.]

§ IV.

40. *Arnaldus de Villanova :* Thesaurus thesaurorum, vel rosarium philosophorum. p. 662.

41. *Arnaldus de V. :* Novum lumen. p. 676.

42. *Arnaldus de V. :* Perfectum magisterium et gaudium et flos florum et thesaurus omnium incomparabilis et margarita. p. 679.

43. *A. d. V. :* Epistola super alchemia ad regem Neapolitanum. p. 683.

44. *A. d. V. :* Speculum alchymia. p. 687.

45. *A. d. V. :* Carmen. p. 698.

46. *A. de V. :* Questiones tam essentiales quam accidentales ad Bonifacium octavum cum suis responsionibus. p. 698.

47. *A. d. V. :* Semita semitæ. p. 702.

48. *A. d. V. :* Testamentum. p. 704.

§ V.

49. *Lullius, Raymundus :* Testamentum. p. 707.

50. *L., R. :* Practica super lapide philosophico. p. 763.

51. *Gerhardus, Joannes :* Analysis partis practicæ Raymundi Lulli in testamento.

52. *Lullius, R. :* Compendium animæ transmutationis artis metallorum, Ruperto Anglorum regi transmissum. p. 780.

53. *L., R. :* Testamentum novissimum regi Carolo dicatum. p. 790.

54. *L., R. :* De practica testamenti novissimi. p. 806.

55. *L., R. :* Elucidatio testamenti. p. 823.

56. *L , R. :* Lux mercuriorum. p. 824.

57. *L., R. :* Experimenta. p. 826.

58. *L., R. :* Liber artis compendiosæ, quem Vade mecum nuncupavit. p. 849.

59. *L., R. :* Compendium animæ transmutationis artis metallorum, Ruperto Anglorum regi transmissum. p. 853. [See p. 780 for the same title and contents.]

60. *L., R. :* Epistola accurationis, missa olim Roberto Anglorum regi. p. 863.

61. *L., R. :* Potestas divitiarum ; in quo libro optima expositio testamenti Hermetis continetur. p. 866.

62. *L., R. :* Clavicula quæ apertorium dicitur. p. 872.

63. *L., R. :* Compendium artis alchimiæ et naturalis philosophiæ secundum cursum. p. 875.

64. *L., R. :* De lapide et oleo philosophorum. p. 878.

65. *L., R. :* Codicillus, seu Vade mecum aut cantilena. p. 880.

66. *Braceschus, Johannes :* Lignum vitæ, seu dialogus ex Italico in

MANGET, JEAN JACQUES. [Cont'd.]

Latinum versus a Georgio Gratarolo physico, quo Raymundi Lullii scripta explicantur. p. 911.

67. *Mutus liber :* in quo tamen tota philosophia hermetica figuris hieroglyphicis depingitur. p. 939. 15 plates.

VOLUMEN II, LIBER III.

SECTIO PRIMA.

§ I.

68. *Petrus Bonus Lombardus, Ferrariensis :* Margarita pretiosa novella correctissima, exhibens introductionem in artem chemiæ integram, ante annos plus minus trecento septuaginto composita. p. 1.

69. *Rupescissa, Joannes de :* Liber de confectione veri lapidis philosophorum. p. 80.

70. *R., J. de :* Liber lucis. p. 84. Plate.

§ III.

71. *Rosarium* philosophorum. p. 87.

72. *Toletanus :* Alterum exemplar Rosarii. p. 119.

73. *Rosarium* abbreviatum e manuscripto vetustissimo. p. 133.

§ IV.

74. *Montanor, Guido de :* Scala philosophorum. p. 134.

§ V.

75. *Clangor buccinæ :* tractatus mirabilis de lapide philosophorum. p. 147.

§ VI.

76. *Correctio fatuorum :* tractatus satis perutilis et authenticus. p. 165.

§ VII.

77. *Ficinus, Marsilius :* Liber de arte chemica. p. 172.

§ VIII.

78. *Calid filius Jaichi :* Liber secretorum artis ex Hebræo in Arabicum et ex Arabico in Latinum translatus. p. 183.

§ IX.

79. *Kalid Rachaidibus :* Liber trium verborum. p. 189.

§ X.

80. *Merlinus :* Allegoria profundissimum philosophici lapidis arcanum perfecte continens. p. 191.

§ XI.

81. [*Efferarius Monachus*] : Thesaurus philosophiæ. p. 192.

MANGET, JEAN JACQUES. [Cont'd.]

§ XII.

82. *Zadith :* Aurelia occulta philosophorum. p. 198.

83. *Zadith :* Tractatus de chemia. p. 216. Plate.

§ XIII.

84. *Consilium coniugii,* seu de massa solis et lunæ Libri III veri aurei et incomparabilis, ex Arabico in Latinum sermonem redacti. p. 235.

SECTIO SECUNDA.

§ I.

85. *Richardus Anglus :* Libellus utilissimus περὶ Χημείας, cui titulum fecit Correctorium. p. 267.

§ II.

86. *Ripley, George :* Liber duodecim portarum. p. 275.

§ III.

87. *Norton, Thomas :* Tractatus Crede mihi, seu ordinale. p. 285.

§ IV.

88. *Daustenius, Joannes :* Rosarium arcanum philosophorum secretissimum comprehendens. p. 309.

§ V.

89. *Dialogus* inter naturam et filium philosophiæ. p. 326.

§ VI.

90. *Zacharius, Dionysius :* Opusculum chemicum, cum adjuncto Nicolai Flamelli in illum librum commentario, item ejusdem Flamelli summario philosophico. p. 336.

§ VII.

91. *Augurellus, Joannes Aurelius :* Chrysopœia et vellus aureum, seu chrysopœia major et minor, ad Leonem pontificem maximum. p. 371.

92. *Albineus, Nathan :* Carmen aureum. p. 387.

§ VIII.

93. *Bernardus Trevisanus :* Liber de secretissimo philosophorum opere chemico. p. 389.

94. *B. T. :* Responsio ad Thomam de Bononia. p. 399.

§ IX.

95. *Valentinus, Basilius :* De magno lapide antiquorum sapientum. p. 409.

96. *V., B. :* Duodecim claves. p. 413. Two plates.

97. *V., B. :* De prima materia lapidis philosophici. p. 421.

MANGET, JEAN JACQUES. [Cont'd.]

MANGET, JEAN JACQUES. [Cont'd.]

§ VI.

117. *Barnaudus Nicolaus*, a Crista Arnaudi Delphinatis ; in ænigmati-
cum quoddam epitaphium Bononiæ studiorum ante multa
secula marmoreo lapidi insculptum commentariolus sequun-
ter. p. 713.

118. Extractum è *Caroli Cæsaris Malvasii* tractatu, super eodem
epitaphio conscripto. p. 717.

§ VII.

119. Anonymus sub nomine *Pantaleon :* Bifolium metallicum, seu
medicina duplex pro metallis et hominibus infirmis, sive
lapis philosophicus. p. 718.

120. *Pantaleon :* Tumulus Hermetis opertus. p. 728.

121. *Pantaleon :* Examen alchimisticum quo, ceu Lydio lapide,
adeptus à sophista et verus philosophus ab impostore
dignoscuntur. p. 736.

122. *Pantaleon :* Disceptatio de lapide physico, in quo tumbam Semiramidis ab anonymo phantastice non hermetice, sigillatam
ab anonymo reclusam . . . p. 744.

123. *Pantaleon :* Tumba Semiramidis hermetice sigillata. p. 760.

§ VIII.

124. *Comitibus, Ludovicus de :* Tractatus de liquore alchaest et lapide
philosophorum, et de sale volatile tartari. p. 764.

125. *C., L. de :* Metallorum ac metallicorum naturæ operum, ex orto-
physicis fundamendis recens elucidatio. p. 781.

126. *C., L. de :* Appendix symbolicæ crucis aliqualem illustrationem
exhibens. p. 840.

§ IX.

127. *Claudius Germanus :* Icon philosophiæ occultæ, sive vera methodus componendi antiquorum philosophorum lapidem. p. 845.

128. *Balduinus, Christianus Adolphus :* Aurum superius et inferius,
auræ superioris et inferioris Hermeticum. p. 856. Plate.

129. *Melchior, Friben :* Epistola sive brevis enumeratio hactenus à se
in chemia actorum. p. 875.

130. *D. F. B. :* De spiritu mundi positiones aliquot. p. 876.

131. *Cnoeffellius, Andreas :* Responsum ad positiones de spiritu
mundi, quod in se continet Referationem tumbæ Semiramidis. p. 880.

132. Trames facilis et planus ad auream Hermetis arcem rectâ producens. p. 886.

§ X.

133. *Stolcius de Stolcenberg, Daniel :* Hortulus Hermeticus flosculis

MANGET, JEAN JACQUES.   [Cont'd.]

> philosophorum cupro incisis conformatus et brevissimis versiculis explicatus. p. 895. Six plates.

---

* Bibliotheca chemico-curiosa enucleata ac illustrata.   Das ist Kern und Stern der vornehmsten chymisch-philosophischen Schrifften die in Mangeti Bibliotheca chemico-curiosa befindlich seyend.   Welche mit sonderbaren Anmerckungen allerseits erläutert daraus auch die vornehmste chymische Denck-Sprüche und bewährteste Experimenta exerpiret oder kürtzlich, jedoch aber mit sonderbaren nutzbringenden Fleiss zusammengetragen, auch also in drey Classes abgetheilet und herausgegeben durch Conrad Horlachern.   Franckfurt, zu finden bey Wolffgang Michahelles und Johann Adolph. 1707.   pp. [xxxvi]–422–[xxiv], 12mo.

MARIÆ PROPHETISSÆ PRACTICA.

> *See* Artis auriferæ ; *also* R[ichebourg], J. M. D. ; *also* Arnaldus de Villa Nova : Schrifften.

MARSCIANO, FRANCISCUS ONUPHRIUS DE.   *See* A B C.

MARTINUS DE DELLE.   *See* Quadratum alchymisticum.

MATTHEW, RICHARD.

* The unlearned Alchymist his Antidote, or a more full and ample explanation of the use, virtue and benefit of my pill, entituled : An effectual diaphoretick, diuretick purgeth by sweating, urine. Whereunto is added, sundry cures and experiences, with particular direction unto particular diseases and distempers ; also sundry plain and easie receits, which the ingenious may prepare for their own health.   London, printed for Joseph Leigh at the upper end of Bazingball-street, near the Nagshead-Tavern, and are there to be sold together with this Pill, and by Giles Calvert at the west-end of Pauls, 1662.   pp. [xi]–204, 16mo.

MAUGUIN DE RICHEBOURG.   *See* R[ichebourg], J. M. D.

* MEDICINISCH-CHYMISCH UND ALCHEMISTISCHES ORACULUM, darinnen man nicht nur alle Zeichen und Abkürzungen welche so wohl in den Recepten und Büchern der Aerzte und Apotheker als auch in den Schriften der Chemisten und Alchemisten vorkommen findet, sondern deme auch ein sehr rares chymisches Manuscript eines gewissen Reichs . . . beygefüget.   Ulm, 1772, bey August Lebrecht Stettin.   pp. [vi]–71, 8vo.   Plates.

> A collection of over 2000 signs, symbols and secret characters used by alchemists to designate substances, apparatus and processes.

MEHUN, JOHANN VON.

* Der Spiegel der Alchymie. Aus dem lateinisch-französischen übersetzt. Dem noch beygefüget worden Ben Adams Traum-Gesichte
   durch Floretum a Bethabor ; nebst Friedrich Galli Reise nach der
   Einöde St. Michael. Ballenstädt und Bernburg, in der Biester-
   feldischen Hof-Buchhandlung, 1771. pp. 48, 16mo.

MEHUNG, JOHANN. [MÉHUN, MESUNG, MÉUN, MEUNG, MUNG.]
   See Musæum hermeticum ; also Métallique (La) transformation ; also Wasserstein der Weisen.

M[EIDINGER], VON.

* Die Richtigkeit der Verwandlung derer Metalle aus der wahrhaften
   Begebenheit welche sich im Jahr 1761 auf der kurfürstlichen trie-
   rischen Münzstatt zu Koblenz mit einem Adepten Namens Georg
   Stahl zugegetragen hat. Beschrieben von dem damaligen kurtrie-
   rischen Münzdirector K. K. wirklichem Hofrathe. Leipzig, bey
   Adam Friedrich Böhmen, 1783. pp. 66, 16mo.

MELCHIOR, FRIBEN. See Aureum Vellus ; also Manget, J. J.

MELVOLODEMET, FRANCISCUS SEBASTIANUS FULVUS. See A B C.

MENENS, GUILLELMUS. See Theatrum chemicum.

MERCURII ARBEIT (EIN). See Aureum Vellus.

MERCURIUS REDIVIVUS, oder Unterricht wie man den philosophischen Stein machen soll.
   See Dreyfaches hermetisches Kleeblatt ; also Besondere Geheimnisse.

MERLINUS. See Artis auriferæ ; also Manget, J. J.

MESUNG, JOHANN VON. [MEHUNG.] See Wasserstein der Weisen.

* MÉTALLIQUE (LA) TRANSFORMATION contenant trois anciens traictez
   en rithme Françoise, à sçavoir ; La fontaine des amoureux de
   science : Autheur J. de la Fontaine. Les remonstrances de Nature
   a l'Alchymiste errant : avec la responce dudict Alchymiste par J.
   de Mung. Ensemble un traicté de son Romant de la Rose, concernant ledict art. Le Sommaire Philosophique de N. Flamel.
   Avec la deffense d'iceluy art, et des honestes personnages qui y vac-
   quent, contre les efforts que J. Girard met à les outrager. Dernière
   édition. À Lyon, chez Pierre Rigaud, ruë Merciere, à l'Enseigne
   de la Fortune, 1618. ff. 88, 24mo.

METAMORPHOSIS LAPIDUM IGNOBILIUM IN GEMMAS QUASDAM.
   See Dreyfaches hermetisches Kleeblatt ; also Besondere Geheimnisse.

MICHELIUS, JOSEPH.
  Apologia chemica adversus invectivas Andreæ Libavii calumnias.   Mittelburg, 1597.   8vo.

MICRERIS.   *See* Theatrum chemicum.

MIRANDULA, JOHANNES FRANCISCUS.   *See* Theatrum chemicum ; *also* Manget, J. J.

* MISCELLANEA CHYMIÆ ET METALLURGIÆ, oder hundert und fünf und funfzig wahre Experimenta, aus denen hinterlassenen Schriften eines berühmten Chymici, verbotenes gezogen mit allen von demselben angemerkten Handgriffen und Productis ; ächten Liebhabern zur Nachahmung und Belustigung vorgelegt, besonders denen Herren Bergofficianten zu Erkänntniss und Untersuchung verschiedener unansehnlicher verachteter Steine und Erden.   Hof, bey Johann Gottlieb Vierling, 1766.   pp. 152, 16mo.   Illustrated title-page.

MODUS PROCEDENDI IN PRÆPARATIONE LAPIDIS PHILOSOPHICI.   *See* Aureum Vellus.

MONTANOR, GUIDO DE.   *See* Manget, J. J. ; *also* Theatrum chemicum.

MONTE RAPHAIM, JOANNES DE.   *See* Roth-Scholtz, Fr. (Deutsches Theatrum chemicum).

MONTE-SNYDERS, JOHANN DE.
  * Chymische Schriften.   Frankfurt und Leipzig, 1773.   pp. 208.

[*Contents :*]

  1. Tractatus de Medicina Universali ; das ist : Von der Universal-Medicin, wie nemlich dieselbe in denen dreyen Reichen der Mineralien, Animalien und Vegetabilien zu finden und daraus zuwege zu bringen durch ein besonders Universal-Menstruum, welches auf- und zuschliessen, und jedes Metall in Materiam primam bringen kann auch wie dadurch das fixe, unzerstörliche Gold in ein wahrhafftes Aurum potabile zu bringen, so sich nimmermehr wieder in ein fix Gold Corpus reduciren lässet. Gott zu Ehren und dem menschlichen Geschlecht zu sonderbarem Trost und Nutzen anjetzo wiederum zum Druck befördert, und mit einer kurzen gründlichen Erklärung, auch beygefügeten spagyrischen Grund-Regeln illustrirt durch A. Gottlob B[londel].   Frankfurt und Leipzig, zu finden bey Johann Paul Kraus in Wien, 1773.

  2. Spagyrische Grund-Regeln aus des vortreflichen Joh. de Monte Snyders Tractat de Medicina Universali verfasset, und dem geneigten Leser zu erspiesslichen Nutzen vorgestellet von A. Gottlob B[londel], 1773.

MONTE-SNYDERS, JOHANN DE.   [Cont'd.]

* Metamorphosis Planetarum ; das ist : Eine wunderbarliche Veränderung der Planeten und metallischen Gestalten in ihr erstes Wesen,
mit beygefügtem Process und Entdeckung der dreyen Schlüssel, so
zu Erlangung der drey Principien gehörig und wie das Universale
Generalissimum zu erlangen in vielen Oertern dieses Büchleins
beschrieben.   Anjetzo wiederum zum Druck befördert durch A.
Gottlob B[londel].   Franckfurt und Leipzig, zu finden bey Johann
Paul Kraus in Wienn, 1774.   pp. 173–[iii], 12mo.

MORGENSTERN, PHILIPPUS ISLEBIENSIS.

* Turba philosophorum ; das ist das Buch von der güldenen Kunst,
neben andern Authoribus, welche mit einander 36 Bücher in sich
haben.   Darinn die besten urältesten Philosophi zusammengetragen
welche tractiren alle einhellig von der Universal Medicin, in zwey
Bücher abgetheilt, und mit schönen Figuren gezieret.   Jetzundt
newlich zu Nutz und Dienst allen waren Kunstliebenden der Natur
(so der lateinischen Sprach unerfahren) mit besonderen Fleisz, Mühe
unnd Arbeit trewlich an Tag geben.   Zu Basel, in Verlegung Ludwig
Königs, 1613.   2 vols., pp. [xiv]–560, 12mo.

   Volume II has the title :

Das ander Theil der güldinen Kunst die sie sonst Chymia nennen,
welches in sich hellt die Schrifften Morieni Romani, von den
metallischen Dingen, und von der verborgenen und höchsten
Artzney der alten Philosophorum, mit andern Authoribus, die da
auff dem nachfolgenden Blatt angezeigt werden.   Gedruckt zu
Basel, bey Johann Schröter, 1613.   pp. [vi]–455, 12mo.   Ill.

   For the contents, see " Artis auriferæ," of which this is a translation.

MORHOF, DANIEL GEORG.

* De metallorum transmutatione ad virum nobilissimum et amplissimum
Joelem Langelottum, serenissimi principis Cimbrici archiatrum
celeberrimum, epistola.   Hamburgi, ex officina Gothofredi Schult-
zen prostant, et Amsterodami, apud Joannem Janssonium à Waesberge, 1673.   pp. 168, 16mo.

   Also in Manget, J. J. : Bibliotheca chemica curiosa ; *also* Roth-Scholtz, F.
   (Deutsches Theatrum chemicum).

   *Vom Goldmachen, oder physikalisch-historische Abhand-
   lung von Verwandlung der Metalle.   Aus dem lateinischen.   Bayreuth, 1674.   pp. 136, 12mo.

   The author was Professor of History at Kiel ; born 1639, died 1691.   He
   demonstrates by historical evidence the verity of alchemy.   Schmieder
   calls him an unprejudiced historian.   *Cf.* Langelot, Joel.

MORESINUS, THOMAS.

Liber novus de metallorum causis et transubstantione, editus per T. M., Aberdonanum Scotum ; in quo chimicorum quorundam inscitia et impostura philosophicis medicis et chimicis rationibus retegitur et demonstratur, et vera iis de rebus doctrina solide asseritur. Francofurti, 1593. pp. 131, 12mo.

> The dedication to James VI is dated 1593. The author, a physician of Aberdeen, contends that the pretensions of hermetic philosophers are vain and their reasoning false.

MORIENUS ROMANUS.

* De transfiguratione metallorum et occulta summaque antiquorum philosophorum medicina, libellus ; Morieno Romano, quondam eremita Hierosolymitano, auctore. Accessit huic nunc primum, $XPY\Sigma OPPHM\Omega N$, sive de arte chymica dialogus, qui præclarissimas huius scientiæ actiones rationi ac naturalibus principiis consentaneas esse demonstrat ; impostorum vero et sycophantarum somnia et nugas retegit. Hanoviæ ad Mœnum, apud Guilielmum Antonium , 1593. pp. 79, sm. 8vo.

> Bound with Duval, Robert : De veritate et antiquitate artis chemicæ. This edition of Morien is unknown to Professor Ferguson.

Morieni Romani, quondam eremitæ Hierosolymitani, de re metallica, metallorum transmutatione, et occulta summaque antiquorum medicina libellus, præter priorem editionem accuratè recognitus. Item, nunc primum in lucem prodit Bernardi Trevirensis responsio ad Thomam de Bononia, Caroli Regis octavi medicum, de mineralibus et elixiris seu pulveris philosophici compositione, quæ par est secretioris phisicæ, scholiis aliquot per Robertam Vallensen Rugl. illustrata. Ad calcem adduntur tabulæ breves ab eodem R. Vallensi conscriptæ quæ antiquorum intentionem de pulveris philosophici compositione, abstrusis eorum scriptis et ænigmatibus involutam, declarant. Cum indice copiosissimo. Parisiis, 1564. ff. [ii]–66–[iv], 4to.

> *See* Manget, J. J. ; *also* Artis auriferæ.

MÜLLER, THEOPHILUS.

De oleis variisque ea extrahendi modis, et de quibusdam alchymiæ ortum et progressum illustrantibus. Hamburgicæ, 1688. 12mo.

MÜLLNER, LEONHARD. *See* Roth-Scholtz, F. (Deutsches Theatrum chemicum).

MUFFETUS, THOMAS ANGLUS. *See* Theatrum chemicum.

MUNDAN, THEODOR. *See* S[chröder, Fr. J. W.].

* MUSÆUM HERMETICUM, omnes sophospagyricæ artis discipulos fidelissime
erudiens, quo pacto summa illa veraque medicina, qua res omnes,
qualemcunque defectum patientes, instaurari possunt (quæ alias
benedictus Lapis Sapientum appellatur), inveniri ac haberi queat ;
continens tractatus chymicos novem præstantissimos, quorum nomina et seriem versa pagella indicabit ; in gratiam filiorum doctrinæ quibus Germanicum idioma ignotum, in Latinum conversum,
ac juris publici factum.   Francofurti, 1625.   pp. [xvj]–446–[ii]–35.
Plates.

> A second and much more complete edition was issued under the title :

*Musæum hermeticum reformatum et amplificatum, omnes sophospagy-
ricæ artis discipulos fidelissimè erudiens, quo pacto summa illa
veraque lapidis philosophici medicina, qua res omnes qualemcunque defectum patientes, instaurantur, inveniri et haberi queat ;
continens tractatus chimicos XXI præstantissimos, quorum
nomina et seriem versa pagella indicabit ; in gratiam filiorum
doctrinæ, quibus Germanicum idioma ignotum est, Latina lingua
ornatum.   Francofurti, apud Hermannum a Sande, 1678.   pp.
[xii]–863, sq. 8vo.   Illustrated title-page.   Ill., and 4 folding
plates.

> A collection of twenty-one alchemical treatises, of which some bear on his-
> tory.   The plates are curious.

[Contents :]

1. [*Trismegistus, Hermes*] : Tractatus aureus de lapide philospho-
rum ab anonymo vero tamen lapidis possessore conscriptus.
p. 1.

2. *Madathanus, Henricus :* Aureum seculum redivivum, quod nunc
iterum apparuit, suaviter floruit, et odoriferum aureumque
semen peperit.   p. 53.

3. Hydrolithus sophicus, seu Aquarium sapientum, hoc est opusculum chymicum, in quo via monstratur, materia nominatur, et
processus describitur, quomodo videlicet ad universalem tincturam perveniendum, hactenus nondum visum.   p. 73.

4. *Mehung, Joannes a :* Demonstratio naturæ quam errantibus chymicis facit, dum de sophista et stolido spiratore carbonario
conqueritur.   p. 145.

5. *Flamellus, Nicolaus :* Tractatus brevis sive summarium philosophicum.   p. 172.

6. [*Valentinus, Basilius*] : Via veritatis unicæ, hoc est elegans, peru-
tile et præstans opusculum, viam veritatis aperiens.   p. 181.

7. [*Barcius. (J. v. Sternberg)*] : Gloria mundi, alias Paradysi tabula,
hoc est : Vera priscæ scientiæ descriptio quam Adam ab ipso

Musæum hermeticum. [Cont'd.]

Deo didicit ; Noe, Abraham et Salomo, tamquam summorum divinorum donorum unum, usurparunt ; omnes sapientes, omnibus temporibus, pro totius mundi thesauro habuerunt, et solis piis post sese reliquerunt ; nimirum de lapide philosophico. p. 203.

8. Tractatus eximius de lapide philosophico, a Germanico quodam philosopho subsequente titulo factus et conscriptus : Vera philosophiæ doctrina de generatione metallorum, veraque illorum origine, luce publice donata. p. 305.

9. Perbreve opusculum de lapide philosophico quod "liber Alze" nuncupatum fuit, nunc vero in lucem editum. p. 323.

10. *Lambsprinck :* De lapide philosophico e Germanico versu Latine redditus, per Nicolaum Barnaudum Delphinatem, hujus scientiæ studiosissimus. p. 337. Plates.

11. Tripus aureus, hoc est tres tractatus chymici selectissimi.

    a. Valentinus, Basilius : Practica una cum duodecim clavibus et appendice ex Germanico. p. 373. Plates.

    b. Norton, Thomas : Crede mihi, seu Ordinale, ante annos centum et quadraginta ab authore scriptum, nunc in Latinum translatum ex Anglicano manuscripto, phrasi cujusque authoris ut et sententia retenta. p. 433.

    c. Cremerus : Testamentum, hactenus nondum publicatum : nunc in diversarum nationum gratiam editi et figuris cupro affabre incisis ornati. p. 533.

12. *Sendivogius :* Novum lumen chemicum e naturæ fonte et manuali experientia depromptum ; cui accessit tractatus de sulphure. Auctoris anagramma : Divi Leschi genus amo. p. 545.

13. Novi luminis chemici tractatus alter de sulphure. Authoris anagramma : Angelus doce mihi jus. p. 601.

14. *Philalethæ :* Introitus apertus ad occlusum regis palatium. p. 647.

15. *Meier, Michæl :* Subtilis allegoria super secreta chymiæ, perspicuæ utilitatis et jucundæ meditationis. p. 701.

16. *Philalethæ :* Tractatus tres. p. 741.

    a. Metallorum metamorphosis.

    b. Brevis manuductio ad rubinum cœlestem.

    c. Fons chymicæ veritatis.

17. *Helvetius, Joannes Friedericus :* Vitulus aureus, quem mundus adorat et orat, in quo tractatur de rarissimo naturæ miraculo transmutandi metalla, nempe quomodo tota plumbi substantia vel intra momentum ex quavis minima lapidis veri philosophici particula in aurum obryzum commutata fuerit Hagæ Comitis. p. 815.

Musæum, hermeticum. [Cont'd.]

    18. Janitor pansophus, seu Figura aenea quadripartita, cunctis musæum hoc introeuntibus superiorum ac inferiorum scientiam Mosaico-Hermeticam analytice exhibens. p. 864. 4 plates.

      For an English translation see Addenda.

Mutus liber. *See* Liber mutus.

* Mysteries of the Rosie Cross, or the History of that curious Sect of the Middle Ages, known as the Rosicrucians ; with examples of their pretensions and claims as set forth in the writings of their leaders and disciples. London, 1891. pp. vi–134, 8vo.

Mysterium occultæ naturæ. *See* Manget, J. J. ; *also* Theatrum chemicum.

Nedagander. *See* A B C.

Neue Sammlung von einigen alten und sehr rar gewordenen philosophisch und alchymistischen Schriften, in welcher anzutreffen sind : 1. *Lullius, Raymundus :* Codicill oder Vade Mecum. 2. *L., R. :* Allgemeine Ausübung des grossen Werks von der Quint-Essenz. 3. *L. R. :* Kleiner Schlüssel, worinnen alles, was zur Alchymie-Arbeit erfordert wird, eröfnet und erkläret ist. 4. *Ventura, Laurentius :* Liber unus de lapide philosophorum, oder Beweis das die Kunst der Alchymie gewiss und wahr seye. Welche sämmtliche Werke nicht nur an und vor sich Selbsten vollständig sind, sondern auch als eine neue Fortsetzung des bekannten deutschen Theatri chymici angesehen und gebraucht werden können. Erster Theil. Frankfurt und Leipzig, zu finden im Kraussischen Buchladen, 1767. pp. [viii]–436–[iii], 8vo.

      Cf. Roth-Scholtz, Fr. : Deutsches Theatrum chemicum.

Nodus sophicus enodatus. *also* A B C.

Norton, Samuel. *See* Dreyfaches hermetisches Kleeblatt.

Norton, Thomas. *See* Manget, J. J. ; *also* Musæum hermeticum ; *also* Ashmole, E. : Theatrum chemicum Britannicum

Novi luminis chemici tractatus alter de sulphure. *See* Musæum hermeticum.

Nuisement, De.

    * Poeme philosophic de la vérité de la phisique minéralle, où sont refutées les objections que peuvent faire les incredules et ennemis de cet art. Auquel est naïfvement et véritablement depeinte la vraye matière des philosophes. Dedié au très-haut, très-puissant et très-vertueux Prince, Monseigneur le Duc de Lorraine et de Bar, etc.

NUISEMENT, DE. [Cont'd.]

À Paris, chez Jeremie Perier et Abdias Buisard à la place Dauphine, prés le Palais au Bellerophon, 1620. pp. 80, 12mo.

* Traitté du vray sel, secret des philosophes et de l'esprit général du monde ; contenant en son intérieur les trois principes naturels, selon la doctrine de Hermes. Oeuvre très-utile et nécéssaire à quiconque désire arriver à la parfaitte prattique de ce prétieux élixir ou médecine universelle, tant célébrée des anciens recogneuë et expérimentée. À Paris, chez Jeremie Perier et Abdias Buizard, en la cour du palais vers les Horlogers, 1621. pp. [xxviii]–332–[ii], 12mo.

> Nuisement was Receveur-Général du Comté de Ligny en Barrois. Lenglet du Fresnoy says of him : "On pretend que cet Auteur est plagiaire. He ! que n'importe qu'il le soit, pourvû qu'il me dise des choses curieuses et peu connues, cependant le malheur est, qu'il n'en dit pas plus que les autres."

* NÜTZLICHE VERSUCHE UND BEMERKUNGEN AUS DEM REICHE DER NATUR, allen Erz- und Naturkündigen, Bergwerksverwandten, wie auch denen Liebhabern der Alchimie zum Gebrauch und Nutzen herausgegeben. Nürnberg, Verlegts Georg Bauer, 1760. pp. [xvi]–214–[ii], 12mo. Ill.

NUYSEMENT, DE.

> See A B C ; also Dreyfaches hermetisches Kleeblatt ; also Besondere Geheimnisse ; also Nuisement in this Section.

ODOMARUS. See Theatrum chemicum.

OLEUM ANTIMONII. See Aureum Vellus.

OLEUM VITRIOLI. See Aureum Vellus.

OLIFFE, CHARLES.

* Les Alchimistes d'autrefois. Paris, 1842. pp. xvi–291, 64mo.

OPUS AD ALBUM. See Theatrum chemicum.

OPUS MULIERUM ET LUDUS PUERORUM.

> See Artis auriferæ ; also Arnaldus de Villa Nova : Schrifften.

ORTHELIUS. See Manget, J. J. ; also Theatrum chemicum.

ORTHOLANUS. See Theatrum chemicum.

OSTEN, HANS VON.

* Eine grosse Herzstärkung für die Chymisten nebst einer Dose voll gutes Niesepulver für die unkundigen Widersprecher der Verwand-

OSTEN, HANS VON.   [Cont'd.]
lungskunst der Metalle, im Kloster zu Oderberg, seit Anno 1426,
aufbehalten durch Hans v. Osten ; welche vor wenigen Monathen
von einem Maurergesellen daselbst gefunden worden.   Begleitet
mit einer Zuschrift an die Chymisten, und einer wahrhaften Nachricht dieser Geschichte nebst dem dazu gehörigen Kupfer.   Auf
Kosten des Verfassers.   Berlin, in Commission bey dem Antiquarius
Johann Friedrich Vieweg, 1771.   pp. [xiv]–108, 12mo.   Frontispiece and plates.

PANTALEON.   *See* Collectanea chymica ; *also* Manget, J. J. ; *also* Roth-Scholtz, Fr.
      (Deutsches Theatrum chemicum).

PANTALEON.
  * Alchimistische Tractätlein : 1. Das eröffnete Hermetische Grab, vom
      Philosophischen Quecksilber.   2. Alchimistische Prüffung eines
      wahren Philosophi und betrügerischen Sophistens.   3. Metallisches
      Zweyblat, vom Stein der alten Weisen.   Welche erstlich vom
      Autore lateinisch beschrieben und absonderlich gedruckt, nun-
      mehr, auf Begehren unterschiedlicher Liebhaber der edlen Chimie,
      ins teutsche übersetzet und zusammengedruckt herausgegeben
      Christophorus Victorinus Artis filius.   Nürnberg, zu finden bey
      Paul Fürstens, Kunst- und Buchh. Seel. Witben und Erben ; ge-
      druckt daselbst bey Christoff Gerhard, Anno 1677.   pp. 175, 12mo.

[PANTALEON.] H. V. D.
  Tumba Semiramidis hermetice sigillatæ, quam si sapiens aperuerit, non
      Cyrus ambitiosus, avarus regum, ille thesauros divitiarum inexhaus-
      tos quod sufficiat inveniet.   Norimbergæ, apud hæredes Johannis
      Danielis Tauberi, 1717.   pp. 20–[43–63], 16mo.

PANTHEUS, JOANNES AUGUSTINUS.   *See* Theatrum chemicum.

PAPUS.
  * La pierre philosophale, preuves irréfutables de son existence.   Paris,
      1889.   pp. 29, 12mo.   Ill.
          The author seriously undertakes to establish the existence of the philosophers'
          stone, and this in A.D. 1889.

PARACELSUS, THEOPHRASTUS.   *See* Aureum Vellus.

PARACELSUS.   [PHILIPPUS AUREOLUS THEOPHRASTUS PARACELSUS BOM-
      BAST VON HOHENHEIM].
  * Of the chymical Transmutation, Genealogy and Generation of Metals
      and Minerals.   Also of the Urim and Thummim of the Jews.
      With an Appendix of the Vertues and Use of an excellent Water

PARACELSUS. [Cont'd.]

made by Dr. Trigge. The second Part of the Mumial Treatise. Whereunto is added Philosophical and Chymical Experiments of that famous Philosopher, Raymund Lully; containing: The right and due Composition of both Elixirs. The admirable and perfect way of making the great Stone of the Philosophers, as it was truely taught in Paris, and sometimes practised in England, by the said Raymund Lully, in the time of King Edward III. Translated into English by R. Turner. London, printed for Rich. Moon at the seven Stars, and Hen. Fletcher at the three gilt Cups in Paul's Church-yard, 1657. pp. [viii]–166, 12mo.

PATER SAPIENTIÆ. *See* Ashmole, E. : Theatrum chemicum Britannicum.

PAULI, JOHANN.

* Chymisch-medicinische Abhandlung von denen harnichten Salzen und Geistern. Kopenhagen, 1770, bey Johann Gottlob Rothen, Königl. dänischen Hof- und Universitäts-Buchhändler. pp. 72, 12mo.

PAYKULL, OTTO ARNOLD.

* Problema chymicum oder chymischer Process, wodurch nach Proportion eines Quentleins præparirten Sulphuris Antimonii, anderthalb Loth Bley in das schöneste und feineste Gold verwandelt worden. Allen der wahren Chymie Liebhabern und rechtschaffenen Philosophis mitgetheilet, und zu dero genauen Untersuchung aufgegeben, nebst beygefügter Vorrede, in welcher sowohl die Ursache der Publication, als auch dasjenige, was in selbigem Process ermangele und annoch zu erforschen sey, kürtzlich eröffnet und angezeiget wird. Berlin, bey Johann Christoph Papen, 1719. pp. [viii]–20, 4to.

> Published over initials only—O. A. P. The career of this Swedish alchemist, his imprisonment and offer to make one million crowns annually for the state, his successful projection before witnesses, and his exposure by the chemist Hjärne are narrated by Güldenfalk, Schmieder and other historians.

PEARCE THE BLACK MONKE. *Sse* Ashmole, E : Theatrum chemicum Britannicum.

PENOTUS, BERNARDUS GEORGIUS. *See* Theatrum chemicum.

PERBREVE OPUSCULUM QUOD " LIBER ALZE" DICITUR. *See* Musæum hermeticum.

PERNETY, ANTOINE-JOSEPH.

* Dictionnaire mytho-hermétique dans lequel on trouve les allégories fabuleuses des poétes, les métaphores, les énigmes et les termes barbares des philosophes hermétiques expliqués. À Paris, chez

PERNETY, ANTOINE-JOSEPH.   [Cont'd.]

Delalain l'aîné, Libraire, rue Saint-Jaques, No. 240, 1787.   Avec approbation et privilége du Roi.   pp. xxiv–546–[ii], 8vo.

> The author appropriately opens his preface thus : "Jamais science n'eut plus besoin de dictionnaire que la Philosophie Hermétique."

* Les Fables Égyptiennes et Grecques dévoilées et réduites au même principe, avec une explication des hiéroglyphes et de la guerre de Troye.   À Paris, Quai des Augustins, chez Bauche, Libraire, à Saint Jean dans le Désert.   Avec approbation et privilége du Roi. 1758.   2 vols., pp. xvi–580–[iv] ; [iii]–627–[v], 8vo.

> The author attempts to give a hermetic signification to Egyptian and Greek mythology. Cf. Maier, Michael.

PETRÆUS, BENEDICTUS NICOLAUS.

See Roth-Scholtz, Fr. (Deutsches Theatrum chemicum).

PETRUS BONUS FERRARIENSIS.

Introductio in divinam chemiæ artem integra magistri Boni Lombardi Ferrariensis Physici.   Basileæ, 1572.   4to.

* Pretiosa margarita novella de thesauro ac pretiosissimo philosophorum lapide ; artis huius divinæ typus et methodus : collectanea ex Arnaldo, Rhaymundo, Rhasi, Alberto, et Michaele Scoto ; per Jannum Lacinium Calabrum nunc primum cum lucupletissimo indicein lucem edita.   Cum privilegio Pauli III Pont. Max. et Senatus Veneti ad annos decem, 1546.   ff. [xxi]–202–[xv], 12mo. Ill.   [Apud Aldi filios, Venetiis], 1546.

> The "Pretiosa margarita novella" was written about 1330. Of this Lenglet du Fresnoy remarks : "Belle édition très-estimée d'un livre recherché des curieux."

> * Pretiosa Margarita, oder neuerfundene köstliche Perle, von dem unvergleichlichen Schatz und höchst-kostbahren Stein der Weisen.   In sich haltend den eigendlichen Grund-Riss und Lehr Arth dieser Göttlichen Kunst : in gleichen andere aus dem Arnaldo, Rhaimundo, Rhasi, Alberto und Michaele Scoto zusammen gelesene Schrifften, durch Janum Lacinium aus Calabria.   Zum erstenmahl in lateinischer Sprache, mit Freyheit Pabsts Pauli Tertii und des Raths zu Venedig, Anno 1546 herausgegeben.   Anjetzo aber um seiner Fürtrefflichkeit Willen in das teutsche übersetzet und ans Licht gestellet, von Wolfgang Georg Stollen, Liebhabern der edlen Chymie.   Nebst einem vollständigen Register.   Leipzig, Verlegts Johann Friedrich Braun, 1714.   pp. [xx]–468–[xxxvi], 4to. Plates.

> Cet auteur qui vivoit au XIVe siècle passe pour un des meilleurs sur la philosophie hermétique.—Lenglet du Fresnoy.

> Cf. Manget, J. J. ; also Theatrum chemicum.

PHILALETHA, EIRENÆUS.   *See* Philaletha, Eugenius ; *also* Ripley, George.

PHILALETHA, EUGENIUS.

    See A B C ; *also* Manget, J. J. ; *also* Musæum hermeticum ; *also* Roth-
    Scholtz, Fr. (Deutsches Theatrum chemicum ; *also* R[ichebourg], J. M.
    D. ; *also* Collectanea chymica.

PICATOSTE, FELIPE.

    La alquimia en nuestros dias.   *In* Museo Universal, vol. v, 1861.   pp.
    250 *et seq.*

PICUS MIRANDULA, JO. FRANCISCUS.

    Io. Francisci Pici Mirandulæ et Concordiæ Domini de auro libri tres ;
    opus sane aureum in quo de auro tum æstimando, tum conficiendo,
    tum utendo ingeniosè et doctè disseritur ; cum explicatione peru-
    tili et perjucunda complurium tam philosophorum quam facultatis
    medicæ arcanorum.   Venetiis, 1586.   pp. [viii]-131, 8vo.

PLATO.   *See* Theatrum chemicum.

PLATT, SIR HUGH.   *See* Collectanea chymica.

PLUSIUS, EDUARD.

    * Spiegel der heutigen Alchimie, das ist wohlgegründeter Bericht was
    von der so beruffenen Goldmacher-Kunst zu halten, und wie man
    sich darinnen behutsam zu erzeigen habe ; allen aufrichtigen Lieb-
    habern derselben zu Dienste aus dem lateinischen wegen seiner
    Nutzbarkeit in das deutsche übersetzt.   Budiszin und Görlitz, bey
    David Richtern, Buchhändler, 1725.   pp. 80, 16mo.

PLYTOFF, G.

    * Les sciences occultes.   Divination, calcul des probabilités, oracles et
    sorts, songes, graphologie, chiromancie, phrénologie, physiogno-
    monie, cryptographie, magie, kabbale, alchimie, astrologie, etc.
    Paris, 1891.   pp. 320, 12mo.   Ill.
        Bibliothèque scientifique contemporaine.

POISSON, ALBERT.

    * Collection d'ouvrages relatifs aux sciences hermétiques.   Théories et
    symboles des alchimistes.   Le Grand-œuvre.   Suivi d'un essai sur
    la bibliographie alchimique du XIXe siècle.   Ouvrage orné de 15
    planches représentant 42 figures.   Paris, 1891.   pp. xii-184.
        Contains a " dictionnaire des symboles hermétiques." The bibliography
        enumerates the writings of 48 authors, all of the nineteenth century.

    * Collection d'ouvrages relatifs aux sciences hermétiques.   Cinq traités
    d'alchimie des plus grands philosophes, Paracelse, Albert le Grand,

POISSON, ALBERT. [Cont'd.]

Roger Bacon, R. Lulle, Arnauld de Villeneuve. Traduits du Latin en Français. Précédés de la Table d'Émeraude, suivis d'un glossaire. Bibliothèque Chacornac. Paris, 1890. pp. viii–134, 8vo.

[Contents :]

1. Table d'Émeraude. p. 1.
2. Notice biographique sur Arnauld de Villeneuve. p. 7.
3. *A. de V. :* Le chemin du chemin [Semita Semitæ]. p. 9.
4. Notice biographique sur R. Lulle. p. 25.
5. *Lulle, Raymond :* La clavicule. p. 27.
6. Notice biographique sur Roger Bacon. p. 53.
7. *Bacon, Roger :* Le Miroir d'alchimie. p. 55.
8. Notice biographique sur Paracelse. p. 77.
9. Le trésor des trésors des alchimistes. p. 79.
10. Notice biographique sur Albert le Grand. p. 91.
11. *Albert le Grand :* Le composé des composés. p. 93.
12. Glossaire. p. 128.

POLEMAN, JOACHIM.

* Novum lumen medicum ; wherein the excellent and most necessary Doctrine of the highly-gifted Philosopher Helmont concerning the great Mystery of the Philosophers' Sulphur is fundamentally cleared. Out of a faithful and good intent to those that are ignorant and straying from the truth as also out of compassion to the sick. Written by the author in the German tongue and now Englished by F. H., a German. London, printed by J. C. for J. Crook at the sign of the Ship in St. Paul's Church-yard, 1662. pp. [viii]–206, 16mo.

Cf. Theatrum chemicum.

PORDÄDSCH, JOHANN. *See* Roth-Scholtz, Fr. (Deutsches Theatrum chemicum).

PORTA, JOHANNES BAPTISTA.

* Magia naturalis, sive de miraculis rerum naturalium libri iv. Coloniæ, apud Johanne Birckmannum et Wernerum Richvuinum, 1562. pp. [xvi]–307–[xv], 24mo.
* Libri viginti magiæ naturalis in quibus scientiarum naturalium divitiæ et deliciæ demonstrantur jam de novo, ab omnibus mendis repurgati, in lucem prodierunt ; accessit index, rem omnem dilucide repræsentans, copiosissimus. Librorum ordinem qui in hoc opere continentur versa pagina indicabit. Hanoviæ, typis Wechelianis, impensis Danielis ac Davidis Aubriorum et Clementis Schleichii, 1619. pp. [xxxii]–622, 12mo.

PORTA, JOHANNES BAPTISTA. [Cont'd.]

Magiæ naturalis qui extant libri sunt :

  1. De mirabilium rerum, causis.
  2. De variis animalibus gignendis.
  3. De novis plantis producendis.
  4. De augenda supellectili.
  5. De metallorum transmutatione.
  6. De gemmarum adulteriis.
  7. De miraculis magnetis.
  8. De portentosis medelis.
  9. De mulierum cosmeticis.
  10. De extrahendis rerum essentiis.
  11. De myropœia.
  12. De incendiariis ignibus.
  13. De raris ferri temperaturis.
  14. De miro conviviorum apparatu.
  15. De capiendis manu feris.
  16. De invisibilibus literarum notis.
  17. De catoptricis imaginibus.
  18. De staticis experimentis.
  19. De pneumaticis.
  20. Chaos.

> Natural Magick in twenty books, wherein are set forth all the riches and delights of the natural sciences, with engravings. London, 1658. Sm. folio.

>> John Baptist Porta, of Naples, was born in 1538 and died in 1615. The first edition of this work, comprised in four books, was published in 1553 when the author was only 15 years of age. This first edition is said not to be extant ; a much enlarged edition was published in 1589, which was followed by many others. Porta's description of the Camera obscura has led historians to attribute to him its invention, but it was known nearly one hundred years earlier to Leonardo da Vinci. In the first edition, as well as in that of 1562, the camera obscura is described in an imperfect form, without the lens. The latter important addition first appears in the issue of 1589. *Cf*. Libri, *Histor. math.*, IV.

POYSELIUS, ULRICHUS. *See* Aureum Vellus.

PRACTICE (THE) OF LIGHTS, or an excellent and ancient treatise of the Philosophers' Stone. *See* Collectanea chymica.

PRÉCEPTES ET INSTRUCTIONS DU PÈRE ABRAHAM À SON FILS. *See* R[ichebourg], J. M. D.

PROCESSUS PRO TINCTURA AUFF DEN MERCURIUM SOLIS ET LUNÆ. *See* Aureum Vellus.

PROPOSITIONES SIVE MAXIMÆ IN QUIBUS VERITAS TOTIUS ARTIS CHEMICÆ COMPREHENDI-TUR. *See* Artis auriferæ ; *also* Theatrum chemicum.

PRUGGMAYR, MARTIN MAXIMILIAN.

Scrutinium philosophicum de vero elixire vitæ, seu genuino auro potabili philosophico, quo non solum omnes humani corporis morbi quondam sanabantur, verum et immunda ac leprosa corpora metallorum curabantur ; opus non minus utile quam necessarium omnibus artis Hermeticæ filiis, in quo docetur, quid scire, quidque vitare debeat versus philosophiæ chemicæ studiosus ; ubi quoque exacte potissima enucleantur, quæ circa præparationem auri potabilis philosophici necessario sunt observanda, neque quidquam dictum reperitur, quin perspicue, succincte, ad oculum ex genuinis philosophorum adeptorum textibus comprobetur . . . Salisburgi, 1687. pp. (xvi)–146 and Index, 12mo.

> * Philosophische Untersuchung des wahrhaften Lebenselixieres oder des echten philosophischen Trinkgoldes, wodurch ehemals nicht nur alle Krankheiten des menschlichen Leibes geheilet, sondern auch die unreinen und aussätzigen Körper der Metallen zu ihrer vollkommenen Reinigkeit gebracht wurden. Aus dem lateinischen übersetzt und mit Anmerkungen versehen ; mit einem hieroglyphischen Kupfer das ganze philosophische Werk vorstellend. Leipzig, bey Adam Friedrich Böhmen, 1790. pp. 312, 12mo. Plates.

> Contains a list of hermetic works that ought to be read by a student of alchemy (Kap. III). The subject is treated historically and with customary credulity.

PSEAUTIER (LE) D'HERMOPHILE. *See* R[ichebourg], J. M. D.

* PYROTECHNICAL DISCOURSES, being—1. An experimental confirmation of chymical philosophy; treating of the several principles in the animal, vegetable, and mineral kingdoms ; with a perspective against chymical non-entities ; written by John Kunkel, Chymist to the Elector of Saxony. 2. A short discourse on the original of metallick veins ; by George Ernest Stahl which may serve as an answer to Dr. Woodward's theory of the earth and was a forerunner to : 3. The grounds of pyrotechnical metallurgy, and metallick essaying ; by John Christian Fritschius of Schwartzburg. All faithfully translated from the Latin and useful for all such as are any ways concern'd in medicine or metals. London, printed and sold by B. Bragg in Avemary-Lane, 1705. pp. x–268, 8vo.

* QUADRATUM ALCHYMISTICUM : Das ist : Vier auserlesene rare Tractätgen vom Stein der Weisen ; Speculum Sapientiæ, in welchem sowol die Sonnenklarheit von Jesu Christo, als auch die wahre

QUADRATUM ALCHYMISTICUM. [Cont'd.]

Tinctur der Weisen gelehret wird. Centrum Naturæ concentratum, welches von dem wiedergebohrnen Saltze der Philosophorum handelt. Discursus de Universali, worin viel geheimnissvolle Excerpta von der Universal Tinctur, und Medicin gesammlet worden. Abyssus Alchymiæ Explorata, in welchem die Verwandlung der Metallen handgreiflich und leichte von Thoma de Vagan, abgehandelt wird. Zum Dienst der Kunst und Weissheit-liebenden Practicorum, itzo herausgegeben von einem Liebhaber verborgener Künste. Hamburg, Verlegts Christian Liebezeit, Druckts Philipp Ludwig Stromer, 1705.

[Contents :]

I. Speculum Sapientiæ ; Das ist : Ein Buch dés Geheimnisses vom Anfang der Welt genannt : Der Himmlischen Sonnen-Klahrheit und Geheimniss von unserm Herrn und Heiland Jesu Christo mit dem Anhang der goldenen Practica de Tinctura Lapidis Physicorum, welches beschrieben im Jahr 1672 den 27 Martii. Hamburg, Verlegts Christian Liebezeit, Druckts Philipp Ludwig Stromer, 1705. pp. 54.

II. *Ali Puli :* Centrum Naturæ concentratum, oder ein Tractat von dem wiedergebohrnen Saltz, insgemein und eigendlich genandt : Der Weisen Stein. In arabischen geschrieben von Ali Puli, einem asiatischen Mohren, darnach in portugisische Sprache durch H. L. V. A. H. und ins hochteutsche versetzt, und herausgegeben von Joh. Otto Helbig, Rittern, Chur-Fürstlichen Pfaltzischen Rath. Gedruckt im Jahr 1682. pp. 38.

III. *Martinus de Delle :* Discursus de Universali. pp. from 39–80.

IV. *Vagan, Thomas de :* Abyssus Alchymiæ exploratus : oder die lang gesuchte und nunmehro glücklich gefundene Verwandlung der Metallen vermöge des Steins der Weisen als des grössesten Geheimnüsses und Wunderwercks der natürlichen Kunst und künstlichen Natur. Zum Nutzen der Philosophorum, zu Erweiterung der Metallurgie, und zum Trost derer, die da Knechte dieses Erbes seyen sollen gezeiget und beschrieben. pp. [viii]–113–[v].

QUATTRAMO, FRANCESCO.

* La vera dichiaratione di tutte le metafore, similitudini, e enimmi degl'antichi filosofi alchimisti, tanto Caldei e Arabi, come Greci e Latini, usati da loro nella descrittione, e compositione dell' oro potabile, elissire della vita, quinta essenza e lapis filosofico. Ove con un breve discorso della generatione de i metalli e di quasi tutte l'opere di natura, secondo i principii della filosofia, si mostra l'errore,

QUATTRAMO, FRANCESCO. [Cont'd.]

e ignoranza (per non dir l'inganno) di tutti gl'alchimisti moderni. Con licenza de i Superiori.   In Roma, appresso Vincentio Accolti, in Borgo novo, 1587.   pp. [xxiv]–230–[xxiv], 4to.

QUERCETANUS.   *See* Duchesne.

QUERCETANUS, JOSEPHUS ARMENIACUS.   *See* Theatrum chemicum.

RACHAIDIBUS.   *See* Kalid Rachaidibus Veradianus.

* RECHTE (DER) WEG ZU DER HERMETISCHEN KUNST, vor die Lehrbe-gierigen Schüler und Liebhaber dieser Wissenschaft. Nebst ver-schiedenen Anmerkungen über das betrügliche Verfahren der sogenannten Sophisten und ihrer Irrwege, herausgegeben von Anonymo. Franckfurt und Leipzig, bey Johann Georg Fleischer, 1773. pp. 104, 12mo.

     Cf. A B C.

* RECONDITORIUM AC RECLUSORIUM opulentiæ sapientiæque numinis mundi magni, cui deditur in titulum *Chymica Vannus*, obtenta qui-dem et erecta auspice mortale cœpto ; sed inventa pro authoribus immortalibus adeptis, quibus conclusum est, sancitum et decretum, ut anno hoc per mysteriarcham Mercurium velut viocurium seu medicurium. Amstelodami, apud Joannem Janssonium à Waesberge, et Elizeum Weyerstraet, 1666.   pp. 292–[ii], 4to.   Plates.

          My copy contains also :

Commentatio de pharmaco catholico ; quomodo nimirum istud in tribus illis naturæ regnis, mineralium, animalium ac vegetabilium reperi-endum atque exinde conficiendum, per excellentissimum univer-sale menstruum vi pollens recludendi occludendique, tum metallum quodlibet in primam sui materiam reducendi ; insuper qualiter per idipsum (supple menstruum) alias fixum illud indestructibile aurum redigendum sit in verum et inculpatum aurum potabile, quod nullo se imposterum artis stratagemate in solidum iterum aureum corpus patitur reduci ; cum primis ad honorem Dei glorios et excelsi, deinde in singulare solatium et emolumentum generis humani propalata inque publicam data lucem, atque Londini in Anglia, ab uno eodemque paraphraste, qui usqueadhuc Chymi-cum Vannum instituit, celeriter, sed tamen fideliter, è Germanismo in Latinismum trajecta. Æra christiana millesmia sexcentesima sexagesima quinta, kalendis octobribus.   pp. 76, 4to.   Ill.

REDMANN, W.   *See* Ashmole, E.: Theatrum chemicum Britannicum.

RESPONSUM AD FRATRES ROSACEÆ CRUCIS.
> *See* Roth-Scholtz, Fr. (Deutsches Theatrum chemicum).

REYHER, SAMUEL.
* Dissertatio de nummis quibusdam ex chymico metallo factis. Kiliæ
  Holsatorum, 1692. Typis Joachimi Reumanni, Acad. Typgr. pp.
  [viii]–141–[iv], 4to. Ill.
  > *Cf.* Bolton, Henry Carrington.

RHASES. *See* Petrus Bonus Ferrariensis : Pretiosa margarita.

RHENANUS, JOANNES.
* Solis e puteo emergens, sive dissertationes chymiatechnica libri tres ;
  in quibus totius operationis chymicæ methodus practica, materia
  lapidis philosophici et nodus solvendi eius, operandique ut et
  clavis operum Paracelsi, qua abstrusa explicantur, deficientia supplentur ; cum præfatione chymiæ veritatem asserente. Libri tres.
  1613. I, pp. [xxiv]–80 ; II, pp. 31 ; III, pp. 24, 4to. Illustrated
  title-page. Ill.

RHODARGIRUS, LUCAS EUTOPIENSIS. *See* Theatrum chemicum.

RICHARDUS ANGLICUS. *See* Theatrum chemicum ; *also* Manget, J. J.

R[ICHEBOURG], J. M[AUGUIN] D[E]. [Editor.]
* Bibliothèque des philosophes chimiques. Nouvelle édition, revue,
  corrigée et augmentée de plusieurs philosophes, avec des figures et
  des notes pour faciliter l'intelligence de leur doctrine. Paris, 1740–
  54. 4 vols, 12mo. Plates. Vol. I, pp. cxliv–384–[40] ; vol. II, pp.
  564 ; vol. III, pp. 522–[4].

  > Vol. IV under the title :

  Bibliothèque des philosophes alchimiques ou hermétiques contenant
  plusieurs ouvrages en ce genre très curieux et utiles, qui n'ont
  point encore parus, précédés de ceux de Philalethe, augmentés et
  corrigés sur l'original Anglois et sur le Latin. Tome quatrième.
  Paris, 1754. pp. viii–590, 12mo.

VOL. I.

1. *Trismegiste, Hermes :* La table d'émeraude, avec le commentaire
   de l'Hortulain. p. 1.
2. *Trismegiste, Hermes :* Les sept chapitres. p. 16.
3. *Le dialogue de Marie* et d'Aros. p. 77.
4. *Geber :* La somme de la perfection ou l'abrégé du magistère parfait de *Geber.* Divisé en deux livres. p. 85.

R[ICHEBOURG], J. M[AUGUIN] D[E].   [Cont'd.]

VOL. II.

1. [*Arisleus*] : La tourbe des philosophes ou l'assemblée des disciples
   de Pythagoras, appellée code de vérité. p. 1.

2. *Entretien du roi Calid* et du philosophe Morien. Sur le magistère
   d'Hermes, rapporté par Galip, esclave de ce Roi. p. 56.

3. *Artefius :* Un livre qui traite de l'art secret, ou de la pierre
   philosophale. p. 112.

4. *Synesius :* Un livre sur l'œuvre des philosophes. p. 175.

5. *Flamel, Nicolas :* Un livre contenant l'explication des figures
   hyérogliphiques qu'il a fait mettre au cimetière des SS. Innocens à Paris. p. 195.

6. *Flamel, Nicolas :* Petit traité d'alchymie intitulé "le sommaire
   philosophique." p. 263.

7. *Flamel, Nicolas :* Le désir désiré. p. 285.

8. *Trévisane, Bernard de la Mordu :* Le livre de la philosophie naturelle des métaux. p. 325.

9. *Trévisane, B. :* La parole délaissée, traité philosophique. p. 400.

10. [*Trévisane, B. :*] Le songe verd, véridique et véritable parce qu'il
    contient vérité. p. 437.

11. *Zachaire,* [*Denys*]: Opuscule de la philosophie naturelle des métaux.
    p. 447.

VOL. III.

1. *Valentin, Basile :* Les douze clefs de philosophie. p. 1.

2. *Valentin, Basile :* L'azoth ou le moyen de faire l'or caché des
   philosophes. p. 84.

3. *L'ancienne guerre* des chevaliers, ou le triomphe hermétique. p. 181.

4. *La lumière sortant* par soi-mesme des ténèbres, poëme sur la composition de la pierre des philosophes, traduit de l'Italien, avec
   un commentaire. p. 322.

VOL. IV.

1. *Philalèthe* (l'amateur de la vérité) : Traité de l'entrée ouverte du
   palais fermé du roi. p. 1.

2. *Philalèthe :* Explication de ce traité par lui-même. p. 121.

3. *Philalèthe :* Expériences sur la préparation du mercure des sages
   pour la pierre, par le régule de mars ou fer, tenant de l'antimoine, et étoilé, et par la lune ou l'argent. p. 138.

4. *Philalèthe :* Explication de la lettre de Georges Ripley à Édouard
   IV Roi d'Angleterre. p. 148.

5. *Philalèthe :* Principes pour diriger les opérations dans l'œuvres
   hermétiques, traduits de l'Anglais. p. 174.

R[ICHEBOURG], J. M[AUGUIN] D[E].   [Cont'd.]

6. *Traité du secret* de l'art philosophique, ou l'arche ouverte, autrement dite la cassettte du petit paysan.   Commenté par
   Valachius, corrigé & élucidé par Ph . . . Ur . . . Amateur
   de la sagesse.   p. 186.

7. *Rouillac, Philippe :* Abrégé du traité du grand œuvres des philosophes.   p. 234.

8. *Lulle, Raimond :* L'élucidations ou l'éclaircissement du testament
   de Raimond Lulle.   p. 297.

9. *Explication très-curieuse* des énigmes et figures hyéroglifiques,
   physiques, qui sont au grand portail de l'église Cathédrale et
   Métropolitaine de Nôtre-Dame de Paris, pas Esprit Gobineau
   de Montluisant, Gentilhomme Chartrain, Amateur et inter-
   préte des vérités hermétiques, avec une instruction préliminaire
   sur l'antique situation et fondation de cette église, et sur l'état
   primitif de la Cité.   p. 307.

10. *Le pseautier d'Hermophile*, envoyé à Philalèthe.   p. 394.

11. *Traité d'un philosophe* inconnu sur l'œuvre hermétique.   p. 461.

12. *Lettre philosophique* de Philovite à Héliodore.   p. 511.

13. *Préceptes et instructions du père Abraham* à son fils, contenant la
    vraie Sagesse hermétique, traduits de l'Arabe.   p. 552.

14. *Lavinius Vincelas de Moravie :* Traité du ciel terrestre.   p. 566.

15. Dictionnaire abrégé des termes de l'art hermétique.   p. 570.

> Lenglet du Fresnoy catalogues the 13 treatises of the first three vols. (being
> all published at the date he wrote) and adds : " There is no doubt that
> this skilful author and his wife possessed the Philosophers' Stone."
> (*Hist. philos. hermétique*, III, 46.)   Some copies of vol. I bear the date
> 1741, and have a somewhat different ornament on the title-page from
> those dated 1740.   *Cf.* Salmon, William : Bibliothèque.

RICHTER, SAMUEL.   *See* Sincerus Renatus.

RINVILLUS BRISZOLLENSIS.   *See* Dreyfaches hermetisches Kleeblatt.

RIPLEY, GEORGE.
> *See* Manget, J. J. ; *also* Theatrum chemicum ; *also* Collectanea chymica ;
> *also* Ashmole, E. : Theatrum chemicum Britannicum.

RIPLEY, GEORGE.
* Chymische Schrifften, darinnen von dem gebenedeyten Stein der
  Weisen und desselben kunstreichen Præparation gründlich ge-
  handelt wird.   Nach der lateinisch- und englischen Edition Herrn
  William Salmon ins teutsche übersetzt durch Benjamin Roth-
  Scholtzen.   Zu finden bey Johann Paul Krauss, Buchhändler in
  Wienn, 1756.   pp. 233, 16mo.   Illustrated title-page.

RIPLEY, GEORGE. [Cont'd.]

[*Contains also :*]

Artephius : Geheimer Haupt-Schlüssel zu dem verborgenen Stein der
    Weisen. Aus der lateinisch und englischen Edition Herrn
    William Salmon ins teutsche übersetzet durch Benjamin Roth-
    Scholtzen. p. 105.

Das eröffnete philosophische Vatter-Herz an seinen Sohn, welches er,
    wegen hohen Alters, nicht länger wollte vor ihm verschlossen
    halten, sondern zeigete und erklärte demselben alles das, was zu
    der völligen Composition und Bereitung des Steins der Weisen
    vonnöthen war. Sonst in französischer, nun aber in teutscher
    publicirt durch Benjamin Roth-Scholtzen. p. 153.

Compound (The) of Alchymy, or the ancient hidden art of alchemie,
    containing the right and perfectest meanes to make the philosophers'
    stone, aurum potabile, with other excellent experiments, divided in
    twelve gates, first written by the learned and rare philosopher of our
    nation, George Ripley, sometime Chanon of Bridlington in York-
    shyre, and dedicated to King Edward IV, whereunto is adjoyned
    his Epistle to the King, his vision, his wheele, and other his works,
    never before published, with certaine brief additions of other
    notable writers concerning the same, set forth by Ralph Rabbards,
    Gentleman, studious and expert in alchemical artes. London,
    imprinted by Thomas Orwin, 1591. Wood-cut border around title
    and a portrait of Queen Elizabeth. 4to.

* Philalethes, Eirenæus : Ripley Reviv'd or, An Exposition upon Sir
    George Ripley's Hermetico-Poetical Works. Containing the plainest
    and most excellent discoveries of the most hidden secrets of the
    Ancient Philosophers, that were ever yet published. London,
    printed by Tho. Ratcliff and Nat. Thompson, for William Cooper
    at the Pelican in Little-Britain, 1678. pp. [xvi]–47–[i]–389–[v]–
    10–[i]–28–[iv]–25–[iii], 12mo. Illustrated title-page.

ROBINSON, THOMAS. *See* Ashmole, E. : Theatrum chemicum Britannicum.

ROCHAS, HENRICUS DE. *See* Theatrum chemicum.

ROSARIUM ABBREVIATUM E MANUSCRIPTO VETUSTISSIMO. *See* Manget, J. J.

ROSARIUM PHILOSOPHORUM. *See* Artis auriferæ ; *also* Manget, J. J.

ROSENCREUTZ, CHRISTIAN.

The Hermetick Romance : or the chymical wedding. Written in High
    Dutch by C. Rosencreutz. Translated by E. Foxcroft. [London,]
    1690. pp. 226, 8vo.

    Cf. Andreä, Johann Valentin.

Rosinus. *See* Artis auriferæ.

Roth-Scholtz, Friedrich. [Editor].

* Deutsches Theatrum chemicum, auf welchem der berühmtesten Philosophen und Alchymisten Schrifften die von dem Stein der Weisen, von Verwandlung der schlechten Metalle in bessere, von Kräutern, von Thieren, von Gesund- und Sauerbrunnen, von warmen Bädern, von herrlichen Artzneyen und von andern grossen Geheimnüssen der Natur handeln, welche bisshero entweder niemahls gedruckt, oder doch sonsten sehr rar worden sind. Drei Theile. Nürnberg, 1728–1732. 3 vols., 12mo. Vol. i, pp. 26–[ii]–680–350 ; ii, pp. 22–936 ; iii, pp. 46–960. Portrait of Edward Kelley and plates.

Contains 53 alchemical treatises as follows :

1. *Buddeus, Joannes Franciscus :* Historisch und politische Untersuchung von der Alchemie und was davon zu halten sey. Aus dem lateinischen ins teutsche übersetzt. p. 1.

2. *Nenter, Georg Philipp :* Bericht von der Alchemie ; darinnen von derselben Ursprung, Fortgang und besten Scriptoribus gehandelt, auf alle Einwürffe der Adversariorum geantwortet, und klar bewiesen wird, dass warhafftig durch die Alchemie der rechte Lapis Philosophorum als eine Universal-Medicin könne bereitet werden. p. 147.

3. *Schröder, Wilhelm von :* Unterricht vom Goldmachen, denen Buccinatoribus oder so sich selbst nennenden Fœderatis Hermeticis auf ihre drey Episteln zur freundlichen Nachricht. p. 219.

4. Treuhertzige Warnungs-Vermahnung an alle Liebhaber der wahren naturgemässen Alchemiæ Transmutatoriæ von einem Liebhaber der Wahrheit aufgesetzt. p. 289.

5. *Müllner, Leonhard :* Gründlicher Bericht von der Generation und Geburt der Metallen. Wie solche durch des Himmels Einfluss in dem Erdreich gewürcket wird ; samt der Verwandlung der Schlechten in Bessere. p. 313.

6. *Morhof, Daniel Georg :* Wahrhaffter und in der Natur gegründeter Bericht von der Generation und Regeneration der Metallen. p. 331.

7. *Hautnorthon, Josaphat Friederich :* Dritter Anfang der mineralischen Dinge oder vom philosophischen Saltz nebenst der wahren Præparation Lapidis et Tincturæ Philosophorum. p. 339.

8. *Chrysogonus de Puris :* Das pontische oder Mercurialwasser der Weisen aus philosophischen Schrifften denen Söhnen der Kunst ordentlich vorgestellet. p. 391.

ROTH-SCHOLTZ, FRIEDRICH. [Cont'd.]

9. *Philaletha, Eugenius :* Euphrates, oder die Wasser vom Aufgang, welches ist ein kurtzer Bericht von dem geheimen Brunnen, dessen Wasser aus dem Feuer quillet, und bey sich die Strahlen der Sonnen und des Mondes führet. Aus dem englischen übersetzet von Johann Langen. p. 415.

10. *Helvetius, Joannes Friedericus :* Vitulus aureus, oder güldenes Kalb, welches die ganze Welt anbetet und verehret ; in welchem das rare und wundersame Werck der Natur in Verwandelung derer Metallen historisch ausgeführet wird. Wie nemlich das gantze Wesen des Bleyes in einem Augenblick durch Hülffe eines sehr kleinen Körnleins des Steins der Weisen, zu dem allerbesten Gold vom obgenandten Herrn Helvetio im Haag gemacht und verwandelt worden. p. 481.

11. *Pordädsch, Johann :* Ein gründlich philosophisch Sendschreiben vom rechten und wahren Steine der Weisheit. Worinnen der ganze Process des philosophischen Wercks, oder wie man das Werck der wahren Wiedergeburt recht anfangen, darinnen glücklich fortgehen, und es zum vollkommenen und seeligen Ende bringen soll, gründlich angewiesen und ausgeführt wird. Geschrieben zum Unterricht und Warnung an eine gute Seele, die nach der ersten Materie dieses herrlichen Steins der göttlichen Tinctur zwar mit grossen Ernst gesucht und gegraben, dieselbe auch würcklich gefunden und geschmeckt gehabt ; aus Ermanglung genugsamen Lichts aber ihn vollkömmlich zu besitzen, und zu völliger Ruhe gekommen zu seyen, ihr allzu frühzeitig eingebildet gehabt. Aus dem englischen übersetzt und gedruckt im Jahr Christi 1694. p. 557.

12. *Monte Raphaim, Joannes de :* Vorbothe der am philosophischen Himmel hervorbrechenden Morgen-Röthe ; sammt einem Anhang etlicher Lehr-Sätze vor die Schüler der Weissheit ans Licht gestellt durch F. Roth-Scholtzen. p. 597.

13. *Valentinus, Basilius :* Triumph-Wagen des Antimonii ; nebst Theodori Kerckringii gelehrten Anmerckungen darüber. Darauf folget Herrn Georg Wolffgang Wedels in einem Programma vom B. Valentino ertheilte Nachricht und Recommendation. p. 653.

VOL. II.

1. *Petræus, Benedictus Nicolaus :* Critique über die alchymistischen Schrifften und deren Scribenten, ihren neuen Projections-

ROTH-SCHOLTZ, FRIEDRICH. [Cont'd.]

Historien, wie auch von der Materia Prima philosophica
und andern Dingen mehr handlend ; sowohl mit einigen
Anmerckungen vermehret und von neuem ans Licht gestellet
durch Fr. Roth-Scholtzen. p. 1.

2. Bedencken über die Frage : ob die Transmutatio Metallorum
möglich. Nebst einem Responsio einer berühmten Juristen-
Facultät : Da sich ein Ehemann belehren lässet : Ob ihm
das seiner Frauen in Gold transmutirte silberne Gefässe
nicht zukomme ? oder doch wenigstens der usus fructus
davon. p. 87.

3. *Sperberus, Julius :* Isagoge, das ist Einleitung zur wahren Erkänntniss des drey-einigen Gottes und der Natur ; worinn
auch viele vortreffliche Dinge von der Materia des philosophischen Steins und dessen gar wunderbaren Gebrauch
enthalten sind. Aus dem lateinischen ins deutsche übersetzet. p. 119.

4. *Pantaleon :* Tumulus Hermetis apertus, oder das eröffnete Hermetische Grab, in welchem sonnenklar zu sehen sind, der
uralten Weisen verborgene natürliche Wahrheiten, und etlicher neuen irrige Meinungen von dem hochberühmten
Wasser, dem philosophischen Quecksilber. Aus dem lateinischen ins teutsche übersetzet. p. 197.

5. *Pantaleon :* Examen Alchymisticum, oder alchimistische Prüffung,
mit welcher, als mit einem Probierstein, ein Besitzer der
Tinctur von einem Sophisten und ein wahrer Philosophus
von einem Betrüger könne unterschieden werden. Aus dem
lateinischen ins teutsche übersetzet. p. 259.

6. *Pantaleon :* Bifolium Metallicum, das ist metallisches Zweyblatt
oder zweyfache Artzney für mangelhaffte Metallen und
Menschen, welche von den Besitzern der Hermetischen
Kunst unterm Namen des Steins der Weisen, erfunden, verfertiget, und der Nachwelt mitgetheilet worden. Aus dem
lateinischen ins deutsche übersetzet. p. 313.

7. *Langelott, Joel :* Sendschreiben an die hochberühmte Natura
Curiosos : Von etlichen in der Chemie ausgelassenen
Stücken, durch welcher Anleitung nicht geringe, bissher aber
für unwesentliche gehaltene Geheimnüssen wohlmeinend
eröffnet, dargethan und erwiesen werden ; sambt denen darzu
nöthigen Kupffern, zum Druck befördert durch F. Roth-
Scholzen. p. 381.

8. *Rudolff, Johann Heinrich :* Elementa amalgamationis, oder Gründlicher Unterricht worinnen die Amalgatio bestehe nebst

ROTH-SCHOLTZ, FRIEDRICH. [Cont'd.]

denen Hülffs-Mittlen und führenden Endzweck, sambt beygefügter Praxi einiger diese Jahre her, durch diesen Modum
et Motum untersuchter Ertze, Vitriolen und Mineren auch
hierzu dienlichen Processen und Handgriffen. Wegen seiner
Seltenheit von neuem zum Druck befördert durch F. Roth-
Scholtzen. p. 407.

9. *Rudolff, Johann Heinrich :* Extra-Ordinair Bergwerck durch die
Amalgamation mit Quecksilber, auff die Metalla, Mineras,
Vitriola, und deren anhängigen Erden und Letten gerichtet.
Wodurch man mit wenigen Unkosten und kurtzer Zeit,
Capital mit Capital gewinnen kan, sambt beygefügten An-
merckungen und hierzu dienlichen Processen. p. 431.

10. *Garlandus, Johannes seu Hortulanus :* Compendium alchimiæ,
oder Erklärung der smaragdischen Tafel Hermetis Trismegisti. Aus dem lateinischen ins teutsche übersetzet. p. 499.

11. *Arnaldus de Villa Nova :* Erklärung über den Commentarium
Hortulani. p. 533.

12. *Faber, Peter Joannes :* Send-Schreiben an einen guten Freund
von der Vortrefflichkeit der chymischen oder vielmehr
philosophischen Schrifften, worinn zugleich zulängliche Anleitung zur Transmutation oder Verwandelung der Metallen,
auch wie der Lapis Philosophorum zu machen in dem Fabro
angewiesen wird. p. 551.

13. *Korndorffer, Bartholomäus :* Beschreibung wie die Edelgesteine
nicht allein von ihren gifftigen Influentien corrigiert, sondern auch wie sie nach geschehener Correction zu Nutz vor
vielerley Kranckheiten dem Menschen adhibiret werden
können. p. 567.

14. *Becher, Joh. Joachim :* Oedipus Chymicus, oder chymischer Rätseldeuter, worinnen derer Alchymisten dunckelste Redens-
arten und Geheimnisse offenbahret und aufgelöset werden.
Aus dem lateinischen ins deutsche übersetzet. p. 619.

15. *D'Espagnet, Johannes :* Das geheime Werck der Hermetischen
Philosophie worinnen die natürlichen und künstlichen Geheimnüsse der Materie des philosophischen Steins, wie auch
die Art und Weise zu arbeiten richtig und ordentlich offenbahret sind. p. 823.

VOL. III.

1. *Baco, Rogerius :* Radix mundi, oder Wurtzel der Welt ; ver-
deutsch nach dem englischen von William Salmon. Mit
Anmerckungen versehenen Exemplar. p. 21.

ROTH-SCHOLTZ, FRIEDRICH. [Cont'd.]

2. *B., R.:* Medulla Alchemiae, darinnen vom Stein der Weisen und von den vornehmsten Tincturen des Goldes, Vitriols und Antimonii, gehandelt wird. Item eine alchymische Epistel, so Alexandro zugeschrieben worden. p. 73.

3. *B., R.:* Tractat vom Golde, oder gründlicher Bericht von der Bereitung des philosophischen Steins, so aus dem Golde gemacht wird. Theoretice und practice klar und deutlich beschrieben, und nicht allein durch natürliche Ursachen demonstrirt und bewiesen, sondern auch mit theologischen Exemplen nach der Natur wol erkläret. p. 103.

[Only the title-page and the preface are to be found at p. 103 ; for " Tractat vom Golde," *see* No. 5, p. 130.]

4. *B., R.:* Spiegel der Alchemie. p. 105.

5. *B., R.:* Tractat vom Golde. p. 130.

6. *B., R.:* Tractat von der Tictur und Oel des Vitriols ; welchen er als ein Edelköstlich und allergewisseste Secretum und Medicin der Menschen und Metallen seinem geliebten Bruder Wilhelmo communicirt und überschicket. p. 179.

7. *B., R.:* Tractat von der Tinctur und Oel des Antimonii, von der wahren und rechten Bereitung des Spiessglases, menschliche Schwachheiten und Krankheiten dadurch zu heilen, und die imperfecten Metallen in Verbesserung zu setzen. p. 205.

8. *Epistel oder Send-Brief* des Kaysers Alexandri, welcher zuerst in Griechenland und Macedonien regieret hat, auch ein Kayser der Persianer gewesen, darinnen der Stein der Weisen durch ein Gleichnüss und Parabel sehr lustig und wohl beschrieben erkläret wird. Jetzo den Philochemicis zum Besten in öffentlichem Druck publiciret. p. 227.

9. *Baco, Rogerius:* Send-Schreiben von geheimen Würckungen der Kunst und der Natur und von der Nichtigkeit der falschen Magiæ. p. 245.

10. *Baco, Rogerius:* Epistola de secretis operibus artis et naturæ et de nullitate magiæ ; opera Johannis Dee, Londinensis, e pluribus exemplaribus castigata olim, et ad sensum integrum restituta ; nunc vero cum notis quibusdam partim ipsius Johannis Dee partim edentis. p. 287.

11. Responsum ad fratres rosaceæ crucis illustres. Heus leo cruce fides, lux sathodie. Nam quando fide curris, onus propulsans ecclesiæ vigebit. p. 349.

12. [*Barcius.* (*Johann v. Sternberg*)] : Gloria mundi, sonsten Paradeiss-Taffel : Das ist Beschreibung der uralten Wissenschaft welche Adam von Gott selbst erlernet, Noa, Abraham und

ROTH-SCHOLTZ, FRIEDRICH.   [Cont'd.]

Salomon, als eine der höchsten Gaben Gottes gebraucht, Weisen zu jederzeit vor den Schatz der ganzen Welt gehalten, und den Gottsfürchtigen allein nachgelassen haben nemlich ; de lapide philosophorum authore anonymo. p. 357,

13. Ein ander Tractätlein gleiches Inhalts mit dem vorigen sehr nützlich zu lesen. p. 511.

14. *Alethophilus :* Philosophische Betrachtung von Verwandlung der Metallen.   Aus dem lateinischen ins teutsche übersetzet. p. 537.

15. Warnungs-Vorrede wider die Sophisten und Betrüger welche ein Anonymus Anno 1670 und Anno 1691, in Hamburg, dess Johannis Ticinencis Anthonii de Abbatia und Edovardi Kellæi chymischen Schrifften vorgesetzet hat. p. 561.

16. *Ticinensis, Johannes :* Chymische Schrifften, oder Process vom Stein der Weisen, in lateinisch und teutscher Sprache von neuem zum Druck befördert. p. 607.

17. *Anthonius de Abbatia :* Send-Schreiben von dem Stein der Weisen und von Verwandlung der Metallen. Aus dem lateinischen ins teutsche übersetzet. p. 651.

18. *A. d. A. :* Epistolæ duæ scrutatoribus artis chymicæ mandatæ ; accessit arcanum a quodam philosopho anonymo deductum. p. 681.

19. *Kellæus, Edvvardus :* Buch von dem Stein der Weysen. An den Römischen Kayser Rudolphum II, anno 1596, in lateinischer Sprache geschrieben.   Nun aber nebst einer Vorrede von dem Leben und Schrifften Kellæi zum Druck befördert von F. Roth-Scholtz. p. 733. [Portrait of Kelley.]

20. *Kellæus, Edvvardus :* Fragmenta quædam ex ipsius epistolis excerpta. p. 799.

21. *K., E. :* Via humida, sive discursus de menstruo vegetabili Saturni, e manuscripto. p. 801.

22. *N . . ., S. :* Aula lucis, oder das Haus des Lichts. In englischer Sprache beschrieben und anno 1690, in das deutsche übersetzt durch Johann Langen. p. 855.

Cf. Neue Sammlung von Schriften.

ROUILLAC, PHILIPPE.   *See* R[ichebourg], J. M. D.

RUDOLF, JOHANN HEINRICH.   *See* Roth-Scholtz, Fr. (Deutsches Theatrum chemicum).

RUESENSTEIN, VON.

* Auserlesene chymiche Universal und Particular Processe, welche Herr Baron von Ruesenstein auf seinen zweyen Reisen mit sechs

RUESENSTEIN, VON. [Cont'd.]

Adepten als Gualdo, Schulz, Fauermann, Koller, Fornegg und Monteschider, erlernet, auch viele selbst davon probirt und mit eigener Hand im Jahr 1664, zusammengetragen hat, und wovon die Originalien in seinem Schloss in einer Mauer gefunden worden sind. Frankfurt und Leipzig, zu finden bey Peter Conrad Monath, 1754. pp. [xvi]-284-[iv], 16mo.

RULAND, MARTIN.

* Lexicon alchemiæ, sive Dictionarium alchemisticum, cum obscuriorum verborum et rerum hermeticarum, tum Theophrasto-Paracelsicarum phrasium, planam explicationem continens. Francofurtensium Repub., 1612. pp. [viii]-471 ; [487], sq. 8vo.

> By an error in pagination the numbers 465 to 471 appear twice, following page 480.

[Another edition.] Frankfurt, 1661. pp. 471-[487].

> The same error occurs. The Preface is dated April, 1611. The author was one of the court physicians of Rudolph II of Germany. This lexicon is very full, less mystical and more practical than some later ones. Useful in explaining early terminology.

RUPESCISSA, JOANNES DE.

> De confectione veri lapidis phil. *See* Manget, J. J., II ; *also* Theatrum chemicum, III.

SACHS A LEVVENHEIMB, PHILIP JACOB. *See* Manget, J. J.

SALMON, WILLIAM.

Bibliothèque des philosophes chimiques, ou Recueil des auteurs les plus approuvés qui ont écrit de la pierre philosophale. Paris, 1672-78. 2 vols., 12mo.

> Of this a second and enlarged edition was published by Richebourg, *q. v.*

* Dictionaire hermétique contenant l'explication des termes, fables, énigmes, emblèmes et manières de parler des vrais philosophes. Accompagné de deux traitez singuliers et utiles aux curieux de l'art. Par un Amateur de la Science. À Paris, chez Laurent d'Houry, ruë Saint Jaques, devant la Fontaine Saint Severin au Saint Esprit, avec privilége du Roy, 1695. pp. [xii]-216, 16mo.

> *Cf.* Pernety, Antoine Joseph.

SALTZTHAL, SONINUS. *See* Theatrum chemicum.

SAMMLUNG DER NEUESTEN und merkwürdigsten Begebenheiten welche sich mit unterschiedlichen vermutlich noch lebenden Adepten zugetragen. Hildesheim, 1780.

SATURNUS SATURATUS DISSOLUTUS ET CŒLO RESTITUTUS.
  See Dreyfaches hermetisches Kleeblatt ; *also* Besondere Geheimnisse.

SCALA PHILOSOPHORUM.   *See* Artis auriferæ.

SCOTUS, MICHAEL.   *See* Petrus Bonus Ferrariensis : Pretiosa margarita.

SCHENKEN.   *See* Aureum Vellus.

SCHÖNER TRACTAT (EIN) VOM PHILOSOPHISCHEN STEINE.   *See* Grasshof, Johann.

SCHÖNES TRACTÄTLEIN (EIN).   *See* Aureum Vellus.

* SCHREIBEN AN DIE GOLD-BEGIERIGEN LIEBHABER DER CHYMIE UND
    ALCHYMIE, worinnen ihnen wohlmeinend durch ein und andere in
    der gesunden Vernunft und Experience gegründeter Beweiss,
    Ursachen, und Widerlegungen abgerathen wird, dieser Kunst nicht
    länger nachzuhangen, um sich nicht in das äusserste Elend zu
    stürzen, durch einen wahren Verehrer der Wahrheit, und aufrichti-
    gen Liebhaber seines Nächsten. Wer das Abysinische Alphabet
    kennet, kennet meinen Namen ሕ. ሎ. ሕ. Frankfurt und Leipzig,
    zu finden im Kraussischen Buchladen, 1770. pp. 190, 12mo.

S[CHRÖDER, FRIEDRICH JOSEPH WILHELM].
  * Neue alchymistische Bibliothek für den Naturkundigen unsers Jahr-
      hunderts ausgesucht und herausgegeben. Frankfurt und Leipzig,
      1772. 2 vols., 12mo.

[*Contents :*]

VOL. I, PART I.

1. *Dickinson, Edmund :* Schreiben an Herrn Theodor Mundan von
    der Goldkunst oder Quintessenz der Filosofen, aus dem Ox-
    forder lateinischen Exemplar übersetzt. p. 1.

Innhalt der Fragen.

a. Was ist der Merkur der Filosofen ?
b. Was ist die Materie des physischen Steins ?
c. Was ist das geheime Feuer der Filosofen ?
d. Was ist das Gold der Filosofen ?
e. Was sind die Gebirge der Filosofen ?
f. Was ist das Meer der Filosofen ?
g. Was ist das Lebenswasser der Filosofen ?
h. Was ist der Filosofen Diana ?
i. Kann der Stein oder die Quintessenz der Weisen durch men-
    schliche Untersuchung gefunden werden ?
j. Kann die Chemie ein allgemeines Arztneymittel dargeben ?
k. Haben die Patriarchen ihr Leben durch dieses Mittel verlängert ?

S[CHRÖDER, FRIEDRICH JOSEPH WILHELM]. [Cont'd.]

2. *Mundan, Theodor:* Antwort auf vorhergehendes Schreiben. p. 115.

3. *Elias der Artist:* Eine Abhandlung von der künstlichen Metall-verwandlung. p. 181.

PART II.

1. *Schwärzer, Sebald* und andere : Die Metallverwandlungskünste aus den sichersten Urkunden hervorgezogen und mit allgemeinen filosofischen Betrachtungen begleitet von W——. p. 1.

2. *Ferrarius :* Chymische Abhandlung für den Pabst . . . p. 159.

3. Das Buch von den Anfängen der Natur und chymischen Kunst. p. 237.

4. *Drebbel, Kornelius :* Abhandlung von der Quintessenz. Herausge-geben von Joachim Morsius im Jahre 1621. p. 291.

VOL. II, PART I.

1. *Clauder, Gabriel :* Eine Abhandlung von dem Universalsteine, wo insbesondere gegen den Pater Athanasius Kircher die Wirk-lichkeit des Steins der Weisen behauptet wird. p. 1.

2. *Germain, Claude :* Abbildung der geheimen Filosofie ; eine ächte Vorschrift den Stein der alten Weisen zu machen. Nihil est tam absconditam, quod non aliquando manifestum fiat. p. 281.

3. Zwey alte Denkmale deutscher Filosofen von der Alchymie, mit Anmerkungen von R . . . aufs neue herausgegeben. Der erste deutsche Tractat vom Jahre 1423 : Eine wahrhaftige Lehre der Filosofie von Gebährung der Metalle und ihrem rechten Beginne. p. 345.

4. Der andere Tractat. Eines alten deutschen Filosofen poetische Belustigung in Reimen von der geheimen Filosofie der Chy-misten. Mit Anmerckungen herausgegeben von R——. p. 379.

PART II.

1. *Westphalus, Josephus :* Von der Goldtinctur der Weisen. p. 1.

2. *Zalento, Petrus de :* Vom metallischen Kunststücke der Weisen. Aus dem lateinischen übersetzt und mit Anmerkungen be-gleitet von J . . . p. 131.

3. *Baco, Rogerius :* Alchymeyspiegel, oder kurzgefasste Abhandlung der Alchymey. p. 167.

4. *Avicenna :* Kleines Büchelchen vom mineralischen Steine, nebst dem Anfange der Erläuterung desselben vom fysischen Steine. p. 193.

SCHRÖDER, FRIEDRICH JOSEPH WILHELM.

Neue Sammlung der Bibliothek für die höhere Naturwissenschaft und Chemie. Leipzig, 1775–76. 2 vols., 12mo. Vol. II in 2 parts. I, pp. [xii]–748 ; II, pp. 286–438–[xi].

> Of vol. I a second edition was issued in 1779. The volume contains 17 treatises, including the following of interest to the historian :

Hermann Boerhaave, Schutzschrift für die Alchemie. Vol. I, pp. 1–88. Schröder's Geschichte der ältesten Chemie und Filosofie, oder sogenannten Filosofie der Egyptier. Vol. I, pp. 89–412.

SCHRÖDER, WILHELM VON.   See Roth-Scholtz, Fr. (Deutsches Theatrum chemicum).

SCHRÖDER, WILHELM VON.

* Fürstliche Schatz- und Rent-Kammer, nebst seinem nothwendigen Unterricht vom Goldmachen.   Zu finden in Leipzig bey Thomas Fritschen, 1704.   pp. [xxviii]–474–[lx], 12mo.
* Nothwendiger Unterricht vom Goldmachen denen Buccinatoribus oder so sich selbst nennenden Foederatis Hermeticis auf ihre drey Episteln zur freundlichen Nachricht, 1705.   pp. [vi]–70, 12mo.

SCHULZ, ELIA.

* Optische Erquickungen in welchen zu finden allerhand ergötzliche optische Maschinen noch von keinem Author beschrieben, vermöge welcher man erstaunende fast unglaubliche Effecten zuwegen bringet, und die mehresten durch ganz geringe Unkosten angeschaffet werden können.   Allen Kunstliebenden zur Ergötzlichkeit und mehrerer Erkänntnuss der Natur aus selbst eigener Erfahrung verfasset und herausgegeben.   Frankfurt und Leipzig, bey Peter Conrad Monath, 1754.   pp. [x]–30.   Plates.

SCHWÄRZER, SEBALD.   See S[chröder, F. J. W.].

SCHWEITZER, JOHANN FRIEDRICH.   See Helvetius, J. F.

SEILERUS, WENCESLAUS.   See Becher, John Joachim.

SEMLER, JOHANN SALOMON.

Hermetische Briefe wider Vorurtheile und Betrügereien.   Erste Sammlung.   Leipzig, 1788.   pp. 144, 12mo.
* Schreiben an Hrn. Baron Hirschen zur Vertheidung des Luftsalzes, als Anhang zu den drei Stücken von hermetischer Arzenei, worin ein Zeugniss eines königlichen preussischen Officiers.   Leipzig, bey Georg Emanuel Beer, 1788.   pp. 47, 16mo.
Unpartheiische Sammlung zur Geschichte der Rosenkreutzer.   Leipzig, 1786–88.   4 parts, 8vo.
* Von ächter hermetischer Arzenei.   An Herrn Leopold Hirschen in Dresden.   Wider falsche Maurer und Rosenkreuzer.   Leipzig, bey Georg Emanuel Beer, 1786.   pp. 84, 12mo.

Sᴇɴᴅɪᴠᴏɢɪᴜs Pᴏʟᴏɴᴜs, Mɪᴄʜᴀᴇʟɪs.

    See Manget, J. J. ; *also* Theatrum chemicum ; *also* Musæum hermeticum ; *also* Hollandus, Joh. Isaacus : Die Hand der Philosophen.

Sᴇɴᴅɪᴠᴏɢɪᴜs, Mɪᴄʜᴀᴇʟ. [Dɪᴠɪ Lᴇsᴄʜɪ ɢᴇɴᴜs ᴀᴍᴏ.]

  * A new Light of Alchymie, taken out of the fountaine of Nature and Manuall Experience. To which is added a Treatise of Sulphur ; also Nine Books of the Nature of Things, written by Paracelsus, viz., of the Generations, Growths, Conservations, Life, Death, Renewing, Transmutation, Separation, Signatures of Natural Things. Also a Chymical Dictionary explaining hard places and words met withall in the writings of Paracelsus and other obscure Authors. All which are faithfully translated out of the Latin into the English tongue by J. F. M. D. London, Printed by Richard Cotes, for Thomas Williams at the Bible in Little-Britain, 1650. Sendivogius, pp. [xvi]-147-[iii]. Paracelsus, pp. [viii]-145. Chymical Dictionary, pp. [lii]. Sm. 4to.

  * Chymische Schriften ; darinnen gar deutlich von dem Ursprung, Bereit- und Vollendung des gebenedeiten Steins der Weisen gehandelt wird. Nebst einem kurtzen Vorbericht ans Licht gestellet durch Friedrich Roth-Scholtzen. Nürnberg, bey Johann Daniel Taubers seel. Erben, 1718. pp. 250, 16mo. Illustrated title-page.

      * [Another edition :] Verlegts Joh. Paul Krauss, Kayser und Königl. privilegirter Niederlags-Verwandter, Buchhändler in Wienn, 1749. 12mo.

    The edition of 1749 *contains :*

  1. S., M. : Processus super centrum universi seu sal centrale, wie solcher in Hrn Johann Joachim Becher's chymischen Glückshafen zu finden ; nun aber zu Complirung der Sendivogianischen Schrifften hiermit beygefüget worden durch F. Roth-Scholtzen. p. 25.

  2. S., M. : Gespräch zwischen dem Mercurio, einem Alchymisten und der Natur gehalten. p. 113.

  3. S., M. : Ein philosophischer Tractat von dem andern Anfang der natürlichen Dinge dem Schwefel. p. 137.

  4. S., M. : Anhang eines gleichförmigen Gesprächs des Geistes Mercurii mit einem Closter-Philosopho gehalten. Hiehero wegen gleichlautender Materie und zu Ergäntzung des Tractätleins aus einem alten Buch beygefüget. p. 221.

  5. S., M. : Epistolæ quinquaginta quinque quæ, a Johanno Jacobo Mangeto in Bibliotheca curiosa chymiæ amatoribus primo donatæ, nunc Sendivogianis operibus annectuntur curæ Friderici Roth-Scholtzen. p. 237.

SENDIVOGIUS, MICHAEL. [Cont'd.]

6. *Synesius :* Chymische Schrifften von dem gebenedeyten Stein der
Weisen uud dessen Bereitung wie solche ehemals aus der
Kayserlichen Bibliothek sind communicirt, nun aber zum
Druck befördert worden durch Friedrich Roth-Scholtzen.
p. 349.

7. *Valentinus, Basilius :* Via veritatis, oder der einige Weg zur Warheit, wie er solchen ehemals beschrieben hinterlassen, nun
aber um dessen Fürtrefflichkeit willen denen Liebhabern der
wahren Weissheit zu Dienste zu den Sendivogianischen
Schrifften mit beygefüget durch Friedrich Roth-Scholtzen.
p. 375.

*Fünf und funfzig Briefe, den Stein der Weisen betreffend. Aus dem
lateinischen übersetzt. Frankfurt und Leipzig, in Johann Georg
Fleischers Buchhandlung, 1770. pp. 152, 12mo.

SERMO IN TURBAM PHILOSOPHORUM. *See* Manget, J. J.; *also* Theatrum chemicum.

SETONIUS SCOTUS, ALEXANDER.

*Les Oeuvres du Cosmopolite divisez en trois traitez, dans lesquels sont
clairement expliquer les trois principes des philosophes chymiques,
sel, soûfre et mercure. [Paris, 1691.] pp. [xiv]–333–238, 16mo.

[*Contents :*]

Cosmopolite ou nouvelle lumière chymique pour servir d'éclaircissement
aux trois principes de la nature exactements décrits dans les trois
traitez suivans. Le I Traité : Du Mercure ; le II : Du Soûfre, et
le III : Du vray sel des philosophes. Dernière édition, revûë et
augmentée. Des lettres philosophiques du mesme auteur. À
Paris, chez Laurent d'Houry, rüe S. Jaques, devant la Fontaine S.
Séverin, au S. Esprit, 1691. Avec privilége du Roy. pp. [xiv]–333.

Traitez du Cosmopolite nouvellement découverts ; où après avoir
donné une idée d'une société de philosophes on explique dans
plusieurs lettres de cet autheur la théorie et la pratique des véritez
hermétiques. À Paris, 1691. pp. 238, 16mo.

Published anonymously.

SIEBMACHER, JOHANN. *See* A B C ; *also* Wasserstein der Weisen.

SILENTO, PETRUS DE. (ZALENTINO.) *See* Theatrum chemicum ; *also* S[chröder, F. J. W.].

SINCERUS RENATUS. [RICHTER, SAMUEL.]

* Sämtliche philosophisch- und chymische Schrifften, als :

1. Die wahrhaffte und vollkommene Bereitung des philosophischen
Steins [etc.].

SINCERUS, RENATUS. [Cont'd.]

2. Theo-Philosophia Theoretico-Practica, oder der wahre Grund Göttlicher und natürlicher Erkänntniss.

3. Goldene Quelle der Natur und Kunst, bestehend in lauter Experimentis und chymischen Handgriffen, etc. Leipzig und Bresslau, Verlegts Michael Hubert, 1741. pp. [xv]–750, 12mo.

* SOUFFEURS (LES) OU LA PIERRE PHILOSOPHALE D'ARLEQUIN. Comédie nouvelle comique et satirique. Suivant la copie de Paris. À Amsterdam, chez Adrian Braakman, Marchand Libraire dans le Beurs-Straat, prés Le Dam à l'enseigne de la ville d'Amsterdam, 1695. pp. 128, 24mo. Illustrated title-page.

> The copy in my possession bears a MS. note stating that the author is Michel Chilliat.

SPECULUM SAPIENTIÆ. See Quadratum Alchymisticum.

SPERBERUS, JULIUS. See Roth-Scholtz, Fr. (Deutsches Theatrum chemicum).

SPIEGEL DER PHILOSOPHEY. See Aureum Vellus.

SPIESS. See A B C.

SPLENDOR SOLIS MIT SEINEN FIGUREN. See Aureum Vellus.

SPONHEIM, TRITHEMIUS DE.

* Güldenes Kleinod oder Schatzkästlein. Aus dem lateinischen, um seiner Unschätzbarkeit willen, ins deutsche übersetzt von Fr. Basilio Valentino, zum erstenmale herausgegeben, nebst zwoen andern forne mit angedruckten seltenen Handschriften von Jamimah KoranhapuCH. Leipzig, bey Paul Gotthelf Kummer, 1782. pp. 135, 16mo. Ill.

> According to Schmieder this is a spurious work by an unknown author.

STAHL, GEORG ERNST. See Pyrotechnical Discourses.

STANDING OF THE GLASS (THE). See Ashmole, E. : Theatrum chemicum Britannicum.

STARKEY, GEORGE. [Or STIRK.]

* Pyrotechny asserted and illustrated to be the surest and safest means for arts triumph over natures infirmities. Being a full and free discovery of the medicinal mysteries studiously concealed by all artists, and onely discoverable by fire. With an Appendix concerning the nature, preparation and virtue of several specifick medicaments, which are noble and succedaneous to the great arcana. London, printed by R. Daniel, for Samuel Thomson at the White Horse in S. Paul's Church-yard, 1658. pp. [xviii]–172, 12mo.

STARKEY, GEORGE. [Cont'd.]

>George Stirk was born in the Bermudas, was graduated at Harvard College in 1646. He went to England and became noted for his chemical medicines. For a list of ten treatises by him, see J. L. Sibley's "Biographical Sketches of Graduates of Harvard," vol. 1, Cambridge, 1873. One of these treatises is found in "Collectanea Chymica," *q. v. Cf.* Wilson, George.

STEINER, PETER. *See* A B C.

STOLCIUS DE STOLCENBERG, DANIEL.

* Viridarium chymicum figuris cupro incisis adornatum et poeticis picturis illustratum ; ita ut non tantum oculorum et animi recreationem suppeditet, sed et profundiorem rerum naturalium considerationem excitet, adhæc forma sua oblonga amicorum alio inservire queat. Francofurti, sumptibus Lucæ Jennissi, 1624. ff. [cxi]– 107 pl. Oblong 16mo.

>One hundred and seven plates, each with a verse from Michael Maier's works and other sources. *Cf.* Manget, J. J.

STREIT (EIN) UND GESPRÄCH DES GOLDS UND MERCURII WIDER DEN STEIN DER WEISEN. *See* Aureum Vellus.

STRUVE, FR. GOTTLIEB.

De jure alchemiæ ; vom Recht der Alchymisten. Jenæ, 1717. 4to.

SUCHTEN, ALEXANDER VON. *See* A B C.

SUCHTEN, ALEXANDER VON.

* Chymische Schrifften alle, so viel deren vorhanden zum erstenmahl zusammengedruckt mit sonderbarem Fleiss von vielen Druckfehlern gesäubert, vermehret und in zwey Theile als die teutschen und lateinischen verfasset. Franckfurt am Mayn, in Verlegung Georg Wolffs, Buchh. in Hamburg, Druckts Johann Görlin, 1680, pp. [xiv]–486–[viii], 12mo. Frontispiece. [Edited by Dagitza.]

[Contents :]

1. Concordantia chymica, id est : Eine Vergleichung etlicher philosophischen Schrifften von Bereitung dess philosophischen Steins, wie solche Würckung mit der Natur übereinstimmet und sich damit vergleichet. Zu besserer Explication Manualis, Tincturæ Philosophorum, Apocalypsis et libri vexationum Theophrasti Paracelsi an Tag gegeben. p. 1.

2. Colloquia chymica, das ist : Allerhand freundliche, lustige und hochnützliche Gespräche, so nicht allein den Lapidem philosophicum betreffen, sein recht Fundament, Gebrauch und Nutzen, sondern auch alle andere Arcana und Mysteria auslegen, klärlich darthum und demonstriren. p 161.

SUCHTEN, ALEXANDER VON. [Cont'd.]

3. Vom Antimonio oder Spiessglass, des edlen und hochgelährten Alexanders von Suchten. p. 229.

4. De Antimonio vulgari, an den edlen Johann Baptista von Seebach geschrieben. p. 267.

5. Dialogus introducens duas personas interlocutrices, Alexandrum et Bernhardum. p. 305.

6. De tribus facultatibus. p. 357.

7. Explicatio Tincturæ Physicorum Theophrasti Paracelsi. p. 383.

8. De vera medicina. p. 458.

9. Elegia. p. 487.

* Mysteria gemina Antimonii, das ist: Von den grossen Geheimnüssen dess Antimonii. In zwey Tractat abgetheilt, deren Einer die Artzeneyen zuaufallenden menschlichen Kranckheiten offenbaret. Der ander aber, wie die Metallen erhöhet, und in Verbesserung übersetzet werden. Mit mancherley künstlichen und philosophischen beyderseits derselbigen Bereitungen, exempelweise illustrirt und zu Vindicirung seines Lobs und Ruhms publicirt worden durch Johann Thölden Hessum. Anjetzo auffs neue übersehen mit einem vollständigen Register vermehret. Mit Röm. Kaiserl. Majest. und Chur-Fürstl. sächsischem Privilegio ; Nürnberg, in Verlegung Paul Fürsten Kunst- und Buchh. S. Wittib und Erben, *n. d.* [1680?]. pp. [vi]-380-[xxviii].

SUGGESTIVE (A) ENQUIRY INTO THE HERMETIC MYSTERY, with a dissertation on the more celebrated of the alchemical philosophers, being an attempt towards the recovery of the ancient experiment of nature. London, 1850. 8vo.

SUPER TRACTATULUM MER : fugi dum bibit. *See* Theatrum chemicum.

SYNESIUS. *See* R[ichebourg], J. M. D. ; *also* Sendivogius, Michaelis : Schriften ; *also* Flamellus, N. : Chymische Werke ; *also* Trois traictez de la philosophie.

TACHENIUS, OTTO.

* Clavis to the ancient Hippocratical Physick or Medicine made by manual experience in the very fountains of nature, whereby through fire and water, in a method unheard of before, the occult mysteries of nature and art are unlocked and clearly explained by a compendious way of operation. London, printed by Tho. James, and are to be sold by Nath. Crouch in Exchange-Alley over against the Royal Exchange in Corn-Hill, 1677. pp. [xiv]-120-[xiii], sm. 4to.

* Hyppocrates Chymicus, discovering the ancient foundation of the late viperine salt, with his clavis thereunto annexed. Translated by J.

TACHENIUS, OTTO. [Cont'd.]

W. London, printed and are to be sold by Nath. Crouch at the George at the lower end of Cornhill over against the Stocks Market, 1677. pp. [xxii]–122–[ix], sm. 4to.

* TASCHENBUCH FÜR ALCHEMISTEN, Theosophen und Weisensteinsforscher die es sind und werden wollen. Leipzig, bey Christian Gottlob Hilscher, 1790. pp. [xvi]–342, 16mo.

* TEXTE (LE) D'ALCHYMIE ET LE SONGE-VERD. À Paris, chez Laurent d'Houry, ruë S. Jacques, devant la Fontaine S. Séverin au Saint Esprit, 1695. Avec privilége du Roy. Quatre parties. pp. 115, 16mo. Plate.

THEATRUM CHEMICUM, præcipuos selectorum auctorum tractatus de chemiæ et lapidis philosophici antiquitate, veritate, jure, præstantia et operationibus continens ; in gratiam veræ chemiæ et medicinæ chemicæ studiosorum (ut qui uberrimam inde optimorum remediorum messem facere poterunt) congestum et in quatuor partes seu volumina digestum ; Singulis voluminibus suo auctorum et librorum catalogo primis pagellis, rerum vero et verborum indice postremis, annexo. Argentorati, sumptibus Lazari Zetzneri, 1613–22. 5 vols., 12mo. Vol. I, pp [viii]–869 and index ; vol. v, pp. [viii]–1009 and index.

Second edition under the title :

* Theatrum chemicum, præcipuos selectorum auctorum tractatus de chemiæ et lapidis philosophici antiquitate, veritate, jure, præstantia et operationibus, continens ; in gratiam veræ chemiæ et medicinæ chemicæ studiosorum (ut qui uberrimam inde optimorum remediorum messem facere poterunt) congestum, et in sex partes seu volumina digestum ; singulis voluminibus suo auctorum et librorum catalogo primis pagellis, rerum vero et verborum indice postremis, annexo. Argentorati, sumptibus Heredum Eberh. Zetzneri, 1659–61. 6 vols., 8vo. Vol. I, [vi]–794–[xxx] ; II, 549–[vii] ; III, 859–[xiii] ; IV, [viii]–1014–[xxxiii] ; v, [viii]–912–[xxix].

Vol. VI *bears the title :*

Theatri chemici volumen sextum, theologis, medicis et tam vulgaribus quam hermeticæ chemiæ studiosis utilissimum, præcipuos selectorum auctorum huius seculi tractatus de chemia et lapidis philosophici antiquitate, veritate jure, præstantia et operationibus continens, ex Germanica et Gallica lingua in Latinam translatum per Johannem

Theatrum chemicum. [Cont'd.]
 Jacobum Heilmannum. Argentorati, sumptib. Hære-
 dum Eberhardi Zetzneri, 1661. pp. [xviii]–772[xxv], 8vo.

[*Contents :*]

VOL. I.

[1] *Vallensis, Robertus :* De veritate et antiquitate artis chemicæ. p. 7.
[2] *V., R. :* De pulveris seu medicinæ philosophorum ac auri potabilis, mira, naturali, occulta ac divina vi, actione et operatione
 in tria rerum genera, animale, vegetale et minerale. p. 22.
[3] *Arnaldus à Villa Nova :* Testamentum. p. 28.
[4] *Fanianus, Chrysippus Joannes :* De arte metallicæ metamorphoseos. p. 33.
[5] *F., C. J. :* De jure artis alchemiæ, hoc est : Variorum authorum
 et præsertim jurisconsultorum judicia et responsa ad quæstionem quotidianam : An alchimia sit ars legitima. p. 49.
[6] *Muffetus, Thomas Anglus :* De jure et præstantia chemicorum
 medicamentorum. p. 70.
[7] *M., Th. A. :* Quinque epistolæ medicinales ad medicos conscriptæ. p. 89.
[8] *Hoghelande, Theobaldus de, Mittelburgensis :* De alchimiæ difficultatibus ; in quo libro docetur, quid scire quidque vitare
 debeat veræ chemiæ studiosus ad perfectionem aspirans.
 p 109.
[9] *Dornæus, Gerardus :* Clavis totius philosophiæ chemisticæ, per
 quam potissima philosophorum dicta referantur. p. 192.
 a. Libri artis chemisticæ. p. 195.
 b. Philosophia speculativa. p. 228.
 c. De artificio supernaturali. p. 277.
[10] *D., G. :* Tractatus de naturæ luce physica ex genesi desumpta.
 p. 326.
 a. Physica genesis. p. 331.
 b. Physica Trismegisti. p. 362.
 c. Physica Trithemii. p. 388.
 d. Philosophia meditiva. p. 399.
 e. Philosophia chemica. p. 418.
[11] *D., G. :* Tractatus de tenebris contra naturam, et vita brevi.
 p. 457.
[12] *D., G. :* Congeries Paracelsicæ chemiæ de transmutationibus
 metallorum. p. 491.
[13] *D., G. :* De genealogia mineralium ex Paracelso. p. 568.
[14] *Penotus, Bernardus Georgius :* Tractatus varii de vera præparatione et usu medicamentorum chemicorum. p. 592.

THEATRUM CHEMICUM. [Cont'd.]

[15] *Trevisanus, Bernardus :* Liber de alchemia.   p. 683.

[16] *Zacharius, Dionysius :* Opusculum philosophiæ naturalis metallorum.   p. 710.

[17] *Flammellus, Nicolaus :* Annotationes.   p. 748.

VOL. II.

[1] *Gasto, Claveus :* Apologia argyropœiæ et chrysopœiæ.   p. 6.

[2] *Ægydius de Vadis :* Dialogus inter naturam et filium philosophiæ.   p. 85.

[3] *Æ. d. V. :* Tabula, diversorum metallorum vocabula, quibus usi sunt veteres ad artem celandam, explicans.   p. 108.

[4] *Ripleus, Georgius :* Duodecim portarum axiomata philosophica.   p. 110.

[5] *Albertus Magnus :* Compendium de ortu et metallorum materia supra quam spagyricus radicalia principia fundet.   p. 123.

[6] *Hollandus, Isaacus :* Fragmentum de lapide philosophorum.   p. 126.

[7] *Penotus, Bernardus :* Questiones et responsiones philosophicæ.   p. 129.

[8] *P., B. :* Regulæ seu canones philosophici quinquaginta septem.   p. 133.

[9] *P., B. :* Vera extractio mercurii sive argenti vivi ex auro.   p. 137.

[10] *P., B. :* Chrysorrhoas, sive de arte chemica dialogus, qui præclarissimas hujus scientiæ actiones rationi ac rationalibus principiis consentaneas esse demonstrat, impostorum vero et sycophantarum somnia et nugas detegit.   p. 139.

[11] *Quercetanus, Josephus Armeniacus :* Brevis responsio ad Jacobi Auberti Vendonis de ortu et causis metallorum contra chemicos explicationem ; ac primum ad ejus epistolam convitiatoriam, qua Paracelsicorum, quos vocat, nonnulla remedia evertere conatur.   p. 150.

[12] *Dee, Joannes, Londinensis :* Monas hieroglyphica, mathematice, magice, cabalistice explicata.   p. 192.

[13] *Ventura, Laurentius :* Liber de ratione conficiendi lapidis philosophici.   p. 215.

[14] *Mirandula, Joannes Franciscus :* Opus aureum de auro tum æstimando, tum conficiendo, tum utendo.   p. 312.

[15] *Baco, Rogerius :* Libellus de alchemia, cui titulum fecit, speculum alchemiæ.   p. 377.

[16] *Richardus Anglicus :* Libellus utilissimus περὶ χημείας, cui titulum fecit, Correctorium.   p. 385.

[17] *R. A. :* Rosarius minor de rerum metallicarum cognitione.   p. 406.

Theatrum chemicum. [Cont'd.]

[18] *Albertus Magnus :* De alchemia. p. 423.

[19] *Pantheus, Joannes Augustinus :* Ars et theoria transmutationis metallicæ, cum Voarchadumia proportionibus, numeris et iconibus rei accommodis illustrata. p. 459.

VOL. III.

[1] *Incertus auctor :* De magni lapidis compositione et operatione. p. 5.

[2] *Incertus auctor :* De eadem materia capita aliquot quæ in secunda editione ob incuriam ad calcem volumini tertii rejecta hic debito loco restituuntur. p. 53.

[3] *Aristoteles :* De perfecto magisterio. p. 76.

[4] *Arnaldus de Villa Nova :* Lumen luminum seu flos florum. p. 128.

[5] *A. de V. N.:* Practica. p. 137.

[6] *Efferarius Monachus :* De lapide philosophorum secundum verum modum formando. p. 143.

[7] *E. M.:* Thesaurus philosophiæ. p. 151.

[8] *Lullius, Raymundus :* Praxis universalis magni operis. p. 165.

[9] *Odomarus :* Practica. p. 166.

[10] *O. :* Historiola antiqua de argento in aurum verso. p. 170.

[11] *O. :* Tractatus de marchasita. p. 173.

[12] *O. :* De arsenico. p. 177.

[13] *O. :* De sale alchali. p. 180.

[14] *O. :* An lapis philosophorum valeat contra pestem. p. 181.

[15] *O. :* Vetus epistola de metallorum materia et artis imitatione. p. 187.

[16] *Rupescissa, Joannes de :* De confectione veri lapidis philosophorum. p. 189.

[17] *Augurellus, Joannes Aurelius :* Chrysopœia carmine conscripta. p. 197.

[18] *A., J. A.:* Carmen ad Petrum Lipomanum (Geronticon). p. 244.

[19] *Aquinas, Thomas :* Secreta alchemiæ magnalia : De corporibus supercœlestibus, quod in rebus inferioribus inveniantur, quoquomodo extrahantur : De lapide animali et plantali. p. 267.

[20] *A., Th. :* Tractatus in arte alchemiæ fratri Reinaldo. p. 278.

[21] *Rupescissa, Johannes de :* Liber lucis. p. 284.

[22] *Lullius, Raymundus :* Clavicula quæ apertorium dicitur in, quo omnia quæ in opere alchemiæ requiruntur venuste declarantur, et sine quo, ut ipse testatur Lullius, alii sui liberi intelligi nequeunt. p. 295.

THEATRUM CHEMICUM. [Cont'd.]

[23] *Hollandus, Isaacus :* Opus minerale sive de lapide philosophica. p. 304.

[24] *Vogelius, Evaldus Belga :* Liber de lapidis physici conditionibus ; quo abditissimorum auctorum Gebri et Raymundi Lullii methodica continetur explicatio. [Septem capita.] pp. 515–649.

[25] *Balbian, Justus à :* Tractatus septem de lapide philosophorum. p. 649.

[26] *Grevverus, Jodocus :* Secretum nobilissimum et verissimum. p. 699.

[27] *Alanus philosophus :* Dicta de lapide philosophico. p. 722.

[28] *Barnaudus, Nicolaus :* Commentariolum in quoddam epitaphium Bononiæ studiorum, ante multa secula marmoreo lapidi insculptum. p. 744.

[29] *B., N. :* Processus chemici aliquot. p. 755.

[30] Triga chemica.
  a. Libellus *Lambsprinck* nobilis Germani philosophi. p. 765.
  b. Antiqus philosophus Gallus : Liber secreti maximi totius mundanæ gloriæ. p. 774.
  c. Extractum ex cimbalo aureo, antiquissimo libro manuscripto ad rem nostram faciens. p. 781.

[31] Quadriga aurifera.
  a. Prima rota : Tractatus de philosophia metallorum. p. 79¹.
  b. Secunda rota : *Georgei Riplei* liber duodecim portarum, nequaquam mutilus sed integer. p. 797.
  c. Tertia rota : *Georgii Ripleii* liber de mercurio et lapide philosophorum. p. 821.
  d. Quarta et ultima rota : Cum tabula de cœlo philosophorum. p. 828.

[32] Auriga chemicus sive theosophiæ palmarium. p. 834.

[33] Epistola de occulta philosophiæ. p. 859.

[34] Paucula dicta sapientum. p. 859.

VOL. IV.

[1] *Lullius, Raymundus :* Theorica. p. 1.

[2] *L., R. :* Practica. p. 135.

[3] *L., R. :* Compendium animæ transmutationis artis metallorum. p. 171.

[4] *Artefius :* Clavis majoris sapientiæ. p. 198.

[5] *Helias, Artista :* Nova disquisitio. p. 214.

[6] *Zanetus, Hieronymus :* Conclusio et comprobatio alchymiæ qua dispositioni et argumentis Angeli respondetur. p. 247.

THEATRUM CHEMICUM. [Cont'd.]

[7] *Happelius, Nicolaus Niger :* Cheiragogia Heliana, de auro philosophico, necdum cognito. p 265.

[8] *Lavinus, Wenceslaus Moravus :* Tractatus de cœlo terrestri. p. 288.

[9] *Happelius, Nicolaus Niger :* Aphorismi Basiliani sive canones Hermetici, de spiritu, anima et corpore medio, majoris et minorls mundi. p. 327.

[10] *Brentzius, Andreas :* Variæ philosophorum sententiæ perveniendi ad lapidem benedictum. p. 333.

[11] *Gasto, Claveus (Dulco):* De triplici præparatione auri et argenti. p. 372.

[12] *G., C. (D.)* : De recta et vera ratione progignendi lapidis philosophici, seu salis argentici et aurifici. p. 388.

[13] [*Sendivogius*] : Tractatus duodecim de lapide philosophico. p. 420.

[14] [*Sendivogius :*] Ænigma philosophorum ad filios veritatis. p. 442.

[15] Dialogus Mercurii, alchymistæ et naturæ. p. 447.

[16] *Aureliæ* occultæ philosophorum partes duæ. p. 462.

[17] *Arnaldus de Villa Nova :* Speculum alchymiæ. p. 515.

[18] *A. de V. N. :* Carmen. p. 542.

[19] *A. de V. N. :* Quæstiones essentiales quam accidentales ad Bonifacium octavum. p. 544.

[20] *Anonymos :* Tractatus de secretissimo antiquorum philosophorum arcano. p. 554.

[21] Viginti duo propositiones sive maximæ, in quibus veritas totius artis chemicæ brevissime comprehenditur. p. 577.

[22] *Lasnioro, Joannes de :* Tractatus aureus de lapide philosophorum. p. 579.

[23] *Trithemius, Joannes :* Tractatus chemicus nobilis. p. 585.

[24] *Trismegistus, Hermes :* Tractatus aureus de lapidis physici secreto in capita septem divisus, nunc vero a quodam anonymo scholiis illustratus. p. 592.

[25] *Lagneus, David :* Harmonia seu consensus philosophorum chemicorum magno cum studio et labore in ordinem digestus. p. 718.

[26] *Albertus Magnus :* De concordantia philosophorum in lapide philosophico. p. 809.

[27] *A. M. :* Compositum de compositis. p. 825.

[28] *A. M. :* Liber octo capitulorum de lapide philosophorum. p. 841.

[29] *Avicenna :* Epistola ad Hasen regem de re recta. p. 863.

[30] *Avicenna :* Declaratio lapidis physici filio suo Aboali. p. 875.

[31] *Avicenna :* De congelatione et conglutinatione lapidum. p. 883.

THEATRUM CHEMICUM. [Cont'd.]

[32] *Ticinensis, Guillelmus :* Lilium de spinis evulsum. p. 887.

[33] *Ortholanus :* Practica vera alchymiæ Parisiis probata et experta sub anno Domini millesimo trecentesimo quinquagesimo octavo. p. 912.

[34] Lumen juvenis experti novum, id est Tractatus in quo nominat lapidem philosophorum. p. 934.

[35] *Valentinus :* Opus præclarum ad utrumque quod protestamento dedit filio suo adoptivo, qui etiam istum tractatulum propria manu scripsit. p. 941.

[36] *Incertus autor :* Tractatulus super hæc verba : Studio namque florenti. p. 955.

[37] Opus ad album. p. 957.

[38] *Aquinas, Thomas :* Liber Lilii benedicti. p. 960.

[39] *Innominatus author :* Super tractatulum mer : fugi dum bibit. p. 974.

[40] *Anonymus :* Breve opus ad rubeum cum sole per aquas fortes. p 984.

[41] *Silento, Petrus de :* Opus. p. 985.

[42] *Anonymus :* Tractatus ad album et rubrum. p. 1001.

[43] *Eck de Sultzbach, Paulus :* Clavis philosophorum. p. 1007.

VOL. V.

[1] *[Arislæus] :* Turba philosophorum ex antiquo manuscripto codice excerpta, qualis nulla hactenus visa est editio. p. 1.

[2] *Anonymus :* Sermo unus in turbam philosophorum. p. 52.

[3] Allegoriæ sapientum, et distinctiones XXIX supra librum turbæ ex eodem manuscripto. p. 52.

[4] *Micreris :* Tractatus suo discipulo Mirnefindo. p. 57.

[5] *Plato :* Quattuor libri, cum commento Hebuhabes Hamed explicati ad Hestole. p. 101.

[6] *Kalid Rachaidibus Rex :* Liber trium verborum. p. 186.

[7] *Zadith Senior filius Hamuelis :* Tabula chimica, marginalibus adaucta. p. 191.

[8] *Meneus, Guilielmus :* Libri tres aurei velleris. p. 240.

[9] *Anonymus :* Consilium conjugii, seu de massa solis et lunæ. p. 429.

[10] *Petrus Bonus Lombardus :* Margarita pretiosissima. p. 507.

[11] *Scotus, Michael :* Quæstio curiosa de natura solis et lunæ. p. 713.

[12] *Rhodargirus, Lucas Eutopiensis :* Pisces Zodiaci inferioris, vel de solutione philosophica, cum ænigmatica totius lapidis epitome. p. 723.

[13] *Alphonsus, Rex Castellæ :* Liber philosophiæ occultioris, præcipue metallorum, profundissimus. p. 766.

THEATRUM CHEMICUM.   [Cont'd.]

cognitione aquarum mineralium et de illarum qualitatibus et virtutibus antehac incognitis, et de spiritu universali. p. 716.

" Les curieux et les amateurs conviennent que s'il y a beaucoup de bons traités dans ce théatre chimique, il s'en trouve aussi un grand nombre qui sont de peu de conséquence, et où l'on voit des procédés sophistiques."—*Lenglet du Fresnoy.*

THEATRUM CHIMICUM, ofte geopende deure der chymische verborgentheden.   Ontsloten van de vermaartste Autheuren, die in de chymische stoffe gelaboreert en geschreven hebben, met groote olyd door een liefhebber der chymie by een gesamelt.   Als Schroderus, Angelus Sala, Rolfinkius, le Febure, Crollius, Charras, Beguinus, als meer andere hedendaagse schryvers.   Met een vervolg over de chymische verborgentheden aangaande de verandering en verbetering der metalen en gesteenten.   Door den Ridder K. Digby. Amsterdam, 1693. pp. [xvi]–490–[xxii]–170–[vi], 8vo. With an additional engraved title-page.   The author signs the preface N. N.

* THEORETISCH-PRAKTISCHES HANDBUCH DER HÖHEREN CHIMIE, in welchem alle zu wissen nöthige Lehrsätze der Philosophen sistematisch vorgetragen und mit nützlichen Erläuterungen versehen sind.   Herausgegeben von einem Liebhaber und Schüler der geheimen Weisheit. Hof, in der Vierlingeschen Buchhandlung, 1784. pp. [vi]–271, 12mo.

* THEORETISCH UND PRAKTISCHER WEGWEISER ZUR HÖHERN CHEMIE. Ausgefertiget von einem Liebhaber der geheimen Physik und chemisch-physikalischer Wahrheiten.   Bresslau und Leipzig, bey Christian Friedrich Gutsch, 1773. pp. [xviii]–206.

[*Contents :*]

1. Eines unbekannten Philosophi und wohlerfahrnen Artisten hinterlassene gründliche Beschreibung von denen Particular- und Universal-Tincturen, aus den wahren Fundamenten der Natur nebst übereinstimmigen Zeugniss etlicher wahren Philosophen, ausführlich zusammengetragen und verfasset im 128sten Jahre seines Alters. N. de Tr. E. ad S. Michaël. Anno 1590. Nunmehro aber zum Druck ausgefertiget, und mit einigen Anmerkungen erläutert. p. 1.

2. Das zweite Buch de via universali, wie sowohl im trocknen Wege als auch im nassen Wege die grosse Mineral-Tinctur zu erlangen. p. 67.

3. *Gallus, Friedrich :* Reise nach der Einöde Sanct Michael und wie er sich daselbst Anno 1602 mit einem Adepto in Unterredung eingelassen. p. 121.

THURNEISSER ZUM THURN, LEONHARD. *See in Section V.*

TICINENSIS, GUILLELMUS. *See* Theatrum chemicum.

TICINENSIS, JOANNES. *See* Roth-Scholtz, Fr. (Deutsches Theatrum chemicum).

TIFFEREAU, C. THÉODORE.

Les métaux sont des corps composés. La production artificielle des métaux précieux est possible, est un fait avéré. Suivi de Paracelse et l'alchimie au XVIe siècle par M. Franck. Vaugirard, Paris, 1855. pp. xxii–114, 12mo.

Deuxième édition revue et corrigée. Paris, 1856.

> In six brief memoirs presented to the French Academy of Sciences in 1854–55 the author claims to have discovered a method of converting silver into gold. He had made his experiments in Mexico at great expense, supporting himself meanwhile by taking daguerreotypes. His process was repeated at the Imperial Mint, Paris, before the assayer M. Levol, but with little success. The operations are fully described. Tiffereau is probably the latest claimant in a public way to the hermetic mystery. In the second part M. Franck defends Paracelsus from the charge of charlatanism and regards him as one of the greatest geniuses of an age abounding in men of superior intellectual powers. *Cf.* Lermina, Jules, for a new edition of Tiffereau's work.

TINCTURA PHYSICA. *See* Aureum Vellus.

TINCTUR (EIN) EINES UNBEKANDTEN PHILOSOPHI. *See* Aureum Vellus.

TOELTIUS, J. G.

* Cœlum reseratum chymicum, oder philosophischer Tractat worinne nicht allein die Materien und Handgriffe, woraus und wie der Lapis Philosophorum in der Vor- und Nach-Arbeit zu bereiten, sondern auch wie aus allen vier Reichen der Natur, als astral-, animal-, vegetabil- und mineralischen Reiche, vortreffliche und unschätzbare Tincturen und Medicamenta sowohl zu Erhaltung der Gesundheit und des Lebens als auch Verbesser- und Transmutirung der unvollkommenen Metallen zu verfertigen, offenhertzig gezeiget wird. Mit Figuren denen Liebhabern der wahren Hermetischen Philosophie zu Liebe ausgefertiget von einen Kenner derselben. Franckfurt und Leipzig, Druckts und Verlegts Carl Friedrich Jungnicols hinterlassene Wittwe in Erffurt, 1737. pp. [xiv]–337, 12mo. Ill.

TOLETANUS. *See* Manget, J. J.

TRACTÄTLEIN (EIN) WELCHES VON DEN PHILOSOPHISCHEN SCHRIFFTEN HANDELT. *See* Dreyfaches hermetisches Kleeblatt ; *also* Besondere Geheimnisse.

TRACTATULUS AD ALBUM ET RUBRUM. *See* Theatrum chemicum.

TRACTATULUS SUPER HAEC VERBA : STUDIO NAMQUE FLORENTI. *See* Theatrum chemicum.

TRACTATUS, DARINNEN DAS GANZE SECRET DER ALCHYMEY VON DEM STEIN DER WEISEN
     BEGRIFFEN IST. *See* Aureum Vellus.

TRACTATUS EXIMIUS DE LAPIDE PHILOSOPHICO. *See* Musæum hermeticum.

* TRAITÉ DE CHYMIE PHILOSOPHIQUE ET HERMÉTIQUE, enrichi des opéra-
     tions les plus curieuses de l'art. À Paris, rue de la Harpe, chez
     Charles-Maurice d'Houry seul Imprimeur de Monseigneur le Duc
     d'Orléans, 1725. Avec approbation et privilége du Roy. pp. 292–
     [iv], 16mo.

TRAITÉ D'UN PHILOSOPHE INCONNU. *See* R[ichebourg], J. M. D.

TRAITÉ DU SECRET DE L'ART PHILOSOPHIQUE OU L'ARCHE OUVERTE, autrement dite la
     cassette du pétit paysan. *See* R[ichebourg], J. M. D.

TRAMES FACILIS et planus ad auream Hermetis arcem recta producens. *See* Manget, J. J.

TREUHERTZIGE WARNUNGS-VERMAHNUNG.
     See Roth-Scholtz, Fr. (Deutsches Theatrum chemicum).

TREVIRENSIS, BERNARDUS. *See* Trevisanus, Bernardus.

TREVISANUS, BERNARDUS.
     See Manget, J. J. ; *also* Theatrum chemicum ; *also* R[ichebourg], J. M. D.;
     also Artis auriferæ ; *also* Collectanea chymica ; *also* Hermetischer
     Rosenkranz ; *also* Divers traitez de la philosophie naturelle ; *also* Trois
     anciens traictez de la philosophie naturelle.

TREVISANUS, BERNHARDUS.
   * Chymische Schrifften von dem gebenedeyten Stein der Weisen ; aus
     dem lateinischen ins teutsche übersetzt, ingleichen mit des Joachim
     Tanckens und anderer Gelehrten Anmerckungen ans Liecht ge-
     stellet durch Caspar Horn. Nürnberg, verlegts Johann Paul
     Krauss, Buchhändler in Wien nächst der Kayserl. Burg, 1747. pp.
     [xcviii]–390–[iv], 8vo. Frontispiece.
          Contains also : Dicta Alani ; and Metallurgia, von einem Philosopho Her-
          metico beschrieben und publiciret durch J. Tanckium.

TRIPUS AUREUS. *See* Musæum hermeticum.

* TRIPUS CHIMICUS SENDIVOGIANUS, dreyfaches chimisches Kleinod ;
     das ist zwölff Tractätlein, von dem philosophischen Stain der
     alten Weisen, in welchem desselbigen ursprung, beraitung, und
     vollendung, so hell und klaar, aus dem Licht der Natur erwisen
     und dargethan wirdt, desgleichen von keinem Authoren vorgehends
     jehmahlen beschehen. II. Ein artlich und sinnreiches Gespräch

TRIPUS CHIMICUS SENDIVOGIANUS. [Cont'd.]

eines Alchymisten, mit dem Mercurio, und der Natur, darinnen, das aller verborgenste gehaimniis dess Stains, der Philosophen Mercurius mit eigentlichen bekandtlichen Farben abgemahlet und aussgestrichen wird. III. Ein Tractat und Gespräch von Schwefel, dem anderen Hauptstuck der Tinctur, welches die aller haimlichsten Mysterien der Natur entdecket, und offenbaret. Erstlichen von einem hocherleuchten, sehr gelehrten unnd wohlerfarnen Philosopho lateinisch beschrieben. Nun aber teutscher Nation zu Ehr, den Kunst Liebhabern zu Lehr, den irrenden zu Wehr unnd abkehr, zum erstenmal fleissig verteutschet durch *Hisaiam sub cruce, Ath.* Strassburg, in Verlegung Lazari Zetzners Seligen Erben, im Jahr 1628. pp. [xxii]-190, 16mo.

TRISMEGISTUS, HERMES.

See A B C ; *also* Theatrum chemicum ; *also* R[ichebourg], J. M. D. ; *also* Manget, J. J. ; *also* Musæum hermeticum ; *also* Bacon, Roger : Le miroir d'alquimie ; *also* Trois anciens traictez de la philosophie naturelle.

TRISMEGISTUS, HERMES.

* Einleitung ins höchste Wissen, von Erkenntniss der Natur und des darin sich offenbarenden grossen Gottes. Begriffen in siebenzehn Büchern nach griechischen und lateinischen Exemplaren ins deutsche übersetzt, nebst Nachricht von der Person des Hermetis dessen Medicin, Chemie, Natur- und Gottesgründe ; mit der Egyptier wundervoller Weisheit und Enthüllung der Geheimnisse der smaragdinischen und bembinischen Tafel. Verfertiget von Alethophilo, 1768. Stuttgart, 1855. pp. 256, 16mo.

* Wahrer alter Naturweg, oder Geheimniss wie die grosse Universaltinctur ohne Gläser auf Menschen und Metalle zu bereiten. Herausgegeben von einem ächten Freymäurer J. C. H. Mit vier Kupfern. Leipzig, bey Adam Friedrich Böhme, 1782. pp. [viii]-100, 12mo. Plates.

TRISMOSINUS, SOLOMON. *See* Aureum Vellus.

TRITHEMIUS, JOANNES. *See* Theatrum chemicum.

* TROIS ANCIENS TRAICTEZ DE LA PHILOSOPHIE NATURELLE. I. Les sept chapites dorez, ou bien les sept sceaux Égyptiens et la table d'esmeraude d'Hermes Trismegiste. II. La response de Messire Bernard Conte de la Marche Trevisane, à Thomas de Boulongue, médecin du Roy Charles huictiesme. III. La Chrysopée de Jean Aurelle Augurel, qui enseigne l'art de faire l'or. Les deux premiers n'ont encore esté traduits en François et le troisiesme est corrigé

TROIS ANCIENS TRAICTEZ, etc. [Cont'd.]

des fautes survenuës en la précédente impression, par Gabriel Joly. À Paris, chez Charles Hulpeau, demeurant à la ruë Dauphine à l'Escharpe royale et en sa boutique sur le Pont-neuf proche les Augustins, 1626. Avec privilége du Roi. pp. [viii]–89–130, 12mo.

The third treatise has an independent title-page :

Augurel, Jean Aurelle : Les trois livres de la chrysopée, c'est a dire de l'art de faire l'or : contenant plusieurs raisons et choses naturelles. Traduict de Latin en François par F. Habert de Berry. Revu et corrigé de nouveau. Paris, 1626. pp. 130.

* TROIS TRAICTEZ DE LA PHILOSOPHIE NATURELLE, non encore imprimez. Sçavoir : Le secret livre du très ancien philosophe Artephius, traictant de l'art occulte et transmutation métallique Latin François. Plus les figures hierogliphiques de Nicolas Flamel, ainsi qu'il les a mises en la quatriesme arche qu'il a bastie au Cimetière des Innocens à Paris, entrant par la grande porte de la ruë S. Denys, et prenant la main droite, avec l'explication d'icelles par iceluy Flamel. Ensemble le vray livre du docte Synesius, Abbé Grec, tiré de la Bibliothèque de l'Empereur sur le mesme sujet, le tout traduict par P. Arnauld, sieur de la Chevallerie Poictevin. À Paris, chez Jaques D'Allin, ruë saint Jaques, à l'Image Saint Estienne, 1659. pp. 98, sm. 4to. Plate.

TRUE LIGHT OF ALCHEMY, containing (1) A correct edition of the " Marrow of Alchemy "; (2) The Errors of a late Tract called " A Short Discourse of the Quintessence of Philosophers "; (3) Method and Materials composing the Sophick Mercury and Transmuting Elixir. London, 1709. 16mo.

TURBA PHILOSOPHORUM.

See Artis auriferæ, vol. 1 ; *also* Manget, J. J., vol. 1 ; *also* Theatrum chemicum, vol. v.

There are two treatises under the above title, one ascribed to Arisleus, who lived about 1140 ; both are found in Artis auriferæ and in Manget.

TURBÆ PHILOSOPHORUM ALIUD EXEMPLAR. *See* Manget, J. J. ; *also* Artis auriferæ.

ULSTADIUS, PHILIPPUS (NIERENBERGENSI).

Cœlum philosophorum, seu de secretis naturæ liber. [Colophon :] Argentorati, arte et impensa Joannis Grienynger anno a virgineo partu MDXXVIII in die sancti Egidij. Fol. 64, 4to.

URALTE RITTERKRIEG UND HERMETISCHE TRIUMPH. *See* A B C.

VALENTINUS, BASILIUS.

See Manget, J. J. ; *also* Theatrum chemicum ; *also* R[ichebourg], J. M. D.; *also* Musæum hermeticum ; *also* Roth-Scholtz, Fr. (Deutsches Theatrum chemicum) ; *also* Sendivogius : Schriften ; *also* Wasserstein der Weisen ; *also* Grasshof, Johann.

VALENTINUS, BASILIUS.

* Chymische Schriften, aus einigen alten Manuscripten aufs fleissigste verbessert, mit vielen Tractaten, auch etlichen Figuren vermehret und nebst einem vollständigen Register in drey Theile verfasset: Samt einer neuen Vorrede von Beurtheilung der alchymistischen Schriften und dem Leben des Basilii begleitet, von Benedicto Nicolao Petræo. Sechste Edition. Leipzig, verlegts Joh. Paul Krauss, Buchhändler in Wien, 1769. pp. [clviii]-1133-[civ], 16mo. Portrait.

* Letztes Testament, darinnen die geheimen Bücher vom grossen Stein der uralten Weisen und andern verborgenen Geheimnüssen der Natur. Auss dem Original, so zu Erffurt in dem hohen Alter, unter einem Marmorsteinen Täfflein gefunden, nachgeschrieben; und nunmehr auff vielfältiges Begehren, den Filiis doctrinæ zu gutem, neben angehengten XII Schüsseln, und in Kupffer gebrachten Figuren, dessen Innhalt nach der Vorrede zusehen, zum Andernmahl ans Licht gebracht. Strassburg, in Verlägung Caspari Dietzels, im Jahr 1651. pp. [xvi]-224, 16mo.

* Of Natural and Supernatural Things, also of the first Tincture, Root and Spirit of Metals and Minerals, how the same are conceived, generated, brought forth, changed, and augmented. Translated out of High Dutch by Daniel Cable. Whereunto is added Alex. Van Suchten of the Secrets of Antimony. Translated out of High Dutch by D. C., a Person of great skill in chymistry. London, printed and are to be sold by Moses Pitt at the White Hart in Little Britain, 1670. Valentinus, pp. 238. Alexander van Suchten, pp. [viii]-80, 16mo.

> The treatise on Natural and Supernatural Things was the first of this author's writings to appear in print; it was published in German by Johann Thölde at Eisleben in 1603.

* Triumphant (The) Chariot of Antimony, being a conscientious discovery of the many real transcendent excellencies included in that mineral. Faithfully Englished and published for the common good, by I. H. Oxon. London, printed for W. S. and are to be sold by Samuel Thomson at the Bishops Head in Pauls Church-Yard, 1661. pp. [iv]-175, 16mo.

* Triumphant (The) Chariot of Antimony with annotations of Theodore Kirkringius. With the true Book of the learned Synesius a Greek Abbot taken out of the Emperour's Library, concerning the Philosophers' Stone. London, printed for Dorman Newman at the Kings Arms in the Poultry, 1678. pp. [xvi]-176, 16mo. Ill.

* Von dem grossen Stein der Uhralten, daran so viel tausend Meister anfangs der Welt hero gemacht haben. Neben angehängten Tractät-

VALENTINUS, BASILIUS.   [Cont'd.]

lein derer Inhalt nach der Vorrede zu finden.   Den Filiis doctrinæ
zu gutem publicirt und jetzo von neuen mit seinen zugehörigen
Figuren in Kupffer ans Licht gebracht.   Strassburg, in Verlegung
Caspari Dietzel, im Jahr 1651.   pp. [viii]–156–[iv], 16mo.   Ill.

> Notwithstanding the many works ascribed to this famous German alchemist,
> doubts are expressed as to his existence.   The first edition of the col-
> lection by Petraeus was published at Hamburg in 1717.   The earliest
> Latin collected works bear date 1700, also at Hamburg.   For numerous
> German editions consult Schmieder's Geschichte der Alchemie.

VALLENSIS, ROBERTUS.   *See* Theatrum chemicum.

VARIA PHILOSOPHICO.   *See* Aureum Vellus.

VAUGHAN, THOMAS.

> *See* Philaletha, Eugenius ; *also* A B C ; *also* Quadratum Alchymisticum.

VAZQUEZ JABLADA, G.   [BISHOP OF OVIEDO.]

Chrysopoiea : being a dissertation on the Hermetical Science.   Wherein
is proved by undeniable arguments, the possibility of making gold
by art, in favour of the Alchymists.   The probability of it, learnedly
discussed and refuted, and such cautions given to those who are
inclined to this study, as will, if diligently attended to, prevent their
falling into fatal errors.   Dedicated to the Venerable Brethren of
the Laudable Order of R. C.   London, 1745.   pp. [vii]–43–[iv], 8vo.

> Stated in the Preface to be written by the Bishop of Oviedo.

VENTURA, LAURENTIUS.

> *See* Theatrum chemicum ; *also* A B C ; *also* Neue Sammlung, etc.

VENUS VITRIOLATA IN ELIXIR CONVERSA.

> *See* Dreyfaches hermetisches Kleeblatt ; *also* Besondere Geheimnisse.

VERÆ ALCHEMIÆ SCRIPTORES, ETC.   *See* Gratarolus, Gul.

VIGENERUS, BLASIUS.   *See* Theatrum chemicum.

VIGILANTIUS DE MONTE CUBITI.   *See* Dreyfaches hermetisches Kleeblatt.

VILLAIN, ABBÉ.   *See* Flamel, Nicolas.

VILLANOVA, ARNALDUS DE.   *See* Arnaldus de Villa Nova.

VOGELIUS, EVALDUS BELGA.   *See* Theatrum chemicum.

VOM ANTIMONIO PHILOSOPHORUM.   *See* Aureum Vellus.

VON OFFENBARUNG DER PHILOSOPHISCHEN MATERIEN UND DINGEN.   *See* Aureum Vellus.

WAITE, ARTHUR EDWARD.

* The real History of the Rosicrucians founded on their own manifestoes
and on facts and documents collected from the writings of initiated
brethren. London, 1887. pp. viii–446, 8vo. Ill.

WARNUNGS VORREDE WIDER DIE SOPHISTEN UND BETRÜGER.

See Roth-Scholtz, Fr. (Deutsches Theatrum chemicum).

WASSERBERG, F. AUGUST VON.

* Chemische Abhandlung vom Schwefel. Wien, bey Johann Paul Krauss,
1788. pp. 375, 12mo.

* WASSERSTEIN DER WEISEN, oder chymisches Tractätlein, darinn der
Weg gezeiget, die Materia genennet und der Process beschrieben
wird, zu dem hohen Geheimniss der Universal-Tinctur zu kommen ;
dabey auch zwey sehr nützliche andere Büchlein der Gleichförmig-
keit und Concordantz wegen angehängt : 1. Johann von Mesung
[*sic*]. 11. Via veritatis, Weg der einigen Wahrheit. Vormahlen
durch Lucas Jennis ausgegeben, nunmehro aber wiederum neu auf-
gelegt, und noch dabei gefüget zwey Responsa von dem F. R. C. so
an etlichen ihro zugethanen abgefertiget. Frankfurt und Leipzig, in
der Fleischerischen Buchhandlung, 1760. pp. 228, 12mo. Frontispiece.

[Contents :]

1. Kurtze Erklärung des Wunder fürtrefflichen Wassersteins der
Weisen, sonsten Lapis Philosophorum genandt. p. 20.

2. *Mesung, Johann von :* Beweiss der Natur, welchen sie den irrenden
Alchymisten thut, indeme sie sich über den Sophisten und
thörichten Kohlenbläser beschweret. p. 135.

3. Gegen-Antwort des Alchymisten, die er der Natur, neben Erkennung seiner Fehl, mit Abbittung und Dancksagung thut. p. 158.

4. *Flamel, Nicolas :* Ein kurzer Tractat genannt summarium philosophicum. p. 177.

5. [*Valentinus, Basilius*] : Via veritatis. Ein sehr schönes, nützliches
und herrliches Tractätlein. p. 189.

6. Prima Responsio des F. R. C. an etliche ihnen zugethanen abge-
lassen. p. 219.

7. Secunda Responsio des B_2. des R. C. p. 227.

My copy contains a manuscript note stating that the Wasserstein der Weisen
was written by Joh. Ambrosius Siebmacher of Nürnberg in the year
1612. *Cf.* A B C.

WEDEL, GEORG WOLFFGANG.

* Introductio in alchimiam. Sumptibus Johannis Bielkii ; Jenæ, litteris
Christophori Krebsii, 1706. pp. [iv]–60, 4to.

WEDEL, GEORG WOLFFGANG.   [Cont'd.]
* Einleitung zur Alchimie aus dem lateinischen ins teutsche
    übersetzet.   Berlin, bey Christoph Gottlieb Nicolai,
    1724.   pp. [xii]-100, 12mo.
Programata de tabula Hermetis smaragdina.   Jenæ, 1704.   4to.
* Vernünfftige Gedancken vom Gold-Machen nebst einer Vorrede
    Christian Gottfried Stentzels.   Zweyte Auflage.   Wittenberg, bey
    Carl Siegemund Henningen, 1734.   pp. [xxvi]-100, 12mo.

WEI PIH YANG.
    Chen tung chi.
        The earliest Chinese work on alchemy, written about the middle of the second
        century A.D.   (Communicated by John Fryer, LL.D., of Shanghai.)

WEIDENFELD, JOHANNES SEGERUS.
* Four Books concerning the secrets of the Adepts or of the use of
    Lully's spirit of wine : A practical work.   With very great study
    collected out of the ancient as well as modern fathers of adept
    philosophy, reconciled together by comparing them one with an-
    other, otherwise disagreeing, and in the newest method so aptly
    digested, that even young practitioners may be able to discern the
    counterfeit or sophistical preparations of animals, vegetables and
    minerals, whether for medicines or metals from true, and so avoid
    vagabond impostors and imaginary processes, together with the
    ruine of estates.   [The first Book.]   London, printed by Will. Bonny,
    for Tho. Howkins in George-Yard in Lombard-Street, 1685.   pp.
    [xlviii]-380, 4to.
        Dedicated to the Hon. Robert Boyle.   Only the first book was translated
        into English.

WELLING, GEORG VON.   See A B C.

WERCK (EIN) DER ALCHEMEY.   See Aureum Vellus.

WESTPHALUS, JOSEPHUS.   See S[chröder, Fr. J. W.].

WIEGLEB, JOHANN CHRISTIAN.
* Historisch-kritische Untersuchung der Alchemie, oder der einge-
    bildeten Goldmacherkunst ; von ihrem Ursprunge sowohl als Fort-
    gange, und was nun von ihr zu halten sey.   Weimar, 1777.   pp.
    [xxii]-437, 12mo.
        A severe attack on the claims of alchemical philosophers which provoked a
        defense by Kortum, K. A., q. v.   Wiegleb gives a short bibliography
        of works written against alchemy (pp. 372-377).
* Onomatologia curiosa artificiosa et magica oder natürliches Zauber-
    Lexicon, in welchem vieles nützliche und angenehme aus der

WIEGLEB, JOHANN CHRISTIAN. [Cont'd.]

Naturgeschichte, Naturlehre und natürlichen Magie nach alphabetischer Ordnung vorgetragen worden. Dritte Auflage, verbessert und mit vielen neuen Zusätzen vermehrt. Nürnberg, auf Kosten der Raspischen Buchhandlung, 1784. pp. [viii]–1743, 8vo.

WILLIS, TIMOTHY.

The Search of Causes, containing a theophysicall investigation of the possibilitie of transmutatorie Alchemie. J. Legatt. London, 1616. 8vo.

ZACHARIAS, DIONYSIUS.

See Manget, J. J.; also Aureum Vellus; also Theatrum chemicum; also R[ichebourg], J. M. D.

ZADITH. See Manget, J. J.; also Theatrum chemicum.

ZALENTO, PETRUS DE. See Silento, Petrus de.

ZANETUS, HIERONYMUS. See Theatrum chemicum.

ZIMPEL, CHARLES F.

* Bemerkungen über den Stein der Weisen (Lapis philosophorum) als Universal-Heilmethode zur möglichen Verhütung des Todes. Göppingen, Würtemberg, 1879. pp. 24, 8vo.

ZWEY ALTE DENKMALE DEUTSCHER FILOSOFEN VON DER ALCHYMIE.

See S[chröder, Fr. Jos. W.].

ZWO TINCTUREN AUFF ROTH UND WEISS. See Aureum Vellus.

SECTION VII.

PERIODICALS.

Titles are alphabetized under the first word, articles and "new" excepted ; with cross references from Editors and in the case of Societies from cities.

EXPLANATION OF SIGNS.

+ Following a date signifies current at the date in question.
‖ Following a date signifies publication discontinued.

ACADÉMIE DES SCIENCES, PARIS.
  See Comptes-rendus hebdomadaires des séances [etc.].

ACADÉMIE IMPÉRIALE DES SCIENCES DE ST. PÉTERSBOURG.
  See Mélanges physiques et chimiques [etc.].

ACHENBACH. *See* Schwäbische (Der) Bierbrauer ; *also* Süddeutsche (Der) Gerber.

1. AFHANDLINGAR I FYSIK, KEMI OCH MINERALOGI. Utgifne af W. Hisinger och J. Berzelius. 6 vols., 8vo. Stockholm, 1806–'18. Vol. I, 1806 ; II, 1807 ; III, 1810 ; IV, 1815 ; V, VI, 1818.

2. AGENDA DU CHIMISTE. 16 vols., 16mo. Paris, 1877–92+

3. AGRICULTURAL SCIENCE. Editor, William Frear. Published by the Editor, State College, Pennsylvania, U. S. A. 6 vols., 8vo. I–VI, 1887–92+

AGRICULTURE, U. S. DEPARTMENT OF. Bulletins of the Division of Chemistry.
  For check-list, see Bulletins of the Division of Chemistry [etc.].

AKADEMIE DER WISSENSCHAFTEN . . . WIEN. *See* Monatshefte für Chemie.

ÅKERMANN, JOACHIM. *See* Jern-Kontorets Annaler.

ALBERTONI, P. *See* Rivista di chimica e farmaceutica.

ALCAN, MICHEL. *See* Moniteur des fils et tissus.

ALCAN, TRÉLAT, etc. *See* Coloration (La) industrielle.

4. ALCOOL (L') ET LE SUCRE. Organe des industries agricoles distilleries, sucréries, féculeries, par un Comité d'Ingénieurs et de spécialistes. Revue mensuelle illustrée. 4to. Paris, 1892+

ALESSANDRI, P. E. *See* Selmi (Il), Giornale [etc.].

5. ALLGEMEINE BAYERISCHE HOPFEN-ZEITUNG. Organ der bayerischen Bierbrauerei. 2 vols., 4to. Nürnberg, 1861–'62.
> *Continued under the title :*

[a] Allgemeine Hopfen-Zeitung. Landwirthschaftliches Blatt für Oeconomen, Brauer und Hopfenhändler (Organ der bayerischen Bierbrauerei). Redacteur : J. Carl. 8 vols. (III–X), 4to. Nürnberg, 1863–'70.
> *Continued under the title :*

[b] Allgemeine Hopfen-Zeitung. Blätter für Hopfenbau und Hopfenhandel, Brauwesen und Landwirthschaft. Organ der deutschen Bierbrauerei. Redigirt von J. Carl. 2 vols. (XI–XII), 4to. Nürnberg, 1871–'72 [|| ?].

ALLGEMEINE BRAUER- UND HOPFEN-ZEITUNG. *See* Allgemeine Hopfen-Zeitung.

6. ALLGEMEINE CHEMISCHE BIBLIOTHEK DES NEUNZEHNTEN JAHRHUNDERTS. Herausgegeben von J. B. Trommsdorff. 5 vols., 8vo. Erfurt, 1801–'05.

7. ALLGEMEINE CHEMIKER-ZEITUNG. Central-Organ für Chemiker, Techniker, Ingenieure, Maschinenbauer, Fabrikanten chemischtechnischer Apparate. Correspondenzblatt chemisch-technischer und Gewerbe-Vereine. Chemisches Central-Annoncenblatt. Herausgegeben von G. Krause. 2 vols., 4to. Cöthen, 1877–78.
> *Continued under the title :*

[a] Chemiker-Zeitung. Central-Organ für Chemiker, Techniker, Fabrikanten, Apotheker, Ingenieure. Herausgegeben von G. Krause. 16 vols., 4to. Cöthen, 1879–92+ Each vol. in 2 parts.

8. ALLGEMEINE HOPFEN-ZEITUNG. Organ des deutschen Brauerbundes, des deutschen Hopfenbau-Vereins und des badischen Brauerbundes. Redigirt von J. Carl. 42 (?) vols., fol. Nürnberg, 1861–'81.
> *Continued under the title :*

[a] Allgemeine Brauer- und Hopfen-Zeitung. Officielles Organ des deutschen Brauerbundes . . . und des deutschen Hopfenbau-Vereins. Fol. Nürnberg, 1882 [+ ?].

ALLGEMEINE HOPFEN-ZEITUNG.   *See* Allgemeine bayerische Hopfen-Zeitung.

ALLGEMEINE NORDISCHE ANNALEN DER CHEMIE.
   See Nordische Blätter für Chemie.

9. ALLGEMEINE ÖSTERREICHISCHE CHEMIKER- UND TECHNIKER-
    ZEITUNG. Organ des allgemeinen technischen Vereines in
    Wien. Central-Organ für Petroleum-Industrie. Fach-Organ
    der Bohrtechniker. Sammt dem Beiblatte "Die Oel- und Fett-
    Industrie." Herausgeber, Hans Urban. 10 vols., 4to. Wien,
    1883-92+

10. ALLGEMEINE ZEITSCHRIFT FÜR BIERBRAUEREI UND MALZ-FABRI-
    KATION. Unter Mitwirkung der tüchtigsten Fachmänner
    herausgegeben von Franz Fasbender. 18 vols., 4to. Wien,
    1873-90+

    ALLGEMEINE ZEITSCHRIFT FÜR SPIRITUS- UND PRESSHEFE-INDUSTRIE.
       See Populäre Zeitschrift für Spiritus- und Presshefe-Industrie.

11. ALLGEMEINE ZEITSCHRIFT FÜR TEXTIL-INDUSTRIE. Populär-
    wissenschaftliches Fachblatt für Spinnerei, Weberei, Wirkerei,
    Färberei, Druckerei, Bleicherei, Appretur und verwandte
    Industrie-Zweige. Herausgegeben unter Mitwirkung hervor-
    ragen der Fachmänner und Industrieller von Ph. Zalud. 7
    vols., 4to. Chemnitz und Leipzig, 1879-85[‖ ?].

12. ALLGEMEINE ZEITUNG FÜR DIE GESAMMTE SPIRITUS-INDUSTRIE.
    Herausgegeben unter Mitwirkung hervorragender Fachautori-
    täten und anerkannt bewährter Fachmänner. 1 vol., 4to.
    Berlin, 1890-'91.

13. ALLGEMEINES JOURNAL DER CHEMIE. Herausgegeben von Alex.
    Nic. Scherer. 10 vols., 8vo. Leipzig, 1798-1803.
        Continued under the title:

    [a] Neues allgemeines Journal der Chemie, von Klaproth, Hermb-
         städt, Scherer, J. B. Richter, J. B. Trommsdorff, heraus-
         gegeben von Ad. Ferd. Gehlen. 8 vols., 8vo. Leipzig,
         1803-'06.
             Continued under the title:

    [b] Journal für die Chemie, Physik [*from* vol. IV] und Mineralogie,
         von Bucholz, Crell, Hermbstädt, Klaproth, Richter, Ritter,
         Trommsdorff, herausgegeben von A. F. Gehlen. 9 vols.,
         8vo. Berlin, 1806-'10.

ALLGEMEINES JOURNAL DER CHEMIE.   [Cont'd.]

Continued under the title :

[c]   Journal für Chemie und Physik, in Verbindung mit J. J. Bern-
      hardi, J. Berzelius, C. F. Bucholz, L. von Crell, A. F.
      Gehlen [*and others*], herausgegeben von J. S. C. Schweigger.
      69 vols., 8vo.   Nürnberg, 1811–'33.

Changes in the title as follows :

1st Series, vols. I–XXX, 1811–'20, *also under the title*, Beiträge zur
      Chemie und Physik.

2d Series, vols. XXXI–LX, 1821–'30, *also under the title*, Jahrbuch der
      Chemie und Physik herausgegeben von Schweigger und Meinecke.

3d Series, vols. LXI–LXIX, 1831–'33, *also under the title*, Neues Jahr-
      buch der Chemie und Physik. *From* 1829, *edited by* Fr. W.
      Schweigger-Seidel.

United in 1834 *with the* Journal für technische und oekonomische
Chemie *and continued under the title :*

[d]   Journal für praktische Chemie, herausgegeben von Otto Linné
      Erdmann, F. W. Schweigger-Seidel (und R. F. Marchand).
      [*From* 1853, *edited by* O. L. Erdmann *and* Gustav Werther.]
      108 vols., 8vo.   Leipzig, 1834–'69.

[e]   Neue Folge. Herausgegeben von Hermann Kolbe. [*From*
      1879 *by* H. Kolbe *and* Ernst von Meyer, *later by* Ernst von
      Meyer *alone.*] 46 vols., 8vo.   Leipzig, 1870–92+

      Sachregister zu den drei Jahrgängen 1823, 1824, und 1825 oder
      Band VII–XV des Jahrbuches der Chemie und Physik.
      Halle, 1826, 8vo.

      Sach- und Namen-Register zu Band I–XXX.   Leipzig, 1844.

      [*The same*], Band XXXI–LX.   Leipzig, 1854.

      [*The same*], bearbeitet von Friedrich Gottschalk, Band LXI–XC.
      Leipzig, 1865.

      [*The same*], Band XCI–CVIII.   Leipzig, 1871.

ALLISON, WILLIAM O.   *See* Oil, Paint and Drug Reporter.

14.   ALMANACCO DI CHIMICA AGRICOLA.   Di A. Selmi.   6 vols., 16mo.
      Milano, 1873–'78 [+ ?].

15.   ALMANACH DE LA CHIMIE, par H. du M.   8 vols., 18mo.   Rouen et
      Paris, 1854–'61.

16.   ALMANACH FÜR SCHEIDEKÜNSTLER UND APOTHEKER.   Herausge-
      geben von J. F. A. Göttling.   23 vols., 8vo.   Weimar, 1780–
      1802.

ALMANACH FÜR SCHEIDEKÜNSTLER UND APOTHEKER.   [Cont'd.]
    Continued under the title :

[a]   Taschenbuch für Scheidekünstler und Apotheker, herausgege-
    ben von Ch. Fr. Bucholz.   17 vols. (XXIV–XL), 8vo.   Wei-
    mar, 1803–'19.
    Continued under the title :

[b]   Trommsdorff's Taschenbuch für Chemiker und Pharmaceuten.
    10 vols. (XLI–L), 8vo.   Jena, 1820–'29 ||

Register, 1780–1803.   1 vol., 8vo.

Edited from 1780–1802, *by* J. F. A. Göttling ; 1803–'05, C. F. Bucholz ;
    1806–'17, C. F. Bucholz *with* Wilh. Meissner ; 1818, C. F.
    Bucholz, *with* Rud. Brandes ; 1819, Rud. Brandes ; 1820–'29,
    J. B. Trommsdorff.

ALMSTRÖM, P. O.   *See* Tekno-kemisk Journal.

AMERICAN ANALYST.   *See* New York Analyst (The).

17.   AMERICAN (THE) BREWER AND DISTILLER.   Published by J. P.
    McDonnell and J. W. Romaine, Paterson, N. J., 1891+

18.   AMERICAN (THE) BREWERS' GAZETTE AND MALT AND HOP
    TRADERS' REVIEW.   Edited by John Flintoff.   12 vols., 4to.
    New York, 1871–'82 [+ ?].

AMERICAN CHEMICAL REVIEW.   *See* Chemical Review and Journal [*a*].

19.   AMERICAN CHEMICAL JOURNAL.   Edited, with the aid of chemists
    at home and abroad, by Ira Remsen.   14 vols., 8vo.   Balti-
    more, Md., 1879–'92+

AMERICAN CHEMICAL SOCIETY.   See Journal of the American Chemical Society.

20.   AMERICAN (THE) CHEMIST.   A monthly journal of theoretical,
    analytical, and technical chemistry.   Edited by Chas. F.
    Chandler and W. H. Chandler.   6 vols. and 7 nos., 4to.   New
    York, 1870–'77 ||

21.   AMERICAN DRUGGISTS' CIRCULAR AND CHEMICAL GAZETTE.   A
    practical journal of chemistry as applied to pharmacy, arts and
    sciences, and general business organ for druggists, chemists
    and apothecaries.   Edited by Henry Bridgeman.   [*From* 1858
    by F. N. Newton.]   9 vols., 8vo.   New York, 1857–'65.
    Continued under the title :

[a]   Druggists' Circular and Chemical Gazette.   A practical journal
    of chemistry as applied to pharmacy, arts and sciences, and
    general business organ for druggists, chemists and apothe-
    caries.   Edited by F. N. Newton.   36 vols., 8vo.   New York,
    1866–92+

22. AMERICAN GAS-LIGHT JOURNAL. Devoted to light, water - supply and sewerage. 4 vols., 4to. New York, 1859–'63.

> *Continued under the title :*

[a] American Gas-Light Journal and Mining and Petroleum Standard. Edited by J. W. Bryant. 4 vols. (V–VIII), 4to. New York, 1864–'67.

> *Continued under the title :*

[b] American Gas-Light Journal and Chemical Repertory. Devoted to the interests of illumination, heating, ventilation and sanitary improvement, domestic economy and general science. [Published by M. L. Callender & Co.] 47 vols. (IX–LV), 4to. New York, 1868–'92+

> *Vols.* VI *and* VII *were also issued under the title :* Mining and Petroleum Standard and American Gas-Light Journal.

23. AMERICAN (THE) GLASS WORKER. Edited by J. H. Leighton. 1 vol., fol. Ottawa, Illinois, 1882–'83 [+?].

24. AMERICAN (THE) LABORATORY. A bi-monthly journal of the progress of chemistry, pharmacy, medicine, recreative science, and the useful arts. 4to. Boston, 1875 ‖

25. *AMERICAN MINERALOGICAL JOURNAL. Being a collection of facts and observations tending to elucidate the Mineralogy and Geology of the United States of America. Together with other information relating to mineralogy, geology and chemistry, derived from scientific sources. Conducted by Archibald Bruce. 1 vol., 8vo. New York, 1814.

> [All published.]

26. AMERICAN (THE) TANNER AND JOURNAL OF THE HIDE AND LEATHER TRADE. 4to. Buffalo, 1883 [+?].

27. AMERIKANISCHE (DER) BIERBRAUER. Monatliche Zeitung. Revue des Hopfen-, Malz- und Getreide-Handels. Officielles Organ des Haupt-Brauer-Vereins in den Vereinigten Staaten. Herausgeber und Redakteur, H. Schwarz. 25 vols., 4to. New York, 1868–92+

> *With a Supplement :* Der Praktische Bierbrauer, *q. v.*

28. AMERICANISCHE ANNALEN DER ARZNEIKUNDE, CHEMIE UND PHYSIK. Herausgegeben von Johann Abraham Albers. 3 nos. Bremen, 1802–3 ‖

29. ANALYST (THE), including the proceedings of the "Society of Public Analysts." A monthly journal of analytical chemistry.

ANALYST (THE). [Cont'd.]
>   Edited [in 1882] by G. W. Wigner and J. Muter [*later by* John Muter, *alone*]. 22 vols., 8vo. London, 1876-'92+

30. ANALYST'S (THE) ANNUAL NOTE-BOOK, 1874. By Sidney W. Rich. pp. iv-59, 12mo. London, 1875[+ ?].

>   ANNALEN DER CHEMIE UND PHARMACIE. *See* Annalen der Pharmacie.

>   ANNALEN DER CHEMISCHEN LITERATUR.
>   *See* Bibliothek der neuesten physisch-chemischen . . . Literatur.

31. ANNALEN DER PHARMACIE. Eine Vereinigung des Archives des Apotheker-Vereins im nördlichen Teutschland, B. XL; und des Magazins für Pharmacie und Experimentalkritik, B. XXXVII. Herausgegeben von Rudolph Brandes, Ph. Lorenz Geiger und Justus Liebig. 10 vols., 8vo. Lemgo und Heidelberg, 1832-'34.

>   *Continued (from vol.* XI, 1834) *under the title :*

[a] Annalen der Pharmacie. Vereinigte Zeitschrift des Neuen Journals der Pharmacie für Aerzte, Apotheker und Chemiker, Band XXVIII; des Archivs des Apothekervereins im nördlichen Deutschland, Band XLIX; und des Magazins für Pharmacie und Experimental-Kritik, Band XLVI. Von Johann Bartholomä Trommsdorff, Rudolph Brandes, Philipp Lorenz Geiger und Justus Liebig. 22 vols. (XI-XXXII), 8vo. Heidelberg, 1834-'39.

>   Vols. XVII-XXII, *edited by* J. B. Trommsdorff, Justus Liebig, *and* Emanuel Merck. Vols. XXIII-XXVI, *edited by* Justus Liebig, Emanuel Merck *and* Friedrich Mohr. Vols. XXVII-XXXII, herausgegeben, unter Mitwirkung der HH. Dumas in Paris und Graham in London, von Friedrich Wöhler und Justus Liebig.

>   *Continued under the title :*

[b] Annalen der Chemie und Pharmacie. Unter Mitwirkung der HH. Dumas in Paris und Graham in London, herausgegeben von Friedrich Wöhler und Justus Liebig. 136 vols. (XXXIII-CLXVIII), 8vo. Heidelberg, 1840-1873.

>   *From* vol. XLI *the names* Dumas and Graham *are dropped. From* vol. LXXVII (1851), *edited by* Friedrich Wöhler, Justus Liebig *and* Hermann Kopp; neue Reihe Band I. *From* vol. CLIX (1871), *edited by the same together with* E. Erlenmeyer *and* J. Volhard.

>   *Continued under the title :*

[c] Justus Liebig's Annalen der Chemie und Pharmacie. Herausgegeben von Friedrich Wöhler, Hermann Kopp, Emil Er-

ANNALEN DER PHARMACIE. [Cont'd.]
lenmeyer, Jacob Volhard. 4 vols. (CLXIX–CLXXII), 8vo. Leipzig und Heidelberg, 1873.

[d] Justus Liebig's Annalen der Chemie. Herausgegeben von Friedrich Wöhler, Hermann Kopp, Emil Erlenmeyer, Jacob Volhard [*later by the same with* A. W. Hofmann, Aug. Kekulé]. 104 vols. (CLXXIII–CCLXXVI), 8vo. Leipzig und Heidelberg, 1874–92+

Supplement-Band I, 1861 ; II, 1862–'63 ; III, 1864–'65 ; IV, 1865–'66 ; V, 1867 ; VI, 1868 ; VII, 1870 ; VIII, 1872.

Autoren- und Sach-Register zu den Bänden I–C (Jahrgang, 1832–1856) der Annalen der Chemie und Pharmacie. Bearbeitet von G. C. Wittstein. 8vo. Leipzig und Heidelberg, 1861.

Autoren- und Sach-Register zu den Bänden CI–CXVI (Jahrgang, 1857–1860) der Annalen der Chemie und Pharmacie. Bearbeitet von G. C. Wittstein. 1 vol., 8vo. Leipzig und Heidelberg, 1861.

Autoren- und Sach-Register zu den Bänden CXVII–CLXIV und den Supplementbänden I–VIII (1861–1872), der Annalen der Chemie und Pharmacie. Bearbeitet von Friedrich Carl. 1 vol., 8vo. Leipzig und Heidelberg, 1874.

General-Register zu den Bänden CLXV–CCXX (1873–1883), von Liebig's Annalen der Chemie (früher Annalen der Chemie und Pharmacie), bearbeitet von Friedrich Carl. 1 vol., 8vo. Leipzig, 1885.

Cf. Magazin für Pharmacie.

ANNALEN DER PHYSIK UND CHEMIE. Poggendorff.
See Journal der Physik. Gren.

32. ANNALES DE CHIMIE, ou Recueil de mémoires concernant la chimie et les arts qui en dépendent [*from* vol. XXXIII], et spéciale-ment la pharmacie. Par de Morveau, Lavoisier, Monge, Ber-thollet, de Fourcroy, de Dieterich, Hassenfratz et Adet. 96 vols., 8vo. Paris, 1789–1816.

Vols. I, II *and* III *were reprinted at Paris in* 1830.
Tables des matières. 3 vols. Paris, 1801, 1807, 1821.
Continued under the title :

[a] Annales de chimie et de physique, par Gay-Lussac et Arago. Deuxième série. 78 vols., 8vo. Paris, 1817–'40.
Tables des matières. 3 vols., 8vo. Paris, 1831–'41.
Continued under the title :

[b] Annales de chimie et de physique, par Arago, Chevreul, Dumas, Pélouze, Boussingault, Regnault. Avec une revue des

ANNALES DE CHIMIE.　[Cont'd.]
travaux de chimie et de physique publiées à l'étranger, par Wurtz et Verdet. Troisième série. 69 vols., 8vo. Paris, 1831–'63.

Tables des matières. 2 vols., 8vo. Paris, vols. I–XXX, 1851 ; vols. XXXI–LXIX, 1866.

Continued under the title :

[c]　Annales de chimie et de physique, par Chevreul, Dumas, Pélouze, Boussingault, Regnault, avec la collaboration de Wurtz. Quatrième série. 30 vols., 8vo. Paris, 1864–'73.
Table des matières. Vols. I–XXX. Paris, 1874.

[d]　Cinquième série. Par Chevreul, Dumas, Boussingault, Wurtz, Berthelot, Pasteur, avec la collaboration de Bertin. 30 vols., 8vo. Paris, 1874–'83.

[e]　Sixième série. Par Berthelot, Pasteur, Friedel, Becquerel, Mascart. 22 vols., 8vo. Paris, 1884–92+
Tables de la cinquième série des Annales de chimie (1874–1883) dressées par Gayon. 1 vol., 8vo. Paris, 1885.

33.　ANNALI DI CHIMICA [*from* vol. IV] e storia naturale, ovvero raccolta di memorie sulle scienza, arti e manufatture ad esse relative, di L. Brugnatelli. 21 vols., 8vo. Pavia, 1790–1802 ‖
Vol. XXI contains an index.

ANNALI DI CHIMICA APPLICATA ALLA MEDICINA. Polli.
See Giornale di farmacia, chimica e scienze accessorie.

ANNALI DI CHIMICA MEDICO-FARMACEUTICA. *See* Giornale di farmacia.

ANNALI DI FISICA. *See* Raccolta fisico-chimica italiana.

34.　ANNALI DI FISICA, CHIMICA E MATEMATICHE, col bullettino dell' industria, meccanica e chimica, diretti dall' ingegnere G. A. Majocchi. 28 vols., 8vo. Milano, 1841–'47.
Continued under the title :

[a]　Annali di fisica, chimica e scienze affini, col bollettino di farmacia e di tecnologia, redatti da G. A. Majocchi e F. Selmi [*from* vol. III], e P. A. Boscarelli. Seconda serie. 4 vols., 8vo. Torino, 1850 ‖

ANNALI DI FISICA, dell' Abbate F. C. Zantedeschi.
See Raccolta fisico-chimico-italiano.

35.　ANNALS OF CHEMICAL MEDICINE, including the application of chemistry to physiology, pathology, therapeutics, pharmacy, toxicology and hygiene. Edited by J. L. W. Thudicum. 2 vols., 8vo. London, 1879–'81 ‖

36. ANNALS (THE) OF CHEMICAL PHILOSOPHY [etc.]. By W. Maugham. 2 vols., 8vo. London, 1828-'29.

37. ANNALS OF CHEMISTRY [etc.], by de Morveau, Lavoisier [*and others*]. Translated from the French. 1 vol., 8vo. London, 1791.
> Translated from the fifth volume of the Annales de chimie.

38. ANNALS (THE) OF CHYMISTRY AND PRACTICAL PHARMACY. Being a weekly summary of the discoveries of philosophers, chiefly continental and transatlantic, in their applications to the chemistry of medicine, agriculture, manufactures, and to the several branches of physics, electricity, galvanism, photography, etc. 1 vol., 8vo. London, 1843.

39. ANNALS OF PHARMACY AND PRACTICAL CHEMISTRY. Edited by W. Bastick and W. Dickenson. 3 vols., 8vo. London, 1852-'54 ‖

40. ANNALS OF PHILOSOPHY; or Magazine of Chemistry, Mineralogy, Mechanics, Natural History, Agriculture, and the Arts. By Thomas Thomson. 16 vols., 8vo. London, 1813-'20.
[a] New series. [Edited by Richard Phillips.] 12 vols., 8vo. London, 1821-'26.
> *United in* 1827 *with the* Philosophical Magazine and Journal. *See* Philosophical Magazine.

41. ANNUAIRE À L'USAGE DU CHIMISTE, DU MÉDECIN, DU PHARMACIEN . . . Par P. J. Hensmans. 1 vol., 8vo. Bruxelles, 1843.

ANNUAIRE DE CHIMIE . . . par Laurent et Gerhardt.
> *See* Comptes rendus mensuels des travaux chimiques, etc.

42. ANNUAIRE DE CHIMIE, comprenant les applications de cette science à la médecine et à la pharmacie, ou répertoire des découvertes et des nouveaux travaux en chimie faits dans les diverses parties de l'Europe, par E. Millon et J. Reiset, avec la collaboration de F. Hoefer et de Nicklès. 7 vols., 8vo. Paris, 1845-'51 ‖

43. ANNUAIRE DE LA CHIMIE INDUSTRIELLE ET DE L'ÉLECTROCHIMIE. Rédigé par Donato Tommasi. Paris, 1888-90+

44. ANNUAIRE DE LA DISTILLERIE FRANÇAISE, contenant tous les renseignements techniques et commerciaux relatifs aux alcools. Paris, Redacteur : Fritsch. 18—.

45. ANNUAIRE DES EAUX MINÉRALES DE FRANCE, 1830 et 1831. Rédigé
     par Longchamp. 18mo. Paris, 1830–32.

46. ANNUAIRE DES FABRICANTS DE SUCRE, distillateurs et liquoristes
     des départements du Nord, du Pas-de-Calais, de l'Aisne et de
     la Somme. 8vo. Douai, 1862.

47. ANNUAIRE DES PRODUITS CHIMIQUES, de la droguerie et de l'épicerie
     en gros, contenant la liste complète des fabricants, etc., de
     France, de l'Italie, de Belgique, et de la Suisse. 4 vols., 8vo
     Paris, 1874–'78.

48. ANNUAIRE DES SCIENCES CHIMIQUES, ou Rapport sur les progrès des
     sciences naturelles présenté à l'académie Stokolm [sic]. Par
     Berzelius. Supplément à son Traité de chimie. Traduit en
     Français par H. D. 8vo. Paris, 1837.
         See Rapport annuel sur les progrès des sciences ; also Årsberättelse
         om Framstegen i Fysik och Kemi.

49. ANNUAIRE GÉNÉRAL DE L'INDUSTRIE de l'éclairage et du chauffage
     par le gaz. Par Émile Durand et Paul Durand. 7 vols.,
     18mo. Paris, 1874–'80 [+ ?].

50. ANNUAL REPORT OF THE PROGRESS OF CHEMISTRY and the allied
     sciences, physics, mineralogy and geology ; including the
     application of chemistry to pharmacy, the arts and manufac-
     tures. By Justus Liebig and H. Kopp, with the co-operation
     of H. Buff, Frederick Knapp, Ernest Dieffenbach, Charles
     Ettling, Henry Will, Frederick Zamminer. Edited by A. W.
     Hofmann and Warren de la Rue, 1847–'53. 7 vols., 8vo.
     London, 1849–'55 ‖
         See Jahresbericht über die Fortschritte der reinen . . . Chemie.
         Giessen.

51. ANNUARIO ALMANACCO PEI CHIMICI, FARMACISTI E MEDICI ITALIANI,
     redatto per cura del farmacista Ign. Cugusi-Persi da Cagliari.
     5 vols., 16mo. Milano, 1874–'79.

52. ANNUARIO CHIMICO ITALIANO dell' anno 1845, diretto da Francesco
     Selmi e compilato dal medesimo in compagnia dei Signori
     Giuseppe Parmeggiani e Giovanni Giorgini. 1 vol., 8vo
     Modena, 1846.

53. ANNUARIO DEL LABORATORIO DI CHIMICA generale e tecnologica
     della R. Accademia Navale di Livorno, a cura di G. Bertoni.
     Anno I : 1890. 8vo. Milano, 1891.

54. ANNUARIO DELLE SCIENZE CHIMICHE E NATURALI. 1 vol., 8vo. Verona, 1840.

55. ANNUARIO DELLE SCIENZE CHIMICHE, FARMACEUTICHE, E MEDICO-LEGALI ad uso dei farmacisti e medici, in continuazione del Supplemento al trattato di farmacia del Sig. Virey; della Gazzetta eclettica di farmacia e chimica. 1 vol., 8vo. Mantova, 1840.

Continued under the title :

[a] Annuario delle scienze chimiche, farmaceutiche e medico-legali, contenente tutte le scoperte relative a queste scienze, la relazione di lavori chimici e naturali, delle riunioni degli scienziati italiani e stranieri, di quelli particolari di J. J. Berzelius, e la traduzione della chimica organica di J. Lie-big. Redattore G. B. Sembenini. 9 vols., 8vo. Mantova, 1841-'49.

Cf. Gazzetta eclettica di farmacia e chimica.

56. ANNUARUL LABORATORULUI DE CHIMIE ORGANICA pe anul bugetar 1888-1889. Isstrati, C. J. Volumul I. Bucuresci, 1889.

57. ANTI-ADULTERATION REVIEW. — vols. London, 1871-'80+

APPLETON, JOHN HOWARD. *See* Laboratory (The) Year-Book.

58. ARCHIV DER AGRICULTURCHEMIE für denkende Landwirthe. Her-ausgegeben von Sig. F. Hermbstädt. 7 vols., 8vo. Berlin, 1803-'18 ||

59. ARCHIV FOR PHARMACI, redigeret af S. M. Trier. 3 vols., 8vo. Kjøbenhavn, 1844-'46.

Continued under the title :

[a] Archiv for Pharmaci og technisk Chemi med deres Grundvid-enskaber. Redigeret af S. M. Trier. Det technisk-chemiske Afsnit redigeret af P. Faber. 45 vols. (IV-XLV), 8vo. Kjøbenhavn, 1847-'90+

Index. vols. I-XV, 1844-'58. Kjøbenhavn, 1859.

ARCHIVES DE PHYSIOLOGIE. *See* Journal de la physiologie.

ARCHIV FÜR CHEMIE UND METEOROLOGIE.
See Archiv für die gesammte Naturlehre.

60. ARCHIV FÜR DIE GESAMMTE NATURLEHRE. In Verbindung mit Bischoff, Förstmann, C. G. Gmelin, Grischow, F. W. von Paula Gruithuisen, Hallaschka, Pl. Heinrich, A. von Humboldt, John

ARCHIV FÜR DIE GESAMMTE NATURLEHRE. [Cont'd.]
Kleefeldt, Lichtenberg, Marx, Olbers, Pleischl, Prechtl, Schmidt, Schön, Späth, Wollner und Zimmerman, herausgegeben von C. W. G. Kastner. 27 vols., 8vo. Nürnberg, 1824–'35.

Vols. XIX–XXVII *also under the title :*

[a] Archiv für Chemie und Meteorologie, in Verbindung mit mehreren Gelehrten herausgegeben von C. W. G. Kastner. 9 vols. (I–IX), 8vo. Nürnberg, 1830–'35.

61. ARCHIV FÜR DIE THEORETISCHE CHEMIE. Herausgegeben von Alex. Nic. Scherer. 1 vol., 8vo. Jena und Berlin, 1800–'02.

62. ARCHIV FÜR DIE THIERISCHE CHEMIE. Herausgegeben von Johann Horkel. 1 vol., 8vo. Halle, 1800–'01.

ARCHIV FÜR PHYSIOLOGISCHE UND PATHOLOGISCHE CHEMIE.
See Beiträge zur physiologischen und pathologischen Chemie.

63. ÅRSBERÄTTELSE OM FRAMSTEGEN I PHYSIK OCH KEMI till Kongl. Vetenskaps-Akademien afgifven af Jac. Berzelius. 1821–'40. 20 vols., 8vo. Stockholm, 1822–'41.

Continued under the title :

[a] Årsberättelse om Framstegen i Kemi och Mineralogi afgifven af Jac. Berzelius. 1841–'47. 7 vols., 8vo. Stockholm, 1841–'48.

Followed by :

[b] Årsberättelse om Framstegen i Kemi till Kongl. Vetenskaps-Akademien afgifven af L. F. Svanberg. 1847–'49. 3 vols., 8vo. Stockholm, 1849–'51 ‖

Sak- och Namn-Register öfver alla af Berzelius . . . afgifvna Årsberättelser (1821–'47). På Kongl. Vetenskaps-Akademiens föranstaltande utgifvet af A. Wiemer. 1 vol., 8vo. Stockholm, 1850.

Cf. Rapport annuel sur les progrès des sciences physiques et chimiques.

ARTUS, W. *See* Jahrbuch für ökonomische Chemie; *also* Vierteljahresschrift für technische Chemie.

64. ASSOCIATION OF OFFICIAL AGRICULTURAL CHEMISTS, Proceedings at the Annual Conventions, Raleigh, [later] Washington. 10 vols., 8vo. 1884–'93+

[Published at first independently and afterwards as Bulletins of the Division of Chemistry, U. S. Department of Agriculture.]

Detailed titles as follows :

ASSOCIATION OF OFFICIAL AGRICULTURAL CHEMISTS. [Cont'd.]

[No. 1] Proceedings of the Convention of Agricultural Chemists at Atlanta, May 15 and 16, 1884. Edited by Charles W. Dabney, Secretary. Published by the members of the Convention. Raleigh, 1884. pp. 60, 8vo.

[No. 2] Proceedings of the Convention of Agricultural Chemists and First Annual Meeting of the Association of Official Agricultural Chemists at Philadelphia, September 8 and 9, 1884. From the Monthly Report for October 1, 1884, of the Department of Agriculture of South Carolina. Columbia, S. C., 1884. pp. 8, 8vo.

[No. 3] Proceedings of the [Second Annual Convention] Association of Official Agricultural Chemists [held at Washington], September 1 and 2, 1885. Methods of Analysis of Commercial Fertilizers. Bulletin No. 7, Division of Chemistry, Department of Agriculture. [Edited by Charles W. Dabney.] Washington, 1885. pp. 49, 8vo.

[No. 4] Proceedings of the Third Annual Convention . . . August 26 and 27, 1886. Bulletin No. 12, Division of Chemistry, Department of Agriculture. Methods of Analysis of Commercial Fertilizers. Washington, 1886. pp. 59, 8vo.

[No. 5] Methods of Analysis of Commercial Fertilizers, Feeding Stuffs, and Dairy Products adopted at the Fourth Annual Convention . . . August 16, 17 and 18, 1887. Edited by Clifford Richardson, Secretary. Bulletin No. 16, Division of Chemistry, U. S. Department of Agriculture. Washington, 1887. pp. 80, 8vo. Ill.

[No. 6] Methods of Analysis of Commercial Fertilizers, Cattle-Foods, Dairy Products, Sugar and Fermented Liquors adopted at the Fifth Annual Convention of the Association of Official Agricultural Chemists, held at the U. S. Department of Agriculture, August 9 and 10, 1888. Edited by Clifford Richardson, Secretary of the Association. Bulletin No. 19, Division of Chemistry, U. S. Department of Agriculture. Washington, 1888. pp. 96, 8vo.

[No. 7] Proceedings of the Sixth Annual Convention . . . held . . . Sept. 10, 11 and 12, 1889 . . . Edited by Harvey W. Wiley. Bulletin No. 24, Division of Chemistry, U. S. Department of Agriculture. Washington, 1890. pp. 235, 8vo.

[No. 8] Proceedings of the Seventh Annual Convention . . . held . . . August 28, 29 and 30, 1890. Edited by Harvey

ASSOCIATION OF OFFICIAL AGRICULTURAL CHEMISTS. [Cont'd.]
    W. Wiley, Secretary of the Association. Bulletin No. 28,
    Division of Chemistry, U. S. Department of Agriculture.
    Washington, 1890. pp. 238, 8vo.

 [No. 9] Proceedings of the Eighth Annual Convention . . .
    held at the Columbian University, Washington, August
    13, 14 and 15, 1891. Methods of Analysis of Com-
    mercial Fertilizers, Foods and Feeding Stuffs, Dairy
    Products, Fermented Liquors and Sugars. Edited by
    Harvey W. Wiley, Secretary of the Association. Pub-
    lished by authority of the Secretary of Agriculture.
    Bulletin No. 31, Division of Chemistry, U. S. Depart-
    ment of Agriculture. Washington, 1891. pp. 253, 8vo.
    Ill.

 [No. 10] Proceedings of the Ninth Annual Convention . . . Bulle-
    tin No. 35, Division of Chemistry, U. S. Department of
    Agriculture. Washington, 1893.

  ATWATER, W. O. *See* Experiment Station Record.

65. AUSWAHL ALLER EIGENTHÜMLICHEN ABHANDLUNGEN UND BEO-
   BACHTUNGEN IN DER CHEMIE, mit einigen Verbesserungen und
   Zusätzen. Herausgegeben von Lorenz Crell. 5 vols., 8vo.
   Leipzig, 1786–'87 ||
    Cf. Crell, Lorenz.

66. AUSWAHL VORZÜGLICHER ABHANDLUNGEN AUS DEN SÄMMTLICHEN
   BÄNDEN DER FRANZÖSISCHEN ANNALEN DER CHEMIE zur vollständigen Benutzung derselben durch Ergänzung der von
   ihrem Anfange an den chemischen Annalen einverleibten Aufsätzen für deutsche Scheidekünstler von Lorenz von Crell.
   1 vol., 8vo. Helmstadt, 1801 ||
    Cf. Crell, Lorenz.

  BARTHEL, KARL. *See* Chemisch-technischer Central-Anzeiger.

  BARTHOLOMEW, GEORGE. *See* Oil and Drug News.

  BASTICK (W.) AND DICKENSON (W.). *See* Annals of Pharmacy and Practical
   Chemistry.

  BAUMANN, J. *See* Correspondenzblatt des Vereins . . . Zuckertechniker.

  BAUMGARTEN, P. *See* Jahresbericht über die Fortschritte in der Lehre von den
   pathogenen Mikroorganismen.

67. BAYERISCHE (DER) BIERBRAUER. Unter Mitwirkung der angesehensten Theoretiker und Praktiker ; redigirt von Karl
   Lintner. 12 vols., 8vo. München, 1866–77.

BAYERISCHE (DER) BIERBRAUER.   [Cont'd.]

Continued under the title :

[a] Zeitschrift für das gesammte Brauwesen.   Organ der wissen-
schaftlichen Station für Brauerei in München.   Herausge-
geben von Karl Lintner und L. Aubry und redigirt von
George Holzner.   13 vols., roy. 8vo.   München, 1878–90+
Repertorium zu den Jahrgang 1866–77.   Angefertigt von Geo.
Holzner.   Roy. 8vo.   1878.

BEIBLÄTTER ZU DEN ANNALEN DER PHYSIK UND CHEMIE.   *See* Journal der Physik.

BEILSTEIN'S ZEITSCHRIFT FÜR CHEMIE.   *See* Kritische Zeitschrift für Chemie.

68.   BEITRÄGE ZUR CHEMIE in Uebersetzung oder vollständigen Auszügen neuer chemischer Abhandlungen, sammt einigen neuen
Aufsätzen.   Herausgegeben von F. A. X. von Wasserberg.
1 vol., 8vo.   Wien, 1791.

BEITRÄGE ZUR CHEMIE UND PHYSIK, von J. S. C. Schweigger.   *See* Allgemeines
Journal der Chemie.

69.   BEYTRÄGE ZUR PHYSIK, OECONOMIE, CHEMIE, TECHNOLOGIE, MIN-
ERALOGIE UND STATISTIK, besonders der russischen und
angränzenden Länder, von B. F. Hermann.   3 vols., 8vo.
Berlin, 1786–'88.

70.   BEITRÄGE ZUR PHYSIOLOGISCHEN UND PATHOLOGISCHEN CHEMIE
und Mikroscopie in ihrer Anwendung auf die praktische
Medicin, unter Mitwirkung der Mitglieder des Vereins für
physiologische und pathologische Chemie und anderer Gelehrten, herausgegeben von Franz Simon.   1 vol., 8vo.   Berlin,
1843.

Continued under the title :

[a] Archiv für physiologische und pathologische Chemie und Mikro-
scopie in ihrer Anwendung auf die praktische Medicin.
Organ für die Fortschritte der gesammten medicinischen
Chemie im In- und Auslande.   Unter Mitwirkung mehrerer
Gelehrten des In- und Auslandes als Fortsetzung der von
Franz Simon in Berlin gegründeten Zeitschrift " Beiträge,
etc.," herausgegeben und redigirt von J. F. Heller.   4 vols.
(I–IV), 8vo.   Wien und Berlin, 1844–'47.

Continued under the title :

[b] Archiv für physiologische und pathologische Chemie und
Mikroscopie mit besonderer Rücksicht auf die medicinische
Diagnostik und Therapie.   Herausgegeben von Joh. Florian
Heller.   Neue Folge.   2 vols. (V, VI), 8vo.   Wien, 1852–'54 ‖

BENNEWITZ, PAUL. *See* Deutscher Chemiker - Kalender ; *also* Technisch-chemischer Kalender.

BERGÉ, HENRI. *See* Chimiste (Le).

71. BERICHT ÜBER DIE NEUESTEN FORTSCHRITTE IN DER CHEMISCHEN UND PHYSIKALISCHEN TECHNIK. Herausgegeben von dem technischen Vereine zu Carlsruhe. 1 part, 8vo. Carlsruhe, 1865 ‖

72. BERICHTE DER DEUTSCHEN CHEMISCHEN GESELLSCHAFT. 25 years. [*From* 1873, Redacteur H. Wichelhaus ; *from* 1883, Ferd. Tiemann ; *from* 1874 *each volume in two parts, from* 1884 *in three parts*], 8vo. Berlin, 1868–'92 +
Generalregister über die ersten zehn Jahrgänge (1868–1877) der Berichte der deutschen chemischen Gesellschaft zu Berlin. Bearbeitet von C. Bischoff. 1 vol., 8vo. Berlin, 1880.
Generalregister über die zweiten zehn Jahrgänge (1878–1887) der Berichte der deutschen chemischen Gesellschaft. Redacteur : Ferd. Tiemann. Stellvertretender Redacteur : F. von Dechend. pp. 1,636, 8vo. Berlin, 1888.

BERLIN, DEUTSCHE CHEMISCHE GESELLSCHAFT.
See Berichte der deutschen chemischen Gesellschaft.

73. BERLINISCHES JAHRBUCH DER PHARMACIE. 1 vol., 12mo. Berlin, 1795.
Continued under the title :
[a] Berlinisches Jahrbuch für die Pharmacie und für die damit verbundenen Wissenschaften. 42 vols. (II–XLIII), 12mo. Berlin, 1796–1840 ‖
From 1803–'10 *published with the additional title :* Neues Berlinisches Jahrbuch für die Pharmacie [*edited by* F. A. Gehlen *and* Val. Rose]. *From* 1811–'14, *with the additional title :* Neues Jahrbuch der Pharmacie, herausgegeben von W. Döbereiner. *From* 1815–'29, *with the additional title :* Deutsches Jahrbuch für die Pharmacie. *Edited from* 1818–'20 *by* C. W. G. Kastner ; 1821–'25, G. H. Stoltze ; 1826–'29, Wilhelm Meissner ; 1830, '31, A. Lucas ; 1833–'40, Lindes.

BERZELIUS, J. J.
See Afhandlingar i Fysik, Kemi [etc.] ; *also* Årsberättelse om Framstegen i Physik och Kemi ; *also* Rapport annuel sur les progrès des sciences physiques et chimiques.

BERZELIUS' JAHRESBERICHT.
See Jahresbericht über die Fortschritte der physischen Wissenschaften.

BESSELICH, N. *See* Zeitung für das Gas- und Wasserfach ; *also* Organ für den Oel- und Fetthandel.

BIBLIOTECA DI FARMACIA. *See* Giornale di farmacia . . . Cattaneo.

74. BIBLIOTHEK DER NEUESTEN PHYSISCH-CHEMISCHEN, METALLUR-GISCHEN, technologischen und pharmaceutischen Literatur. Herausgegeben von S. F. Hermbstädt. 4 vols., 8vo. Berlin, 1788–'95.

> *Continued under the title :*

[a] Annalen der chemischen Literatur. Herausgegeben von Wolf. 1 vol., 8vo. Berlin, 1802.

BIEDERMANN, R. *See* Centralblatt für Agriculturchemie ; *also* Chemiker-Kalender ; *also* Technisch-chemisches Jahrbuch.

75. BIERBRAUER (DER). Monatsbericht über die Fortschritte des gesammten Brauwesens. Herausgegeben von G. E. Habich [*later by* Conrad Schneider ; Neue Folge in 1870]. 23 vols., roy. 8vo. Leipzig und Halle a. S., 1859–92+

76. BIERBRAUEREI (DIE). Milwaukee, 1875.

BLONDEAU, P. F. *See* Corps (Les) gras industriels ; *also* Journal de l'exploitation des corps gras industriels.

77. BÖHMISCHE BIERBRAUER. Organ des Brau-Industrie-Vereines im Königreich Böhmen. Herausgegeben und redigirt von Ant. St. Schmelzer. 15 vols., 8vo. Prag, 1874–88 [|| ?].

BÖRNSTEIN, E. *See* Farben-Industrie (Die).

BON, EDUARD. *See* Deutsche Muster-Zeitung.

BORDET, G. *See* Sucre (Le).

78. BOSTON JOURNAL OF CHEMISTRY. Devoted to chemistry, as applied to medicine, agriculture and the arts. [*From* vol. v, devoted to the science of home life, the arts, agriculture and medicine.] Edited by Jas. R. Nichols. 14 vols., 4to. Boston (July), 1866–'80.

> *Continued under the title :*

[a] Boston Journal of Chemistry and Popular Science Review. Devoted to chemistry, pharmacy, geology, agriculture, astronomy, hygiene, medicine, practical arts, home science [etc.]. 2 vols. (XV, XVI), 4to. Boston, 1881–'82.

> *Continued under the title :*

BOSTON JOURNAL OF CHEMISTRY. [Cont'd.]

[b] Popular Science News and Boston Journal of Chemistry. A journal of useful knowledge for all classes—house-keepers, farmers, mechanics, physicians, druggists, dentists, chemists, lawyers, etc. 10 vols. (XVII–XXVI), 4to. Boston, 1883–'92+

BOURGEOIS, P. *See* Journal des fabricants de sucre ; *also* Moniteur de la sucrerie.

BRANDES, RUDOLPH. *See* Annalen der Pharmacie.

79. BRASSEUR (LE). Organe special des intérêts de la brasserie. 15 vols. Namur, 1866–'80 [+?].

80. BRAUER (DER) UND MALZER. Ein Journal dem Interesse der Brau- und Malz-Industrie, dem Hopfen- und Gersten-Handel gewidmet. Herausgeber : Eugene A. Sittig. 4to. Chicago, 1882+

81. BRENNEREI-ZEITUNG (NEUE). Praktische Mittheilungen der bei der Spiritus-Fabrikation vorkommenden Erfahrungen und wichtigen Erfindungen sowie 'der neu erfundenen Maschinen, Geräthe und Apparate. Herausgegeben von L. Gumbinner. 14 vols., 8vo. Berlin, 1872–'85 ||

BRESLAUER. *See* Deutsche Chemiker-Zeitung.

82. BREWER (THE). 4to. St. Louis, 1878[+?].

83. BREWERS' (THE) GUARDIAN. A first-class paper devoted to the protection of brewers' interests in licensing, legal and parliamentary matters, review of the malt, hop, and sugar trades, and wine and spirit trades' record. The official organ of the Country Brewers' Society. Edited by Thomas Lampray. 20 vols., 4to. London, 1871–90+

84. BREWERS' JOURNAL (THE) AND HOP AND MALT TRADES' REVIEW. A Monthly Trade Journal. 28 vols., 4to. London, 1869–92+

85. BREWERS' (THE) AND DEALERS' JOURNAL. Edited by P. Henry Doyle. 4to. Philadelphia, 1881–'82[+?].

86. BREWING (THE) WORLD. Edited by Carl Fabian Falk. 1 vol., 4to. Milwaukee, 1883[+?].

BREWSTER'S JOURNAL OF SCIENCE.
     See Edinburgh Journal of Science, *and cf*. Philosophical Magazine.

87. BROD (DAS). Organ des wirthschaftlichen und wissenschaftlich-technischen Vereins für Fortbildung des Backwesens und des

Brod (Das). [Cont'd.]
Backgewerbes und Sprechsaal für die einschlagende Technik sowie für die Nahrungsfrage überhaupt. Herausgegeben von dem Vereins-Ausschusse und redigirt von Mor. Freiherrn von Eberstein. 2 vols., 8vo. Leipzig, 1869–70 ||

Brown-Séquard, F. *See* Journal de la physiologie.

Bruce, Archibald. *See* American Mineralogical Journal.

Brugnatelli, L. *See* Annali di chimica ; *also* Giornale di fisica, chimica e storia naturale.

Bryant, J. W. *See* American Gas-Light Journal.

Bucholz, Christian Friedrich.
See Almanach für Scheidekünstler [*a*] ; *also* Beiträge zur Erweiterung . . . der Chemie ; *also* Taschenbuch für Scheidekünstler.

Buchner's Repertorium. *See* Repertorium für die Pharmacie.

88. Bulletin commercial des industries tinctoriales et textiles.
— vols. Paris, 18—'80[+?].

Bulletin de la Société chimique de Paris.
See Bulletin des Séances de la Société chimique de Paris.

89. Bulletin de la Société de chimie industrielle de Paris.
5 vols. Paris, 1888–92+

90. Bulletin de l'association des chimistes de sucrerie et de distillerie de France et des colonies. 8vo. Paris, 1883[+?].

91. Bulletin de pharmacie [*from* vol. vi] et des sciences accessoires. Rédigé par Parmentier, C. L. Cadet, L. A. Planche, P. F. G. Boullay, J. P. Boudet et P. R. Destouches. 6 vols., 8vo. Paris, 1809–'14.
Continued under the title :
[a] Journal de pharmacie et des sciences accessoires. Rédigé par C. L. Cadet, L. A. Planche, P. F. G. Boullay, J. P. Boudet, J. J. Virey, J. Pelletier et A. Vogel. Deuxième série. 27 vols. (i–xxvii), 8vo. Paris, 1815–'41.
Continued under the title :
[b] Journal de pharmacie et de chimie contenant une revue de tous les travaux publiés en France et à l'étranger sur les sciences physiques, naturelles, médicales et industrielles, ainsi que le bulletin des travaux de la Société de pharmacie de Paris. Troisième série. 46 vols. (i–xlvi), 8vo. Paris, 1842–'64.
Continued under the title :

BULLETIN DE PHARMACIE.   [Cont'd.]

[c] Journal de pharmacie et de chimie, par Boullay, Bussy, Soubeiran, Henry, F. Boudet, Cap, Boutron-Charlard, Fremy, Guibourd, Barreswil, Buignet, Gobley et Léon Soubeiran, contenant une revue médicale par Le Vigla et une revue des travaux chimiques publiés à l'étranger par J. Nicklès. Correspondants : Durand, Girardin, Morin, Sobrero, C. Calvert, J. Liebig, Taddei, Vogel, Redwood, Malaguti, Persoz, de Vrij, Christison. Quatrième série. 30 vols. (I–xxx). 1865–'79.

[d] Cinquième série. Journal de pharmacie et de chimie rédigé par Bussy, Fremy, L. Soubeiran, Regnauld, Lefort, Planchon, Riche, Coulier, Jungfleisch et Mialhe, contenant les travaux de la société de pharmacie de Paris. Une revue médicale par Vulpian, une revue des travaux de pharmacie publiés à l'étranger par M. Méhu, et une revue des travaux de chimie publiés à l'étranger par Jungfleisch. 23 vols. (I–xxiii). Paris, 1880–'91+

Table analytique des auteurs cités et des matières contenus dans les tomes I–xvi (1809–'30) du bulletin de pharmacie et des sciences accessoires. 8vo. Paris, 1831.

Table analytique des auteurs cités et des matières contenus dans les tomes xvii–xxvii (1831–'41) du journal de pharmacie et des sciences accessoires. 8vo. Paris, 1842.

92. [A] BULLETIN DES SÉANCES DE LA SOCIÉTÉ CHIMIQUE DE PARIS, publié par Adolphe Wurtz et Felix Le Blanc. 1858–62. 3 vols., 8vo. Paris, 1861–62.

   And simultaneously :

[B] Répertoire de chimie pure et appliquée. Compte rendu des progrès de la chimie pure en France et à l'étranger. Par Adolphe Wurtz, avec la collaboration de Chas. Friedel, Girard, LeBlanc et A. Riche, pour la France ; Williamson, pour l'Angleterre ; Lieben, pour l'Allemagne ; L. Schischkoff, pour la Russie ; Rosing, pour les pays Skandinaves ; Frapolli, pour l'Italie. 4 vols., 8vo. Paris, 1858–'62.

   Simultaneously with the above a section devoted to applied chemistry was published under the title :

[C] Répertoire de chimie pure et appliquée. Compte rendu des applications de la chimie en France et à l'étranger. Par Ch. Barreswil, avec la collaboration de Daniel Koechlin, Hervé Mangon, Em. Kopp, de Clermont, pour la France ; Knapp, Boettger, Sobrero, Rosing, Boutlerow, pour l'étranger. 5 vols., 8vo. Paris, 1858–'63.

BULLETIN DES SÉANCES DE LA SOCIÉTÉ CHIMIQUE DE PARIS. [Cont'd.]
   [A] *united with* [B] *in* 1863 *and with* [C] *in* 1864, *forming* [D].

[D] Bulletin de la Société chimique de Paris, comprenant le compte rendu des travaux de la Société et l'analyse des mémoires de chimie pure et appliquée publiés en France et à l'étranger, par Ch. Barreswil, J. Bouis, Ch. Friedel, E. Kopp, F. LeBlanc, A. Scheurer-Kestner et Ad. Wurtz, avec la collaboration de C. G. Foster, A. Girard, A. Lieben, A. Riche, A. Rosing, Thoyot, A. Vée et E. Willm. Nouvelle série. 18 vols., 8vo. 1864–1872.
   Continued under the title :

[E] Bulletin de la Société chimique de Paris, comprenant le procés-verbal des séances, les mémoires présentés à la Société, l'analyse des travaux de chimie pure et appliquée publiés en France et à l'étranger, la revue des brevets, etc. Comité de rédaction : J. Bouis, Ph. de Clermont, P. T. Clève, G. Daremberg, P. P. Dehérain, Ch. Friedel, Ch. Girard, A. Henninger, F. de Lalande, F. LeBlanc, A. Riche, G. Salet, P. Schutzenberger, G. Vogt, E. Willm, A. Wurtz. Nouvelle série. 32 vols. (XIX–L). Paris, 1873–1888.
   Continued under the same title :

[F] Troisième série. Secrétaire de la rédaction : M. Hanriot. 8 vols. Paris, 1889–92+

Table analytique des matières contenues dans le Bulletin de la Société chimique 1re et 2e Séries, 1858 à 1874, et dans les Répertoires de chimie pure et de chimie appliquée ; suivie de la Table alphabétique des auteurs dressés par Ed. Willm. 1 vol., 8vo. Paris, 1876.

Tables des années 1875 à 1888 du Bulletin de la Société chimique (Table analytique, Table alphabétique des auteurs) dressés par Th. Schneider. 1 vol., 8vo. Paris, 1890.
   Part II not issued in 1892.

93. BULLETIN DES SCIENCES MATHÉMATIQUES, ASTRONOMIQUES, PHYSIQUES ET CHIMIQUES. Rédigé par Saigey. Première section du bulletin universel des sciences et de l'industrie, publié sous la direction du Baron de Férussac. 16 vols., 8vo. Paris, 1824–'31 ‖

94. BULLETIN DES SCIENCES PHYSIQUES ET NATURELLES EN NÉERLANDE. Rédigé par F. A. W. Miquel, G. J. Mulder et W. Wenckebach. 1 vol., 4to, and 2 vols., 8vo. Leyde, Rotterdam et Utrecht. 1838–'40.

95. BULLETIN DES SUCRES FRANÇAIS ET ÉTRANGERS.   Paris, 1837-'38.

96. BULLETIN DU LABORATOIRE DE CHIMIE DU CAIRE.   [Dirigé par Albert Ismalun.]   Année 1881–82.   pp. 53, 12mo.   Caire, Décembre, 1882.

> The only number I have seen bears the additional title : Compterendu des travaux faits au laboratoire ; rapport présenté à son Excellence Chérif Pacha.

97. BULLETIN OF THE CHEMICAL SOCIETY OF WASHINGTON.   8 nos., 1884-'92.   Washington [D. C.], 1886-'93+

98. BULLETIN (THE) OF THE SCIENTIFIC LABORATORIES OF DENISON UNIVERSITY.   Edited by C. L. Herrick.   Granville, Ohio, 1886.

99. BULLETINS OF THE DIVISION OF CHEMISTRY, U. S. DEPARTMENT OF AGRICULTURE.   Washington, D. C., 1883-'93.

> For full titles *see* names of Authors and Editors *in Section V.*

No. 1. Composition of American Wheat and Corn; edited by Clifford Richardson.   1883.

No. 2. Diffusion applied to Sugar Cane and Sorghum in 1883 ; edited by H. W. Wiley.   1883.

No. 3. The Northern Sugar Industry ; edited by H. W. Wiley.   1883.

No. 4. Composition of American Wheat and Corn, second report ; edited by Clifford Richardson.   1884.

No. 5. The Sugar Industry of the United States ; edited by H. W. Wiley.   1885.

No. 6. Experiments with Diffusion and Carbonatation at Ottawa, Kansas, in 1885 ; edited by H. W. Wiley.   1885.

No. 7. Proceedings of the Second Meeting of Association of Official Agricultural Chemists ; edited by Chas. W. Dabney.   1885.

No. 8. Methods and Machinery for the Application of Diffusion ; edited by H. W. Wiley.   1886.

No. 9. Chemical Composition and Physical Properties of American Cereals, Wheat, Oats, Barley and Rye ; edited by Clifford Richardson.   1886.

No. 10. Principles and Methods of Soil Analysis ; edited by Edgar Richards.   1886.

No. 11. Report on Manufacture of Sugar at Magnolia Plantation, Lawrence, La., 1885–86 ; edited by G. L. Spencer.   1886.

No. 12. Proceedings of the Third Meeting of Association of Official Agricultural Chemists, 1886 ; edited by Clifford Richardson.   1886.

No. 13. Foods and Food Adulterants.

BULLETINS OF THE DIVISIONS OF CHEMISTRY. [Cont'd.]

Part First. Dairy Products ; edited by H. W. Wiley. 1887.

Part Second. Spices and Condiments ; edited by Clifford Richardson. 1887.

Part Third. Fermented Beverages ; edited by C. A. Crampton. 1887.

Part Fourth. Lard and Lard Adulterants ; edited by H. W. Wiley. 1889.

Part Fifth. Baking Powders ; edited by C. A. Crampton. 1889.

Part Sixth. Sugar, Molasses, Sirup, Confections, Honey and Beeswax ; edited by H. W. Wiley. 1892.

Part Seventh. Tea, Coffee and Cocoa Preparations ; edited by Guilford L. Spencer. 1892.

No. 14. Record of Experiments at Fort Scott, Kansas, in the Manufacture of Sugar from Sorghum and Sugar Canes ; edited by H. W. Wiley. 1886.

No. 15. Report on Manufacture of Sugar at Magnolia Plantation, 1886–87 (third report) ; edited by G. L. Spencer. 1887.

No. 16. Proceedings of the Fourth Meeting of Association of Official Agricultural Chemists ; edited by Clifford Richardson. 1887.

No. 17. Manufacture of Sugar from Sorghum and Sugar Cane, 1887–88 ; edited by H. W. Wiley. 1888.

No. 18. Sugar-Producing Plants : Experiments at Fort Scott, Rio Grande and Magnolia Plantation, 1887–88 ; edited by H. W. Wiley. 1888.

No. 19. Proceedings of the Fifth Meeting of Association of Official Agricultural Chemists ; edited by Clifford Richardson. 1888.

No. 20. Manufacture of Sugar from Sorghum at Rio Grande, Kenner, Conway Springs, Douglass and Sterling ; edited by H. W. Wiley. 1889.

No. 21. Manufacture of Sugar by Diffusion at Magnolia Plantation, 1888–89 ; edited by G. L. Spencer. 1889.

No. 22. Experiments at Des Lignes, 1888 ; edited by C. A. Crampton. 1888.

No. 23. Experiments at Calumet Plantation, 1889 ; edited by Hubert Edson. 1889.

No. 24. Proceedings of the Sixth Meeting of Association of Official Agricultural Chemists ; edited by H. W. Wiley. 1889.

BULLETINS OF THE DIVISION OF CHEMISTRY. [Cont'd.]

No. 25. Extent and Character of Food Adulterations ; edited by A. J. Wedderburn. 1890.

No. 26. Experiments in Production of Sugar from Sorghum in 1889 ; edited by H. W. Wiley. 1889.

No. 27. The Sugar Beet Industry ; edited by H. W. Wiley. 1889.

No. 28. Proceedings of the Seventh Meeting of Association of Official Agricultural Chemists ; edited by H. W. Wiley. 1890.

No. 29. Records of Experiments with Sorghum in 1890 ; edited by H. W. Wiley. 1891.

No. 30. Experiments with Sugar Beets in 1890 ; edited by H. W. Wiley. 1891.

Farmers' Bulletin No. 3. Culture of the Sugar Beet ; edited by H. W. Wiley. 1891.

No. 31. Proceedings of the Eighth Meeting of Association of Official Agricultural Chemists ; edited by H. W. Wiley. 1891.

No. 32. Special Report on the Extent and Character of Food Adulterations ; edited by Alex. J. Wedderburn. 1892.

No. 35. Proceedings of the Ninth Annual Convention of Association of Official Agricultural Chemists. 1893.

CAIRO, LABORATORY AT. *See* Bulletin du laboratoire de chimie du Caire.

CARL, J. *See* Allgemeine Hopfen-Zeitung.

CARLSBERG LABORATORIET. *See* Meddelelser fra Carlsberg Laboratoriet.

100. ČASOPIS CHEMIKŮ ČESKÝCH. Redaktor : Karel Otakar Čech. Roy. 8vo. v Praze, 1869.

Continued under the title :

[a] Časopis chemiků českých. Spolu organ spolku cukrovárníků východních Čech. Redaktor : Karel Otakar Čech. Hlavní spolupracovník : Frant. Štolba. 5 vols., roy. 8vo. v Praze, 1870-'74 ||

A Supplement to Průmyslník, *q. v.*

101. ČASOPIS CUKROVARNICKÝ. Redaktor : K. Preis. 3 vols. v Praze, 1872-74 ||

102. ČASOPIS PRO PRUMYSL CHEMICKÝ. Redaktor : Stolba a Bélohoubek. 1 vol., 4to. v Praze, 1891+

CATTANEO, ANTONIO. *See* Giornale di farmacia, chimica e scienze accessorie.

ČECH, KAREL OTAKAR. *See* Časopis chemiků českých ; *also* Průmyslník.

103. Centralblatt der gesammten chemischen Grossindustrie. Herausgegeben von A. Engelhardt, redigirt von H. Krätzer. 4to. Augsburg, 1891.

104. Centralblatt für Agricultur-Chemie und rationellen Wirthschaftsbetrieb. Referirendes Organ für naturwissenschaftliche Forschungen in ihrer Anwendung auf die Landwirthschaft. Herausgegeben von R. Biedermann. [*Edited in* 1875 *by* W. Detmer ; *in* 1876–'79 *by* R. Biedermann, unter Mitwirkung von Mor. Fleischer und Bernh. Tollens.] 18 vols., 8vo. Leipzig, 1872–'80.

Continued under the title :

[a] Biedermann's Centralblatt für Agrikultur-Chemie und rationellen Landwirthschafts-Betrieb. Referirendes Organ für naturwissenschaftliche Forschungen in ihrer Anwendung auf die Landwirthschaft. Fortgesetzt unter der Redaction von M. Fleischer und unter Mitwirkung von W. Borgmann, O. Kellner, A. König [*and others*]. 12 vols., 8vo. Leipzig, 1881–'92+

105. Centralblatt für Branntweinbrennerei. Die Vervollkommnungen in der Spiritus-Fabrikation und den Nebenzweigen. Herausgegeben von E. Kreplin. 4to. Leipzig, 1866–'69.

106. Centralblatt für die Textil-Industrie. Organ für die Gesammt-Interessen der Wollen-, Baumwollen-, Flachs- und Seiden-Industrie, Spinnerei, Weberei, Färberei, Druckerei, Bleicherei und Appretur. Organ des deutsch-oesterreichischen Webschullehrer-Verbandes. Redacteur : C. Sonntag [*later*, F. Stöpel, Friedrich Schulze]. 23 vols., 4to. Berlin, 1869–'92+

107. Český sladek. Red.: K. Suk. 4 vols. v Praze, 1878–'81 ‖

Chandler, C. F., and W. H. Chandler. *See* American Chemist.

108. Chemical (The) Gazette ; or, Journal of Practical Chemistry in all its applications to Pharmacy, Arts and Manufactures. Conducted by William Francis and Henry Croft. 17 vols., 8vo. London, 1843–'59.

Followed by :

[a] Chemical (The) News [*from* vol. iii, and Journal of Physical Science], with which is incorporated the "Chemical Gazette." A journal of practical chemistry, in all its applications to pharmacy, arts and manufactures. Edited by William Crookes. 66 vols., sm. 4to. London, 1860–'92+

109. CHEMICAL NEWS (THE) AND JOURNAL OF PHYSICAL SCIENCE, with which is incorporated the "Chemical Gazette." A journal of practical chemistry in all its applications to pharmacy, arts and manufactures. Edited by William Crookes. *Authorized American reprint.* 6 vols., 4to. New York, 1867–'70 ||
> *Followed by the* American Chemist, *q. v.*

CHEMICAL RECORD. *See* Pharmacist and Chemical Record.

110. CHEMICAL (THE) RECORD AND DRUG PRICE CURRENT. 1 vol., 4to. London, 1851, '52 ||

111. CHEMICAL REVIEW and Journal for the Spirit, Vinegar and Sugar Industry. Published by J. E. Siebel. 1 vol., 4to. Chicago, 1881.
> *Continued under the title :*

[a] American Chemical Review and Journal for the Spirit, Vinegar and Sugar Industry. Devoted to the interests of the arts of applied chemistry. 14 vols., 4to. Chicago, 1882–'91+

112. CHEMICAL (THE) REVIEW. A monthly journal for manufacturing chemists and druggists, dyers, printers, bleachers, sizers, paper-makers and stainers, leather-dressers, &c. 22 vols., 4to. London, 1871–'92+

CHEMICAL (THE) TIMES AND JOURNAL OF PHARMACY. *See* Pharmaceutical Times.

113. CHEMICAL TRADES JOURNAL (THE). Edited by George E. Davis. 3 vols., sm. 4to. Manchester, 1887–1890+

114. CHEMIKER-KALENDER. Herausgegeben von Rudolph Biedermann. 14 vols., 16mo. Berlin, 1880–'93+
> *Accompanied by a* " Beilage."

CHEMIKER-ZEITUNG. *See* Allgemeine Chemiker-Zeitung.

115. CHEMIKER (DER) UND DROGIST. Haupt-Organ für Chemiker, Drogisten, Gewerbtreibende, etc., [*later*] Correspondenzblatt des Vereins deutsche Berufs-Chemiker, Dresden. Herausgegeben von H. Krätzer. 8 vols., 4to. Leipzig, 1885–'92+

116. CHEMISCHE (DER) ACKERSMANN. Naturkundliches Zeitblatt für deutsche Landwirthe, herausgegeben von J. Adolph Stöckhardt. 21 vols., 8vo. Leipzig, 1855–'75 ||

117. CHEMISCHE ANNALEN FÜR DIE FREUNDE DER NATURLEHRE, Arzneigelahrtheit, Haushaltungskunst und Manufacturen von Lorenz Crell.   40 vols. (2 vols. *per annum, not numbered*), 8vo. Helmstädt und Leipzig, 1784–1803.

> *Accompanied by :*

Beiträge zu den chemischen Annalen von Lorenz Crell.   6 vols., 8vo.   Leipzig und Dessau, 1785–'99.   Vol. I, 1785–'86 ; II, 1787 ; III, 1788 ; IV, 1789 ; V, 1794 ; VI, 1799.

Vol. IV *also under the title :* Beiträge zur Erweiterung der Chemie.

> *Cf.* Chemisches Archiv ; *also* Chemisches Journal.

118. CHEMISCHE EN PHYSISCHE OEFENINGEN VOOR DE BEMINNAARS DER SCHEI- EN NATUURKUNDE.   Door P. J. Kastelyn, vervolgt door Bondt en J. R. Deiman.   3 vols.   Amsterdam en Leyden, 1788.

> *Followed by :*

[a] Nieuwe chemische en physische oefeningen.   1 part, 8vo.   1797.

119. CHEMISCHE (DIE) INDUSTRIE.   Monatsschrift [*later*, Zeitschrift], herausgegeben vom Verein zur Wahrung der Interessen der chemischen Industrie Deutschlands.   Redacteur : Emil Jacobsen.   15 vols., 4to.   Berlin, 1878–'92+

120. CHEMISCHES ARCHIV.   Herausgegeben von L. F. F. von Crell.   2 vols., 8vo.   Leipzig, 1783.

> *Continued under the title :*

[a] Neues chemisches Archiv.   Herausgegeben von L. F. F. von Crell.   8 vols., 8vo.   Leipzig, 1784–'91.

> *Followed by :*

[b] Neuestes chemisches Archiv.   Herausgegeben von L. F. F. von Crell.   1 vol., 8vo.   Weimar, 1798 ||

> *Cf.* Chemische Annalen ; *also* Chemisches Journal.

CHEMISCHES CENTRALBLATT.   *See* Pharmaceutisches Centralblatt.

121. CHEMISCHES JOURNAL FÜR DIE FREUNDE DER NATURLEHRE, Arzneygelahrtheit, Haushaltungskunst und Manufakturen.   Entworfen von Lorenz Crell.   6 vols., 12mo.   Lemgo, 1778–'81.

> *Continued under the title :*

[a] Entdeckungen (Die neuesten) in der Chemie ; gesammelt von Lorenz Crell.   13 vols., 8vo.   Leipzig, 1781–'86 ||

> *Cf.* Chemische Annalen für die Freunde der Naturlehre ; *also* Chemisches Archiv.

122. CHEMISCH-PHARMACEUTISCH ARCHIEF ; uitgegeven door J. E. de Vrij, B. Eickma en A. F. van der Vliet.   2 vols.   Schoonhoven, 1840–'41.

CHEMISCH-PHARMACEUTISCHES CENTRALBLATT.
See Pharmaceutisches Centralblatt.

123.   CHEMISCH-TECHNISCHE ZEITUNG.   Herausgegeben von Otto Prinz.
6 vols.   Leipzig, 1883–88 ‖
> *Accompanied by a* " Beiblatt " : Chemisch-technischer Central-Anzeiger, *q. v.*

124.   CHEMISCH-TECHNISCHEN MITTHEILUNGEN (DIE) DER NEUESTEN
ZEIT ; ihrem wesentlichen Inhalte nach alphabetisch zu-
sammengestellt.   Herausgegeben von L. Elsner [*later,* fortge-
führt von F. Elsner ; *later,* von Ch. Heinzerling].   1846–'86.
37 vols., 8vo (4to).   Berlin, Halle a. S., 1849–'86+
Sach-Register zu den ersten acht Heften der chemisch-tech-
nischen Mittheilungen der neuesten Zeit ihrem wesentlichen
Inhalte nach alphabetisch zusammengestellt von L. Elsner.
Die Jahre 1846–'59 enthaltend.   8vo.   Berlin, 1860.
Sach-Register zu den bisher erschienen zwanzig Heften [etc.].
Die Jahre 1846–'71 enthaltend.   8vo.   Berlin, 1873.

125.   CHEMISCH-TECHNISCHER CENTRAL-ANZEIGER.   Fach- und Handels-
blatt für Chemiker, Techniker, Drogisten, Apotheker, Fabrikanten.   Central-Insertions-Organ für die gesammten chemischen
Industriezweige und deren Hilfsindustrieen.   Verantwörtlicher
Redacteur : Karl Barthel [*later,* Otto Prinz, *also* Gumpert].
10 vols., sm. fol.   Leipzig, 1883–'92+
> *Begun as a* " Beiblatt " *to* chemisch-technische Zeitung, *q. v.*

126.   CHEMISCH-TECHNISCHES REPERTORIUM.   Uebersichtlich geordnete
Mittheilungen der neuesten Erfindungen, Fortschritte und Ver-
besserungen auf dem Gebiete der technischen und industriellen
Chemie mit Hinweis auf Maschinen, Apparate und Literatur.
Herausgegeben von Emil Jacobsen.   1862–'91.   30 vols., 8vo.
Berlin, 1863–'92+
General-Register zu Jahrg. I–V (1862–66).   Berlin, 1867.
General-Register zu Jahrg. VI–X (1867–71).   Berlin, 1873.
General-Register zu Jahrg. XI–XV (1872–76).   Berlin, 1879.
General-Register zu Jahrg. XVI–XX (1877–81).   Berlin, 1884.
General-Register zu Jahrg. XXI–XXV (1882–86).   Berlin, 1889.

127.   CHEMIST (THE).   [Edited by Mongredieu ?]   2 vols., 8vo.   London,
1824–'25 ‖

128.   CHEMIST (THE) ; or, Reporter of Chemical Discoveries and Improvements and Protector of the Rights of the Chemist and

CHEMIST (THE). [Cont'd]

> Chemical Manufacturer. Edited by Charles Watt and John Watt. 6 vols., 8vo. London, 1840–'45.
>
> *Continued under the title :*

[a] Chemist (The) ; or, Reporter of Discoveries and Improvements in Analytical, Manufacturing, and Agricultural Chemistry. Edited by John Higgs Newton. 1 vol. (VII), 8vo. London, 1846–'48.

> *Continued under the title :*

[b] Chemist (The). A monthly journal of chemical philosophy and of chemistry, applied to the arts, manufactures, agriculture, and medicine, and record of pharmacy. Edited by John and Charles Watt. New series. 4 vols., 8vo. London. 1849–'53.

> *Continued under the title :*

[c] Chemist (The). A monthly journal of chemical and physical science. Edited by John and Charles Watt. New series. 5 vols., 8vo. London, 1854–'58 ||

129. CHEMIST (THE) AND DRUGGIST. A monthly trade circular. 12 vols., 8vo. London, 1859–'71.

> *Continued under the same title :*

[a] 31 vols. (XIII–XXXI), 4to. London, 1872–'90+

130. CHEMIST (THE) AND METEOROLOGICAL JOURNAL. July 8th to Dec. 9th, 1826. John R. Cotting, Editor. Amherst, Mass. 8vo. *n. d.*

131. CHEMISTS' AND DRUGGISTS' ADVOCATE. London, 1873–'80.

132. CHEMISTS' (THE) DESK COMPANION for 1865[–'66]. The Year-Book of Pharmacy. A practical summary of researches in pharmacy, materia medica and pharmaceutical chemistry. Edited by Chas. Wood and Chas. Sharp. 2 vols., 8vo. London, 1865–'66.

> *Cf.* Year-Book of Pharmacy.

133. CHEMISTS' (THE) JOURNAL. 6 vols. London, 1880–'82+

CHEVREUL, E. *See* Annales de chimie et de physique. *Also* Journal de chimie médicale.

CHIAPPERO, F. *See* Giornale di farmacia, chimica [etc.].

CHICAGO COLLEGE OF PHARMACY. *See* Pharmacist and Chemical Record.

CHEVALIER, FEE, GUIBOURT, JULIA FONTENELLE, LAUGIER, ORFILA [etc.].
See Journal de chimie médicale.

134. CHIMISTE (LE). Journal de chimie appliqué aux arts, à l'industrie et à l'agriculture ; publié par Henri Bergé. 5 vols., 8vo. Bruxelles, 1865-'69.

135. CHIMISTE (LE), JOURNAL DES DISTILLATEURS. Organe spécial de la chimie appliquée à la distillation et à la conservation des vins et spiritueux. Rédacteur-en-chef : Simon. 4to. Paris, 1859, '60.

CHODOUNSKY, F. *See* Pivovarnické Zprávy.

COLLEGE (ROYAL) OF CHEMISTRY. *See* Reports.

136. COLORATION (LA) INDUSTRIELLE. Journal spécial de la teinture et de l'apprêt des étoffes, de la production et de la préparation des matières tinctoriales, de l'impression et de la fabrication des tissus et des papiers peints, et en général de toutes les substances colorantes appliquées à l'industrie et aux arts. [Edited by Lecouturier.] 4to. Paris, 1857-'61.

Continued under the title :

[a] Coloriste (Le) industriel. Journal de la teinturerie parisienne, du blanchiment et de l'apprêt des étoffes, de la production et de la préparation des matières colorantes, de l'impression et de la fabrication des tissus. Par Alcan, Trélat, Péligot, Wolowski, Payen, Persoz, Chevreul. Deuxième série. 4to. Paris, 1862-'66.

Continued under the title :

[b] Moniteur de la teinture et de la fabrication des tissus. Journal scientifique industriel et commercial, spécialement consacré à l'apprêt ; à la teinture et à l'impression des étoffes ; à la fabrications des papiers peints ; à la production et à la préparation des matières tinctoriales ; en général à tous les produits colorantes et matières textiles employés dans l'industrie et dans les arts. Par A. Félix Gouillon. [From 1872 by the same with P. F. Blondeau.] Troisième série. 4to. Paris, 1867-'92+

[Vol. XXXVI in 1892.]

COLORISTE (LE) INDUSTRIEL. *See* Coloration (La) industrielle.

COLUMBIAN CHEMICAL SOCIETY. *See* Memoirs of the Columbian Chemical Society.

137. COMPTES-RENDUS HEBDOMADAIRES DES SÉANCES DE L'ACADÉMIE DES SCIENCES, publiés conformément à une décision de l'Aca-

COMPTES-RENDUS HEBDOMADAIRES, etc. [Cont'd.]
démie en date du 13 Juillet, 1835, par MM. les Secrétaires
perpétuels. 115 vols., 4to. Paris, 1835–92+

Supplément aux Comptes-Rendus hebdomadaires des séances de
l'Académie des Sciences, publiés conformément à une décision de l'Académie en date du 13 Juillet, 1835, par MM. les
Secrétaires perpétuels. 2 vols., 4to. Paris, 1856 *and* 1861.

Table générale des Comptes-Rendus hebdomadaires des séances
de l'Académie des Sciences, publié par MM. les Secrétaires
perpétuels, etc. Tomes I–XXXI, 3 Août 1835 à 30 Décembre
1850. 1 vol., 4to. Paris, 1853.

138. COMPTES-RENDUS MENSUELS DES TRAVAUX CHIMIQUES de l'étranger,
ainsi que les laboratoires de Bordeaux et de Montpellier,
rédigés, avec la collaboration de A. Laurent, par Charles Ger-
hardt. 7 vols., 8vo. Montpellier, 1845–'51 ||

> *The paper cover also bears the title :* Annuaire de chimie.

CORPS (LES) GRAS INDUSTRIELS.
> *See* Journal de l'exploitation des corps gras industriels.

139. CORRESPONDENZBLATT DES VEREINS AKADEMISCH-GEBILDETER
ZUCKERTECHNIKER. Redigirt von J. Baumann. 8vo. Berlin,
1891+

140. COUNTRY BREWERS' GAZETTE. 4 vols. London, 1876–88+

CRELL, LORENZ VON.
> *See* Auswahl aller eigenthümlichen Abhandlungen ; *also* Auswahl
> vorzüglicher Abhandlungen ; *also* Chemische Annalen für die
> Freunde der Naturlehre ; *also* Chemisches Archiv ; *also* Che-
> misches Journal.

141. CRELL'S CHEMICAL JOURNAL. Giving an account of the latest
discoveries in chemistry, with extracts from various foreign
transactions ; translated from the German, with occasional
additions. 3 vols , 8vo. London, 1791–'93.
> *An English edition of* Chemische Annalen für die Freunde der
> Naturlehre, *q. v*

CROOKES, WM. *See* Chemical Gazette [*b*].

DABNEY, CHARLES W. *See* Association of Official Agricultural Chemists.

DAVIS, GEORGE E. *See* Chemical Trades Journal.

DELBRÜCK UND HAYDUCK. *See* Wochenschrift für Brauerei.

DEITE, C.   *See* Seifenfabrikant (Der).

DENZER'S GEWERBEBLATT.   *See* Technisch-chemisches Gewerbeblatt.

DETMER, W.   *See* Jahresbericht der Agriculturchemie.

142.   DEUTSCHE AMERICANISCHE BRAUER-ZEITUNG.   O. Mega, Editor.
  Fol.   New York, 1874, '75.

143.   DEUTSCHER BRAUER-KALENDER.   Unter Mitwirkung der ersten
  Fachmänner, herausgegeben von Carl Homann.   2 vols., 16mo.
  Nürnberg, 1877, '78 ‖

144.   DEUTSCHE (DER) BRAUEREI-KALENDER.   Eine gedrängte Sammlung der wichtigsten Tabellen, Berechnungen, Hilfsmittel und
  täglichen Controll-Listen bei Führung der Brauerei [etc.],
  nebst einem Notizbuch zum praktischen Gebrauch.   Herausgegeben von Alb. Hayn.   1 vol., 16mo.   Dresden, 1877.
         Continued under the title :
  [a] Deutsche (Der) Brauerei-Kalender für die Brauerei-Campagne
      1877.   Eine gedrängte Sammlung [etc.].   Herausgegeben
      von Alb. Hayn unter Mitwirkung von Henry T. Böttinger.
      3 vols., 16mo.   Dresden [*later*, Frankfurt a. O.], 1877–'80.

DEUTSCHE BRAU-INDUSTRIE.   *See* Norddeutsche Brauer-Zeitung.

145.   DEUTSCHE CHEMIKER-ZEITUNG.   Centralblatt für die chemische
  Praxis und öffentliche Gesundheitspflege.   Unter Mitwirkung
  bewährter Fachgelehrten herausgegeben und redigirt von
  Breslauer.   7 vols., 4to.   Berlin, 1886–92+

DEUTSCHE CHEMISCHE GESELLSCHAFT.   *See* Berichte der ――.

DEUTSCHE FÄRBER-ZEITUNG.   *See* Thüringer Muster-Zeitung.

146.   DEUTSCHER CHEMIKER-KALENDER.   Jahrbuch und Notizbuch für
  den theoretischen und praktischen Chemiker, Fabrikanten,
  Bierbrauer, Branntweinbrenner, Zuckerfabrikanten.   Herausgegeben von Paul Bennewitz.   [*Later*, von H. von Gehren.]
  3 vols., 16mo.   Dresden, 1875–'77 ‖

147.   DEUTSCHE GERBER-ZEITUNG.   Zeitung für Lederfabrikation und
  Lederhandel.   Organ des Vereins deutscher Gerber.   Redigirt
  von F. A. Günther.   4 vols., fol.   Berlin, 1865–'68.
         Vol. IV, 1868, is styled the II. Jahrgang.
         Continued under the title :
  [a] Deutsche Gerber-Zeitung.   Organ des (ersten) Vereins deutscher

Deutsche Gerber-Zeitung. [Cont'd.]
Gerber. Redacteur : F. A. Günther [und] A. Schöniger. 3 vols. (xii–xiv), fol. Berlin, 1869–'71.

Continued under the title :

[b] Deutsche Gerber-Zeitung. Deutsche Sattler- und Wagenbau-Zeitung. Organ für Leder-Industrie und Lederhandel. Organ des (ersten) deutschen Gerber-Vereins. Redacteur : F. A. Günther [und] A. Schöniger. [Vol. xvii, by J. W. Dust.] 3 vols. (xv–xvii). Berlin, 1872–'74.

Continued under the title :

[c] Deutsche Gerber-Zeitung. Organ für Leder-Industrie, Leder-handel und Leimfabrikation. Organ des (ersten) deutschen Gerber-Vereins. Redacteur : J. V. W. Dust. Mit Beiblatt : Sorgenfrei ; Wochenschrift für Unterhaltung und Belehrung. 1 Jahrgang. 1 vol. (xviii), fol. Berlin, 1875.

Continued under the title :

[d] Deutsche Gerber-Zeitung. Organ für Leder-Industrie, Leder-handel und Leimfabrikation, des (ersten) deutschen Gerber-Vereins und des von der Westdeutschen Versicherungs-Aktien-Bank in Essen garantirten Feuer-Versicherungs-Verbandes für die deutsche Leder-Industrie. Redaction-Commission : F. A. Günther, A. Schöniger. Mit Beiblatt : Sorgenfrei ; Wochenschrift für Unterhaltung und Belehrung. 2–3 Jahrgang. 2 vols. (xix, xx), fol. Berlin, 1876, '77.

Continued under the title :

[e] Deutsche Gerber-Zeitung. Organ des Centralverbandes der deutschen Lederindustriellen, des von der West-deutschen Versicherungs-Actien-Bank in Essen garantirten Feuer-Versicherungs-Verbandes für die deutsche Leder-Industrie und des von der "Victoria" zu Berlin allgemeine Ver-sicherungs-Actien Gesellschaft garantirten Lebens-Ver-sicherungs-Verbandes für die deutsche Lederindustrie. Mit Beiblatt : Sorgenfrei ; Wochenschrift für Unterhaltung und Belehrung. 4–18 Jahrgang. Redaction-Commission : F. A. Günther, H. Schöniger. 15 vols., fol. Berlin, 1878–'92+

148. Deutsche Muster-Zeitung für Färberei, Druckerei, Blei-cherei, Appretur. Darstellung verwandter technischer und chemischer Erzeugnisse. Herausgegeben von Eduard Bon und Wilhelm Grüne, jun. 5 vols., 4to. Berlin, 1850–'54.
Neue Folge. Redigirt von Hermann Reidel [from 1859, von R. Engels]. 11 vols. (i–xi), 4to. Berlin, 1857–'67.

DEUTSCHE MUSTER-ZEITUNG, etc. [Cont'd.]

Continued under the title :

[a] Muster-Zeitung für Färberei, Druckerei, Bleicherei, Appretur, Darstellung verwandter technischer und chemischer Erzeugnisse. Redacteur : M. Reimann und E. Wolffenstein. 3 vols. (XVII–XIX), 4to. Berlin, 1868–'70.

Continued under the title :

[b] Muster-Zeitung für Färberei, Druckerei, Bleicherei, Appretur von Gespinnsten, Geweben, Papieren, etc., und für die gesammte Farbenanwendung unter besonderer Berücksichtigung der Spinnerei, Weberrei, etc. Redigirt von Wilh. Grüne und Hermann Grothe. 2 vols. (XX, XXI), 4to. Berlin, 1871, '72.

Continued under the title :

[c] Muster-Zeitung. Zeitschrift für Färberei, Druckerei, Bleicherei, Appretur von Gespinnsten, Geweben, Papieren, etc., und für die gesammte Farbenanwendung. Redacteur : Ferd. Springmühl. 2 vols. (XXII, XXIII), 4to. Berlin, 1873, '74.

Continued under the title :

[d] Muster-Zeitung des Färbers. Zeitschrift für Färberei, Bleicherei, Druckerei, Appretur von Gespinnsten, Geweben, Papieren, etc., und Farbenfabrikation. Redacteur : Ferd. Springmühl. 1 vol. (XXIV), 4to. Berlin, 1875.

Continued under the title :

[e] Muster-Zeitung. Zeitschrift für Färberei, Bleicherei, Druckerei, Appretur von Gespinnsten, Geweben, Papieren, etc., und Farbenfabrikation. [From 1877] Central-Organ für Veröffentlichung neuer Erfindungen und Verbesserungen auf dem Gesammtgebiete der Färbereien, Zeugdruckereien, Kattunfabriken, etc. Redakteur : Ferd. Springmühl. 7 vols. (XXV–XXVIII), 4to. Leipzig, 1876–'79.

Continued uuder the title :

[f] Färberei-Muster-Zeitung. Wochenschrift für Färberei, Bleicherei, Druckerei, Appretur und Farbenfabrikation. Centralorgan sämmtlicher Färber-Vereine für Veröffentlichung neuer Erfindungen und Verbesserungen auf dem Gesammtgebiete der Farbenchemie für Färbereien, Zeugdruckereien, Kattunfabriken, etc. Herausgegeben von einem Verein von Fachmännern. 11 vols. (XXIX–XXXIX), 4to. Ill. Leipzig, 1880–'90.

Continued under the title :

[g] Leipziger Färber-Zeitung. Wochenschrift für Färberei, Bleicherei, Druckerei, Appretur, Farben- und Chemikalien-

DEUTSCHE MUSTER-ZEITUNG, etc. [Cont'd.]
Fabrikation und Handel. Central - Organ sämmtlicher
Färber-Vereine für Veröffentlichung neuer Erfindungen und
Verbesserungen auf dem Gesammtgebiete der Farbenchemie
für Färbereien, Zeugdruckereien, Kattunfabriken, Farben-
und Chemikalien-Fabriken, Farbenhändler, Chemiker, Coloristen, Bunt- und Teppichwebereien, Buntpapier- und Tape-
tenfabriken, Appreteure und Bleicher. 2 vols., 4to. Leipzig,
1891–92+

149. DEUTSCHE (DIE) ZUCKER-INDUSTRIE. Wochenblatt für Landwirthschaft, Fabrication und Handel. Redacteur : Wilh.
Herbertz. 16 vols., 4to. Berlin, 1876–'91+

Accompanied by :

Wöchentlicher Markt-Bericht für den internationalen Zucker-
Handel. Berlin, 1886–'91.

DEUTSCHES JAHRBUCH FÜR DIE PHARMACIE.
See Berlinisches Jahrbuch der Pharmacie.

150. DEUTSCHES UND AMERIKANISCHES BRAUER-JOURNAL UND GERSTE-,
MALZ- UND HOPFEN-REVUE. A. E. J. Tovey, Redacteur.
5 vols., 4to. New York, 1876–'80 [+?].

Cf. German (The) and American Brewers' Journal.

151. DIAMANT (DER). Technisch-komerziell-soziale Fachzeitschrift für
die gesamte Glasindustrie. Organ der Glaser-Innungen, des
Glashandels, der Glasmalerei, Glasschleiferei, etc. Redacteur :
Alexander Duncker. 4 vols., 4to. Leipzig, 1879–'82.

DINGLER, J. G. *See* Journal für Zitz-, Katun- und Indiennendruckerei.

DINGLER'S POLYTECHNISCHES JOURNAL. *See* Polytechnisches Journal.

DIVISION OF CHEMISTRY, U. S. DEPARTMENT OF AGRICULTURE.
See Bulletins of the Division of Chemistry [etc.].

DOYLE, P. HENRY. *See* Brewers' (The) and Dealers' Journal.

DRUG, PAINT AND OIL TRADE. *See* Oil, Paint and Drug Reporter.

152. DRUGGIST (THE) AND CHEMIST. Edited by C. C. Vanderbeck.
8vo. Philadelphia, 1878–'79.

DRUGGISTS' CIRCULAR AND CHEMICAL GAZETTE.
See American Druggists' Circular and Chemical Gazette.

153. Dublin Journal of Medical and Chemical Science. Exhibiting a comprehensive view of the latest discoveries in medicine, surgery, chemistry and the collateral sciences. 28 vols., 8vo. Dublin, 1832–'45.

> *Continued under the title :*

[a] Dublin Quarterly Journal of Medical Science. [Exclusively medical.]

Duncker, Alexander. *See* Diamant (Der).

Durand, Émile et Paul. *See* Annuaire général de l'industrie de l'éclairage.

Dust, J. V. W. *See* Deutsche Gerber-Zeitung.

154. Dyer (The), Calico-Printer and Textile-Review. London, 1881–'90.

Eberstein, Mor. Freiherr von. *See* Brod (Das).

155. Éclairage (L') public. Organe spécial de l'industrie du gaz. Commerce, hygiène, fabrication. Fol. Paris, 1862.

156. Edinburgh (The) Journal of Science. Exhibiting a view of the progress of discovery in natural philosophy, chemistry, mineralogy, geology, botany, zoölogy, comparative anatomy, practical mechanics, geography, navigation, statistics, antiquities and the fine and useful arts. Conducted by David Brewster ; with the assistance of John MacCulloch for geology, chemistry, etc. ; W. Jackson Hooker for botany ; John Fleming for natural history ; Will. Haidinger for mineralogy ; Robt. Knox for zoölogy and comparative anatomy ; Sam. Hibbert for antiquities and geology. 10 vols., 8vo. Edinburgh, 1824–'29.

[a] Second series. 6 vols., 8vo. Edinburgh, 1829–'32.

> *United in* 1832 *with the* " Philosophical Magazine, or Annals " [etc.],
> *forming the* " London and Edinburgh Philosophical Magazine."
> *See* Philosophical Magazine.

157. Elsässische Hopfen- und Brauer-Zeitung. — vols. Hagenau, 18— –'80 [+?].

Elsner, L. (and F.). *See* Chemisch-technische Mittheilungen.

Elwert, J. C. *See* Magazin für Apotheker, Chemisten [etc.].

Engelhardt, Alwin. *See* Seifensiederei-Zeitung (Neue).

Entdeckungen (Die neuesten) in der Chemie. *See* Chemisches Journal.

Erdmann, Otto L. *See* Allgemeines Journal der Chemie [*d*].

ERDMANN'S JOURNAL.   *See* Journal für technische und ökonomische Chemie.

ERLENMEYER, E.   *See* Kritische Zeitschrift für Chemie.

158. EXPERIMENT STATION RECORD. Published by authority of the Secretary of Agriculture, U. S. Department of Agriculture. Office of Experiment Stations, W. O. Atwater, director.   3 vols., 8vo.   Washington, 1889–92+

159. FACH-ZEITSCHRIFT FÜR DIE CHEMISCHE SEITE DER TEXTIL-INDUS-TRIE. (Mittheilungen des technologischen Gewerbe-Museums.) 1 vol.   Wien, 1883.

FÄRBER-ZEITUNG.   *See* Reimanns Färber-Zeitung.

FÄRBER-ZEITUNG MIT MUSTERN.   *See* Muster-Zeitung für den Färberstand.

FÄRBEREI MUSTER-ZEITUNG.   *See* Deutsche Muster-Zeitung.

FAHDT, JULIUS.
       See Glashütte (Die) und Keramik ; *also* Monatsbericht der Glashütte.

FALK, CARL FABIAN.   *See* Brewing (The) World.

160. FARBEN-INDUSTRIE (DIE). Vierteljahrsschrift über die Leistungen auf dem Gebiete des Steinkohlentheers, der Chemie der aromatischen Verbindungen, des künstlichen Farbestoffe, der Färberei, Bleicherei des Zeugdruckes und der Appretur, in Verein mit fachkundigen Mitarbeitern herausgegeben von E. Börnstein.   6 pts., 4to.   Berlin, 1888–90 ||

FASBENDER, FRANZ.   *See* Oesterreichische Zeitschrift für Bierbrauerei ; *also* Oesterreichisch-ungarischer Brauer- und Mälzer-Kalender ; *also* Allgemeine Zeitschrift für Bierbrauerei.

FECHNER, GUSTAV THEODOR.   *See* Repertorium der neuen Entdeckungen in der unorganischen Chemie ; *also* Repertorium der organischen Chemie.

FÉRUSSAC, BARON DE.
       See Bulletin des sciences mathématiques . . . et chimiques.

FISCHER, F.   *See* Zeitschrift für angewandte Chemie ; *also* Zeitschrift für die chemische Industrie.

FLINTOFF, JOHN.   *See* American (The) Brewers' Gazette.

161. FORTSCHRITTE (DIE) AUF DEM GEBIETE DER TECHNISCHEN CHEMIE. 1874–'76.   1 no., 8vo.   Leipzig, 1877.
       Reprinted from Vierteljahres-Revue der Naturwissenschaften.

162. FORTSCHRITTE (DIE) AUF DEM GEBIETE DER THEORETISCHEN
CHEMIE. 1 no., 8vo. Leipzig, 1874.
> *Reprinted from* Vierteljahres-Revue der Naturwissenschaften.

163. FORTSCHRITTE (DIE) DER CHEMIE. Herausgegeben von H. J.
Klein. 6 nos., 8vo. Köln, 1879–'85.
> *Reprinted from* Vierteljahres-Revue der Naturwissenschaften.

FRANCIS (WM.) AND CROFT (HENRY). *See* Chemical Gazette.

FRANK, M. *See* Textile (The) Colorist.

FREAR, WILLIAM. *See* Agricultural Science.

FRESENIUS' ZEITSCHRIFT. *See* Zeitschrift für analytische Chemie.

164. GÄHRUNGSTECHNISCHES JAHRBUCH. Bericht über die wissenschaftlichen und gewerblichen Fortschritte auf dem Gebiete
der Brauerei, Brennerei, Presshefenfabrikation, Weinbereitung,
Essigfabrikation, Molkerei, Kälteerzeugung, Stärke-, Dextrin-
und Stärkezuckerfabrikation. Herausgegeben von A. Schrohe.
1891. 8vo. Berlin, 1892+

GAJANI, MARIANO. *See* Giornale di farmacia, chimica [etc.].

165. GAMBRINUS BRAUER-ZEITUNG. Herausgegeben und redigirt von
Sigmund Spitz und Adolf Lichtblau unter Mitarbeitung hervorragender Fachmänner. 17 vols., 4to. Wien, 1874–'90+

166. GAS AND WATER. London, 1884–86.
> *Continued under the title :*

[a] Gas World. London, 1886–90.

167. GAS ENGINEER. Birmingham, 1884–87.
> *Continued under the title :*

[a] Gas Engineer's Magazine. Birmingham, 1888–89.

168. GAS-KALENDER. Zum Gebrauche für Gasanstalts-Dirigenten, Gas-
Techniker, sowie Gas- und Wasser-Installateure. Bearbeitet
von G. F. Schaar. 3 vols., 16mo [later, 12mo]. Leipzig,
1878–80.
> *Continued under the title :*

[a] Kalender für Gas- und Wasserfach-Techniker. Zum Gebrauche
für Dirigenten und technische Beamte der Gas- und Wasser-
werke, sowie für Gas- und Wasser-Installateure. Bearbeitet
von G. F. Schaar. [IV–XVI], 12mo. München, 1881–'92.

GAUGER, C. *See* Repertorium für Pharmacie und praktische Chemie.

GAULTIER DE CLAUBRY, HENRI FRANÇOIS. *See* Répertoire de chimie.

GAY-LUSSAC ET ARAGO. *See* Annales de chimie.

169. GAZ (LE), Journal des consommateurs des gaz d'éclairage et de chauffage. [*Later.*] Organe des intérêts de l'industrie de l'éclairage et du chauffage par le gaz. 34 vols., 4to. Paris, 1857-'90+

170. GAZZETTA CHIMICA ITALIANA. Ed. da M. Paterno. 22 vols., roy. 8vo. Palermo, 1871-'92+

> *Since* 1884 *accompanied by :*

> APPENDICE diretto da E Paterno. 6 vols., 8vo. Palermo, 1884–88.

> Devoted to extracts from foreign chemical periodicals.

171. GAZZETTA ECLETTICA DI CHIMICA TECNOLOGICA. Red. : Sembenini. — vols., 8vo. Verona, 1833-'34.

172. GAZZETTA ECLETTICA DI FARMACIA E CHIMICA MEDICA. Red. : Sembenini. — vols., 8vo. Verona, 1831-'34.

> *United with* Gazzetta eclettica di chimica tecnologica *and continued under the title :*

[a] Gazzetta eclettica di chimica farmaceutica medica tecnologica e di rispettiva letteratura e commentario della conversazione chimico-farmaceutica. Red. : Sembenini. — vols., 8vo. Verona, 1835-'37.

Serie terza. — vols. 1838-'39 [+ ?].

> *Cf.* Annuario delle scienze chimiche, farmaceutiche e medico-legali.

173. GAZZETTA DI FARMACIA E DI CHIMICA che da prima pubblicavasi in Este. 8vo. Venezia, 1855-'57.

GEHLEN, A. F. *See* Allgemeines Journal der Chemie [*a*] ; *also* Repertorium für die Pharmacie.

GEIGER, PH. LORENZ. *See* Annalen der Pharmacie.

GEIGER'S MAGAZIN. *See* Nordische Blätter für Chemie [*b*].

174. GERBER (DER). Wien, 1887–90.

GERBER-ZEITUNG. *See* Deutsche Gerber-Zeitung.

175. GERBER-ZEITUNG. Zeitung für Lederfabrikation und Lederhandel. Organ des Vereins deutscher Gerber [*later*] und des Versicher-ungs–Verbandes deutscher Lederindustrieller. Redigirt von G. Kerst. [From 1876 by G. Lewinstein, *later* with Kampff-meyer.] 35 vols., fol. Berlin, 1858–'92+

176. GERMAN (THE) AND AMERICAN BREWERS' JOURNAL and Barley, Malt, and Hop Traders' Reporter. Published in the German and English Languages. Edited by A. E. J. Tovey. 15 vols., fol. New York, 1876–'90+

   Cf. Deutsches und Amerikanisches Brauer-Journal.

GEYER, J. C. G. *See* Thüringer Muster-Zeitung.

GILBERT'S ANNALEN. *See* Journal der Physik. Gren.

177. GIORNALE DI CHIMICA E VETERINARIA. Di Pozzi. 8vo. Milano, 18— (?).

178. GIORNALE DI FARMACIA, CHIMICA E MATERIA MEDICA APPLICATA ANCHE ALLA VETERINARIA. Diretto da Mariano Gajani. 4to. Ancona, 1861–'62.

179. GIORNALE DI FARMACIA, CHIMICA E DI SCIENZE AFFINI. Diretto da F. Chiappero ; pubblicato dalla società di farmacia di Torino. 29 vols., 12mo. Torino, 1852–'80+

180. GIORNALE DI FARMACIA, CHIMICA E SCIENZE ACCESSORIE ; ossia raccolta delle scoperte, ritrovati e miglioramenti fatti in far-macia ed in chimica. Compilato da Antonio Cattaneo. 19 vols., 8vo. Milano, 1824–'34.
   Continued under the title :

   [a] Biblioteca di farmacia, chimica, fisica, medicina, chirurgia, tera-peutica, storia naturale [etc.] ; compilato da Antonio Cat-taneo. 23 vols. (I–XXIII), 8vo. Milano, 1834–'45.
   Continued under the title :

   [b] Annali di chimica applicata alla medicina, cioè, alla farmacia, alla tossicologia, all' igiene, alla fisiologia, alla patologia ed alla terapeutica, compilati da Giovanni Polli. 80 vols. (I–LXXX), 8vo. Milano, 1845–'84+
      United with Rivista di chimica medica e farmaceutica *and continued under the title :*

   [c] Annali di chimica medico-farmaccutica e di farmacologia. Diret-tori, P. Albertoni, I. Guareschi ; Condirettori, A. Pavesi, G. Colombo. 1 vol. Milano, 1885.

GIORNALE DI FARMACIA, etc. [Cont'd.]

Continued under the title :

[d] Annali di chimica e di farmacologia. 5 vols. Milano, 1886–90. Supplemento agli annali di chimica medico-farmaceutica e di farmacologia. Direttori P. Albertoni e J. Guareschi ; condirettori A. Pavesi e G. Colombo. Milano, 1885.

181. GIORNALE DI FISICA, CHIMICA ED ARTI. Di Majocchi. 8vo. Milano, 1839.

182. GIORNALE DI FISICA, CHIMICA E STORIA NATURALE, ossia raccolta di memorie sulle scienze, arti e manufatture ad esse relative. Di L. Brugnatelli. 10 vols., 4to. Pavia, 1808–'17.

Continued under the title :

[a] Giornale di fisica, chimica, storia naturale, medicina ed arti, di Brugnatelli, Brunacci e Configliachi, compilato da Gaspare Brugnatelli. Decade seconda. 10 vols. (I–X), 4to. Pavia, 1818–'27 ||

183. GIORNALE FISICO-CHIMICO ITALIANO, ossia raccolta di scritti risguardanti la fisica e la chimica degl' italiani di Francesco Zantedeschi. 8vo. Venezia, 1846–'48 [–'51 (?)].

184. GLASHÜTTE (DIE) UND KERAMIK. Technisch-kommerzielle Fachzeitschrift für die gesammte Glas-, Porzellan- und Thonwaaren-Industrie und den Handel. Herausgegeben und redigirt von Julius Fahdt in Dresden. 16 vols., 4to. Dresden und Leipzig, 1871–'86 ||

GÖRZ, J. *See* Wöchentlicher Marktbericht für den internationalen Zuckerhandel.

GÖTTLING, J. F. A. *See* Almanach für Scheidekünstler.

GOUILLON, A. FÉLIX. *See* Coloration (La) industrielle.

GRAF, EMIL. *See* Rundschau für die Interessen der Pharmacie.

GREN, F. A. C. *See* Journal der Physik.

GRETSCHEL, H. *See* Jahrbuch der Erfindungen.

GRINDEL, DAV. H. *See* Russisches Jahrbuch der Pharmacie.

GRÜNE, WILH. *See* Deutsche Musterzeitung.

GUARESCHI, ICILIO. *See* Supplemento annuale alla enciclopedia di chimica.

GUMBINNER, L. *See* Brennerei-Zeitung (Neue).

GÜNTHER, F. A.   *See* Deutsche Gerber-Zeitung.

HAAXMANN, P. J.   *See* Tijdschrift voor schei- en artsenijbereidkunde.

HABICH, G. E.   *See* Bierbrauer (Der).

HAMEL-ROOS, G. F. VAN.
  See Revue internationale scientifique et populaire des falsifications.

HART, EDWARD.   *See* Journal of Analytical Chemistry.

HASSE, T. S.   *See* Magazin für Eisen-, Berg- und Hüttenkunde.

HATSCHEK, MOR.   *See* Zeitschrift (Neue) für die österreichisch-ungarische
  Spiritus-Industrie.

HAYN, ALB.   *See* Deutsche (Der) Brauerei-Kalender.

185.   HELFENBERGER ANNALEN, herausgegeben von der chemischen
  Fabrik E. Dieterich in Helfenberg bei Dresden.   8vo.   Berlin, 1892+

HELLER, J. F.   *See* Beiträge zur physiologischen und pathologischen Chemie.

HENSMANS, P. J.   *See* Annuaire à l'usage du chimiste ; *also* Répertoire de
  chimie, pharmacie [etc.].

HERBERTZ, WILH.   *See* Deutsche (Die) Zucker-Industrie.

HERMANN, B. F.   *See* Beiträge zur Physik.

HERMBSTÄDT, SIG. F.
  See Archiv der Agricultur-Chemie ; *also* Bibliothek der neuesten
  physisch-chemischen . . . Literatur ; *also* Magazin für Färber.

186.   HERMETISCHES JOURNAL zur endlichen Beruhigung für Zweifler
  und Sucher von der hermetischen Gesellschaft.   1 No.   Cam-
  burg, 1802.

   Followed by :

[a]   *Hermes,* eine Zeitschrift in zwanglosen Heften zur endlichen
  Beruhigung für Zweifler und Sucher herausgegeben von
  L. F. von Sternhagen in Karlsruhe.   [Karlsruhe ?] 1805.

HIGGINS, BRYAN.   *See* Minutes of the Society for Philosophical Experiments.

HILGER, DR.   *See* Mittheilungen aus dem chemischen Laboratorium.

HILGER, A.   *See* Mittheilungen aus dem pharmaceutischen Institute . . . Erlangen ; *also* Vierteljahresschrift über die Fortschritte auf dem
  Gebiete der Chemie der Nahrungs- und Genussmittel.

HIRZEL, H.   *See* Jahrbuch der Erfindungen.

HOFFMANN'S JAHRESBERICHT.
    See Jahresbericht über die Fortschritte der Agriculturchemie.

HOFMANN (A. W.), AND DE LA RUE.
    See Annual Report ot the Progress of Chemistry.

HOLZNER, GEORGE.   *See* Zeitschrift für das gesammte Brauwesen.

HOMANN, CARL.   *See* Deutscher Brauer-Kalender ; *also* Hopfenlaube (Die).

187.   HOPFENLAUBE (DIE).   Fach- und Handels-Zeitung für das deutsche
    Brauwesen.   Officielles Organ des badischen Brauerbundes.
    Herausgeber und Chef-Redacteur : C. Homann.   [Edited in
    1878 by Alb. Sertz.]   3 vols., fol.   Nürnberg, 1876–'78.
        Continued under the title :

[a]   Hopfenlaube (Die).   Officielles Organ des Bezirks-Vereins vom
    allgemeinen deutschen Brauerbunde in Leipzig und des
    badischen Brauerbundes.   Redacteur : Alb. Sertz.   7 vols.
    (IV).   Nürnberg, 1879–82.
        Continued under the title : Hopfen-Kurier, 1883–92 +

HOPPE-SEYLER, FELIX.   *See* Zeitschrift für physiologische Chemie ; *also* Medi-
    cinisch-chemische Untersuchungen.

HORKEL, JOH.   *See* Archiv für die thierische Chemie.

188.   HUMPHREY'S PAINT AND OIL TRADE AND WHOLESALE DRUGGIST.
    Fol.   New York, 1872–'74 [–'77 (?)].

D'HURCOURT, ED. R.   *See* Moniteur (Le) du gaz.

189.   INCORAGGIAMENTO (L').   Giornale di chimica e di scienze affini,
    d'industrie e di arti.   8vo.   Napoli, 1865.

190.   INDUSTRIELLES WOCHENBLATT FÜR GLASINDUSTRIE.   — vols.,
    Coblenz, 1870, '71.

191.   INSTITUTE OF BREWING, TRANSACTIONS OF.   5 vols., 8vo.   London,
    1887, '92.

192.   INTRODUCTION AUX OBSERVATIONS SUR LA PHYSIQUE, SUR L'HIS-
    TOIRE NATURELLE ET SUR LES ARTS.   Par l'abbé Rozier.   18
    vols., 12mo.   Paris, 1771–'72.
        A second edition in 2 vols., 4to, was published in 1777.   Vol. II *has
        also the title :*
    Tableau du travail annuel de toutes les académies de l'Europe ; ou
        observations sur la physique, sur l'histoire naturelle et sur les
        arts et métiers.   Par Jean Rozier.   1 vol., 4to.   1777.

INTRODUCTION AUX OBSERVATIONS SUR LA PHYSIQUE, etc.   [Cont'd.]
          Continued under the title :

[a]   Observations et mémoires sur la physique, sur l'histoire naturelle
      et sur les arts et métiers.   Par l'abbé Rozier.   1 vol., 4to.
      Paris, 1773.
          Continued under the title :

[b]   Observations sur la physique, sur l'histoire naturelle et sur les
      arts.   Par l'abbé Rozier [*from* 1779] et J. A. Mongez
      [*from* 1785] et de la Méthérie.   42 vols. (II–XLIII), 4to.
      Paris, 1773–'94.
          Continued under the title :

[c]   Journal de physique, de chimie, d'histoire naturelle et des arts.
      Par Jean Claude Laméthérie [*from* 1817] et H. M. Ducro-
      tay de Blainville.   [*From* vol. LXXXV *by* de Blainville *alone.*]
      53 vols. (XLIV–XCVI), 4to.   Paris, 1794–1822 ||
      Suppléments.   2 vols., 4to.   Paris, 1778, 1782.
      Vol. X *contains :* Table des articles contenus dans les volumes
      in 4to de ce recueil imprimés depuis le commencement de
      1773 et dans les 18 vols. in 12mo imprimés depuis Juillet
      1771, jusqu' à la fin de 1772, actuellement imprimés en 2
      vols. in 4to sous le titre d'Introduction aux observations
      sur la physique, etc.
      Vol. XXIX, 1786, *contains :* Table générale des articles contenus
      dans les vingt volumes de ce journal depuis 1778.
      Vol. LV, 1802, *contains :* Table générale des articles contenus
      dans les vingt-six derniers volumes du Journal de physique,
      depuis 1787 jusqu' en 1802, pour faire suite à celle qui est
      imprimée à la fin du second volume de l'année 1786.   Par
      L. Cotte.

ISMALUN, ALBERT.   *See* Bulletin du laboratoire de chimie du Caire.

JACOBSEN, EMIL.
          See Chemische Industrie ; *also* Chemisch-technisches Repertorium.

JAHN, RICH.   *See* Oesterreichisch-ungarischer Brennerei-Kalender.

JAHRBUCH DER CHEMIE UND PHYSIK ; Schweigger.
          See Allgemeines Journal der Chemie.

193.   JAHRBUCH DER CHEMIE ; Bericht über die neuesten und wichtig-
       sten Fortschritte der reinen und angewandten Chemie.   Unter
       Mitwirkung von H. Beckurts (Braunschweig), R. Benedikt
       (Wien), C. A. Bischoff (Riga), L. Bühring (Halle), E. F.
       Dürre (Aachen), J. M. Eder (Wien), C. Häussermann (Stutt-

JAHRBUCH DER CHEMIE.   [Cont'd.]
gart), G. Krüss (München), M. Märcker (Halle), W. Nernst (Göttingen), F. Röhmann (Breslau), E. Valenta (Wien), herausgegeben von Richard Meyer (Braunschweig). 1 Jahrgang, 1891. Frankfurt am Main, 1892+

194. JAHRBUCH DER ERFINDUNGEN UND FORTSCHRITTE AUF DEN GEBIETEN DER PHYSIK UND CHEMIE, der Technologie und Mechanik, der Astronomie und Meteorologie. Herausgegeben von H. Hirzel und H. Gretschel [*later, by* H. Gretschel *and* G. Wunder]. 28 vols., 8vo. Leipzig, 1865–'92+
Register. Vols. I–V, 1869 ; vols. VI–IX, 18— ; vols. X–XI, 18— ; vols. XII–XXIV, 18—.

JAHRBUCH (NEUES) DER PHARMACIE.   *See* Berlinisches Jahrbuch der Pharmacie.

195. JAHRBUCH FÜR FABRIKANTEN. — vols. Prag, 1839–'44.

196. JAHRBUCH FÜR ÖKONOMISCHE CHEMIE UND VERWANDTE FÄCHER. Eine Sammlung des Wichtigsten aus der landwirthschaftlichen Chemie, den mit derselben verwandten landwirthschaftlichen Gewerben und der Hauswirthschaft. Herausgegeben von Willibad Artus. 3 vols., 8vo. Leipzig, 1847–'49.

197. JAHRESBERICHT DER AGRICULTURCHEMIE. Separat-Ausgabe des Centralblattes für Agriculturchemie und rationellen Wirthschaftsbetrieb. Herausgegeben von W. Detmer. 2 vols., 8vo. Berlin, 1875 ||
Cf. Centralblatt für Agriculturchemie.

198. JAHRESBERICHT DES CHEMISCHEN UNTERSUCHUNGSAMTES DER STADT BRESLAU für das Jahr 1890–92, erstattet von B. Fischer. 8vo. Ill. Breslau, 1891–'92+

199. JAHRESBERICHT ÜBER DIE FORTSCHRITTE AUF DEM GEBIETE DER REINEN CHEMIE. Bearbeitet im Verein mit mehreren Fachgenossen und herausgegeben von Wilhelm Staedel. 1873–'81. 9 vols., roy. 8vo. Tübingen, 1874–'83 ||

200. JAHRESBERICHT ÜBER DIE FORTSCHRITTE DER AGRICULTURCHEMIE, mit besonderer Berücksichtigung der Pflanzenchemie und Pflanzenphysiologie. Herausgegeben von Robert Hoffmann. 1838– '64. 6 vols., 8vo. Berlin, 1860–'65.
Continued under the title :

[a] Jahresbericht über die Fortschritte auf dem Gesammtgebiete

JAHRESBERICHT ÜBER DIE FORTSCHRITTE, etc. (HOFFMANN.) [Cont'd.]
der Agriculturchemie. Fortgesetzt von Edward Peters.
[*From* 186–, *edited by* Th. Dietrich, H. Hellriegel, J. Flittbogen, H. Ulbricht.] 14 vols. (VII–XX), 8vo. Berlin,
1864–'77.

Neue Folge. von A. Hilger. 14 vols. (XXI–XXXIV). Berlin,
1878–'92+.

From 1874, annual volumes are divided into portions having indepen-
dent pagination and bearing the following titles :

Jahresbericht über die Fortschritte der Chemie des Bodens, der
Luft und des Düngers. Herausgegeben von Th. Dietrich.

Jahresbericht über die Fortschritte der Chemie der Pflanze.
Herausgegeben von J. Flittbogen.

Jahresbericht über die Fortschritte der Chemie der Thiernährung und der chemischen Technologie der landwirthschaftlichen Nebengewerbe. Herausgegeben von J. König und
A. Hilger.

General-Register über Jahrg. I–XX [1858–'77]. Unter Mit-
wirkung von E. von Gerichten, C. Krauch, E. von Raumer,
W. Rössler und O. H. Will. Herausgegeben von A. Hilger.
I vol., 8vo. Berlin, 1879.

JAHRESBERICHT ÜBER DIE FORTSCHRITTE DER CHEMIE DES BODENS.
See Jahresbericht über die Fortschritte der Agriculturchemie.

JAHRESBERICHT ÜBER DIE FORTSCHRITTE DER CHEMIE DER PFLANZE.
See Jahresbericht über die Fortschritte der Agriculturchemie.

JAHRESBERICHT ÜBER DIE FORTSCHRITTE DER CHEMIE DER THIERNÄHRUNG.
See Jahresbericht über die Fortschritte der Agriculturchemie.

201. JAHRESBERICHT ÜBER DIE FORTSCHRITTE DER CHEMISCHEN TECH-
NOLOGIE für Fabrikanten, Hütten- und Forstleute, Cameralisten,
Chemiker und Pharmaceuten. Herausgegeben von Joh. Rud.
Wagner. 6 vols., 8vo. Leipzig, 1855–'60.

Continued under the title :

[a] Jahresbericht über die Fortschritte und Leistungen der chemischen Technologie und technischen Chemie. Herausgegeben von J. R. Wagner. 9 vols. (VII–XV), 8vo. Leipzig,
1861–'69.

Continued under the title :

[b] Jahresbericht über die Fortschritte und Leistungen der chemischen Technologie und technischen Chemie, (*from* vol.
XVIII,) mit besonderer Berücksichtigung der Gewerbesta-

JAHRESBERICHT ÜBER DIE FORTSCHRITTE, etc. (WAGNER.) [Cont'd.]

    tistik. Herausgegeben von J. R. Wagner. Neue Folge.
    11 vols. (XVI–XXVI), 8vo. Leipzig, 1870–'80.

      Continued under the title :

[c] Wagner's (R. von) Jahresbericht über die Leistungen der chemi-
    schen Technologie, mit besonderer Berücksichtigung der
    Gewerbestatistik für das Jahr 1880–'91. Herausgegeben
    von F. Fischer. 12 vols. [XXVII–XXXVIII], 8vo. Leipzig,
    1881–'92+

    General-Register über Band 1 bis 10, bearbeitet von Fr. Gott-
    schalk. 1 vol., 8vo. Leipzig, 1866.

    General-Register über Band 11 bis 20, bearbeitet von Fr. Gott-
    schalk. 1 vol., 8vo. Leipzig, 1876.

202. JAHRESBERICHT ÜBER DIE FORTSCHRITTE DER GESAMMTEN PHAR-
    MACIE UND PHARMACOLOGIE IM IN- UND AUSLANDE. Von
    Dierbach und Martius. Separat-Abdruck für Pharmaceuten
    aus Canstatt's Jahresbericht über die Fortschritte der gesamm-
    ten Medicin in allen Ländern. 1 vol., roy. 8vo. Erlangen,
    1841.

      Continued under the title :

[a] Jahresbericht über die Fortschritte der Pharmacie in allen Län-
    dern. Herausgegeben von Dierbach, Martius, Scherer und
    Simon. [*In* 1843 *edited by* Siebert, Martius *and* Scherer ;
    in 1844 *and* 1845 *edited by* Scherer *and* Wiggers ; *from*
    1846–'49 *by* Scherer, Heidenreich *and* Wiggers.] Separat-
    Abdruck für Pharmaceuten aus Canstatt's und Eisenmann's
    Jahresbericht über die Fortschritte der gesammten Medicin
    in allen Ländern. [This addition is dropped after 1845.]
    8 vols. (II–IX), roy. 8vo. Erlangen, 1842–'49.

      Continued under the title :

[b] C. Canstatt's Jahresbericht über die Fortschritte in der Pharma-
    cie in allen Ländern im Jahre 1849. Herausgegeben von
    Wiggers, Scherer und Heidenreich. 1 vol. (x), roy. 8vo.
    Erlangen, 1850.

      Continued under the title :

[c] Canstatt's Jahresbericht über die Fortschritte in der Pharmacie
    und verwandten Wissenschaften [*from* 1853] in allen Län-
    dern. Verfasst von Martel, Frank, Heidenreich, Löschner,
    Scherer und Wiggers. [*In* 1852 *edited by* Eisenmann,
    Falk, Klencke, Löschner, Ludwig, Scherer *and* Wiggers ;
    from 1853, '54 *by* Scherer, Virchow *and* Eisenmann ; *from*

JAHRESBERICHT ÜBER DIE FORTSCHRITTE, etc. (CANSTATT.) [Cont'd.]
1855, '56 *by the same*, unter Mitwirkung von Friedreich ;
after 1857 *edited by* Scherer, Virchow *and* Eisenmann.]
Neue Folge. 14 vols. (XI–XXV), roy. 8vo. Würzburg,
1851–'65.

> *Continued under the title :*

[d]   Jahresbericht über die Fortschritte der Pharmakognosie, Pharmacie und Toxicologie. (Neue Folge des mit Ende 1865
abgeschlossenen Canstatt'schen pharmaceutischen Jahres-
berichtes. Herausgegeben von N. Wulfsberg, G. Dragendorff und W. Marmé ; [*later by*] Heinrich Beckurts.
1866–'89. 24 Jahrgang (49 der ganzen Reihe). 8vo.
Göttingen, 1867–'90.

> *Continued under the title :*

[e]   Jahresbericht der Pharmacie, herausgegeben vom deutschen
Apothekerverein unter Redaction von Heinr. Beckurts.
Neue Folge ; 1890–'91. 2 vols. (XXV–XXVI). Göttingen,
1891–'92+

203.   JAHRESBERICHT ÜBER DIE FORTSCHRITTE DER PHYSISCHEN WISSEN-
SCHAFTEN. Von Jacob Berzelius. Aus dem schwedischen
übersetzt von C. G. Gmelin [*later*,] und F. Wöhler. 20 vols.,
8vo. Tübingen, 1822–'41.

> *Continued under the title :*

[a]   Jahresbericht über die Fortschritte der Chemie und Mineralogie.
Eingereicht an die schwedische Akademie der Wissen-
schaften. Von Jacob Berzelius. [*From* 1849–'51, *edited by*
L. F. Svanberg.] Aus dem schwedischen übersetzt von F.
Wöhler. 10 vols. (XXI–XXX), 8vo. Tübingen, 1842–'51 ‖
Register. Vols. I–XVII, 1839 ; I–XXV, 1847.

> *Cf.* Årsberättelse om Framstegen i Fysik och Kemi ; *also* Annuaire
> des sciences chimiques ; *also* Rapport annuel sur les progrès des
> sciences physiques et chimiques.

204.   JAHRESBERICHT ÜBER DIE FORTSCHRITTE DER REINEN, PHARMA-
CEUTISCHEN UND TECHNISCHEN CHEMIE, Physik, Mineralogie
und Geologie. Unter Mitwirkung von H. Buff, E. Dieffen-
bach, C. Ettling, F. Knapp, H. Will, F. Zamminer. Herausgegeben von Justus Liebig und Hermann Kopp. 1847–'56. 9
vols., 8vo. Giessen, 1849–'57.

> *Continued under the title :*

[a]   Jahresbericht über die Fortschritte der Chemie und verwandter
Theile anderer Wissenschaften. Von Hermann Kopp und

JAHRESBERICHT ÜBER DIE FORTSCHRITTE DER . . . CHEMIE. [Cont'd.]
Heinrich Will. [*In* 1861, unter Mitwirkung von Th. En-
gelbach, W. Hallwachs, A. Knop, herausgegeben von Hermann Kopp und Heinrich Will. *In* 1864, unter Mitwirkung
von C. Bohn und Th. Engelbach, herausgegeben von Heinrich Will. *In* 1868, unter Mitwirkung von Th. Engelbach,
Al. Naumann, W. Städel, herausgegeben von Adolph
Strecker. *In* 1870, unter Mitwirkung von K. Birnbaum,
W. Dittmar, F. Hoppe-Seyler, A. Laubenheimer, A.
Michaelis, F. Nies, Th. Zincke, K. Zöppritz, herausgegeben
von Alexander Naumann. *In* 1875, unter Mitwirkung von
K. Birnbaum, C. Boettinger, C. Hell, H. Klinger, A. Laubenheimer, E. Ludwig, A. Michaelis, A. Naumann, F. Nies,
H. Salkowski, Zd. H. Skraup, K. Zöppritz, herausgegeben
von F. Fittica. *In* 1882, unter Mitwirkung von A. Bornträ-
ger, A. Elsas, E. Erdmann, C. Hell, H. Klinger, E. Ludwig,
A. Naumann, F. Nies, H. Salskowski, G. Schultz, herausge-
geben von F. Fittica. *From* 1884, II Theile : *In* 1885,
Begründet von. J. Liebig und H. Kopp, unter Mitwirkung
von A. Bornträger, A. Elsas, H. Erdmann, C. Hell, H.
Klinger, C. Laar, E. Ludwig, A. Naumann, F. Nies, W.
Roser, H. Salkowski, W. Sonne, W. Suida, herausgegeben
von F. Fittica. *In* 1888, unter Mitwirkung von O. T.
Christensen, A. Elsas, W. Fahrion, A. Fock, C. Hell, A.
Kehrer, F. W. Küster, C. Laar, E. Ludwig, F. W. Schmidt,
W. Sonne, W. Suida, A. Weltner, herausgegeben von F.
Fittica.] 1857-'89. 33 vols. (x–XLIII). Giessen [*from*
1886 (1889) Braunschweig], 1858–92+

The volume for 1889 is not complete in 1892.

Since 1857 *also under the title :*

Jahresbericht über die Fortschritte der reinen, pharmaceutischen
und technischen Chemie, Physik, Mineralogie und Geologie.
Bericht über die Fortschritte der Chemie und verwandter
Theile anderer Wissenschaften.

Accompanied by :

Jahresbericht über die Fortschritte der Physik, von Friedrich
Zamminer. Für 1857. 1 vol., 8vo. Giessen, 1858.

Register zu den Berichten für 1847 bis 1856. 1 vol., 8vo.
Giessen, 1857.

Register zu den Berichten für 1857 bis 1866. 1 vol., 8vo.
Giessen, 1868.

Register zu den Berichten für 1867 bis 1876. 1 vol., 8vo.
Giessen, 1880.

205. JAHRESBERICHT ÜBER DIE FORTSCHRITTE DER THIERCHEMIE. Herausgegeben von Richard Maly. 1871–'89. 19 vols., 8vo. Wien, 1872–'90+

Sach- und Autoren- Register zu Band 1–x, bearbeitet von Rudolph Andreasch. 8vo. Wiesbaden, 1881.

206. JAHRESBERICHT ÜBER DIE FORTSCHRITTE IN DER LEHRE VON DEN PATHOGENEN MIKROORGANISMEN, umfassend Bacterien, Pilze und Protozoën. Unter Mitwirkung von Fachgenossen bearbeitet und herausgegeben von P. Baumgarten. 6 vols., 8vo. 1885–90. Braunschweig, 1886–92+

207. JAHRESBERICHT ÜBER DIE UNTERSUCHUNGEN UND FORTSCHRITTE AUS DEM GESAMMTGEBIETE DER ZUCKERFABRIKATION. Herausgegeben von C. Scheibler und K. Stammer. 29 vols., 8vo. (1863–89). Braunschweig, 1863–90+

JETTEL, WLADIMIR. *See* Zeitschrift für Zündwaarenfabrikation.

208. JERN-KONTORETS ANNALER. 29 vols., 8vo. Stockholm, 1817–'45.
　　　　　　　　Continued under the same title :

[a] Ny Serie. En Tidskrift för Svenska Bergshandteringen. Redigerad af Joachim Åkermann. [From 1855–'57 by Jonas Samuel Bagge och Gustav Svedelius ; from 1858 by Kunt Styffe, J. S. Bagge och G. Svedelius.] 43 vols., 8vo. Stockholm, 1846–85+

Register öfver Årgångarne 1881–85 af A. Ekelund. Stockholm, 1886. 8vo.

209. JERN OCH STÅL. Tidskrift för Bergsmanna-Sällskapet. 8vo. Stockholm, 1879–'80 [+?].

JOHANNESSON, B. *See* Norddeutsche Brauer-Zeitung.

210. JOURNAL DE CHIMIE ET DE PHYSIQUE, ou Recueil périodique des découvertes dans les sciences chimiques et physiques, tant en France que chez l'étranger. Par J. B. van Mons. 6 vols., 12mo. Bruxelles, An x [1801]–XII [1804].
　　　　　Cf. Journal de chimie pour servir [etc.].

211. JOURNAL DE CHIMIE MÉDICALE, DE PHARMACIE ET DE TOXICOLOGIE. Rédigé par Chevallier, Fée, Guibourt, Julia-Fontenelle, Laugier, Orfila, Payen, Gabrielle Pelletan, Lassaigne, A. Richard, Robinet, Segalas d'Etchepare. 10 vols., 8vo. Paris, 1825–'34.
　　　　　　　Continued under the title :

[a] Journal de chimie médicale, de pharmacie et de toxicologie.

JOURNAL DE CHIMIE MÉDICALE. [Cont'd.]

Revue des nouvelles scientifiques nationales et étrangères Par les membres de la Société de chimie médicale. Deuxième série. 10 vols. (I–x), 8vo. Paris, 1835–'44.

Table générale des matières et des auteurs de la deuxième série du Journal de chimie médicale, de pharmacie et de toxicologie ; de 1835–'44. 1 vol., 8vo. Paris, 1845.

Continued under the title :

[b] Journal de chimie médicale, de pharmacie, de toxicologie et revue des nouvelles scientifiques nationales et étrangères. Par les membres de la Société de chimie médicale, Béral, Chevallier, Dumas, Fée, Guibourt, Lassaigne, Orfila, Payen, E. Péligot, G. Pelletan, Pelouze, A. Richard, S. Robinet. Troisième série. 10 vols. (I–x), 8vo. Paris, 1845–'54.

Continued under the title :

[c] Journal de chimie médicale, de pharmacie, de toxicologie et revue des nouvelles scientifiques nationales et étrangères. Publié sous la direction de A. Chevallier. Quatrième série. 10 vols. (I–x), 8vo. Paris, 1855–'64.

Continued under the title :

[d] Journal de chimie médicale, de pharmacie, de toxicologie et revue des nouvelles scientifiques nationales et étrangères. Revue industrielle. Publié sous la direction de A. Chevallier. Cinquième série. 12 vols. (I–XII), 8vo. Paris, 1865–'76 ‖

United in March, 1876, *with the* Répertoire de pharmacie *and continued under the title :* Répertoire de pharmacie et Journal de chimie médicale réunis, *q. v.*

212. JOURNAL DE CHIMIE pour servir de complément aux Annales de chimie et d'autres ouvrages périodiques français de cette science. Rédigé par J. B. van Mons. 6 vols., 8vo. Bruxelles, 1792–1804.

Cf. Journal de chimie et de physique.

JOURNAL DE LA SOCIÉTÉ CHIMIQUE RUSSE.

See Zhurnal russkova khimicheskova (etc.).

213. JOURNAL DE L'ÉCLAIRAGE AU GAZ. Organe spécial et pratique de l'industrie de l'éclairage et du chauffage. [*From* 186– *with the additional words :*] du service des eaux et de la salubrité publique. 39 vols., 4to. Paris, 1852–'90+

Tables des matières *in* vol. XIII. 1864, '65.

214. Journal de l'exploitation des corps gras industriels. Paris, 1874–77.

> *Continued under the title :*

[a] Corps (Les) gras industriels. Journal des fabricants d'huiles, savons, suif, bougies, etc., et du matériel de ces industries. Dirigé par P. F. Blondeau. 8 vols., 4to. Paris, 1878–'90+

215. Journal de la physiologie de l'homme et des animaux. Publié sous la direction de F. Brown-Séquard. 6 vols., 8vo. Paris, 1858–'63.

> *Continued in 1868 under the title :*

[a] Archives de physiologie normale et pathologique (suite du Journal de la physiologie de l'homme et des animaux). Dirigées par Brown-Séquard, Charcot et Vulpian. Deuxième série. 44 vols. (I–XLIV), 8vo. Paris, 1869–'92+

Journal de pharmacie.   *See* Bulletin de pharmacie.

Journal de physique, de chimie   . . .   Laméthérie.
See Introduction aux observations sur la physique [*c*].

216. Journal de teinture. Organe de la teinture, de l'impression, du blanchiment, des apprêts [etc.]. Par Reimann. 17 vols., 8vo. Paris, 1874–'90+

> *See* Reimann's Färber-Zeitung, of which this is a translation.

Journal der Chemie.   *See* Allgemeines Journal der Chemie.

217. Journal der Pharmacie für Aerzte und Apotheker. Herausgegeben von Joh. B. Trommsdorff. 26 vols., 8vo. Leipzig, 1793–1816.

> *Continued under the title :*

[a] Neues Journal der Pharmacie für Aerzte, Apotheker und Chemiker. Herausgegeben von Joh. B. Trommsdorff. 27 vols., 8vo. Leipzig, 1817–'34.
> *United in 1834 with the* Annalen der Pharmacie, *q. v.*

218. Journal der Physik. Herausgegeben von Fr. Albrecht Carl Gren. 8 vols., 8vo. Halle und Leipzig, 1790–'94.

> *Continued under the title :*

[a] Neues Journal der Physik. Herausgegeben von Fr. Albrecht Carl Gren. 4 vols. (I–IV), 8vo. Halle und Leipzig, 1794–'98.
> *Continued under the title :*

[b] Annalen der Physik. Angefangen von Fr. Albr. Carl Gren, fortgesetzt von Ludwig Wilhelm Gilbert. [*From* vol. IV,

JOURNAL DER PHYSIK. [Cont'd.]

  herausgegeben von Ludwig Wilhelm Gilbert.] 30 vols.
  (I–XXX), 8vo. Halle, 1799–1808.

  Neue Folge. 30 vols. (I–XXX *and* XXXI–LX), 8vo. Halle,
  1809–'18.

  Supplement zu Band XII. Halle, 1803.

    Continued under the title :

[c] Annalen der Physik und der physikalischen Chemie. Herausgegeben von Ludwig Wilhelm Gilbert. Dritte Reihe. 16
  vols. (I–XVI *and* LXI–LXXVI), 8vo. Halle, 1819–'24.

    Vol. LXI, 1819, *is also entitled :* Neueste Folge. Band 1.
    Continued under the title :

[d] Annalen der Physik und Chemie. Herausgegeben von J. C.
  Poggendorff. 30 vols. (I–XXX *and* LXXVII–CVI), 8vo. Berlin, 1824–'33.

  Zweite Reihe. 30 vols. (I–XXX *and* CVII–CXXXVI). 1834–'43.
  Dritte Reihe. 30 vols. (I–XXX *and* CXXXVII–CLXVI). 1844–'53.
  Vierte Reihe. 30 vols. (I–XXX *and* CLXVII–CXCVI). 1854–'63.
  Fünfte Reihe. 30 vols. (I–XXX *and* CXCVII–CCXXVI). 1864–'73.
  Sechste Reihe. 10 vols. (I–X *and* CCXXVII–CCXXXVI). 1874–'77.

    Continued under the title :

[e] Annalen der Physik und Chemie. [*From* vol. XV,] Begründet
  und fortgeführt durch F. A. C. Gren, L. W. Gilbert, J. C.
  Poggendorff. Neue Folge. Unter Mitwirkung der physikalischen Gesellschaft in Berlin und insbesondere des
  Herrn H. Helmholtz. Herausgegeben von G. Wiedemann.
  47 vols. (I–XLVII and CCXXXVII–CCLXXXIII), 8vo. Leipzig,
  1877–'92+

  Ergänzungs-Bände : Bd. I, 1843 ; II, 1848 ; III, 1853 ; IV, 1854 ;
  V, 1871 ; VI, 1874 ; Jubelband, 1874.

  Indexes : Vols. I–XXX, 1833 ; XXXI–XLII, 1837 ; XLIII–LI,
  1842 ; LI–LX, 1844 ; LXI–LXIX, 1846 ; LXI–LXXV, 1848 ;
  LXXVI–LXXXIV, 1851 ; LXI–XC, 1854.

  Namen-Register zu Band I bis CL, Ergänzungsband I bis VI
  nebst Jubelband, und Sach-Register zu Band CXXI bis CL,
  Ergänzungsband V und VI nebst Jubelband, bearbeitet von
  W. Barentin, nebst einem Anhange von J. C. Poggendorff
  enthalten Verzeichniss der verstorbenen Auctoren, und
  Zeittafel zu den Bänden. 8vo. Leipzig, 1875.

    From 1877 *the* Annalen *are accompanied by :*

[f] Beiblätter zu den Annalen der Physik und Chemie. Herausgegeben unter Mitwirkung befreundeter Physiker von J. C.

JOURNAL DER PHYSIK.   [Cont'd.]
> Poggendorff, [*later*] von G. und E. Wiedemann.   16 vols., 8vo.   Leipzig, 1877–'92+

JOURNAL DER RUSSISCHEN PHYSISCH-CHEMISCHEN GESELLSCHAFT.
> *See* Zhurnal russkova khimicheskova [etc.].

219.   JOURNAL DES BRASSEURS, ANNALES DE LA BRASSERIE.   Fol.   Paris, 1857–'71.

220.   JOURNAL DES BRASSEURS DU NORD DE LA FRANCE.   Organe spécial du syndicat des brasseurs des départements du Nord, Pas-de-Calais, Oise, Somme et Ardennes.   Fol.   Lille, 1871.

221.   JOURNAL DES FABRICANTS DE SUCRE ET DES DISTILLATEURS.   Organe de la sucrerie indigène et coloniale.   Directeur : P. Bourgeois [*later*, Durau].   33 vols., folio.   Paris, 1860–92+

222.   JOURNAL DES USINES À GAZ.   Organe de la Société technique de l'industrie du gaz en France.   17 vols., 4to.   Paris, 1877–'92+

223.   JOURNAL DU GAZ ET DE L'ÉLECTRICITÉ.   Organe spécial aux intérêts financiers des usines à gaz et des Sociétés d'électricité.   12 vols., 4to.   Paris, 1881–'92+

JOURNAL FÜR DIE CHEMIE, PHYSIK [etc.].
> *See* Allgemeines Journal der Chemie [*b*].

JOURNAL (NEUES) FÜR DIE INDIENNENDRUCKEREI.
> *See* Journal für Zitz-, Kattun- und Indiennendruckerei.

224.   JOURNAL FÜR GASBELEUCHTUNG UND VERWANDTEN BELEUCH-TUNGSARTEN.   [*From* vol. v,] Organ des Vereins von Gas-fachmännern Deutschlands [*from* vol. XII] und seiner Zweig-Vereine, sowie des Vereins für Mineralöl-Industrie.   [*From* vol. XIII,] Mit einem Anhang über Wasserversorgung.   Monatschrift von N. H. Schilling und A. Schels.   [*From* vol. III, *by* Schilling *alone*.]   13 vols., 4to.   München, 1858–'70.
> *Continued under the title :*

[a]   Journal für Gasbeleuchtung und Wasserversorgung.   Organ des Vereins von Gas- und Wasserfachmännern Deutschlands mit seinem Zweig-Vereinen und des Vereins für Mineralöl-Industrie.   Von N. H. Schilling [from vol. XIX] und H. Bunte.   22 vols. (XIV–XXV), 4to.   [*From* vol. XIX, roy. 8vo.].   München, 1871–'92+.

Generalregister zu Jahrgang 1858 bis einschliesslich 1873.   Bearbeitet von L. Diehl.   4to.   München, 1875.

225. JOURNAL FÜR GAS-INDUSTRIE. Organ des Vereins der Gas-Industriellen in Oesterreich-Ungarn. Herausgeber: Carl Pataky. 4 vols., sm. fol. Wien, 1881-'84 [+?].

226. JOURNAL FÜR PHYSIK UND PHYSIKALISCHE CHEMIE DES AUS-LANDES. In vollständigen Uebersetzungen, herausgegeben von A. Krönig. 3 vols., 8vo. Berlin, 1851 ‖

JOURNAL FÜR PRAKTISCHE CHEMIE. *See* Allgemeines Journal der Chemie.

227. JOURNAL FÜR TECHNISCHE MITTHEILUNGEN AUS DEM GEBIETE DER ZUCKERFABRIKATION. Herausgegeben von Mitgliedern des Vorstands des Raffinerie-Vereins und des technischen Vereins für Zuckerfabrikation in Halberstadt. 8vo. Braunschweig, 1877[+?].

228. JOURNAL FÜR TECHNISCHE UND ÖKONOMISCHE CHEMIE. Heraus-gegeben von Otto Linné Erdmann. 18 vols., 8vo. Leipzig, 1828-'33.
Register. 1837.

Also under the title :

Forschungen (Die neuesten) im Gebiete der technischen und ökonomischen Chemie.

United in 1834 *with the* Journal für Chemie und Physik *and continued under the title :* Journal für praktische Chemie. *See* Allgemeines Journal der Chemie.

229. JOURNAL FÜR ZITZ-, KATTUN- UND INDIENNENDRUCKEREI. Herausgegeben von J. G. Dingler. 2 vols. Leipzig, 1806-'07.

Continued under the title :

[a] Neues Journal für die Indiennen- und Baumwollendruckerei. Herausgegeben von J. G. Dingler. 4 vols. Augsburg und Leipzig, 1815-'17.

Followed by :

[b] Magazin für die Druck-, Färbe- und Bleichkunst [etc.]. Herausgegeben von J. G. Dingler. 3 vols. Leipzig, 1818-'20.

230. JOURNAL (THE) OF ANALYTICAL CHEMISTRY. Edited by Edward Hart. Associate editors, P. W. Shimer [and] John Eyerman. With the assistance of the following specialists : F. C. Blake, H. C. Bolton, F. W. Clarke, Isaac Ott, V. C. Vaughan, H. W. Wiley. 4 vols., 8vo. Easton, Pa., 1887-'90.

Continued under the title :

[a] Journal (The) of Analytical and Applied Chemistry. Edited

JOURNAL (THE) OF ANALYTICAL CHEMISTRY. [Cont'd.]
>by Edward Hart. Associate editors : Stuart Croasdale, P.
W. Shimer ; with the assistance of A. A. Blair, F. C. Blake,
H. Carrington Bolton, F. W. Clarke, C. W. Marsh, I. A.
Palmer, V. C. Vaughan, A. H. Welles, J. E. Whitfield and
H. W. Wiley. 2 vols., 8vo. Easton, Pa., 1891–'92+.

231. JOURNAL (THE) OF APPLIED CHEMISTRY. Devoted to chemistry
as applied to the arts, manufactures, metallurgy, and agriculture. 10 vols., 4to. New York, Philadelphia, and Boston,
1866–'75 ‖ .

232. JOURNAL (THE) OF GAS-LIGHTING, WATER-SUPPLY AND SANITARY
IMPROVEMENT. 56 vols., 4to. London, 1849–'90+.

233. JOURNAL (A) OF NATURAL PHILOSOPHY, CHEMISTRY AND THE
ARTS. Illustrated with engravings. By William Nicholson.
5 vols., 4to. London, 1797–1801.
Second series. 36 vols. (I–XXXVI), 8vo. London, 1802-'13 ‖
>*United in* 1814 *with the* Philosophical Magazine, *q. v.*

JOURNAL OF THE AMERICAN CHEMICAL SOCIETY.
>*See* Proceedings of the American Chemical Society.

JOURNAL OF THE LONDON CHEMICAL SOCIETY.
>*See* Proceedings of the Chemical Society of London.

JOURNAL OF THE RUSSIAN PHYSICAL AND CHEMICAL SOCIETY.
>*See* Zhurnal russkova khimicheskova [etc.].

234. JOURNAL (THE) OF THE SOCIETY OF CHEMICAL INDUSTRY. A
monthly Record for all interested in chemical manufactures.
Edited by Watson Smith. 11 vols., 4to. London, [*later*]
Manchester, 1882–'92+

235. JOURNAL (THE) OF THE SOCIETY OF DYERS AND COLOURISTS, for
all interested in the use of manufactures of colours and in
calico printing, bleaching, etc. 8 vols. Manchester, 1884–'92 +

KALENDER FÜR GAS- UND WASSERFACH-TECHNIKER. *See* Gas-Kalender.

236. KALENDARZ DLA UŻYTKU FARMACEUTÓW I CHEMIKÓW 1881. Redactor : Dunin v. Wąsowicz. 12mo. Lwów, 1880.

237. KALENDER FÜR TEXTIL-INDUSTRIE. Eine Sammlung der wichtigsten Regeln, Notizen und Resultate aus der Praxis der Spinnerei, Weberei, Appretur, Bleiche und Färberei. Unter Mitwirkung von Fachmännern, herausgegeben von W. H. Uhland.
11 vols., 12mo. Leipzig, 1880–'90+

KASTELYN, P. J.   *See* Chemische en physische oefeuingen.

KASTNER, C. W. G.   *See* Archiv für die gesammte Naturlehre.

KEKULÉ, A.   *See* Kritische Zeitschrift für Chemie.

238. KEMISKA NOTISER.   Organ for Kemistsamfundet i Stockholm. Redaction : J. Landin, E. Petersen, K. Sondén.   Utgifvare K. Sondén.   Stockholm, 1887–'88.
> *Continued under the title :*

[a]   Svensk Kemisk Tidskrift (Fortsättning af Kemiska Notiser). Utgifvare K. Sondén.   3 vols.   Stockholm, 1889–'92+

KERST, G.   *See* Gerber-Zeitung.

KLAPROTH, M. H.   *See* Beiträge zur chemischen Kenntniss der Mineralkörper.

239. KLEINE PHYSIKALISCH-CHEMISCHE ABHANDLUNGEN.   Herausgegeben von Joh. F. Westrumb.   8 vols., 8vo.   Leipzig, 1785–'97.

KLETZINSKY, V.   *See* Mittheilungen aus dem Gebiete der reinen und angewandten Chemie.

KOLBE, H.   *See* Allgemeines Journal der Chemie [*e*].

KOPP, H.   *See* Jahresbericht über die Fortschritte der reinen . . . Chemie.

KRÄTZER, H.   *See* Central-Blatt der gesammten chemischen Grossindustrie.

KREPLIN, E.   *See* Centralblatt für Branntweinbrennerei.

240. KRITISCHE ZEITSCHRIFT FÜR CHEMIE, PHYSIK UND MATHEMATIK. Herausgegeben in Heidelberg von A. Kekulé, F. Eisenlohr, G. Lewinstein, M. Cantor.   1 vol., 8vo.   Erlangen, 1858.
> *Continued under the title :*

[a]   Kritische Zeitschrift für Chemie, Physik, Mathematik und die verwandten Wissenschaften und Disciplinen, als Pharmacie, Technologie, Agriculturchemie, Physik und Mineralogie. Unter Mitwirkung von Fachmännern, herausgegeben von E. Erlenmeyer und G. Lewinstein.   1 vol. (II), 8vo.   Erlangen, 1859.
> *Continued under the title :*

[b]   Zeitschrift für Chemie und Pharmacie.   Correspondenzblatt, Archiv und kritisches Journal für Chemie, Pharmacie und die verwandten Disciplinen.   Unter Mitwirkung von Fachmännern, herausgegeben zu Heidelberg von E. Erlenmeyer und G. Lewinstein.   4 vols. (III–VII), 8vo.   Erlangen [*later*, Heidelberg], 1860–'64.

KRITISCHE ZEITSCHRIFT FÜR CHEMIE, etc. [Cont'd.]
Continued under the title :

[c] Zeitschrift für Chemie. Archiv für das Gesammtgebiet der Wissenschaft. Unter Mitwirkung von F. Beilstein und Rud. Fittig, herausgegeben von H. Hübner. Neue Folge. 7 vols. (VIII–XIV), 8vo. Göttingen, 1865–'71 ||

KRÖNIG, A. *See* Journal für Physik und physikalische Chemie.

KRÜSS, GERHARD. *See* Zeitschrift für anorganische Chemie.

241. KURZER VIERTELJAHRESBERICHT ÜBER DIE FORTSCHRITTE DER ZUCKERFABRIKATION von M. Märcker und Albert Jena. Berlin, 1878. 8vo, 4 parts.
Reprinted from Zeitschrift für das chemische Grossgewerbe.

242. KVAS. Časopis spolků pro průmysl v kralovství Českém. Red. : A. Schmelzer. 19 vols., 8vo. v Praze, 1873–'91+

243. LABORATORIUM (DAS). Eine Sammlung von Abbildungen und Beschreibungen der besten und neuesten Apparate, zum Behuf der practischen und physikalischen Chemie. 44 nos., 4to. Weimar, 1825–'40.

244. LABORATORY (THE). A monthly journal of the progress of chemistry, pharmacy, medicine, recreative science, and the useful arts. By J. F. Babcock. 2 vols., 4to. Boston, 1874–'76 ||

245. LABORATORY (THE). A weekly record of scientific research. 1 vol., 8vo. London, April to October, 1867 ||

LABORATORY (THE). *See* American Laboratory.

246. LABORATORY (THE) YEARBOOK. Edited by John Howard Appleton. 10 parts, 16mo. Providence, 1883–'92+

LA CAVA, P. *See* Metamorfico (II).

LAIWITZ, GEBRÜDER. *See* Muster-Zeitung für den Färberstand.

LAMPRAY, THOMAS. *See* Brewers' (The) Guardian.

LARTIGUE, A. *See* Répertoire de pharmacie.

LASSING, HENRY. *See* New York (The) Analyst.

LAURENT, A., ET GERHARDT, CH.
See Comptes-rendus mensuels des travaux chimiques.

LECOUTURIER,　*See* Coloration (La) industrielle.

LEIGHTON, J. H.　*See* American (The) Glass Worker.

LEIPZIGER FÄRBER-ZEITUNG.　*See* Deutsche Muster-Zeitung.

247.　LICHT'S (F. O.) WOCHENBERICHTE [*later*, MONATSBERICHTE] ÜBER
　　　DIE RÜBENZUCKER-INDUSTRIE. – vols. Magdeburg, 18——-91+

　　　LIEBIG AND KOPP'S JAHRESBERICHT.　*See* Jahresbericht über die Fortschritte der
　　　reinen, pharmaceutischen und technischen Chemie.

　　　LIEBIG'S ANNALEN.　*See* Annalen der Pharmacie [*c*].

　　　LINTNER, KARL.　*See* Bayerische (Der) Bierbrauer.

248.　LISTY CHEMICKÉ.　Organ spolku chemiků českých.　Red. : K.
　　　Preis a A. Bélohoubek [1875–78]; Preis a K. Kruis [1878–83];
　　　Preis a B. Raÿman [1883–91]; K. Preis [1891].　16 vols., roy.
　　　8vo.　v Praze, 1875–'91+

249.　LISTY CUKROVARNICKÉ.　Red. : K. Preis a M. Nevole.　9 vols.
　　　v Praze, 1883–'91+

250.　LISTY PEKAŘSKÉ.　Red.: T. Soehor.　[*From* 1890, Red.: A. Liegert.]
　　　5 vols., 4to.　v Praze, 1887–'91+

　　　LÖFFLER, CARL.　*See* Zeitung für Zuckerfabrikanten.

　　　LÖWIG, C.　*See* Repertorium für organische Chemie.

　　　LONDON CHEMICAL SOCIETY.　*See* Proceedings of the London Chemical Society.

251.　LONDON WATER SUPPLY (REPORTS).　London, 1882–'92+

　　　MAANDBLAD VOOR TOEGEPASTE SCHEIKUNDE.　*See* Toegepaste scheikunde.

　　　MACKENSIE, COLIN.　*See* Mémorial pratique de chimie manufacturière.

252.　MAGAZIN FÜR APOTHEKER, CHEMISTEN UND MATERIALISTEN.
　　　Herausgegeben von Joh. Casp. Elwert.　3 parts, 8vo.　Nürn-
　　　berg, 1785–'87.
　　　　　　Continued under the title :

　[a]　Repertorium für Chemie, Pharmacie und Arzneimittelkunde.
　　　Herausgegeben von Joh. Casp. Elwert.　8vo.　Hildesheim,
　　　1790.　[*Second edition in* 1796].

253.　MAGAZIN FÜR DIE HÖHERE NATURWISSENSCHAFT UND CHEMIE.
　　　2 vols., 8vo.　Tübingen, 1784–'87.

MAGAZIN FÜR DIE NEUESTEN ERFAHRUNGEN . . . DER PHARMACIE.
See Nordische Blätter für Chemie [*b*].

254. MAGAZIN FÜR EISEN-, BERG- UND HÜTTENKUNDE. Herausgegeben von J. L. Jordan und T. S. Hasse. 1 vol., 8vo. Berlin, 1802–'07.

255. MAGAZIN FÜR FÄRBER, ZEUGDRUCKER UND BLEICHER. Herausgegeben von S. F. Hermbstädt. 6 vols., 8vo. Berlin, 1802–'07.

MAJOCCHI, G. A. *See* Annali di fisica, chimica e matematiche ; *also* Giornale di fisica, chimica ed arti.

MALY, RICH. *See* Jahresbericht über die Fortschritte der Thierchemie.

MARKL, A. *See* Vzorný Pivovar.

MARTIN, CH. *See* Répertoire de chimie, de physique [etc.

MAUGHAM, W. *See* Annals of Chemical Philosophy.

256. MECHANIC AND CHEMIST. 8 vols. London, 1836–'42.

257. MEDDELELSER FRA CARLSBERG LABORATORIET. Udgivet ved Laboratoriets Bestyrelse. 3 vols., 8vo. Kjøbenhavn, 1885–'91 +

258. MEDICINISCH-CHEMISCHE UNTERSUCHUNGEN AUS DEM LABORATORIUM FÜR ANGEWANDTE CHEMIE ZU TÜBINGEN. Herausgegeben von Felix Hoppe-Seyler. 6 vols., roy. 8vo. Berlin, 1866–'71 ‖

MEIJLINK. *See* Schei-artsenijmeng- en natuurkundige bibliotheek.

259. MÉLANGES PHYSIQUES ET CHIMIQUES tirés du Bulletin physico-mathématique de l'Académie Impériale des sciences de St. Pétersbourg. 1846–'89. 13 vols., 8vo. St. Pétersbourg, 1854–'91+

MEMOIRS AND PROCEEDINGS OF THE CHEMICAL SOCIETY OF LONDON.
See Proceedings of the Chemical Society of London.

260. *MEMOIRS OF THE COLUMBIAN CHEMICAL SOCIETY OF PHILADELPHIA. Published by Isaac Peirce, No. 3 S. Fourth St. pp. xiv-221, 8vo. [Philadelphia], 1813.

261. MÉMORIAL PRATIQUE DE CHIMIE MANUFACTURIÈRE. Par Colin Mackensie. 3 vols., 8vo. Paris, 1824.

262. MEMORIE DELLA SOCIETÁ DEGLI SPETTROSCOPISTI ITALIANI. Palermo, 1872–'90 +

263. MERCK'S BULLETIN. A periodical record of new discoveries, introductions or applications of medicinal chemicals. Vol. I–IV, 8vo. New York [Darmstadt and London], 1887–'91.

264. METAMORFICO (IL). Giornale di chimica, farmacia e scienze affini, redatto da P. La–Cava. 3 vols., 8vo. Napoli, 1845–'47.

MÉTHÉRIE, DE LA. *See* Introduction aux observations sur la physique [etc.].

MEYER, RICHARD. *See* Jahrbuch der Chemie.

265. MILCHWIRTHSCHAFTLICHES TASCHENBUCH. 15 vols., 16mo. Danzig [*later*, Bremen], 1877–'91+

266. MILCH-ZEITUNG. Organ für die gesammte Viehhaltung und das Molkereiwesen. Unter Mitwirkung von Fachmännern, herausgegeben von C. Petersen. 21 vols. (I–XXI), 4to. Bremen, 1872–'92+

267. MILK (THE) JOURNAL. A monthly review of the dairy (etc.). 4to. London, 1871–'73.

MILLON, E. ET REISET, J. *See* Annuaire de chimie.

MINING AND PETROLEUM STANDARD. *See* American Gas-Light Journal.

268. MINUTES OF THE SOCIETY FOR PHILOSOPHICAL EXPERIMENTS . . . edited by Bryan Higgins. London, 1794.
[a] Protokolle der Verhandlungen einer Privat-Gesellschaft in London über die neuern Gegenstände der Chemie. Geführt unter der Anleitung von Bryan Higgins. Herausgegeben von Alex. Nic. Scherer. Halle, 1803. Ill.

269. MISCELLANEA DI CHIMICA, FISICA E STORIA NATURALE. 8vo. Pisa, 1843.

270. MITTHEILUNGEN AUS DEM CHEMISCHEN LABORATORIUM DER UNIVERSITÄT INNSBRUCK. 19 nos., roy. 8vo. Wien, 1867–'73.

271. MITTHEILUNGEN AUS DEM CHEMISCHEN LABORATORIUM VON DR. HILGER. Herausgegeben von Hilger und F. Nies. 8vo. Würzburg, 1873.

272. MITTHEILUNGEN AUS DEM GEBIETE DER REINEN UND ANGEWANDTEN CHEMIE. Für Fachchemiker, Aerzte, Oeconomen, etc. Als Jahresbericht des Laboratoriums für 18—. Veröffentlicht von V. Kletzinsky. 4to. Wien, 1865–'66.

273. MITTHEILUNGEN AUS DEM LABORATORIUM DER ALLGEMEINEN
      CHEMIE an der Kais. Kön. technischen Hochschule zu Brünn.
      3 nos., 8vo. Wien, 1876.
         Reprinted from Sitzungsberichte der K. Akademie der Wissenschaften.

274. MITTHEILUNGEN AUS DEM PHARMACEUTISCHEN INSTITUTE und
      Laboratorium für angewandte Chemie der Universität Er-
      langen. Redigirt von A. Hilger. Erlangen, 1889–'90+

275. MITTHEILUNGEN AUS DER AMTLICHEN LEBENSMITTEL-UNTER-
      SUCHUNGS-ANSTALT UND CHEMISCHEN VERSUCHSSTATION ZU
      WIESBADEN, über die geschäftliche und wissenschaftliche
      Thätigkeit in dem Betriebsjahre 1884. Herausgegeben von
      Director Schmitt. 8vo. Berlin, 1885.

276. MITTHEILUNGEN DER K. K. CHEMISCH-PHYSIOLOGISCHEN VERSUCHS-
      STATION für Wein- und Obstbau in Klosterneuburg bei Wien.
      Herausgegeben von L. Rössler. Wien, 1888+

277. MONATH-SCHRIFT VON NÜTZLICHEN UND NEUEN ERFAHRUNGEN
      AUS DEM REICHE DER SCHEIDEKUNST und andern Wissen-
      schaften, von J. A. Weber. 8vo. Tübingen, 1773.
         Only the Erster Monath, 32 pp., seems to have been published.

278. MONATSBERICHT DER GLASHÜTTE. Allgemeines Organ für die
      Glasindustrie. Mittheilungen merkantilischen, technischen
      und historischen Inhalts über alle die Glasindustrie betreffende
      Vorkommnisse des In- und Auslandes. Redigirt und heraus-
      gegeben von Julius Fahdt. 4to. Dresden, 1873.

279. MONATSHEFTE FÜR CHEMIE UND VERWANDTE THEILE ANDERER
      WISSENSCHAFTEN. Gesammelte Abhandlungen aus den Sit-
      zungsberichte der K. Akademie der Wissenschaften. 13 vols.,
      8vo. Wien, 1880–'92+

280. MONITEUR (LE) DE LA BRASSERIE ET DU COMMERCE DES VINS ET
      DES SPIRITUEUX. 31 vols. Bruxelles, 1860–'90+

281. MONITEUR DE LA BRASSERIE. 34 vols., fol. Paris, 1859–'92+
         Identical with preceding ?

282. MONITEUR DE LA CÉRAMIQUE, de la verrerie et des industries qui
      s'y rattachent. Publié sous la direction d'une Société d'ingé-
      nieurs, de professeurs et de fabricants. [Edmond Rousset,
      directeur-gérant.] 24 vols., 4to. Paris, 1869–'92+

283. MONITEUR DE LA SUCRERIE ET DE LA DISTILLERIE. 3 nos., fol. Paris, 1859.

Continued under the title :

[a] Moniteur de la sucrerie et des industries agricoles. Rédigé par P. Bourgeois. Fol. Paris, 1859–'60.

MONITEUR DE LA TEINTURE. *See* Coloration (La) industrielle.

284. MONITEUR DE L'INDUSTRIE DU GAZ. Organe des abonnés et des usiniers. 8 vols., 4to. Paris, 1875–'82 [+?]

285. MONITEUR DES FILS ET TISSUS, des apprêts, de la teinture et du matériel de ces industries. Fondé sous la direction de Michel Alcan. Par une Société d'ingénieurs, d'industriels et de négociants. Edmond Rousset, directeur-gérant. 23 vols., 4to. Paris, 1870–'92+

286. MONITEUR (LE) DES PRODUITS CHIMIQUES, pour l'industrie, les sciences et les arts, et du matériel de ces industries. Publié par une Société de chimistes et d'industriels. Edmond Rousset, gérant. 23 vols, 4to. Paris, 1870–'92+

287. MONITEUR (LE) DU GAZ. Journal de l'éclairage [etc.]. Publié sous la direction de Ed. R. d'Hurcourt. 4to. Paris, 1864–'65.

MONITEUR SCIENTIFIQUE DU CHIMISTE [etc.].
See Revue scientifique et industrielle.

MONS, J. B. VAN. *See* Journal de chimie et de physique ; *also* Journal de chimie pour servir [etc.].

288. MONTHLY (THE) MAGAZINE OF PHARMACY, CHEMISTRY, MEDICINE, etc. 5 (?) vols. London, 1876–'80+

MULDER, G. J. *See* Natuur- en scheikundig archief ; *also* Scheikundige onderzoekingen.

MUSTER-ZEITUNG. *See* Deutsche Muster-Zeitung.

289. MUSTER-ZEITUNG FÜR DEN FÄRBERSTAND. Monatsschrift für Färberei, Druckerei und Farbwaarenkunde. Herausgegeben von Gebrüder Lairitz. 18 nos., fol. Leipzig, 1847–'48.

Continued under the title :

[a] Färber-Zeitung mit Mustern. Monatsschrift für Farbwaarenkunde, Färberei und Druckerei. 1849. 1 vol., 4to. Leipzig, 1850 ‖

MUTER, J.   *See* Analyst (The).

290. NATURHISTORISCHE UND CHEMISCH-TECHNISCHE NOTIZEN, nach den neuesten Erfahrungen zur Nutzanwendung für Gewerbe, Fabrikwesen und Landwirthschaft.   11 vols., 8vo.   Berlin, 1854-'59.
Neue Folge.   4 vols. (XII–XV), 8vo.   Berlin, 1860–'62 ||

291. NATUUR- EN SCHEIKUNDIG ARCHIEF.   Door G. J. Mulder [en W. Wenckebach].   6 vols., 8vo.   Rotterdam en Leijden, 1833–'38 ||

NEUE, NEUERE, NEUESTE, NEW [etc.].
These adjectives have been omitted in alphabetization.   *See* next succeeding word of title.

NÉVOLE, M.   *See* Zeitschrift für Zucker-Industrie in Böhmen.

NEWCASTLE-UPON-TYNE CHEMICAL SOCIETY.   *See* Transactions of.

292. NEW YORK ANALYST (THE).   Devoted to the interests of sanitary science, food, medicine and the suppression of adulteration. Edited by Henry Lassing.   18 nos., 4to.   New York, 1885.
[This is successor to an American reprint of the ANALYST published in London ; hence the above is styled : " New Series."]
From No. 19, Oct. 1st, 1885, *continued under the title :*

[a]   American Analyst.   A popular semi-monthly [*later*, weekly] review devoted to industrial progress, sanitation and the chemistry of commercial products.   H. Lassing, editor and publisher.   New series.   Vol. II–VIII, 4to.   New York, 1885–'92+

NEW YORK DRUG BULLETIN.   *See* Oil, Paint and Drug Reporter.

NEW YORK DRUGGISTS' PRICE CURRENT.   *See* Oil, Paint and Drug Reporter.

NICHOLS, JAS. R.   *See* Boston Journal of Chemistry.

NICHOLSON'S JOURNAL.   *See* Journal of Natural Philosophy.

293. NORDDEUTSCHE BRAUER-ZEITUNG.   Fach Zeitschrift für Bierbrauerei und Malzfabrikation.   Herausgegeben von B. Johannesson in Berlin.   12 vols., 8vo.   Berlin, 1876–'87.
Continued under the title :

[a]   Deutsche Brau-Industrie.   Offizielles Organ des deutschen Braumeister-Vereins.   Redactor : B. Johannesson.   5 vols., 8vo.   Berlin, 1887–'92+

294. NORDISCHE BLÄTTER FÜR CHEMIE. Herausgegeben von Alex. Nic. Scherer. 1 vol., 8vo. Halle, 1817.

Followed by :

[a] Allgemeine nordische Annalen der Chemie für die Freunde der Naturkunde und Arzneiwissenschaft insbesondere der Pharmacie, Arzneimittellehre, Physiologie, Physik, Mineralogie und Technologie im russischen Reiche. Herausgegeben von Alex. Nic. Scherer. 7 vols. (II–VIII), 8vo. St. Petersburg, 1819–'22.

Followed by :

[b] Magazin für die neuesten Erfahrungen, Entdeckungen und Berichtungen im Gebiete der Pharmacie, mit Hinsicht auf physiologische Prüfung und practisch bewährte Anwendbarkeit der Heilmittel, vorzüglich neuentdeckter Arzneistoffe in der Therapie. Herausgegeben von G. F. Hänle [*from* 1829, *by* P. L. Geiger]. 36 vols., 8vo. Carlsruhe, 1823–'31 ||

> *From* 1829–'31 *also under the title :* Magazin für Pharmacie und Experimental-Kritik. *United in* 1832 *with* Archiv des Apotheker-Vereins im nördlichen Teutschland, *forming the* Annalen der Pharmacie, *q. v.*

OBSERVATIONS ET MÉMOIRES SUR LA PHYSIQUE [etc.].
See Introduction aux observations sur la physique.

295. OESTERREICHISCHE (DER) BIERBRAUER. Illustrirte Monatsschrift für das gesammte Brauwesen. Herausgegeben und redigirt von Fernand Stamm. 1 vol., fol. Wien, 1864–'65 ||

296. OESTERREICHISCHE ZEITSCHRIFT FÜR BIERBRAUEREI. Unter Mitwirkung der tüchtigsten Fachmänner, herausgegeben von Franz Fasbender und redigirt von Ludwig Berger. 2 vols., roy. 8vo. Wien, 1873–'75 ||

297. OESTERREICHISCH-UNGARISCHER BRAUER- UND MÄLZER-KALENDER. Redigirt von Franz Fasbender. 3 vols., 16mo. Wien, 1878–'80.

298. OESTERREICHISCH-UNGARISCHER BRENNEREI-KALENDER für das Jahr 1878. Herausgegeben von Rich. Jahn. 2 vols., 16mo. Wien, 1878–'79 ||

OESTERREICH-UNGARISCHE ZEITSCHRIFT FÜR ZUCKERINDUSTRIE. *See* Organ des Vereins für Rübenzucker-Industrie.

OFFICIAL AGRICULTURAL CHEMISTS.
    See Association of Official Agricultural Chemists.

299. OIL AND DRUG NEWS [*later*] AND PAINT REVIEW. A weekly journal devoted to the oil, paint, drug and allied trades. George Bartholomew, editor. 3 vols., 4to. New York, 1880–'82 ||

300. OIL, PAINT AND DRUG REPORTER. Embracing the " Drug, Paint and Oil Trade," " New York Druggists' Price Current," and " New York Drug Bulletin." William O. Allison, proprietor. 40 vols. (2 vols. per annum), fol. New York, 1871–'91+

O'NEILL, CHARLES. *See* Textile Colorist (The).

OPWYRDA, R. J. *See* Toegepaste scheikunde.

301. ORGAN DES VEREINS FÜR RÜBENZUCKER-INDUSTRIE in der oesterreichisch-ungarischen Monarchie. Zeitschrift für Landwirthschaft und technische Fortschritt der landwirthschaftliche Gewerbe, vorzugsweise für Rübenzucker-Industrie. Redigirt von Otto Kohlrausch. 25 vols., 4to. Wien, 1863–'87.
    Continued under the title :

[a]   Oesterreich-ungarische Zeitschrift für Zuckerindustrie und Landwirthschaft. Herausgegeben vom Centralverein für Rübenzucker-Industrie in der oesterreichisch-ungarische Monarchie. Redigirt von F. Strohmer. 3 vols., roy. 8vo. Wien, 1888–'92+

302. ORGAN FÜR DEN OEL- UND FETTHANDEL. Die Seifen- und Lichte-Fabrication, die Oel-, Harz- und Theerproduction-Industrie, sowie die Wachswaaren- und Parfümerie-, die Mineralöl- und Paraffin-, die Degras- und Schmiermaterialenfabrikation. Herausgeber : N. Besselich. 21 vols., fol. Trier, 1868–'88+

303. OROSI (L'). Bollettino [*later*, Giornale] di chimica, farmacia e scienze affini. Pubblicato per cura dell' associazione chimico-farmaceutica fiorentina. Antonio Bernardi, editore. 13 vols., 8vo., 1878–'90 Firenze, 1878–'91 +

OSANN, F. *See* Stahl und Eisen.

OSTWALD, WILHELM. *See* Zeitschrift für physikalische Chemie.

PARIS, SOCIÉTÉ CHIMIQUE DE. *See* Bulletin des séances de la Société [etc.].

PATAKY, CARL. *See* Journal für Gas-Industrie.

PATERNO, M. *See* Gazzetta chimica italiana.

PETERSEN, C. *See* Milch-Zeitung.

304. PENNY MECHANIC AND CHEMIST. 8 vols. London, 1836-'42.

305. PHARMACIST (THE) AND CHEMICAL RECORD. A monthly journal
    devoted to pharmacy, chemistry and the collateral sciences.
    Published by the Chicago College of Pharmacy. Editor, N.
    Gray Bartlett ; associate editor, Albert E. Ebert. [*From* vol.
    III, *edited by* E. H. Sargent.] 5 vols., 8vo. Chicago, 1868-'72.
        Continued under the title :

[a] Pharmacist (The). 6 vols. (VI-IX). Chicago, 1873-'78.
        Continued under the title :

[b] Pharmacist (The) and Chemist. Published by the Chicago Col-
    lege of Pharmacy. [*Conducted by* Robert H. Cowdrey.] 6
    vols. (XII-XVII). Chicago, 1879-'84.
        Continued under the title :

[c] Pharmacist (The). 1 vol. (XVIII). Chicago, 1885.
        In 1886 *merged in :*

[d] Western (The) Druggist. Chicago, 1879-'92+

306. PHARMACEUTICAL (THE) TIMES. A journal of chemistry applied
    to the arts, agriculture and manufactures. 3 vols., 4to. Lon-
    don, 1847-'48.
        Continued under the title :

[a] Chemical (The) Times and Journal of Pharmacy, Manufactures
    [etc.]. [*Edited by* G. M. Mowbray.] 2 vols , 4to. London,
    1848-'49.

307. PHARMACEUTISCHES CENTRALBLATT. [*Edited from* 1830-'38 *anony-*
    mously ; from 1840-'44, *by* A. Weinlig ; 1845-'47, *by* R. Buch-
    heim ; 1848-'49, *by* W. Knop.] 20 vols., 8vo. Leipzig,
    1830-'49.
        Continued under the title :

[a] Chemisch-pharmaceutisches Centralblatt. Redacteur : W. Knop.
    5 vols. (XXI-XXV), 8vo. Leipzig, 1850-'55.
        Continued under the title :

[b] Chemisches Centralblatt. Repertorium für reine, pharmaceu-
    tische, physiologische und technische Chemie. Redaction :
    W. Knop. [*From* 1862, Red. : Rud. Arendt.] Neue
    Folge. 14 vols. (XXVI-XXXIX), 8vo. Leipzig, 1856-'69.

[c] Dritte Folge. Redigirt von Rudolph Arendt. 19 vols. (XL-LIX),
    8vo. Leipzig, 1870-'88.

PHARMACEUTISCHES CENTRALBLATT. [Cont'd.]

[d] Chemisches Centralblatt. Vollständiges Repertorium für alle
     Zweige der reinen und angewandten Chemie. Vierte Folge.
     Redigirt von Rudolph Arendt. 4 vols. (LX–LXIII, *of the
     entire series*). Hamburg und Leipzig, 1889–'92+

     General-Register zum chemischen Centralblatt, III Folge,
     Jahrgang I–XII. 1870–'81. Redigirt von Rud. Arendt.
     1 vol., 8vo. Leipzig, 1882–'83.

PHILADELPHIA. *See* Memoirs of the Columbian Chemical Society.

308. PHILOSOPHICAL (THE) MAGAZINE. Comprehending the various
      branches of science, the liberal and fine arts, agriculture, manu-
      factures, and commerce. By Alexander Tilloch. 42 vols.,
      8vo. London, 1798–1813.

      United in 1814 *with the* Journal of Natural Philosophy, by William
      Nicholson, *and continued under the title :*

[a] Philosophical (The) Magazine and Journal. Comprehending
     the various branches of science, the liberal and the fine arts,
     geology, agriculture, manufactures, and commerce. By Al-
     exander Tilloch [*from* 1824 *by* Alexander Tilloch *and*
     Richard Taylor.] 26 vols. (XLIII–LXVIII), 8vo. London,
     1814–'26.

      United in 1827 *with the* Annals of Philosophy, or Magazine of
      Chemistry, by Richard Phillips, *and continued under the title :*

[b] Philosophical (The) Magazine ; or Annals of Chemistry,
     Mathematics, Astronomy, Natural History, and General
     Science. New and united series of the Philosophical Maga-
     azine and Annals of Philosophy. By Richard Taylor and
     Richard Phillips. 11 vols. (I–XI), 8vo. London, 1827–'32.

      United in 1832 *with the* Edinburgh Journal of Science, by David
      Brewster, *and continued under the title :*

[c] London and Edinburgh Philosophical Magazine and Journal of
     Science. Conducted by David Brewster, Richard Taylor,
     and Richard Phillips. New and united series of the Philo-
     sophical Magazine and Journal of Science. 37 vols.
     (I–XXXVII), 8vo. London, 1832–'50.

      Continued under the title :

[d] London, Edinburgh, and Dublin Philosophical Magazine and
     Journal of Science. Conducted by David Brewster, Rich-
     ard Taylor, Richard Phillips, Robert Kane, and William
     Francis. Fourth series. 50 vols. (I–L), 8vo. London,
     1851–'75.

PHILOSOPHICAL (THE) MAGAZINE.   [Cont'd.]
> Fifth series. Edited by R. Kane, W. Thomson, and W.
> Francis. 32 vols. (I–XXXII), 8vo. London, 1876–'92+

309. PHYSIKALISCH–CHEMISCHES MAGAZIN FÜR AERZTE, CHEMISTEN
> UND KÜNSTLER. Von J[acob] A[ndreas] Weber. Berlin,
> 1780. 2 vols., 8vo.

310. PHYSIKALISCHE REVUE, herausgegeben von L. Graetz. 8vo. Stutt-
> gart, 1892+

311. PIRIA (IL). Giornale di scienze chimiche. Napoli, 1875.

312. PIVOVARNICKÉ ZPRÁVY. Red. F. Chodounský. 2 vols., 4to. v
> Praze, 1890–'91+

313. PIVOVARSKI LISTY. Red. K. Tiller. 9 vols., 4to. v Praze, 1883–
> '91+

POGGENDORFF'S ANNALEN. *See* Journal der Physik.

POLLI, GIOVANNI. *See* Giornale di farmacia, chimica e scienze accessorie.

314. POLYTECHNISCHES JOURNAL. Herausgegeben von Johann Gottfried
> Dingler. [*From* 1831, vol. XXXIX, *edited by the same and*
> Emil Maximilian Dingler ; *from* 1834–'40, vols. LI–LXXVII, *by*
> *the above and* Julius Hermann Schultes ; *from* 1841, *by* Johann
> Gottfried Dingler und Emil Maximilian Dingler ; *from* 1855,
> vol. CXXXVI, *by the last named alone.*] 211 vols., 8vo. Augs-
> burg, 1820–'73.
>> *Continued under the title :*

[a] Dingler's polytechnisches Journal. Herausgegeben von Johann
> Zeman und Ferd. Fischer. [*From* 1887, *by* J. Zeman und
> H. Kast. *From* 1888, *by* A. Hollenberg und H. Kast.]
> 67 vols. (212–278). 8vo. Augsburg, [*later*] Stuttgart.
> 1874–'90.
>> Vol. 248 is entitled : Atlas zu Dingler's polytechnischem Journal,
>> Jahrgang 1883, and consists wholly of plates.
>> *Changed to small folio and continued under the title :*

[b] Dingler's polytechnisches Journal. Unter Mitwirkung von C.
> Engler herausgegeben von A. Hollenberg und H. Kast.
> 8 vols. (279–287), sm. fol. Stuttgart, 1891–'92+

POPULAR SCIENCE NEWS. *See* Boston Journal of Chemistry.

315. POPULÄRE ZEITSCHRIFT FÜR SPIRITUS- UND PRESSHEFE-INDUSTRIE. Herausgegeben von Alois Schönberg, unter Mitwirkung hervorragender Fachautoritäten und anerkannt bewährter Fachmänner. 6 vols., 4to. Wien, 1880–'85.

Continued under the title :

[a] Allgemeine Zeitschrift für Spiritus- und Presshefe-Industrie, Cognac-, Liqueur- und Essig-Fabrikation. Herausgegeben von Heinrich Blechner. Unter Mitwirkung hervorragendster Fachautoritaten und erster Practiker. 4 vols. (VI–IX), 8vo. Wien, 1885–'88+

316. PORTEFEUILLE FÜR GEGENSTÄNDE DER CHEMIE UND PHARMACIE. 1 no., 8vo. Hamburg, 1784.

POSKE, F. *See* Zeitschrift für den physikalischen und chemischen Unterrricht.

POST, JULIUS. *See* Zeitschrift für das Grossgewerbe.

317. PRAKTISCHE (DER) BIERBRAUER. 8vo. New York, 1870 [?]–'92+

A Supplement to the Amerkanische Bierbrauer, *q. v.*

PREIS, K. A M. NEVOLE. *See* Listy cukrovarnické.

PREIS, K. *See* Zeitschrift für Zucker-Industrie.

318. PROCEEDINGS OF THE AMERICAN CHEMICAL SOCIETY. 1876–'78. 1 vol., 8vo. New York, 1877.

Continued under the title :

[a] The Journal of the American Chemical Society. 14 vols., 8vo. New York, 1879–'92+

319. PROCEEDINGS OF THE CHEMICAL SOCIETY OF LONDON. 1841–'43. 1 vol., 8vo. London, 1843.

Continued under the title :

[a] Memoirs and Proceedings of the Chemical Society of London. 1841–'48. 3 vols., 8vo. London, 1843–'48.

Continued under the title :

[b] Quarterly Journal of the Chemical Society of London. 14 vols. (I–XIV), 8vo. London, 1849–'62.

Continued under the title :

[c] The Journal of the Chemical Society of London. 1 vol. (XV), 8vo. London, 1862.

[d] New series. 13 vols. (I–XIII). (Entire series, vol. XVI–XXVIII). London, 1863–'75.

PROCEEDINGS OF THE CHEMICAL SOCIETY OF LONDON. [Cont'd.]
> *Continued under the title :*

[e]   Journal of the Chemical Society, containing the papers read before the Society, and abstracts of chemical papers published in other journals. [Third Series.] 2 vols. (I–II). [Entire Series, XXIX–XXX.] London, 1876.

[f]   [Fourth Series.] 2 vols. (I–II). [Entire Series, XXXI–XXXII.] London, 1877.

[g]   [Fifth Series.] The Journal of the Chemical Society. 30 vols., Vol. XXXII–LXII. London, 1878–'92+

> From 1878 to date two vols. per annum, one with the sub-title "Transactions," and one with the sub-title "Abstracts."
> In 1878 the editors gave to the preceding and current volumes the numbers of the entire series, disregarding the individual series.

PROCEEDINGS OF THE CONVENTION OF AGRICULTURAL CHEMISTS.
> *See* Association of Official Agricultural Chemists.

320.   PRŮMYSLNÍK. List věnovaný zájmým domácího průmyslu, spolu organ cukrovárníků českých. Majetník a redaktor : Karel Otakar Čech. S přilohou : "Časopis chemiků českých." 3 vols., 8vo. v Praze, 1869–'72.
> *Cf.* Časopis chemiků českých.

PUBLIC ANALYSTS, SOCIETY OF.   *See* Analyst (The).

QUARTERLY JOURNAL OF THE CHEMICAL SOCIETY OF LONDON.
> *See* Proceedings of the Chemical Society of London.

QUESNEVILLE.   *See* Revue scientifique et industrielle.

321.   RACCOLTA FISICO-CHIMICA ITALIANA. Ossia collezione di memorie originali edite ed inedite di fisici, chimici e naturalisti italiani dell' Ab. Francesco Zantedeschi. 3 vols., roy. 8vo. Venezia, 1846–'48.
> *Followed by :*

[a]   Annali di fisica, dell' Abbate F. C. Zantedeschi. 1 vol., roy. 8vo. Padova, 1849–'50.

322.   RAPPORT ANNUEL SUR LES PROGRÈS DES SCIENCES PHYSIQUES ET CHIMIQUES présenté . . . à l'académie royale des sciences de Stockholm par J. Berzelius. Traduit du Suédois par Ph. Plantamour. 4 vols., 8vo. Paris, 1841–'44.
> *Continued under the title :*

[a]   Rapport annuel sur les progrès de la chimie, présenté . . . à

RAPPORT ANNUEL, etc.   [Cont'd.]
l'académie royale des sciences de Stockholm par J. Berzelius. Traduit du Suédois par Ph. Plantamour. 2 vols., 8vo. Paris, 1845–'46.

> *Cf*. Annuaire des sciences chimiques ; *also* Årsberättelse om Framstegen i Physik och Chemi ; *also* Jahresbericht über die Fortschritte der physischen Wissenschaften.

323. RECHERCHES PHYSICO-CHIMIQUES. 3 nos., 4to. Amsterdam, 1792–'94.

324. RECUEIL DES MÉMOIRES LES PLUS INTÉRESSANS DE CHIMIE ET D'HISTOIRE NATURELLE, contenus dans les actes de l'académie d'Upsal et dans ceux de l'académie royale des sciences de Stockholm. Publiés depuis 1720 jusqu' en 1760. Traduit du Latin et de l'Allemand. 2 vols., 12mo. Paris, 1764.

325. RECUEIL DES TRAVAUX CHIMIQUES DES PAYS-BAS. Par W. A. van Dorp, A. P. N. Franchimont, S. Hooge-Werff, E. Mulder et A. C. Oudemans, Jr. 11 vols., roy. 8vo. Leide, 1882–'92+

REIMANN, M. *See* Deutsche Musterzeitung ; *also* Journal de teinture.

326. REIMANN'S FÄRBER-ZEITUNG. Organ für Färberei, Druckerei, Bleicherei, Appretur, Farbwaaren und Buntpapierfabrication, Droguenhandel, Spinnerei und Weberei. Redacteur und Herausgeber : M. Reimann. [*From* 1878,] Organ des "Allgemeinen Färber- und Fachgenossenvereins," und der "Färber-Akademie" zu Berlin. 21 vols., 8vo. Berlin, 1870–'90+
See Journal de teinture de Reimann.

REMSEN, IRA. *See* American Chemical Journal.

327. RÉPERTOIRE DE CHIMIE, pharmacie, matière pharmaceutique et de chimie industrielle. Par P. J. Hensmans. 2 vols., 8vo. Louvain, 1828–'30.
   [a]   Nouvelle répertoire [etc.]   1 vol., 8vo. Louvain, 1831.

328. RÉPERTOIRE DE CHIMIE, DE PHYSIQUE, ET D'APPLICATIONS AUX ARTS. Contenant les traductions ou extraits des travaux qui se publient sur ces matières dans les pays étrangers, et de plus un résumé rapide des mémoires parus en France. Rédigé par Ch. Martin, sous la direction de Gaultier de Claubry. 1 vol., 8vo. Paris, 1837.

RÉPERTOIRE DE CHIMIE, etc.   [Cont'd.]
> *Continued under the title :*

[a]   Répertoire de chimie scientifique et industrielle.   Contenant les traductions ou extraits des travaux qui se publient sur cette matière dans les pays étrangers, et de plus un résumé des mémoires les plus intéressants parus en France.   Rédigé par Ch. Martin, sous la direction de Gaultier de Claubry.   4 vols., 8vo.   Paris, 1837–'38.

> *Continued under the title :*

[b]   Répertoire de chimie.   Mémorial des travaux étrangers.   Rédigé par Gaultier de Claubry et Ch. Gerhardt.   Deuxième Série.   1 vol., 8vo.   Paris, 1839 ‖

RÉPERTOIRE DE CHIMIE PURE ET APPLIQUÉ.
See Bulletin des séances de la Société chimique de Paris.

329.   RÉPERTOIRE DE PHARMACIE, DE CHIMIE, DE PHYSIQUE, D'HYGIÈNE PUBLIQUE, de la médecine légale et de thérapeutique ; réimpression générale des ouvrages périodiques publiés en France sur ces sciences.   1 vol., 8vo.   Bruxelles, 1842.

330.   RÉPERTOIRE DE PHARMACIE.   Recueil pratique.   Rédigé par A. Lartigue.   [*From* vol. III, *by* Bouchardat.]   29 vols., 8vo. Paris, 1844–'73.
     Nouvelle série.   3 vols., 8vo.   Paris, 1874–'76.

> *United with the* Journal de chimie médicale, *and continued under the title :*

[a]   Répertoire de pharmacie et Journal de chimie médicale réunis. Dirigé par Eug. Lebaigue.   (?) vols., 8vo.   Paris, 1876–'91+
     Cf. Journal de chimie médicale.

331.   RÉPERTOIRE PRATIQUE DES VINS, vinaigres, boissons fermentées, alcools et spiritueux, tartres et acides tartriques.   3 vols., 8vo. Paris, 1888–'91+

332.   REPERTORIUM DER ANALYTISCHEN CHEMIE für Handel, Gewerbe und öffentliche Gesundheitspflege.   Redigirt von J. Skalweit. 7 vols., 8vo.   Hannover [*later*,] Hamburg, 1881–'87+

> *Incorporated with* Zeitschrift für die Chemische Industrie, *and continued under the title :* Zeitschrift für angewandte Chemie, *q. v.*

333.   REPERTORIUM DER CHEMIE UND PHARMACIE.   Herausgegeben von Swittau.   8vo.   St. Petersburg, 1837 [+ ?].

334. REPERTORIUM DER ORGANISCHEN CHEMIE. Herausgegeben von Gustav Theodor Fechner. 2 vols. in 5 parts. Leipzig, 1826–'28.

Continued under the title :

[a] Repertorium der neuen Entdeckungen in der organischen Chemie. Herausgegeben von Gustav Theodor Fechner. 2 vols., 8vo. Leipzig, 1829–'33.

335. REPERTORIUM DER NEUEN ENTDECKUNGEN IN DER UNORGANISCHEN CHEMIE. Herausgegeben von Gustav Theodor Fechner. 2 vols., 8vo. Leipzig, 1826–'27.

REPERTORIUM FÜR CHEMIE. ELWERT. *See* Magazin für Apotheker.

336. REPERTORIUM FÜR DIE PHARMACIE. Angefangen von Adolph Ferdinand Gehlen und fortgesetzt in Verbindung mit C. F. Bucholz, Rink und Anderen, von Johann Andreas Buchner. [*From* vol. v, unter Mitwirkung des Apotheker-Vereins in Baiern, herausgegeben von Johann Andreas Buchner.] 50 vols., 12mo. Nürnberg, 1815–'34.
Zweite Reihe. 50 vols., 12mo. Nürnberg, 1835–'48.
Dritte Reihe. 10 vols., 12mo. Nürnberg, 1849–'51.

Continued under the title :

[a] Repertorium (Neues) für Pharmacie. Unter Mitwirkung von Alb. Frickhinger, C. F. Hänle, J. E. Herberger, X. Landerer, Th. W. Ch. Martius, W. Mittenheimer, Friedrich Mohr, Max Pettenkofer, A. Schnizlein, F. L. Winkler, herausgegeben von J. A. Buchner. 25 vols., 12mo. Nürnberg, 1852–'76 ‖
Ergänzungsband. 1 vol., 12mo. Nürnberg, 1816.

337. REPERTORIUM FÜR ORGANISCHE CHEMIE. Herausgegeben von C. Löwig. 3 vols., 8vo. Zürich, 1841–'43 ‖

REPERTORIUM FÜR PHARMACIE. BUCHNER. *See* Repertorium für die Pharmacie.

338. REPERTORIUM FÜR PHARMACIE UND PRAKTISCHE CHEMIE IN RUSSLAND ; oder, Zusammenstellung des Wichtigsten und Wissenswerthesten aus den neuesten Entdeckungen im Gebiete der Pharmacie und Chemie mit vorzüglicher Rücksicht auf das russische Reich. Red. : C. Gauger. 8vo. St. Petersburg, 1842.

339. REPORTS OF THE ROYAL COLLEGE OF CHEMISTRY. London, 1849.

340. Résumé du Comte-rendu des Travaux du Laboratoire de Carlsberg. Vol. I, 8vo. Copenhague, 1891.

See Meddelelser fra Carlsberg Laboratoriet.

341. Review of Gas and Water Engineering. London, 1880–'83.

Continued under the title :

[a] Gas and Water Review and Journal of Electric Lighting. London, 1884–'90+

342. Revue de chimie industrielle. Rédacteur : A. M. Villon. 2 vols., 4to. 1890–'91+

343. Revue des bières, des vins et des alcools. Organe officiel de l'Association générale des brasseurs belges. Directeur : J. P. Roux-Matignon. Bruxelles, 1874.

344. Revue des industries chimiques et agricoles. 14 vols., 8vo. Paris, 1878–'91+

345. Revue hebdomadaire de chimie scientifique et industrielle, publiée sous la direction de Ch. Mène. 7 vols., 8vo. Paris, 1869–'75.

346. Revue internationale scientifique et populaire des falsifications des denrées alimentaires. Rédigée par G. F. Van Hamel-Roos. 5 vols., 4to. Amsterdam, 1887–'92+

347. Revue scientifique et industrielle des faits les plus utiles et les plus curieux observés dans la médecine, l'hygiène, la physique, la chimie, la pharmacie, l'économie rurale et domestique, l'industrie nationale et étrangère. Sous la direction de Quesneville. 16 vols. (I–XVI), 8vo. Paris, 1840–'44.

Deuxième série. 15 vols. (I–XV), 8vo. 1844–'47.

Troisième série. 9 vols. (I–IX), 8vo. 1848–'51.

Quatrième série. 1 vol., 8vo. 1852.

Followed by :

[a] Moniteur (Le) scientifique du chimiste et du manufacturier. Livre-Journal de chimie appliqué aux arts et à l'industrie. Spécialement consacré à la chimie générale pure et appliquée, par Quesneville. 5 vols. (I–V), 4to. Paris, 1857–'63.

Continued under the title :

[b] Moniteur (Le) scientifique. Journal des sciences pures et appliquées. Deuxième série. 7 vols. (VI–XII), 4to. Paris, 1864–'70.

REVUE SCIENTIFIQUE ET INDUSTRIELLE.   [Cont'd.]
   Continued under the title :

[c]   Moniteur scientifique—Quesneville.   Journal des sciences pures
       et appliquées, compte rendu des académies et sociétés
       savantes et revue des progrès accomplis dans les sciences
       mathématiques, physiques et naturelles.   Photographie,
       chimie, pharmacie, médecine, revue des inventions nouvelles
       et industrie manufacturière des arts chimiques.   Journal
       fondé et dirigé par Quesneville.   Troisième série.   16 vols.
       (XIII–XXVIII), 4to.   Paris, 1871–'86.
          Continued under the same title :

[d]   Quatrième série.   5 vols. (XXIX–XXXIII), 4to.   Paris, 1887–'91+

348.   REVUE UNIVERSELLE DE LA BRASSERIE ET DE LA DISTILLERIE.
       Industrie, commerce, finance, examen des questions politiques,
       fiscales et économiques.   Directeur-gérant : J. Paul Roux.   18
       vols., fol.   Paris et Bruxelles, 1873–'90+

349.   REVUE UNIVERSELLE DES MINES, de la métallurgie, des travaux
       publics, des sciences et des arts appliqués à l'industrie.   Sous
       la direction de Ch. de Cuyper.   40 vols., 8vo.   Paris et Liége,
       1857–'76.
          Continued under the title :

[a]   Revue universelle des mines, de la métallurgie, des travaux
       publics, des sciences et des arts appliqués à l'industrie.
       Annuaire de l'Association des ingénieurs sortis de l'école
       de Liége.   Sous la direction de Ch. de Cuyper et A. Habets.
       Deuxième série de la Revue universelle des mines ; troisi-
       ème série de l'Annuaire de l'Association des ingénieurs.
       36 vols., 8vo.   Paris et Liége, 1877–'92+
       Tables des matières.   Vols. I–XII, 1863 ; vols. I–XXVI, 1870.

350.   REVUE UNIVERSELLE DES PROGRÈS DE LA FABRICATION DU SUCRE.
       Gemblaux, 1887–'90+

351.   REVUE UNIVERSELLE DES PROGRÈS DE LA FABRICATION DU SUCRE
       pour l'année 1883–'84.   Rédigée par F. Sachs, Le Docte et A.
       Raeymaeckers.   Vol. I, 8vo.   Paris, 1885.

   RICHARDSON, CLIFFORD.   *See* Association of Official Agricultural Chemists.

   RICHTER, J. B.   *See* Ueber die neueren Gegenstände in der Chemie.

352.   RIVISTA DI CHIMICA MEDICA E FARMACEUTICA, TOSSICOLOGIA, FAR-
       MACOLOGIA E TERAPIA.   Diretta da P. Albertoni e J. Guareschi.
       3 vols.   Torino, 1883–'85 ||
          United in 1885 *with* Annali di chimica applicata [etc.].   *See* Gior-
          nale di farmacia, chimica e scienze accessorie . . . Milano.

353. RIVISTA SCIENTIFICA ; organo dell' associazione chimico-farmaceutica provinciale. Messina, 1880+

RÖSSLER, L. *See* Mittheilungen der K. K. chemisch-physiologischen Versuchsstation.

ROUSSET, EDMOND. *See* Moniteur de la céramique.

ROUX, J. PAUL. *See* Revue universelle de la brasserie.

ROUX-MATIGNON, J. P. *See* Revue des bières, des vins et des alcools.

ROZIER'S JOURNAL. *See* Introduction aux observations sur la physique [etc.].

354. RUNDSCHAU FÜR DIE INTERESSEN DER PHARMACIE, CHEMIE, UND
DER VERWANDTEN FÄCHER. Redacteur : Emil Graf [*later*,
und A. Vomácka]. 11 vols., roy. 4to. Leitmeritz, 1875–'85.
[*New series*] 3 vols. (XII–XVIII), 4to. Prag, 1886–'92+

355. RUSSISCHES JAHRBUCH DER PHARMACIE. Herausgegeben von
Dav. H. Grindel. 6 vols., 12mo. Riga, 1803–'08.

Continued under the title :

[a] Russisches Jahrbuch für die Chemie und Pharmacie. Herausgegeben von Fd. Giese und D. H. Grindel. 2 vols. (VII,
VIII), 12mo. Riga, 1809–'10.

ŠAFAŘIK, V. *See* Zprávy spolku chemiku českých.

SAIGEY. *See* Bulletin des sciences mathématiques.

SAKURAI, J. *See* Tokyo Kagakkai Kaishi.

356. SAMMLUNG AUSERLESENER ABHANDLUNGEN ÜBER DIE INTERESSAN-
TESTEN GEGENSTÄNDE DER CHEMIE. Aus dem lateinischen
mit einigen Anmerkungen begleitet. Redigirt von Hochheimer.
1 vol., 8vo. Leipzig, 1793.

357. SAVONNERIE (LA). Revue spéciale de l'huilerie, la savonnerie, la
parfumerie et la stéarinerie. Rédigé par A. Engelhardt. 5
vols., 4to. Zürich, 1887–'92+

SCHAAR, G. F. *See* Gas-Kalender *also* Kalender für Gas- und Wasserfach-
Techniker.

358. SCHEI- ARTSENIJMENG- EN NATUURKUNDIGE BIBLIOTHEEK. Uitgegeven door Meijlink. 18 vols., 8vo. Deventer, 1827–18—.

SCHEIBLER, C. *See* Zeitschrift des Vereins für Rübenzuckerindustrie ; *also* Zeitschrift (Neue) für Rübenzucker.

359. SCHEI- EN HUISHOUDKUNDIG MAGAZIJN. Uitgegeven door Daniel Craanen. 8vo. Amsterdam, 1809.

360. SCHEIKUNDIGE BIBLIOTHEEK, waarin de voornaamste nieuwe ont-dekkingen en verbeteringen, welke in de scheikunde in ons vaderland, doch wel meest in andere landen van tijd tot tijd gedaan worden, kortelijk worden voorgedragen. Door een gezelschap van beminaaren dezer wetenschap. 2 vols., 8vo. Delft, 1790–'98.

Continued under the title :

[a]   Scheikundige (Nieuwe) bibliotheek.   3 vols., 8vo.   Amsterdam, 1799–1802.

361.   SCHEIKUNDIGE BIJDRAGEN.   — vols.   Amsterdam, 1867.

362.   SCHEIKUNDIGE ONDERZOEKINGEN GEDAAN IN HET LABORATORIUM DER UTRECHTSCHE HOOGESCHOOL. Uitgegeven door G. J. Mulder. 6 vols., 4to. Rotterdam, 1845–'52.

Followed by :

[a]   Scheikundige verhandelingen en onderzoekingen uitgegeven door G. J. Mulder.   3 vols., 8vo.   Rotterdam, 1857–'64.

Followed by :

[b]   Scheikundige aanteekeningen uitgegeven door G. J. Mulder. 1 vol., 8vo.   Utrecht, 1865–'67.

Followed by :

[c]   Scheikundige onderzoekingen gedaan in het physiologisch labor-atorium der Utrechtsche Hoogeschool.   Nieuwe serie.   3 vols., 8vo.   Rotterdam, 1867–'71.
Derde serie.   4 vols., 8vo.   1871–'76 [+ ?].

SCHEIKUNDIGE VERHANDELINGEN.   Rotterdam.
See Scheikundige onderzoekingen.

SCHERER.   *See* Higgins Protokolle.

SCHERER, ALEX. NIC.   *See* Allgemeine nordische Annalen der Chemie ; *also* Allgemeines Journal der Chemie ; *also* Archiv für die theoretische Chemie ; *also* Nordische Blätter für Chemie

SCHILLING, N. H.   *See* Journal für Gasbeleuchtung.

363.   SCHLESISCHE BRAUER-ZEITUNG. Freisinniges Fachblatt für den deutschen Brauerstand.   Redacteur: P. Sitte.   3 vols., 4to. Myslowitz, 1877–'79.

SCHMELZER, ANT. ST. *See* Böhmische (Der) Bierbrauer ; *also* Kvas. Časopis spolků pro průmysl.

SCHMIDT, A. *See* Skandinaviens kemisk-tekniske Centralblad.

SCHÖNBERG, ALOIS. *See* Populäre Zeitschrift für Spiritus- und Presshefe-Industrie.

364. SCHOOL (THE) OF MINES QUARTERLY. Published by the chemical and engineering societies of the School of Mines, Columbia College, New York. [*Later,*] A Journal of Applied Science. 13 vols., 8vo. New York, 1879–'92+

SCHROHE, A. *See* Gährungstechnisches Jahrbuch.

365. SCHWÄBISCHE (DER) BIERBRAUER. Organ für die gesammten Interessen der Bierbrauerei, sowie für den würtembergischen Brauerbund. Herausgegeben von Achenbach. 19 vols., 4to. Waldsee, 1872–'90+

SCHWARZ, H. *See* Amerikanische (Der) Bierbrauer.

SCHWARZWÄLLER, UDO. *See* Zeitschrift für deutsche Spiritus-Fabrikation.

SCHWEIGGER, J. S. C. *See* Allgemeines Journal der Chemie [*c*].

366. SCHWEIZERISCHE MILCHZEITUNG. Organ des Ostschweizerischen Käsevereins. 8 vols., sm. fol. Schaffhausen, 1875–'82 [+?]

367. SEIFENFABRIKANT (DER). Zeitschrift für Seifen-, Kerzen- und Parfümerie-Fabrikation, sowie verwandte Geschäftszweige. Organ des Verbandes der Seifenfabrikanten. Herausgegeben von C. Deite. 12 vols. Berlin, 1881–'92+

368. SEIFENSIEDER - ZEITUNG. Central-Organ der Seifenfabrikanten. Die neuesten Fortschritte in der Seifenfabrikation und den damit verwandten Geschäftszweigen. Herausgegeben von Alwin Engelhardt. Mitredacteur : Hermann Krätzer. 19 vols., sm. fol. Leipzig, 1874–'92+

SELMI, FRANCESCO. *See* Annuario chimico italiano.

369. SELMI (IL). Giornale di chimica e fisica applicate allo studio delle alterazioni e falsificazioni. Diretto da P. E. Alessandri. 3 vols., 8vo. Pavia, 1890–'92+

SEMBENINI, G. B. *See* Annuario delle scienze chimiche [etc.] ; *also* Gazzetta eclettica di chimica tecnologica ; *also* Gazzetta eclettica di farmacia e chimica medica.

SERTZ, ALB.   *See* Hopfenlaube (Die).

SIEBEL, J. E.   *See* Chemical Review and Journal.   Chicago.

SIMON, FRANZ.   *See* Beiträge zur physiologischen und pathologischen Chemie.

SITTE, P.   *See* Schlesische Brauer-Zeitung.

SITTIG, EUGÈNE A.   *See* Brauer (Der) und Malzer.

SKALWEIT'S REPERTORIUM.   *See* Repertorium der analytischen Chemie.

370. SKANDINAVIENS KEMISK-TEKNISKE CENTRALBLAD for Danmark, Sverige, Norge og Finland. Redigeret af G. A. Schmidt. 10 vols., 8vo. Kjøbenhavn, 1882–'91+

SOCIÉTÉ DE CHIMIE INDUSTRIELLE DE PARIS.   *See* Bulletin de la Société [etc.]

SOCIETIES, CHEMICAL.
>   *American*, *see* Proceedings of the American Chemical Society.
>   *Chemical Industry*, *see* Journal of the Society of Chemical Industry.
>   *Columbian*, *see* Memoirs of the Columbian Chemical Society.
>   *French*, *see* Bulletin de la Société de chimie industrielle de Paris.
>   *French*, *see* Bulletin des séances de la Société chimique.
>   *German*, *see* Berichte der deutschen chemischen Gesellschaft.
>   *Japanese*, *see* Tokyo Kagakkai Kaishi.
>   *London*, *see* Minutes of the Society for Philosophical Experiments.
>   *London*, *see* Proceedings of the Chemical Society of London.
>   *Public Analysts*, *see* Analyst (The).
>   *Russian*, *see* Zhurnal russkova khimicheskova [etc.].
>   *School of Mines*, *see* School of Mines Quarterly.
>   *Washington*, *see* Bulletin of the Chemical Society of Washington.

SOCIETIES.   Verein für die Rübenzuckerindustrie im Zollverein.   *See* Zeitschrift des Vereins [etc.].
>   Verein für Rübenzucker-Industrie in der oesterreichisch-ungarischen Monarchie.   *See* Organ des Vereins [etc.].
>   Versuchs- und Lehranstalt für Brauerei in Berlin.   *See* Wochenschrift für Brauerei.

SOCIETY OF DYERS AND COLOURISTS.   *See* Journal of the Society of Dyers [etc.].

SOEHOR, T., A A LIEGERT.   *See* Listy pekařské.

SONDÉN, K.   *See* Kemiska Notiser.

SONNTAG, C.   *See* Centralblatt für die Textil-Industrie.

SORGENFREI.   *See* Deutsche Gerber-Zeitung,

371. SORGO (THE) JOURNAL AND FARM MACHINIST. Devoted to Northern cane and sugar-beet culture, improved farm machinery, and progressive husbandry. 8vo. Cincinnati, 1863.

SPITZ, SIGMUND. *See* Gambrinus.

SPRINGMÜHL, FERD. *See* Deutsche Muster-Zeitung.

STAEDEL, W. *See* Jahresbericht über die Fortschritte auf dem Gebiete der reinen Chemie.

372. STAHL UND EISEN. Zeitschrift des Vereins deutscher Eisenhütten-leute. Redigirt von F. Osann. [*Later*, Zeitschrift der nord-westlichen Gruppe des Vereins deutscher Eisen- und Stahl-industrieller und des Vereins deutscher Eisenhüttenleute. Herausgegeben von den Vereins-Vorständen. Redigirt von den Geschäftsführern beider Vereine.] 11 vols., roy. 8vo. Düsseldorf, 1881-'91+

STAMM, FERNAND. *See* Oesterreichische (Der) Bierbrauer.

STAMMER, K. *See* Taschenkalender für Zuckerfabrikanten.

STÖCKHARDT, J. A. *See* Chemische Ackersmann.

ST. PETERSBURG. *See* Zhurnal russkova khimicheskova [etc.].

STROHMER, F. *See* Oesterreich-ungarische Zeitschrift für Zuckerindustrie.

373. SUCRE (LE). Journal de la sucrerie indigène et exotique, industriel, commercial et financier. [Edited by G. Bordet.] Fol. Paris, 1874.

374. SUCRERIE (LA) BELGE. Organe publié par la Société générale des fabricants de sucre de Belgique. 20 vols., 4to. Liége, 1872-'91+

375. SUCRERIE (LA) INDIGÈNE. Revue périodique fondée en Mai 1866 à Valenciennes, sous le patronage du comité des fabricants de sucre des arrondissements de Valenciennes et d'Avesnes. Organe des intérêts des fabricants du sucre et des distillateurs. Technologie, commerce, économie politique, agriculture, statis-tique, législation et jurisprudence industrielles, bibliographie. H. E. Tardieu, directeur-gérant. 40 vols., 8vo. Paris, 1866-'92+

376. SÜDDEUTSCHE (DER) GERBER. Wochenblatt für Leder-Industrie und Leder-Handel. Herausgegeben von Achenbach [later, Reisenbichler]. 18 vols., 4to. Waldsee, 1875-'92+

377. SUGAR BEET (THE). Devoted to the cultivation and utilization of the sugar beet. 12 vols., fol. Philadelphia, 1880-'91+

378. SUGAR CANE (THE).   A monthly magazine devoted to the interests of the sugar cane industry.   22 vols., 8vo.   Manchester, 1869–'90+

379. SUGAR CANE (THE).   — vols.   Newcastle-on-Tyne, 1878–'80+
         Same as the preceding [?].

380. SUGAR PLANTER.   Franklin, Massachusetts, 1880+

SUK, K.   *See* Český sladek.

SVANBERG, L. F.   *See* Årsberättelse om Framstegen i Physik och Chemi [*b*].

SVENSK KEMISK TIDSKRIFT.   *See* Kemiska Notiser.

SWITTAU'S REPERTORIUM.   *See* Repertorium der Chemie und Pharmacie.

TABLEAU DU TRAVAIL ANNUEL (etc.).
         See Introduction aux observations sur la physique [etc.].

TARDIEU, H.   *See* Sucrerie (La) indigène et coloniale.

381. TASCHENBUCH FÜR SCHEIDEKÜNSTLER UND APOTHEKER.   Herausgegeben von Ch. F. Buchholz.   8vo.   Weimar, 1803–'19.
         Cf. Almanach für Scheidekünstler und Apotheker.

382. TASCHEN-KALENDER FÜR ZUCKERFABRIKANTEN.   Herausgegeben von K. Stammer.   16 vols., 16mo (1876–'92).   Berlin, 1877–'93+

TAYLOR, RICHARD.   *See* Philosophical Magazine.

383. TECHNISCH-CHEMISCHER KALENDER FÜR OESTERREICH-UNGARN.   Jahrbuch und Notizbuch für den theoretischen und praktischen Chemiker, Fabrikanten, Bierbrauer, Branntweinbrenner, Zuckerfabrikanten.   Herausgegeben von Paul Bennewitz.   2 vols., 16mo.   Wien, 1875, '76 ‖

384. TECHNISCH-CHEMISCHES GEWERBEBLATT.   Gesammelte Vorschriften und Erfahrungen in der technischen Chemie.   Zum Nutzen der Gewerbtreibenden und Fabrikanten herausgegeben von L. F. Denzer.   1 vol., 8vo.   Berlin, 1849–'50 ‖

385. TECHNISCH-CHEMISCHES JAHRBUCH.   Herausgegeben von Rudolph Biedermann.   12 vols., 16mo.   Berlin, 1880–'91+
         Vols. I to III *were issued as a Supplement to the* Chemiker-Kalender ;
         after that it forms an independent journal with the sub-title :
         Ein Bericht über die Fortschritte auf dem Gebiete der chemischen Technologie.

386. TECHNISCH-CHEMISCHES RECEPT-TASCHENBUCH. Redigiert von E. Winckler. 9 vols. Leipzig, 1875–'83.
Continued under the title:

[a] Technisch-chemisches Recept-Taschenbuch, enthaltend in circa 10,000 Recepten und Mittheilungen die nützlichsten Entdeckungen aus dem Gebiete der technischen Chemie und der Gewerbskunde. 8vo. Halle, 1884.

387. TECHNISCHES CENTRALBLATT. Allgemeines Repertorium für mechanische und chemische Technik. Herausgegeben von G. Behrend unter Mitwirkung von Ch. Heinzerling. No. 1, Januar. 4to. Halle, 1884.

388. TEINTURIER (LE) PRATIQUE. Journal pour teinturiers, imprimeurs, blanchisseurs, apprêteurs, filateurs, droguistes, fabricants de papier, etc. Seul rédacteur, Max Singer. 21 vols., 4to. Tournay, 1872–'92+

389. TEINTURIER (LE) UNIVERSEL, ou L'Echo des applications des matières colorantes aux arts et à l'industrie. Journal spécial de la teinture et de l'apprêt des étoffes, de la production et de la préparation des matières tinctoriales, etc. Publié sous la direction de Jacob. 4to. Paris, 1860–'67 [+?]

390. TEKNO-KEMISK JOURNAL. P. O. Almström. Stockholm, 1847–'48.

391. TEXTILE (THE) COLORIST. A journal of bleaching, printing, dyeing, and finishing textile fabrics, and the manufacture and application of coloring matters. Edited by Charles O'Neill. — vols. London, 1876–'77.

392. TEXTILE (THE) COLORIST. A monthly journal devoted to practical dyeing, bleaching, printing and finishing; dyes, dyestuffs and chemicals, as applied to dyeing. M. Frank, manager. 12 vols., 4to. Philadelphia, 1879–'90+

THOMSEN, A. (AND J.). *See* Tidsskrift for Physik og Chemi.

THOMSON'S ANNALS. *See* Annals of Philosophy.

THUDICHUM, J. L. W. *See* Annals of Chemical Medicine.

393. THÜRINGER MUSTER-ZEITUNG FÜR FÄRBEREI UND DRUCKEREI. Redigirt von J. C. H. Geyer. 4 vols., 4to. Berlin, 1865–'68.
Continued under the title:

[a] Deutsche Färber-Zeitung. Organ für alle Zweige der Färberei, Druckerei, Bleiche und Appretur. Redigirt von J. C. G. Geyer und A. Stubenrauch. 24 vols. (V–XXVIII), 4to. München, 1869–'92+

394. TIDSSKRIFT FOR ANVENDT CHEMI, for Fabrikanter, Chemikere,
      Pharmaceuter og Handlende. Udgivet af T. Holm og A.
      E. M. Schleisner. 1 vol., 8vo. Kjøbenhavn, 1869–'70.

395. TIDSSKRIFT FOR PHYSIK OG CHEMI SAMT DISSE VIDENSKABERS AN-
      VENDELSE. Udgivet af A. og J. Thomsen. 9 vols., 8vo.
      Kjøbenhavn, 1862–'70.
      Anden Række. 21 vols., 8vo. 1871–'91+

396. TIJDSCHRIFT VOOR SCHEI- EN ARTSENIJBEREIDKUNDE. Door P. J.
      Haaxmann. 2 vols., 8vo. Leiden, 1844–'45.

397. TIJDSCHRIFT VOOR WETENSCHAPPELIJKE PHARMACIE, benevens
      mededeelingen over chemie, pharmacie en pharmacognosie van
      het planten-, dieren- en delfstoffelijk rijk. Geredigeerd door
      P. J. Haaxmann. 5 vols., 8vo. Voorburg, 1849–'53.
      Tweede serie. 5 vols., 8vo. 's Gravenhage, 1854–'58.
            Continued under the title :
      [a] Tijdschrift voor wetenschappelijke pharmacie. Geredigeerd
            door P. J. Haaxmann, bevattende de mededeelingen der
            Nederlandsche maatschappij ter bevordering der pharmacie.
            Derde serie. 6 vols., 8vo. Gorinchem, 1859–'64.
            Nieuwe serie. 9 vols., 8vo. Gorinchem, 1865–'73.

      TILLER, K. *See* Pivovarski Listy.

      TILLOCH, ALEX. *See* Philosophical Magazine.

398. TOEGEPASTE SCHEIKUNDE. Tweemaandelijksch tijdschrift, bevat-
      tende mededeelingen uit het gebied der toegepaste scheikunde
      voor het algemeen. Onder redactie van R. J. Opwyrda. 5
      vols., 8vo Vlaardingen, 1865–'69.
      Nieuwe serie. 4 vols., 8vo. 1870–'75.
            Continued under the title :
      [a] Maandblad voor toegepaste scheikunde, bevattende mededeelin-
            gen uit het gebied der toegepaste scheikunde voor het
            algemeen. Redacteur : R. J. Opwyrda. Derde serie.
            5 vols., 8vo. Amsterdam, 1876–'80 [+?]

399. TOKYO KAGAKKAI KAISHI. [Editor] J. Sakurai. 6 vols., 8vo.
      Tokyo, June, 1880–'86+
            Tokyo Chemical Society's Journal, Tokyo, Japan.

      TOMMASI, DONATO. *See* Annuaire de la chimie industrielle.

      TOVEY, A. E. J. *See* Deutsches und Amerikanisches Brauer-Journal ; *also*
            German (The) and American Brewers' Journal.

400. TRANSACTIONS OF THE LABORATORY CLUB. London, 1887–'90.

401. TRANSACTIONS OF THE NEWCASTLE-UPON-TYNE CHEMICAL SOCIETY. Newcastle, 1868–'82.

> *For continuation see* Journal of the Society of Chemical Industry.

TRIER, S. M. *See* Archiv for Pharmaci.

TROMMSDORFF, J. B. *See* Allgemeine chemische Bibliothek ; *also* Almanach für Scheidekünstler [*b*] ; *also* Annalen der Pharmacie ; *also* Journal der Pharmacie für Aerzte und Apotheker.

TROMMSDORFF'S TASCHENBUCH FÜR SCHEIDEKÜNSTLER. *See* Almanach für Scheidekünstler.

402. UEBER DIE NEUEREN GEGENSTÄNDE IN DER CHEMIE. Herausgegeben von J. B. Richter. 11 parts, 8vo. Breslau, Hirschberg und Lissa, 1791–1802 ‖

UHLAND, W. H. *See* Kalender für Textil-Industrie.

UNITED STATES DEPARTMENT OF AGRICULTURE, DIVISION OF CHEMISTRY. Bulletins. *For check list, see* Bulletins of the Division of Chemistry [etc.].

403. UNTERSUCHUNGEN AUS LIEBIG'S LABORATORIUM. 1 vol., 8vo. Wien, 1872.

VANDERBECK, C. C. *See* Druggist (The) and Chemist.

404. VEGYTANI LAPOK. [Edited by] Rudolf Fabinyi. 5 vols. Kolozsvár, 1883–'88 ‖

VEREIN FÜR DIE RÜBENZUCKER-INDUSTRIE. *See* Zeitschrift des Vereins, etc.

405. VERMISCHTE ABHANDLUNGEN DER PHYSISCH-CHEMISCHEN WARSCHAUER GESELLSCHAFT zur Beförderung der praktischen Kenntnisse in der Naturkunde, Oekonomie, Manufacturen und Fabriken, besonders in Absicht auf Polen, Warschau und Dresden. 1 vol., 8vo. Ill. 1768 ‖

406. VIERTELJAHRESSCHRIFT FÜR TECHNISCHE CHEMIE, landwirthschaftliche Gewerbe, Fabrikwesen und Gewerbtreibende überhaupt. Unter Mitwirkung mehrerer Gelehrten, Fabrikanten und Techniker herausgegeben von Wilibad Artus. 10 vols., 8vo. Quedlinburg, 1859–'69 ‖

407. VIERTELJAHRESSCHRIFT ÜBER DIE FORTSCHRITTE AUF DEM GEBIETE DER CHEMIE DER NAHRUNGS- UND GENUSSMITTEL, der Gebrauchsgegenstände, sowie der hierher gehörenden Indus-

VIERTELJAHRESSCHRIFT ÜBER DIE FORTSCHRITTE, etc. [Cont'd.]
triezweige. Unter Mitwirkung von Degener, Hochstetter, P.
Lohman, Benno Martiny, Paack, Proskauer, Würzburg, L.
Aubry, R. Sendtner, H. Will, von Peters, Weigmann, J. Mayr-
hofer, E. von Raumer, Röttger, herausgegeben von A. Hilger,
R. Kayser, J. König, E. Sell. 7 vols., 8vo. 1886–92. Berlin,
1887–'93+

VILLON. *See* Revue de chimie industrielle et agricole.

408. VZORNÝ PIVOVAR. Red.: A. Markl. 2 vols., 8vo. v Praze, 1876–
'77 ‖

WAGNER'S JAHRESBERICHT.
         See Jahresbericht über die Fortschritte der chemischen Technologie.

WĄSOWICZ, DUNIN V. *See* Kalendarz dla uzytku farmaceutów i chemików.

WASSERBERG, F. A. X. *See* Beiträge zur Chemie.

409. WATER, GAS AND STEAM. 4to. Philadelphia, 1878.

WATT, CHARLES (AND JOHN). *See* Chemist (The).

WEBER, J. A. *See* Monath-Schrift . . . der Scheidekunst. *Also* Physikalisch-
chemisches Magazin. . . .

WEIN, E. *See* Zymotechnisches Centralblatt.

WESTRUMB, J. F. *See* Kleine physikalisch-chemische Abhandlungen.

WIEDEMANN'S ANNALEN. *See* Journal der Physik [*e*].

WIGNER, G. W. *See* Analyst (The).

WILEY, HARVEY W. *See* Association of Official Agricultural Chemists.

410. WOCHENSCHRIFT DES CENTRALVEREINS FÜR RÜBENZUCKER-
INDUSTRIE IN DER OESTERREICH-UNGARISCHEN MONARCHIE.
Redigirt von E. Kutschera. 30 vols. Wien, 1863–'92+

411. WOCHENSCHRIFT FÜR BRAUEREI. Eigenthum der Vereins-Ver-
suchs- und Lehranstalt für Brauerei in Berlin. Herausgegeben
von M. Delbrück und M. Hayduck. 9 vols., 4to. Berlin,
1884–92+

412. WOCHENSCHRIFT FÜR ZUCKERFABRIKATION. — vols. Braunschweig, 18— –'81 [+?]

WÖCHENTLICHER MARKT-BERICHT. *See* Deutsche (Die) Zucker-Industrie.

413. WÖCHENTLICHER MARKTBERICHT FÜR DEN INTERNATIONALEN ZUCKERHANDEL. (Auch als Gratisbeilage zum Wochenblatt : Die deutsche Zuckerindustrie.) Redigirt von J. Görz in Berlin. 6 vols., 4to. Berlin, 1886–'92+

WÖHLER, FRIEDRICH. *See* Annalen der Pharmacie [*b*].

WOOD, CHARLES H., AND SHARP, CHARLES.
See Chemist's Desk Companion ; *also* Year-book of Pharmacy.

WURTZ, ADOLPHE. *See* Bulletin des séances de la Société chimique de Paris.

414. YEAR-BOOK OF PHARMACY. A practical summary of researches in pharmacy, materia medica, and pharmaceutical chemistry [*in* 1881, and Transactions of the Pharmaceutical Conference]. Edited by Charles H. Wood and Charles Sharp. 26 vols , 8vo. London, 1865–'90+
Cf. Chemist's (The) Desk Companion.

ZALUD, PH. *See* Allgemeine Zeitschrift für Textil-Industrie.

ZANTEDESCHI, FRANCESCO. *See* Giornale fisico-chimico italiano ; *also* Raccolta fisico-chimica italiana.

415. ZEITSCHRIFT DES VEREINS FÜR DIE RÜBENZUCKERINDUSTRIE IM ZOLLVEREIN. Herausgegeben durch C. Scheibler [and others]. *From* vol. 14, Neue Folge. *From* vol. 23, Industrie des deutschen Reichs. 41 vols. Berlin, 1850–'91+
Register, vols. i–x, *in* vol. x.

ZEITSCHRIFT FÜR ANGEWANDTE CHEMIE.
See Zeitschrift für die chemische Industrie.

416. ZEITSCHRIFT FÜR ANALYTISCHE CHEMIE. Herausgegeben von C. Remigius Fresenius [*later*], unter Mitwirkung von Heinrich Fresenius. 31 vols, 8vo. Wiesbaden, 1862–'92+
Autoren- und Sach-Register zu den Bänden i–x (1862–'71). 1 vol., 8vo. Wiesbaden, 1872.
Autoren- und Sach-Register zu den Bänden xi–xx (1873–'71). Bearbeitet von Heinr. Fresenius, unter Mitwirkung von Wilh. Lenz. 8vo. Wiesbaden, 1882.

417. ZEITSCHRIFT FÜR ANORGANISCHE CHEMIE. Unter Mitwirkung von M. Berthelot, C. W. Blomstrand . . . und anderen Fachgenossen herausgegeben von Gerhard Krüss. Hamburg, 1892+

ZEITSCHRIFT FÜR CHEMIE UND PHARMACIE. *See* Kritische Zeitschrift für Chemie.

418. ZEITSCHRIFT FÜR DAS CHEMISCHE GROSSGEWERBE. Kurzer Bericht über die Fortschritte der chemischen Grossindustrie. Unter Mitwirkung von angesehenen Technologen und Technikern, sowie von F. Frerichs, J. Landgraf, K. Plostorff, P. Wagner, H. Wiesinger, F. Wunderlich, herausgegeben von Jul. Post. 7 vols., 8vo. Berlin, 1876–'82 ‖

ZEITSCHRIFT FÜR DAS GESAMMTE BRAUWESEN. *See* Bayerische (Der) Bierbrauer.

419. ZEITSCHRIFT FÜR DEN PHYSIKALISCHEN UND CHEMISCHEN UNTERRICHT. Unter Mitwirkung von E. Mach und B. Schwalbe, herausgegeben von F. Poske. 6 vols., 4to. Berlin, 1887–'93+

ZEITSCHRIFT (NEUE) FÜR DEUTSCHE SPIRITUSFABRIKANTEN.
See Zeitschrift für deutsche Spiritusfabrikation.

420. ZEITSCHRIFT FÜR DEUTSCHE SPIRITUSFABRIKATION. Redigirt von U. Schwarzwäller. 8 vols., 8vo. Leipzig, 1859–'66.

Continued under the title :

[a] Neue Zeitschrift für deutsche Spiritusfabrikanten. Organ des Vereins der Spiritusfabrikanten in Deutschland von Udo Schwarzwäller. 11 vols., 8vo. Leipzig, 1867–'77.

Continued under the title :

[b] Neue Zeitschrift für deutsche Spiritusfabrikanten. Organ des Vereins und der Versuchsstation der Spiritusfabrikanten in Deutschland. (Zweite Folge der Schwarzwällerschen Zeitschrift für deutsche Spiritusfabrikanten.) Unter Mitwirkung von M. Märker, herausgegeben von M. Delbrück. Neue Folge. (I–II), [*whole number*, XII–XIII], 4to. Berlin, 1878–'79.

Continued under the title :

[c] Zeitschrift für Spiritusindustrie. Organ des Vereins und der Versuchsstation der Spiritusfabrikanten in Deutschland. (Zweite Folge der Schwarzwällerschen Zeitschrift für deutsche Spiritusfabrikanten.) Unter Mitwirkung von M. Märker herausgegeben von M. Delbrück. Neue Folge. 13 vols. (III–XV), [*whole number*, XV–XXVII]. Berlin, 1880–'92+

421. ZEITSCHRIFT FÜR DIE CHEMISCHE INDUSTRIE, mit besonderer Berücksichtigung der chemisch-technischen Untersuchungsverfahren. Herausgegeben von Ferdinand Fischer. 2 vols., roy. 8vo. Berlin, 1887.

ZEITSCHRIFT FÜR DIE CHEMISCHE INDUSTRIE. [Cont'd.]
Continued under the title:

[a] Zeitschrift für angewandte Chemie. Organ der deutschen
Gesellschaft für angewandte Chemie. Herausgegeben von
Ferd. Fischer. 6 vols., 4to. Berlin, 1887–'92+

422. ZEITSCHRIFT (NEUE) FÜR DIE ÖSTERREICHISCH-UNGARISCHE SPIRI-
TUS-INDUSTRIE. Unter Mitwirkung anerkannt tüchtiger Kräfte,
redigirt von Mor. Hatschek. 1 vol., 4to. Wien, 1879 ‖

423. ZEITSCHRIFT FÜR MINERALWASSERFABRIKATION. Herausgegeben
von Lohmann. Berlin, 1890.

424. ZEITSCHRIFT FÜR NAHRUNGSMITTEL-UNTERSUCHUNG UND HY-
GIENE. Eine Monatschrift [*later*, Halb-Monatsschrift] für
chemische und mikroscopische Untersuchung von Nahrungs-
und Genussmitteln und Gebrauchsgegenständen. Beiblatt der
Wochenschrift "Pharmaceutische Post." Herausgegeben und
redigirt von Hans Heger in Wien. 7 vols. Wien, 1886–'92+

425. ZEITSCHRIFT FÜR PHYSIKALISCHE CHEMIE, STÖCHIOMETRIE UND
VERWANDSCHAFTSLEHRE. Herausgegeben von Wilh. Ostwald
[*later*] und J. H. Van't Hoff. 10 vols., 8vo. Riga und Leipzig,
1887–'93+

426. ZEITSCHRIFT FÜR PHYSIOLOGISCHE CHEMIE. Unter Mitwirkung
von E. Baumann, Gähtgens, Hüfner [etc.], herausgegeben von
F. Hoppe-Seyler. 16 vols., 8vo. Strassburg, 1877–'92+
Sach- und Namen-Register zu Band I–IV. Strassburg, 1882.
Sach- und Namen-Register zu Band V–VIII. Strassburg, 1888.

427. ZEITSCHRIFT (NEUE) FÜR RÜBENZUCKER-INDUSTRIE. Wochenblatt
für die Gesammtinteressen der Zuckerfabrication. Herausgegeben von C. Scheibler. 29 vols., 4to. Berlin, 1876–'92+
Register (vols. I–XX).

ZEITSCHRIFT FÜR SPIRITUSINDUSTRIE.
See Zeitschrift für deutsche Spiritusfabrikation [*c*].

428. ZEITSCHRIFT FÜR ZUCKER-INDUSTRIE. Organ des Vereins zur
Hebung der Zuckerfabrikation im Königreich Böhmen. Unter
Mitwirkung von Aug. Weiler, redigirt von K. Preis. 3 vols.,
8vo. Prag, 1872–'74 ‖

429. ZEITSCHRIFT FÜR ZUCKER-INDUSTRIE IN BÖHMEN. Redigirt von
M. Névole. 15 vols., roy. 8vo. Prag, 1877–'90+

430. ZEITSCHRIFT FÜR ZÜNDWAARENFABRIKATION. Organ des Vereins deutscher Zündwaarenfabrikanten. Redacteur: Wladimir Jettel. 1 vol., 8vo. Clausthal, 1874 ||

431. ZEITUNG FÜR DAS GAS- UND WASSERFACH, die Klein- und Feinmechanik, die Kunst- und Bauschlosserei [*later*, das gesammte Heizungs- und Beleuchtungswesen, den Pumpen- und Spritzenbau und die gesammte Hydraulik]. Herausgeber: N. Besselich. 19 vols., fol. Trier, 1870–'88+

432. ZEITUNG FÜR ZUCKERFABRIKANTEN. Organ für Rübenbau, Zuckerfabrikation, Maschinenbau [etc.]. Redigirt von Karl Löffler. 2 vols., 4to. Dresden, 1863–'64 ||

433. Журналъ русскаго химическаго и физическаго общества при С.-Петербургскомъ университетѣ. С.-Петербургъ, 1869—86. 18 vols 8º

> Zhurnal russkova khimicheskova i fizicheskova obschetsva pri St. Peterburgskom Universitetye.
>
> RUSSIAN CHEMICAL AND PHYSICAL JOURNAL OF THE ST. PETERSBURG UNIVERSITY. 18 vols., 8vo. St. Petersburg, 1869–'86+

434. ZPRÁVY SPOLKU CHEMIKŮ ČESKÝCH. Rediguje: V. Šafařik. 2 vols., 8vo. v Praze, 1872–'76 ||

435. ZYMOTECHNISCHES CENTRALBLATT. Referirendes Organ über die neuesten wissenschaftlichen Forschungen und praktischen Erfahrungen auf dem Gesammtgebiete der Bierbrauerei, Spiritusbrennerei, Malz- und Hefefabrikation, redigirt von E. Wein. 1 vol., 8vo. München, 1892+

436. ZYMOTECHNISK TIDSSKRIFT FOR BRYGGERI, Brænderidrift og Gjærfabrikation i Danmark, Norge, Sverige og Finland. Redigeret af A. Jörgensen. 8 vols., 8vo. Kjøbenhavn, 1884–'92+

ABBREVIATIONS OF TITLES OF CHEMICAL PERIODICALS.

This list is an extension of a similar one prepared by the Committee on Indexing Chemical Literature of the American Association for the Advancement of Science. (Proc. Am. Assoc. Adv. Sci., XXXVI, p. 45, 1887.)

The numbers refer to the titles in Section VII.

1. Afh. Fys. Kemi.
2. Agenda chim.
3. Agric. Sci.
4. L'Alcool.
5. Allg. bayr. Hopfen-Ztg.
5a. Allg. Hopfen-Ztg.
6. Allg. chem. Bibl. (Trommsdorff).
7. Allg. Chem.-Ztg.
7a. Chem.-Ztg.
8. Allg. Hopfen-Ztg. (Carl).
8a. Allg. Brauer- u. Hopfen-Ztg.
9. Allg. österr. Chemiker-Ztg.
10. Allg. Ztschr. Bierbrauerei.
11. Allg. Ztschr. Textil-Ind.
12. Allg. Ztg. Spiritus-Ind.
13. Allg. J. Chem. (Scherer).
13a. N. allg. J. Chem. (Gehlen).
13b. J. für Chem. (Gehlen).
13c. J. für Chem. (Schweigger).
13d. J. prakt. Chem.
14. Alm. di chim. agric.
15. Alm. de chim.
16. Alm. Scheid. Apoth.
16a. Taschenb. f. Scheid.
16b. Taschenb. f. Chem. (Trommsdorff).
17. Am. Brewer and Dist.
18. Am. Brewers' Gaz.
19. Am. Chem. J.
20. Am. Chem.
21. Am. Drugg. Circ.
21a. Druggists' Circ.
22. Am. Gas-Light J.
23. Am. Glass-Worker.
24. Am. Lab.
25. Am. Min. J.
26. ·Am. Tanner.

27. Am. Bierbrauer.
28. Am. Ann. Arzneik.
29. Analyst.
30. Analyst's Note-Book.
31. Ann. der Pharm.
31b. Ann. Chem. (Liebig).
32. Ann. de chim.
32a. Ann. chim. phys.
33. Ann. di chim. (Brugnatelli).
34. Ann. fis. chim. (Majocchi).
35. Ann. Chem. Med.
36. Ann. Chem. Phil.
37. [Refer to the original.]
38. Ann. Chym. Pract. Pharm.
39. Ann. Pharm. (Bastick).
40. Ann. Phil. (Thomson).
41. Annuaire du chim. (Hensmans).
42. Annuaire chim.
43. Annuaire chim. indust.
44. Annuaire dist. Française.
45. Annuaire eaux min. France.
46. Annuaire fabric. sucre.
47. Annuaire prod. chim.
48. Annuaire sci. chim.
49. Annuaire indust. éclairage.
50. Annual Rep. Chem.
51. Anno. alm. pei chim.
52. Anno. chim. ital.
53. Anno. lab. chim. Livorno.
54. Anno. scienze chim. nat.
55. Anno. scienze chim. farm.
56. Annuarul Lab. (Bucuresci).
57. Anti-Adult. Rev.
58. Arch. der Agr.-Chem.
59. Arch. for Pharm. (Trier).
60. Arch. ges. Naturl.

60a. Arch. für Chem. (Kastner).
61. Arch. für theor. Chem.
62. Arch. thier. Chem.
63. Årsb. Phys. Kemi.
63a. Årsb. Kemi.
64. Proc. Assoc. Off. Agr. Chemists.
65. Ausw. Abh. Chem. (Crell).
66. Ausw. Ann. Chem. (Crell).
67. Bayer. Bierbr.
67a. Ztschr. ges. Brauwesen.
68. Beitr. Chem. (Wasserberg).
69. Beitr. Phys. (Hermann).
70. Beitr. physiol. Chem.
70a. Arch. für physiol. Chem.
71. Ber. Fortschr. chem. Technik.
72. Ber. d. chem. Ges. [or Ber.]
73. Berl. Jahrb. Pharm.
74. Bibl. phys. Lit. (Hermbstädt).
74a. Ann. chem. Lit.
75. Bierbrauer.
76. Bierbrauerei (Milwaukee).
77. Böhm. Bierbrauer.
78. Boston J. Chem.
78b. Pop. Sci. News.
79. Brasseur.
80. Brauer u. Malzer.
81. N. Brennerei-Ztg.
82. Brewer (St. Louis).
83. Brewers' Guardian.
84. Brewers' J. (London).
85. Brewers' J. (Phil.)
86. Brewing World.
87. Brod.
88. Bull. industr. tinct.
89. Bull. Soc. chim. industr.
90. Bull. Assoc. chim. sucrerie.
91. Bull. de pharm.
91a. J. de pharm.
92A. Bull. séances Soc. chim. (Paris).
92B. Rép. chim. pure.
92C. Rép. chim. appl.
92D. Bull. Soc. chim. Paris.
93. Bull. math. chim. (Férussac).
94. Bull. sci. phys. Néerlande.
95. Bull. sucres français.
96. Bull. lab. chim. Caire.
97. Bull. Chem. Soc. Washington.
98. Bull. Lab. Denison.
99. Bull. Chem. U. S. Dept. Agr.
100. Casopis chem.

101. Casopis cukrov.
102. Casopis pro prumysl chem.
103. Centrbl. chem. Grossind.
104. Centrbl. Agr.-Chem.
105. Centrbl. Branntweinbr.
106. Centrbl. Textil-Ind.
107. Cesky sladek.
108. Chem. Gaz.
108a. Chem. News.
109. Chem. News, Am. Repr.
110. Chem. Record.
111. Chem. Review (Chicago).
111a. Am. Chem. Review.
112. Chem. Review (London).
113. Chem. Trades J.
114. Chem. Kal.
115. Chemiker u. Drogist.
116. Chem. Ackersmann.
117. Chem. Ann. (Crell).
118. Chem. en phys. oefeningen.
119. Chem. Ind. (Jacobsen).
120. Chem. Archiv. (Crell).
121. Chem. J. (Crell).
121a. Neuest. Entdeck. Chem. (Crell).
122. Chem.-pharm. archief.
123. Chem.-techn. Ztg.
124. Chem.-techn. Mitthl.
125. Chem.-techn. Central-Anz.
126. Chem.-techn. Repert.
127. Chemist (Mongredieu).
128. Chemist (Watt).
129. Chemist and Drug.
130. Chemist Meteorol. J. (Amherst).
131. Chemists' Advocate.
132. Chemists' Desk Comp.
133. Chemists' J.
134. Chimiste. (Bruxelles).
135. Chimiste. (Paris).
136. Coloration ind.
136a. Coloriste ind.
136b. Monit. teinture.
137. Comptes rend. [or C. R.]
138. C. R. Montpellier.
139. Correspondenzbl. Zuckertechn.
140. Country Brewers' Gaz.
141. Crell's Chem. J. (Lond.)
142. D. Amerik. Brauer-Ztg.
143. D. Brauer-Kal.
144. D. Brauerei-Kal.
145. D. Chem. Ztg.

146. D. Chem. Kal.
147. D. Gerber-Ztg.
148. D. Muster-Ztg.
148a. Muster-Ztg.
148f. Färberei-Muster-Ztg.
148g. Leipz. Färber-Ztg.
149. D. Zucker-Ind.
149a. Wöchentl. Markt-Ber.
150. D. Am. Brauer-J.
151. Diamant.
152. Drugg. Chemist. (Phila.).
153. Dublin J. Med. Sci.
153a. Dublin Q. J. Med. Sci.
154. Dyer, Calico Printer (London).
155. Éclairage publ.
156. Edinb. J. Sci.
157. Elsäss. Hopfen-Ztg.
158. Exp. Station Rec. (Washington).
159. Fach-Ztschr. Textil-Ind.
160. Farben-Industrie.
161. Fortschr. techn. Chem.
162. Fortschr. theoret. Chem.
163. Fortschr. Chem. (Klein).
164. Gährungstechn. Jahrb.
165. Gambrinus.
166. Gas and Water.
166a. Gas World.
167. Gas Engineer.
167a. Gas Engineer's Mag.
168. Gas-Kalender.
168a. Kalender Gas-Techn.
169. Le Gaz. (Paris).
170. Gazz. chim. ital.
171. Gazz. chim. tecn. (Sembenini).
172. Gazz. farm. chim. (Sembenini).
172a. Gazz. chim. farm. (Sembenini).
173. Gazz. farm. chim. (Venezia).
174. Gerber (Wien).
175. Gerber-Ztg.
176. German Am. Brewers' J.
177. Giorn. chim. veterin.
178. Giorn. farm. chim. (Gajani).
179. Giorn. di farm. (Chiappero).
180. Giorn. di farm. (Cattaneo).
180a. Bibl. di farm.
180b. Ann. di chim. appl. (Polli).
180c. Ann. di chim. med. farm.
180d. Ann. di chim. farmacol.
181. Giorn. di fis. (Majocchi).
182. Giorn. di fis. (Brugnatelli).

183. Giorn. fis.-chim. ital.
184. Glashütte.
185. Helfenberger Ann.
186. Hermetisches J.
186a. Hermes (Karlsruhe).
187. Hopfenlaube.
188. Humphrey's Paint and Drug.
189. Incorragiamento.
190. Industrielles Wochenbl.
191. Trans. Inst. Brewing.
192. Introd. obs. phys. (Rozier).
192a. Obs. sur phys. (Rozier).
192c. J. de phys.
193. Jahrb. Chem.
194. Jahrb. Erfind.
195. Jahrb. Fabrik.
196. Jahrb. ökon. Chem.
197. Jsb. Agr.-Chem. (Detmer).
198. Jsb. Unters. Breslau.
199. Jsb. rein. Chem.
200. Jsb. Agr.-Chem.
201. Jsb. chem. Techn. [or Wagner's Jsb.].
202. Canstatt's Jsb.
202e. Jsb. Pharm.
203. Jsb. phys. Wiss. [or Berzelius' Jsb.].
204. Jsb. Chem.
205. Jsb. Thierchem.
206. Jsb. pathog. Mikroorganismen.
207. Jsb. Zuckerfabr.
208. Jern-Kont. Ann.
209. Jern och Stål.
210. J. chim. phys. (Van Mons).
211. J. chim. méd.
212. J. chim. (Van Mons).
213. J. éclair. gaz.
214. J. corps gras.
214a. Corps gras. industr.
215. J. physiol. homme.
215a. Archives physiologie.
216. J. teinture (Reimann).
217. J. der Pharm. (Trommsd.).
217a. N. J. der Pharm. (Trommsd.).
218. J. der Phys. (Gren).
218a. N. J. der Phys. (Gren).
218b. Ann. der Phys. (Gren).
218c. Ann. der Phys. (Gilbert).
218d. Ann. der Phys. (Pogg.).
218e. Ann. der Phys. (Wied.).
218f. Beibl. Ann. der Phys.
219. J. des brasseurs.

220. J. des brasseurs du Nord.
221. J. fabricants sucre.
222. J. usines à gaz.
223. J. du gaz.
224. J. für Gasbeleucht.
225. J. für Gas-Ind.
226. J. Phys. (Krönig).
227. J. techn. Mitthl. Zuckerfabr.
228. J. techn. Chem.
229. J. Zitz-Druckerei.
230. J. Anal. Chem.
230a. J. Anal. Appl. Chem.
231. J. Appl. Chem.
232. J. Gas-Lighting.
233. Nicholson's J.
234. J. Soc. Chem. Ind.
235. J. Soc. Dyers.
236. Kalend. farm. (Wasowicz).
237. Kalend. Textil-Ind.
238. Kemiska Notiser.
238a. Svensk Kem. Tidskr.
239. Kl. phys.-chem. Abh.
240. Ztschr. Chem.
241. Vjsb. Zuckerfabr.
242. Kvas.
243. Laboratorium.
244. Laboratory (Boston).
245. Laboratory (London).
246. Laboratory Yearbook.
247. Licht's Ber. Rübenzucker-Ind.
248. Listy chem.
249. Listy cukrovar.
250. Listy pekarské.
251. Lond. Water-Reports.
252. Mag. f. Apoth. (Elwert).
252a. Repert. f. Chem. (Elwert).
253. Mag. f. höh. Chem.
254. Mag. f. Hüttenk.
255. Mag. f. Färber.
256. Mech. and Chem.
257. Meddelelser, Carlsberg.
258. Med.-chem. Unters. Tübingen.
259. Mél. phys. chim.
260. Mem. Columbian Chem. Soc.
261. Mém. prat. chim.
262. Mem. Soc. spettroscopisti.
263. Merk's Bull.
264. Metamorfico.
265. Milchw. Taschenb.
266. Milch-Ztg.

267. Milk J.
268. Min. Soc. Phil. Exp.
268a. Protokolle (Higgins).
269. Misc. chim. fis. (Pisa).
270. Mitthl. chem. Lab. Innsbruck.
271. Mitthl. chem. Lab. (Hilger).
272. Mitthl. Chem. (Kletzinsky).
273. Mitthl. Lab. Chem. Brünn.
274. Mitthl. pharm. Inst. Erlangen.
275. Mitthl. Versuchsst. Wiesbaden.
276. Mitthl. Versuchsst. Klosterneuburg.
277. Monathschr. Scheidek.
278. Monatsber. Glashütte.
279. Monatsh. Chem.
280. Monit. brasserie (Bruxelles).
281. Monit. brasserie (Paris).
282. Monit. céramique.
283. Monit. sucrerie.
284. Monit. ind. gaz.
285. Monit. fils tissus.
286. Monit. prod. chim.
287. Monit. gaz.
288. Month. Mag. Pharm.
289. Muster-Ztg. Färberst.
289a. Färber-Ztg. Mustern.
290. Naturh. chem. Notiz.
291. Natuur. scheik. archief.
292. N. Y. Analyst.
292a. Am. Analyst.
293. Nordd. Brauer-Ztg.
293a. D. Brau-Ind.
294. Nord. Blätter Chem.
294a. Allg. nord. Ann. Chem.
294b. Mag. für Pharm. (Geiger).
295. Oesterr. Bierbrauer.
296. Oesterr. Ztschr. Bierbrauer.
297. Oesterr.-ung. Brauer-Kal.
298. Oesterr.-ung. Brennerei-Kal.
299. Oil and Drug News.
300. Oil, Paint, Rep.
301. Organ Ver. Rübenzucker-Ind.
301a. Oesterr.-ung. Ztschr. Zuckerind.
302. Organ Oel-Handel.
303. Orosi.
304. Penny Mech.
305. Pharmacist (Chicago).
306. Pharm. Times.
306a. Chem. Times.
307. Pharm. Centrbl.
307a. Chem. Centrbl.

308. Phil. Mag.
309. Phys. chem. Mag. (Weber).
310. Physikal. Revue.
311. Piria.
312. Pivovarn. Zpr.
313. Pivovarski L.
314. Polyt. J. (Dingler).
314a. Dingl. pol. J.
315. Pop. Ztsch. Spiritus-Ind.
315a. Allg. Ztschr. Spiritus-Ind.
316. Portefeuille f. Chem.
317. Prakt. Bierbrauer (New York).
318. Proc. Am. Chem. Soc.
318a. J. Am. Chem. Soc.
319. Proc. Chem. Soc. (Lond.).
319b. Q. J. Chem. Soc. (Lond.).
319c. J. Chem. Soc. (Lond.).
320. Prumyslník.
321. Raccolta fis. chim.
321a. Ann. fis. (Zantedeschi).
322. [Refer to No. 63.]
323. Rech. phys. chim. (Amsterdam)
324. Recueil mém. chim. Upsal.
325. Recueil trav. chim. Pays-Bas.
326. Reimann's Färber-Ztg.
327. Rép. chim. pharm. (Hensmans).
328. Rép. chim. phys. arts.
328a. Rép. chim. scientif. ind.
328b. Rép. chim. (Claubry).
329. Rép. pharm. chim. (Bruxelles).
330. Rép. pharm. (Bouchardat).
331. Rép. prat. vins.
332. Rep. anal. Chem.
333. Rep. Chem. Pharm. (St. P.).
334. Rep. organ. Chem. (Fechner).
335. Rep. Entdeck. Chem.
336. Rep. für Pharm. (Buchner).
337. Rep. organ. Chem. (Löwig).
338. Rep. Pharm. Russland.
339. Reports Roy. Coll. Chem.
340. Résumé trav. lab. Carlsberg.
341. Rev. Gas-Engineering.
341a. Gas, Water Rev.
342. Rev. chim. ind. (Villon).
343. Rev. des bières.
344. Rev. ind. chim. agric.
345. Rev. hebd. chim.
346. Rev. intern. falsifications.
347. Rev. sci. (Quesneville).
347a. Monit. sci. (Quesneville).

347c. Monit. sci. (Quesneville).
348. Rev. univ. brasserie.
349. Rev. univ. mines.
350. Rev. univ. fabric. sucre (Gemblaux).
351. Rev. univ. fabric. sucre (Paris).
352. Rivista di chim.
353. Rivista sci.
354. Rundschau Pharm. Chem.
355. Russ. Jahrb. Pharm.
356. Samml. Abh. Chem. (Hochheimer).
357. Savonnerie.
358. Scheik. bibl. (Meijlink).
359. Scheik. mag.
360. Scheik. bibl.
361. Scheik. bijdr.
362. Scheik. onderzoek (Mulder).
362a. Scheik. verhandel. (Mulder).
362b. Scheik. aanteek (Mulder).
362c. Scheik. onderzoek (Utrecht).
363. Schles. Brauer-Ztg.
364. School Mines Q.
365. Schwäb. Bierbrauer.
366. Schweiz. Milch-Ztg.
367. Seifenfabr.
368. Seifensieder-Ztg.
369. Selmi.
370. Skand. kem. Centralblad.
371. Sorgo J.
372. Stahl und Eisen.
373. Sucre (Le).
374. Sucrerie Belge.
375. Sucrerie indigène.
376. Südd. Gerber.
377. Sugar Beet (Philadelphia).
378. Sugar Cane (Manchester).
379. Sugar Cane (Newcastle).
380. Sugar Planter.
381. Taschenb. Scheidek.
382. Taschen-Kal. Zuckerf.
383. Techn.-chem. Kal.
384. Techn.-chem. Gewerbebl.
385. Techn.-chem. Jahrb.
386. Techn.-chem. Receptb.
387. Techn. Centralbl.
388. Teinturier prat.
389. Teinturier univ.
390. Tekno-Kemisk J.
391. Textile Colorist (London).
392. Textile Colorist (Philadelphia).
393. Thüringer Muster-Ztg.

393a. D. Färber-Ztg.
394. Tidssk. anv. Chemi.
395. Tidssk. Phys. Chemi.
396. Tijdschr. scheik.
397. Tijdschr. wet. pharm.
398. Toeg. scheik.
398a. Maandbl. toeg. scheik.
399. Tokyo Kag. Kaishi.
400. Trans. Lab. Club.
401. Trans. Newcastle Chem. Soc.
402. N. Gegenst. Chem. (Richter).
403. Unters. Liebig's Lab.
404. Vegytani L.
405. Verm. Abh. Warschauer Ges.
406. Vjschr. techn. Chem.
407. Vjschr. Nahrungsmittel.
408. Vzorny Pivovar.
409. Water, Gas and Steam.
410. Wochenschr. Rübenzuckerind.
411. Wochenschr. Brauerei.
412. Wochenschr. Zuckerfabr.
413. Wöchentl. Marktber. Zuckerhandel.
414. Yearbook Pharm.
415. Ztschr. Rübenzuckerind.

416. Ztschr. anal. Chem.
417. Ztschr. anorgan. Chem.
418. Ztschr. chem. Grossgew.
419. Ztschr. physikal. Unterricht.
420. Ztschr. Spiritusfabr.
420a. N. Ztschr. Spiritusfabr.
420c. Ztschr. Spiritusind.
421. Ztschr. chem. Ind.
421a. Ztschr. angew. Chem.
422. Ztschr. oesterr. Spiritusind.
423. Ztschr. Mineralwasser Fabrik.
424. Ztschr. Nahrungsm. Unters.
425. Ztschr. physikal. Chem.
426. Ztschr. physiol. Chem.
427. N. Ztschr. Rübenzucker-Ind.
428. Ztschr. Zucker-Ind.
429. Ztschr. Zucker-Ind. Böhmen.
430. Ztschr. Zündwaarenfabr.
431. Ztg. Gas Wasserfach.
432. Ztg. Zuckerfabrikanten.
433. Zhurnal russkova khim.
434. Zprávy chem. ceskych.
435. Zymotechn. Centrlbl.
436. Zymotechnisk Tidsskrift.

ADDENDA.

Chiefly of works published while the foregoing pages were in press. [Add.] following a name indicates an addition to a title in the given Section.

SECTION I.—BIBLIOGRAPHY.

Bay, J. Christian.
Materials for a Monograph on Inuline. Trans. Acad. Sci., St. Louis. Vol. VI, p. 151. March, 1893.
> Contains a bibliography of inuline.

Bolton, Henry Carrington. [Add.]
A Bibliography of Analytical and Applied Chemistry for the year 1892. J. Anal. Appl. Chem., vol. VII, p. 19. (Jan., 1893.)

Eiloart, Arnold.
An Index to the Literature of Stereochemistry. J. Am. Chem. Soc., vol. XIV, pp. 241–286. (Oct., 1892.)

Kukula, R. [Add.]
Bibliographisches Jahrbuch der deutschen Hochschulen. Vollständig umgearbeitete neue Auflage des Allgemeinen deutschen Hochschulen Almanachs. Innsbruck, 1892. 8vo.

Munroe, Charles E. [Add.]
Index to the Literature of Explosives. Part II, pp. 43–195. Baltimore, 1893. 8vo.

Tuckerman, Alfred.
Index to the Literature of Thermodynamics. Smithsonian Miscellaneous Collections. Washington, 1890. pp. v–239, 8vo.

SECTION II.—DICTIONARIES.

Dammer, O., und F. Rung.
Chemisches Handwörterbuch. Zweite Auflage. Stuttgart, 1892. 8vo.

FOURTIER, H.

Dictionnaire pratique de chimie photographique contenant une étude méthodique des divers corps usités en photographie. Paris, 1892. 8vo.

FREMY, EDMONDE. [Add.]

Encyclopédie chimique. Tome v. Partie 2. Section ii. Traitement des minérais auro-argentifères, par E. Cumenge et E. Fuchs. Paris, 1892. 8vo.

Tome vii. 5e Fascicule. 3e Section. Acides à fonction complexe (Acides à six equivalents d'oxygène), par Bourgoin. Paris, 1892. 8vo.

Tome ix. 2e Fascicule. 2e Partie. Chimie des liquides et des tissus de l'organisme, par Garnier, Lambling et Schlagdenhauffen. Paris, 1892. 8vo.

Tome x. La conservation des substances alimentaires, par Urbain. Paris, 1892. 8vo. Ill.

VILLON, A. M.

Dictionnaire de chimie industrielle, contenant toutes les applications de la chimie à l'industrie, la métallurgie, l'agriculture, la pharmacie, la pyrotechnie, les arts et métiers. Avec la traduction russe, anglaise, allemande, italienne, espagnole de la plupart des termes techniques. Paris, 1892. Sm. 4to.

　　　To be completed in three volumes.

SECTION III.—HISTORY.

BOLTON, HENRY CARRINGTON. [Add.]

Progress of Chemistry as depicted in Apparatus and Laboratories. Abstract Transactions N. Y. Academy of Sciences, vol. xii, p. 128. (Feb., 1893.)

CARO, H.

Ueber die Entwickelung der Theerfarben-Industrie. Ber. d. chem. Ges. xxv, pp. 955-1105. (1892.)

DAUBENY, CHARLES GILES B. · [Add.]

An Introduction to the Atomic Theory. Supplement. London, 1840.
　　　Second edition, enlarged. Oxford, 1850.

FOORD, G.

Lecture on Alchemy. Chem. News, xlviii, 93 (1883).

Hallopeau et Campredon.
La métallurgie (fonte, fers, aciers) à l'Exposition universelle de 1889.
Paris, 1892. 8vo. Ill.

Hoefer, Ferdinand. [Add.]
Histoire de la physique et de la chimie depuis les temps les plus reculés
jusqu'à nos jours. Nouvelle édition. Paris, 1892. 12mo.

Hoffmann, Friedrich.
Arabesken aus der ältesten Geschichte der Chemie. Pharmaceutische
Rundschau (New York), vol. III, p. 112 and p. 160.

Luzi, W.
* Das Ende des Zeitalters der Alchemie, und der Beginn der iatro-
chemischen Periode. Berlin, 1892. pp. 33, 8vo.
    Sammlung populärer Schriften herausgegeben von der Gesellschaft Urania
    zu Berlin. Nro. 13.

Gerland, E.
Geschichte der Physik. Leipzig, 1892. 8vo. Ill.

Klaproth, J.
Sur les connoissances chimiques des Chinois dans le VIIIe siècle. Mém.
Acad., St. Pétersb., vol. II, 1810.

Lorscheid, J.
Aristoteles' Einfluss auf die Entwickelung der Chemie. Münster, 1872.

Lübimof, N. A.
Istorija [etc.] [History of Physics in Russian language]. St. Petersburg,
1892. [In progress.]

Mercier, P.
Virages et fixages. Traité historique théorique et pratique. Partie I,
Notice des historique virages aux sels d'or. Paris, 1892. 8vo.

Ott, Adolph.
History of Chemical Nomenclature prior to Lavoisier. J. Franklin
Inst., August and November, 1869

Prescott, Albert B.
The Immediate Work in Chemical Science. J. Am. Chem. Soc., XIV,
p. 190. (Sept., 1892.)
    Address as Retiring President of the American Association for the Advance-
    ment of Science, at Rochester, August, 1892.

RODWELL, GEO. F.
On the Supposed Nature of Air prior to the Discovery of Oxygen.
Chem. News, VIII, 113 (1863).

SMITH, J. D.
On Early Egyptian Chemistry.   Phil. Mag. [4], IV, 143 (1852).

SWANK, J. M.
History of the Manufacture of Iron in all ages and particularly in the
United States from colonial times to 1891 ; also a short history of
early Coal Mining in the United States and a full account of the
influences which long delayed the development of all American
Manufacturing Industries.   Second edition thoroughly revised and
greatly enlarged.   Philadelphia, 1892.   8vo.

SECTION IV.—BIOGRAPHY.

BERZELIUS UND LIEBIG.
* Ihre Briefe von 1831–1845 mit erläuternden Einschaltungen aus gleich-
zeitigen Briefen von Liebig und Wöhler, sowie wissenschaftlichen
Nachweisen herausgegeben mit Unterstützung der Kgl. bayer.
Akademie der Wissenschaften von Justus Carrière.   München und
Leipzig, 1893.   pp. viii–279, 8vo.   Portraits of Liebig and Berzelius.

BOERHAAVE, H.
* Versuch über den Character des grossen Arztes, oder kritische Lebensbeschreibung, nebst einem Verzeichnisse der Boerhaavischen Schriften.   Aus dem französischen übersetzt.   Leipzig, 1748.   pp. [xxxii]
–160, 12mo.

BOISSAT.
DEBANS, C.   Boissat, chimiste.   Paris, 1892.   8vo.

CAHOURS, AUGUSTE THOMAS.
ÉTARD, A.   Notice sur la vie et les travaux de A. T. C.   Bull. Soc. chim.
[3], VII, i–xii (Nov., 1892).   Bibliography and portrait.

CARRIÈRE, JUSTUS.   *See* Berzelius und Liebig, ihre Briefe .   .   .

DITTMAR, WILLIAM.
Obituary.   Chem. News, LXV, 117 (March 4, 1892).

HARE, ROBERT.
Sketch of R. H.   Pop. Sci. Mon., vol. XLII, p. 695 (March, 1893).   Portrait.

HOFMANN, AUGUST WILHELM VON.
    MACQUENNE, L. Notice sur A. W. von H., avec portrait. Bull. Soc.
    chim. [3], IX, 1 (April 5, 1893).
    Obituary. Chem. News, LXV, 237 (May 13, 1892).

JOHNSON, SAMUEL WILLIAM.
    Biographical Sketch and Portrait. Pop. Sci. Mon., vol. XLIII, p. 117
    (May, 1893).

KOENE AND STAS.
    Chemistry at Brussels in 1840–1860. By T. L. PHIPSON. Chem. News,
    LXVII, p. 51 (Feb. 3, 1893).

KOPP, CHARLES ÉMILE.
    CABOT, S., JR. [Biographical sketch of] Charles Émile Kopp. Am.
    Chem., vol. vi, p. 211 (Dec., 1875).

PHIPSON, T. L. *See* Koene and Stas.

SCHEELE, CARL WILHELM.
    * Nachgelassene Briefe und Aufzeichnungen. Herausgegeben von. A.
    E. NORDENSKJØLD. Stockholm, 1892. pp. xliii–491–6, roy. 8vo. Ill.

        This sumptuous volume contains a view of Scheele's statue, a bibliography
        of works by and about him, and fac-similes of a letter and of pages from
        his laboratory note-book. There is also an edition in the Swedish
        language.

    THORPE, T. E. [Biographical sketch of] Scheele. Nature, vol. 47,
    p. 152 (Dec. 15, 1892).

SCHORLEMMER, CARL.
    SPIEGEL, ADOLF. Obituary and portrait. Ber. d. chem. Ges. XXV,
    pp. 1107–1123 (1892).

SEELEY, CHARLES A.
    Obituary. New York Daily Tribune, November 7, 1892.

STAS, JEAN SERVAIS.
    MALLET, J. W. Stas Memorial Lecture. (Abstract.) Chem. News,
    vol. LXVII, p. 19 (Jan. 13, 1893).
    Manifestation en l'honneur de J. S. Stas à l'occasion du 50 anniversaire
    de sa nomination comme membre titulaire de la classe des sciences
    de l'Académie royale de Belgique. Bruxelles, 1891. 8vo.
    MORLEY, EDWARD W. Jean Servais Stas [Biography and chemical
    work]. J. Am. Chem. Soc, XIV, 173 (Sept., 1892).

Wöhler, Friedrich.
    Thorpe, T. E.   The chemical work of Wöhler.   Proc. Roy. Inst. of
    Great Britain, vol. x, p. 477 (1884).

Young, Thomas.
    Tyndall, John.   Thomas Young.   Proceedings Roy. Inst. of Great
    Britain, vol. xi, p. 553 (1887).

SECTION V.—CHEMISTRY, PURE AND APPLIED.

Ackroyd, William.
    Elementary Chemical Analysis.   Halifax, 1892.

Allen, Alfred H.   [Add.]
    Commercial Organic Analysis, vol. iii, Part ii.   Amines and Ammonium
    Bases, Hydrazines, Bases from Tar, Vegetable Alkaloids.   London,
    1892.   8vo.

Arche, A.
    Praktische Chemie.   Anleitung zum Gebrauche der einfachen chemi-
    schen Geräthe und Reagentien in ihrer Anwendung zur Unter-
    suchung der Körper auf trockenem und nassem Wege.   Triest,
    1892.   8vo.

Arendt, R.   [Add.].
    Grundzüge der Chemie . . . Vierte vermehrte Auflage.   Hamburg,
    1892.   8vo.   Ill.

Autenrieth, W.
    Kurze Anleitung zur Auffindung der Gifte und stark wirkender Arznei-
    stoffe, zum Gebrauche in chemischen Laboratorien.   Freiburg,
    1892.   8vo.   Ill.

Barlet, A.
    Cours de la physique resolutive, vulgairement dite chymie.   Paris, 1653.
    4to.   34 plates.

Bartlett, Edwin J.
    Laboratory Exercises in Chemistry, with simple Apparatus.   Hanover,
    N. H., 1889.   pp. 22, 8vo.   Third edition, 1892.

Béguin, Jean.   [Add.]
    *Les élémens de chymie, revuez, notez, expliquez et augmentez par I.
    L. D. R. B. IC. E. M.   En ceste dernière édition ont esté adjoustées

BÉGUIN, JEAN. [Cont'd.]

plusieurs explications obmises aux précédentes impressions, et plusieurs préparations de remedes tirés de la dernière édition Latine. À Lyon, chez Pierre et Claude Rigaud, en ruë Merciere à la Fortune. 1666. pp. [xvi]–445–[xl], 12mo. Ill.

BEILSTEIN, F. [Add.]

Anleitung zur qualitativen chemischen Analyse. Siebente umgearbeitete Auflage. Leipzig, 1892. 8vo.

BENDER, A., UND H. ERDMANN.

Chemische Präparatenkunde. Band 1. Anleitung zur Darstellung anorganischer Präparate, von A. Bender. Stuttgart, 1893. 8vo. Ill.

BERTHELOT, MARCELLIN.

Explosives and their powers. Translated and condensed by C. N. Hake and W. MacNab. With preface by J. P. Cundill. London, 1892. 8vo. Ill.

BERZELIUS, J. J.

Versuch die bestimmten und einfachen Verhältnisse aufzufinden, nach welchen die Bestandtheile der unorganischen Natur verbunden sind (1811–12). Herausgegeben von W. Ostwald. Leipzig, 1892. 8vo.
Ostwald's Klassiker der exakten Wissenschaften.

BIELER, K., UND W. SCHNEIDEWIND.

Die agriculturchemische Versuchsstation Halle a. S., ihre Einrichtung und Thätigkeit. Berlin, 1892. 8vo. Ill.

BÖCKMANN, F. [Add.]

Chemisch-technische Untersuchungsmethoden der Grossindustrie, der Versuchsstationen und Handelslaboratorien. Herausgegeben unter Mitwirkung von C. Bischof, R. Nietzki, K. Stammer, A. Stutzer und A. Dritte vermehrte und umgearbeitete Auflage. Berlin, 1893. 2 vols., 8vo. Ill.

BOYLE, ROBERT.

[Addendum to Birch's edition of Boyle's Complete Works.]
[Another edition.] Edited by T. Birch. London, 1772. 6 vols., 4to. Portrait.

BREYER, F.

Gewinnung (Die) von sterilem Wasser in grösster Menge auf dem kalten Wege der Filtration. Zweite vermehrte Auflage. Wien, 1892. 8vo. Ill.

BREYER, TH., AND H. SCHWEITZER.
Chemists' Pocket-book of Ready Reference.   New York, 1893. pp.
viii-[56]-115, 24mo.

BUNSEN, R., UND H. E. ROSCOE.
Photochemische Untersuchungen (1855–59).   Herausgegeben von W.
Ostwald.   Leipzig, 1892.   8vo.
Ostwald's Klassiker der exakten Wissenschaften.

CASTNER, J.
Das Schiesspulver in seinen Beziehungen zur Entwickelung der ge-
zogenen Geschütze.   Berlin, 1892.   8vo.   Ill.

CHAPEL, E.
Le caoutchouc et la guttapercha.   Paris, 1892.   8vo.   Ill.

CHEMISTRY.   Part I : Inorganic.   Part II : Inorganic and Organic.   Catechism Series.   Edinburgh, 1892.
Published anonymously.

CHEMISTS' (THE) LEGAL HAND-BOOK.   By a Barrister.   London, 1892.
This title is misleading, as the hand-book is intended solely for pharma-
ceutists or apothecaries, who are misnamed " chemists " in England.

CHIMIE à l'usage des ingénieurs, des manufactures de l'État, directeurs et
contremaitres d'usines chimiques, professeurs, pharmaciens, industrielles, employés des régies.   Paris, n. d.   32mo.
Agenda Dunod Nro. 4 ; Arts et manufactures.   Anonymous.

CHRISTENSEN, C.
Uorganisk Kemi.   Femte udgave.   Kjøbenhavn, 1892.   8vo.

CHRISTENSEN, O. T.
Grundtraek af den organiske Kemi.   Anden udgave.   Kjøbenhavn,
1892.   8vo.

CHURCH, A. H.
Researches on Turacin, an animal pigment containing copper.   London,
1892.   Roy. 4to.   Ill.

CLARKE, FRANK WIGGLESWORTH.
The Elements of Chemistry.   New York, 1884.   pp. VII–369.   8vo.   Ill.

CLASSEN, A.   [Add.]
Quantitative chemische Analyse durch Electrolyse.   Dritte vermehrte
und verbesserte Auflage.   Berlin, 1892.   8vo.   Ill.

CONINCK, OECHSNER DE.
Cours de chimie organique. Paris, 1892. 8vo. Ill.
To be completed in two volumes.

DEFERT, R.
Guide pratique d'analyse qualitative par voie humide. Paris, 1892.
12mo.

DELMART, A.
Stück- (Die) und Kammgarnfärberei in ihrem ganzen Umfange.
Reichenberg i. B., 1892. 8vo
With dyed samples.

DENIGÈS, G.
*Esposizion elemental de los prinzipios fundamentales de la teoría
atómiqa. Bersion qastellana de Manuel A. Délano qon autoriza-
tion espezial del autor. Paris, 1893. pp. 38, 8vo.

DITTE, ALFRED.
Leçons sur les métaux professées à la Faculté des sciences de Paris.
Paris, 1891. 2 vols., 4to.

DOBBIN, L., AND J. WALKER.
Chemical Theory for Beginners. London, 1892. 12mo.

DÜRRE, E. F.
Die Praxis des Chemikers bei Untersuchung von Nahrungsmitteln und
Gebrauchsgegenständen, Handelsproducten, Luft, Boden, Wasser,
bei bacteriologischen Untersuchungen, sowie in der gerichtlichen
und Harnanalyse. Fünfte umgearbeitete und vermehrte Auflage.
Hamburg, 1892. 8vo. Ill.

DUHEM, P.
Introduction à la mécanique chimique. Gand, 1892. 8vo.

ÉGASSE, E. ET GUYENOT.
Eaux minérales naturelles autorisées de France et de l'Algérie ; leur
analyse et leurs applications thérapeutiques. Avec une préface par
Dujardin-Beaumetz. Deuxième édition. Paris, 1892. 8vo.

EGGERTZ, V. [Add.]
Om kemisk profning af jern, jernmalmer, bränmaterialer, och bränsle-
gaser. Andra upplaga tillökad och utgifven af C. G. Särnström.
Stockholm, 1892. 8vo. Ill.

EILOART, ARNOLD.
 A Guide to Stereochemistry, based on lectures delivered at Cornell
  University with an Index to the Literature and an Appendix:
  Models for use in teaching organic chemistry. Reprinted with
  additions from scientific periodicals. New York, 1893. 8vo. Ill.
     Contains a full bibliography

ELLIOTT, ARTHUR H.
 A System of Instruction in Qualitative Chemical Analysis. New York,
  1892. 8vo.

ELSNER, FRITZ. [Add.]
 Praxis (Die) des Chemikers . . . Fünfte umgearbeitete und vermehrte
  Auflage. Hamburg und Leipzig, 1892. 8vo. Ill.

ENGEL, R. [Add.]
 Nouveaux éléments de chimie . . . Quatrième édition revue et corrigée.
  Paris, 1892. 8vo. Ill.

ENGELHARDT, A. [Add.]
  Chemisch-technische Herstellung täglicher Bedarfsartikel.
       Second edition under the title :
  Chemisch-technisches Recept-Taschenbuch, enthaltend 1800 Vorschriften und Fabrikationsverfahren aus dem Gebiete der chemisch-
  technischen Industrie und Gewerbekunde. Zweite vollständig
  neubearbeitete Auflage von chemisch-technischer Herstellung täglicher Bedarfsartikel u. s. w. Leipzig, 1892. 8vo.

FABRE, J. HENRI.
 Notions de chimie à l'usage de l'enseignement secondaire spécial.
  Deuxième année. Métalloïdes. Paris, 1884. 12mo.
       [*The same.*] Quatrième année. Chimie organique. Paris,
       1886. 12mo.
 Notions de chimie organique à l'usage de l'enseignement secondaire des
  jeunes filles. Paris, 1880. 12mo.
 Notions élémentaires de chimie à l'usage des écoles primaires et des
  classes élémentaires. Paris, 1881. 12mo.
 Notions préliminaires de chimie à l'usage de l'enseignement secondaire
  des jeunes filles. Paris, 1880. 12mo.

FALL, DELOS.
 Introduction (An) to Qualitative Chemical Analysis by the Inductive
  Method. A Laboratory Manual for colleges and high-schools.
  Boston and New York, *n. d.* [1892]. pp. vii–85, 8vo.

FINGER.
Die Erklärung der Isomerie chemischer Verbindungen nach den heutigen Ansichten. Rappoltsweiler, 1891. 4to.

FISCHER, E. [Add.]
Exercises . . . with a preface by W. Dittmar. Glasgow, 1892. 8vo.

FISCHER, F. E.
Das Gesammtgebiet der Glasätzerei, Aetzen der Tafelgläser, Hohlgläser, Beleuchtungsartikel u. s. w. Braunschweig, 1892. 8vo. Ill

FLEISCHMANN, WILHELM. [Add.]
Industrie (L') laitière au point de vue scientifique et pratique. Traduit sur l'édition allemande revue et complétée par l'auteur par G. Brélaz et J. Oettli. Paris, 18—. Ill.

GRAEBE, C.
Guide pratique pour l'analyse quantitative. Paris, 1892. 8vo.

GREY, ANDREW.
Aids to Chemical Science, for teachers and students ; more especially for those preparing for " Class D " examination of the New Zealand Education Department. [London ?], 1892.
The author dates his preface from Remuera, Auckland.

GUARESCHI.
Introduzione allo studio degli alcaloidi vegetali. Torino, 1892.

GUTTMANN, O.
Handbuch der Sprengarbeit. Braunschweig, 1892. 8vo. Ill.

HAHN, E., UND J. HOLFERT.
Specialitäten und Geheimmittel mit Angabe ihrer Zusammensetzung. Eine Sammlung von Analysen, Gutachten und Literaturangaben. Fünfte völlig umgearbeitete, vermehrte und verbesserte Auflage. Berlin, 1893. 8vo.

HARTMANN, E.
Chemie für das Tentamen physicum. Dritte Auflage. Leipzig, 1892. 8vo.

HERZFELD, J. [Add.]
Das Färben und Bleichen der Baumwolle . . . Theil III. Die Praxis der Färberei. Berlin, 1893. 8vo. Ill.

HIGGINS, BRYAN.

* Experiments and Observations made with the view of improving the
art of composing and applying calcareous cements and of preparing
quicklime.   Theory of these arts, and specification of the Author's
cheap and durable cement for building, incrustation or stuccoing
and artificial stone.   London, 1780.   pp. xi–233, 8vo.

HIORNS, A. H.

Metal Colouring and Bronzing.   London, 1892.   8vo.

HIRSCHLER, J.

Methoden zur Bestimmung von Näherungswerten der Moleculargrösse.
Marburg, 1891.   4to.

HOFF, J. H. VAN'T.   [Add.]

Stéréochimie.   Nouvelle édition de " Dix années dans l'histoire d'une
théorie."   Rédigée par W. Meyerhoffer.   Paris, 1892.   8vo.

HUARD, AUGUSTE.

Traité comparé de chimie organique théorique et pratique, servant à
l'étude et à l'enseignement de la chimie organique.   Paris, 1852.
12mo.

HURST, GEORGE H.

Painters' Colours, Oils and Varnishes.   A practical manual.   London,
1892.   8vo.   Ill.

Ильинъ, А.

Краткій повторительный курсъ неорганической химіи.   С.-Петербургъ, 1875.

IL'IN, A.   Short Course of Inorganic Chemistry.   St. Petersburg, 1875.

Жакинъ, Іосифъ.

Начальныя основанія всеобщей и врачебной химіи.   С.-Петербургъ, 1808.

JACQUIN, JOSEPH.   First Principles of General and of Medical Chemistry.
St. Petersburg, 1808.

JAMMES, L.

Aide-mémoire de pharmacie chimique.   Paris, 1892.   12mo.   Ill.

JOHNSTON, J. F. W.   [Add.]

Catechism of Agricultural Chemistry, from the edition of Chas. A.
Cameron, revised and enlarged by C. M. Aikman.   Ninety-second
thousand.   Edinburgh and London, 1892.

JOLY, A.

Cours élémentaire de chimie ; notation atomique.   Paris, 1892.   12mo.

JOYNSON, F.
Iron and Steel Maker, being detailed descriptions of the various processes for the conversion of the ores of iron into pig-iron and cast-
iron, wrought or malleable iron and the different qualities of steel.
London, 1892. 8vo. Ill.

JUNGFLEISCH, E. [Add.]
Manipulations . . . Deuxième édition, revue et augmentée. Paris,
1892. 8vo. Ill.

KIMMINS, C. W.
The Chemistry of Life and Health. London, 1892. 8vo.

KLETWICH, JOHANN CHRISTOPH.
* Dissertatio de phosphoro liquido et solido . . . Nov. 9, 1688. Fran-
cofurti, *n. d.* pp. [56], 4to.
 Contains the history of the discovery of phosphorus.

KNAB, L.
Traité des alliages et des dépôts métalliques. Paris, 1892. 8vo. Ill.

KNAPP, F. L.
Mineralgerbung mit Metallsalzen und Verbindungen aus diesen mit
organischen Substanzen als Gerbemittel. Braunschweig, 1892. 8vo.

KNECHT (DR.), CHR. RAWSON, AND R. LOEWENTHAL.
Manual (A) of Dyeing, for the use of practical dyers, manufacturers
and students. London, 1892. 8vo. Ill.
 Specimens of dyed fabrics.

KOLLER, F.
Die Surrogate ; ein Handbuch der Herstellung der künstlichen Ersatz-
stoffe für den praktischen Gebrauch von Industriellen und Technikern. Frankfurt a. M., 1893. 8vo. Ill.

KOLLER, THEODOR.
Wasserstoffsuperoxyd (Das) in seiner technischen, industriellen und
öconomischen Bedeutung. Hamburg, 1892. Sm. 8vo.

KURZES REPETITORIUM . . . [Add.]
Abtheilung II, Quantitative Analyse. Wien, 1892.

LAINER, ALEXANDER.
Anleitung zu Laboratoriumsarbeiten mit besonderer Rücksicht auf die
Bedürfnisse des Photographen. Halle a. S., 1892. Ill.

LANGLOIS, PAUL.
  Le lait.   Paris, 1892.   12mo.

LEHNE, ADOLF.
  Tabellarische Uebersicht über die künstlichen organischen Farbstoffe
    und ihre Anwendung in Färberei und Zeugdruck.   Mit Ausfär-
    bungen jedes einzelnen Farbstoffes und Zeugdruckmustern.   Ber-
    lin, 1893.   Sm. 4to.
        In progress.

LUDWIG, E.
  Chemie und Rechtspflege.   Wien, 1892.   8vo.

LUDWIG, H.
  Kleine Gelegenheitsschriften.   Leipzig, 1892.   2 parts, 8vo.
        Part I : Petroleumfarben und sogenannte Harzölfarben.
        Part II : In Zimmerluft nassbleibendes, an der Sonne verdun-
        stendes Petroleum.

LUNGE, G.   [Add.]
  Taschenbuch für die Soda . . .   Zweite umgearbeitete Auflage.   Ber-
    lin, 1892.   8vo.   Ill.

MALY, RICHARD, UND K. BRUNNER.
  Anleitung zu pharmaceutisch-medicinisch-chemischen Uebungen.   Wies-
    baden, 1892.   Ill.

MAMY, J.
  Notions générales sur les matières colorantes organiques artificielles.
    Paris, 1892.   12mo.

MARCO, F.
  Elementi di chimica secondo il sistema periodico.   Ottava edizione
    riveduta ed aumentata.   Torino, 1892.   8vo.   Ill.

MATTHEY, E.
  On the Liquation of Metals of the Platinum Group.   London, 1892.   4to.

MAUMENÉ.
  Manuel de chimie photographique.   Paris, 1892.   12mo.   Ill.

MENSCHUTKIN, N.   [Add.]
  Analytische Chemie . . .   Dritte umgearbeitete Auflage.   Leipzig,
    1892.   8vo.

MILLS, E. J. [Add.]
Destructive Distillation. Fourth edition. London, 1892.

MOLINARI, F.
Combustili industriali e petrolio. Milano, 1892. 16mo.

MORITZ, ED. RALPH, AND G. H. MORRIS.
Text-book of the Science of Brewing, based upon a course of lectures
delivered by E. R. M. at the Finsbury Technical College of the
City and Guilds of London Institution. London, 1891. 8vo.

MOSEBACH, O.
Die Rohstoffe der Lackfabrikation. Zwickau, 1892. 8vo.

MÜLLER, C. [Add.]
Anleitung zur Prüfung der Kuhmilch. Sechste vollständig umgearbeitete
Auflage von E. Müller. Bern, 1892. 8vo. Ill.

NEWTH, G. S.
Chemical Lecture Experiments. Non-metallic Elements. London and
New York, 1892. 8vo.

OST, H. [Add.]
Lehrbuch der technischen Chemie. Zweite verbesserte Auflage. Berlin,
1893. 8vo. Ill.

OSTWALD, WILHELM. [Editor.]
Klassiker (Die) der exakten Wissenschaften, herausgegeben von Wilhelm
Ostwald. Leipzig, 18— -'92.
>No. 3 and No. 4. *See* page 712.
>No. 8. AVOGADRO, A., UND AMPÈRE. Abhandlungen zur Molekular-
theorie. (1811 und 1814). Ill.
>No. 9. HESS, H. Thermochemische Untersuchungen. (1839-1842.)
>No. 15. SAUSSURE, THÉODORE DE. Chemische Untersuchungen über
die Vegetation. (1804.) Uebersetzt von A. Wieler.
2 parts.
>No. 22. WÖHLER UND LIEBIG. Untersuchungen über das Radikal
der Benzoesäure. (1832.) Herausgegeben von Hermann Kopp.
>No. 26. LIEBIG, JUSTUS. Ueber die Constitution der organischen
Säuren. (1838.) Herausgegeben von Hermann Kopp.
>No. 27. BUNSEN, ROBERT. Untersuchungen über die Kakodylreihe.
(1837-1843.) Herausgegeben von Adolf von Baeyer.
>No. 28. PASTEUR, L. Ueber die Asymmetrie bei natürlich vorkom-
menden organischen Verbindungen. (1860.) Uebersetzt und herausgegeben von M. und A. Ladenburg.

OSTWALD, WILHELM. [Coht'd.]

No. 29. WILHELMY, LUDWIG. Ueber das Gesetz nach welchem die Einwirkung der Säuren auf den Rohrzucker stattfindet. (1850.)

No. 30. CANNIZZARO, S. Abriss eines Lehrganges der theoretischen Chemie vorgetragen an der Königlichen Universität Genua. (1858.) Uebersetzt von Arthur Miolati aus Mantua. Herausgegeben von Lothar Meyer.

No. 34. BUNSEN, ROBERT, und H. E. ROSCOE. Photochemische Untersuchungen. (1855–1859.) Ill. 2 parts. (No. 38.)

No. 35. BERZELIUS, JACOB. Versuch die bestimmten und einfachen Verhältnisse aufzufinden nach welchen die Bestandtheile der unorganischen Natur mit einander verbunden sind. (1811–1812.)

No. 39. PASTEUR, L. Die in der Atmosphäre vorhandenen organisirten Körperchen. Prüfung der Lehre von der Urzeugung. (1862.) Uebersetzt von A. Wieler. Ill.

No. 40. LAVOISIER, A. L., und P. S. DE LAPLACE. Zwei Abhandlungen über die Wärme. (Aus den Jahren 1780 und 1784.) Herausgegeben von J. Rosenthal. Ill.

The omitted numbers treat of purely mathematical and physical topics. The series is in progress.

PASCAL, T.
Tintura della seta ; studio chimico-tecnico. Milano, 1892. 12mo.

PREYER, W.
Genetische (Das) System der chemischen Elemente. Berlin, 1893. 8vo. Ill.

RAINS, GEO. W.
Interesting Chemical Exercises in Qualitative Analysis for ordinary schools. New York, *n. d.* [1879]. pp. 59, 12mo.

RICHE ET ROUME.
Rapport sur la production, l'industrie et le commerce des huiles minérales aux États-Unis d'Amérique. Paris, 1892. 4to.

RICHTER, V. VON. [Add.]
Lehrbuch der anorganischen Chemie. Siebente Auflage neu bearbeitet von H. Klinger. Bonn, 1892. Ill.

RIZZATTI, F.
Le analisi al canello ferruminatorio. Manuale pratico pel mineralogista e pel chimico. Seconda edizione. Torino, 1892. 8vo. Ill.

SARRAU, E.
Introduction à la théorie des explosifs. Paris, 1893. 8vo. Ill.

SCHREITMAN, C.
Probierbüchlein ; frembde und subtile Kunst, vormals in Truck nie gesehen, von Wage und Gewicht, auch von allerhandt Proben auff Ertz, Golt, Silber und andere Metall. Franckfurt, Chr. Egen-[olphus] Erben, 1578. 12mo. Ill.

SIMON, WILLIAM. [Add.]
Manual of Chemistry . . . Fourth edition. Philadelphia, 1893. 8vo. Ill.

SÖRENSEN, S. P. L.
Kortfattet Vejledning i kvalitativ Analyse af simple uorganiske Blandinger. Kjøbenhavn, 1892. 8vo.

SPEZIA, G.
Sull' origine del solfo nei giacimenti solfiferi della Sicilia. Torino, 1892. 8vo. Ill.

SUPINO, R.
Veleni e contraveleni. Milano, 1892. 12mo.

TAYLOR, R. L.
Practical Chemistry. New edition, revised. London, 1892. 12mo.

TROOST, L. [Add.]
Précis de chimie. Vingt cinquième édition. Paris, 1893. 12mo. Ill.

VEITH, A.
Das Erdöl (Petroleum) und seine Verarbeitung. Gewinnung, Untersuchung, Verwendung und Eigenschaften des Erdöles. Braunschweig, 1892. 8vo. Ill.

SECTION VI.—ALCHEMY.

* HERMETIC (THE) MUSEUM, restored and enlarged, most faithfully instructing all the disciples of the sopho-spagyric art how that greatest and truest medicine of the Philosophers' Stone may be found and held. Now first done into English from the rare Latin original published at Frankfurt in the year 1678. Containing twenty-two most celebrated tracts. London, 1892. 2 vols., 4to. Vol. I, pp. xi–357 ; plates. Vol. II, pp. [iv]–322 ; plates.

> The name of the translator is not given ; the preface is signed Arthur Edward Waite. Edition limited to 250 copies.

* CHANG PIH TUAN.
    Wu-chen-pien.   2 vols., 8vo.   Vol. I, pp. [99] ; II, pp. [156].

> Written about 1075 A.D.   There are three Commentaries on this work pub-
> lished together.   A few extracts from Wu-chen-pien will be found in
> Dr. W. A. P. Martin's " The Chinese," New York, 1881, pp. 182, 183.
> My copy is a reprint issued during the reign of the present Emperor of
> China, Kuang Hsü.

* TS'AN T'UNG CH'I.   pp. [165], 8vo.

> The earliest Chinese work on alchemy now extant.   My copy is a reprint
> issued during the reign of the present Emperor of China, Kuang Hsü.
> The clearest Commentary of recent times is by a scholar of the Yuen
> dynasty, named Ch'en Chih Hsu.
> For these Chinese alchemical works I am indebted to the courtesy of John
> Fryer, LL.D., of Shanghai.

SECTION VII.—PERIODICALS.

CHEMISCH-ÖKONOMISCHES TASCHENBUCH auf 1804, 1805 . . . 1812.
    Herausgegeben von Carl Friedrich Richter. — vols. Chemnitz,
    1805–'13.

MÉMORIAL DES POUDRES ET DES SALPÊTRES, publiés par soins du Service
    des poudres et salpêtres, avec l'autorisation du Ministre de la
    guerre.   3 vols. (III–V), 8vo.   Paris, 1890–'92+

> The Mémorial was changed to a periodical in 1890.   Vol. I was issued in
> 1882, II in 1889.

SUBJECT-INDEX.

This index is limited to special topics and does not include works of an encyclopedic character.

Section IV being self-indexed, and Section VI being confined to a single topic, are not included.

Entries are made under specific words so far as possible. Roman numbers refer to Sections ; Arabic to pages, except in Section VII in which the numbers refer to titles. In a few instances dates have been inserted for reasons that are obvious.

ANALYTICAL CHEMISTRY. [Cont'd.]
Leyde, 621 ; Ludwig, 642 ; McGill,
645 ; Madan, 646 ; Marlin, 654 ; Mason, 657 ; Matcovich, 658 ; Medicus,
660 ; Meinecke, 661 ; Mendeléeff, 664 ;
Menschutkin, 664 ; Methods, 665 ;
Metzner, 665 ; Meyer, 666 ; Meylink,
668 ; Miller, 671 ; Mills, 671 ; Mitchell, 672 ; Mojon, 676 ; Müller, 685 ;
Müntz, 685 ; Munner, 689 ; Musaio,
690 ; Muspratt, 690 ; Muter, 690 ;
Nayudu, 694 ; Nilson, 701 ; Noad,
701 ; Normandy, 702 ; Northcote,
703 ; Noyes, 703 ; O'Brine, 704 ;
Orfila, 707 ; Orosi, 711 ; Palm, 717 ;
Parmentier, 720 ; Parnell, 720 ; Pasqualini, 720 ; Patchett, 721 ; Pauly,
722 ; Pavesi, 723 ; Peñuelas, 730 ;
Petersen, 733 ; Pfaff, 735 ; Pilley,
740 ; Pink, 741 ; Pisani, 742 ; Posselt,
748 ; Post, 748 ; Pott, 749 ; Preis,
752 ; Prescott, 752 ; Prestinari, 752 ;
Pritchett, 755 ; Prunier, 756 ; Purgotti,
756 ; Quadrat, 756 ; Qualitative, 757 ;
Rammelsberg, 759 ; Reactions, 762 ;
Redtel, 762 ; Reynolds, 772 ; Ribaucourt, 773 ; Richardson, 773 ; Richter,
775 ; Righini, 777 ; Rivot, 779 ; Rodi-
guez, 780 ; Rose, 784 ; Rouelle (1774),
788 ; Rude, 790 ; Rüdorff, 790 ; Sage,
795 ; Salomon, 797 ; Sbriziolo, 800 ;
Schatz, 802 ; Schema, 805 ; Schlickum,
808 ; Schmidt, 811, 813 ; Schorn, 817 ;
Schramm, 817 ; Schweizer, 824 ; Scivoletto, 824 ; Scoffern, 825 ; Scott-
White, 825 ; Sella, 826 ; Semple, 828 ;
Sexton, 830 ; Siegel, 832 ; Silvestrini,
833 ; Simmons, 834 ; Slater, 835 ; Slatter, 836 ; Snively, 838 ; Sobrero, 839 ;
Sonnenschein, 841 ; Sousa Pinto, 842 ;
Spencer, 844 ; Spica, 844 ; Spirgatis,
845 ; Städeler, 845 ; Stahl, 848 ; Stévart, 841 ; Stoddard, 852 ; Strasser,
856 ; Swartz, 861 ; Tableaux, 862 ;
Tassinari, 864 ; Taylor, 865 ; Terreil,
865 ; Thaulow, 867 ; Thénard, 868 ;
Thomson, 871 ; Thorpe, 872 ; Tornöe,
876 ; Torres, 876 ; Treadwell, 877 ;
Trimble, 877 ; Trommsdorff, 878 ;
Tuttle, 880 ; Ulbricht, 881 ; Valentin,
883 ; Venable, 885 ; Villiers, 888 ;
Vogel, 891 ; Waage, 894 ; Wacken-
roder, 894 ; Wackenroder, 895 ; Wal-
lach, 901 ; Wartha, 904 ; Weinzierl,
910 ; Weselsky, 912 ; White, 914 ;
Wicke, 914 ; Wiecke, 915 ; Wiley,
917 ; Will, 920, 921 ; Wills, 923 ;
Windisch, 924 ; Winkelblech, 925 ;
Winkler, 925 ; Wöhler, 930 ; Wuttig,
936 ; Zenneck, 939 ; Zenoni, 939 ;
Zettnow, 940.
VII. 29, 39, 64, 230, 275, 292, 332,
416.
V. (Add.) Ackroyd, 1170 ; Allen, 1170 ;
Beilstein, 1171 ; Böckmann, 1171 ;
Classen, 1172 ; Defert, 1173 ; Dürre,
1173 ; Eggertz, 1173 ; Elliott, 1174 ;
Fall, 1174 ; Graebe, 1175 ; Kurzes,
1177 ; Menschutkin, 1178 ; Rains,
1180 ; Sörensen, 1181.

ANILIN DYES. (*See also* COAL-TAR.)
V. Bersch, 304 ; Dépierre, 397 ; Heu-
mann, 529 ; Hödl, 536 ; Kertész, 574 ;
Krieg, 590 ; Nölting, 702 ; Reimann,
767 ; Rosenstiehl, 786 ; Soxlet, 842 ;
Vallhonesta, 884 ; Vogel, 892.

ANIMAL CHEMISTRY. (*See also* PHYSI-
OLOGY.)
I. Jahresbericht, 18.
II. John, 63 ; Berzelius, 90 ; Johnson,
123.
V. Arends, 271 ; Barral, 283 ; Baumé,
287 ; Berzelius, 308, 310 ; Du Ménil,
416 ; Fleck, 442 ; Hemmer, 522 ;
Herpin, 528 ; Heusel, 529 ; John,
556 ; Jones, 562 ; Juch, 563 ; Kapp,
568 ; Kingzett, 575 ; Kletzinsky, 578 ;
Kühn, 593 ; Liebermann, 624 ; Liebig,
624, 627, 630, 632 ; Luigi, 642 ; Matteucci, 658 ; Odling, 704 ; Schlossberger, 809 ; Schmidt, 810 ; Thomson,
871.
VII. 62, 205.

ANTHRACENE.
V. Auerbach, 275 ; Hödl, 536.

ANTIDOTES. (*See also* POISONS.)
V. Bandlin, 280 ; Bunsen, 346 ; Cox,
380 ; Craanen, 380 ; Dyrenfurth, 420 ;
Navier, 694 ; Plenck, 744 ; Sage, 795 ;
Supino, 1181.

ATOMIC PHILOSOPHY. [Cont'd.]

Macneven, 645 ; Mann, 651 ; Ostwald, 712, 1179 ; Petzold, 734 ; Phipson, 737 ; Poleck, 746 ; Ponci, 747 ; Purgotti, 756 ; Raynant, 762 ; Reinsch, 768 ; Schlegel, 808 ; Tarugi, 864 ; Wurtz, 935.

ATOMIC VOLUMES.

V. Avogadro, 276 ; Graham, 493 ; Schröder, 818 ; Vlenten, 890 ; West, 912.

ATOMIC WEIGHTS.

I. Becker, 3.

II. Dulk, 48 ; Clarke, 45 ; Meyer und Seubert, 69.

III. Mulder, 140 ; Newlands, 142 ; Oudemans, 143.

V. Baumhauer, 288 ; Baup, 288 ; Berzelius, 307 ; Cooke, 377 ; Errera, 431 ; Grothuss, 501 ; Hartsen, 518 ; Kessler, 574 ; Kreusler, 590 ; Krüss, 591 ; Meyer and Seubert, 667 ; Mosmann, 682 ; Pizzi, 742 ; Sebelien, 825 ; Stas, 849 ; Strecker, 857 ; Turner, 880 ; Vogel, 891 ; Wackenroder, 894 ; Wandtafeln, 902 ; Weber, 906 ; Wilde, 917.

ATOMICITY. (*See* VALENCE.)

AZOTE. (*See* NITROGEN.)

B.

BACTERIOLOGY.

VII. 206.

BAKING POWDERS.

V. Crampton, 381, 919.

BARIUM AND ITS COMPOUNDS.

V. Bucholz, 344.

BEER, EXAMINATION OF.

I. Gracklauer, 15.

III. Grässe, 115 ; Reiber, 149.

V. Boulin, 331 ; Dannehl, 388 ; Hopff, 542 ; Mulder, 686 ; Pasteur, 721 ; Placotomus (1551), 742 ; Reischauer, 768 ; Stierlin, 852 ; Vogel, 891 ; Wackenroder, 895 ; Wernerus (1567), 912 ; Zenneck, 939.

BEESWAX. (*See* WAXES.)

BEET-SUGAR.

I. Zeitschrift, 36.

III. Scheibler, 154 ; Stöckhardt, 159 ; Ware, 164.

V. Achard, 261 ; Anleitung, 269 ; Barruel, 284 ; Birnbaum, 314 ; Bley, 318 : Braumüller, 337 ; Cadet, 349 ; Claudius, 370 ; Clemandot, 370 ; Congrès, 375 ; Crookes, 383 ; Dennstedt, 397 ; Dombasle, 404 ; Dubrunfaut, 408, 409 ; Dureau, 419 ; Erxleben, 431 ; F., 432 ; Gautier, 471 ; Girard, 480 ; Göttling, 487 ; Graftian, 492 ; Grant, 495 ; Grauvogel, 495 ; Grebner, 496 : Hartmann, 517 ; Heidenkampf, 520 ; Hermann,, 525 ; Hermbstädt, 526 ; Hlubek, 535 ; Jacobi, 551 ; Joulie, 563 ; Juch, 563 ; Kauzmann, 569 ; Kirchhof, 576 ; Koppi, 587 ; Krause, 589 ; L——, 596 ; Lampadius, 599 ; Leuchs, 619 ; Lüdersdorff, 642 : Meidinger, 661 ; Meyer, 666 ; Michaelis, 669 ; Nebbien, 695 ; Nikolai, 701 ; Noldechen, 702 ; Otto, 714 ; Payen, 726 ; Péligot, 727 ; Pexa, 735 ; Pickny, 738 ; Poutet, 751 ; Preuss, 753 ; Rad, 758 ; Rath, 761 ; Regner, 765 ; Rössig, 781 ; Scheibler, 804 ; Schmidt, 812 ; Schubarth, 818 ; Schulz, 821 ; Stadiot, 845 ; Sykora, 862 ; Ueber, 881 ; Walkhoff, 900 ; Ware, 903 ; Weinrich, 910 ; Wiley, 919 ; Zdeborsky, 939.

VII. 247, 301.

BENZENE. (Benzole.)

V. Mansfield, 651.

BENZOIC ACID.

V. Wöhler, 930, 1179.

BERLIN SCIENTIFIC INSTITUTIONS.

III. Guttstadt, 116.

BERYLLIUM. (*See* GLUCINUM.)

BILE.

V. Trifanovsky, 877.

BISMUTH AND ITS COMPOUNDS.

V. Jaeger, 552 ; Viluef, 889.

BITUMEN.

V. Halleck, 510 ; Kirwan, 577 ; Payen, 724.

BLASTING POWDER.

V. Bolley, 326 ; Guttmann, 1175.

BLEACHING. (*See also* DYEING.)

V. Curaudau, 384 ; Dosne, 405 ; Gillet, 479 ; Hermbstädt, 526 ; Higgins, 530 ; Home, 541 ; Joclét, 555 ; Kurrer, 596 ; Romen, 782 ; Rouget, 788 ; Stein, 850 ; Westrumb, 912, 913.

DISSOCIATION.

V. Sainte-Claire-Déville, 795 ; Société (1864), 840.

DISTILLATION OF LIQUORS.

I. Gracklauer, 15.

V. Balling, 279 ; Barbet, 281 ; Bolley, 325 ; Brevans, 338 ; Briem, 338 ; Brunschwick (1512), 342 ; Demachy, 396 ; Duplais, 418 ; Duportal, 418 ; Eidherr, 424 ; Elsholtz (1674), 426 ; Fischer, 440 ; French (1651), 454 ; Fritsch, 462 ; Gaber, 466 ; Hermbstädt, 526 ; Keller, 570, 571 ; Krüger, 591 ; Maercker, 646 ; Moewes, 674 ; Nicolas, 700 ; Otto, 715 ; Payen, 725 ; Poncelet, 747 ; Roux, 789 ; Sala, 796 ; Schmidt, 811 ; Schubert, 819 ; Sell, 826 ; Ulbricht, 882 ; Westrumb, 912 ; Wilfert, 920.

VII. 12, 44, 81, 105, 111, 135, 283, 298, 315, 348, 420, 422.

DÖBEREINER'S LAMP.

V. Böttger, 323 ; Döbereiner, 404.

DUTCH CHEMICAL BIBLIOGRAPHY.

I. Bidstrup, 4 ; Bierens de Haan, 4 ; Holtrop, 16, 17 ; Meulen, 22 ; Mourick, 23.

DYEING.

I. Gracklauer, 15 ; Leuchs, 21.

II. Hochheimer, 61 ; O'Neill, 71.

III. Bischoff, 91 ; Kielmeyer, 125 ; Kurrer, 129.

V. Angaryd, 269 ; Beckers, 294 ; Berthollet, 306 ; Bird, 314 ; Centner, 359 ; Chaptal, 361 ; Chevreul, 365 ; Cooper, 377 ; Crace-Calvert, 380 ; Crookes, 383 ; Delmart, 396 ; Dépierre, 397 ; Dollfus-Ausset, 404 ; Dumas, 416 ; Dyer, 420 ; Fiedler, 439 ; Fol, 445 ; Garcia-López, 468 ; Gardner, 468 ; Gibson, 478 ; Grison, 500 ; Gülich, 502 ; Haserick, 518 ; Hellot, 521 ; Hermbstädt, 527 ; Herzfeld, 528 ; Hummel, 547 ; Joclét, 555 ; Kurrer, 596 ; Lau, 604 ; Lauber, 604 ; Levaux, 619 ; Macquer, 646 ; Martin, 655 ; Moyret, 683 ; Napier, 692 ; O'Neill, 706 ; Parnell, 720 ; Persoz, 732 ; Piéquet, 739 ; Pörner, 745 ; Polleyn, 746 ; Ponci, 747 ; Reimann, 767 ; Reisig,

768 ; Riffault, 777 ; Romen, 782 ; Rudolf, 791 ; Runge, 791, 792 ; Sadtler, 794 ; Sammlung, 798 ; Sansone, 799 ; Schmidt, 812 ; Schützenberger, 820 ; Smith, 837 ; Soxlet, 842 ; Stein, 850 ; Tassart, 864 ; Trommsdorff, 878 ; Tschelnitz, 879 ; Vinant, 889 ; Vitalis, 890.

VII. 11, 88, 112, 136, 148, 154, 159, 160, 216, 229, 235, 237, 255, 285, 289, 326, 388, 389, 391, 392, 393.

V. (Add.) Delmart, 1173 ; Herzfeld, 1175 ; Knecht, 1177 ; Pascal, 1180.

DYESTUFFS.

II. Schultz, 75.

III. Lightfoot, 133 ; Noelting, 142 ; Wurtz, 169.

V. Bancroft, 280 ; Bolley, 325 ; Gentele, 473 ; Goppelsroeder, 489 ; Hollunder, 541 ; Keller, 571 ; Kopella, 586 ; Leuchs, 618 ; Mierziński, 670 ; Möhlau, 674 ; Mühlhäuser, 683 ; Nietzki, 700 ; Nölting, 702 ; Ostermayer, 712 ; Pelouze, 728 ; Renard, 770 ; Schmidt, 811, 813 ; Schreger, 817 ; Schreiberey, 817 ; Slater, 836 ; Société [Schützenberger], 840 ; Société [Nœlting], 840 ; Springmühl, 845 ; Stein, 851 ; Stöckhardt, 855 ; Villon, 889 ; Koller, 1177 ; Lehne, 1178 ; Mamy, 1178.

DYNAMICAL CHEMISTRY.

I. Warder, 33.

III. Urech, 163.

V. Lutoslawski, 643 ; Prieto, 754 ; Urech, 883.

DYNAMITE. (See also NITRO-GLYCERIN.)

V. Dumas-Guilin, 416 ; Eissler, 424 ; Mowbray, 682 ; Ruggieri, 791 ; Trauzl, 877.

E.

EGYPTIANS, ANCIENT, CHEMISTRY OF.

III. Berthelot, 89 ; Conring, 101 ; Herapath, 117 ; Schaefer, 154 · Smith, 1168 ; Tollius, 162.

V. Fischer, 442.

ELECTRICITY.

III. Becquerel, 88 ; Boullet, 96 ; Cozzi, 102 ; Faulwetter, 108 ; Sigaud de la Fond, 158.

HISTORY OF ALCHEMY.

III. Almqvist, 85 ; Amelung, 85 ; Bauer, 87 ; Bégin, 88 ; Berthelot, 90 ; Beytrag, 90 ; Bodenstein, 92 ; Bolton, 93 ; Borrichius, 95 ; Buddeus, 98 ; Christmas, 100 ; Conring, 101 ; Dewar, 104 ; Duval, 105 ; Edelgeborne, 106 ; Eaton, 106 ; Eyssenhardt, 108 ; Ferber, 108 ; Figuier, 109 ; Gericke, 112 ; Gildemeister, 113 ; Gladstone, 113 ; Güldenfalk, 115 ; Henkel, 117 ; Herverdi, 118 ; Hirsching, 118 ; Hoffmann, 1167 ; Hofmann, 120 ; Hoghelande, 120 ; K, 123 ; Kircher, 125 ; Kirchmaier, 125 ; Kopp, 127 ; Kortum, 128 ; Leemans, 131 ; Lenglet du Fresnoy, 131 ; Lenz, 132 ; Lewinstein, 132 ; Libavius, 132 ; Luzi, 1167 ; Mackay, 135 ; Marchand, 135 ; Meidinger, 136 ; Mikowec, 138 ; Murr, 140 ; Nenter, 141 ; Olliffe, 143 ; Osten, 143 ; Rhamm, 150 ; Rosa, 152 ; Schaefer, 154 ; Schmieder, 155 ; Schröder, 157 ; Svátek, 160 ; Tonni-Bazza, 162 ; Tresor, 162 ; Vlasto, 164 ; Vulpius, 164 ; Wedel, 165 ; Weech, 165 ; Wiegleb, 167 ; Wild, 167.

HISTORY OF CHEMISTRY.

I. Bolton, 6.

III. Béchamp, 87 ; Bégin, 88 ; Bernardus, 89 ; Berthelot, 89 ; Bley, 91 ; Blomstrand, 92 ; Bolton, 93, 94, 1166 ; Borrichius, 95 ; Brande, 96 ; Buchner, 97 ; Buckley, 98 ; Büchel, 97 ; Buff, 98 ; Caillot, 98 ; Casali, 99 ; Chevreul, 100 ; Chronologische Tabelle, 100 ; Cibot, 100 ; Cozé, 101 ; Cuvier, 102 ; Davy, 103 ; Deventer, 104 ; Draper, 104 ; Dupont, 105 ; Duval (1561), 105 ; Ferguson, 109 ; Fownes, 110 ; Frazer, 110 ; Fremy, 111 ; Freund, 111 ; Gadd, 111 ; Gautier, 112 ; Gerding, 112 ; Gerhard, 112 ; Gigli, 113 ; Gish, 113 ; Gmelin, 113, 114 ; Goebel, 114 ; Hebenstreit, 117 ; Hilger, 118 ; Hinterberger, 118 ; Hoefer, 119, 1167 ; Hoffmann, 119, 1167 ; Hofmann, 120 ; Isambert, 122 ; Jackson, 122 ; Jacob, 122 ; Jagnaux, 123 ; Kastner, 124 ; Klaproth, 1167 ; Knyazeff, 126 ; Koninck, 126 ; Kopp, 127 ; Lacroix, 129 ; Ladenburg,

129 ; Lamarck, 129 ; Lemoine, 131 ; Lense, 132 ; Lusson, 154 ; Meyer, 137 and 138 ; Micé, 138 ; Mitchell, 138 ; Napier, 141 ; Oersted, 143 ; Ohsson, 143 ; Paets van Troostwijk, 144 ; Picton, 147 ; Poppe, 147 ; Post, 148 ; Pott, 148 ; Proceedings, 148 ; Reuvens, 150 ; Rodwell, 151 ; Romegialli, 152 ; Routledge, 153 ; Rüdorff, 153 ; Savérien, 153 ; Schreger, 156 ; Schröder, 156 ; Schubarth, 157 ; Schützenberger, 157 ; Sennert, 157 ; Siebert, 157 ; Silvestri, 158 ; Smith, 158, 1168 ; Spica, 159 ; Sprengel, 159 ; Stöckhardt, 159 ; T., 160 ; Thomson, 161 ; Tieboel, 162 ; Trommsdorff, 163 ; Troost, 163 ; Wagner, 164 ; Weidner, 165 : Weihrich, 166 ; Weitz, 166 ; Whewell, 166 ; White, 167 ; Wiegleb, 167 ; Winterl, 168 ; Wisser, 168 ; Wittstein, 168 ; Wohlwill, 168 ; Woltter, 168 ; Wright, 169 ; Wurtz, 169 ; Wurzer, 170 ; Yeats, 170.

V. Stölzel, 855.

HONEY.

I. Colby, 11 ; Wiley, 35.

III. Venturi, 163.

V. Wiley, 919.

HOPS. (*See also* BREWING.)

VII. 5, 8, 187.

HYDROCHLORIC ACID.

III. Wehrle, 165.

V. Nagy, 690.

HYDROCYANIC ACID.

I. Preyer, 26.

V. Grindel, 499 ; Henry, 523 ; Ittner, 551 ; Kobert, 581 ; Magendie, 648 ; Preyer, 753 ; Société [Gautier], 840.

HYDROFLUORIC ACID.

V. Scheele, 802.

HYDROGEN.

V. Cormier, 377 : Heldt, 521.

HYDROGEN, PEROXIDE OF.

I. Leeds, 21.

V. Koller, 1177.

I.

ICE.

V. Petterson, 734 ; Ritter, 778 ; Tyndall, 881.

MILK-ANALYSIS.

V. Adam, 262 ; Becke, 293 ; Bouchardat, 330 ; Dietzsch, 400 ; Esbach, 431 ; Gerber, 474 ; Herz, 528 ; Meigs, 661 ; Monnier, 678 ; Müller, 684, 1179 ; Pfeiffer, 735; Quesneville, 757; Sartori, 799 ; Schrodt, 818 ; Vieth, 887 ; Vogel, 891 ; Wanklyn, 902.

MINERAL WATERS (*cf.* POTABLE WATERS).

II. Hoffmann, 62.

V. Ankum, 269 ; Ash, 273 ; Bailly, 278 ; Bayen, 288 ; Berzelius, 310 ; Bley, 318 ; Blum, 319 ; Bottler, 330 ; Bouquet, 331 ; Brandes, 336 ; Bucholz, 344 ; Cadet, 348 ; Costel, 379 ; Daquin, 388 ; Darapsky, 388 ; Demachy, 396 ; Döbereiner, 403 ; Duclos, 410 ; Dupasquier, 418; Egasse, 1173; Eimbke, 424 ; Ekeberg, 424 ; Fischer, 441 ; French (1652), 454 ; Fresenius, 454, 455 ; Garnett, 468 ; Gesner (1553), 477 ; Glover, 484 ; Goldberg, 488 ; Granville, 495 ; Gunning, 504 ; Hagen, 507 ; Harmens (1734), 515 ; Henry, 523 ; Herpin, 528 ; Hirsch, 532 ; Hirschfeld, 532 ; Hofmann, 540 ; Jervis, 554 ; John, 556 ; Jones, 562 ; Julia-Fontenelle, 565 ; Kersten, 574 ; Kielmayer, 575 ; Kitaibel, 577 ; Klaproth, 578 ; Lagrange, 598 ; Lancellotti, 600 ; Le Givre (1654), 611 ; Lersch, 617 ; Liebig, 627 ; Limbourg, 632 ; Löwig, 638 ; Maret, 653 ; Marteau, 655 ; Mauro, 658 ; Melandri, 662 ; Mönch, 674 ; Mojon, 676 ; Monheim, 678 ; Monnet, 678 ; Monro, 679 ; Moorman, 679 ; Morell, 680 ; Morus, 681 ; Nicolas, 700 ; Nivet, 701 ; Opoix, 707 ; Osann, 711, Osburg, 711 ; Patissier, 721 ; Peale, 726 ; Piepenbring, 739 ; Poggiale, 745 ; Poleck, 746 ; Quincke, 757 ; Rahn, 758 ; Raulin, 762 ; Redtenbacher, 763 ; Reichardt, 766 ; Remler, 769 ; Reuss, 771; Saunders, 800 ; Sauvages, 800 ; Scherer, 806 ; Schweitzer, 824 ; Sestini, 829 ; Steele, 850 ; Suckow, 859 ; Than, 866 ; Vanuccini, 884 ; Westrumb, 913 ; Wiggers, 916 ; Wurzer, 836.

VII. 45.

MINERAL WATERS ; LOCALITIES.

Aargau, Switzerland, 638 ; Acqui, 676 ; Aix en Savoie, 388 ; Aix-la-Chapelle, 273, 627, 678 ; Alach, 711 ; Alais, 800; Algiers, 1173 ; Aumale, 655 ; Austria, 330 ; Bagnères de Luchon, 288 ; Ballston, N. Y., 850 ; Bardfeld, 577 ; Berlin, 455 ; Bertrich, 455 ; Bilin, 763 ; Birresborn, 454 ; Biskirchen, 455 ; Boisse, Borcette, 678 ; Brohl-Thal, 455 ; Budapest, 455, 577 ; Burtscheit, 678 ; Chili, 388 ; Denmark, 902 ; Driburg, 454, 455 ; Eilsen, 454 ; Ems, 454 ; England, 484, 495 ; Europe, 532, 711 ; Ferdinandsbrunn, 574 ; Flinsberg, 746 ; Forges, 655 ; France, 410, 1078, 1173 ; Franzenbad, 771 ; Geismarch, 674 ; Germany, 62, 318, 330, 477 (1553), 495 ; Godesberg, 936 ; Harrowgate, 468 ; Holland, 269, 504 ; Homburg, 454 ; Homburg vor der Höhe, 454, 681 ; Hungary, 866 ; Italy, 454 ; Jena, 859 ; Karlsbad, 578 ; Kolberg, 455 ; Kreuzbrunn, 574 ; Lamotte-les-Bains, 278 ; Lamscheid, 455 ; Langenschwalbach, 824 ; Liebenstein, 766 ; Loreto, 829 ; Lorraine, 700 ; Mecklenburg, 932 ; Meinberg, 336 ; Nassau, Duchy of, 455 ; Neudorf, Bohemia, 454 ; Oelheim, 455 ; Oldesloensis, 424 ; Offenbach am Main, 455 ; Orezza, 745 ; Padua, 663 ; Paris barracks, 745 ; Passy, 348, 396 ; Pont de Vesle, Dijon, 653 ; Pougues, 379 ; Pozzuoli, 600 ; Provins, 611, 707, 762 ; Puy de Dôme, 701 ; Pyrmont, 454, 739, 913, 916 ; Rivers, 418 ; Rome, 658 ; Russia, 806 ; Saidschütz, 771 ; Salzbrunn, 441, 455 ; Saratoga, 850 ; Schinznach, Switzerland, 638 ; Seewen, Switzerland, 638 ; Seine, 745 ; Selters, 771, 913 ; Sondershausen, 344 ; Spa, 273, 562, 632, 678 ; Stachelberg, 575 ; St. Diez, 700 ; Stettin, 455; Switzerland, 477 (1553), 680 ; Tatonhausen, 336 ; Thuren, Prussia, 507 ; United States of North America, 726 ; Verberie, 396 ; Vichy, 331 ; Voltri, 676 ; Werne, Westphalen, 455 ; Wiesbaden, 455 ; Yorkshire Spas, 454.

O.

OILS. (*See also* FATS.)

V. Askinson, 273 ; Bornemann, 329 ; Brannt, 337 ; Cameron, 350 ; Carl, 353 ; Gorini, 489 ; Gusenberger, 504 ; Hager, 508 , Julia-Fontenelle, 565 ; Maier, 648 ; Mierzinski, 670 ; Mulder, 686 ; Popper, 747 ; Schaedler, 801 ; Schmidt, 813 ; Schweitzer, 824 ; Wright, 934.

VII. 214, 299, 300, 302.

OLEO-MARGARINE. (*See* BUTTER, ARTIFICIAL.)

OPIUM.

V. Descharmes, 398 ; Hirzel, 533 ; Stratingh, 857.

ORGANIC ANALYSIS.

V. Allen, 267 ; Chevreul, 365 ; Ellis, 426 ; Fleury, 443 ; Horn, 543 ; Kopfer, 586 ; Liebig, 624 ; Prescott, 752 ; Schiel, 807 ; Vejledning, 885 ; Vortmann, 893 ; Walker, 900.

ORGANIC CHEMISTRY.

I. Wolff, 35.

II. Weltzien, 82.

III. Cahours, 98 ; Remsen, 150 ; Schorlemmer, 156 ; Wenghöffer, 166.

V. Alexéyeff, 266 ; Butlerof, 348 ; Chastaing, 363 ; Claus, 370 ; Clements, 371 ; Cleve, 371 ; Coninck, 1173 ; Delffs, 395 ; Doering, 404 ; Dumas, 414 ; Erlenmeyer, 430 ; Fischer, 440 ; Franchimont, 451 ; Friedel, 461 ; Gerhardt, 475 ; Girard, 480 ; Girardin, 480 ; Gregory, 496 ; Hétet, 529 ; Hjelt, 534 ; Hofmann, 539 ; Huard, 1176 ; Kekulé, 570 ; König, 582 ; Kolbe, 584 ; Lassar-Cohn, 604 ; Lavén, 605 ; Levy, 620 ; Leyde, 621 ; Liebig, 624–632 ; Limpricht, 633 ; Löwig, 638 ; Machado, 646 ; Marchand, 653 ; Masserotti, 657 ; Masure, 657 ; Meyer, 668 ; Oechsner, 705 ; Paupaille, 722 ; Payen, 724 ; Pennetier, 730 ; Piazza, 738 ; Pinner, 741 ; Piria, 741 ; Puerta, 756 ; Remsen, 769 ; Repetitorium, 770 ; Richter, 776 ; Rideal, 776 ; Roscoe and Schorlemmer, 785 ; Rosenberg, 786 ; Rossignon, 787 ; Schaedler,

801 ; Scheurer-Kestner, 807 ; Schiel, 807 ; Schorlemmer, 816 ; Schramm, 817 ; Seelig, 826 ; Société, 839, 840 ; Streatfield, 857 ; Tauba, 864 ; Thomson, 871 ; Wagner, 895 ; Watts, 905 ; Wenghöffer, 911 ; Wislicenus, 926 ; Witt, 926 ; Wöhler, 928, 929.

VII. 56, 334, 337.

OSMOSE. (*See* DIFFUSION.)

OXYGEN.

III. Bromeis, 97 ; Déhérain, 103 ; Draper, 105 ; Olschanetzky, 143 ; Rodwell, 151 ; Yeats, 170.

V. Beschreibung, 311 ; Böttcher, 323 ; Fremy, 453 ; Heldt, 521 ; Lvof, 644 ; Mayow (1674), 659 ; Meissner, 662 ; Philipps, 736 ; Pictet, 739 ; Ramaer, 759 ; Schurer, 822 ; Stamm, 848 ; Valon, 884 ; Wroblewsky, 935.

OZONE.

I. Leeds, 21.

III. Dachauer, 102 ; Engler, 107 ; Leeds, 131 ; Moigno, 439 ; Odling, 143.

V. Andrews, 269 ; Bellucci, 299 ; Chappius, 361 ; Fox, 451 ; Fremy, 453 ; Hammerschmied, 512 ; Meissner, 662 ; Moigno, 675 ; Odling, 705 ; Oppio, 707 ; Ozone, 716 ; Schönbein, 815, 816 ; Scoutetten, 825 ; Stabel, 845 ; Stamm, 848 ; Wolf, 931 ; Zinno, 940.

P.

PAINTS. (*See* PIGMENTS.)

PALLADIUM.

V. Wollaston, 933.

PAPER.

V. Bolley, 326.

PARACELSUS' TERMINOLOGY.

II. Onomastica duo, 71 ; Chymicall Dict., 45 ; Dorneus, 47 ; Johnson, 63.

PARAFFIN.

V. Albrecht, 265 ; Mills, 671 ; Perl, 732 ; Uhlenhuth, 881.

PERFUMERY.

V. Askinson, 273 ; Debay, 392 ; Deite, 394 ; Morfit, 680 ; Perutz, 732 ; Piesse, 740.

PERIODIC LAW.

III. Newlands, 142.

V. Huth, 548 ; Rydberg, 792.

PHOSPHORUS. [Cont'd.]
    326 ; Branchi, 335 ; Buttatz, 348 ; Cel-
    lio (1680), 358 ; Cohausen (1717), 372 ;
    Elsholtz (1671), 426 ; Kletwich, 1177 ;
    Koning, 586; Kunckel (1678), 594 ;
    Lehmann, 613 ; Munk, 688 ; Reinlein,
    767 ; Scherer, 806 ; Schrötter, 818 ;
    Société [Moissan], 840 ; Spallanzani
    (1796), 842 ; Stratingh, 857 ; Wed-
    ding, 908 ; Wilson, 923.
PHOTOGRAPHIC CHEMISTRY. (*See also*
    LIGHT, CHEMICAL ACTION OF.)
    I. Zuchold, 37.
    II. Bollmann, 42 ; Fourtier, 1166.
    III. Daguerre, 102 ; Eder, 107 ; Harri-
    son, 117 ; Martin, 136 ; Niepce, 142 ;
    Petersen, 146 ; Schiendl, 155 ; Talbot,
    161 ; Tissandier, 162 ; Vogel, 164 ;
    Werge, 166.
    V. Barreswil, 283 ; Belloc, 298 ; Eder,
    422 ; Hardwich, 514 ; Hunt, 547 ;
    Köhler, 582 ; Lainer, 598, 1177 ;
    Maumené, 1178 ; Niepce, 700 ; Ogo-
    nowski, 706 ; Pizzighelli, 742 ; Talbot,
    863 ; Toifel, 875 ; Villon, 889 ; Vogel,
    891, 892.
PHYSICS.
    I. Annalen, 1 ; Bibliotheca, 4 ; Jahres-
    bericht, 18 ; Reuss, 27 ; Rohr, 27.
    II. Carnelley, 44 ; Clarke, 45 ; Göttling,
    57 ; Handwörterbuch, 59 ; Jehan, 63 ;
    Landolt, 67 ; Pozzi, 72 ; Zenneck, 84.
    III. Boisselet, 93 ; Bostock, 95 ; Boullet,
    96 ; Buckley, 98 ; Cronsaz, 102 ; Cu-
    vier, 102 ; Fischer, 109 ; Gerland,
    1167 ; Heinecken, 117 ; Heller, 117 ;
    Hoefer, 119, 1167 ; Libes, 132 ; Libri,
    132 ; Loys, 134 ; Mair, 135 ; Lübimof,
    1167 ; Marie, 135 ; Murhard, 140 ;
    Poggendorff, 147 ; Quetelet, 149 ; Reg-
    nault, 149 ; Rosenberger, 152 ; Schnei-
    der, 156 ; Sturmius, 160 ; Valentini,
    163.
    V. Clarke, 369 ; Cooke, 376 ; Cutler,
    385 ; Erxleben, 431 ; Gänge, 466 ;
    Gonzalez Valledor, 488 ; Grothuss, 501 ;
    Grove, 501 ; Kremers, 590 ; Sage, 795 ;
    Schönbein, 815 ; Sertürner, 829 ; Waals,
    894 ; Wittwer, 928 ; Yarkovski, 937 ;
    Youmans, 937 ; Zantedeschi, 938.

    VII. 34, 63, 69, 93, 94, 108*a*, 181, 183,
    192, 194, 203, 210, 218, 226, 240, 249,
    310, 321, 323, 328, 395, 419, 425, 433.
PHYSIOLOGICAL CHEMISTRY.
    I. Zeitschrift, 36.
    III. Laugel, 130.
    V. Arends, 271 ; Bibra, 312 ; Bruylants,
    343 ; Bunge, 346 ; Charles, 362 ; Chit-
    tenden, 366 ; Dey, 392 ; Denis, 396,
    397 ; Eberle, 421 ; Ebermaier, 421 ;
    Folwarczny, 445 ; Frey, 460 ; Fried-
    rich, 461 ; Funke, 464 ; Gamgee, 468 ;
    Gautier, 471 ; Geubel, 477 ; Gmelin,
    486 ; Gorup-Besanez, 489, 490, 491 ;
    Hales, 510 ; Halliburton, 510 ; Hand-
    buch, 512 ; Hardy, 514 ; Hoppe-Seyler,
    542 ; Hünefeld, 546 ; Jones, 562 ;
    Khodnef, 575 ; Kingzett, 575 ; Kohl-
    rausch, 584 ; Krukenberg, 591 ; Kühne,
    593 ; Larkin, 603 ; Le Cat, 608 ; Leh-
    mann, 611 ; L'Heritier, 621 ; Liebig,
    624–632 ; Loew, 638 ; Magendie, 648 ;
    Marchand, 653 ; Maschi, 657 ; Mayow
    (1674), 659 ; Miahle, 668 ; Moleschott,
    676 ; Moore, 679 ; Moser, 682 ; Mulder,
    687 ; Nysten, 704 ; Penrose, 730 ;
    Pettenkofer, 734 ; Plenck, 744 ; Pog-
    giale, 745 ; Quinquand, 757 ; Ralfe,
    758 ; Robin, 779 ; Schlossberger, 809 ;
    Schützenberger, 820 ; Schwarz, 823 ;
    Simon, 834 ; Stenberg, 851 ; Thudi-
    chum, 872 ; Vaughan, 885 ; Watson,
    905 ; Weyl, 913 ; Wurtz, 936.
    VII. 70, 215, 426.
PICRIC ACID.
    V. Rymsza, 792.
PIGMENTS.
    V. Bersch, 304 ; Church, 367 ; Hurst,
    1176 ; Lefort, 611 ; Schmidt, 813 ;
    Villejean, 888.
    VII. 188.
PLANT-ANALYSIS AND PLANT CHEMISTRY.
    II. John, 63.
    V. Abbott, 258 ; Claus, 370 ; Cramer,
    380 ; Crügner, 384 ; Decandolle, 393 ;
    Demel, 396 ; Dragendorff, 406 ; Draper,
    407 ; Fechner, 436 ; Flückiger, 444 ;
    Fremy, 453 ; Fromherz, 462 ; Geubel,
    477 ; Grischow, 499 ; Hales, 509 ;
    Hempel, 522 ; Hermbstädt, 526 ;

R.

REACTIONS, CHEMICAL.
II. Heppe, 60.
V. Claus, 370; Drechsel, 407; Flückiger, 444; Genth, 473; Gustavson, 504; Hagemann, 507; Horstmann, 544; Klencke, 578; Nechaeff, 695.

REAGENTS, CHEMICAL.
II. Lassaigne, 67.
V. Accum, 260; Buchner, 344; Curtman, 385; Fischer, 441; Krauch, 589; Mamone, 651; Moldenhauer, 676; Payen, 725; Prestinari, 752; Schulze-Montanus, 821; Schweinsberg, 824.

RESINS.
V. Cameron, 350; Campari, 350; Thenius, 868; Wiesner, 916.

RESPIRATION.
V. Andral, 268; Archer, 270; Regnault, 764.

RHODIZONIC ACID.
V. Heller, 521; Meusel, 666.

ROMANS, CHEMISTRY AND TECHNOLOGY OF THE.
III. Blümner, 92.

ROYAL INSTITUTION.
V. Davy, 391.

ROYAL SOCIETY.
III. Birch, 91; Sprat, 159; Thomson, 161; Weld, 166; Yearbook, 170.

RUBIDIUM.
V. Société (1863), 839; Wurtz [Grandeau], 936.

RUSSIAN CHEMICAL BIBLIOGRAPHY.
I. Krebel, 20.

RUSSIAN CHEMICAL TERMINOLOGY.
II. Scherer, 74; Severghin, 76.

S.

SACCHARIC ACID.
V. Afzelius, 263; Heintz, 520.

SACCHARIMETRY.
V. Balling, 279; Clerget, 371; Landolt, 601; Moigno, 675; Pellet, 728.

SACCHARINE.
V. Marquisan, 655; Stutzer, 859.

SALAMMONIAC.
V. Alberti, 265; Casartelli, 355; Gött-

ling, 487; Jaeger, 552; Leblanc, 607; Löscher, 637.

SALT, COMMON.
III. Boddy, 92; Ratton, 149; Schleiden, 155.
V. Balthasar, 280; Branche, 335; Brownrigg (1748), 341; Cochrane (1785), 372; Grube (1703), 501; Harmens (1748), 515; Karsten, 568; Licht (1749), 623; Phipson, 737; Ratton, 761.

SALTPETER.
V. Becker, 293; Chaptal, 362; Clarcke (1670), 368; Conring (1672), 376; Cornette, 378; Fiedler, 439; Hare, 515; Harmens, 515; Jaeger, 552; Reuss, 771; Riffault, 777; Schelhammer, 805; Stahl, 847; Wallerius, 902; Weber, 906.
VII. Mémorial, 1182.

SALTS.
II. Hoffmann, 62; Gruner, 57.
V. Brix, 339; Dalton, 386; Fehling, 436; Fick, 439; Gay Lussac, 471; Göttling, 487; Hagen, 508; Harmens, 515; Köhler, 582; Mansfield, 652; Pialat, 737; Remer, 768; Remler, 769; Richardson, 774; Rothe, 787; Stahl, 846; Trommsdorff, 878; Wallerius, 902; Wiegleb, 915.

SANITARY CHEMISTRY.
V. Cheyne, 365; Fleck, 442; Fox, 451; Gautier, 471; Hankel, 513; Heinzerling, 520; Kimmins, 1177; Lambotte, 599; Liebermann, 623, 624; Lindes, 633; Nichols, 699; Poiré, 745; Schmid, 810; Selmi, 827; Spataro, 843; Verver, 887; Zaborowski, 937.
VII. 292.

SCANDINAVIAN CHEMICAL TERMS.
II. Oersted, 70.

SEA-WATER.
V. Meerten, 661; Mulder, 688; Report . . . Challenger, 770; Sage, 795; Salzer, 798; Schmelck, 809; Simons, 835; Tornöe, 876.

SECRET PREPARATIONS, ANALYSIS OF.
II. Capaun, 43, 352.
V. Hahn, 1175; Thein, 867.

SEIGNETTE'S SALT.
V. Teichmeyer, 865; Wilkens, 881.